T0204168

BIODESIGN
The Process of Innovating Medical Technologies

A practical guide to the new era of global opportunity and value-based innovation in medical technology

This step-by-step guide to medical technology innovation, now in full color, has been rewritten to reflect recent trends of industry globalization and value-conscious health-care. Written by a team of medical, engineering, and business experts, the authors provide a comprehensive resource that leads students, researchers, and entrepreneurs through a proven process for the identification, invention, and implementation of new solutions.

- Nearly 70 case studies on innovative products from around the world explore successes and failures, provide practical advice, and enable readers to learn from real projects.
- "Getting Started" sections for each chapter encourage readers to take action and apply what they've learned to their own work.
- A collection of nearly 300 videos, created for the second edition of the book, expand upon critical concepts, demonstrate essential activities within the process, and bring the innovation experience to life.
- A wealth of additional material supports the book, including active links to external websites and resources, supplementary appendices, and timely updates.
- New to this edition, two opening sections highlight the importance of globalization and cost-effective healthcare in the medtech industry, themes which are carried throughout the book.

Readers can access videos and additional materials quickly, easily, and at the most relevant point in the text within the ebook, or on the companion website at ebiodesign.org, alongside instructor resources.

"Biodesign is on the forward edge of one of the most exciting new frontiers of healthcare. This impressive and engaging work provides a thorough look at the innovation process. But this is certainly not just for the scientific innovators: it is a must-read for anyone in any aspect of healthcare today."

Alex Gorsky, *Chairman and CEO, Johnson & Johnson*

"I can't think of a more important place to turn creativity loose than in designing the future of healthcare. But it's a complicated scene – and it's easy to get lost in the maze of stakeholders, regulation, and financing. Biodesign lays out a clear and logical map to find and pursue opportunities for real innovation. One of the core messages in this new edition is that, by placing the need for affordability up front in design process, innovators can more explicitly create technologies that bring value to the healthcare system. This is design thinking at its best!"

David Kelley, *Founder, Hasso Plattner Institute of Design at Stanford University, Founder, IDEO*

"A must-to-read textbook for anyone in academia or industry, in any country, who wants to innovate and deliver value to patients and health systems around the world."

Koji Nakao, *Chairman of Terumo and the Japanese Federation of Medical Device Associations*

"If you want to know how to come up with a both innovative and transformative technology in medicine, there isn't a better resource than this book by Paul Yock and his colleagues at Biodesign. Over 13 years ago, the program at Stanford brought together trans-disciplinary innovators – engineers, physicians and business experts – to not only design their formidable program, but to teach all the rest of us how to do it."

Eric J. Topol, *Director, Scripps Translational Science Institute*

"This book on biodesign will be invaluable for any inventor or entrepreneur. It contains very useful information on such critical areas as design principles, regulatory issues, clinical trial strategies, intellectual property, reimbursement strategies, and funding- and it backs them up with interesting real-life experiences and case studies".

Robert Langer, *David H. Koch Institute Professor, MIT*

"This practical but comprehensive resource is keeping up with the rapid developments affecting medical device innovation. The authors draw on their own extensive experiences and insights, as well as diverse case studies, to present the full range of strategic and operational considerations to bring valuable new therapies to patients in the US and around the world."

Mark McClellan, *Director, Health Care Innovation and Value Initiative, Brookings Institution*

BIODESIGN

The Process of Innovating Medical Technologies

EDITORS
Paul G. Yock
Stefanos Zenios
Josh Makower
Todd J. Brinton
Uday N. Kumar
F. T. Jay Watkins

PRINCIPAL WRITER
Lyn Denend

SPECIALTY EDITOR
Thomas M. Krummel

WEB EDITOR
Christine Q. Kurihara

ebiodesign.org

University Printing House, Cambridge CB2 8BS, United Kingdom

One Liberty Plaza, 20th Floor, New York, NY 10006, USA

477 Williamstown Road, Port Melbourne, VIC 3207, Australia

314-321, 3rd Floor, Plot 3, Splendor Forum, Jasola District Centre, New Delhi 110025, India

103 Penang Road, #05-06/07, Visioncrest Commercial, Singapore 238467

Cambridge University Press is part of the University of Cambridge.

It furthers the University's mission by disseminating knowledge in the pursuit
of education, learning, and research at the highest international levels of excellence.

www.cambridge.org
Information on this title: www.cambridge.org/9781107087354

First published 2010
Second edition 2015
10th printing 2022

Printed in the United Kingdom by Bell and Bain Ltd, Glasgow

A catalogue record for this publication is available from the British Library

Library of Congress Cataloguing in Publication data
Biodesign : The process of innovating medical technologies / editors, Paul G. Yock, Stefanos Zenios,
Joshua Makower, Todd J. Brinton, Uday N. Kumar, F. T. Jay Watkins ; principal writer, Lyn Denend ;
speciality editor, Thomas M. Krummel ; web editor, Christine Kurihara. – 2.
      p. ; cm.
Includes bibliographical references and index.
ISBN 978-1-107-08735-4 (Hardback)
I. Yock, Paul G., editor.
[DNLM: 1. Biomedical Engineering–organization & administration.   2. Biomedical
Technology.   QT 36]
R856
610.28–dc23   2014025957
ISBN 978-1-107-08735-4 Hardback

Additional resources for this publication at ebiodesign.org
Cambridge University Press has no responsibility for the persistence or accuracy of
URLs for external or third-party Internet websites referred to in this publication,
and does not guarantee that any content on such websites is, or will remain,
accurate or appropriate.

To innovators – past, present, and future – and the patients who inspire them . . .

. . . and in tribute to Wallace H. Coulter, a pioneer in developing affordable healthcare technologies with a global impact.

Contents

See **ebiodesign.org** for videos, online appendices, and active web links to the resources listed in each chapter.

Preface

There is no greater satisfaction than seeing a patient being helped by a technology that you've had a hand in creating. And thanks to continuing advances in science and technology, healthcare is more open for innovation than at any time in history.

Despite this promise, however, medical technology innovators face significant hurdles – especially in the new era of cost containment. If not managed skillfully, patents, regulatory approval, reimbursement, market dynamics, business models, competition, financing, clinical trials, technical feasibility, and team dynamics (just to name a few of many potential challenges) can all prevent even the best idea from reaching patient care.

So, where should you begin as an innovator? What process can you use to improve your chances of success? What lessons can you learn from the inventors, engineers, physicians, and entrepreneurs who have succeeded and failed in this endeavor before? This book delivers practical answers to these important questions.

Who should read it and why?

Biodesign: The Process of Innovating Medical Technologies provides a comprehensive roadmap for identifying, inventing, and implementing new medical devices, diagnostics, and other technologies intended to create value for healthcare stakeholders. It has been written to be approachable for engineering, medical, and business students at both the undergraduate and graduate level, yet comprehensive and sophisticated enough to satisfy the needs of experienced entrepreneurs and medtech executives. For instructors, it provides a proven approach for teaching medical technology innovation that begins pre-idea and extends through preparing for commercialization. It is ideally suited to support team-oriented, project-based learning experiences in academic and industry settings.

The text describes the biodesign innovation process, which we initially developed to support the biodesign innovation and fellowship programs at Stanford University. Over 13+ years, the process has been built and refined based on:

- Presentations and mentoring by more than 200 industry leaders who have participated in our training programs
- Our experience advising more than 150 project teams that have applied the process to their work
- Feedback from those who have learned the process through our executive education courses, as well as input and suggestions from students, fellows, instructors at other universities, and industry representatives using the first edition of the book
- Extensive field-based research

Our confidence that the process is effective is based on the results of the students and fellows trained at Stanford and through our university-based partnerships in India and Singapore. Already over 30 of these projects have been converted to externally funded companies that have raised an aggregate of over $250 million. More importantly, even though these are young companies, over 250,000 patients have already been treated by the technologies invented by our trainees. We have also been encouraged by the positive feedback we received on the process following the release of the first edition of the text.

What's new and important in the second edition of the biodesign book?

We initially wrote the *Biodesign* book because there was no comprehensive text that described the complete innovation process with a focus on the medical technology sector. Many excellent books address entrepreneurship generally or pieces of the device development process, but our goal was (and is) to provide a definitive, comprehensive resource for the medtech community.

Since the first edition of *Biodesign* was published in 2010, however, the medical technology industry and, more broadly, the healthcare ecosystem has experienced tumultuous change. As healthcare costs escalate on an unsustainable trajectory, a high priority is being placed on medical technologies that deliver *value* – that is, good outcomes at an affordable cost. In parallel with these forces, the global medical technology landscape is evolving rapidly, with large-scale demand for improved healthcare and a new focus on frugal innovation for developing economies. In this changing environment, veteran medtech innovators may feel as though they are treading unfamiliar new ground, and aspiring inventors and entrepreneurs are faced with navigating an even more complex and challenging landscape.

Besides the need to update the text in response to these major environmental changes, we felt a personal imperative to create the second edition. Over the past several years we have learned more about how to teach the biodesign innovation process. We've had the chance to use the text with students, fellows, entrepreneurs, and executives, and gather feedback from instructors at other universities around the world who are using it in their courses. Through these interactions, we realized that there were messages that we could clarify and some that we should emphasize more strongly. As a result, we have revised the text substantially for the second edition to address three critical factors:

1. **Value orientation** – The healthcare industry has become increasingly competitive, with the primary customers of medical technologies – governments, private payers, provider groups, and patients – focusing intensely on the cost of medical technologies and related services. In this environment, it is more essential than ever for products and related services to demonstrate measurable value to their intended users. The second edition of *Biodesign* more explicitly recognizes the importance of value generation in healthcare and includes guidance to better address this imperative in all phases of our process. Be sure to read the section "Focus on Value" in the pages that follow the preface for more context on value and how it is treated within the text.

2. **Going global** – The first edition of the text was largely US-centric, but in the second edition we devote significantly more attention to describing the changes in the process of medtech innovation resulting from the growing importance of markets, clinical opportunities, and sources of innovation outside of the United States. We focus on key strategic considerations for operating in a more global healthcare environment and share substantially more examples from medtech innovators working around the world. To dig more deeply into some key issues, we have added a section on "Global Perspectives," in which we spotlight six regions that present interesting medtech opportunities.

3. **Better ways to teach and learn** – While the fundamental biodesign innovation process remains the same in the new version of the text, we have rewritten a number of sections to provide more focus and clarity; and we offer more examples and case material in areas that are best understood experientially. One important take-away is that our approach appeared too linear in the first edition, and we have made concerted effort to explain within the chapters when and why a more iterative method is necessary. We have also captured a number of important lessons in the "Process Insights" section that follows the preface. Readers will significantly increase their effectiveness if they take these key themes to heart and keep them in mind as they work through the chapters within each major section of the text.

Our core belief remains that innovation is both a process and a skill that can be learned. We hope that the new edition of *Biodesign* will help to better equip aspiring and experienced innovators alike to be successful in the dynamic medtech industry. Tumultuous changes notwithstanding, the dynamics of the emerging healthcare burden around the world demand continued innovation, and technology innovators will continue to be central to this mission.

How to maximize the benefit of this book: a user's guide

The steps in the biodesign innovation process build on each other and, in this respect, it makes sense for readers to work their way through the text in chapter order. Taking this approach provides innovators with the most complete understanding of the biodesign innovation process and the most valuable overall learning experience. We have heard of many medtech innovators using the text as a roadmap for their projects, starting at the beginning and following the process to help drive their progress.

That said, each chapter is sufficiently robust to support alternate approaches to the content. For instance, instructors can pick and choose the chapters most relevant to their specific courses (e.g., some of the chapters in the Implement section may be a bit advanced for undergraduates, but they are ideally suited to graduate-level innovation or business planning classes). And experienced device executive and entrepreneurs can use the book as a reference as they encounter specific challenges on their way to market with a new technology.

In terms of organization, we present the biodesign innovation process in:

- three distinct **phases**, Identify, Invent, and Implement;
- that are divided into two **stages** each (six in total);
- which are supported by 29 core **activities**, with a chapter on each one.

Figure P1 summarizes the overall process. Keep in mind that it's not nearly as linear in practice as it appears in this depiction. The iterative and cyclical nature of the process is further explained throughout the text.

As you navigate *Biodesign*, we encourage you to pay attention to a series of different features that

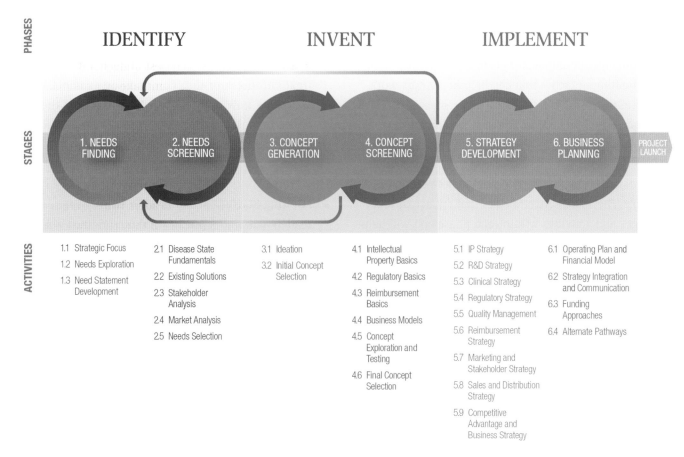

FIGURE P1
The biodesign innovation process.

have been designed to help you optimize the value you receive from the text.

As you begin – Immediately following the preface, you'll find relevant information that expands upon the three primary reasons we created the second edition of the book. These materials set a context for understanding and applying the content of the chapters.

- **Focus on value** – The medtech industry is in the midst of a transition to a stronger value orientation, in which the improvement a technology offers relative to its price is an essential ingredient of success. This section explores the forces behind this shift and their implications to innovators as they design, develop, and prepare to commercialize new products and services.
- **Global perspectives** – An introduction to factors driving the globalization of the medtech industry and changes in how innovators source, develop, and sell their technologies. We also profile six regions, Africa, China, Europe, India, Japan, and Latin America, providing background on these geographies, highlighting potential barriers to medtech innovation, and outlining tactics that can help innovators work more successfully in these areas.
- **Process insights** – Through feedback and our teaching experience, we have identified a series of key themes that you should keep top-of-mind while reading the chapters within each major section of the book. These are core strategies that cut across the stages and activities within each phase and will help you to keep on track as you proceed with the process. Instructors that emphasize these points in their teaching and readers who embrace this information will be able to navigate the biodesign innovation process more effectively.

Throughout the book – You should also be on the look-out for a few categories of information that have been added or broadened in the second edition.

New

- **Videos** – The second edition of *Biodesign* is supported by a brand new collection of nearly 300 videos on topics spanning the complete biodesign innovation process. These clips, which include expert presentations and advice, interviews with innovators, demonstrations, and other exercises, are available to all readers in the video library at ebiodesign.org. Those reading the electronic version will find select videos embedded in the book directly where they are most relevant.

Expanded

- **"From the Field" case studies** – These short stories, which provide real-world examples of how innovators, teams, and companies have tackled important challenges in the biodesign innovation process, were one of the most popular features of the first edition. Accordingly, we increased the number of case studies by more than 50 percent. Look for 36 new and/or rewritten stories in the second edition of the text, many of which spotlight groups developing innovative medtech solutions outside of the US. At the end of each stage, we present a case study on Acclarent, maker of a device to treat chronic sinusitis. This running example spotlights how one real company executed the entire biodesign innovation process, from need finding to commercialization.

Updated

- **"Getting Started" sections** – For each chapter, readers will find a practical, action-oriented guide that they can follow to execute every step in the biodesign innovation process when working on an actual project. To make these sections more useful in the electronic version of the text, they have been populated with active web links to take readers directly to essential references and resources. In the print version, the key steps for getting started are listed, with the complete, interactive guides accessible at ebiodesign.org.

Enhanced

- **ebiodesign.org** – To better support the second edition of *Biodesign*, we have completely redesigned ebiodesign.org to be more user friendly and content rich. In addition to the video library and interactive getting started sections, ebiodesign.org includes additional content in the form of online appendices for many chapters. This is also where we'll post important updates, new videos, and other learning materials as they become available. Instructors can access our course syllabus, select presentation slides, and exam questions/answers via the Instructor Resources section of the site.

Focus on Value

What do we mean by "value" and why is it so important?

The escalation of healthcare costs is one of the major economic and political issues of our time. The problem is most apparent in the United States, where healthcare as a share of the economy has more than doubled over the past 35 years. Spending on health accounted for 7.2 percent of the nation's gross domestic product (GDP) in 1970, expanded to 16 percent in 2005, and is projected to be as high as 20 percent of GDP by 2015.[1]

Simply put, the US economy cannot sustain this spending trajectory, which has outpaced GDP growth for years (see Figure V1).[2] The problem is not just straining the federal budget: state and local governments have been forced to reduce support for education, infrastructure, and other critical expenditures as they struggle to fund Medicaid and other health programs. In the private sector, the cost of employment-based health insurance is one of the main reasons workers have seen their wages stagnate.[3]

Despite the fact that the US spends two-and-a-half times more per capita on health than most developed countries,[4] it does not necessarily provide the best care to its citizens. In 2000, when the World Health Organization ranked the health systems of its 191 member states for the first time ever, the US found itself in 37th position.[5] In a more recent study that compared the US to Australia, Canada, Germany, the Netherlands, New Zealand, and the United Kingdom on measures of quality, efficiency, access to care, equity, and the ability of citizens to lead long, healthy lives, America occupied last place. As the report pointed out, "While there is room for improvement in every country, the US stands out for not getting good value for its healthcare dollars."[6]

Against this backdrop, economists, researchers, and policy makers alike have pointed to medical technology as a dominant factor driving increased health expenditures in the US. Their estimates of the impact of technical innovation on accelerating costs vary considerably, but some argue that new technologies and the procedures that accompany them account for one-third to one-half of real long-term spending growth in healthcare.[7] To be sure, many of these technologies have provided major advancements in health and longevity, ranging from

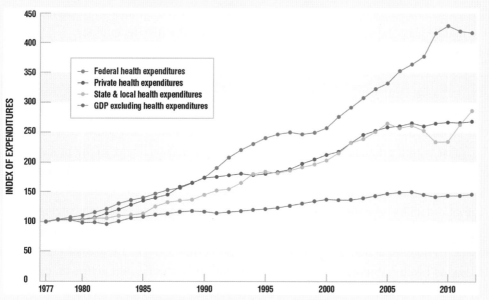

FIGURE V1

Indexes of US health expenditures and GDP (excluding health expenditures), per capita, adjusted for inflation, 1977–2007 (compiled based on National Health Expenditure data, CMS.gov).

diagnostic breakthroughs such as CT and MRI scanning to life-saving surgical and interventional therapies for the heart and brain. Increasingly, however, even revolutionary developments such as these are being weighed against the unsustainable rise in healthcare costs.

Since the birth of the modern medtech industry in the mid-twentieth century, the majority of medical technology companies pursued a philosophy that has been described as "progress at any price."[8] Innovators and companies were focused on developing new products that resulted in improved clinical outcomes, almost regardless of their associated cost. In some cases, this meant simply making marginal enhancements in order to sell a next-generation technology at a higher price. These strategies were successful for many years because the fee-for-service payment system in the US largely uncoupled the providers, who make the treatment decisions, from the payers, who bear the costs of their choices. In this way, the market forces that operate in other sectors of the economy have not been effective in maximizing the value of health technologies and services. By spending trillions of dollars on new innovations, the US fueled the growth of the medical technology industry and helped to foster a view that complex and expensive technology was the hallmark of superior healthcare.

While the US has been hardest hit by uncontrolled health spending, it certainly is not alone. The countries in the European Union and Japan, which together with the US account for 75 percent of all medtech sales today,[9] have also been wrestling with how to manage mounting healthcare costs. Moreover, as the middle class expands in developing countries such as India, China, and Brazil, these patients are demanding increased access to more advanced healthcare, potentially initiating the same spiral of escalating health expenditures. In fact, these issues are already emerging, with medical device sales growing two- to five-times faster in these markets than in developed countries.[10]

Together these forces have launched a fundamental shift in the healthcare sector. The affordability of care relative to its quality is now a primary focus in both developed and developing markets. "Progress at any price" is no longer a tenable strategy as health systems universally place increasing emphasis on ensuring a good value for the healthcare dollars they spend. In developed countries such as the US, providers, hospitals, clinics, and (in some cases) payers are consolidating to achieve economies of scale and organization. Value-based payment models are emerging. And purchasing managers and executives are playing a more central role in deciding which medical technologies to adopt, with physicians influencing, rather than dictating, those choices. In developing countries, health systems recognize they are facing increased demand for medical technologies but are actively pursuing more affordable, cost-effective products and services designed specifically to address the needs of patients and providers in settings with fewer resources. In other words, around the world, the need for medical technologies that deliver clear *value* to their intended users has never been more imperative.

The concept of value is widely understood in general terms, but is more difficult to articulate as a concept to be considered throughout the biodesign innovation process. Here are a few key points that resonate with us about value and value creation:

- Value is an expression of the improvement(s) a new technology and its associated services offer relative to the incremental cost. Just because a new technology provides an improvement doesn't mean it will create value.

- Importantly, value is not realized unless the cost/ improvement equation is compelling enough – that is, has enough marginal benefit over other available solutions – to cause decision makers to change their behavior and adopt the new technology.

- We are in a period of transition with respect to who the key decision makers are in the healthcare field. In particular, purchasing power is shifting from individual physicians to integrated health systems and patients are becoming more knowledgeable and active healthcare consumers. In the process, both of these audiences are demanding greater cost transparency.

- In parallel, the assessment of value is evolving from being product specific to outcomes oriented. Stated another way, decision makers are increasingly evaluating total solution offerings across an episode of care rather than focusing on an individual technology or service. Within this context, new types of value-based offerings and innovative business models are emerging.

Understanding what we mean by value is important because it has a major impact on how you approach the biodesign innovation process. In short, while medtech companies used to strive to produce products that delivered optimal improvement (without undue attention to cost), we are now seeing purchasers demand offerings that drive cost as low as possible. In certain situations there will be willingness to sacrifice some degree of performance for a better price (see Figure V2). Amidst the uncertainty of today's value-oriented environment, technologies that significantly – not incrementally – generate measurable savings while providing acceptable (or better) quality will be the ones with the clearest path forward.

So how can innovators practically address value in the design, development, and commercialization of their medtech offerings? There are multiple steps in the biodesign innovation process where opportunity exists to create and deliver value (as you navigate the book, you will see substantial attention to value in almost every chapter). But there are three critical points at which value should be a primary focus:

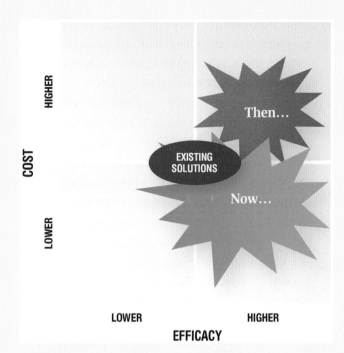

FIGURE V2

The medtech landscape – then and now.

- **Value exploration** – Early in the biodesign innovation process (see chapters 1.1 and 1.2), innovators should begin scanning for problems and opportunities that are ripe for value realization. This means actively seeking *need areas* where improved economic outcomes can potentially be generated. As they perform research, observations, and interviews, innovators have traditionally watched for what we call *practice-based value signposts*; for example, opportunities to address problems such as keeping patients out of the hospital, shortening the length of hospital stays, and reducing procedure time. But in the new environment, they should take a more explicit plunge into investigating *budget-based value signposts*, such as big line items on facility budgets, negative outliers in the cost-effectiveness of existing treatments, and extreme variations in treatment costs across geographies. These and other economic signals will guide the next generation of medtech innovators to promising areas to begin needs finding.

- **Value estimate** – Once promising *needs* have been identified, innovators dive deeper into understanding the potential to create and deliver value through the needs screening stage of the process (especially chapters 2.4 and 2.5). Quantifying value in this stage of the process can be tricky since no specific solutions have yet to be defined. However, innovators can still develop directional estimates of the value associated with their needs in order to ensure it is worth moving forward into concept generation. These estimates are based broadly on understanding who the real decision makers are with respect to adoption/purchasing decisions in each need area, how significant they perceive the need to be, to what degree available solutions are effectively addressing the need, and therefore how much margin there is to offer a new technology with a different improvement/cost equation. The insights gleaned from explicitly considering value at this early stage can save innovators from investing time, resources, and energy in developing solutions that ultimately will not offer a significant enough value proposition (see below) to drive decision makers to adopt them.

- **Value proposition** – As the *solution* to a promising need begins to take shape, innovators can begin thinking about value in more concrete, concept-specific terms. A value proposition describes the net impact of the cost/improvement equation associated with a new offering in terms that are meaningful to decision makers and sufficiently convincing to elicit a change in their behavior. Value propositions form the core of a company's sales and marketing activities and become a source of its competitive advantage and differentiation (see chapters 5.7 and 5.9). Importantly, value propositions must be backed by strong evidence that resonates with decision makers and the influencers that surround them. In the new healthcare environment, value propositions increasingly require the company to share the risk of ensuring that the promised improvements and desired outcomes are realized at the stated cost.

These mechanisms for anchoring the biodesign innovation process on value are broad and directional. We are still in the early stages of what is clearly a profound shift in the way medical technology innovation will address the economics of healthcare. But we hope that these initial ideas, as well as the discussion of value that permeates the text, will serve as a useful starting point for innovators as they embrace this new paradigm in device innovation.

As with any major economic and social transformation, there are tremendous opportunities for those who can position themselves to understand and take advantage of the changes. And the wonderful part about this particular technology sector is that the innovators who are able to make the transition may have the opportunity to benefit millions of patients around the globe.

The biodesign working group on value

Laurence Baker
Chief of Health Services Research, Stanford University

Aaron (Ronnie) Chatterji
Associate Professor, Duke University

Former Senior Economist at the White House Council of Economic Advisers

Victor Fuchs
Professor of Economics and Health Research and Policy (emeritus), Stanford University

John Hernandez
Vice President, Health Economics and Outcomes Research, Abbott Vascular

Doug Owens
Director of the Center for Health Policy, Stanford University

Jan Pietzsch
President and CEO, Wing Tech Inc.
Consulting Associate Professor, Stanford University

Bob Rebitzer
Consultant to the Clinical Excellence Research Center, Stanford University

Gordon Saul
Executive Director of Biodesign, Stanford University

Christopher Wasden
Managing Director, US Healthcare Strategy and Innovation Practice, PricewaterhouseCoopers

NOTES

1 "Snapshots: How Changes in Medical Technology Affect Health Care Costs," Henry J. Kaiser Family Foundation, March 2, 2007, http://kff.org/health-costs/issue-brief/snapshots-how-changes-in-medical-technology-affect/ (March 25, 2014).

2 Victor R. Fuchs, "New Priorities for Biomedical Innovation," *New England Journal of Medicine*, August 19, 2010, http://www.nejm.org/doi/full/10.1056/NEJMp0906597 (March 25, 2014).

3 Ibid.

4 "Why is Health Spending in the United States so High?," *Health at a Glance 2011: OECD Indicators*, Organization for Economic Cooperation and Development, http://www.oecd.org/unitedstates/49084355.pdf (March 25, 2014).

5 "Health Systems: Improving Performance," *The World Health Report*, 2000, http://www.who.int/whr/2000/en/whr00_en.pdf (March 25, 2014).

6 "U.S. Ranks Last Among Seven Countries on Health System Performance Based on Measures of Quality, Efficiency, Access, Equity, and Healthy Lives," The Commonwealth Fund, June 23, 2010, http://www.commonwealthfund.org/News/News-Releases/2010/Jun/US-Ranks-Last-Among-Seven-Countries.aspx (March 25, 2014).

7 For an example, see Sheila Smith, Joseph P. Newhouse, and
Mark Freeland, "Income, Insurance, and Technology: Why
Does Health Spending Outpace Economic Growth," *Health
Affairs*, September/October 2009, http://content.healthaffairs.
org/content/28/5/1276.full (April 29, 2014).

8 Fuchs, op. cit.

9 "Medical Device Growth in Emerging Markets: Lessons from
Other Industries," *In Vivo*, June 2012, file:///C:/Users/Lyn/
Downloads/
Medical_device_growth_in_emerging_markets_InVivo_1206%
20(3).pdf (March 25, 2014).

10 Ibid.

Global Perspectives

A world of opportunity ...

Although the United States and Europe remain global leaders in medical technology innovation, the story of the medtech sector has become much more diverse in recent years as healthcare has become a global priority. Inventors and companies in countries around the world are playing an increasingly important role in sourcing

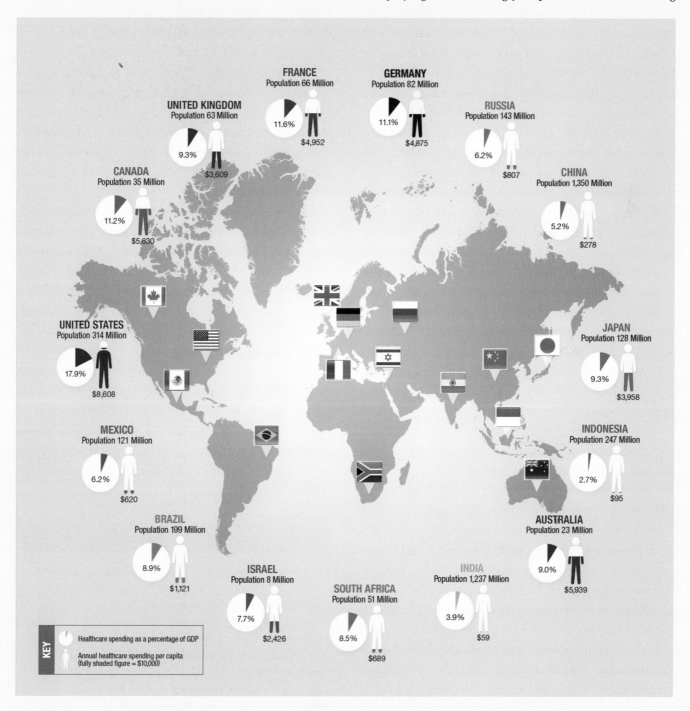

FIGURE G1

A snapshot of health and health-related spending in select countries around the world (compiled from The World Bank data, 2011).

ideas, designing and developing them into viable products and services, and introducing them into patient care. In parallel, device sales in developing countries are expanding at a rapid pace. As the US and Europe both sustain growth rates in the low single digits, medtech revenues in countries such as India and China are forecast to increase at a compound annual growth rate of 14 percent and 26 percent, respectively.[1]

The global transformation of the medtech sector has been driven by multiple, interrelated factors. In developed markets, health systems are actively seeking to slow health spending associated with medical technologies as they become more cost conscious and attuned to the value these products deliver. Moreover, as the time, expense, and complexity of developing new solutions in environments like the US continues to increase, innovators are moving offshore and creating new innovation hubs in locations around the world.[2]

In developing markets, disease profiles are shifting from infectious to chronic conditions, which makes diagnostic and device solutions a more important part of efforts to meet the healthcare needs of patients. Governments and private healthcare providers alike are increasing health-related spending (see Figure G1). And innovators and companies in low-resource settings are becoming leaders in inventing more affordable solutions that enable care delivery in any setting and reduce (rather than increase) its cost.[3]

Medtech innovators can certainly find compelling opportunities in both environments. They can also benefit from thinking more globally about how – and where – they source, develop, and sell their new solutions. While many innovators historically used a single market as their base, got established, and then expanded into new markets in a serial manner, they can now take a more global approach from the very beginning of the biodesign innovation process. Various regions in the world are moving into prominence in different parts of the medtech innovation process. To take just a few examples: Israel is home to over 700 medical device companies and leads the world in the medtech patents filed per capita.[4] It has become a hotbed of invention and incubation of medical technologies, with a robust start-up scene. Argentina, Brazil, and Chile have become leaders in conducting high-quality, yet affordable clinical trials for pharmaceutical and medical device companies

from around the world.[5] Ireland has developed into a prominent medtech manufacturing center, serving eight of the top 20 medtech multinationals[6] and attracting new enterprises of all sizes.

Of course, each region has its own unique challenges and opportunities. In the pages that follow, we have tried to give innovators a flavor for this range of issues and possibilities by profiling six important medtech markets. Europe and Japan represent geographies outside of the US with well-established device industries; India, China, Latin America, and Africa represent those in which the sector is still emerging. The purpose of these profiles is to provide a context for healthcare innovation in these locations, highlight some of the barriers that innovators may encounter in working there, and share tactics they can utilize to increase their chances of success. We're grateful to the experts who worked with us to develop this valuable content.

Additionally, innovators will find significantly more global content through the remainder of the *Biodesign* text. While the book is still grounded in what's required to identify, invent, and implement a new medical technology in the US, we expanded our treatment of other markets through the inclusion of more global guidance, as well as case studies that feature innovators and companies working across the globe.

Global expansion in the medtech sector can make it possible for patients traditionally underserved by medical devices to benefit from advanced technologies in new and different ways. With the global medtech market on its way to $440 billion by 2018,[7] a world of opportunity truly awaits medtech innovators and the patients they are committed to helping.

NOTES

1 "Global Market for Medical Devices, 4th Edition," *Kalorama Report*, 2013, http://www.kaloramainformation.com/Global-Medical-Devices-7546398/ (March 10, 2014).

2 "Medical Technology Innovation Scorecard: The Race for Global Leadership," PricewaterhouseCoopers, 2011, http://download. pwc.com/ie/pubs/2011_medical_technology_innovation_ scorecard_the_race_for_global_leadership_jan.pdf (March 10, 2014).

3 Ibid.

4 "How Did Israel Become a Hotbed for Medical Devices?," Fierce Medical Devices, August 14, 2013, http://www.fiercemedicaldevices.com/story/how-did-israel-become-hotbed-medical-devices/2013-08-14 (January 24, 2014).

5 "Climate Change in Latin America Makes for Successful Clinical Trials," MEDPACE, http://www.medpace.com/pdf/conductingtrialsinlatinamerica.pdf (March 10, 2014).

6 "Business in Ireland," IDA Ireland, http://www.idaireland.com/business-in-ireland/life-sciences-medical-tec/ (January 27, 2014).

7 "Medtech Market to Achieve Global Sales of $440B by 2018," Evaluate press release, October 12, 2012, http://www.evaluategroup.com/public/PressReleases/Medtech-Market-to-Achieve-Global-Sales-of-$440-Billion-by-2018.aspx (January 27, 2014).

Africa

Background

Africa is on the rise. The twenty-first century has been called the "African Century" due to the continent's potential for increased economic development in the coming decades.[1] From 2000–2012, economic growth averaged more than 5 percent per year,[2] driven by the recovery of commodity prices, government economic and policy reforms, and restoration of international donor confidence and aid.[3] Africa's collective gross domestic product (GDP) topped US$1.7 trillion in 2012 (making it nearly comparable to Russia or Brazil),[4] and its middle class expanded to more than 34 percent of the continent's 1 billion people.[5]

Poverty is declining, yet Africa still has the highest poverty rate in the world with 47.5 percent of the population living on less than US$1.25 a day.[6] The continent also accounts for 25 percent of the global disease burden.[7] Maternal health, child health, HIV, tuberculosis, and malaria continue to be the continent's greatest health challenges. What may be surprising is that over the next 10 years, Africa will experience the largest increase in deaths from cardiovascular disease, cancer, respiratory disease, and diabetes of any continent in the world.[8] For instance, the World Health Organization estimated that in 2008 the prevalence of hypertension was highest in its Africa region, with nearly half of the population affected,[9] and this figure is on the rise.

Generalities are difficult to apply across this diverse continent. It is a massive, highly fragmented mosaic of more than 50 countries, with an estimated 2,000 languages spoken and thousands of distinct ethnic groups. The continent's diverse population is expected to double by 2050, from 1 billion to more than 2 billion.[10] Africa is endowed with more than 30 million square miles of varied geography and could fit China, India, the United States, and most of Europe within its physical boundaries.[11] Across this great expanse, the continent's healthcare infrastructure is evolving. African governments are working to expand healthcare delivery systems through public and private investment,[12] but in the meantime, millions of people must travel vast distances to receive basic medical care. As access to care improves, it is estimated that Africa will still require at least 800,000 additional doctors and nurses to adequately meet the healthcare needs of its population.[13]

However, advances are under way with the potential to improve healthcare delivery. Low-cost broadband mobile phones and Internet connections are reaching new populations and accelerating Africa's economic development. Mobile phone penetration surpassed 80 percent in late 2013.[14] Approximately 16 percent of people on the continent are now online, and that number is rapidly growing.[15] In the health sector, access to these technologies is expected to enable greater use of remote diagnosis, treatment, and education – extending the reach of scarce physician and nursing resources. Applying technology to improve healthcare in Africa is estimated to improve productivity, reduce costs, and deliver financial gains to the economy of US$84–$188 billion by 2025.[16]

While Africa has great potential for economic growth, the medical device industry is in its earliest stages of development. Combined sales of medical device and equipment across African countries are just over US$3.2 billion,[17] with most medtech products imported from Asia, Europe, and North America. Medical products imports expanded at a compound annual rate of 7.5 percent from 2006–2010, with the fastest growth seen in western and northern Africa.[18] Two key factors have prevented a stronger growth rate in medtech sales to date. First, Africa currently has insufficient buying power for high-end technologies. Second, in some countries there is not a medical technology ecosystem in place that can support adoption through the consistent and effective sale, distribution, and service of complex medtech products as well as the training of healthcare providers in their use. South Africa, Nigeria, and several North and East African countries represent the largest opportunities for medtech companies, both for adoption and local manufacture of medtech products. It is anticipated that medium-sized African economies such as Kenya, Ethiopia, and Tanzania will become significant medtech growth drivers for the continent in the coming decades.

Total annual health expenditure in the continent was estimated at US$117 billion in 2012, with roughly half of this amount funded by African governments and the other half provided by private sources, including charitable/aid organizations and out-of-pocket payments.[19] Although this spending is dramatically uneven across countries (e.g., South Africa accounts for nearly 30 percent of the total), there is substantial room for the medtech industry to grow.[20] Currently, many medtech innovations are aimed at the "bottom-of-the-pyramid" population. This massive group is likely to see benefits from the growing community of non-governmental organizations (NGOs), governments, and entrepreneurs devoting resources to address their health needs. To be well equipped for the future, innovators also need to prepare for Africa's rising middle class, as this growing population will lead to a bigger consumer market. Africa is on the cusp of transformative change, enabled, in part, by innovations that improve access and quality at an affordable cost.

Challenges

Medtech innovators should be mindful that many African markets are smaller, riskier, and therefore less attractive for private companies (such as venture capitalists) to invest in, especially without special incentives. Additionally, since advanced medical technologies have been largely absent in many countries, innovators must demonstrate the long-term value of their technologies before adoption will be considered. Innovators and companies working in Africa should expect to devote significant time and resources to market-development activities, such as awareness raising, demand generation, comprehensive introduction strategies, and training of healthcare providers.

Another potential barrier is linked to the complex interplay of healthcare stakeholders and decision makers in Africa. Often, "the people who choose, the people who use, and the people who pay the dues" for medical technologies are distinct and not always aligned. This can result in products failing to achieve widespread or sustained adoption. For example, the Ministry of Health within an African nation might decide to mandate the use of auto-disable syringes within its public health centers, and a large multilateral organization or NGO may agree to fund the initiative. However, successful scale-up may depend on getting buy-in from in-country healthcare providers treating patients in public hospitals. Healthcare provider input is critical to the long-term sustainability and use of the technology. Misalignment among healthcare stakeholders can also lead to products being procured that are not appropriate for the local setting. For instance, healthcare providers and patients in Africa can benefit from high-throughput diagnostic technologies that can be deployed in central labs as well as rugged point-of-care tests that can be used in clinics and healthcare centers in remote, rural areas. The challenge is to make sure that the equipment funded, procured, and deployed is appropriate for the setting of use, underscoring the importance of decision makers being in tune with local needs and requirements.

Perhaps the greatest challenge of working in many parts of Africa is related to its limited physical infrastructure, which can hinder productivity and add considerable expense to medtech applications. Specifically, supply chain issues such as transportation and power are critical barriers to overcome. Where available, the power supply can be unreliable, prone to chronic outages, and expensive.[21] Similarly, despite being the main mode of transport for goods, roads are scarce; only one in three rural Africans has access to an all-season road.[22] Also, transportation costs in Africa can be costly; basic services can be twice as much as the world's average.[23] Air travel, essential to develop regional markets, is constrained by insufficient capacity. In fact, in many instances, flying between African countries may involve a connecting flight via the Middle East or Europe. The implications of these infrastructure challenges can be considerable, ranging from product stock-outs and bottlenecks to lifesaving interventions failing to reach patients in a timely manner.

The regulatory infrastructure for medical products is also nascent in Africa, with regulatory processes and requirements varying from country to country. Few African countries have national regulatory agencies for medical devices[24] although many have drug regulatory agencies. Some countries accept the regulatory approvals of Europe and the United States for devices. In most instances, international approvals do not replace African country policies; however, these approvals can often allow for faster approval of a product for in-country use. It is important to note that the European CE mark and US Food and Drug Administration approval are intended for products used in environments within Europe and the United States. The settings of use in Africa may be drastically different, and innovators should expect to conduct in-depth needs assessments as well as in-country clinical studies to ensure that products are appropriate for the place of use and acceptable for the people who will use them.

Tactics

Although Africa can be a challenging place to work, opportunities abound for those interested in applying appropriate, affordable medtech solutions to the continent's vast health problems. As they tackle product development and commercialization, innovators will benefit from the following guidelines:

1. **It's all about providing a complete solution**. Medical devices that are affordable, robust, easy to use, and low maintenance are needed in Africa. Innovators can potentially make a huge impact by introducing fundamental – yet disruptive – health technologies that are appropriately designed for Africa's remote and low-resource settings. Innovators must learn the unique regional needs within the African context and address the ones where essential clinical, infrastructure, economic, and ongoing support requirements can be met. (See 1.2 Needs Exploration, 2.5 Needs Selection, and 4.6 Final Concept Selection.)
2. **Partnerships are key**. Novel, creative collaborations can provide medtech innovators with new pathways to success and scale in Africa. Conditions are primed for governments, NGOs, and private businesses to work together. Increasingly, these public–private partnerships are being formed to tackle health and social issues in Africa. Each of the partners brings

something unique and critical to the solution. Historically, NGOs have provided knowledge of the communities and technical capacity with minimal cost to the private-sector partner. The involvement of private-sector participants enhances the financial viability of the innovation and can lead to more sustainable results, sometimes tapping into market forces to improve access and affordability. Governments offer broad decision-making authority and the ability to align stakeholders around the activities of the partnership. While public–private partnerships require diligence, cross-sector cooperation is critical to creating the momentum needed to advance medical technologies in Africa. (See 2.3 Stakeholder Analysis and 6.4 Alternate Pathways.)

3. **Funding models require innovative thinking, too**. Innovations, even affordable ones, require substantial investment. Admittedly, traditional medtech funding sources such as venture capital are rare in Africa. But the climate for stimulating investment in an African-based manufacturer to produce medical device products for Africa is improving. International agencies are increasingly shifting their support to in-country, on-the-ground ventures and transitioning from aid to investment.[25] And governments, global donors doing in-country work, and corporations seeking to enter or expand into Africa are playing interesting new roles. For instance, a recent medtech collaboration in Africa involved a large medtech multinational providing technology and product development support, an NGO assisting with product validation and business model development, a local government agency committing co-funding to the project, and a local entrepreneur leading the management of the venture. "Funding," in other words, can come in many different forms. (See 6.3 Funding Approaches.)

Good luck! Bahati nzuri! Sterkte! Nasiib wacan! ‏حظا سعيدا‏

Anurag Mairal
Program Leader, Technology Solutions, PATH

Rachel Seeley
Information and Communications Specialist, PATH

NOTES

1 "Thabo Mbeki's Victory Speech," *BBC News*, June 3, 1999, http://news.bbc.co.uk/2/hi/world/monitoring/360349.stm (March 20, 2014).

2 "Main Drivers of Africa's Economic Performance," *African Development Report 2012*, African Development Bank Group, http://www.afdb.org/fileadmin/uploads/afdb/Documents/Publications/African%20Development%20Report%202012%20-%20Main%20Drivers%20of%20Africa%E2%80%99s%20Economic%20Performance.pdf (March 20, 2014).

3 "Hospital Purchasing and Reimbursement for Medical Devices in Key Sub-Saharan African Markets," Frost & Sullivan, 2007, http://www.frost.com/prod/servlet/frost-home.pag (July 16, 2014).

4 "Annual Development Effectiveness Review," African Development Bank, 2012, http://www.afdb.org/fileadmin/uploads/afdb/Documents/Project-and-Operations/ADER%202012%20(En).pdf (March 20, 2014).

5 Mthuli Ncube, Charles Leyeka Lufumpa, and Steve Kayizzi-Mugerwa, "The Middle of the Pyramid: Dynamics of the Middle Class in Africa," African Development Bank, April 20, 2011, http://www.afdb.org/fileadmin/uploads/afdb/Documents/Publications/The%20Middle%20of%20the%20Pyramid_The%20Middle%20of%20the%20Pyramid.pdf (March 20, 2014).

6 "Africa Development Indicators," The World Bank, 2013, https://openknowledge.worldbank.org/bitstream/handle/10986/13504/9780821396162.pdf?sequence=1 (March 20, 2014).

7 "Hospital Purchasing and Reimbursement for Medical Devices in Key Sub-Saharan African Markets," op. cit.

8 Ama de-Graft Aikins, Nigel Unwin, Charles Agyemang, Pascale Allotey, Catherine Campbell, and Daniel Arhinful, "Tackling Africa's Chronic Disease Burden: From the Local to the Global," Globalization and Health, 2010, http://www.globalizationandhealth.com/content/6/1/5 (March 20, 2014).

9 "Raised Blood Pressure: Situation and Trends," World Health Organization, http://www.who.int/gho/ncd/risk_factors/blood_pressure_prevalence_text/en/index.html (March 20, 2014).

10 Claire Provost, "Nigeria Expected to Have Larger Population than U.S. by 2050," *The Guardian*, June 13, 2013, http://www.theguardian.com/global-development/2013/jun/13/nigeria-larger-population-us-2050 (January 29, 2014).

11 "The True Size of Africa," *The Economist*, 2010, http://www.economist.com/blogs/dailychart/2010/11/cartography (March 20, 2014).

12 "Hospital Purchasing and Reimbursement for Medical Devices in Key Sub-Saharan African Markets," op. cit.

13 Ibid.

14 "Naziha Bagui, Mobile Money: The Best Route to the African Consumer," Infomineo.com, January 14, 2014, http://blog.infomineo.com/2014/01/14/mobile-money-the-route-to-the-african-consumer/#more-530 (March 20, 2014).

15 "Lions Go Digital: The Internet's Transformative Potential in Africa," McKinsey & Company, November 2013, http://www.mckinsey.com/insights/high_tech_telecoms_internet/lions_go_digital_the_internets_transformative_potential_in_africa (March 20, 2014).

16 Ibid.

17 "African Medical Device Market: Facts and Figures 2012," ReporterLinker.com press release, December 13, 2012, http://www.prnewswire.com/news-releases/african-medical-device-market-facts-and-figures-2012-183352861.html (March 20, 2014).

18 Ibid.

19 Ibid.

20 Ibid.

21 "Fact Sheet: The World Bank and Energy in Africa," The World Bank, http://go.worldbank.org/8VI6E7MRU0 (March 20, 2014).

22 "Transforming Africa's Infrastructure," The World Bank, November 12, 2009, http://go.worldbank.org/NGTDDHDDB0 (March 20, 2014).

23 Ibid.

24 "National Regulatory Agencies for Medical Devices: Africa Region," World Health Organization, http://www.who.int/medical_devices/safety/NRA_Africa_Region.pdf (March, 2014).

25 Bekele Geleta, "Investing in Africa: A Sustainable Means to End Aid Dependency," Devex, October 19, 2012, https://www.devex.com/en/news/investing-in-africa-a-sustainable-means-to-end-aid/79494 (March 20, 2014).

China

Background

China is perhaps the most impressive economic development story in modern history. Sustaining annual growth rates upwards of 9 percent for more than two decades,[1] the country's gross domestic product (GDP) reached US$8 trillion in 2012 (second only to the United States at US$16 trillion).[2] This remarkable expansion has lifted hundreds of millions of Chinese out of poverty and created a new middle class that is larger than the entire US population.[3]

With more than 1.35 billion people, China has the largest citizenry in the world.[4] In 2011, the country's urban population surpassed its rural population for the first time, with close to 700 million people living in China's cities.[5] Population growth in China has decreased steadily over the last 20 years due to the controversial one-child policy (from approximately 1.2 percent to less than half of one percent)[6] and is expected to continue to decline. The country's median age is just 35 years, compared to nearly 40 years in more developed countries.[7] However, as a whole, the population is aging rapidly; senior citizens will account for as much as 35 percent of the Chinese people by 2053.[8]

One of the most important challenges facing China in the twenty-first century is how to allocate healthcare resources for its massive population. Despite progress in the country's economic transformation, China significantly lags the developed world in its ability to provide even basic health services to the vast majority of its people.[9] The Chinese government spent approximately 5 percent of GDP on healthcare in 2011, compared to roughly 18 percent spent in the US[10] and 9 percent on average in the OECD countries.[11] Per capita spending on medical technologies is just US$12[12] in China versus US$399 in the US.[13]

China's centrally planned economy provides health insurance coverage to approximately 90 percent of the population under three primary programs (an employer-based system, one for urban residents, and another covering the rural population).[14] These insurance schemes are largely inadequate to cover basic care but rather focus on protecting patients from catastrophic health events. As a result, the Chinese typically pay for basic health services out-of-pocket, causing many individuals to delay diagnosis and treatment until they are critically ill. For those who do seek care, access and quality are dramatically uneven between urban and rural settings, and highly dependent on one's ability to pay. Shortages of physicians, facilities, and other resources further complicate China's ability to provide adequate care.[15]

The government is working to reform China's healthcare system, with a goal of making basic care available across the country by 2020.[16] However, with several hundred million people entering the healthcare system over the past decade, the Chinese government, as a single payer, is critically concerned about managing healthcare costs. Generally, this translates into intense price competition (especially from indigenous manufacturers), lower reimbursement for medical devices, and a dramatic need for innovations that can facilitate adequate (not cutting-edge) care for large numbers of patients at affordable rates. For medical technologies, the Ministry of Health oversees the bidding and tendering system used in public hospitals to purchase new medical equipment. The tender process sets prices, which are subject to a ceiling in most parts of China. They also decide which medical device manufacturers can engage with hospital purchasing departments. Of course, a growing segment of the Chinese population has discretionary capacity to pay, which can potentially be tapped through direct-to-consumer products – especially imported health-related goods that are perceived to be of higher quality than local alternatives.

Medical devices sales took off in China during the last decade, growing roughly 20 percent per annum[17] and making China the world's 4th largest medtech market behind the US, Japan, and Germany. The industry is currently estimated to be worth roughly US$17 billion.[18] Imported medical devices, primarily advanced technologies such as imaging equipment and implants that are targeted at top-tier hospitals in urban settings, account for over 60 percent of the market.[19] Among the top 10 medical technology manufacturers in China, seven are foreign firms or joint ventures. Domestic players have tended to function on a regional basis, selling lower-tech devices in markets outside the major cities. However, notable exceptions are emerging in certain product categories. Coronary stents were first introduced on a large scale in China by some of the top five multinational medtech companies, and they rapidly captured 70–80 percent market share.[20] Within a few years, however, local companies launched mid-tier alternatives that they offered at prices 30–40 percent lower than their multinational competitors. As a result, they quickly took over the market. Today, local stent manufacturers dominate the market on a national scale.[21] Ultrasound machines provide another example of a product category where majority market share is rapidly moving from multinationals to local Chinese manufacturers.[22] On the whole, local companies are increasingly consolidating their operations, augmenting their product pipelines with new and acquired products, and making inroads into mid-tech device sectors.

Challenges

Compared to the growth prospects they contend with in other markets, many healthcare companies consider China to be a "bright spot" in the global healthcare landscape.[23] Indeed the country is rich with opportunities, but it is not a market to be entered without considerable thought and planning.

One of the most challenging aspects of working in China is navigating the formal rules set forth by regulators and other government agencies in parallel with the informal norms that are integral to making progress in the country. Informal requirements and the precedents set by other companies on their way to market are powerful forces with which innovators must contend. However, it can be difficult for innovators to know what customary behaviors are expected and how to balance them against more formal rules and guidelines. As one innovator described, very little related to doing business in China is "black and white," but is instead characterized by "shades of gray."

The most effective way to understand formal and informal systems within China, and how they interact, is by developing a strategic network of relationships within the country. For example, when seeking regulatory approval for a device in China, the official requirements of the China Food and Drug Administration (CFDA) stipulate that a company's first meeting with the agency take place *after* it makes its submission. However, informal pre-submissions meetings with CFDA officials, which help to clarify clinical requirements and streamline the review process, can be arranged for companies with the right relationships. Such connections are built through years of time and effort. Companies entering China for the first time almost always must hire consultants and other experts who can lend subject matter expertise as well as the right relationships to a project.

Product distribution is another challenging area, where companies are rarely successful without extensive relationship building. China's diverse and distributed population must be accessed province by province. And each area has unique government policies, adaptation of the centralized tender process, and other local requirements that make the notion of a national sales and distribution model completely impractical in China. Moreover, contracts are awarded, sales are made, and products are adopted based on the relationships that exist between distributors and the facilities and physicians they serve. As some multinational companies have learned the hard way, replicating these relationships is not only cost prohibitive, but virtually impossible. As a result, it's not uncommon for large medtech companies seeking the broad dissemination of its products to partner with as many as 2,000 regionally focused distributors – a daunting and costly necessity of doing business in the country.

When developing in-country relationships, companies should expect to find competing priorities and conflicting signals from the stakeholders in their networks. While this is common in any geography, the types of tensions that arise in China, at times, can be more sensitive and potentially difficult to reconcile. For example, in the country, there is a history of payments that flow directly from drug and device companies to physicians and hospitals.[24] In the wake of a high profile scandal involving a multinational drug company, the central government launched a new anti-corruption campaign that it hopes will contribute to changes in the behavior of multinational corporations (which are at risk of greater scrutiny relative to in-country and global anti-bribery laws), as well as the activities of local drug and device companies.[25] Efforts such as these may help eliminate some of the tensions that firms experience when doing business in China.

On a different note, the protection of intellectual property (IP) and trade secrets in China poses significant concern for medtech companies. According to the US Embassy in Beijing, China has one of the world's highest piracy rates, with counterfeit goods accounting for over 20 percent of products sold in the country.[26] Inadequate enforcement of international laws governing IP rights and a protectionist instinct by the government combine to hinder efforts to reduce IP infringement.[27] Foreign companies traditionally have not had much success seeking redress for infringement in Chinese courts. That said, enforcement and reparations are starting to improve as Chinese companies increasingly find themselves the victims of IP theft. Innovators should be wary of taking easily copied innovations into the Chinese market, unless they can erect other barriers to protect their assets (as described below).

Tactics

China is the proverbial 800-pound gorilla that cannot be ignored when considering global markets. The continuation of economic and demographic trends, health-related reform, improvements in infrastructure, and significant interest in innovation all provide real opportunities to medtech companies.[28] Successful execution can be challenging but rewarding.

Given the many unknowns associated with working in the Chinese market, innovators and companies with an interest in China are advised to remember that:

1. **Relationships are key**. As described, having an active and extensive network is essential to successfully conducting business in the country. Innovators interested in entering China will need

strong relationships with distributors, hospital administrators, CFDA officials, and municipal and central government officials in order to understand and mitigate important risks and reduce the time and cost of getting to market in China. If innovators do not already have useful connections that can help them build a network, they should make it a priority to partner with local experts and/or consulting firms that are experienced at navigating the medical device development and commercialization process in China. (See 2.3 Stakeholder Analysis and 5.7 Marketing and Stakeholder Strategy.)

2. **Proactively erect competitive barriers**. Start-up companies often focus on intellectual property protection as the main barrier to entry for their competition. In China, however, patents are not easily enforced. Successful companies create novel barriers to protect their competitive position. For instance, the CFDA allows companies seeking regulatory approval for novel, innovative products to generate their own product testing standard, which can become the established standard for other companies that may seek to develop "me-too" products. Innovators can use this opportunity to erect regulatory barriers to entry to slow fast followers. Distributor partnerships and successful tender bids can also function as barriers to entry. Innovators in China must think carefully about their potential positional and capability-based advantages and use them to protect against competition. (See 5.9 Competitive Advantage and Business Strategy.)

3. **Consider the pros and cons of being an outsider**. Innovators working in China sometimes perceive that they are at a disadvantage to domestic competitors. For example, local firms may enjoy an advantage in the Chinese tender process, and some observers believe they are favored in the Chinese courts when it comes to patent infringement litigation. Domestic products can also follow a separate pathway for regulatory and reimbursement approval in China that may be faster and less complex than the pathways available to foreign manufacturers. On the other hand, products from foreign multinational brands are often perceived as

being higher quality than domestic products, which allows them to command higher prices. Innovators should appreciate the advantages and disadvantages to being an outsider working in China and take these factors into account in constructing a business strategy. (See 5.9 Competitive Advantage and Business Strategy.)

Good luck! 好运

Christopher Shen

Executive Director, Singapore-Stanford Biodesign
Consulting Assistant Professor of Medicine, Stanford
School of Medicine

NOTES

1 "GDP Growth (Annual%)," The World Bank, 2012, http://data. worldbank.org/indicator/NY.GDP.MKTP.KD.ZG (March 14, 2014).

2 "China: Health Data," The World Bank, http://data.worldbank. org/indicator/NY.GDP.MKTP.CD (January 10, 2014).

3 Helen W. Wang, "The Biggest Story of Our Time: The Rise of China's Middle Class," *Forbes*, December 21, 2011, http://www. forbes.com/sites/helenwang/2011/12/21/the-biggest-story-of-our-time-the-rise-of-chinas-middle-class/ (January 15, 2014).

4 "China: Data," The World Bank, http://data.worldbank.org/ country/china (January 15, 2014).

5 "China Urban Population Exceeds Rural for First Time," *Bloomberg News*, Jan 17, 2012 (January 10, 2014).

6 "Population Growth Rate," The World Bank, 2012, hhttp://data. worldbank.org/indicator/SP.POP.GROW?page = 6 (March 14, 2014).

7 "Median Age of the Population in China, India, Europe, and USA from 1950–2100," China Profile Data, June 12, 2011, http:// www.china-profile.com/data/fig_WPP2010_Median-Age.htm (January 10, 2014).

8 "China's Aging Population to Double by 2053," *China Daily*, October 23, 2012, http://www.chinadaily.com.cn/china/2012-10/23/content_15837794.htm (January 10, 2014).

9 Bradley Blackburn, "'World News' Gets Answers on China: Health Care," *ABC News*, November 18, 2010, http://abcnews. go.com/International/China/health-care-china-trails-developed-countries-world-news/story?id = 12171915&singlePage = true (January 13, 2014).

10 "Health Expenditure Total (% of GDP)," World Bank, http:// data.worldbank.org/indicator/SH.XPD.TOTL.ZS (January 15, 2014).

11 OECD Health Data 2013 http://www.oecd.org/unitedstates/ Briefing-Note-USA-2013.pdf (January 15, 2014).

12 "Medical Device Market: China," PRWeb, November 25, 2013, http://www.prweb.com/releases/2013/11/prweb11367826. htm (January 15, 2014)

13 "Medical Device Market: USA," Espicom, February 19, 2014 (March 16, 2014).

14 Blackburn, op. cit.

15 Ibid.

16 "China's New Health Plan Targets Vulnerable," *Bulletin of the World Health Organization*, January 2010, http://www.who. int/bulletin/volumes/88/1/10-010110/en/ (January 14, 2014).

17 Jamie Hartford, "The Medical Device Market in China," Medical Device and Diagnostic Industry, June 18, 2013 http://www. mddionline.com/article/medical-device-market-china (January 15, 2014).

18 "Medical Device Market: China," Reportbuyer.com press release, November 25, 2013, http://www.prweb.com/releases/ 2013/11/prweb11367826.htm (January 15, 2014).

19 Hartford, op. cit.

20 Nicholas Donoghue et al., "Medical Device Growth in Emerging Markets: Lessons from Other Industries," *In Vivo*, June 2012, http://www.elsevierbi.com/publications/in-vivo/30/6/ medical-device-growth-in-emerging-markets-lessons-from-other-industries (February 12, 2014).

21 Ibid.

22 Ibid.

23 Franck Le Deu, Rajesh Parekh, Fangning Zhang, and Gaobo Zhou, "Health Care in China: Entering 'Uncharted Waters,'" McKinsey & Company, November 2012, http://www.mckinsey. com/insights/health_systems_and_services/ health_care_in_china_entering_uncharted_waters (January 14, 2014).

24 Andrew Jack and Patti Waldmeir, "GSK China Probe Flags Up Wider Concerns," *The Financial Times*, December 17, 2013, http://www.ft.com/intl/cms/s/0/ba26aa2c-6648-11e3-aa10-00144feabdc0.html#axzz2qPYpCP7c (January 14, 2014).

25 Ibid.

26 "Intellectual Property Rights in China," American International Education Foundation, http://www.aief-usa.org/ipr/ipr_facts/ index.htm (January 10, 2014).

27 "Intellectual Property Rights in China," loc. cit.

28 Le Deu, Parekh, Zhang, and Zhou, op. cit.

Europe

Background

Europe, in geographic terms, comprises 47 independent countries that jointly can be considered the largest economy on earth.[1] The European Union (EU), as an economically and politically integrated group of member states, includes 28 countries,[2] with 18 of these sharing the euro as their common currency.[3] The EU member states have a total gross domestic product (GDP) of more than US$16 trillion, with a per capita GDP of roughly US$34,000.[4] In terms of medical devices, the EU is often referred to as the "European market" because of its common device regulation under the CE mark. However, innovators should appreciate that the European market extends beyond the EU and includes such non-member states as Switzerland and Norway. Russia, which geographically belongs to both Europe and Asia, is also commonly considered part of the larger European medical device market, as most of its economy and population is located in the western portion of the country.

Europe has a population of nearly 740 million people, approximately 7 percent of the global population (with the current 28 EU member states accounting for 69 percent of the total).[5] Compared to other parts of the world, population growth in Europe is rather slow and the median age comparatively high. Nine of the top 10 countries with the highest median age, worldwide, are European countries, with only Japan having an older population.[6]

Spending on healthcare as a percentage of GDP ranges widely across European countries. France, Germany, the Netherlands, and Denmark commit more than 11 percent of GDP to health, while Romania and Cyprus spend less than 6 percent.[7] In 2010, health expenditures as a percentage of GDP dropped in the EU for the first time since 1975. From an annual average growth rate of 4.6 percent between 2000 and 2009, growth in health spending per capita fell to –0.6 percent in 2010[8] and has been stagnant in many countries ever since.[9] Among EU member states, those with higher average income levels per person generally spend more on health-related products and services.[10] When considering absolute amounts of healthcare spending per capita, the variation in healthcare spending is even more evident, ranging from about US$5,000 in France to merely US$500 in Romania.[11]

Medical technology plays an important role in Europe, both in terms of its use in clinical practice, as well as R&D and manufacturing. In fact, approximately 7.5 percent of total healthcare spending can be attributed to medical technologies.[12] In 2012, Europe accounted for approximately 30 percent of total global sales in medical technology.[13] In core countries of the European Union, including Germany, the United Kingdom, France, and Italy, utilization of innovative device technologies is often comparable to the United States. In fact, the EU often leads the US with earlier medtech market introductions that are a result of different regulatory systems in the two regions. Due largely to different clinical data requirements for CE marking, innovative devices are often commercialized in Europe first, receiving EU regulatory clearance years ahead of FDA (US Food and Drug Administration) approval, with resulting delays in US market introduction.

More than 60 percent of total EU medical technology sales come from its four largest countries (Germany 27 percent, France 16 percent, Italy 10 percent, and the UK 11 percent).[14] Germany, Ireland, Sweden, Finland, the Netherlands, and Belgium have a positive trade balance, exporting more medical technology than they import.[15] Ireland has evolved into a major medical device hub in Europe and hosts manufacturing sites for eight of the top 20 medtech multinationals.[16] Government tax incentives for large corporations, a technically trained workforce, and a budding start-up ecosystem are major contributors to Ireland's success in medtech manufacturing.

Germany

Germany is Europe's largest economy, with a GDP of roughly US$3.4 trillion.[17] It is the second most populous country in the region, behind Russia, with 82 million people.[18] Health insurance is compulsory for everyone living in Germany.[19] For those earning less than approximately US$68,000 per year, insurance is provided by the public statutory health insurance scheme (SHI), known in Germany as Gesetzliche Krankenversicherung (GKV). The rest of the population has the option of purchasing private health insurance plans, although a full 85 percent opts to remain with SHI.[20] In Germany, a strict separation exists between payers (insurances/sickness funds) and healthcare service providers, with many hospitals and all doctors' offices privately owned and operated. However, service fees are determined via bargaining processes between the major healthcare institutions.[21] Germany is also Europe's largest medical device market (at US$27 billion) and the third largest in the world behind the US and Japan.[22] Medical technology is a key industry in Germany, with substantial employment, the highest total sales among European countries, and a significant export rate that continues to grow at approximately 12 percent per year.[23]

France

France is Europe's second-largest economy, with a GDP of US$2.6 trillion.[24] Similar to a number of other countries in Europe, the government bears the majority of the healthcare expenditure, with private payments accounting for less than 24 percent in 2013.[25] Universal medical coverage (couverture maladie universelle, or CMU) was introduced in 2000, and all residents receive publicly financed healthcare.[26] Ninety-two percent of the population also has access to complementary or supplementary health insurance through employers or the government.[27] The French healthcare system has been lauded for providing high-quality care at less than half the per-capita health spending level as the United States.[28] The French medical device market, which is the second biggest in Europe (at US$15 billion), is the fifth-largest medtech market worldwide.[29] However, despite its attractive size, France is known to be challenging when it comes to device commercialization. Domestically, French medical technology companies excel in producing highly advanced devices such as implants.[30]

United Kingdom

The UK is Europe's third largest economy by GDP, at approximately US$2.4 trillion. It is home to the one of the largest healthcare systems in the world, the National Health Service (NHS), a universal coverage, single payer, integrated healthcare delivery system.[31] Through the NHS, residents of the UK automatically receive healthcare that is largely free at the point of use. The NHS, via its trusts, operates hospitals, doctor's offices, and other related health services delivery channels, and doctors, nurses, and other care providers are directly employed by the agency.[32] The medical device market in the UK is valued at about US$11 billion, making it the third largest in Europe.[33] The UK market for medical devices is projected to increase by 7.3 percent per annum to about US$14 billion by 2018.[34] The market is predominantly import-led, with only 25 percent of domestic demand met through in-country manufacturing.[35] Overall, spending cuts in healthcare and an increasing focus on value-based pricing put substantial pressure on manufacturers of devices and pharmaceuticals to drive down costs.

Challenges

While the EU's regulatory system for medical devices has long been touted as innovation-friendly, decisions related to reimbursement and payment for new devices can be more challenging. Coverage must be negotiated separately with payers in each country. Structured and centralized processes exist in a number of countries, including the UK, France, and Germany. In other countries, coverage is often decided at the regional level (e.g., in Italy) or is handled through less formalized negotiation processes between manufacturers and payers. This can be burdensome to medtech companies, especially start-ups, and also requires a clear strategic and tactical focus.

Medical device reimbursement is still considered favorable in a number of European countries, and some useful pathways exist for "innovation" or add-on payments for devices. For instance, Germany's NUB (Neue Untersuchungs-und Behandlungsmethoden) system provides a mechanism for hospitals to receive reimbursement for some newly introduced devices,[36] even though it can be challenging to obtain a positive decision. However, there is a growing trend across countries towards more cost-conscious decision making and higher barriers for reimbursement. This change is evidenced by the significant growth of health technology assessment

(HTA) programs across Europe that focus on balancing the two major health system objectives of outcome improvement and cost control.[37] The UK launched this movement by establishing the National Institute for Health and Care Excellence (NICE) in 1999. NICE has since implemented methodologically rigorous assessment processes that inform reimbursement decision making based on cost-effectiveness assessments. For medtech companies, this means an early focus is necessary to appreciate and collect the clinical and cost evidence that is required to win a favorable reimbursement decision. This is costly, and also leads to the exclusion of technologies that do not demonstrate sufficient "value" in terms of the specific healthcare system's willingness-to-pay.

Another challenge is the strain the recent financial crisis has put on the economies and financial budgets of many European countries since 2008. This has led, as noted, to a number of European countries reducing and/ or slowing expenditures on healthcare. These cuts directly impact available spending on medical technology and have already led to substantial pressures on medical device sales prices.

Further, historical, political, and socio-economic factors, as well as national laws that govern healthcare system design, are explicitly excluded from harmonization per the EU treaty. This exclusion has contributed to maintained differences between the structures of healthcare delivery systems. As a result, innovators must appreciate that healthcare delivery and medical practices vary among European countries, with implications to the use of medical technology. In addition to these structural differences, pronounced variations exist in cultures, attitudes, and languages across European countries, which add complexity to the implementation of a comprehensive market entry strategy.

Tactics

In addition to its substantive market size, Europe is especially attractive to innovators because many of its countries have highly advanced healthcare systems with experienced clinicians that tend to be open to innovation and commonly are early adopters of new medical technologies. This is further supported by Europe's regulatory system for medical devices, which often facilitates earlier market access of new technologies compared to the US, based on different regulatory requirements; the focus in Europe is on the demonstration of safety and performance, as opposed to safety and clinical effectiveness in the United States. Europe's rapidly aging population and its distinct clinical needs present another significant opportunity for medtech innovation. Finally, economic growth in a number of European countries that had limited healthcare resources in the past is creating new market opportunities. For instance, the Russian medical device market has gained increasing attention by multinational medtech companies in recent years because of its size and growth potential, and the willingness of a portion of the population to pay out-of-pocket for innovative new devices and procedures. Device sales in Russia, currently at US$6 billion, are estimated to experience a six-fold increase by 2020.[38]

In preparing to tackle the European market, innovators and companies are advised to devote considerable attention to the following issues:

1. **Appreciate country-specific differences**. While Europe is often seen and referred to as one market, it in fact is not. Each country has its own healthcare delivery system and payment systems, mostly governed by national laws. This leads to a variety of differences between individual countries' healthcare systems, ranging from differences in qualifications and responsibilities of healthcare staff, to variations in patient referral and flow patterns, clinical practice, and the medtech value chain. In addition, patient preferences and specific needs may differ based on cultural and historic distinctions among member states. Innovators should therefore be prepared to conduct a thorough stakeholder and clinical needs analysis, starting with the major individual medtech markets in Europe, including Germany, France, the UK, and Italy. Also, innovators should anticipate dealing with a diversity of languages and local regulations, which can be burdensome and requires careful planning. (See 1.2 Needs Exploration, 2.3 Stakeholder Analysis, 2.4 Market Analysis, and 2.5 Needs Selection.)

2. **Be prepared to demonstrate value**. As noted, Europe has been at the forefront of health technology assessment efforts for the last two decades. As a result, the focus on cost-effectiveness and true value contribution of new technologies is much stronger in Europe than it is in the United States. Innovators should expect these assessments and proactively seek to understand the technology assessment processes in their countries of interest and the types of clinical and economic evidence that is likely needed for their technologies. (See 5.3 Clinical Strategy, 5.6 Reimbursement Strategy, and 5.7 Marketing and Stakeholder Strategy.)

3. **Leverage European activities for global market entry**. The current regulatory system in Europe, as has been outlined, frequently facilitates earlier market entry than in the US. Innovators should weigh the benefits such early market entry could provide for the global commercialization of their technologies. Among these benefits are potential first revenues that can help a company's bottom line. But, more importantly, early market entry provides opportunities to gain commercial experience with new products that can help to further improve and streamline the product offering. In addition, similar to FDA approval, European CE marking is seen as a stamp of approval that can be highly useful when entering emerging markets. In fact, a number of countries, including India and some nations in Latin America, provide substantially lower regulatory hurdles for products that already have obtained the CE mark, or may even waive any further regulatory requirements. The value of early European activities for further US and foreign commercialization should therefore be considered in strategic decision making. (See chapters 4.2 Regulatory Basics, 5.2 R&D Strategy, 5.4 Regulatory Strategy, and 6.1 Operating Plan and Financial Model.)

Good luck! Bonne chance! Viel Glück! Buona fortuna! Buena suerte!

Jan B. Pietzsch
President and CEO, Wing Tech Inc.
Consulting Associate Professor, Stanford University

NOTES

1 "EU Position in World Trade," European Commission, http://ec.europa.eu/trade/policy/eu-position-in-world-trade/ (March 16, 2014).
2 "List of Countries," European Union, http://europa.eu/about-eu/countries/index_en.htm (January 31, 2014).
3 "The Euro," European Commission, Economic and Financial Affairs, http://ec.europa.eu/economy_finance/euro/ (January 31, 2014).
4 "European Union," *CIA World Factbook*, Central Intelligence Agency, https://www.cia.gov/library/publications/the-world-factbook/geos/ee.html (February 14, 2014).
5 "European Population Compared with World Population," EuroStat November 2012, http://epp.eurostat.ec.europa.eu/statistics_explained/index.php/European_population_compared_with_world_population (January 31, 2014).
6 Ibid.
7 "Healthcare Statistics," European Commission, Eurostat, September 2012, http://epp.eurostat.ec.europa.eu/statistics_explained/index.php/Healthcare_statistics (February 16, 2014).
8 "Health Spending in Europe Falls for the First Time in Decades," OECD Newsroom, November 6, 2012, http://www.oecd.org/newsroom/healthspendingineuropefallsforthefirsttimeindecades.htm (February 16, 2014).
9 "Health Spending Continues to Stagnate, Says OECD," OECD Newsroom, June 27, 2013, http://www.oecd.org/els/health-systems/health-spending-continues-to-stagnate-says-oecd.htm (February 16, 2014).
10 "Healthcare Statistics," op. cit.
11 "Health Expenditure per Capita," The World Bank, 2013, http://data.worldbank.org/indicator/SH.XPD.PUBL (February 13, 2014).
12 "The European Medical Technology Industry in Figures," MedTech Europe, 2013, http://www.eucomed.org/uploads/Modules/Publications/the_emti_in_fig_broch_12_pages_v09_pbp.pdf, (January 31, 2014).
13 "Medtech Industry in Europe," Eucomed, http://www.eucomed.org/uploads/Modules/Publications/medtech_graphic_a2_130912_landscape.pdf (March 17, 2014).
14 "The European Medical Technology Industry in Figures," op. cit.
15 Ibid.
16 "Business in Ireland," IDA Ireland, http://www.idaireland.com/business-in-ireland/life-sciences-medical-tec/ (January 27, 2014).

17 "Germany," The World Bank, http://data.worldbank.org/
country/germany (February 13, 2014).

18 Ibid.

19 David Green, Benedict Irvine, Emily Clarke, and Elliot Bidgood,
"Healthcare Systems: Germany," Civitas, 2013, http://www.
civitas.org.uk/nhs/download/germany.pdf (February 13, 2014).

20 Ibid.

21 Ibid.

22 "What is Germany's Secret? How the World Can Learn from a
Thriving Medtech Industry," MMDI Online, May 30, 2012,
http://www.mddionline.com/article/what-germany%E2%
80%99s-secret-how-world-can-learn-thriving-medtech-industry
(February 13, 2014).

23 "Industry Report Medtech 2013," BVMed, March 2013, http://
www.bvmed.de/themen/medizinprodukteindustrie-1/CE-
Kennzeichnung/article/2013-03-branchendarstellung-medtech-
2013.html (February 13, 2014).

24 "France," The World Bank, http://data.worldbank.org/
country/france (February 16, 2014).

25 "France: Health Expenditure," The World Bank, 2013, http://
data.worldbank.org/indicator/SH.XPD.PUBL (February
13, 2014).

26 "International Profiles of Healthcare Systems 2013,"
Commonwealth Fund, 2013, http://www.commonwealthfund.
org/~/media/Files/Publications/Fund%20Report/2013/Nov/
1717_Thomson_intl_profiles_hlt_care_sys_2013_v2.pdf
(February 13, 2014).

27 Ibid.

28 "Healthcare Lessons from France," National Public Radio, 2008,
http://www.npr.org/templates/story/story.php?
storyId=92419273 (February 13, 2014).

29 "Medical Device Market: France," Espicom, 2014, http://www.
espicom.com/france-medical-device-market (February
13, 2014).

30 "France: Medical Device Industry," Emergo Group, http://
www.emergogroup.com/resources/market-france (February
21, 2014).

31 "The U.K. Healthcare System," The Commonwealth Fund,
2013, http://www.commonwealthfund.org/Topics/
International-Health-Policy/Countries/United-Kingdom.aspx
(February 13, 2014).

32 Ibid.

33 "The Global Market for Medical Devices, 4th Edition,"
Kalorama Information, May 2013, http://www.
kaloramainformation.com/Global-Medical-Devices-7546398/
(March 10, 2014).

34 "Medical Device Market: United Kingdom," Espicom, 2014,
http://www.espicom.com/uk-medical-device-market.html
(February 13, 2014).

35 Ibid.

36 "ISPOR Global Healthcare Systems Roadmap," International
Society for Pharmacoeconomics and Outcomes Research, April
2011, http://www.ispor.org/htaroadmaps/germanymd.asp#4
(January 31, 2014).

37 "Health Technology Assessment and Health Policy Making in
Europe," European Observatory on Health Systems and Policies
Report, Studies Series No. 14, http://www.euro.who.int/
__data/assets/pdf_file/0003/90426/E91922.pdf (January
31, 2014).

38 "Executive Guide to Doing Business in Russia," Emergo Group,
January 2013, http://www.emergogroup.com/resources/
market-russia (February 21, 2014).

India

Background

South Asia is generally considered to include Afghanistan, Bangladesh, Bhutan, India, Maldives, Nepal, Pakistan, and Sri Lanka. Over the past 20 years, the region has experienced robust economic growth, averaging 6 percent per year.[1] As a result, poverty rates have declined, with the percentage of South Asians living on less than US$1.25 per day decreasing from 61 percent to 36 percent between 1981 and 2008. While the region is still home to approximately 44 percent of the developing world's poor, growth and development in South Asia are expected to continue.[2]

The largest and most influential country in the region is India. With approximately 1.3 billion people, India is the fourth largest global economy by purchasing power parity (PPP).[3] India's gross domestic product (GDP) reached nearly US$2 trillion in 2012,[4] and it is expected to continue increasing at a healthy rate as the country further integrates into the global economy. Growth will also be driven by increased domestic demand as India's burgeoning middle class expands from roughly 50 million in 2007 to almost 600 million people between by 2025.[5]

India's healthcare system is plagued by low spending levels. Healthcare expenditure per capita was only US$59 in 2011.[6] The country's private and public sector combined spent only about 4 percent of GDP on healthcare in 2011,[7] although the government is planning to increase its share from 1.4 percent to 2.5 percent of GDP over the next five years.[8] In the past half-century, India's public sector has steadily given up market share to the private sector in providing healthcare.[9] Accordingly to one study, the private sector accounted for over 90 percent of all hospitals, 85 percent of doctors, 80 percent of outpatient care, and almost 60 percent of inpatient care.[10]

Fortunately, India's private sector has been responsible for some remarkable innovations in healthcare delivery. Several major hospital systems in the country are able to deliver high-quality outcomes at a fraction of the cost of care in developed country settings. For instance, one cardiac care center offers open-heart surgery for less than US$2,000 per patient, with outcomes similar to those at US-based centers where the price tag can exceed US$100,000.[11] Similar examples exist for ophthalmology, oncology, nephrology, and OB-GYN specialty hospitals in India.[12] Some of the successful strategies employed by these healthcare centers include generating high volumes of patients, aggressively trimming procedure costs, and shifting tasks to lower-skilled care providers.[13]

Health insurance coverage is still relatively uncommon in India, but its availability is improving. Estimates vary, but as much as 25 percent of the population now has some form of health insurance,[14] although a much smaller percentage has full or substantial coverage. Both government and private insurers are working to increase access to insurance. Analysts estimate that almost half the population will enjoy some level of health insurance coverage by 2020.[15] The National Rural Health Mission (NRHM), which the Indian government rolled out in 2005, will account for some of this increase. NRHM is an ambitious and wide-ranging public health program that seeks to improve healthcare delivery in rural India.[16] The Rashtriya Swasthya Bima Yojna (RSBY, translated as National Health Insurance Program) also strives to increase health insurance access for families below the poverty line.[17]

The recent increase in individual purchasing power is important given the relative lack of health insurance coverage in India. Patients make approximately 70 percent of total healthcare payments in the country.[18] Accordingly, they tend to be highly sensitive to both the cost and value of the medical interventions they receive. Many Indians are willing to commit their family savings to high-impact, life-saving medical interventions such as the implantation of pacemakers or stents, but may only be willing to spend minimally to address health issues and chronic conditions that they perceive to be "optional" or non-life-threatening.[19]

India's medical device market is conservatively worth more than US$3 billion.[20] It is forecast to continue expanding at a compounded annual growth rate of over 15 percent through 2016,[21] far better than the 2 to 3 percent growth anticipated for the sector in the United States and Europe. As a result, many global medical technology companies view India as one of the most promising emerging markets for direct investment.[22] Several of the largest multinational companies in medical technology have invested in large product development centers in India to develop solutions suited for the local market. These product development centers are also creating examples of reverse innovation. For instance, some of these locally developed products, such as inexpensive blood glucose meters, have been launched in developed markets with great success.

Another factor affecting the demand for medical technologies in India is the growing prevalence of chronic diseases, linked to increased longevity, greater urbanization, and shifting lifestyle choices within the population. Communicable diseases such as malaria and tuberculosis and tropical diseases such as Japanese encephalitis and dengue fever traditionally represented a large proportion of India's disease profile. However, coronary heart disease, diabetes, asthma, and other chronic non-communicable diseases are significantly increasing in prevalence. For example, analysts predicted that Indians would account for some 60 percent of the world's heart patients by 2010.[23] While this trend poses challenges for the country's healthcare system, it also presents significant opportunities for medical device companies with products that treat these conditions. However, Indian patients will have to be convinced of the value of paying for treatments to address such chronic conditions.

Challenges

Despite the promise India offers as an emerging medical technology market, there are still relatively few examples of innovative medical technologies that have been adopted on a large scale across the country. Imports from medical technology companies dominate the medical technology sector, accounting for approximately 80 percent of the value of all devices sold within India.[24] Some companies have created products that are simplified versions of products sold in Western markets that can be produced less expensively and offered to Indian customers at more affordable prices.[25]

To date, many multinational companies have focused largely on making capital equipment, such as imaging equipment and incubators, available within India. In contrast, local medical technology companies have traditionally concentrated on low-cost offerings such as medical supplies and consumables (sutures, catheters) that allow them to take advantage of inexpensive labor and manufacturing costs but do not require extensive research and development.[26]

State-of-the-art Indian secondary and tertiary care institutions, which attract both domestic patients who can afford their world-class services as well as hundreds of thousands of medical tourists each year,[27] are benefitting from the innovative medical technologies that are imported into the country. However, most of the Indian population is served by healthcare facilities without adequate resources, staff, or capacity to access these products. And, unfortunately, the vast majority of imported medical technology products may not appropriately address their needs. Some are too expensive to be made widely available. Others do not function dependably in areas with unreliable power or other infrastructure challenges. Still others may be too technically complex or resource-intensive to operate or maintain by healthcare workers who are under-trained relative to staff in top-tier facilities.

Although medical technologies are still largely under-utilized across the country, India has developed a large, well-established clinical trial industry led by the pharmaceutical industry. Clinical testing in the country can be as little as one-twentieth the cost of conducting trials elsewhere.[28] However, at the time of this writing, new restrictions put in place by the Indian Supreme Court in 2013 have stalled most clinical trial activity in the country.[29] These restrictions enforce stricter monitoring and what critics perceive to be unreasonable requirements for compensating patients for research injuries or death.[30] The long-term effects of these changes remain to be seen, but some observers of the medical technology and pharmaceutical industries anticipate that they will significantly reduce the number of trials conducted in India and, in turn, the availability of new treatments.[31]

Distribution is another variable in scaling the adoption of innovative medical technologies beyond India's premium healthcare settings in large urban centers. Distribution networks for medical products are fragmented by region, medical specialty, and product category. The distributors serve an important role in that they often have deep relationships with healthcare providers, especially in areas of the country where sales representatives of the medical device companies do not have relationships. Further, distributors may extend credit with favorable terms to smaller hospitals or physicians operating community clinics and surgery centers.

Equally as important as figuring out how to physically sell and distribute a product in India, is devising a way to do so on a sustainable basis. Given the extreme requirement for more affordable medical technologies in the country, innovators and companies often struggle if they rely on traditional business models for generating revenue. Accordingly, business model innovation – or coming up with new and different ways to engage with stakeholders in the healthcare value chain, align their incentives, and realize a financial return – is becoming paramount to success in India's medical technology sector. The problem is that business model innovation is difficult and can add considerably to the resource requirements for a medical technology innovation project.

Tactics

The current market characteristics and barriers combine to make the Indian market a distinctive opportunity for medical technology innovation. Not only is the country's large and growing population in need of more inventive, appropriate solutions to common medical problems, but products and services that work in India may also be relevant in other markets such as Eastern Europe, the Middle East, and Africa.

Innovators and companies choosing to target this market will face challenging conditions, to be sure. There are a few key issues related to the biodesign innovation process that deserve special emphasis:

1. **Search for needs in country**. Don't try to import needs (or their solutions) into the market. On-the-

ground clinical immersion is essential to understanding India's heterogeneous nature and what is truly required to more fully address the needs of segments of its diverse population. Investigate problems and opportunities across geographic regions (north, south, east, and west) – needs can be considerably different depending on the area. The same applies for urban versus rural settings, public versus private centers, and different socioeconomic classes. (See 1.2 Needs Exploration, 2.4 Market Analysis, and 2.5 Needs Selection.)

2. **Go deep on stakeholder analysis**. India's stakeholder landscape is significantly different than what innovators traditionally encounter in more developed markets. For instance, a low-skilled health worker or a family caregiver in India may perform procedures usually performed by a physician or skilled nurse in a developed market. Carefully understand the interests of all those involved in the cycle of care, flow of money, and medical technology ecosystem to identify advocates and anticipate resistance. Pay attention to the many different levels of care providers, the multi-faceted medical technology value chain, and to the extensive role of patients and their families in making care decisions. (See 2.3 Stakeholder Analysis.)

3. **Keep innovating beyond the technology**. With few rules or precedents to follow, innovators have no choice but to become creative; not just with the products they design and develop, but with the business strategies they craft to support their commercialization. Use your deep understanding of the need and the relevant stakeholders to enable successful business model innovation, unique partnerships, and other non-traditional approaches with the potential to overcome common barriers. And don't be afraid to experiment. In India, creativity is a necessity that has spawned many advances, such as financing schemes for more expensive interventions and mobile clinics and transport solutions to increase access to essential medical services in remote rural areas, to name a few. (See 4.4 Business Models and 5.8 Sales and Distribution Strategy.)

Good luck! गुड लक
Rajiv Doshi
Executive Director (US), Stanford-India Biodesign
Consulting Associate Professor, Stanford University

NOTES

1 "South Asia Overview," The World Bank, http://www. worldbank.org/en/region/sar/overview (January 3, 2014).

2 Ibid.

3 "Country Comparison: GDP (Purchasing Power Parity)," *The World Factbook*, Central Intelligence Agency, 2012, https://www.cia.gov/library/publications/the-world-factbook/rankorder/2001rank.html?countryname = India&countrycode = in&regionCode = sas&rank = 4#in (January 3, 2014).

4 "GDP (Current US$)," The World Bank, 2012, http://data.worldbank.org/indicator/NY.GDP.MKTP.CD (January 3, 2014).

5 "Taking Advantage of the Medtech Market Potential in India," PricewaterhouseCoopers, 2012, pg. 4 http://www.pwc.com/mx/es/industrias/archivo/2012-09-taking-advantage-india.pdf (January 3, 2014).

6 "Health Expenditure Per Capita" The World Bank, 2012, http://data.worldbank.org/indicator/SH.XPD.PCAP (March 14, 2014).

7 "Health Expenditure, Total (% of GDP)," The World Bank, 2012, http://data.worldbank.org/indicator/SH.XPD.TOTL.ZS (February 14, 2014).

8 "Healthcare Spending to Rise to 2.5 Percent," *Indian Express*, March 1, 2012, http://www.indianexpress.com/news/healthcare-spend-to-rise-to-2.5-of-gdp/918380 (March 7, 2014).

9 Ramya Kannan, "More People Opting for Private Healthcare," *The Hindu*, August 1, 2013, http://www.thehindu.com/sci-tech/health/policy-and-issues/more-people-opting-for-private-healthcare/article4967288.ece (March 7, 2014).

10 "Private Sector in Healthcare Delivery in India," National Commission on Macroeconomics and Health, 2005, pg. 5 http://www.who.int/macrohealth/action/Report%20of%20the%20National%20Commission.pdf (March 7, 2014).

11 Ketaki Ghokhale, "Heart Surgery in India for $1,583 Costs $106,385 in U.S.," Bloomberg News, July 28, 2013, http://www.bloomberg.com/news/2013-07-28/heart-surgery-in-india-for-1-583-costs-106-385-in-u-s-.html (February 12, 2014).

12 Vijay Govindrajan and Ravi Ramamurti, "Delivering World-Class Healthcare Affordably," *Harvard Business Review*, November 2013, http://www.aravind.org/aravindcontentmanagement/file/MF00000053.pdf (February 12, 2014).

13 Ibid.

14 "Government Sponsored Health Insurance in India: Are You Covered?," The World Bank, October 11, 2012, http://www. worldbank.org/en/news/2012/10/11/government-sponsored-health-insurance-in-india-are-you-covered (February 22, 2013).

15 "Indian Pharma 2020: Propelling Access and Acceptance, Realizing Potential," McKinsey and Company, 2010, p. 18 http://www.mckinsey.com/ ~ /media/mckinsey/dotcom/client_service/Pharma%20and%20Medical%20Products/PMP%20NEW/PDFs/778886_India_Pharma_2020_Propelling_Access_and_Acceptance_Realising_True_Potential.ashx (March 7, 2014).

16 Deoki Nandan, "National Rural Health Mission: Turning into Reality," *Indian Journal of Community Medicine*, vol. 35, no. 4, 2010, pp. 453–4. http://www.ncbi.nlm.nih.gov/pmc/articles/PMC3026119/ (March 7, 2014).

17 Nagesh Prabhu, "Rashtriya Swasthya Bima Yojana for BPL families too," *The Hindu*, July 8, 2013, http://www.thehindu.com/news/national/karnataka/rashtriya-swasthya-bima-yojana-for-bpl-families-too/article4892098.ece (March 7, 2014).

18 "Health Financing: Private Expenditures on Health as a Percentage of Total Expenditures on Health," World Health Organization, 2011, http://gamapserver.who.int/gho/interactive_charts/health_financing/atlas.html?indicator = i2&date = 2011 (October 2, 2013).

19 "Taking Advantage of the Medtech Market Potential in India," op. cit.

20 Ibid.

21 "Medical Devices Market in India May Grow to $5.8 Billion by 2014: Report," *BioSpectrum*, October 3, 2013, http://www.biospectrumasia.com/biospectrum/news/197400/medical-devices-market-india-grow-usd58-billion-2014-report#.UxoIqPldV8E (March 7, 2014).

22 "Taking Advantage of the Medtech Market Potential in India," op. cit.

23 David Kohn, "Getting to the Heart of the Matter in India," *The Lancet*, August 16, 2008, http://www.thelancet.com/journals/lancet/article/PIIS0140-6736(08)61217-9/fulltext (January 3, 2014).

24 "India Medical Device Consulting," Pacific Bridge Medical, http://www.pacificbridgemedical.com/business-services/medical-device-consulting/india/ (January 3, 2014).

25 "Taking Advantage of the Medtech Market Potential in India," op. cit.

26 "The Medical Device Market: India," op. cit.

27 "Stricter Rules Driving Away Medical Tourism from India," *The Economic Times*, August 15, 2013, http://articles.economictimes.indiatimes.com/2013-08-15/news/41413559_1_apollo-hospitals-prathap-c-reddy-overseas-patients (October 3, 2013).

28 Kenan Machado, "New Restrictions Stall Drug Trials in India," *Wall Street Journal India*, January 28, 2014, http://blogs.wsj.

com/indiarealtime/2014/01/28/new-restrictions-stall-drug-trials-in-india/ (February 14, 2014).

29 Dinsa Sachan, "Supreme Court Ruling Brings Clinical Trials to a Halt in India," *Chemistry World*, October 15, 2013, http://www.rsc.org/chemistryworld/2013/10/supreme-court-ruling-clinical-trials-halt-india (February 14, 2014).

30 Jeremy Sugarman, Harvey M. Meyerhoff, Anant Bhan, Robert Bollinger, Amita Gupta, "India's New Policy to Protect Research Participants," *British Medical Journal*, July 2013, http://www.bmj.com/content/347/bmj.f4841 (January 2, 2014).

31 S. Seethalakshmi, "Foreign Companies Stop Clinical Trials in India After Government Amends Rules on Compensation," *The Times of India*, August 1, 2013, http://articles.timesofindia.indiatimes.com/2013-08-01/bangalore/40960487_1_clinical-trials-iscr-suneela-thatte (February 14, 2014).

Japan

Background

At nearly US$6 trillion, Japan has the third largest gross domestic product (GDP) in the world, after the US and China.[1] Real GDP grew at about 1.9 percent in 2012, and it is projected to expand at roughly one percent per year through 2020.[2] The Japanese economy was badly hit by the global recession in 2008–2009 and the massive tsunami in 2011, and it continues to suffer from persistent deflation. Japan's economy has been led by the advanced manufacturing sector, which generates strong export activity. However, the country is looking to bolster domestic demand in order to drive increased growth.[3]

The population of Japan, at more than 128 million people, is decreasing about 0.2 percent each year.[4] Trends indicate that the country's total inhabitants will decline by almost 30 percent by 2060 due to a low birthrate, limited immigration, and an aging population.[5] Approximately 23 percent of the Japanese people were over 65 years of age in 2012; by 2060, more than 40 percent of the population will be senior citizens.[6] As the population ages, cancer, heart disease, and pneumonia have become the country's leading causes of death.[7]

The Japanese universal healthcare system, known as *kaihoken*, has been lauded for increasing the quality of life of the Japanese people and is cited as a key reason the Japanese have the longest life expectancy in the world.[8] However, increasing healthcare costs, in combination with the country's rapidly aging population and slow-growth economy, are creating the need for reforms and cost cutting.[9] Japanese patients tend to visit physicians more frequently than their counterparts in the US (13.2 versus 3.9 appointments per person per year). And their hospitals stays are significantly longer (18.8 versus 5.5 days). Japan also has three times as many acute care hospital beds per 1,000 people (8.1 versus 2.7).[10] Moreover, Japanese expenditures on medical devices are the highest in Asia at US$165 per capita, compared to just US$10 per capita in China.[11] To date, Japan has maintained one of the most technologically advanced healthcare systems in the world. For instance, more than 70 percent of the country's hospitals have CT scanners and Japan has the most MRI machines among the Organization for Economic Cooperation and Development (OECD) countries.[12] Still, the country commits 9.3 percent of its GDP to healthcare compared to 17.9 percent in the US.[13]

Japan finances the delivery of healthcare through a universal health insurance system that has three primary parts: (1) Employees' Health Insurance for employed individuals and their families; (2) National Health Insurance for the self-employed and poor; and (3) Late-Stage Medical Care System for individuals over 75 years of age. The result is that nearly all residents are covered by insurance. People insured under the first two categories are responsible for copayments equal to 30 percent of their care, up to a maximum limit. The elderly pay 10 percent, up to a maximum limit (unless their income is equivalent to an active worker). Expenses over the defined limits are paid for by the government. In total, roughly 82 percent of health spending is funded by public sources.[14] The government uses regulation of the country's hospitals, which are mostly private, to ensure that access and the quality of care remains universal and egalitarian.[15] And although the government carefully regulates healthcare financing and the country's insurance system, patients enjoy great freedom of choice in which doctors they see, and physicians and other medical professional are generally in control of the delivery of care.

Japan is the second largest medical technology market in the world, behind only the US. It is the third largest importer of medical equipment (after the US and Germany) and the eighth largest medical device exporter in the world.[16] At US$31 billion,[17] Japan accounts for about 10 percent[18] of global medtech sales. Some multinational medical device companies have a presence in Japan, and many Japanese companies are aggressive players in the global medtech market themselves. Many of these companies, such Toshiba, Hitachi, and Fuji Film have entered the medical field from the high-technology and electronics sectors. Japanese firms tend to be stronger in diagnostic devices, particularly imaging, while most innovative therapeutics are imported.

Challenges

Although Japan is a large, stable market that is receptive to advanced technology, it is frequently considered to be the most difficult Asian market to enter.[19] Many innovators are challenged by the language barrier since not all forms and guidelines are readily available in other languages. They also may find certain processes to be complex and laden with "hidden" costs. For instance, to help them navigate the path to market, foreign companies must engage with a Marketing Authorization Holder (MAH). In addition to assisting with the regulatory process, the MAH helps facilitate distribution, which typically involves a primary distributor that works through a network of secondary distributors. Each of these external parties must be managed, and also requires a commission on product sales. Similarly, when conducting clinical trials in Japan, physicians tend to be less directly involved in data collection than their counterparts in the US and Europe, and experienced clinical research coordinators are scarce. Accordingly, companies often must depend on contract research organizations (CROs) to play a much more active, hands-on role in their Japanese trials, incurring significantly higher costs in the process. These costs are offset to some extent by the relatively high reimbursement rates traditionally authorized by Japan's Ministry of Health, Labour, and Welfare. However, this entity is working to bring reimbursement in Japan into closer alignment with reimbursement levels in the US. It has also has begun reevaluating payment levels for medical devices every two years as part of its cost reduction efforts. These activities will diminish this advantage to companies over time.

Device regulation through the country's Pharmaceuticals and Medical Devices Agency (PMDA) provides another example of where companies often struggle. This agency traditionally has been known for its rigid standards, lengthy approval processes, and requirements for extensive documentation. The PMDA recently enacted reforms focused on encouraging medtech innovation and decreasing the time to approval for novel devices. The number of reviewers at the PMDA has been tripled since 2009, aspects of the approval process have been clarified, and more frequent consultation on regulatory submissions is now allowed. However, despite these changes, PMDA remains understaffed. In addition, review times for priority devices are only just catching up to the US FDA, while non-priority devices can take up to 2 years longer.[20] Such barriers have led to a well-documented "device lag" that delays product launches in the country and prevents Japanese patients from benefiting from the world's most advanced diagnostics and devices in a timely manner.

An additional factor that can make Japan a difficult environment for medtech innovation is that the country does not have a strong history of entrepreneurship in the healthcare sector. Japan's leading examples of entrepreneurship exist almost exclusively in high-technology. Culturally, Japanese society is still not widely accepting of mistakes and failure, which discourages the risk-taking behavior that is required to create start-up companies. The country is admired for its commitment to research and development, devoting a higher percentage of GDP to this activity than all other countries except Israel.[21] Yet, relatively few Japanese innovations are transformed into viable businesses.[22] Of those individuals and teams that do decide to launch new companies, most have a tendency to "play it safe" by pursuing incremental improvements and "me too" technologies. Regardless, they often have difficulty recruiting engineering talent, as few people are willing to leave large, stable organizations to join a start-up. As a result, in-country medtech innovators have trouble finding experienced mentors to help guide them. Innovators and companies seeking to enter Japan with innovative new medical products also feel the effects of this problem. With few start-ups founded within the country and many larger organizations able to leverage partnerships and resources developed in the high-tech and electronic sectors, the medtech ecosystem in Japan is still in its infancy.

Tactics

A survey conducted by the American Medical Devices and Diagnostics Manufacturers' Association (AMDD) revealed that 85 percent of Japanese people desire faster access to the world's most advanced medical technologies, and 66 percent of respondents indicated that they favor these technologies even if they cost slightly more.[23] The government is also supportive, having recently established an Office for Health Care and Medical Strategy that is focused on helping drive the development and commercialization of more medical technologies from the country's investment in R&D. The prime minister has further designated the medtech industry as one of three strategic sectors that will revitalize the country's economy. Against this backdrop, Japan is ripe for medtech innovation from outside and within.

In preparing to tackle the Japanese market, innovators and should pay specific attention to these factors:

1. **Credibility matters**. In Japan's hierarchical society, it can be difficult for innovators and start-up companies to make progress on multiple fronts without the support or involvement of a well-known doctor, leading academic connection, or strong corporate and/or government backing. The repercussions of this issue are felt throughout the biodesign innovation process, from gaining access to hospitals and physicians, through identifying sources of funding. Innovators with prestigious in-country connections should actively seek to leverage them. While there is no easy solution for those without, they should anticipate potential resistance and think creatively about workarounds until they are able to build desired relationships. (See 2.3 Stakeholder Analysis and 5.7 Marketing and Stakeholder Strategy.)

2. **When it comes to regulation, plan ahead**. Given the Japanese device lag, begin thinking about an in-country regulatory strategy relatively early. Choosing the right MAH is essential to the regulatory process, so innovators should seek referrals and screen these prospective partners carefully. Additionally, innovators are advised to take advantage of the many consultation sessions offered by PDMA. These meetings can be costly (up to US$28,000 for a two-hour meeting), but the feedback is reliable, making the sessions a valuable investment.[24] Although medtech regulation in Japan can be a challenge, the good news is that review times are getting shorter and

almost all innovative devices are able to secure local reimbursement within three to six months following PMDA approval. (See 4.2 Regulatory Basics and 5.4 Regulatory Strategy.)

3. **Getting down to business can be burdensome**. Medtech specific activities such as filing a patent, running clinical trials, and seeking regulatory approval can all require additional effort and expense when working in Japan. In addition, Japan is ranked 122 out of 189 economies when it comes to the time and cost involved in launching a business in the country.[25] In combination, innovators must take these factors into account as they develop their plans and timelines to get to market, and also as they prepare to raise the funding necessary to support their in-country efforts. (See 6.1 Operating Plan and Financial Model and 6.3 Funding Approaches.)

Good luck! 幸運
Fumiaki Ikeno
Research Associate, Cardiovascular Medicine, Stanford University
Global Product Development Partnership (PDP) Liaison, Stanford Biodesign

NOTES

1 "Japan Indicators," The World Bank, 2012, http://data.worldbank.org/country/japan (January 14, 2014).
2 "Japan GDP Growth Forecast 2013–2015," Knoema, 2013 http://knoema.com/igsdjtg/japan-gdp-growth-forecast-2013-2015-and-up-to-2060-data-and-charts (January 14, 2014).
3 "Japan Overview," *Encyclopedia of the Nations*, http://www.nationsencyclopedia.com/economies/Asia-and-the-Pacific/Japan-OVERVIEW-OF-ECONOMY.html (January 14, 2014).
4 "Japan Indicators," The World Bank, 2012, http://data.worldbank.org/country/japan (January 14, 2014).
5 "Japan's Population Logs Record Drop," *CBC News*, January 2, 2013, http://www.cbc.ca/news/world/story/2013/01/02/japan-population-record-decline.html (January 14, 2014).
6 Ibid.
7 "Causes of Death," Japanese Ministry of Health Data, http://www.mhlw.go.jp/toukei/saikin/hw/jinkou/geppo/nengai11/kekka03.html#k3_2 (January 14, 2014).
8 "Healthcare in Japan," *The Economist*, September 12, 2011, http://www.economist.com/node/21528660 (January 14, 2014).

9 Meredith Milnick, "Japanese Longevity – How Long Will It Last?," *Time*, September 5, 2011, http://healthland.time.com/2011/09/05/japanese-longevity-%E2%80%94-how-long-will-it-last/ (March 14, 2014).

10 Hideki Hashimoto, Naoki Ikegami, Kenji Shibuya, Nobuyuki Izumida, Haruko Noguchi, Hideo Yasunaga, Hiroaki Miyata, Jose M. Acuin, and Michael R. Reich, "Japan: Universal Health Care at 50 Years," *The Lancet*, August 30 2011, http://www.thelancet.com/journals/lancet/article/PIIS0140-6736(11)60987-2/abstract (March 17, 2014).

11 "Medical Device Market: Japan," Pacific Bridge Medical, http://www.pacificbridgemedical.com/business-services/medical-device-consulting/japan/ (January 14, 2014).

12 "OECD Health Data 2013: How Does Japan Compare?," Organization for Economic Cooperation and Development, June 2013, http://www.oecd.org/els/health-systems/Briefing-Note-JAPAN-2013.pdf (January 14, 2014).

13 "Health Expenditure, Total (as% of GDP), The World Bank 2011, http://data.worldbank.org/indicator/SH.XPD.TOTL.ZS (March 14, 2014).

14 "OECD Health Data 2013: How Does Japan Compare?," op. cit.

15 Kavitha A. Davidson, "The Most Efficient Health Care Systems in the World," *The Huffington Post*, August 28, 2013, http://www.huffingtonpost.com/2013/08/29/most-efficient-healthcare_n_3825477.html (January 14, 2014).

16 "Medical Device Market in Japan," Espicom Reports, 2012, http://www.espicom.com/japan-medical-device-market (January 14, 2014).

17 "The Global Market for Medical Devices, 4th Edition," Kalorama Information, May 2013, http://www.kaloramainformation.com/Global-Medical-Devices-7546398/ (March 10, 2014).

18 Miki Anzal, "Japan's Medical Device Market is Getting Better," *European Medical Device Technology*, November 2012, http://www.emdt.co.uk/article/japan%E2%80%99s-medical-device-market-getting-better (January 14, 2014).

19 Ames Gross, "PMDA Consultation Sessions for Medical Device Registration in Japan," Pacific Bridge Medical, April 4, 2013, http://www.pacificbridgemedical.com/publications/pmda-consultation-sessions-for-medical-device-registration-in-japan/ (January 29, 2014).

20 Ibid.

21 "Medical Technology Innovation Scorecard: The Race for Global Leadership," PricewaterhouseCoopers, January 2011, http://download.pwc.com/ie/pubs/2011_medical_technology_innovation_scorecard_the_race_for_global_leadership_jan.pdf (January 14, 2014).

22 Michael Fitzpatrick, "Japan: Where Medical Miracles Are Waiting to Get Out of the Lab," *CNN Money*, April 8, 2013, http://tech.fortune.cnn.com/2013/04/08/where-medical-miracles-are-just-waiting-to-get-out-of-the-lab/ (January 14, 2014).

23 "AMDD Announces Japan Advanced Medical Device and Diagnostics Public Opinion Survey Results," American Medical Devices and Diagnostics Manufacturers' Association, December 17, 2010, http://www.amdd.jp/en/technology/press101217.html (January 14, 2014).

24 Gross, op. cit.

25 "Starting a Business in Japan," International Finance Corporation and The World Bank, http://www.doingbusiness.org/data/exploreeconomies/japan/starting-a-business/ (January 17, 2014).

Latin America

Background

Latin America includes approximately 20 countries in North, South, and Central America and the Caribbean. In contrast to most of the United States and Canada, Latin languages – mainly Spanish and Portuguese – are primarily spoken in the region.

While economic growth varies substantially across Latin America, the gross domestic product (GDP) growth in the region as a whole increased by an average of 5 percent from 2000 to 2008,[1] and decreased to closer to 3 percent subsequently.[2] Brazil and Mexico are the largest economies, accounting for roughly 65 percent of region's combined GDP.[3] Globally, Brazil and Mexico have the 7th and 14th largest GDPs, respectively.[4]

Economic gains have stimulated increases in consumption, population, and longevity, as well as more demand for adequate healthcare by a growing middle class. This, in turn, has led to enhanced opportunities in the public and private healthcare markets and increased investment in healthcare access and infrastructure. Concurrently, Latin America has experienced a rise in the prevalence of chronic diseases across the region. Mortality due to cardiovascular diseases alone is predicted to increase by 145 percent between 1990 and 2020.[5] Healthcare expenditure per capita in the region hovers around US$661,[6] compared to approximately US$8,600 in the United States,[7] suggesting future room for expansion .

As a whole, Latin America has a medical technology industry valued at more than US$8 billion, which makes it one of the larger global markets.[8] Moreover, medical device sales have been expanding at a rate of more than 10 percent per year in the region.[9] Based on their size, Brazil and Mexico comprise the most important medical device markets in Latin America and, accordingly, will be covered in more detail below. However, Colombia has the fastest growing medtech market, with a projected 2013–2018 compound annual growth rate of 13.3 percent. This market is forecast to reach US$2.2 billion in 2018.[10] Chile and Peru are among the region's top economic performers,[11,12] but both of these countries represent small markets for medical devices. Argentina and Venezuela have demonstrated a higher demand for medical technologies, yet both countries are experiencing serious macroeconomic challenges that are hindering continued medtech expansion.[13]

Brazil

Brazil is the region's heavyweight, with the largest population (199 million people), the biggest geographic footprint, and largest economy (US$2.2 trillion in 2012).[14] The country's economic growth has created a large middle class, with the number of people living below the national poverty line declining from 21 percent in 2003 to 11 percent in 2009.[15] The disease profile of population is dominated by chronic diseases, such as cancer and cardiovascular disease.[16]

The Brazilian government provides universal healthcare coverage to approximately 75 percent of its citizens under the Unified Health System (Sistema Unico de Saúde – SUS)[17] and spends about 9 percent of GDP (or US$1,121 per capita) on healthcare.[18] In 2010, the Ministry of Health launched the "More Health" (Mais Saúde) initiative, a healthcare program that targets the strengthening of the SUS by extending healthcare coverage and improving quality and access. Alongside the SUS, Brazil also has the second largest private health insurance sector in the Americas. Over 1,200 insurers provide supplementary medical coverage to approximately 25 percent of the population, with services typically purchased by middle and upper income households.[19] The Agência Nacional de Saude Suplementar (ANS) regulates the supplementary healthcare sector and healthcare plans have a mandatory obligation to pay for inpatient drugs and medical devices that are part of the statutory list, but are not required to cover drugs and medical products dispensed by retail pharmacies.

Brazil is the largest medical device market in Latin America, with revenues of about US$6 billion in 2012.[20] The country has a relatively well-established medical technology industry that includes both local and multinational companies. The local companies mostly manufacture low-to-mid complexity, less expensive medical devices, while most high-end, expensive devices are imported. Medical devices are regulated by the Brazil national health surveillance agency called Agência Nacional de Vigilância Sanitária (ANVISA), which has requirements similar to those found in the European Union.

Mexico

Mexico is the largest Spanish-speaking country in the world, with a population of about 121 million people.[21] With a GDP of US$1.1 trillion, it has the second largest economy in the region.[22] Its proximity to the United States makes it an attractive market for many medtech companies. Health expenditure is low compared to other Latin American countries – the World Bank estimates that Mexico spends 6.3 percent of GDP on healthcare,[23] with public expenditure accounting for about 50 percent of the total.[24] Most private spending is out-of-pocket, as private insurance companies represent a small proportion of the healthcare market.[25]

Healthcare provision varies widely across Mexico. The largest public hospitals and private facilities are generally well equipped and staffed, with private hospitals in Mexico City catering to upper-middle class locals and medical tourists. However, the country's overall hospital infrastructure is underdeveloped, with only one hospital bed per 1,000 people in 2013, less than half the rate of provision in Argentina, Brazil, or Chile.[26] Smaller, more remote facilities are in great need of upgraded and expanded equipment.

Mexico is the second largest medical device market in Latin America, at about US$4 billion.[27] It has become an important medical device manufacturing base for multinationals due to its Maquiladora program, under which manufacturers can bring in components, parts, or even capital equipment from the US free of import duties due to the North American Free Trade Agreement (NAFTA).[28] Geographical proximity to the United States and a less expensive cost of labor encourages many American medical device companies to set up manufacturing facilities in Mexico or use third-party manufacturing services provided by local companies. Maquiladora activity concentrates along the US border in the state of Baja California. The domestic medtech industry in Mexico is geared towards exports, with the United States as the dominant destination. The local markets, on the other hand, are predominantly supplied by imports. In combination, these factors make Mexico both the leading medical device exporter and importer in Latin America.[29] The Mexican regulatory agency, Subsecretaría de Regulación y Fomento Sanitario of the Secretaría de Salud (SSA), through a division known as COFEPRIS, follows similar guidelines to the US FDA and works in cooperation with that agency.[30]

Challenges

Economic growth and development in Latin America has been undeniably robust over the last decade, making it a compelling geographic target for medtech innovators. However, as with any emerging market, the region is not without its challenges. For one, the countries that comprise Latin America are incredibly diverse, offering significantly different levels of opportunity, stability, productivity, and competitiveness to the companies that do business within them. On the World Economic Forum's 2013–24 Global Competitiveness Index, Chile was a top performer, ranked 34th out of 148 countries, while Venezuela received the lowest ranking at 134th (Mexico and Brazil earned the 55th and 56th positions on the report).[31]

Despite this wide-ranging performance, many countries in Latin America struggle with common factors that limit their competitiveness by international standards. For example, multiple countries within the region, including Brazil and Mexico, have persistent problems related to the overall quality of their physical infrastructures. Government efficiency, corruption, and security are also common concerns.[32] Another important issue relates to the smooth, transparent functioning of many major institutions, including the regulatory bodies that play a critical role in the medtech field. Brazil and Mexico have the most mature and stable regulatory systems in the region, but requirements for market entry can still be somewhat confusing and excessively bureaucratic. For instance, gaining regulatory approval in Brazil through ANVISA can take from six months to two years, with unexplained delays often slowing licensing, registration, and review processes.[33] According to one report, bureaucracy and corruption costs Brazil over US$40 billion each year.[34]

Intellectual property (IP) protection is another area where Latin America's institutions have room for continued improvement. In contrast to the United States and Europe, Latin America does not have a longstanding tradition of protecting IP rights. In recent years, the region has made significant strides in adopting legal reforms and aligning its IP policies with those advocated by international agencies such as United States Patent and Trademark Office, the European Patent Office, and the World Intellectual Property Organization. However, piracy and enforcement remain ongoing challenges.[35] In some countries, other hurdles exist. For example, in Brazil, any patent application for products affecting "public health" must first be approved by ANVISA before it can be examined on its merits by the in-country patent office. This extra requirement, known as "prior consent," can add significant time, cost, and risk to the patent applications of pharmaceutical and medical technology companies.[36] Other Latin American countries such as Paraguay also follow prior consent rules.[37]

Finally, Latin America suffers from uneven quality in its educational systems, which contributes to a scarcity of skilled workers in high technology fields such as medical device development. This, in turn, limits the extent of innovation being generated from within the region. For instance, despite a thriving medical device manufacturing sector in Mexico,[38] few domestic companies are developing innovative medical solutions specifically for the Latin American market. Similarly, entrepreneurship is an important driver of the economy – small businesses employ over half of all workers. But these businesses often fail to grow into large, sustainable enterprises.[39]

Tactics

Large multinational corporations have been selling medical technologies in Latin America for more than 50 years, but relatively few start-ups have targeted the region with their offerings.[40] Latin America is large, growing, and full of opportunity. And, according to some, it may be a less complicated, more approachable market for young medtech companies to enter than other emerging markets, such as China.[41] Innovators are well served to consider Latin America as a potential business destination, and they should keep these three interrelated factors in mind when doing so:

1. **Conduct an in-market experiment**. Innovators who are uncertain whether Latin America is the right market to enter with a product should consider gaining some experience in the region before making up their minds. For example, Argentina, Brazil, and Chile have developed vibrant industries focused on conducting clinical trials for pharmaceutical and medical device companies from around the world. The advantages of conducting trials in the region include lower costs, faster enrollment, rates strong patient retention, competent and enthusiastic investigators, and compliance with Good Clinical Practices (GCPs).[42] Similarly, locations such as Mexico are well known for their medtech contract manufacturing capabilities. Baja California alone has more than 65 facilities devoted manufacturing ISO, FDA, and CE-mark certified medical devices.[43] Costa Rica is another manufacturing hub, with medtech exports expanding at a compound annual growth rate of 24 percent since 1998. Products produced in the country range from high-tech devices for multinationals to low-end disposables.[44] Starting with a small trial or manufacturing project can be an effective way to gain exposure to Latin America and begin cultivating relationships in markets that may be of interest at a later date. (See 5.2 R&D Strategy and 5.3 Clinical Strategy.)

2. **Establish a beachhead from which to expand**. When innovators are ready to tackle Latin America as a market, thinking about the region as a whole can be intimidating. A more effective approach is to establish a foundation in a single market from which the company can grow. Brazil or Mexico, with their large economies and more established medical device markets, can serve as an excellent starting point. For US-based innovators, Mexico offers great proximity, regulatory requirements that are becoming harmonized with those of the FDA, and a base from which to enter other Spanish-speaking countries.[45] Brazil encourages companies to establish operations in the country by offering a

government procurement preference for goods manufactured locally, even when their prices are up to 25 percent higher.[46] The country is also a member of MERCOSUR,[47] an economic and political agreement among eight South American nations to promote the free movement of goods, services, and people among member states. As MERCOSUR participants tend to follow Brazil's lead when it comes to the regulation of technologies such as medical devices, Brazil can be an effective springboard into these neighboring countries.[48] (See 5.4 Regulatory Strategy and 5.9 Competitive Advantage and Business Strategy.)

3. **Capitalize on government incentives.** As they experience sustained economic growth, the countries in Latin America are focused on improving access to and the quality of healthcare for their citizens. For instance, Mexico established the Sistema de Protección Social en Salud (SPSS) in 2003, introducing popular health insurance for those below the poverty line, in addition to all employed citizens, their dependents, and government employees.[49] With this move, the government dramatically expanded healthcare spending and created opportunities for the introduction of new treatments and procedures. Some countries are going further to stimulate growth and innovation in the healthcare sector. The Brazilian government, for one, launched the Investment Program for the Health Industrial Complex or Programa para o Desenvolvimento do Complexo Industrial da Saúde (PROCIS) with the objective of encouraging the manufacture of drugs, vaccines, and medical devices in the country. Similarly, the Brazilian bank Banco Nacional de Desenvolvimento Economico e Social (BNDES) provides initial funding of healthcare start-ups.[50] Innovators are encouraged to investigate programs such as these to determine how they might be useful in underwriting the development of the product and/or its purchase once on the market. (See 4.3 Reimbursement Basics and 6.3 Funding Approaches.)

Good luck! Buena suerte! Boa sorte!

Robson Capasso

Clinical Assistant Professor, Otolaryngology, Stanford School of Medicine

Lecturer, Stanford Biodesign Program (global course)

Fernanda O. Machado

Associate Vice President, Global Strategy and Analysis, Advamed

NOTES

1 "Latin America Economic Outlook," Organization for Economic Cooperation and Development, 2012, http://www.oecd.org/dev/americas/48965859.pdf (February 5, 2014).

2 "Regional Economic Update – Latin America and the Caribbean," International Monetary Fund, October 2013, https://www.imf.org/external/pubs/ft/reo/2013/whd/eng/pdf/wreo1013.pdf (February 6, 2014).

3 "Latin America and Caribbean Indicators," The World Bank, 2013, http://data.worldbank.org/region/LAC (February 6, 2014).

4 "GDP Ranking," The World Bank, 2013, http://databank.worldbank.org/data/download/GDP.pdf (February 10, 2014).

5 S. Yusuf, S. Reddy, S. Ounpuu, and S. Anand, "Review Global Burden of Cardiovascular Diseases: Part I: General Considerations, The Epidemiologic Transition, Risk Factors, and Impact of Urbanization," *Circulation*, November 27, 2001, http://circ.ahajournals.org/content/104/22/2746.full (February 6, 2014).

6 "Healthcare Expenditure and Financing in Latin America," Pan American Health Organization, December 2012, http://www.paho.org/hq/index.php?option = com_docman&task = doc_download&gid = 20057&Itemid = 270&lang = en (February 5, 2014).

7 "Health Indicators," The World Bank, 2012, http://data.worldbank.org/indicator/SH.XPD.PCAP (February 5, 2014).

8 Fred Aslan, "Huge Markets for Devices Emerging in Brazil, Latin America," *IN VIVO: The Business and Medicine Report*, October 2012, http://www.wsgr.com/news/medicaldevice/pdf/latam.pdf (February 10, 2014).

9 Ibid.

10 "Colombia Medical Devices Report," Espicom, December 2013, http://store.businessmonitor.com/colombia-medical-devices-report.html (February 5, 2014).

11 "GDP per Capita: Chile," The World Bank, 2013, http://data.worldbank.org/indicator/NY.GDP.PCAP.CD (February 5, 2014).

12 "Peru: Latin America's Economic Performer," *IMF Survey Magazine*, February 22, 2013, http://www.imf.org/external/pubs/ft/survey/so/2013/car022213d.htm (February 5, 2014).

13 "Latin America Medical Device Market Report," Reportlinker. com press release, February 3, 2014, http://www.prnewswire. com/news-releases/latin-america-medical-device-market-reports-243317201.html (February 6, 2014).

14 "Brazil Country Profile," The World Bank, 2013, http://www. worldbank.org/en/country/brazil (February 6, 2014).

15 "Brazil Overview," The World Bank, 2013, http://www. worldbank.org/en/country/brazil/overview (February 6, 2014).

16 "Global Burden of Disease Profile: Brazil," GBD, 2013, http:// www.healthmetricsandevaluation.org/sites/default/files/ country-profiles/GBD%20Country%20Report%20-%20Brazil. pdf (February 6, 2014).

17 "Brazil's March Toward Universal Coverage," *Bulletin of the World Health Organization*, September 2010, http://www. who.int/bulletin/volumes/88/9/10-020910/en/ (February 6, 2014).

18 "Health Indicators," The World Bank, 2013, http://data. worldbank.org/indicator/SH.XPD.PCAP (February 5, 2014).

19 "Brazil's March Toward Universal Coverage," op. cit.

20 "The Global Market for Medical Devices, 4th Edition," Kalorama Information, May 2013, http://www. kaloramainformation.com/Global-Medical-Devices-7546398/ (March 10, 2014).

21 "Mexico Country Profile," The World Bank, 2013, http://www. worldbank.org/en/country/mexico (February 6, 2014).

22 Ibid.

23 "Health Expenditure, Total (% of GDP)," The World Bank 2012, http://data.worldbank.org/indicator/SH.XPD.TOTL.ZS (March 16, 2014).

24 "Health Expenditure, Public (% of Total Health Expenditure)," The World Bank 2012, http://data.worldbank.org/indicator/ SH.XPD.PUBL (March 16, 2014).

25 "Democratization of Health in Mexico," *Bulletin of the World Health Organization*, July 2009, http://www.who.int/bulletin/ volumes/87/7/08-053199/en/ (February 6, 2014).

26 "Hospital Beds (per 1000 people)," The World Bank 2012, http://data.worldbank.org/indicator/SH.MED.BEDS.ZS (March 16, 2014).

27 "Medical Device Market: Mexico," Espicom, 2013, http://www. espicom.com/mexico-medical-device-market (February 6, 2014).

28 "Medical Device Manufacturing in Mexico," Maquila Reference, http://www.maquilareference.com/2013/03/medical-device-manufacturing-in-mexico/ (February 6, 2014).

29 "Medical Device Market: Mexico," op. cit.

30 Undersecretary of Health Regulation and Development: Mexico, http://www.nl.gob.mx/?P = s_regulacion (February 06, 2014).

31 Klaus Schwab and Xavier Sala-i-Martin, "The Global Competitiveness Report 2013–2014," The World Economic Forum, 2013,http://www3.weforum.org/docs/ WEF_GlobalCompetitivenessReport_2013-14.pdf (February 10, 2014).

32 Ibid.

33 Pablo Halpern and Benny Spiewak, "Medical Devices in Brazil – Problem, Challenge or Opportunity?," *European Medical Device Technology*, October 2, 2013, http://www.emdt. co.uk/article/medical-devices-brazil-oportunity (February 10, 2014).

34 Stewart, A. "Brazil: Corruption Costs $41 Billion," *Latin Business Chronicle*, October 5, 2010, http://www. latinbusinesschronicle.com/app/article.aspx?id = 4550 (February 7, 2014).

35 Álvaro Ramirez Bonilla, David Switzer, Danny G. Pérez y Soto, "The State of Intellectual Property in Latin America," B&R Latin America IP, October 2012, http://gallery.mailchimp.com/ 146754dfefdb520a9b0fbd631/files/The_State_of_Intellectual_ Property_in_Latin_America_PDF_High_Res.pdf (February 10, 2014).

36 Ryan O'Quinn and Sanya Sukduang, "Drug Patents Under Fire in Brazil," PharmaExec.com, June 4, 2013, http://blog. pharmexec.com/2013/06/04/drug-patents-under-fire-in-brazil/ (February 10, 2014).

37 Roy Whalen, "IP in Latin America: Growing Recognition of the Importance of IP to Innovation," Patently Biotech, May 29, 2013, http://www.biotech-now.org/public-policy/ patently-biotech/2013/05/ip-in-latin-america-growing-recognition-of-the-importance-of-ip-to-innovation-2 (February 7, 2014).

38 "Medical Devices Sector: Mexico," ProMexico Trade and Investment, http://embamex.sre.gob.mx/kenia/images/ stories/pdf/medical_services.pdf (February 7, 2014).

39 "Latin America: Entrepreneurs' Lack of Innovation Curbs Creation of Quality Jobs," *World Bank News*, December 5, 2013, http://www.worldbank.org/en/news/feature/2013/12/05/ latin-america-many-entrepreneurs-little-innovation-growth (February 6, 2014).

40 Aslan, op. cit.

41 Ibid.

42 "The Top Reasons for Conducting a Clinical Trial in Latin America," Estern Medical, http://www.esternmedical.com/ information/article21.php (February 10, 2014).

43 "Medical Device Manufacturing in Mexico," op. cit.

44 Penny Bamber and Gary Gereffi, "Costa Rica in the Medical Devices Global Value Chain: Opportunities for Upgrading," Duke University Center on Globalization, Governance, and Competitiveness, August 2013, http://www.cggc.duke.edu/ pdfs/2013-08-20_Ch2_Medical_Devices.pdf, (February 12, 2014).

45 "Medical Device Manufacturing in Mexico," op. cit.

46 "Brazil," United States Trade Representative, http://www.ustr.gov/sites/default/files/2013%20NTE%20Brazil%20Final.pdf (February 10, 2014).

47 MERCOSUR, http://www.mercosur.int/msweb/portal%20intermediario/ (February 7, 2014).

48 "The Outlook for Medical Devices in Latin America," *PRNewswire*, July 31, 2012, http://www.prnewswire.com/news-releases/the-outlook-for-medical-devices-in-latin-america-164407766.html (February 7, 2014).

49 Mariana Barraza-Lloréns, Stefano Bertozzi, Eduardo González-Pier, and Juan Pablo Gutiérrez, "Addressing Inequity in Health and Health Care in Mexico," *Health Affairs*, May 2002, http://content.healthaffairs.org/content/21/3/47.full?sid=b3b6de23-b597-4dff-99cf-0fbbcbfca739 (February 6, 2014).

50 "Industrial Policy," ABDI, http://www.abdi.com.br/Paginas/politica_industrial.aspx (February 6, 2014).

IDENTIFY

The purpose of the *Identify* phase (shown in Figure I1) is to gather a number of unmet medical needs through observation and then screen this list down to a promising few, based on information about the key clinical, stakeholder, and market characteristics.

The output is a small set of carefully formulated need specifications that frame truly promising opportunities for invention.

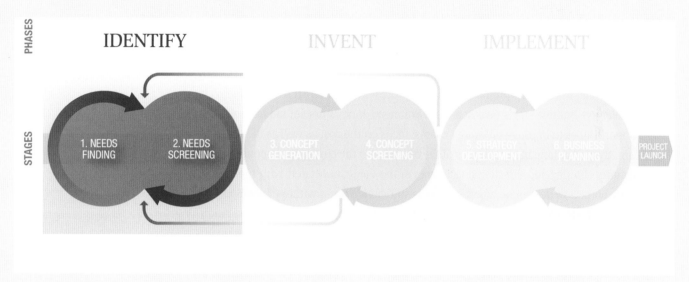

Key themes for the Identify phase, which span both stages of needs finding and needs screening

FIGURE I1

The Identify phase kicks off the biodesign innovation process.

→ **Biodesign innovation is driven by a compelling need.** Medical technology innovators in both business and university environments often follow a "technology push" strategy: they discover or invent a potentially useful technology and then go searching for a clinical application to commercialize it. This model is especially common in biotech and pharmaceutical development, where scientific discoveries such as a new molecule or pathway ultimately lead to the creation of new drugs. In medtech, too, there are important examples of technology push innovations (for example, medical lasers and surgical robotics). And this is unquestionably a productive route for product development. But the distinctive characteristic of the medical technology field is that it's almost always possible to start *purely with an important unmet clinical need* and then invent a technology that will help solve it. This needs-driven approach is not well understood or practiced by either industry or academia. And this, in a nutshell, is the key reason for writing this text. Our "mantra" is that *a well-characterized need is* *the DNA of a good invention.* Stated another way, the Identify phase of recognizing and understanding needs is the cornerstone of the entire biodesign innovation process. Once you get the need wired, you have a solid foundation for moving on to invention.

→ **Even if you start with a technology, you should still figure out the need.** Suppose you're given a technology to develop and bring to market in your role at a medtech company. You will still benefit from following the biodesign innovation process. In fact, it is critically important to go back to the Identify phase and carefully evaluate the underlying assumptions that have been made about the need behind the new technology. By pausing to deeply understand and clearly articulate the need, you will be able to perform an instant WOMBAT check (is this a Waste of Money, Brains, and Time?). By researching the need further and defining the key criteria that any solution must address, you will create a basis for evaluating whether the new approach is on track or whether a different solution might, in fact, be better.

→ **Pay attention to value up front.** The core strategy in exploring needs is to dive into the clinical environment, looking for suboptimal patient outcomes, recurring complications, frustrations on the part of care providers, or other signs of problems in care delivery. But in today's healthcare environment, innovators should actively look for opportunities to improve on *value* – that is, focus on the cost as well as the outcome of care. By researching the economics of the disease and the current treatment options, innovators can develop directional estimates of the value associated with their need. The key point is to determine whether there is sufficient margin within the competitive landscape to develop a new technology with a better improvement/cost equation. Technologies, procedures, and systems that drive costs down (or provide major benefits at low incremental costs) are ripe areas for innovation.

→ **Let needs compete to survive.** The Identify phase begins with an expansion of possible opportunities as you collect a relatively large set of potential needs. Then, in the need screening stage, you progressively cut down the list, learning just

IDENTIFY

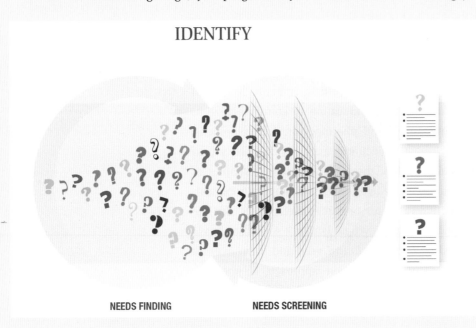

NEEDS FINDING   NEEDS SCREENING

FIGURE I2

In the Identify phase, innovators first expand their range of possible opportunities and then narrow their focus to the most interesting and attractive needs.

enough about the needs areas to be able to jettison those needs that are less promising compared to others in the group (see Figure I2).

There is some psychology at play here: it's hard to "kill" needs. All of them stem from a real problem in clinical care and, in that sense, they all seem important. The biodesign innovation process provides an efficient and relatively painless mechanism to do the hatchet job. Innovators feel empowered to kill a pretty good need because, by comparison, the next one on the list is *really* good. And it turns out that it's possible to do this prioritization without getting into paralyzing detail about every aspect of every need. The process of letting the needs compete allows you to find out *just enough* information, *just in time* to appreciate that one need is better than another. This process continues until a handful of needs emerge at the top of the list. As a bonus, by going through the process of filtering, the innovators will have accumulated enough information to create a robust specification for their surviving needs.

→ **Be prepared to multitask and iterate.** While the activities in the Identify phase may appear somewhat linear, the reality is that they are highly iterative and often conducted in parallel (as indicated by the cyclic arrows in the biodesign innovation process diagram). As you tackle needs finding and needs screening, get ready to loop back and forth between these stages. After sourcing and filtering your first set of needs, you may find it's necessary to perform more observations, re-scope certain needs, and/or conduct additional due diligence. In the most effective projects, innovators move fluidly between needs finding and needs screening activities.

→ **Start with a sketch.** As you cycle through these activities, your understanding of your needs will progressively deepen, as will your confidence in your ability to decide which needs to take forward. Resist the temptation to go too deep into detail too quickly. The progression in the Identify phase is like painting a portrait. The artist's ultimate goal is to produce an extraordinary work, rich in form, detail,

FIGURE I3
An innovator's understanding of a need grows progressively detailed, just as a great portrait moves from a sketch to a completed work.

and color. But it starts with a sketch. Through an iterative, progressive process, the artist adds elements to the canvas – colors, textures, shadows, background. It's the same with needs characterization – the masterpiece emerges with time (see Figure I3). Your "sketch" at this point in the process is a rough idea of the potential clinical demand and market uptake. You will only have a full, nuanced picture of the complete opportunity much later in the biodesign innovation process.

INVENT

The purpose of the *Invent* phase (shown in Figure I4) is to devise solutions to one or more defined needs, taking advantage of creative ideation techniques, prototyping and testing methods, and a filtering process that is based on objective risk criteria.

The output of this phase is a final concept that you will advance into strategic planning for implementation, with the goal of bringing the invention forward into patient care.

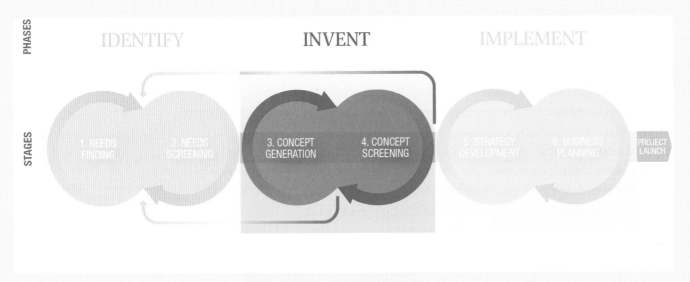

FIGURE I4

The Invent phase covers activities in the middle of the biodesign innovation process.

Key themes for the Invent phase, which span both stages of concept generation and concept screening

→ **What you learned during Identify will guide you through Invent.** Each of the needs you identify and screen in the Identify phase is made actionable through a detailed need specification. Once you have a great "need spec," you're poised to invent innovative solutions to address it. Although ideation is a creative exercise, it's not completely free-form. The need spec provides a map of the areas where you should go looking for new ideas – and, if used effectively, it makes concept generation much more productive. Then, when it comes to concept screening, the need spec provides an objective set of criteria against which you can evaluate competing solutions to ensure that your final concept truly addresses the most important aspects of the need.

→ **Keep cycling.** Just as when you're working with needs, the process of generating and screening concepts is iterative, not linear. Occasionally, teams go through ideation only once and never look back. But more often than not, innovators return to brainstorming after conducting preliminary concept screening. Sometimes they use these sessions to refine their initial concepts based on newly available information. At other times, they go back to the drawing board for an entirely different set of solutions. It's a matter of instinct and practice when to revisit ideation, but it's certainly worth keeping this as a possibility throughout the Invent phase. Building prototypes as early and as quickly as possible typically helps to speed up the process of cycling. There is something almost magical about how making a rough model brings into focus the possibilities and/or limitations of a given concept.

→ **Never change a need to suit a concept.** At first glance, nothing looks as pretty as your own invention! Unfortunately, you may find your solution to be so attractive that you lose site of the fact that it doesn't really suit the need. Use the need specification to help you stay clear-headed. In short, it's always a mistake to change your need specification to suit an invention. Your need spec tells you clearly which essential features your invention must have to be useful – stick with these. If you're having trouble generating concepts that really address a need, the answer isn't to change the specification. Instead, try setting the need aside and shifting to a different need that stimulates a richer set of concepts. If necessary, go back to the Identify phase and pick up a need you didn't take forward initially – there are undoubtedly some good ones you left behind that will perform better in the Invent phase.

→ **Again, let the concepts compete.** As with Identify, the Invent phase relies on a process of generating and then filtering a number of possibilities – this time, it's concepts – using a set of ranking criteria (see Figure I5). These criteria are different from the needs screening filters used in the Identify phase and they are specific to the *solutions* being evaluated: intellectual property (IP), regulatory and reimbursement pathways, technical feasibility, and business model. These are "meaty" categories, to be sure.

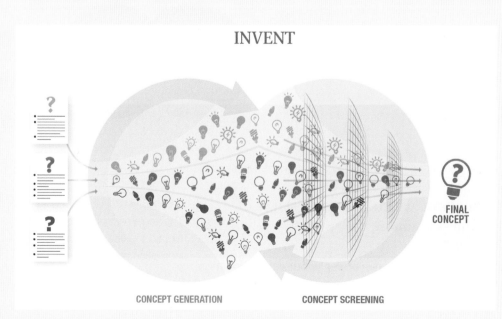

INVENT

CONCEPT GENERATION   CONCEPT SCREENING

FINAL CONCEPT

FIGURE I5
In the Invent phase, innovators generate many solution ideas and then select a final concept to take forward through a progressively in-depth screening process.

It may seem impossible that you'll be able to filter dozens, if not hundreds of solutions down to a single concept within any reasonable period of time. For example, a detailed IP analysis on one particular concept could, by itself, take many weeks to complete. The key, as described in Invent, is to get *just enough* information, *just in time* to be able assess concepts without spending huge amounts of time on any one issue. You are still working from a "sketch," though now you will be adding more detail (Figure I3). Your goal is to learn enough about your concepts to be able to drop the ones that have major flaws in terms of these primary filters. Then, you can further explore those that survive the preliminary screens. Through increasingly detailed probes into available information, you'll gravitate to the final concept that objectively wins out over all others. Later on, you'll learn much more about that concept as you take it forward into the Implement phase. But, having gone through the concept screening process, you can be assured that your investment of additional time and effort into the solution is worthwhile.

IMPLEMENT

The purpose of the *Implement* phase (shown in Figure I6) is to create a multi-year plan for developing a concept into a real product that is safe and effective for patients and attractive to providers and payers.

The focus of this phase is on developing and integrating core strategies for the launch of a new business or a new program in an existing business. Execution comes next, when you use the output of this phase as a roadmap for product development and market initiation.

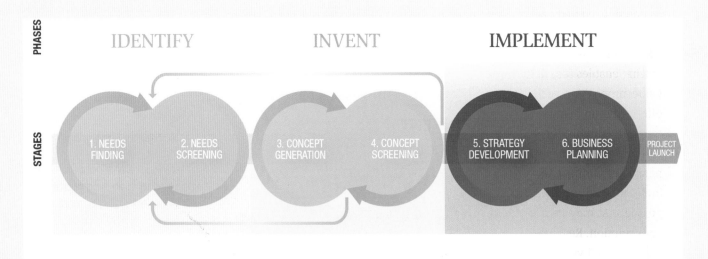

Key themes for the Implement phase, which span both stages of strategy development and business planning

FIGURE I6

The Implement phase comes toward the end of the biodesign innovation process as innovators prepare for project launch.

→ **The end is only the beginning.** The final phase of the biodesign innovation process is where all of the critical factors involved in developing and

44

commercializing a new medical technology come together into a unified plan for delivering your product or service to the market. In strategy development, your focus is on leveraging your knowledge to design the path for all critical activities (IP, R&D, clinical regulatory, reimbursement, etc.) and anticipate the risks involved in each. Integration is then about prioritizing and sequencing the time, resources, and funding requirements of these essential activities into a cohesive and actionable plan. By the end of the Implement phase, you will be on a clear trajectory to launch your new product or service.

→ **Prioritize your risks: identify the key questions.** The idea behind defining key questions is to consider all of the risks facing a company and then prioritize them, with an emphasis on determining the one or two most critical issues that must be addressed *first* in order for the project to remain viable. These risks are often technical in nature, in which case the key question may be something along the lines of, "Can we build a working prototype that successfully performs the function in question?" Sometimes the key question is clinical and the basic issue is, "Will this work in humans?" In this scenario, the priority is to find a path to first-in-human testing that is as fast and inexpensive as possible. Getting the key question(s) right is the single most important aspect of planning for a new business or project in that it shapes many important decisions (e.g., how much funding to raise, who to hire, what space/equipment to activate, etc.).

→ **Maintain a value-orientation.** During needs finding and screening, savvy innovators will pursue needs where there is the promise to drive down costs or provide major benefits at a low incremental cost. During the Implement phase, it's time to figure out how to make good on this promise. In today's cost-conscious environment, new technologies must deliver strong value in the eyes of the customers. Stated another way, the improvement/cost equation must be compelling enough to motivate decision makers to change their behavior and adopt something new. The way that a team's functional strategies come together is what enables it to deliver on a meaningful value proposition. Even with the best value proposition, achieving reimbursement (or another form of payment) will often wind up being a killer risk. But the clearer and more compelling the value of the new solution, the greater its likelihood of commercial success.

→ **There's more than one way to deliver a new technology to patients.** Starting a business to bring a new medtech product to market is *one way* that innovators can build and capture the value of their ideas. But there are other pathways you can pursue that may be a better fit given the context of your need and the nature of your solution. Regardless of whether you intend to start a company or, for example, license or sell your solution to an existing one, you need to anticipate and prepare for the requirements of commercialization. The more effectively you can demonstrate that your product has a clear path to market, the easier it will be for investors, licensors, partners, or acquirers to appreciate the opportunity enabled by your offering. This means that *all* innovators need to think about the complete set of activities addressed in this stage, even though they may not have to address them at an equal level of detail depending on their path.

→ **Don't underestimate the importance of "telling and selling" your story.** Once you have a comprehensive plan for getting your technology to market, it's time to assume the role of storyteller. A well-crafted story is essential for justifying the time, capital, and other resources you'll need to execute your plan and convince key stakeholders – your boss within a medtech company, a potential funder for your start-up, or a prospective employee who is considering joining your project – to support the effort. These days, your story may take many different forms (e.g., a business plan, pitch, or other type of presentation). But the framing should always be customized for the intended audience. Effective storytellers understand what the target audience values most and positions their core messages to address those specific interests.

→ **Get comfortable making assumptions.** Medtech innovators work in a field dominated by science but, when it comes to developing strategies for implementation, they are asked to make countless assumptions. One of the most difficult aspects of doing this, especially for first-time innovators, is achieving confidence that your predictions are "accurate." Benchmarking, proxy analysis, and engaging experts are all tools that can help you develop credible assumptions. You will be required to make educated guesses, but they must be directionally correct. Learn to make and work with reasonable assumptions until better information becomes available, otherwise you risk becoming paralyzed by uncertainty.

→ **Ask for help.** Fundamentally, there's no substitute for experience. Throughout the Implement phase, recognize when and where you'll benefit from professional expertise – and then plan to go out and get it. Whether you tap mentors and advisors, hire consultants, or add strategic hires to your team, the input of the right experts at the right time can be invaluable to your progress. A good team will effectively pivot as it encounters roadblocks; an inexperienced team will flounder. That's why experienced investors almost always bet on a great team – and they will expect you to know who you need to add to yours.

By keeping these important themes in mind as you explore the chapters ahead, it's our hope that you'll be better prepared to understand the biodesign innovation process and more effective in applying it. Each of the phases requires some difficult decision making. But the process will guide you to pay sufficient attention to critical considerations while not getting bogged down by trying to get too much information, too early in the sequence. In short, you can trust the process to help you to make tough choices. As we have seen in project after project, the rigor and discipline of the biodesign innovation process will help you exercise sound judgment. And, ultimately, it will improve your odds of getting your new medical technology into patient care. Even in a challenging environment, unlimited opportunities exist to create new medical technologies that can deliver cost-effective care to patients around the world. Now it's time to go out and get started!

IDENTIFY ▸ Needs Finding

PHASES　　　**STAGES**　　　**ACTIVITIES**

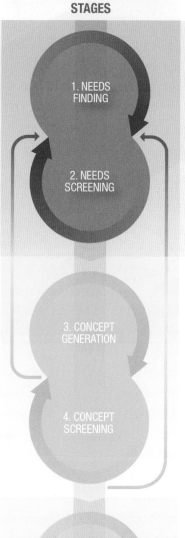

IDENTIFY

1. NEEDS FINDING

2. NEEDS SCREENING

1.1　Strategic Focus
1.2　Needs Exploration
1.3　Need Statement Development

2.1　Disease State Fundamentals
2.2　Existing Solutions
2.3　Stakeholder Analysis
2.4　Market Analysis
2.5　Needs Selection

INVENT

3. CONCEPT GENERATION

4. CONCEPT SCREENING

3.1　Ideation
3.2　Initial Concept Selection

4.1　Intellectual Property Basics
4.2　Regulatory Basics
4.3　Reimbursement Basics
4.4　Business Models
4.5　Concept Exploration and Testing
4.6　Final Concept Selection

IMPLEMENT

5. STRATEGY DEVELOPMENT

6. BUSINESS PLANNING

5.1　IP Strategy
5.2　R&D Strategy
5.3　Clinical Strategy
5.4　Regulatory Strategy
5.5　Quality Management
5.6　Reimbursement Strategy
5.7　Marketing and Stakeholder Strategy
5.8　Sales and Distribution Strategy
5.9　Competitive Advantage and Business Strategy

6.1　Operating Plan and Financial Model
6.2　Strategy Integration and Communication
6.3　Funding Approaches
6.4　Alternate Pathways

PROJECT LAUNCH

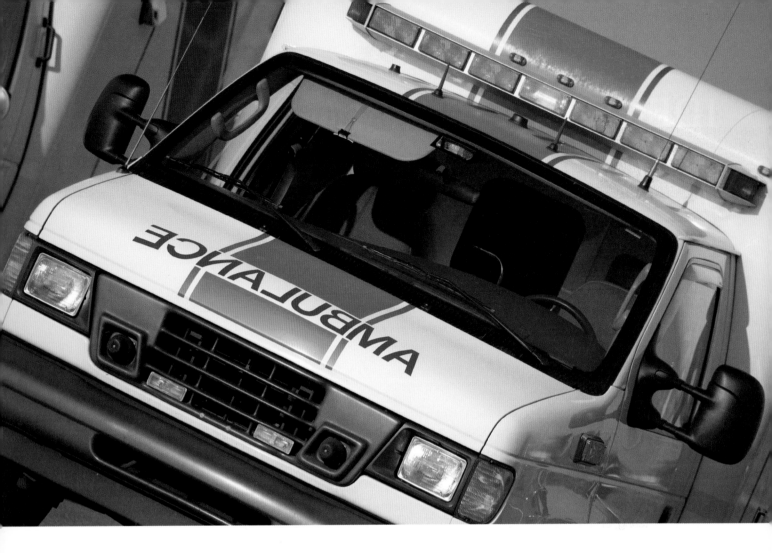

If you want to have good ideas you must have many ideas.

Linus Pauling[1]

If I had asked my customers what they wanted, they would have said a faster horse.

Henry Ford[2]

1. NEEDS FINDING

IDENTIFY

Both Pauling and Ford offer great insights into this *most important* starting point. Identifying a compelling clinical need may seem simple and obvious, but it is not. Get it right and you have a chance; get it wrong and all further effort is likely to be wasted.

Identifying needs involves first a broad screening survey, which we call needs finding. The follow-on process, needs screening, is covered in Stage 2. By way of analogy, needs finding is akin to snorkeling; needs screening is more like a deep dive.

Needs finding is a simple and yet profound process. All of the diagnostic and therapeutic workings of the healthcare system offer fertile ground to search for unsolved problems. From the back of an ambulance to the OR, then ICU and the outpatient clinic, real problems abound. The principle is to observe real people and real-life situations in order to fully understand clinical procedures and techniques, *as they are currently practiced*. The observer then looks for difficulties that providers, patients, or other healthcare stakeholders are encountering, and major obstacles or technical barriers that may be modified. Look for what might be missing (Henry Ford). The essential task is to identify the real clinical challenges and problems that impose a significant medical burden.

One of the biggest challenges is the high cost of healthcare. Think about what is driving the costs of the procedure or treatment. The hospital administrator or purchasing agent may be more useful than a doctor or nurse in identifying an opportunity to create economic benefit while still preserving the quality of care.

This is neither an armchair exercise nor an isolated epiphany. Rather, thoughtful observation of clinical encounters with "fresh eyes" is most likely to identify substantial unsolved problems. It may be a spoken need, such as a surgeon asking for a "third hand"; it may be the unspoken need, only appreciated when clinical troubles or complications are the expectation of the treating team. When an untoward clinical outcome or complication is met with the retort, "Oh, we see this" – *pay attention*. This should be a great stimulus for the innovator to ask: "*Why* do you see this?" "Should you see this?" "Is this inevitable?"

This sequential and iterative process from early need statement to final need specification produces real clarity. A well-characterized need becomes the DNA of the invention/innovation to follow.

NOTES

1 As quoted by Francis Crick in his presentation "The Impact of Linus Pauling on Molecular Biology," 1995.
2 Unsourced quotation widely attributed to Henry Ford.

1.1 Strategic Focus

INTRODUCTION

An engineer with a needle-phobic mother decides to design an alternate method for administering the daily insulin she takes to control her diabetes. A spinal surgeon, frustrated with the limitations of the implants she uses to treat vertebral compression fractures, starts working on improvements to the device. A business student observing a birth at a hospital in Africa is surprised by the extent of blood spray during the process and becomes concerned about protecting healthcare workers when the mother is infected with HIV. A resident studying oncology becomes passionate about understanding the disease more fully and commits himself to cancer research and the pursuit of a cure. While all of these paths are worthwhile, they are not universally appealing. The course that excites one innovator may be uninteresting or overwhelming to another. But, the one thing that these paths have in common is that they are compelling to the people undertaking them. By aligning these passions with their unique competencies and using these factors to create criteria for selecting or rejecting specific projects, innovators increase their likelihood of choosing a successful path that will also keep them motivated and engaged as they navigate the many challenges that await them in the biodesign innovation process.

One of the first, most important steps in the biodesign innovation process is for innovators to discover and commit themselves to the strategic focus area that stimulates their personal enthusiasm. By explicitly deciding in what areas to focus, innovators accept different risks, challenges, and potential rewards (e.g., working on heart problems is much different from working on wrinkle removal). To make an effective, meaningful decision about a strategic focus, innovators must ask themselves questions about why they want to pursue this path, what they hope to accomplish, and how their strengths and weaknesses may affect their efforts. Additionally, a high-level assessment of the characteristics of the focus area and the environmental factors affecting it should be taken into account relative to these goals. Ultimately, the most rewarding and successful biodesign projects are those that achieve a high degree of alignment

OBJECTIVES

- Understand that innovators must explicitly choose their strategic focus.
- Appreciate the importance of achieving alignment between the mission and strengths and weaknesses of the individual and/or team and the strategic focus area that is chosen.
- Recognize the steps involved in choosing a strategic focus.

between the values and competencies of the innovators and the defining characteristics of the strategic focus area that is chosen.

 See ebiodesign.org for featured videos on choosing a strategic focus.

STRATEGIC FOCUS FUNDAMENTALS

As Mir Imran, CEO of InCube Labs and founder of more than 20 medical device companies, said:[1]

I knew once I found a problem, I could solve it. The biggest challenge for me was which problem to solve.

The most productive way to launch the biodesign innovation process is to choose a strategic focus – an area to pursue that matches the innovator's core competencies and personal or organizational mission. Determining a strategic focus involves:

- Deciding what the innovator or organization values or wants to achieve, independent of any specific vehicle or project for accomplishing it.
- Accurately assessing what competencies the innovator or the organization have (or do not have) that will affect the ability to realize those goals.
- And then translating these insights into criteria that can be used to objectively evaluate opportunities and decide which problems or focus areas to pursue.

If one thinks of the innovation process as a journey – from discovering medical needs to developing and commercializing new medical technologies to solve those needs – then the selection of a strategic focus is analogous to charting a course.

Developing a strategic focus

The notion that innovators typically create new inventions in a spontaneous stroke of genius is a myth. For most medtech innovators, ideas do not just happen – they are the result of an intentional decision to go out and seek opportunities and problems in a specific area.

Making this decision is not always easy. But using a structured approach can help (see Figure 1.1.1). Choosing a strategic focus begins with performing a personal inventory. Explicitly defining a mission and understanding their strengths and weaknesses helps innovators define "acceptance criteria" – conditions that they will used to determine whether a project is a good fit. Acceptance criteria are also shaped by factors in the external environment that may affect the ability of innovators to act on their interests and realize their goals. Once they are comfortable with their acceptance criteria, innovators can begin evaluating a variety of opportunity areas to arrive at a strategic focus.

Conducting a personal inventory

The personal inventory should be performed before the innovator begins thinking about any particular opportunity or problem area. The purpose of the inventory is to identify the mission of the individual or team, as well as their strengths and weaknesses.

Conducting a personal inventory is equally important for individual innovators, academics/researchers, small teams, young companies, and large corporations, in that it helps ensure that the person (or people) undertaking the innovation process are enthusiastic about the strategic focus that is chosen and that they have the necessary capabilities to pursue it. The issues and priorities that emerge as a result of the inventory will be different based on the constituency performing it; however, the value of the exercise will be the same.

Determining a mission Innovators must be unambiguous about their mission. A mission is a broad, directional aspiration that defines what an individual or group wants to accomplish. Articulating a mission sets a desired destination for an innovation project and provides clarity about the ultimate goal(s) the individual or group hopes to achieve.

To define a mission, individuals and groups should think about their priorities, beginning with questions

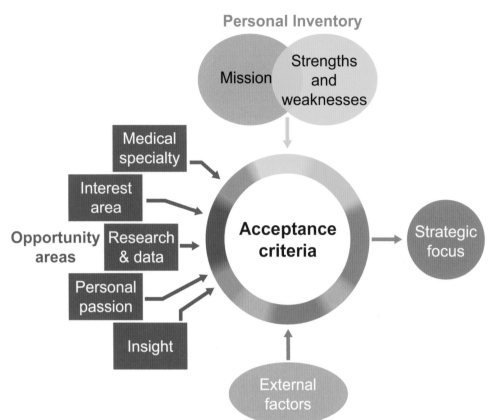

Personal Inventory

Mission

Strengths and weaknesses

Opportunity areas

Medical specialty

Interest area

Research & data

Personal passion

Insight

Acceptance criteria

Strategic focus

External factors

FIGURE 1.1.1

Using a structured approach that takes internal and external factors into account can help lead innovators to a strategic focus that provides a good fit.

about what is most important to them (or, conversely, not important to them). For example, a priority for someone pursuing a career in research or academia might be to engage in an exceptionally compelling research project that, if successful, would have a dramatic impact on healthcare worldwide. While such a long-term mission might take an entire career to achieve, the magnitude of the potential outcome would be large enough to make that commitment worthwhile to someone with this goal. Getting involved in a project with a less significant outcome might take less time and effort to achieve, but would be less interesting to the individual due to the misalignment with his/her mission.

In companies and other established organizations, the mission sometimes takes the form of what is commonly known as a mission statement. The Medtronic example that follows illustrates how a corporate mission statement might look.

Large corporations may also choose to define specific missions for their divisions or groups. At this level, other priorities may surface as they approach the innovation process. With established portfolios of products to leverage (and protect), a division might not always be interested in finding the biggest innovation. Instead, it may focus on driving incremental improvements in existing product lines that enable it to stay ahead of the competition. Or, with more extensive resources at its disposal, a company might be willing to make slightly larger, longer-term investments with the intent of leap-frogging competitors over time.

The missions of aspiring entrepreneurs or young start-up companies may be different still. These individuals and teams do not necessarily have to create mission statements that are as formal or expansive as those of a large company. As long as the mission is clearly articulated, it can be significantly more informal (although it is still advisable to put it in writing). Additionally, the mission might be somewhat more practical or applied. For example, without the resources to support a vast, long-term research program, two innovators working together on a shoestring budget might decide that an important aspect of their mission is to identify a solution that is readily achievable (within two to three years) and compelling enough from a business perspective to raise

FROM THE FIELD ▷ MEDTRONIC

Defining a meaningful mission statement

Medtronic was founded in 1949 by Earl Bakken and his brother-in-law Palmer Hermundslieco as a medical equipment repair shop. The fledgling company quickly expanded into services and then into device design, development, and manufacturing.[2]

During the early years, Bakken was moved by the emotional response patients had to the company's products. Many were overjoyed to regain mobility, to feel better, and sometimes even to be alive as a result of Medtronic's work (see Figure 1.1.2).[3]

Inspired by their stories and the desire to make this type of human benefit the purpose of the organization's efforts, he and the board of directors created the Medtronic Mission, which remains an integral part of the company's culture and the driving force behind every project that it undertakes. This Mission guides the

FIGURE 1.1.2
Earl Bakken with a young Medtronic patient (courtesy of Medtronic).

company's day-to-day work and keeps employees focused on the goal of changing the face of chronic disease for millions of people around the world.

Medtronic's Mission is:[4]

- To **contribute to human welfare** by application of biomedical engineering in the research, design, manufacture, and sale of instruments or appliances that alleviate pain, restore health, and extend life.
- To **direct our growth in the areas of biomedical engineering where we display maximum strength and ability**; to gather people and facilities that tend to augment these areas; to continuously build on these areas through education and knowledge assimilation; to avoid participation in areas where we cannot make unique and worthy contributions.
- To **strive without reserve for the greatest possible reliability and quality in our products**; to be the unsurpassed standard of comparison and to be recognized as a company of dedication, honesty, integrity, and service.
- To **make a fair profit** on current operations to meet our obligations, sustain our growth, and reach our goals.
- To **recognize the personal worth of employees** by providing an employment framework that allows personal satisfaction in work accomplished, security, advancement opportunity, and means to share in the company's success.
- To **maintain good citizenship** as a company.

As William Hawkins, former CEO of Medtronic explained while he was at the helm of the organization, "The mission is our moral compass. It is the glue that binds all of our businesses together. It underpins everything we do. In good times and tough times, the one constant in our business model is our core values. We use the mission to ensure that we work on the right things and that we strive to do things right."[5]

financial support. Unlike the researcher or aspiring academic, these innovators would be more focused on near-term opportunities that are meaningful, but not too expensive to pursue.

Identifying strengths and weaknesses In addition to thinking about a mission, individual innovators, academics/researchers, small teams, young companies, and large corporations will all benefit from assessing their strengths and weaknesses. Specifically, they should evaluate what they do well, and how they can capitalize on these strengths. They should also consider in what areas they are less experienced, competent, or confident, and how they can compensate for these relative weaknesses.

Some people can be successful in leading the innovation process (especially in its early stages) on their own. However, many individuals *and* groups recognize, after they assess their strengths and weaknesses, that they will benefit from collaborating with others who offer different, complementary skill sets. For example, if an innovator is a strong clinician, but not an engineer, it might be helpful to partner with an engineer if the mission is to develop a device technology. Or, if that same innovator is interested in developing a business plan to pursue a concept, s/he might want to consider collaborating with someone with business training or experience to help construct and execute that plan. Wildly creative types are best paired with grounded detail-oriented types, and so on. Fundamentally, the most important objective of this step is to identify where certain competency gaps and opportunities exist so that the innovator can address them when the time is right. It is rare for one person to embody all the talents necessary to identify, invent, develop, and commercialize a technology all alone. However, innovators are aware of areas where help may be necessary, they can begin building a team with the strengths that complement known weaknesses, and can make sure that team expands as requirements for more diverse skills increase.

Considering external factors
Finding the right fit from a personal (or internal) perspective is essential but, as the medtech field becomes increasingly complex, it is also important to evaluate what is happening in the external environment. Over the past several years, factors out of the innovator's direct control have shifted dramatically. In the US, for example, regulatory and reimbursement pathways have become significantly less predictable and more time and resource intensive. In parallel, access to capital is more uncertain and difficult to obtain. A similar shift is being felt in Europe where macroeconomic conditions are contributing to the adoption of more conservative policies with the potential to slow the pace of innovation. Against this backdrop, patients, physicians, facilities, and payers alike are looking more critically at the affordability of healthcare and the economic value associated with new innovations.

In combination with completing a personal inventory, innovators should think about how these external factors may affect their goals and/or play to their strengths and weaknesses. Factors in the external environment can add significantly more risk to a project by increasing the amount of time, money, and resources required to achieve results. Innovators will have different "appetites" for risk in the external environment. Some will be driven to pursue their goals at any cost, while others may prefer a more moderate level of challenge with the hope of increasing their chances of success. Understanding this at a conceptual level is an important input to subsequently defining project acceptance criteria.

Defining acceptance criteria
At their most basic, acceptance criteria are conditions that must be met to make an innovation project attractive to the innovator. These criteria are defined based on what the innovators have learned about themselves by conducting the personal inventory and considering the external environment. Innovators apply their acceptance criteria to choose an area of strategic focus, as well as to help assess the needs they discover in the early stages of the biodesign innovation process.

Of course, there is no single set of acceptance criteria that works for every individual or team. However, most acceptance criteria are built around common themes that, when customized by the innovator, become

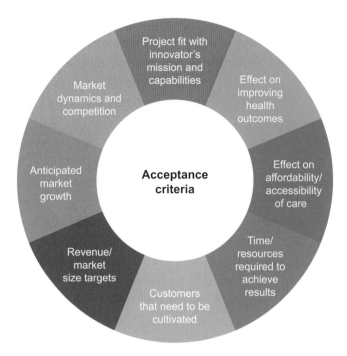

FIGURE 1.1.3

Acceptance criteria are frequently built around some combination of these common themes.

requirements that an innovation project must meet (see Figure 1.1.3).

For example, suppose that a large corporation has a mission to develop a product that expands its portfolio into a new clinical area that drives increased growth within the company. Before defining its acceptance criteria, the company would have to think about what strengths and weaknesses it has that would enable it to achieve this goal. The availability of resources (staff, funding, time) would certainly be an enabler. However, the way in which its existing sales force is deployed (i.e., which types of doctors it already calls on) could be a strength or a weakness, depending on the specific area of focus that is chosen. After performing an assessment, the corporation might decide to engage in a project only if it meets the following acceptance criteria:

- The clinical practice area is new to the company and is growing at a minimum of 10 percent per year and/ or can generate a minimum of $100 million in revenue per year.
- Technologies in this space have a relatively well-understood regulatory pathway and clear clinical trial

requirements so they can be brought to market within two to three years.
- The company's established sales force already calls on these same customers, so the commercial fit is good.
- A new solution can potentially be cost-neutral or cost-reducing to key healthcare stakeholders.

In the mid-1990s, American Medical Systems (AMS) had two primary products: an implantable urinary sphincter and a penile prosthetic line. Its mission was to become a well-rounded urology company by broadening its focus to include other urological products. As the company began to think about its acceptance criteria for new opportunities, the list included the following: (1) technologies that could be sold to the same customer or at the same "call point"; (2) technologies that were more mechanical in function than biological; and (3) opportunities/ areas that could grow at greater than 20 percent per year to add to the company's revenue growth.[6] Under different circumstances (for instance, if the company had saturated its existing customer base), the corporation might have eliminated the criterion to stay within the same customer group. While this would have made a wider cross-section of potential projects attractive to the company, it might not have allowed it to achieve certain economies of scale by offering the same customers a wider line of products through the existing sales force. In this respect, the acceptance criteria defined by the AMS appropriately reflected its priorities at the time and capitalized on the perceived strength of its established sales arm.

Without any limitations imposed by a preexisting business, an innovator or young company might define acceptance criteria around the magnitude of the impact its solutions can have on peoples' lives. In this scenario, with a mission to improve important outcomes for patients on a major scale, the acceptance criteria might require a project that:

- Has a total potential market of $1 billion or more.
- Will be cost-neutral or cost-reducing (so it gets adopted).
- Will be attractive to investors (so it gets adequate financial support).
- Results in an innovation that has a significant impact on patients' **quality of life** (as opposed to an

innovation that makes a device cheaper, faster, or easier to use).

- Has platform potential so that the benefits from one medical specialty can be rapidly leveraged to affect patients in other practice areas.
- Is focused on a patient segment where head-to-head competition can be avoided, especially if the company is concerned about its ability to compete with entrenched firms.

The acceptance criteria above are similar to those used by medtech **incubators** such as ExploraMed, The Foundry, The Innovation Factory, or Coridea. Such criteria enable these organizations to continually deliver powerful innovations in a number of diverse fields. However, these particular criteria represent just one approach. Given the variety of players active in the medtech field, there are other incubators, as well as other organizations and innovators whose acceptance criteria would look completely different. For instance, a more philanthropically oriented organization might have acceptance criteria that guide them to projects that:

- Benefit individuals living on less than $2 per day.
- Significantly expand access to healthcare in areas where treatments exist but have previously been unaffordable.
- Have dual-market potential so that by charging customers "market rates" in wealthier settings the organization can subsidize costs in low-resource settings.
- Are aligned with the priorities of certain foundations and/or non-governmental organizations to help attract funding for product development.

Evaluating opportunities against acceptance criteria
Once specific acceptance criteria have been defined, innovators can start exploring different interest areas for a good fit. The idea is to screen each potential opportunity area against the acceptance criteria, setting aside the ones that do not provide a good match and looking more deeply at those that do. Innovators are encouraged to consider a broad range of possibilities, keeping in mind that deep expertise in a field is not necessarily required. All too often, people who are deeply immersed in a field fail to see the opportunities and problems that surround them because they have been indoctrinated into a certain way of doing things. Individuals and teams who bring diverse experiences and different backgrounds to a field can sometimes be more successful in uncovering opportunities and problems because they are more willing to question the status quo.

One approach is to start by evaluating problems and opportunities related to a personal interest or passion. For instance, someone might become committed to working in the breast cancer field after losing a loved one to the disease. While this is certainly a valid method for choosing a strategic focus, innovators must still conduct research in that area to determine the extent to which opportunities and problems in the space meet their defined acceptance criteria at more than just a superficial level. They must also be prepared to "walk away" if significant gaps are uncovered between the acceptance criteria and the area being explored. It can also be helpful to get more specific about the strategic focus area. For instance, would it be a better fit to embark on a long-term research-based path to cure the disease, or to pioneer near-term improvements in the effectiveness of breast cancer treatment? Innovators can use their other acceptance criteria to define a concentration within the desired field that is most likely to lead to a fulfilling experience and outcome.

If innovators do not have a specific passion for a particular opportunity or problem area, another way to begin the process of screening potential focus areas against their acceptance criteria is to examine high-level data related to a practice area or disease state (note that more in-depth research will be performed in subsequent steps of the biodesign innovation process). Statistics to consider include the number of people affected by a condition, the clinical impact of the disease, the outcomes and costs of existing treatments, the profitability of existing treatments, and the rate at which spending is growing (see Table 1.1.1). Innovators can also glean insights from the total revenue realized each year in a particular medical field (see Figures 1.1.4 and 1.1.5).

The more thoughtful this evaluation process, the better. However, even a cursory evaluation of different clinical areas (and their subspecialties) will potentially help to narrow one's focus. For example, an innovator or

Table 1.1.1 Data such as total expenses for selected conditions and percent distribution by type of service, as shown in the table, can be an interesting source of ideas regarding areas that might meet the innovator's acceptance criteria (Agency for Healthcare Research and Quality, "Total Expenses and Percent Distribution for Selected Conditions by Type of Service: United States, 2010," Medical Expenditure Panel Survey Household Component Data, generated interactively (September 19, 2013).

United States, 2010 Conditions	Total expenses (millions)	Percent distribution by type of services				
		Hospital outpatient of office-based visits	Hospital inpatient stays	ER visits	Prescribed medicines	Home health
Heart conditions	107,186.40	18.0	62.9	5.2	9.2	4.7
Trauma-related disorders	82,303.57	43.2	38.0	13.3	1.8	3.7
Cancer	81,734.62	50.2	36.8	0.3	8.3	4.4*
Mental disorders	73,060.24	24.1	15.1	1.4	45.3	14.0
COPD, asthma	63,782.99	23.5	27.2	4.4	37.7	7.1
Osteoarthritis and other non-traumatic joint disorders	62,362.98	40.3	31.4	1.2	17.4	9.7
Diabetes mellitus	51,310.57	21.9	22.1	1.2	48.0	6.7
Hypertension	42,943.38	30.4	12.5	2.0	47.4	7.8
Back problems	39,259.66	56.6	28.4	2.4	9.4	3.2*
Hyperlipidemia	37,174.19	25.7	2.7*	0.1*	69.1	2.3*
Normal birth/live born	34,945.69	23.6	72.8	2.3*	1.0*	0.2*
Systemic lupus and connective tissues disorders	30,836.17	48.1	29.5	2.2*	12.3	7.9*
Other central nervous system disorders	25,898.88	44.0	42.6	2.8	7.5	3.1
Disorders of the upper GI	23,457.37	20.3	18.5	4.5	53.7	2.9*
Kidney disease	22,967.52	47.7	33.8	8.2	6.8	3.5*
Other circulatory conditions arteries, veins, and lymphatics	22,678.64	32.3	57.4	2.0	6.3	2.1*
Gallbladder, pancreatic, and liver disease	22,646.35	15.7	72.4	8.1*	2.7*	1.2*
Other endocrine, nutritional and immune disorder	22,097.34	52.4	25.3*	2.5*	15.4	4.3*
Infectious diseases	21,909.62	21.6	32.0	4.1*	39.1	3.2*
Cerebrovascular disease	20,576.60	12.4	59.1	8.3*	5.9	14.3*

* Relative standard error equal to or greater than 30 percent; total percentages do not always add to 100 due to rounding

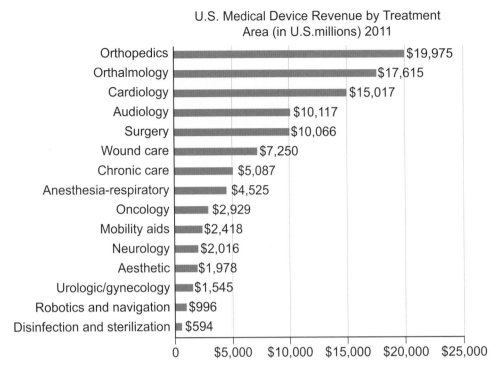

U.S. Medical Device Revenue by Treatment Area (in U.S.millions) 2011

Orthopedics	$19,975
Orthalmology	$17,615
Cardiology	$15,017
Audiology	$10,117
Surgery	$10,066
Wound care	$7,250
Chronic care	$5,087
Anesthesia-respiratory	$4,525
Oncology	$2,929
Mobility aids	$2,418
Neurology	$2,016
Aesthetic	$1,978
Urologic/gynecology	$1,545
Robotics and navigation	$996
Disinfection and sterilization	$594

FIGURE 1.1.4
Information about medical device revenues by major medical segment can be helpful in choosing a strategic focus (from "U.S. Medical Devices Market Outlook," Frost & Sullivan, 2012; reprinted with permission).

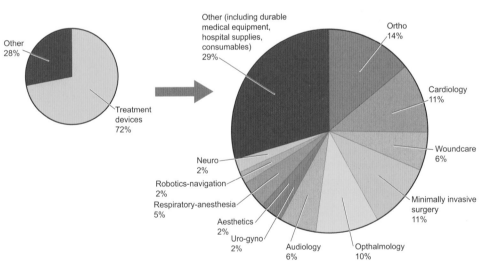

Global Medical Devices Market: Segmentation by Product Class, 2012

Other 28%
Treatment devices 72%

Other (including durable medical equipment, hospital supplies, consumables) 29%
Ortho 14%
Cardiology 11%
Woundcare 6%
Minimally invasive surgery 11%
Opthalmology 10%
Audiology 6%
Uro-gyno 2%
Aesthetics 2%
Respiratory-anesthesia 5%
Robotics-navigation 2%
Neuro 2%

FIGURE 1.1.5
Market segmentation by product class can vary by geography, so innovators may wish to consider data for different locations. Note: all figures are rounded; the base year is 2012 (from "Global Medical Devices Market Outlook," Frost & Sullivan, 2013; reprinted with permission).

company seeking a large business opportunity might review certain statistics and other data (as shown in Table 1.1.1) and immediately become interested in the cardiovascular field. Yet, the fact that this is a relatively well-established, mature field may conflict with some of the other acceptance criteria that the innovator has defined. If the innovator is committed to new opportunities and problems that have not yet been defined or where innovation has not occurred for quite some time, another field outside of cardiology might be a better fit (e.g., respiratory medicine or urology). In an area with a well-defined market opportunity, there may be intense competition and a great deal of pressure to be first to market with technology that could set the new

standard of care. In less popular areas, the advantages of weaker competition are balanced by greater uncertainties – both regarding the ability to attract investment and motivate behavior change among physicians who are entrenched in the old ways of treating patients. This is where the innovator's acceptance criteria (and how they are prioritized) can help to resolve inherent conflicts and facilitate effective trade-offs, which become clearer when evaluating these different risks and rewards.

As innovators evaluate opportunity and problem areas against their acceptance criteria, a strategic focus or a few acceptable focus areas should begin to emerge. Regardless of the specific area that is chosen, a strong sense of "the right fit" is essential to anyone embarking on the biodesign innovation journey.

The following story from ExploraMed describes how one innovator worked through the process of choosing a strategic focus.

FROM THE FIELD ⟩ **EXPLORAMED**

Applying acceptance criteria in evaluating a strategic focus

Making an explicit decision about the strategic focus to be pursued is an essential exercise for individual innovators, teams, companies, and incubators alike. According to Josh Makower, founder and CEO of medical device incubator ExploraMed, "Choosing what is not a fit is as important as determining what is."[7] ExploraMed has embedded this step in its process for identifying, creating, and developing new medical device businesses. When Makower initiates a new business, he and his team spend time assessing their relative strengths and weaknesses and articulating the acceptance criteria against which they will screen potential opportunities.

ExploraMed's defined mission is to "focus on clinical needs where there is an opportunity to dramatically improve outcomes and build freestanding businesses." As Makower explained, "I get excited about working on things that are going to have a major impact on medicine. We want to work on projects that make a substantial contribution, can potentially change the direction of healthcare, and affect outcomes for thousands or millions of patients. If a large number of people are affected by a problem and currently have poor outcomes from the existing set of treatments, it could be a hot area for us to investigate." Recognizing their own strengths and weaknesses, Makower and

team further constrain their efforts to medical device opportunities, leaving drugs, diagnostics, and other healthcare technologies to a different set of innovators and entrepreneurs. Finally, they specifically like the idea of being "contrarians." "We like to go where others haven't gone and where people believe there aren't reasonable opportunities. You can create a competitive advantage for yourself by being the first to go in another direction. The other thing that we're trying to do is create big enterprises. To do this, we almost always have to be willing to go into a space where there aren't a lot of other players. A little fish can grow to be pretty big if he finds himself in a big pond all by himself. I like that a lot better than trying to establish a foothold in an already crowded market."

These defined acceptance criteria are routinely used by ExploraMed to evaluate which opportunities to pursue, as the following example demonstrates. Early in the company's history, when the team was actively investigating new projects, Makower's elderly aunt fell and broke her hip. "Before the accident, she was energetic, vibrant, and active. After she fell, her life changed dramatically. She had trouble with her daily activities, as well as doing the things she loved, like seeing her children and grandchildren. Suddenly, she was an old lady, when that wasn't how she lived before." With a new passion to address problems and opportunities in this area, Makower and Ted Lamson, an ExploraMed project creator at the time (see Figure 1.1.6),

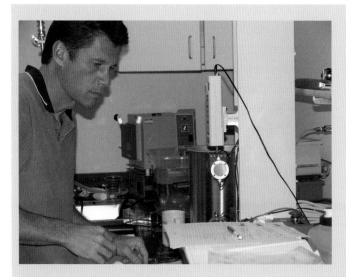

FIGURE 1.1.6
Lamson working on a device test (courtesy of ExploraMed).

began to investigate the space. What they quickly learned was that hip fractures represent a sizable problem and account for 350,000 hospital admissions and 60,000 nursing home admissions each year. More than 4 percent of hip fracture patients die during their initial hospitalization, and a full 24 percent die within a year of the injury. 50 percent lose the ability to walk.[8] "It's a shocking mortality rate," noted Makower. "We speculated that there was a need for a less invasive alternative to hip replacement, and that the size of the incision or **morbidity** from the operation itself was the key to the problem. Eventually, we discovered that this guess was wrong and that the real problem was not in the surgery, but in post-surgery recovery. If we hadn't been following a defined process for researching the space, we easily could have become biased towards a solution early on that would have sent us in the wrong direction," he emphasized.

ExploraMed's first acceptance criterion – the size and severity of the problem – was the first screen the team applied to the problem, and hip fractures appeared to be a promising market. Makower and Lamson conducted further preliminary research to understand what companies and innovations were active in the space. It turned out that numerous advancements had been made in hip surgery and the devices used to support it,

but that few new technologies existed to improve post-operative care and recovery. "If you get the patient up immediately post-procedure, and you effectively manage their pain locally so they can walk around and never waste any of their muscles, then their outcomes are fantastic. But if they stay in bed more than they should because their pain is not managed well, they do terribly. What happens is that they lose muscle mass, they get sick or become depressed, and then they die of pneumonia or some other complicating condition." With few individuals or companies working to address the non-surgical issues associated with hip fractures, the field appeared to be wide open to ExploraMed.

Unfortunately, as Makower and Lamson got further and further into their exploration, they identified a conflict with one of their important acceptance criteria. After weeks and weeks of interviewing patients and doctors and researching the space, they became concerned that the most compelling opportunities in the space might not be best addressed with device technology. "We discovered that the most pressing problems were related to improving local pain management to help patients ambulate more quickly," recalled Makower. "It was a big opportunity in an open market, but we realized it would probably be best addressed by a drug solution. However, we didn't have the right technology, skills, or resources to take on a drug project. We really wanted to figure it out, but we realized that we weren't the right guys to do it. Regardless of your passion for an area, you have to be honest with yourself about you and your team's strengths and weaknesses."

Wanting to be sure before abandoning hip fractures as an opportunity area, the team confirmed its hunch through additional research and consultations with experts in the field. "Upon further investigation we actually discovered systems to do exactly what we wanted to do already existed, but were not being utilized because of healthcare management constraints or cost. This was very discouraging . . . the answer was there and doctors were actually aware of it, but they were not using it for one reason or another," Makower commented. Eventually, the team decided to reject the project and continue their search elsewhere. "You have to be willing to accept a lot of failure," he said, reflecting on the experience. "But you've got to keep on trying – and failing if necessary – in order to understand the parameters that will make you successful and

ultimately enable you to choose the right path." Later, Makower and Lamson redirected their focus to an entirely different clinical area and, after several months of investigation, found a compelling opportunity that met all their criteria and became a company called NeoTract, Inc. (see 5.2 R&D Strategy for more information about NeoTract).

Global consideration in choosing a strategic focus

The fundamental process of choosing a strategic focus is the same whether the innovator wishes to work in the United States, Europe, China, India, or any other geographic market. Yet, as the epicenter of medical device innovation continues to gradually shift away from the US, innovators are encouraged to make their geographic focus an explicit part of their internal and external analysis before making a decision.

In 2012, the worldwide medical device market reached $331 billion.[9] Although the United States, Europe, and Japan still represent approximately 75 percent of that total, medical device sales in emerging markets are growing two to five times faster than in developed countries.[10] In the coming years, China, India, and Brazil are forecast to experience the sharpest increase in total and per capita healthcare spending. China is expected to be the world's third largest healthcare market by 2020, closing in on Japan in second place.[11] Health expenditures in these countries will be fueled by the demand for value-driven, lower-cost technologies to meet the needs of their vast populations. In contrast, Israel, Japan, and some European countries will continue to drive up per capita spending on a combination of value and premium products (see Figure 1.1.7).

Although the vast majority of medtech innovation is currently centered in the West, this will continue to shift over the next two decades. These changes are already underway as device manufacturers, seeking relief from the external factors noted above, establish new or move their operations elsewhere. Accordingly, greater numbers of innovators from inside and outside

the US will find their focus drawn to opportunities that span the globe. Whether or not problems and opportunities in emerging countries provide a good fit depends on the innovators' acceptance criteria. For example, countries such as China, India, and Africa have large groups of patients with dire needs for effective, affordable solutions (see Figure 1.1.8). But mechanisms to pay for new medical technologies are still being developed, especially when it comes to treating low-income populations. Innovators with strong acceptance criteria around helping others and having a major impact on large numbers of prospective patients may choose to gravitate toward these opportunities, while those with a stronger interest in optimizing their financial return may choose other projects.

Innovators exploring strategic focus areas in emerging markets also face an increased level of risk. In places like China, for example, intellectual property (IP) protection and regulatory processes are still under development. This additional uncertainty prevents some companies and innovators from moving into these markets and creates additional challenges for those who do. Many of these types of risks will almost certainly be reduced or resolved over time. But it will take the efforts of many motivated and committed innovators to make this happen.

Another source of risk is related to the innovator's familiarity with a given market. A critical aspect of the biodesign innovation process is the ability to gain a deep understanding and a true sense of **empathy** for the patients being targeted, no matter where they are located. Anytime US-based innovators focus on opportunities overseas, or overseas innovators focus on problems in the US, they should be aware that extra time,

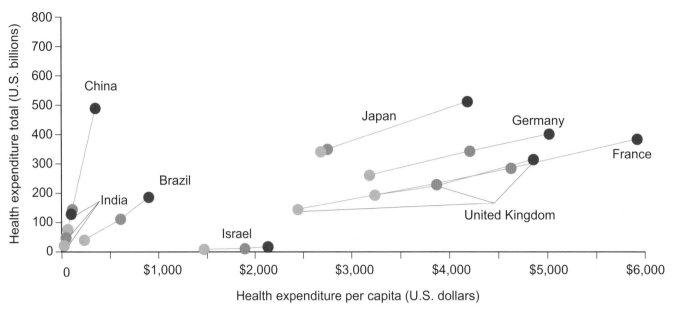

The World Bank, World Health Organization, and PwC analysis

FIGURE 1.1.7

This global trend analysis highlights how countries vary in their appetite for cost- versus outcome-driven innovation per capita. As a benchmark, US per capita spending on healthcare was $7,285 in 2007, compared to per capita spending shown in the chart, which ranges from $40 in India to $4,209 in France for the same time period. Total US spending on healthcare in 2007 was $2,159 (in U.S. billions). ("Medical Technology Innovation Scorecard: The Race for Global Leadership," PricewaterhouseCoopers, January 2011; reprinted with permission.)

effort, and resources will be required to access, research, and truly understand the problems and opportunities, the people experiencing them, and the context within which they exist. This work is certainly not impossible, but it must not be overlooked and rarely can be done from afar.

Ethics in the biodesign innovation process

Choosing a strategic focus is among the first of many steps in the biodesign innovation process where innovators may face ethical dilemmas. The potential for ethical conflicts exists at nearly every stage of an innovator's journey. Ethics are the intentional choices that people make and the basic moral principles that are used to guide these decisions. Ethics do not provide a specific value system for making choices, but rather a set of basic principles that can be followed to guide decision making.[12] Often, there is not one "right" choice. Stated another way, ethics provide the rules or standards that guide (but do not determine) the conduct of a person or the members of a profession. In developing and bringing

new medical technologies to market, the need to maintain the highest ethical standards extends to everyone involved in the process.

At the heart of most ethical issues are conflicts of interest, which arise when one person's interests are at odds with another's. For example, confidentiality (or the practice of discerning what is privileged information and rigorously protecting it) is an important principle in the medical field. If one party has an incentive to disclose confidential information about another party, a conflict of interest may arise. Because there are so many individuals and groups in the development and commercialization of any medical innovation, conflicts of interest are inevitable. Realistically, the objective of any innovator should not be to avoid such conflicts, but to minimize their occurrence and ethically address and resolve them when they arise. In particular, in any scenario where conflicts of interests involve patients and the care they receive, innovators have a special obligation to act ethically because of the potential to both improve and harm human lives.

399 black men from 1932 to 1972. These patients, who were mostly poor and illiterate, had late-stage syphilis but were not informed from what disease they were suffering. The doctors involved in the experiment had no intention of curing the men – instead, their objective was to collect scientific data from their autopsies. Over years, the patients experienced tumors, heart disease, paralysis, blindness, insanity and, eventually, death.[13] Since then, significant strides have been made, internal and external to the profession, in enforcing a strong code of ethics across the medical community. Moreover, every member of the medical community, medtech innovators included, has an important role to play in promoting and adhering to ethical behavior.

Early in the biodesign innovation process in particular, innovators often struggle with the tension between altruistically addressing important medical problems and the imperative to do so in such a way that the solution has a viable chance of reaching the market for which it is intended. An inspirational new therapy cannot, in most cases, reach patients without the necessary capital to develop it; yet capital will be provided by commercial investors primarily if they feel that a reasonable profit can be obtained. It can be frustrating to innovators to identify problems and opportunities only to discover that the market or profit potential for a solution is too small or risky to attract funding. Although capital can be obtained from government grants, non-governmental organizations (**NGOs**), or beneficent donors, there are inherent limitations associated with this type of funding that can prevent an innovation from achieving its full potential.

To develop a truly sustainable solution, most innovators have to strike a balance among satisfying the requirements of the target audience, optimizing patient benefit, and satisfying the interests of investors. Protecting investors is known as a fiduciary duty. A fiduciary is any individual or group that has the legal responsibility for managing somebody else's money.[14] As a fiduciary, the innovator has an obligation to carry out the responsibility of managing others' funds with the utmost degree of "good faith, honesty, integrity, loyalty, and undivided service of the beneficiary's interest."[15] At the most basic level, this means that

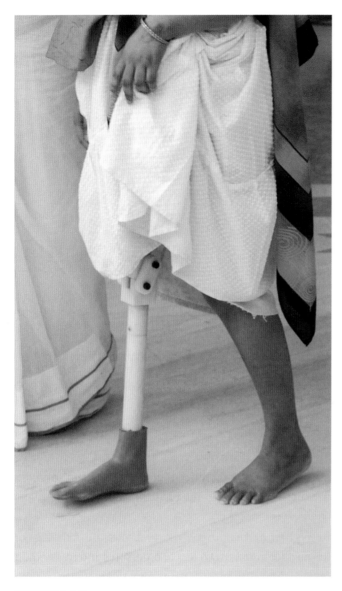

FIGURE 1.1.8
The Jaipur artificial knee (as shown above) is one example of an effective, affordable solution targeted at low-income patients with a limited ability to pay. The product costs $35–40 compared to $6,000–35,000 for similar technology in the US (courtesy of the Stanford–Jaipur Knee team from Professor Tom Andriacchi's mechanical engineering course: L. Ayo Roberts, Joel Sadler, Angelo Szychowski, and Eric Thorsell).

Ethics in the medical field have a difficult past, with trials, such as the Tuskegee syphilis experiment, creating issues of fear and distrust between medical providers and the patients they are meant to serve. In this particular case, the US Public Health Service ran an experiment on

Table 1.1.2 An example of a medical ethics (from the American Medical Association's "Principles of Medical Ethics"; reprinted with permission).

American Medical Association's Principles of Medical Ethics
A physician shall be dedicated to providing competent medical care, with compassion and respect for human dignity and rights.
A physician shall uphold the standards of professionalism, be honest in all professional interactions, and strive to report physicians deficient in character or competence, or engaging in fraud or deception, to appropriate entities.
A physician shall respect the law and also recognize a responsibility to seek changes in those requirements which are contrary to the best interests of the patient.
A physician shall respect the rights of patients, colleagues, and other health professionals, and shall safeguard patient confidences and privacy within the constraints of the law.
A physician shall continue to study, apply, and advance scientific knowledge, maintain a commitment to medical education, make relevant information available to patients, colleagues, and the public, obtain consultation, and use the talents of other health professionals when indicated.
A physician shall, in the provision of appropriate patient care, except in emergencies, be free to choose whom to serve, with whom to associate, and the environment in which to provide medical care.
A physician shall recognize a responsibility to participate in activities contributing to the improvement of the community and the betterment of public health.
A physician shall, while caring for a patient, regard responsibility to the patient as paramount.
A physician shall support access to medical care for all people.

innovators have a duty not to favor anyone else's interests (including their own) over those of the beneficiary.[16] If the fiduciary violates this responsibility, they may be subject to legal liability, which is another reason why ethical behavior is so important throughout the innovation process.

Striking an appropriate balance can be difficult when conflicting interests arise. The important thing to remember is that the "right" solution may vary for every innovator based on their individual ethical compass. By openly acknowledging the fact that "gray areas" exist and taking time for self-reflection, innovators can more readily determine the approach that is most closely aligned with their values.

Regarding other ethical conflicts in the innovation process, innovators are generally advised to maintain a primary focus on the best interests of patients in resolving issues. Seeking input and advice from objective third parties can be an invaluable resource for resolving conflicts. However, more often than not, innovators must rely on their own codes of personal and professional

ethics. The following four principles are widely accepted as ethical standards in the medical field.[17,18]

Respect for autonomy

Respect for autonomy refers to others' rights to make their own choices. This means, for example, that all parties with an interest in a new innovation must be informed about its risks and benefits, any potential conflicts of interests among those involved in its development and delivery, and about any other factors that could conceivably affect their choice. Ultimately, the patient has the right to refuse the offer to participate in an investigative study.

Beneficence

Beneficence is the practice of doing good. Medical personnel are often taught, "First, Do No Harm," but there is usually the possibility of some harm if medical devices either provoke complications of their use or they malfunction. In the field of medical innovation, this mandate extends to maximizing benefits while seeking to minimize potential harm.

Non-maleficence

The mandate of non-maleficence also is captured by the phrase "First, Do No Harm." Often beneficence and non-maleficence cannot be separated. In the process of providing a medical benefit, healthcare providers may also expose patients to risk. For instance, in clinical trials, patients are exposed to risks for the sake of others, by making it possible for life-saving devices to reach the market. The Hippocratic oath taken by many physicians essentially combines the principles of beneficence with non-maleficence, by stating that the obligation of healthcare professionals is to provide the greatest net medical benefit at minimal risk.[19]

Justice or fairness

All those in the medical field have an obligation to fairly decide among competing concerns and interests. At a minimum, this requires recognizing potential conflicts of interest and objectively determining, sometimes with third-party assistance, how they should be resolved. This principle also extends to fairness in dealing with the subjects of clinical trials; and to the reporting of all data from such trials, including negative findings, so as to benefit others in general and specifically to prevent repetition of trials without benefit for patients.

Because so many interactions in the biodesign innovation process involve clinicians, innovators should become familiar with the specific ethical codes developed by relevant medical professional societies. For instance, it may be helpful to familiarize oneself with the World Medical Association's Physician's Oath as defined in the Declaration of Geneva in 1948.[20] Table 1.1.2 summarizes the American Medical Association's Principles of Medical Ethics, the foundation of the **AMA** Code of Medical Ethics, one of the most well-known and widely practiced codes of ethics in the medical field. The complete code, consisting of these Principles and the opinions of the Council on Ethical and Judicial Affairs, is available online.[21]

When conflicts must be resolved or difficult decisions must be made, a strong code of ethics can be used by innovators as an essential guide.

Online Resources

Visit www.ebiodesign.org/1.1 for more content, including:

 Activities and links for "Getting Started"
- Take inventory
- Articulate a strategic focus

 Videos on strategic focus

 An appendix that lists professional associations for select medical conditions

CREDITS

The editors would like to thank William Hawkins and Richard L. Popp for their contributions to this chapter.

NOTES

1 From remarks made by Mir Imran as part of the "From the Innovator's Workbench" speaker series hosted by Stanford's Program in Biodesign, April 28, 2004, http://biodesign.stanford.edu/bdn/networking/pastinnovators.jsp. Reprinted with permission.

2 "Our History," Medtronic.com, http://www.medtronic.com/about-medtronic/our-story/garage-years/index.htm (September 11, 2013).

3 "Our Mission," Medtronic.com, http://www.medtronic.com/about-medtronic/our-mission/index.htm (September 11, 2013). Reprinted with permission.

4 Ibid.

5 From an exchange with William Hawkins, CEO of Medtronic, Fall 2008. Reprinted with permission.

6 From an exchange with Thom Gunderson, Medical Device Analyst for Piper Jaffray, Fall 2008. Reprinted with permission.

7 All quotations are from interviews conducted by the authors, unless otherwise cited. Reprinted with permission.

8 "Hip Fractures in Seniors: A Call for Health System Reform," American Academy of Orthopaedic Surgeons and American Association of Orthopaedic Surgeons, 2008, http://www.aaos.org/about/papers/position/1144.asp (September 12, 2013).

9 Mary Mosquera, "Global Medical Device Market Increases Just 3 Percent in 2012," *Healthcare Finance News*, May 28, 2013,

http://www.healthcarefinancenews.com/news/medical-devices-grow-3-percent-2012 (September 11, 2013).

10 Nicholas Donoghoe, "Medical Device Growth in Emerging Markets: Lessons from Other Industries," *InVivo*, The Business and Medicine Report, June 2012, http://www.elsevierbi.com/publications/in-vivo/30/6/medical-device-growth-in-emerging-markets-lessons-from-other-industries (September 11, 2013).

11 "Medical Technology Innovation Scorecard: The Race for Global Leadership," PricewaterhouseCoopers, January 2011, http://download.pwc.com/ie/pubs/2011_medical_technology_innovation_scorecard_the_race_for_global_leadership_jan.pdf (September 19, 2013).

12 R. J. Devettere, *Practical Decision Making in Health Care Ethics* (Georgetown University Press, 2000).

13 "The Tuskegee Syphilis Experiment," Infoplease.com, http://www.infoplease.com/ipa/A0762136.html (September 18, 2008).

14 Jerry Sais Jr. and Melissa W. Sais, "Meeting Your Fiduciary Responsibility," Investopedia, http://www.investopedia.com/articles/08/fiduciary-responsiblity.asp (September 12, 2013).

15 Errold F. Moody, Jr., "Fiduciary Responsibility," EFMoody.com, http://www.efmoody.com/arbitration/fiduciary.html (September 12, 2013).

16 Ibid.

17 T.L. Beaucham and J.F. Childress, *Principles of Biomedical Ethics* (Oxford University Press, 1989).

18 "The Belmont Report: Ethical Principles and Guidelines for the Protection of HumanSubjects of Research," The National Commission for the Protection of Human Subjects of Biomedical and Behavioral Research, April 18, 1979, http://www.hhs.gov/ohrp/humansubjects/guidance/belmont.html (November 26, 2013).

19 R. Gillon, "Medical Ethics: Four Principles Plus Attention to Scope," *British Medical Journal*, July 16, 1994, p. 184.

20 See The World Medical Association's Physician's Oath, Declaration of Geneva, 1948, http://www.cirp.org/library/ethics/geneva/ (September 12, 2013).

21 See "Council on Ethical and Judicial Affairs," American Medical Association, http://www.ama-assn.org/ama/pub/physician-resources/medical-ethics/code-medical-ethics.page? (September 12, 2013).

1.2 Needs Exploration

INTRODUCTION

Two aspiring innovators are looking for ways to improve sternotomy[1] procedures, and they both contact a leading clinician. One simply asks the clinician for an interview. The other requests permission to follow patients through surgery and post-op care, and to talk with an administrator who can help identify the associated costs. Both innovators are likely to glean important insights from their investigations. But the one who actually sees the procedure performed, follows the patient into recovery, and digs into the actual costs associated with episode will learn dramatically more about the opportunities for improvement in care and where real value can be created.

Before the development of any new solutions can actually take place, innovators must first identify and understand the opportunities that are associated with their chosen strategic focus area. The process of identifying opportunities requires innovators to utilize a combination of background research, first-hand observations, and interviews to find new ways of looking at medical processes, procedures, events, costs, and resource allocation. The well-observed problems that emerge through these activities are at the heart of defining a need – the fundamental building block of the biodesign innovation process.

 See ebiodesign.org for featured videos on needs exploration.

NEEDS EXPLORATION FUNDAMENTALS

Needs exploration is all about understanding various elements of a problem that a new technology or solution may be able to address. To understand a **need** fully, it is useful to consider it in three dimensions. First, there is the core *problem* – the basic issue that is somehow limiting the quality and/or affordability of care somewhere in the continuum of healthcare delivery. While some problems are obvious, others have not yet been recognized, even by those closest to them (see Figure 1.2.1). It may be that the same issue, or a version of it, will be observed in a variety of healthcare settings. The second dimension of the need is the *population* affected by the problem. The relevant population could be a subgroup of patients, a set of providers in a particular specialty, a type of hospital with a certain cost issue,

FIGURE 1.2.1
Direct observations help ensure that important clinical problems (and the associated needs) are not overlooked.[2]

Background research

At this early stage of the process, innovators perform background research to prepare for more detailed, first-hand data collection in their interest areas. In chapter 1.1, it was only necessary to look at high-level information to directionally understand the focus areas under consideration. For example, innovators investigated factors such as the size of the total population affected by a disease, the nature of the disease burden, annual expenditures on diagnosis and treatments, the general effectiveness of available technologies, and how crowded the space is with competitors. During needs exploration, it is time to go deeper to better understand the disease state (pathophysiology, patient demographics, and key terminology), existing solutions (companies linked to available technologies and how/where available diagnostics and treatments are delivered), **stakeholders** (the range of participants in core processes, procedures, and related interactions), and market factors (major expenditures in the space, what are the big ticket items) (see Figure 1.2.2). Primed with this background, innovators will better understand what is being said and done as they conduct observations and interviews. Chapters 2.1 through 2.4 can be used as a guide for deciding which factors to research at this stage in the biodesign innovation process. These chapters outline detailed approaches to conducting research in each of these areas. Innovators should begin their research at a relatively

or an entire healthcare system. The third dimension is the desired *outcome* – that is, the positive change or improved end result that would be experienced by the population if the problem is appropriately solved. Eventually, these three types of information come together in a **need statement**, as outlined in more detail in 1.3 Need Statement Development.

The three most common techniques for performing needs exploration are background research, observations, and interviews, with research typically completed first to help innovators prepare for observations and interviews. This chapter describes effective approaches for conducting this work.

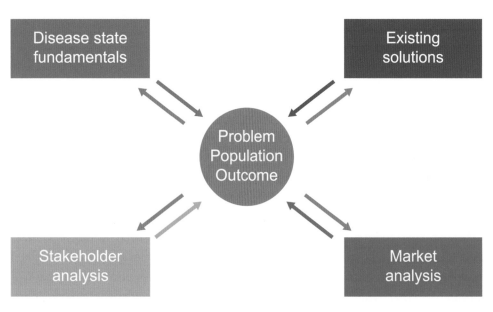

FIGURE 1.2.2
Maximizing the value of needs exploration requires a fundamental understanding of disease, existing solutions, stakeholder, and market factors.

high level with the expectation that it will become more detailed and in-depth as they become more focused on specific needs.

As innovators become more knowledgeable about their focus areas through their research, they can use the acquired information to plan for observations and craft questions to direct their interviews. In doing so, it is essential to explicitly recognize the different perspectives of all involved stakeholders throughout the **cycle of care**, including patients, physicians, nurses, and the many other representatives of the healthcare system. Problems often are uncovered when the inadequacies or limitations of current approaches are identified. But something that is problematic to one stakeholder may be viewed as perfectly acceptable and routine to another. To be sure that nothing is missed, it is essential to research and then explore how processes work from the point of view of every participant. By outlining questions or inquiries from multiple points of view, innovators ensure a broad understanding of potential issues and opportunities across a focus area. As a guide, the questions in Figures 1.2.3–1.2.5 can be used to frame observations from multiple perspectives for a patient undergoing a hospital-based surgical procedure.

Consider colonoscopy screening for colon cancer as an example. Background research in this area should highlight at least two primary stakeholders to consider when preparing for observations and interviews – physicians and patients – and should reveal that their points of view may differ substantially across the cycle of care. If innovators plan only to observe the procedure and interact with the physician, important patient-related insights will be lost. For instance, when physicians comment on the most difficult aspects of the procedure, they often refer to the technical challenges of reaching the furthest regions of the bowel with the colonoscope. In contrast, most colonoscopy patients report that, by far, the biggest problem with the procedure is the "bowel preparation" the day prior to treatment. Patients are asked to drink large quantities of a special preparation that causes severe diarrhea, often accompanied by bloating, flatulence, and other forms of discomfort. Yet this problem is not observable during the procedure. Even when asked directly about patient discomfort, physicians often

The Patient

○ What did the patient have to undergo in terms of pre-operative tests, appointments, etc.?
○ What time did the patient have to get up to prepare for the procedure?
○ Was s/he allowed to eat the night before?
○ What sort of preparation was required?
○ Did the preparation have any negative or unintended side effects?
○ What did the patient experience when s/he arrived at the hospital?
○ How long did s/he have to wait?
○ Was the patient taken to the operating room in a wheel chair or on a gurney?
○ How long did the procedure take?
○ What were the steps of the procedure and how long did each one take?
○ Did the procedure require a general anesthetic?
○ How much pain (or discomfort) did the patient experience during the procedure? Post-operatively? After discharge?
○ What was involved in the post-operative process?
○ What sort of bandage did the patient receive?
○ How often was the bandage changed/wound drained?
○ Was a urinary catheter required?
○ Was intravenous (IV) access required?
○ Were there any complications that resulted from these procedures?
○ How long was it before the patient could discontinue the drain, catheter, or IV?
○ Are there any variations in the ways patients are prepared for, treated during, or cared for after a procedure, depending on the environment?
○ Did the patient need to stay in the hospital overnight? For how many nights?
○ Did the patient need any assistance after hospital discharge?
○ What was the plan for post-operative medications to address infection and pain control?
○ What was the time required before the patient could resume normal activities?

FIGURE 1.2.3

A checklist of illustrative questions for exploring the patient's perspective across all aspects of their care.

describe colonoscopy as relatively easy to tolerate. Their point of view is based on the fact that they see only limited patient discomfort when the scope is in the bowel and tend to forget about or minimize the discomfort that many patients experience when preparing at home.

The Provider

○ What training and certification is required to perform the procedure?

○ Who prepares the patient for the procedure?

○ How many people are present in the operating room?

○ What are their various roles?

○ Does the same person perform the procedure from start to finish?

○ Are practitioner staffing levels and roles the same across different environments?

○ Why is work allocated across practitioners in this way?

○ How long has this been the standard of care?

○ How was the procedure performed before the current approach became standard?

○ What are the accepted primary limitations or difficulties associated with the current procedure?

○ Do the devices (or other tools used in the procedure) perform as the providers want/need them to?

○ How does the provider use the device?

○ Does the provider appear confident using the device?

○ Did the provider have difficulties using the device? Operating it? Implanting it?

○ How many hands were required to operate/implant/use the device properly (i.e., did the provider need assistance operating the device)?

○ Did the provider make any errors while using the device?

○ Was there any evidence of operator fatigue or distraction during the case?

○ How much follow-up is required of the surgical provider(s) after the procedure?

○ What are the most common complications associated with the procedure?

○ Who treats the complications?

○ How (and where) are they treated?

FIGURE 1.2.4

A checklist of illustrative questions for understanding the provider's perspective across all aspects of patient care.

Others in the Healthcare System

Facility, Payer, etc.

○ How much is billed for the procedure?

○ At what rate is the procedure reimbursed?

○ Does reimbursement for the procedure differ depending on the payer?

○ Is the procedure profitable?

○ What factors are most likely to drive up (or down) costs?

○ How long does the procedure take to perform?

○ What aspect(s) of the procedure take the longest to complete?

○ How many resources are tied up as the procedure is being performed?

○ What facilities (e.g., rooms) are tied up as a result of the procedure?

○ What devices, equipment, or supplies are required to support the procedure?

○ How much do the devices, equipment, and supplies cost?

○ To what extent do they affect the profitability of the procedure?

○ Is the procedure performed in only one setting (e.g., operating room) or can it be performed in other venues (e.g., outpatient procedure or radiology lab)?

○ If there are complications to the procedure, who bears this cost?

FIGURE 1.2.5

A checklist of illustrative questions for probing other perspectives during observation and interviews.

Background research can alert innovators to important "disconnects" like this one, and can serve as a guide for planning comprehensive observations and interviews.

When using research to prepare for observations and interviews, innovators must be thoughtful about how they frame their questions, since the way something is asked can influence or **bias** the answer. For example,

when balloon angioplasty (a technique for inserting a catheter into a blocked or narrowed artery in order to inflate a balloon to reopen the vessel) was in its early stages of development, many cardiac surgeons were asked about the **value** of a technology that did not require a sternotomy for treating patients with coronary artery disease. Most responded that they simply could not envision the potential benefits of a procedure that did not allow them to visually access and directly repair the arteries by opening the chest, stopping the heart, and engaging cardiopulmonary bypass. However, when the question was reframed and cardiac surgeons were asked about the value of this approach for high-risk patients who could not tolerate coronary artery bypass grafting (CABG) surgery, most saw the potential for angioplasty to be useful as an alternative.

Observations

Clinical problems, populations, and desired outcomes come to life through direct observations. Typically an observation centers on a singular event that the innovator witnesses. In order to qualify as a real problem, meriting further attention in the biodesign innovation process, the issue raised in an observation should involve an insight about *recurring* situations in which doubt, uncertainty, difficulty, inadequacy, and/or undue cost are encountered. Consider an example that demonstrates how an observation can lead to the identification of a problem, as well as a population and desired outcome:

Observation: A medical resident in training struggles to intubate a patient (place a breathing tube into a patient's trachea) in the emergency room, leading to a drop in the patient's oxygen levels.
Problem: Difficulty placing the endotracheal breathing tube in an emergency setting.
Population: Untrained/unskilled practitioners.
Outcome: Ability to place an endotracheal tube in a timely manner without a dangerous drop in oxygen saturation.

Innovators must carefully assess their observations to ensure an appropriate opportunity is identified. For instance, in the example above, the innovator should ask whether the problem might be a concern for a larger population than just residents. The existence of a more widespread problem is possible, but that determination should only be made with further observations and data collection (i.e., the innovator may need to observe intubations performed by experienced physicians, paramedics, and other care providers). Further, all assumptions made in identifying the problem through an observation should be validated and tested. Again, in the example, the problem noted is based on the assumption that the resident's lack of skill led to the requirement of extra time to place the tube. If this proves to be incorrect (perhaps the problem is instead caused by certain types of patients with challenging anatomy and may not change with improved skill), the innovator could potentially invest time, effort, and money in pursuing a need that does not exist. Notice also that, in this case, reduced time is defined as the core of the problem. However, time may

not be as important to patient care as a more specific clinical endpoint, such as minimizing oxygen saturation changes during the procedure.

A second example reinforces the relationship between observations, problems, populations, and outcomes:

Observation: When an elderly patient was discharged from the hospital after treatment of a cardiac arrhythmia (abnormal heart rhythm), his previous medications were modified and a new medication was added for treatment of the arrhythmia. When seen in the clinic for follow-up in a week later, there is confusion about the medications he is taking and concern about potential interactions between the medications that could be life threatening.
Problem: Directions for medication usage after discharge from hospital.
Population: Elderly patients discharged after hospitalization.
Outcome: Clearly defined instructions for medication use at hospital discharge that results in a reduction in hospital readmissions due to medication interactions.

In both examples, innovators should repeatedly validate the variables related to an issue before naming the problem, population, and outcome, and then subsequently translating it into a need statement (as described in chapter 1.3). This is especially important and potentially difficult for latent problems that have not previously been described.

Setting up observations

Innovators find that it can sometimes be challenging to gain access to appropriate clinical settings to perform observations. One of the main reasons that a large number of **medtech** inventions come from physician inventors is that their work allows them to directly observe relevant problems on a regular basis. Creating the opportunity to perform observations for non-clinician innovators can be difficult, especially given the many safeguards in place to protect patient privacy (see A Note on Ethics and Observations later in this chapter). One of the most effective access strategies is for innovators to leverage their personal networks (and often their extended networks, i.e., friends of friends, distant family members, and introductions gained through casual acquaintances) to conduct observations in a diversity of relevant facilities. Another approach is to partner with a physician or medical professional to address the access issue. If these two strategies are not effective, innovators can make "cold calls" to facilities to request permission to make site visits (an approach that requires substantial patience and perseverance).

When thinking about access, keep in mind that it is important to make observations that span the entire timeline of care. When innovators watch only a few minutes of a surgery, for example, they are almost certain to miss important insights. Observers must understand what is involved in the preparation, procedure, and post-operative care to truly understand potential problems (and corresponding needs) in a focus area. For this reason, it is not enough to just gain access to the operating or examining room. Innovators should also make arrangements to observe waiting areas, laboratory areas, and even administrative spaces. Whenever possible, innovators should also explore environments where patients present with preliminary symptoms and receive follow-up care. To understand how a procedure or interaction is paid for or reimbursed, innovators should further seek access to the finance department of a hospital and/or the accountant in a doctor's office. Here the main focus is to gather high-level information about how the provider is paid and who covers the charges. In many cases, separate permissions may be needed to access each of these target areas.

Conducting observations

The next step is for innovators to immerse themselves in the clinical situation of interest. The process of observation is linked to an approach called **ethnographic research**. The basic ethnographic method involves the researchers becoming immersed in the activities of the people that they want to study with the goal of gaining the in-depth perspectives of that group, including clues about what they think, feel, and may need.[3] In the biodesign innovation context, this means trying as much as possible to assimilate with the group being studied to understand the perspective of the "insiders."

A core feature of modern ethnographic research is to devote considerable attention to establishing **empathy** for the people being studied – that is, sharing their experiences and feelings.[4] This requires that innovators figuratively step into another person's shoes and allow themselves to truly consider what it is like to think, feel, and experience everything that person encounters in a given interaction. One way to "get inside someone else's head" is to get creative about ways to better understand their perspective, above and beyond what is directly observed. For example, when exploring the needs of above-knee amputees in rural India, one team of students wanted a way to better understand the psychology of patients in the target population. To do so, they acquired a specialized device that one could strap to a bent knee to simulate a prosthesis. By walking on this device, particularly over uneven surfaces that would be common in remote parts of India, each engineer experienced first-hand what it would be like to depend on an artificial limb. "It was terrifying," recalled Joel Sadler, one of the team members. The insight they gathered from this experience was that many of the inexpensive prostheses available in the market were scary to use and that amputees could benefit from an alternative that would help them feel more confident walking with the device.[5]

Another important technique for gaining empathy is to "adopt a beginner's mindset." Discovery is partially about being in the right place at the right time. However, it is also about being receptive to new ideas and opportunities when they arise. If innovators go into an observation thinking, "I've already seen this before," it is unlikely that they will be in the state of mind necessary to pick up on the subtle insights that often lead to great ideas. Innovators' own assumptions may actually be misconceptions or stereotypes that can restrict the amount of real empathy that they can build. Adopting a beginner's mindset, by following the points listed below, can help innovators set aside their biases so they can approach observations with fresh eyes:[6]

- **Don't judge.** Try to observe and engage with stakeholder without the influence of value judgments upon their actions, circumstances, decisions, or any other issues that surface.

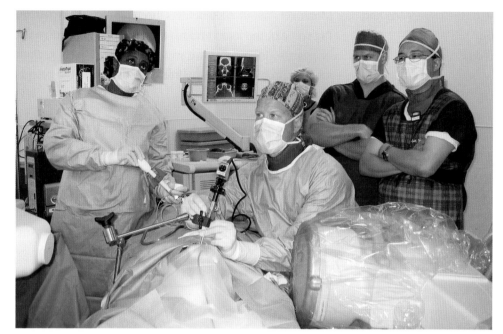

FIGURE 1.2.6
Two leaders of an ENT start-up company observing sinus surgery. Ideally, they will observe not just the procedure, but the complete episode of care that surrounds the procedure (courtesy of Acclarent).

- **Be truly curious.** Strive to assume a posture of questioning and curiosity, especially in circumstances that seem either familiar or uncomfortable.
- **Question everything.** Keep a running list of new inquiries stimulated by different observations.
- **Find patterns.** Look for interesting threads and themes that emerge across interactions with stakeholders.
- **Listen. Really.** Let go of any competing agenda and let the scene soak into the psyche. Absorb what stakeholders do, what they say, and how they say it, without thinking about anything else.

To be considered part of the team being studied, observers need to be willing to commit substantial time and energy to their observations. For example, if innovators are seeking to identify the problems associated with the use of a device in a certain surgical procedure, they need to arrive at the hospital when the surgeon does, watch several unrelated cases, and then observe the entire procedure and the post-operative routine (see Figure 1.2.6). If, instead, the innovators arrive just before the device is used and leave the room as soon as it is put away, they will miss important learning opportunities and potentially leave with an incorrect impression. The surgeon, nurses, and other assisting professionals are unlikely to share feedback and opinions that might reveal problems if they feel that the observers are not really engaged in the **episode of care**. But if the observers stand as long as the team members stand, rest only when they do, and join the team in the cafeteria after the case, people are far more likely to open up. These behaviors are just as likely to result in relevant insights regarding problems as watching the device in use. Usually this is not due to any *formal* statements that are made, but to information gleaned from watching members work and participating in their *informal* discussions. As noted, having some understanding of the medical situation being observed (through background research and a review of the medical literature) will also be viewed favorably and will help innovators present themselves in a professional, educated, interested manner.

One other issue to consider is known as the "observer effect." This refers to changes in the phenomenon being observed that are caused by the mere act of observation. For instance, physicians and other providers may perform tasks or respond to problems differently when they are being watched. Patients, too, may modify their behavior (e.g., their response to pain) when observers are present. As the observers become part of the team, the observation effect can often be diminished because members feel more comfortable with their presence and less conspicuous about their actions.

Performing repeated observations across different settings is also an imperative. While common clinical problems exist throughout the healthcare system, there is tremendous variability in how similar problems are handled in different healthcare environments. For instance, innovators exploring hospital-related problems are likely to uncover significantly different issues in university teaching hospitals compared to community care hospitals. To illustrate this point, consider the process of closing a skin flap after plastic surgery, a practice that requires suturing by hand and can often take as long as three hours to complete. In a university hospital, the surgeon would complete a procedure and then turn the patient over to a resident to close the flap. With many residents on staff, all eager for experience, this time-consuming, labor-intensive process is not viewed as a problem. In a community hospital, however, there are no "extra" resources to complete these kinds of tasks. Surgeons close their own flaps, which ties up their time, potentially limits the number of cases they can manage on any given day, affects their ability to deliver other forms of patient care, and may lessen the overall amount of money that the physician and the facility can earn. Without this type of differential insight, an innovator could potentially miss a problem and an important driver of the related need.

Finally, during the observation process, certain types of events and behaviors can signal innovators about significant problems. These clues (as shown in Table 1.2.1) are often specific to one particular perspective. When these signals are observed, they should be investigated further as they can lead to the identification of opportunities. Although each type of clue is linked to a different perspective, too often the physician's perspective is given top priority, especially during observations, when there is much to be gained from taking another point of view. For example, instead of considering how to find *a faster way to cut during surgery*, the patient's perspective might well be to eliminate the need for cutting altogether. It can be equally helpful to consider problems by making observations from the **third-party payer's** perspective (public and/or private health insurance providers) and the facility's perspective (hospitals, outpatient clinics, etc.).

Moreover, as all stakeholders become more focused on the *value* of the interventions they receive, perform, and/or pay for, innovators are well served to actively seek to identify problems and opportunities related to the affordability of care.

Documenting observations

Thoroughly documenting observations is as important as conducting the observations themselves. Innovators often use an **innovation notebook** to capture their observations so that they can later perform follow-up analysis to identifying problems and insights. The sooner documentation is completed (i.e., during or immediately after the observation), the more likely it is that the data captured will be accurate and allow for key facts not to be forgotten or influenced by innovator biases. Moreover, although the innovators are not detailing any inventions during this early stage of the biodesign innovation process, they are establishing a pattern of documentation that may be useful when they eventually seek to protect their work (see 4.1 Intellectual Property Basics). In particular, capturing observations in the early pages of an innovation notebook helps innovators tell a holistic story of how their ideas came to fruition, which can be helpful if the invention is ever contested.

Importantly, while there is no specific or "correct" way to document observations, innovators should be detailed in their notes (see Table 1.2.2) and record only what is seen. For example:

> *The patient was laid flat on the table. The physician's assistant sterilized the groin area. Then, the doctor tried to gain vascular access through the groin. This took multiple attempts. The doctor mentioned that this was because the vessels were deep and nonpalpable. The patient seemed to experience pain each time the needle was inserted and the physician became increasingly frustrated.*

Ideally, innovators will complete their documentation on an ongoing basis (e.g., at the end of each observation session). In making notes, avoid the temptation to editorialize. Do not begin filtering or classifying information

Table 1.2.1 When conducting observations, innovators should pay special attention to these clues, which may signal opportunities.

The patient	
Pain	Watch patients throughout the cycle of care to identify any points at which they experience any suffering, which may range from mild discomfort to excruciating pain. Sometimes pain is caused by an issue that can be easily corrected (e.g., the patient is under medicated), but in other scenarios a larger problem with the current procedure or device may exist.
Complications	Complications take many forms, ranging from minor incidents that have a limited effect on patients to serious issues that may (in a worst case scenario) result in their death. Anytime a complication is observed, investigate how frequently it occurs and consider if it may be a preventable problem. Pay attention to big incisions, lots of blood, poor healing, and/or infections.
Stress	Stress refers to physical, mental, or emotional strain or tension. Watch patients throughout the cycle of care and seek to determine what aspects of the procedure create anxiety for them. It is also important to watch physicians and other members of the care team to identify when they experience visible stress (and the extent to which the timing or activities correspond to patient stress). Seek to understand what problem(s) might be causing the tension.
Time and convenience	It is not only the physicians' time that matters in a healthcare encounter. As patients exert increasing power as informed consumers (and purchasers) of healthcare, their time and convenience will become more influential factors. Pay attention to how long health encounters take and the time patients must spend away from work (often without pay) as a potentially important need area.
The provider	
Risk	Risk is exposure to the chance of injury or loss. Generally, in their quest to "do no harm," physicians seek to minimize risks when delivering care. If a physician (or other provider) advocates a treatment alternative with higher perceived risk, understand what problems have necessitated the riskier approach.
Malfunction	Whenever a device or other piece of equipment malfunctions, look closely at what caused the problem and how stakeholders respond. Consider the results of the malfunction, including the complications it created and any stress or pain that it caused. And think about whether inadequate training with the device, or a high degree of complexity in using the device, contributed to the malfunction.
Uncertainty	In addition to looking for stress, watch for instances in which a provider is unsure or indecisive about how to proceed. If there is a discussion at this point, listen carefully to identify what the core issue is – why this particular situation is different or particularly challenging. These occurrences may point to problems that have yet to be solved.
Dogma	Dogma refers to settled or established opinions, principles, or beliefs that may or may not represent optimal behavior. If an observer asks why a procedure is performed in a certain way and the provider says, "Because that's how I was trained to do it," or "This is always how it is done," this type of response may be a good indicator that the practice area or procedure may not have been evaluated critically in quite some time.

Table 1.2.1 (*cont.*)

Others in the healthcare system (facility, payer, etc.)	
Inefficiency	Consider the treatment process from the perspectives of the patient, the provider, and the system when seeking to identify problems of inefficiency. For example, in what instances must patients be held overnight while they await test results? Or when is additional staff required to perform only a small part of a procedure? View these issues from the perspective of what a consumer-friendly business outside the healthcare sector would provide to its customers.
Information gaps	The transfer of information and activities from one individual or group to another can be a source of error and stress. Watch particularly for handoffs of information, paying special attention to which parties are involved, why the transfer of information is required, and how much time and energy is needed to complete the transition.
Cost	Cost typically cannot be directly observed. However, innovators can observe certain factors that are drivers of cost, including staffing levels and the skills levels of care providers involved in a procedure, the venue in which it occurs, resource utilization, waste, etc. Watch to understand the role that each participant plays and how the setting and other resources contribute (or not) to the effectiveness of the work being performed. (See the section "Value Exploration" for a more detailed treatment of this topic.)

at this point. And, do not risk trying to interpret information before adequate data is collected. Stay focused on capturing raw data for analysis and interpretation at a later date. (More information about documentation is provided in chapter 4.1.)

Table 1.2.2 The following types of information should be routinely recorded as part of the observation process.

Documentation guidelines
Date, time, and place of observation.
Who was present (name of doctor, number of nurses and other staff members by type, etc.).
Specific facts, numbers, details of what happens at the site.
Sensory impressions: sights, sounds, textures, smells.
Personal responses to the fact of recording field notes (i.e., did someone comment when this particular effect was noted?).
Specific words, phrases, summaries of conversations, and insider language.
Timing of various steps of a process, procedure, or interaction; often good to have a stopwatch available.
Questions about people or their behaviors to be investigated later.

If, during observations, innovators get ideas regarding an invention, they should capture them briefly. Then, it is time to leave these ideas alone until later in the biodesign innovation process (3.1 Ideation). More than anything, innovators should not anchor on any particular solution until more work has been done to understand the real problem – or need – that must be addressed.

A note on ethics and observations

When scheduling and performing observations, it is essential to remain professional at all times and respectful of the approach/limitations of key contacts. People seek medical care due to illness and, therefore, the medical environment is fraught with fear of the unknown and the possibility of impairment or death. Patients and families are fragile during these periods, and providers are ethically obligated to provide them with a safe, respectful environment when delivering care. For these reasons, innovators must remain sensitive to privacy-related issues while working in the medical environment and also the boundaries that providers and healthcare facilities may put into place to protect their patients.

For instance, under the Privacy Rule set forth by the Health Insurance Portability and Accountability Act

(**HIPAA**), patients are provided with comprehensive federal protection for the privacy of their personal health information.[7] This rule, which took effect in 2003, establishes regulations for the use and disclosure of an individual's protected health information (**PHI**) and has resulted in a climate of caution with respect to sharing patient information, including granting admission to observers in the clinical environment. Any individual seeking to perform observations that involve patients and/or patient data should have a thorough understanding of HIPAA regulations and demonstrate sensitivity to the healthcare provider's constraints and limitations under the law. Most facilities and providers require an observer to become HIPAA certified, which is accomplished through a several-hour long training session. Others may request that the observer get written patient consent. As a rule, innovators should be responsive and resourceful in responding to these requests in an effort to increase their likelihood of gaining permission to conduct observations. When in doubt, they should always seek guidance before entering a patient care environment.

Remember that it is a privilege for an innovator to gain access to a healthcare team and patients to conduct observations. The people on the healthcare team are providing real medical care to patients in need during observations. As a result, the innovator's purpose or agenda must always be secondary to allowing the normal pace and manner of healthcare delivery to occur. The innovator must gauge where and when it is appropriate to be present and ask questions. This can be determined by talking with the healthcare team during more informal, less critical periods to gain a better understanding of the team's expectations of the innovator during observations.

Interviews

Once a critical mass of observations have been conducted and documented, innovators can begin organizing and reviewing their data to begin thinking about interesting problems and related insights. In doing so, they should be looking for data points that are particularly memorable or thought-provoking. If working in a team, one way to accomplish this is to do a "story share-and-capture." Ask each member of the group to explain what stood out most about all the things they saw and heard during their fieldwork. Even if all team members were present during the same observations, different problems and insights are sure to emerge from their varying points of view. By listening and probing for more information, team members can draw out more nuance and meaning from the experience than they may have initially realized, which starts the synthesis process. Capture each headline, quote, surprise, or other interesting bits of information on sticky notes somewhere the team can continue to reflect on the information.[8] Over time, innovators can prioritize what has been learned and identify the most compelling problem/insight to pursue. Importantly, though, innovators should seek to synthesize their findings rather than accepting one team member's view over another's. In particular, innovators should be careful not to place more value on the observations of team members with medical training over those from other backgrounds. They should also watch out for information that is incomplete, contradictory, or confusing. These issues may signal the need for additional observations as well as interviews to clarify what is really going on.

Interviews are an essential part of needs exploration, but this activity comes with a few potential pitfalls that are worth noting. In particular, what people *say* about who, what, when, where, how, and why they do something can be somewhat misleading. As Thomas Fogarty, inventor of the embolectomy balloon catheter, as well as dozens of other medtech devices, said:[9]

> *Innovators tend to go out and ask doctors what they want rather than observe what they need. When you talk to physicians, as well as others involved in the delivery of care, you've got to learn the difference between what they say, what they want, what they'll pay for, and what they actually do.*

For this reason, innovators should not rely on interviews in isolation of observations. Instead, they should think about the two techniques as working best in combination.

To prepare for interviews, innovators should go back to their research and the questions they outlined, then update and modify them based on what has been learned through observations. Then they should prepare a unique interview guide for each of the specific people they will be talking with, making sure it includes an appropriate number of questions for the allotted time, as well as those that are best aligned with the perspectives and expertise of the interviewee.

When conducting the actual interviews, keep these guidelines in mind:[10]

- **Ask why.** Even if they think they know the answer, innovators should ask people why they do or say things as the answers can sometime be surprising. Let a conversation started from a single question go on as long as it needs to.
- **Never say "usually" when asking a question.** Instead, ask about a specific instance or occurrence, such as "tell me about the last time you _____."
- **Encourage stories.** Whether or not the stories people tell are true, they reveal how they think about the world.
- **Look for inconsistencies.** Again, what people say and what they do can be different. Watch for these inconsistencies as they can often hide interesting insights.
- **Pay attention to nonverbal cues.** Be aware of body language and emotions.
- **Do not be afraid of silence.** Interviewers often feel the need to ask another question when there is a pause. If innovators allow for silence, a person can reflect on what they've just said and may reveal something deeper.
- **Do not suggest answers to questions.** Even if interviewees pause before answering, don't help them by suggesting an answer. This can unintentionally get people to say things that agree with your expectations.
- **Avoid binary questions.** Binary questions can be answered in a word; you want to host a conversation built upon stories.
- **Be prepared to document.** Always interview in pairs or use a voice recorder. It is nearly impossible to properly engage a **user** and take detailed notes at the same time.

Knowing when to stop and transition to the next step

The process of exploring needs is inherently unpredictable and inefficient. Innovators may have to watch dozens (or even hundreds) of procedures before any significant issues are revealed. In some cases, even that might not be enough. There are certainly instances of smart people exploring interesting focus areas without uncovering any meaningful needs. This potentially means that there is a mismatch in the fit between the focus area and the innovator. Rather than pursuing one strategy indefinitely, there may be times when one is better served by going back and reevaluating the chosen focus area (see 1.1 Strategic Focus).

Because it is difficult to provide an estimate of how much time innovators should devote to needs exploration, an example may be helpful. In the Stanford Biodesign Fellowship (which usually spans about one year), innovators spend approximately two months doing background research, performing observations, conducting interviews, and validating what they have learned. In general, they commit roughly three weeks of this time to a preliminary round of research, observations, and interviews; they then spend the remaining weeks more deeply exploring particular areas of interest (which frequently involves a return to the clinic for more observation and conversation). After that, the flow of ideas tends to taper off. A sign of this may be that there are fewer and fewer *new* observations or insights and a few others that come up repeatedly. If that is not the case, and the innovators are still searching for problems that they perceive to be clinically important, it may be time to move on to another focus area.

When preparing to take the next step and begin translating problems, opportunities, and outcomes into needs (as outlined in chapter 1.3), be certain to maintain good relationships with the patients, providers, and representatives of the system who have been observed. Once a need statement has been developed and additional research performed, it will be necessary to return to the clinical environment to validate the need before concept generation begins. Having these relationships to leverage in the validation process is extremely helpful.

The following story about a multidisciplinary innovation team at the University of Cincinnati provides an example of how needs exploration can effectively be performed.

FROM THE FIELD ▶ UNIVERSITY OF CINCINNATI MEDICAL DEVICE TEAM

Observing problems as part of needs exploration

Mary Beth Privitera, an associate professor of biomedical engineering, is always on the lookout for problems in the medical field. As the co-developer of the University of Cincinnati's Medical Device & Entrepreneurship Program, she is responsible for bringing together students in their senior year from biomedical engineering, industrial design, and the business honors program, dividing them into multidisciplinary teams, and assigning them real-world medical issues to investigate as part of a year-long innovation process. Each academic year, these projects are sponsored by companies and/or physician researchers in the medical device field and guided by experienced faculty from the colleges of design, art, architecture and planning, engineering, medicine, and business.

In the fall of 2006, Privitera was approached by Respironics, a medical device company with a focus on sleep and respiratory solutions, to identify the problems and needs of sleep apnea patients. Obstructive sleep apnea (OSA) is a condition that causes an individual to stop breathing repeatedly during sleep because the airway collapses. The most common symptoms of OSA are loud snoring and restless sleep, but it can also cause headaches, forgetfulness, depression, and anxiety, as well as other mood changes. In severe cases, sleep apnea causes pressure on the heart that can ultimately lead to heart failure or stroke. Approximately one in five Americans suffers at least minor sleep apnea.[11] Although some patients undergo surgical procedures to modify the airway mechanics, most patients opt for a non-invasive solution. Continuous positive airway pressure (CPAP) is a mask-like device worn during sleep, which supplies a constant stream of pressurized air to prevent the airway from collapsing (see Figure 1.2.7).

While CPAP is effective in preventing the symptoms of sleep apnea, many patients find the device

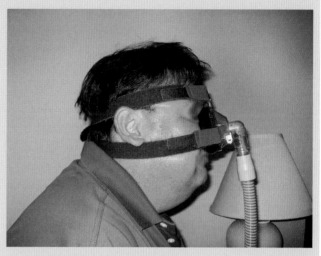

FIGURE 1.2.7

A patient with obstructive sleep apnea demonstrates how he wears the CPAP device (courtesy of Mary Beth Privitera and the University of Cincinnati; note that the patient is disguised to protect his privacy).

uncomfortable and difficult to use. As a result, "Respironics was particularly concerned with patient compliance and promoting a more positive patient experience during device use," recalled Privitera.[12]

Privitera assembled a team of biomedical engineers and industrial designers to better understand the problem. Despite the availability of sleep clinics that provide an environment in which a patient's sleep patterns can be carefully monitored, Privitera, along with Respironics and a team of faculty guided the students to interview and observe patients in their homes. "I'm a firm believer that the problems people have don't happen in a lab," she said. While carefully controlled experiments have their place in observing problems, Privitera felt that the student observers would learn more by being able to see how patients interact with their CPAP devices in the environment where they use them. "I always try to send students to the location where the problems really occur," she said, emphasizing that this approach is consistent with the focus on ethnographic research that she advocates in her courses.[13]

To gain a clinical perspective on the problems associated with non-invasive treatment of sleep apnea, the faculty team suggested that students should also meet with the specialists treating sleep apnea patients in the sleep centers as part of the observation process. To prepare for these interactions, the students researched sleep apnea and acquainted themselves with the current CPAP devices. Privitera highlighted the importance of prior research before beginning to interview doctors: "We don't expect a physician to talk in laymen's terms. I want the student to speak in the language of who they're interviewing, and who they're observing."

The students also developed extensive interview protocols to ensure consistency in the observation and data collection processes. "With multiple people on the team, we needed to do everything we could to achieve consistency across interviews," Privitera said. "The research protocols outlined specific questions they would ask, and then specific activities that they would observe. For example, in patient interviews the team would ask what they liked and disliked about the device, what improvements they would recommend, and how they used the apparatus. They would also observe the patient using the equipment, cleaning it, and performing other common behaviors.

When this background work was done, the next challenge was to identify patients and physicians willing to participate in the observation process. "I'm a firm believer that students need to learn how to make these contacts themselves," explained Privitera. Working with nothing more than a list of sleep centers in the area, the students made "cold calls" to the clinics to schedule appointments and gain access to lists of patients who might be willing to participate.

Reflecting on common pitfalls associated with the observation process, Privitera noted that knowing when to stop observing can be a challenge. "Typically you have enough information when you start to recognize repeat patterns of behavior," she said. "It might be after

12 hours, or it might be after 30 hours, but you have to stop at the point when the same situation repeats itself a couple of times." She also cautioned observers to take their time and let the patients [or physicians] do the talking. In her experience, observers are often eager to volunteer information to show their knowledge. They can also have a tendency to anticipate people's answers or misinterpret a response if it differs from what they are expecting.

As with every needs exploration process, the students on the sleep apnea project identified problems that they did not initially foresee. Through careful observation, it became clear that sleep apnea patients were actually quite diverse in the issues they faced and the extent to which these problems affected their compliance with their recommended treatment regimen. This led the team to develop a series of different "personas" to help differentiate patients and their needs. Privitera explained, "When we put forth our plans for the quarter, we didn't expect to develop the personas, but it happened. We saw different interrelations that led us to these six categories of users, which did not necessarily reflect one person, but were combinations of people that had some of the same sensitivities and were like-minded" (see Figure 1.2.8).

These personas helped the team define much more detailed patient problems, populations, and desired outcomes that varied by segment. For example, the Hipster persona was young, single, socially motivated, and concerned about appearances. To address the specific concerns of this group, a solution would have to be quiet (so as not to disturb roommates), compact and easily camouflaged for communal apartment and/or dorm room living, non-institutional in appearance, and customizable in its fit and style. In contrast, the Metro persona was health conscious, spiritual, concerned with personal fulfillment, and interested in enhanced experiences. The desires of this segment would be driven by a serene user experience and could include criteria such as a built-in sleep mask (to block out light) and integrated audio (for "white noise" or other soothing sounds). The Dude persona had a completely different

FIGURE 1.2.8
The six personas identified by the team: Nomad, Hipster, Metro, Dude, Gramps, and Trucker (courtesy of Mary Beth Privitera and the University of Cincinnati).

perspective. Members of this segment were relatively unhygienic, unconcerned with their appearance, and motivated by convenience above all else. This led to an interest in a disposable contact interface, a rugged and durable device, self-cleaning functionality (or some mechanism that provided automatic feedback when it was time to clean or replace it), and an exceptionally easy user experience.

As the biodesign innovation process progressed, the team used this information to develop individual need statements and, later, specific need criteria for each persona. They presented this information to the company, giving Respironics the opportunity to potentially develop unique solutions for the segments of greatest interest.

A focus on value exploration

A central theme throughout this text is the critical importance of maximizing economic value in developing new medical technologies. A strong value orientation begins during needs exploration. Although delivering a clinical improvement at a reasonable cost has always been a goal for innovators, healthcare stakeholders of all types are placing increased weight on the cost of new products and services in the value equation. Today, it is not unusual to hear investors or companies say that they are only really interested in technologies that can actually *reduce* cost (not "merely" deliver more health at a

reasonable cost). This profound shift in emphasis creates a new set of opportunities and challenges for innovators – and a new emphasis for needs exploration. As innovators conduct their observations, they should be actively searching for waste, inefficiency, and other sources of undue costs that can be eliminated. This activity, which can be thought of as "**value exploration**," involves scanning for need areas where costs potentially can be reduced while holding quality/outcomes steady, or even reducing outcomes to a small, acceptable degree.

Some of the target areas outlined in Table 1.2.2 can help innovators identify value-related opportunities.

However, there is a more specific set of practice-based markers or value "signposts" that innovators can observe during their clinical immersion, as shown in Table 1.2.3.

Many of the examples included within this text follow the direction of at least one of these value signposts. The development of coronary angioplasty described earlier in this chapter is a classic example. Angioplasty gradually replaced much of bypass surgery in certain patient populations due to its positive patient outcomes and the improved value it offered through cost reduction. Savings were realized in a number of ways, as summarized in Table 1.2.4.

Value exploration is an even greater priority for innovators working in severely resource-constrained environments. As is the case for all types of needs exploration, it is important for the innovator to have a direct experience of the value environment – as illustrated in the story of a team's experience in India (see next page).

Looking back at the practice-based value signposts, the team in this example was able to significantly reduce the resource use associated with limb immobilization. By making a device that was simple and intuitive to use correctly, they also were able to equip care providers to more effectively perform their work and prevent more highly trained professionals from having to intervene as they sometimes were required to do with more complex splints.

Opportunities for greater value realization can be found in almost any environment. Again referring to the story, the emergency department at Stanford was most likely just as ripe with possibilities for reducing costs; however, the "clues" were more apparent in the Indian setting. The key is to think critically about the **standard of care** and question why activities are undertaken in the prevailing manner. For instance,

Table 1.2.3 These observable value exploration signposts can point innovators to opportunities to reduce the cost of care.

Practice-based value signposts
Potential to keep a patient out of the hospital and/or emergency department.
Potential to change the location of care to a less expensive venue.
Potential to shorten a patient's length of hospital stay.
Potential to reduce the procedure time or resource use for a given intervention.
Potential to shift a procedure or service to a lower-cost provider.
Potential to reduce the number of staff and/or intensity of labor necessary to administer a given intervention.
Potential to diagnose a condition earlier to reduce complications and/or slow/prevent disease progression.

Table 1.2.4 The transition from CABG to angioplasty created economic value for healthcare stakeholders in multiple ways.

Practice-based value signposts	Example of value realized through the shift from CABG surgery to angioplasty
Potential to change the location of care to a less expensive venue.	Procedure moved from the operating room to catheterization suite. Patients discharged home or to ward rather than to the ICU or post-anesthesia recovery unit.
Potential to shorten a patient's length of hospital stay.	Over time, elective angioplasty became primarily an outpatient procedure.
Potential to reduce the procedure time or resource use for a given intervention.	Both resource usage and procedure time were reduced with the elimination of cardiopulmonary bypass.
Potential to reduce the number of staff and/or intensity of labor necessary to administer a given intervention.	Angioplasty is typically performed by a single cardiologist with a nurse and technician assistant, compared to a full surgical team (including multiple OR nurses and a pump perfusionist).

FROM THE FIELD ▷ HICARE LIMO

Needs exploration with a focus on value

Darshan Nayak and Pulin Raje were Stanford-India Biodesign fellows when they conducted clinical immersion at the Stanford University Medical Center. The focus area for their fellowship was emergency medicine, and they spent several weeks observing clinical practice in the Stanford emergency room and with its ambulance service looking for unaddressed problems and opportunities. When asked to summarize some of the key take-aways from these observations, Nayak commented. "The settings in which we performed our observations were relatively well-resourced, with highly trained staff members, state-of-the-art equipment, and an abundance of supplies. Everybody seemed to have what they needed."

When Raje and Nayak returned to India, they had another chance to conduct observations in a variety of emergency medicine environments. Through their mentors at the All India Institute of Medical Sciences (AIIMS), they gained access to five different settings: the AIIMS emergency department, the AIIMS trauma center, a local ambulance service, rural district hospitals, and primary care centers. The two men were struck by the differences they observed. In stark contrast to their US experience, ERs in India were overcrowded and severely under-staffed. Nayak recalled, "There could be multiple patients arriving at the ER for treatment every few minutes, with just a small number of doctors responsible for delivering care. It is extremely difficult to triage and treat patients in this scenario, and clear protocols often do not exist or are not followed." Additionally, they found that staff members were often less well-trained. And in most settings, especially outside of tertiary-level urban centers, facilities lacked necessary equipment, supplies, and other resources necessary to provide high-quality care.

Raje explained how fragmentation among the players in emergency medicine further contributed to suboptimal care: "Unlike the US where everyone knows to call 911 and there is an unbroken chain of care between the ambulance and the ER, in India the patient's first contact after a trauma maybe the rickshaw puller who brings him to the hospital." One other factor complicating the landscape is that health insurance is practically non-existent, with most Indian patients paying out of pocket for medical procedures and devices. In this environment, both the patients and the providers are extremely cost conscious. According to Nayak, everyone is thinking, "Is this really needed or not?"

Nayak stressed that the system in India, though under-resourced, was not "bad," but "simply different." "We had to set aside what we saw at Stanford and start with a blank slate, rather than comparing what we saw here to what was observed in a different setting," added Raje. Importantly, they believed the environment was rich with opportunities to add value by improving care quality while bringing down costs.

One problem that caught the team's attention stemmed from multiple observations. Nayak and Raje first saw a patient come in to the district level hospital in Vallabhgarh, outside New Delhi, after a road accident. "He was taken off the rickshaw and carried in by three people, who were not hospital staff. He had a broken limb, which had not been supported or immobilized since the accident. The first responder in the hospital was a ward boy, followed by a junior resident who checked the patient's airway and circulation," Raje described. Since the hospital did not have an orthopedic specialist, the patient was informed that he could not be treated at this facility and that he would have to get himself to a different center. This time, to help immobilize the injured leg during transport, hospital staff strapped a wooden plank to his leg using gauze and bandages. Later, at the AIIMS trauma center, a tertiary-level facility where patients like the man with the leg injury are often referred, Nayak and Raje observed the same problem from a different point of view. Patients routinely arrived at the trauma center with dangling limbs or make-shift splints attached to their broken limbs. A relatively small percentage of patients arrived on an ambulance with a proper splint. However, the team noticed that these devices were often not used correctly. Subsequent investigation revealed that many care providers in emergency medicine did not understand basic concepts of limb immobilization and found available splints difficult to fit to injured patients.

Nayak and Raje spoke with ambulance attendants to get their point of view on the issue of limb immobilization. Through these discussions, they uncovered important information. As Nayak explained, "Ambulance services are accountable for the equipment they own. Even though they have access to existing splint technologies, the attendants often won't deploy them because of the high cost associated with the splint." The team also observed scenarios in which an ambulance attendant would fit the patient with a splint during transport but then retrieve the device before leaving the patient at the hospital. Unfortunately, the removal of the splint often aggravated the patient's injury. After further investigation, Nayak and Raje discovered that many hospitals and primary care centers were similarly reluctant to part with the splints they had in inventory.

Sensing that they had hit on an area where they could add value for patients and providers alike, the team became interested in improving limb immobilization (problem) in trauma patients (population) to prevent the aggravation of their wounds during transfer (outcome). "The key insight was that the splint had to be affordable enough to give away, while also being effective and easy-to-use," said Raje. To gain a better understanding of what would be required to change existing patterns of behavior, Nayak and Raje again spoke with ambulance attendants, as well as procurement teams at ambulance companies. "We learned that available leg splints cost about Rs. 1200 (approximately US$20) and that an ambulance attendant would try to use one device three to five times before discarding it or leaving it with a patient," Raje stated. This information indicated that if they could devise a better solution that cost roughly Rs. 300 (US$5), they would potentially be able to stimulate adoption.

Raje and Nayak went on to develop a new lower-limb splint that is effective, easy to use, and inexpensive

enough for providers to leave with patients after transfer. The device is manufactured using a special pressed paper, coated with protective plastic layer. It is longitudinally reinforced to provide excellent structural support for immobilizing the limb. The device conforms to multiple limb sizes, as well as both the left and right limbs, and it is fitted to the leg using a series of simple Velcro straps (see Figure 1.2.9).

The composition of the splint is entirely radiolucent, so the device does not have to be removed during X-ray imaging (preventing further exacerbation of the wound). Finally, the device is easily stacked for space-efficient storage in an ambulance or ER setting, and it is disposable after use. In 2013, they licensed the technology to a major Indian manufacturing company called HLL Lifecare Limited. Now available in the market under the brand name HiCARE LIMO, the manufacturer hoped to capture market share based on the value delivered by the innovative and affordable solution.

Reflecting on the important role of observations and needs exploration in their experience, Nayak and Raje emphasized that it is critical to view an interesting problem from multiple points of view. "You don't want to only think about problems from only the clinical perspective," noted Raje. By looking beyond care delivery processes to the purchasing and cost factors that drive then, the team was able to identify a meaningful need. They also noted that sometimes significant value can be generated by addressing "a simple problem that requires a simple solution."

FIGURE 1.2.9

A patient using the HiCARE LIMO splint as part of a clinical trial (courtesy of Pulin Raje, Darshan Nayak, and Stanford-India Biodesign).

when observing a doctor diagnosing a patient in an office visit, watch for the presence (or absence) of factors that make that face-to-face interaction essential. Later, when it is time to begin thinking about possible solutions, innovators can question whether the in-office diagnosis could potentially be replaced with a less expensive encounter – or perhaps some other form of communication.

Beyond these practice-based value signposts, innovators should be aware of a second set of value indicators that are emerging in the medtech field (see Table 1.2.5). These budget-based signposts are not found by direct observation in the clinical setting, but instead are identified as part of the background research that innovators complete in preparation for clinical immersion, as well as the interviews they conduct to clarify potential problems and opportunities. Given the economic nature of these signposts and the fact that relevant information can sometimes be buried deep within facility budgets and/or institutional documentation and reports, identifying budget-based signposts can require a certain level of in-depth investigation. Contacts within the finance departments of hospitals and other care facilities can be useful in this exercise.

By exploring these signposts through different forms of inquiry, innovators will potentially uncover additional problems which, in turn, may be worthy of direct observation. For example, a report by the US Institute of Medicine found that the single greatest driver of geographic variation in Medicare spending is post-acute care expenditures (i.e., the use of home health services, skilled nursing facilities, rehabilitation facilities, long-term care hospitals, and hospices).[14] This is a budget signpost, meaning that innovators may discover opportunities to improve economic value by focusing on post-acute care (for example, care for patients following myocardial infarction, congestive heart failure, or hip fracture). To begin to dig into the issues, innovators could conduct observations in facilities that fall above the average spending level, as well as those that fall below it. They should develop an understanding of the major drivers of cost at these facilities and understand what accounts for key differences. Importantly, they should also understand how evolving policies from

Table 1.2.5 Budget-based value signposts may not be directly observable, but they can point innovators to areas that are ripe for innovation and where observations can be performed.

Budget-based value signposts
Diagnostics, treatments, or procedures that are outliers in terms of their cost-effectiveness.
Diagnostics, treatments, or procedures that represent big line items in health system, payer, facility, or physician practice budgets.
Diagnostics, treatments, or procedures that are routinely unprofitable.
Diagnostics, treatments, or procedures that are significantly less expensive in other geographies.
Diagnostics, treatments, or procedures in which technology is a high percentage of the total cost.
Conditions for which the life-long cost of care is especially high.
Areas where providers are challenges to achieve quality measures linked to shared savings goals for Accountable Care Organization (see chapter 2.3 for an introduction to ACOs).
"Never events" – complications or outcomes where a reimbursement penalty has been (or will be) implemented.

Medicare or other payers may impact the way caregivers and facilities are reimbursed for this kind of care.[15]

Another example of how a budget-based signpost can lead to an interesting need is found in David Green's experience conducting needs exploration in India. When he was working with the SEVA Foundation, whose mission is to prevent blindness and restore sight worldwide, he became intrigued by the work of Dr. Govindappa Venkataswamy. "Dr. V," as he is known, mortgaged his home to establish the Aravind Eye Hospital and provide free and low-cost cataract surgery to Indians who otherwise would not be able to afford treatment. The hospital performed an impressive 5,000 surgeries in its first year.[16] However, studying the model, Green realized that the number of surgeries Dr. V could complete was constrained by the high cost of the replacement lenses required for each patient. At up to $150 per pair from US manufacturers,[17] Aravind was limited in the total units it could purchase without jeopardizing the sustainability of its model. The value signpost here was the fact that the technology accounted for such a high proportion of the total cost for the episode of care. In response, Green invented a new manufacturing solution that allowed him to produce comparable lenses for just $10 per pair,[18] significantly reducing the cost of the technology as a percentage of the overall procedure. Due in part to Green's work, the cost of cataract care in an Aravind center is now sustainably provided at a fraction of the cost of surgery in a Western facility. This striking geographic variation (for outcomes that do not appear to be substantially compromised) could now potentially signal another round of innovation for improved value realization in developed markets.

New opportunities related to the budget-based value signpost associated with **reimbursement** penalties are also gaining increased attention in the US. In fiscal year 2013, the Centers for Medicare and Medicaid Services (CMS) implemented its Hospital Readmissions Reductions Program. In the first year, this program withheld up to one percent of regular reimbursements for hospitals that exceed pre-defined patient readmission rates within 30 days of discharge for three medical conditions: heart attack, heart failure, and pneumonia. At one percent, these penalties amounted to $280 million. However, at the time of this writing, the maximum penalty was scheduled to increase and the program was expected to expanded to include readmissions for other medical conditions.[19] Innovators like Xiangwen Zang, founder of AirCare, are capitalizing on these penalties (and the demand they create among healthcare stakeholders seeking to avoid them) by developing technologies and systems to improve patient monitoring post-discharge and facilitate more targeted interventions that keep patients healthy and out of the hospital. Zang described his needs exploration process in an interview: "I went about looking at how to fix the healthcare industry, and realized readmissions is such a big issue for hospital systems – it's one of the top priorities for any hospital executive, and that's because the readmissions penalty

alone has the potential to wipe out a third of hospitals' profits."[20] The solution that his company ultimately devised allows patients to fill out a checklist of symptoms and indicate whether they have taken their medication as prescribed using an Internet and tablet-based application. Nurses can review this information without having to place outbound calls to the patients, which saves them significant time. The technology also provides them with a list of at-risk patients so that they receive timely follow up. To date, the company is primarily targeting the reduction of heart-failure readmission rates.[21]

Global considerations in exploring needs

The fundamental aspects of needs exploration are the same in any geographic location. However, there are some important considerations to take into account when innovators are not originally from the region in which they intend to work. Before conducting any observations or interviews, it is imperative to understand the dominant cultural norms in the environment where they will take place. By adapting their approach based on relevant customs, innovators can dramatically increase their effectiveness. For example, something as simple as knowing how to dress appropriately and the level of formality to use when interacting with physicians, nurses, patients, and other healthcare stakeholders can help make an interview or a day of clinical immersion much more successful and productive.

It is also important to take factors unique to the geography into account when planning research, observations, and interviews. One graduate from the Stanford-India Biodesign Innovation Fellowship, Ritu Kamal, shared her experience conducting needs exploration in India to highlight the types of issues that should be considered. "Healthcare facilities across India are incredibly heterogeneous, so it's critical to conduct your clinical immersions at a wide variety of sites in order to fully observe clinical practice," she said. Healthcare delivery varies by state and region, but also between tertiary, secondary, and primary care facilities, public and private institutions, and healthcare providers in rural and urban settings. For instance, in leading tertiary care facilities in urban areas, such as the Narayana Hrudayalaya hospitals and clinics, innovators encounter state-of-the-art technology and healthcare professionals whose training is on par with physicians in the US and Europe. At the other end of the spectrum, primary care clinics in more remote and poorer locations might depend on a single healthcare provider (who may or may not be a licensed doctor) doing his/her best to care for community members with little more than a stethoscope and a blood pressure cuff in the way of devices (pharmaceutical products are much more widely available than medical technologies). Moreover, while clear standards of care exist in higher-end facilities, they may be lacking or absent in public and rural settings, making it significantly more difficult for innovators to confidently discern routine procedures and processes. The cost of care can also vary dramatically, along with the ability of patients to pay for diagnosis and treatment. These extreme differences contribute to dramatic inconsistencies from setting to setting when it comes to the problems being experienced, the populations they affect, and the improved outcomes that are needed.

Mark Bruzzi, director of the BioInnovate medical device training program at the National University of Ireland, underscored a similar theme when innovators are working in Europe. It is important, he explained, to know enough about how healthcare is delivered in the target geography to pursue the right settings for observations. For instance, while innovators sometimes overlook primary health settings for clinical observations in the US, that would be a mistake in parts of Europe. "In Ireland and the UK, primary care physicians often act as the gatekeepers to hospital and specialty services. It's often not possible to self-refer to specialists, so these physicians play an important role in directing treatment. And their decisions have a tremendous impact on how the patient is managed from presentation through post-operative care," he said.

Another issue raised by Kamal had to do with access. "In India, there are very few rules governing things like patient privacy, so if a physician agrees to take you around to perform observations you can basically go anywhere and see everything," she commented. This scenario allows for tremendous learning during needs exploration, but requires innovators to act responsibly, self-monitoring to ensure that patients are treated with dignity and respect and that sensitive information is treated confidentially. "You also have to be careful that people don't mistake you for a doctor," Kamal added. "Doctors are so revered and their time is so difficult to get that people may approach anyone in a lab coat or with an identification badge to ask for their medical advice." If this happens, she advised, innovators should quickly redirect the patient to someone in the facility who is authorized to help.

In Ireland, gaining access to clinical settings to perform observations also relies heavily on doctors. As Bruzzi explained, "With respect to the Irish system, people are very open to innovators going in and observing, but it's very much led by the clinician." However, he added, "I would emphasize that clinical nurse managers have significant power and influence regarding what goes on in the wards and the operating theaters, so getting to know these individuals can be very useful."

Other stakeholders may play unexpected roles in unfamiliar regions. Again using India as an example, Kamal said, "Patients often travel long distances to get to the hospital or clinic, and it's not unusual for them to bring their entire family with them. So you'll see big groups of family members camped out in waiting areas or at the bedside. And they're much more involved in helping deliver care at the hospital and in the home." Accordingly, in some settings, innovators may be well served to consider families as a key stakeholder group and observe their involvement in the delivery of patient care. However, cautioned Kamal, "Be sure to ask permission of the physician who's hosting you before directly approaching a patient's family. It may seem strange to have to worry about this since observers are generally granted such broad access, but there's something different about engaging with family members and you need to be sure it's acceptable to your host."

A final issue worth noting has to do with understanding how medical technologies are regarded in different environments. Innovators should not assume that stakeholders are receptive or comfortable with devices, or that their presence necessarily corresponds to their use. In some low-resource settings, innovators will be struck by the lack of devices, or that devices manufactured for a single-use are being sterilized and re-used on different patients. In other locations, they may discover that devices have been made available to address important needs, but they are not being used appropriately (or at all). For instance, when team members at D-Rev became interested in the problem of infant jaundice, they initiated a detailed assessment of the phototherapy landscapes in India and Nigeria. From that work, D-Rev confirmed that jaundice was a challenge in rural areas, where equipment to treat the condition was virtually non-existent. But the team discovered that the situation was also problematic in urban hospitals and clinics in those countries. In these settings, phototherapy equipment was typically available, but a full 90 percent of the devices evaluated by D-Rev (in collaboration with the Stanford School of Medicine) were ineffective. The vast majority of solutions designed for these low-resource settings failed to meet international quality standards and offered suboptimal performance. Some healthcare providers even relied on homemade solutions that were not only unproductive but dangerous to the infants that they were intended to treat. Another factor that rendered phototherapy solutions ineffective in these settings was the cost to operate and maintain them. In D-Rev's research, one out of three phototherapy devices had at least one bulb missing or burned out, with many healthcare providers unable to reliably procure or afford replacements.[22]

Bruzzi provided a contrast to this scenario in describing the environment in Ireland. "In general," he said, "the vast majority of doctors are very interested in new innovations and in engaging with innovators. They're eager for new and better ways of doing things and open to trying new technologies." The point is that stakeholder receptivity to device-based tests and treatments can vary significantly, as can the skill levels of healthcare providers to use complex technology. Issues such as these must be explored when conducting a clinical immersion.

 Online Resources

Visit www.ebiodesign.org/1.2 for more content, including:

 Activities and links for "Getting Started"
- Perform background research
- Set up observations
- Conduct observations
- Document observations
- Refine problems/insights through interviews

Videos on needs exploration

CREDITS

The editors would like to acknowledge Steve Fair and Asha Nayak for their help in developing the original chapter as well as Mark Bruzzi, Ronnie Chatterji, Krista Donaldson, Raj Doshi, Ritu Kamal, and Greg Lambrecht for adding their insights to the updated version. Many thanks also go to Darshan Nayak and Pulin Raje of HiCARE LIMO, as well as Mary Beth Privitera, Respironics, and the University of Cincinnati student team: Laurie Burck (Biomedical Engineering Leader), Nate Giraitis (Industrial Design Leader), Celina Castaneda, Christina Droira, Adam Feist, Tom Franke, Christine Louie, Bryan Porter, Nicole Reinert, and Rebecca Robbins.

NOTES

1 In a sternotomy, a vertical incision is made along the sternum and it is cracked open to access the heart or lungs during surgery.
2 All cartoons by Josh Makower, unless otherwise cited.
3 "Understand Mixtape: Discovering Insights Via Human Behavior," Hasso Plattner Institute of Design, Stanford University, 2012, http://dschool.stanford.edu/wp-content/uploads/2012/02/understand-mixtape-v8.pdf (September 26, 2013).
4 "Empathy," *Merriam-Webster*, http://www.merriam-webster.com/dictionary/empathy (September 26, 2013).
5 "Global Health Innovation Guidebook," Stanford University Graduate School of Business, August 2013, http://csi.gsb.stanford.edu/sites/csi.gsb.stanford.edu/files/GlobalHealthInnovationGuidebook_2.pdf (September 26, 2013).

6 "Understand Mixtape: Discovering Insights via Human Behavior," op. cit.

7 See United States Department of Health and Human Services, "Health Information Privacy," http://www.hhs.gov/ocr/hipaa/ (September 26, 2013).

8 "Understand Mixtape: Discovering Insights via Human Behavior," op. cit.

9 From remarks made by Thomas Fogarty as part of the "From the Innovator's Workbench" speaker series hosted by Stanford's Program in Biodesign, January 27, 2003, http://biodesign.stanford.edu/bdn/networking/pastinnovators.jsp. Reprinted with permission.

10 Ibid.

11 A.S. Shamsuzzaman, B.J. Gersh, V.K. Somers, "Obstructive Sleep Apnea: Implications for Cardiac and Vascular Disease," *Journal of the American Medical Association*, October 2003, pp. 1906–14.

12 All quotations are from interviews conducted by the authors, unless otherwise cited. Reprinted with permission.

13 According to Privitera, ethnography is a research method completed through in-depth user interviews and directed observations in the context of people and tasks targeted with the design problems. Its primary advantages are that the approach: (1) helps uncover the differences between what people say and what they do; and (2) enables the researcher to describe what a device needs to do in context.

14 Joseph P. Newhouse and Alan M. Garber, "Geographic Variation in Medicare Services," *The New England Journal of Medicine*, April 18, 2013, http://www.nejm.org/doi/full/10.1056/NEJMp1302981 (February 25, 2014).

15 Robert Mechanic, "Post-Acute Care – The Next Frontier for Controlling Medicare Spending, *New England Journal of Medicine*, February 20, 2014, http://www.nejm.org/doi/full/10.1056/NEJMp1315607 (February 25, 2014).

16 "The New Heroes: Dr. Govindappa Venkataswamy ('Dr. V') & David Green," PBS.org, http://www.pbs.org/opb/thenewheroes/meet/green.html (February 27, 2014).

17 Ibid.

18 Ibid.

19 "CMS: The 2,225 Hospitals That Will Pay Readmissions Penalties Next Year," Advisory.com, August 5, 2013, http://www.advisory.com/daily-briefing/2013/08/05/cms-2225-hospitals-will-pay-readmissions-penalties-next-year (February 27, 2014).

20 "AirCare Aims to Fix Readmissions Problems," *Region's Business*, October 24, 2013, http://philadelphia.regionsbusiness.com/print-edition-news/aircare-aims-to-fix-readmissions-problems/ (February 27, 2014).

21 Ibid.

22 Lyn Denend, Julie Manriquez, Stefanos Zenios, "Brilliance I: From Prototype to Product Company," *Global Health Innovation Insight Series*, June 2012, http://csi.gsb.stanford.edu/brilliance-i-prototype-product-company (September 26, 2013).

1.3 Need Statement Development

INTRODUCTION

Innovators are usually able to quickly identify problems during observation. The greater challenge is in understanding the associated clinical need and in translating problems into a meaningful need statement. For example, after observing the difficulties some physicians have when cutting the sternum for thoracic surgery, a well-intentioned innovator may define a need for "a more effective cutting device to perform a sternotomy." This need will lead her to investigate multiple solutions for cutting the skin and bone during thoracic procedures. However, if someone else is simultaneously exploring the need for "a way to access the chest to perform procedures on organs of the thorax," the options for innovative solutions will be much broader. In fact, the second need statement could lead to minimally invasive thoracotomy, which the first need statement would not. Ultimately, the first innovator may find that she faces a significantly diminished demand for her new cutting tool if the broader need is met by the second innovator's solution.

Needs correspond to opportunities for innovation. They are characterized by defining an outcome that currently is unmet for a problem in a particular population, which helps direct the opportunity. Too often, clever innovations fail because they have not been developed to address "real" customer and/or market needs. Creating explicit needs statements is a powerful way to prevent this mistake. By clearly and concisely articulating the needs they have observed, innovators will be in a much better position to then determine which ones represent the most compelling opportunities.

 See ebiodesign.org for featured videos on need statement development.

OBJECTIVES

- Learn how to translate problems, populations, and desired outcomes identified through observations into clinical need statements that are accurate, descriptive, and solution-independent.

- Recognize the importance of need scoping and its role in improving the effectiveness of need statements.

- Understand common pitfalls in developing need statements, the impact of these mistakes, and how to avoid them.

NEED STATEMENT FUNDAMENTALS

Once interesting problems have been identified through research, observations, and interviews (as outlined in 1.2 Needs Exploration), an innovator's next challenge is to translate what has been learned into a set of meaningful **need statements**. Because need

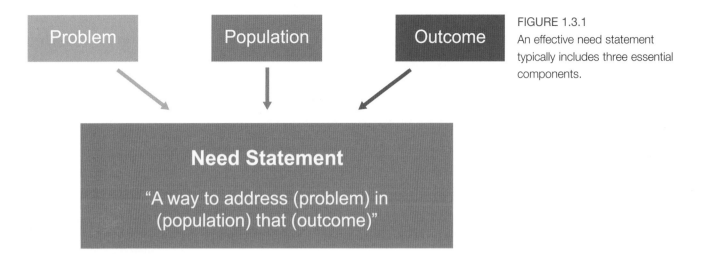

FIGURE 1.3.1
An effective need statement typically includes three essential components.

statements provide the foundation for all further steps in the biodesign innovation process, innovators should expect to invest significant time and energy in their careful construction.

Shaping a need statement has been described by seasoned innovators as something of an art form. It is an iterative exercise that starts with creating rough prototype need statements that progressively become more descriptive and refined, first through needs scoping and then through needs validation.

As previewed in 1.2 Needs Exploration, well-constructed need statements have three essential components: (1) the problem; (2) the effected population; and (3) the targeted change in outcome (see Figure 1.3.1).

The *problem* communicates the health-related dilemma that requires attention. The *population* clarifies the group that is experiencing the problem (and potentially foreshadows the market for the solution). The *outcome* specifies the targeted change in outcome, against which solutions to the problem will be evaluated.

Many innovators find that it is intuitive to include the problem and population in a needs statement, but not as natural or automatic to specify an outcome. However, being clear about the outcome up front is important in that it keeps innovators focused on the results that their eventual solutions must deliver to satisfy their target audience. For example, consider the following need statement, which is based on one of the observations described in chapter 1.2: *a way to reduce the time required for unskilled medical practitioners to place*

endotracheal tubes in an emergency setting. In this statement, the specified outcome measure is the reduction in time associated with the procedure. However, further observation and research might lead an innovator to define the outcome differently, as *a way for unskilled medical practitioners to place an endotracheal tube in an emergency setting without a drop in oxygen saturation.* Both of these need statements are "correct," but they target somewhat different outcomes and therefore have the potential to lead to different solutions. The process of creating a need statement brings these differences into focus and allows the team to make a disciplined and informed decision about the opportunity it wants to pursue.

As this example illustrates, outcomes should be stated objectively so that they can easily and effectively be measured. Table 1.3.1 provides a sample of some of the desired outcomes associated with medical need statements and recommendations for how they can be assessed.

It is worth noting that a **need** may ultimately be solved in such a way that it results in multiple benefits (or a number of improved outcomes). For example, the development of a surgical procedure that can be performed in the physician's office as opposed to the operating room will likely result in measurably lower costs (primary benefit), but may also be significantly more convenient for patients and physicians (secondary benefits). In situations like this, a need statement is typically crafted to include just the primary outcome measure, rather than listing all outcomes that may be positively affected. This

Table 1.3.1 Common changes in outcomes and how they are measured in need statements.

Desired outcomes	As measured by ...
• Improved clinical outcome	• Treatment success rates in clinical trials
• Increased patient safety	• Rate of adverse events in clinical trials
• Reduced cost	• Total cost of care relative to available alternatives
• Improved physician/facility productivity	• Time and resources required to perform procedure
• Improved physician ease of use	• Elimination of complex workarounds and/or the simplification of workflow
• Improved patient convenience	• Frequency and occurrence of required treatment, change in treatment venue (inpatient versus outpatient, physician's office versus home), etc.
• Accelerated patient recovery	• Length of hospital stay, recovery period, and/or days out of work

approach can be less cumbersome to innovators and keep them focused on the most important result. It also helps eliminate the perception that a need has been successfully addressed only if it performs well against *all* outcome measures (e.g., improved efficacy and safety, reduced cost, etc.) – a difficult and often unattainable challenge. Ultimately, the targeted outcome foreshadows the value that a solution would bring if it appropriately addresses the need. For this reason, innovators should make their primary focus the outcome with the opportunity to have the greatest impact.

All of this said, innovators may find that in some circumstances the outcome is implied rather than explicitly included within the need statement. For instance, consider the simplified need: *a way to prevent stroke in patients with atrial fibrillation*. In this example, stroke prevention is both the problem and the desired outcome, so restating the outcome may seem obvious or redundant. The key is to be sure the team carefully evaluates the problem, the population, *and* the target outcome when defining and assessing any need.

When thinking about how problems, populations, and outcomes come together in a need statement, it is important to carefully evaluate every word that is chosen because specific wording can potentially lead to dramatically different solutions. For example, consider the differences among the following simplified need statements:

- *A way to prevent hip dislocation in high-risk patients ...*

- *A way to prevent recurrent hip dislocations in high-risk patients ...*
- *A way to prevent recurrent hip dislocations in patients after surgical treatment of a first hip dislocation ...*

All three statements address the same general clinical issue (hip dislocation in high-risk patients). Yet, each one identifies certain existing conditions (past dislocation or previous surgery for past dislocation) that would cause innovators to target different patient populations and that could potentially send them in an entirely different direction in terms of what they attempt to accomplish.

A slightly more elaborate example further illustrates this point. Imagine that a team of new innovators has observed a problem with long-term urinary catheters causing infections in patients in the intensive care unit (ICU), and members are discussing the need statement with a mentor:

INNOVATORS: "I think I've finally defined a need statement for *a catheter that will not track infection*."

MENTOR: "Are you talking about all catheters or just urinary catheters?"

INNOVATORS: "Just urinary catheters."

MENTOR: "OK. Do you intend to address all urinary catheters?"

INNOVATORS: "No, just long-term catheters that are used for more than two weeks."

MENTOR: "Great. But is catheterization the only possible solution to this problem?"

INNOVATORS: "Well, I suppose we might be able to develop a number of other approaches. Maybe something that allows for the evacuation of urine without keeping a catheter in place at all times. Or perhaps we could come up with some sort of an implant that releases localized antibiotics in the area of the urethra . . . ."

MENTOR: "Good. Try to set aside any specific solution and focus on the need. Maybe what you're trying to address is more appropriately defined as *a way to reduce the incidence of urinary tract infections in ICU patients*. What do you think?"

INNOVATORS: "Yes. That's it."

MENTOR: "So, what value is created by reducing urinary tract infections in the ICU? Selecting a targeted outcome will better describe the value that the project can deliver.

INNOVATORS: "How about *a way to reduce the incidence of urinary tract infections in ICU patients that reduces hospital stay* . . . ."

The last iteration of the need statement expresses with much greater clarity the true nature of the problem, the metric necessary to achieve the desired outcome, and the potential value that would be created by the solution that meets the outcome. This variation is also more focused on the target audience. If the innovators were to adopt the first need statement, a number of potential (non-catheter based) solutions never would be considered. Without a specific, clearly identified target user, potential solutions could have been developed for users to whom the need was not truly applicable. The example also highlights the importance of defining an appropriate scope for the need. The goal is to establish the need as broadly as possible while keeping it linked to a specific, validated problem. More information about scoping needs is provided later in this chapter.

Drafting preliminary need statements

Drafting the first version of a need statement can seem a little daunting. The thing to keep in mind is that this version does not have to be perfect. Instead, think of it as a crude prototype. One strategy for getting started is to treat the exercise like a game of Mad Libs. Mad Libs is a word game in which one player prompts another for a list of words to substitute for blanks in a template. In this case, the team or innovator would use the need statement template (*a way to address [problem] in [population] that [outcome]*) and try substituting a variety of different words related to the observations they have made to create a cohesive need statement for each interesting problem.[1] Different variations can be tried using sticky notes or a whiteboard, for example, so it is easy to make modifications and experiment with diverse word combinations. Initially, speaking the "language" of need statements may seem awkward, but it will get easier with practice.

Ultimately, the team can select a version of the need statement that seems to most accurately, completely, and compellingly capture the need based on the current knowledge. Then, as their knowledge of the need area deepens through the activities described in chapters 2.1–2.4, they can modify and refine each need statement.

Need scoping

After innovators prototype their preliminary need statements, the next step is to begin actively testing and refining them through an exercise called need scoping. Need scoping allows innovators to further explore the problem, the population, and the desired outcome – and the interaction between these three components – through a series of thought experiments that will lead to a description for each of the components that is "just right." The point of scoping is to systematically try out different levels of focus or specificity for each of the components of the need statement while remaining centered in the general area of the need. Starting with the draft needs statement, the innovators ask themselves questions such as:

- Is the problem just the one outlined in the draft statement (e.g., from the example in the chapter introduction, cutting through the sternum) or could it in fact be broadened (e.g., gaining access to the thorax)?
- Is this issue actually relevant to a larger population than initially described (e.g., not just patients with

urinary catheters, but all patients with urinary tract infections)?

- Conversely, upon closer inspection, is this need actually most relevant and important when applied to a smaller subset of the population?
- Is the outcome described in the needs statement really the most essential one, or is there another outcome that is more compelling?
- Is the need, as scoped, consistent with the team's strategic focus?

This type of scoping exercise allows the innovators to methodically revisit the assumptions they have made in developing the needs statement in a way that results in the optimal framing for the need, so it is detailed and actionable without being too limiting.

Consider another example, starting from the draft need statement: *a way to decrease the incidence of infections associated with hip implants in the elderly in order to reduce hospital stays.* Even though this particular need has been observed in the elderly, the innovators should ask if it might be generalizable to a broader segment of the population. Through research and additional observations, they may determine that the need to decrease the incidence of infections associated with hip implants actually applies to all recipients, not just to those over a certain age. A next step, again through research and observation, would be to explore whether the need applies to other types of joint implants (e.g., artificial knees). The result could be that the potential target market is significantly larger and, thus, more compelling than originally estimated. It is also worth probing whether reduction in hospital stays is really the most important and measurable outcome for this need. Perhaps the reduction in morbidity (e.g., suffering associated with the infection rate) is really the most compelling outcome, provided it can be measured well. Finally, the innovators may want to ask whether there is, in fact, a broader problem that warrants consideration. In the example, infections associated with hip implants may be a need worth addressing. However, it is also part of a larger need to find *a better way to treat osteoarthritis.* The innovators should at least consider whether they would be well served to work on this "higher level" need.

The closer a need is to addressing the fundamental aspects of a disease state, the less likely it is that the need will be displaced by a **superseding need**; that is, a need that is proximal or upstream of the need under consideration and, if solved, would make this need superfluous. For instance, take the case of atrial fibrillation (AF), a disease in which the irregular heartbeat causes clots to form in the heart that can potentially dislodge and travel to the brain, causing a stroke. An innovator might choose to focus on how to prevent a thrombus from leaving the heart to travel to the brain (one approach would be to seal off an out-pouching of the heart called the atrial appendage, where clots frequently form). In scoping this need the innovator should consider whether it could be superseded by a way to prevent clots from forming (as would be provided by a better blood thinning mediations). Even this need, though, could be superseded by a way to prevent AF from occurring in the first place (maintaining sinus rhythm through medications, surgical or catheter-based ablation). Figure 1.3.2 shows these options as progressive branches in a tree.

The further away innovators work from the trunk of the tree, the more likely it is that the branch where their innovation exists could be cut off (or superseded by another invention). As Mir Imran, serial inventor, entrepreneur, and founder of InCube Labs, summarized:[2]

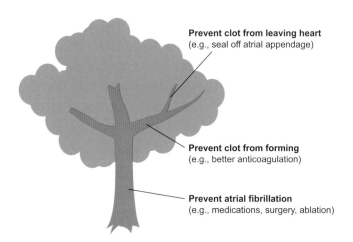

Prevent clot from leaving heart
(e.g., seal off atrial appendage)

Prevent clot from forming
(e.g., better anticoagulation)

Prevent atrial fibrillation
(e.g., medications, surgery, ablation)

FIGURE 1.3.2

An example of superseding needs and related solutions.

One of the things that device company executives worry about most is that a technology they've worked very hard on for many years will do everything they want it to do and solve exactly the clinical problem that they figured, but by the time they actually get it to the market, it's been passed by another technology, or the clinical problem has been solved in some other way.

The point is that there is a cascade of events that create a need, and each event within the series may be associated with its own unique need. This creates a hierarchy of related needs that directly affects the risk profile associated with the issue the innovator is seeking to address. In general, broad needs (e.g., that seek to cure, eliminate, or prevent a disease) often have the potential to supersede other needs. In contrast, needs focused on changes to existing treatments are often at risk of being superseded.

Of course, there is also a risk that a need can be framed *too* broadly. In an effort to avoid unnecessary constraints, innovators sometimes over-generalize a need by making the assumption that it applies to a broader population when, in fact, it does not. In the need discussed above, *a way to decrease the incidence of infections associated with hip implants in the elderly*, the scoping exercise would, as mentioned, cause the innovators to consider whether the need should be broadened to include other joints (e.g., knee implants). However, with a bit of research and perhaps more observations, it may become clear that the nature of the infections in the two joints are dissimilar in important ways due, for instance, to differences in susceptibility, different effects of the infection on the joints, and/or different mechanisms of healing. Broadening the focus to include knees may cause the innovator to overlook that one unique insight about infections in the hip that would provide the direction for a novel solution.

Embedded solutions and other need statement pitfalls

Beyond the pitfalls of framing needs either too broadly or too narrowly, a few other problems in generating needs statements deserve mention (see Figure 1.3.3). The trickiest of these for first-time innovators is the tendency to embed a solution within the need. At the most fundamental level, a need statement should address *what* change in outcome is required to resolve a stated problem, not *how* the problem will be addressed. Too often innovators incorporate elements of a solution into their need statements because they quickly envision ideas to solve the problems they observe. This is especially tempting when a respected figure – a key opinion leader (**KOL**), for example – offers a solution for how s/he would approach the need area in question. Sometimes this occurs blatantly, sometimes subtly. In either case, embedding a solution into a need statement seriously reduces the range of possible opportunities that are explored, constrains the creativity of the team, and places unnecessary boundaries on the potential market. More importantly, it can lead to a need statement that does not truly represent the actual clinical problem and, thus, may result in solutions that do not effectively address the need.

One young company, for example, focused on a problem with stents (a mesh-like tubular scaffold that can be deployed in blood vessels to expand a narrowed region). It noted that although stents are beneficial in holding open arteries, during deployment they can cause a shower of emboli (debris that becomes dislodged, travels through the bloodstream, and potentially creates blockages by lodging in other smaller blood vessels). Centering on this problem, the company framed a need for *a coronary stent that could prevent vessel wall material from embolizing* (the implied outcome in this need is to minimize the risk of stroke). The members of the design team surmised that the relatively large gaps between the struts of the stent could allow fragments of atherosclerotic plaque or thrombus to dislodge from the vessel wall and pass through, resulting in distal embolization. They decided to develop a "covered" stent incorporating a material that would stretch over the holes and prevent the emboli from breaking free. However, after development and testing, they found that the covering prevented the natural blood vessel surface from reforming around the stent after the procedure – a phenomenon that could create other serious complications, including more embolization. Ultimately, the team failed to deliver a product to the market and the company was shut down.

Pitfall	Problematic Example	Improved Example
NEED IS TOO GENERAL	*A way to improve outcome of spine surgery* ● Not clear which surgery, initial diagnosis, or how to improve	*A way to reduce risk of re-herniation after lumbar discectomy for sciatica to reduce re-operation* ● Clear about procedure, diagnosis, and complication
NEED IS TOO SPECIFIC	*A way to treat bifurcation lesions in the left main coronary artery to reduce recurrence rates* ● No reason to limit to this population—many patients have bifurcation lesions in other main coronary vessels	*A way to treat coronary bifurcation lesions to reduce recurrence rates* ● Increases the patient population at least several-fold ● Likely that same solution will work across all types/patients
NEED IS STUCK IN CURRENT PRACTICE	*A way to close sternotomy without risk of sternal-wire breaking* ● Focuses on sternal-wire; closes out other approaches ● Focuses on part of procedure that doesn't deliver result	*A way to close a sternotomy following CABG quickly and securely that reduces wound dehiscence* ● No reference to current solution, targets procedure goal
NEED HAS AN EMBEDDED SOLUTION	*A way to ultrasonically weld suture in surgery* ● Completely limits to one approach	*A way to secure an aortic valve prosthesis with minimal or no on-pump time to reduce cognitive dysfunction after surgery* ● Identifies specific procedure and problem while leaving solution open
NEED IS BUILT ON A NEGATIVE	*A way to not have infections related to dialysis catheters* ● Focuses negatively on one specific issue of one solution	*A way to provide long-term, high-flow vascular access for hemodialysis with reduced risk of infection* ● Not so negative; focuses on the goal of dialysis catheters rather than the specific solution and is therefore less limited

FIGURE 1.3.3
Potential pitfalls in writing needs statements.

Another company took a different approach to solving the problem of emboli after stenting. This company framed the basic need as *a way to prevent the consequences of emboli in patients undergoing coronary interventional procedures*. Notice that there is no solution embedded in the need itself; this need statement leaves open a number of different potential directions to pursue. The team decided that rather than focusing on stopping the emboli from being generated, it might be more effective to catch any emboli that were created by means of a basket deployed downstream from the site of intervention. After development and successful clinical testing of a basket device, this company was acquired by one of the major medical device companies. By defining the need independent of any particular solution, the team avoided the inherent limitations of a stent-based approach and opened up a more diverse range of possibilities. Both companies were staffed with talented and

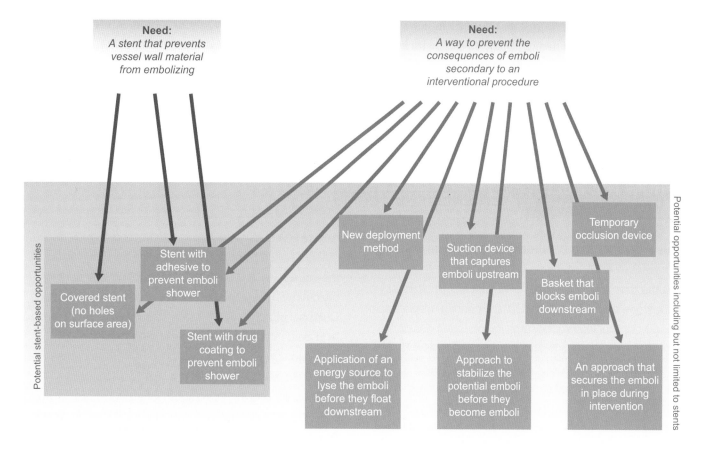

FIGURE 1.3.4

In this example, the first company restricted itself to a need with an embedded solution (a better stent) and thereby limited itself to a relatively small set of opportunities (small dotted box to the left). Removing the embedded solution of the stent and focusing on the outcome to be prevented (consequences of emboli) opened up a much broader set of possibilities.

creative engineers. The first, however, was at a disadvantage because of the solution **bias** embedded within the way the need was framed. Ultimately, taking an approach that anticipated a particular type of solution imposed artificial constraints on the team and prevented it from considering more feasible and effective approaches (see Figure 1.3.4).

Two other pitfalls in need construction deserve brief mention. The first is sticking too closely to current medical practice in formulating the need. In other words, the problem here is letting the prevailing approach or technology shape the need statement. In one of the problematic examples shown in Figure 1.3.3, the need assumes that sternotomy closure requires a sternal wire and so precludes creative thinking about other ways of accomplishing the same outcome. In a way, this problem is a variant of embedding a solution in a

need in that it accepts a particular aspect of current medical practice and builds it into the need statement. It is important to point out that, in some cases, a team will (by virtue of its strategic focus) be specifically motivated to improve an existing technology rather than make a breakthrough discovery. As long as the team is clear that its goal is to create an incremental improvement, maintaining an existing paradigm is not an issue. Instead, it becomes an accepted constraint on the need statement.

The remaining pitfall is a subtle one: formulating a need in a negative way. Since a need describes a problem, the natural tendency is to frame it in a way to eliminate the problem (*a way **not** to...* or *a way to **avoid**...*). The challenge with a negative need statement is that it tends to constrain the open-mindedness that leads to the most creative solutions. Whenever possible, innovators should try to restate the need in a way that optimizes a positive outcome. Doing so can be tricky, but a positive need statement can lead to more open and constructive ideation sessions.

The following story, about a biomedical engineering team from Northwestern University, illustrates some of the challenges related to need statement development.

FROM THE FIELD ▶ NORTHWESTERN UNIVERSITY BIOMEDICAL ENGINEERING TEAMS

Navigating the challenges of needs finding

Some time ago the leaders of Northwestern University's capstone course in biomedical engineering, David Kelso and Matt Glucksberg, became interested in design issues associated with global health problems. "The challenge with the equipment and devices used to address health issues in developing countries was not that they were poorly designed, but that they were not designed for the environment in which they would be used," Kelso said.[3] "If people began designing devices specifically for resource-poor settings, they could come up with much better solutions."

Motivated to make a difference, Kelso and Glucksberg initiated a program that gave senior students the opportunity to design solutions that specifically addressed medical needs in developing parts of the world. While their goal was to have students work on "real" projects for "real" end users, they initially launched the program targeting health-related issues identified by the World Health Organization (WHO) or other universities around the world. In one case, they read about a project initiated by engineers at the Massachusetts Institute of Technology (MIT) to develop an incubator that would help address the high rate of infant mortality in developing nations. In countries such as Bangladesh, the area where Kelso and a team of five students decided to focus, as many as 30 percent of all births were premature, a figure that translated into approximately 3,500 premature babies a day.

The team committed to developing *a better incubator that would be designed for local conditions to help reduce infant mortality*. Early in the process, they also defined specific **need criteria** for the solution, including the capacity of the system to operate without electricity, maintain a baby's temperature at a constant 37 degrees Celsius, help protect the infant from infections, contain a high percentage of local material, and have low manufacturing and operating costs. Over the course of the 10-week academic quarter, the team networked extensively, seeking input about incubator design from contacts with experience in healthcare delivery in South Asia as well as those with prenatal education and neonatal baby care. After developing an initial prototype in a plastic laundry basket (see Figure 1.3.5), the team decided to make the container from jute so it could be sourced and manufactured at low cost in Bangladesh.

The phase-change material they used to control temperature in the new model worked just as well as

FIGURE 1.3.5
A photograph of the team's early prototype (courtesy of David M. Kelso).

electrically powered devices. "We were really excited about this," noted Kelso, who immediately began seeking ways to take the project beyond prototype and into production.

Tapping into a Northwestern study abroad program, Kelso formed a second team of students located in Capetown, South Africa and began working with this group to make the incubator relevant for the South African market. Targeting the most prominent neonatal intensive care unit in the area, they scheduled a meeting with the head of that department at Karl Bremer hospital. Kelso and team brought with them photographs and storyboards that described their incubator project. "But as we got off the elevator, we saw incubators piled up in the corner," he recalled. "They were not at all interested in our incubator solution, but invited us to come in and see how they care for premature babies. There were 30 mothers in the neonatal intensive care unit, all caring for their newborns with something called Kangaroo Mother Care."

Kangaroo Mother Care (KMC) had been pioneered in 1978 in Bogotá, Colombia to overcome the inadequacies of neonatal care in developing countries. The basic idea is to place the infant (without clothes, except for a diaper, cap, and booties) upright between the mother's breasts.[4] The baby is held inside the mother's blouse by a pouch made from a large piece of fabric. The method promotes breastfeeding on demand, thermal maintenance through skin-to-skin contact, and maternal–infant bonding. While few large-scale studies have been conducted or published in mainstream medical journals, KMC is believed to help babies stabilize faster and to provide more protection from infection (from the antibodies gained through frequent breastfeeding) compared to babies isolated in incubators.[5] Many believe it also leads to reduced mortality rates among premature and low birth weight infants, although these results are still being studied. "Basically, it provides superior results at no cost," summarized Kelso.

"During our early discussions, we heard about Kangaroo Mother Care as a method they were trying to teach in Bangladesh," he remembered. "But the concept didn't affect the team's design." Kelso continued, "The right way to specify a design challenge is to do it in solution-independent form. By saying we would develop an incubator, we had over-constrained the need. Otherwise, almost all of the need criteria on our list were spot-on." It just so happened that KMC also met these need criteria, while offering other benefits to the infants as well as the mothers who preferred not to be separated from their babies.

Recognizing that the incubator solution was no longer appropriate for this environment, Kelso and his team quickly revisited the needs finding process. One of the associated needs they uncovered was a way to identify apnea in neonates (a problem that could have tragic consequences). Ultimately, they developed an innovative monitor that was appropriate for babies and could be used in conjunction with KMC.

As the Northwestern story illustrates, developing effective need statements is highly iterative and experiential – many innovators master this skill "the hard way" (by making mistakes and learning from them). This can be a costly process, since a poorly defined need statement usually is not discovered until the solution for that need statement misses the mark much later in the biodesign innovation process, after significant time, money, and effort have been invested.

Categorizing needs

Once defined and scoped, needs can be organized into three general categories: **incremental**, **blue sky**, and **mixed**. These three primary need categories exist upon a continuum (with incremental needs on one end, blue-sky needs on the other, and mixed need in between) based on the extent to which they operate within existing treatment paradigms. See Figure 1.3.6.

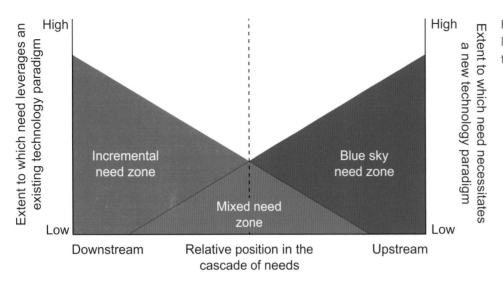

FIGURE 1.3.6
Different types of needs carry with them different benefits and risks.

Categorizing needs is useful for several reasons. First, this exercise provides a checkpoint for the team's alignment with its strategic focus – specifically, does the timeframe and budget of a given type of need fit with the priorities of the group? One team may be focused on launching a project that will support a company's goal to introduce new products within a two-year horizon (and so would concentrate on incremental needs); another group may be looking to create the biggest possible impact in the field of cancer (a blue-sky need). Second, organizing the needs in this manner can also help innovators appreciate the dependence of the potential solution on existing technology. Blue-sky needs typically are wide open for innovation, whereas many incremental needs and some mixed needs build on established practices or paradigms. Finally, by categorizing need statements at this step in the biodesign innovation process, innovators can better understand the range of needs that stemmed from their observations. Later, in 2.5 Needs Selection, they may again refer to these categories as one of several screening factors used to help them select which needs to take forward into invention. Each category is described in more detail in the sections that follow.

Incremental

An incremental need is focused on addressing issues with or making modifications to an existing solution, such as the function of a device or other technology.

For example, *a way to prevent clogging of a device used to remove tumors during neurosurgery to reduce surgical time* would be considered an incremental need. It is important to understand that incremental needs typically assume that underlying treatment paradigms or technologies will continue to be used and applied. This is not the same as saying that solutions are built into the need, but rather that solutions are constrained by being further downstream in the cascade of needs. As noted, incremental needs are generally approachable, but run the risk of being superseded when new technology paradigms are introduced.

Blue sky

Blue-sky needs, on the other hand, require solutions that represent a major departure from currently available alternatives and address needs that are further upstream in the cascade of needs. As a result, they may be difficult to define. Blue-sky needs are often focused on curing, eliminating, or preventing various disease states and, therefore, are more focused on physiology and mechanisms of action than existing treatments or solutions. For example, rather than concentrating on improvements to existing machinery or procedures, a blue-sky need might be something like *a way to prevent the spread of colon cancer to improve survival*. Blue-sky needs, if solved, will often supersede most other related needs within a treatment area. A delicate balance exists in determining whether a blue-sky need is approachable at the current point in time or if it requires more study. For a blue-sky need to be entertained as something that could be solved, it is important that at least some of the underlying disease mechanisms are understood in the medical community. This differentiates a blue-sky need from a "science experiment," or an exercise for which the fundamental building blocks of a solution are not yet known and are unlikely to be solved.

Mixed

A mixed need exists somewhere in between an incremental and blue-sky need on the continuum. With a mixed need, most of the problem may be defined, yet the solution requires expansive thinking. *A better way to surgically remove breast cancer at the time of surgery to ensure all of it has been eliminated from the site while minimizing breast tissue loss* is an example of a mixed need.

Importantly, the scope of the problem or type of need does not necessarily correspond to the size of the business opportunity or potential market. Incremental needs can be solved by solutions that can result in sizable business opportunities if undertaken at the right time. Conversely, blue-sky needs may result in solutions that are direction-changing in the industry but may not necessarily translate into significant commercial opportunities.

Early needs validation

After actively scoping each need statement, it is essential to gather more information by talking directly with potential users, customers, and other stakeholders about the problem that has been observed and the need(s) associated with it. These discussions are important because the change outlined by a need *must* be driven by what the target audience wants and/or requires. If innovators seek to solve a problem that is not important to the target population, then the innovation may not be widely adopted. If the problem identified through observation is an issue about which the target population is not readily aware, it can be more challenging for the innovators to validate the need.

Consider an example. In approximately 5–20 percent of colonoscopy procedures, the cecum (the pouch at the base of the ascending colon) is never reached due to the difficulty in navigating the endoscope through the entirety of the colon.[6] As a result, some instances of colon cancer that exist deep within the colon are not detected. Using current technology, it is difficult for physicians to know with certainty whether or not they have reached the cecum. However, despite the fact that published colonoscopy completion rates vary substantially, when endoscopists are interviewed many say that this is not a problem that they personally experience. In cases such as this, innovators may be successful initiating a productive dialog if they identify the problem in a generalized manner, based on research, without asking if a physician has personally ever experienced it. With this approach, physicians can acknowledge and discuss the need for more reliable colonoscopy completion results, whether or not they feel comfortable admitting any personal familiarity with the problem. Moreover, if a solution is introduced that makes the procedure more failsafe for all physicians, with no additional cost or risk, most would be likely to adopt it.

In talking with members of the target population, it is essential to ask them exactly *what* results they would want, not *how* to achieve them – at least at this stage. The point is to keep the discussion focused on the need and not potential solutions. (*How* comes later, when team members decide they want to include an expert in ideation or they want to bring concepts or prototypes to stakeholders for feedback and/or talk with them about product features.) Deconstruct the problem, breaking it down to each component to ensure that it is understood at every level. Make sure to understand any possible interactions between the various components of the problem and develop hypotheses for the root causes of each component that can be validated or refuted by the target audience. Then, using input from the target population, seek to identify the key elements that an ideal solution would have to include to satisfy them (ideally, these elements should be linked back to the root causes they are likely to address). Individuals can be asked to not only identify these elements, but also prioritize them in order of importance. Keep in mind that experts may have different requirements than "common users," but that common users often represent the greater market. They may also have different biases based on their own experience and perspectives that should be considered when gathering feedback from them about an observed clinical problem.

Needs validation begins right after need statement development but is repeated in an iterative fashion throughout needs screening. Efforts to validate needs play an essential role in refining the need statements and also in needs selection (see chapter 2.5 for more information).

Need criteria

Need criteria are an essential component of the need specifications that the team will develop as part of 2.5 Needs Selection. When the innovators begin to feel relatively confident in their preliminary characterization of need statements, they can begin to think more deeply about the need criteria that any solution must meet to address the need as defined. Need criteria should be based on a team's research and observations, as well as information collected in interviews and discussions with providers, patients, and other stakeholders. For instance, for the need *to reduce the incidence of urinary tract infections in ICU patients to reduce hospital stay*, the need criteria might include the following:

- Whatever the solution may be, it must be deployable by personnel that are already available within the ICU.
- It has to last for at least two weeks (since this is how long the average patient spends in the ICU). And if it is limited to two weeks, it needs to be repeatable, if necessary, with no adverse consequences.
- It must have a similar safety profile to existing treatment (traditional urinary catheters) so as not to introduce the risk of a consequence more dangerous than a urinary tract infection.
- The cost of the solution should be comparable to the cost of a traditional catheter (or only slightly more, based on the incidence of urinary tract infection and the costs associated with its treatment in those who become infected) in order for it to be commercially viable.

Although the most important need criterion is integrated into the need statement itself in the form of the desired outcome, innovators will have learned a great deal about other supporting principles that are of value to the target population. These principles might include "softer" solution attributes, such as ease of use, speed, patient convenience, etc. After conducting observations and working through the development of need statements, innovators may be able to start a list of potential need criteria to support some of their needs. Just like the need statements themselves, these criteria will be refined over time as additional information is gathered. The key is to make note of any essential insights about preliminary requirements associated with a need to ensure they do not get overlooked.

Importantly, if at any point the need criteria cause the innovators to consider modifying the need statement, it is essential that they reconfirm (with data and additional observations) that the revised need statement is still valid. In this way, the early need criteria can provide innovators with additional "boundary conditions" that can be used to stimulate further investigation during need screening.

Because needs emerge from observations, innovators should keep in mind that the need criteria may vary based on where the observations were performed. For instance, when exploring needs in low-resource facilities in areas with emerging healthcare ecosystems, innovators may find a greater imperative for solutions that are:

- Inexpensive.
- Locally manufactured (using relatively simple manufacturing methods).
- Able to withstand tough environmental conditions (dust, humidity).
- Operational despite inadequate infrastructure (irregular power supply, poor maintenance).
- Usable with minimal specialized skills or training.
- Easily repaired (with accessible replacement parts).

However, across all settings, need criteria will become more specific and actionable as innovators perform additional research.

A final word on insight

As mentioned in several places in this text, the core "mantra" of the biodesign innovation process is that *a well-characterized need is the DNA of a great invention.* One of the wonderful characteristics of the medtech field is that if an innovator or team is able to identify a truly promising need, there is a high likelihood that they can find a solution to bring into patient care to address it. In fact, there are many times when the real *insight* behind an important technology innovation is in the recognition of the need, not in the solution. The case study of Acclarent that follows this chapter provides a good example of this point. The founding team recognized that a whole category of sinus surgery could be performed with catheter-based tools that are similar to those developed to treat vascular blockages (balloon catheters, guides, etc.). Developing the tools for this application was by no means trivial, but the core insight lay in appreciating the need for this less invasive approach.

Sometimes the key insight behind an important need is camouflaged by years of medical practice (an opportunity like this is called a "latent need"). For instance, in the early days of coronary angioplasty (catheter-based opening of arterial blockages), patients were subjected to 20–30 minutes of a "groin hold" following a procedure (i.e., manual pressure applied by a doctor or nurse to help seal the femoral artery site where the catheter had been inserted). Patients routinely complained that the hold was by far the worst part of the procedure from the standpoint of pain and discomfort. But, presumably because physicians were so focused on the dramatic outcome of opening the coronary arteries, this complaint did not register as a significant problem. Finally, some 15 years after angioplasty began, the medtech community woke up to the insight that managing the entry site in the femoral artery was an important need. A number of different approaches were invented to seal the arteriotomy site and a major new sector of the industry was born.

Unfortunately, innovators cannot solely count on finding a need where, within the need itself, there is a radical or transforming insight. But by taking a systematic approach to creating a need statement, innovators will clarify their understanding of what they have observed and be able to identify the best possible opportunity residing within the need area. Time spent on crafting effective, meaningful need statements is an invaluable investment that will pay off throughout the biodesign innovation process. And, it represents a discipline that, with some practice, will become an essential part of the innovator's skill set.

↘ Online Resources

Visit www.ebiodesign.org/1.3 for more content, including:

 Activities and links for "Getting Started"
- Translate problems, populations, and outcomes into need statements
- Confirm that needs are solution independent
- Scope each need
- Perform early needs validation
- Categorize needs and define need criteria

 Videos on need statement development

CREDITS

The editors would like to acknowledge Asha Nayak for her help in developing the original chapter. Many thanks also go to David M. Kelso and the student teams at Northwestern University for sharing their story.

NOTES

1 "Point-of-View Mad Lib," Stanford Institute of Design, http://dschool.stanford.edu/wp-content/themes/dschool/method-cards/point-of-view-madlib.pdf (September 30, 2013).
2 From remarks made by Mir Imran as part of the "From the Innovator's Workbench" speaker series hosted by Stanford's Program in Biodesign, April 28, 2004, http://biodesign.stanford.edu/bdn/networking/pastinnovators.jsp (September 30, 2013). Reprinted with permission.
3 All quotations are from interviews conducted by authors, unless otherwise cited. Reprinted with permission.
4 A.-M. Bergh, "Kangaroo Mother Care to Reduce Morbidity and Mortality in Low-Birth-Weight Infants," World Health

Organization, Reproductive Health Library, http://apps.who.int/rhl/newborn/cd002771_bergham_com/en/ (September 30, 2013).

5 "What Is KMC?," Kangaroo Mother Care, http://www.kangaroomothercare.com/what-kmc-is.aspx (September 30, 2013).

6 Jane Neff Rollins, "Many New Colonoscopic Devices are in Pipeline," *Internal Medicine News*, August 1, 2006, http://www.thefreelibrary.com/Many+new+colonoscopic+devices+are+in+pipeline.-a0171953276 (September 30, 2013).

Acclarent Case Study

Throughout the biodesign innovation process, innovators face a continual stream of interconnected challenges and opportunities as they move from needs finding to integration, and then on to commercial launch. New information becomes available at every stage of the process, which can require them to revisit previous decisions, address new risks, and consider complicated trade-offs. Moreover, the biodesign innovation process takes place within the competitive medtech field, against the backdrop of the increasingly demanding and complicated domestic and global healthcare environments.

Nothing demonstrates the difficult, ever-changing, yet potentially rewarding nature of this process better than a real-world example. The following case study tells the story of a company called Acclarent, Inc. as it moves through each stage of the biodesign innovation process toward the commercial launch of its innovative new technology.

STAGE 1: NEEDS FINDING

After completing the sale of his most recent company, TransVascular, to Medtronic, Josh Makower was at a crossroads. With Makower's leadership, TransVascular had pioneered the development of a proprietary catheter-based platform to facilitate existing and emerging intravascular procedures. The new technology could be used to bypass occluded vessels in the coronaries and peripheral vasculature, rescue failed attempts to navigate total occlusions, and deliver therapeutic agents (e.g., cells, genes, and drugs) to precise locations within the vascular architecture.[1] One potential application for the system was to repair the damaged heart tissue that resulted from the more than 1.5 million heart attacks suffered annually.[2] In September 2003, Medtronic acquired substantially all of TransVascular's assets for a deal valued up to $90 million, leaving Makower in a position to decide what he wanted to do next. While the sale to Medtronic was a positive financial outcome for the investors and employees, it fell short of the much higher expectations the TransVascular team had for the business when its members set out to "pioneer the vascular highway" in 1996, the year in which the company was founded.

1.1 Strategic Focus

"Anytime you have an opportunity to stop, step back, and reassess, it's just a great time to check in on your priorities and the things that you want to try to accomplish in life to make sure that you're heading in the right direction," Makower said.[3] Recognizing the chance to define a fresh strategic focus in his career, he initiated a personal inventory and started by thinking about his mission. "For me, it was about trying to make sure that I learned from the mistakes that I had made in the past. And I also wanted to stay true to what I initially set out to do, which was to work on medical problems of a magnitude that, if solved, would result in a significant improvement of quality of life for thousands, if not millions of patients," he recalled.

In terms of his strengths, Makower, who holds a SB in mechanical engineering from MIT, an MD from the New York University School of Medicine, and an MBA from Columbia University, had now delivered successful liquidity events for investors from his first two companies. This gave him not only a valuable educational background, but the battle-scars of real-life experience on which to depend. In addition, "I felt that a skill I could rely upon was the ability to sift out important things to work on, and create projects that were compelling enough to draw extremely talented people together. I also knew that I had selling skills that would help me raise money and communicate enough enthusiasm to others that they would see the vision, commit themselves to a project, and join me in giving it all we have," he commented. Makower further recognized that he had an advantage in his close relationship with venture capital firm NEA, which had invested in TransVascular and his prior start-up, EndoMatrix.[4] His ties with this entity bolstered his belief that he would be able to secure enough funding to get his next idea off the ground. Based on his past experiences (with start-ups as well as Pfizer's Strategic Innovation Group), his education and, most importantly, his mastery of the biodesign innovation process, Makower also felt confident in his ability "to venture with a blank sheet of paper into any clinical field and come up with something that was meaningful to improve the lives of patients."

As far as assessing his weaknesses, Makower tried to be brutally honest with himself. "One weakness," he recalled, "was that I knew that I did not want to be a CEO again, at least not for a long period of time. While I understood the skills required for management, appreciated its value, and probably could do it, it just wasn't fun for me. I knew managing large groups of people was not for me and it was not until I brought a new CEO into TransVascular, Wick Goodspeed, that I started enjoying my role again. I liked being a part of finding the solution, being a problem-solver, and providing a vision of the future. But I didn't enjoy having to manage people by milestones, conduct performance reviews, and all the other things that good managers do to manage and lead a company." This brought him to the conclusion that he preferred working with small teams during their start-up phase and growing the business to the point where he could reasonably hire a CEO to lead the dozens or even hundreds of employees that might come afterwards.

In addition, "I realized that while I deeply enjoy pushing the edge of medicine and exploring completely new concepts in medical areas that are not well understood, basic research is not a good place to operate a venture-backed company," he said. Due to the highly theoretical nature of this kind of work and the extreme levels of uncertainty that innovators face, Makower recalled, "I had strong feelings about not wanting to get people – employees and investors – on board with a vision and then have them be disappointed because our theory was

wrong after so much good effort and hard work. We had been there before with TransVascular, and I just didn't want to do that to myself and the people that I worked with again. I wanted to look for opportunities that were much more concrete and could be realized commercially in a reasonable time frame."

After evaluating his strengths and weaknesses, Makower thought about specific project acceptance criteria that would make a new project attractive to him. "I viewed EndoMatrix and TransVascular as good learning experiences, but not tremendously successful. I wanted to have the opportunity to deliver on a project that was very successful. I felt like it was time to take what I had learned and really apply it." This led him to focus on opportunities with a reasonably high chance of success. "No more science experiments" became a mantra of sorts as he and his eventual team began evaluating possibilities.

Another key acceptance criterion was Makower's desire to work on problems that affected a large number of people. Finding a compelling market – opportunities with the potential to reach millions of people and achieve $1 billion in revenue – was another important factor. "We knew we needed to create a company that within a 10-year timeframe could have $100–$200+ million in annual revenues with a reasonable growth rate to achieve our investor's return expectations," he recalled.

Finally, he decided to commit himself to projects that would not involve patient deaths. According to Makower, "At TransVascular, we worked on a technology targeted at critically ill patients. Our interventions risked patient lives in an effort to try to save them. This kind of project requires a certain level of intestinal fortitude and a willingness to accept dire consequences for miscalculating the unknowns. I respect it. I've done it. But I didn't want to do it again, at least not as my next big thing. It's just too much emotion and stress, worrying about the patients."

With these (and a handful of other) acceptance criteria defined, Makower set out to identify one or more specific strategic focus areas that would meet his requirements. To help accomplish this, he restarted medical device incubator ExploraMed. ExploraMed I, which was originally founded by Makower in 1995, spawned EndoMatrix

and TransVascular. In its new form, ExploraMed II (as it would be called) was intended to become a platform for launching two to three new medical device businesses.

Makower's first move was to secure trusted team members in key roles within the incubator. Karen Nguyen signed on to oversee the finances and Maria Marshall agreed to continue on this new ExploraMed venture as his executive assistant. His first technical hire was John Chang, a seasoned R&D veteran who had been a core part of the engineering team at TransVascular (see Figure C1.1). In fact, Makower accelerated his plans to restart ExploraMed in an effort to help Chang avoid having to accept another job. "I think one of the important parts of my model is an emphasis on people," he said. "At the end of the day, the value of a business is in the people. Ideas are great, but the people who make it all work are the reason why you're successful. I wanted to work with John again because he's just the most positive, energetic, happy, hard working, smart, dedicated, loyal, and trustworthy guy anyone would ever want to have on a team. So I restarted ExploraMed sooner than I wanted to, so we could have a chance to work together again."

Together, Makower and Chang decided to explore four key areas. Two of these areas were orthopedics and respiratory disease. Another was focused on trying to find a niche within the congestive heart failure (CHF) arena that would meet Makower's acceptance criteria. This field was interesting to the two men because of their prior experience at TransVascular, but they needed to define a scope and focus that would be more applied. "The thought was, 'Is there anything that we can do that would be simpler than what we were working on at TransVascular – something that's not going to require us to create new science?'" Makower recalled. They started to focus on pulmonary edema associated with CHF (the effect that causes patients to become starved for oxygen during the night and unable to sleep). They noticed it was an important side effect that dramatically affected patients' quality of life, yet it seemed to have some interesting mechanical implications.

The fourth potential focus area was in the ear, nose, and throat (ENT) specialty – a space in which Makower had already performed some preliminary research and generated some ideas. As someone who suffered from chronic sinusitis, a condition involving the recurring inflammation of the cavities behind the eyes and nose commonly caused by bacterial or viral infections,[5] Makower was intimately familiar with the inadequacies of existing treatment alternatives for this condition. Patients usually were treated with over-the-counter and prescription medications, including antibiotics, when severe infections occurred. In fact, sinusitis was the fifth most common condition for which antibiotics were prescribed in the US.[6] Steroids were another type of therapy employed when these other treatments failed to produce or sustain results. In a relatively small number of the worst cases, chronic sinusitis patients qualified for surgical procedures, the most common of which was functional endoscopic sinus surgery (FESS). Frustrated with the efficacy and nature of his treatment alternatives, Makower had informally begun exploring new approaches in this area. When Chang joined ExploraMed, the time was right to more formally evaluate the strength of potential opportunities in ENT. "We set aside my previous work for the time being to try to treat this like any of the other projects," said Makower. "We needed to start with the basic clinical problem, develop a deep understanding of the real clinical needs, and do our due diligence to see if we would arrive at the same general conclusions and ideas that I had going in."

1.2 Needs Exploration

According to Chang, the two men developed a plan to spend their first few weeks collecting general information about what was going on in the areas of orthopedics, respiratory disease, CHF, and ENT. "I scheduled a number of meetings with physicians I knew, and I attended several conferences," he said. In these meetings, Chang remembered, "We said, 'So, tell us about what you do. Tell us about some of the patients you see. Tell us about the challenges you face as a physician. What are some of your greatest needs?'" Sometimes the physicians had specific ideas to share but, more often than not, they simply shared their experiences and discussed whatever frustrations were giving them problems. The team knew that the real insights would come from observing physicians in the operating room, with their patients in the clinic, and at clinical meetings as they debated and discussed current therapy.

Accordingly, the next step was to schedule first-hand observations. Makower explained: "We needed to find clinicians that would allow us to see a large volume of cases, get a lot of patient experience, and quickly come up to speed on the space." Makower had already begun teaching at Stanford University's Program in Biodesign and, as a result, had a rich network at the university's medical center to tap into for contacts in certain specialties. However, in some fields, such as ENT, he and Chang had to "cold call" the hospital. "We called and asked, 'Who's the rhinologist at this hospital?'" Makower remembered.

Once they made the appropriate connections, a formal observation process was launched. Again, in the ENT space, "We spent a couple of days following a surgeon around in clinic, getting an appreciation for his day-to-day routine – what patients were coming in, and why. Was it an initial visit, a post-operative follow-up, or some other type of appointment?" said Chang. "We also spent time in the OR asking 'dumb' questions like, 'Hey, I notice you did that four times. Is there a reason why you have to do that?' The idea was to keep the eyes and the mind open."

"We went to clinic, we watched cases, we talked to patients, and we observed surgeries," Makower reiterated. "And basically, the more we heard and saw, the more we began to feel that we really had something here." FESS surgery to treat severe cases of chronic sinusitis became the dominant form of sinus surgery in the mid-1980s, led by the introduction of the endoscope to the field. This approach involved the insertion of a glass-rod optic, called an endoscope, into the nose for a direct visual examination of the openings into the sinuses. Then, under direct visualization, several cutting and grasping instruments were used to remove abnormal and obstructive tissues in an effort to open the sinus drainage pathways. In the majority of cases, the surgical procedure was performed entirely through the nostrils (rather than through incisions in the patient's face, mouth, or scalp, as was previously necessary).[7] "Conceptually, the specialty made this huge leap forward 20 years ago from large open incisions to FESS, which was considered to be minimally invasive and atraumatic," commented Chang. "But when observing a FESS procedure, the video image coming off the endoscopic camera was often a sea of red. As an outsider, you just think, 'Gosh, maybe that's better than it used to be, but that doesn't seem so atraumatic to me. Ouch!"

According to Makower, one of the reasons the process was so bloody was because a significant amount of bone and tissue was being removed in every procedure. As an example, he explained:

The uncinate process is a bone that sits at the edge of the maxillary sinus. It is never diseased, yet it is completely removed in almost every conventional sinus surgery. To me, it was amazing that they were removing a structure solely because it was in the way. I had to ask the question two or three times: "So, the only reason why you're taking it out is because you can't see around it?" "Yes." "But it's not diseased?" It was kind of incredible. And that was the beginning of us coming to the realization that there was a lot more cutting involved in the procedure than ideally needed to be done. We used to joke that it was analogous to someone deciding that they needed to make their bedroom door a little bit bigger and then choosing the method of driving a bulldozer through the entry doorway, through the living room, and demolishing the kitchen along the way just to get to the bedroom door, because all those things were just "in the way."

Other major problems identified through their observations included post-operative scarring, mostly related to the trauma imparted during surgery that often led to suboptimal outcomes and a need for repeated procedures solely to address the recurrence of their scars. Additionally, "There were other potential complications, like cerebrospinal fluid (CSF) leaks, as well as a high level of complexity of the procedure, not to mention the significant post-operative pain and bleeding," Makower added. The team perceived that these factors, in combination, presented an opportunity to create significant value in the space by offering a better solution.

1.3 Need Statement Development

Using the information gleaned from their initial observations, Makower and Chang defined a need statement and a preliminary list of need criteria. Summing up their takeaways, Makower said, "We saw a need for a minimally invasive approach to treating chronic sinusitis that had less bleeding, less pain, less bone and tissue removal, less risk of scarring, and that was faster, easier, and safer to perform. We tried hard to put aside the ideas that we had come up with already and stay true to the process, but our excitement about the opportunity was clearly building."

NOTES

1 "Medtronic Completes Transaction with TransVascular, Inc.," *BusinessWire*, September 24, 2003, http://www.businesswire.com/news/home/20030924005389/en/Medtronic-Completes-Transaction-TransVascular (September 16, 2013).

2 "Medtronic Agrees to Acquire Assets of TransVascular, Inc., Maker of Next-Generation Vascular Devices," *BusinessWire*, August 11, 2003, http://www.businesswire.com/news/home/20030811005356/en/Medtronic-Agrees-Acquire-Assets-TransVascular-Maker-Next-Generation (September 16, 2013).

3 All quotations are from interviews conducted by the authors, unless otherwise cited.

4 EndoMatrix was a medical device company focused on the treatment of incontinence and gastro-esophageal reflux. It was acquired by C. R. Bard in July, 1997.

5 "Balloon Therapy," *Forbes*, May 22, 2006, p. 82.

6 Carol Sorgen, "Sinus Management Innovation Leads to an Evolution in Practice Patterns," *MD News*, May/June 2007, http://www.clevelandnasalsinus.com/webdocuments/Acclar-Cleveland-md-news.pdf (September 16, 2013).

7 "Fact Sheet: Sinus Surgery," American Society of Otolaryngology – Head and Neck Surgery, http://www.entnet.org/HealthInformation/SinusSurgery.cfm (March 22, 2014).

IDENTIFY ▶ Needs Screening

PHASES	STAGES	ACTIVITIES

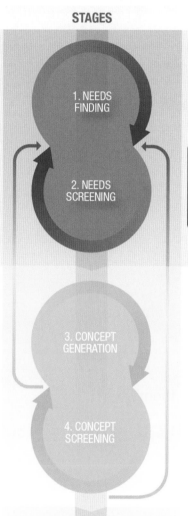

IDENTIFY

1. NEEDS FINDING

2. NEEDS SCREENING

1.1 Strategic Focus
1.2 Needs Exploration
1.3 Need Statement Development

2.1 Disease State Fundamentals
2.2 Existing Solutions
2.3 Stakeholder Analysis
2.4 Market Analysis
2.5 Needs Selection

INVENT

3. CONCEPT GENERATION

4. CONCEPT SCREENING

3.1 Ideation
3.2 Initial Concept Selection

4.1 Intellectual Property Basics
4.2 Regulatory Basics
4.3 Reimbursement Basics
4.4 Business Models
4.5 Concept Exploration and Testing
4.6 Final Concept Selection

IMPLEMENT

5. STRATEGY DEVELOPMENT

6. BUSINESS PLANNING

5.1 IP Strategy
5.2 R&D Strategy
5.3 Clinical Strategy
5.4 Regulatory Strategy
5.5 Quality Management
5.6 Reimbursement Strategy
5.7 Marketing and Stakeholder Strategy
5.8 Sales and Distribution Strategy
5.9 Competitive Advantage and Business Strategy

6.1 Operating Plan and Financial Model
6.2 Strategy Integration and Communication
6.3 Funding Approaches
6.4 Alternate Pathways

PROJECT LAUNCH

Successful entrepreneurs do not wait until "the muse kisses them" and gives them "a bright idea": they go to work . . . . Those entrepreneurs who start out with the idea that they'll make it big – and in a hurry – can be guaranteed failure.

Peter Drucker[1]

I find out what the world needs. Then, I go ahead and invent it.

Thomas Edison[2]

2. NEEDS SCREENING

IDENTIFY

After collecting many needs, a rigorous, follow-on process of screening and specification is required before you begin inventing – the deep dive. This is not an intuitive skill for most; typically, bright people will encounter a clinical need and proceed directly to devising solutions without first validating whether that need is *really* the most important one to take on.

In fact, careful scrutiny of all facets of the need is essential. While serial innovators may do this intuitively, a formal process is highly useful for those with less experience. The iterative process of "walking around the problem" may be enhanced by an occasional cooling off period. There is a perfectly natural human tendency to fall in love with a need and remain anchored to it. Dispassionate review and reflection can prevent a WOMBAT experience (Waste of Money, Brains, and Time).

By the end of this deep dive, the innovation team should have become absolutely expert on the problem, with a detailed specification of the need including clinical characteristics, market dynamics, competitors and their current solutions, and stakeholder requirements – both "must-haves" and "nice-to-haves." The all-important headline is the need statement, a single sentence that contains the essential features of the need: the problem, the population affected, and the outcome desired for the new solution. This is the genetic code for the entire project to come.

NOTES

[1] Peter Drucker, *Innovation and Entrepreneurship: Practice and Principles* (HarperCollins, 2006).
[2] Robert A. Wilson and Stanley Marcus, *American Greats* (PublicAffairs, 1999).

2.1 Disease State Fundamentals

INTRODUCTION

In the excitement of having identified one or more compelling needs, innovators' instincts may compel them to quickly jump ahead and begin inventing. However, establishing a detailed knowledge of the relevant disease state, with a particular focus on its mechanism of action, is fundamental to validating any need and understanding how it can best be addressed. Disciplined disease state research is an essential part of the biodesign innovation process and an invaluable activity for clinician and non-clinician innovators alike.

Understanding disease state fundamentals involves researching the epidemiology, anatomy and physiology, pathophysiology, symptoms, outcomes, and economic impact of a disease. This information is pertinent to the process of finding a clinical need or in validating a need that has already been established. The process also provides innovators with a critical level of knowledge about a condition so they can be credible when speaking to external healthcare stakeholders, such as physicians or other experts in the field.

 See ebiodesign.org for featured videos on disease state fundamentals.

OBJECTIVES

- Understand the importance and role of disease state analysis.

- Know what factors to investigate as part of this research.

- Appreciate how to effectively search for and summarize this information to aid the needs screening process.

DISEASE STATE FUNDAMENTALS

Performing disease state research is iterative. It begins at the highest level when choosing a strategic focus (chapter 1.1), preparing for observation (chapter 1.2), and creating **need statements** (chapter 1.3). It becomes even more important once need statements have been created. Disease state research serves a critical role in forming the basis for screening multiple **needs** against one another later in the biodesign innovation process (2.5 Needs Selection). Additionally, innovators often underappreciate that a team's understanding of the disease anatomy, physiology, **pathophysiology**, and **mechanism of action** provides the foundation for **concept** generation during the Invent phase of the biodesign innovation process.

Disease state research is first performed using general scientific resources such as medical textbooks or medical information websites, then transitions over time into a more comprehensive, in-depth review of historical and current medical literature. This approach allows the innovator to begin by developing a general understanding of a disease and then become increasingly knowledgeable about aspects of the condition that are most relevant to the need. Obtaining an understanding of a condition's mechanism of action – or the science behind how the disease works from a biologic or physiologic perspective – is especially important. Some disease states are well understood and, therefore, needs in the field are more readily approachable. Disease states in which the mechanism of action is unclear may pose a significant challenge, and needs in these areas may not be selected for projects for just this reason.

Because disease state research can be tedious, innovators may be tempted to skip this step. Those with a medical background may figure that they already know enough to understand the disease state associated with a need. In contrast, innovators from business or engineering backgrounds may have a tendency to shortcut the research in their enthusiasm to evaluate the market or other factors that will help determine if an opportunity is promising. However, underinvesting in this process is almost always shortsighted. Disease research not only provides a foundation for understanding the underlying disease state, but lends valuable knowledge that aids in the investigation of existing treatments, the current market, and important **stakeholders**. It also helps with ideation – without a sufficient disease state knowledge, brainstorming can stall within the first few minutes of a session. Later in the biodesign innovation process, this information can be used again to assess the clinical, technical, and commercial feasibility of any solution concept that will eventually be developed.

The following example, which references one of the great **medtech** success stories of the 1990s, illustrates that even the most experienced innovators and companies, regardless of their prior experiences and training, can realize significant value from disease state research and should regard the analysis as indispensable.

FROM THE FIELD ▸ JOHNSON & JOHNSON

Understanding disease state fundamentals as part of the needs screening process

Johnson & Johnson (J&J), through its subsidiary Cordis, was an early pioneer in the market for bare metal stents, small mesh-like tubular scaffolds which can be used to open narrowed heart arteries. The company dominated the treatment space after the introduction of its Palmaz-Schatz® coronary stent in 1994. J&J held a firm leadership position until 1997 when competition from other medical device manufacturers began to intensify, particularly with the launch of Guidant's Multi-Link® bare metal stent. Seeking a way to regain the company's leadership position while further reducing the need for repeat procedures in patients with coronary artery disease, Bob Croce, J&J's group chairman of Cordis Corporation at the time, went back to the drawing board with his team to reexamine disease state fundamentals as part of the need screening process.

Coronary artery disease occurs when plaque, a mixture of cholesterol and other substances, accumulates over time within the arterial wall, through a process called atherosclerosis (see Figure 2.1.1). This causes a reduction in the available area for blood flow.

FIGURE 2.1.1

A blood vessel narrowed by atherosclerosis (developed by Yasuhiro Honda; reprinted with permission).

The restriction of blood flow to the heart can result in angina (chest pain) or lead to a myocardial infarction (heart attack), depending on the severity of the narrowing. Nearly 7 million people suffer from angina,[1] and 1.2 million people experience new or recurrent heart attacks each year in the United States (the company's primary market at the time).[2] Approximately 40 percent of these heart attacks are fatal,[3] making cardiovascular disease the leading cause of death in the US. The treatment of coronary artery disease is a major contributor to the roughly $15 billion market for cardiology devices.[4]

Angioplasty, an interventional procedure in which a physician inserts a balloon-tipped catheter into a narrowed artery to increase blood flow, revolutionized the treatment of coronary artery disease in 1977 by providing a less invasive, lower-risk alternative to coronary artery bypass surgery.[5] However, there was often a recoiling effect of the arterial wall, which meant that the artery remained only partially open after the balloon catheter was removed. In addition, there was scarring within the artery as a response to the injury from the balloon – called restenosis – that occurred in 30–40 percent of patients within 6 months of the angioplasty as the body sought to heal the artery.[6] For these reasons, many patients required repeat angioplasty procedures or bypass surgery, resulting in increased risk for the patient and added cost for the healthcare system.

Bare metal stents were incorporated into the balloon angioplasty procedure to address these issues. When the balloon was inflated at the site of the blockage, a stent – a small mesh-like tubular scaffold – was expanded and locked into the wall of the artery (see Figure 2.1.2).

The stent physically held the artery open and prevented it from recoiling once the balloon was extracted. As a result, the number of repeat procedures declined and patient restenosis rates dropped to approximately 20–25 percent. While bare metal stents were widely considered to be a major breakthrough, "The statistics weren't that great," said Croce.[7] "The stents corrected one problem, the retracting of the arterial wall, and they improved outcomes compared to using the balloon alone. Unfortunately, they also caused the re-narrowing of the arteries through neointimal growth." Neointimal growth was the formation of scar tissue within the stent as a result of the trauma involved with the insertion of the stent and the body's reaction to it. Thus, through a different mechanism, the arteries could still eventually become narrowed.

(a)

(b)

FIGURE 2.1.2
A balloon is used to deploy a stent within the arterial wall (developed by Yasuhiro Honda; reprinted with permission).

To better understand the **need criteria** that any new solution would have to satisfy, Croce and his team spent significant time revisiting the physiology of the coronary arteries and the pathophysiology associated with neointimal growth. One of the most significant insights from this research was that the original disease state had shifted. The need was not just to address atherosclerosis, the build-up of plaque in the vessel wall, but ultimately, the new disease state of restenosis caused by neointimal growth. The fundamentals of the disease state were generally known, so "It wasn't like the cycle of neointimal growth in the arteries was a brand new discovery," recalled Croce. However, "Many smart people in the area had not been trained in neointimal growth for a long time and in some cases they never did understand it since it wasn't important to them in their practice," he continued. By conducting a thorough study of the disease state, Croce and team increased their understanding, as well as their confidence that no opportunities would be overlooked. "No matter how experienced you are, you can't go into this process assuming that you know everything. It's essential to stay open-minded and force yourself to analyze all the different aspects of the disease," he said.

An in-depth understanding of the pathophysiology of the disease was particularly important in this situation because of the solution that was eventually chosen – a combination of drug and device – which eventually became the drug-eluting stent. Only through revisiting the underlying need and studying neointimal growth, a fundamental aspect of the disease state, did the team determine that certain drugs could be used to prevent the problem linked to stent placement. However, the marriage of medical device developers with pharmaceutical scientists was not an easy one. Medical device companies had a **bias** toward engineering a better stent that would not scar the arteries, while pharmaceutical companies were predisposed toward small molecule and biotech solutions, and did not necessarily want or know how to consider device development. "Before drug-eluting stents, there were no major drug-device combination products. So, there was a lot of hesitation on all sides of this project," remembered Croce. In addition, physicians were skeptical that such a novel concept could produce results. Nonetheless, the science behind the disease provided a uniting factor around which all parties could converge and, therefore, served as a critical building block for the effort.

After years of development, J&J's drug-eluting stent moved into **clinical trials**. In its first-in-human studies, the Cypher stent demonstrated in-stent restenosis rates of 0–3 percent and in-segment or vessel restenosis rates of up to only 9 percent, compared to 33 percent in the baremetal stent arm.[8] When Cypher received **FDA** approval in April 2003, J&J decisively regained its leadership position in the treatment of coronary artery disease.

As this case illustrates, understanding a disease state can be a dynamic exercise. In some scenarios, new physiological issues can arise in response to an existing, widely adopted treatment. As a result, innovators must be certain they stay abreast of new developments over time.

An approach to disease state analysis

Disease state research is best approached in a systematic manner, particularly if the need is related to a single, specific disease area. Six key areas, outlined in Table 2.1.1, should be addressed to ensure a thorough understanding of a disease state.

For needs that cross more than one disease area, the innovator should establish a clear understanding of each interrelated condition. Realistically, it may be necessary to take a somewhat broader perspective, paying close attention to those aspects of the various disease areas that are most directly related to the need. In these cases, the anatomy, physiology, and even pathophysiology that are studied may not be for a specific organ or system, but instead for a fundamental biologic process that is shared across the multiple disease areas.

Throughout the remainder of this chapter, atrial fibrillation (AF), a disease in which the heart has an abnormal

Table 2.1.1 Six key areas of disease state analysis.

Focus area	Description
Epidemiology	Describes the causes, distribution, and control of disease in the population.[9]
Anatomy and physiology	Describes the normal anatomy and/or function of the organ system, which may include various organs or areas of the body affected by the need.
Pathophysiology	Describes the disturbance of normal anatomy and physiology caused by a disease or other underlying physical, mechanical, electrical, or biochemical abnormality.
Clinical presentation	Profiles the patient state and clinical status associated with a disease. These include the symptoms (what the patient feels and experiences) and signs of the disease (what one might find on a clinical exam or with lab testing).
Clinical outcomes	Profiles the most common outcomes experienced by patients as a result of having the disease.
Economic impact	Outlines the cost of the disease to the healthcare system.

rhythm, is used as an example to illustrate the types of disease state analysis innovators should perform. Importantly, the evaluation of disease state fundamentals is distinct from understanding the existing solutions used to address the disease. An in-depth review of existing solutions is covered in a separate chapter (see 2.2 Existing Solutions). However, the analysis of available diagnostics, therapies, and management tools may lead to a refined understanding of disease state fundamentals and vice versa.

Epidemiology

Review of disease **epidemiology** is one of the most efficient ways to gain an understanding of the breadth and impact of a particular disease state. This data is extremely helpful when trying to make early decisions in the needs selection process. It also serves as essential inputs to performing market analysis (see chapter 2.4). Understanding the extent and severity of a disease also can be useful in refining a need statement and selecting which needs to take forward.

Effective epidemiology evaluation must be detailed and specific. Innovators should seek data for the disease as a whole, as well as the most relevant patient subsegments. Additionally, they should try to find information about disease dynamics, such as its growth rate, to illustrate how the disease will impact society in the future.

A thorough assessment of epidemiology addresses the *incidence* of a disease, which is the rate at which it occurs (i.e., number of new cases diagnosed per year). It will also include *prevalence* data, or a measurement of all people afflicted with the disease at a given point in time.

For the AF example, innovators should start by capturing incidence and prevalence data for the overall disease state and its most meaningful subgroups (e.g., paroxysmal AF). They should also understand how incidence and prevalence rates are changing.

Working Example
Epidemiology of atrial fibrillation

Estimates of the diagnosed incidence and prevalence of AF in the US vary widely in medical literature. These inconsistencies are attributable to differences in study design, covered time period, birth cohort, and temporal effects, as well as improvements in AF diagnosis.[10] The objective of a study published in 2012 was to estimate and project the incidence and prevalence of diagnosed AF among adults in the US from 2010 through 2030. Researchers used data from a large health insurance claims database for the years 2001 to 2008 to represent a geographically diverse 5 percent of the target population. The trend and growth rate in AF incidence and prevalence was then projected by a dynamic age-period cohort simulation progression model that included all diagnosed AF cases in future prevalence projections regardless of follow-up treatment, as well as those cases expected to be chronic in nature.[11] The model showed that AF incidence was 1.2 million cases in 2010 and expected to double to 2.6 million cases in 2030.

Prevalence, in turn, was forecast to increase from 5.2 million in 2010 to 12.1 million cases in 2030.[12]

This dramatic increase in prevalence is due, in part, to the aging of the general population. Additionally, improvements in medical care are leading to increased longevity in patients with coronary artery disease, hypertension, and heart failure, which are all chronic cardiac conditions that increase an individual's risk for AF.[13]

Anatomy and physiology

Obtaining a basic working knowledge of the normal anatomy and physiology of the organ, system, or structure of the body that is affected by a need is important because it establishes a baseline against which abnormalities are understood. While some diseases affect a specific organ or system, other disease states affect multiple organs or systems within the body. Through research of the normal anatomy and physiology, innovators should quickly be able to determine whether narrowing their focus to one organ or area is appropriate. This research also provides innovators with an understanding of important vocabulary and context as they delve into further research.

The disease will be much easier to comprehend if the anatomy of the affected organ or organ system is clearly understood and can be visualized. For example, an innovator with an engineering mindset will likely be assisted in understanding the disease by knowing the position, size, and proximity of the affected organ or system in relation to other systems. In addition, both gross and cellular anatomy must be evaluated. Gross anatomy refers to the study of the anatomy at a macroscopic level (through dissection, endoscopy, X-ray, etc.). Cellular anatomy, also called histology, refers to the study of the body using a microscope. It is often most effective to start with gross anatomy, as this is generally easier for most innovators to grasp. Then, with knowledge of the gross anatomy serving as context, innovators can more effectively tackle cellular anatomy.

Physiology, or the way in which biologic tissues function, is often better understood after innovators establish a working knowledge of the normal anatomy. As with anatomy, physiology should be investigated at both the gross and cellular levels. Once innovators learn about normal patterns of function within an affected area, they have a basis for understanding how the disease functions (as described in the next section). Biologic and physiologic processes can be evaluated in terms of their mechanical, electrical, and chemical mechanisms. Later in the biodesign innovation process, this information can serve as a basis for brainstorming new concepts that act on these mechanisms of action.

Using the AF example, innovators would begin by determining that AF is a disease of the electrical system of the heart, which is part of the cardiovascular system. As the heart is the primarily affected organ, they could then focus on investigating the basic gross anatomy of the heart and its normal function. Understanding the heart's size, location, and position in relation to other structures quickly establishes a baseline context for exploring more complex concepts and interactions, such as how the electrical system of the heart establishes a rhythm that affects the organ's ability to mechanically contract. While the appropriate level of detail to capture varies significantly with each specific need and its associated disease state, the Working Example below is representative of the detail that is appropriate for a *preliminary* disease state assessment.

Working Example
Normal anatomy and physiology of the heart

The pumping action of the heart depends on precise electrical coordination between the upper loading chambers (atria) and lower pumping chambers (ventricles) as shown in Figure 2.1.3.

The contraction of the atria and ventricles is regulated by electrical signals. During normal sinus rhythm[14] the sinoatrial (SA) node (often referred to as the "pacemaker"), which is located in the high right atrium, releases an electrical discharge that causes the atria to contract. The electrical signal then propagates through the atria to the atrioventricular (AV) node, which is located between the atria and the ventricles to help regulate the conduction of electrical activity to the ventricles. Electrical signals are conducted from the AV node through Purkinje fibers[15] into the ventricles, causing the ventricles to contract.

The rate of the electrical impulses discharged by the SA node determines the heart rate. At rest, the frequency of discharges is low, and the heart typically beats at a rate of 60 to 80 beats per minute. During periods of exercise or excitement, an increase in the heart rate is mediated by the input of the central nervous system onto the SA node, which subsequently discharges more rapidly.

The heart's mechanical and electrical coupling is the result of the organ's fundamental cardiac cellular physiology. While the heart is composed primarily of connective tissue, cardiac muscle tissue is responsible for the electro-mechanical coupling of electrical signals and mechanical pumping. Muscle contraction is essentially the result of changes in the voltage of a cell due to the movement of charged ions across the cell's surface. This initial voltage change is the result of ions flowing from cell to cell, and is usually initiated by pacemaker cells, such as the cells of the SA node, which intrinsically cycle through voltage changes. This voltage change then triggers the movement of other ions within the cardiac muscle cell to cause changes in mechanical structures that result in contraction. The majority of cardiac muscle contracts due to depolarization, which is a change in voltage caused by the influx of sodium ions and the outflux of potassium ions.

This flow of ions results in changes in the heart's baseline voltage, which causes both the influx of calcium ions and the release of internal calcium stores. Calcium, in turn, results in the interaction of various cellular components, bringing about a contraction in the mechanical filaments of the muscle cell (see Figure 2.1.4).

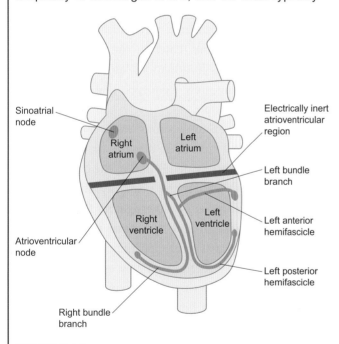

FIGURE 2.1.3
The heart's electrical system is one aspect of normal anatomy and physiology that an innovator must understand when initiating an investigation of AF (reproduced from Steve Meek and Francis Morris, "ABC of Clinical Electrocardiography: Introduction," *British Medical Journal*, 324, 415–18, 2002; reprinted with permission from BMJ Publishing Group Ltd.).

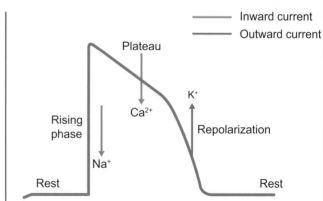

FIGURE 2.1.4
Depolarization is the result of the inward flux of sodium and the outward flux of potassium ions. Changes in the baseline voltage due to depolarization results in the inward flow and intracellular release of calcium and muscle contraction.

This process propagates throughout the heart from cell to cell, such that the result of all the cellular filaments contracting is the temporally and spatially coordinated contraction of the heart muscle. Once contraction is complete, various cellular components are activated to reset the filament structure and ionic balance so that the process can begin again.

Pathophysiology

Once an understanding of anatomy and physiology of the relevant organ in a healthy individual is established, then innovators can examine how the disease disturbs the normal structure and function. It is critically important to take into account the fact that most diseases are not homogeneous. Stated another way by some medical educators, "diseases don't read textbooks," which underscores the point that the description of a disease in a book or other reference only represents one example of its presentation. Different subtypes of a disease often exist and the heterogeneous nature of patient populations can result in a broad range of effects for any given disease state.

When investigating pathophysiology, the first step is to better understand how the disease works from a biologic and physiologic perspective and then how this affects the normal function of the organ or system. The second step is to identify the risk factors and causal associations (e.g., genetics, age, associated diseases, lifestyle) that characterize the disease. Finally, innovators can seek to understand the disease progression. Disease progression examines the rate (e.g., days, weeks, or years) at which the disease leads to abnormal function. This includes the peak age of the effect and the types of changes that occur at each stage of the disease.

In the AF example, innovators would explore how the heart might be structurally altered, leading to abnormal function, and whether or not the condition can cause structural changes in the organ. They should also look at the common causes of AF, its primary risk factors, and how AF progresses. Time should be spent understanding the different types of AF and the unique characteristics of each variation of the disease. This might include looking at which type of AF is most common among different groups of patients, whether all AF patients progress in the same way (or if progression is more directly affected by other factors such as coexisting conditions), and how likely patients are to progress from one type of AF to another.

Working Example
Pathophysiology of atrial fibrillation

Disease function

In a normally functioning heart, the rates of contraction for the atria and ventricles are typically equal and result in a regular heartbeat. However, during AF, ventricular and atrial contractions become irregular and unsynchronized. Instead of electrical discharges being regularly generated solely by the SA node, rapid and irregular discharges come from other areas in the atria. Since these other areas are discharging so fast, the SA node's slower, more regular rate is suppressed. There are several "trigger points" for this electrical activity, which create a pattern of rapid, chaotic electrical activity that is characteristic of AF. The majority of these focal sources (approximately 94 percent) are located in areas around the four pulmonary veins, which are connected to the left atrium. Other less common areas include the superior vena cava, right and left atrium, and the coronary sinus.[16] Though not fully understood, causal factors (see below) may result in inflammation and injury to the heart, causing alterations in cell structure and predisposing it to abnormal electrical discharges that can initiate and maintain AF.

As a result of these irregular discharges, the atria contract between 300 and 600 times per minute.[17]

However, the atria do not actually contract as a whole – the rapid contractions of parts of the atria may be better thought of as a quiver, or fibrillation, rather than regular beating. This results in improper filling and ejection of blood, as well as a decreased efficiency of the heart's pumping process. Since all electrical activity from the atria can typically only get to the ventricle via the AV node, the AV node is able to filter many of the irregular electrical discharges associated with AF, preventing the rapid rate of the atrial beat from being conducted into the ventricles. However, not all of the signals are blocked and AF is often accompanied by irregular ventricular beating, at 50 to 150 per minute.[18]

Causal factors

The most common causes of AF are advanced age, abnormalities in the heart's structure, uncontrolled hypertension (e.g., high blood pressure), thyroid disease (e.g., an overactive thyroid or other metabolic imbalance), and acute exposure to heart stimulants (e.g., alcohol).

Disease progression

According to the American Heart Association, AF can be classified into three clinical subtypes: paroxysmal, persistent, and permanent. In the case of AF, the subtypes parallel disease progression with one subtype transitioning to the next over several years in a large majority of patients. These subtypes are defined by the ease with which episodes of AF terminate. Paroxysmal AF refers to recurrent or lone[19] episodes that spontaneously self-terminate after a relatively short period of time. Persistent AF requires pharmacological or electrical cardioversion[20] (e.g., giving medicines or an electric shock) to restore regular sinus rhythm. In patients with permanent AF, regular sinus rhythm cannot be restored and the irregular heartbeat becomes the accepted rhythm.[21] These subtypes stand in stark contrast to AF associated with reversible causes (e.g., thyrotoxicosis,[22] electrolyte abnormalities) and the occurrence of AF secondary to acute myocardial infarction, cardiac surgery, or acute pulmonary disease. These conditions are considered separately since the AF is unlikely to recur once the precipitating condition has been treated.

Clinical presentation

Research of clinical presentation focuses on the impact of the disease on the patient. It emphasizes the symptoms (what patients say they experience) and the signs (what the astute healthcare provider identifies or observes during the patient examination) of a disorder or disease. Gaining an understanding of clinical presentation is important because it is often the target for improved care and the development of new therapies that address identified needs. When evaluating clinical presentation, describe what patients complain about when they see a clinician and how they feel. Note that patients with the same disease may present differently based on a number of factors, such as age, gender, ethnicity, and coexisting conditions. Since every individual is different, each is likely to experience symptoms slightly differently. Ultimately, clinical presentation may manifest itself in the signs/symptoms that result from the primary effect of the disease or from the long-term consequences of having and managing the disease over time.

When researching AF, innovators would seek to understand the most common symptoms for patients with the disease, how they feel with AF, and the signs most commonly observed by physicians in patients with the disease. They should also consider whether all AF patients are affected by the same symptoms and what factors have the greatest impact on symptoms presented (e.g., age, coexisting conditions). For example, young patients are much more likely to report symptoms of palpitations with AF than older ones. This may directly impact the goal of therapy for different age groups.

One strategy that may be helpful in evaluating clinical presentation is to take the perspective of the healthcare delivery system (e.g., insurance company or hospital). From a provider's perspective, what symptoms or comorbidities bring patients in for clinical care? From an insurance company's perspective, what types of bills are submitted from the providers who first see a patient with the disease and what is the frequency of care?

Working Example
Clinical presentation of atrial fibrillation

While some patients do not experience noticeable symptoms due to AF, others have fatigue, weakness, lightheadedness, shortness of breath, or chest pain. Palpitations – sensations of a racing, uncomfortable, or irregular heartbeat – are also quite common. Symptomatic AF is widely recognized as leading to reduced patient **quality of life**, functional status, and cardiac performance.[23]

Importantly, one of the most common presentations for AF is stroke. In fact, AF is the heart condition that most commonly causes stroke.[24] Because the atria are fibrillating and not contracting, the flow of blood in the atria can become sluggish, especially in certain parts of the left atrium. This blood can coagulate leading to the formation of a clot. If a clot is dislodged and pumped out of the heart to the brain, it can cause a stroke. As a result, many patients with AF are treated by physicians for stroke using medicines that prevent the blood from clotting easily. Since these medicines can lead to the side effect of bleeding, a consequence is that physicians sometimes occasionally need to treat patients for a side effect of the stroke treatment itself.

In general, younger patients tend to have more "palpitations symptoms" which cause them to seek medical care. Older patients tend to have few (or no) symptoms of palpitations, but may be more compromised by fatigue. Patients with preexisting cardiac disease, such as heart failure, in which the heart does not function well at baseline, can become severely ill if they develop AF, sometimes resulting in the need for acute hospital care.

Clinical outcomes

Importantly, clinical outcomes are different from symptoms. Outcomes generally refer to hard data points associated with a disease that can be measured. The two most important types of clinical outcomes to consider are **morbidity** and mortality. Morbidity refers to the severity of the disease and its associated complications. Measures of morbidity may be evaluated using quality-of-life questionnaires, or they can be assessed by more specific endpoints such as distance walked in six minutes, hospital admissions, or a clinical event which does not cause immediate death (e.g., stroke, heart attack). Mortality refers to the death rate associated with a disease. Clinical outcomes are particularly important as they often serve as endpoints for clinical trials, since they can be assessed more easily and objectively than symptoms and have a direct impact on cost.

In the AF example, key clinical outcomes to address are the morbidities associated with AF, their likelihood of occurrence, and what factors have the greatest impact on them (e.g., age).

Working Example
Clinical outcomes associated with atrial fibrillation

Morbidities

One of the most frequent reasons that patients come to the emergency room for evaluation is palpitations. Although a number of rhythm disturbances can cause palpitations, AF is one of the most common. AF not only causes an irregular heartbeat, but can result in a rapid heart rate of up to 180 beats per minute, which can make patients feel nauseated and short of breath. These symptoms often improve with treatment by intravenous medications that slow AV conduction and reduce the heart rate. Episodes of AF are extremely scary and have a major impact on quality of life. They also result in a large number of emergency room visits each year.

In addition to acute symptoms with accelerated ventricular rates, AF can lead to a four- to five-fold increase in the risk of stroke.[25] The risk of stroke due to AF increases with age, rising from 1.5 percent for patients in their 50s to 23.5 percent for those in their 80s.[26] Overall, the annual risk of stroke in patients with AF ranges from 3–8 percent per year, depending on associated stroke risk factors[27] – a rate that is roughly five times higher than the rate of stroke in patients without AF.[28] As a result of this risk, one of the most common reasons for hospitalization each year is the need to anti-coagulate the patient and reduce the rate of stroke, driving huge impact on patients and the healthcare system.

Beyond the risk of stroke, AF is widely believed to reduce the heart's pumping capability by as much as 20–30 percent. As a result, AF (combined with a rapid heart rate over a sustained period of time) can lead to congestive heart failure (CHF). More directly, patients with existing heart failure often decompensate when they develop AF, requiring prolonged hospitalization.

Mortality

AF is also associated with an increased risk of death. According to the Framingham Heart Study, AF leads to a doubling of mortality in both sexes. After making adjustments for comorbidities, the risk remains 1.5 times higher in patients with AF. This increased rate of mortality is mainly due to strokes, progressive ventricular dysfunction and heart failure, and increased mortality from coronary events.[29]

Economic impact

At this stage of the biodesign innovation process, the focus of economic research should be on understanding the overall costs of the disease on the system at large, including the annual cost of treatment, hospitalization, and lost productivity due to absenteeism from work. Consider these costs at the system level, not necessarily for individual treatment alternatives. More detailed analysis of costs and healthcare payments will be performed as part of 2.2 Existing Solutions, 2.4 Market Analysis, and 4.3 Reimbursement Basics.

Be diligent in trying to understand the distribution of costs. Is the primary expenditure for acute or chronic medications, a device-based treatment, or a major surgery? Does the treatment of symptoms require hospitalization, or is this type of care mainly provided in the outpatient setting (which can be more cost effective)? Remember to take into consideration the life-long costs of care, as well as those associated with episodes such as hospitalization. The answers to these questions may reinforce the presence of **value** signposts (as described in 1.2 Needs Exploration), which signal that innovators may be able to create increased value for the stakeholders affected by AF in addressing the defined need. The potential value associated with a need is explored further in 2.4 Market Analysis, with the economic data gathered as part of disease state analysis acting as an important input.

For AF, innovators should look at the aggregate, system-level cost of AF on an annual basis, the treatment-related annual cost of AF, the annual cost of hospitalization, and the annual cost of lost productivity from absenteeism due to AF.

Working Example
Economic cost of atrial fibrillation

According to study published in 2011, AF costs the US healthcare system up to $26 billion each year. This retrospective, observational cohort study, based on administrative claims from the MarketScan Commercial and Medicare Supplemental research databases from 2004 to 2006, estimated that more than 460,000 hospitalizations per year cite atrial fibrillation as the primary diagnosis. The authors also concluded that AF contributes to 80,000 annual deaths.[30]

Just under three-quarters of total annual AF costs are associated with patient hospitalization. Another 23 percent is spent on outpatient care and testing. And 4 percent goes to outpatient prescription drugs.[31] Hospital costs are high among AF patients because they often require readmission. As the US population ages, costs associated with AF are expected to increase significantly.

Optimizing disease state research

The approach to disease state research described in this chapter is applicable no matter where in the world innovators intend to work. Just keep in mind that certain aspects, such as epidemiology, pathophysiology, clinical outcomes, and the economic impact of a disease, can vary dramatically by geographic setting. For example, a global registry of more than 14,000 patients spanning North America, Latin America, Europe, the Middle East, Africa, India, China, and other parts of Asia, demonstrated that hypertension was the most common risk factor for AF, present in 62 percent of patients worldwide. But its prevalence ranged from 41.5 percent in India to 80 percent in Eastern Europe. Rheumatic heart disease was present in only 2.1 percent of North American AF patients, compared to 15.3 percent in both the Middle East and China, 22.0 in Africa, and 30.9 percent in India. In Africa, more than 5 percent of AF cases were associated with pericarditis or endomyocardial fibrosis, while these conditions were present in less than 1 percent of patients in the other locations.[32] Innovators are advised to understand the leading research in the field at large, as well as disease information that is specific to the location they are targeting.

As innovators perform disease state research, they also should be careful not to lose sight of the human side of the disease. Studying clinical presentation as part of this recommended research approach provides a good start; but even this line of inquiry can tend to be rather scientific. To gain a full appreciation of a disease, it can be helpful for innovators to understand the emotional toll it takes on patients, their families, and the providers who care for them. The following quote from an article by one AF patient underscores how significant the disease burden can be:[33]

I never knew when an episode would strike – while washing the dog, walking, talking on a conference call, sitting in a meeting – so I was always afraid. My heart was like a flopping fish inside my chest. I would get so dizzy and lightheaded that I thought I would pass out. I was paralyzed and scared. When it was over, I was so wiped out that all I could do was crash. Life with atrial fibrillation wasn't normal . . . . My family was scared [too] and wouldn't let me out of their sight. We traveled together in the motorhome [for vacation]. . .. We planned our route to be near hospitals, and I knew every hospital along the way. You can't imagine, unless you've lived through it, the toll that atrial fibrillation takes. It takes a huge physical toll and a huge emotional toll, not just on you, but on your whole family, too. Worst of all is the financial toll – huge medical bills, inability to get insurance once you have atrial fibrillation, lost time from work and lost income, and for some people, lost jobs and careers and even lost houses and life savings.

The human aspects of a disease may not be relevant to every need. However, in some cases, it can provide important insights about need criteria that potential solutions should addresses. It also expands the innovators' understanding of key stakeholder (see chapter 2.3) and promotes a stronger sense of **empathy** for those affected by the disease.

In terms of the mechanics of performing disease state research, it is best to give priority to professional medical resources (textbooks, **peer-reviewed** medical journals, and websites targeted toward physicians and backed by accredited medical institutions). Peer-reviewed medical journals are usually the most up-to-date resources. However, not all medical journals are necessarily equal – just because an article is published in a medical journal does not make it fact. The higher the quality of the medical journal, the more likely the research design, process, and conclusions are accurate. One method to evaluate the credibility of a journal is to review its "impact factor." The impact factor, a measure of the frequency with which the "average article" in a journal has been cited in a particular year or period, is often used as an indicator of the importance of the journal to its field.[34] A "citation impact" can be used to measure the significance of an individual work or author. Impact factor and citation impact information is available on the Thomson Reuters Web of Knowledge, in its annual *Journal Citation Reports*.[35] Other tools, such as Google Scholar, also provide some information on citation frequency. No matter how credible the journal or its authors, innovators are advised not to accept any information blindly. Always search back for any original references that are available and try to triangulate data via multiple sources.

When searching, it is often valuable to start with a series of general searches (e.g., on "atrial fibrillation") using the sources outlined in the Getting Started section. Then, specific gaps in available information can be addressed through additional, increasingly directed data searches (e.g., annual cost of hospitalization for atrial fibrillation) until innovators have a thorough understanding of the space. In addition to moving from general to more specific inquiries, another helpful approach for completing disease state research is to look up references cited in some of the most informative documents and then to find and review the listed papers, especially the peer-reviewed ones. Lastly, analyst reports can be invaluable for understanding hard-to-find economic impact data, which is usually important for discerning the economic impact of a disease.

Summarizing the data

When summarizing what has been learned about a disease state, innovators should strive to keep the target audience in mind. Write the overview in an appropriate manner (i.e., not too technical if intended for potential investors, but adequately scientific if targeted to clinicians). Additionally, make sure to cite the sources of all statistics, study results, and clinical outcomes, as well as the source of interviews with physicians or other experts. Unless this information is sourced, the credibility of the research is subject to question. If conflicting information is uncovered during the research process, give priority to data from peer-reviewed medical journals or other similar resources, as noted above.

The case example on The Foundry and Ardian, Inc. describes how one team approached the challenge of disease state research.

FROM THE FIELD ▶ THE FOUNDRY AND ARDIAN

Using disease state research as a building block for an innovative therapy

As part of its efforts to identify its next new project, medical device incubator The Foundry routinely looks at disease states not adequately served by existing technologies. The company identifies clinical needs internally and also evaluates ideas presented by other innovators. Years ago, as the team debated its next focus area, serial entrepreneurs Howard Gelfand and Mark Levin came to The Foundry with a novel idea. Levin, a heart failure cardiologist, and Gelfand, a biomedical engineer, had studied the interactions between the kidney and central nervous system that help the body regulate blood pressure and fluid balance. They believed that blocking the activity of the renal nerves[36] could positively affect three major disease states: heart failure (HF), hypertension (HT), and chronic kidney disease (CKD). In particular, having previously invented a dialysis-like device to remove excess fluid from patients in congestive heart failure, Gelfand and Levin hypothesized that preventing the transmission of signals along the renal nerves could help HF patients offload the fluid that builds up in the lungs and causes these individuals to be repeatedly hospitalized with shortness of breath. The entrepreneurs had even proposed a solution – an implantable neurostimulator or drug pump that could block renal nerve activity. They dubbed their concept renal denervation.

While the need for better solutions for fluid-overloaded HF patients was compelling, the complex physiology involved in targeting the renal nerves, combined with an implantable solution, sounded too much like "a science project" to Foundry partner Hanson Gifford. Partner Mark Deem, however, was not dissuaded. Intrigued by the idea of manipulating the renal nerves to influence other body systems, Deem set aside Gelfand and Levin's possible solution and dove into disease state research to better understand the mechanisms of action related to the need.

Rather than starting with an investigation of heart failure or hypertension, Deem focused on the neurophysiology of the kidney and the integrated processes that help the body control blood pressure and fluid to maintain

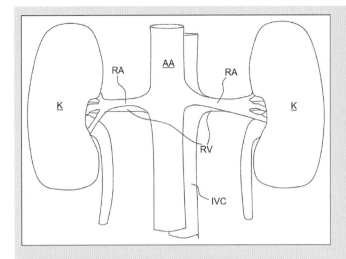

FIGURE 2.1.5
The first drawing from a seminal Ardian patent reflects the fact that the team's focus was grounded squarely in the disease state (from US utility patent 7,653,438-14).

homeostasis (see Figure 2.1.5). "I wanted to understand how the renal nerves play into this thing," Deem said. He began by accessing relevant journal articles via Medline. Describing his process, Deem explained, "When I'm working in body systems that I don't know well, I keep an excel spreadsheet of keyword searches as I work. The first column contains my exact search term, complete with quotes and Booleans. The second column is the number of hits, and the third is the number of relevant hits. Then I go one row per citation with title and author columns and a notes box I use to capture a three line synopsis that summarizes what I got out of the paper." According to Deem, this system keeps his research organized, prevents him from re-visiting papers he has already reviewed as he follows interesting citations, and helps him identify the best keyword strings. In the journal articles, Deem also highlights passages and uses margin notes to summarize key points. "I highlight what I think is important initially, but I often need to go back when I realize that actually different information is important. The margin notes help me quickly find the most relevant information in the paper."

Ultimately, Deem's Medline searches led him to papers written by Dr. Gerald F. DiBona, professor and vice chairman of the Department of Internal Medicine at the

University of Iowa College of Medicine. "He was the guy who has produced the largest body of work on renal neurophysiology in the world," Deem recalled. "And there was this one 'magnum opus' of a paper that had hundreds of citations. And for easily a month to two months, I just sat with that one paper. I'd read that paper and then I'd hit a brick wall. And there'd be a citation there. So, I'd go back to this huge bibliography, and I'd go pull that paper. And so I had the floor of an office basically covered with Gerry's paper and all of the citations from Gerry's paper, trying to build a fundamental understanding of the neurophysiology of the kidney and how that influenced hypertension and heart failure."

Through his research, Deem learned that in response to a decrease in blood pressure, a drop in sodium, or signals from the sympathetic nervous system, special cells in the kidneys generate a chain of neurohormonal responses that raise blood pressure by constricting the blood vessels, stimulating the heart to pump harder and faster, and causing the body to retain water and salt.[37] Although designed to maintain homeostasis, these regulatory mechanisms are chronically hyperactivated in some patients, causing the sustained high blood pressure, fluid overload, and dangerous sequelae that are the hallmarks of heart failure, hypertension, and chronic kidney disease.[38,39] Accordingly, blocking renal sympathetic activity could interrupt this cycle and have a dramatic effect on these disease states.

Following this deep dive into the literature, Deem was ready to test what he had learned by talking to doctors in the field. "I tend not to just go and try to sit down with docs and just ask them a bunch of questions," he explained. "Personally, I find it more useful to wait until I am very conversant about the anatomy and physiology that we're talking about. Otherwise, we can't really have a discussion, and I can't challenge them. And a huge amount of us being successful at what we do is being able to challenge the docs on the dogma of the clinical practice in the area that we're in. We say, 'Why is it that way?' and 'Couldn't it be this way?' And you can't do that unless you understand the system you are questioning." In line with this approach, Deem did not reach out to Gerry DiBona until he had read "everything I could find," he recalled. Describing their conversation, he continued, "We got pretty deep into the physiologic details, and he finally asked me who I had studied with. When I told him I'd just been reading up on it, he kind of laughed. I thought it was a pretty cool compliment."

Subsequent to speaking with DiBona, Deem also mentioned the concept of renal denervation as a treatment for heart failure to a practicing nephrologist. This conversation led him to an important historical precedent that served as early proof of concept and helped validate the proposed mechanism of action. Specifically, he was guided to a 1956 study of a now outdated procedure in which portions of the sympathetic nerve chain controlling the kidneys were surgically removed. While high risk, the surgery had been shown to slow progression of heart failure, resolve congestive symptoms, reduce blood pressure in 30–50 percent of cases, and increase survival. Deem also found more recent studies that concluded that surgical denervation (severing the renal nerves, as in kidney transplantation) was well-tolerated, increased urine output, and made it nearly impossible to induce or maintain a hypertensive state.

While Deem's research validated the role of the renal nerves in all three disease states, The Foundry team initially focused on heart failure. "Although kidney disease is a terrible disease, it was unclear that there would be a significant market there. So that was out from the get-go. Between heart failure and hypertension – well, we really went back and forth on that a lot," Deem remembered. Ultimately the preclinical focus by Levin and Gelfand on heart failure, and the fact that it was a huge medical opportunity that The Foundry had considered before, biased the team to focus on this space.

Gifford, who had worked in cardiovascular devices prior to forming The Foundry, provided the relevant disease state research on heart failure. With this condition, the patient's heart muscle cannot pump enough blood to meet the body's needs or keep up with the return of blood to the heart. Because of this, patients feel tired or weak. They also experience fluid build-up and swelling in the abdomen and extremities due to ineffective circulation. Fluid also backs up in the lungs, causing shortness of breath, especially upon exertion or when lying down. Doctors categorize heart failure patients as Class I through Class IV according to the severity of the symptoms,[40] and patients in the more advanced stages of the disease tend to be unstable, suddenly worsening (acute decompensated heart failure) and then improving temporarily after treatment with drugs and diuretics in the hospital.

With regard to clinical outcomes, 40 percent of Class IV patients are hospitalized at least four times a year, most often because of shortness of breath due to fluid overload. Late-stage heart failure has a 50 percent 5-year mortality.[41] Epidemiological research at the time indicated that there were five million HF patients and one million HF-related hospitalizations in the US each year.[42] The economic impact was significant; heart failure is one of the most resource-intensive conditions with direct and indirect costs in the United States estimated at $39.2 billion in 2010.[43]

While the compensatory hormonal responses that Deem had studied were meant to improve perfusion in HF patients, they ultimately put more stress on the heart and caused it to become even less efficient.[44] As noted, medications could help, but as heart failure advances these drugs often become ineffective.[45] Based on their improved understanding of the disease area, The Foundry team refined its direction, focusing on a way to offload fluid longer term than drugs to get congestive HF patients into a more stable, more medically manageable state in order to reduce the length and frequency of hospitalizations. Deem and his colleagues founded Ardian, named for the phonetic pronunciation of RDN for **re**nal **den**ervation, in January of 2005 to formally pursue this need.

Over time, as the team progressed, the solution evolved from an implantable device to a catheter-based treatment that would quiet the renal nerves using energy. Although the team believed the approach was technically sound, as they moved closer to their first-in-human milestone CEO Andrew Cleeland and other members of the team began to worry about the fragility and unpredictability of heart failure patients as a study group. Deem explained: "The big problem with heart failure is that you have patients who are Class II, and then they progress rapidly to Class IV. You get them in the hospital and pump them full of meds, you get the water off, and they go back to Class III. And then they go to II, and then they're back to IV. I mean, it's called 'unstable heart failure' for a reason. These patients are really hard to characterize, and it's difficult to build a study around that." Denise Zarins, Ardian's VP of R&D added, "We were worried that if we took them when they were in a decompensated state, with all that fluid on board, even having them lie down for a procedure could be difficult."

Going back to Deem's deep research and the effect of the renal nerves on all three disease states, Cleeland and the team realized that they could change the trial design for the first-in-human study to treat patients with HT resistant to drugs rather than heart failure. Describing the decision as "an evolution of thought," Cleeland explained, "Our [initial] goal was just to show that we could safely denervate a human being – not that we were going to cause a specific physiologic response. So we thought, maybe we should begin with the HT population because they are relatively healthier and more robust. Hypertension leads to heart failure, it leads to stroke, it leads to everything else, and it is the precursor." On the other hand, hypertension can be managed with safe, inexpensive drugs in many patients, so it wasn't clear that the risks of an interventional procedure would be justified. If the company was going to focus on treating hypertension, the procedure would need to be extraordinarily safe.

The results of the first trial not only demonstrated safety and proof of denervation, but were highly encouraging, as HT patients in the study experienced significant blood pressure reductions, with very few complications. Accordingly, while the team had planned to continue its HT study, and also start pilot studies in two HF groups (chronic and acute decompensated), the HF study arms "dropped by the wayside pretty quickly when the team started seeing 20–30 point reductions in blood pressure," recalled Deem. Convinced that they were onto something, the Ardian team decided to change its primary indication to HT.

Ardian's first pivotal trial, called Symplicity HTN-2, supported the decision to pursue hypertension. This study involved more than 100 patients in 24 centers in Europe, Australia, and New Zealand. The primary endpoint was office blood pressure as measured six months post treatment. Patients randomized to renal denervation therapy plus antihypertensive medications achieved a significant reduction in mean blood pressure (–32/–12 mmHg) at six months, whereas patients in the control group (randomized to receive antihypertensive medications alone) had blood pressures that did not vary from baseline (+1/0 mmHg).[46] In addition, over 20 percent of treated patients were able to decrease their hypertension medications or reduce their dosage at the six-month point when physicians were allowed to change their prescription regimens.[47] Based on the results of this study, Ardian received a CE mark for its technology and began commercializing the product in Europe. Shortly thereafter, the company was acquired by Medtronic.

As part of the strategy to bring the technology to the US, Medtronic worked with the FDA to design and launch another pivotal trial, known as HTN-3. In this blinded, randomized controlled study of more than 500 treatment-resistant hypertension patients, renal denervation met its primary safety endpoint, but did not show a significant difference from a sham intervention for lowering office systolic blood pressure through 6 months among patients who continued taking their anti-hypertensive medications.[48]

These results underscore the fact that a complicated disease process, such a sympathetic overactivity, can be unpredictable and difficult to study. In this case, several of the key issues affecting the outcome of the trial were directly related to disease factors, including the impact of medication roll-in and demographic mix on treatment results, as well as the extent of renal artery sympathetic nerve response to ablation energy.

Following the announcement of the HTN-3 results, Medtronic reconfirmed its commitment to the technology and indicated that it would continue working with the FDA to prepare for additional US trials.[49] In response to the HTN-3 data, one of the team's top priorities was to return to disease state research to expand its understanding of hypertension in different populations and the most effective way to study the link between the disease and renal denervation.

⬎ Online Resources

Visit www.ebiodesign.org/2.1 for more content, including:

Activities and links for "Getting Started"
- Assess anatomy and physiology
- Understand the pathophysiology of the disease
- Understand clinical presentation
- Assess clinical outcomes
- Gather epidemiology data
- Evaluate the economic impact
- Assess and summarize the information

Videos on disease state research

CREDITS

The editors would like to acknowledge Robert Croce for his assistance with the J&J case example, as well as Andrew Cleeland, Mark Deem, and Hanson Gifford for sharing The Foundry/Ardian story. Additional thanks go to Darin Buxbaum, Steve Fair, and Stacey McCutcheon for their help with the chapter.

NOTES

1 "What is Angina?," National Institutes of Health, National Heart, Lungs and Blood Institute, http://www.nhlbi.nih.gov/health/health-topics/topics/angina/ (October 16, 2013).

2 "State Heart Disease and Stroke Prevention Program Addresses Heart Attack Prevention," Centers for Disease Control and Prevention, http://www.cdc.gov/dhdsp/data_statistics/fact_sheets/fs_state_heartattack.htm (October 16, 2013).

3 Ibid.

4 "U.S. Medical Device Market Outlook," Frost & Sullivan, 2010, http://www.frost.com/prod/servlet/frost-home.pag (July 16, 2014).

5 In coronary artery bypass surgery, a patient's chest is opened and an artery or vein from another part of the body is attached to the diseased artery such that it allows blood to flow around, or bypass, the narrowing or blockage.

6 Matthew Dodds, Efren Kamen, and Jit Soon Lim, "DES Outlook: Adding the Wild Cards to the Mix," *Citigroup*, September 27, 2005, p. 4.

7 All quotations are from interviews conducted by the authors, unless otherwise cited. Reprinted with permission.

8 Robert J. Applegate, "Drug-Eluting Stents: The Final Answer to Restenosis?," Wake Forest University Medical Center, blogs.epfl.ch/stents/documents/Featured%20Article.doc (October 16, 2013).

9 "Epidemiology," Dictionary.com, http://dictionary.reference.com/search?r=2&q=epidemiology (October 16, 2013).

10 S. Colilla, A. Crow, W. Petkun, D.E. Singer, T. Simon, and X. Liu, "Estimates of Current Future Incidence and Prevalence of Atrial Fibrillation in the U.S. Adult Population," *American Journal of Cardiology*, 2013, pp. 1142–7.

11 Ibid.

12 Ibid.

13 J. S. Steinberg, "Atrial Fibrillation: An Emerging Epidemic?," *Heart*, 2004, p. 239.

14 Sinus rhythm is the term used to refer to the normal beating of the heart.

15 Purkinje fibers form a network in the ventricular walls and rapidly conduct electric impulses to allow the synchronized contraction of the ventricles.

16 Michel Haissaguerre, Pierre Jais, Dipen Shaw, Atsushi Takahashi et al., "Spontaneous Initiation of Atrial Fibrillation by Ectopic Beats Originating in the Pulmonary Veins," *New England Journal of Medicine*, September 3, 1998, p. 659.

17 "What is Atrial Fibrillation?," Cleveland Clinic: Heart and Vascular Institute, http://my.clevelandclinic.org/heart/atrial_fibrillation/afib.aspx (October 16, 2013).

18 Ibid.

19 In a "lone" episode, AF occurs in a heart that seems to be otherwise structurally and functionally normal.

20 Cardioversion is a method to restore a rapid heart beat back to normal.

21 Maurits A. Allessie et al., "Pathophysiology and Prevention of Atrial Fibrillation," *Circulation*, 2001, p. 769.

22 Thyrotoxicosis, or hyperthyroidism, is a condition in which the thyroid gland produces excess thyroid hormone (thyroxine) which affects the whole body.

23 Alan S. Go, Elaine M. Hylek, Kathleen A. Phillips et al., "Prevalence of Diagnosed Atrial Fibrillation in Adults," *JAMA*, vol. 285, no. 18, 2001, p. 2370.

24 "Advanced Imaging Can ID More Causes of Stroke Before They Strike," *Science Daily*, March 22, 2007, http://www.sciencedaily.com/releases/2007/03/070320084738.htm (October 16, 2013).

25 Gregory Y. H. Yip and Hung Fat-Tse, "Management of Atrial Fibrillation," *The Lancet*, 2007, p. 612.

26 "ACC/AHA/ESC 2006 Guidelines for the Management of Patients with Atrial Fibrillation – Executive Summary," American College of Cardiology Foundation, the American Heart Association, and the European Society of Cardiology, 2006, p. 866.

27 Ibid.

28 "Atrial Fibrillation," American Heart Association, http://www.americanheart.org/presenter.jhtml?identifier=4451 (October 16, 2013).

29 Lloyd-Jones, Wang, Leip et al., "Lifetime Risk for Development of Atrial Fibrillation," *Circulation*, 2004, pp. 1042–6.

30 M. H. Kim, S. S. Johnston, B. C. Chu, M. R. Dalal, and K. L. Schulman, "Estimation of Total Incremental Health Care Costs in Patients with Atrial Fibrillation in the United States," *Circulation: Cardiovascular Quality and Outcomes*, 2011, pp. 313–20.

31 Matthew R. Reynolds and Vidal Esseberg, "Economic Burden of Atrial Fibrillation: Implications for Intervention," *The American Journal of Pharmacy Benefits*, 2012, pp. 58–65.

32 Jeff S. Healey et al., "Global Variation in the Etiology and Management of Atrial Fibrillation: Results from a Global Atrial Fibrillation Registry," *Circulation*, 2011, http://circ.ahajournals.org/cgi/content/meeting_abstract/124/21_MeetingAbstracts/A9174 (January 15, 2014).

33 Mellanie True Hills, "Patient Perspective: Living with Atrial Fibrillation," StopAFib.org, September 2008, http://www.stopafib.org/downloads/News109.pdf (January 15, 2014).

34 "The Thomson Reuters Impact Factor," *Web of Knowledge*, Thomson Reuters, http://wokinfo.com/essays/impact-factor/ (October 16, 2013).

35 "Journal Citation Reports," *Web of Knowledge*, Thomson Reuters, http://wokinfo.com/products_tools/analytical/jcr/ (January 15, 2014).

36 The renal nerves are the sympathetic nerves to and from the kidneys.

37 *Cardiovascular Physiology Concepts*, April 19, 2007, http://www.cvphysiology.com/Heart%20Failure/HF003.htm (November 17, 2012).

38 "High Blood Pressure Dangers: Hypertension's Effects on Your Body," Mayo Clinic, January 21, 2011, http://www.mayoclinic.com/health/high-blood-pressure/HI00062 (October 3, 2012).

39 Markus P. Schlaich, Paul A. Sobotka, Henry Krum et al., "Renal Denervation as a Therapeutic Approach for Hypertension: Novel Implications for an Old Concept," *Hypertension*, vol. 54, 2009, pp. 1195–201.

40 "NYHA Classification–The Stages of Heart Failure," Heart Failure Society of America, December 5, 2011, http://www.abouthf.org/questions_stages.htm (January 31, 2013).

41 C. D. Kemp and J. V. Conte, "The Pathophysiology of Heart Failure," *Cardiovascular Pathology*, vol. 21, no. 5, 2012, pp. 356–71, http://www.ncbi.nlm.nih.gov/pubmed/22227365 (November 18, 2012).

42 Ardian, Series B Financing Presentation, 2004.

43 D. Lloyd-Jones, R. J. Adams, T. M. Brown et al., "Heart Disease and Stroke Statistics, 2010 Update; A Report from the American Heart Association," *Circulation*, vol. 121, no. 7, 2010, pp. e46–215.

44 Richard E. Klabunde, "Pathophysiology of Heart Failure," *Cardiovascular Physiology Concepts*, April 19, 2007, http://www.cvphysiology.com/Heart%20Failure/HF003.htm (November 17, 2012).

45 L. De Bruyne, "Mechanisms and Management of Diuretic Resistance in Congestive Heart Failure," *PostGraduate Medical Journal*, vol. 79, 2003, pp. 268–71, http://pmj.bmj.com/content/79/931/268.full (November 30, 2012).

46 Symplicity HTN-2 Investigators, "Renal Sympathetic Denervation in Patients with Treatment-Resistent Hypertension," *The Lancet*, December 2010, http://www.thelancet.com/journals/lancet/article/PIIS0140-6736(10)62039-9/abstract (January 7, 2014).

47 Murray Esler, presentation at the American Cardiology Conference, 2012.

48 Todd Neale, "Medtronic's Renal Denervation System Fails," *MedPage Today*, January 9, 2014, http://www.medpagetoday.com/Cardiology/Hypertension/43715 (January 22, 2014).

49 "Medtronic Releases Results of SYMPLICITY HTN-3," Medtronic press release, March 29, 2014, http://newsroom.medtronic.com/phoenix.zhtml?c = 251324&p = irol-newsArticle&ID = 1913401&highlight = (April 25, 2014).

2.2 Existing Solutions

INTRODUCTION

The team is about to take its second-generation prototype into the animal lab for testing. The pressure is on because the initial seed round of $200,000 is almost gone. Then, the bad news arrives: another group has just presented an abstract at the European College meetings describing the first eight patients treated with an approach that will most likely render this technology obsolete. The team now recalls one of their early clinical advisors mentioning another technology under development in Germany. If only they had tracked down this lead at the time, they might have saved money, time – and maybe the company.

The goal of any solution is to improve outcomes for patients with a particular disease or disorder. The analysis of existing solutions involves detailed research to understand what established *and* emerging products and services are available for diagnosing, treating, and managing a condition, how and when they are used, how and why they work, their effectiveness, their costs, and the overall value they deliver. Through the process of investigating existing solutions and creating a comprehensive profile of how a condition is typically addressed, areas for improvement and new opportunities may become apparent. This analysis also helps provide innovators with an understanding of the clinical, patient-related, and economic requirements that any new solution must meet to be considered equivalent or superior to existing alternatives. It further establishes a baseline of knowledge against which the uniqueness and other merits of new solutions can eventually be evaluated.

 See ebiodesign.org for featured videos on researching existing solutions.

OBJECTIVES

- Appreciate the importance of understanding the full range of existing and emerging solutions for diagnosing, treating, and managing a given disease state.

- Know how to effectively search for and summarize information about existing solutions into a useful format.

- Understand how to perform a gap analysis that can lead to the identification of opportunities to create value within the landscape of existing solutions.

EXISTING SOLUTION FUNDAMENTALS

Existing solutions can only be evaluated after gaining a working knowledge of the disease state (see chapter 2.1). As with disease state analysis, initial research of existing solutions is performed broadly at first, and then again in much greater detail as part of validating or screening promising **needs**.

The primary goal of researching existing solutions is to learn what alternatives are already available for diagnosing, treating, and managing a disease, as well as to identify where the most compelling opportunities to address unmet needs may exist. Solution research also provides a platform for better understanding patient, provider, and system-related requirements that will ultimately become a vital component of the **need specification** (as described in 2.5 Needs Selection).

Types of solutions to consider

Innovators are encouraged to perform a comprehensive search of existing solutions, being careful not to overlook any relevant diagnostics, interventional or surgical therapies, or management tools and services. For example, a team of innovators may have a primary interest in developing device-related therapies, but their research should include all diagnostics, treatment alternatives that extend beyond devices (e.g., drugs, surgery), and products and services related to managing a patient's condition (e.g., rehabilitation, monitoring). Table 2.2.1 outlines the full range of solutions to assess.

Other solution considerations

In addition to being exhaustive about the types of existing solutions associated with a given disease state, innovators should carefully evaluate the type of provider that delivers each one. The mapping of different solutions to their practitioners will be a useful input to **stakeholder** analysis (see 2.3 Stakeholder Analysis). It also provides a construct to think about the capabilities of each provider type. For instance, colonoscopy requires dexterity and finesse on the part of the physician to avoid patient discomfort. Gastroenterologists who routinely perform this procedure are more likely to have the skills required for other delicate or sensitive procedures using this approach, while other types of physicians working in the space might be less suited to deliver new solutions of this nature.

The skills or requirements that each solution places on patients should also be taken into account. For example, orthopedic surgical procedures are usually associated with rehabilitation in order to optimize the surgical result. Clearly, compliance varies among different patient populations based on age, functional status, and even economic background. These issues are often tied to the likelihood of procedural success or follow-up complications. More generally, the degree of compliance with existing solutions (be it rehabilitation, medication, diet, exercise, or other lifestyle factors) is a fertile area to survey in assessing the needs landscape. A great deal of the current excitement surrounding the mobile health (**mHealth**) "revolution" is based on the expectation that compliance gaps can be narrowed, in

Table 2.2.1 Be careful not to overlook any solution types when researching existing solutions.

Type of solution	Description
Diagnostics	Determination of whether (and to what extent) patients are affected by a given condition.
Behavioral and lifestyle treatments	Patient-driven solutions such as modifying one's diet and exercise.
Pharmacologic or biologic treatments	Chemical or biologic agents; usually injected or orally delivered.
Percutaneous treatments	Therapies administered via a catheter.
Minimally invasive treatments	Procedures performed using small incisions (e.g., laparoscopic procedures to access the abdomen or a joint; pacemaker placement).
Open surgery	Procedures that require the surgeon to cut larger areas of skin and tissues to gain direct access to the structures or organs involved.
Services	Human-centered interventions such as physical therapy and respiratory therapy.
Disease management	Products or services to monitor a patient's condition and guide therapy.

Table 2.2.2 Several different types of gaps can exist in the landscape of existing solutions, each leading to different types of opportunities.

Solution gaps	Description
Gap between the desired outcomes and the outcomes achieved by existing solutions	As existing solutions are researched and evaluated, think about ways in which they fail to meet the desired outcomes defined in the preliminary need criteria and what may be the cause of their shortcomings. For instance, gaps may be caused by applying an inappropriate solution to a clinical need. Or, they may stem from solutions that embody the correct approach, but need further development or specific refinements to achieve improved results.
Gap between what specific subsegments of patients need and what is offered by existing/emerging solutions	Diseases and disorders often have different manifestations depending on the stage or severity of the condition. Particular subgroups, such as the elderly or patients with certain comorbidities (e.g., diabetes) may also respond differently to differing solution. Accordingly, gaps may exist for specific patient subtypes and/or stages of a disease.
Gap between the outcomes/effectiveness that existing or emerging solutions deliver and their cost	Value gaps take into account the cost of a solution relative to its effectiveness. In an increasingly cost-sensitive environment, where facility administrators and purchasing managers are playing a more dominant role in adoption decisions for new medical technologies, value gaps must be addressed by driving up effectiveness, bringing down costs, or ideally accomplishing both.

part, by new technologies and services. Accordingly, innovators should think about how the various solutions in the disease area of interest are likely to align with the interests, motivations, and capabilities of patients.

Finding gaps to identify opportunities

The desired outcome of this analysis is to find a gap (or gaps) in the landscape of existing solutions that represents an opportunity to address the stated need. There are at least three types of gaps to consider, as shown in Table 2.2.2.

When performing a gap analysis, it is important to think about how the solution landscape will look several years in the future, not just how it appears today. Solutions represent a moving target, so the analysis of existing and emerging solutions needs to be referenced to the time frame of the team's entry into the market. If the group anticipates a long development cycle (some devices can take 10 years or more to get into patient

care), there is a much bigger time frame for other technologies to enter the market and eliminate or substantially change the opportunity. Without considering this temporal aspect to the improvement of existing solutions and the development of new ones, innovators may focus only on the current opportunity gap, not realizing that it is already diminishing or even on the path to closing altogether.

Just as looking forward is important, a great deal can be learned from looking backward. If there appears to be a major gap in the solution landscape, an innovator would be well served to do some research on prior diagnostics, therapies, and management tools that sought to address that gap but failed. Studying the unsuccessful experiments of other innovators can highlight important pitfalls, risks to be avoided, and fundamental learnings that can be leveraged to accelerate efforts moving forward. The example below, focused on another company from The Foundry, called Emphasys Medical, illustrates this approach.

FROM THE FIELD ⟩ THE FOUNDRY AND EMPHASYS MEDICAL

Understanding existing solutions as part of the needs screening process

Medical device incubator The Foundry has learned that, early in the innovation process, a thorough examination of all available options for diagnosing, treating, and managing a disease can spark ideas that others may have missed.

In the late 1990s, managing partners Hanson Gifford and Mark Deem were interested in pursuing opportunities related to the treatment of emphysema (see Figure 2.2.1). As part of their early research, they performed a detailed assessment of established and emerging solution alternatives in the space. At the time, Gifford pointed out, "The only real treatment for emphysema was lung volume reduction surgery (LVRS)."[1] LVRS is a highly invasive procedure in which the surgeon opens a patient's chest and removes roughly 30 percent of the diseased tissue in order to increase the flow of oxygen to the remainder of the lungs. The operation is difficult to perform and extremely painful for the patient. Moreover, the associated mortality rate ranges from 6 to 10 percent, among the highest for any elective procedure.[2] "We spent a lot of time trying to understand how we might be able to do the surgery better. We did extensive

FIGURE 2.2.1
Mark Deem and Hanson Gifford of The Foundry (courtesy of Stanford Biodesign).

literature research, talked with many surgeons and pulmonologists, and really explored the disease state, including its natural history, the cellular degradation of the lungs, and so on," explained Gifford. Besides disease state analysis, they also proactively and exhaustively researched existing and emerging solutions by cold-calling inventors, companies, and experts working in the field and networking with other entrepreneurs. Through this process, they realized that emerging technologies were predominantly focused on incrementally improving LVRS and almost no research was being done to look for an alternative, non-surgical procedure.

Recognizing that there was a huge gap in the solution landscape for emphysema (and driven by a desire to develop a less painful and invasive option for patients), Gifford and Deem decided to refocus their solution research on non-surgical alternatives. "One of our major breakthroughs came when we decided that no matter what kind of a device we came up with, what we really needed to do was make this a non-surgical procedure," recalled Gifford. "Everybody was looking at surgery, which became just this little corner of the solution landscape for us," added Deem. "There was this huge area outside of surgery where nobody was working that was full of potential opportunities."

The absence of other innovators or companies pursuing non-surgical treatments for emphysema left the field wide open, but it also created a host of difficult challenges that they would potentially face in pioneering a new procedure and/or device. Their research had shown that pulmonologists (physicians specializing in the lungs) saw emphysema patients on a regular basis, but it was surgeons who performed LVRS. In order for a solution to be non-surgical, pulmonologists would have to be able to deliver treatment. At the time, "Pulmonologists didn't really perform therapeutic procedures," said Deem. "In some ways, we would have to develop a new field of medicine, or at least expand a traditional specialty to make a non-surgical treatment for emphysema possible." To evaluate the feasibility of such a shift, they invested significant effort in examining the current referral patterns among doctors, how equipment was procured and funded in hospitals, and whether pulmonologists had the right skills, resources, and physical space to perform therapeutic procedures.

They also evaluated economics in the solution area. "We took a fairly broad brush toward the financials at that point," said Gifford. "We looked at the overall cost to the doctor, the hospital, and the healthcare system for comparable procedures because – right or wrong – if these entities are already used to paying 'X' dollars for the treatment of a disease, you're more likely to be able to get that same amount." He added, "LVRS cost in the range of $20,000 to $30,000. If we could come up with a therapeutic procedure that only costs a few thousand dollars, then there would be room for a reasonably well-priced device."

Despite the many challenges in the field, Gifford and Deem saw the potential for a breakthrough product. More than 3.7 million patients have been diagnosed with emphysema in the United States alone.[3] While a relatively small number of LVRS procedures are performed each year, the market for a less invasive treatment alternative would be significant, particularly given the mortality rates associated with the current LVRS procedure and the pain and suffering to patients associated with both the surgery and the disease. According to Deem, "The standard line that physicians would hear from emphysema patients was, 'Cure me or kill me. I don't care which one, but I can't stand being perpetually short of breath.' "

Ultimately, The Foundry decided to move forward with this project, collaborating with John McCutcheon and Tony Fields to found Emphasys Medical in 2000. Emphasys developed a minimally invasive procedure utilizing removable valves that could be inserted into a patient in as little as 20 minutes.

The valves could control air flow in and out of the diseased portions of the lungs to help healthier portions function normally. The **concept** was to collapse the diseased portions of the lungs without having to remove the tissue surgically. More about Emphasys Medical can be found in chapters 5.3 Clinical Strategy and 5.4 Regulatory Strategy.

Approach to existing solution analysis

A complete analysis of existing solutions should include information in the core areas outlined below:

- **Overview of solution options:** Provide a high-level description of relevant solutions in the field, including a summary of each alternative, how it is typically used in practice, and the skills required by the **user**. This provides a foundation for both gap analysis and the refinement of needs criteria.
- **Clinical solution profile:** Describe the clinical rationale for why and when each solution is used. In particular, outline the clinical **mechanism of action**, clinical evidence on safety and effectiveness, indications, and patient segments. This information can be useful for performing gap analysis and refining needs criteria.
- **Economic solution profile:** Outline the cost associated with each solution, including both direct and ancillary costs and who or what entity is incurring them. Begin to understand the **value** of available treatment options by exploring costs relative to their effectiveness. This information is useful in performing a more in-depth market analysis of the solution landscape (see 2.4 Market Analysis).
- **Utilization solution profile:** Describe how each solution is used in clinical practice, by whom, and where. Utilization is another important component of the market analysis.
- **Emerging solution profile:** Capture new products and procedures for the diagnosis, treatment, and management of a condition likely to affect the solution landscape within the next 3 to 10 years. Diligent research of possible emerging technologies

helps prepare the gap analysis, refine the needs criteria, and provide support for the technical and clinical feasibility of potentially new solutions.

- **Summary of the solution landscape:** Develop a cohesive assessment of potential opportunities within the current and emerging solution environment, focusing on the gap between these alternatives and the defined need criteria.

The remainder of this chapter examines each of the areas above, using atrial fibrillation (AF) as an example. In this case, detailed analysis is provided for one existing therapy (pulmonary vein isolation) and one emerging therapy (left atrial appendage occlusion). To better understand the example, refer back to the information given in 2.1 Disease State Fundamentals as it provides an overview of AF, which is relevant to understanding existing solutions. The focus on only two example treatments in this chapter is for the purpose of illustration. In reality, innovators must examine all available solution and solution types within their area of interest.

As with disease state analysis, the appropriate level of detail for assessing existing solutions varies significantly based on the number of alternatives that exist within a particular field, as well as the number of different needs being evaluated. Innovators should remember to take an iterative approach, exploring each solution in progressively more depth, as their direction becomes increasingly clear. They should also revisit this analysis as additional information is gathered via other steps of the biodesign innovation process (such as 4.3 Reimbursement Basics). The example below has been prepared at a relatively high level as a simplified illustration of the approach.

Overview of existing solutions

The first step in performing this analysis is to investigate and summarize the current solutions for the disease. Developing an overview of the solution landscape may be best accomplished by categorizing the options based on common or shared features (see the Working Example "Overview of Existing Solutions for Atrial Fibrillation" for a sample). These common features are usually unique to the particular clinical need area, but may include patient populations, technology platform, or even mechanism of action. Remember not to limit this analysis to any specific solution type, but to include everything from diagnostics and lifestyle modification to open surgical therapies and disease management tools. For each one, explain the objective(s) of the solution, how it is typically applied in practice (e.g., does the solution have progressive steps?), the person delivering the solution, and the skills required by the user. Organize the information in such a way that it provides a complete sense of the solution alternatives that are currently used in practice. It is also important to elaborate on the relative strengths and weaknesses of each approach.

Working Example
Overview of existing solutions for atrial fibrillation

Atrial fibrillation is a disease in which an irregular, typically rapid, and chaotic heart rhythm replaces sinus rhythm (normal heart rhythm) leading to a variety of symptoms as well as blood clot formation that can result in a stroke. Available treatments for AF typically seek to accomplish one of three different objectives: (1) the restoration and maintenance of normal sinus rhythm; (2) the control of the ventricular rate; or (3) the reduction of the risk of forming and dislodging a clot, or thromboembolic risk (to be explained in more detail below). Therefore, it is natural to classify the different treatments into one of these three categories according to their objective. Within each category, a variety of pharmacological, surgical, and device therapies are utilized. Depending on the type of atrial fibrillation (paroxysmal, persistent, or permanent) and symptoms, an appropriate therapy can be selected. Typically, a pharmacological approach is used first for treating AF, along with lifestyle changes such as limiting the intake of alcohol and caffeine. When medications do not work, are not tolerated, or lose their effectiveness over time, surgical and/or device therapy may be required.[4]

Restoration and maintenance of sinus rhythm

One treatment strategy for patients with AF, termed rhythm control, is to reestablish and maintain normal sinus rhythm since this can improve symptoms, correct atrial function and structure, reduce the risk of blood clot formation (thereby reducing stroke risk), and potentially reduce the need for long-term treatment with anticoagulants. Regardless of whether underlying heart disease is present in a patient, restoring sinus rhythm is associated with improved oxygen utilization, lifestyle improvements, and increased exercise capacity.[5]

The primary advantages of rhythm control treatments are improved cardiac function and reduced symptoms in some patients. While a rhythm control strategy can help reduce the frequency of AF, it may not eliminate it all together. As a result, most patients using rhythm control treatments are still required to take anticoagulants on a long-term basis to reduce their risk of blood clots and strokes.

Cardioversion and pharmacologic rhythm maintenance

The most common way to return the heart to sinus rhythm is through direct current cardioversion – the restoration of the heartbeat to normal function by electrical countershock. Another approach is chemical cardioversion, which involves the use of antiarrhythmic drugs to convert the heart rhythm to normal.

Regardless of whether an electrical or pharmacological approach to cardioversion is used, patients are generally also required to take anticoagulants for some period after the cardioversion to prevent blood clots from forming while the atria recover from the stunning of the cardioversion procedure. Some patients are also required to take such medication before the cardioversion to reduce the risk of stroke due to the cardioversion itself. This is separate from the need for long-term treatment with anticoagulants due to the possibility of AF recurrence.

The relatively high rate of AF recurrence using a cardioversion and antiarrhythmic medication strategy for rhythm control is one of the key disadvantages of this treatment approach. After successfully being returned to a normal sinus rhythm, only 20–30 percent of patients remain in sinus rhythm after one year.

Although this percentage can be increased to 40–80 percent through the sustained use of antiarrhythmic drugs,[6] medications that affect the electrical properties of cells in the heart to help prevent the occurrence of AF, the overall risk of recurrence remains significant. Antiarrhythmic medications also have serious potential side effects, including the development of new, abnormal heart rhythms.[7]

Catheter-based pulmonary vein isolation

Catheter ablation is a minimally invasive procedure used to terminate AF by eliminating and/or "disconnecting" the pathways supporting the initiation and maintenance of AF. Catheters introduced into the heart direct energy to destroy tissue in specific areas (mostly located in or around the pulmonary veins) that are the source of AF, and to prevent it from initiating or conducting any type of electrical impulse to the rest of the heart, thereby allowing normal sinus rhythm to continue. While there are several variations of catheter ablation procedures currently in use to address AF, they all seek to either electrically isolate or eliminate the pulmonary vein triggers of AF and/or eliminate any other areas in the left or right atrium capable of initiating or maintaining AF. For the sake of simplicity, all of these varied procedures will be categorized together under the term "pulmonary vein isolation" in this example.

Cox–Maze surgery

The primary surgical approach to restoring sinus rhythm in patients with AF is the Cox–Maze procedure, during which a series of precise incisions in the right and left atria are made to interrupt the conduction of abnormal electrical impulses and to direct normal sinus impulses to the atrioventricular (AV) node, as in normal heart function.[8] Because this procedure is very invasive and complex, it is usually offered only to patients with a high risk of stroke and who are already undergoing another form of cardiac surgery.

Implantable atrial defibrillators

Another method of restoring sinus rhythm is through an implantable atrial defibrillator. This device delivers small electrical shocks via leads placed in the heart to convert abnormal rhythms to sinus rhythm. Patients can turn them on and off to treat AF when episodes occur, or they can be set to operate automatically. The device is inserted by a cardiologist in a cardiac catheterization laboratory using X-ray guidance. The primary limitations of atrial defibrillators are their relatively large size and, more importantly, the fact that the shocks they deliver can be quite painful. As such, they are not currently used to treat AF on a widespread basis.

Ventricular rate control

In patients who receive ventricular rate control treatments, no attempt is made to cure or eliminate AF. Rather, these treatment alternatives are focused on slowing the conduction of electrical impulses through the AV node, the part of the heart's conduction system through which all impulses from the atria typically need to pass before activating the ventricles. By controlling the rate at this junction point, the ventricular heart rate can be brought back into the normal range and thereby potentially mitigate symptoms due to the rapid pumping of the heart.[9] Because AF continues, anticoagulants are recommended to prevent blood clot formation and strokes, typically on an indefinite basis. Another disadvantage associated with these treatment options is the fact that it can be difficult to adequately control the heart rate and relieve the symptoms on a long-term basis.[10]

Pharmacological rate control

Rate therapy using various medications leverages the gatekeeper properties of the AV node to reduce the ventricular rate to 60 to 90 beats per minute during AF. Such a change does not eliminate the irregular heartbeat but, by slowing the heart rate, reduces the workload of the heart and potentially the symptoms that a patient may be experiencing.

Catheter ablation of the AV node

Catheter ablation of the AV node is typically reserved as a last resort for patients who have failed other treatment options.[11] This procedure, in which a catheter is inserted into the heart and energy is delivered to destroy the AV node, thereby disconnecting the electrical pathway between the atria and the ventricles. Without an atrial source to drive the contraction of the ventricles, which are responsible for pumping blood to the lungs and body, a permanent pacemaker needs to be implanted at the time of the procedure to restore and sustain regular ventricular contractions. The pacemaker is a device that sends electrical pulses to the heart muscle, causing it to contract at a regular rate. Even though the atria continue to fibrillate, the symptoms of AF are reduced in many patients.

Reduction of thromboembolic risk

Preventive treatment to reduce the risk of blood clots and strokes (thromboembolic risk) in patients with AF is an important consideration in any AF treatment regime. Anticoagulant or antiplatelet therapy medications are commonly used on both a short-term (e.g., before and after electrical cardioversion) and long-term (e.g., in conjunction with ventricular rate control treatments) basis. While anticoagulant and antiplatelet therapies carry a bleeding risk, their use is warranted in patients where the risk of thromboembolic events is greater than the risk of bleeding complications. Key risk factors for stroke in AF patients include previous history of ischemic attack or stroke, hypertension, age, diabetes, rheumatic, structural, or other heart disease, and ventricular dysfunction.[12]

Pharmacological therapy

For many years, the most common drugs used to treat the risk of thromboembolic events were aspirin and warfarin. Aspirin is an over-the-counter medication that is an antiplatelet therapy; platelets are blood constituents that play an important role in the clotting cascade. By affecting the adherence of circulating platelets to one another, aspirin can reduce the likelihood of clot formation. However, aspirin is usually only effective in AF patients who are young and who do not have any significant structural heart disease. Warfarin (Coumadin) is an anticoagulant used to prevent blood clots, strokes, and heart attacks. Warfarin is a vitamin K antagonist which reduces the rate at which several blood clotting factors are produced. The metabolism and activity of warfarin can be affected by various other medications, foods, and physiologic states. As such, the dosage of warfarin needs to be monitored closely and usually necessitates frequent blood tests to check the degree to which it is anticoagulating a patient's blood. Over-anticoagulation with warfarin can lead to a significantly higher bleeding risk for a patient.

The development of direct thrombin inhibitors (DTIs) for prevention of thromboembolic risk has been a major medical advancement. For years, numerous companies sought to develop a therapy to reduce the risk of thromboembolism without the elevated risk of bleeding. The direct thrombin inhibitors are a class of anticoagulants that includes vitamin K antagonists and Factor Xa inhibitors. These medications were initially developed for the prevention of deep vein thrombosis. However, their safety profile made them ideally suited for use in patients with atrial fibrillation not due to valvular heart disease. Dabigatran (Pradaxa®), the commercially available DTI, received US Food and Drug Administration (FDA) approval in October 2010[13] for prevention of stroke and systemic embolism in patients with non-valvular AF. The American Heart Association/American College of Cardiology Foundation/Heart Rhythm Society recently gave dabigatran a Class I indication recommendation[14] for use as an alternative pharmacologic agent to warfarin based on the results of two large studies. The PETRO study evaluated 502 patients with non-valvular atrial fibrillation and increased risk of thromboembolism. Study results demonstrated non-inferiority to warfarin.[15] The RE-LY study then evaluated two different does of dabigatran in 18,113 patients compared to warfarin and demonstrated non-inferiority and superiority to warfarin depending on the dose.[16] The second DTI rivaroxaban (Xarelto®) was approved by the FDA for prevention of stroke in patients with non-valvular AF in November 2011.[17] This was the result of the ROCKET AF study which evaluated 14,264 patient with non-valvular AF and increased risk of stroke. The results demonstrated non-inferiority to warfarin therapy, but a statistically significant reduction in intracranial bleeding without a mortality benefit.[18] This second DTI offers the advantage of being a once-a-day medication compared to dabigatran which must be taken twice a day. Both of these therapies have been widely adopted because they do not require regular monitoring of blood thinning effects as warfarin does.

Clinical solution profile

With the universe of existing solutions defined, the next step is to assess the clinical rationale for why and when each one is used. This includes researching the following areas:

- **Mechanism of action:** Review what has been learned about the disease in terms of the steps or sequence of events that occurs – also called its mechanism(s) of action. Then, identify which disease mechanisms are targeted by a given solution option and how each one seeks to affect the disease.

- **Indications:** Identify the patient populations for which each solution is indicated or contraindicated.

For drugs and devices, consider the specific indications approved by the FDA or other regulatory bodies outside the US. For any type of invasive procedure, determine whether there are specific patient segments for which the solution is recommended and understand why.

- **Efficacy:** Define the benefits for each solution. Ideally, these should be measurable benefits (e.g., reduction in mortality), which are best demonstrated by **clinical trials** for relevant solution types.

- **Safety:** Describe the risks of each solution, including precautions and adverse reactions.

Working Example
Clinical solution profile for pulmonary vein isolation

Pulmonary vein isolation seeks to prevent abnormal electrical impulses that initiate and maintain AF from reaching the atria. The procedure is focused on destroying, or ablating, the abnormal "triggers" that originate in and around the pulmonary veins, or creating lesions to effectively isolate them such that they can no longer electrically communicate with the rest of the heart.

The procedure is performed using conscious sedation (intravenous medications for pain and anesthesia) or general anesthesia, depending on the complexity and length of the planned procedure. Patients also receive an injection of local anesthetic to the groin where catheters are inserted (there is usually minimal patient discomfort). The procedure is performed by a cardiac electrophysiologist and can take between three to six hours, depending on the number of areas treated (although estimates range from as few as two to as many as 10 hours). Careful monitoring and a series of tests are completed following the procedure. If there are no complications, patients can be discharged from the hospital after approximately 24 hours. Patients generally resume their normal activities after two or three days.

Mechanism of action

During the pulmonary vein isolation procedure, multiple catheters are advanced through the blood vessels and positioned in various locations of the heart chambers. The catheters are used to electrically stimulate the heart and intentionally trigger AF, after which the catheters record and/or map the heart's electrical activity in and around the pulmonary veins and atria. Using this data, the tissue responsible for the abnormal electrical impulses causing AF can usually be identified. A different type of catheter then can be used to apply energy to destroy, or ablate, this tissue. Radiofrequency energy can be used to heat the tissue to create the lesions, or cryothermy can be used to create the lesions through freezing. Ultrasound and laser techniques are also under development.

Ablation lesions are placed at the interface between the atrial tissue and pulmonary veins to effectively create continuous encircling lesions which electrically isolate the pulmonary veins so that any abnormal impulses originating in them cannot reach the rest of the heart and initiate AF (see Figure 2.2.2). Ablating too deep in the pulmonary veins can cause narrowing, which will cause long-term complications for the patient. The lesions heal within four to eight weeks, forming scars around the pulmonary veins. The use of this technique can "cure" AF in many patients.

Indications

Technologies used by physicians to treat atrial fibrillation using pulmonary vein isolation can be separated into diagnostic mapping systems and therapeutic ablation

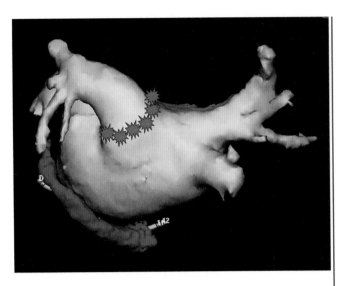

FIGURE 2.2.2

A three-dimensional image of the left atrium and pulmonary veins prior to pulmonary vein isolation (obtained using NavX™ navigation and visualization technology by St. Jude Medical). The star-shaped marks indicate the targeted region for pulmonary vein isolation; the structure beneath the left atrium illustrates the position of the coronary sinus, which wraps around the atria (courtesy of Amin Al-Ahmad and Paul Wang, Stanford University).

systems, including radiofrequency ablation products and cryoballoon catheter ablation devices.[19] In the US, the FDA has approved both technologies for clinical use in patients with paroxysmal atrial fibrillation. The same technologies are also used to perform the procedure in patients with persistent or permanent atrial fibrillation, although these indications are not FDA-approved and so technically are considered "**off-label**."[20,21] Guidelines released by the American College of Cardiology, the American Heart Association, and the European Cardiology Society in 2012 specified that the procedure has a Class I indication recommendation for patients with paroxysmal atrial fibrillation who are symptomatic and refractory to medications. It also has a 2A indication recommendation (slightly lower but still recommended indication)[22] for patients with paroxysmal atrial fibrillation who are symptomatic with limited or no drug therapy. Lastly, it has a 2A indication for patients with persistent atrial fibrillation who are symptomatic. It is not indicated for patients with permanent atrial fibrillation or that will soon be identified as long-standing persistent atrial fibrillation.[23]

Based on the approach, pulmonary vein isolation is most successful for patients with AF originating in the pulmonary vein(s), although continued variations in the procedure have additionally incorporated the identification and ablation of non-pulmonary vein triggers, such as those in the atria. However, since it is nearly impossible for clinicians to determine the origin of the arrhythmia without intracardiac mapping, which is too invasive and resource intensive to use on all patients as a screening tool to determine whether they need subsequent pulmonary vein ablation, clinicians typically need to rely on the clinical pattern and patient profile to determine whether a patient will be a good candidate for the procedure.

Efficacy

While there have been numerous non-controlled studies demonstrating the efficacy of varied types of pulmonary vein isolation in patients primarily with paroxysmal AF, there are several **randomized controlled trials** comparing pulmonary vein isolation with **standard treatments**. While each of the studies has limitations, they all report differences in favor of an ablation strategy in terms of relevant outcomes.

Wazni et al. (2005) conducted a small, randomized, unblinded trial that compared the procedure with a rhythm-control strategy. In this study, the magnitude of benefit in reducing AF recurrence was large at one year (13 percent recurrence in pulmonary vein isolation versus 63 percent recurrence in the medical group). There was also an improvement in quality that was considered clinically significant.[24]

Pappone et al. (2006) conducted a larger randomized trial to compare pulmonary vein isolation with antiarrhythmic drug therapy in patients with paroxysmal AF. This study concluded that pulmonary vein isolation is more successful than drug therapy with few complications. 86 percent of patients receiving pulmonary vein isolation were free from recurrent atrial tachyarrythmias versus 22 percent of patients receiving antiarrhythmic drug therapy.[25]

Wilber et al. (2010) conducted a multi-center controlled, 2:1 randomized trial to investigate the effectiveness of the Thermacool system (Biosense Webster). This study evaluated 167 patients and demonstrated that 66 percent of those treated with the catheter technique versus 16 percent treated with anti-arrhythmic drugs were free from atrial fibrillation at the nine-month primary endpoint.[26] This study led to the approval of the first combined diagnostic mapping and therapeutic ablation system for atrial fibrillation.

Table 2.2.3 Common complication rates for pulmonary vein isolation (from Atul Verma and Andrea Natale, "Why Atrial Fibrillation Should be Considered for First Line Therapy," *Circulation*.

Complication	Rate of occurrence
Transient ischemic stroke	0.4 percent
Permanent stroke	0.1 percent
Severe PV stenosis (greater than 70 percent, symptomatic)	0.3 percent
Moderate PV stenosis (40–70 percent, asymptomatic)	1.3 percent
Tamponade/perforation	0.5 percent
Severe vascular access complication	0.3 percent

Overall, these studies built the foundation for the first approved system for treatment of atrial fibrillation using pulmonary vein isolation. As a result, several ongoing studies are now broadening the experience with pulmonary vein isolation in an effort to gain approval for additional systems for treatment. Ultimately, long-term survival data will be important to demonstrate the value of such technologies in the scheme of therapies for AF.

Safety

While pulmonary vein isolation is generally considered safe when performed by experienced doctors, there are serious risks to consider, especially in patients in whom the likelihood of success may be low due to various factors such as structural heart disease. Complications from the procedure include those shown in Table 2.2.3 (based on pooled data from six studies involving more than 1,000 patients).

The procedure also can result in valvular injury, esophageal injury, and proarrhythmia. Complications appear to be declining with modifications to the procedure, new technology, and greater clinician experience. Advocates for the procedure assert that it is safe and may reduce **morbidity** and mortality associated with medical therapy.[27] Critics acknowledge that the short-term safety of newer ablation procedures has improved, but maintain that serious life-threatening complications do exist and that long-term safety is relatively unknown.[28]

Economic solution profile

For each existing solution, it is important to understand the financial impact at the individual level, as well as at the level of the healthcare system. The innovator should seek to identify the costs of providing the solution (diagnosis, drug cost, procedure, hospital stay, rehabilitation, ongoing management, etc.), as well as potential cost savings for using one particular solution versus another. This information will be considered in combination with the efficacy or effectiveness of the solutions later in the chapter.

Remember to determine the total cost of a solution across the entire **episode of care**. A long-term treatment, such as medical therapy, should be evaluated for its long-term beneficial potential for the estimated remaining lifetime of a patient versus the benefits of one-time surgical or device therapy. It is also important to take into account how cost-effective each solution is perceived to be by the various stakeholders in the healthcare system. **Reimbursement** is another critical consideration; however, it will be evaluated in more detail later in the biodesign innovation process (see 4.3 Reimbursement Basics).

Working Example
Economic solution profile for pulmonary vein isolation

The cost of pulmonary vein isolation procedures varies and limited data is available. However, in a Canadian study designed to compare the cost of medical therapy to catheter ablation in patients with paroxysmal AF, the cost of the procedure was found to range from $16,278 to $21,294, with a median estimate of $18,151. Follow-up costs ranged from $1,597 to $2,132 per year.[29]

In another study, Verma and Natale found that the initially high cost associated with pulmonary vein isolation would offset the ongoing costs of antiarrhythmic medication in year three, assuming a 72 percent ablation cure rate after 1.5 procedures. After four years, ablation becomes more cost-effective than treatment with antiarrhythmic medication.[30]

Additional studies have attempted to evaluate the economics of catheter ablation with an eye to selecting the best patient candidates for the procedure. One European study compared the costs of medical therapy for atrial fibrillation, with its accompanying need for long-term management and monitoring, with the costs of ablation. The authors concluded that catheter ablation was a cost-effective procedure primarily in patients with paroxysmal atrial fibrillation who are refractory to antiarrhythmic medications, especially if the success of the procedure and accompanying benefit in **quality of life** persist for more than five years, and the complication rate is low. (The authors noted that the lower complication rates experienced in high-volume centers would further increase the cost-effectiveness of the procedure.) However, for patients whose quality of life is not significantly affected by atrial fibrillation, or those whose poor quality of life is attributable to other health conditions besides atrial fibrillation, catheter ablation was unlikely to be cost effective.[31]

As noted earlier, the lack of long-term outcome data on quality of life and potential reduction of adverse outcomes (e.g., stroke reduction) hampers definitive statements about cost-effectiveness. A recent systematic review and meta-analysis sought to address this issue by reviewing studies that described outcomes at three years post-treatment and beyond, with a mean follow-up of more than 24 months after the index procedure. Data extracted from 19 studies found 53.1 percent of patients undergoing a single ablation procedure were free of atrial arrhythmia at long-term follow up and that, with multiple procedures (the average number per patient was 1.51), the long-term success rate was 79.8 percent.[32] Recognizing the importance of the long-term results to patient decision-making, as well as the determination of cost-effectiveness, the authors concluded, "The data presented in the current study suggest that long-term freedom of atrial arrhythmia can be achieved in the majority of AF cases, taking into account the need for multiple procedures in a significant proportion of patients."[33]

Utilization solution profile

Next, an innovator should seek to provide an overview of how each solution is currently used in clinical practice. This may include when (and why) certain solutions are used and the frequency of procedures. Physician organizations disseminate best practices for treatment regimens via guideline documents, which may be helpful. However, keep in mind that physicians may deviate from or modify these guidelines (as well as FDA regulations) in certain medically appropriate circumstances. Physicians are legally obliged to provide standard of care to their patients. As medical practice evolves, it sometimes outpaces regulatory approvals. This is particularly relevant in the area of pulmonary vein isolation devices since tens of thousands of procedures are performed annually using existing ablation devices on an off-label basis. This type of usage highlights the importance of gaining insight into how diagnostics, drugs, surgical procedures, devices, and other services are used by the majority of physicians in everyday practice, not just what usage they are cleared for in the market.

Emerging solution profile

Clinical solutions change over time as companies develop new products and physicians (and other care providers) develop new techniques. For some disease states, new solutions are introduced at a rapid rate and have the potential to dramatically change the landscape. Researching emerging solutions provides innovators with a working understanding of new diagnostics, treatments, and management tools potentially on the horizon, the areas they are targeting, and their timeline for their development and/or entry into the market. Although available information may be limited, try to cover as many of the topics outlined under the clinical solution profiles section as possible.

When investigating emerging solutions, look for leads by talking with experts in the field who might be aware of what technologies are under development. Clinicians

> **Working Example**
> Utilization profile for pulmonary vein isolation
>
> Without access to costly analyst reports, getting detailed information about the number of catheter ablation procedures performed each year can be challenging. In 2011, approximately 50,000 AF ablations were performed in the US, with another 60,000 performed in Europe.[34] The vast majority of these AF ablations were pulmonary vein isolation. Use of the AF ablation is growing at a rate of roughly 15 percent per year.[35]
>
> Traditionally, pulmonary vein isolation had been used as a therapy after a patient has failed at least one (if not two or more) antiarrhythmic drugs or had an intolerance or contraindication to antiarrhythmic therapy. As a result, the Centers for Medicare and Medicaid Services (CMS), as well as many private insurance plans, covered pulmonary vein isolation as a second-line therapy for specific patient groups. However, with changes in the guidelines in 2012, first line therapy is now broadly covered. (See 2.3 Stakeholder Analysis for more information about the role of public and private insurance companies – or **payers** – in the **medtech** field.)

may be aware of new products and services being tested in clinical trials, and investors may have information about new products and companies seeking funding. In terms of useful secondary research, innovators can learn a great deal by looking carefully at the abstracts of the relevant clinical meetings as soon as they are published to get late-breaking information about new technologies. Clinical trial databases are another source of information about products under development. For each emerging solution, seek to understand which mechanism or symptom of the disease it addresses and how it works. Evaluate the hypothesized efficacy and safety of the solution, as well as its anticipated time to market.

Working Example

Emerging solution profile for left atrial appendage occluder

As much as 90 percent of the embolisms, or clots that dislodge, associated with AF originate from the left atrial appendage (LAA), a small pouch-like sac attached to the left atrium.[36] Traditionally, chronic anticoagulation therapy has been used to manage this risk; however, these drugs can present safety and tolerability problems, particularly in patients older than 75 years of age (the group accounting for approximately half of AF-related stroke patients). Unfortunately, antiplatelet medications, such as aspirin, are less effective compared to drugs such as warfarin. Long-term warfarin therapy has additional downsides in that it requires costly and inconvenient patient monitoring, can have unpredictable drug and dietary interactions, and can be difficult to administer due to the frequent dosage adjustments needed to keep the risks of clotting and bleeding events appropriately balanced.

Since most clots in AF form in the LAA, one option for preventing clot formation would be to occlude or eliminate the LAA through surgical means. However, the Left Atrial Appendage Occlusion Study (LAAOS), which evaluated LAA occlusion performed at the time of coronary artery bypass grafting, showed that complete occlusion was achieved in only 45 percent of cases using sutures and in 72 percent of cases using a stapler.[37]

These observations laid the foundation for the development of device-based approaches to address the shortcomings of surgery and traditional drug therapy by controlling embolisms that come from thrombus in the LAA to reduce the risk of stroke. These devices work by mechanically preventing communication between the appendage and left atrium, thereby isolating or occluding the LAA from blood flow that would prevent harmful-sized emboli that may form in this location from exiting into the blood stream and traveling to the brain, resulting in a stroke.[38] The LAA occluders typically have a nitinol wire cage with a material coating that promotes endothelialization on the atrial side such that approximately four months following implantation, the device becomes part of the atrial wall. These devices have been designed to be placed surgically during open heart surgery, through minimally invasive approaches, or percutaneously.[39]

Appriva Medical (later acquired by ev3, Inc.) was the first company to develop a device to be placed percutaneously. The device received CE mark approval in Europe for its nitinol-based PLAATO™ (Percutaneous Left Atrial Appendage Transcatheter Occlusion) system in 2002 (see 4.2 Regulatory Basics for more information about CE marking and other forms of regulatory clearance).[40] Ultimately, the device was not commercialized in the US.

However, the WATCHMAN™ device (by Atritech, later acquired by Boston Scientific) is actively being targeted to the US market. In 2009, an FDA review panel initially voted 7:5 to recommend approval of the device, yet the agency opted to not grant its approval in 2010, citing concerns over safety from procedural complications in the initial 800-patient PROTECT AF study, as well as other issue with the conduct of the trial.[41] The initial study showed significantly lower hemorrhagic stroke in the device compared to treatment with warfarin. However, more safety events occurred in the device group.[42] The follow up PREVAIL study demonstrated non-inferiority comparing the device to warfarin for prevention of ischemic stroke and systemic embolization,[43] as well as a lower rate of procedural complications. Based on these results, a 2013 panel of FDA advisors voted 13:1 to recommend approval for WATCHMAN.[44] The FDA still requested additional data, and a third panel meeting was convened a year later. As of November 2014, a final decision from the FDA was still pending.

Lastly, the Amplatzer device (by AGA Medical, later acquired by St. Jude Medical), which was originally designed for the closure of atrial septal defects, has also been used to close the appendage but has not received regulatory approval for this indication in the US.

Solution landscape

Once a comprehensive set of data has been gathered, innovators should synthesize all of the findings into a comprehensive framework that summarizes the solution landscape. Such a summary should include an overview of what is known about a disease's mechanisms of action, which mechanisms are targeted by what solutions, and which are not currently targeted.

After creating an overarching summary, innovators should assess the more granular information they have gathered in a variety of different ways in an effort to better understand where potential solution gaps exist. Experiment with several of these approaches, choosing the ones that seem most likely to highlight opportunities in the solution landscape.

Causes versus consequence Certain solutions attack the cause of the disease while others ameliorate symptoms. In the case of atrial fibrillation, rhythm control works to target the disease mechanism and rate control targets the improvement of symptoms associated with rapid ventricular rates. It might be helpful to draw a tree diagram, with the causes shown nearer to the trunk and symptoms branching out from them. Place each solution above the location in the diagram that it targets to understand the hierarchy and where new solutions might supersede existing alternatives.

Mechanism of action frequency tally Typically, a number of solutions may target a single mechanism of action. Based on the clinical need area, more than one mechanisms may be targets. Count the treatments in each area to see which spaces are more crowded than others. In AF, for example, it would be instructive to count the number of therapies directed at treatment of primary arrhythmia versus the control of ventricular rate.

Patient segment frequency plot Different solutions are indicated and utilized for different patient populations. Plotting how many options are available to each segment may uncover underserved populations. In AF,

for example, elderly patients tend to have better outcomes with a rate control strategy as opposed to a rhythm control strategy; this is not same for younger patients.[45]

Risk versus benefit Summarize the risks and benefits of each solution. In doing so, think about the inherent trade-offs associated with each one and where opportunities exist for lowering risks while improving a wide range of benefits.

Cost versus effectiveness Cost-effectiveness analysis puts a slightly different twist on risk–benefit analysis by specifically describing the trade-offs between the cost of the various solutions and their clinical efficacy and/or the outcomes they deliver. Performing this analysis for existing (and emerging) solutions is the first step in initiating a market analysis for the needs under consideration and estimating their potential value to the stakeholders they affect (see 2.4 Market Analysis for a more detailed exploration of this topic). The classic way of presenting this analysis is to create a grid with effectiveness on one axis and cost on the other and then place solutions within that construct, based on what is known about their performance and price. By viewing the solution landscape in this way, innovators can potentially identify gaps that may be associated with an opportunity to create value by developing a solution with a better combination of cost and effectiveness.

Once a variety of analyses have been performed, innovators may find it useful to summarize their key takeaways in an overall gap analysis of available solutions.

Working Example
Summary of the solution landscape for atrial fibrillation

AF is the most common sustained cardiac arrhythmia, affecting millions of people in the US each year, yet there is not a clearly defined and agreed-upon strategy for the treatment of the disease.[46] Treatments to address the three categories of sinus rhythm maintenance, rate control, and thromboembolic risk are often

administered according to the general guidelines that exist in the field. Typically, clinicians seek to achieve rhythm control through the use of cardioversion and antiarrhythmic drugs. If that approach fails, the next step is to try ventricular rate control therapy or AV node ablation with the insertion of a pacemaker. Throughout the treatment process, thromboembolic risk is treated on an as-needed basis, typically with anticoagulants. A visual representation of these treatment options is presented in Figure 2.2.3. (Without delving into all of

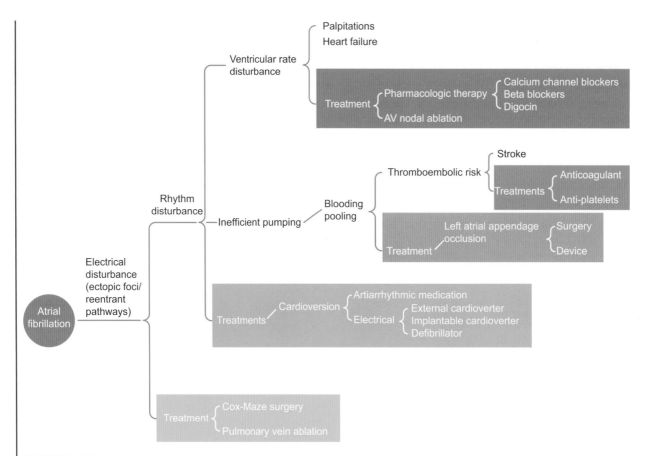

FIGURE 2.2.3

A high-level summary of select AF treatment options.

the solutions considered throughout this chapter, innovators can consider the debate regarding rhythm and rate control as an example of how to compare solution options.)

Until recently, there was little evidence that supported the traditional, somewhat sequential approach that favored rhythm control over rate control as a first-line therapy. In the AFFIRM trial, which included 4,060 patients, researchers compared the effects of long-term rhythm management and rate management treatment strategies to determine whether one approach offered significant advantages over the other in terms of benefits and risks to patients. Over a mean follow-up time of 3.5 years, the results of the study showed that there were no clear advantages to rhythm versus rate control. In fact, there was a trend toward increased mortality in the rhythm control group. The patients in the rhythm control group were also significantly more likely to be

hospitalized and have adverse drug effects than those in the rate control group.[47] The conclusion of the researchers was that rate control therapy should be considered a primary approach to the treatment of AF and that rhythm control should be abandoned early if it is not fully satisfactory.[48] However, this study focused on an older patient population so the generalizability of the study needs to be approached cautiously. As well, this trial did not include catheter ablation strategies for achieving rhythm control.

This information highlights the fact that while the study is an important first step in starting to evaluate one type of treatment option against another in the management of AF, more research is needed before conclusive guidelines that address all patient variations and the use of newer therapies are in place.

At a high level, the benefits and risks of the primary treatment alternatives for AF are summarized in online

Appendix 2.2.1. A gap analysis is provided in online Appendix 2.2.2. As can be seen, despite the significance and consequences of AF, many areas remain open for further research and development. Some examples include higher efficacy ablation procedures, improved medications for rhythm or rate control, better devices

for the treatment of the thromboembolic complications of AF, and safer and less invasive surgical procedures. By understanding the entire treatment landscape, gaps and inadequacies in current treatment methods can be used to help identify where promising clinical needs may exist that require innovative solutions.

Online Resources

Visit www.ebiodesign.org/2.2 for more content, including:

Activities and links for "Getting Started"
- Develop an overview of solution options
- Evaluate clinical solution profiles
- Analyze economic solution profiles
- Explore utilization solution profiles
- Investigate emerging solution profiles
- Summarize the solution landscape

Videos on existing solutions

Sample appendices that demonstrate
- A summary of risks and benefits for atrial fibrillation
- A gap analysis matrix for atrial fibrillation

CREDITS

The editors would like to acknowledge Mark Deem and Hanson Gifford of The Foundry for their assistance with the case example. Many thanks also go to Darin Buxbaum and Steve Fair for their help in developing the original chapter, as well as Stacey McCutcheon for her assistance with the updates.

NOTES

1 All quotations are from interviews conducted by the authors, unless otherwise cited. Reprinted with permission.
2 "Lung Volume Reduction Surgery," Thoracic Surgery Division, University of Maryland Medical Center, http://umm.edu/programs/thoracic/services/lvr (October 16, 2013).
3 "Understanding Emphysema?," American Lung Association, http://www.lung.org/lung-disease/emphysema/understanding-emphysema.html (October 16, 2013).
4 "What is Atrial Fibrillation?," Cleveland Clinic: Heart and Vascular Institute, http://my.clevelandclinic.org/heart/atrial_fibrillation/afib.aspx (October 16, 2013).
5 Nicholas S. Peters, Richard J Schilling, Prapa Kanagaratnam, and Vias Markides, "Atrial Fibrillation: Strategies to Control, Combat, and Cure," *The Lancet*, 2002, p. 596.
6 Leonard Ganz, "Patient Information: Atrial Fibrillation (Beyond the Basics)," Up-to-Date, http://www.uptodate.com/contents/atrial-fibrillation-beyond-the-basics (October 16, 2013).
7 Ibid.
8 "What is Atrial Fibrillation?," Cleveland Clinic: Heart and Vascular Institute, http://my.clevelandclinic.org/heart/atrial_fibrillation/afib.aspx (October 16, 2013).
9 Ganz, op. cit.
10 Ibid.
11 Peters, Schilling, Kanagaratnam, Markides, loc. cit.
12 Ibid., p. 600.
13 "FDA Approves Pradaxa to Prevent Stroke in People with Atrial Fibrillation," U.S. Food and Drug Administration, October 19, 2010, http://www.fda.gov/newsevents/newsroom/pressannouncements/ucm230241.htm (March 27, 2014).
14 A Class I indication recommendation means that there is evidence and/or general agreement that the treatment should be utilized as indicated. As the consensus to use the treatment increases, its class of indication is improved. Class I is the highest recommendation.
15 M. D. Ezekowitz, P. A. Reilly, G. Nehmiz, T. A. Simmers, R. Nagarakanti, K. Parcham-Azad, K. E. Pedersen, D. A. Lionetti, J. Stangier, and L. Wallentin, "Dabigatran with or without Concomitant Aspirin Compared with Warfarin Alone in Patients with Nonvalvular Atrial Fibrillation (PETRO Study)," *American Journal of Cardiology*, November 1, 2007, http://www.ncbi.nlm.nih.gov/pubmed/17950801 (March 27, 2014).
16 Connolly et al., "Dabigatran versus Warfarin in Patients with Atrial Fibrillation," *New England Journal of Medicine*, September 17, 2009, http://www.nejm.org/doi/full/10.1056/NEJMoa0905561 (March 27, 2014).
17 "FDA Approves Xarelto to Prevent Stroke in People with Common Type of Abnormal Heart Rhythm," U.S. Food and

Drug Administration, November 4, 2011, http://www.fda.gov/ NewsEvents/Newsroom/PressAnnouncements/ucm278646. htm (March 27, 2014).

18 Patel et al., "Rivaroxaban versus Warfarin in Nonvalvular Atrial Fibrillation," *New England Journal of Medicine*, September 8, 2011, http://www.nejm.org/doi/full/10.1056/NEJMoa1009638 (March 27, 2014).

19 Jeffrey Mandell, Frank Amico, Sameer Parekh et al., "Early Experience with the Cryoablation Balloon Procedure for the Treatment of Atrial Fibrillation by an Experienced Radiofrequency Catheter Ablation Center," *Journal of Invasive Cardiology*, June 2013, http://www.ncbi.nlm.nih.gov/ pubmed/23735354 (April 10, 2014).

20 Oussama Wazni, Bruce Wilkoff, and Walid Saba, "Catheter Ablation for Atrial Fibrillation," *New England Journal of Medicine*, December 15, 2011, http://www.nejm.org/doi/pdf/ 10.1056/NEJMct1109977 (April 10, 2014).

21 If a device is used in a way that deviates from the indication on the FDA-approved label, it is called "off-label" usage. While physicians may elect to engage in off-label use, the manufacturer of the device is not allowed to promote any methods of use other than those cleared by the FDA.

22 A Class 2A indication means that the weight of evidence/ opinion is in favor of the treatment's usefulness/efficacy as indicated.

23 Calkins et al., "2012 HRS/EHRA/ECAS Expert Consensus Statement on Catheter and Surgical Ablation of Atrial Fibrillation," EP Europace, March 27, 2012, http://europace. oxfordjournals.org/content/early/2012/02/29/europace. eus027 (March 27, 2014).

24 Wazni et al., "Radiofrequency Ablation versus Antiarrhythmic Drugs as First-Line Treatment of Symptomatic Atrial Fibrillation," *JAMA*, 2005, pp. 634–40.

25 Pappone et al., "A Randomized Trial of Circumferential Pulmonary Vein Ablation versus Drug Therapy in Paroxysmal Atrial Fibrillation," *Journal of the American College of Cardiology*, 2006, pp. 2340–7.

26 Wilber et al., "Comparison of Antiarrhythmic Drug Therapy and Radiofrequency Catheter Ablation in Patients with Paroxysmal Atrial Fibrillation," *Journal of the American Medical Association*, January 27, 2010, http://jama.jamanetwork.com/ article.aspx?articleid=185277 (March 27, 2014).

27 Ibid., p. 1219.

28 Benzy J. Pandanilam and Eric N. Prystowsky, "Should Ablation Be First-Line Therapy and for Whom: The Antagonist Position," *Circulation*, August 23, 2005, p. 1227.

29 Khaykin et al., "Cost Comparison of Catheter Ablation and Medical Therapy in Atrial Fibrillation," *Journal of Cardiovascular Electrophysiology*, September 2007, p. 909.

30 Atul Verma and Andrea Natale, "Why Atrial Fibrillation Ablation Should Be Considered First-Line Therapy for Some

Patients," *Circulation*, http://circ.ahajournals.org/content/ 112/8/1214.full (March 27, 2014).

31 Josef Kautzner, Veronika Bulkova, and Gerhard Hindricks, "Atrial Fibrillation Ablation: A Cost or an Investment?," *Europace*, European Society of Cardiology, 2011, http:// europace.oxfordjournals.org/content/13/suppl_2/ii39.full.pdf +html (April 10, 2014).

32 Anand N. Ganesan, Nicholas J. Shipp, Anthony G. Brooks et al., "Long-Term Outcomes of Catheter Ablation of Atrial Fibrillation: A Systematic Review and Meta-Analysis," *Journal of the American Heart Association*, March 18, 2013, http://jaha. ahajournals.org/content/2/2/e004549.full.pdf (April 10, 2014).

33 Ibid.

34 Andrew D'Silva and Matthew Wright, "Advances in Imaging for Atrial Fibrillation Ablation," *Radiology and Research Practice*, 2011, http://www.hindawi.com/journals/rrp/2011/714864/ (March 27, 2014).

35 Patrick P. Kneeland and Margaret C. Fang, "Trends in Catheter Ablation for Atrial Fibrillation in the United States," *Journal of Hospital Medicine*, September 2009, http://www.ncbi.nlm.nih. gov/pmc/articles/PMC2919218/ (March 27, 2014).

36 B. Meier, I. Palacios, S. Windecker, M. Rotter, Q. L. Cao, D. Keane, C. E. Ruiz, and Z.M. Hijazi, "Transcatheter Left Atrial Appendage Occlusion with Amplatzer Devices to Obviate Anticoagulation in Patients with Atrial Fibrillation," *Catheter and Cardiovasc Intervention*, vol. 60, no. 3, 2003, pp. 417–22.

37 Jeff S. Healey, Eugene Crystal, Andre Lamy, Kevin Teoh, Lloyd Semelhago, Stefan H. Hohnloser, Irene Cybulsky, Labib Abouzahr, Corey Sawchuck, Sandra Carroll, Carlos Morillo, Peter Kleine, Victor Chu, Eva Lonn, and Stuart J. Connolly, "Left Atrial Appendage Occlusion Study (LAAOS): Results of a Randomized Controlled Pilot Study of Left Atrial Appendage Occlusion During Coronary Bypass Surgery in Patients at Risk for Stroke," MedScape, September 9, 2005.

38 Ibid.

39 A procedures that is performed by entering the left atrium via the right atrium across the septum, which is the wall that separates them.

40 "APS: Appriva Medical Receives Approval to Commercialize PLAATO," ANP Pers Support, http://www.perssupport.anp.nl/ Home/Persberichten/Actueel?itemId=40776&show=true (October 17, 2006).

41 David Pittman, "Watchman Afib Device Wins FDA Panel Nod," *MedPage Today*, December 11, 2003, http://www. medpagetoday.com/Cardiology/Strokes/43380 (March 27, 2014).

42 David R. Holmes and Robert S. Schwartz, "Controversies in Cardiovascular Medicine," *Circulation*, https://circ. ahajournals.org/content/120/19/1919.full (March 27, 2014).

43 "Final Results of Randomized Trial of Left Atrial Appendage Closure versus Warfarin for Stroke/ Thromboembolic

Prevention in Patients with Non-Valvular Atrial Fibrillation (PREVAIL)," American Heart Association and American Stroke Association, http://my.americanheart.org/idc/groups/ ahamah-public/@wcm/@sop/@scon/documents/ downloadable/ucm_449994.pdf (March 27, 2014).

44 Pittman, op. cit.

45 Philip A. Wolf, Emelia J. Benhamin, Albert J. Belanger et al., "Secular Trends in the Prevalence of Atrial Fibrillation: The Framingham Study," *American Heart Journal*, vol. 131, 1996, pp. 790–5. Cited by Craig T. January, Samuel Wann, Joseph S.

Alpert et al., in the "2014 AHA/ACC/HRS Guideline for the Management of Patients with Atrial Fibrillation," *Journal of the American College of Cardiology*, 2014, accepted manuscript. http://content.onlinejacc.org/article.aspx?articleid = 1854231 (April 10, 2014).

46 The AFFIRM Writing Group, "A Comparison of Rate Control and Rhythm Control in Patients with Atrial Fibrillation," *New England Journal of Medicine*, December 5, 2002, pp. 1825–33.

47 Ibid.

48 Ibid.

2.3 Stakeholder Analysis

INTRODUCTION

A clinical need often begins with patients, their symptoms, and an underlying medical problem. But that is just the tip of the iceberg. Think about the physician and the nurses involved in the patient's care. Also, somewhere in the back office, there is a facility manager crunching numbers to decide whether or not to invest in the necessary equipment and infrastructure to support the patient's treatment. And then, perhaps a thousand miles away, is an insurance administrator or government official who decides whether or not to pay for the care that has been delivered. All of these are stakeholders – individuals and groups who are touched by the need and have a stake in how it is ultimately addressed.

In stakeholder analysis, the innovator systematically examines the direct and indirect interactions of all parties involved in financing and delivering care to the patient. The purpose of this analysis is to understand how these entities are affected by the need and to determine their requirements (or their stake) in how it is addressed. Stakeholders have different perspectives – for instance, some will benefit if the need is addressed, but others may be adversely affected. Uncovering these points-of-view and any potential conflicts is critical to shaping and refining the need statement and preliminary need criteria that have been identified. It also allows the innovator to anticipate resistance, as well as to define and prioritize the requirements that will shape the eventual solution to maximize its chance of adoption among the most important and influential stakeholders – often referred to as decision makers. For these reasons, stakeholder analysis should begin early in the biodesign innovation process, while needs are being identified and assessed. It can then be expanded as more information becomes known and progress is made toward a solution.

This is the first of two chapters focused on stakeholders. The output from the basic stakeholder analysis described here informs 5.7 Marketing and Stakeholder Strategy. The chapter is also closely linked to 2.4 Market Analysis. Among other topics, market analysis focuses on the assessment of competitors (i.e., businesses offering competing products) and other suppliers of products that address a given need. Competitors obviously have an important stake in the need and could, as a result, technically be considered stakeholders. However, because they will always resist new solutions proposed by competing innovators, they are excluded from traditional stakeholder analysis and considered among the other market forces that can create barriers to the adoption of a new idea.

 See ebiodesign.org for featured videos on stakeholder analysis.

STAKEHOLDER ANALYSIS FUNDAMENTALS

The need for **stakeholder** analysis is based on the multifaceted nature of healthcare systems and the fact that multiple groups and individuals – not just a single customer – drive the adoption of health-related products and services. Richard Stack, physician, inventor and investor, succinctly summarized the complex nature of medical innovation and reinforced the need for in-depth stakeholder analysis:[1]

You have to know who your customer is. Certainly you have to know what the patient wants . . .. But the person actually buying the product, you have to know that psychology very, very well.

One of the early steps in performing a stakeholder analysis is to identify the many different parties involved in delivering and financing care related to the **need**.

There are two primary methods for identifying stakeholders in the medical field, which should be performed in conjunction. The first focuses on stakeholders involved in the **cycle of care** – patient diagnosis and the delivery of treatment. With this approach, innovators study how patients move through their healthcare experience, making note of all of the different players, their roles, and interests. The second method is concerned with stakeholders involved in financing patient care. In this analysis, the innovator follows the **flow of money** from one entity to the next as charges and payments are made. The results from these two methods can be triangulated by referring back to the observations made during needs exploration (see chapter 1.2), as well as the data collected as part of the disease state and existing solutions analysis (see chapters 2.1 and 2.2).

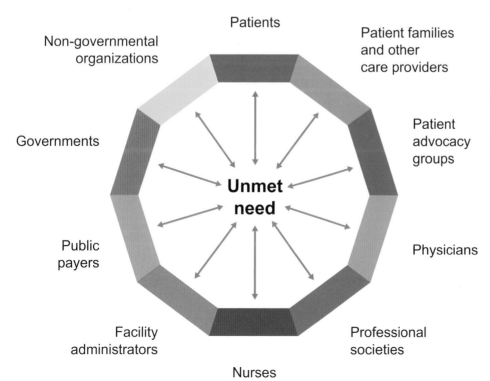

FIGURE 2.3.1

All stakeholders have the capacity to embrace or resist new medical technologies. While some exert more influence than others, all should be considered, particularly in a preliminary stakeholder analysis.

Figure 2.3.1 provides a detailed representation of the many different stakeholders with a potential interest in a new medical technology to address a defined need.

Cycle of care analysis

Cycle of care analysis is based on how patients interact with the medical system. The focus of this form of assessment is specifically on understanding the patient's diagnosis and treatment (not yet worrying about payments). Innovators should investigate who diagnoses the condition, who provides preliminary treatment, who provides next-level treatment if the condition progresses, what parties are involved in the ongoing management of the disease, and the role that patients themselves play in their care. It is especially important to pay attention to what different medical specialties are involved in this cycle, what referral patterns exist, and whether there is any tension as the patient moves between practice areas. The Working Example illustrates the cycle of care for end-stage renal disease.

Working Example
The cycle of care for end-stage renal disease

End-stage renal disease (ESRD) is characterized by chronic failure of the kidneys. The traditional cycle of care for this condition resembles the flow outlined in Figure 2.3.2. It starts with the patient developing certain symptoms that may trigger a visit to a primary care physician (PCP) or, in extreme cases, to the emergency room (ER). In both situations, a series of laboratory tests are performed that will be used by the attending physician to make a diagnosis. When ESRD is confirmed, the patient is referred to a nephrologist (specialist in kidney disease).

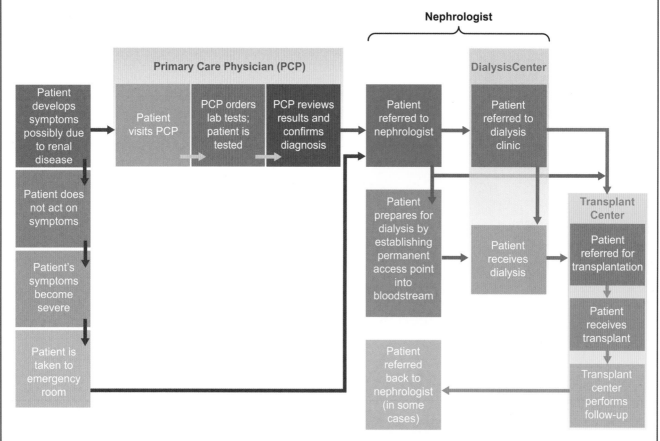

FIGURE 2.3.2
The cycle of care for ESRD can be relatively complex, involving multiple provider and facility types.

The nephrologist evaluates the patient and determines whether dialysis is needed. If so, s/he refers the patient to a dialysis clinic, as well as a vascular surgeon to prepare the patient's access for dialysis. (In dialysis, a patient's blood stream is accessed with a needle through a permanent access point established by a vascular surgeon. The blood flow is diverted through a filter in the dialysis machine to clear the toxins that the failing kidneys cannot eliminate. Dialysis involves three lengthy treatments per week.) Simultaneously, the nephrologist determines whether the patient is a good candidate for a kidney transplant (a surgical procedure in which an ESRD patient receives a new organ from a donor). If so, s/he refers the patient to a transplant center and a transplant surgeon. If the patient receives a transplant, the transplant center is initially involved in follow-up care with the involvement of a transplant nephrologist. Eventually, the patient may be referred back to the original nephrologist for long-term follow-up care.

Once the cycle of care has been mapped, as in the ESRD example, the innovator's objective is to examine who interacts with the patient, the nature of their relationships with the patient, and the duration and timing of the interactions. All of the individuals and groups in the cycle should be considered stakeholders in the process, including the patient. However, innovators should note that each stakeholder has unique needs, requirements, and interests. For example, patients almost certainly have a requirement to get a life-threatening medical condition under control, but may also have an interest in minimizing the effect of the disease on their **quality of life**. While other stakeholders involved in the delivery of care may share the first objective, they may not be as concerned with the second since it does not affect them in the same way and they routinely deal with patients who have made the life-changing move to dialysis.

Other stakeholders from the ESRD example include multiple clinical specialists (primary care physician, emergency care physician, nephrologist, transplant nephrologist, vascular surgeon, and transplant surgeon), different nursing specialties, and numerous facility types (doctor's office, ER, dialysis clinic, and transplant hospital). Not all of these stakeholders will be intimately involved with every need considered in the ESRD cycle of care. However, by using a method for identifying everyone with even a remote stake in a need (and how it is solved) innovators ensure that no one is overlooked.

In conducting this analysis, it is important to be aware that referral patterns in the cycle of care can be a source of potential conflicts between stakeholders, especially in cases where multiple specialties are involved. In the ESRD example, nephrologists lose their patients at the point when transplantation occurs, with only some patients referred back to them for long-term care (the remaining patients are cared for by the transplant center). A similar conflict exists between dialysis centers and transplant centers due to the loss of revenues to dialysis providers after patients receive transplants. So, it follows that if a new breakthrough becomes available that allows more patients to receive transplants, nephrologists and dialysis clinics might mount some resistance since the change could potentially lead to a substantial loss in their patient-care revenue.

This particular point of view is not entirely speculative. A study published in the *New England Journal of Medicine* confirmed that the likelihood of being placed on the waiting list for a renal transplant was lower for patients treated at for-profit dialysis centers than non-profit ones, which led to increased mortality in this patient group.[2] While this study and the previous discussion appears to suggest that healthcare providers and physicians may not always have the best interests of their patients in mind, the real message in the story is more nuanced. In the medical field, choosing the proper course of treatment for each patient requires the careful balance of the risks and benefits associated with the treatment options. Yet, this equation can be ambiguous and perspectives can vary from provider to provider. Transplantation, for example, may be perceived by some providers to be too risky for certain patients (based on their age, coexisting conditions, or other factors). The equation is further complicated when a provider has a financial incentive that makes one course of treatment preferable to another (dialysis versus transplantation). These situations create conflicts of interest that can affect the likelihood of adoption of a particular treatment for certain high-risk patients. Such resistance can sometimes be overcome by understanding and addressing the motivations and concerns of the involved stakeholders. For instance, if ESRD patients are explicitly referred back to their initial nephrologists for ongoing care after transplantation (even though the transplant centers are capable of administering such services), then this may reduce resistance to high-risk transplantation among nephrologists and stimulate more transplant referrals.

To succeed in global markets, innovators need to recognize that the results of a stakeholder analysis will differ considerably depending on the country where treatment occurs. Importantly, they should not assume that a US-based stakeholder analysis will be adequate for understanding other markets. Detailed analysis of the countries of interest must be performed to ensure that subtle (and not so subtle) distinctions in stakeholder attitudes, preferences, and perceptions are understood. The principles and approach to the analysis will be the same as those described within this chapter, but the output will likely be different. For instance, the cycle of care for a group of patients would be remarkably different in China than in the US, Europe, or other Western nations. Services that are routinely part of healthcare delivery in one country may be considered non-health-related in another. For example, a heart failure patient might be prescribed ACE inhibitors in the US (to control high blood pressure), but not in some Southern European nations where such treatments are not as widely used. In China, treatments could differ even further, with a patient receiving traditional Chinese medicines in lieu of Western solutions.

In India, vast disparities in the quality and availability of care characterize the healthcare system and affect the stakeholder experience. State-of-the-art Indian secondary and tertiary care institutions attract both domestic patients who can afford their world-class services and approximately 350,000 medical tourists each year.[3] However, healthcare facilities without adequate supplies, staff, or capacity to provide affordable care are responsible for serving a large majority of the population. A significant percentage of the existing infrastructure, both public and private, is unreliable.[4] In rural settings, where the penetration of healthcare facilities has been low, patients often resort to traditional healers for healthcare.[5]

In China, the Ministry of Health administers public hospitals. Additionally, this organization cooperates with local governments to oversee the urban healthcare system, which traditionally has been hospital based. Hospitals are classified into Tier 1, 2, or 3, each with three sub-levels: A, B, and C.[6] The most sophisticated and well-equipped hospitals are awarded the highest rating: 3A. Tier 2 hospitals are usually found at the district level; and Tier 1 hospitals are small, community-based centers.[7] Outpatient services attached to these hospitals are the first point of care for most patients, for even minor ailments, which creates major bottlenecks in the system. Long lines and overcrowded waiting rooms are common.[8]

As this basic information suggests, each geography has unique factors that directly affect healthcare delivery and the cycle of care for any given condition. Such factors can introduce a diversity of stakeholders into the equation (some of which US-based innovators may not initially anticipate). They also underscore the importance of researching the cycle of care across multiple settings to ensure that stakeholder attitudes, preferences, and perceptions are comprehensively understood.

Flow of money analysis

A flow of money analysis identifies stakeholders who directly or indirectly finance the cycle of care. It focuses on who pays for the services and procedures performed in diagnosis and treatment of patients. In the process, it highlights all of the entities with a direct financial stake related to the need, including the patients and their families. Often this analysis is used by innovators to identify those stakeholders who are most likely to be decision makers.

When it comes to analyzing stakeholders in the flow of money, the simplest model is one in which the patient is also the one who pays for the procedure. This "**out-of-pocket**" payment structure can be found in every country in the world to varying degrees. Alternatively, payments can be made on the patient's behalf through public or private insurance programs (see 4.3

Reimbursement Basics for more information). The National Health Service (NHS) in the UK and Medicare in the US [administered by the Centers for Medicare and Medicaid Services (**CMS**)] are both examples of public health insurance programs financed by taxpayer money. Examples of private health insurers in the US include companies such as Blue Cross Blue Shield (**BCBS**), United Healthcare, and Aetna. Such companies provide insurance on behalf of their subscribers, who pay a premium in exchange for health insurance **coverage**. Many, but not all, procedures delivered to patients by physicians are reimbursed under these plans. However, in some countries with **third-party payer** systems, there is a strong trend toward shifting the burden of payment to the patient through deductibles and exclusions for elective procedures. Common exclusions include aesthetics (within dermatology) and certain reconstructive surgery procedures (within plastic surgery and dentistry), which are generally not covered unless they are deemed medically necessary. Otherwise, patients are required to foot the bill for these "elective" procedures.

The Working Example outlines some key points about how the healthcare financing systems works in the US.

Working Example
Overview of the US healthcare financing system

The US healthcare financing system is undeniably complex, involving both public and private **payers**. A simplified view of the stakeholders involved in financing the US healthcare system and their interactions with the stakeholders delivering and receiving care is depicted in Figure 2.3.3.

The US has two public insurance programs (Medicare for the elderly and disabled, and Medicaid for the poor), as well as a multitude of private insurers. Medicare and Medicaid are funded by individual and corporate taxpayer money and administered by the Centers for Medicare and Medicaid Services (CMS), with a significant state involvement for Medicaid. In many instances, both

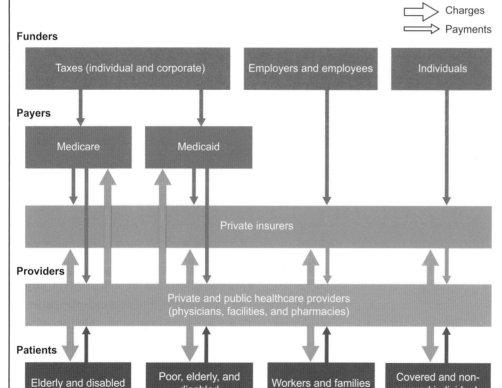

FIGURE 2.3.3
There are many interrelated entities involved in the flow of money in the US healthcare system (based on N. Sekhri, *Bulletin of World Health Organization*, vol. 78, no. 6 (2000): 832; reprinted with permission).

Medicare and Medicaid subcontract with private insurers for the administration of the benefits they cover.

Private insurers collect premiums from individual subscribers or from employers who provide health insurance benefits to their employees (historically, most of the employed non-elderly obtain health insurance through their employer, while some may purchase individual insurance).

Insurers (both private and public) then pay healthcare providers (facilities and physicians) for the services they provide to the individuals they insure. In many instances, the payments made by the insurers do not cover all of the charges made, such that the individuals receiving treatment must pay the balance (called a copayment).

This is becoming increasingly common, with private insurance plans developing hybrid approaches that require individuals to bear a larger portion of their total costs through deductibles, copayments, and/or limits on coverage.

Until the passage of the Affordable Care Act (**ACA**), individuals without any insurance coverage had to pay for all of their healthcare services out-of-pocket. As of 2014, however, the individual mandate stipulates that most adults will have to purchase health insurance or face a financial penalty. For more information about healthcare financing and **reimbursement** in the US, see 4.3 Reimbursement Basics.

Despite the involvement of multiple players and the many handoffs between participants, the flow of money in many treatment areas tends to follow a standard path in the US. For instance, most patients who receive an implantable cardiac defibrillator for the treatment of heart rhythm disorders are covered by private insurance until they turn 65 years of age (or become disabled), at which time they are covered by Medicare. However, because there are always idiosyncrasies and variations within the system, innovators should be cautioned about making assumptions regarding the flow of money. For example, many ESRD patients without private insurance are covered by Medicare three months after they begin dialysis and continue to be covered until three years after transplantation, regardless of their age. Most ESRD patients with private insurance are covered under their private health insurance policy for the first three years of treatment, but then convert to Medicare (again regardless of age).[9] In this scenario, it would be imperative for an innovator working in the ESRD space to understand the role of Medicare in the flow of money for the treatment area and then focus on this group as a primary stakeholder, even if the innovator's solution targets a subset of the population under the age of 65.

A recent shift in the US healthcare financing landscape is worth noting because of its effect on the financial incentives of various stakeholders. An increasing number of provider/payer networks have emerged which combine the provider (a direct participant in the cycle of care) and the payer (the direct source of payment in the flow of money). One of oldest and most well-known examples of this type of collaboration is Kaiser Permanente, which is an integrated delivery network (IDN) that includes the Kaiser Foundation Health Plan, Kaiser Foundation Hospitals, and the Permanente Medical Group (which represents the physicians). Kaiser serves eight regions in the US and has become significant in large markets like California where the organization holds a 40 percent share of individual and employer health insurance customers.[10] However, more recently, in response to the Affordable Care Act, voluntary consortiums of *independent* physician groups, hospitals, and insurers have been developed as Accountable Care Organizations (**ACOs**). These groups agree to share the responsibility for caring for a defined population of Medicare beneficiaries over a defined period of time. In the process, they can earn incentives for saving money through more coordinated care that avoids duplicate or unnecessary procedures and tests.[11] The proliferation of these provider/payer networks is important because it changes the interests of the involved stakeholders as well as their receptivity to certain types of innovations. For instance, because the participants of an integrated delivery network or ACO are jointly accountable for the patient's longer-term cost of care, they may be more amenable to preventative care technologies or solutions that incur a higher near-term cost but promise savings over a greater time horizon. Integrated networks also tend to be more focused on innovations that can lower treatment costs while maintaining outcomes. Moreover, they are more motivated by **value** than by volume.

It is also important to note that many of the world's largest corporations have recently begun to function in much the same way as the integrated delivery networks and ACOs. Gaining control of their rapidly rising healthcare costs has become a necessity in order to preserve their overall cost competitiveness. Because they often carry the burden of insuring the long-term health of hundreds of thousands of employees and their retired former employees, these companies are motivated to make different decisions surrounding wellness and preventive care than stand-alone providers or payers might make.[12]

In other countries, particularly in the developing world, the health expenditures may look significantly different based, in part, on variability in public versus private payments as well as the role of health plans versus out-of-pocket health expenditures (see Figure 2.3.4). For example, in India, the role of third-party public and private payers is significantly diminished and patients bear a much larger portion of the healthcare payment burden. In 2011, just under 70 percent of all healthcare payments in the country were made by private individuals.[13] In contrast, government expenditures on healthcare accounted for approximately 25 percent of total spending. On a per capita basis, government spending on healthcare in India was only $56 compared to $208 in China and $964 in Brazil in 2010 (all figures in US dollars).[14]

The availability of health insurance is rare but growing in India. In 2003, only 55 million people were covered by a health insurance policy but, by 2010, this figure had increased to 300 million people, mostly below the poverty line. Still, only 3–5 percent of individuals in the country have full or substantial coverage.[15] Both government and private insurers are working to address this problem, and analysts estimate that about half the population will enjoy some level of health insurance coverage by 2020.[16]

In China, health insurance coverage is more widespread. Roughly 87 percent of China's population had some form of health insurance coverage as of 2008;[17] by 2010, this figure had grown to 90 percent, with the government aiming for 100 percent coverage by 2020.[18] Despite these advances, out-of-pocket payments made by patients are still the predominant source of private healthcare financing, accounting for 44 percent in private payments in 2011.[19] Insured individuals' out-of-pocket payments remain high in China because insurance plans do not adequately cover large healthcare expenditures.[20] For example, copayments for inpatient care can be as much as 10–35 percent of the total cost of care.[21] As a result, patients are cost-conscious and exercise considerable influence over the medical devices used in their procedures.[22]

Stakeholder interests
Once all key stakeholders with an interest in a need have been identified, innovators can next dive in to understanding the barriers that might cause a stakeholder to resist the adoption of a new innovation, as well as the improvements or benefits that may drive their acceptance (chapter 2.4 includes a discussion of how improvements can subsequently be translated into value for key stakeholders). An effective analysis assesses the factors that *directly* and *indirectly* affect stakeholder behaviors. Direct factors may include a loss/gain in revenue, decrease/increase in profitability, decrease/increase in time away from work. Indirect factors may include impacts on reputation, ease of use, and especially **opportunity costs** (which are defined as the cost and benefits of giving up one alternative to pursue another).[23] Understanding the drivers of stakeholder behavior is essential to being able to influence stakeholder actions. The following sections articulate some of the common drivers of stakeholder behavior for four primary groups: patients, physicians, facilities, and payers.

Patients
Patients are the ultimate gatekeepers as to whether or not they will agree to undergo a specific test or treatment. Traditionally, their decisions have been made based on

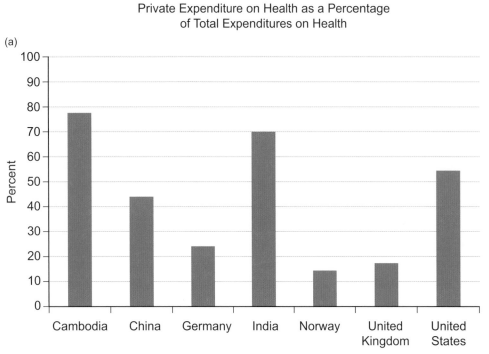

(a)

Private Expenditure on Health as a Percentage
of Total Expenditures on Health

FIGURE 2.3.4
Innovators can expect to see great variation in the relative prominence of public, private, and out-of-pocket health payments in different countries around the world. (compiled from "Health Financing: Health Expenditure Ratios by Country," Global Health Observatory Data Repository, World Health Organization).

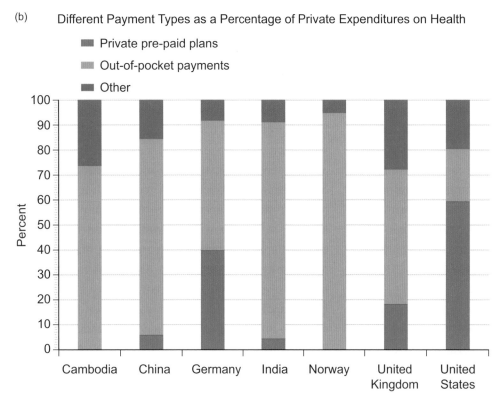

(b) Different Payment Types as a Percentage of Private Expenditures on Health

■ Private pre-paid plans
■ Out-of-pocket payments
■ Other

information and advice received from physicians. However, patients are now more empowered than ever and have access to vast quantities of medical information directly from other resources (e.g., advertisements, online knowledge bases and blogs, bulletin boards, and discussion and support groups). Accordingly, they may play a more active role in making health-related choices based on the information available to them (even though not all available data may be credible). In the US, 50 percent of hospitals and 40 percent of physicians in ambulatory practices offer some sort of Internet portal for patient use.[24] In addition, pharmaceutical companies, medical device manufacturers, and health insurance companies are increasingly seeking to influence patient behavior through financial means (copayment requirements), as well as non-financial mechanisms (direct-to-consumer advertising).

In locations like India, patient power may extend even further. For example, in many non-tertiary healthcare settings, patients requiring a stent in their coronary arteries might be informed by the doctor of the product choices. These patients would then go to a local vendor of medical products to evaluate the available stents, which typically include options made in-country as well as those manufactured by multinational corporations. Medical products made by the multinationals tend to be perceived as higher quality, but are significantly more expensive. More often than not, patients purchase the most expensive product they are able to afford and then bring it with them on the day of the procedure to be placed. In this way, they exercise an unparalleled level of control over their treatment. Cost tends to be the primary driver of the decision, with the recommendations of the physician and/or vendor having a secondary effect on patient behavior.

A patient's family is another related stakeholder group to carefully consider. Particularly in developing countries, family members tend to be much more involved in delivering care, both at home and directly in the hospital. They also may be the ones to decide about health-related expenditures. Especially in environments with high patient-pay requirements, it is often the patient's son or daughter who must take on a second job or make other sacrifices to fund the patient's treatment.

In more developed environments, innovators should also evaluate relevant patient advocacy groups. These entities have the ability to influence patient opinions. They are also frequently sponsored by major corporations whose interests may be served or threatened by the new innovation. Consider the sponsors' interests alongside the interests of the advocacy groups and the patients they are intended to serve.

Patient behavior with respect to accepting a certain treatment is often driven by the following *direct* factors:

Clinical outcomes Patients are interested in the treatment that will best resolve their primary problem. However, they are also concerned with the elimination of symptoms and the avoidance of unintended consequences from the treatment. The order of importance of these factors is likely to be different for each patient. For instance, when considering a patient's reaction to a new treatment alternative that may have clear benefits, it is still necessary to think about the amount of pain the patient will experience, whether or not the patient's appearance will be altered, and/or other potential side effects that might be associated with one treatment but not another. Also, the innovator should evaluate the benefit of living longer if a treatment helps delay mortality.

Safety While procedures and their associated risks may be considered routine from a physician's perspective, the idea of undergoing certain treatments can be traumatic for a patient. Patients must consider whether the "cost" of living with a disease is higher than the risk of being treated, based on the safety profile of the treatment and a patient's own individual preferences.

Economic impact In the case of new technologies, patients are often required to cover a larger portion of the total treatment cost. Or, in some cases, they are required to cover all treatment expenses. Determine what out-of-pocket expenses a patient should expect to incur relative to other treatment alternatives and evaluate this cost against the anticipated change in clinical outcome. Keep in mind that roughly 100 million people each year are pushed under the poverty line because of healthcare payments, with more than 90 percent of these individuals living in low-income countries.[25]

Convenience The impact that a new treatment may have on a patient's life can vary from inconvenient to life-changing (in a positive or negative way). In deciding on a treatment, patients often think about whether the treatment is available nearby, how easy it is to schedule, what impact it will have on days off work, and the long-term implications on their quality of life.

Indirect factors influencing patient behavior include:

Opportunity cost Innovators should consider what patients could do with their time, money, and energy if they elected to have one solution over another. Think about this question on both a near-term and long-term basis. Also, remember to include a patient's choice to do nothing about the problem as one potential alternative.

Perceived risk Sometimes perceived risk can be a major factor in a patient's decision-making process, even if the actual risk associated with the treatment is relatively low. Particularly for experimental treatments, consider how the perceived risk is likely to affect a patient's behavior. It is also important to take into account the psychological effect that the new treatment is likely to have on the patient relative to established treatments.

Importantly, not all patients facing the same medical need will perceive it the same way. It can be useful for innovators to differentiate between patient types by developing a series of patient profiles. Within these profiles, patients will view a need similarly and have comparable reactions to different treatment alternatives. However, across profiles, patient perceptions of the need and how it is treated will be distinct. To understand how these profiles can be developed, refer back to the case example in 1.2 Needs Exploration, which describes how a biodesign team from the University of Cincinnati created different patient personas to better understand needs in sleep apnea. Identifying these types of profiles will also prove helpful in 2.4 Market Analysis, where one of the goals is to define clear segments of patients with uniform perceptions towards the need.

Physicians

Because physicians are the primary individuals recommending patient treatment, they are critical stakeholders for almost every medical treatment option. While physicians are first and foremost driven by the desire to provide patients with the best possible treatment, they also face the need to earn a living. Innovations that make new procedures possible can often help physicians achieve both of these desired outcomes.

Yet the influence of physicians on adopting new technologies can vary significantly and depends on the type of organization in which they practice medicine (e.g., a private practice, integrated delivery network such as Kaiser Permanente, or non-profit, community-based hospital or clinic). Historically, in large markets like the US, physician preferences primarily drove device usage. However, as hospitals face increased pressure to contain costs, many have adopted **value analysis committees** (VACs), also known as technology assessment committees, to assist with decision-making. In a survey of 4,500 US hospitals, nearly 75 percent of respondents either had a strategy for standardizing physician preference items (**PPI**) or were working on developing one. Additionally, 64 percent of hospitals were using value-analysis teams to evaluate and select PPIs and other supplies.[26] This strong trend toward more centralized purchasing, which is almost certain to continue with the expansion of ACOs and a system-wide focus on cost containment, can be expected to further erode the individual influence of the physicians on purchasing decisions, even while they remain on the front line of patient care.

In assessing and anticipating physician behavior, keep in mind the potential for conflicts to arise. For instance, if a technology shifts patients from one specialty to another, this may cause "turf wars" between physicians and create serious obstacles to the technology's adoption. Another scenario that might raise a conflict is one in which a new technology requires the skill set of one specialty, but this specialty is not currently involved in the cycle of care for the given disease state. As physicians become increasingly specialized within a single field, these kinds of conflicts are becoming more common. Guy Lebeau, a physician and businessman who led the growth of Cordis Corporation's cardiology, endovascular, neurovascular, and electrophysiology businesses, commented on the benefits of this trend, using the field of cardiology as an example:[27]

I think the fact that we are no longer going to have one cardiologist with one set of knowledge, but probably 10 different types of cardiologists who are going to focus their energy on treating one type or one part of the disease, is excellent because this creates a situation where the learning and the competency of physicians is going to be higher.

The downside is that, as new technologies disrupt referral patterns, they create "winners" and "losers," particularly among physicians within these narrowly defined subspecialties. When considering physicians as stakeholders, the innovator must be on the lookout not just for the relevant specialties, but the potential subspecialties that exist within them. Then the interests of all such stakeholders need to be taken into account.

Another issue that may emerge is related physician willingness to try new devices and the procedures they enable. Certain specialists (e.g., cardiologists) are known for their receptivity to new technologies. But other physician groups may be more risk averse and, as a result, less willing to adopt new ways of working. This conservative stance toward technology can be particularly pronounced in certain geographies. For instance, Stanford Biodesign Fellows working in India sometimes express frustration that it can be difficult to find physicians who are open to trying new devices. These innovators report that they must expend considerable time and energy to identify one or two key doctors who will act as early adopters of an invention.

The role of professional societies is another important factor to consider. For most well-established medical fields, there are multiple associations that address the area within which a need exists. Physicians involved in these groups are often considered thought leaders in their respective fields and are likely to have strong opinions on the benefits and costs of any new developments in the field.

Physician behavior with respect to the potential utilization of a new treatment is generally driven by the following *direct* factors:

Agency The mandate of physicians is to represent the best interests of their patients. As an agent of the patient, a physician's treatment recommendations must carefully balance the risks and benefits to patients and take into account patient preferences. Medical ethics underlie this relationship (see the section of this chapter entitled "Ethical Considerations in Stakeholder Analysis").

Clinical outcomes Clinical outcomes refer to the manner and degree to which physicians will be able to improve medical results through the use of a new solution. In partnership with agency, this is one of the most persuasive factors for getting physicians to adopt a new technology.

Economic impact The financial impact on the physician of adopting a new treatment is an important consideration. This includes how (and how much) physicians might be reimbursed or otherwise compensated for the treatment and the number of treatments they would perform annually. These calculations must be understood relative to existing technologies and their significance within a physician's practice. For example, resistance may be encountered if a new solution requires substantial training, disrupts existing work flows, requires major equipment purchases, or renders existing (costly) equipment obsolete.

Mobile health (**mHealth**) and remote monitoring technologies provide an example of how the lack of direct financial incentives can slow the adoption of new technologies. Many digital devices and applications, with the promise of improving care while reducing cost, have been introduced in the US. However, they have not been widely reimbursed under the traditional fee-for-service payment scheme. Some physicians have gradually gravitated toward them anyway, justifying the expense based on the efficiencies and quality improvement that the technologies enable. But larger-scale adoption will only be achieved when payment is assured. The move toward value-based reimbursement under ACOs and patient-centered medical homes will potentially act as a catalyst for this shift.[28]

Risks The risks to the physician of adopting a new treatment include any potential increase (or decrease) in clinical uncertainty (e.g., safety risk and side effects for patients). It also covers malpractice liability and/or the liability of not complying with evidence-based guidelines, as well as the impact of the change on the physician's malpractice insurance. In some cases, new innovations can reduce physician risks and malpractice liability.

Physician behavior with respect to the utilization of a new treatment is also driven by a series of *indirect* factors:

Opportunity cost   In determining how a new treatment might fit into a practice, innovators should analyze how the physicians currently use their time. Understand how long current treatment options take to administer, how many providers are involved, and how many procedures are typically performed within a given period of time. This information can then be used as a benchmark as more becomes known about the need and the solution that will eventually address it. Specifically, when physicians adopt a new solution, the revenue they would earn per unit of time should be at least the same as the revenue currently generated from the same unit of time. Otherwise, the opportunity cost associated with adopting the new treatment alternative may be perceived as being too high to make its adoption appealing. In these cases, the clinical benefit would have to be extremely compelling to make the desired change in physician behavior feasible.

Workflow   While all new treatments may not carry with them significant capital expenditures (e.g., investments in new equipment), almost all will require a change in workflow (or the process by which they are used). Consider how disruptive a new treatment may be to established physician practices, or whether it can be integrated relatively seamlessly into common processes and if so how much this may cost. It is important to remember that it is much more difficult to get physicians to adopt a new treatment if it requires a significant change in their accustomed workflow, compared to one that can be integrated easily into their existing daily routine.

Ease of use   Evaluate whether or not a new innovation is likely to make the physician's job easier or harder to perform. While workflow takes into account the process

by which a device is used, ease of use considers how difficult or easy the device is to use at each stage of that process. In terms of costs, the innovator should consider what training might be required to perform a new treatment and any new or special skills that physicians may need to develop. On the benefit side, identify the ways in which an innovation might make it easier for physicians to provide effective treatment. Ease of use can be a potent advantage for a new technology and a factor that can rapidly drive acceptance. This was demonstrated when Guidant Corporation introduced a new bare metal stent to treat coronary artery disease. Because it was so much easier to use (and did not require nearly the same amount of physician experience and skill to effectively place within a patient's artery), it quickly displaced the market leader, a bare metal stent marketed by Cordis Corporation, despite the fact that Cordis was first to market with its product.

Reputation   Consider whether the adoption of a new innovation might be perceived positively or negatively by patients and, in turn, what its effect might be on a physician's standing in the physician community. If physicians are known for being leaders in their field, evaluate the indirect benefits (e.g., visibility) to their practices of adopting the innovation. Conversely, if a physician is risk averse and takes pride in providing proven treatments, consider the reputational cost of adopting an exploratory solution. There are significant differences in perceptions of new technologies and their impact on physician reputation among both physicians and specialties. For example, while interventional cardiologists pride themselves on being quick adopters of new technologies, cardiac surgeons are more conservative in their approach. This may explain why the adoption rates of new technologies vary dramatically between these two specialties.

Facilities
The primary interests of facilities, such as hospitals, surgical centers, laboratories, and other settings where care is delivered, are largely financially driven. However, their perspectives on a need will be heavily influenced by how their operations are organized. If they are

participants in a single-payer system (e.g., Norway), an integrated delivery network (e.g., Kaiser), or an ACO (e.g., Cedars-Sinai Accountable Care), their financial motivations will be more complex than if they are organized as a more traditional fee-for-service facility. Yet, this simple truth will remain constant: innovations that increase procurement costs are most likely to meet resistance from this stakeholder group, while those that reduce procurement costs and are assured of third-party reimbursement are most likely to be accepted.

Innovators should also understand that there can be significant differences in financial motivation and behavior depending on type of setting in which treatment occurs. Economic calculations can differ greatly if an institution is a non-profit versus a for-profit organization; or an academic training center versus a community hospital.

In some geographies, such as the US, facilities also can be sensitive to innovations that shift the location where a treatment, procedure, or test is delivered since the revenue they receive is often adjusted based on the location. As an example, consider point-of-care (**POC**) testing for hemoglobin A1c (HbA1c), a variant of hemoglobin (an oxygen-carrying molecule of blood) that can be used as a marker of glucose control in diabetics. Testing for HbA1c typically requires a patient to go to an outpatient lab. POC testing would change the venue of testing from the lab to the clinic (or the doctor's office). For such tests, the stakeholders include not only the physicians who would perform the POC tests, but also the laboratories that previously provided this service. Following approval by the **FDA** of one such POC test, the Metrika InView, a series of studies were performed (not managed or influenced by the manufacturer) that compared the results from the POC tests to those from tests performed in the lab, with lab tests considered the "gold standard." The studies showed that the correlation between the results was high, but not high enough to make the POC test a substitute for the test in the lab.[29] An editorial accompanying one of these studies, written by a professor of pathology, stated that there are numerous issues to consider when evaluating a new method for HbA1c point-of-care testing, including whether or not the method is NGSP-

certified,[30] how well it performs in a field setting, and if it is free from common interferences.[31]

Such a response demonstrates that pathologists working in labs have a stake in the adoption of POC testing. When their views appear in medical journals, they have the potential to hinder or catalyze the adoption of a new test. Innovators should try to anticipate such viewpoints and develop a strategy to preempt them, for example, by securing appropriate certifications and by carefully designing clinical studies that may go beyond FDA requirements.

While facilities would hope for all of their procedures to be profitable, it is not uncommon for some to be designated as "loss leaders" – procedures billed for less than they cost because they generate business (patient traffic, additional revenue) in other areas of the facility. For example, at many dialysis centers, the delivery of dialysis is performed at a loss since it allows profits to be generated from other services, such as the administration of epogen, a drug necessary to stimulate red blood cell production and help control anemia, which is typically administered while patients receive dialysis. If an innovation will not be profitable for the facility where it is administered or utilized, the innovator should think creatively about related products/services that can be bundled with it, modified, or eliminated such that the innovation results in a net benefit to the facility.

Participation in an ACO is another factor that can influence the procedures, tests, and treatments in which facilities invest. Hospitals and other providers that form an ACO are eligible for the Medicare Shared Savings Program, which rewards the consortium for lowering its aggregate growth in healthcare costs while meeting performance standards on quality of care and "putting patients first."[32] Providers in the ACO continue to receive reimbursement payments under Medicare fee-for-service rules but, if they meet or exceed defined quality standards relative to an established benchmark and achieve savings at or above a Minimum Savings Rate (MSR), they share in the total savings based on their quality scores.[33] Thirty-three quality metrics have been defined using nationally recognized measures in four key domains:[34]

- Patient/caregiver experience (7 measures)
- Care coordination/patient safety (6 measures)
- Preventive health (8 measures)
- At-risk population:
  ○ Diabetes (1 measure and 1 composite consisting of 5 measures)
  ○ Hypertension (1 measure)
  ○ Ischemic Vascular Disease (2 measures)
  ○ Heart Failure (1 measure)
  ○ Coronary Artery Disease (1 composite consisting of 2 measures)

Importantly, this construct begins to provide traditional fee-for-service providers with a sound financial rationale (where one previously did not exist) for shifting their care priorities from a strict focus on volume to improved results. For example, facility executives have indicated increased interest in new programs and technologies to actively manage patients upon discharge to prevent hospital readmissions and reduce emergency room usage; create more robust chronic disease management programs; improve the management of patient care transitions from hospital to home (or other care venues); and experiment with patient-centered medical home models.[35]

As noted, many groups shape the treatment (and purchasing) decisions of a facility, including physicians, facility executives, and purchasing professionals. However, because physicians are evaluated separately, innovators should place their primary focus on understanding how management and purchasing respond to the need during stakeholder analysis.

Direct factors driving facility stakeholder behaviors for adoption of a new treatment include:

Economic impact   Depending on how the costs of a new treatment will be covered, innovators should begin thinking about whether a potential change may increase, decrease, or hold constant the overall cost of treating a given disease state. Since traditional facility payments for treatments typically do not adjust higher for increased costs that may be incurred (i.e., facilities typically receive fixed payments for a treatment), a new innovation must reduce ancillary costs associated with the treatment to decrease overall cost or have a neutral effect on a

facility's budget. For example, reducing a patient's length of stay in the hospital following surgery can provide large financial incentives for a facility to adopt a new treatment if the facility's payment for the surgery is fixed and does not increase with a longer stay (see 4.3 Reimbursement Basics). Also, think carefully about innovations that may change the location where treatment is administered (i.e., takes business away from a facility), as the POC example illustrates. If the facility is part of an ACO, recognize that they are likely to be receptive to needs that, if addressed, will help them not only save money but meet the quality standards that act as a hurdle in the Shared Savings Program.

Risk   Consider the effect of the new treatment on a facility's risk profile. Some procedures may significantly reduce facility risk while others may increase it. An increase in risk can carry with it direct financial costs by affecting liability and insurance.

Indirect factors influencing the behavior of facility representatives regarding the adoption of new treatments include:

Opportunity costs   Facilities have limited resources in terms of their providers, support staff, and physical space in which to provide care. If a new treatment will change the number of procedures performed each year, it may create or consume procedural time in the operating room or other settings. This is time that could be used on other procedures. Therefore, the potential profit that could be generated per unit of procedure time should exceed the profit generated by existing procedures.

Reputation   Being seen as a leader in a certain field can attract patients to a facility. If an innovation serves as a magnet for a facility to draw additional patients, the facility may be willing to make trade-offs in other areas to achieve the benefit of additional patient traffic, especially if the additional patient traffic results in the need for additional ancillary services, such as testing. The DaVinci robot, an innovative surgical robot that can be used to gain improved surgical results and make procedures less invasive, provides a good example of a technology that was used by the hospitals that were early adopters to enhance their reputation.

Payers

If payers grant adequate reimbursement for a medical innovation, it is a powerful force in stimulating adoption in settings where insurance systems dominate the payment landscape. On the other hand, if payers deny, delay, or restrict reimbursement, it can be extremely detrimental to the success of a new treatment unless the treatment is attractive enough to get patients to pay for it directly. In many cases, identifying a path to reimbursement has become equally or more critical to success as developing an approach to gain regulatory clearance. Importantly, innovators and medical device companies must appreciate that the data required for regulatory approval is sometimes not enough to make a compelling case to payers. (4.3 Reimbursement Basics and 5.6 Reimbursement Strategy provide more details on payers, their reimbursement decisions, and how innovators can influence them.)

Historically, many new innovations have been synonymous with increased costs from the payer's perspective. The reason that payers have continued to fund new interventions is the promise of better outcomes, especially when this is coupled with the possibility of lower long-term costs for a given patient (e.g., fewer hospitalizations, surgeries, or other expensive forms of care). However, if the cost burden becomes too great or the perceived clinical benefits are not significant enough, both public and private payers may deny coverage and/or limit the number of patients eligible for a new treatment by dividing the patient population into subgroups and restricting reimbursement to a specific subgroup. Another potential scenario that payers use is implementing step-therapy guidelines, which force physicians to try alternative therapies before utilizing the new treatments.

In general, payers are most likely to cover new medical technologies if they are proven (through robust **clinical trials**) to improve hard clinical endpoints (mortality, **morbidity**) and/or achieve comparable clinical outcomes at significantly lower cost – with this latter factor increasing in importance. Softer endpoints, such as patient convenience, physician convenience, or quality of life, are less likely to gain reimbursement unless the

improvements are shown to be medically necessary (and medical necessity is a rather ambiguous concept that can often be shaped by the innovators as part of their reimbursement and marketing efforts – see 5.6 Reimbursement Strategy and 5.7 Marketing and Stakeholder Strategy). Payers also need to be convinced that new technologies do not add risk to the treatment paradigm.

In some countries, payers already have embraced clear, evidence-based approaches to making decisions about healthcare spending. The National Health Service (NHS), through its National Institute for Health and Care Excellence (**NICE**), provides the most well-known example. Through a formal technology appraisal process, NICE assesses clinical evidence to evaluate how well a new treatment works, along with economic evidence that measures how well it performs relative to its cost. Based on this assessment, a recommendation is made that the NHS is legally obliged to follow.[36] The purpose of these appraisals is to eliminate reimbursement uncertainty and help standardize access to healthcare across the country.[37] However, technology appraisals also allow the NHS to make unambiguous decisions about the most effective use of its finite resources. NICE relies on an internationally recognized method to compare different treatments and measure their clinical effectiveness: the quality-adjusted life years measurement (called the **QALY**). A QALY provides an estimate of how many extra months or years of life of a reasonable quality a person might gain as a result of new treatment. Cost-effectiveness is then determined by calculating how much the treatment costs per QALY. Each intervention is considered on a case-by-case basis, but NICE generally stipulates that if a treatment costs more than £20,000-30,000 per QALY (about $49,000), then it is not considered cost effective.[38]

The trend toward using evidence-based analysis to justify treatment reimbursement is increasingly being embraced by payers and governments around the world. While a comparable approach has yet to be widely adopted in the US, where the idea of "rationing healthcare" has been politically unpopular and vilified in the press, it is clear that the cost-effectiveness of new treatments and technologies – and the value they deliver – is becoming paramount in today's budget-constrained environment. One challenge with these approaches can be the extra time required to conduct a definitive comparative analyses or, alternatively, to gather data from use of a technology in clinical practice to justify favorable reimbursement decisions. A good example is the case of the Guglielmi Detachable Coil (GDC) for catheter-based treatment of brain aneurysms. The device received FDA approval in the US in 1995, but hospitals using it initially lost money on related procedures due to inadequate reimbursement payment levels. It was not until 2003 that the product's manufacturer, Boston Scientific, was able to present analysis of Medicare claims data to CMS that demonstrated the extent to which hospital costs exceeded payments. In 2004, CMS agreed to a change that doubled the average payment level (see 5.6 Reimbursement Strategy for more information).

Payers and manufacturers will sometimes need to use creative approaches to demonstrate value while maximizing patient access to new innovations. For example, consider Genomic Health, a company that developed a high-end genetic test to help determine if women with early-stage breast cancer will benefit from chemotherapy (a commonly prescribed treatment that is effective in only a small percentage of patients). Although the effectiveness of the diagnostic is backed by strong clinical evidence, its value to payers is realized only if women with a negative test result choose *not* to receive chemotherapy (thereby saving money in administering ineffective treatment). However, according to one payer, UnitedHealthcare, too many women were still receiving chemotherapy even if the test suggested they did not need it. For this reason, United entered into a conditional agreement with Genomic Health under which it covered the cost of the test for an 18-month trial period while the outcomes were monitored. If enough women with low scores on the diagnostic did not abstain from chemotherapy, then United had the opportunity to negotiate a lower price with Genomic Health on the grounds that the test was not having the intended impact on actual medical practice. According to Dr. Lee N. Newcomer, senior vice president for oncology at UnitedHealthcare, this arrangement was designed to make the manufacturer more responsible for how its product was used in the medical marketplace.[39] Following the trial period, UnitedHealth extended coverage for the test for patients with estrogen-receptor positive, node-negative carcinoma of the breast[40] (see chapters 2.4 and 5.7 for more information about Genomic Health).

Direct factors driving payer behavior regarding adoption of a new treatment include:

Clinical outcomes Innovators should consider both the near-term medical benefits, as well as the longer-term effects of ongoing treatment in improving outcomes relative to any existing treatment alternatives. The elimination of symptoms and side effects that often require separate treatment can also be significant from a payer's perspective. In the US, the standard for proving clinical outcomes has gradually risen with many payers now requiring two separate **randomized controlled clinical trials** to be published in **peer-reviewed** journals before they will act on a reimbursement decision.

Economic impact Be prepared to evaluate the total cost of any new treatment relative to existing treatment alternatives. When more is known about a potential solution, the innovator can start with the payment per treatment (how much the payer would be willing to reimburse for the new procedure) and then multiply this by the anticipated number of treatments per year. Compare this to data for alternative treatments to calculate by how much the new treatment will increase payer costs. In some cases, an innovator may be able to evaluate whether the innovation can decrease near-term or long-term costs to payers by reducing other services requiring reimbursement, such as hospitalization or additional testing (e.g., blood test or X-rays). Another way to think about the financial impact to payers is the incremental increase or decrease to its cost per member per month. If the per procedure increase is large – but the relative size of the patient pool is small – then payers might be less sensitive to the marginally higher cost of a better outcome. Conversely, if the increase in cost is small but the patient pool is large, the innovator can anticipate resistance. It is also important to consider if a new treatment could potentially expand the market for treatment in such a way that significantly more patients will seek treatment, which can represent a sizable cost increase to payers.

Indirect factors that can also influence payers regarding adoption of a new treatment are:

Competition Payers often move as a group. Consider competitive dynamics among payers in making reimbursement decisions (particularly in the private sector) and think about the benefits (e.g., in terms of market share) and costs of being the first (or last) payer to cover a new solution.

Reputation It can also be helpful for the innovator to think about the effect on the payer's reputation of offering the new treatment. If any new treatment is widely perceived as being ground-breaking, the payer will have a more difficult time justifying a decision not to cover it. Conversely, if a new treatment is marginally effective yet costly relative to available alternatives, a payer will have little incentive to justify reimbursement in a cost-conscious environment where its own internal and external stakeholders would be critical of such a move.

Relative power and linkages between stakeholders

As the forces that drive stakeholder behavior are understood, stakeholders can be classified based on their unique characteristics, motivations, and level of potential impact. Importantly, all decision makers are stakeholders, but not all stakeholders are decision makers when it comes to the adoption of a new solution. It is essential for innovators to identify which stakeholders fundamentally will be the gatekeepers to an adoption decision and which ones will play an influencer role. Generally, more time, effort, and resources should be devoted to understanding and managing the involvement and commitment of decision makers, with a secondary emphasis placed on influencers.[41]

Relationships between key stakeholder groups should also be explored. In the medical environment, no single stakeholder group operates in isolation from the others. For example, purchasing professionals within facilities have their own issues, priorities, and considerations but also must satisfy the needs and demands of their associated physicians and align their efforts with the strategic priorities set forth by executives at the helm of the organization. Patients typically follow the instructions of their physicians, but are increasingly exercising greater control over medical decision making, including treatment alternatives and locations. Physicians and patients may seek to embrace a new innovation but have their adoption hindered by the decision-making process of the payer system. As a result of these types of interconnected issues, the prioritization of stakeholder interests is critical because it can determine the order in which these interests are addressed.

Innovators should keep in mind that the forces that shape stakeholder behavior are dynamic and constantly evolving. Trends like provider/payer consolidation, accountable care, and value-based reimbursement incentives can have a profound impact on stakeholder motivations over the many years that are often required to bring a medical innovation to market. Those innovators who consider how tomorrow's landscape may look are likely be more successful than those who optimize exclusively for today's situation.

The following story of InnerPulse, Inc. demonstrates how stakeholder analysis works in practice and highlights some of the linkages between stakeholder groups with varying degrees of relative power and influence.

FROM THE FIELD ▶ **INNERPULSE, INC.**

Anticipating and managing stakeholder reactions in the innovation process

When it was founded, InnerPulse, Inc. (formerly Interventional Rhythm Management) was focused on developing PICDs™ (percutaneous implantable cardioverter defibrillators), or miniaturized ICDs, that could be placed via a catheter-based approach. ICDs are used to prevent sudden cardiac death by issuing a lifesaving jolt to the heart when a patient suffers sudden cardiac arrest. The company's product is made of a chain of pencil-thin components measuring 56 centimeters in total length. The device can be placed in the vascular system percutaneously (across the skin), using a standard catheter-based approach, in under 10 minutes. PICDs are differentiated from conventional ICDs in two ways: their size and delivery method. Conventional ICDs are roughly the size of a hockey puck, and are usually surgically implanted in the upper chest of a patient. Because of the complexity of ICD devices and the accompanying procedure, implantation is usually only performed by a small group of heart rhythm specialists known as cardiac electrophysiologists (EPs). In contrast, the PICD, due to its less invasive delivery method, can be implanted by EPs and interventional cardiologists (ICs), who typically use catheters to treat blockages in blood vessels. As a consequence, it can be made accessible to a much larger group of defibrillator candidates who would potentially benefit from ICDs (see Figure 2.3.5).

There are two primary groups of patients who benefit from ICDs: (1) **secondary prevention patients** – patients with a prior episode of sudden cardiac arrest; and (2) **primary prevention patients** – patients at high risk of sudden cardiac arrest who have a weakened heart manifested by a left ventricular ejection fraction <35 percent (normal ejection fraction is typically ≥55 percent). The primary prevention market dramatically increased as a result of two clinical trials [MADIT-II (2001) and SCD-HeFT (2004)], which showed that ICDs dramatically reduced the mortality rates of these patients. Based on these results, the indication for ICD implantation expanded to more than 1 million similar individuals at risk for sudden cardiac death. However, the large number of primary prevention patients seeking ICDs is currently overwhelming the 1,800 EPs nationwide. The consequence of this problem is that only 10 to 20 percent of all patients who would potentially benefit from an ICD have received the device. PICDs have the potential to unlock this large, previously untapped market. This under penetration of the market due to the limited number of EPs available to initiate

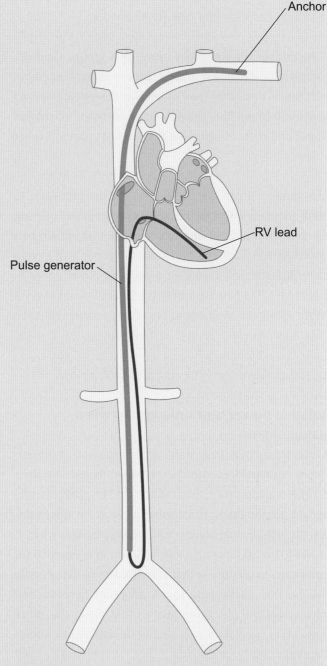

FIGURE 2.3.5
Schematic of the InnerPulse PICD (courtesy of InnerPulse, Inc.).

defibrillator therapy was an important factor in helping understand the role PICDs could play.

Because PICDs can be administered by ICs, instead of just EPs, they raise some important stakeholder issues. When InnerPulse was initially pursuing the *need to develop a less invasive way to prevent sudden cardiac death*, it had to consider the effect a non-surgical implant would have on stakeholders in the field. There is an unwritten conventional rule in medical devices that if a new device "steals" business from one specialty to benefit another specialty, conflicts between the specialties may ensue, creating obstacles to the adoption of the new technology. For example, the tension between cardiac surgeons and ICs that followed the introduction of balloon angioplasty as a minimally invasive alternative to invasive coronary artery bypass surgery was legendary. The team at InnerPulse had to ask itself if the introduction of a less invasive technology implanted by non-EPs would create similar tensions with EPs, who controlled the traditional ICDs market.

Bill Starling, chairman of the InnerPulse board, company co-founder, and an early investor in the venture, believed that turf wars could be avoided in this case if the need for a less invasive solution was focused on the primary (as opposed to the secondary) prevention market. He explained: "The great opportunity here is the primary prevention market. Electrophysiologists have traditionally ignored this market because they do not see these patients. EPs only see the secondary prevention patients who have developed an arrhythmia. Primary prevention patients are seen by the interventionists [ICs]. Primary prevention patients usually have coronary artery disease. The cardiologists put in stents, give them some drugs, and send them home because there is nothing more they can do. With a less invasive solution, the interventionist [IC] would now be able to do something more for the patient. And remember, these are patients that the EPs would not see anyway."[42] Thus, the fact that a patient would typically see a cardiologist or IC for some initial treatment or evaluation prior to being referred to an EP was key in understanding the stakeholder relationships. Without referrals from colleagues in other areas of cardiology, an EP would not have patients requiring ICD implants.

According to InnerPulse, the argument in favor of pursuing the need that eventually led to the PICD was even stronger than this. The company predicted that addressing this need would not only create new business for ICs, but it would expand the market for EPs, as well. Using an analogy shared by the company, PICDs are like an air bag: they prevent death when an accident (sudden cardiac arrest) occurs, but they also lead to additional service and maintenance work. When a patient with a PICD experiences an electric shock, he will visit his IC. Chances are he will then be referred to an EP, since this is what ICs do for their patients with an arrhythmia. Coming back to the airbag analogy, the IC installs the airbag (PICD), but when the airbag inflates (sudden cardiac arrest), the patient – now a secondary prevention patient – is sent to the EP for treatment/repair, potentially with a more complex, traditional ICD.

Despite their early hypothesis that both IC and EPs would respond favorably to the need for a less invasive solution, Starling and cardiologist Richard Stack (the other company co-founder) did not leave anything to chance. Toward the beginning of the biodesign innovation process, they assembled an advisory board of internationally recognized scientific thought leaders. They used this group of five ICs and five EPs to test their basic assumptions. Through this process, they confirmed that both the ICs and the EPs could see that a solution like the PICD would expand the practices and markets for both specialties. Also, it became clear to them that ICs would want to use these devices. Since EPs would be central to the early adoption of the technology, the clinical trials would be largely managed by EPs. This was partly a pragmatic business decision (EPs have deep experience running trials with ICDs, so the trial design would be more readily accepted by the FDA). However, it was also a savvy business strategy that would create a greater sense of buy-in for the device among EPs (by positioning the PICD as a device for them, yet one that could also be used by ICs in patients that the EPs did not normally treat), thereby helping to minimize any potential conflicts.

Beyond that, Starling and Stack were careful in considering all other possible stakeholders in deciding to move forward. A summary of the key stakeholders, their concerns, and how the company ultimately believed each stakeholder would respond to the PICD is shown later in this chapter in Table 2.3.1.

One stakeholder "wild card" that emerged when the company was about five years into development was the reaction of payers to the new technology. While reimbursement codes were in place that would cover PICDs, there was always the risk that once PICDs become widely adopted, major payers would seek to revise the payment levels downwards to physicians and facilities, in order to reflect the lower duration and lower complexity of using and implanting PICDs compared to ICDs. In fact, following the release of MADIT-II, there was widespread informal agreement among major payers that if the technology achieved 100 percent penetration it would have a significant negative effect on their income statements. InnerPulse wondered if payer reactions to the introduction of PICDs could create obstacles to their vision of expanding the market for ICDs.

Starling was convinced that any payer-related issues would be minimal and could be managed by the company. ICDs addressed such an important medical need that efforts by payers to reduce reimbursement or restrict coverage would backfire, he asserted. History seemed to support his belief: Medicare analyzed the results from the MADIT-II and SCD-HeFT trials to argue that not all primary prevention candidates should get ICDs. However, clinician opposition was so strong that Medicare had to backtrack shortly afterwards. Starling maintained that the same dynamics that overcame payer resistance in the past would play to InnerPulse's advantage in its quest to expand this market.

Unfortunately, InnerPulse hit a bump in the road that had little to do with its assessment of the stakeholder landscape. During animal tests, the team uncovered problems with the lead that connects the PICD to the heart. According to Stack, "Addressing the problems with the lead will take a lot of time and a lot of tests,"[43] which has delayed the company's market launch indefinitely.

As noted, the InnerPulse team believed that if there was sufficient interest on the part of the adopting physicians – the interventional cardiologists – it eventually would be able to win the support of Medicare and other payers. However, since the time this case was written, gaining reimbursement has become increasingly challenging. As a result, the importance of payers as stakeholders and critical decision makers should not be underestimated. Innovators should keep in mind that the relative power of stakeholders can change significantly over the extended period of time required to develop and commercialize a new medical device and factor these dynamics into their stakeholder analysis when possible.

A second story about Daktari Diagnostics, Inc. provides another example of important learnings through stakeholder analysis. In this case, the role of another important stakeholder – governments – is emphasized. In some settings, governments are dominant gatekeepers to many activities that affect the adoption of new medical technologies. For instance, in China, the private sector currently does not play a sizable role in healthcare delivery as the government has vastly expanded its role in healthcare delivery over the last decade. Private-sector hospitals account for less than 10 percent of care delivered to patients in urban facilities. Private grassroots clinics treat 50 percent of patients in rural China.[44] The Ministry of Health is responsible for the bidding and tendering system used in public hospitals to purchase new medical equipment. Tenders, which are issued at the province level, set prices that are subject to a ceiling in most parts of China. They also provide medical device manufacturers with a point of entry into hospital procurement departments. Since 1999, China has required a formal tendering process for public health centers to purchase medical equipment. This has increased the transparency of purchases and reduced prices for end **users**. However, this approach raised costs for device makers (through tendering fees and bid bonds), lengthened purchase cycle times, and increased bureaucratic red tape. The tender process favors domestic manufacturers, which benefit from wide distribution networks, cultural affinity with government officials, and highly competitive prices. For instance, Shanghai authorities in recent years fixed the ceiling price for procedures using coronary stents at a level below the price of imported stents. As a result, only patients who can afford to pay the price difference normally choose an imported stent over a locally made one. Similarly, reimbursement rates for imported devices, which vary by locale, may be less than the rates fixed for domestic devices.

In contrast, the role of the government in India is much more fragmented, primarily because the health system is so heavily privatized. As in China, public healthcare facilities use central government or state level tender processes to procure medical supplies. These processes vary from state to state and can be bureaucratic and difficult to break into for new companies. And, unfortunately, the medical device manufacturers that invest the resources to navigate these time-consuming tenders only gain access to a small percent of the total market. Less than 10 percent of care is delivered in public facilities in India. Although exact figures vary by region, the private sector accounts for roughly 90 percent of all hospitals, 85 percent of doctors, 80 percent of outpatient care, and 60 percent of inpatient care.[45] To access the rest of the market, innovators and companies have to find points of entry to the scores of private hospitals and clinics that differ in their procurement practices. As a result, many end up engaging regional distributors that have devoted years to developing relationships with doctors and procurement personnel in private facilities.

In Africa, where Daktari Diagnostics has been working, many of the major markets are truly consolidated, with the government acting as both the primary provider of modern healthcare and sole purchaser of medical technologies. As a result, the government is the key decision maker. If a company is able to convince the Ministry of Health to adopt a new offering, it has access to the vast majority of the country's population. However, these kinds of agreements must be negotiated on a case by case basis across Africa's 55 internationally recognized states. And, as illustrated by the Daktari example below, convincing a government to make an adoption decision can be a complicated undertaking in its own right.

FROM THE FIELD ▶ DAKTARI DIAGNOSTICS

Understanding governments in Africa as a key stakeholder and critical decision maker

Sub-Saharan Africa bears nearly 70 percent of the global HIV burden, with approximately 23 million infected people living in the region.[46] Fueled by the widespread availability of rapid HIV-antibody diagnostic testing, African countries and other low-resource settings have seen significant increases in the number of adults who have had an HIV test and know their status.[47] Support from governments and international organizations has also led to the expanded availability of antiretroviral (ARV) medication to treat HIV+ patients. In 2011, 6.2 million people in the region – or 56 percent of all eligible patients – received ARV therapy (compared to a global average of 54 percent).[48] In less than a decade, access to HIV treatment in sub-Saharan Africa increased more than 100-fold.[49] However, to initiate, stage, and sustain ARV treatment requires careful monitoring of the patient's CD4 antibody count. While flow cytometry diagnostic tests for CD4 cell counting exist in the market, they are expensive, complex, and not broadly accessible to patients and their care providers in low-resource settings.

More than a decade ago, Bill Rodriguez was among the global health pioneers working to expand access to ARVs in a variety of low-income areas, such as Haiti, Vietnam, and South Africa. As an HIV specialist at Harvard Medical School, he advised Ministry of Health officials in several developing countries on how to test and treat HIV+ patients, and ran training programs for healthcare providers there. Later, he became a member of the World Health Organization (WHO)'s HIV Guidelines Committee, which produced global guidelines for HIV care. Through these interactions, he learned first-hand about the barriers associated with CD4 cell count testing, which prevented healthcare providers from delivering adequate care. "I repeatedly heard, 'We need a better test for measuring a CD4 cell count. It needs to be simple and easy to use by doctors and nurses in the field. It needs to deliver results in 15 minutes. And it needs to be inexpensive,'" Rodriguez recalled.

When Rodriguez joined the Clinton Foundation in 2003 as its Chief Medical Officer, he was part of the team that brokered large-scale deals between pharmaceutical companies, diagnostic providers, and Ministries of Health, initially in various African countries, and then throughout the developing world. Even as these discussions progressed well, and affordable ARVs became widely available, he continued to receive the same feedback about CD4 cell count testing, but now it was coming from Ministry officials and representatives from multilateral organizations like the **WHO**, as well as clinicians in the field. "In my role as CMO, I would get emails and phone calls on a daily basis," Rodriguez said. "It was a variety of voices, all saying the same thing, which is that we are desperately in need of new technology for diagnostics." Eventually, he felt the unaddressed need in this space was too great to ignore. Rodriguez left the Clinton Foundation and founded Daktari Diagnostics to develop simple, affordable, and accurate diagnostic tests for developing country settings. The CD4 cell count test is the organization's first project.

Relying on his extensive global health experience, Rodriguez focused on five main stakeholder groups with an interest in the CD4 cell count test. "The patients are the primary stakeholders, because they are the ones who are dying in the absence of adequate care," he explained. While not decision makers about what test will be used, patients are relatively influential given their prominent position in the cycle of care. Patients are often represented by advocacy groups such the Treatment Action Campaign (TAC) in South Africa, or The AIDS Support Organization (TASO) in Uganda. Next are healthcare providers, including physicians, nurses, and community health workers. These individuals would utilize the test to prescribe and manage ARV therapy for their patients. Front-line healthcare providers are the ideal source of information about the clinical need and requirements for a new test. Like patients, they can influence decisions, but are usually not decision makers themselves. In many African countries, most healthcare is delivered through the public sector, particularly to the poorest people. Representatives from the Ministry of Health (MoH) are the key decision makers for the selection and procurement of medical products that are made widely available. Ministries of Finance also have to be persuaded to pay for new technologies, but they generally follow the recommendations of the MOH technical specialists. Finally, global government and international development agencies such as the WHO, the Global Fund for TB, AIDS, and Malaria (GFATM), the Joint United Nations Programme on HIV/AIDS (UNAIDS), Médecins Sans Frontières (Doctors Without Borders) and the US President's Emergency Plan for AIDS Relief (PEPFAR) are also major stakeholders. They tend to be influential in recommending which technologies should be used and how they can be effectively deployed. In some cases, they also provide subsidies to developing world governments, and some non-governmental organizations, to help underwrite the development of new technologies and/or make them more affordable to their intended audience.

Daktari initially focused its information gathering on patients and clinicians, to better understand the need for an improved CD4 cell count test. Rodriguez and his team talked at length with representatives of these stakeholder groups, and began creating use cases to better understand the requirements for the solution Daktari would ultimately develop. Sharing an example, Rodriguez said, "Consider a healthcare worker. Her name is Molly. She packs her bag in the morning with batteries, gloves, and the equipment she needs. The bag weighs 3.5 kilos, and contains her diagnostic device and all the supplies she needs for a day's work. Her work environment varies between 10 and 40 degrees Celsius. She visits 30 patients a day, walking through dusty or rainy roads, or traveling by car or motorbike. And so on." The team tested these use cases with experts in the field, and then leveraged them as a guide for creating detailed design specifications, market requirement documents, and product requirement specifications for the new test.

The Daktari team also began to engage with decision makers at the MoH in several African countries to understand their point of view. As noted, government stakeholders had generally expressed interest in a more cost-effective CD4 cell count test. But when Rodriguez began talking with them in more concrete terms, they voiced an unexpected amount of skepticism that a device could be built at a low enough price point to be widely deployed in the field. Rodriguez highlighted that this could be one of the difficulties faced by young organizations: "They think, 'Why are you, a start-up company, going to be able to solve my problem, when none of the large medical equipment manufacturers have been able to do it?'"

Daktari decided to press forward with product development despite the skepticism. However, the team committed itself to learning more about the unexpected opposition it was getting from key decision makers in government. As Rodriguez and his colleagues gained a more sophisticated understanding of the government's role, they realized that by developing a field-based, point-of-care (POC) test, they would be creating a "major headache" for program managers at the MoH. As he described, "Imagine a Central Program Manager at a MoH, who is used to having every CD4 test happening in a centralized facility near her. She can check the results, see if the test results are accurate, the equipment is

working properly, the technicians are up-to-date on their training, and if the treatment decisions are appropriate. A POC test puts the care provider, the patient, and the test result in the same room at the same time," but reduces the visibility and control of the program managers. "They feel blind," Rodriguez added. In order for a new test to be adopted, the team would have to find a way to give the program managers better visibility and on-demand access to the information they needed to stay actively involved. "We envisioned a dashboard, accessed online or on a mobile phone, that shows where things are going right, and where things are going wrong," he stated. This became an important part of the **need specification** because, without the program managers on board, "there was no product."

Another factor that the Daktari team wrestled with early in the biodesign innovation process was how much value a new test would have to deliver in order to compel the various stakeholders to change their behavior and embrace the new technology. Initially, Rodriguez thought about this issue in terms of a basic cost-effectiveness trade-off. The CD4 cell count tests performed by central labs were extremely accurate, but expensive and slow. A POC diagnostic would be fast and inexpensive, but inherently less accurate. Daktari believed that its test had to strike a perfect balance between accuracy and cost that would satisfy clinicians, decision makers, funders and payers alike. Rodriguez explained, "We framed the tradeoff not as between perfect and imperfect information, but instead between no information and really good information." He and his teammates "shopped around" various scenarios, but ultimately had to trust their own instincts about which value equation would be persuasive enough to MoH representatives to get them to purchase the test, while also meeting the very specific, important needs of patients and clinicians.

When asked for his advice on effectively pursuing a need that involves governments in developing countries as a primary stakeholder, Rodriguez shared a number of ideas. First, he noted that the purchasing process can take some time. In particular, MoH representatives will often want to conduct a local study to generate clinical data specific to their environment. Even if the manufacturer has run clinical trials in a neighboring country, the MoH is likely to conduct an independent trial. "Every country wants its own small study, in their environment, to vet the product," he said. The organization will be asked to provide the product and offer support, yet it may be given little control over how the study is designed. Additionally, Rodriguez stated that having a **CE mark** or FDA regulatory approval is a necessary pre-condition of submitting a government bid for most medical products. ISO 13485 certification is also required. He also mentioned that one way organizations can differentiate themselves in the tender process is by anticipating the government's need for post-market service and maintenance of medical products. "Governments are used to lots of broken, unused equipment. Innovators have to be able to convince the MoH that they have a plan for servicing their product."

Finally, Rodriguez pointed out that in pursuing government customers, "There are a lot of relationships involved. And it's a pretty complicated stakeholder landscape." More than anything, he said, innovators should seek to add individuals to their teams who have significant, first-hand global health expertise and the deep stakeholder relationships that only can be acquired through years of work in the field. "If I were an innovator with a global health technology, I would be very hesitant to go after this without someone experienced, who had worked in global health for a few years. By now, there are many people out there with the requisite experience from

FIGURE 2.3.6
Health workers with the Daktari CD4 in the field (courtesy of Daktari Diagnostics).

working with foundations, multilateral organizations, drug companies, or governments. And I would turn to them for their expertise."

As of late 2013, Daktari had developed a fast, easy-to-use diagnostic designed specifically for low-resource settings that uses microfluidics and electrochemical testing to provide CD4 cell count results in the field (see Figure 2.3.6). The Daktari CD4 test is in late-stage product development and testing, with its first commercial sales expected in mid-2014.

Output from stakeholder analysis

After a stakeholder analysis is complete, an innovator should compile the information into a comprehensive summary of the stakeholder environment. Table 2.3.1 provides a sample of that output for InnerPulse. It summarizes the key stakeholders, identifies the decision maker(s), outlines the primary benefits and costs associated with a new solution to address the defined need, and provides a subjective assessment of the overall net impact of a new technology on each group. Importantly, this example deals with a stakeholder analysis for which the solution is already known. In most cases, at this stage in the biodesign innovation process, the innovator should not yet have defined a solution. As a result, the

Table 2.3.1 The sample stakeholder analysis for InnerPulse demonstrates how the key take-aways from a stakeholder analysis can be summarized in a concise, actionable format.

Stakeholders	Role	Primary benefits	Primary costs	Assessment of net impact
Payers	Decision maker	Expansion of life saving technology to many patients who could benefit from it.	Increased costs.	**Negative**: Total cost of delivering defibrillator therapy will go up. Will they try to reduce reimbursement for ICDs and/or PICDs, given that PICDs can be implanted more quickly?
Physicians: interventional cardiologists (ICs)	Influencer	Expanded practice and market, additional revenue. Allows for retention of primary prevention patients without need for EP referral.	Learning of new procedure.	**Positive**: ICs are typically quick to embrace new technologies that expand their market, especially if combined with an attractive reimbursement.
Physicians: electrophysiologists (EPs)	Influencer	Expanded overall referrals from ICs of primary prevention patients who develop arrhythmias and thus become secondary prevention patients.	Possible loss of primary prevention patient referrals.	**Neutral**: EPs seemingly are interested in using PICDs and, by nature, prefer to focus on complex arrhythmia cases not seen by ICs; but any loss of patient referrals could be perceived as a threat.
Facilities: EP labs and IC catheterization labs	Influencer	Increase device implantation volume for EPs and ICs.	Overall costs, including expensive components, for EPs and ICs become similar.	**Positive**: As long as reimbursement for PICDs remains the same as for ICDs.
Patients	Influencer	Reduced invasiveness compared to traditional ICDs. More convenient, shorter recovery time.	Need for re-implantation after a documented arrhythmia.	**Positive**: Reduced invasiveness expected to increase patient's comfort with the procedure – but ultimately will defer to advice from physician

stakeholder summary would capture the *potential* factors that might stimulate adoption or resistance. As more becomes know about the eventual solution, the stakeholder summary can be made more specific until it resembles the example. At this point, it can serve as the foundation for developing specific stakeholder management strategies and to facilitate decisions about how much time and energy should be invested in winning over each stakeholder group. It can also be leveraged to help forge important relationships and develop key messages appropriately targeted at various stakeholder groups (see 5.7 Marketing and Stakeholder Strategy).

Ethical considerations in stakeholder analysis

As described in 1.1 Strategic Focus, ethics focus on the intentional choices people make and the basic moral principles they use to guide their decisions. They do not provide a specific value system for making choices, but rather a set of basic principles to follow in decision making and in ethically managing the conflicts of interest that these choices may create.[50]

Stakeholder analysis informs the innovator about the types of interests that could be affected by a given decision, and can help identify potential conflicts that may arise. Personal interests (what is best for the individual making the decision), social interests (what is good for a community or society at large), and professional interests (what is good for the company or the patient/client) are factors that usually need to be considered when making almost any decision. At times, these interests are aligned and the level of conflict is low. At other times, they directly conflict and the "right" answer may not be obvious. For example, how will dialysis centers react to an innovation that makes kidney transplants more available?

As a general rule, innovators must make every effort to avoid putting stakeholders in a position that might potentially compromise their ethics. This can be accomplished if the innovator systematically uses a code of well-established ethical principles in interacting with each stakeholder group and making choices available to them (this involves treatment options offered, but also treatment options withheld). As profiled in 1.1 Strategic Focus, key ethical principles for medical device innovators include truthfulness, fairness, beneficence, non-maleficence, respect, and confidentiality.

An appreciation of these principles by innovators not only helps them think through their own choices, but can explain why stakeholders might potentially resist the adoption of a new innovation if it presents an ethical dilemma according to these guidelines (e.g., trying a new treatment if there are questions about its efficacy or safety as compared to other options). For example, this issue was seen in the case of left ventricular assist devices (LVADs), which are mini-pumps implanted into the chest to help a patient's failing heart pump blood. LVADs were initially approved as a temporary "bridge" to help maintain the functioning of the heart while patients with severe heart disease awaited heart transplantation. The FDA eventually approved LVADs for permanent use in terminally ill patients who were not eligible for a heart transplant due to comorbidities or age.[51] However, despite the fact that the LVAD manufacturer had completed FDA approval, achieved relatively strong clinical data that showed improved survival rates compared to drug therapy, and launched an aggressive marketing campaign, physicians resisted widespread adoption of the treatment based, in part, on poor economics (the procedure was reimbursed at a fraction of its actual cost).[52] More importantly, physicians raised concerns about LVAD-related infections, which often were quite serious. Without effective clinical evidence to verify the magnitude of the infection risk, physicians did not perceive the potential benefits to their patients as adequate to justify the increased safety risk under the guideline of beneficence, or the obligation to do no harm. As a result, they were slow to adopt the innovation in large numbers.

While this response alone created significant resistance to the technology, the interpretation of the LVAD example is even more nuanced. With healthcare budgets continually being squeezed, physicians increasingly recognize that expensive technologies strain patient access to care. In the case of LVADs, one could reasonably argue that certain physicians resisted their adoption because they considered it unjust in the face of severe budget constraints. Some questioned the fairness of spending upwards of $200,000 on a device with a questionable benefit and serious safety issues while other patients had difficulties obtaining access to more basic care. Although such megatrends are always in the background of decisions, it is inevitable that individual physician judgments will play a role in decisions that impact patients. Innovators should try to anticipate how such individual value judgments may affect the adoption of a new innovation.

As noted previously, the risk–benefit ratio associated with a new technology ultimately drives its adoption. Medical ethics dictate that physicians carefully balance and discuss the risks and benefits of any treatment with their patients. Regulatory authorities make decisions to approve devices by evaluating evidence on their safety and risk at an overall system level. However, when physicians evaluate the same data one-on-one with a patient, the risk–benefit ratio might look significantly different. Physicians in the field are empowered (and obligated) to make their own decisions, together with their patients. Their own personal experiences in successfully and (perhaps more importantly) unsuccessfully treating a disease may often be a stronger force in their decision making than clinically validated evidence, especially with regard to issues of safety. When the risks are grave, they may not recommend a treatment to a patient even if, for example, it has been shown statistically to prolong the life of a terminally ill patient, as the LVAD example illustrates. While there is a certain group of "early adopters" in any medical field, the majority of medical personnel require evidence of safety and effectiveness demonstrated in a substantial number of patients before implementing a new technology. This extends to including a broad spectrum of patients with any condition to be addressed in studies to demonstrate the benefit of the innovation. These ethical issues need to be applied to considerations of each stakeholder: the patient, the medical personnel applying the technology, those buying the technology, those paying for the technology, and the support groups.

Online Resources

Visit www.ebiodesign.org/2.3 for more content, including:

Activities and links for "Getting Started"
- Identify stakeholders
- Outline benefits and costs for each stakeholder group
- Summarize net impact and key issues for each stakeholder group
- Classify stakeholders and assess trade-offs

Videos on stakeholder analysis

CREDITS

The editors would like to acknowledge William N. Starling for sharing the InnerPulse story and Bill Rodriquez for sharing the Daktari Diagnostics case. Many thanks also go to Darin Buxbaum, Ritu Kamal, Donald K. Lee, and Richard L. Popp for their contributions to the chapter.

NOTES

1 From remarks made by Richard Stack as part of the "From the Innovator's Workbench" speaker series hosted by Stanford's Program in Biodesign, March 6, 2006, http://biodesign.stanford.edu/bdn/networking/pastinnovators.jsp (December 2, 2013). Reprinted with permission.

2 Pushkal P. Garg, Kevin D. Frick, Marie Diener-West, and Neil R. Powe, "Effect of the Ownership of Dialysis Facilities on Patients' Survival and Referral for Transplantation," *New England Journal of Medicine*, November 25, 1999, http://content.nejm.org/cgi/content/abstract/341/22/1653 (October 3, 2013).

3 "Stricter Rules Driving Away Medical Tourism from India," *The Economic Times*, August 15, 2013, http://articles.economictimes.indiatimes.com/2013-08-15/news/41413559_1_apollo-hospitals-prathap-c-reddy-overseas-patients (October 3, 2013).

4 "High Level Expert Group Report on Universal Health Coverage in India," Planning Commission of India, November 2011, p. 189, http://uhcforward.org/publications/high-level-expert-group-report-universal-health-coverage-india (October 3, 2013).

5 Srinath Reddy, K. Srinath Reddy, Vikram Patel, Prabhat Jha, Vinod K. Paul, A. K. Shiva Kumar, and Lalit Dandona, "Towards Achievement of Universal Healthcare in India by 2020: A Call to Action," *Lancet*, vol. 377, 2011, pp.760–68, http://cghr.org/wordpress/wp-content/uploads/Towards-achievement-of-universal-health-care-in-India-by-2020-a-call-to-action-2011.pdf (October 3, 2013).

6 Wang Fan, "Hospital Ratings Reflect a Flawed System," *ChinaNews.com*, August 28, 2012, http://www.ecns.cn/2012/08-28/23744.shtml (October 3, 2013).

7 Pacific Bridge Medical, "China's Hospital Market," January 1, 2011, http://www.pacificbridgemedical.com/publications/chinas-hospital-market/ (October 3, 2013).

8 Karen Eggleston, "Healthcare for 1.3 Billion," *Asia Health Policy Program*, Working Paper 28, APARC Stanford University 2012, p. 6.

9 "Medicare Coverage of Kidney Dialysis & Kidney Transplant Services," Centers for Medicare & Medicaid Services, April 2012, http://www.medicare.gov/Pubs/pdf/10128.pdf (December 2, 2013).

10 Chad Terhune, "Kaiser Tops State Health Insurance Market with 40 Percent Share," *Los Angeles Times*, January 29, 2013, http://articles.latimes.com/2013/jan/29/business/la-fi-mo-health-insure-market-20130129 (October 2, 2013).

11 Jenny Gold, "FAQ on ACOs: Accountable Care Organizations, Explained," Kaiser Health News, August 23, 2013, http://www.kaiserhealthnews.org/stories/2011/january/13/aco-accountable-care-organization-faq.aspx (October 2, 2013).

12 "Employer Wellness Initiatives Grow Rapidly, Nut Effectiveness Varies Widely," National Institute for Health Care Reform, July 2010, http://www.nihcr.org/Employer-Wellness-Initiatives (December 2, 2013).

13 "Health Financing: Private Expenditures on Health as a Percentage of Total Expenditures on Health," World Health Organization, 2011, http://gamapserver.who.int/gho/interactive_charts/health_financing/atlas.html?indicator=i2&date=2011 (October 2, 2013).

14 "Taking Advantage of the Medtech Market Potential in India," PricewaterhouseCoopers, 2012, p. 4, http://www.pwc.com/en_GX/gx/pharma-life-sciences/publications/asia-pharma-newsletter/assets/taking-advantage-of-the-medtech-market-potential-in-india.pdf (October 3, 2013).

15 "Government Sponsored Health Insurance in India: Are You Covered?," World Bank, October 11, 2012, http://www.worldbank.org/en/news/2012/10/11/government-sponsored-health-insurance-in-india-are-you-covered (October 3, 2013).

16 "Indian Pharma 2020: Propelling Access and Acceptance, Realizing Potential," McKinsey & Company, 2010, p. 18, http://www.mckinsey.com/~/media/mckinsey/dotcom/client_service/Pharma%20and%20Medical%20Products/PMP%20NEW/PDFs/778886_India_Pharma_2020_Propelling_Access_and_Acceptance_Realising_True_Potential.ashx (October 3, 2013).

17 Shenglan Tang, "Has China's Health Reform Improved the Affordability of Healthcare for the Rural Population," China Health Policy Report, July 2012, http://sites.duke.edu/chinahealthpolicyreport/2012/07/ (February 11, 2013).

18 William Gallo, "Chinese Healthcare Improves, But More Reforms Needed," VOA News, December 27, 2012, http://www.voanews.com/content/analysts-china-health-care-improving-but-more-reforms-needed/1573022.html (October 3, 2013).

19 "Health Financing: Private Expenditures on Health as a Percentage of Total Expenditures on Health," op. cit.

20 Gallo, loc. cit.

21 "Private Health Insurance in China: Finding the Winning Formula," McKinsey & Company, 2012, https://www.google.com/url?sa=t&rct=j&q=&esrc=s&source=web&cd=1&ved=0CCwQFjAA&url=http%3A%2F%2Fwww.mckinsey.com%2F~%2Fmedia%2Fmckinsey%2Fdotcom%2Fclient_service%2FHealthcare%2520Systems%2520and%2520Services%2FHealth%2520International%2FIssue%252012%2520PDFs%2FHI12_74-84%2520ChinaPrivate

Healthcare_R5.ashx&ei=IpZNUrf2LIznigKW_YHoBg&usg=AFQjCNFbzzSn1nM1R6cGHWRsY00GnINbVg&sig2=wS_FTrWWEKwFdotKjGEMmA&bvm=bv.53537100,d.cGE (October 3, 2013).

22 Celia Deng, Wei Sun, Zhiyi Tong, and Paul Zhang, "Tales of Three Medical Device Markets in China," In Vivo: The Business & Medicine Report, November 2012, p. 57.

23 "Opportunity Cost," Dictionary.com, http://dictionary.reference.com/browse/opportunity%20cost (September 22, 2008).

24 "Market Disruption Imminent as Hospitals and Physicians Aggressively Adopt Patient Portal Technology," Frost & Sullivan press release, September 27, 2013, http://www.prnewswire.com/news-releases/market-disruption-imminent-as-hospitals-and-physicians-aggressively-adopt-patient-portal-technology-225498752.html (October 3, 2013).

25 "The World Health Report, Chapter 1: Where Are We Now?," World Health Organization, 2010, http://www.who.int/whr/2010/10_chap01_en.pdf (October 3, 2013).

26 Ryan Saadi, Nicole C. Ferko, Peter Ehrhardt, and Daniel T. Grima, "Strategies for Medical Device Manufacturers to Address Hospital Value Analysis," Medical Device and Diagnostic Industry, News, May 7, 2013, http://www.mddionline.com/article/strategies-medical-device-manufacturers-address-hospital-value-analysis (October 3, 2013).

27 From remarks made by Guy Lebeau as part of the "From the Innovator's Workbench" speaker series hosted by Stanford's Program in Biodesign, April 5, 2005, http://biodesign.stanford.edu/bdn/networking/pastinnovators.jsp (October 3, 2013). Reprinted with permission.

28 Sara Michaels, "Making Mobile Health Work," PhysiciansPractice.com, January 21, 2011, http://www.physicianspractice.com/articles/making-mobile-health-work-your-medical-practice (October 3, 2013).

29 Laurence Kennedy and William H. Herman, "Glycated Hemoglobin Assessment in Clinical Practice: Comparison of the A1cNow™ Point-of-Care Device with Central Laboratory Testing (GOAL A1C Study)," Diabetes Technology & Therapeutics, vol. 7, no. 6, 2005, pp. 907–12.

30 NGSP (National Glycohemoglobin Standardization Program) is a certification method beyond what was required for FDA approval.

31 Kennedy and Herman, loc. cit.

32 "Shared Savings Program," Centers for Medicare & Medicaid Services, http://www.cms.gov/Medicare/Medicare-Fee-for-Service-Payment/sharedsavingsprogram/index.html?redirect=/sharedsavingsprogram/ (October 3, 2013).

33 "Accountable Care Organization 2013 Program Analysis," Quality Measurement & Health Assessment Group, Centers for

Medicare & Medicaid Services, December 21, 2012, http://www.cms.gov/Medicare/Medicare-Fee-for-Service-Payment/sharedsavingsprogram/Downloads/ACO-NarrativeMeasures-Specs.pdf (October 3, 2013).

34 Ibid.

35 Craig Abbott, "Healthcare Trends: Developing ACOs," Health Dimensions Group, http://www.healthdimensionsgroup.com/pdfs/HDG_WhitePaper_FINAL_print.pdf (October 3, 2013).

36 "About Technology Appraisals," National Institute for Health and Care Excellence, http://www.nice.org.uk/aboutnice/whatwedo/abouttechnologyappraisals/about_technology_appraisals.jsp (October 3, 2013).

37 Ibid.

38 "Measuring Effectiveness and Cost Effectiveness: The QALY," National Institute for Health and Care Excellence, http://www.nice.org.uk/newsroom/features/measuringeffectivenessandcosteffectivenesstheqaly.jsp (October 3, 2013).

39 Andrew Pollack, "Pricing Pills and the Results," *The New York Times*, July 14, 2007.

40 "Does Insurance Cover the Oncotype DX Assay?," OncotypeDX.com, http://www.oncotypedx.com/en-US/Breast/HealthcareProfessionalsInvasive/Insurance/InsuranceInfo (October 3, 2013).

41 "Stakeholder Analysis," NHS Institute for Innovation and Improvement, http://www.institute.nhs.uk/NoDelaysAchiever/ServiceImprovement/Tools/IT145_stakeholder.htm (November 13, 2006).

42 All quotations are from interviews conducted by the authors, unless otherwise cited. Reprinted with permission.

43 Frank Vinluan, "InnerPulse Cuts Jobs, Including CEO's," *Triangle Business Journal*, August 18, 2008, http://www.bizjournals.com/triangle/stories/2008/08/18/story1.html?page=all (October 3, 2013).

44 Eggleston, op. cit., p. 7.

45 "Private Sector in Healthcare Delivery in India," National Commission on Macroeconomics and Health, 2005, http://www.nihfw.org/WBI/docs/PPP_SessionBriefs/PPP%20Course%20sessions/Need%20and%20Scope%20for%20PPP/Private%20Sector%20in%20Health%20Care%20Delivery%20in%20India.pdf (February 22, 2013).

46 UNAIDS Fact Sheet, 2012, http://www.unaids.org/en/media/unaids/contentassets/documents/epidemiology/2012/gr2012/2012_FS_regional_ssa_en.pdf (October 21, 2013).

47 "HIV Diagnosis," UNICEF, January 2008, http://www.unicef.org/supply/files/HIV_DIAGNOSIS_A_Guide_for_Selecting_RDT_Jan08.pdf (October 21, 2013).

48 UNAIDS, op. cit.

49 "HIV Treatment Now Reaching More Than 6 Million People in Sub-Saharan Africa," UNAIDS press release, July 6, 2012, http://www.unaids.org/en/resources/presscentre/pressreleaseandstatementarchive/2012/july/20120706prafricatreatment/ (November 3, 2013).

50 R. J. Devettere, *Practical Decision Making in Health Care Ethics* (Georgetown University Press, 2000).

51 Susan Conova, "FDA Approves Heart Pump for Terminally Ill Patients," *In Vivo*, December 4, 2002, http://www.cumc.columbia.edu/news/in-vivo/Vol1_Iss20_dec04_02/index.html (December 2, 2013).

52 Muriel R. Gillick, "The Technological Imperative and the Battle for the Hearts of America," *Perspectives in Biology and Medicine*, 2007, http://muse.jhu.edu/journals/perspectives_in_biology_and_medicine/v050/50.2gillick.html (October 3, 2013).

2.4 Market Analysis

INTRODUCTION

Not all needs are created equal. For better or worse, innovators must recognize that even seemingly important needs cannot be addressed unless there is a compelling, accessible market to support the effort and expense required to bring forward a new solution. In today's healthcare environment, the size and growth rate of the target market, and the presence and nature of competition in the field all directly affect the attractiveness of the need. In addition, innovators must consider the opportunity to create value in the need area. There must be "room" within the market for the team to develop a solution that can deliver requisite improvements at or below a defined cost threshold in order to the capture the attention of the target market and drive adoption.

Market analysis enables innovators to understand which of their needs are associated with a commercially viable market. Key customers, prospective investors, and the team itself must all see value in addressing a need in order for it to be worth pursuing.

Market analysis begins with a landscaping exercise that is intended to provide a broad understanding of the need area in terms of its total size, range of existing solutions and competitors, and gaps that may exist and indicate opportunities for innovation. Next it focuses on a progressively detailed evaluation of different segments within the overall market to determine which one(s) the innovators should potentially target if they decide to pursue the need. An important part of this analysis is deriving a value estimate that articulates how much and what type of improvements a new solution must deliver, at or below a defined cost threshold, in order to have a reasonable likelihood of being adopted by the target market.

The information gathered through 2.4 Market Analysis is important throughout the biodesign innovation process. It provides essential input into 2.5 Needs Selection. As innovators make the transition from working with needs to solutions, market analysis is repeated and performed in greater depth in conjunction with 5.7 Marketing and Stakeholder Strategy. The output of that chapter then becomes essential input into 6.1 Operating Plan and Financial Model.

 See ebiodesign.org for featured videos on market analysis.

OBJECTIVES

- Understand how to perform broad market landscaping to get a directional sense of a market when working with large numbers of needs.

- Learn to divide a market into segments of homogenous customers that share similar perceptions of and/or responses to medical need.

- Appreciate how to define the market size, growth, and competitive dynamics of each segment.

- Recognize how to determine the extent to which stakeholder needs are currently being addressed within each segment, assess stakeholder willingness to pay for alternate solutions, and bring this information together to create a value estimate for a market segment.

- Recognize key considerations in choosing a target market.

MARKET ANALYSIS FUNDAMENTALS

Market analysis is performed early in the biodesign innovation process to assess whether a potential need being evaluated is associated with a commercially viable market. Importantly, innovators must determine for themselves what constitutes an attractive market using input from potential customers, as well as feedback from possible investors (or other providers of capital). While the perspectives of these three constituencies will vary, customers, investors, and the innovators themselves must all see value in addressing a need in order to make pursuing a solution worthwhile. Fundamentally, the question that must be answered is: Will the gain be worth the pain?

As the cost and complexity associated with developing and commercializing medical devices continues to escalate, many innovators believe that having a large market is one key to success. As Richard Stack, who serves as the president of Synecor Inc., a Silicon Valley-based medical device **incubator**, summarized:[1]

The need has to have a very large market .... There's only so much time in the day, and it's really just as easy to develop a solution for a large market as it is for a small market.

In contrast, other innovators prefer an alternate approach: starting with a smaller market and expanding over time. John Abele, co-founder of Boston Scientific, provided this perspective:[2]

I'm a big fan of niches – of not trying to take on an entire market at once – because there's less resistance to innovation when you do it on a smaller scale.

While innovators may favor different market types, there is widespread agreement regarding the importance of performing market analysis before deciding on which **needs** to take forward into invention. Using market characteristics as factors for choosing which needs to pursue ensures that innovators understand the extent of the **value** that potentially can be generated and captured in a need area. The outcome of market analysis also enables innovators to appropriately refine the **need statement** and develop an expanded **need specification** (see 2.5

Needs Selection) that focuses on the unique needs and requirements of the chosen target market.

Market analysis is a multi-step process that leads the innovator through an increasingly specific investigation, as described in Table 2.4.1. The remainder of this chapter explores each of these important steps in more detail.

Step 1 – Landscape the market

With dozens or even hundreds of needs to evaluate, innovators cannot realistically perform an in-depth market analysis for each one. Often, it makes sense to start with a higher-level approach, focused on understanding the broad market landscape for each need, and then work toward increasing depth and detail as the list of needs is narrowed down through the iterative needs screening process.

A useful way to initiate a market landscape is to start by thinking about the total market for the need area. Innovators can go back to the research performed as part of 2.1 Disease State Fundamentals to refresh themselves on total spending for the disease state. For example, when Moshe Pinto, Dean Hu, and Kenton Fong were students in the Stanford Biodesign program, they began working on the need *to promote the healing of chronic wounds*. As part of the team's research, Pinto and his teammates discovered that the wound care market is immense, accounting for more than $20 billion in annual healthcare costs in the US alone and $566 million in wound care products.[3]

To begin thinking about how well needs are being addressed in the overall market, innovators can next revisit and build on the information they gathered in 2.2 Existing Solutions. Through this assessment, they should already have at least some data on which diagnostics, treatments, and therapies dominate the need area, as well as their approximate cost, utilization, and relative effectiveness. Use this information to create a map of the market from a solution point of view, conducting additional research as needed to fill in missing data. For example, in a simple version of a solution map, innovators place cost on one axis and effectiveness on the other and then plot the comparative position of the major treatment alternatives. In a slightly more advanced version, different sized "bubbles" can be used to convey

Table 2.4.1 The three key steps for performing a high-level market analysis.

Step	Topic	Questions to investigate
1	Market landscape	Thinking about all relevant stakeholders or the "total market," how well are customer needs generally addressed by existing solutions? If no solutions currently exist, then what is the magnitude or importance of the need when viewed from the customer's perspective? *Gaps* How closely aligned are available solutions with the need the innovators are seeking to address? What gap(s) exist where a new solution could potentially add value? How important is addressing this gap viewed from the customer's perspective? *Size and growth* What is the approximate size of the total market for all existing solutions? What is the size of the market by available solution? Is the market expanding or contracting? Where is the most predominant positive or negative growth occurring?
2	Market segmentation	What are the key factors that can be used to divide the total market of all potential customers into distinct segments, in which the population shares common needs and perceptions (e.g., patient characteristics, solution options, provider attributes, and **payer** mix)? Do meaningful differences exist? *Size and growth* What is the size of the market opportunity in each market segment and the potential for growth and expansion? *Competitive dynamics* What are the competitive dynamics in each market segment that a new entrant would face? How intense is the competition? Are new companies created and are they successful? What is the nature of their competitive relationship with existing companies? Are companies acquired in this space? *Needs* What are the unique customer needs of each market segment? *Willingness to pay* Which **stakeholder** could recognize the greatest value created by a new solution? What level and type of value would potentially cause the members of the market segment to change their behavior and embrace a new solution? How willing is each market segment to pay for a new solution and, if applicable, what do customers pay for existing solutions? How motivated is each segment to adopt a new solution?
3	Target market	Which market segment(s) stands to gain the most value from a new solution? Which market segment is likely to create the most value for the: • Customer (the decision maker that will pay for a solution) • Investor (the provider of capital) • Innovator (the provider of intellectual capital and labor)?

the approximate size of the market for each treatment. The rate at which treatment utilization is growing (or contracting) is another factor to consider adding to the landscape.

In some cases, innovators create multiple views of the high-level market landscape – for instance, developing one version that sizes the bubbles for each treatment based on the total number of patients using the intervention and another version that sizes them based on total annual spending or the growth rate in each category. Innovators can also evaluate what treatments are available to patients (and at what cost) at different stages of the disease by replacing effectiveness with disease severity on one axis and creating a completely different map. Or they can evaluate the cost of treatment per patient versus total spending on treatment for patients at each stage of the disease.

After one or more views of the market landscape have been developed, innovators can begin thinking about what gaps exist in the overall market. Gaps often correspond to opportunities to create and deliver value. For instance, a landscape that looks at the cost of available treatments relative to their general effectiveness can help innovators estimate the *value* that any new solution would have to be able to deliver in order to displace available treatments. More information about constructing a **value estimate** is provided later in this chapter.

Figure 2.4.1 shows a market landscape diagram created as part of the wound care project referenced previously. This team, which eventually went on to found a company called Spiracur, identified that a sizable gap existed in the chronic wound care market for a solution with a moderate cost, along with effectiveness equivalent to leading solutions. They also perceived that

the large treatment category of negative pressure dressings, which was the fastest-growing part of the market, was the "solution to beat." If they could provide a lower-cost solution that delivered comparable results, it would be feasible to make inroads into the market.

Figure 2.4.2 provides another example from a team focused on needs related to chronic obstructive pulmonary disease (COPD), a condition that costs $32 billion to treat each year in the United States. These innovators were drawn to the overall need for more effective and affordable solutions for diagnosis and treatment in the disease area. As team member Michael Winlo summarized, "COPD is one of the few chronic diseases that is *increasing* in prevalence globally, and the **morbidity** of the disease is really significant. But there aren't a lot of really great solutions to effectively treat this condition."[4] This group believed that a gap existed in the market for a solution that helped manage the disease that was less expensive yet more effective than basic telemonitoring.

Maps like the ones shown here are sufficient for providing an early, directional sense of where high-value opportunities may exist. However, as they are further developed and become more detailed, they also become more useful. For instance, by further quantifying each axis, innovators can better estimate what level of effectiveness a new solution would have to deliver, within

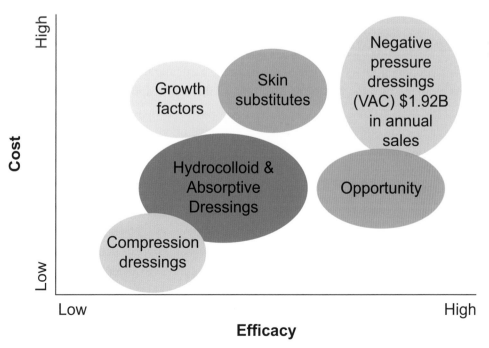

FIGURE 2.4.1

A high-level market landscape for wound care (courtesy of Spiracur, Inc.).

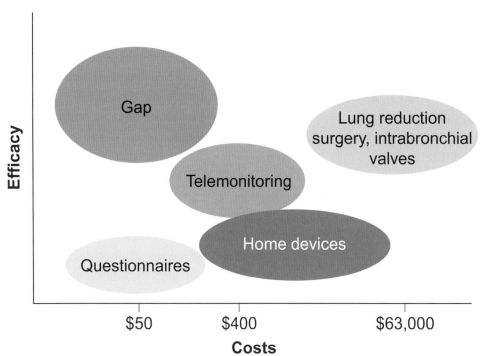

FIGURE 2.4.2
A market landscape for chronic obstructive pulmonary disease (courtesy of Mridusmita Choudhury, Jonathan Hofius, Raymond Hoi, Zubin Huang, Peter Livingston, and Michael Winlo).

what specific price range, in order to be perceived as compelling by the stakeholder(s) that will decide whether or not to adopt it. It is through the more detailed analysis that innovators can gain clearer perspective on exactly how well the need is served by competitive alternatives and how much room remains for an "improved" solution. The risk, as always, is that innovators will overestimate the extent to which an improvement will be significant from the perspective of the customer. As innovators begin narrowing their focus to a manageable set of needs and start thinking about segmenting the market, they can begin layering in this additional information to make their landscape maps more instructive.

Depending on how a particular need has been scoped, it can sometimes be necessary to examine competitive positioning that may exist *within* an existing treatment area, as well. For example, in the COPD case, a closer evaluation of the approaches described as "Home Devices" would result in a map of the market positions occupied by current competitors in the space with devices to diagnose exacerbations (as shown in Figure 2.4.3). This type of competitive landscaping helps to determine how well a given need is currently being met by participants in a specific product category. Additionally, it can highlight gaps in competitive product

offerings and guide the innovator toward more concrete opportunity areas that may be ripe for value creation. In the case of home devices, said Winlo, "What we realized was that none of these tools were very effective. That was clear from the published literature. There were very few alternatives, and the sensitivity and specificity of the portable solutions was very poor."

Like the need itself, the relevant competitive landscape can be scoped "up and down" by looking at different product categories. Innovators should recognize the strong relationship between needs scoping and competitive analysis and be particularly attentive where the solutions identified in this landscaping step are overlapped with respect to cost and efficacy to avoid overlooking a potentially competitive offering.

Step 2 – Segment the market

It is rare that any new solution to address a need will meet the requirements of *all* customers or stakeholders across a broad medical field (e.g., all patients with the heart rhythm disorder known as atrial fibrillation). It is far more likely for a solution to address the needs of a subset of the total population (e.g., patients with paroxysmal atrial fibrillation originating in the pulmonary veins, over the age of 65, treated by an electrophysiologist, and covered

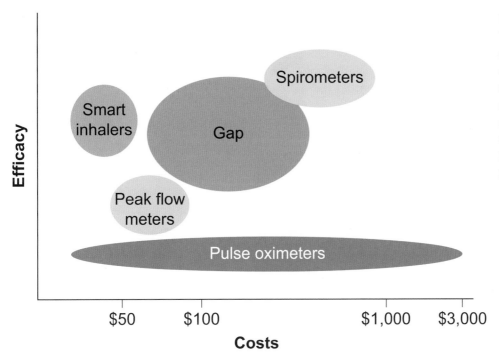

FIGURE 2.4.3
This treatment landscape for home devices used to diagnose an exacerbation provides a directional representation of the cost and relative effectiveness of these devices (courtesy of Mridusmita Choudhury, Jonathan Hofius, Raymond Hoi, Zubin Huang, Peter Livingston, and Michael Winlo).

by Medicare). Innovators can use **market segmentation** to divide patients, payers, and providers (in this context, physicians and facilities) into distinct groups that share similar perceptions about a need. The purpose of this exercise is to recognize and understand the subgroups that comprise a total market (even if those subgroups are somewhat over simplified at this stage of the process).

An effective market segmentation scheme identifies subsets of the population with the greatest similarity while maintaining maximum differences between the groups. Market segments should be measurable, accessible, durable (so as not to change too quickly), and substantial enough to be profitably and/or sustainably served.[5] Innovators will seek to understand the characteristics of customers in these different groupings in order to predict which one or two market segments are most likely to place the highest value on a solution to a given need. In turn, these segments will become the initial target market (read on for more information about target markets).

Because of the complex interactions among the multiple stakeholders that exist within medical markets, segmentation should be thought of as an iterative activity. With each iteration, innovators can use one or more

attributes to divide larger groups into smaller ones. Figure 2.4.4 summarizes a sample of the dimensions that can be used to differentiate between market segments. When conducting a market analysis, recognize that important differences between market segments may not exist across all of these dimensions. However, they all should be considered to ensure that no important sources of distinction between subgroups are overlooked. Furthermore, keep in mind that there may be additional factors to consider based on the unique and distinguishing characteristics of the market being assessed.

One approach to segmentation is to begin with a relatively simple patient-based analysis and then layer in increasingly complex factors (e.g., payers, providers) to eventually account for the many different stakeholders and interactions that need to be considered in the assessment of the market. Four steps can be used to experiment with this method:

1. Segment patients based on symptoms and risk factors.
2. Segment patient subgroups based on treatment.
3. Segment patient/treatment subgroups based on providers.

Patients

• Geography
• Demographics (age, gender,etc.)
• Socioeconomic status
• Clinical risk factors (symptoms, coexisting medical conditions, physiological variables, disease progression/severity)
• Attitude toward health and health care providers
• Existing treatments
• Attitude toward new technology (technophobe versus technophile)

Providers

• Geography
• Demographic and socioeconomic focus
• Specialty and training
• Venue (hospital, multi-specialty group practice, independent provider)
• Attitude towards new technology (technophobe versus technophile)
• Attitude toward evidence-based versus experiential medicine

Payers

• Geography
• Self-pay versus insurance-pay
• National versus regional focus
• Private (for-profit or nonprofit) versus public insurer
• Approach to technology assessment (formal evidence-based versus information relation-based approach)
• Attitude towards medical technology (receptive versus hostile versus agnostic)
• Power within management structure (physicians versus actuaries)
• Types of employers (large, medium, small, self-employed)
• Expectations regarding profitability and reimbursement rates

FIGURE 2.4.4

There are many attributes that can be used to differentiate between market segments. Innovators should consider a wide range of these factors in their efforts to come up with a meaningful segmentation scheme.

4. Segment patient/treatment/provider subgroups based on payers.

The high-level example that follows, which is focused on patients with chronic kidney disease (CKD), illustrates these four steps.

Segment patients based on symptoms and risk factors
First, to segment patients based on symptoms and risk factors, innovators can divide patients into diagnosed and undiagnosed categories, then further refine these groups based on the stage of their disease [stages 1–2 = mild, 3 = moderate, 4 = severe, and 5 = end-stage renal disease (ESRD)]. Most stage 1 patients are undiagnosed, while most stage 5 are diagnosed. Figure 2.4.5 provides a representation of the resulting segments. In this and the other segmentation figures that follow, the area of each

subgroup reflects the size of the segment in terms of the number of patients it includes.

Segment patient subgroups based on treatment
The next step is to further subdivide patients based on the treatment they receive. Common treatment categories for CKD might include self-care management (e.g., dietary and fluid restrictions), drug treatment (glucose control medication for diabetics; blood pressure control therapy for those suffering from high blood pressure; or other treatments for patients with anemia or bone disease); hemodialysis (HD); peritoneal dialysis (PD); or transplantation (Tx). Figure 2.4.6 shows how patients in the diagnosed stage 5 (ESRD) group can be subdivided based on treatment type.

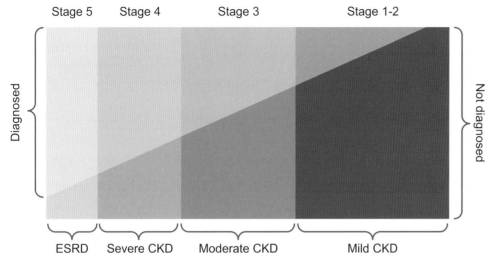

Stage 5   Stage 4   Stage 3   Stage 1-2

Diagnosed

Not diagnosed

ESRD   Severe CKD   Moderate CKD   Mild CKD

FIGURE 2.4.5
Segmentation of CKD patients based on diagnosis and disease stage.

Segment patient/treatment subgroups based on providers

These six treatment segments for stage 5 patients with ESRD can again be refined by overlaying the type of provider administering care to the patient. Typically, in the CKD market, there is a one-to-one correlation between providers and the treatments they deliver. For example, for PD and HD, treatment is delivered by a nephrologist in combination with a dialysis clinic. For transplantation, treatment is administered by a surgeon in collaboration with a transplant program, as shown in Figure 2.4.7.

Segment patient/treatment/provider subgroups based on payers

The final segmentation step is to evaluate payer type relative to the subgroups. To prevent the example from becoming too complex, Figure 2.4.8 seeks to differentiate only between patients covered by private insurance versus public insurance (Medicare in the US).

This simplified CKD example demonstrates steps for performing progressively detailed patient, provider, and payer segmentation using a sequence of Venn-like diagrams. Although the method may seem complex (and the figures difficult to replicate for another disease area), the basic approach provides a structured way for aspiring innovators to begin thinking about market segmentation in the early stages of the biodesign innovation process. More experienced innovators also perform these steps,

although they often do so in a less structured, more instinctive fashion.

Another way to illustrate market segments is by using simple tables alone or in combination with diagrams that resemble decision trees. For example, when the team working on needs related to COPD was thinking about how to segment the total market of 12 million US patients, it experimented with many different approaches, taking into account patient gender, age, risk factors, treatment type, and other factors. The innovators also considered patients based on the severity of their disease, ranging from mild to very severe. Seeking to further differentiate patients in these disease subgroups, they evaluated exacerbation rates associated with each stage of COPD and learned that they steadily increase as the condition worsens. Additional research revealed that exacerbations are major drivers of treatment costs. Across all disease subgroups, 1.2 million patients seek treatment in the emergency department (ED) during a COPD incident, and more than 900,000 are admitted to the hospital. Intrigued by the fact that hospitalizations for COPD account for more than 43 percent of all spending on the disease, the team dove deeper into characterizing the segment of patients hospitalized from COPD. In doing so, they learned that 20 percent of this population is readmitted within 30 days, and 38 percent is readmitted within 18 months.

Key data relevant to this aspect of the team's segmentation analysis are shown in Table 2.4.2 and depicted in Figure 2.4.9.

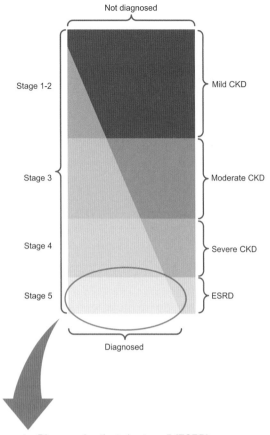

Not diagnosed

Stage 1-2 — Mild CKD

Stage 3 — Moderate CKD

Stage 4 — Severe CKD

Stage 5 — ESRD

Diagnosed

FIGURE 2.4.6
Further segmentation of
diagnosed patients in stage 5
(ESRD) based on type of
treatment.

Segment = Diagnosed patients in stage 5 (ESRD)

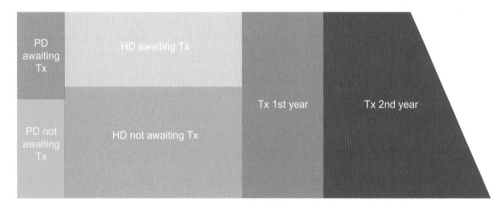

Assess segment size and growth

After defining meaningful market segments, innovators can size the market opportunity associated with each one. This involves addressing a series of important questions that are not always easy to answer (as Figure 2.4.10 illustrates):

1. How many patients are there in each market segment (if not calculated as part of the segmentation exercise), and how can they be reached?

2. What diagnostics, treatments, and/or therapies are currently being used within each segment? What companies provide these solutions? What is their relative market share?

3. What is the total dollar value of each market segment (i.e., total cost of treatment or total medical expenditures)?

4. What is the epidemiologic growth pattern in the segment? Is the segment contracting or expanding?

Table 2.4.2 Exacerbation and hospitalization rates for diagnosed COPD patients by severity of disease (courtesy of Mridusmita Choudhury, Jonathan Hofius, Raymond Hoi, Zubin Huang, Peter Livingston, and Michael Winlo).

Severity of disease	Percentage of all COPD patients	Exacerbation rate/year	% Requiring hospitalization/ year	# of hospitalizations/ year
Mild	40%	0.82	2%	64,800
Moderate	55%	1.17	7%	462,000
Severe	4%	1.67	18%	324,000
Very severe	1%	2.1	33%	118,000
Total				**968,800**

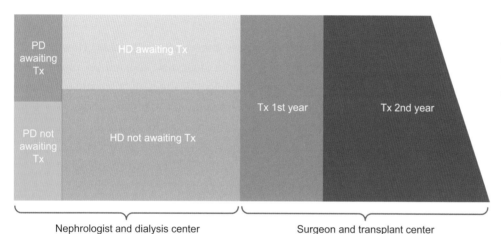

FIGURE 2.4.7
Additional segmentation of diagnosed patients in stage 5 (ESRD) based on the provider administering care.

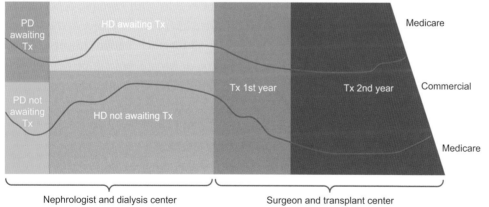

FIGURE 2.4.8
One more layer of segmentation of diagnosed patients in stage 5 (ESRD) based on payer type. The center section, bounded by the lines, indicates the portion of private or commercially insured patients, while the remaining areas, top and bottom, represent those covered by Medicare.

What do growth projections look like in the near, medium, and long term?

Keep in mind that the size of the market segment should not be considered in a vacuum, with bigger always presumed to be better. In some cases, a smaller market that is easier to access may be more attractive than a larger market in which the potential purchasers of an offering are difficult to reach.

Innovators can take either a "**top-down**" or a "**bottom-up**" approach to calculating the dollar value of each

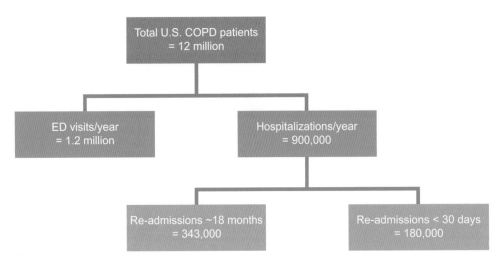

Total U.S. COPD patients
= 12 million

ED visits/year
= 1.2 million

Hospitalizations/year
= 900,000

Re-admissions ~18 months
= 343,000

Re-admissions < 30 days
= 180,000

FIGURE 2.4.9

High-level patient segmentation for US COPD patients hospitalized due to exacerbations (courtesy of Mridusmita Choudhury, Jonathan Hofius, Raymond Hoi, Zubin Huang, Peter Livingston, and Michael Winlo).

FIGURE 2.4.10

Market analysis can provide innovators with confusing or conflicting answers to important questions. But a rigorous assessment approach will help generate less ambiguous information.[6]

segment. A top-down approach begins with the overall spending on the disease state, and then divides total spending into categories based on the percentage of customers in each segment. The problem with the top-down approach is that it essentially assumes the same spending per patient across the various segments. However, its advantage is that it does not demand as much data, which can be helpful when innovators are still evaluating multiple needs. A bottom-up approach derives the total market size by multiplying the number of customers within each market segment by the associated costs of their treatment. It is more precise than the top-down approach but may require data that are not readily available at this point in the biodesign innovation process. Initially, either of these approaches can be used to determine a high-level estimate of market size. However, it may be useful to perform both types of analysis and compare/validate the results. (An example of a bottom-up analysis is provided later in this chapter in the section entitled "Bringing it all together: **patient towers**." More information about top-down and bottom-up market estimates can also be found in 6.1 Operating Plan and Financial Model.)

The Spiracur team used top-down market sizing when performing its early assessment of the chronic wound care market. One of the simple market segments that the innovators looked at was patients suffering from venous ulcers. They were intrigued by this particular segment because they felt that the gap between existing solutions and desired outcomes was so great – most ulcers remained open after 12 weeks of treatment (an unacceptably long period). To size the subgroup, the team estimated that between 0.5 and 1.5 percent of the US population suffers from venous ulcers based on extrapolations of available epidemiologic statistics.

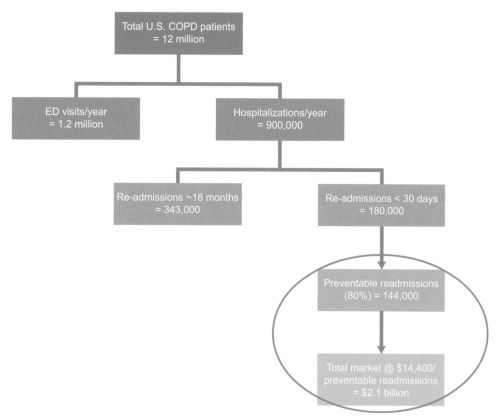

FIGURE 2.4.11

Estimated total market opportunity for the segment of US patients with mild to very severe COPD who are hospitalized (courtesy of Mridusmita Choudhury, Jonathan Hofius, Raymond Hoi, Zubin Huang, Peter Livingston, and Michael Winlo).

Competitive analysis revealed that the aggregate daily cost of the leading product for venous ulcers was $150.[7] To obtain the potential segment value, the team multiplied this amount by the duration of the therapy and by the prevalence of venous ulcers, and then incorporated growth estimates into the calculations to reach a total potential market size of $12 billion by 2009[8] (the total market size included all patients with venous ulcers). At the time of the analysis, that market segment had a penetration rate of approximately 30 percent (i.e., less than one-third of patients with a venous ulcer received some form of treatment). Even without any increased penetration in the market, the team believed that a new solution would have a total potential market opportunity worth $3.6 billion.[9] This top-down approach sizes the market using data from a comparable solution as a benchmark. It assumes full penetration with an alternative solution whose cost to the customer would be equal to (for the purposes of this estimate) the full avoided cost of the prevailing best-in-class solution ($150 per day).

The COPD team derived an early, top-down market size estimate in a slightly different way. For the segment of the US market comprising patients with mild to very severe COPD who are hospitalized, they believed that the total market opportunity could be estimated based on cost savings associated with preventing unnecessary readmissions. In particular, they believed that there was a sizable opportunity linked to preventing readmissions that occur within 30 days of discharge because research revealed that physicians believe as many of 80 percent of these repeat hospitalizations for COPD are avoidable.[10] Cost per admission ranged from $1,500 to $40,000, but they decided to use the average of $14,400 for their preliminary top-down calculations. This data led them to calculate the market opportunity estimate shown in Figure 2.4.11.

In contrast to the comparables approach taken by the Spiracur team, this approach to sizing the market considers the total "solution value" viewed from the purchaser or payer's perspective. Specifically, the team assumed the substitution of an unknown solution that would prevent the total cost of readmission. Stated another way, this top-down market estimate indicates that the solution could be worth up to a maximum of

$14,400 per prevented readmissions to its target customer. Importantly, crude top-down market sizing models often overestimate the size of the opportunity.

Innovators should expect to share potential savings (or other economic benefits) with the stakeholder(s) that will purchase or pay for a solution to give them an incentive to change their behavior and adopt a new offering. This idea is best illustrated by the COPD example. In this case, payers (as a group) most directly stand to benefit from a solution by avoiding up to $2.1 billion in readmission-related costs. But if an innovator or company tries to capture the full $2.1 billion in potential savings through a product that ends up costing payers $2.1 billion to purchase, then the payers arguably could be indifferent to the offering on purely financial terms. In this case, to encourage a change in practice, the innovators would need to leave a portion of the total potential savings with the payers to create a meaningful economic incentive for adoption to occur. To understand the "threshold" at which a behavior change would occur, innovators must assess how important a new solution is likely to be to the buyers and what magnitude of savings would potentially motivate them to adopt it. Although it varies with the size of the market being addressed, a rule of thumb is that innovators will need to consider sharing from 30–50 percent of the savings with purchasers in order for a new solution to be compelling. Note that sometimes, if the savings opportunity allows a facility to redeploy assets (e.g., beds) on more profitable services, then this guideline may be somewhat more flexible.

Evaluate market dynamics

After sizing each market segment, it is important to spend more time understanding the market dynamics within them. For many markets that are characterized by the presence of active competitors, this means figuring who they are, how they compete with one another, and what are their strengths and weaknesses. Innovators sometimes assume that the current market situation will remain static during the many years required to develop and introduce a new solution. Focusing specifically on market dynamics helps innovators anticipate how key market opportunities may change over time, identify possible hurdles, and project the likely responses of competitors in the space. For example, with the COPD example, it would seem that a new solution focused on preventing readmissions would be attractive to payers but controversial to hospitals operating under a fee-for-service **reimbursement** model in that it would potentially reduce their revenue. However, if these hospitals were facing a reimbursement penalty for patients rehospitalized within 30 days, they would be dramatically more receptivite to the new offering. Such factors can only be uncovered and understood through a thorough assessment of the current and expected market landscape. As Winlo of the COPD team described, impending changes made the needs to prevent COPD-related rehospitalization significantly more attractive. "Readmissions for COPD exacerbations were set to be penalized, so that gave hospitals a large incentive to do something proactive," he said. "Another important trend was the shifting of medical care reimbursement towards value-based **payment** models such as accountable care organizations (**ACOs**) that were just beginning to emerge at the time. These organizations would be 100 percent on-the-line for caring and for treating high-risk COPD patients and so they are more motivated to pay for solutions helping prevent unnecessary hospitalizations."

By assessing market and competitive dynamics, innovators can also strengthen their understanding of: (1) how much value (in terms of improved performance and cost) may be necessary to differentiate a new solution; and (2) how much market share they may realistically capture in the early years following a product's launch. These aspects of market analysis are the source of the two most common and costly mistakes that innovators make. It is critical that innovators distinguish between opportunities to create simple, incremental *improvements* and those necessary to create *value sufficient to change customer behavior*. Virtually any product or service can be improved by addressing its specific shortcomings. Incremental improvement opportunities are plentiful. Yet once a stakeholder's basic needs are met, they tend to be resistant to upgrading or accepting new features that incrementally enhance the performance of an offering until those improvements achieve a threshold of innovation meaningful enough to motivate a behavior change. Most solutions fail because the innovators underestimate this threshold within their target market. Innovators can gain insight into what magnitude of improvement may be needed to capture the attention of prospective customers by examining the history of innovation within the segment of interest. Direct customer feedback can also be used to help validate the importance of the need and magnitude of the desire for new solutions. However, as always, innovators should remain as objective as possible and be alert to the **biases** that this feedback can contain.

Innovators may find at least two well-established frameworks helpful for assessing competitors and other dynamics in a market. The first is **Porter's Five Forces**, which can be used to evaluate the overall competitive landscape in a field. The second is **SWOT analysis**, which is focused on understanding the dynamics surrounding individual competitors. This section briefly describes how both of these approaches can be used to evaluate market dynamics. It also includes a Working Example that revisits the InnerPulse story (introduced in 2.3 Stakeholder Analysis) to demonstrate the importance of considering market dynamics.

Porter's Five Forces   Porter's Five Forces framework, developed by Michael Porter of Harvard Business School, identifies the five primary forces that drive competition within an industry:[11]

- The **threat of new competitors** to the market (and the barriers that prevent them from entering).
- The **bargaining power of suppliers** (as measured by the number of suppliers in a market and the costs of switching from one to another).
- The **bargaining power of buyers** (in terms of the extent to which buyers can directly affect profitable sales).
- Pressure from **substitute products or services** (which necessitates differentiation).
- The **intensity of rivalry** among existing competitors.

These five forces interact with one another to shape the profit potential of an industry (and the primary firms within it).[12] By investigating each factor, and how they are interrelated, innovators can objectively characterize the market, the potential for profitability and/or sustainability, and the critical issues that must be managed. The

most common mistakes innovators make when doing this analysis include underestimating competitive R&D pipelines, failing to account for all substitute solutions (including the customer's option to do nothing), and ignoring the extent to which **switching cost** can influence buyer behaviors.

For optimal results, the Five Forces framework should be applied within a single specialty or therapeutic field. An overview of the framework and a **medtech**-specific example is available in online Appendix 2.4.1

SWOT analysis   SWOT analysis, which is credited to Albert Humphrey of Stanford University, focuses on the internal and external factors affecting a project or company (including strengths, weaknesses, opportunities, and threats).[13] Using SWOT analysis to understand what is occurring within the competitive environment requires an innovator to identify the primary competitors in a market and assess each one individually. In doing so, the first step is to identify each competitor's primary objective. Then, the following factors can be considered to determine the relative power of their competitive position:[14]

- **Strengths** – *Internal* attributes of the organization that are helpful in achieving the objective. Identify the company's competitive advantage and what differentiates its products/services.
- **Weaknesses** – *Internal* attributes of the organization that are harmful in achieving the objective. Consider what barriers or hurdles the company is facing.
- **Opportunities** – *External* conditions that are helpful in achieving the objective. In the medtech field, these might include intellectual property (IP) and regulatory issues, partnerships, stakeholder satisfaction, and other economic, social, and technological factors.
- **Threats** – *External* conditions that are harmful in achieving the objective. Explore the same factors listed above, in addition to the activities of the company's direct and indirect competitors.

An assessment of competitors' strengths and weaknesses helps an innovator understand the capabilities and limitations of other players in the market. Strengths represent capabilities that must be overcome (e.g., barriers to entry and/or attributes against which a company must differentiate itself). Weaknesses represent limitations that have the potential to derail a competitor's success. Innovators must be careful to avoid or mitigate the same risks within their own projects (and may be able to exploit the weaknesses of competitors by turning them into opportunities). More information and a sample medtech SWOT analysis can be found in online Appendix 2.4.2.

After performing these types of analyses for the most important competitors within each market segment, innovators can summarize the resulting data and look for trends and issues across products and the organizations that market them. At a minimum, such a summary should include the following information:

- Number of established and/or emerging companies working within the market segment.
- Specific products or services offered.
- Maturity of products or services offered.
- Actual or projected pricing.
- Market penetration.
- Estimated market share.
- Revenue.
- Actual or projected profitability.

Another important factor to consider as part of market dynamics is investor interest in the space. Because almost all medtech innovations require significant funding to reach commercialization, it is essential to understand how compelling a need might be to the providers of capital, whether these "investors" are traditional venture capitalists, the executives responsible for funding projects within large, existing organizations, or foundations that back technologies targeted at addressing the needs of underserved **user** populations. Unless these providers of capital can be convinced that they will realize value from investing in a solution, then the project will go unfunded and the customer need will go unaddressed. In the simplest case, a target market will already have sufficient investor interest that the innovators can feel optimistic about raising the money necessary to fund future development. If so, the challenge for the innovator will not be to persuade the investors that the market has commercially attractive needs (which can sometimes be a difficult challenge if, for example, the team is the first to identify a need). Instead, it will be convincing them that the need identified and the **concept** that is eventually chosen to address it are unique and can be differentiated from existing and future competitive offerings. Ultimately, an investor's question is similar to the customer's: "If I invest in this, how likely am I to get more out of it than I'm required to put in?" Signs that investor funding is available include the creation of investor-funded new businesses in the space, as well as merger and acquisition activity (which can provide a path to profitable liquidity for investors and drive new investments – see 6.3 Funding Approaches).

Understand segment needs

The next challenge is for innovators to understand how the needs of stakeholders from segment to segment may vary and how well those needs are currently addressed by existing solutions. This assessment is closely linked to evaluating segment willingness to pay (as described in the next section). Ultimately, the extent to which the stakeholders in a segment are potentially motivated to adopt a new solution and how much they are presumably willing to pay come together to provide a "*value estimate*" associated with a need. A value estimate helps innovators continued to expand their understanding of how much and what type of improvements a new solution must deliver, at or below a defined cost threshold, in order to have a reasonable likelihood of being adopted. Quantifying value in this stage of the process is difficult since no specific solutions have yet been defined to address the need. However, innovators can develop a directional value estimate for a need area by considering who the real decision makers are with respect to adopting any new solution, how significant they perceive the need to be, to what degree available solutions are effectively addressing their needs, and therefore how much margin there is to offer a new technology with a different improvement/cost equation.

As described in 2.3 Stakeholder Analysis, identifying the decision maker(s) involved in a need area is imperative. Within each market segment, innovators should be sure they understand who makes adoption decisions and how they are typically made. In light of this information, the team can then carefully consider what it has learned about each market segment to discern in what ways the

Working Example
InnerPulse revisited

Porter's Five Forces can be used to examine the decision made by InnerPulse (introduced in 2.3 Stakeholder Analysis) to develop a less invasive implantable cardioverter defibrillator (ICD) for patients at risk of experiencing sudden cardiac death (SCD). In 2007, when InnerPulse was getting started, the existing competitors in the ICD space were Boston Scientific, Medtronic, and St. Jude Medical (Boston Scientific entered the ICD market after its **acquisition** of Guidant in 2006, following a bidding war with Johnson & Johnson). These firms traditionally enjoyed a comfortable oligopoly because of significant barriers to entry, which include the high capital requirements to develop and get approval for new ICD devices, as well as their strong IP positions. Suppliers of components for the industry used to have some power, but it was diminishing as certain manufacturers were independently developing components that they previously would have had to purchase from suppliers. Buyers (payers, such as private insurance companies and Medicare) had significant power, but it had been weakened through clinical and patient advocacy efforts supporting the need for ICDs. There was also some evidence of competitive rivalry (e.g., litigation around IP and the bidding war between Boston Scientific and Johnson & Johnson for the acquisition of Guidant). However, this rivalry could primarily be observed among new entrants, rather than between established firms. Further, alternative or substitute solutions were not available to treat patients who had suffered from secondary prevention or were at risk of primary prevention (i.e., no drugs had been shown to be as effective at reducing the risk of SCD as an ICD). Importantly for InnerPulse, unlike the smaller secondary market, which was well penetrated, significant growth opportunities existed in the larger primary prevention market since the overall market penetration was low (approximately 20 percent). Stated another way, only one in five patients at risk for secondary prevention received an ICD. This was primarily due to the inability of primary prevention patients to be seen by the physicians who implanted ICDs.

This quick analysis suggests that the incumbents are highly profitable (in fact, gross profit margins for ICDs are in excess of 80 percent) and that they will protect their market aggressively. InnerPulse's strategy to focus on the under-penetrated primary prevention market (in which it could fill the major void in the total ICD market as opposed to trying to "steal" market share from incumbents) showed that it recognized the challenges of direct head-to-head competition with the incumbents. By taking this approach, InnerPulse could maximize its likelihood of differentiating its own product from products of the established companies as it entered the market. Because the incumbents were undoubtedly looking into the primary prevention market as a growth opportunity, InnerPulse could eventually become an acquisition target (see 6.3 Funding Approaches).

When possible, innovators who envision building a start-up company may want to initially avoid segments where they will face direct, head-to-head competition with strong entrenched players. If they can focus instead on untapped or under-penetrated market segments with a reasonable market size and need, they can potentially differentiate their solutions more effectively from the products of the incumbents. However, this guidance is never absolute. Recalling Stack's comments in the InnerPulse example in chapter 2.3, sometimes, by examining industry structure in this way and identifying specific gaps in the product offerings of the entrenched competitors, innovators can position themselves for early acquisition. This scenario was played out multiple times in the evolution of the markets for balloon angioplasty and stents, and at the time of this writing was proving to be the case in the early days of renal denervation.

needs or improvements being sought by the decision makers within each subgroup are distinct. Remember, the goal is not necessarily to find the largest segment with at least some interest in the need, but rather the (usually smaller) segment with a strongest possible interest in the need. Refer back to notes from observations and other forms of research gathered through the biodesign innovation process to date to see if they shed light on unique problems, preferences, or other requirements that exist within the segments and how they affect the need. In some cases, it may be necessary to perform additional observations or interviews.

Next, seek to understand how well the needs of key decision makers within each subgroup are currently being met. As part of the broad market landscape, innovators performed this type of gap analysis for the total market. Now they can get more detailed and specific by looking at each market segment. To accomplish this, they can begin by evaluating outcomes for patients within each segment. For example, what is the prognosis for patients over the age of 65 with paroxysmal atrial fibrillation originating in the pulmonary veins? What outcomes should they expect if their condition goes untreated?

Once this information is understood, evaluate how expected outcomes change as a result of the existing solutions that represent the **standard of care** within the market segment. Using the same example, to what extent does the prognosis change when patients over the age of 65 with paroxysmal atrial fibrillation originating in the pulmonary veins are treated by a cardiologist using a minimally invasive procedure such as pulmonary vein ablation (or another relevant treatment)? Be sure to take note of the improvement in outcome, as well as any complications or new risk factors introduced by the treatment, as these can have a significant effect on the overall level of patient satisfaction with the intervention, as well as the satisfaction of the physicians delivering it.

Determining how well existing solutions meet patient needs, along with the needs of physicians, requires some qualitative analysis. There are no "rules" that define how great an improvement in outcome a treatment must provide to satisfy their needs. Ideally, each treatment would provide a long-term "cure" or prevent the consequences of the condition it was developed to address. However, in the absence of a complete cure, innovators must apply their judgment to evaluate how effective existing treatments are in improving patient outcomes *and* the extent to which new complications and risks associated with each treatment are considered an acceptable trade-off by patients and physicians alike.

These factors, in combination, determine how receptive decision makers are likely to be to available solutions. By extension, their level of satisfaction with existing treatments typically corresponds to the level of

perceived need for new treatment alternatives within a market segment. If the level of satisfaction is relatively high, it does not necessarily mean that a new treatment is not needed. It does, however, mean that any new treatment option introduced into the market will have to perform significantly better than available treatments (in terms of improved outcomes and diminished risks/complications) in order to be adopted, or it will have to deliver comparable results at a significantly lower cost.

Of course, the needs of other stakeholders, such as facility purchasing managers and public and private payers, must also be taken into account. The needs of these individuals and groups are often cost-related, but they are also interested in other outcomes such as improving quality and increasing efficiencies. The important thing to keep in mind is that if a representative from a payer or purchasing organization is the primary decision maker for adopting a new solution, then the new solution must offer strong economic benefit along with its other advantages.

In addition to looking at individual treatment alternatives, innovators should review the competitive analysis that they performed when exploring the dynamics within each market segment. The results of this analysis can help innovators identify broad unmet market needs and gaps in competitor offerings that should be taken into account when evaluating overall market needs.

Competitive analysis can also provide innovators with examples of new solutions that were perceived as attractive enough by the key stakeholders in a market segment to cause them to change their behavior and adopt a new approach. Benchmark these examples and evaluate them for insights regarding the extent of the improvements offered by new solutions that achieved significant market penetration.

Importantly, not every segment will be equally receptive to new solutions. In some medical areas, new treatments are introduced and embraced frequently; in others, the same treatment paradigms have been in use for decades. Similarly, certain patient, provider, and payer groups have characteristics that make them more or less willing to adopt new solutions. Do not underestimate the importance of this receptivity and keep in mind that the benefits will have to be significant to get certain segments to change behavior.

Table 2.4.3 Patients, providers, facilities, payers can all affect willingness to pay for a new medical innovation.

Decision makers	Common issues affecting their willingness to pay

Payers	• Is the need being addressed sufficiently important when viewed from the perspective of the payer?
	  ○ Measured against other opportunities, what is the magnitude of the benefit that a new solution to the need must create?
	  ○ What is the total cost of care for patients in each segment with existing solutions, and how might a new solution affect the total cost of care?
	  ○ To what extent must a new solution reduce related expenditures for a defined population (e.g., through prevention and/or eliminating complications)?
	• Is a new solution likely to be covered by existing reimbursement codes or will it require a new code?

Patients	• Is there a precedent for patients paying **out-of-pocket** for solutions to the need? • If so, how much have patients been willing to pay for relevant solutions?

Physicians	• Will the physician be impacted economically by a new solution?
	  ○ Does a new solution have the potential to increase their productivity?
	  ○ What would make physicians potentially willing to cover the cost of a new solution from the reimbursement they receive for the procedure?
	• What clinical benefits would a new solution have to offer relative to existing solutions to get physicians interested? How much have physicians paid for solutions in the past that have provided similar benefits?

Facilities	• How is adoption of a new solution likely to positively or negatively impact overall facility economics?
	  ○ To what extent might a new solution reduce costs for the facility?
	  ○ Could a new solution provide other market benefits for the facility, such as the opportunity to grow its market penetration?
	  ○ Is the profitability of the procedure more or less than a new procedure that might replace it?
	• To what extent could a new solution enable the facility to achieve key quality goals?

Assess willingness to pay

Once market needs are understood, innovators can consider how much the members of a market segment are potentially willing to pay for a new solution, and then bring these two factors together into a value estimate. Note that understanding willingness to pay is not the same as asking how much the segment is currently paying for an existing solution since new solutions may be held to a different standard. In today's value-oriented environment, new solutions are increasingly required to offer substantially better performance or a meaningfully lower cost – or sometimes both – to drive significant adoption.

Like market needs, willingness to pay depends to a large degree on which stakeholders are the decision makers for adopting a new solution – who will be making the payments and how will the new innovation affect them? Remember that some decision makers will be attracted to the improvements offered by a new solution but lack the budget to act on their interest. This scenario is becoming more common as facility purchasing managers become more powerful in making buying decisions at the same time the spending authority of individual physicians wanes. Table 2.4.3 outlines some of the main stakeholders who may act as decision makers and the issues to consider when anticipating their response to a new device technology.

To evaluate and understand the answers to these questions, innovators can use one of the following approaches. Whenever possible, experiment with multiple methods to triangulate the information collected. Remember that this willingness to pay analysis will be "directional" (and, admittedly, somewhat speculative) since a specific solution has not yet been defined to address the need.

1. **Interviews and/or surveys** – Gathering information directly from the decision makers who would purchase a new solution can be extremely helpful. Surveys can be issued in a written format, but similar information is often gathered via interviews or other one-on-one interactions. The idea is to determine the maximum payment a decision maker would be willing to make for different degrees of innovation. A survey should provide a description of the improvement a potential solution might provide and a range of prices. Users can then be asked whether or not they would be willing to use a solution offering those specific improvements for each of the prices presented. Take care to be specific about the advantages a new solution will enable. Sometimes a useful question to ask is, "At what price would you stop being interested in purchasing the product's described benefits?" Customers should be strongly enthusiastic about the prospect of a new solution and offer limited resistance to the possible changes the innovators describe for a new solution to have reasonable chance of adoption. Some stakeholders will not indicate much price sensitivity – especially if they do not actively participate in the purchasing decision. Be aware of the potential for this kind of positive bias and try hard to reach stakeholders that play active roles in relevant purchasing decisions.

2. **Comparables** – Evaluate the prices for technologies across medical fields that deliver benefits similar to those desired by the market segment, could potentially be adapted to address the need, or that are roughly aligned with the solution gap in the segment. Learn as much as possible about pricing for existing solutions within the same medical field, too. These comparables provide a rough benchmark for how high a price the market segment is likely to bear, although (again) new solutions may be held to a different standard. Remember, pricing can be dynamic – be sure to factor in any information that may show evidence of a negative overall trend,

keeping in mind that a new solution may take several years to get to market.

Once innovators have a reasonable sense of a segment's willingness to pay, they can consider this information in combination with the improvements a new solution must deliver to create a value estimate for the need. Remember, the value estimate is intended to provide guidance on how much and what type of improvements a new solution must deliver, at or below a defined cost threshold, in order to have a reasonable likelihood of being adopted. In 2.5 Needs Selection, these data will be translated into **need criteria** that become part of the need specification and guide the development of solutions. Typically, requirements defined by the value estimate translate into "must-have" need criteria – meaning that they must be satisfied by any solution that the team decides to bring forward into invention.

The COPD team did not go deep into creating a value estimate, but background research gave the members a directional sense of the improvement they would need to deliver, and at what cost, in devising a new solution. As Winlo explained, "One of our mentors at the time, who was COO of a major healthcare organization, was very motivated to do something about COPD. His organization had seen some early success in home outreach – hiring nurses to make outbound calls to COPD patients who had recently been discharged to see how they were doing. But these nurses could only reach so many patients in any given day." This care management team would semi-randomly contact patients and ask them a series of questions about their condition. Periodically, they would catch a patient as s/he was experiencing an exacerbation. By swiftly intervening to manage the symptoms before they worsened and get the patient into see a doctor, the nurses were able to prevent some hospital readmissions. "They were literally catching exacerbations by luck," Winlo said. "We figured there had to be a better way." On the other hand, the team was encouraged by the fact that early intervention was effectively preventing exacerbations from progressing to the need for rehospitalization. They also saw the potential for a new solution to improve upon this model. "There's an infrastructure that's already in place that has shown some positive results," he noted,

"How can we make it more effective?" Analysis revealed that it cost healthcare payers approximately $400 per patient for a care management team to follow up periodically during the 30-day post-discharge period. Limited information was available, but a couple of studies indicated that follow up from care management teams could reduce hospital readmissions by up to 50 percent within the 30-day post-discharge period. Based on these directional inputs, the team estimated that it would either have to come up with a solution that more systematically intercepted a substantially greater number of exacerbations during the 30-day follow up (i.e., reduce readmissions by 75 percent) or make the care management team members significantly more productive (i.e., enable them to manage a greater number of patients while achieving comparable results) to drive down the per-patient cost of follow up. In a third scenario, the team could seek to devise a solution that would prevent more readmissions *and* make the care management teams more productive to reduce the per-patient cost of the program. This approach would, of course, be most compelling to the purchasers of the new solution.

Step 3 – Choose the target market

Choosing a target market takes into account all of the information uncovered throughout the market analysis process to focus on the segment with the most promising combination of size, dynamics, needs, and willingness to pay. The ideal target market is the market segment that offers the innovators the greatest opportunity to create value – first, for potential customers, then for prospective investors, and finally for themselves. Clearly the emphasis should be placed on assessing how favorably customers – or purchasers – respond to the possibility of a new solution in the need area. However, unless the innovators are motivated to pursue the opportunity, and the providers of capital are motivated to fund it, then the market will not be commercially viable (see Figure 2.4.12).

Innovators should be especially careful not to mistake the enthusiasm of their potential customers for the

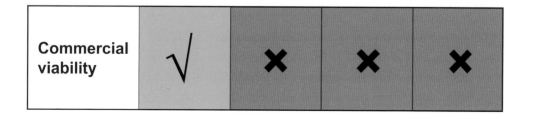

	Segment 1	Segment 2	Segment 3	Segment 4
Customers (stakeholders making the purchasing decision)	+	−	+	+
Investors (providers of capital)	+	+	−	+
Innovator (providers of labor and intellectual capital)	+	+	+	−
Commercial viability	√	✗	✗	✗

FIGURE 2.4.12

If addressing a need for a specific market segment does not have the potential to create value for any one of these key participants – customers, investors, and the innovators – then it is not likely to be commercially viable. Note that some ideas that address an important need may not support the development of a commercial business, but instead may have value in a non-profit context.

Table 2.4.4 Choosing a target market can be challenging, but the following guidelines can help.

	Risks to consider	**Questions to answer**	**Common innovator errors**
Customers (stakeholder making the purchasing decision)	• Value of a solution is too small to change behavior	• Is the need sufficiently important to the customer? • Is gain sufficient to motivate change in behavior?	• Assume improvement is sufficient to change behavior • Assume that a "need" equates to an "opportunity" • Overestimate the role price plays in purchase decision • Underestimate impact of switching costs Unrealistic timeline for adoption
Investors (providers of capital)	• Market is too small to justify effort and investment • Takes too long to develop and commercialize • Competitors may improve their products	• Is the potential market at least 3–4 times the cost of developing a solution? • Are there ways to mitigate risk during development by shortening timelines? • Will a new solution still be competitive when it reaches the market?	• Underestimate competitors ability to improve product offerings over time • Project unrealistic market share
Innovators (providers of labor and intellectual capital)	• Need is too weak to justify time necessary to develop, fund, and introduce a new solution	• How long will a solution take to develop? • What is the **opportunity cost**? • When and how do innovators tell when they have succeeded or failed in the space?	• Losing objectivity • Misreading customer encouragement to mean market validation • Giving up too early • Giving up too late

viability of the market segment. Other risks and pitfalls are outlined in Table 2.4.4. At this stage of the biodesign innovation process, it is difficult to know with certainty the extent of the value that can be created for these participants within any particular market segment. The idea is to objectively consider how the segments compare and then choose the one as a target that seems to hold the greatest overall promise.

Keep in mind that sometimes the greatest opportunities can be asymmetrical – that is, they may be exceptionally strong in one particular dimension while significantly weaker in others. Some innovators create quantitative ratings systems to help them score and prioritize segments when the choice of a target market is not clear; others rely purely on a qualitative assessment. With either approach, they must exercise judgment in making the final determination. Deciding on the best available target market requires innovators to make a series of assumption about decision maker and stakeholder behavior based on limited information. Benchmarking, proxy analysis, and engaging experts are all tools that can be helpful in developing credible assumptions. Make a best effort to gather information and then make an educated choice to avoid becoming paralyzed by uncertainty. Innovators should keep in mind that they are using estimates as an input to screening and, at this early stage, there is still time for doing significant validation work to confirm their assumptions as the biodesign innovation process moves forward.

The rapid development of mobile health (**mHealth**) applications targeted at healthcare consumers provides a good example of how innovators can validate their assumptions. Because the development cost for these technologies is low and the time to market can be rapid compared to traditional medical devices, many innovators are motivated to work in this space. In doing so, they can make use of this ready access to the market to validate their product's features focusing on a concept called the Minimum Viable Product (MVP).[15] The idea is that getting traction with an initial set of customers – at the lowest possible cost to investors and the entrepreneurs – is the best way to prove a need and iterate a product offering, while also refining the target market segment. However, innovators should note that these exceptionally low barriers to entry come with their disadvantages in other aspects of market selection. The competitive landscape in these markets is exceedingly complex and rapidly changing, with roughly 100,000 healthcare apps competing for the attention of the consumer in 2013,[16] and that number expected to increase by 25 percent per year for the foreseeable future.[17] As a result, marketing and promotion have become more important at the same time that

competition has made funding difficult to obtain. Whether innovators undertake rapid development of a mobile app or the longer-term creation of a new implant therapy, the key is to fully understand the characteristics of the market they are targeting through ongoing information gathering and then the validation of specific assumptions.

Admittedly, the process of performing detailed market analysis and identifying a target market involves collecting an immense amount of information and synthesizing it into a coherent story that supports the chosen target market, as shown in the Working Example for chronic kidney disease. The results of a preliminary market analysis also enable innovators to refine their need statements (and ultimately develop a need specification) to appropriately address the needs of the target market. Eventually, this output is also used as the foundation for the development of detailed marketing and sales plans (see chapters 5.7 and 5.8), as well as to support the company's **financial model** and funding strategy (see chapters 6.1 and 6.3).

Working Example
Market analysis for chronic kidney disease

Patients with CKD are often diagnosed late in the disease's progression and are frequently hospitalized. However, early and effective diagnosis of CKD reduces the incidence of hospitalization and also diminishes the rate of disease progression. As a result, there is a need for a device, method, or system that would successfully diagnose patients with CKD earlier in the **cycle of care** to help prevent disease progression.

After landscaping the market, innovators can dive into a patient-based market segmentation, using markers of disease progression to identify basic similarities and differences among patient populations. Recall that, for CKD, the marker is a measure of kidney function which can be used to divide patients into five CKD stages: 1–2 = mild, 3 = moderate, 4 = severe, and 5 = ESRD (end-stage renal disease). (Note that provider and payer considerations will be taken into account later in this example.)

Once these basic segments are defined, primary and secondary research can be conducted to identify the size

of each segment (number of patients), the extent of the medical need in each segment (mortality and morbidity rates, expenses per patient), and segment growth rates. Literature searches can be used as a starting point, but beyond that, public databases can provide relevant information. Two such databases that are widely used are the National Health and Nutrition Examination Survey (NHANES)[18] and the Medical Expenditure Panel Survey (MEPS).[19] These surveys provide **longitudinal data** for large samples of patients regarding medical condition(s) and medical expenditures for families and individuals, and they provide a wealth of information that can be used for any preliminary market analysis. Using data from sources such as these, a bottom-up analysis of the market size for the various segments can be performed.

The NHANES database provides information that innovators can use to calculate the prevalence of each stage of CKD, as well as mortality rates, morbidity rates, and hospitalization expenditures. Figure 2.4.13 presents the number of patients in four segments and the corresponding rates of hospitalization.

These data are presented in a "patient tower" format in which the total patient population is divided in segments

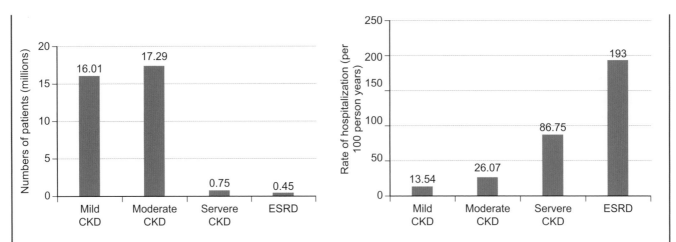

FIGURE 2.4.13

Patient towers for CKD: (a) number of patients by segment and (b) hospitalization rate by segment.

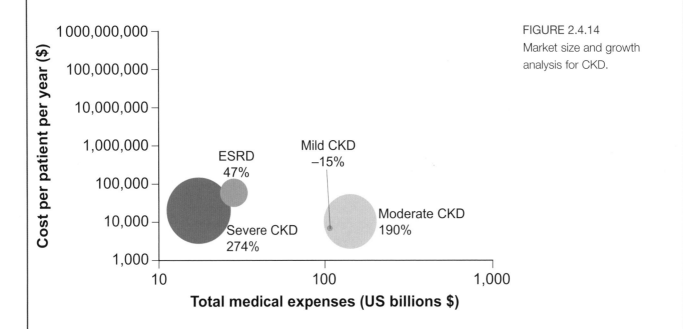

FIGURE 2.4.14
Market size and growth analysis for CKD.

of diminishing size, but increasing medical needs. Reading from the left of the diagrams (mild CKD) to the right (ESRD), the number of patients typically decreases but the severity of the condition increases. While other visual methods can be used to present these data, the patient tower approach has the advantage of visually contrasting the trade-off between the number of patients in each segment and the severity of the medical issues affecting each segment.

To identify a target market, the innovator also needs to get a sense for the cost associated with each stage of the disease. Sources such as NHANES do not provide cost information (even though they provide information on the utilization of medical services). Therefore, some data extrapolation using numbers obtained from a literature search may be required. Another potential strategy is to use data from MEPS, which typically includes expenditure information or data from disease-specific data sources. Figure 2.4.14 summarizes the data (the horizontal axis gives the total medical expenses for each segment, the vertical axis gives the cost per patient in each segment, and the size of the bubble indicates the expected growth rate).

The data in the figure were based on expenditure data provided in the US Renal Data System report (the annual medical cost per prevalent case of severe CKD was

$20,784 and for ESRD patients it was $59,412).[20] To obtain cost estimates for the other two segments (where no data were reported), an assumption can be made. In this case, the CKD cost for earlier stages was scaled by taking the cost for stage 4 and multiplying the ratio of the hospitalization rate in each stage, divided by the hospitalization rate for beyond stage 4 [cost for stage 1 (or 2) = cost of stage 4 × hospitalization rate in stage 1 (or 2)/ hospitalization in stage 4]. Using this approach, the total annual medical cost for each stage was estimated; then the growth in the prevalent population (estimated from NHANES) was used to obtain the growth in total costs for each stage.

With this information, innovators can define the characteristics of the ideal market segment: large total medical expense (all segments satisfy this), high expenses per patient (which makes the total number of patients relatively small and easier to access), and a high growth rate. Using these characteristics as a filter, the ESRD and severe CKD segments emerge as most attractive.

However, before the analysis is complete, the innovator needs to study whether all patients who have the condition are actually aware of it (if not, they are unlikely to seek treatment, which will negatively affect the total potential market size). In the example, data provided via NHANES included a response to a survey question that asked patients whether or not they were aware of diminished kidney function. This survey showed that only 25 percent of the patients with serious CKD were aware of it, and all those who were aware of it had other symptoms that helped make the diagnosis (e.g., elevated albumin to creatinine ratio, ACR). This information suggests that patients who are correctly diagnosed are often patients with high ACR. As a result, the target market could be defined in even more specific terms: patients with severe CKD and high ACR.

To some, this definition of the target market may appear to be surprising (after all, these are the patients who are already diagnosed). However, what makes this patient segment attractive is that their disease is already recognized by the healthcare system and, therefore, this segment can be accessed by the innovator. For this reason, addressing the need for better diagnosis and/or control of complications for these patients is more likely to be clinically feasible than targeting patients who are not currently diagnosed. The latter segment (of undiagnosed patients) could potentially provide an innovator with an opportunity for expansion, once the preliminary target market has been addressed, but may not be as easy to target.

A word of caution

The most common mistake in market analysis is to make assumptions that lead to overly optimistic market estimates. Overestimating a market opportunity can give innovators unrealistic expectations and result in a business plan that is unachievable and/or unsustainable. It can also hurt their credibility with investors and other stakeholders if they present an idea that is based on overly aggressive estimates and/or impractical assumptions.

To avoid this common pitfall, watch out for these four sources of error:

1. Calculating market size based on the total market, not the market segment(s) most likely to adopt an innovation. This error can be driven by failure to understand the needs of specific segments and their different adoption patterns.

2. Underestimating how much of an improvement is necessary for early adopters to change behavior. A strong value estimate is needed to understand how much benefit must be delivered, at what cost, to drive adoption.

3. Failing to recognize that not all people with a need will take steps to address it (e.g., seek treatment). This could be the consequence of relying on "sanitized" data obtained from well-controlled studies. When obtaining estimates for number of patients in a treatment area, it is important to confirm that the estimates reflect community practices and not well-controlled academic environments.

4. Overestimating the amount of market share that a new entrant can capture in an established market or the rate at which a new innovation will be adopted in

an emerging market. This error is often driven by the tendency to underestimate: (1) the effect of competition; (2) the capital requirements needed to sell a product; and (3) the time required to establish a sales or distribution capability. In most cases, innovators should not expect to gain more than 1 percent market share in the first year, increasing to a maximum of 15–30 percent based on market conditions.

Remember that market analysis can be complicated and difficult, potentially requiring a certain amount of trial and error. An iterative, increasingly detailed approach to market analysis is recommended. Innovators should be sure to revisit and adjust their preliminary market analysis as more becomes known about the need, as well as potential solutions. The story of Genomic Health demonstrates how one company tackled the challenge of market analysis.

FROM THE FIELD ▸ GENOMIC HEALTH INC.

Market analysis for a revolutionary product

Identifying the target market segment and sizing the overall market can be especially challenging for innovators considering a need with ground-breaking potential, but no proven market. Genomic Health Inc., founded by Randy Scott, Joffre Baker, and Steve Shak, is one company that faced this challenge.

In his previous role as the co-founder of Incyte Corporation, Scott and his team had made massive databases of genomic information available to major pharmaceutical companies to aid their research and development efforts in identifying targeted therapies at the molecular level. However, as the cost of genomic information came down (due to what Scott called a "Moore's Law effect" in biotechnology), he recognized that it would become possible to analyze genomic information on a patient-by-patient basis and develop truly personalized regimes to treat disease. Passionate about helping make this happen, he set out to pursue *the need for high-value, information-rich diagnostics based on patient-level (gene expression) genomic testing to enable more personalized treatment decisions*.

Scott initially pitched the idea to Incyte, but the need was met with mixed reviews. "One of my more controversial views with Incyte at the time," recalled Scott, "was that drugs would ultimately be commodities, and that greater value would come from genomic information about

disease. I believed that there was more power in the information than in the solution because what's more important than understanding exactly, precisely what the molecular cause of a disease is? Once you have that information, there will be 10 potential therapies developed to treat it. These drugs will ultimately become generic, but the value in the diagnosis of that information will hold." Scott positioned this as a paradigm shift in the healthcare industry that would occur gradually over the next 30 years. When Incyte decided it was not interested in taking this direction, Scott decided to pursue the idea himself, recruiting Baker and Shak to work with him.

In terms of analyzing the market opportunity related to the need, Scott remembered that they had an initial interest in focusing on cancer patients. However, with so many other life-threatening conditions affecting relatively sophisticated patients who would understand the benefits of personalized, genomic-based medicine, more in-depth research of the potential market segments was required. To narrow the scope of their segmentation analysis, they identified several fields (such as oncology, inflammation, cardiovascular, infertility) that would potentially benefit from personalized medicine and ranked them according to several clinical and market criteria, including market size; potential for genomic information to predict disease progression; drug development pipelines (and whether the percentage of responders for the new drug would be high); patient involvement in treatment choices; and physician willingness to adopt new treatment paradigms. After performing this analysis, "The bottom line," said Scott "was that we just kept coming back to cancer." Unlike other diseases, such as cardiovascular, where lifestyle factors played a big role in disease progression, cancer is mostly a genome-based disease. Further, response to existing drugs (or drugs under development) was considered to be variable and uncertain, and a strong history of patient advocacy showed that patients would drive the adoption of new technologies. Beyond that, the total market for oncology drugs was in the multibillions of dollars, with most drug manufacturers developing several new therapeutics.

While identifying cancer as a focus area was a good first step, it was only the beginning of their segmentation analysis. The next challenge was to determine which type of cancer to target. Ultimately, they decided to focus on breast cancer for five primary reasons: (1) *prior experience*: Shack led Genentech's clinical development program for Herceptin® (a novel, genomic-based cancer therapeutic[21]) and had a vast network of contacts, as well as a strong reputation in that clinical area; (2) *top four*: breast cancer is one of four most prevalent cancers (the others being colon, prostate, and lung cancer) and the team wanted to "do something big"; (3) *market accessibility*: breast cancer is characterized by a large system of patient advocacy groups, readily accessible education channels that make it easy for a new company to reach the market segment, and patients that tend to be highly involved in their treatment decisions; (4) *likelihood of adopting new technology*: both physicians and patients are likely to embrace new treatments in this clinical area; (5) *clinical knowledge*: there is a deep and broad body of knowledge on the genetic basis of breast cancer and a vast library of breast cancer tissue samples that make it technically feasible to develop and clinically validate any new test.

Within breast cancer, the founders focused on further refining the target market segment. The team examined the various drugs available to breast cancer patients, and their effectiveness or ineffectiveness at various stages of the disease. What they learned is that for late-stage

cancer patients, treatment choices are unambiguous and highly aggressive. Patients and physicians do not want to "give up hope" and they will try everything to stop or slow the disease. In contrast, decisions about what treatments to pursue (beyond surgery) for early-stage patients were relatively subjective and varied significantly from physician to physician, making it difficult for patients to determine the appropriate course of treatment to prevent recurrence. Most early-stage patients do well with only surgery and hormone therapy. However, chemotherapy is commonly prescribed to minimize a woman's chance of repeat tumors even though only a small fraction of early-stage breast cancer patients benefit from it (approximately 4 percent). This led to an "a-ha" moment for the team: predicting distant recurrence of breast cancer for early-stage patients would likely help them make better decisions about what treatment to pursue. For example, patients with a low risk of distant recurrence could potentially be counseled to forego chemotherapy, a physically stressful and disruptive therapy, since their cancer was unlikely to return. Taking the process one step further, based on what was known about the disease, the team defined its preliminary target market segment as patients with early stage, node-negative (N–), estrogen receptor positive (ER+) breast cancer. They also further refined their need statement based on the result of their market analysis to address requirements for *high-value, information-rich diagnostics based on patient-level (gene expression) genomic testing to predict the recurrence of early stage, N–, ER+ breast cancer and enable personalized treatment decisions*.

With the market segment identified and need refined, the next challenge was to determine the size of the segment in order to determine if the market potential would justify the high anticipated cost of clinical development and the anticipated risk to prospective investors. According to the company's estimates at the time, there were approximately 200,000 new breast cancer cases diagnosed in the US each year, of which roughly 50 percent were early-stage, N–, ER+ cases. Quantifying the total market size (in dollars) was challenging, though, because of the lack of comparable products available in any clinical field. Any solution that would address Genomic Health's defined need would be classified as a diagnostic test. However, Scott and team determined that diagnostic companies traditionally charge between $25 and $50 for their tests, commanding margins of just 5 to 10 percent. Using a $50 price for the new test, the total potential market size (assuming full penetration) would be just $5 million per year. The results of this analysis clearly illustrated that the success of the venture would necessitate a completely different pricing paradigm to support the heavy investment in research and development required to make genomic testing practical.

Two types of analysis suggested that a price in the range of $1,000 to $7,000 per test could be viable. The first was comparables analysis. Kim Popovits, COO of Genomic Health, recalled, "There was another diagnostic in the marketplace at that time, a genetic test that looked at the mutation of the BRAC-1 and -2 genes to assess a woman's hereditary risk of breast cancer." This test was priced around $3,000 and was on its way to being reimbursed on a relatively broad scale. The second was based on the value estimate for the need. Over time, Genomic Health's test had the potential to save money for the overall healthcare system and could, thus, shift the pricing power from therapeutics to diagnostics. Specifically, the total cost of chemotherapy for early-stage breast cancer patients was conservatively $15,000. If the test cut the number of patients undergoing chemotherapy by 50 percent (by predicting low recurrence risk), then the total savings to the healthcare system would be roughly $7,000 per patient. This meant that Genomic Health could command a price of up to $7,000 per test, making the total potential market for kits sold to the healthcare system $700 million.

In fact, this estimate brought Scott and team back to an earlier market estimate they developed when they raised their preliminary funding. Back then, Genomic Health's business plan was not as precise about the market segment and size (highlighting the iterative nature of

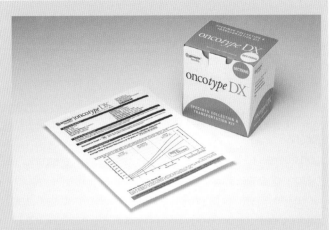

FIGURE 2.4.15

The Oncotype DX specimen collection and transportation kit, along with a results report (courtesy of Genomic Health).

market analysis as part of the biodesign innovation process). The original vision was that consumers would pay for genomic test results out-of-pocket. As a result, the team felt that a reasonable price would be somewhere in the four figures range (primarily because consumers pay analogous amounts for cosmetic surgery and other elective medical procedures funded out-of-pocket). Using similar projections regarding the total number of consumers willing to pay for the test, the founders reached a total market estimate in the billion dollar range and raised more than $30 million in investments. About a year later, they focused their market segment and further developed the company's value forecast.

The Genomic Health team went on to develop the Oncotype DX® test to predict the risk of distant recurrence for early-stage breast cancer patients (see Figure 2.4.15). Multiple clinical studies established that the test is not only effective in predicting the likelihood of recurrence, but also a patient's response to chemotherapy. Priced at $3,650, in line with the team's initial value forecast, the test is now included in the treatment guidelines set forth by the American Society of Clinical Oncology and the National Comprehensive Cancer Network.

Market analysis in emerging economies

The approach to market analysis described in this chapter can be effectively utilized in any geographic setting. However, innovators working in emerging economies are likely to face a much greater challenge in implementing it. This is particularly true when innovators seek to work in a location where they do not yet have a deep, personal knowledge of cultural, business, and healthcare customs. In these cases, fundamental differences in healthcare stakeholders, how healthcare is financed and delivered, and important variations in standards of care must be understood as a precursor to assessing the market. Similarly, significant attention should be paid to understanding the unique aspects of local competition, which can vary dramatically from what innovators may be used to seeing in the US or Europe.

Another important distinction about working in less developed markets is that reliable market data can be much less widely available. Research reports and formal data sets rarely exist, which can make learning about the market difficult, especially from afar. Even in areas where market research exists, the information can be inaccurate, dated, or misleading. Understanding how data has been collected can help innovators assess its relevancy and reliability. (For more information about taking the lack of available secondary data into account during needs screening, see 2.5 Needs Selection.)

More often than not, innovators have to perform their own market research in the space where their need exists. However, there are certain challenges to relying solely on primary data collection when assessing a market. Intentionally or not, stakeholders have a tendency to apply their own lens to a situation, especially when they have a vested interest in either maintaining the status quo or altering it. Innovators must consider the source for all information gathered and to remember that not everyone will necessarily want to be helpful.

Additionally, stakeholders may be tempted to provide innovators with the answer that they believe they want to hear, instead of what they really think or feel (although this can happen in any geography). This is especially true if a perceived power differential exists between the person being interviewed and the person conducting the interview. According to one innovator working on health solutions in Kenya, "You have to be really careful how you frame your questions and sometimes take people's answers with a grain of salt." One way to validate the information provided in interviews is to observe actual user behavior whenever possible. Even if no product or service exists to address a particular need, find a proxy offering against which to benchmark stakeholder actions. For example, when trying to understand willingness to pay, this innovator continued, "I find it far more valuable to see what people are actually buying. Look at what's really going on as opposed to what people report they would hypothetically spend their money on." John Anner, President of the East Meets West Foundation, elaborated on that point in an article: "Listening is tricky. Without multiple ways to verify what people really want, it can be hard to cross the barriers of language and culture to truly understand what people need and want (good or bad). This is why the best way to figure it out is to provide a range of options and let people vote with their own money."[22]

Another issue is that primary research takes time and often cannot effectively be conducted from a remote location. This can be a challenge for US-based innovators with an interest addressing global health needs in that it requires them to spend significant time in the field. As Krista Donaldson, CEO of the non-profit product design company D-Rev described:

You can spend a lot of time just getting to the people you want to talk to. That might mean a lengthy car trip on rural roads or several hours sitting in traffic. Then there's navigating the hospital to find the right person. And sometimes that means tea with the medical superintendent first. Not infrequently, you could spend all day waiting to talk to one person, particularly in busy hospitals where, rightly, we are a relatively low priority. Innovators often want to get in and get out – that's very Western treatment. But you just can't do that. This type of work takes a lot of patience and relationship building. And it also takes the ability to go with the flow and recognize that you may spend a week trying to visit hospitals and get nowhere, but you're still learning a lot. You just might not appreciate that until you're able to step back and think about it.

Donaldson also noted that it can be a challenge for innovators to know when they have enough information to stop and move on. While the answer varies from case to case, she offered this advice: "Once you start hearing the same thing over and over again, you know that you've done good research."

The following story, from a team that ultimately became Consure Medical, highlights some of the challenges involved in conducting market analysis in an emerging economy.

FROM THE FIELD ▸ CONSURE MEDICAL PVT. LTD.

Redefining a need statement through market analysis

Nish Chasmawala and Amit Sharma, both participants in the Stanford-India Biodesign Program, spent six months in California learning and applying the biodesign innovation process. Early in their Fellowship, one of their classmates went back to India to help care for his mother, who had become very ill. When that colleague returned to Stanford, Chasmawala and Sharma tried to distract him from his worries by asking if he had identified any needs while he was away. "And the need that he threw out on the table that day was fecal incontinence," Chasmawala recalled. "It's pretty embarrassing and nobody talks about it. But it's also a genuine need, and there didn't seem to be any good solutions for it."

The second half of the yearlong Fellowship took place in India, where Chasmawala and Sharma decided to further investigate opportunities and problems related to fecal incontinence (FI). Patients with FI are unable to control their bowel movements, which can result in the involuntary excretion of liquid, mucous, or solid stool from the rectum. Ranging in severity from minor soiling to a complete loss of bowel control, FI is a psychologically distressing condition that significantly affects **quality of life** both within and outside institutional settings.

For six weeks, the team conducted observations in a variety of medical settings, including a large tertiary care hospital, a new, state-of-the-art trauma center, and rural and community public health centers. While both men were born and raised in India, they were still surprised by what they saw when observing the cycle of care for FI. Absorbent pads (or adult diapers) were the standard of care for most FI patient. Within home settings, Sharma described a typical situation: "Imagine a 65–70 year old bed-bound male with fecal incontinence. In India, generally speaking, a female family member will provide the home care. So, you might have a daughter-in-law who has to change her father-in-law's diaper, clean the area, and so on, which is undignifying for both of them." In addition, he pointed out, "The diaper is not even a good solution." Diapers delivered mixed results. They prevented bed soiling, but caused skin breakdown. They also were uncomfortable to wear, and unpleasant and time-consuming to change. "If you look at the current gold standard, you realize that it is not at all 'gold'," Sharma noted.

In the hospital setting, the situation was equally as troubling. Overworked nurses often relied on family members to address the problem of FI, even when patients were under the hospital's care. Chasmawala explained, "In rural community hospitals in India, there are usually eight beds in a room with limited ventilation and inadequate waste disposal systems [see Figure 2.4.16]. Family members usually try to form a human shield to do an absorbent pad change and clean the patient." Summing up the observation experience, he said, "Spending time in these settings created a sense of pathos in us, but also motivated us to develop a solution for this unaddressed problem."[23] Based on what they saw, one of the need statements they defined was *for a better way to manage FI in resource-constrained environments*.

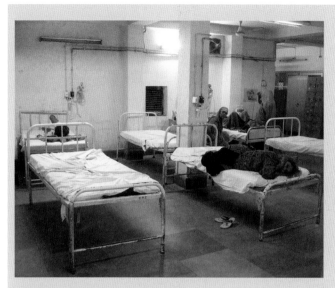

FIGURE 2.4.16
A typical rural community hospital in India, where patients experiencing fecal incontinence would be treated (courtesy of Consure Medical Pvt. Ltd.).

In parallel with disease research, an assessment of existing solutions, and stakeholder analysis, Chasmawala and Sharma dedicated themselves to understanding the market for FI. However, their search for information confirmed that conducting market research in India could be difficult, with reliable numbers hard to find. Sizing the market for FI was particularly challenging because, in most cases, it is a secondary condition or comorbidity to a primary disease state, such as stroke or other condition that damages the nervous system. This makes it somewhat more problematic to study. Prevalence studies of fecal incontinence in the general population are rare, and the results can differ significantly. As one literature review revealed, the estimated prevalence of fecal incontinence in selected studies varied from 0.4 to 18 percent.[24]

In terms of spending and competitive activity in the FI space, the team thought about the per-day treatment burden rather than the cost of competitive solutions themselves. Diapers were the most widely used product, and also the least expensive at roughly US$10 per day. Collection devices, which were slightly more effective in containing feces, caused more clinical complications and were available at a higher cost of roughly US$15 per day. In-dwelling catheters were significantly more expensive at US$60 per day. These devices could only be placed by a trained care provider and were typically used only in the intensive care unit (ICU) setting. According to Sharma, they were also not widely adopted because they were suboptimal in many ways: "Products that existed to address other needs have been scaled in their size and used for FI. For instance, most in-dwelling catheters are essentially a larger version of Foley's catheter [used for urinary incontinence]. Nobody has looked at the physiology of the rectum and designed an appropriate solution for fecal incontinence." For chronic fecal incontinence due to neuromuscular degradation, surgery was also an alternative. However, this treatment was exceedingly expensive for the majority of patients in India. In fact, the team believed that cost was a critical factor in each treatment category since most patients in India covered their medical expenses out of pocket. Taking all of this information into account, Chasmawala and Sharma determined that there was a significant gap in the market landscape and they had significant room to add value (see Figure 2.4.17). Their primary focus would be on seeking to displace the adult diaper, FI collection devices, and in-dwelling catheters with a new solution.

To segment the market, they started by evaluating a variety of different subgroups within the total FI patient population. Specifically, they explored patient characteristics across five different dimensions: (1) acute versus chronic condition; (2) pediatric versus adult; (3) ambulatory versus non-ambulatory; (4) major versus minor incontinence; and (5) hospital versus home setting. This assessment revealed important insights about the need. For instance, in patients with acute FI (e.g., linked to a stroke), the function of the bowels is temporarily compromised. In contrast, patients with chronic FI have experienced neuromuscular degradation or other changes to the apparatus of the bowels from which they will not recover. Patients with acute FI, who receive care in a hospital setting, have access to skilled healthcare providers capable of assisting with treatment (beyond

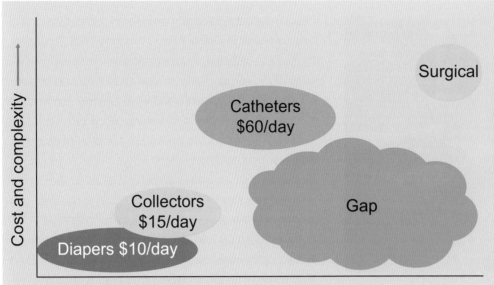

absorbent pads). However, patients with chronic FI are forced to manage this condition over the long term at home and must rely on family members without any medical training to assist them.

From this analysis, the team defined two primary market segments that seemed attractive as potential targets. The first segment included ambulatory patients who suffer from chronic minor incontinence outside the hospital, often linked to old age, diabetes, traumatic childbirth in women, and other such conditions. "These patients experience some leakage of feces when they cough, sneeze, or laugh loudly. This is not a major clinical problem, but it is extremely embarrassing for patients and can become a traumatic issue that isolates and depresses them," Sharma explained. The second market segment focused on non-ambulatory acute patients suffering from major incontinence, who start in the hospital setting but then move to the home setting for long-term care.

Thinking further about their two potential market segments, Chasmawala and Sharma believed the basic market needs were similar across both groups. For example, a new solution should be easy to use to meet the needs of both segments. For the home, the solution would have to be so simple that unskilled, untrained,

possibly illiterate family members could administer it. In low-resource hospital settings, a solution that could be delivered by a nurse or an assistant would be preferred. Both segments also required a more effective and more dignified solution.

One important difference that emerged between the two segments was how the team would potentially access them. The minor incontinence, home-setting segment of the market would primarily be accessed by selling products over the counter to patients via medical product shops, whereas the major incontinence, hospital segment would be reached through hospital procurement departments and doctor recommendations. Further, Sharma mentioned, "The need in the hospital setting was bigger. From the clinical point of view, the under-management of fecal incontinence can lead to a higher rate of comorbidities in the patient. Thus, the hospital setting had a great 'pull' factor associated with it, as opposed to the home setting where we would have to push the device." In combination, these factors drove the team to pursue the hospital market segment as its first priority.

To develop a more specific understanding of the size of this segment, they conducted additional literature searches and also performed primary research to gather anecdotal market estimates from doctors in the field. Eventually, due to the paucity of available information, they investigated the number of the non-ambulatory patients suffering from major FI as a secondary condition within institutional settings in the US. Acknowledging that it was an imperfect approach, they determined that they could use US data as a rough proxy for market size with the assumption that the US population is roughly one-third the size of the Indian population, but in India only one-third of people can afford regular clinical care. After validating a series of rough estimates with in-country clinical and medical technology experts, they felt comfortable that they were directionally on target. This research suggested that their target segment included approximately 16 million patients in India each year. They sized the total worldwide market at 100 million patients.

In terms of willingness to pay, Chasmawala and Sharma thought more deeply about the value that any new solution would have to deliver to get physicians and other stakeholders to adopt it. In terms of effectiveness, the bar was set relatively low and they felt that patients and care providers would be unanimously receptive to an alternative to diapers and clumsy collection devices if it was affordable. Using diapers as a comparable benchmark, they determined that if they could create a considerably better approach that costs approximately US$10 per day, their solution would be compelling enough to get patients and providers to change their behavior.

Based on the outcome of their market research, Chasmawala and Sharma were able to refine their need statement. As Sharma stated, "The new need is for *a better way to manage fecal incontinence in order to improve clinical outcomes and reduce costs to hospitals by minimizing hospital-acquired infections and decreasing nursing time*." Over a number of months, the team, which founded a company called Consure Medical, developed a device similar to a short-term implant to address this need. The Consure solution diverted all fecal material from the rectum to an external collection device without interfering in peristaltic movements and other physiological functioning of the anorectal apparatus. It also had a slick insertion mechanism that allowed it to be placed in a clean and hygienic manner and was explicitly designed to be simple enough for a nurse or family member to use without physician involvement. Additionally, it would be priced to be competitive with adult diapers. As of 2013, Consure had completed a 10-patient safety study and another 10-patient efficacy study in India, which produced promising results and no adverse events. The company had also raised a Series A round of funding and was on its way to achieving regulatory clearance for the device. Although India remained the company's preliminary geographic focus, the team was already working with sites in the United States and the United Kingdom that were enthusiastic about the device.

 Online Resources

Visit www.ebiodesign.org/2.4 for more content, including:

 Activities and links for "Getting Started"
- Landscape the market
- Segment the market
- Size the segments
- Understand the dynamics of each segment
- Assess market needs
- Evaluate willingness to pay
- Choose a target market
- Bring it all together

 Videos on market analysis

Sample appendices that demonstrate
- Porter's Five Forces analysis for drug eluting stents
- SWOT analysis for drug eluting stents

CREDITS

The editors would like to acknowledge Randy Scott and Kim Popovits for the Genomic Health story, as well as Amit Sharma and Nish Chasmawala for the Consure Medical example. Further thanks go to the Spiracur team (Moshe Pinto, Dean Hu, and Kenton Fong) and the COPD team (Mridusmita Choudhury, Jonathan Hofius, Raymond Hoi, Zubin Huang, Peter Livingston, and Michael Winlo) for providing valuable input to the chapter, along with Ross Jaffe of Versant Venture for sharing his perspective on market analysis and Greg Lambrecht for his thoughts on solution value. Darin Buxbaum, Ritu Kamal, and Donald K. Lee also contributed to the material.

NOTES

1 From remarks made by Richard Stack as part of the "From the Innovator's Workbench" speaker series hosted by Stanford's Program in Biodesign, March 6, 2006, http://biodesign.stanford.edu/bdn/networking/pastinnovators.jsp. Reprinted with permission.

2 From remarks made by John Abele as part of the "From the Innovator's Workbench" speaker series hosted by Stanford's Program in Biodesign, February 2, 2004, http://biodesign.stanford.edu/bdn/networking/pastinnovators.jsp. Reprinted with permission.

3 "U.S. Medical Device Market Outlook," Frost & Sullivan, 2008, http://www.frost.com/prod/servlet/frost-home.pag (July 16, 2014).

4 All quotations are from interviews conducted by the authors, unless otherwise cited. Reprinted with permission.

5 "Market Segmentation," www.QuickMBA.com, http://www.quickmba.com/marketing/market-segmentation/ (March 29, 2014).

6 Cartoons created by Josh Makower, unless otherwise cited.

7 KCI 10-K form for FY2006 and analyst reports revealed that the VAC device is rented at a monthly rate of $2,000. Adding to that the cost of the disposable elements of the therapy (screens, dressings, etc.) and the cost of the health care professional changing the dressings, the total cost was estimated at $150 per day.

8 It should be noted that ulcers larger than 5 cm^2 take longer to heal (68 days) but, to be conservative, it was assumed that the average healing time is 52 days, as it is for smaller venous ulcers.

9 Estimates from Spiracur analysis.

10 Estimate based on team interviews with practicing physicians.

11 Michael E. Porter, "How Competitive Forces Shape Strategy," *Harvard Business Review*, March–April 1979.

12 Ibid.

13 "SWOT Analysis," www.wikipedia.org, http://en.wikipedia.org/wiki/Swot_analysis (March 29, 2014).

14 Ibid.

15 "Methodology," The Lean Startup, http://theleanstartup.com/principles (March 29, 2014).

16 Chantal Tode, "Mobile Health App Marketplace to Take Off, Expected to reach $26B by 2017," Mobile Marketer.com, March 25, 2013, http://www.mobilemarketer.com/cms/news/research/15023.html (April 22, 2014).

17 "Health Policy Brief: mHealth and FDA Guidance," *Health Affairs*, Robert Wood Johnson Foundation, December 5, 2013, http://www.rwjf.org/content/dam/farm/reports/issue_briefs/2013/rwjf409203 (April 22, 2014).

18 National Center for Health Statistics, Center for Disease Control and Prevention, http://www.cdc.gov/nchs/nhanes.htm (March 29, 2014).

19 Medical Expenditure Panel Survey (MEPS), Agency for Healthcare Research and Policy, http://www.ahrq.gov/data/mepsix.htm (March 29, 2014).

20 "Cost of CKD and ESRD," 2007 U.S. Renal Data System, 2007, http://www.usrds.org/2007/pdf/11_econ_07.pdf (March 29, 2014).

21 Only those patients who over-expressed the genetic alteration for HER2 responded to the drug Herceptin. By screening patients with a genetic diagnostic for this marker, physicians could make more personalized, effective treatment decisions and prescribe Herceptin to only those patients with the greatest probability of benefiting from its effects.

22 Jonathan C. Lewis, "Product Paternalism," *Stanford Social Innovation Review*, June 10, 2013, http://www.ssireview.org/blog/entry/product_paternalism (March 29, 2014).

23 Stacey McCutcheon, Lyn Denend, and Stefanos Zenios, "Consure Medical I: Translating a Need Into an Actionable Path Forward," *Global Health Innovation Insight Series*, February 2013, http://csi.gsb.stanford.edu/consure-medical-i-translating-need-actionable-path-forward (September 26, 2013).

24 A. K. Macmillan, A. E. Merrie, R. J. Marshall, and B. R. Parry, "The Prevalence of Fecal Incontinence in Community-Dwelling Adults: A Systematic Review of the Literature," *Diseases of the Colon and Rectum*, August 2004, pp. 1341–9.

2.5 Needs Selection

INTRODUCTION

To aspiring biodesign innovators, the process of selecting the right needs to bring forward into invention can feel a lot like comparing apples and oranges (and peaches, pears, bananas . . .) that all look good on the surface. One unmet need that applies to millions of patients might offer the opportunity to do little more than alleviate certain symptoms of their underlying condition. Another unmet need involving the fundamental mechanics of curing a disease might require years of research to affect just thousands of sufferers. Yet another could address a vital need common to a vast population, but in a geographic market where patients have little or no ability to pay for the remedy. With all of these trade-offs and challenges to evaluate, innovators require a systematic process for deciding which needs will provide the greatest opportunity to create and deliver value.

Needs selection is highly dependent on all of the data collected to this point in the biodesign innovation process. The goal of this activity is to identify, from the vast list of potential needs originally under consideration, a smaller subset of needs (or sometimes a single need) that warrants further investigation. The process of needs selection is inherently subjective in that it seeks to determine which needs are in greatest alignment with the interests and priorities of the innovators, as defined by their strategic focus. However, to maximize the innovator's chances of success, needs selection also must ensure that critical data about important objective factors are considered and given appropriate weight. The information gathered through iterative and increasingly in-depth research into various aspects of each need, such as an understanding of the disease state, solution landscape, stakeholders, and market, helps to drive innovators toward needs for which significant risks can be most effectively mitigated.

There is no one "correct" way to perform needs selection. However, significant value is realized by letting needs directly compete against one another, with a combination of both subjective and objective factors used to evaluate their relative merit. This type of comparative approach, complemented by increasingly detailed due diligence as the total

OBJECTIVES

- Understand how to develop a needs ranking system.
- Learn how to utilize the needs ranking system iteratively, through the incorporation of additional research, to reduce the number of needs in a stepwise fashion.
- Recognize when and how to interact with key stakeholders to validate needs.
- Learn how to create a need specification for the highest-scoring needs.

number of needs gets smaller and smaller, allows innovators to ultimately decide on a few high-potential needs to pursue.

The output of needs selection is the need specification – a document that synthesizes all of the important data gathered through observations, research, and needs evaluation. The "need spec" also outlines the criteria that any solution must meet in order to satisfy the need. This information is then used as the starting point for generating preliminary solution concepts.

 See ebiodesign.org for featured videos on needs selection.

NEEDS SELECTION FUNDAMENTALS

Needs selection requires the innovator to systematically compare one **need statement** against another to identify the most promising opportunities before moving into concept generation. Since most innovators transition from the needs finding stage into the needs screening stage with an abundance of **needs** and no clear sense of which ones hold the greatest potential, a formal process to select needs helps ensure that sound, unbiased decisions are made using the data gathered from 2.1 Disease State Fundamentals through 2.4 Market Analysis. Through this process, innovators can typically narrow the list of opportunities they are pursuing to between 1 and 10 top-priority needs.

The process of needs selection can be performed in many different ways, but a rigorous, structured approach usually involves six essential steps:

1. Select factors to consider in making a decision.
2. Assign ratings for each factor for each need.
3. Combine values to produce a score that can be used to rank the needs.
4. Select a smaller set of needs to be investigated further based on the rankings.
5. Perform additional research on the smaller set of remaining needs and repeat steps 1 through 4 until only a few top-priority needs remain, interspersing step 6 (below), as the number of needs decreases.
6. Engage key **stakeholders** to validate that the needs are meaningful.

Each of these steps is described in more detail later in this chapter.

Depending on how many needs innovators start with, the iterative process outlined in steps 1–4 may be repeated several times in order to get to a focused set of 1–10 top-priority needs. Some innovators move decisively through the process, revisiting these steps just once or twice, while others take a more progressive, cyclical approach, looping back through the steps multiple times and making measured cuts with each iteration.

The key to needs selection is to gather just enough information with each pass through the process to make informed choices about ranking needs without allowing the research and analysis cycle to become so burdensome that it impedes project progress. Since the biodesign innovation process fundamentally leads innovators to identify dozens (or even hundreds) of needs, it is unrealistic to think that they can spend weeks investigating each one. As noted, research, as described in chapters 2.1 through 2.4, should initially be kept at a relatively high level. As needs are screened through an iterative process and the list of those being considered becomes progressively smaller, innovators should perform increasingly more detailed and thorough research on the more focused list of needs that remain.

One technique that can help innovators appropriately limit how much time they spend on research is to set deadlines regarding when data collection for a given round of needs selection must be complete. For example, if innovators give themselves two weeks to investigate 100 needs, the research will be relatively high level by necessity. If they then take another two weeks to explore the 40 needs that "survived" the preliminary selection process, they will be able to dive deeper into understanding them, and so on.

Some innovators worry that if they do not exhaustively research each need, they might inadvertently eliminate a high-potential opportunity. But if the biodesign process is followed, innovators almost always find that the needs selection process leads them to a few excellent needs that are well aligned with their strategic focus and fully supported by the increasingly detailed needs research performed up to that point. That said, if new research uncovers information pertinent to a need that has been eliminated, innovators can always re-score the need and potentially reconsider it. Of course, this should be done empirically and not to "save" needs for which innovators have an affinity. But if innovators understand this is an option, they may be less fearful about eliminating promising needs based on their preliminary research. Re-scoring needs can even happen during the creation of a **need specification**, which occurs quite late in the needs selection process. The point is that research happens throughout the full process of selecting needs and any relevant information that is uncovered can and should be given serious consideration, even if it means adjusting prior decisions.

Again, while there is no single way to perform needs selection, Figure 2.5.1 provides a summary of how one team worked through its needs selection exercise. As the figure illustrates, this team was quite systematic in its approach to needs selection. Accordingly, this team's experience is used as an example in describing the six steps of needs selection. However, other innovators may work in a slightly less structured manner, as the two From the Field examples illustrate later in this chapter. Many different approaches can be successful as long as they: (1) allow for the rigorous evaluation of needs against a combination of subjective and objective factors that are important to the team; and (2) position needs to compete against one another, especially when screening many needs in a relatively short period of time.

Step 1 – Choose selection factors

The first step in the needs selection process is to clearly identify which factors to evaluate in comparing the needs under consideration. Each team must figure out which factors (and how many) make the most sense based on the types of needs being assessed and the interests of its members. However, choosing multiple objective factors and not more than one or two subjective factors (e.g., the team's level of enthusiasm for the need or the extent to which the need is life-saving) should be sufficient to understand and compare all of the needs on a level playing field.

# of needs   **Selection approach and criteria**

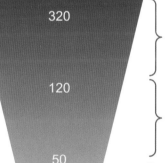

320
- Approach: Make quick decision based on minimal research and intuition
- Major factors: (1) does need duplicate another? (2) big market? (3) real need? (4) team interest?

120
- Approach: Use weighted formula based on moderate research and intuition
- Factors: (1) market size; (2) patient impact; (3) provider impact
- Rely on intuition if need does not "fit"

50
- Approach: Use weighted formula based on major research and intuition
- Factors: (1) market size; (2) patient impact; (3) provider impact; (4) need met? (5) feasibility of solving
- Rely on intuition if need does not "fit"

10

FIGURE 2.5.1

An example of the needs selection approach used by a multidisciplinary team from Stanford Biodesign. The team started with a set of 320 needs, which were cut to 120, then to 50, and finally to 10 target needs. Increasingly detailed criteria (developed through more thorough research) were used at each step in the selection process (with permission from Uday N. Kumar; developed with John White, Kityee Au-Yeung, and Joseph Knight).

Some of the most common objective factors that innovators use to evaluate one need against another are outlined in Table 2.5.1. Background research should enable the innovator to answers the questions associated with each factor (summarized in the table) for each need under consideration.

In addition to the objective factors shown in the table, innovators often find it helpful to consider the type of need that is being evaluated – blue sky, **mixed**, or incremental (see 1.3 Need Statement Development). Some innovators specifically want to evaluate a diversity of need types, while others prefer only to investigate those

Table 2.5.1 Objective factors that can be chosen to structure the needs selection process.

Factor	Question(s)	Issues/implications
Disease state mechanisms (see 2.1 Disease State Fundamentals)	Are the mechanisms of the disease state associated with the need well understood?	• Needs that are focused on poorly understood disease states may be more challenging to solve, especially with devices (e.g., autoimmune deficiencies). • Needs focused on multi-factorial disease states (e.g., diabetes) may not yield simple/direct solutions if not targeted to a specific aspect of the disease state. • Needs focused on disease states with large numbers of affected people are favorable since information tends to be available and can be discovered quickly (e.g., gastroesophageal reflux disease). • Patients in developing and emerging markets may present with later stages of a given disease state, compared to presentations in developed nations, and thus there may opportunities to shift the timing of care through earlier diagnosis and/or prevention.
Solution landscape (see 2.2 Existing Solutions)	To what extent is the need currently being addressed? Is there a gap between the need and existing treatments (i.e., a "white space" characterized by no solutions)?	• If treatment options currently exist that address the need (or could be "tweaked" to provide a solution), there may be significant competition in the market. • If established treatments exist but the need is still unmet, understand their limitations or shortfalls as a source of information. • If the shortfall or limitation of established treatments may be cost-related or due to complexity, there may be opportunities to solve the need in more **value**-oriented ways, which may also be important for emerging markets. • Pay attention if no treatment options exist to address the need. This can signal both opportunities and risks.
Stakeholder impact (see 2.3 Stakeholder Analysis)	How significant would the benefits to patients be if the need is addressed? How would physicians and facilities be affected? How likely are decision makers to pay for a solution that addresses the need?	• Any perceived or real ethical issues relating to solving a need can have an impact on how that need is received by key stakeholders. • Needs focused on improving patient outcomes are generally considered particularly important. • In addition to the benefits realized by some facilities and/or physicians from addressing the need, keep in mind that a solution might adversely affect others (e.g., by taking away business or shifting care from one specialty or facility to another).

Table 2.5.1 (*cont.*)

Factor	Question(s)	Issues/implications
		• If **payers** are the decision maker, they are more likely to provide **reimbursement** if the need is related to improved safety or efficacy and/or cost reduction and increased efficiency rather than increased convenience or a better **user** experience. • Other stakeholders, including facility administrators and patients, are increasingly playing the role of decision makers for products and services not covered by insurance. In developed and emerging markets, these customers are interested in both effectiveness and affordability.
Market size and competitive dynamics (see 2.4 Market Analysis)	How large is the market for a solution to the need? Which players (and how many of them) offer or are developing therapies in the need space? What magnitude of funding will be required to develop and commercialize a solution in the need area? Is the market large enough to justify this requirement? How substantial is potential value (in terms of clinical and economic outcomes) for the need?	• Market size helps quantify the potential impact that an innovator can have by addressing the need. • Market size is also an important factor in determining the likelihood that funding can be raised. • **Blue-sky needs** tend to have greater financing requirements so they require a larger market to support them. • **Incremental needs** may have lower funding requirements and may be feasible with a somewhat smaller market. • The level and nature of competition in an area of need has significant implications for how easy (or difficult) it will be to penetrate the market. • Regardless of their type, needs that lend themselves to driving favorable economic outcomes of a reasonable scale will be more likely to drive adoption in developed and emerging markets. • Sound market data can be more difficult to amass in emerging markets and will require more time and effort from innovators.

in a particular category. Generally, the decision of what type (or types) of needs to pursue is linked to the risk–reward equation that the innovators desire. Blue-sky needs typically involve significant risks, but can lead to greater rewards; the opposite is usually true for incremental needs. For this reason, innovators sometimes chose to categorize all needs by type and then determine how many in each category to pursue based on their preferences.

When choosing objective factors to use in the needs selection process, it is essential to think about how much value can be generated by solving each need under consideration. While including one or more value-related factors in the selection process may have been "optional" years ago, it is now imperative. In cost-sensitive healthcare environments, for example, certain incremental needs that can drive favorable economic outcomes for stakeholders in decision-making roles may gain more traction than blue-sky needs that deliver greater clinical benefits at a higher cost. Regardless of need type, innovators must identify the value of their needs in both clinical and economic terms and then use this information as an objective factor in needs selection.

Innovator preferences should also be taken into account through a series of subjective factors that link back to the strategic focus decided upon at the outset of the biodesign innovation process. As described in chapter 1.1, **acceptance criteria** are the mechanism through which an innovator's mission, strengths, weaknesses, and tolerance for risks in the external environment are woven together into a list of requirements that an innovation project should meet to provide a good fit. By translating the most importance acceptance criteria into selection factors, innovators can ensure that the projects emerging from the needs selection process are, in fact, aligned with their personal priorities. For teams of innovators working together, a personal interest rating provided by each team member can be used to collect and incorporate everyone's feedback on the most important subjective factors to evaluate.

The selection factors that innovators decide to use in the needs selection process can be tracked in many different ways. Some experienced innovators do little more than keep a mental list, while others record them with their notes from other steps in the biodesign innovation process. However, for most innovators, especially those who are just starting out, it can be extremely helpful to construct a spreadsheet or database to perform this tracking function. An electronic tool provides great flexibility in organizing, manipulating, and sorting information, especially when there are many needs to track. Figure 2.5.2 provides an example of the selection factors chosen by one team, as well as the method used to compile and manage this information in a database. Keep in mind that the data shown in the screen shot were collected incrementally, over multiple iterations through the selection process. Some needs were eliminated based on information related to a few key selection factors early in the process, while others that made it further in the selection process were evaluated based on research from the entire set of selection factors.

Step 2 – Assign ratings for each factor

Once a set of selection factors has been chosen, innovators next assign ratings to those factors for each need. Through this rating process, the needs under consideration begin to empirically compete against one another.

Before assigning ratings, it is necessary to decide on a rating scale that can be applied consistently across needs. Since the ratings ultimately lead to scores that should help separate strong needs from ones that may not be as compelling, innovators and their teams must be careful about choosing a rating scale. Too often, when innovators define a scale with an odd number of choices, they tend to "play it safe" by assigning ratings at the midpoint. This results in clusters of needs with scores that are similar. To help spread scores across a wider range of ratings, it may be useful to use a scale with an even number of choices so that innovators must assign scores that tip to one side of the midpoint or the other. For example, a scale of 1 to 4, where 4 is the best possible score and 1 is the lowest, often works well.

Next, the rating scale must be further defined in meaningful terms for each factor. For instance, for an objective factor related to *patient impact*, the innovator might define the scale shown in Table 2.5.2. For an objective factor related to *treatment landscape*, the scale in Table 2.5.3 might be used. For a value-related factor such as the cost/benefit of a need area, a sample rating scale is shown in Table 2.5.4.

Specific rating scales should also be defined for the subjective factors that the innovators have chosen. For example, if one of the acceptance criteria from the strategic focus is to work on projects that can have a positive effect on a large patient population, innovators may outline a factor and its rating scale as shown in Table 2.5.5.

The most difficult part of assigning ratings is that innovators must do so based only on what is known about the *need*. Because they should still be thinking in solution-independent terms, it is essential not to take into account information or **biases** related to any particular solution. Innovators have a tendency to jump ahead – for example, by rating the market size associated with a new type of catheter, rather than the size of the market associated with a better way to perform interventional procedures. The effect of this bias is to skew the rating process based on preconceived notions (e.g., "Catheter-based solutions are really hot right now, so I'm going to rate the need high," or "Medication pumps are ubiquitous commodities, so I'm going to rate the need low"). The best way to prevent this type of bias is to stay focused on rating the impact of *addressing the need* rather than the impact of a specific type of solution (see 1.3 Need Statement Development for more information about the pitfalls associated with embedding solutions within needs).

As ratings are assigned to each factor for a given need, they should be captured in the database or spreadsheet

Need ID#	Bucket	Type of Need
245	EPS & Ablation	Incremental

☐ Eliminate (50 to 12)

Personal Favorite: Kityee ☐ John ☑ Uday ☑ Joe ☐

Validation Owner: UK

Need Owner:

Need A faster, more effective way to access and deliver therapy to the area of the pulmonary veins to improve the success of atrial arrhythmia procedures

Est Market Size
Really only an issue in AF ablations, so market should be a 1 (but could become much larger)

Patient Impact
This should make the procedure faster and possibly reduce re-do rate since often don't isolate RLPV

Incidence_old:

Procedures Performed: 12000

AF ☑ Coronary Artery Disease ☐
Heart Failure ☐ Congenital Heart Disease ☐
SVT ☐ Bradycardia ☐
VT ☐
Sudden Cardiac Death ☐ Other_disease ☐ Specify:

Need met?
Current method is using adjustable ablation catheter; believed to work most of the time

Provider Impact
Could give provider improved ability to maneuver and provide full lesion set

Maximum Addressable VC Market per year

Prevalence (US)	2,200,000

prevalence of AF = 2.2M

new cases of AF = 300,000
% addressed = 15%
total = 45,000
- cost could include cost of "system" to get from groin to RLPV
- rough number of AF ablation per years,

Incidence	45,000
Device Cost - Variable Portion	750
New devices / patient per year	2
Max Market Total	67,500,000

Ideas
rapido; incorporate into transseptal

Provider Opinion (quotes are good)
"sometimes the doctors just give up getting there."

Total Fixed Cost Opportunity

Locations	
Devce Cost - Fixed Portion	
Devices / Location	
Fixed Cost Total	

Comments
- # procedures performed = current # of ablations performed
- while other ablation procedures are performed in the LA, most don't target the posterior LA; even if these were included, the vast majority of procedures in the LA would be for AF

Current Market
There is currently no specific catheter/sheath designed to specifically get to RLPV

Number of treatment / year	12,000
% Number of treatment / year	100.00%
Number of treatment per year	12,000
Cost of current device	0
Current devices / patient	0
Current Market Total	0

Feasibility Index	3 ⌄

FIGURE 2.5.2

A screen shot from a database developed by a Stanford Biodesign team to support the needs selection process (with permission from Uday N. Kumar; developed with John White, Kityee Au-Yeung, and Joseph Knight).

used to manage the selection process, as shown in Figure 2.5.3.

When evaluating a large number of needs, it can be a good idea to "test" the rating system before scoring every need. Choose a small, diverse sample of the needs under consideration, assign ratings, and then combine values to produce a single score for each need, as described in Step 3 below. If the scores that result from the ratings seem reasonable and are not clustered too closely, then go ahead and apply the rating system to the remaining needs. Alternatively, if there is limited differentiation between scores and they are clustered into a few dominant groupings, innovators may want to rethink the rating scale, definition of each rating, or even the factors chosen and then try rating the needs again before pressing ahead.

Table 2.5.2 A sample rating scale for the objective factor: patient impact.

Rating	Description
4	Addressing the need would be life-saving to patients.
3	Addressing the need would reduce **morbidity** and/or eliminate the risk of serious complications.
2	Addressing the need would not have an impact on morbidity, but would positively affect **quality of life** by eliminating undesirable symptoms of the disease.
1	Addressing the need would not have a significant impact on patients.

Table 2.5.3 A sample rating scale for the objective factor: treatment landscape.

Rating	Description
4	There are no existing treatments available to address the need and the field.
3	Treatments exist to address the need but have *serious* deficiencies that must be overcome.
2	Treatments exist to address the need but have *minor* deficiencies that must be overcome.
1	Treatments exist to address the need that are generally well accepted by the target user population and address the need well.

After ratings have been made for every need, be sure to look across needs, factor by factor, to confirm that the values have been assigned consistently, fairly, and without undue bias.

Step 3 – Combine values to produce a score

The next step in the selection process is to figure out how to combine ratings in such a way that each need ends up with a single numeric score. These scores should reflect the relative strengths and weaknesses of each need to allow for their direct comparison and prioritization.

Table 2.5.4 A sample rating scale for the objective factor: potential to add value.

Rating	Description
4	Positive clinical benefit with major cost savings.
3	Positive clinical benefit with incremental cost savings.
2	Positive clinical benefit with no effect on cost.
1	Positive clinical benefit that increases cost.

Table 2.5.5 A sample rating scale for the subjective factor: desire to positively impact a large patient population.

Rating	Description
4	Need directly affects more than 1 million people.
3	Need directly affects 100,000 to 1 million people.
2	Need directly affects 10,000 to 100,000 people.
1	Need directly affects fewer than 10,000 people.

One approach is to simply add or average the ratings assigned to each factor to come up with a score for each need. If the right mix of objective and subjective factors has been developed, this simple approach to calculating an overall score may be appropriate. Another method is to assign different weights to the various factors and then calculate a total score or weighted average. Weighting the factors is important if the innovator believes that certain criteria are significantly more essential than others. For example, if an innovator is driven to build a thriving business more than anything else, it might be appropriate to assign a greater weight to a factor such as market size. On the other hand, if an innovator is motivated above all else to improve outcomes for patients, a factor such as patient impact might warrant greater weight.

The process of weighting factors is inherently subjective – there really is no right way to do it. Innovators must trust their instincts in determining the best possible approach. The weightings used to calculate scores for the example team are shown in Figure 2.5.4. Note that, in this case, team members varied their weightings by

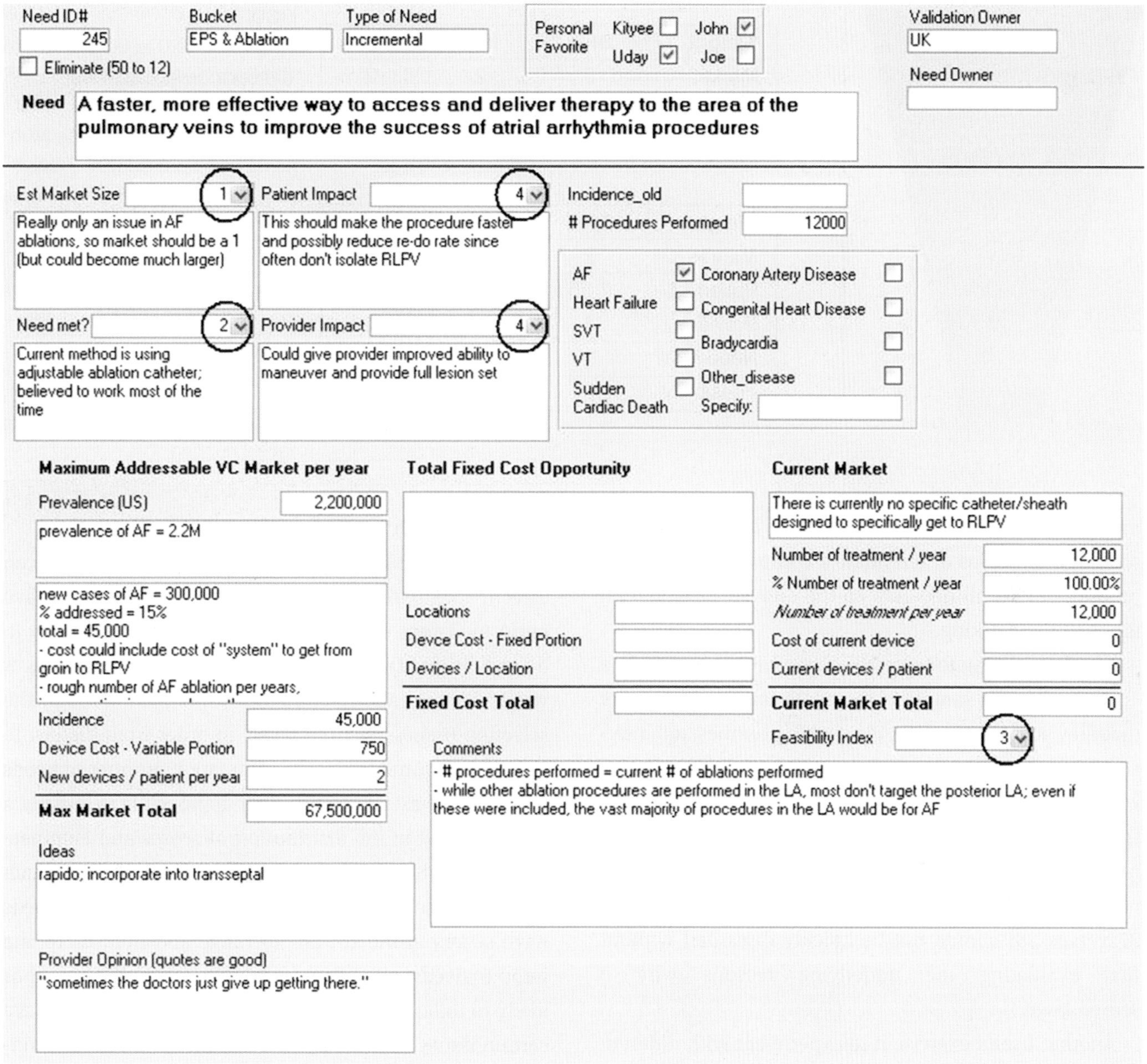

Need ID# 245
Bucket: EPS & Ablation
Type of Need: Incremental
☐ Eliminate (50 to 12)

Personal Favorite: Kityee ☐ John ☑ Uday ☑ Joe ☐

Validation Owner: UK
Need Owner:

Need A faster, more effective way to access and deliver therapy to the area of the pulmonary veins to improve the success of atrial arrhythmia procedures

Est Market Size: 1
Really only an issue in AF ablations, so market should be a 1 (but could become much larger)

Patient Impact: 4
This should make the procedure faster and possibly reduce re-do rate since often don't isolate RLPV

Incidence_old:
Procedures Performed: 12000

Need met? 2
Current method is using adjustable ablation catheter; believed to work most of the time

Provider Impact: 4
Could give provider improved ability to maneuver and provide full lesion set

AF ☑ Coronary Artery Disease ☐
Heart Failure ☐ Congenital Heart Disease ☐
SVT ☐ Bradycardia ☐
VT ☐
Sudden Cardiac Death ☐ Other_disease ☐ Specify:

Maximum Addressable VC Market per year

Prevalence (US): 2,200,000
prevalence of AF = 2.2M

new cases of AF = 300,000
% addressed = 15%
total = 45,000
- cost could include cost of "system" to get from groin to RLPV
- rough number of AF ablation per years,

Incidence: 45,000
Device Cost - Variable Portion: 750
New devices / patient per year: 2
Max Market Total 67,500,000

Ideas
rapido; incorporate into transseptal

Provider Opinion (quotes are good)
"sometimes the doctors just give up getting there."

Total Fixed Cost Opportunity

Locations:
Devce Cost - Fixed Portion:
Devices / Location:
Fixed Cost Total

Comments
- # procedures performed = current # of ablations performed
- while other ablation procedures are performed in the LA, most don't target the posterior LA; even if these were included, the vast majority of procedures in the LA would be for AF

Current Market

There is currently no specific catheter/sheath designed to specifically get to RLPV

Number of treatment / year: 12,000
% Number of treatment / year: 100.00%
Number of treatment per year: 12,000
Cost of current device: 0
Current devices / patient: 0
Current Market Total 0

Feasibility Index: 3

FIGURE 2.5.3
A screen shot from a database developed by a team in Stanford's Program in Biodesign showing assigned ratings (with permission from Uday N. Kumar; developed with John White, Kityee Au-Yeung, and Joseph Knight).

need type (blue-sky, mixed, or incremental) to help adjust for the natural bias of certain teams towards blue-sky needs. These needs may often gain the most points simply because they are so broad and can offer the greatest clinical and market rewards if successfully addressed. However, they may not be the optimal needs to pursue if, for example, a team has a strong desire to bring a tangible product to market in a relatively short period of time or prefers not to take on the significant risk that can sometimes accompany blue-sky needs. Of note, innovators in different geographies may value needs differently based on local market factors and the overall healthcare climate in the area where they are working. Along those lines, in increasingly cost-sensitive environments, incremental needs might not be valued as highly by payers if they offer limited clinical benefit. Yet, they

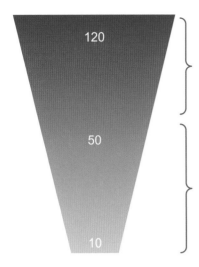

Weight	Blue	Mix	Incr
Market	2	2	2
Patient	1.25	1.5	1
Provider	1	1.5	2

Weight	Blue	Mix	Incr
Market	2	2	3
Patient	1.25	1.5	1
Provider	1	1.5	2
Need Met?	1	1	0
Feasibility	2	2	2

FIGURE 2.5.4
To get from 120 to 50 and then to 10 needs, weightings were assigned to increasingly specific filtering criteria by need type (with permission from Uday N. Kumar; developed with John White, Kityee Au-Yeung, and Joseph Knight).

may be valued by potential corporate acquirers if they solving them helps differentiate a company's product or product line. It all depends on the context in which an innovator is working.

After calculating the overall scores for each need, do a "gut check" on a sample of the final scores to make sure that they accurately reflect the team's priorities and interests, as well as the more objective factors that characterize a strong opportunity. It is important to keep in mind that the scores represent estimates of the strength of each need, but are not perfectly precise. This means that a need with a slightly lower score than another need (e.g., 27 versus 28) should not be quickly discarded without first evaluating the underlying reasons why it scored lower.

Another useful exercise it to experiment with differential weightings to see how this may alter the scores. Doing this might allow the impact of key factors to be magnified, which can help spread out scores. If certain needs consistently rise to the top of the list using different scoring scenarios, these are likely to be strong needs to pay attention to as the needs selection process progresses. Additionally, as described earlier, if significant clustering of scores remains, it may be worthwhile to revisit the rating scales, rating definitions, and selection factors.

Just as with collecting data for each need, a database or spreadsheet can be used to capture the approach to weighting certain factors (if appropriate) and to calculate the overall scores for each need as shown in Figure 2.5.5.

Step 4 – Select needs

With scores assigned for each need, innovators can directly compare them and select the ones that have performed best through the process so far. One way to do this is for the team to decide how many needs it would like to take forward from this pass through the selection process, list the needs in order from highest to lowest score, and then select the target number of needs with the highest scores. Another approach is to look for a natural break in the distribution of scores and eliminate any that fall below that point. The natural break is sometimes relatively obvious – for example, if many needs have scores in the 28, 27, 26 range and then there is a jump to needs with scores of 23, 22 21, etc. Such a gap is likely to indicate that the needs below a score of 26 have certain issues or risks that may be more difficult to overcome, making 26 a natural cut-off point for this round of needs selection. Another way to identify an appropriate cut-off point is to plot the scores for each need on a graph, as shown in Figure 2.5.6.

Step 5 – Iterate the selection process

As emphasized, the need selection process should be highly iterative. The first time through the process, an innovator might cut the number of possible need statements in half, complete additional research, adjust certain ratings and scores, and then work through this process again to reduce the list by another 50 percent. This approach ensures that more and more time is

Need ID.	Need	Est Market Size	Patient Impact	Provider Impact	Type of Need	Degree to Which the Need is Met	Feasibility Index	Personal Favorite-Kit	Personal Favorite-Uday	Personal Favorite-Joe	Personal Favorite-John	Total	Rank
INCREMENTAL		3	1	2	0	0	2	0	0	0	0		
116	A method to ablate more tissue, over a larger area, with each application of energy	2	2	2	1	3	3	0	0	0	0	18	1
84	A way to change the RF energy output location along the length of the catheter without moving the catheter at all	2	1	2	1	4	3	0	0	0	0	17	3
123	A cheap, simple method with which to accurately and easily enter the arterial and venous systems	3	1	1	1	2	3	0	0	0	0	18	1
51	A better way to stabilize device lead in coronary sinus	2	1	2	1	2	3	0	0	0	0	17	3
152	A way to quickly map and ablate within a 3D structure	2	2	2	1	3	2	0	0	0	1	16	5
245	A better way to access the right lower pulmonary vein. (changed better "catheter design" to better way)	1	2	2	1	2	3	0	1	0	1	15	9
159	A way for one cath lab operator to perform all necessary equipment movements for ablation and EPS procedures	3	1	2	1	4	1	0	0	0	0	16	5
70	Lighter x-ray shielding outfits	1	1	3	1	3	2	0	0	0	0	14	11
21	A way to avoid esophageal injury during ablation	1	1	2	1	2	3	0	0	0	0	14	11
56	A way to increase the battery life in the average implanted device	3	3	1	1	3	1	0	0	0	0	16	5
52	A more effective way to close an implanted device pocket	1	1	2	1	2	3	0	0	0	0	14	11
321	A way to determine the settings of an implanted device without a programmer available	3	1	2	1	4	1	0	0	0	0	16	5
40	A way to prevent electrophysiology catheter cables from building up torque outside of the patient	1	1	2	1	4	3	0	0	0	0	14	11
191	A way to prevent thrombosis/vegetation on device leads	3	2	1	1	2	1	0	0	0	0	15	9
24	A cost effective way to remotely monitor many inpatients	3	1	1	1	2	1	0	0	0	0	14	11
22	A better handle for ablation catheters	2	1	2	1	2	0	0	0	0	0	11	16
MIXED		2	1.5	1.5	0	1	2	0	0	0	0		
300	A way to remove the left atrial appendage from within the left heart	3	3	2	2	4	1	0	0	0	0	19.5	1
119	A method to assess the depth of the lesion created during ablation	2	2	3	2	4	2	1	1	0	0	19.5	1
325	A way to access the pericardial space	2	2	2	3	3	3	0	0	0	0	19	3
76	A better, more reliable way to determine if an arrhythmia that was present is now gone	2	2	3	2	3	2	0	0	0	0	18.5	4
149	A better way to determine/estimate ablation site prior to the actual procedure	2	2	3	2	3	2	0	0	0	0	18.5	4
144	A way to improve and angioplasty balloon	3	2	2	2	2	2	1	0	0	0	18	6
57	A way to recharge an implanted device without removing the entire generator	3	3	1	2	4	1	0	0	0	0	18	6
226	A way to anchor the wire / lead / catheter once you are in a vessel you are interested in.	2	1	2	2	3	3	0	0	0	0	17.5	8
118	A method or a device/suite of devices that would allow one person to perform all the functions necessary to perform a complete ablation	2	1	3	2	3	2	0	0	0	0	17	9
155	A way to determine the appropriate ablation "dosage" to account for tissue variability	2	2	2	2	3	2	1	0	0	0	17	9
279	A better way to more effectively deliver energy during defibrillation	3	2	1	2	2	2	0	0	0	0	16.5	11
284	A way to change the stiffness of a balloon (complaint to non-compliant)	3	1	2	2	2	2	0	0	0	0	16.5	11
180	A better way to non-invasively monitor arrhythmias long-term	3	2	1	3	2	2	0	0	0	0	16.5	11
248	A way to treat vessel perforation during a procedure	1	3	1	2	4	2	0	0	0	0	16	14
73	A better way to keep an ablation catheter against the wall of the heart	2	1	2	2	3	2	0	0	0	0	15.5	15
298	A way to make an ablation catheter assume any shape	2	1	2	2	3	2	0	0	0	0	15.5	15
47	A better way to access the coronary sinus	2	1	2	2	2	2	0	0	0	0	14.5	17
195	A way for an EP recording system to interpret signals, not just display them.	2	1	2	2	2	2	0	0	0	0	14.5	17
314	A way to guide a catheter to the previously stored location	2	1	2	2	2	2	0	0	0	0	14.5	17
307	A way to turn different parts of a catheter (REMOVED " without having to torque at the handle")	2	1	2	2	2	2	0	0	0	0	14.5	17
131	A way to alter the properties (stiffness, softness, slickness) of the wire/stylet without exchanging it	2	1	2	2	1	2	0	0	0	0	13.5	21
86	A way to include depth perception on a fluoro image	2	1	2	2	3	1	0	0	0	0	13.5	21
99	A way to eliminate putting 12 individual ECG patches on the patient	2	1	1	2	0	3	0	0	0	0	13	23
299	A way to place permanent leads in the left heart without an increased risk of thrombus	2	1	1	2	2	2	0	0	0	0	13	23
67	A way to prevent bleeding locally in the device pocket.	1	1	2	2	2	2	0	0	0	0	12.5	25
143	A way to change the visibility of equipment on the imaging system	2	1	2	2	2	1	0	0	0	0	12.5	25
305	A way to outline the borders of the endocardium for the duration of a procedure	1	1	2	2	3	1	0	0	0	0	11.5	27
301	A way to image the left atrial appendage without using a transesophageal approach	1	1	2	2	2	1	0	0	0	0	10.5	28
163	An easier way to compare/characterize/determine the direction of waveform propagation and its origin	1	1	2	2	2	1	0	0	0	0	10.5	28
BLUE SKY		2	1.25	1	0	1	2	0	0	0	0		
326	A way to improve the contractility of the heart	3	3	2	3	4	1	0	0	0	0	17.75	1
122	An alternative to fluoroscopy that will allow visualization of internal structures in real-time which would also allow real-time procedures to be performed	3	2	3	3	4	1	0	0	0	0	17.5	2
107	A method to non-invasively image coronary and peripheral arterial plaques susceptible to rupture	3	3	2	3	3	1	0	0	0	0	16.75	3
14	A way to treat a stenosis	3	3	1	3	2	2	0	0	0	0	16.75	3
8	A method to identify the likelihood of malignant arrhythmias in patients who are currently not symptomatic from the arrhythmia	3	3	1	3	4	1	0	0	0	0	16.75	3
112	A non-invasive method to ablate internal cardiac tissue	2	3	3	3	4	1	0	0	0	0	16.75	3
210	A way to cause scar (not iatrogenic fibrosis e.g. from a lead placement) tissue to conduct again	3	3	1	3	4	1	0	0	0	1	16.75	3
324	A better way to treat AF	3	3	1	3	3	1	0	0	0	0	15.75	8
124	A method with which to assess electrical and mechanical ventricular dyssynchrony	3	2	2	3	3	1	0	0	0	0	15.5	9
322	A way to prevent arrhythmias in post-MI patients with low EF	2	3	1	3	4	1	0	0	0	0	14.75	10

FIGURE 2.5.5

A screen shot from a spreadsheet developed by a Stanford Biodesign team showing the assigned ratings, weightings (in gray), and total scores for numerous needs. Note: the highlighted numbers indicate the weighting used for each factor (with permission from Uday N. Kumar; developed with John White, Kityee Au-Yeung, and Joseph Knight).

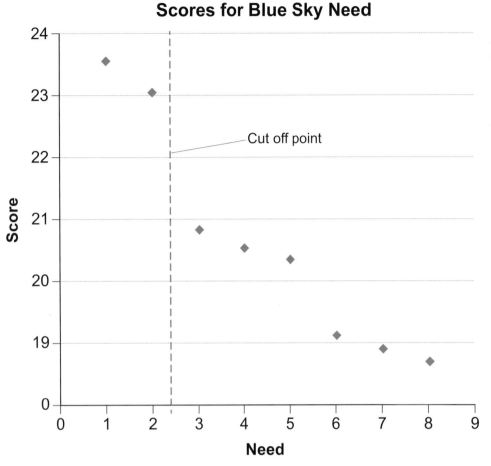

Scores for Blue Sky Need

FIGURE 2.5.6

By plotting the score of nine different needs, one team was able to identify the two most promising needs to take forward (courtesy of Rena Dharmawan, Prusothman Sina Raja, Benjamin Chee Keong Tee, and Cecilia Yao Wang).

invested in researching only the most promising needs that continue to "make the cut."

Typically, the quantitative scores developed through this process provide an excellent way to make the "easy" decisions about which needs to eliminate. For example, some needs clearly will not be as compelling as others in terms of the potential opportunities they represent, and their scores will cause them to be quickly set aside. However, most innovators find that, as the list of needs gets smaller and smaller, the quantitative scores become less helpful and they must rely on more qualitative decision-making criteria, potentially building off some of the subjective criteria used earlier to reduce the list to 1–10 top-priority needs. This is a natural part of the process, demonstrating that there is fundamentally no substitute for an innovator's professional judgment. But, if the process is followed consistently, this will help maintain some level of structure and objectivity to the exercise.

Step 6 – Validate needs with stakeholders

An important step that can help innovators arrive at a list of 1–10 top-priority needs is to reengage stakeholders for their input. Validating needs with key influencers and decision makers (as initially described in 1.3 Need Statement Development) helps ensure that the needs are, in fact, interesting and important. Traditionally, innovators have predominantly validated their needs with the physicians and other care providers they observed during needs exploration. However, it is essential to revisit a team's stakeholder analysis and validate needs with decision makers (often payers or purchasing managers), as well as influencers (like physicians).

Innovators can choose to validate specific aspects of their needs with stakeholders at any time during the needs selection process. However, they usually get good results when they do so on a smaller set after significant cuts have been made (e.g., from 320 to 50 needs). Asking stakeholders to consider dozens of needs is far more manageable than asking them to consider hundreds. On the other hand, if the list gets too small, there may not be enough choices among the needs to enable stakeholders to compare/contrast in order to identify the ones they truly find most compelling.

When approaching stakeholders for feedback on a list of needs, it is often advisable to quickly go through each one (or perhaps provide the list ahead of time for review) so that the full set of needs is understood before the stakeholder provides input. If each need is reviewed sequentially, invariably more time and effort is focused on the first few that are discussed. Another potential issue is that needs presented early may seem artificially interesting in the absence of information about more compelling needs that show up later on the list.

Another useful technique is to provide stakeholders with a few relevant selection factors and their associated rating scales and then ask the reviewers to objectively compare the needs. Importantly, these factors and their rating scales should be carefully customized to the perspective of each stakeholder. For example, a patient impact or physician impact selection factor might be appropriate when speaking with physicians, whereas a cost-effectiveness selection factor would be more appropriate for a payer or purchasing administrator. In addition to generating interesting insights, this approach can provide an efficient way for innovators to directly populate ratings for a stakeholder-oriented rating factor that can be used in the next round of needs selection.

Once stakeholders have validated the needs, this step usually does not need to be repeated in getting to the final list of top-priority needs since stakeholder opinions are unlikely to change significantly as the list of needs gets smaller.

Although stakeholder validation is a valuable technique and an effective way to keep needs selection linked to the interests of key stakeholders, innovators must strike a careful balance. It is absolutely critical that the input provided from stakeholders is used only as one data point in the ongoing needs selection process. All stakeholders have their own biases, and no single individual or group should be allowed to unduly influence the selection process. Ultimately, the team should choose its top-priority needs based on *all* of the input it has gathered and the research it has performed through the biodesign innovation process so far. Depending on the relationship between the innovators and the people they approach to validate their needs, it may be helpful to explain this in advance so that there are no misunderstandings if a need that appeals to a stakeholder is eventually set aside.

More needs selection examples

Given all the different ways to execute the six steps described above, there are countless variations of the needs selection process when it is put into practice. The factors, rating scales, weights, and scores chosen by each team will be unique. For example, one team from the Stanford-India Biodesign (SIB) Program followed the basic process when screening needs in pediatric medicine and maternal health, but ended up with an approach that was distinctive from the one shown in Figures 2.5.2 through 2.5.5 in several notable ways. Figure 2.5.7 summarizes the SIB team's needs selection approach.

This team entered the needs screening stage of the biodesign innovation process with more than 370 needs and went through four iterative selection rounds to eventually get to three top-priority needs. For each round, they used different factors, each with a specialized rating scale, as a mechanism for eliminating some needs and selecting others for additional research and evaluation. The SIB team did not explicitly assign weightings to the factors it used in each round. However, the rating scale chosen to support each factor implied a prioritization of the factors. For instance, in round 2, a need could earn up to five points for patient impact, but only a maximum of three points for the number of affected patients, understanding of the disease state and problem area, and treatment landscape. Such a system inherently assigned greater weight to the patient impact factor. Similarly, a mixed or incremental need was rated a one while a blue-sky need was given a zero, effectively weighting the selection process against blue-sky needs based on the team's preferred focus.

Stage 2: Needs Screening

370+ Need Statements

Round 1 : Gut Check Screening						
Factor 1	Team passionate about need	Yes = 1	No = 0			
Factor 2	Need type (blue-sky, mixed, incremental)	Mixed/ Incremental = 1	Blue-Sky = 0			
If Factor 1=0 or Factor 2=0, need is eliminated						

~175 Need Statements

Round 2 : Screening Based on Moderate Research					
Factor 1	Number of affected patients	Large = 3	Moderate = 2	Small = 1	
Factor 2	Understanding of disease state and problem area	Clear understanding of a well-focused problem area = 3	Reasonable understanding of a somewhat defined problem area = 2	Limited understanding of an ambiguous problem area = 1	
Factor 3	Treatment landscape	No good options available to address need = 3	A few available options but need is still largely unmet = 2	Numerous options available to address need although they may be may be suboptimal in some way = 1	
Factor 4	Patient impact	Life saving = 5	Eliminates serious complications = 4	Eliminates undesirable outcomes but no impact on morbidity = 3	Improves patient experience, but has no clinical impact =2	No major impact = 1
Factor 5	Positive feedback from informal advisors	Yes = 1	No = 0		
Rank order needs based on total score and take top 50 needs					

~50 Need Statements

Round 3: Screening Based on More In-Depth Research and Expert Validation					
Factor 1	Market size	Greater than INR 50 Cr (USD $10 million) per year = 3	Between INR 50 Cr (USD $10 million) and INR 10 Cr (USD $2 million) per year = 2	Less than INR 10 Cr (USD $2 million) per year = 1	
Factor 2	Expert opinion	High preference = 3	Mid preference = 2	No preference = 1	
Factor 3	Provider value (cost savings to hospitals, reduced procedure time, lower skill requirements)	Changes 2 or more factors = 3	Changes at least 1 factor = 2	Changes no factors = 1	
Rank order needs based on total score and take top 15 needs					

~15 Need Statements

Round 4: Final Screening					
Factor 1	Competitive landscape	Intense = 1	Moderate = 2	Low = 3	
Factor 2	Industry/partner opinion	Yes = 0.5	No = 0		
Factor 3	Department of Biotech (and other govt. agencies) opinion	Yes = 0.5	No = 0		
Identify natural breaking point, then discuss remaining choices as a team to choose final set of top 3 needs					

FIGURE 2.5.7

An overview of the factors, rating system, and general approach used by a team from the Stanford-India Biodesign program to move from 370 to 3 top-priority needs (with permission from Ritu Kamal; developed with Nitin Sisodia and Pushkar Ingale).

Another interesting aspect of this approach is that the team validated needs with four different types of external experts at three different times during the selection process. In emerging markets like India, which do not have mature medical device industries, it is difficult to gather relevant and reliable market data. Accordingly, multiple levels of feedback from a diverse set of experts can help teams understand the relative merits of different needs. In round 2, SIB team members gathered feedback about a relatively sizable list of needs from informal advisors such as former Fellows and the physicians they interacted with during observations. In round 3, they validated their top 50 needs with department heads in pediatric medicine and other key opinion leaders. In round 4, the innovators gathered feedback on their final 15 needs from prospective industry partners, such as representatives from major **medtech** firms and foundations working in India. They also validated these needs with the Indian Department of Biotechnology and representatives from other government agencies to gauge their interest in having specific needs addressed. This input into the needs selection process reflects the desire of innovators in countries like India to partner with industry and government to increase their chances of success in developing new medical technologies. While there are certainly no guarantees, if a need is aligned with the interests of a prospective partner and/or the priorities of the government, it can potentially eliminate some barriers to funding and commercialization. In this case, one of the industry partners, Siemens Healthcare, went on to provide seed-stage funding for the SIB team's top need (*a better way to screen neonates for hearing loss in resource-constrained settings*) as it was closely aligned with the company's area of focus.[1]

Generally speaking, the needs selection process as described in this chapter works in any geographic environment. However, as the SIB example demonstrates, the factors, rating scale, weightings, and approach toward validation may vary depending on where the work is being performed. The team at Indian medtech accelerator InnAccel, which includes Jagdish Chaturvedi, Siraj Dhanani, and A. Vijayrajan, emphasized

two main differences innovators should anticipate in working in a low-resource setting. First, they underscored the lack of available secondary data to use when conducting needs selection. As Chaturvedi, a former SIB Fellow and clinician, explained, "Gathering data in India can be very unpredictable, and many times we get information from research that is unreliable or irrelevant to the Indian setting." Accordingly, innovators at InnAccel use a factor in their needs selection activities that allows them to take into account the caliber of the data used to evaluate each need. If the need is supported by information that is India-specific and credible, meaning that it comes from **peer-reviewed** journals or reputable analyst reports, it is given a score of 1. Data that is not unique to India or comes from less reliable sources (or is mixed) is scored with a 0. In this way, data quality is lightly reflected in the overall scores for each need.

The InnAccel team also stressed that the needs selection process should be weighted toward factors linked to the cost of potential solutions and stakeholders' willingness and ability to pay for them. Given that most healthcare in India is funded using a self-pay model and, as a result, nearly all stakeholders in this environment are price sensitive, innovators must prioritize high value solutions that can be delivered at an affordable cost in order for them to have a reasonable chance of being adopted. Solutions that affordably deliver desired improvements are becoming increasingly important in all markets, but they are an absolute necessity in resource-constrained settings. InnAccel always includes a factor in its needs selection process that assesses the capacity of key stakeholders to pay. Another factor that the group sometimes evaluates is the extent to which needs can potentially be solved by shifting care to lower skilled care providers, since this is one way to effectively reduce cost and make solutions more affordable.

Independent of geography, a great deal can be learned from understanding the different ways that teams of innovators work through the needs selection process. The following story describes the approach used by the US-based medical device incubator, The Foundry.

An incubator's approach to selecting needs

As described in chapters 2.1 and 2.2, The Foundry is a medical device incubator that helps innovators "rapidly transform their **concepts** into companies."[2] Because of the large number of needs the company evaluates each year, Hanson Gifford managing partners and Mark Deem are constantly in the needs selection process as they decide which projects to pursue and which opportunities will translate into the most promising solutions.

Gifford and Deem are inundated with new ideas every day. Some of these needs arise organically from related projects they are working on, while others are brought to them by entrepreneurs and inventors. With limited resources and even more limited time, Gifford and Deem are forced to make tough and relatively quick decisions about which needs should proceed to concept generation and which should not. While Gifford and Deem appreciate the merits of a well-defined, clearly structured needs selection process, they admit that their approach is only partially systematic. As Deem commented, "I'd like to say our project analysis is completely calculated, but each opportunity is different. At the end of the day, you have to take the facts around each of them and make a judgment call. Sometimes, as we move along, one need starts to make more sense. We've got better ideas and information in one area than we do in another and it becomes the lead horse," he explained.[3]

Despite their informal characterization of The Foundry's process for selecting needs, further discussion revealed that Gifford and Deem do, in fact, evaluate many of the objective and subjective factors described within this chapter. While they may not formally assign quantitative ratings and explicitly weight the factors via rigid analysis to come up with an overall score, they do much more than depend simply on their gut instincts. When pressed for information on how The Foundry makes decisions regarding which needs to pursue, Gifford highlighted the main factors that they consider: "The market opportunity, the clinical benefit, and the overall investment landscape – is the need attractive to investors? Are we going to be able to get there within our lifetime and with less cost than the national debt? We also carefully consider how likely something is to really succeed in the market." These criteria, consistently applied across needs, provide a basis of comparison that lends objectivity to the selection process. If there is only a small market for a product, or if the intellectual property is already substantially owned by others, The Foundry may be inclined to pass on the opportunity. Likewise, Gifford and Deem may steer away from a need if the referral and reimbursement patterns are unduly complex, or if prior research suggests clinical feasibility may be low.

Continuing to describe the factors they take into account in selecting needs, they emphasized the importance of stakeholder analysis, abiding by the saying that "A new therapy needs to be attractive to patients, physicians, and payers." Typically, the Foundry investigates the **value proposition** to all three groups before making a decision.

Gifford explained that it is also important to take the innovator's interests and motivations into account in the needs selection process: "This isn't just a one-time decision about whether projects A, B, or C are interesting to work on. It's a decision followed by several years of blood, sweat, and tears to make it happen. We have to take a good, hard look at ourselves and ask which opportunity we want to commit a portion of our lives to." That is not to say that they always stick with what they know and like best – as Deem pointed out, The Foundry has taken on highly diverse projects over the course of its history. Rather, he said, they must be excited about any need they decide to pursue. Over time, they have learned that a project only succeeds if the people involved want to work on it and not simply because it looks financially promising. If they are not enthusiastic about and truly interested in addressing the need, Gifford and Deem are more than willing to shelve it and move on. They recognize that it is this sense of passion that has kept them excited about coming to work every day through many start-up experiences since 1998.

One of The Foundry's former companies, called Cierra, Inc., provides an example of how the group's needs selection criteria were applied to a new opportunity. Cierra was founded to develop a novel approach for the treatment of patent foramen ovale (PFO). This condition results from the incomplete closure of the septal wall between the right and left atria (upper chambers) of the heart, an opening that exists before birth to allow oxygenated blood to circulate throughout the fetus without having to pass through its lungs. In 75 percent of those with the condition, the foramen ovale closes naturally after birth. In the remaining 25 percent, it does not seal completely, allowing blood to flow directly, under certain conditions, from the right atrium to the left atrium.[4] As a result of the existence of a PFO, the natural filtration that the lungs provide is partially bypassed. Consequently, blood clots and other agents in the blood can go directly into the arterial system, potentially causing paradoxical embolism, cryptogenic stroke, and right to left nitrogen embolism in severe decompression illness.

Early in the innovation process, when Gifford and Deem first began looking at PFO, they did so at the urging of cardiologists in their network who anticipated that this was going to become an increasingly important need. At the time, there was no clearly defined **standard of care**. Many physicians believed there was no reason to treat for PFO unless a patient experienced cryptogenic stroke (a stroke with unknown causes). In these cases, patients often were prescribed chronic anticoagulation therapy. The condition could also be addressed through open surgery or a transcatheter intervention that used an implantable closure device to address the problem. Two transcatheter devices had received approval from the **FDA** for the treatment of PFO under a humanitarian device exemption (**HDE**), a category of FDA clearance that applies to devices designed to treat a population of less than 4,000 patients.

After initially studying the disease and evaluating stroke-related PFO opportunities against their selection criteria, Gifford and Deem were somewhat less than enthusiastic about the need. The market opportunity was relatively small, the attractiveness to investors was dubious, and the clinical benefits were not compelling relative to the amount of effort and investment required to conduct a stroke trial (which were notorious for being costly, lengthy, and difficult to enroll). "We just couldn't get excited about signing up for a project where we were going to have do a one-thousand patient **clinical trial** to try to show an improvement in stroke rates from one single digit number to another single digit number," recalled Deem. "We were about to shelve it," he continued, "when we came across some of the first articles that were being published on the migraine-PFO association." As Erik Engelson, who later became CEO of Cierra, described, "There were observations and some single-arm studies that were published. In deep-sea divers who were having PFO treated for decompression illness and in stroke patients who also had migraines and had their PFO closed, the observation was that the frequency of migraine decreased after the procedure." "That caught our interest, and we started studying it a little bit more," said Deem. "It was still a relatively unproven association, but we felt like there was probably something there that could transform this into a huge opportunity." Although there was still a fair amount of uncertainty surrounding the linkage between migraines and PFO closure, as well as the amount of time and cost required to develop a solution, Gifford and Deem became passionate about pursuing the need.

When Gifford and Deem reapplied their selection criteria to the migraine-related opportunity, it was more appealing, in part, because it seemed that the market could potentially be more easily and quickly accessed than the stroke market. While the symptoms of this larger population of migraine sufferers were not life threatening, they could be severely debilitating so Gifford and Deem perceived the clinical benefit of a solution to be significant in terms of improving quality of life. Moreover, Gifford and Deem anticipated being able to more effectively get investors interested due, in part, to recent activity in the field. "All of a sudden this market was hot," recalled Engelson. "There was this public company called NMT Medical, the market cap of which doubled on the buzz of upcoming completion of their UK-based migraine study [the MIST-I study]."

The apparent attractiveness of the opportunity was also enhanced by the novel concept of closing PFO with a non-implant solution (rather than leaving behind an implant in the heart). As a result, The Foundry team refined its **need criteria** to include this design requirement. "An implant is all well and good if you're a 75-year old patient who has already had one or two strokes and you're trying to prevent another one," Deem said. "But if you're a 25-year old migraine sufferer who is otherwise reasonably healthy, having a metal implant in your heart for the rest of your life is a completely different value proposition due to known [and yet-to-be known] complications associated with some of the PFO implant devices. The notion of a non-implant solution played positively to cardiologists, patients, neurologists, and investors. So, we decided that we needed a solution that left nothing behind in the heart." It would have been a lot easier and faster to simply pursue a better implant. But, as Deem summarized, "We felt like the benefit of leaving absolutely nothing behind versus having the next best clip warranted the decision to go forward there."

When Engelson was recruited to head Cierra, the company had 12 employees. The team had evaluated clips, snaps, staples, sutures, and patches to build a broad IP position, but ultimately decided on an implant-free system that closed the PFO using suction and "tissue welding" performed percutaneously using radiofrequency (RF) energy (see Figure 2.5.8).

Under Engelson's leadership, Cierra refined its product, completed animal testing, treated eight human patients in Germany (with an initial 75 percent PFO closure rate), and raised $21 million in next-round funding (incremental to the previous $8 million). The team also had a series of promising meetings with the FDA. In order for the PFO-migraine solution to be successful, the company needed

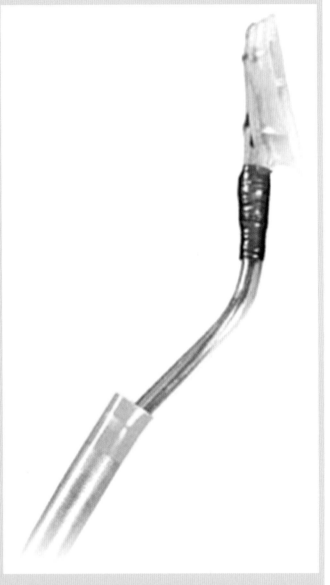

FIGURE 2.5.8
The Cierra device (courtesy of Cierra, Inc.).

to be able to design a clinical study that would address the agency's safety concerns related to applying RF energy to the heart. "The FDA really seemed to like the non-implant solution. They gave us a verbal thumbs-up on our safety data," Engelson recalled.

Shortly thereafter, however, Cierra faced a reversal of fortune. "When we formally submitted our **clinical protocol**, the written feedback we received seemed to differ significantly from the verbal feedback on the conceptual study design we had previously discussed with the FDA," said Engelson. The FDA's written feedback (questions and suggestions) directed the company to conform its study design to that of the other migraine studies, which were not enrolling. Around the same time, the company found itself achieving mixed technical results. "We had learned to segment by size and we were getting good PFO closure in smaller PFOs, but still had work to do in the larger ones. Closing the large PFOs was becoming an insurmountable challenge with the RF technology," noted Engelson. Additionally, when NMT released partial results of the MIST I trial in spring 2006, the buzz around PFO-migraine opportunities diminished. In that study of 147 patients, there was no difference in headache cure between migraine sufferers receiving PFO closure and those receiving the sham procedure.[5] Despite these challenges, Engelson noted, "We believed in this." "We had signed up and we were fighting this war together. So we carried on." The company's efforts continued until late 2007, when it finally discontinued its operations. Interestingly, NMT Medical announced that it was halting its MIST II (US-based migraine) trial one month later, citing difficulty in enrolling patients in the study and the need to redirect the millions of dollars necessary to sustain their stroke investigation.[6]

Reflecting on the Cierra experience, Engelson commented that some of the challenges faced by the company could be traced back to steps in the needs selection process. While the team's market analysis demonstrated that migraine was, in fact, a bigger and potentially more easily accessible market segment, it later became clear that neurologists, not cardiologists, were the key stakeholder group since they would be the ones to control study enrollment. And, unfortunately, they were generally not supportive of PFO-migraine studies. "It wasn't a cardiology deal," said Engelson. "It was a neurology deal. Compared to cardiologists, neurologists tend to be less aggressive – they were not enthusiastic to try new, interventional technologies. They're used to using drugs to manage their patients," he noted. While initiating an interventional procedure to address a quality of life issue might have been broadly accepted in cardiology, the concept was viewed skeptically in neurology.

The definition of a need criterion to develop a non-implantable solution also may have unnecessarily constrained the company's efforts and eventually contributed to its difficulties in overcoming the technical challenges associated with closing large PFOs. According to Engelson, "We spent a lot of time and money developing a totally non-implantable solution when, in hindsight, I'm not sure that this was as critical as initially thought. We may have been better served by developing a succession of products for PFO closure: first a very small implant such as a clip, which would be much smaller than the existing implants, followed by the non-implant technology. This technology-product tradeoff decision (or product portfolio planning) process might have been more effective had we had a visionary and inventive physician as part of our internal thinking process."

Overall, summarized Engelson, "It was the FDA's high study-design hurdle, the lack of support by migraine neurologists, and the remaining technical challenge in closing larger PFOs without an implant that collectively led to our shut-down decision." Similar issues had proven to be challenging for numerous other device-based companies working in the field. Ultimately, the combination of having to demonstrate the safety and effectiveness of the intervention in parallel with addressing basic science questions about the PFO-migraine link that remained unanswered presented too much risk for the company to proceed.

Need specification

Once a small set of top-priority needs has emerged from the needs selection process, a need specification should be created for each one. A need specification is a detailed, but succinct stand-alone document that (1) presents the need statement, (2) summarizes the data gathered through the needs screening process (chapters 2.1 through 2.5), and (3) outlines the need criteria that any solution must address in order to satisfy the need. These criteria should be organized into "must-haves" and "nice-to-haves," with typically 4–6 key requirements in each category.

The information in the need specification serves as guiding principles for the next stage of the biodesign innovation process – concept generation. For this reason, it is essential that each specification is prepared thoroughly. However, this should not be viewed as an exhaustive task. Innovators should apply roughly the same level of effort and rigor to creating a need specification as they would to developing a research paper.

Keep in mind that the need specification is not a static document, presentation, or outline, but rather evolves and changes as the data that goes into it is updated and revised. Ultimately, the need specification has to be finalized prior to entering the concept generation stage but, as with many parts of the biodesign innovation process, getting to that final document is itself iterative.

In creating a need specification, innovators may also uncover information that throws the needs selection process into a new light. For example, when compiling more detailed data about various treatment options, they may realize that one solution, previously thought to be unimportant, is actually of significance. If elements about competitive landscape were included in the selection process, new information could suggest that certain ratings may need to be adjusted. Although minor adjustments are not likely to have a significant impact on the overall ranking of needs, it is important to confirm this by revisiting the needs selection process. Major adjustments in rankings will typically only occur if important information was somehow missed in the initial screening process or if the research performed at each stage of needs selection was kept too superficial, even as the number of needs was being reduced. This interaction of the needs selection process with the creation of the need specification once again highlights the iterative nature of the biodesign innovation process.

Sometimes innovators may try to shortcut need specification by preparing one background document for multiple needs that overlap, particularly when they all focus on a given disease state. However, in doing this, the innovator runs the risk of creating confusion about the key drivers of each need, especially if those are not clearly separated. Additionally, since a need specification provides a key input into upcoming ideation sessions, it is undesirable for information about competing or alternate needs to distract participants or bias their thinking if it is included in a combined need specification. Additionally, even with only a few needs remaining, some will continue to fall off the list of the innovator's top priorities as more information becomes known. If each need has its own specification, this can more easily be accomplished.

In terms of content, a need specification provides a summary of the most important and relevant information gathered in support of the need. It is often organized into the sections outlined below, although the document does not have to follow this outline exactly. Innovators have the freedom to experiment with different headings and ways of organizing their need specification, as long as all of the most relevant and informative content is included. See online Appendices 2.5.1 and 2.5.2 for two real-world examples of need specifications that have quite different forms, but cover most of the same fundamental information. The format of the need specification can also vary. Need specifications usually are created as either several-page papers, slide presentations, or bulleted outlines.

One final point is that a need specification should incorporate quantitative data that support the need. In describing the problem area to which the need is related and the market that could be addressed by a new solution, innovators should reference relevant facts and figures to add specificity and credibility to the document. Need criteria are often used to help quantify ideas captured in the need statement. For example, if a need statement refers to a requirement for something smaller, a need criterion might provide a specific size, limit, or target range. Need criteria can also clarify why certain requirements exist; for instance, a solution should be small so it can placed through a specific type of incision.

The need statement
The need statement can be used as originally written (see 1.3 Need Statement Development) or refined based on what has been learned through researching. Sometimes innovators and teams refine their needs progressively, as

research is performed. In other cases, they make adjustments when creating the need specification. Both approaches are acceptable, as long as there is not a fundamental altering of what was actually observed. Refining and clarifying the need helps ensure the team shares a common understanding and can position it for greater success during ideation.

The problem

The summary of the problem should address affected anatomy and physiology, disease mechanisms of action, disease progression, past and current approaches to addressing the problem, outcomes, complication rates, and a gap analysis of existing and emerging treatments, including costs relative to the improvements they deliver (see 2.1 Disease State Fundamentals and 2.2 Existing Solutions).

Market description

The market overview should cover high-level market-related information from the needs research process, including data about the target market, procedure volume, treatment penetration, competitive landscape, and burden on the healthcare system (see 2.3 Stakeholder Analysis and 2.4 Market Analysis). Additionally, important insights related to the potential value of a solution to the need should be included.

Need criteria

As noted, need criteria are the key elements required and/or desired from any new solution by the stakeholders with an interest in the need (e.g., efficacy rates, compatibility with other devices, ease of use). Must-have criteria are essential to addressing the need and should correspond include the core requirements of decision makers. Nice-to-have need criteria are not imperative but would make the solution more attractive. They often correspond to the requirements that are most important to influencers.

A preliminary set of need criteria requirements is usually developed as the need statement is defined (see 1.3 Need Statement Development). However, unlike these initial requirements, which are primarily based on observations, the criteria in the need specification are significantly influenced by the data gathered from research about the problem and the market. These data allow innovators to more deeply understand the key issues that must be met for a need to truly be addressed.

For example, consider the need for *a way to perform testing for skin lesions at the point-of-care to enable accurate, inexpensive diagnosis of malignant melanoma by a dermatologist*. The team members working on this project developed four must-have need criteria that they believed were critical to the adoption of any potential solution, along with a comparable set of nice-to-have criteria that would potentially increase the attractiveness of an offering (as shown in Table 2.5.6).

As this example illustrates, need criteria (particularly the must-haves) should be specific and measurable. Quantitative criteria are especially useful in helping the team be as unambiguous as possible in understanding what is required to adequately address a defined need. Quantitative need criteria can usually be developed relative to benchmarks uncovered through the team's observations and research in the need area. For example, after learning that the national average for the length of a dermatology appointment was approximately 15 minutes, this team determined that dermatologists must be able to administer a new point-of-care diagnostic *and*

Table 2.5.6 Defining strong need criteria is essential for guiding concept generation and selection (with permission from Varun Boriah, Tiffany Chao, and Ryan Krone).

Must-haves	Nice-to-haves
Point of care: < 10 minutes at dermatology clinic	< 1 minute interface time with patient
Accurate: Sensitivity > 98 percent Specificity > 50 percent	Interface component is physically small (can hold in one hand)
Inexpensive: < $190 (which is the current reimbursement rate biopsy + pathology)	Disposable component
Portable	Can be performed by a mid-level provider

deliver the test results comfortably within this time frame. They derived requirements around the sensitivity and specificity of a new solution by assessing the current performance of tests typically performed by primary care physicians and dermatologists, as well as standards set by the state-of-the-art technology available in the field. They then chose accuracy targets that were likely to be viewed favorably relative to the increased convenience and affordability of a new solution, and so on.

Developing need criteria can be more art than science, but it is important that the information gathered through needs research be used to help refine and specify the criteria in such a way that imperative requirements can be distinguished from those that are desirable but not on the critical path. However, recognize that creating too many absolute criteria (more than 6–8) will place too many constraints on concept generation and screening. Innovators should stay focused on the few criteria that are absolutely essential to address the need.

Finally, while some need criteria may specify product attributes (e.g., small, able to be placed via a blood vessel), these more specific attributes should still be kept at a high level. The point is to avoid the implication of a solution while reflecting the accepted constraints for the need (e.g., developing a solution for a problem accepting the fact that it should optimally leverage a catheter-based platform). Need criteria should also continue to evolve and become increasingly concrete as innovators iterate the need specification.

References

All quantitative informative included in the need specification must be cited using commonly accepted conventions for footnotes or endnotes. Quotes from stakeholders should also be cited in the list of references.

The following story describes how one team went through the process of selecting a need and developing a need specification.

FROM THE FIELD ▷ **BLOOD STREAM INFECTION TEAM**

Selecting a need and developing a need specification

As part of their experience in Stanford's Biodesign Innovation course, Eric Chehab, Carl Dambkowski, Jon Fritz, Siddartha Joshi, Brian Matesic, and Julie Papanek began thinking about the incremental need for *a way to reduce catheter-related bloodstream infections*. Recognizing that they would have to refine the need statement to be more actionable and focused, they simultaneously initiated needs scoping and needs research. They started by investigating blood stream infections and the different types of catheters used in the hospital setting, such as dialysis catheters, central lines, PICC lines, chemotherapy ports, and pediatric catheters. After gathering preliminary disease state, treatment, stakeholder, and market data in each area, they defined four preliminary factors to help them compare the opportunities and narrow their focus: (1) infection rate; (2) utilization/volume; (3) the extent to which existing technologies effectively targeted the problem; and (4) the fit with the team's interests.

Through the first selection round, the team converged on pediatric catheters and related blood stream infections (BSIs). In particular, infections rates with pediatric catheters were much higher than in the other areas. "We were drawn toward the highest infections rates where we could really make a difference and where showing an improvement would be easier," said Papanek. They were also struck by the lack of existing technologies designed specifically to address pediatric needs and the fact that few (if any) companies seemed to be working on unique solutions to reduce catheter-related BSIs in neonates.

With this refined focus, they dove back into research to go deeper into pediatric-related disease, treatment, stakeholder, and market characteristics. "Basically," said

Chehab, "we redid our research each time we refined the need." Papanek added that, "As we were pulling in more and more data, our understanding of the needs improved significantly." In total, the team completed two additional rounds of needs scoping and refinement, each time using the same four factors to help them make a decision. Coming out of the second round, they narrowed their focus to catheter-related BSIs in premature babies. The third round led them to umbilical cord catheters and the need for *a way to prevent bacteria from entering the blood stream in neonates with umbilical cord catheters in order to reduce the rate of BSI infections*.

In validating the need statement, many of the experts they spoke with drew attention to the fact that the market for neonatal umbilical catheters was relatively small. The team had uncovered this in its research, but decided that it offered some attractive trade-offs. Although the market was not likely to support a stand-alone business, it was an important need that could have a substantial impact on neonatal health. In addition, they believed a solution could be developed quickly and with limited technical risk because non-neonate-specific devices already existed. The team members did not have to come up with something completely novel, they just had to figure out how to adapt available technologies to the unusual morphology of the umbilical catheter and this unique patient population. For these reasons, it would be a good "first project" for the group to undertake, with the hope of licensing the solution to a larger firm and then moving on to other, potentially more lucrative opportunities.

Another important insight from the team's research and its need validation activities had to do with the value that any solution would have to deliver to the healthcare system. "From the get-go we knew that we were trying to reduce overall cost for the hospital, especially since hospital-acquired blood stream infections are expensive and no longer reimbursed by Medicare," said Chehab. For each infant that contracted a BSI, costs to the hospital increased by approximately $40,000. That said, catheter-related BSIs affected 5–15 percent of babies admitted to the neonatal intensive care unit (NICU). For a

facility like Lucile Packard Children's Hospital, which typically administers umbilical catheters to less than 250 NICU patients a year, the annual cost of the related BSIs was approximately $260,000. As Papanek summarized, "The majority of babies never get a blood stream infection, so a solution essentially has to be a cheap insurance policy for the hospital."

Taking this information into account, the team defined another factor that would affect the success of a solution. To be widely adopted, the members believed that any solution they designed would have to seamlessly and easily integrate with existing catheterization protocols. "We could not expect to reinvent catheter placement or change the current procedure. This solution had to be an add-on to the current catheter placement method – possibly bundled with the current catheter set – to reduce catheter-induced infections," said Dambkowski.

With these factors clearly in mind, the team was able to define clear need criteria as part of developing its need specification. Issues related to value and adoption were placed prominently on the list of "must-have" criteria,

Need criteria

A way to reduce BSI infections caused by umbilical catheters in premature neonates without increasing additional side-effects

Must Haves
- BSI rate <7.1/1k patient days
- No significant increase in small artery elasticity
- Deliver required TNP, drug, and fluid and enable blood monitoring available in competing products
- Can be used immediately after birth
- Can be used in babies with APGAR score <7
- Integrates into existing procedure
- Does not increase net healthcare expenses to system

Nice to Haves
- Safely and effectively utilized for >14 days
- Can be used in very low birth weight infants (<1000g)
- Does not increase antibiotic resistance
- Can be inserted without X-ray confirmation
- Can be legally placed by a nurse
- Simple training to facilitate rapid adoption

FIGURE 2.5.9

Need criteria as defined by the team (with permission from Eric Chehab, Carl Dambkowski, Jonathan Fritz, Siddhartha Joshi, Brian Matesic, and Julie Papanek).

along with a variety of additional clinical and technical requirements, supported by quantitative guidelines. Other requirements to improve the patient and physician experience and make an easy-to-use solution were included among the "nice-to-have" criteria as shown in Figure 2.5.9.

Eager to transition into concept generation, Chehab, Dambkowski, Papanek, and their colleagues committed themselves to using these need criteria as a guide

through ideation and eventually in choosing a solution. A variation of the team's complete need specification is available in online Appendix 2.5.1.

Ultimately, the innovators designed a product called the LifeBubble – a small rigid bubble that is placed around the umbilical catheter insertion site to secure the catheter and reduce BSIs by covering and isolating the area, to prevent the migration of bacteria from the skin to the stump. LifeBubble is still in development.

↘ Online Resources

Visit www.ebiodesign.org/2.5 for more content, including:

 Activities and links for "Getting Started"
- Choose selection factors
- Assign ratings
- Calculate scores
- Validate and select needs
- Create need specification(s)

 Videos on needs selection

 Appendices that provide sample need specifications for two projects

CREDITS

The editors would like to acknowledge Hanson Gifford and Mark Deem of The Foundry and Erik Engelson, formerly of Cierra, as well as Eric Chehab, Carl Damb-kowski, Jon Fritz, Siddartha Joshi, Brian Matesic, and Julie Papanek for their assistance with the case examples. Varun Boriah, Jagdish Chaturvedi, Tiffany Chao, Siraj Dhanani, Rena Dharmawan Pushkar Ingale, Ritu Kamal, Ryan Krone, Prusothman Sina Raja, Nitin Sisodia, Benjamin Chee Keong Tee, A. Vijayrajan, and Cecilia Yao Wang also contributed important insights and examples to this chapter. Further thanks go to Steve Fair for his help in developing the original chapter.

NOTES

1 "About Us," Sohum Innovation Lab, http://www.sohumforall.com/team/ (November 19, 2013).

2 The Foundry, http://www.the-foundry.com/ (November 19, 2013).

3 All quotations are from interviews conducted by authors, unless otherwise cited. Reprinted with permission.

4 "Patent Foramen Ovale," KidsMD, Boston Children's Hospital, http://www.childrenshospital.org/health-topics/conditions/p/patent-foramen-ovale (November 19, 2013).

5 Shelley Wood, "Mixed Results for PFO Closure in Migraine Cloud Interpretation of MIST," TheHeart.org, March 13, 2006, http://www.medscape.com/viewarticle/788296?t = 1 (November 19, 2013).

6 Shelley Wood, "NMT Announces Termination of Its MIST II Trial of PFO Closure for Migraine," *HeartWire*, January 24, 2008, http://www.medscape.com/viewarticle/569169 (November 19, 2013).

Acclarent Case Study

STAGE 2: NEEDS SCREENING

At the same time that Makower and Chang were investigating the chronic sinusitis need, they continued to press forward with investigating needs in orthopedics, respiratory disease, and congestive heart failure (CHF). However, delays in gathering information and gaining access to perform observations allowed chronic sinusitis to become somewhat of a "lead horse" in their efforts to identify which project to work on.

Importantly, they developed more than one need in this area. For instance, in addition to finding a potential alternative to FESS surgery, they also saw the need for improved, more dynamic diagnostics that would allow ENT specialists to more accurately assess a patient's condition. As Chang explained, "In ENT surgery, they use endoscopy and computer tomography (CT) scans to assess symptoms. But the results from these measures don't necessarily correlate. CT scans are static – they provide a snapshot in time, just like endoscopy. So, you can have a picture-perfect CT scan from a week ago, but the patient feels terrible today, and vice versa. So it's pretty obvious that the surgeons are working virtually in the dark and going on their best clinical judgment rather than a single conclusive diagnostic test."[1]

With an understanding of the clinical problems in the area and a working hypothesis for addressing more than one important need, the next step in the process was to perform more detailed research regarding the disease state, existing solutions, stakeholders, and market. Once this was completed, they would be in a much better position to screen the needs and decide which one should become their chief focus.

2.1 Disease State Fundamentals

To understand the disease state, Makower and Chang threw themselves into a thorough literature search. They sought to understand the anatomy and physiology of the sinuses by looking at surgical textbooks, cadaver dissection books, and hundreds of CT scans (see Figure C2.1).

In terms of understanding the pathophysiology of chronic sinusitis, they did more book research and combed peer-reviewed clinical articles, paying particular attention to the disease's mechanism of action. However, as Sharon Lam Wang, a consultant to the ExploraMed II team, pointed out, "Chronic sinusitis is not a given. It's not like we understand what this is. Even the physicians in the field don't agree on the definition. Is it a syndrome? Is it a disease? How do allergies play into it? How does the anatomic makeup of the patient affect the condition? It's a multi-factorial disease that seems to have a life of its own." Lam Wang was an experienced device professional who had worked with Makower and Chang at TransVascular and had played an important early role at both Kyphon and Hansen Medical.

Given the complexity of the disease, Makower and Chang felt that secondary research would not be enough to ensure an adequate understanding of it. Their primary research strategy would involve direct physician input, as well as cadaver dissections. However, they wanted to be highly selective about interfacing with ENT specialists so early in their process. "We needed somebody who we really felt was a safe person, who understood this space very well, but ideally was someone who did not work directly as a rhinologist," explained Makower. "This kind of person would be able to provide great clinical feedback and training for us, but we wouldn't have burned a bridge or potentially mismanaged a relationship with anyone in our target market. We needed someone without a vested interest in what we were doing." The team found such an individual in Dr. Dave Kim, who was an ENT physician, but a specialist in facial plastics for cosmetic and post-traumatic applications. The referral to Dr. Kim came through one of ExploraMed's board members. According to Makower, "We went to him and basically started from scratch. What's a sinus? Where are they located? Why do they remove that structure? We revisited all of the basic physiology, how it functions, and so on. And then we went into the anatomy lab and

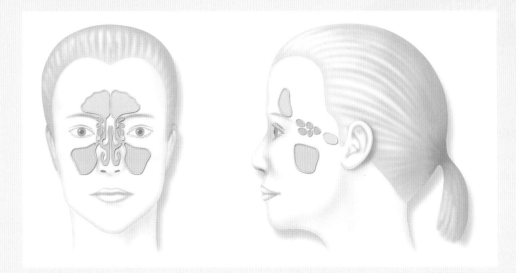

FIGURE C2.1
The sinuses are four bilateral sets of air-filled cavities with ostia that connect to the nose. The frontal sinuses are in the forehead, the ethmoid sinuses are groups of small air pockets located between the eyes, the sphenoid sinuses are behind the eyes, and the maxillary sinuses are in the cheek area. Each sinus has an opening through which mucus drains. The drainage of mucus is a normal process that keeps the sinuses healthy. Mucus moistens the nasal lining and protects the inside of the nose from impurities such as dust and bacteria (provided by Acclarent).

did a detailed dissection with him where he pointed out all the structures, which was very helpful." In terms of a key take-away from the investigation, Makower stated, "For most people it seemed that when you have good flow out of your sinuses, you have healthy sinuses. That seemed to be the bottom line" (see Figure C2.2).

Other important information gleaned from the team's disease research is shown in Table C2.1.[2,3,4]

2.2 Existing Solutions

In parallel with learning about the disease state of chronic sinusitis, Makower and Chang studied existing and emerging solutions in the field. "This is where being a patient myself came back into play," said Makower, "because I had been very frustrated with my choices." He elaborated on the progression of clinical treatment:

> First, there are a variety of over-the-counter drugs, like Sudafed and other allergy medications, to reduce mucus and relieve minor symptoms. At the next level, under doctor's prescription are antibiotics and sprayable steroids, such as Flonase. These always scared me – I never liked the idea of spraying steroids into my nose. Then, before someone would

qualify to get surgery, a lot of doctors prescribed high doses of oral steroids, which can introduce significant side effects. Finally, in the worst cases, functional endoscopic surgery was performed. Over the years, I had seen multiple ENTs regarding the possibility of surgery. But each time, they would look at my CT scans and say, "You know, you just don't have bad enough disease in your sinuses for us to do anything." And that never really made sense to me, because I was really suffering. But, after watching these surgeries, I finally understood why they were steering patients away. They were protecting me from the potential complications of surgery, which are severe when they happen. The surgical tools they use to perform FESS [see Figure C2.3] can sometimes cause scarring, which actually can create a whole new level of disease itself. For a recurrent sinusitis sufferer like me, without surgery, I might have had months of symptoms from swollen sinus that would eventually subside. But after FESS, it would be possible to end up with a permanently scarred ostium that would have had little or no chance of draining, with the possibility of even worse symptoms than I had before.

Table C2.1 Chronic sinusitis is a serious medical condition that affects a significant number of patients.

Sinusitis facts
Sinusitis affects approximately 39 million people each year (or 14 percent of the adult US population), making it one of the most common health problems in the US.
It is more prevalent than heart disease and asthma and has a greater impact on quality of life than chronic back pain, diabetes, or congestive heart failure.
Common symptoms, which can affect patients physically, functionally, and emotionally, include facial pain, pressure and congestion, discharge and drainage, loss of one's sense of smell, headache, bad breath, and fatigue.
Direct healthcare expenditures due to sinusitis total more than $8.1 billion per year.
Patients with chronic sinusitis have symptoms that last more than 12 weeks (patients with acute sinusitis usually have their symptoms resolved in less than four weeks).
Approximately 1.4 million patients are medically managed for chronic sinusitis each year.
Patients with chronic sinusitis make 18–22 million annual office visits per year and have twice as many visits to their primary care doctors and five times as many pharmacy fills as those who do not.

(a)

(b)

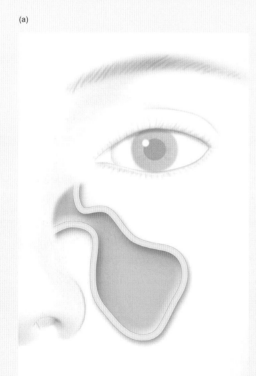
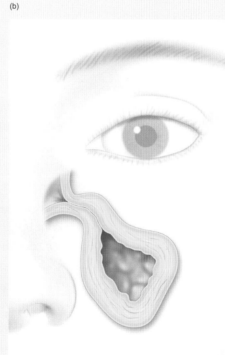

FIGURE C2.2

Sinusitis is an inflammation of the sinus lining most commonly caused by bacterial, viral, and/or microbial infections, as well as structural issues such as blockage of the sinus opening (ostium). If the ostium becomes swollen shut, normal mucus drainage may not occur. This condition may lead to infection and inflammation of the sinuses (provided by Acclarent).

Of the 1.4 million people who are medically managed for chronic sinusitis each year, approximately 970,000 are considered candidates for FESS surgery. The goal of this procedure is to remove bone and tissue to enlarge the sinus opening and restore drainage. Yet, a surprisingly small number of the qualified surgical candidates – just 500,000 per year – undergo the FESS procedure.[5] Some of the most common reasons patients declined the surgery are summarized in Table C2.2.[6]

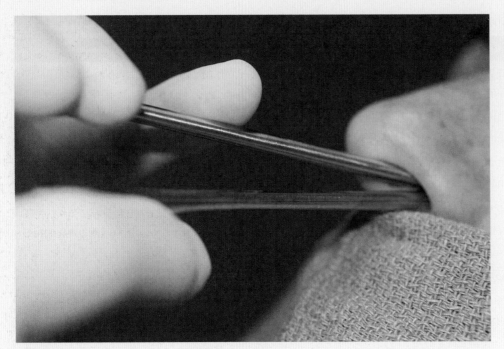

FIGURE C2.3
Functional endoscopic surgery uses straight, rigid tools in the tortuous sinus anatomy. As a result, surgeons are often forced to remove healthy tissue and bone simply to reach the infected sinus(es) (courtesy of Acclarent).

After investigating these risk factors, Makower noted, "It was clear that this procedure had not been completely fixed yet." Chang added, "Patients described feeling terrible for a week or two post-operatively, as if they were recovering from being hit in the face with a baseball bat. This was often accompanied by swelling and the occasional black eye. But, even worse is the fact that there can be a significant revision rate. That's horrible! How do you ask patients to go through such a painful surgery and recovery with an almost one-in-four chance that they will need to go through it again?"

2.3 Stakeholder Analysis

Understanding the risks involved with FESS helped provide Makower and team with insights into how the patients felt about the procedure. As a result, they spent the majority of their time during their preliminary stakeholder analysis evaluating the perspectives of physicians. Research was also performed on payers and facilities, but this work will be described in more detail in the stage 4 section of this case (under reimbursement).

Lam Wang performed much of this physician research, joining the team to determine if the group could realistically make a business in this area. When she got involved in the project, she believed that ENT specialists would respond positively to new innovations in the area of chronic sinusitis:

I thought that they were open to new technology, primarily because of the FESS revolution that had occurred in the 1980s. Before then, sinus surgery was done open through cuts on the face or through the gums to get to your sinuses. They would literally peel away the skin on your face to get to your sinuses. With the advent of the endoscope, physicians switched over to accessing the sinus through the nose. It was almost like an overnight switch, and they adopted the new approach with no clinical data indicating that it worked better. Even when we were researching the problem, there was very little data out there. And so that told me these folks are willing to be practical and open-minded to new ways of doing things.

To confirm this point of view, she conducted interviews with four or five well-respected doctors in the ENT specialty – those physicians who routinely performed FESS. "In these initial discussions, nothing led me to believe that they would be resistant," she recalled.

Table C2.2 FESS procedures can be fraught with pain and risk.

About FESS
FESS is successful in approximately 76 to 98 percent of all procedures[7] (with revision rates up to 24 percent).
Post-operative visits to remove debris, clots, and scars from the surgical site are often needed and can be painful.
Rigid steel surgical instruments are used to perform the procedure through the patient's nostrils, which creates the need to remove bone and tissue so that the physician can reach target sinus and open the ostium to facilitate drainage.
The procedure is conducted under general anesthesia.
Given the proximity of the sinuses to the skull and eyes, there is a small chance with every surgery (one in 200) that the surgeon will inadvertently remove a piece of thin bone between the sinus and the brain, which can cause a spinal fluid leak and can introduce the risk of meningitis.
Blindness is another possible complication from the procedure if the wall of the sinus against the eye socket is breached.
Due to the highly vascular nature of the nose and sinuses, the nasal cavity can sometimes require packing with gauze to absorb bleeding from the procedure, and its subsequent removal can be painful.
Although the procedure is usually performed on an outpatient basis, it is often described by patients as one of the worst experiences of their lives.
Following surgery, it can take weeks for the ancillary effects (e.g., swelling) to subside in order to determine whether the procedure was successful in opening the ostium.

From Chang's perspective, however, there were some mixed signals coming from their physician stakeholders. He explained:

On the one hand, we'd talk to them and they'd claim to be "gadget guys." They'd tell us how much they loved new toys, like powered instrumentation and surgical navigation. This gave us the feeling they might be excited to adopt new technology that did things a little better. On the other hand, as we peeled back the onion, we recognized that most of what they were talking were about incremental innovations taken from other specialties – powered shavers from orthopedics and surgical navigation for neurosurgery. Procedurally, they were still cutting and removing bone, where the shaver and navigation system allowed them to do it more efficiently and safely. These were certainly improvements, but not huge leaps forward in their approach to surgery. They were also very proud of keeping the same instrument set for the last 12 years. They would handle these steel instruments with the utmost care because they didn't want to break them and subsequently have to pay for a new set. This did concern us.

Despite the uncertainty raised by the conflicting feedback gathered by Lam Wang and Chang, the team felt reasonably comfortable pressing forward with their investigation of the needs in this space.

2.4 Market Analysis

The next step was to perform a more detailed assessment of the market. The team believed that to get a truly accurate estimate of the market, it would not be enough to rely on the broad-brush estimates published in the available literature. So, they started from scratch. To determine the incidence of people affected by chronic sinusitis each year, Lam Wang researched the clinical literature and found a study that used ICD-9 diagnosis codes to estimate the true number of adults diagnosed with the condition.

Next, she worked through a cycle of care analysis to identify the "low hanging fruit" – people who do not benefit from drug therapy, are considered candidates

Estimated Market Potential for Exploramed

U.S. Market Model for 2004

294.2	M	U.S. population 2004
1.96%		annual incidence rate of CRS as indicated by ICD-9
5.8	M	individuals diagnosed with CRS
87.0%		non-pediatric
5.0	M	adults diagnosed with CRS in 2004

X 90%		$\geq$ 12 wk symptoms		X 10%		recurrent acute
4.5	M			0.50	M	
X 20%		refractory to medical therapy		X 30%		due to unfortunate anatomy/ostial obstruction
0.9	M	referred to ENT for surgery consult		0.15	M	candidates for surgery
X 90%		positive CT scan		- 0.03	M	FESS on recurrent acute pts today
0.8	M	really have CRS				
- 0.3	M	FESS procedures on CRS pts today				
0.5	M	refractory to meds but don't get surgery today		0.12	M	recurrent acute pts with unmet clinical need
515,705		CRS pts with unmet need for Exploramed		117,501		recurr acute pts with unmet need for Exploramed

Potential initial target market	117,501	Recurrent acute pts with ostial obstruction but no surgery today
Potential target market	515,705	CRS pts refractory to meds and no polyps but no surgery today
Potential expansion populations	330,000	Cannibalize FESS Pediatric Undiagnosed pts who do not seek treatment

TOTAL Potential Market	963,206

FIGURE C2.4
The market analysis for chronic sinusitis pointed to a large unmet medical need (provided by Acclarent).

for surgery, but elect not to undergo the FESS procedure. After diving into all the numbers, validating them with practicing physicians in the field, and testing all the assumptions in the model through an iterative process, she had a preliminary answer. "It turns out it was a phenomenally high number, like 900,000 patients per year, who might be candidates for an alternate procedure," Lam Wang recalled. "So that spoke to a huge unmet need." Importantly, this figure included the more than 300,000 patients per year currently undergoing FESS (assuming they might eventually move toward an alternate procedure). It also took into account a certain type of chronic sinusitis called recurrent acute sinusitis, which met the definition of sinusitis but was "considered a different animal." According to Lam Wang, "These are folks who have frequent episodes, possibly because of their anatomy – a narrowing in their passageway. Getting

at these numbers required a lot of iteration with the doctors." See Figure C2.4 for the basic model.

The fact that there were "tons and tons of possible patients, but not a lot of procedures being performed," indicated to Lam Wang, Makower, and Chang that the current and future opportunity was highly attractive. If they could create a less invasive solution that could be performed in the specialist's office rather than in the operating room, they had the potential to create even greater value by reducing costs in the process of providing an improved solution.

2.5 Needs Selection

By the time the team members had determined that the total potential market for an alternative technology for FESS surgery could reach nearly a million patients per year, they had all the information necessary to confidently commit to this opportunity in ongoing innovation efforts. Enough was known about the disease state to make new solutions feasible, the solution landscape was ripe for innovation, patients were eager for an alternative technology for FESS, and physicians seemed to be generally open to considering new ideas. It was also a strong fit with the defined strategic focus and Makower's related acceptance criteria.

While the team was still investigating needs related to orthopedics, respiratory disease, and CHF (as well as other chronic sinusitis needs), the need for *a minimally invasive approach to treating chronic sinusitis that had less bleeding, less pain, less bone and tissue removal, less risk of scarring, and that was faster, easier, and safer to perform* had strongly overtaken any of the other opportunities being considered. As Makower explained, rather than performing all of their needs screening activities at one point in time, the team had been filtering its needs on more of a rolling basis as new information became available:

We did our filtering along the way. When we teach the biodesign innovation process, we focus on isolating the fundamentals of how people innovate to make each step clear, but in doing so, it can make the process seem like all of these activities occur in sequence. In practice, they can happen almost simultaneously. There's an analogy that I like to use with my project architects: Getting to the finish line with a winning project is like a horse race. You can put the need areas or ideas (horses) in at any time during the race, and they can get knocked out at any

time, too. Some horses start at the beginning of the race and go very far before being knocked out by other screening criteria. Others don't make it very far at all. Eventually there's one horse that keeps on going and gets out ahead of the others, clearing each screening hurdle with ease. Once you see one getting far ahead of the others, you solely focus on that one horse and you try to ride it to the finish line.

As a seasoned innovator and entrepreneur, Makower was comfortable leaving some opportunities behind when others seemed more promising. "You can only win by eventually focusing on one thing at a time. If you spread your efforts too thin, you'll be too distracted to execute well on any one need, and you'll never get there."

NOTES

1 All quotations are from interviews conducted by the authors, unless otherwise cited. Reprinted with permission.
2 "Sinusitis Overview," Balloonsinuplasty.com, http://www.balloonsinuplasty.com/learn-about-sinusitis (September 16, 2013).
3 Carol Sorgen, "Sinus Management Innovation Leads to an Evolution in Practice Patterns," *MD News*, May/June 2007, http://www.clevelandnasalsinus.com/webdocuments/Acclar-Cleveland-md-news.pdf (September 16, 2013).
4 Stephen Levin, "Acclarent: Can Balloons Open Sinuses and the ENT Device Market?," *In Vivo*, January 2006.
5 Ibid.
6 Ibid.
7 R.S. Jiang and C.Y. Hsu, "Revision Functional Endoscopic Sinus Surgery," *Annals of Otology, Rhinology and Laryngology*, February 2002, pp. 55–59.

INVENT → Concept Generation

PHASES	STAGES	ACTIVITIES

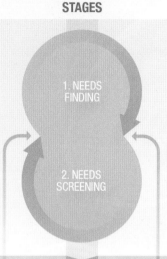

IDENTIFY

1. NEEDS FINDING

1.1 Strategic Focus
1.2 Needs Exploration
1.3 Need Statement Development

2. NEEDS SCREENING

2.1 Disease State Fundamentals
2.2 Existing Solutions
2.3 Stakeholder Analysis
2.4 Market Analysis
2.5 Needs Selection

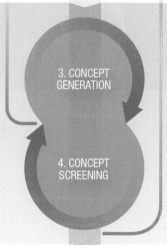

INVENT

3. CONCEPT GENERATION

3.1 Ideation
3.2 Initial Concept Selection

4. CONCEPT SCREENING

4.1 Intellectual Property Basics
4.2 Regulatory Basics
4.3 Reimbursement Basics
4.4 Business Models
4.5 Concept Exploration and Testing
4.6 Final Concept Selection

IMPLEMENT

5. STRATEGY DEVELOPMENT

5.1 IP Strategy
5.2 R&D Strategy
5.3 Clinical Strategy
5.4 Regulatory Strategy
5.5 Quality Management
5.6 Reimbursement Strategy
5.7 Marketing and Stakeholder Strategy
5.8 Sales and Distribution Strategy
5.9 Competitive Advantage and Business Strategy

6. BUSINESS PLANNING

6.1 Operating Plan and Financial Model
6.2 Strategy Integration and Communication
6.3 Funding Approaches
6.4 Alternate Pathways

PROJECT LAUNCH

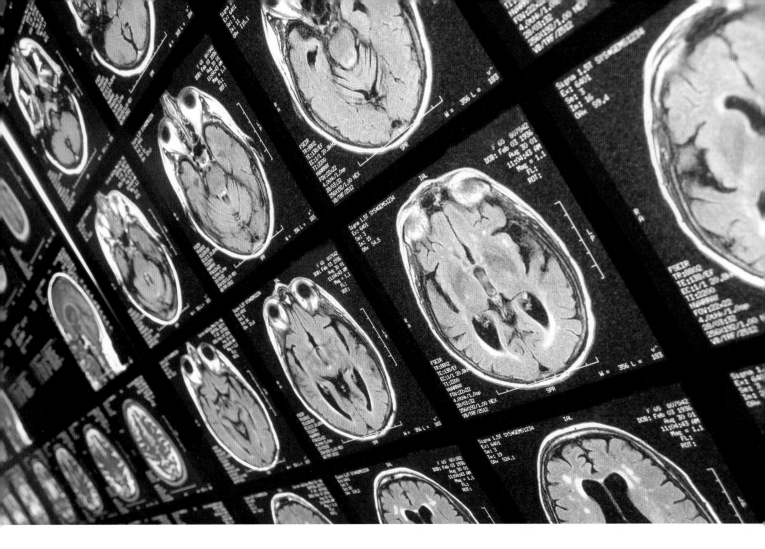

Innovation is now recognized as the single most important ingredient in any modern economy.

The Economist[1]

The devil's advocate may be the biggest innovation killer . . .

Tom Kelley[2]

3. CONCEPT GENERATION

If invention/innovation is so important, why are there so many devil's advocates? Can we banish them . . . at least early on?

Once an important clinical need is clearly identified, it's time to have some fun. It's time to invent. The recurring theme of constant iteration developed in Phase 1 of the biodesign innovation process (Identify) continues as an essential component, during both the concept generation and concept screening stages.

Concept generation, getting the ideas, begins with ideation. One common approach is brainstorming, which originated half a century ago in Alex Osborn's *Applied Imagination*[3] and launched the study of creativity in business development. Its premise is clear. There are three things to work with – facts, ideas, and solutions; *each* deserves quality time. The natural tendency is to leap from facts to solutions, skipping over the play and exploration that should be at the heart of finding new ideas. Most of us are experienced with fact finding; it's a consequence of contemporary education's preoccupation with facts. We're also familiar with solutions; most of us like to solve problems and move on. Idea finding may seem childlike (and it should be) but at its core is the exploration of possibilities *free from as many constraints as possible*. If nothing revolutionary, weird, or goofy surfaces during concept generation, then this stage has failed. The vibe should be upbeat – a chance to try things out, to free associate, and to challenge the wisdom of the present.

NOTES

[1] As cited in Tom Kelley, *The Ten Faces of Innovation* (Doubleday Business, 2005).
[2] Ibid.
[3] Alex F. Osborn, *Applied Imagination* (Charles Scribner's Sons, 1957).

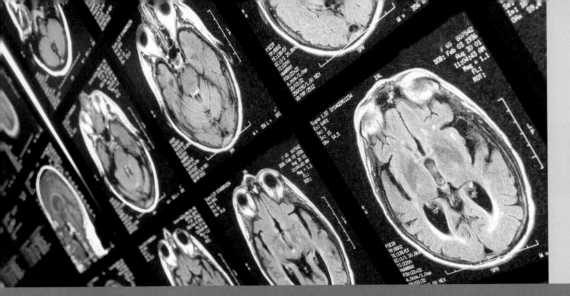

3.1 Ideation

One compelling medical need. Six knowledgeable participants with a range of backgrounds and experiences. Four empty whiteboards. 60 minutes. Plenty of candy. And a goal to develop 75 new ideas – one of which just may be the key to finally delivering a practical solution to a pressing medical need.

"Ideation," which refers to the process of creating new concepts or ideas, is useful in the biodesign innovation process whenever new solutions are required to address well-defined needs. There are various approaches to ideation, developed for either individuals or groups. The most familiar of these is a form of group brainstorming in which participants are asked to suspend their instinct to criticize new ideas and open their minds to a rapid flow of creative possibilities. Other forms of ideation may take a more structured approach to generating ideas that allows for constructive conflict rather than minimizing it. Innovators may wish to experiment with different techniques in their efforts to come up with solutions to address their needs.

 See ebiodesign.org for featured videos on ideation and brainstorming.

OBJECTIVES

- Understand the role of ideation in the context of the biodesign innovation process.

- Learn the basic methods of brainstorming and how to plan and execute a session.

- Consider other ideation approaches and tips that are specific to biomedical technology inventing.

IDEATION FUNDAMENTALS

Since the time of Thomas Edison, the stereotypic picture of the invention process has been the "aha" moment – the burst of inspiration from the brilliant mind of the inventor that is portrayed by a light bulb flashing on. Edison himself, however, made it clear that inventing is a disciplined process that involves patience and hard work, saying, "None of my inventions came by accident. I see a worthwhile **need** to be met and I make trial after trial until it comes. What it boils down to is 1 percent inspiration and 99 percent perspiration."[1] Although he became an icon as an individual inventor, the key to Edison's productivity was that he developed a multidisciplinary *team* of innovators in his Menlo Park, New Jersey laboratory. He also knew that the *quantity* of ideas was important, that *prototyping* was an integral part of invention, and that the *failure* of many potential solutions was inevitable – for example, the lab tried several thousand different filaments before finding a stable material for the first successful light bulb.[2]

Ideation in the medtech field

Invention of medical devices can be more involved and complex than invention in many other technology sectors. In contrast to the situation with many consumer products, healthcare typically involves multiple **stakeholders**, often with competing interests, as well as unique hurdles related to regulation and **reimbursement**. Against this backdrop, innovators may have an "aha" moment at some point, but it will likely come only after investing a lot of hard work and diligence. John Simpson, a pioneering medical device inventor and founder of multiple companies in the cardiovascular space, has a wry but insightful way of describing his inventing process:[3]

> *I know that there have been people who have had these visions and suddenly the design of a catheter pops into their mind and they make it that afternoon and then the following day they achieve enormous wealth. Every morning, I get up, I check my mind to see if it's in there and if it's not, then I go to work. And so far, I haven't missed a day of work yet.*

A few general guidelines related to ideation in the **medtech** field are worth emphasizing at the outset. First, it is important to understand that ideation calls for a mindset that is different from what is required at other stages of the biodesign innovation process. While critical filtering is necessary and important at other points in the process, it can be counterproductive when first considering solutions. Inventors need to open their minds to a creative flow of ideas, set aside their preconceived notions, and look beyond the solutions that they may have been consciously or subconsciously forming during needs screening. It is not important to get the solution completely "right" at this stage in the process. *The need for perfection is the enemy of ideation.*

A second main theme is the importance of cross-pollination in the ideation process. In medtech, abundant opportunities exist to look across specialties to adapt the technologies and approaches from one area to another. Acclarent, a company profiled in this text, is a case in point: it developed a new, minimally invasive approach to functional endoscopic sinus surgery to treat chronic sinusitis by adapting the equipment and procedures of coronary angioplasty that were originally developed for the heart. Medtech is rich with similar opportunities for cross-pollination – between different medical specialties; between physicians and engineers; and even between medical and non-medical technologies. Julio Palmaz, inventor of the balloon-expandable coronary stent, thought of the metal lattice approach when looking at the "chicken wire" used for masonry support.[4] Thomas Fogarty, one of today's most prolific medtech innovators, likes to walk the aisles of a hardware store to look at different materials and tools when he is struggling for new approaches to a medical need.[5]

A third general point is that ideation approaches can be applied at many different stages in the biodesign innovation process. For instance, ideation is particularly useful after the inventor has developed a **need specification** and is ready to begin thinking about different solution **concepts** to address the need. But there are many other circumstances that may be appropriate for this activity, including when an inventor or team is exploring a new or more specific direction for a solution (e.g., finding a delivery method to place a device in a particular spot) or when a team is refining its approach to an already accepted solution (e.g., exploring ways to modify the design of a device to improve its function). It can also be used to address different types of problems, including technical, clinical, or market-related challenges. In fact, solutions to many of the strategic issues that will need to be addressed later in the biodesign innovation process can be developed using an ideation approach.

Finally, it is important to realize that ideation is part of an iterative or cyclic approach. New information and new circumstances crop up at all stages of the innovation process and may require the team to go back into ideation mode. There is an especially important feedback loop between prototyping and ideation: good **prototypes** provide powerful stimuli for new ideas.

A guide to brainstorming

As noted, there are a number of different ideation techniques and a fair amount of controversy about which of these are most effective.[6] For example, some studies suggest that individuals can generate more ideas more rapidly on their own versus in a group.[7] The counterpoint to this view is that group ideation may provide ideas that have more potential for success because they tap into the motivation and capabilities of the team that will be responsible for carrying the concepts forward.[8] Another source of controversy is related to the issue of constructive conflict versus uncritical acceptance of ideas as a guiding principle for group **brainstorming** (more about this later). In the end, as with many other steps in the biodesign innovation process, innovators should experiment with different methods and evolve toward the approach that they personally find most productive and fulfilling.

Framing the question

No matter which approach to ideation an individual or team pursues, an essential first step is to identify the core questions that need to be answered. In the biodesign innovation process, this framing of questions should be based on the need specification – in particular, the core need statement and need criteria (see 2.5 Needs Selection). In order to transition from the need specification to a more actionable process of ideation, it is useful to employ a tool from design thinking called "How Might We" (HMW) questions.[9] The basic idea is to convert the need statement and criteria into a set of focused questions to jumpstart the ideation process using the format "How might we . . .?" for each question (see the Working Example on this topic). Typically, there will be at least several of these HMWs for a given need. The HMWs should probe different areas based on the different components of the need statement (problem, population, outcome) and/or the most important need criteria. Developing the HMWs is really a first step in the ideation process itself and it lends itself well to some of the approaches outlined in the remainder of this chapter. In particular, it is a useful approach to helping the team shift from focusing on needs to thinking about solutions. To apply the technique, team members should generate a number of HMW questions and then select the handful among them that seem most promising or inviting for full-fledged ideation. As a next step, the team could schedule a series of brainstorming session focused on each question and use the HMWs as a prompt to get the creative process going.

Working Example
Generating "how might we" questions

Consider a hypothetical need statement related to the problem of atrial fibrillation: *a way to identify the onset and duration of atrial fibrillation in patients with recurrent AF in order improve the expediency and efficiency of physician visits.* Suppose the must-have need criteria for this need statement are:

- Easy to implement at home or on the road.
- Comfortable and unobtrusive.
- Sensitivity greater than 90 percent and specificity greater than 80 percent.
- Can be read/deciphered in any physician office.

Based on this information, some sample HMW questions could be:

- *HMW make a tiny recorder that the patient will not even notice?* This could be a good start to a brainstorm. There are likely some fun (if impractical) solutions that would come up in ideation.
- *HMW record a heart rhythm at home without touching the patient?* This is a "stretch" question that carries a challenge to consider some blue-sky opportunities.
- *HMW use a cell phone to record rhythm?* This could be a more practical field for ideation, leading to thoughts about ubiquitous technologies that could be "borrowed" for this purpose.
- *HMW employ online resources to read the EKG for the doctor/nurse?* This focuses on another component of the problem: how the information is transmitted to the physician.

FIGURE 3.1.1

A relaxed space for a company brainstorm (courtesy of Cordis Corporation).

and innovation firm IDEO. A more detailed treatment can be found in *The Art of Innovation*, by Tom Kelley, IDEO's general manager.[11] (Other references are listed in the online Getting Started section for this chapter.)

Working Example

IDEO's seven rules for effective brainstorming

1. **Defer judgment** – Don't dismiss any ideas.
2. **Encourage wild ideas** – Think "outside the box."
3. **Build on the ideas of others** – No "buts," only "ands."
4. **Go for quantity** – Aim for 100 ideas in 60 minutes!
5. **One conversation at a time** – Let people have their say.
6. **Stay focused on the topic** – Keep the discussion on target.
7. **Be visual** – Take advantage of your space. Use objects and toys to stimulate ideas.

Although HMW questions can be a fruitful way to initiate ideation and move a team into a "solution mindset," innovators should be cautious not to let this approach limit the range of ideas they ultimately consider. For instance, in the Working Example, using a cell phone to record heart rhythm is only one way to measure heart rate at home or on the road. After "warming up" on this approach, the team should be sure to explore other potential solutions as ideation progresses.

The "classic" approach to team brainstorming

Many of the core concepts behind what we now think of as brainstorming – and in fact the term "brainstorming" itself – originated with the work of an advertising executive from Buffalo, NY named Alex Osborn. Osborn outlined a basic set of ground rules that anticipated much of what we now recognize as brainstorming (though the methods have continued to evolve over time through research and practice in both university programs and commercial design firms).[10] Currently, the most widely recognized list of rules for brainstorming (outlined in the Working Example) comes from the design

1. **Defer judgment** Deferring judgment is perhaps the most counterintuitive and difficult to follow of all the brainstorming rules. The point is to suspend any critical thoughts or commentary until later in the innovation process (well after the brainstorming session). The purpose of brainstorming is to open up both individual creativity and the group's creative process. One good way to make this happen is to accept any new idea – even those suggestions that seem at first to be impractical or silly – and move on quickly to the next concept.

Learning to defer judgment can be challenging. If some of the participants in a brainstorming session are new to the process, it is especially important to explain this first rule and to practice it with an informal warm-up. Another effective way to make sure that this first rule is followed is to be careful when selecting participants for the brainstorm. There are certain people who are extremely valuable as critics and will be essential later in the biodesign innovation process (e.g., during concept screening), but they do not belong in an early-stage brainstorming session because they may negatively affect the flow of creative ideas.

2. Encourage wild ideas When brainstorming, participants should do more than suspend their critical filters. They should actually practice thinking in new and creative ways to generate ideas that are true outliers, meaning that they are different enough to offer the possibility of a real breakthrough. Fun (or even silly) ideas serve an important purpose in that they can stimulate collective creativity by building a connection to something far removed from the current conversation. This, in turn, can lead to new inspirations. As one former member of the IDEO team shared informally, "Great ideas hide right behind the goofy ones."[12] In an effective brainstorming session there is a rhythm or momentum that builds up, and this can be stimulated by group members offering far-fetched ideas. Additionally, wild ideas help keep the energy in the session upbeat and thought-provoking, which is essential to the creative process (see Figure 3.1.2).

3. Build on the ideas of others Building on the ideas of others means leveraging one idea as a foundation from which to make another suggestion. The power of this method is seen as one participant's idea stimulates other participants to come up with solution enhancements, novel connections, and even new ideas they would not have thought of otherwise.

One technique for helping participants with this behavior is to encourage them to explicitly say the words, "Building on [this person's] idea, what if ...." This approach acknowledges the contribution of the individual with a preceding idea, while also offering a way to enhance or improve it without being critical. This works especially well in brainstorming sessions that include participants across different functional areas or disciplines.

4. Go for quantity A successful brainstorming session builds a flow that breaks through the usual inhibitions of a group. One way to achieve this desired effect is to set a target goal for the group to create a large number of concepts without regard to how "good" the ideas are. A typical brainstorming session lasts for about an hour (more than 90 minutes of intense brainstorming can be exhausting and unproductive). Within this timeframe, a team might expect at least 60 new ideas to be generated, with a "stretch" goal of 100 or more. Clearly, developing so many ideas within such a short period of time requires the group to move quickly from idea to idea rather than dwelling on any single suggestion.

FIGURE 3.1.2
A playful warm-up to a brainstorm helps to jumpstart the flow of wild ideas (by Steve Castillo, courtesy of the Stanford Graduate School of Business).

5. One conversation at a time The critical concept underlying this rule is that listening can be as important as talking during the creative process. Enforcing this rule usually falls to the facilitator (see below), although each participant should help keep the group focused on one discussion. As with all of the brainstorming rules, the need for one conversation at a time should be established as an expectation going into the session so that the facilitator has an accepted, non-threatening basis for holding people accountable to the rule if a problem arises. It can be helpful to have the brainstorming rules posted in the room, both as a reminder for everyone and to make it a little bit easier for someone with a concern to point to the rule and say, "Remember: one conversation at a time."

6. Stay focused on the topic Even the most disciplined participants may have a tendency to let their conversations "wander" in a brainstorming session. While these digressions sometimes result in valuable ideas and information, they tend to have a negative effect on both the productivity and the flow of the meeting. To help a group stay focused, try to avoid distractions, side conversations, real-time analysis, and filtering of ideas (which comes later). One useful strategy to minimize the occurrence of these distractions is to have a special flip chart or section of a whiteboard where unrelated suggestions or questions can be set aside until another time (this is sometimes called the "parking lot"). If the momentum of the meeting begins to slow down, the facilitator should try to jump to a new area of brainstorming within the general topic of the session. If the meeting is seriously stalling, the group should consider taking a short break or stopping altogether for the day.

Keep in mind that wild ideas are not digressions, but valuable components of the brainstorming process (even though they may seem somewhat off-topic at the time). However, there is no clear rule for distinguishing one from the other. As a result, the facilitator and participants should not get too concerned about policing this distinction. Remember that brainstorming is an iterative process. If a brainstorming area is important, it will be worth conducting more than one session. With this approach, today's digression may turn into tomorrow's great idea.

7. Be visual Being visual begins with the physical space in which brainstorming occurs. It is useful to have blackboards, whiteboards, large flip pads, sticky notes and/or other means of drawing and writing that encourage an open, fast, and unlimited flow of ideas. Everyone needs to be able to see what is being written and it is useful not to have to erase or cover up pages (that is, have a lot of space to write). It is typical to have a scribe – sometimes two, if the idea flow is rapid – to make sure that all ideas are captured and made visible to participants throughout the session. Another approach is to have everyone write down their own ideas and then take turns briefly presenting them to the group as they post them in the space. Giving everyone a pen and paper can help ensure that people do not forget their ideas, and it can also help ensure that one person talks at a time. When documenting, encourage people to use as few words as possible to describe the idea clearly. Catchy labels or "headlines" are helpful, as long as everyone will remember what they mean.

Cartoon-like doodles can increase the speed and economy of communication. The point is not to create great art, but to stimulate the flow of ideas. This is similar to the game "Pictionary" in which players must use drawings (not words) to help their team guess a word, saying, or concept. It is not the most skilled artist who wins, but rather the player who can most quickly use rudimentary drawings to communicate critical information.

Another strategy for making a brainstorming session more visual is to make various artifacts available (toys and props) since they can be useful in prompting new connections and inspiring insights. Simple props, such as blocks, pipe cleaners, clay, balls, tubes, or Lego, can be used to help create an interactive and fertile mood during the session. They also can be used to visually stimulate ideas and demonstrate three-dimensional (3D) concepts. Tape and staplers help support the assembly of more complex (although still quick and crude) mock-ups. As an example, in one brainstorming session focused on addressing common problems with traditional otolaryngology tools, a whiteboard marker taped to a film canister became the key prop for stimulating ideation (see Figure 3.1.3).[13] Miming, role playing, or physically

FIGURE 3.1.3
A quick prototype from a
brainstorm session led to a
breakthrough in conceptualizing a
new ENT dissection tool (courtesy
of IDEO).

simulating the use of a device can also be a good way to express a 3D concept to others, in order to provide rapid understanding and provoke new ideas.

Tracking ideas as the number reaches 100 or more can be challenging, but keeping the ideas visually displayed in the working space is important. A typical brainstorm can fill multiple whiteboards or flip chart pages, so be prepared with plenty of space for capturing ideas. Planning the spatial locations where ideas will be written down (or where notes are placed) in the room is also important for staying oriented and organized, and should be given careful thought prior to the session. It is often useful to cluster ideas into different "regions" of the room if they seem to be related (sticky notes can be particularly convenient when it comes to clustering ideas). Showing a flow of ideas (i.e., one idea building on the other) with arrows is a different way to increase the visual aspects of a brainstorming session. Another simple strategy is to number all ideas as they are generated so that they can be quickly and easily referred to when building on previous concepts. Some groups like to use computer-based concept maps (see 3.2 Initial Concept Selection) as a way of tracking ideation in real time during a brainstorming session.

Whether and how to use computers and the Internet during a brainstorm depends on the participants and purpose of the session. Online connectivity can be useful to quickly pull up images, search through devices, and watch procedure videos, but the opportunity for distraction and breaking the rhythm of the session is obviously quite high. In general, it is probably best to avoid the Internet while brainstorming unless someone has been designated as the online searcher and has had the chance to prepare ahead of time.

It is not necessary to have a space specifically designed for brainstorming (see Figure 3.1.4) – productive sessions have been conducted in nearly every type of space using flips pads, whiteboards, or sticky notes on the walls. Once the session is underway, check in to confirm whether or not the space and configuration is working for the group. If not, it may be worthwhile to interrupt the session and try a different approach. It is also a good practice to change venues for different brainstorming sessions – go to another room or building, someone's living room, or an inspiring outdoor space. A fresh location may bring a new spark and some new associations into the group's ideation process.

Special considerations for medtech brainstorming
Before identifying participants and a facilitator for a brainstorming session, it is important for the individual or group hosting the session to clearly define the topic to be addressed. For medtech brainstorming, the need specification (see 2.5 Needs Selection) generally should be used as a starting point for defining the scope of the meeting. The idea is to address the core problem

FIGURE 3.1.4

An immersion room designed for brainstorming offers the fun of being able to write directly on the walls! But almost any space can be made effective for ideation with a little advanced preparation (by Anne Knudsen, courtesy of the Stanford Graduate School of Business).

outlined in the need statement while leaving enough room for participants to be able get creative and think beyond existing solutions. Accordingly, the topic for a given session should not be too broad or too restrictive.[14] In practice, it is not always easy to dial in the right "focal zone" in advance of the session, but it is possible to avoid major errors. For example, "reducing the pain and disability of arthritis" is too general a topic to provide traction in a brainstorm, but a session on "slowing or reversing the progression of early cartilage deterioration in the knee" has a good chance at being productive. If the given problem is particularly broad in scope, consider dividing it into multiple topics that can be addressed through a series of brainstorming sessions (e.g., one session on ways to address key aspects of the disease state, another in identifying gaps in the treatment landscape). Sessions can also be divided up according to potential engineering solution sets (e.g., electrical, mechanical, chemical, biologic) or potential business models (disposable, reusable, implantable, capital equipment, etc.). Good topic choices come with practice. Do not worry too much about the topic at first; remember, the group can always schedule another brainstorming session with a different focus or even make a change in the middle of a session when the focus becomes clearer.

Selecting participants Choosing the team for medtech brainstorming can be a challenge, given that domain experts in the medical device field are typically physicians and engineers who, by training, can be reflexively critical when it comes to new ideas (including their own). Medical training is based on the maxim of "doing no harm." The deeply ingrained value of avoiding damage to patients tends to make many experts in the field conservative when it comes to innovation or change. This factor is influential enough that an inventor or team may want to think carefully whether or not to invite experts in the problem area to participate in the brainstorming session, particularly early in the biodesign innovation process. Again, guidance from these physicians and engineers may be more appropriate later in the process, when screening and improving ideas become the focus.

A related challenge with involving experts who are currently active in the field is that they may be disproportionately focused on near-term improvements that can make their work safer or more efficacious. For example, some physicians will be full of good ideas for the next generation of a device (one to two years in the future). However, they may not have spent as much time thinking about breakthrough concepts or radical new ideas that will revolutionize a practice area in the long term (a 10- to 20-year vision). Similarly, executives, managers, and engineers within medical device firms may have a wealth of knowledge to share in the brainstorming process, but bring with them certain preferences or **biases** based on the products their companies are investing in and/or currently marketing.

Among all these cautions about the "expert problem" in brainstorming, it is important to mention that there are many exceptions to this profile – that is, there are many wonderfully innovative physicians, engineers, and executives who have no trouble whatsoever contributing to freewheeling brainstorms. Finding these individuals can supercharge a brainstorming session and lead to impressive results.

With or without domain experts, it can be helpful to include participants with little or no medical expertise, but who bring a unique perspective, or whose background can cross-pollinate the brainstorming session. For example, in addition to one or two physicians with relevant clinical experience in a practice area, it may be useful to invite specialists from other areas unrelated to the one in which the need exists. The contributions of these individuals are often based on technologies or procedures they know from their own fields. Similarly, it is helpful to include different types of engineers in the process (e.g., mechanical, electrical, and chemical engineers). It can also be useful to involve a "maker" in the group – an expert at prototyping devices who can conceptualize how a device might look and perhaps make a quick mock-up in the session.

In developing the list of participants to invite to a brainstorming session, try to anticipate different areas that potentially will come into play in designing and developing a solution (of course, this can be hard to do since it is impossible to predict exactly where the solution will lie). It is also important to find people who understand the field of interest and existing technologies, but equally essential that they have the ability see past their own knowledge and biases so as not to steer the group toward a particular type of solution. For example, if a team is brainstorming new solutions for visualizing the gastrointestinal (GI) tract, which is currently performed using an endoscope, it may want to enlist cross-functional representation from individuals who understand fluids, light, electronics, displays, optics, and the mechanics of scopes. In finding a person with an understanding of endoscope mechanics, the group should target someone with an open mind and the ability to think beyond the known mechanics of today's endoscopes to fruitfully brainstorm new solutions for visualizing the GI tract.

In the same example, it could be helpful to include physicians from the clinical area in which the scope would be used, as well as experts who could represent how other body corridors are assessed (e.g., catheters for blood vessels). The team should not overlook the possibility of including "non-technical" individuals within the group, too. Business people from various functional backgrounds (finance, operations, marketing, sales) can all make strong contributions to the creative process, particularly since they may overlook (or overcome) conventional practices entrenched in the status quo. In the Stanford Biodesign course several years ago, one device concept that went forward to form a start-up company was proposed by an MBA with a finance background as he brainstormed with two PhD engineers and a cardiologist.

Choosing a facilitator The facilitator's job is to run the session, enforce the rules of brainstorming, and make sure the process works smoothly. Facilitators need to be able to stimulate ideas, prevent lingering too long on any one idea, and move the group in fresh directions when idea generation lags (see Figure 3.1.5). Often, it is best to have facilitators refrain from being participants so that they can be completely focused on managing the effectiveness of the session.

Especially for new inventors and teams, it may be helpful to avoid positioning an expert clinician or engineer as the session leader as these individuals may naturally wind up in an authority position within the group, which could stifle less experienced participants from sharing ideas. Depending on the dynamics of the team, the participants, and the topic, having a neutral third party act as facilitator of the session can be effective. The facilitator must, of course, understand the area well enough to effectively navigate the participants' contributions and stimulate their thinking. However, it is

FIGURE 3.1.5
A skilled facilitator can help make a
brainstorming session productive
and fun (by Steve Castillo,
courtesy of the Stanford Graduate
School of Business).

equally important for the facilitator to have the ability to keep the session balanced and well-paced, and serve as a referee regarding adherence to the brainstorming rules.

Although the model of one facilitator per brainstorm is a good starting point, the team may want to consider other alternatives. It can be useful to switch facilitators during the course of a brainstorm to keep the session fresh, to use the talents of the facilitator as a participant, or to make other adjustments in response to a change in the group dynamic that becomes obvious during the session. The point is to stay alert regarding the effectiveness of the session and be flexible about the facilitator's role.

In advance of the session   Once participants and a facilitator have been chosen, the individual or group hosting the brainstorming session should think about what background information can be provided before the session to aid the idea generation process. For brainstorming sessions focused on creating solutions to needs, usually the need specification can be the foundation of this background information (see chapter 2.5). A disease-state fundamentals report (see chapter 2.1), an overview of existing solutions (see chapter 2.2), or other data relevant to the creation of the need specification may also be

provided. However, the aim is to achieve a balance between offering ample information for effective brainstorming without constraining creativity by restrictive framing of background information about the need itself. This is another reason to mix in participants who are not specific experts in the medical or technical areas under consideration.

Medical props   As mentioned, it is useful to make props available to participants in a brainstorm. For medtech sessions, consider having a few real medical tools among the generic blocks and Lego on hand in the brainstorming room. Simple instruments – clamps, retractors, trocars, basic balloon catheters, etc. – can be mixed in with the other artifacts since they can be useful in quickly conveying a concept or inspiring a connection to a new idea (see Figure 3.1.6).

Anatomical drawings and models are other tools that can be useful in a medical-related brainstorming session, as long as the discussion does not get bogged down in detail. Many companies sell large anatomical charts of the different organ systems that are both helpful and visually stimulating. Simple pictures or drawings of a particular operation or procedure can also help motivate ideas. Even better are 3D models of the human anatomy (or portions directly related to the brainstorm focus area,

FIGURE 3.1.6
Medtech brainstorming tools – a creative jumble of anatomy aids, devices and toys (by Anne Knudsen, courtesy of the Stanford Graduate School of Business).

e.g., the GI tract, the heart, the skeleton). These plastic models are typically expensive, but can be purchased online or borrowed from a medical school or specialty practice office. Brainstorming with actual tissue parts is another idea to consider, especially when the physical characteristics of the organ are important to understanding the need and developing potential solutions (e.g., looking at a pig stomach when exploring GI needs – see case example). Just be sure to understand the bio-safety regulations in the chosen setting if any actual tissues will be brought to the session.

Innovators should also think beyond the medical arena for props and gadgets that can be useful in stimulating ideas during medical ideation. For example, one physician noted how he had got inspiration for new surgical tools from devices as diverse as bicycle components and ponytail clamps. At IDEO, the professional designers maintain a collection of interesting devices and toys with different mechanisms. As Tom Kelley described, "The complex and extremely reliable mechanisms of rock-climbing camalots or the simple mechanism of a Japanese folding razor can get people thinking in new ways as they hold the those items in their hands." Think creatively about materials and parts that might make a fruitful addition to the ideation process and make a variety of tools available to participants.

Sugar, caffeine, and brainstorming endorphins It is part of brainstorming ritual to provide participants with candy, cookies, and cola to help ensure a high energy level. Whether or not a medtech group thinks a metabolic "buzz" is desirable (or healthy!), it can certainly help create a positive, engaging, and upbeat environment. The props and the set-up all contribute to the right tone and feeling for the session. It can also be a good idea for the facilitator to encourage participants at the outset of the meeting by pointing out that the attendees have been chosen for their talent and creativity, and conveying that it is an honor to have been invited. The positive reinforcement of ideas also goes a long way and can be helpful as the group is just getting going.

Managing the session It is essential to be clear with participants that the brainstorming session will require 60 to 90 minutes of uninterrupted attention. Physicians are especially vulnerable to being phoned or texted by the hospital. If possible, get them when they are not on call or are able to "sign out" to someone else during the session. It is appropriate to ask participants to turn off their communication devices during the meeting. If an important interruption occurs, the participant should leave the session to take the call. The idea is to focus on the topic, avoid interruptions, and maintain the momentum that the group has built.

At the beginning of the session, invest a few minutes in doing some kind of quick warm-up exercise (the design thinkers call this "stoking"[15]). This is particularly valuable when dealing with participants who are new to the brainstorming process. An unrelated ice breaker can help in making this transition. Finding an "easy" medical-related exercise that participants should be relatively familiar with can be an effective way to get started. For example, the facilitator might ask the group to come up with as many suggestions as possible within two minutes for improving a hospital bed. By making it clear that s/he is looking for creative and even wacky suggestions delivered at a rapid rate, the facilitator can bring the group up to brainstorming speed without spending too much time on a warm-up. Another approach is to coax a warm-up need from the group. Ask a simple question like, "What was the most irritating technical problem you had today?" Make a quick list of responses on the whiteboard and then choose the one that seems to resonate most with the group. Dedicate just a few minutes to coming up with solutions, with the facilitator encouraging the group to think of quick, interesting, and goofy responses. Some groups may appreciate a non-medical warm-up. Try brainstorming a topic such as "How could you prove there is life on Mars?" Another strategy is to watch a short video, for example, with a far-out science fiction scene full of cool inventions. It can also be good to get participants up and out of their chairs for a short game. For instance, for a round of "category, category, out!" the facilitator asks participants to line up, names a category (breakfast cereals, vegetables, animals, car manufacturers, etc.), and then points at each person in rapid succession, skipping around the group. Each player has to name something in the category. If someone cannot, everyone yells "out!" and that player sits out for the next round.[16] The point of stoking is to briefly engage everyone, build a rhythm of rapid responses, and have some fun!

One of the most challenging issues for both the facilitator and the group can be dealing with the spectrum of personality types among participants. One of the criticisms of the classic brainstorming approach is that it is designed for extroverts and creates an environment that can be difficult for introverts (who may be equally or more creative).[17] If the facilitator discovers that there is a reticent or quiet member or two in the group, it is important to be encouraging but not heavy-handed in inviting participation. It is absolutely not the case that everyone in a brainstorm must contribute equally – in fact, one of the best ways to kill ideation is to "cold call" on participants to make sure everyone is getting equal say.

On the other end of the spectrum is the overbearing, controlling, or dominating participant. Here the facilitator, joined by any brave members of the group, needs to help direct this participant toward a more positive contribution. The brainstorming rules generally provide a useful framework for making these suggestions. If this is not successful, the facilitator might choose to declare a break and speak to the participant privately.

A less extreme version of this same issue is a brainstorming session in which there is a hierarchy of participants. This problem can arise in business or academic settings and requires both a skilled facilitator and the willingness of the group to overcome. Different cultural contexts can also lead to this particular challenges. For example, in reflecting on his experience brainstorming in India, Consure Medical co-founder Amit Sharma remarked, "Not everyone will speak up with the same frequency. For instance, sometimes the more experienced, senior people think that their ideas should be given more weight compared to the other people in the room. And the other people may defer to those who are more senior and experienced. The emphasis on the true merit of the ideas can occasionally be missing." Here the challenge for the moderator is considerable, requiring a blend of deep cultural knowledge and leadership skills.

Capturing the results It is essential to capture and preserve the results of every brainstorm with as much accuracy and richness as possible. If large paper or sticky notes have been used, these should be gathered up and saved. It is worth paying attention to the spatial order in collecting them, because remembering where a particular note was physically made in the room may aid more detailed recollection later on. If a follow-on brainstorm is conducted, it can be useful to place the sheets back in the position they were in at the end of the first session. When whiteboards are used, digital photographs of the boards are convenient. The photos can be projected later for a second session and are also easy to archive electronically. It is important that some method of recording is performed immediately at the end of the brainstorm, even if that requires the brute force method of simply copying down the output of the session. Again, it is most useful to do this in a way that is spatially close to the original organization. Remember to include any numbering scheme that was used in the session. Further recommendations for collecting and organizing the results of a brainstorm are presented in 3.2 Initial Concept Selection.

In capturing the results of a brainstorm it is important to assess whether all of the solutions are sufficiently well thought out to potentially take forward into concept screening and prototype development, or whether one or more ideas may still be too general and in need of further brainstorming. For instance, the idea to "capture a clot in the blood vessel" is representative of a general approach that is not yet actionable as a potential solution. In comparison, the idea to "use a stent with integrated mesh to a capture clot in carotid artery" is far more concrete. Of course, not all general concepts will yield workable solutions. It is a matter of judgment to decide whether the team should invest in additional brainstorming on the approach, or whether to set aside the general concept and move on.

Intellectual property ownership Brainstorming creates new ideas, and new ideas are the heart of new intellectual property (IP – see 4.1 Intellectual Property Basics). The inventor or team needs to be clear with the members of a brainstorming group regarding the implications of their participation in the session with respect to IP. If the participants are all from the same company or university, the ownership of an important idea may stay within that organization (and, thus, ownership is

relatively uncomplicated). If the team employs a facilitator or other participants who are not part of the company or university, it may be reasonable to ask them to sign a non-disclosure agreement (see chapter 4.1).

Even in a situation where all participants are from the same entity, recognize that each participant may become an inventor as a result of ideas discussed in the session and that ownership should be decided in a fair and equitable manner. It is impossible to give specific guidance on how to do this, as each situation is unique. In practice, it will only be clear later in the biodesign innovation process whether an idea coming from a brainstorm session will move forward into a **patent**. Basically, as described in chapter 4.1, an inventor is someone who materially contributes to one or more claims of a patent. In some cases it will be clear that a single individual from the group came up with the key concept that was patentable, while in other cases it will be some combination of individuals or the whole group. One way to clarify this from the outset is to jot down the initials of the key contributor(s) alongside the idea, in real time, as the concepts are generated. The disadvantage of this approach is that the participants may get distracted by concerns about which ideas they do or do not "own." A more subtle method is to ask each participant in the session to use different colored sticky notes so that ideas can easily be linked to individuals. Ultimately, it is the responsibility of the team leader (the person or people responsible for driving the project) to make a determination of IP ownership in consultation with the group's members. It is, of course, best to clarify who is an inventor and what his/her relative contribution is, as early as possible in the patent filing process.

The case example demonstrates how all of these considerations come together in planning and executing a medtech brainstorming session.

Other approaches to ideation

Although classic brainstorming may be the most popular form of ideation being taught at universities, schools, and corporations around the world as the core of "design thinking," it is certainly not the only effective technique for generating concepts.

FROM THE FIELD ⟩ A TEAM IN THE STANFORD BIODESIGN PROGRAM

Brainstorming early concepts in the obesity space

As part of a biodesign innovation course, Darin Buxbaum, a first-year Stanford MBA student, and his team collaborated to address the need for *a less invasive way to help morbidly obese people lose weight*. In addition to Buxbaum, who had a business background in the medical device field, the team included one postgraduate resident who had completed a rotation in general surgery and was working on a specialty in plastic surgery, two students pursuing master's degrees in bioengineering with a specialty in biomaterials, and a PhD student in bioengineering with a background in mechanical engineering (see Figure 3.1.7).

After investigating the disease state of morbid obesity, understanding the current solution landscape, performing stakeholder and market analyses, and developing a need specification, the team was ready to begin the process of concept generation. "The need was so big to begin with," recalled Buxbaum, "that we agreed to focus our first brainstorming session on coming up with broad mechanisms for losing weight, like increasing energy expenditure or reducing caloric intake. And then we would devote a session to exploring the categories that seemed to be the most promising."

All of the team members were relatively new to brainstorming, so they decided to bring in an outside facilitator to lead the first session. "When we were first getting together, there was no formal organization," noted Buxbaum. "Bringing in an outside facilitator helped impose some structure on the brainstorm – there was a designated facilitator and everyone else on the team was an equal contributor." With limited funds at their disposal during this early stage, the team called in a favor with a friend who worked at a design firm, asking him to lead the session. However, according to Buxbaum, it would have been worth paying a junior design consultant to facilitate based on the value they extracted from the role. Because the chosen facilitator was a personal friend, they did not ask the individual to sign a non-disclosure agreement. However, Buxbaum advised other teams to be more cautious. "You just never know who's going to be the one to come up with something really interesting in this kind of session," he said.

FIGURE 3.1.7
Buxbaum (standing by laptop) and his team at an early brainstorming session (by Anne Knudsen, courtesy of the Stanford Graduate School of Business).

The facilitator started the session by leading a couple of short, fast-paced warm-up exercises to get the team going, for example asking the participants to name as many different animals as they could in one minute. "This is something we would have completely glossed over if we hadn't used a facilitator," commented Buxbaum. "But it really helped people get used to the approach of throwing out ideas as fast as they could and building on the ideas of others."

Where the facilitator really demonstrated value, however, was in helping the group manage a team member who had difficulty allowing for wild and crazy ideas without passing judgment. According to Buxbaum, "Any time a really far-fetched idea would be thrown out like, 'Let's put a black hole in the stomach,' this person would say, 'That's not possible.' He complained about any idea that wasn't physically feasible. In the end, he shut down and stopped giving suggestions. It was really frustrating for the team, and must have been painfully frustrating for him. If it wasn't for the facilitator, we probably would have had a team argument about what we should and shouldn't be able to brainstorm." However, the facilitator was able to mediate the situation, enforce the rules, and keep the session going. Over time, the problematic team member grew increasingly comfortable with the brainstorming approach. "In subsequent sessions, he loosened up and eventually came up with some of our most creative ideas," noted Buxbaum. In later sessions, the team was able to facilitate the brainstorms on its own. "The norms became so ingrained in us that we were able to self-police," he said.

Because the team members came from a diversity of backgrounds, they decided that some advance preparation was required to get everyone up to speed before each session. "One of the things we worked incredibly hard on was looking at the mechanisms by which the human body works," said Buxbaum. "The physician on our team would create a presentation before each meeting and teach us everything about the relevant anatomy. For instance, if we wanted to talk about mechanical ways for making people eat less, he went over all the different parts of the stomach and how

they interact. One time before a session, he presented the entire gastrointestinal system, how the chemical pathways work, and how the hormones are released to stimulate satiety. Having that depth of knowledge led to some really neat ideas. For example, one of our more interesting concepts was to put a stent in the intestines where it would elute fatty acids to activate a hormonal pathway to make people feel full and slow down the entire digestive tract." However, Buxbaum admitted that the concept of advance preparation could be taken too far. "One way that we may have over-prepared was to look at IP early in the process. While it seemed like a good idea to understand what concepts were already being worked on, this may have actually clouded our judgment and put unnecessary constraints on our thinking," he said.

Another technique that proved to be invaluable to the team was using drawings, crude prototypes, and medical props during the sessions, including pig stomachs in a few sessions (see Figure 3.1.8). In fact, Buxbaum credited one of the team's major breakthroughs to its use of the stomachs in a brainstorming session: "We were playing with a pig stomach in one of our meetings, literally just manipulating

FIGURE 3.1.8
Having a pig stomach available in the brainstorm provided the key insight for the team's invention (courtesy of Jennifer Blundo, Darin Buxbaum, Charles Hsu, Ivan Tzetanov, and Fan Zhang).

it while we were thinking, and it just flopped into the right geometry. We never would have imagined that the stomach could take this particular shape unless we had a real piece of tissue in front of us." Acquiring the pig stomach was an adventure in itself. First the team tried a traditional butcher shop. "They made a really big deal about it," remembered Buxbaum. "They said they would have to call the place where they got their meat and convince them to stop the line, pull out the stomach, inspect it, and package it for us. They charged us $30 and it took about two weeks for us to get it." He continued: "In the meantime, a couple of our team members happened to be at a Chinese grocery and saw a whole stack of pig stomachs for $1.99." These stomachs ended up working just as well, at much lower cost and greater convenience to the team.

In total, the team held approximately 12 brainstorming sessions that typically ranged from 60 to 120 minutes. "Usually, we wouldn't go longer than 90 minutes," said Buxbaum. "Our longest session was probably two hours but that was pretty draining. If we went over our time limit, we would table the process and pick it up in another session." In these sessions, the team's goal was to generate two to three ideas a minute, sometimes walking out with as many as 200 ideas from a single meeting.

Reflecting back on the process, he noted, "We have a lot of fond memories of brainstorming. It was a real coming together for the team." In fact, Buxbaum recommended using brainstorming as a technique for fueling progress beyond concept generation. "Whenever you're dealing with an undefined space, brainstorming can really help you fill it out. Our team didn't brainstorm to figure out how to design the device. To our detriment, we sat around tables at coffee shops or in people's living rooms instead. The space wasn't as good and we didn't use all the rules. As a result, our ideas weren't as creative and we didn't get anywhere near the same quality."

One alternate approach focuses on constructive conflict as a mechanism for stimulating effective ideation. This more critical or debate-oriented style of ideation sidesteps the cardinal rule of brainstorming to suspend judgment. Yet, it has been shown in some situations to produce an advantage in creative output when compared to traditional brainstorming.[18] Proponents for this approach assert that dissent opens the mind, allowing for a broader consideration of different possibilities. They also believe that competition can be effective in stimulating creativity.[19] However, there are some important conditions to making this approach work. As one professor summarized, "Whenever you're fighting about ideas ... it's important that you're engaging in the 'right fight,' criticizing another person's ideas and not the person himself. This type of conflict, what researchers call 'intellectual' or 'task' conflict, must be done in an atmosphere of mutual respect and must be based on the factual information available."[20]

Team members at The Foundry, a successful medical technology incubator in Menlo Park, California, have embraced the creative conflict approach. As managing partner Mark Deem explained, "When we brainstorm at The Foundry, we feel comfortable challenging each other. We've been together a long time – we trust each other and we respect the different perspectives each of us brings. So we're not afraid to mix it up a bit. There *are* bad ideas and wrong assumptions, and when we're in a brainstorm, we challenge those. Not, 'that's a stupid idea,' but 'that's not going to work because of this.' So it's about being factual, not critical."

Another variation on ideation seeks to take advantage of the deep knowledge of one or more experts by abandoning the "horizontal" or non-hierarchical philosophy of traditional brainstorming. In particular, medical device companies often use a format where they invite in a physician key opinion leader (**KOL**) to visit the company and participate in a group ideation session. While this type of facilitated discussion may draw on many of the same aspects of the brainstorming process described in this chapter (e.g., having different types of people from engineering, sales and marketing, etc. in attendance), it differs in one important way. The purpose of these working sessions is usually to "uncork" the expert's mind and stimulate interesting ideas. Unlike a brainstorming session in which all participants are encouraged to contribute more equally, in these situations the group is asymmetrical, with the expert's ideas being given the most attention. Despite this important distinction, the presence of the group and the upbeat, thought-provoking nature of the interaction can help stimulate powerful ideas.

A substantially different approach to ideation is represented by the TRIZ method. Developed by a Soviet inventor (Genrich Altshuller) and his colleagues from the 1940s through 1980s, TRIZ advocates a highly structured, pattern-driven approach to generating solutions (the acronym is Russian for the Theory of Inventive Problem Solving). Like brainstorming, it is meant to create a leap forward in ideation, especially when there is some form of barrier or blockage impeding progress. However, TRIZ does not rely on the collective intuition of the team. Instead, the core philosophy is that it is possible to tap into ideas from other inventions, particularly those in other industries, to find solutions to the current need or problem. In short, "Somebody someplace has already solved this problem (or one very similar to it) …. Creativity involves finding that solution and adapting it to this particular problem."[21]

To help innovators tap into the solutions of other inventors, the TRIZ creators analyzed three million patents to identify the "contradictions" or trade-offs that they believe are at the root of most problems. These can be engineering trade-offs (e.g., in order to provide more diagnostic functionality in a pacemaker, the complexity increases and battery life decreases) or physical contradictions (e.g., in order to create a lower catheter profile, thinner balloon material is required, which means the balloon will rupture more easily after inflation). They then codified common problem-solution patterns into 40 Principles of Problem Solving, which are available in a knowledgebase along with a collection of tools for applying them in a structured way.[22] Importantly, the TRIZ approach may be particularly useful when ideation is applied to concept exploration and testing.

A final note on ideation

Consistent across all approaches to ideation is the fact that the process of coming up with new solutions to fundamental problems is a challenging one. Even professionals in design firms who routinely brainstorm with their expert colleagues continue to develop and learn new techniques to help them more effectively address the challenges of this work. These experts also point out that there is an important element of "staying in shape" for ideation, both for the individual and the team. Like other forms of creative and intellectual activity, practicing the skill helps maintain a level of ongoing fitness for the undertaking.

↘ Online Resources

Visit www.ebiodesign.org/3.1 for more content, including:

 Activities and links for "Getting Started"
- Understand basic ideation and/or brainstorming concepts
- Define the topic
- Identify participants
- Choose a facilitator
- Prepare for the session
- Conduct the session and capture output

 Videos on ideation and brainstorming

CREDITS

The content of this chapter is based, in part, on lectures in Stanford University's Program in Biodesign by Dennis Boyle, David Kelley, Tom Kelley, George Kembel, and Tad Simons. The editors would also like to acknowledge Darin Buxbaum, Asha Nayak, and Tad Simons for their help in developing the original chapter, as well Tom Kelley for reviewing the updated material. Additional thanks go to Darin Buxbaum, Jennifer Blundo, Charles Hsu, Ivan Tzvetanov, and Fan Zhang for contributing to the case example, and to Amit Sharma for providing examples.

NOTES

1 "Thomas Edison," Wikiquote.org, http://en.wikiquote.org/wiki/Thomas_Alva_Edison (October 2, 2013).

2 Paul Israel, *Edison: A Life of Invention* (John Wiley & Sons, 1998).

3 From remarks made by John Simpson as part of the "From the Innovator's Workbench" speaker series hosted by Stanford's Program in Biodesign, March 3, 2003, http://biodesign.stanford.edu/bdn/networking/pastinnovators.jsp (October 2, 2013). Reprinted with permission.

4 From remarks made by Julio Palmaz as part of the "From the Innovator's Workbench" speaker series hosted by Stanford's Program in Biodesign, February 10, 2003, http://biodesign.stanford.edu/bdn/networking/pastinnovators.jsp (October 2, 2013). Reprinted with permission.

5 From remarks made by Thomas Fogarty as part of the "From the Innovator's Workbench" speaker series hosted by Stanford's Program in Biodesign, January 27, 2003, http://biodesign.stanford.edu/bdn/networking/pastinnovators.jsp (October 2, 2013). Reprinted with permission.

6 Scott G. Isaksen, "A Review of Brainstorming Research: Six Critical Issues for Inquiry," Monograph #3-2, Creative Problem Solving Group, June 1998, http://www.cpsb.com/resources/downloads/public/302-Brainstorm.pdf (November 16, 2013).

7 Donald W. Taylor, Paul C. Berry, and Clifford H. Block, "Does Group Participation When Using Brainstorming Facilitate or Inhibit Creative Thinking?," *Administrative Science Quarterly*, June 1958, pp. 23–47.

8 Warren Bennis and Patricia Ward Biederman, *Organizing Genius: The Secrets of Creative Collaboration* (Addison-Wesley, 1997).

9 Warren Berger, "How Might We: The Secret Phrase Top Innovators Use," Harvard Business Review Blog, September 17, 2012, http://blogs.hbr.org/2012/09/the-secret-phrase-top-innovato/ (February 6, 2014).

10 Alex Osborn, *Your Creative Power* (C. Scribner's Sons, 1948).

11 Tom Kelley, *The Art of Innovation* (Currency/Doubleday, 2001).

12 All quotations are from an interview conducted by the authors, unless otherwise cited. Reprinted with permission.

13 "Diego Powered Dissector System for Gyrus ACMI, ENT Division," IDEO.com, http://www.ideo.com/work/diego-powered-dissector-system/ (November 6, 2013).

14 Tom Kelley, *The Ten Faces of Innovation* (Doubleday, 2005), pp. 151–152.

15 "Ideate Mixtape," Hasso Plattner Institute of Design, Stanford University, 2012, http://dschool.stanford.edu/wp-content/uploads/2012/02/ideate-mixtape-v8.pdf (November 16, 2013).

16 Ibid.

17 Susan Cain, "The Rise of New Groupthink," *The New York Times*, January 13, 2013, http://www.nytimes.com/2012/01/15/opinion/sunday/the-rise-of-the-new-groupthink.html?pagewanted=all&_r=0 (November 20, 2013).

18 Charlan J. Nemeth, Bernard Personnaz, Marie Personnaz, and Jack A. Goncalo, "The Liberating Role of Conflict in Group Creativity: A Study in Two Countries," *European Journal of Social Psychology*, July/August 2004, http://onlinelibrary.wiley.com/doi/10.1002/ejsp.210/pdf (October 1, 2013).

19 David Burkus, "Why Fighting Over Our Ideas Makes Them Better," 99u.com, http://99u.com/articles/7224/why-fighting-for-our-ideas-makes-them-better (October 1, 2013).

20 Ibid.

21 Katie Barry, Ellen Domb, and Michael S. Slocum, "TRIZ – What Is TRIZ?," *TRIZ Journal*, http://www.triz-journal.com/archives/what_is_triz/ (October 2, 2013).

22 "TRIZ 40 Principles," TRIZ40.com, http://www.triz40.com/aff_Principles.htm (October 2, 2013).

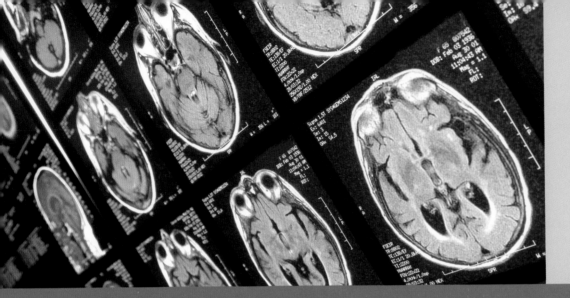

3.2 Initial Concept Selection

INTRODUCTION

A "leap of faith" is an act of accepting or trusting in something that cannot readily be seen or proved.[1] More than any other step in the biodesign innovation process, initial concept selection qualifies as a leap of faith. Based only on information about the need, innovators choose a handful of promising solution concepts to develop further from dozens or even hundreds of ideas coming out of their ideation sessions. At this point, they have little or no hard information about which solutions may ultimately best solve the need – just a host of questions. Is the solution truly technically feasible? Is enough understood about the underlying mechanisms of the disease that the solution targets? Is there sufficient insight into critical interactions between the different engineering elements to make the solution viable? How much time, effort, and money will it take to determine the practicality of the idea? How will doctors and other important stakeholders respond to the concept? In the face of these unknowns, innovators leap forward, setting aside the majority of their ideas to investigate the few that appear most likely to succeed.

OBJECTIVES

- Understand how to cluster and organize the output of ideation so it can be analyzed in order to determine whether additional ideation is needed.

- Learn to objectively compare solution concepts against the criteria in the need specification to determine which concepts to pursue.

Being selective about which concepts to take forward and further investigate from the many possibilities generated through ideation is an important step of the biodesign innovation process. Similar to the needs selection process, performing initial concept selection well increases an innovator's efficiency by helping triage concepts so that only the most favorable few move on to more detailed investigation. Spending lots of time researching many concepts in detail is not worthwhile as the bulk will be eliminated.

Initial concept selection involves comparing all of the ideas generated against the need statement and need criteria defined in the need specification to evaluate how well each solution may (or may not) solve the need. As part of this process, innovators usually organize their ideas into related groups to identify potential gaps or biases in the proposed solutions. If significant gaps are found or certain biases are discovered that unnecessarily constrain the solutions, additional ideation may be required to generate new solutions that directly target these issues. Ultimately, the output of initial concept selection is a few promising concepts to take into the concept screening stage of the biodesign innovation process.

 See ebiodesign.org for featured videos on initial concept selection.

INITIAL CONCEPT SELECTION FUNDAMENTALS

As described in 3.1 Ideation, one of the key rules of brainstorming is to "go for quantity," encouraging as many ideas as possible. This is important to ensure that many varied solutions are considered before an innovator settles on any particular **concept**. However, generating a vast number of ideas can create a challenge in that it is unrealistic for the innovator to research and develop them all. At the most basic level, the process of initial concept selection can be thought of as an objective, comprehensive method for organizing and evaluating the information generated through ideation in order to go from many concepts to a few.

Transitioning from ideation to initial concept selection

Before beginning initial concept selection, innovators should first carefully review the raw output from their ideation sessions to make sure that all ideas are fully and accurately captured and clearly understandable (see Figure 3.2.1). If they realize that many of the ideas they generated still represent general approaches rather than concrete solutions, they may be well served to go back to ideation to expand the ideas into clearer solutions before going further with initial concept selection.

Ideally, reviewing the raw output of brainstorming should be done immediately following the session so that the innovator can seek clarification from the participants

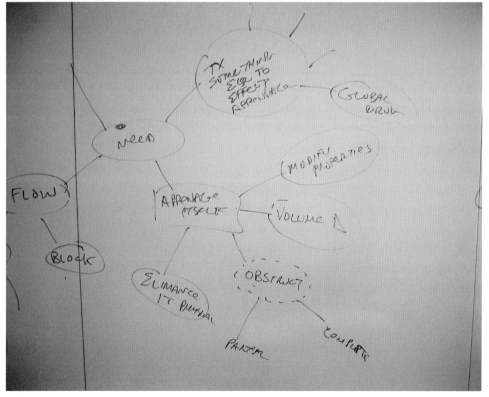

FIGURE 3.2.1

Example of the raw output from a brainstorming session (provided by Uday N. Kumar; developed with John White, Kityee Au-Yeung, Joseph Knight, and Josh Makower).

if something is unclear. Misinterpretation can be a source of error and may inadvertently lead to the elimination of potentially good ideas. If each idea is linked to the participant(s) that contributed it immediately following (or during) the brainstorm, as suggested in 3.1 Ideation, it should be relatively easy for the innovator to follow up.

Remember that it is important to consider all of the notes, pictures, drawings, diagrams, and three-dimensional mock-ups that were created during the brainstorm, and not just the documentation produced by the scribe. The goal is to review all ideas, in their entirety, so that they can be adequately assessed.

Once innovators feel comfortable that they have accurately identified and interpreted all of the raw data from ideation, and that the ideas seem to represent well-defined concepts, the information can be recorded in some meaningful and easily understandable way. One approach is to tag each idea with a short phrase or name to make it quickly identifiable. By standardizing the way ideas are described, an innovator can more efficiently compile and organize the concepts. For example, in the obesity space, an idea to put an implant in the stomach that elutes a substance to simulate satiety, or the sensation of feeling full, might be named "drug-eluting implant."

Grouping and organizing ideas

Once the raw data from ideation have been reviewed, "cleaned up," and labeled, innovators can group and organize the concepts. There are many methods for accomplishing this, but the most effective techniques usually involve clustering the ideas according to some organizing construct, such as the mechanism of action or presumed technical feasibility, and then visually organizing them into a **concept map**.

Organizing concepts is important because it allows innovators to identify gaps, biases, and synergies among the ideas. Identifying gaps among groups of solutions is necessary so that additional ideation can be performed to address any opportunities that may have been missed. For example, if an innovator working on the **need** for *a way to prevent strokes due to clots coming from the heart* generated a preliminary set of solutions that did not include any concepts involving the blood vessels that convey clots to the brain, this would be an obvious gap to focus on during subsequent ideation sessions.

The innovator should also watch out for biases in the types of solutions proposed. For example, if all the concepts initially generated to solve the stroke need were heavily focused on mechanical engineering solutions, involving individuals from other engineering disciplines might lead to different approaches and concepts (e.g., chemistry or drug-based solutions).

An effective approach to organizing concepts also allows the innovator to identify commonalities and complementarities between concepts so that they can be merged to create synergistic, combined concepts. For example, for the same stroke need, one group of solutions might focus on meshes to capture clots while another set focuses on medications to dissolve clots. By putting these clusters together, a new set of ideas, centered on a mesh that elutes a substance to dissolve a clot, may emerge. In many cases, the combination of concepts can result in a stronger overall solution.

Clustering

The notion of arranging ideas into groups is simple in theory, but can be difficult to accomplish in practice (requiring more art than science). At its most basic, clustering requires the examination of ideas to identify common themes or similarities and then grouping them based on their related characteristics. For instance, the roughly 100 ideas generated in a brainstorm may fit within 10 general clusters based on the most dominant traits they have in common.

The first step in the grouping process is to identify the primary organizing principle for creating the clusters. Some examples of common ways to group ideas in the **medtech** field are outlined in Table 3.2.1. The challenge to the innovator is to find the most meaningful organizing principle for the solutions being evaluated. Factors like anatomical location and engineering area are simple to understand and provide easy comparisons, but they may not be relevant for some needs. Factors such as mechanism of action may be harder to define and apply, but potentially could be more significant in terms of identifying the similarities and differences between ideas and for assessing the likelihood of success. Innovators

Table 3.2.1 Organizing principles that can be used to help cluster ideas initial concept slection.

Organizing principle	Description	Example
Anatomic location	Group ideas according to the part of the anatomy they pertain to and/or target. Differences between groupings might be small if all solutions are in a highly focused area (e.g., the vertebral discs). Alternatively, groupings might span entire regions/organ systems if the solutions focus on a need pertinent to a significant portion of the body (such as the various places to which emboli dislodged from the heart can travel).	For solutions to address the problem of obesity, ideas might naturally cluster around the mouth, esophagus, stomach, pylorus, small intestines (duodenum, jejunum, and ileum), large intestines, and the various valves in the GI tract.
Mechanism of action	Group ideas according to how the solutions are intended to work.	Increasing energy expenditure, regulating food intake, reducing nutrient absorption, and reducing the motivation to take in energy (or eat) are all different mechanisms of action for reducing weight.
Engineering or scientific area	Group ideas according to the type of engineering or scientific approach underlying the solution.	Solutions could be supported by three main types of engineering: *chemical* (pharmacological weight control), *electrical* (gastric pacing), or *mechanical* (laparoscopic banding, bariatric surgery, liposuction).
Technical feasibility	Group ideas according to their likelihood of coming to fruition. This is based on understanding what is feasible using current engineering and scientific methods, which implies some knowledge of the science behind the solution and/or the engineering development timeline required.	Solutions such as reprogramming fat cells might have low feasibility; a drug-eluting implant might have moderate feasibility; a space-occupying stomach device would have high feasibility.
Funding required	Group ideas around the amount and/or source of funding required to develop them. While this may be difficult before researching the funding landscape, a "best guess" based on prior information (see 2.4 Market Analysis and 2.5 Needs Selection) may suffice.	For obesity, a solution such as filling the stomach with a space-occupying device would likely require less money to develop than a drug-eluting implant.
Appeal to influencers	Group ideas around the interests of important influencers involved with the need, typically the patient or healthcare provider. While this may result in rather general groupings, it can provide insights into which concepts are more likely to be attractive to these **stakeholders** (see 2.3 Stakeholder Analysis).	For obesity, solutions focusing on medications might be more appealing to patients than those requiring surgeries. In contrast, surgeons used to performing bariatric procedures may favor other surgical interventions that are more effective and/or easier to perform but allow them to maintain procedure volumes.

Table 3.2.1 (*cont.*)

Organizing principle	Description	Example
Appeal to decision makers	Group ideas around the interests of decision makers who will potentially pay for a solution to the need, which is often a facility or **payer**. These stakeholders are usually interested in the cost-effectiveness of new solutions relative to available treatments. Remember that in some cases, patients may be the ultimate decision makers if a solution must be paid for out-of-pocket, which can have slightly different implications.	For obesity, solutions requiring a one-time payment may be more attractive than solutions requiring recurrent **payments**. Solutions that reduce costs by moving a procedure out of the operating room and into the clinic or from a specialist to a primary care physician may also be viewed favorably.

are encouraged to experiment with several different organizing methods before choosing the one or two that make most sense and best capture the potential value of the ideas generated.

Another approach for clustering ideas is to create a hierarchy of organizing principles where, for example, concepts are first clustered according to one factor (e.g., anatomical location) and then, within each of the resulting clusters, arranged into subgroups according to a different organizing principle (e.g., mechanism of action). The innovator can then continue this process, incorporating additional organizing principles at deeper and deeper levels. For instance, in the example focused on preventing strokes due to clots from the heart, the first level organizing principle might be anatomical (solutions could be broken down by location: heart, blood vessel, and brain). The concepts within each anatomical location could next be organized by engineering area (mechanical solutions, chemical solutions, or biological solutions). Alternatively, the innovator might start by organizing concepts according to engineering area and then create subgroups based on anatomical location. Again, the key is to try several different approaches, watching for patterns that illuminate important gaps or commonalities among the concepts.

Concept mapping

After one or more organizing principles have been chosen and ideas have been placed into groups, they can be documented in a concept map (also called a mind map). A concept map visually illustrates how ideas relate to one another and to the need itself. As noted, these maps are meant to help the innovator recognize patterns and build connections between solution concepts, as well as between the concepts and the need. Concepts maps are especially useful for highlighting gaps in the solution set – for example, innovators who are well versed in the chosen organizing principle (e.g., all the anatomical locations relevant to the chosen need) can quickly spot what might be missing if it is not represented in the concept map.

When developing a concept map, the need is placed at the center, with the clusters of ideas spanning out in different directions. Figure 3.2.2 shows the concept map generated for a need focused on obesity. In this example, the primary clusters are broken down by various mechanisms of action: (1) regulating food intake; (2) increasing energy expenditure; (3) reducing the motivation to take in energy; (3) reducing nutrient absorption; (4) pharmacologics; and (5) non-physiologic solutions. These clusters are further broken down into subgroupings as they are laid out in the concept map. For instance, the cluster focused on "ways to reduce the motivation to take in energy" is organized into a subgroup called "ways to reduce hunger." This includes solutions that affect volume (space-filling, volume reduction) or reduce hunger through neural means (neural stimulation of the stomach, stimulating the vagus nerve). Using the chosen organizing principle, the actual solutions are then placed where they fit best within these clusters

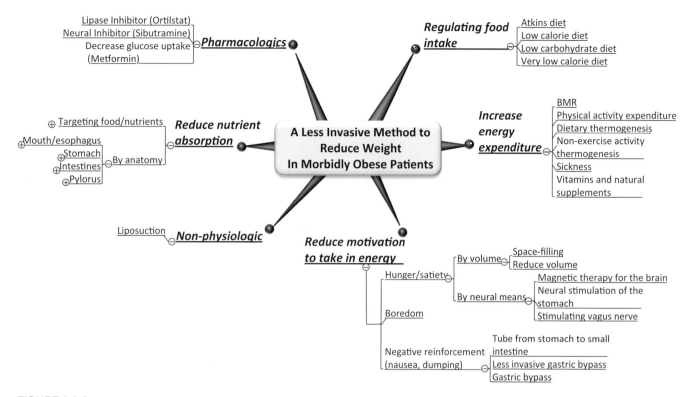

FIGURE 3.2.2

A sample concept map addressing the need for *a less invasive method to reduce weight in morbidly obese patients*, as developed by a team in Stanford University's Program in Biodesign (with permission from Darin Buxbaum; developed with Jennifer Blundo, Charles Hsu, MD, Ivan Tzvetanov, and Fan Zhang).

and subgroupings. In studying the ideas at the edges of the map, it is apparent that some are well-defined concepts while others are mostly general approaches. This could be the result of entering initial concept selection too early, without additional rounds of ideation to flesh out these areas. Alternatively, the innovators may have discovered through the process of creating the concept map that they had many more possible solutions to explore in these clusters. In this scenario, the innovators could choose to focus only on the clusters with well-defined solutions. Or, they could return to ideation to fill out the concept map with more concrete solutions before determining which ones to pursue. Regardless of the path chosen, the exercise of concept mapping plays an invaluable role in helping a team identify its next steps..

Another example focused on the need for *a better way for accessing the pulmonary veins* is shown in Figure 3.2.3.

Reviewing the figure shows that many solutions have been identified to support the cluster called "systems to aim at the PV" (pulmonary vein). However, when looking at the cluster "ablation systems," more brainstorming may be needed to flesh out the ideas since many of these solutions are much less well-defined in scope and are not closely linked to the central need. For their concept maps to be effective, innovators should ensure that all of the clusters have an obvious relationship to the need. They should also check to be sure that all solutions fit comfortably underneath the chosen headings. If an idea appears "forced" to fit within a group, this may indicate that the map is not based on the optimal organizing principle or that another heading is missing. This sometimes occurs if innovators have gone into the brainstorming session with a bias towards certain types of solutions (which places an artificial constraint on the brainstorming process). For example, if the need being

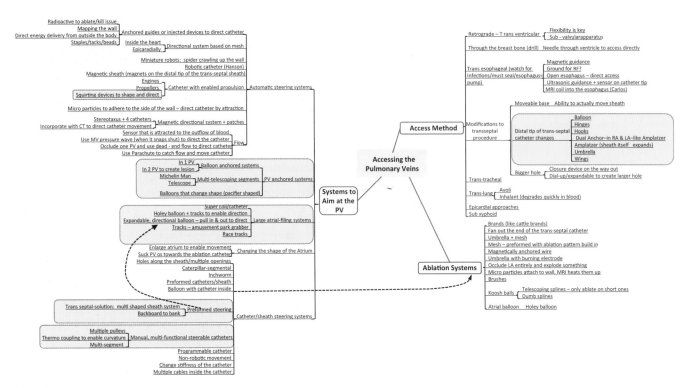

FIGURE 3.2.3

A sample concept map addressing the need for *a better way of accessing the pulmonary veins*, as developed by a team in Stanford University's Program in Biodesign (with permission from Uday N. Kumar; developed with John White, Kityee Au-Yeung, and Joseph Knight).

explored is for *a way to prevent bleeding from an artery*, a group with a mechanical engineering bias may subconsciously focus on mechanical solutions, completely forgetting about chemical or electrical ideas. When such biases are discovered, the team should complete more ideation. Involving people with different backgrounds, expertise, and experiences can help keep ideation balanced and potentially reduce the likelihood of generating significantly biased solutions sets (see chapter 3.1).

These examples show that no two concept maps are exactly alike. The basic layout is similar, but innovators have a good deal of freedom to decide exactly how to visually present the concepts in the most meaningful and useful way. Again, it can be helpful to experiment with multiple concept maps based on various organizing principles before choosing the one that best presents the data from ideation. Additionally, it is important to remember that by creating maps, different gaps in the solution set can be more readily identified. Furthermore, well-defined solutions assigned to different clusters on differently arranged concept maps may help the innovator

better define each concept along a variety of parameters. Both of these outcomes can lead to a more effective assessment of each solution against the need.

There are no specific guidelines for determining at what point a map has reached the optimal presentation of the data. Instead, innovators must rely on making an appropriate judgment call on a map that is logical, balanced, and comprehensive before moving to the next stage of concept screening.

Comparing ideas using the need specification

With a solid understanding of the intent behind the various solutions and a concept map created, innovators are ready to compare the proposed solutions against the need. The goal of this process is to narrow the universe of ideas to a few concepts that seem to best address the need. To accomplish this, innovators first should revisit the **need specification** developed in chapter 2.5 and then evaluate each concept relative to the **need criteria** that have been defined. It is advisable to start by assessing the extent to which proposed solutions satisfy the **need**

statement itself and then the must-have need criteria. Concepts that fail to meet these essential requirements should be set aside. Those that perform well against the must-haves can then be evaluated against the nice-to-have criteria. Often the nice-to-have requirements can be used for "tie breakers" for choosing between multiple solutions that seem equally promising.

As this comparison is performed, it is essential for the innovator to apply the need criteria as originally defined. Modifying the need criteria or compromising in terms of how they are applied to different solutions can tarnish the integrity of the selection process and lead to poor choices, since they will no longer be driven by the pure requirements of the need. Importantly, while it can be tempting to let personal interest in a solution bias the process, this can ultimately be a costly mistake, leading to a solution that does not appropriately address the real need. This, in turn, can undermine the final success of any product emerging from the biodesign innovation process.

That said, the initial selection process is inherently subjective in that innovators are using their best judgment to determine which solution ideas demonstrate the greatest promise in addressing the need. While there is no fail-safe way to eliminate this subjectivity, the chance of successfully identifying the most promising solution concepts can be increased by considering each concept in a consistent manner. As mentioned above, the creation of different concept maps will naturally lead to greater understanding of different parameters along which each solution is aligned. This may also help in comparing these solutions with the need criteria. Involving more individuals in the selection process is another specific tactic that can be used to help minimize personal bias.

One way to perform this assessment is by holding a solution selection meeting. This is similar to a brainstorming session, in that a group of individuals with diverse backgrounds should be invited and a clearly defined protocol should be used. A facilitator should run the meeting with the goal of driving the group to identify the most promising solutions. The facilitator will be in charge of defining the process to select the ideas, as well as ensuring that the selection committee follows the

process. However, unlike the facilitator's role in the brainstorming process, this individual should actively participate in the selection process.

Because initial concept selection requires a different skill set from what is required for brainstorming or other forms of ideation, it makes sense to involve different types of participants. For example, this is a stage where experts in the field can be invaluable. These experts might include physicians, nurses, and/or patients who can help evaluate the clinical usefulness of the various solutions. Also involving various types of engineers might help with understanding the technical feasibility of a solution and/or whether it seems realistic that it could solve the need. Participants with business backgrounds might also be able to help evaluate how needs might meet certain cost, marketing, or **value** criteria that may have been called out in the need specification. Having multiple perspectives can enhance the decision-making process. However, for practical purposes, innovators should seek to limit the number of people from which they solicit feedback. If too many people are involved, it can become impossible to rationalize each person's input, based on their individual preferences and biases.

In the meeting itself, participants can be asked to provide input and potentially even to vote to eliminate or to prioritize ideas. One approach is to give every person approximately five sticky notes and ask them to place one next to each of their top five ideas on the concept map (which can be drawn/copied on to a whiteboard or flip chart). After tallying the votes, the list of ideas can be reduced to roughly 10 leading concepts. Once the list of ideas is shortened to ten, the group can assess them in more detail relative to the need criteria, considering the must-haves and nice-to-haves, until a smaller set of concepts has been agreed upon. When experimenting with this approach, innovators should be sure to remind participants to make their choices based on which proposed solutions they believe will best address need statement and the must-have need criteria. If participants simply pick concepts that "sound cool," this can result in many varied concepts with little connection to those likely to deliver the best results.

At times, it may not be feasible to get all desired participants into a room at the same time and place to participate in the selection process. Another approach is to meet with targeted individuals one-on-one to seek their input. The feedback gathered from the experts can then be used as an input to the selection process that innovators and their teams employ. Regardless of how an innovator collects input from experts in the field, this feedback is essential to initial concept selection.

During this step of the process, it is possible that some proposed solutions will have to be eliminated even though they show potential to address the need statement and satisfy the must-have need criteria. In some cases, certain solution concepts are simply impractical or infeasible. This is often due to technology constraints, but other factors related to patient, provider, and payer concerns may make an idea unrealistic. Although additional primary and/or secondary research may shed light on whether or not an idea is truly impractical, an innovator must usually rely on a "gut check" regarding what can realistically be accomplished within a reasonable timeframe.

Innovators may also eliminate promising solution concepts based on their personal project **acceptance criteria** (as defined in 1.1 Strategic Focus). For example, if a team has conceptualized an electrical engineering solution that meets its need criteria, but the innovators are dedicated to working on mechanical solutions, they may choose to abandon the electrical idea. Another alternative would be to modify their acceptance criteria. While it is perfectly acceptable to make adjustments to acceptance criteria at this stage in the process, remember that the need criteria should remain intact.

As noted, innovators may also end up with different *approaches* to address the need, rather than specific solutions. Sometimes approaches may seem to meet the need criteria more effectively because of their lack of detail – concrete details are often the key factors in determining why a particular concept does not meet a need. For instance, revisiting the example shown in Figure 3.2.3, which focuses on the need of *a better way for accessing the pulmonary veins*, some of the need criteria were: (1) the ability to reach all configurations of the pulmonary veins; (2) easy to use; (3) faster than current methods; (4) low/no **morbidity**; and (5) must enable at least radiofrequency energy delivery. Other factors taken into account included the team's interest in pursuing solutions in line with current catheter-based procedures, the feasibility of the solution, and the expertise of the team. Based on these criteria, this team chose the subgroupings and associated solutions in the outlined and shaded areas on the concept map. Even after these initial choices, the team had to complete additional brainstorming and research to generate more specific concepts from these approaches, as there were still more than could realistically be taken forward. Performing more brainstorming after going through initial concept selection is certainly acceptable. However, it is usually more efficient for innovators to identify if they are working with approaches rather than true concepts before they initiate initial concept selection.

Although it is relatively rare, it is possible for initial concept selection to yield too many genuine concepts that meet the need criteria. In this scenario, the need criteria may be too broad. This can require innovators to revisit the need specification to generate more specific must-have and nice-to-have need criteria (see 2.5 Needs Selection). Importantly, innovators should not consider this a failure of the need specification step, but rather another example of the iterative nature of the biodesign innovation process.

As with so many other elements of the biodesign innovation process, there is no exact way to know when enough solution ideas have been generated and when they have been appropriately screened. Similarly, there is no right number with respect to the final number of concepts to take forward. In most cases, innovators will focus on three to five of the concepts that best meet their need criteria. Many of these may be chosen because they are practical, but it is also acceptable to include an idea that could represent a potential breakthrough and/or one that inherently excites the team. These are the concepts that will undergo more in-depth ideation, usually around their technical details, as well as research and prototyping in sequence or in parallel, in the next stage of the biodesign innovation process: concept screening.

The case example from the Singapore-Stanford Biodesign program illustrates how one team navigated initial concept selection.

FROM THE FIELD ▶ SINGAPORE-STANFORD BIODESIGN TEAM

Organizing and assessing ideas to assist with concept selection

When Tze Kiat Ng, Luke Tay, Justin Phoon, and Pearline Teo were fellows in the Singapore-Stanford Biodesign program, they were focused on solving clinical needs in the ear, nose, and throat (ENT) specialty. After clinical immersion at the National University Hospital in Singapore, the team had about 300 needs, which it narrowed to a set of eight top opportunities. At this point, team members conducted a series of short, intense brainstorm sessions for the eight needs. "We wanted to see which needs we could get most creative with," said Phoon.[2] The innovators also used these initial brainstorms as a way to gauge team interest in the needs so they could use this factor in their final needs selection exercise.

"After the mini-brainstorms, it became quite obvious which needs triggered a lot of interest and creativity within the members of the team, and which ones seemed to be more limited in terms of possible solutions. We went from eight to four needs based on this input," Teo recalled. They then evaluated the remaining needs

against a variety of other factors, to get from four needs to the two that they would officially take forward into concept generation. They also created need specifications and defined clear need criteria to help guide more in-depth ideation.

According to Teo, "When we began brainstorming in earnest, we averaged about two sessions per need." The first meeting usually resulted in several broad approaches and general concepts for solving the problem. For instance, for their need for *a safer way to prevent cerumen impaction at home*, the team came up with ideas such as a soft expandable cleaning tip, a suction device, an ear "stent," and an endoscopic curette. The team relied primarily on whiteboards to capture the results of their brainstorms in real time.

Before the second brainstorming session, the team organized these broad approaches and general concepts into different categories (see Figure 3.2.4). Going back to the cerumen impaction need, the team grouped its ideas into the categories of preventing wax from sticking in the ear canal, improving drainage in the ear, "eating" or dissolving the wax, and safe cleaning instruments. This helped the team members identify

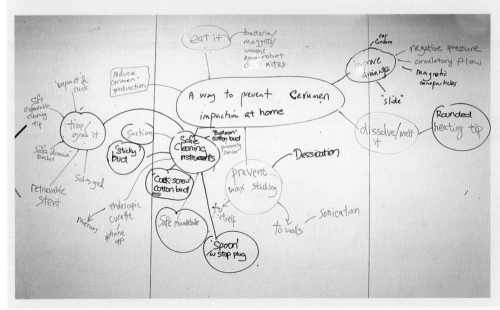

FIGURE 3.2.4

One of the team's early concept maps (courtesy of Tze Kiat Ng, Justin Phoon, Luke Tay, and Pearline Teo).

gaps in their ideas, with areas that they missed serving as a starting point for the second brainstorm meeting. For example, with the cerumen need, they realized that they had not touched upon ways to reduce cerumen production in the ear. As Phoon described, "Our strategy to organize our ideas into categories paid off because it helped us identify gaps which, in turn, drove more focused, productive brainstorms." Teo added, "This approach also made it clear when most of the concepts generated were basically variations of the same approach to solving the problem. This pushed us to expand our thinking into other areas."

Another important objective of the second brainstorming session was to help the team get more specific about their most compelling ideas. According to Teo, "In the case of the more interesting approaches, we challenged ourselves to come up with more detailed and actionable concepts – how exactly would we create something that tangibly embodies the approach." The team used some if its need criteria to help its members stay focused on generating actionable concepts. For example, Teo explained, "We mostly wanted to work on relatively simple mechanical concepts that we could develop in roughly a six-month time frame." With these criteria top of mind, they encouraged one another to come up with increasingly concrete solutions.

The series of second brainstorming sessions resulted in detailed concepts for several of the most interesting approaches. For example, with the cerumen need, the team members generated eight to 10 actionable concepts on which they could potentially work.

To help them move into the next stage of the biodesign innovation process, concept screening, with a reasonable number of ideas, the team decided to use a voting system to eliminate some ideas from consideration. For each need, individual team members were given three votes and asked to assign the first to the concept they thought was most revolutionary, the second to the concept they thought was technically most likely to succeed, and the third to the concept that they were personally most excited to pursue. When the scores were tallied, the most promising three or four concepts emerged. "The concepts with the highest scores were the ones we chose to spend our time on," Teo stated. At that point, the team evaluated each of the top concepts against the need specification, creating matrices to help compare each solution to the must-have and nice-to-have need criteria. The solutions that were best aligned with these important criteria were the first to move into prototyping to assess their technical feasibility. Concurrently, the team conducted more detailed research and evaluation of those ideas.

⬊ Online Resources

Visit www.ebiodesign.org/3.2 for more content, including:

  Activities and links for "Getting Started"
- Review and document raw data
- Cluster ideas

- Develop a concept map
- Assess the concept map
- Compare concepts against the need to complete concept screening

  Videos on initial concept selection

CREDITS

The editors would like to acknowledge Joy Goor and Ritu Kamal for their help in developing the chapter. Many thanks also go to Justin Phoon and Pearline Teo for contributing to the case example.

NOTES

1 "Leap of Faith," Dictionary.com, http://dictionary.reference.com/browse/leap%20of%20faith (November 19, 2013).
2 All quotations are from interviews conducted by the authors, unless otherwise cited. Reprinted with permission.

Acclarent Case Study

STAGE 3: CONCEPT GENERATION

With a well-validated need for *a minimally invasive approach to treating chronic sinusitis that had less bleeding, less pain, less bone and tissue removal, less risk of scarring, and that was faster, easier, and safer to perform*, as well as a firm commitment to finding a less invasive way to treat chronic sinusitis, Makower and Chang began aggressively exploring potential solution concepts. In doing so, they finally returned to some of Makower's preliminary ideas before launching the biodesign innovation process, including the concept of developing "interventional devices" for sinusitis. This approach made even more sense in the context of the team's recent disease state and solution research. "Seeing the three-dimensional anatomy in cadavers and reviewing so many CT scans caused us to appreciate how tortuous the bones and mucosa were in the sinuses. It reinforced the observation that those drainage pathways resembled blood vessels," Makower commented.[1] "So we started to imagine how we could implement cutting, dilating, stenting, energy delivery, and drug delivery across this bony network, utilizing whatever we had learned from our experiences in the cardiovascular space."

3.1 Ideation

To develop a more specific set of potential solution concepts, Makower and Chang began exploring different possibilities in earnest. Makower, who had a bias against disclosing his ideas with a wider audience too early, advocated for a scaled-down, hands-on version of the ideation process. While they did not assemble a cross-functional group of participants to hold formal brainstorming sessions, Chang noted, they still benefitted from multidisciplinary input: "In a way, we did get multiple perspectives, because Josh is an engineer, a doctor, a scientist, and a business person,

and I'm an engineer." He went on to describe their ideation approach:

> Josh and I spent a lot of time brainstorming around what we had learned about the disease and anatomy and from our cadaver work. We dug into how we could use flexible instruments to get around and preserve anatomical structures while gaining access to the sinuses. And we considered what we could do to make access easier. For example, one way we developed this understanding was to section a cadaver head at one of the labs. By sectioning the specimen, we had an unobstructed view of the anatomy with our own eyes, instead of using an endoscope. We laid down malleable wire and got an appreciation for the dimensions of the anatomy, relative locations of the structures, openings, and where instruments would need to travel. It was hands-on brainstorming while we were assessing the anatomy.

Through this process, they developed multiple types of potential solutions. First they thought about ways to dilate the anatomy to gain access to the sinuses, using tools such as lasers, balloons, stents, and what Chang described as a "chomper." They also generated ideas to address a blockage area once it was reached. "We thought about a catheter that cuts, a spinning burr, different kinds of lasers, and balloons. Maybe we could freeze it, or maybe we could use microwave energy. We considered anything that would ablate or cut, that we could deliver on a flexible platform," Chang said (see Figure C3.1).

3.2 Initial Concept Selection

After the team felt that they had more or less exhausted the possibilities for ways to address the need, they started to narrow down the list of concepts. Makower

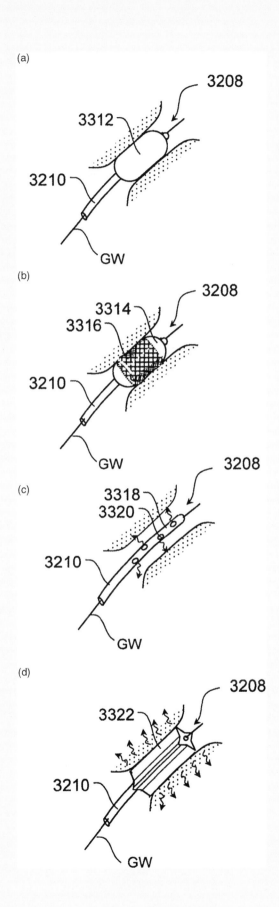

and Chang both had a desire to do something fast and had a preference to try one idea at a time. For these reasons, they saw some logic in starting with one of the concepts that was quickest to prototype. If it worked, they would press forward. If not, they would go back to the drawing board and choose another solution to explore. When they rationalized their list, they decided that the quickest idea to investigate was related to the use of balloons. "We said, 'Look, if we can make it work with a balloon, that'd be great,' " Makower remembered. "It doesn't involve complicated instruments. It doesn't involve bleeding. There's no cutting, no energy, and no hardware. Given our backgrounds, we could also make or obtain balloons to test easily." He continued, "If the balloon works, we're in great shape. If not, we can try energy next, or blades, or maybe even stenting. We knew our mission was to open the sinus drainage pathways in a flexible, less invasive way, but which technology was going to be required was really not clear yet."

NOTES

1 All quotations are from interviews conducted by the authors, unless otherwise cited. Reprinted with permission.

FIGURE C3.1

The team considered many methods of opening sinus pathways, including the use of balloons (a), stents (b), drug delivery (c), and energy-based devices (d) (excerpted from US utility patent 7,462,175 with permission from Acclarent).

PHASES **STAGES** **ACTIVITIES**

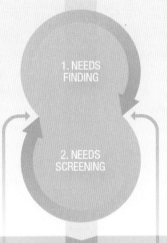

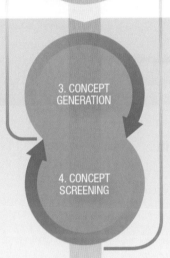

IDENTIFY

1. NEEDS
FINDING

2. NEEDS
SCREENING

1.1 Strategic Focus
1.2 Needs Exploration
1.3 Need Statement Development

2.1 Disease State Fundamentals
2.2 Existing Solutions
2.3 Stakeholder Analysis
2.4 Market Analysis
2.5 Needs Selection

INVENT

3. CONCEPT
GENERATION

4. CONCEPT
SCREENING

3.1 Ideation
3.2 Initial Concept Selection

4.1 Intellectual Property Basics
4.2 Regulatory Basics
4.3 Reimbursement Basics
4.4 Business Models
4.5 Concept Exploration and Testing
4.6 Final Concept Selection

IMPLEMENT

5. STRATEGY
DEVELOPMENT

6. BUSINESS
PLANNING

5.1 IP Strategy
5.2 R&D Strategy
5.3 Clinical Strategy
5.4 Regulatory Strategy
5.5 Quality Management
5.6 Reimbursement Strategy
5.7 Marketing and Stakeholder Strategy
5.8 Sales and Distribution Strategy
5.9 Competitive Advantage and Business
     Strategy

6.1 Operating Plan and Financial Model
6.2 Strategy Integration and Communication
6.3 Funding Approaches
6.4 Alternate Pathways

PROJECT
LAUNCH

Opportunity is missed by most people because it is dressed in overalls and looks like work.

Thomas Edison[1]

Junk can dance.

David Kelley[2]

4. CONCEPT SCREENING

INVENT

After careful screening of many needs, you have a clear statement of a focused problem worth solving. Ideation has generated hundreds of ideas, many whimsical or even outlandish. In the Identify phase, many needs were winnowed down to a focused, clinical needs specification. In this second stage of the Invent phase it's time again to make choices. Some innovators have a tendency to perseverate, to delay because of uncertainty. The plain reality is that no decision is a failure. It is impossible to pursue a dozen concepts simultaneously.

The United States Marine Corps knows something about decision making in the middle of uncertainty. In the classic book *War Fighting*,[3] the message is clear – perfect certainty doesn't exist. It is far better to move expeditiously with the best information available at the moment and to react flexibly as the situation changes and as you learn on the fly.

So, too, choosing a concept that is directionally correct, and acting upon it, will frequently provide new insights and refine your direction. There is no shame in meeting a blind alley and even reversing if necessary. Strong teams know how to backtrack to success.

Concept screening requires an understanding of the rules of the road for medtech innovation. The complex and frequently conflicting interplay between intellectual property, reimbursement, regulatory, and business model options requires sophisticated judgment. There is seldom a single "right" way.

Experienced, serial entrepreneurs will emphasize that time is money ... and more. A balance between thorough deliberation and the opportunity cost of time lost to paralytic analysis is needed.

Just as concept screening is an iterative process, so too is concept exploration and testing. Prototyping, which is at the core of this activity, involves several steps, the first of which is closely linked to ideation. We build to learn. Early on, crude mock-ups are constructed to serve as a "looks like" or "feels like" version of a device made of easily shaped and assembled materials. Access to junk makes this easy and fun. Foam core, plastic, cardboard, outdated surgical instruments, catheters, and endoscopes can be "cannibalized" for early prototyping. (Duct tape rules!) The goal is to fail early, fail cheaply, and ultimately to fail better. Crude can become more refined and the rules of the road revisited.

NOTES

[1] As quoted in John L. Mason, *An Enemy Called Average* (Insight Publishing Group, 1990).
[2] As quoted in Tom Kelley, *The Ten Faces of Innovation* (Doubleday Business, 2005).
[3] U.S. Marine Corps Staff, *War Fighting* (Create Space Independent Publishing Platform, 2012).

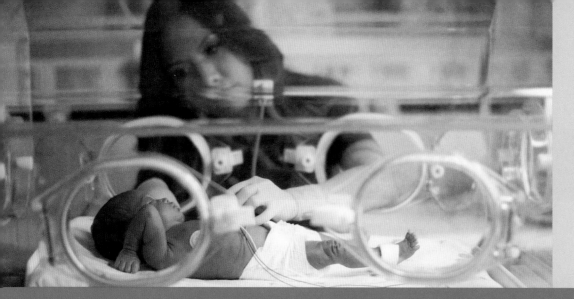

4.1 Intellectual Property Basics

INTRODUCTION

The first months of the project have sped by without any real roadblocks. The team's prototypes are working better than expected in the preliminary animal model. The regulatory pathway appears straightforward. Existing reimbursement codes are in place to cover the new technology. And the project has begun to attract some investor interest. Then an email arrives from the patent attorney: "See attached Patent Cooperation Treaty publication from Israel – just uncovered. Please call immediately."

As soon as a team has identified a promising new concept, it is time to start exploring intellectual property (IP) and developing an approach to patenting. A patent is a legal document that gives an inventor the right to exclude others from commercial use of the invention. The presence of existing patents in the field can complicate or even derail the innovator's ability to launch a new company or R&D program. Alternatively, a strong patent position can add tremendous value to an invention, even at an early stage of development.

This is the first of two chapters on IP. It covers the fundamentals of medtech patents, focusing on the initial steps required to obtain a patent. The subsequent chapter, 5.1 IP Strategy, describes how to build an effective medical device patent portfolio and addresses strategies, tactics, and methods for asserting the inventor's rights.

 See ebiodesign.org for featured videos on intellectual property basics.

OBJECTIVES

- Understand the different types of US patents, including the basic elements of provisional and utility patents.

- Recognize the requirements of patentability, including practical aspects of the filing process for medical devices.

- Develop familiarity with the patent search process.

- Understand the fundamentals of international patent coverage.

- Appreciate how to use intellectual property risks as a screen for prioritizing concepts.

INTELLECTUAL PROPERTY FUNDAMENTALS

IP is defined as any product of the human mind or intellect (e.g., an idea, invention, expression, unique name, business method, or industrial process) which has some value in the marketplace and, ultimately, can be reduced to a tangible form, such as a device, medical method, drug, software program, process, or other invention.[1] IP law governs how individuals and groups may capitalize on innovations by determining who owns the IP, when owners can exclude others from using the invention for commercial purposes, and the extent to which courts will enforce the patent holder's rights.[2]

There are several categories of intellectual property that are of major interest to the medical device innovator – **patents**, trademarks, copyrights, and **trade secrets**.[3] A patent is a grant of **exclusive rights** from the government to make, use, sell, or import an invention. Although it is commonly described as a "monopoly," it is more accurate to say that a patent is the right to *exclude others* from making, using, selling, or importing the technology. Governments are willing to grant this right to the inventors for the life of the patent – generally 20 years from the time of filing – in exchange for the inventors disclosing the details of the invention. Patents benefit society for two main reasons: (1) without this guarantee, innovators and companies would not have the incentive to invest the time and resources needed to bring a technology forward, and (2) the right to exclude others from making or using the invention encourages them to try designing alternatives.

In the United States, the most common type of patent sought for medical devices is the **utility patent**. Utility patents, which are granted by the US Patent and Trademark Office (**USPTO**), describe an apparatus or method wherein the novel invention is *useful* for accomplishing a specific end result. In the **medtech** industry, utility patents cover devices themselves, as well as methods of use and how the devices are made and manufactured. It follows that utility patents are by far the most important category of medical device IP. There are also *design patents* which cover the unique ornamental, visible shape, or design of a non-naturally occurring object. These are particularly important in the consumer products area (one example would be the ornamental features of adhesive bandages). **Provisional patents** are preliminary filings that are generally submitted in order to secure a first filing date for subsequent utility patents (see below).

A *trademark* is a word, phrase, or symbol that is consistently associated with a specific product or medical method and gives the holder exclusive rights to use a word, phrase, or symbol as a brand name, tagline, or logo. Trademarks come into play when devices or medical methods are named or logos are created for companies or products. Trademark rights arise automatically from use of the name or logo in connection with the goods or services (that is, no formal application is required to obtain trademark protection). However, registering a trademark generally strengthens the holder's legal position for asserting the validity of the mark in court and prevent others from using it.[4] Trademarks cannot be the generic name of a product or surgical method. They can be thought of as adjectives in front of the generic product name. For example, Xience® stent, Adapta® pacemaker, or Contiuum® hip replacement.

A *copyright* grants authors and artists of written and graphical materials the right to prevent others from using their original works of expression without permission. In the medtech industry, copyrights are important for software (for example, the programs that run an ultrasound scanner or the algorithms that decipher heart rhythm disturbances in wearable monitors). Copyrights also come into play with advertising and educational materials (written, web, audio, and video). Works are automatically copyrighted when they are placed in some tangible form. As with trademarks, federal registration of a copyright leads to additional procedural rights.

Trade secrets are information, processes, techniques, designs, or other knowledge not generally understood or made public, which provide the holder with a competitive advantage in the marketplace. With some medtech products, trade secrets may be more important than patents. For example, a knee joint device may involve a special manufacturing process that creates smoother contact surfaces. In this case, the company may choose *not* to pursue a patent for its manufacturing process in an effort to keep it a secret and avoid "tipping off" competitors about an innovation that could improve their products. Instead, the company may attempt to protect the process as a trade secret, putting in place various safeguards to keep other companies from finding out anything about this manufacturing know-how. Some strategic considerations regarding trade secrets versus patents are provided in 5.1 IP Strategy.

The remainder of this chapter explores three basic patent considerations: criteria for obtaining a patent, **freedom to operate**, and **prior art** searching. It then provides an overview of provisional and utility patents, identifies important information related to international patenting, and addresses a number of other fundamental IP concepts, such as confidentiality.

Criteria for obtaining a patent

There are three basic parameters for judging the *patentability* of an invention:[5]

1. **Utility** – The invention must do something useful.
2. **Novelty** – It must be new with respect to other patents, products, or publicly available descriptions anywhere in the world (collectively known as the "prior art").
3. **Obviousness** – The invention must not be obvious in light of the prior art to someone of ordinary skill and knowledge working in the given field.

Only if an invention objectively meets these criteria is it considered to be patentable. In practice, the first criterion – utility – is rarely a problem for a new medical device invention. However, an inventor's ability to demonstrate novelty can be trickier. Prior art may exist in the field that would naturally lead to, or anticipate, an inventor's innovation but it may not yet be published (e.g., patent applications that are filed but not published). Beyond this is the world of medical device innovators who are constantly developing new technologies, testing them, and making **public disclosures** in some fashion (sometimes in fairly obscure places). It is not uncommon for an innovator to come up with a new device only to discover, for example, that a scientist working in some other country tested a similar **concept** in animals. In order for a patent to be granted, at least one feature of the device must be novel relative to any previous patents, publications, abstracts, speeches or other presentations – in short, *any public disclosure, from anywhere in the world, in any language, over all time*. The only way to uncover such information is to diligently and patiently search the literature. The advent of the Internet and powerful search engines (such as Google) has made this discovery process easier in recent years (see section "Patent searching" below). Still, this is a daunting task, with over 8 million patents in existence worldwide and a vast store of published information to consider.[6]

Another factor of great practical importance regarding novelty is the timing of the filing of the patent relative to other patent applications. The America Invents Act (**AIA**) of 2011 fundamentally changed the law in the United States from a "first to invent" strategy to a "first to file" approach (making the US similar in this respect to most other countries). This means that the date of the patent filing, also known as the **priority date**, is the critically important timing factor in determining whether the patent can be issued over competing patents.

The point that most often becomes the focus of both patent prosecution and litigation in the medtech field is whether or not the invention is obvious. Obviousness can be complicated, and no inventor is ever completely secure on this point. The criterion that the invention is not obvious to someone "of ordinary skill and knowledge" has a specific legal meaning. As a commonsense guideline, think of this hypothetical person as someone working in the field that uses the invention (say a gastrointestinal endoscope) and who magically has complete knowledge of the prior art in the field of endoscopy. If the invention would be obvious to this hypothetical person, it is not patentable. In practice, the determination of obviousness becomes an issue that the patent examiner decides; later on, it may become a subject of litigation and be determined by the courts. In recent years, the patent office and the courts have tried to base decisions on objective measures of non-obviousness, including commercial success of the invention, prior lack of a solution to a longstanding **need**, and failure of others to come up with the invention.[7] An experienced patent attorney is the inventor's best source of advice on the obviousness of an invention.

Beyond the three essential criteria for patentability, the law specifies that a patent must describe the invention in a clear, unambiguous, and definite way so that a person with knowledge in the field would be able to make and use it. Patent law also requires that the innovator include a description of the *best mode* for the invention, which generally means the "embodiment" or way of practicing that is, in the innovator's opinion, the most effective configuration or version of the invention. This may seem like a difficult requirement but it really just asks the inventors to include the approach they favor at the time of filing regardless of whether a different approach proves to be better later. For example, if the inventor of a particular stent structure described the best mode at the time of filing as the use of stainless steel in the construction of the stent – and, in subsequent development and testing, cobalt-chromium proved to be superior material – the requirement for disclosing best mode would be satisfied. The AIA has softened (and potentially confused) this requirement by stipulating that a patent can no longer be *invalidated* for failure to include best mode – though the original law requiring that the inventor include the best mode in the patent submission is still in force.[8] The resulting uncertainty will likely be resolved through court decisions over time but, for now, it is strongly advisable that innovators disclose the best mode of practicing their inventions in patent applications.

Reduction to practice

Reduction to practice is another important factor for patentability because it shows that the idea is being diligently pursued. With medical devices, this can mean building a **prototype** and performing some kind of bench-top or animal testing. The model making and testing can occur after the patent filing, and there are no strict guidelines about timeliness of this reduction to practice other than what is reasonable for that type of technology. As an alternative to making and testing a model, an invention can also be reduced to practice "constructively" by a careful description in the patent, with sufficient detail and in an "adequately predictive" manner that someone knowledgeable in the field could construct the device. This is particularly useful when the development of a working model will be extremely costly, time intensive, or difficult to accomplish. Constructive reduction to practice via a patent application has the same legal effect as evidence of an actual reduction to practice.[9]

Freedom to operate: an introduction

In addition to patentability, the other key aspect of patent law that determines the commercial usefulness of an invention is freedom to operate (FTO). FTO is often confusing for first-time inventors to understand. The key concept is that *receiving a patent on a device does not guarantee that the inventor is free and clear to make, sell, use, or import the technology (that is, has freedom to operate)*. Despite the fact that the new device may have some new, patentable feature, there may be existing patents that describe other necessary features of the

new device. For example, an inventor could patent a new kind of intraocular lens that has a high resistance to infection – this would be a patentable feature that is not covered by any prior art. However, despite being able to obtain a patent with claims on this new feature of the lens, the inventor may not be free to commercialize this invention if there are current patents in force with claims that cover other important features of the general structure and function of an intraocular lens – features that are also necessary parts of the new device. *A new device has FTO only if the features of the device are free and clear of valid claims from patents that are still in force in the country in question* (generally patents that are 20 years old or less). More specifically, there should be no single claim in any prior patent that describes the features that are included in the device in question *and nothing more*. Another way of saying this is that if there are features in the prior claim that are not part of the new invention, then FTO is preserved (even though there are some features that are comparable). This concept is diagrammed in Figure 4.1.1.

Because it affects whether or not a device can actually be sold, determining FTO for a new invention is equally important as determining patentability and requires a diligent search of the prior patent art (see below). Strategies for dealing with FTO issues are outlined in 5.1 IP Strategy.

Before filing: the prior art search

Understanding the prior art landscape for a new invention is one of the most critically important parts of the biodesign innovation process. Inventors naturally hope to discover that their idea is indeed completely new. However, even if the search reveals that there is troublesome prior art, the team benefits greatly from finding this information as quickly as possible. Such discoveries can potentially save an inventor huge amounts of wasted time and resources in pursuing the wrong idea. There is also the opportunity to modify the invention into an approach that is patentable and has freedom to operate.

Prior art searching is as much an art as a science, and at a later stage it is almost always wise to have a professional search conducted by an attorney or patent agent before the filing of a utility patent (see 5.1 IP Strategy for

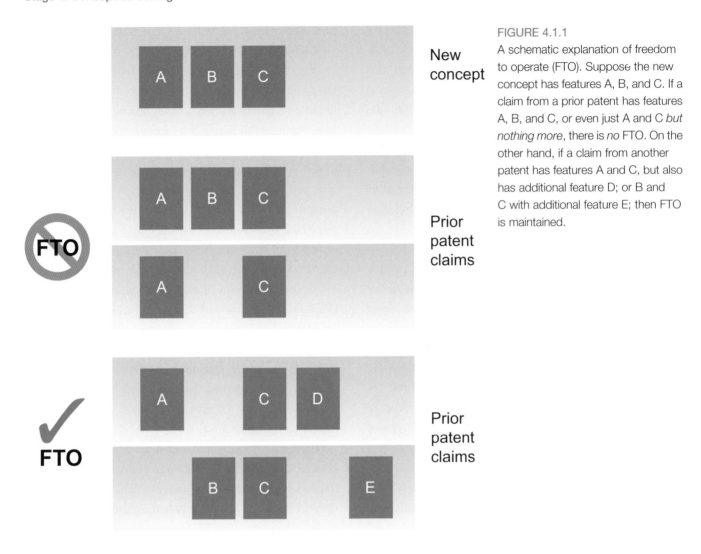

New concept

Prior patent claims

Prior patent claims

FIGURE 4.1.1

A schematic explanation of freedom to operate (FTO). Suppose the new concept has features A, B, and C. If a claim from a prior patent has features A, B, and C, or even just A and C *but nothing more*, there is *no* FTO. On the other hand, if a claim from another patent has features A and C, but also has additional feature D; or B and C with additional feature E; then FTO is maintained.

considerations about hiring an attorney). These experts will usually locate prior art that the team has not uncovered. Still later in the process, it is not uncommon for the patent examiner to find art that the expert attorney has not discovered. Despite the fact that the initial search will not be perfect, it is essential for inventors to conduct their own search early in the process. They will learn a tremendous amount about the field, and this up-front diligence will make the work of the attorney or agent much more effective and economical.

For the inventor just launching a search, the vast scope of the territory for prior art may at first seem intimidating. As mentioned earlier, in the search for patentability the inventor is looking for any previous disclosure of the idea in a public manner – not only patents, but all written descriptions and public presentations at anytime, anywhere in the world. In practice, searching patent databases

is an effective way to start since most commercially significant technologies are patented (especially in the medtech field). One subtlety that is often overlooked is the fact that abandoned and expired patents are also public information and, thus, constitute legitimate prior art.

Given the large scope of the search territory, the prior art search process is unavoidably time-consuming. Depending on the invention, it is not excessive to allocate as much as half (or more) of the team's time to searching during the concept screening stage. Like other steps in the biodesign innovation process, searching should be performed at multiple points in the sequence. This is because there is a continuous flow of potential new prior art coming into public view, both in the form of patents (published in the US every Tuesday) and patent applications (every Thursday), as well as in public disclosures at conferences, in the literature, and in the press.

Additionally, the patent search process is complex, involving many combinations of different keywords and expressions. So searching must be an ongoing process. Looking at patents and publications uncovers new keywords and expressions. This information should be used, in turn, to conduct additional searches.[10]

In the biodesign innovation process, patent searching is usually deferred until after concept generation, the logic being that too much early information about patents may prematurely inhibit the team from developing concepts in an area that might turn out to be highly fruitful. However, a preliminary patent search *should* be completed in relatively short order after initial concept selection to avoid the risk of getting locked into a particular solution before the major features of the patent landscape are understood. As with so many other parts of the biodesign innovation process, identifying and understanding IP is an iterative process, with progressively deeper dives into the material as the concept moves forward. As a potential invention is developed, continued patent searching helps refine the idea and almost always leads to a stronger patent application.

In the US, inventors have a legal requirement to disclose any relevant prior art they uncover to the patent office upon and after filing. Practically speaking, a careful cataloging of prior art can help the patent examiner more quickly assess the technology and review the application.[11] It is a foregone conclusion that any significant prior art will be uncovered at some point – either during the careful examination by the patent office or in litigation once the patent is issued – so it is in the inventor's best interests to find and disclose these materials as early as possible.

Search basics
Before starting a patent search, the inventor must have reached a precise understanding of the invention that will be claimed. This underscores the requirement for the idea to be developed in some detail before an in-depth patent search is performed. A concise summary should be written that captures the most fundamental elements of the invention, including the need or problem that the invention addresses, the structure of the invention, and its function. It can be useful to draft a few

hypothetical claims that help clarify the scope of the invention in advance of the search. These do not need to be written in "legalese" – just a plain language description of what characterizes the invention.

Several conceptual points are important to understand about the actual search process (see the Getting Started section for more practical tips and guidelines). First, there are different types of searches that are conducted depending on the stage of development of the product and the purpose of the search. In the initial stages following an invention, there are two basic types of searches. A search for *patentability* focuses on novelty and obviousness in light of the prior art. This search is intended to locate any information within any of the parts of a patent (most commonly the detailed description and the drawings – see below for more details) that would make the examiner conclude that the current concept is obvious or not novel. A *freedom to operate* search is directed toward finding *claims* of patents currently in force that specifically describe features of the new concept so that it is not possible to commercialize the device without designing around or licensing the patent (see chapter 5.1). Claims, as described more fully below, are the numbered paragraphs at the end of the patent that stake out precisely which aspects of the invention will be exclusively owned by the innovator or company. This is the reason that the claims are the focus of the FTO search.

Second, there are two primary ways to approach searching. The most intuitive method is text searching, also called *keyword* searching. This is a familiar process in the Google era in which keywords are entered into a search engine that looks through a patent database (Google itself now maintains an excellent patent database – Google Patent Search – along with the USPTO and others listed in the Getting Started section). The text search usually starts with keywords that describe the invention, though it can also be productive later on to search for known inventor names in the field, assignees (companies that own an important patent), competitor companies, etc. The precision of a text search can be greatly improved by combining keywords using *Boolean logic operators* such as AND and OR. One other practical piece of advice is to be careful about the fact that there may be many versions of keywords for the same technology. For example, the guidewire used in angio-plasty is variously described as "guidewire," "guide-wire," "guide wire," "guide element," "wire," "wire guide," and probably other terms. The search engines have a feature that helps with this problem called *wild card symbols* or "truncation limiters." These are symbols attached to the root of a keyword that allow the searcher to find any other words that have that same root. For example, by searching "guide*" (the exact wild card symbol depends on the search database), the searcher can look for all keywords with "guide-" as a common root.

The other general category of search is by *classification*. The USPTO and other international patent agencies have systems for grouping and coding patents that are based on the industry, the structure, and function of the invention and its intended use. The database developed by USPTO is called the patent classification (USPC) index. (Note: The US is migrating to a Cooperative Patent Classification system, or CPC, being developed jointly with the European Patent Office.[12] At the time of writing, the use of this system by the USPTO is expected to be mandatory in early 2015.[13]) It can be useful to perform a search by the assigned classes and subclasses, since the keywords prior inventors have chosen to describe their concepts may not be standard or intuitively obvious (such that a text search using "reasonable" keywords may miss some important prior art). Some inventors like to start with a classification search as a quicker way to understand the landscape of their invention than a keyword search. The classification systems used by USPTO and other international agencies are not themselves necessarily intuitive or consistent, so some patience and serendipity may be required. More search tips can be found in the following Working Example.

Working Example
Tips for keyword searching

Start by evaluating three basic questions:[14]

1. What problem does the invention solve (in the context of the biodesign innovation process, what is the basic need)?
2. What is the structure of the invention?
3. What is the function of the invention (what does it do)?

In developing the answers, generate as many relevant terms as possible. These terms will provide the basis for the search. For example, for a new angioplasty catheter, a sample response is shown in Table 4.1.1.

Use these keywords to search either the USPTO or Google Patent website (other patent databases can be used in subsequent rounds). Make use of Boolean logic operators and the advanced search techniques available through the site to refine the searches. For example, search "atherosclerosis AND angioplasty" to find patents using both of these keywords; or "angina OR myocardial infarction" to find patents covering either of these two conditions. To make the search more efficient, use the wild card symbols or truncation limiters to reach all keywords that have a common root. The use of these symbols is usually described on the search site.

Table 4.1.1 An effective way to approach keyword searches (based on an approach outlined by David Hunt, Long Nguyen, and Matthew Rodgers in *Patent Searching: Tools & Techniques* (Wiley, 2007)).

Issue	Possible search terms
Need addressed by the invention	Atherosclerosis, arteriosclerosis, coronary artery disease, coronary stenosis, arterial blockage, arterial stenosis, arterial flow, coronary flow, myocardial infarction, heart attack, angina, chest pain, etc.
Structure of the invention	Catheter, balloon catheter, angioplasty catheter, atherectomy, tube, balloon, flexible member, etc.
Function of the invention	Dilate, expand, open, inflate, pressurize, remove blockage, compress, compact, etc.

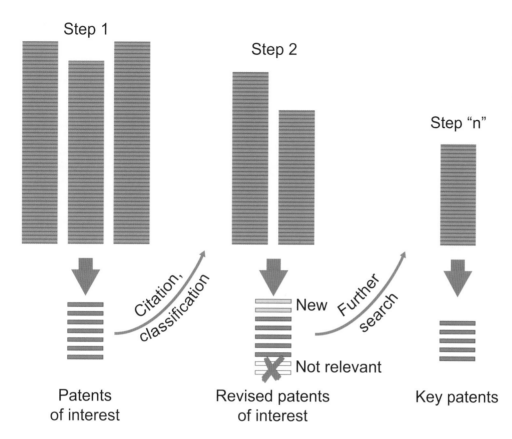

Step 1

Step 2

Step "n"

Citation, classification

New

Further search

Not relevant

Patents of interest

Revised patents of interest

Key patents

FIGURE 4.1.2
The cycle of patent searching: a repeated process of going out broadly into the patent literature, finding patents of interest, and using information from these patents to launch a new, more refined search.

Once a core set of interesting patents is identified by keyword and/or classification searching, it is helpful to use the citation search functions to look forward and backward in time to see what other patents are linked to these core examples. Search backwards in time using the citations (references) that are listed on the face sheets of the key patents that have been identified. For forward searches, start with patents where the drawing and specifications suggest that the patent represents a major advance in the field, then search forward in time using this patent number itself as a search term to see what other, newer patents have referenced this invention.

In practice, combining these different methods leads to a search process that is cyclical in nature – a repeated process of looking out broadly into the patent art and then narrowing it down to progressively more relevant patents. (This iterative process is illustrated in Figure 4.1.2.) The first search may yield many hundreds, or even thousands, of patents.

Through a careful review of these, the innovator can filter the list down to perhaps 10 to 20 key patents. In step 2, these patents of interest are used to create another expanded search, triggered by citations in the patents of interest or classifications of the inventions (or both). This results in another large group of patents to search but, typically, this group is smaller than the patents identified in step 1. Careful review of this art in step 2 produces a few new patents of interest (and probably also leads to the exclusion of some of the initial patents), resulting in a revised set of patents of interest. This process is repeated again, often several more times ("n" times in the figure), until there is a stable group of key patents – perhaps 5 to 15, depending on the area – that continue to be clearly relevant. This is the group of patents to focus on in moving forward.

Once the innovators identify a set of search terms that they believe produces the most relevant results, that refined search should be run repeatedly (i.e., weekly or

Title	Assignee (Company)	Key Claims	Search terms	Patent #	Inventor	Pub Date	File Date
Satiety - Flow Rate							
Gastrointestinal electrical stimulation	UT Austin	Retrograde feedback control of GI action by a stimulating electrode and a detection sensor. Method of external implantation.	"gastric bypass"	6826428	Chen, Jiande; Pasricha, Punkaj Jay	30-Nov-04	11-Oct-00
Adjustable Sphincter System	Ethicon Endo-Surgery, Inc	Artificial sphincter that encircles a passageway that adjusts with fluid. Also discusses invasive and non-invasive manual adjustment methods.		20050272968	Byrum, Randal; Huitema, Thomas; Hassler, William	8-Dec-05	2-Jun-04
Gastric ablation followed by gastric pacing	N/A	Method of ablating pacemaker cells in stomach and then replacing them with electrical stimulation from a pacemaker.		20050240239	Birinder, Boveja and Widhany, Angely	27-Oct-05	29-Jun-05
Satiety - Space Occupying							
Endoscopoic stomach insert for treating obesity and method for use	N/A	Upon release in the stomach, flexible blades for a dome shaped cage, applying pressure to stomach.	"gastric bypass"	5868141	Ellias, Yakub A	9-Feb-99	14-May-97
Endoscopic gastric balloon with transgastric feeding tube	N/A	A gastric balloon that also allows a transgastric feeding tube to go through it and nourishment to pass to the jejunum.		20060025799	Basu, Patrick	2-Feb-06	26-Jul-05
Satiety - By Neural Means							
Treatment of obesity by bilateral vagus nerve stimulation	Cyberonics	Prevent overeating by stimulating vagus nerves to condition the neural stimulus of the stomach.	"gastric bypass"	6587719	Burke, Barrett; Reddy, Ramish K; Roslin, Mitchell S.	1-Jul-03	1-Jul-99
Treatment of obesity by electrical pulses to sympathetic or vagal nerves with rechargeable pulse generator	N/A	Full system (pacemaker, programmer, leads) for stimulating or blocking sympathetic response		20050149146	Birinder, Boveja and Angely	7-Jul-05	31-Jan-05
Methods and devices for the surgical creation of satiety and biofeedback pathways	None	Sensors built into stomach restrictive devices to sense changes in the volume of the restrictive device; sensor then discharges a signal to induce satiety		2005/0267533	Gurtner, Michael	1-Dec-05	15-Jun-05

FIGURE 4.1.3

A sample format for capturing information from a prior art search related to the treatment of obesity as developed by a team in Stanford University's Program in Biodesign. Note that the team's analysis has been removed from this example for the purpose of confidentiality (courtesy of Jennifer Blundo, Darin Buxbaum, Charles Hsu, MD, Ivan Tzvetanov, and Fan Zhang).

monthly) to stay abreast of new information as it is published.

In conducting searches, it is important to develop a system for capturing the output. This is essential because the volume of information is high, the search will likely go on in stages over a long period of time, and the information that is gathered is critically important. Some type of worksheet or spreadsheet is extremely helpful (see Figure 4.1.3 for one example). For the most relevant patents, this worksheet should include the patent number, title, assignee, key claims, inventor(s), publication and filing dates, and agency classification.

A summary of the invention in the words of the searcher and a cut/paste of key claims and figures can also be useful. Additionally, it is a good idea to have a section for comments from the searcher about what is important about this invention and claims (though remember that this information is potentially discoverable in litigation, so it would be wise not to record a comment such as, "This looks identical to our idea . . ."). Beyond the key patents of interest, it is important to keep track of any interesting patent unearthed in a search, since that patent may emerge again in a later search and it is efficient not to have to restudy it. It is also worthwhile to keep track of the search terms used, and the number of "hits" generated by different combinations of terms.

Because the search process is often a complex and time-consuming undertaking, it is wise to plan the work flow ahead of time and divide tasks among the members of an innovator's team. For instance, one team member could start with a classification search while other members divide up the keyword list and begin these inquiries. It is very useful for the different team members to use the same central spreadsheet for recording search results – this can be web-based for convenience.

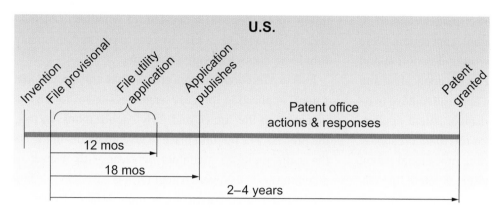

FIGURE 4.1.4

Typical timing associated with US patent filings.

Remember that, in searching for patentability, any publicly disclosed information can serve as prior art. This includes scholarly papers in journals, popular articles, trade show presentations, conference abstracts, newspapers, analyst reports, and many other potential sources. Careful searching with Google, Google Scholar, PubMed, and other appropriate resources is essential. Talk to experts in the field for leads on where to look for public disclosures that may be relevant to the invention but are not yet published. Company booths at medical meetings can also be a great source of information.

A final general comment about searching is particularly important for the first-time inventor: It is essentially *inevitable* that the search process will uncover disclosures that appear to be similar to the new idea, particularly in a dynamic technology domain like medtech. Searching can be an emotional rollercoaster for this reason. Do not be immediately discouraged if something potentially problematic comes to light during a patent search. There are almost always ways to work through the issues, whether by modifying the invention or partnering with owners of existing IP. The important thing is to be thorough and accurate in the search in the first place, so that these issues surface as quickly as possible.

Provisional patents

Medtech inventors frequently file a provisional patent application before pursuing a utility patent in the US. Basically, a provisional patent is an initial description of an invention that establishes a priority date that will

ultimately be used by the patent examiner to help determine patentability, but only after the inventor files a utility patent on the same invention. Provisional patent applications typically require less time and expertise to prepare than a utility patent, and are often written by inventors rather than attorneys (although it can be a good idea to have an attorney at least review a provisional patent application before filing). They are also relatively inexpensive to file with the USPTO, costing approximately $150 for an individual or small business entity. However, the provisional patent is *not itself examined by the USPTO* and can never become an issued patent with enforcement rights. In order to ensure the art contained in the provisional application can become protected IP and that the priority date of the provisional application can be utilized, the inventor must file a nonprovisional (utility) patent that is clearly derived the art described in the provisional application *within 12 months* of filing the provisional application (see Figure 4.1.4).

It is important to note that the provisional patent application is never published, so after filing it is possible to work on the new invention in complete secrecy. It is only after a US utility patent application is filed claiming priority to the provisional application that the invention will ultimately be published (18 months after the provisional filing date).

In order for a utility patent application to claim the priority date benefit of a provisional patent application two criteria must be met. First, the written description and any drawing(s) of the provisional patent application must adequately support the subject matter claimed in the later-filed non-provisional patent application. Note that this does not mean that the written description and any drawings filed in a provisional patent application and a later-filed non-provisional patent application have to be identical. However, the non-provisional patent application is only entitled to benefit from the *common subject matter* disclosed in the corresponding provisional patent application. Second, the non-provisional patent application must have at least one inventor in common with the inventor(s) named in the provisional patent application to claim benefit of its filing date.

Many medtech inventors choose to write their own provisional patents. This is a reasonable approach, provided that the application is thorough and sufficiently detailed to communicate the basic invention and support the claims of a later utility patent. However, it bears repeating that it is wise to have a patent attorney review the application prior to filing, if at all possible. There is no required format for the provisional application. The application should, however, include descriptions of:

1. The need or problem that the invention addresses.
2. Shortcomings of current solutions.
3. Motivation for the new invention.
4. Description of the invention (in enough detail that someone skilled in the field would know how to make and use it).
5. Advantages of this solution to the problem.

In addressing these different parts of the application, the inventor is providing the important details for making a compelling case to the patent examiner for what is novel and non-obvious about the new device. The more that the provisional patent application looks like a utility application, the better the chance that the utility application will be able to successfully claim the benefit of the provisional filing date. Drawings can be very helpful in clarifying an invention and should certainly be included in the application. It is not necessary for a provisional patent application to include claims, although it may be advisable to write claims if they help to clarify the invention (and thereby support the subject matter in a later

utility application). The decision of whether to include claims is made on a case-by-case basis and optimally should be discussed with an attorney.

It is a common misconception that the provisional patent application process was developed to enable inventors to file "quick and dirty" applications. In fact, the process was put into place to give inventors in the US the same rights as inventors overseas with respect to patent term. The 20-year patent term is based on the date a non-provisional patent application is filed. However, inventors outside the US have one year from their ex-US filing date to file a patent application covering the same subject matter in the US. In this way, they gain an early filing date without impacting the term of their US patent (that is, they have the full 20 years after the filing of the US patent). Since, with the provisional patent approach, inventors in the US now have exactly 12 months from the filing date of a provisional patent application to submit a non-provisional patent application, this process creates a US patent term equivalent to the term afforded to those who originally filed first outside the US.

Beyond preserving the term of the eventual utility patent, there are additional advantages to consider with provisional patents. One circumstance where a provisional patent is extremely useful is when an inventor is confronted with an unavoidable and urgent public disclosure – for example, a physician-inventor who is about to give a talk or abstract presentation at a conference. As long as the provisional patent application is received by governmental authorities prior to the presentation, which is verified by obtaining a postmark (a sign that the government postal service has taken possession), the priority date is established and the ability to gain foreign as well as US patent rights is preserved. A more subtle advantage is that the provisional patent application constitutes a legal reduction to practice of the invention without actually building the device (provided there is sufficiently clear and detailed information about construction and use in the application). If there is litigation about the date of the invention later on, the provisional patent is a clear and unambiguous time point regarding both the invention and reduction to practice.

The main disadvantage of the provisional patent alternative is that it can give a false sense of security. Inventors should be aware that filing a last-minute cocktail napkin sketch or a PowerPoint presentation that includes a few high-level ideas about an invention may not provide sufficient support to provide a priority date for the ultimate utility patent. If the inventor does find him/herself in an urgent need-to-file situation, it is important to include as much detailed information as possible in the provisional submission. Many IP attorneys recommend investing nearly as much time in the development of a provisional patent application as a non-provisional patent application. (See online Appendix 4.1.1 for a sample provisional patent application.)

Basic structure of utility patents

A utility patent has three primary parts: a specification, drawings, and claims.

Part 1: Specification

The specification is a description that explains the essential features of the invention. This is where the innovator must provide a detailed enough description to teach a person who is skilled in the field of the invention how to make and use the invention. For a medical device patent, the specification usually begins with a brief background section that clearly describes the need or basic medical problem (including a description of the disease or condition and a discussion of existing approaches to treatment – see 2.1 Disease State Fundamentals and 2.2 Existing Solutions). Within the specification, uncommon medical terms should be explained so that anyone reading the patent, including the examiner, clearly understands the context for the invention, regardless of their level of medical background. The writing should be targeted to the "*Scientific American*" level – intelligible to a bright reader who is not necessarily versed in the field. The specification also must describe in detail what the new device does, how it is to be used (referring to diagrams with labels, as appropriate), and how it is made. This section can describe several different approaches for using the device but, as noted above, the inventor has a legal obligation to describe the best

mode of the invention, as well as sufficient detail to enable others in the field to be able to make and use the invention claimed in the patent.

Part 2: Drawings

Drawings refer to illustrations, typically with a series of reference numbers labeling each of the key features that convey the structure of the invention and how its parts work together. For medical device patents, it is often valuable to include anatomical figures to show the interaction of the device with the organ or system where it is intended to be used. Patent attorneys often employ illustrators who can help render complicated figures based on sketches provided by the inventor.[15] Figure 4.1.5 shows a sample patent drawing next to a

(a)

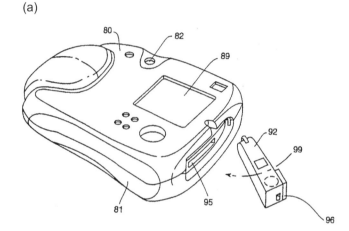

(b)

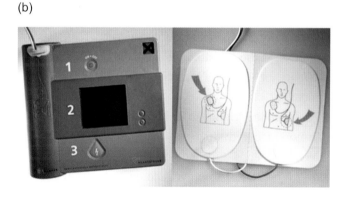

FIGURE 4.1.5
Patent drawing of an external cardiac defibrillator for public use, and the commercial product that was ultimately manufactured and sold (courtesy of Philips Intellectual Property & Standards; reprinted with permission).

photo of the commercial device that ultimately evolved from the invention.

Part 3: Claims

Claims are technically part of the specification, but are presented as a numbered list of points at the end of the patent. Claims are written statements that define the invention and the aspects of that invention that can be legally enforced. As such, they are the real "teeth" of the patent. During prosecution of the patent application, it is the claims that provide the basis for comparison with the prior art to determine patentability of the invention. Upon issuance, the claims provide the basis for a determination of third-party infringement – a ruling that another technology uses features covered by the patent, thereby making it illegal to make, use, sell or import that device without a license. In this sense, the claims are like a deed to a piece of real estate – they specifically describe the boundaries of what is owned. Claims spell out, in precise terms, exactly how an invention or discovery differs from the prior art. Writing claims is such an arcane and expert exercise that inventors are advised to seek the assistance of an IP attorney. However, even when working with an IP expert, the inventor should have a basic familiarity with the art of claims writing in order to effectively understand and assist the work of the attorney.

US utility patents can include two basic types of claims. *Independent claims*, as the name suggests, stand on their own. *Dependent claims* are claims that refer to one or more independent claims and generally express particular embodiments (typically they add additional elements or steps of carrying out the invention).[16] Dependent claims are used to clarify the language of an independent claim, so each dependent claim is more narrowly focused than the independent claim upon which it depends. Independent claims are typically written in broad terms to prevent competitors from circumventing the claim by modifying some aspect of the basic design. However, when a broad term is used, it may raise a question as to the scope of the term itself. For example, if there is any question about whether or not a "base" described in a patent application includes a "set

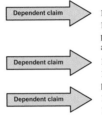

13. An expandable intraluminal vascular graft, comprising:

a tubular shaped member having first and second ends and a wall surface disposed between the first and second ends, the wall surface being formed by a plurality of intersecting elongate members, at least some of the elongate members intersecting with one another intermediate the first and second ends of the tubular shaped member;

the tubular-shaped member having a first diameter which permits intraluminal delivery of the tubular-shaped member into a body passageway having a lumen; and the tubular-shaped member having a second, expanded diameter, upon the application from the interior of the tubular-shaped member of a radially, outwardly extended force, which second diameter is variable and controlled by the amount of force applied to the tubular-shaped member, at least some of the elongate members being deformed by the radially, outwardly extended force, to retain the tubular shaped member with the second, expanded diameter, whereby the tubular-shaped member may be expanded to expand the body passageway and remain therein.

14. The expandable intraluminal vascular graft of claim 13, wherein the plurality of elongate members are a plurality of wires, and the wires are fixedly secured to one another where the wires intersect with one another.

15. The expandable intraluminal vascular graft of claim 14, wherein the plurality of elongate members are a plurality of tantalum wires.

16. The expandable intraluminal vascular graft of claim 13 wherein the plurality of elongate members are a plurality of thin bars which are fixedly secured to one another where the bars intersect with one another.

FIGURE 4.1.6

Independent claims can stand alone, whereas dependent claims rely on a parent claim (from US utility patent 4,733,665).

of legs," a dependent claim that included the phrase, "wherein said base comprises a set of legs," would clarify that, in at least some variations, it does.[17] While the independent claim broadly describes the invention, each dependent claim may describe specific aspects or variants that build on the embodiment or invention captured in the independent claim (frequently, there is a string of dependent claims attached to an important independent claim). Strategically, dependent claims provide a safety margin in the case where the independent claim is invalidated by some prior art that comes to light after the filing (the dependent claim, when combined with the independent claim, leads to something more specific, which may still be valid). Figure 4.1.6 shows how independent and dependent claims work together in the classic Palmaz patent for the intravascular stent.

Another way to think about claims is to categorize them based on what they cover. *Product or apparatus claims* cover a physical entity, such as a material, system, or device. *Method claims* cover an activity, such as a process or method of use or manufacture – for example, the way a medical device is deployed or used in the body. Note that certain method claims covering surgical or diagnostic procedures are not allowed in most countries outside the US (see more about international patenting below).

In addition to the major components in the body of the patent, the *face* or front page includes important summary information that is very useful for searching (see Figure 4.1.7).

Key information to review includes:

- **Patent number** – Assigned by the USPTO.
- **Filing date** – The date the inventors submitted the application and the start of the 20-year period of coverage.
- **Issue date** – The date in which the patent is officially granted.
- **Assignee** – The legal entity that has rights to the patent, often a company.
- **References cited** – The patents and literature cited by the inventor(s) and/or examiners in obtaining the patent. Note that this is not the full list of patents or references that could be related to the patent, as would have been uncovered during the patent search process, but specifically the ones referenced in the patent itself.
- **Abstract** – A brief description of the invention.
- **Classes** – A list of the classes and subclasses assigned to the invention by the patent office.

Preparing a utility patent

Utility patents are the "workhorse" form of IP in the medical technology industry. As such, it is important for the inventor to understand them well. For practical purposes, drafting of the patent should almost always be performed by an expert attorney or patent agent. However, the inventor can save considerable time and expense by providing the attorney or agent the detailed background required for the specification, drafts of the drawings, and the results of the prior art search with

analysis and comments. It is important for the inventors to make sure that all of the alternative ways of making the device are described, and all of the alternative devices they can think of that can be used to perform the method are also described. The point is to stake a claim to as broad an IP position as possible, as a defense against competitors who might pursue the same general concept with a different embodiment or version of the technology.

The inventors will also want to provide a detailed review and editing of the application that is drafted by the attorney or agent prior to filing. For most inventors, the most alien and difficult part of this process is trying to understand the claims. When reviewing claims, an inventor should seek to ensure that they cover all commercially relevant aspects of the invention while not being too limiting or narrow in scope. This includes an invention's physical form, materials, and methods of use. Too often, inventors emphasize the structural elements of the device without paying adequate attention to the clinical significance or the important functionality of the invention. For example, it would not be uncommon for an inventor to submit a claim such as, "A device comprising X, Y, Z ....." However, a much more valuable and strategic approach would be to think about the clinical advantages that the device enables (e.g., it is faster, easier to use, less expensive, etc.) and reflect in the patent the functionality that reveals those unique advantages. For example, a device that enables faster surgical times because it deploys elements simultaneously, as opposed to individually and sequentially, may read, "A device for use in X procedure wherein the device enables simultaneous deployment of elements ...." Similarly, a device that is easier for a surgeon to use because it requires only one hand may read, "A device for use in X procedure wherein the device may be fully operated with a single hand ...." In addition to protecting the invention in a way that is more difficult for competitors to circumvent, this approach to drafting claims also protects the inventor as the device evolves through ongoing product development.

When developing claims, it is important to try to understand the IP activities of the inventor's competitors (other individuals and companies working in the same

United States Patent [19]

Yock

[11] **Patent Number:** 5,061,273

[45] **Date of Patent:** Oct. 29, 1991

[54] **ANGIOPLASTY APPARATUS FACILITATING RAPID EXCHANGES**

[76] Inventor: **Paul G. Yock,** 1216 San Mateo Dr., Menlo Park, Calif. 94025

[21] Appl. No.: **548,200**

[22] Filed: **Jul. 5, 1990**

Related U.S. Application Data

[63] Continuation of Ser. No. 361,676, Jun. 1, 1989, abandoned, which is a continuation of Ser. No. 117,357, Oct. 27, 1987, abandoned, which is a continuation of Ser. No. 852,197, Apr. 15, 1986, abandoned.

[51] Int. Cl.⁵ A61M 25/00
[52] U.S. Cl. 606/194; 604/96
[58] Field of Search 604/96, 101, 102; 606/192, 193, 194

[56] **References Cited**

U.S. PATENT DOCUMENTS

2,043,083	6/1936	Wappler	128/303.11
2,687,131	8/1954	Raiche	128/349
2,883,986	4/1959	de Luca et al.	128/351
2,936,760	5/1960	Gants	128/349
3,731,962	5/1973	Goodyear	128/351
3,769,981	11/1973	McWhorter	128/348
3,882,852	5/1975	Sinnreich	128/4
4,195,637	4/1980	Gruntzig et al.	128/348.1 X
4,198,981	4/1980	Sinnreich	128/344
4,289,128	9/1981	Rusch	128/207
4,299,226	11/1981	Banka	128/344
4,367,747	1/1983	Witzel	128/344
4,468,224	8/1984	Enzmann et al.	604/247
4,545,390	10/1985	Leary	604/95
4,569,347	2/1986	Frisbie	128/344
4,610,662	9/1986	Weikl et al.	128/348.1 X
4,616,653	10/1986	Samson et al.	128/344
4,619,263	10/1986	Frisbie et al.	128/344
4,652,258	3/1987	Drach	604/53

FOREIGN PATENT DOCUMENTS

591963 4/1925 France .
627828 10/1978 U.S.S.R. .

OTHER PUBLICATIONS

A New PTCA System with Improved Steerability, Contrast Medium Application and Exchangeable Intracoronary Catheters, Tassilo Bonzel, PTCA Proc. Abstract, Course 3, Center for Cardiology, University Hospital, Geneva, Switzerland, Mar. 24–26, 1986.
Nordenstrom, ACTA Radiology, vol. 57, Nov. 1962, pp. 411–416.
Nordenstrom, Radiology, vol. 85, pp. 256–259 (1965).
Diseases of the Nose and Throat, at pp. 776–794–797 (S. Thomson) (6th Ed., 1955).
Achalasia of the Esophagus, at pp. 122–147 (F. Ellis, Jr. et al.) (1969) (vol. IX in the Series Major Problems in Clinical Surgery, J. Dunphy, M.D., Ed.)

Primary Examiner—Michael H. Thaler
Attorney, Agent, or Firm—Fulwider, Patton, Lee & Utecht

[57] **ABSTRACT**

Apparatus for introduction into the vessel of a patient comprising a guiding catheter adapted to be inserted into the vessel of the patient and a device adapted to be inserted into the guiding catheter. The device includes a flexible elongate member and a sleeve carried by the flexible elongate member near the distal extremity thereof and extending from a region near the distal extremity to a region spaced from the distal extremity of the flexible elongate element. The device also includes a guide wire adapted to extend through the sleeve so that the guide wire extends rearwardly of the sleeve extending alongside of and exteriorly of the flexible elongate element into a region near the proximal extremity of the flexible elongate element.

6 Claims, 3 Drawing Sheets

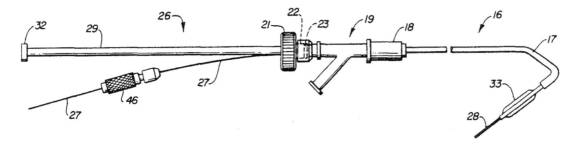

FIGURE 4.1.7

The face page of a patent includes a summary of important information (from US utility patent 5,061,273; the complete patent is shown in online Appendix 4.1.2).

field). Actively review what information is being published and what patents are being issued in the field and evaluate this information against the inventor's established and desired claims. More information about developing an ongoing patent strategy is presented in 5.1 IP Strategy.

Filing and review of utility patents in the US

A patent application can be filed by mail or electronically. The USPTO manages a large volume of patent applications (the office received 576,763 utility patents in 2012[18]) and has been struggling to reduce a substantial backlog, estimated at more than 600,000 applications.[19] Several new programs have been piloted to deal with the backlog, including "Track One," which provides a prioritized examination timetable for an additional fee ($4,000 as of 2014[20]). Basically, while there are some strategic limitations with this path (described in more detail in chapter 5.1), the Track One option allows inventors to move to the front of the line and receive a final action within one year of filing. For regular filings, it can take anywhere from 18 months to up to three years for the patent examiner to pick up the patent application for initial review.

When a utility patent application is submitted, it is opened, sorted, given a serial number, scanned, and checked for completion by the Office of Initial Patent Examination (**OIPE**). A USPTO drawing inspectors also reviews the quality of the figures for clarity and conformity with guidelines. Complete applications (that meet all USPTO submission guidelines) are then routed to a "technology center," which is the organizational unit of USPTO responsible for that type of invention. For medtech, the main technology center is 3700 (Mechanical Engineering, Manufacturing and Products) which includes 37D (Medical and Surgical Instruments, Treatment Devices, Surgery and Surgical Supplier), 37E (Medical Instruments, Diagnostic Equipment, Treatment Devices), and 37F (Body Treatment, Kinesthery, Exercising). If the application is missing information or otherwise incomplete, the OIPE issues a notice to the filer and sets a time period for submitting the missing information. Failure to submit missing or inaccurate information during the time period will result in abandonment of the application.

Within the technology centers, utility patent applications are assigned to a patent examiner who is considered an expert in the subject area. This person researches previous patents and available technical literature to determine whether a patent should be granted based on the parameters outlined above.

Utility applications are made public after 18 months from their earliest priority date, usually the filing date of the provisional patent application on which the nonprovisional utility patent application is based. Inventors can file a "Nonpublication Request" to the USPTO to prevent having their inventions published; however, this process requires an inventor to forego any foreign patent protection for the invention.

Once a thorough examination has been performed, the USPTO issues a first **Office Action** (OA) to the filer. A patent is almost never approved in the first OA. The letter from the examiner will instead outline his/her objections to the application (e.g., it may cite prior art that, in the examiner's opinion, renders the new concept obvious or not novel; it may list the claims that are rejected; or it could call out any problems in the specification and drawings). The filer is given three months to respond to the office action in writing (an additional three months will be granted if an extension fee is paid). In the response, the filer can either argue against the rejections and objections stated in the office action and/or make the required changes in a document called an amendment, which modifies the claims language in a direction intended to make the claims acceptable in light of the examiner's analysis. The filer has an option to meet with the examiner to discuss the application, which can be a highly productive and efficient way of understanding and addressing the examiner's concerns. The office action/amendment process is often repeated a number of times before the patent application is in condition for allowance – or is finally rejected. On balance, there are two points in the patent application process where a patent application may be allowed or rejected. If a **Notice of Allowance** is granted, the USPTO issues the patent upon receipt of required Issue Fee. The complete process from the initial filing to an allowance usually takes between two and four years. For applicants who ultimately do not get the approval they seek, there is an appeals process through the USPTO Patent Trial and Appeal Board, or they can refile the patent application (request for continuing examination). Decisions from the Board may be further appealed to the US Court of Appeals for the Federal Circuit.

One confusing feature of an issued utility patent is the priority date, which is effectively the date against which prior art challenges to a patent will be adjudicated. For international filings, any prior art published before the priority date can be used to help invalidate the patent. In the US, non-patent publications by the inventor that are dated within a year before the priority date do not invalidate the patent filing. For example, if a provisional patent was filed on January 1, 2015, any publication by the inventor in the prior year (subsequent to January 1, 2014) would not be considered as prior art. In practice, this means that an inventor who publishes a paper has a year in the US to obtain patent protection.

If an inventor has filed a provisional patent and then follows with a utility patent, the date of filing the provisional patent is the priority date. If the original filing is made as a utility patent, the utility patent filing date becomes the priority date.

There are several layers of costs associated with utility patents. Attorney costs are reviewed in more detail in chapter 5.1 but, in general, will typically run from $10,000 to $20,000 for the filing of an uncomplicated medtech utility patent. The basic filing fees paid to the USPTO depend on the number of claims in the patent and some other factors, but are approximately $1,600.[21] If the patent is granted, there is an additional USPTO issue fee followed by three maintenance fees at 3–3.5 years, 7–7.5 years, and 11–11.5 years. The total of the issue and maintenance fees are on the order of $13,000.[22]

International patenting

The protection provided by a patent issued by the USPTO stops at the US borders. To obtain patent protection outside the US, an inventor must file patent applications in each country in which patent protection is desired. Under a treaty called the Paris Convention, most countries in the world (with the exception of Taiwan) give patent applicants one year to file a corresponding patent application in their patent office and use the US priority date as the effective date of the foreign filing. In addition, through the international Patent Cooperation Treaty (**PCT**), an inventor can file a unified patent application to seek patent coverage in each of a large number of contracting countries. As of 2013, there were 148 countries participating in PCT, representing most major nations (excluding Iraq, Pakistan, and Taiwan among other countries).[23] The PCT is administered under the auspices of the World Intellectual Property Organization (**WIPO**), a specialized agency of the United Nations, which manages this program and over 20 other international patent treaties. As with other Paris Convention filings, international inventors have one year from filing an international application under PCT to file a utility patent application in the US. (This timing is parallel to that for the US inventor who files a provisional patent, which means that the inventor will often be filing US utility cases and foreign and/or PCT cases simultaneously on or before the one-year anniversary of the provisional filing date.)

Some patent applicants file a PCT first and enter the USPTO through the PCT in the national stage (discussed below). If the PCT is the first patent application filed, the international filing date becomes the priority date in the US.

Under the PCT, a single filing of an international application (called a PCT or international application) is made with a Receiving Office in a single language (the receiving offices are typically the patent offices of the PCT contracting states). An International Searching Authority (**ISA**) performs an extensive international patent search and delivers a written opinion regarding the patentability of the invention.[24] Importantly, this is a non-binding opinion. There is no collective "international patent" that results from PCT process – patents must be reviewed and granted individually by patent agencies in each of the countries in which coverage is sought. What the PCT provides is a way of starting the application process in an efficient and relatively economical way (typically a few thousand dollars) and receiving a preliminary opinion on patentability from an expert international agency. In some cases, the applicant can further request a preliminary examination by an International Preliminary Examining Authority, whose opinion can supersede that of the ISA.

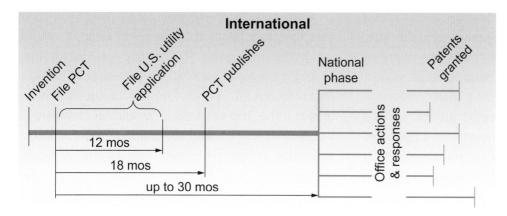

FIGURE 4.1.8
Typical timing associated with international patent filings.

The process of pursing country-by-country patent coverage is called the national phase or stage. Depending on the number of countries selected, this can be an expensive step, consuming hundreds of thousands of dollars. Some strategies for selecting countries for filing are reviewed in 5.1 IP Strategy. The PCT allows inventors to delay national phase filing in PCT member countries for up to 30 months from the priority date, which can allow for substantial development and testing prior to making this major financial commitment (see Figure 4.1.8).

The international application is published by the International Bureau of WIPO, 18 months after the filing date. The WIPO patent database can be an invaluable source of search information about patents in process, particularly those being pursued outside the US.

Although the national phase must generally proceed country-by-country, the European Patent Office (**EPO**) constitutes an important exception.[25] The EPO provides a uniform application procedure for individual inventors and companies to be granted patent coverage in up to 37 European countries through a single application process that is consistent with PCT guidelines. However, the European patent must be validated in each of the European countries the inventor wishes to pursue and usually must be translated into the local language, which represents additional expense. Obtaining the European patent is still an advantage to the inventor because it allows a delay of expenditure for national coverage until the European patent issues and there is more clarity around the value of the claims received. There is movement to establish a Unified Patent Court among a number of the

EU countries, which would create a central authority for deciding matters of validity and infringement of European patents.[26]

Regardless of the specific country in which international patent coverage is being sought, innovators should keep in mind a few important considerations. For US inventors who file a PCT application, the USPTO may serve as the searching agency and will render the patentability opinion. (Some inventors choose to use the European Patent Office, Russian Patent Office, or the Korean Patent Office as the search authority in order to have someone other than a US patent examiner perform the search in an effort to save some money and uncover different prior art.) As noted above, under the PCT, an inventor can wait until one year from the first patent filing in the US to file in other countries and still use the US filing date as the effective filing date on the non-US patent applications.[27]

IP and patenting in India and China
Because of their relatively large size and emerging economic importance, India and China are of particular interest in international patenting. Both countries have had a history of relatively weak IP rights and patent enforcement, which has been a barrier to foreign interest in many technology sectors, including medtech. While both countries are improving, they remain on the Priority Watch List of the Office of the US Trade Representative (along with eight others).[28] The activities of the nations on this list are carefully scrutinized for the adequacy of the IP protection and enforcement rights provided to inventors.[29]

In India, patenting is managed by the Indian Patent Office (IPO). The state of IP enforcement is in flux after an unusual history. Before the Indian Parliament passed the Indian Patents Act of 2005, India enforced product patents but not process patents. As a result, companies could legally sell reverse-engineered products in the Indian market. Although this phenomenon helped fuel the growth of a large generic pharmaceutical industry in India, it also contributed to lax IP protection, a problem that persists.

The 2005 Indian patent law recognized utility patents, copyrights, and trademarks as protected IP, similar to other World Trade Organization member countries.[30] India follows a first to file rule, with patent protection extending 20 years from the priority date. Innovators can take one of two primary paths for pursuing patent coverage in India. The first is to file a patent application directly with the IPO. Indian patent offices are located at Delhi, Kolkata, Mumbai, and Chennai, and inventors must file with the appropriate office based on their location. After preparing and making a filing, the inventors have 48 months to submit a request for examination in order for the patent office to take up the application. Typically, an initial examination report is issued and the inventors are given the opportunity to address any objections raised by the patent examiner. They must comply with the requirements before the stated deadline, otherwise the application will be treated as though it was abandoned. If/when all patentability requirements are met, the patent is granted and recorded in the Patent Office Journal.[31] It can take as much as three to four years to obtain a patent from the IPO.

The second approach is to file a PCT application since India is a contracting state to the Patent Cooperation Treaty. This approach allows the priority date or date of patent application to be recognized in other countries.[32]

Despite recent progress, India is still criticized for the unpredictable enforcement of its patent law, the growing backlog in the patent office, a slow judiciary process, and the theft of proprietary information.[33] Patent protection in pharmaceuticals has been particularly contentious, with several high-profile patent denials of oncology drugs by the Indian Supreme Court.[34] Yet, the indigenous pharmaceutical industry is increasingly making progress in creating and protecting IP as legitimate source of revenue in the Indian healthcare market. The Indian medical technology industry is less well developed than the pharmaceutical industry, and there is not yet a dominant pattern of patent activity.

In China, the State Intellectual Property Office (SIPO) governs patenting, as well as the country's membership in the PCT. In 2011, the SIPO overtook the USPTO to become the patent office that receives the most patent applications in the world (see Figure 4.1.9).[35]

Using a first to file system, the SIPO issues three types of patents: invention patents, utility model (UM) patents, and design patents. The invention patent is the closest form of protection to a US utility patent in China. It protects "any new technical solution relating to a product, a process or improvement"[36] for a period of 20 years from the date of filing. Typically, invention patents take three to five years to be issued. The UM patent protects "any new technical solution relating to the shape, the structure, or their combination, of a product, which is fit for practical use,"[37] and cannot be used to cover methods and processes. UM patents are somewhat easier and faster (5 to 10 months) to obtain because, unlike invention patents, they are not substantively examined. These patents are valid for 10 years from the filing date.[38] Design patents in China are similar to design patents in the US, providing protection for "any new design of the shape, the pattern, or their combination, or the combination of the color with shape or pattern, of a product, which creates an aesthetic feeling and is fit for industrial application."[39] Design patents, like UM patents in China, are not substantively examined (issuing in roughly four to nine months) and have a term of 10 years from the filing date. Although foreign companies working in China overwhelmingly file invention patents, they are increasingly being advised to more fully consider the benefits provided by all three types of patents available under Chinese law.[40] In contrast, the majority of domestic

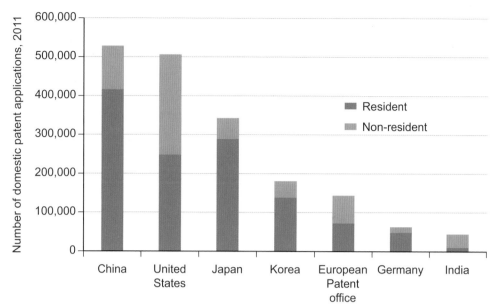

FIGURE 4.1.9
China's SIPO now receives more patent applications each year than the USPTO (compiled from "World Intellectual Property Indicators," WIPO, 2012).

companies and Chinese individuals file for utility model and design patents.

As with India, China moved to strengthen its IP position following its ascension to the WTO in 2001. The country has successfully improved its legal framework and amended its IP laws and regulations, but critics charge that China has been remiss in enforcing these new rules, that the support requirements applied to foreign entity invention patent applications are too strict, and that the flood of unexamined utility model and design patents in the country has created (and continues to create) a large mass of low-quality patents. They further charge that these issues have the consequence of making it difficult to obtain important invention patents and enforce IP rights. Another consequence is that in this environment it is easier for non-inventors to potentially secure patents on technologies that are not their own.[41] In particular, because UM and design patents are granted without a full investigation, they can be perceived as providing an unfair advantage to those who obtain them without actually inventing the technology or process in question, since the burden of proof is on the inventors or company challenging a patent to show that the innovation in question rightfully belongs to them.[42] Moreover, the Chinese government provides incentives, such as financial remuneration and tax breaks, in exchange for filing patents, which has been blamed for the focus on patent quantity rather than quality.[43]

In general, the Chinese patent landscape has become significantly more litigious recently, with more patent lawsuits filed in China than the US.[44] A number of cases have been brought by multinational corporations, alleging that Chinese companies have appropriated foreign technologies covered by the multinational corporation's patents in China. However, given the relatively unpredictable results they achieve, some foreign entities are hesitant to file patent lawsuits within China.[45] Foreign companies are also finding themselves as defendants in patent infringement actions. As one article put it, "Foreign companies face a growing risk that Chinese entities may unscrupulously patent foreign technology in China and demand a toll to do business there."[46] Such cases often move through the courts rapidly, and foreign defendants often consider themselves to be at a significant disadvantage against Chinese plaintiffs due, in part, to uncertainties associated with litigation procedures.

In most regions around the world, there are significant challenges associated with working in an area with a crowded IP landscape, as the story about CoreValve and its transcatheter aortic valve replacement technology illustrates.

FROM THE FIELD ▸ COREVALVE

Assessing intellectual property in a competitive space

One of the major technical advances in cardiac therapy in recent years is transcatheter aortic valve replacement or implantation (which is known as TAVR or TAVI). TAVR was pioneered as an alternative for treating aortic stenosis, a relatively common and deadly heart condition (particularly in the elderly). Aortic stenosis is a degenerative disease in which the leaflets of the aortic valve become fibrotic, calcified, and stiffened. As a result, the valve does not open fully, restricting blood flow from a patient's heart into the aorta and onward to the rest of the body. Aortic stenosis is life threatening because it weakens the heart, ultimately causing heart failure. This final deterioration occurs relatively quickly – after the onset of symptoms, which can include chest pain, shortness of breath, and dizziness, half of untreated patients die within an average of two years.

Historically, aortic stenosis has been treated primarily via surgical valve replacement in patients who could tolerate open-chest surgery and cardiopulmonary bypass. However, a significant portion of the population suffering from the condition is considered inoperable or too high-risk to receive open surgery. Balloon aortic valvuloplasty, in which a balloon catheter is used to expand the narrowed aortic valve, is available to some of these patients but typically does not provide a lasting solution.

Physicians had been intrigued by the idea of minimally invasive valve replacement for decades. Danish cardiologist Henning Rud Andersen was the first to experiment with a valve-stent device in animals. Interventional cardiologist Alain Cribier performed the first transcatheter aortic valve replacement in a human patient in France. Cribier then co-founded a start-up company in Israel, Percutaneous Valve Technologies (PVT), to develop and commercialize the experimental technology.[47]

Shortly thereafter, another physician, Jacques Seguin, began working on a minimally invasive solution when he was a professor of heart surgery at Paris University. As he explained in an article, "My focus was on heart valve therapy and while performing such surgeries, I just knew that what I was doing every day – basically, opening the sternum, arresting and opening the heart, and changing out valves – should be able to be done in a less invasive way. That's the reason why I started design work and experimentation on animals. Eventually, I left the university and my practice to create CoreValve in 2001 to pursue this vision of developing a technology to allow for a heart valve replacement by a catheter-based technique."[48]

The fundamental procedure developed by Cribier, Seguin, and others worked by first dilating the diseased aortic valve with a balloon to make room for the new valve. The prosthetic valve, attached to a wire stent, was then guided by catheter (a thin, flexible tube) to the heart. Once in the proper position, the stent was expanded to hold open the old valve tissue and secure the new valve into place, thus allowing the new aortic valve to open and begin pumping blood. Because the treatment was delivered without the use of a heart-lung machine and did not require sternotomy (opening of the chest cavity), it promised to be appropriate for patients contraindicated for standard valve replacement surgery and potentially for other patients for whom surgery represented a significant risk.

In Sequin's case, he went on to develop a first-generation version of what he initially called a ReValving system (see Figure 4.1.10). After successful **first-in-human** implantations, he expanded his engineering team and decided to pursue the development of the solution in earnest. In parallel with his early experimentation, Sequin approached Antoine Papiernik of French investment firm Sofinnova Partners to see if he would be interested in helping fund the project. At first, Papiernik was skeptical about the need for an alternative to traditional valve surgery, but he and his team started looking into the idea. "We learned that another French physician had just done

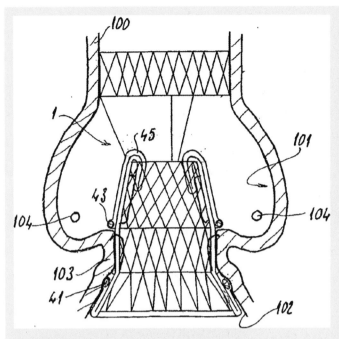

FIGURE 4.1.10

A drawing of the valve from a patent application filed by Jacques Seguin in 2002 (from US utility patent 20,050,043,790 A1).

the first few cases with an Israeli start-up company called PVT (Percutaneous Valve Technologies)," Papiernik recalled, referencing Cribier, "so we realized that Seguin wasn't the only one who saw an opportunity."[49] Intrigued, Sofinnova initiated a formal due diligence process.

"PVT was a little bit ahead," Papiernik said. Cribier and his company had filed multiple patents in the space, so intellectual property was a primary focus during due diligence. "We went to a specialized patent lawyer with strong litigation experience," he explained, "and sought his guidance on the potential risks. We wanted to know if this was going to be a problem for CoreValve. We concluded there were strong positions that supported going forward with the investment." Papiernik continued: "If you are clinically successful with a disruptive technology, somebody almost always will come after you." Entering into a complicated IP landscape was certainly not Papiernik's first choice, but he had become convinced that CoreValve was going after an important clinical need that could translate into a blockbuster market opportunity. He decided to invest.

"At least we went in with our eyes open," he said. "It was a leap of faith."

Sofinnova Partners offered CoreValve a term sheet and ultimately became the sole investor in the company's Series A round of fundraising. Even before the deal was finalized, PVT was acquired by US-based Edwards Lifesciences. Now with the backing of a much larger company, Papiernik anticipated that PVT/Edwards would act aggressively to defend its position.

The first lawsuits were filed within just a few years, first in Germany and the United Kingdom, and then eventually in the US. And, as one article described, "Medtronic [which acquired CoreValve in 2009] and Edwards have been battling one another in the courts [ever since], with each side accusing the other of patent infringement related to TAVR."[50] "The litigation between the two companies is . . . notable because of the sheer number of cases filed against one another," stated Nathan Lowenstein, a partner at Goldberg, Lowenstein & Weatherwax LLP, who was quoted in the same article.[51]

Although they perform roughly the same function, the two technologies are different in multiple ways. For example, the Edwards device is slightly larger, made out of stainless steel, and is deployed when a heart specialist inflates a balloon to position it. In contrast, the frame of the CoreValve device is made of nitinol that automatically expands, giving structure to the valve, when the heart surgeon releases it.[52] However, the specifics of the patent lawsuits come down to the details outlined in the patent claims. As Papiernik put it, "You can win or lose on a single word." For example, CoreValve won early suits in the Germany and the UK based largely on the use of the term "cylindrical." According to press coverage on the UK decision, the Edwards patent in question included "a requirement that the stent be 'cylindrical' and it is this integer that was the subject of the dispute. The Court rejected Edwards' submission that confining 'cylindrical' to that which is approximately geometrically cylindrical is to give the claim a purposeless limitation. Although the Court accepted that in some cases a patentee may use a geometric term without requiring mathematical precision, the result of Edwards' purposive construction was effectively to strike out the requirement for cylindricality altogether, and this, the Court held, was not acceptable. Having considered the relevant passages in the specification, the Court held that the word 'cylindrical', as used in the claims, should be given its ordinary acontextual meaning. Although it is circular in cross section, the CoreValve device varies significantly in diameter across its length. The Court of Appeal agreed with the first instance decision that this device is not cylindrical."[53]

Since then, both Medtronic (CoreValve) and Edwards had experienced wins and losses in the courts. "I have a stack of IP documents that are probably three meters high," Papiernik said. "And we are still not at the end of this story."

When asked for advice about working in a competitive IP space, Papiernik said, "Hire the best IP attorney you can afford as early as possible, and make sure it's someone who will tell you the hard truth. Then work with that person to try to kill the idea. If you can't kill it, then it may have legs to stand on." A strong IP attorney can also help innovators devise ways to strengthen their patent position. For example, he said, "You may be able to modify your design or license key patents" in an effort to mitigate the risk of IP litigation.

Importantly, he emphasized that the idea must be compelling and the market substantial enough to justify a litigious (and costly) situation. In this case, the heart valve market in the US alone is expected to reach $1.5 billion by 2016, with TAVR driving the majority of that growth.[54]

Papiernik further recommended that innovators think globally about their IP position. "Protecting your technology in the US and EU used to be enough, but not anymore. You can't be everywhere as a small company, but reach as far as possible." With TAVR, CoreValve filed patents in the US, EU, and strategic markets in Asia and South America.

More than anything, Papiernik concluded, "You must believe you have a truly defensible position to justify moving forward in a crowded space. But even then, there are no guarantees."

Other important IP concepts

In addition to understanding issues around patents per se, there are a number of basic points about IP that the innovator should understand well.

Documentation

To help to develop a maximally effective patent filing, inventors should begin keeping an **innovation notebook** early in the innovation process (see 1.2 Needs Exploration). Julio Palmaz, inventor of the coronary stent, emphasized that his early documentation proved to be critically important in litigation surrounding the stent, advising innovators to, "Get in the habit of putting down in writing what is in your mind."[55] The contents of an innovation notebook can help innovators tell a holistic story of how their ideas came to fruition, which can be helpful if the invention is ever contested. When performing this documentation, follow these specific guidelines to ensure that the entries are legally defensible:

- **Format** – Choose a *bound* notebook with *numbered* pages. Never tear out or add pages.
- **Process** – Date and sign each page. Never back-date any entry. Even if a problem was observed (but not noted) on the previous day, always date each entry with the actual day it was written. An explanation for the delay in making an entry can be made, but do not falsify any information as this can constitute fraud if an issue ever goes to court.
- **Authentication** – Have a non-innovator act as a witness, signing and dating each page. This can be done every few weeks – it does not need to be done daily.
- **Additions** – Any added material that is pasted in must be "signed over" (this means pasting in the page and

then signing/dating the addition in such a way that the signature is partially on the original page and partially on the addition).

- **Blank space** – Cross out blank spaces to ensure that no retroactive entries can be made.
- **Deletions** – Never white out anything. Use a single line to strike through any errors and initial the corrections.

Inventorship and ownership

Any patent application must include all of the inventors who contributed to the creation of an innovation, as defined by the claims of the patent. To be considered an inventor, an individual must have contributed to the conception of the idea. If someone adds to any part of an invention as it is being claimed, then they are a joint inventor. From the standpoint of patent law regarding inventorship, it is irrelevant whether an inventor was instrumental to conception of a core concept or only conceived of some detail. However, if someone provided only labor, supervision of routine techniques, or other non-substantive contributions, then they are not to be considered an inventor.[56] With medical devices, this issue often surfaces when a prototyping engineer is involved in building the first versions of the device. If the builder of the device is operating entirely under the inventor's instructions, the builder is not an inventor. On the other hand, it is frequently the case that the builder adds a key insight. For example, when one of the editors of this textbook was developing the Smart Needle™ concept (a needle with a Doppler transducer), the engineer who was building the prototype pointed out that moving the transducer to a different position could improve the signal. This was an important addition to the design, and the engineer became an inventor by virtue of this contribution.

It is important to be aware that in the absence of any employment or consulting agreement obligating inventors to assign their inventions to another entity (such as a university or an employer), each inventor has both an ownership interest in the patent and the right to license, commercially develop, or sell the invention *as an individual,* with or without the knowledge or permission of the other inventors. This is true even if the inventor contributed to only a single issued claim. While most inventors prefer to work together to maximize the value of their patent in any licensing deal, the ability of each inventor to act alone is something to consider when involving multiple individuals in the innovation process.

Understanding the implications of ownership for an invention is essential prior to initiating a collaboration. For example, in the case of medical devices, it is common for the inventors of a new technology to seek out a physician expert to validate their early ideas. It can be irresistible for the physician to add some potential improvements to the core idea. These suggestions may be useful and important, but in the process of receiving the feedback the original team may have added a new inventor – at least for some of the claims of the patent. This is particularly problematic if the new inventors have a special relationship with a company that presents a conflict later on, or if they are a faculty member of a university that will automatically assume ownership of the IP (see below).

If it is discovered retroactively that the inventorship on a patent is inaccurate, and the mistake was made in bad faith (for example, if a valid inventor was purposely not included on the patent), the patent can be invalidated.[57] While this happens infrequently, litigators challenging a patent will often start with inventorship as possible grounds for invalidating it.

Student, faculty, and employee inventors

Inventors working in an academic setting must keep in mind that if an invention is conceived and/or developed using significant university resources, the university may assert its rights in taking ownership of the invention. Generally the interpretation of "significant resources" does not include desks, computers, the use of conference rooms, or the use of other commonly available equipment or facilities. University policy regarding IP ownership comes out of the institutions' need to avoid situations in which public resources (such as grants from the NIH) are used to support private enterprise without some oversight regarding the fairness and best deployment of the technology. Among all science and technology areas, the biomedical sciences are, in fact, the area of highest concern for government regulators with respect to university inventors. Unfortunately, there have been some high-profile and tragic examples of conflicts of interest related to ownership of biomedical innovations (e.g., the 1999 Jesse Gelsinger case).[58] As a result, universities and government agencies have become extremely attentive to issues of IP ownership of medical technologies.

Some universities have language in their policies that states, in effect, that inventions in the faculty member's areas of expertise also belong to the university. For student inventors, policies vary across academic institutions. In some universities, students are treated exactly like faculty; in others, there are no IP regulations. A wide range of policies exists between these two extremes. It is important for any university-based inventor to have a clear understanding of the relevant rules. If the university guidelines are not clear, it can be useful to enlist the assistance of an outside lawyer who has experience with that particular university and its inventorship policies (more information about licensing back one's own invention from a university's Office of Technology Licensing can be found in 6.4 Alternate Pathways).

The situation where students are inventing together as a team warrants special mention. If an invention is poised to go forward to commercialization, teams sometimes find that misunderstandings arise regarding the respective contributions of each inventor to the IP. Conflicts can develop when one of the team members feels that s/he has "come up with the idea" in the context of a group ideation process but other team members feel they have made an important contribution in framing the concept. Because of this issue, students in the Stanford Biodesign fellowship and classes create a "memo of understanding" at the beginning of their project about how IP will be divided. The teams typically decide to split the IP evenly, or develop some algorithm for deciding how to value the contribution of a lead inventor. The important thing is that the team has an appreciation of the issues involved upstream of any actual conflict.

Within a corporate context, it is common for an employee to be obligated to assign ownership of inventions to the company, provided that the inventions are within the employer's business or if the inventors used the employer's resources in developing the invention. The company may have an incentive or reward scheme for the inventors, but this varies from organization to organization. For anyone working in more than one company, or for a student or faculty member working for a company part-time as an employee or as a consultant, it is important to understand whose ownership rights apply to any inventions.

Public disclosure

Many inventors underestimate the importance of confidentiality in the patenting process. In the early phases of a new project it is natural to want to share the excitement of the new approach – or present the concept to academic colleagues. As noted earlier, in the US inventors have a grace period of one year to file a patent application from the date that they first publicly disclose the concept, as shown in Figure 4.1.11 (i.e., before the

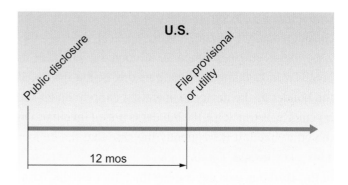

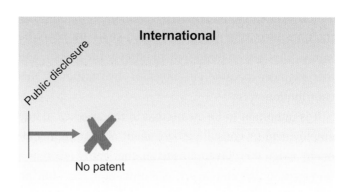

FIGURE 4.1.11

Public disclosure has different implications for patenting in US and international markets.

inventors' own disclosures become prior art that can be used against the patentability of their inventions).

Public disclosure can include publication, presentation, and announcement, as well as commercial use, offerings, or sales. Be aware that casual conversations with colleagues are similarly considered to be a form of disclosure. Medical technology inventors (especially academic engineers or physicians) run a particular risk of submitting an abstract or manuscript that describes their invention assuming that the "confidential" editorial review process protects their patent rights. This situation is at best in the gray zone for public disclosure, and it is wise to file at least a provisional patent application before sending in a manuscript or abstract.

Internationally, there is no grace period. If the innovator publicly discloses the invention prior to filing a patent application, the disclosure is likely to preclude patent protection outside the US, even if the disclosure is within the one-year grace period provided by US law. As a result, experts in the field strongly suggest filing a patent application before making any public disclosures.[59] For medical devices, issues with public disclosure have led many inventions to have coverage in the US only. Depending on the device, this has not necessarily been a significant problem in the past, given the historically large size of the American device market compared to that of other countries. With the expansion of the global markets, however, international coverage is becoming increasingly important, making issues related to public disclosure a serious matter.

Non-disclosure agreements

To keep an idea or invention confidential, inventors are strongly encouraged to use **non-disclosure agreements** (**NDAs**), alternatively known as confidential disclosure agreements. NDAs are legally binding documents that enable an entity to record the terms under which confidential information is exchanged (thus preventing the interaction from qualifying as a public disclosure). Such agreements can be useful anytime an individual or a company must disclose the details of its technical secrets, including (but not limited to) discussions with potential investors, partners, third-party suppliers, employees, contractors, and consultants. For students, the need for

confidentiality can extend to faculty members and fellow students, as well as other mentors, family, and friends. In discussions with an IP attorney, confidentiality is legally protected (with or without an NDA) under attorney–client privilege.

The key issue to address in an NDA is *use* (how the party to whom the confidential information is disclosed can use the information that is shared). It is not enough to ask someone to keep quiet about an invention or other confidential information. A strong NDA must explicitly restrict the ways in which the information can be used.

An example of an NDA is provided in online Appendix 4.1.3. In general, an NDA will have several components:

1. It specifies who the legal parties are in the agreement.
2. It defines the content and scope of the confidential information.
3. It provides for some reasonable exclusions from the obligation of confidentiality (for example, when the receiving party can prove they already knew about the information – or when it was public knowledge).
4. It details what the nature of the obligation is for the receiving party – that is, prevent disclosure and specify how the information can and cannot be used (e.g., restrict access of employees to the information; do not publish or otherwise disclose the information).
5. It specifies a time period that confidentiality is in force (typically for three to five years, or in some cases, until the information becomes public).

NDAs are used to protect disclosures made in multiple scenarios – individual to individual; individual to company; company to individual; and company to company. Typically, innovators and companies are discouraged from using NDAs that require the discloser to document in writing what was disclosed orally in the meeting. Although it is always a good idea to summarize what information was shared, the innovator or company should not be required to do so within a specific time frame and should not be constrained by what may or may not have been included within a summary document.

An NDA does not guarantee that a breach of confidence will not occur. However, it does give the company or innovator a stronger legal position if a problem arises. Some entities will not want to sign an NDA and the inventors must decide how much they are willing to disclose without protection. In some situations, disclosure may be still be warranted. For example, venture capital firms routinely abstain from signing NDAs in favor of a generally accepted principle that sensitive information will be kept confidential. In this case, innovators are advised to mark all documents confidential (including each slide of PowerPoint presentations) and disclose only what is absolutely necessary to get the firm interested. In other scenarios, where the level of trust is less certain, it is best not to reveal confidential information without an NDA.

When in doubt, inventors are encouraged to "file first, disclose later," using NDAs to help control the sharing of any confidential information before and after the filing date. Whenever innovators disclose information before filing a provisional or utility patent application, the risk exists that the person(s) they have disclosed to might develop a competing invention based on the revealed information and worse, file a patent application before them, precluding their patent rights. Using a "file first" strategy before disclosing can help minimize this risk.

The following case example on iRhythm Technologies, Inc. demonstrates how the essential information presented in this chapter can be brought together to help innovators and their teams understand the IP landscape for a concept under consideration and begin to establish a strong IP position.

FROM THE FIELD ▸ IRHYTHM TECHNOLOGIES, INC.

Assessing the IP landscape and making preliminary filings

When Uday N. Kumar, John White, Kityee Au-Yeung, and Joseph Knight were fellows in Stanford University's Program in Biodesign, they began working on the need for *a better way to detect potential rhythm disturbances in non-hospitalized patients with suspected arrhythmias*. Arrhythmias are abnormal heart rhythms (atrial fibrillation, described at length in 2.1 Disease State Fundamentals, is one common type). Although some arrhythmias can have few negative health consequences, others can lead to serious heart disease, stroke, or sudden cardiac death.

According to Kumar, a cardiologist specializing in cardiac arrhythmias, "The vast majority of people who have symptoms that could be due to arrhythmia first go to their primary care physician or an emergency room doctor. But physicians in those settings are not enabled to fully diagnose the problem since current cardiac rhythm monitoring technologies are complicated and have many drawbacks that limit their use by these doctors. So, many people never get assessed until they have a very severe presentation, such as passing out or experiencing a cardiac arrest. Unless something relatively serious happens, patients are often told, 'You had palpitations two days ago. But you look fine now. Let's wait and see if it happens again.'" In contrast, if these patients were referred to a cardiologist, advanced monitoring technologies could be used to diagnose the problem and facilitate more proactive treatment, in many cases, before more serious symptoms occurred. "We figured out that enabling better diagnosis through more accessible and simple approaches was the key to getting more people into treatment earlier in the process," said Kumar.

Once the team understood the parameters of the need and the key **need criteria** (as articulated in the **need specification**), its members held a series of **brainstorming** sessions. Together, they generated dozens of potential solutions and screened them against the need. "After we had a solution that we thought would best meet the need spec," Kumar recalled, "we had to figure out if people had done this before so we conducted IP searches. Our focus was on patents that had already been issued as well as pending patent applications," he added. Importantly, the team did not spend too much time looking at medical journals and other general sources since, as a field, cardiac rhythm monitoring had not experienced a great deal of innovation in recent years. "There just weren't many recent publications out there about talking about cardiac rhythm monitoring," Kumar recalled.

The initial patent searches were meant, in part, to establish the utility, novelty, and non-obviousness of the team's idea. Demonstrating the utility of a device designed to perform outpatient cardiac rhythm monitoring was not a primary concern. However, the team was intent on clearly identifying that its solution was novel and non-obvious. Kumar and his teammates performed many searches using a wide variety of search terms related to the description of the device concept, its functionality, appearance, interaction with the body, and links to the disease. Patents on existing cardiac rhythm

monitoring therapies were also searched for relevant information. The team used different combinations of search terms to help ensure that the scope of the research was not too narrow and that nothing was inadvertently overlooked. "You have to be broad in your searches," said Kumar. "The information you uncover really depends on how you search. I used a thesaurus to figure out different ways to describe the same thing. People use different approaches, and you can miss a whole patent if you don't expand your thinking about search terms." The team also realized that being too broad in its search strategy led to many irrelevant results. Understanding the limitations of being too narrow or too broad allowed the team to develop a search strategy that gave them a good sense of what was out there.

As the team reviewed patent documents it became increasingly clear that their solution had several important elements that made it unique. To address the non-obviousness of the device, they spent significant time and energy figuring out where the cardiac rhythm monitoring field was trending. "If you understand where the field is heading, you can determine how you fit in,"

FIGURE 4.1.12
The team with its first provisional patent filing (courtesy of Uday N. Kumar).

Kumar commented. "If we're going in one direction and everyone else is going another, it shows that we're on a different path, which makes it difficult for the patent office to say that what we're doing is really obvious."

At that point, in parallel with other activities in the biodesign innovation process, the team was eager to seek input and opinions about its solution from other people in the field, "so we filed a provisional patent application," said Kumar. "It wasn't just a paper napkin with a sketch. It had a detailed background section and drawings." It also included a single claim the team believed captured the device's utility. Filing a provisional application secured a priority date in the patent office and protected the team from third-party disclosure (see Figure 4.1.12).

As biodesign fellows during this preliminary IP assessment, the team did not have the financial means to engage an IP attorney. However, upon completing the program, "I made the decision to start a company," said Kumar. "That's when I knew it was time to get a patent attorney involved." To prepare for his interactions with an IP lawyer, Kumar revisited the team's early IP assessment and initiated a more detailed review of the prior art. "I spent a lot of time going back to the original patents we had examined. I also did more searching, brushed up on what else was out there, and dived much more deeply into the analysis of specific claims," he remembered. Kumar and team had started a claims analysis but realized they needed outside expertise to complete it. For this reason, they decided to file the provisional application and come back to claims analysis later, if any of them decided to pursue the project beyond graduation. "At the end of the day, the claims are what matters," Kumar commented. "In the first instance, it can be discouraging to read descriptions and see figures in another patent that look similar. But 'similar' doesn't mean 'the same.' You have to closely examine and interpret the claims, especially the independent claims, to determine whether a potential problem exists." Kumar looked at both device and method claims, since his product and business model were directed towards a monitoring device and an approach for using the monitoring technology.

Based on his preliminary claims analysis, Kumar narrowed the list of patents to those that were closest and most relevant to its proposed solution. "Then, I went back to look at the text to understand the arguments that were made to support the claim sets. This also helped me think about and refine my own arguments for why the solution was novel and non-obvious compared to what was already out there."

With his own analysis complete, Kumar was ready to engage an attorney. As a fellow, he had been introduced to Ben Glenn, who regularly volunteered time to be an "IP coach" to teams in the Program in Biodesign. "I recalled meeting Ben and the positive experience our team had in talking with him. He had deep knowledge and experience in medtech, so I sought his help," said Kumar. He shared his assessment with Glenn as the first step in establishing a close and highly collaborative relationship. Over time, "I spent countless hours with Ben, really explaining exactly what we were doing and bringing him up to speed. This way he could help direct how I could be most helpful as we parsed up the work. Even with a great IP lawyer, the innovator really has to stay involved and understand what's happening because you know better than anyone else what the technology is intended to do." In addition to validating the team's assessment of the IP landscape, Glenn orchestrated a conversion of the provisional application that emphasized the differences and advantages of the device and the method of use. The result was an IP foundation with three utility patent applications (covering the device and associated methods), as well as a PCT application. "In the end, it is hard to overstate the importance of having an experienced IP lawyer such as Ben on my side. He definitely spent a great deal of time really trying to understand the space so that he could thoughtfully put together claims and send them to me. He'd ask, 'Is this really what you mean?' and we'd go back and forth, trying to define different elements or combinations of elements to distinguish over the prior art. Being pushed and questioned by Ben was the key to developing claims that really got to the essence of what was patentable about the solution," Kumar explained.

One important tool that Kumar recommended in developing a preliminary patent position was the USPTO Patent Application Information Retrieval system (or PAIR database) which allows users to access the status of current patent applications. "It gives you all of the history on what's gone on between a patent applicant and the patent examiner," he explained. "By reading these interactions, you can see where applications have been rejected by an examiner, and for what reason. It shows you what claims were rejected, the prior art used in the rejection, and the rationale behind it. With Ben's help, I was also able to appreciate which of the many documents in the file for a given patent application really were significant. This information helped me to think more specifically about where Ben and I might expect

push-back in the examination of our own applications, and it allowed us to realistically address these potential issues."

The company that Kumar founded around the IP that he worked on and licensed from Stanford became an important foundational element for iRhythm Technologies, Inc., based in San Francisco, California. In addition to developing the device and establishing the methods to use it, the company has continued to monitor the IP landscape to stay aware of salient new developments in the field. In addition to being granted numerous utility patents stemming from the early IP work done by Kumar and Glenn, the company has also filed additional patent applications.

A final note: using IP to screen and eliminate concepts

There is a practical problem to address head-on in screening concepts from the standpoint of IP. As is clear from the chapter, the amount of work required to perform a thorough patentability and FTO analysis on just one idea is formidable. So, from the standpoint of workflow alone, it is a major challenge for the team to research the IP landscape on multiple concepts. The key, as in many other steps of the biodesign innovation process, is to gather just enough information in a stepwise fashion to make the screening process both effective *and* efficient. Innovators should plan to conduct a series of progressively deeper searches, eliminating concepts with killer risks as they go. At this stage, a "killer risk" is just that – a problem so severe that it makes clear that the project cannot proceed. In the IP space, for example, this could be a major FTO problem where the technology is owned by a company that would have no incentive to license or co-develop a new product.

For concepts not affected by a killer risk, innovators should gather data about IP (along with regulatory, reimbursement, business models – as described in chapters 4.2, 4.3, and 4.4) that enable them to create and use a risk scoring matrix on their way toward a final solution.

This exercise is described in more detail in 4.6 Final Concept Selection. But, by carefully understanding the risks related to their concepts on a rolling basis, innovators will position their team to efficiently move from multiple concepts to one final solution.

In the end, the team should only do a detailed patentability and FTO search on, at most, a small set of leading concepts that survive the full screening process described in chapter 4.6. In the interim, the following tips can help make the IP screening process manageable:

1. **Red, green, yellow** – To resist the urge to get too deep into IP too early, set a timeline to do an initial screen of all concepts and, using the best information available in that short time, group the concepts into rough categories based on the initial promise of the IP landscape. Given the time constraint, innovators should start with a patent database like Google (and not try to research the world's literature). Use a simple system for categorizing such as: red (troublesome IP problems), green (looks OK), and yellow (some issues raised).

2. **Accentuate the positive** – For the time being, defer any further work on the concepts that appear to have

major IP problems (remember, if it turns out there are compelling reasons to pursue one of these ideas later on, there may be ways of dealing with the IP issues). Broaden the search on the promising concepts, looking into the biomedical literature. Set a timeline for these searches also – do as much as you can in the defined time period, understanding that there is much more searching ahead. Some of the previously green concepts may move to yellow or red through this activity.

3. **Prioritize based on other screens** – Since the next round of IP searching will be extremely time-consuming, use the screens from the other categories to triage where the team will invest its time going forward. If a concept has serious regulatory, reimbursement, or business model issues, or prototyping is raising questions regarding its technical feasibility, do not waste the team's time with a painstaking IP search.

One last point: remember that at this stage in the biodesign innovation process it is common for innovators to identify problematic patents related to their ideas. In certain cases, these issues constitute killer risks and warrant setting aside a potential solution. However, if the concept looks promising relative to all other screens and is an idea the team is excited about, it may be worth seeking an expert opinion before abandoning the idea. As noted, skilled patent counsels can sometimes identify ways to successfully work within the existing patent landscape that may not be readily obvious to innovators without extensive experience in the field.

↘ Online Resources

Visit www.ebiodesign.org/4.1 for more content, including:

 Activities and links for "Getting Started"
- Compile background information
- Search the prior art

- Identify relevant prior art for patentability
- Prepare a patent application
- File a patent application

 Videos on intellectual property basics

 Appendices that provide examples for:
- A provisional patent application
- A USPTO utility patent
- A standard non-disclosure agreement

CREDITS

This chapter was based on lectures and mentoring in Stanford University's Program in Biodesign by Tom Ciotti and Mika Mayer of Morrison & Foerster LLP and Jim Shay and Ben Glenn of Shay Glenn LLP. The editors would like to acknowledge Jessica Hudak of Hudak Group LLC for her careful editing of this material, as well as Antoine Papiernik of Sofinnova Partners for sharing the CoreValve story. Eb Bright of ExploraMed, Nish Chasmawala of Consure Medical, Geoffrey Lin or Ropes & Gray LLP, Jeffrey Schox of Schox PLC, and Jim Shay of Shay Glenn LLP also made valuable contributions. Many thanks also go to Darin Buxbaum, Ritu Kamal, and Asha Nayak for their help with the chapter.

NOTES

1 David Pressman, *Patent It Yourself* (Nolo Press, 2008).
2 Ibid.
3 Ibid.. The following information is drawn from Chapter 1 of *Patent It Yourself*.
4 William H. Eilberg, "Frequently Asked Questions About Trademarks," http://www.eilberg.com/trademarkfaq.html (March 18, 2014).
5 Pressman, p. 5/3.
6 "8 Million Patents and Going," USPTO, http://www.uspto.gov/about/ipm/8_Millionth_Patent.jsp (November 25, 2013).

7 Pressman, 2008, op. cit., p. 160.

8 Erik Combs, "Uncertainty Surrounds the Best Mode Requirement in the Wake of the American Invents Act," Mondaq, February 19, 2013, http://www.mondaq.com/unitedstates/x/222416/Patent/Uncertainty±Surrounds ±The±Best±Mode±Requirement±In±The±Wake±Of± The±America±Invents±Act (November 25, 2013).

9 2138.05 "Reduction to Practice" [R-5]," USPTO, http://www.uspto.gov/web/offices/pac/mpep/s2138.html (November 25, 2013).

10 Patent Search Tutorial," op. cit., http://www.stanford.edu/group/biodesign/patentsearch/goals.html (November 25, 2013).

11 Ibid., http://www.stanford.edu/group/biodesign/patentsearch/benefits.html (November 25, 2013).

12 Cooperative Patent Classification, European Patent Office and United States Patent and Trademark Office, http://www.cooperativepatentclassification.org/index.html; jsessionid = 1nmesnh2n8a3s (December 23, 2013).

13 Bruce Kisliuk, "Introduction to the Cooperative Patent Classification (CPC) EPO and USPTO Bi-Laterla Classification System," Patent Pubic Advisory Committee Meeting presentation, September 27, 2012, http://www.uspto.gov/about/advisory/ppac/120927-09a-international_cpc.pdf (April 23, 2014).

14 Based on the approach outlined by David Hunt, Long Nguyen, and Matthew Rodgers in *Patent Searching: Tools & Techniques* (Wiley, 2007).

15 Inventors can also produce these drawings themselves using resources such as Jack Lo and David Pressman's *How to Make Patent Drawings Yourself* (Nolo Press, 1999).

16 "Claim (Patent)," www.wikipedia.org, http://en.wikipedia.org/wiki/Claim_patent)((November 25, 2013).

17 Ibid.

18 "U.S. Patent Statistics Chart, Calendar Years 1963–2012," USPTO, http://www.uspto.gov/web/offices/ac/ido/oeip/taf/us_stat.htm (November 25, 2013).

19 "Performance and Accountability Report," Fiscal Year 2012, USPTO, http://www.uspto.gov/about/stratplan/ar/USPTOFY2012PAR.pdf (November 25, 2013).

20 "Revised Fee Schedule Effective January 1, 2014," USPTO, http://www.uspto.gov/web/offices/ac/qs/ope/fee010114.htm (November 25, 2013).

21 Ibid.

22 Ibid.

23 "PCT Contracting States," World Intellectual Property Organization, October 4, 2013, http://www.wipo.int/pct/guide/en/gdvol1/annexes/annexa/ax_a.pdf (November 25, 2013).

24 Applicants are sometimes given the choice of having a search done on the invention at a patent office other than at the Receiving Office. Furthermore, not all Receiving Offices are authorized to act as International Searching Authorities.

25 See The European Patent Office's website at www.epo.org.

26 Unified Patent Court, European Patent Office, http://www.epo.org/law-practice/unitary/patent-court.html (December 23, 2013).

27 The full text of the Paris Convention for the Protection of Industrial Property can be found at http://www.wipo.int/treaties/en/ip/paris/trtdocs_wo020.html. However, the treaty is difficult to understand. Consult a patent attorney for practical advice about how to obtain international patent protection.

28 "2013 Special 301 Report," Office of the United States Trade Representative, 2013, http://www.ustr.gov/about-us/press-office/reports-and-publications/2013/2013-special-301-report (December 3, 2013).

29 "USTR Issues 2008 Special 301 Report," Office of the U.S. Trade Representative, April 25, 2008, http://www.ustr.gov/Document_Library/Press_Releases/2008/April/USTR_Issues_2008_Special_301_Report.html (October 2, 2008).

30 "Note on India's Intellectual Property Regime," Embassy of India, https://www.indianembassy.org/press_detail.php?nid = 37 (December 4, 2013).

31 "Frequently Asked Questions," Controller General of Patents and Design Trademarks, http://ipindia.nic.in/ipr/patent/patents.htm (December 21, 2013).

32 "PCT – The International Patent System," World Intellectual Property Organization, http://www.wipo.int/pct/en/ (December 4, 2013).

33 Eric S. Langer, "Understanding India's New Patent Laws," BioPharma International, April 1, 2008, http://biopharminternational.findpharma.com/biopharm/India±Today/Understanding-Indias-New-Patent-Laws/ArticleStandard/Article/detail/507465 (December 4, 2013).

34 Tim Smedley, "Patent Wars: Has India Taken on Big Pharma and Won?," *The Guardian*, May 14, 2013, http://www.theguardian.com/sustainable-business/patent-wars-india-takes-on-big-pharma (December 4, 2013).

35 Lee Chen Yee, "China Tops U.S, Japan to Become Top Patent Filer," Reuters, December 21, 2011 http://www.reuters.com/article/2011/12/21/us-china-patents-idUSTRE7BK0LQ20111221 (December 4, 2013).

36 "Definition," China Patent Trademark Office, http://www.chinatrademarkoffice.com/index.php/ptreg (December 4, 2013).

37 Ibid.

38 Peng Li, Kenneth X. Xie, and David T. Yang, "Patent Procurement and Enforcement in China: A Field Guide," Association of Corporate Council, November 18, 2011, http://www.lexology.com/library/detail.aspx?g = d564bca4-3fdd-454a-b0a2-a30b9388a332 (December 4, 2013).

39 "Definition," op. cit.

40 Li, Xie, and Yang, op. cit.

41 Thomas S. Babel, "Patents in China – Is There Any Real Protection?," IP Frontline, April 30, 2008, http://www.ipfrontline.com/depts/article.aspx?id = 18723&deptid = 3# (December 4, 2013).

42 Ibid.

43 Chris Neumeyer, "China's Great Leap Forward in Patents," IPWatchdog, April 4, 2013 http://www.ipwatchdog.com/2013/04/04/chinas-great-leap-forward-in-patents/id = 38625/ (December 4, 2013).

44 Ibid.

45 Li, Xie, and Yang, op. cit.

46 Neumeyer, op. cit..

47 Brian Buntz, "TAVR: Still the Next Big Thing in Cardiology?," MDDI Online, June 29, 2012, http://www.mddionline.com/article/tavr-still-next-big-thing-cardiology (February 9, 2014).

48 "Dr. Jacques R. Seguin," The Wall Street Transcript, August 24, 2007, http://www.twst.com/interview/24705 (February 7, 2014).

49 All quotations are from interviews conducted by the authors, unless otherwise cited. Reprinted with permission.

50 Brian Buntz, "TAVR on the Global Stage," MDDI Online, June 30, 2012, http://www.mddionline.com/article/tavr-global-stage (February 7, 2014).

51 Ibid.

52 Barry Meier, "Two Medical Devices, Two Different Methods," The New York Times, September 30, 2009, http://www.nytimes.com/2009/10/01/business/01valveside.html?_r = 1& (February 7, 2014).

53 "UK – Medtronic CoreValve v. Edwards Lifesciences," EPLAW Patent Blog, June 30, 2010, http://www.eplawpatentblog.com/eplaw/2010/06/uk-medtronic-corevalve-v-edwards-lifesciences.html (February 7, 2014).

54 Arundhati Parmar, "TAVR Will Drive U.S. Heart Valve Market to Reach $1.5 Billion in 2016," MedCity News, June 27, 2012, http://medcitynews.com/2012/06/tavr-will-drive-u-s-heart-valve-market-to-reach-1-5-billion-in-2016/ (February 7, 2014).

55 From remarks made by Julio Palmaz as part of the "From the Innovator's Workbench" speaker series hosted by Stanford's Program in Biodesign, February 10, 2003, http://biodesign.stanford.edu/bdn/networking/pastinnovators.jsp (November 25, 2013). Reprinted with permission.

56 "Patent Search Tutorial," Stanford Biodesign Program, http://www.stanford.edu/group/biodesign/patentsearch/inventor.html (November 25, 2013).

57 Pressman, 2008, op. cit., p. 16/3.

58 Gelsinger was the first person publicly identified as having died in a clinical trial. An FDA investigation concluded that the scientists leading the trial took shortcuts that may have contributed to this outcome. In this case both the scientists and their university had equity in the company conducting the trial. See Kristen Philipkoski, "Perils of Gene Experimentation," Wired, February 21, 2003, http://www.wired.com/techbiz/media/news/2003/02/57752 (November 25, 2013).

59 Pressman, 2008, op. cit., p. 5/2.

4.2 Regulatory Basics

INTRODUCTION

Without regulatory approval or clearance by the FDA (or the equivalent agency abroad), even the most innovative and important breakthrough in medical technology will never reach patients. The issues involved in determining the safety and effectiveness of a new technology are often complex – and the data on which decisions are made are never perfect. Innovators can lose patience with a process that seems vague, arbitrary, and interminable, while FDA reviewers can lose sleep over the prospect of approving a device that may someday do unexpected harm to patients.

Because of the critically important role that regulatory issues play in the ultimate success of a new technology, understanding the regulatory landscape early in the biodesign innovation process is essential. In practice, innovators almost always employ an expert to write and manage their regulatory submissions. However, given the extent to which regulatory requirements affect product design, development, and commercialization, innovators must have at least a general understanding of regulatory requirements, options, and nomenclature in order to provide effective leadership in the biodesign innovation process.

This chapter provides an overview of regulatory terminology and a primer on basic regulatory pathways. A second chapter, 5.4 Regulatory Strategy, describes the more nuanced implications of regulatory requirements and how they can become a source of competitive advantage (or disadvantage) to an innovator or company.

 See ebiodesign.org for featured videos on regulatory basics.

OBJECTIVES

- Understand the basic goals of the FDA and how the agency is organized.

- Learn about the US medical device classification system and how it relates to the two main regulatory pathways for medical devices: 510(k) and PMA.

- Develop a basic understanding of requirements for regulatory approval outside the US.

- Appreciate how to use regulatory risks as a screen for prioritizing concepts.

REGULATORY FUNDAMENTALS

The US Food and Drug Administration (**FDA**) is a regulatory, scientific, and public health agency with a vast jurisdiction, overseeing products that account for roughly 25 percent of all consumer spending in the US.[1] Products under FDA jurisdiction include most foods (other than meat and poultry), human and animal drugs, therapeutic agents of biological origin, radiation-emitting products, cosmetics, animal feed and, of course, medical devices.[2] The FDA is the lead regulatory agency in the world, although other important device markets either have or are developing robust regulatory systems (regulatory approaches in countries outside the US are addressed later in this chapter).

FDA background

The modern era of the FDA began in 1906 with the passage of the Federal Food and Drugs Act, which created the regulatory authority for the agency. This law was replaced in 1938 by the Food, Drug & Cosmetic Act, which focused primarily on drug safety.[3] Remarkably, devices were essentially not regulated until the Medical Device Amendments Act of 1976. The device amendments were stimulated, in part, by a therapeutic disaster in which thousands of women were injured by the Dalkon Shield intrauterine device.[4] The new law provided for three classes of medical devices based on risk, each requiring a different level of regulatory scrutiny (see below for details).

At the most basic level, the FDA's goal is to protect the public health by assuring the *safety* and *effectiveness* of the products under its supervision.[5] A strong focus on safety is at the core of the FDA's mission and culture. In most cases, the laws that have shaped the regulatory authority of the agency have come in response to high-profile incidents involving unintended, harmful effects of drugs or devices. Even though the FDA's mission also refers to the goal of "advancing the public health by helping to speed innovation," the mandate to protect the public health takes precedence. It is important for innovators to understand this priority and how it influences the reviewers who make key decisions about new technologies. These reviewers are government employees who share a deeply held motivation to serve the public by spotting problems before they get to patients. It follows that there is less incentive to approve submissions quickly than there is to be as certain as possible about the safety of the device.

To understand what the FDA means by effectiveness, innovators should realize that the end result of a successful FDA submission is that the agency clears or approves the marketing and sale of a device for certain, specific clinical indications. For example, a pacemaker might be approved to treat symptomatic bradyarrhythmias (slow heart rates causing dizziness and other symptoms). The exact language FDA approves for the use of the device is reviewed in great detail by the agency and results in a statement of "indications for use" (**IFU**) that is included in the product packaging and advertising. To judge effectiveness, the FDA must decide that the device functions as specified by the IFU. Exactly what kind of evidence is required by the FDA to prove effectiveness depends on the risk associated with the device. While a device with minimal risk will be exempt from any type of FDA premarket clearance (no evidence is required), a device that treats a life-threatening condition will typically require a large-scale, **controlled clinical trial** for approval. It is important to understand that the criteria for clearing or approving a device are not fixed, but evolve with time in response to a number of factors, including new clinical science and accumulating experience in the marketplace with medical technologies. As mentioned, the general tenor of regulation can be strongly influenced by major safety failures that attract media and public attention. For instance, issues with pacemaker leads[6] and metal-on-metal hip implants[7] are two relatively recent examples of incidents that have contributed to a climate of particular caution and scrutiny at the agency. Because device regulation was implemented relatively recently (compared to other areas under the FDA's authority), and because of the complexity of medical devices, the agency's practices and policies continue to evolve as it tries to keep up.

There are two broad and important areas over which the FDA does not exert regulatory control. First, *cost-effectiveness has no part in the FDA's assessment of new technologies*. Data about cost are not part of any submission and there is no mandate for the FDA to be involved in the determination of prices or **reimbursement**.[8] In the federal government, reimbursement levels for medical technologies are determined by the Centers for Medicare and Medicaid Services (**CMS**) – see 4.3 Reimbursement Basics. CMS generally awaits FDA approval before making a positive reimbursement decision, but this is not an absolute requirement. Second, *the FDA does not regulate or otherwise monitor the practice of individual physicians*. Once a device is cleared or approved for sale in the US, physicians can use it as they see fit.[9] If something goes wrong, a physician may be sued for malpractice if the device has been used in a manner that is not in accordance with the clinical **standard of care** in the medical community. However, the FDA has no jurisdiction in the matter. The agency only has jurisdiction over the device manufacturers and how those companies promote and sell their products.

The FDA is headquartered in Silver Spring, Maryland, in the White Oak Federal Center. The agency has nearly 11,000 permanent employees with approximately 4,000 of them working for Center for Devices and Radiological Health (**CDRH**), the main unit overseeing medical devices.[10] In 1980, the FDA became part of the Department of Health and Human Services. The agency is periodically reauthorized by Congress and is subject to Congressional oversight which, in practice, is distributed across a large number of committees. The proposed FDA operating budget for fiscal year 2013 was approximately $4.5 billion.[11] The FDA commissioner is nominated by the President and confirmed by the US Senate. For the most part, however, the FDA has relatively few political appointees compared to other government agencies, so it is less subject to internal staff changes with turnover in political administrations.

FDA's Center for Devices and Radiological Health

The FDA is organized into several centers according to types of products they regulate. As noted, CDRH provides oversight for devices and also regulates radiation emitting products (including X-ray and ultrasound instrumentation). The Center for Drug Evaluation and Research (**CDER**) regulates pharmaceuticals, while the Center for Biologics Evaluation and Research (**CBER**) oversees biologics (e.g., vaccines, blood products and biotechnology-derived products). Within CDRH, the Office of Device Evaluation (ODE) is the entity responsible for review and approval of most devices, while primarily in vitro diagnostic technologies fall under the Office of In Vitro Diagnostics and Radiological Health (OIR). The Office of Combination Products was established in 2002 to help triage submissions and manage drug-device, drug-biologic, and device-biologic therapies.

The FDA defines a medical device as:[12]

An instrument, apparatus, implement, machine, contrivance, implant, in vitro reagent, or other similar or related article, including a component part, or accessory which is … intended for use in the diagnosis of disease or other conditions, or in the cure, mitigation, treatment, or prevention of disease, in man or other animals, or intended to affect the structure or any function of the body of man or other animals, and which does not achieve any of its primary intended purposes through chemical action within or on the body of man or other animals and which is not dependent upon being metabolized for the achievement of any of its primary intended purposes.

The key to decoding this complicated description is to understand that chemical action and metabolic change are hallmarks of drugs and biologics. Therefore this definition says, in effect, that if the product is not a drug or biologic, it *is* a device. Of course, there are therapies that fall in the gray zone of this definition (e.g., drug-eluting stents). If there is ambiguity about whether or not a new product will gain access to the market via a drug or biological application or device application, the company can file a request for designation (RFD) and propose a recommendation.[13] However, the FDA will ultimately make the determination. The Office of Combination Products is now making the assignments in most of these ambiguous situations.

Device classification

Once it is clear that a new product is properly characterized as a medical device, the next major consideration is to determine its risk profile based on the current three-tier safety classification system (see Table 4.2.1). **Class I** devices are those with the lowest risk, while **Class III** includes those with the greatest risk.[14] This categorization serves as the basis for determining the regulatory pathway that the device must take before being cleared or approved for human use (the "premarket" stage), as described in more detail below.

Innovators developing devices are initially required to make a "best-guess" selection of the appropriate

Table 4.2.1 Device classification has direct implications on the number and complexity of the requirements imposed by the FDA.

Class	Examples	Description	FDA requirements
I	Bandages, tongue depressors, bedpans, examination gloves, hand-held surgical instruments	Class I devices present minimal potential harm to the person they are being used on and are typically simple in design.	With Class I devices, most are exempt from premarket clearance. There is no need for **clinical trials** or proof of safety and/or efficacy since adequate predicate experience exists with similar devices. However, they must meet the following "general controls": • Registration of the establishment with the FDA. • Medical device listing. • General FDA labeling requirements. • Compliance with quality system regulation (**QSR**), with the exception of design controls, unless specifically called out in the regulation.
II	X-ray machines, powered wheelchairs, surgical needles, infusion pumps, suture materials	**Class II** devices are often non-invasive, but tend to be more complicated in design than Class I devices and, therefore, must demonstrate that they will perform as expected and will not cause injury or harm to their **users**.	Class II devices are generally cleared to market via the **510(k)** process, unless exempt by regulation. They must meet all Class I requirements, in addition to the "special controls" which may include: • Special labeling requirements. • Mandatory performance standards. • Design controls. • Post-market surveillance.
III	Replacement heart valves, silicone breast implants, implanted cerebellar stimulators, implantable pacemakers	Class III devices are high-risk devices. These are typically implantable, therapeutic, or life-sustaining devices, or high-risk devices for which a predicate does not exist.	Class III devices must generally be approved by the **PMA** regulatory pathway, although a small number are still eligible for 510(k) clearance. (FDA has begun the process of requiring PMAs for all of these.) Class III devices must meet all Class I and II requirements, in addition to stringent regulatory approval requirements that necessitate valid scientific evidence to demonstrate their safety and effectiveness, before they can be used in humans.

classification of their device in consultation with their expert regulatory consultants. The FDA offers device classification panels and codes that can be referenced to help with this determination.[15] The classification will be reviewed by the branch of CRDH that evaluates the technology. Roughly half of all medical devices fall within Class I, 40–45 percent in Class II, and 5–10 percent in Class III.[16] If the technology does not fit into one of the existing device regulation intended uses (published in the Code of Federal Regulation (CFR) database) or if the technology is so novel that it raises new questions of safety or effectiveness such that the innovator cannot judge the likely classification, then informal or formal discussions can be pursued with the FDA to clarify the classification. Be advised, this can significantly impact the time and cost of device development.

Through CDRH, the FDA regulates more than 100,000 medical devices ranging from simple thermometers, tongue depressors, and heating pads to pacemakers and kidney dialysis machines. These devices are organized into 1,700 different categories of technology which are managed within ODE. Currently there are seven divisions with a total of 33 branches that are based on medical specialties (see Table 4.2.2).[17] Note that ODE frequently reorganizes its structure as part of efforts to improve the review process.

Each of these branches has separate teams of reviewers who are experts in that area. The primary reviewer will typically have at least an undergraduate degree in engineering or one of the biomedical sciences. Reviewers work on a dozen or more submissions at a time. As mentioned, in vitro diagnostic devices are evaluated under a separate Office of In Vitro Diagnostic and Radiological Health, which has five divisions (Chemistry and Toxicology Devices, Immunology and Hematology Devices, Microbiology Devices, Radiological Health, and Mammography Quality Standards).

Innovators can choose which branch to target for their device submission based on the intended use (although the FDA will ultimately determine which branch reviews the device). The process for making the initial selection of a branch is described in more detail in the Getting Started section of this chapter, but basically involves either searching the FDA's classification database or browsing the regulations for device precedents to determine where other similar devices have been assigned. Using these tools, innovators are able to determine an appropriate classification by finding the description that best matches their own device. For instance, if a team created a new type of steerable colonoscope, members could go to the classification database and do a search on "colonoscope." Alternatively, they could access the device classification panels, choose Gastroenterology Devices, select Diagnostic Devices, and then review the description for Endoscope and Accessories.

Regulatory pathways

There are three major pathways for medical device regulation by CDRH, which are based on the three-level risk classification (although, unfortunately, there is not a one-to-one correspondence between the classification and the pathway). Which pathway the device takes is extremely important to the innovator and any company developing the technology because the effort, time, and cost associated with these different alternatives vary significantly (see Table 4.2.3). Note that the regulatory pathways for pharmaceuticals and biologics, which are overseen by CDER and CBER, respectively, are different than the ones described here (see the websites of those centers for further information).

Relatively speaking, only a few medical devices are required to receive premarket approval (PMA). For example, in fiscal year 2012 the ODE received over 4,000 510(k) submissions, but only 33 original PMA applications or panel track PMA supplements.[18]

Exempt devices

Roughly three-quarters of Class I devices are exempt, meaning that they do not require FDA clearance to be marketed.[19] Examples of exempt Class I devices include elastic bandages, tongue depressors, bedpans, and surgical gloves (see Table 4.2.1). A much smaller number of Class II devices are exempt (less than 10 percent), based on the agency's determination that they represent a minor safety risk. No Class III devices qualify for exemption. In addition to devices that pose little or no risk to patients, there are some other special circumstances under which an exempt classification is given, such as

Table 4.2.2 The ODE is currently organized into seven major divisions with 33 branches that are based on medical specialties (compiled from the US Food and Drug Administration's CDRH Management Directory by Organization).

Office of Device Evaluation (ODE)	
Division	**Branch**
Division of Neurological and Physical Medicine Devices	• Neurostimulation Devices • Neurodiagnostic and Neurosurgical Devices • Physical Medicine Devices
Division of Orthopedic Devices	• Joint Fixation Devices Branch One • Joint Fixation Devices Branch Two • Restorative and Repair Devices • Anterior Spine Devices • Posterior Spine Devices
Division of Surgical Devices	• General Surgery Devices Branch One • General Surgery Devices Branch Two • Plastics and Reconstructive Surgery Devices One • Plastics and Reconstructive Surgery Devices Branch Two
Division of Cardiovascular Devices	• Cardiac Diagnostics Devices • Cardiac Electrophysiology Devices • Circulatory Support Devices • Interventional Cardiology Devices • Implantable Electrophysiology Devices • Peripheral Interventional Devices • Structural Heart Device Branch • Vascular Surgery Devices
Division of Opthalmic and ENT Devices	• Contact Lenses and Retinal Devices • Diagnostic and Surgical Devices • Intraocular and Corneal Implants • Ear, Nose, and Throat Devices
Division of Reproductive, Gastro- Renal and Urological Devices	• Obstetrics/Gynecology Devices • Urology and Lithotripsy Devices • Renal Devices • Gastroenterology Devices
Division of Anesthesiology, General Hospital, Infection Control and Dental Devices	• Anesthesiology Devices • General Hospital Devices • Respiratory Devices • Infection Control Devices • Dental Devices

Table 4.2.3 The three regulatory pathways for medical devices vary in their requirements based on the level of risk associated with the device (compiled by authors from the FDA website).

Pathway	Description
Device exemption	These are devices for which the risk is so low that they are exempt from regulatory clearance. Most Class I devices take this pathway.
510(k)	This is the largest category of medical device applications, in which clearance is based on a device being similar (or substantially equivalent) to existing, predicate devices in clinical use. Some Class I devices and most Class II devices take this pathway.
Premarket approval (PMA)	This is the most stringent pathway, used for devices that are significantly different from existing technologies and/or represent the highest risk to patients. The vast majority of Class III devices take the PMA pathway, although a few still remain eligible for 510(k) clearance (these will ultimately be eliminated).

finished devices that are not sold in the US or custom devices (one-off devices made for a specific patient or application).[20]

Even if a device is determined to be exempt, it still must comply with a minimum set of FDA requirements called "general controls" that also apply to the other two regulatory pathways. These requirements oblige the company to register their facility or establishment with the FDA, fill out a form listing the device and its classification, comply with general FDA labeling and packaging requirements, and adhere to the FDA's Quality Systems Regulation (QSR) – a set of guidelines for safe design and manufacturing (see 5.5 Quality Management). However, exempt devices are not subject to the Design Control Regulations of the QSR unless specifically noted in the regulations. A limited number of Class I exempt devices are also exempt from other QSR requirements.

The 510(k) pathway

The 510(k) review process applies to devices of moderate risk where there is some similarity to an existing technology already in use. This is the pathway required for most Class II devices. A device that passes FDA scrutiny by 510(k) is said to be *cleared* and to have achieved *premarket notification* (in contrast, a Class III device that follows the PMA pathway is *approved* by the FDA). The company making a submission for either a 510(k) or a PMA is referred to as the *sponsor* of that submission.

The "510(k)" nomenclature refers to the section of the 1976 Medical Device Amendments (MDAA) to the Federal Food, Drug & Cosmetic Act that describes this pathway. In creating a new system for medical device approval, the MDAA took into account the fact that there were a number of existing, moderate-risk devices that were already widely and safely in use. These devices were essentially grandfathered by the act and are now described as *pre-amendment devices*. The MDAA also enacted a mechanism for clearing new devices based on similarities to these existing pre-amendment devices that had been "road tested" prior to 1976. A further provision allows for new devices to be cleared based on comparison with other devices that have received 510(k) clearance subsequent to 1976. In any of these cases, the preexisting device to which the new device is compared is called the *predicate device* (see Figure 4.2.1).

To support the 510(k) pathway, the innovator must demonstrate how the device is *substantially equivalent* to the predicate(s) to allow the FDA to compare the new device to these existing devices. In essence, in order for a new device to be found substantially equivalent to a predicate device, it must: (1) have the same indication for use; (2) have technological characteristics that are similar to the existing device; and (3) not raise any new questions of safety and effectiveness in those areas where there are differences with the predicate device.[21] Specifically, this means that the device must be comparable to the predicate device in terms of its intended use,

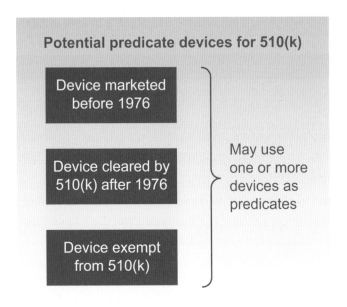

Potential predicate devices for 510(k)

Device marketed before 1976

Device cleared by 510(k) after 1976

Device exempt from 510(k)

May use one or more devices as predicates

FIGURE 4.2.1

510(k) clearance can be obtained based on either a pre-amendment predicate device, a post-amendment device that has already been cleared via the 510(k) pathway, or a 510(k) exempt device.

design, energy used or delivered, materials, chemical composition, manufacturing process, performance, safety, effectiveness, labeling, biocompatibility, standards, and other characteristics, as applicable.[22] Substantial equivalence does not mean that the new and existing devices are identical. However, it does require that the new device provides a relevant comparison (a concept that is ultimately decided by the FDA) and is at least as safe and effective as the predicate device.

Another important aspect of the 510(k) pathway is that a sponsor may choose *more than one predicate device* to make the argument of substantial equivalence. However, the FDA will not allow "split predicates," that is, a situation in which a sponsor is attempting to split the 510(k) decision-making process by demonstrating that a new device has the same intended use as one marketed device and the same technological characteristics as a second, different type of marketed device.

Substantial equivalence for 510(k) clearance usually can be demonstrated on the basis of bench and animal testing. However, although the FDA only requires clinical data for approximately 10 percent of annual 510(k)

submissions,[23] that figure is gradually increasing. The FDA will advise innovators and companies whether clinical data collection is necessary. If required, the clinical studies to gather such data are typically much smaller, faster, and less expensive than the trials required for the PMA pathway (see 5.3 Clinical Strategy).

In recent history, the FDA's application of the 510(k) process has been under scrutiny from the media and Congress. An Institute of Medicine (IOM) report criticized the FDA's application of the 510(k) process as overly liberal, allowing too many products to be cleared via this pathway with minimal clinical testing as opposed to the more rigorous PMA pathway.[24] This, along with continued media attention, has led to continued debate surrounding devices cleared via the 510(k) pathway. There is a general trend for more extensive data requirements for 510(k)s, which in some cases may approach the type of filing more typically required for a PMA.

De novo 510(k) clearance In 1997, a new category of 510(k) clearance was introduced called the de novo pathway. This alternative was further modified in 2013 by Congress in the FDA Safety and Innovation Act (FDA-SIA). The de novo pathway is intended for devices that do not have the major risks of a Class III device, but for which no predicates exist – or for products that raise different questions of safety and effectiveness than those for a legally marketed predicate. The de novo 510(k) generally requires a higher level of proof of efficacy than a standard 510(k), but less evidence than for a PMA. One example of a de novo 510(k) clearance is the Given Imaging Pillcam™ endoscopic capsule. This capsule contains a tiny video camera that transmits images from inside the intestines as the capsule works its way through the gastrointestinal tract. The company provided convincing data to the FDA that the capsule provided acceptable images to aid in diagnostic evaluation, as an adjunct to standard endoscopy procedures. The FDA did not regard the device as sufficiently similar to any existing predicate to support 510(k) clearance, but was willing to grant the de novo 510(k) because of the favorable safety profile of the capsule and the demonstration of effectiveness in imaging. This pathway historically has been used mostly for in vitro diagnostic devices. However, it may prove to be a useful tool as more complex devices are developed that do not rise to the level of high risk but for which no predicate exists or which raise different questions of safety and effectiveness versus a legally marketed predicate.

Two other categories of 510(k) clearance are worth a brief mention. A company can pursue a special 510(k) when it has modified its own device and is seeking clearance for this modification. An abbreviated 510(k) can be used when the company is able to certify compliance with an FDA recognized special standard – a published document that lists explicit requirements regarding the characteristics or performance of the device. In this case, the company submits a declaration that the device is in conformance with these standards and does not need to submit the detailed test reports required for a traditional 510(k).

Mechanics of 510(k) submissions   In almost all circumstances an innovator will use an expert regulatory consultant to prepare a 510(k) submission. The necessary documentation can be many hundreds of pages long (or more), and the approach to choosing predicates and arguing substantial equivalence requires experience.

While there is no standard 510(k) application, the requirements for a 510(k) submission are relatively well defined.[25] The heart of the submission is a section in which the new device is compared to the predicate(s). This requires a detailed and scientific comparison that includes device performance characteristics, data from bench testing and, in some cases, the results of animal and clinical tests. A second important section is the indications for use – a list of the clinical indications for which clearance is sought. The submission also includes a copy of all draft printed material and labeling to be distributed with the device or provided to patients. The sponsor may also submit sample advertising and educational materials. If biocompatibility or shelf-life/stability data are required, these results are also provided. Finally, the submission includes a 510(k) summary or 510(k) statement, which is a public statement that will be posted on the FDA website if the device is cleared.

Numerous device-specific guidance documents are available via the FDA's guidance document database.

These guidance documents contain detailed information regarding the FDA's current expectations in order to determine substantial equivalence for the new device,[26] as well as the format to be used,[27] and how to be sure that nothing is overlooked.[28]

510(k) review process and timeline   As noted, each 510(k) submission received by the FDA is assigned to one of the agency's primary divisions for review. Not all divisions have the same approach to doing business or working through the review/approval process and there may be benefit in trying to direct the submission to a particular group.

The FDA is required to review a traditional 510(k) submission within 90 days of its receipt. This does not necessarily mean that the FDA must issue a decision within 90 days, but it is obligated to provide feedback within that period. If the FDA cannot make the determination based on the information that has been submitted, the sponsor will receive a request for additional information. There may be multiple such requests and, each time the FDA requests information, the clock may be stopped until the manufacturer submits the requested information. Ultimately, the FDA issues either a substantially equivalent (**SE**) determination or a not substantially equivalent determination (**NSE**). The notification of substantial equivalence comes in the form of a letter from the FDA stating that the device can be marketed in the US. An NSE decision puts the company "back to the drawing board" and is a major liability not only for that device but for the company in general, since it can take a significant amount of time to either file a PMA or rework the 510(k) in order to allow it to be cleared (e.g., change the intended use, collect additional data, etc.).

Although the statutory review timetable for 510(k) clearance is 90 days and there are target review timelines issued by Congress, the actual time for review of a 510(k) can be unpredictable. For the most part, a straightforward clearance can be obtained in several months. Some 510(k)s can drag on for more than a year, based on the complexity of the analysis and number of requests for additional information. The process can also be slowed by substandard preparation of the submission on the part of the sponsor or poor communication between the sponsor and the reviewer. The FDA now conducts an initial review upon receipt of the submission to ensure that all the required information is present before beginning the 510(k) review process.

The PMA pathway

The PMA pathway is required for devices that represent the highest risk to patients and/or are significantly different from existing technologies in use within a field. There are also certain strategic reasons for an innovator or company to pursue a PMA versus a 510(k); these are reviewed in chapter 5.4. A PMA is based on a determination by the FDA that sufficient, valid scientific evidence exists to assure that a device is safe and effective for its intended use(s) before it is made commercially available. Approval of a PMA device is made based on the merits of that device alone, regardless of any similar Class III devices that may exist. The large majority of PMA applications are submitted for Class III devices. Although there are certain Class III pre-amendment devices that can still be cleared by the 510(k) pathway, the FDA is working to complete classification actions for these remaining devices.[29]

Pursuing a PMA is considerably more complicated than pursuing a 510(k), primarily because the sponsor needs to provide valid scientific evidence to support safety and effectiveness and this often means clinical data from a **pivotal study**. Such studies are typically large, multi-center, **randomized clinical trials** and often represent the single largest expense – and the biggest risk – in the entire biodesign innovation process (see 5.3 Clinical Strategy). As a rough rule of thumb, these studies involve hundreds of patients and cost millions of dollars – or tens of millions. In turn, PMA submissions often reach thousands of pages in length (see Figure 4.2.2).

If a device is life-sustaining, it will require PMA submission whether or not there are similar devices that are already approved. For example, coronary stents are a type of Class III device. Even though other coronary stents exist in the market, all new stents must follow the PMA pathway. A device that was previously classified as Class III and entered the market via premarket approval *cannot* be used as a predicate device for 510(k) regulatory clearance.

FIGURE 4.2.2

Informally, the length of a PMA submission is often measured in feet rather than by pages (by Joan Lyons, courtesy of Emphasys Medical).

Mechanics of PMA submissions The PMA submission begins with a summary of the safety and effectiveness data. The core of the submission is the clinical study report, which includes the study design and protocol, patient enrollment and exclusion data, **primary** and **secondary endpoints** of the study, data from all patients entered into the trial, and detailed statistical analysis of the results. There are also major sections on technical data, which include biocompatibility and non-clinical (animal or biological) testing, as well as non-clinical laboratory testing that encompasses bench-top testing results and data on stress and fatigue, shelf-life, sterilization, software validation and verification, and other relevant non-clinical tests to support the safety or effectiveness of the device. The proposed labels and IFU are included in the submission. The sponsor may also provide a physician training plan. There is a section on the manufacturing of the device and compliance with

QSR for the FDA to use in preparation of a preapproval facility inspection. There is also a section that outlines a negotiated agreement of additional studies that will be required following approval of the device (called post-market surveillance or post-approval studies). Regulations and guidance documents clearly outline the FDA's expectations and requirements for the contents and format for a PMA submission.[30] Guidance documents for similar products across divisions or for other devices within the division can be helpful to understanding the FDA's current thinking or expectations for a given device.

PMA applications are reviewed by a special advisory panel, a group of 5 to 15 physicians, statisticians and other experts (all non-FDA employees) who serve a three-year term. In addition to the core experts, the panel can add topic experts on a case-by-case basis and also has nonvoting industry and consumer members. After the PMA is submitted and has undergone FDA review, the panel convenes to hear presentations from the company sponsor, its expert consultants, and from an FDA review team. The panel votes to answer questions related to the safety and effectiveness of the device. The recommendations are non-binding, but generally carry great weight with the FDA in making its determination whether to approve or disapprove a new device. The final decision is based on the analysis of the CDRH branch team, subject to the approval of the director of the Office of Device Evaluation.

There are two different types of PMA routes.[31] The traditional PMA is an all-in-one submission that is typically used when the clinical testing has already been completed or when the bench/non-clinical testing or other information will not be completed ahead of the clinical study. The modular PMA is increasingly being used by sponsors in the US. With this approach, the complete contents of a PMA are broken down into well-delineated components (or modules) and each component is submitted to the FDA as soon as the sponsor has finished it, compiling a complete PMA over time. The FDA reviews each module separately, as it is received, allowing companies to gain timely feedback during the review process. This approach may lead to a quicker approval, though this is not guaranteed. The caveat is that the clinical module must be the last one submitted.

The **product development protocol (PDP)** method is an alternative to the PMA process, and is essentially a contract between the sponsor and FDA that describes the agreed-upon details of design and development activities, the outputs of these activities, and acceptance criteria for these outputs. Ideal candidates for the PDP process are those devices for which the technology is well established in the industry. Note that although the PDP mechanism has existed for years, it has not been widely used, and the FDA is currently exploring ways to make the process more efficient and more attractive to sponsors.[32]

PMA review process and timeline The formal FDA review period for PMA submissions is 180 days. As with 510(k) applications, the FDA will approve, deny, or request additional information from the company upon review of its application. There are usually at least two cycles of requests and responses before a decision is made. If the sponsor submits new information on its own initiative or at the request of the FDA, the agency can extend the review period up to 180 days. Final approval comes in the form of a letter from the FDA and represents, in effect, a private license granted to the applicant for marketing a particular medical device (see online Appendix 4.2.1 for a sample). There is no reliable approval time for PMA submissions, but it is by no means unusual for it to take a year or longer.

Investigational device exemptions

In order to begin human testing in advance of regulatory clearance or approval, official permission must be granted to the innovator, either by the FDA, the supervising institutions where studies will be conducted, or both. For a device that is low risk, the innovators can apply for approval of a study to the Institutional Review Board (IRB) of one or more hospitals (see 5.3 Clinical Strategy for more information about IRBs). If the IRB(s) agree that the study involves a non-significant risk, there is no requirement for the FDA or other agency to review the study before patient enrollment commences. However, if even one of the IRBs does not approve the study, the study may not proceed at *any* hospital without FDA involvement.

For significant risk devices that are headed for a 510(k) or PMA submission, clearance to begin clinical testing must be granted by the FDA via an Investigational Device Exemption (**IDE**). No patients in the US can be enrolled in one of these studies before an IDE is approved by the FDA. Unlike the submissions required for 510(k)s and PMAs, the IDE application process is manageable enough that the innovators who are interested in testing the device may make a submission without using an expert regulatory consultant (although a review by an expert is advisable).[33] Importantly, an IDE does not allow a company to market the device; it is a legal exemption to ship the device for investigation under well-defined and carefully controlled circumstances. Companies can charge for investigational devices under an IDE.

To obtain IDE approval by the FDA, sufficient data must be presented to demonstrate that the product is safe for human clinical use; this may require mechanical, electrical, animal, biocompatibility, or other supportive testing. In addition, the patient consent form (also called the **informed consent** form) to be used in the study must be approved by the FDA. IRB approval is also required for all centers in which testing will be performed. Sponsors have the opportunity to meet with the FDA in an informal pre-submission meeting, during which the **clinical protocol** and any preclinical studies can be reviewed. Strategies for approaching these and other FDA meetings are described in chapter 5.4 Regulatory Strategy. Following submission of an IDE application, the FDA has 30 days to respond. The application is approved, disapproved, or conditionally approved. If the response is a conditional approval, the sponsor has 45 days to respond to the FDA with the revised device and/or clinical trial proposal.

Humanitarian device exemptions

The FDA recognizes that certain devices have a limited application in terms of numbers of patients affected, but still are important medically. For these devices, termed Humanitarian Use Devices (HUDs), there is a special approval pathway, the Humanitarian Device Exemption (**HDE**). The agency defines a HUD as a device that would

be used in 4,000 or fewer patients in the US per year. The first step in pursuing a HDE is for the sponsor to apply for a HUD designation for its device from the Office of Orphan Products Development. The sponsor then must have IRB approval before submitting an application for an HDE. Because clinical data for these devices are so difficult to obtain, the HDE pathway does not require the same type or size of trials as for a PMA (in general safety must be assured, but effectiveness requires a lower standard of proof than for a typical PMA). The sponsor must also make a convincing argument that it cannot develop the product except by using the HDE pathway and that no existing device can be used as effectively for the same clinical purpose.

The review period for an HDE is 75 days. An example of a device receiving HDE approval is the Amplatzer® PFO occluder for the treatment of patent foramen ovale (PFO). This condition results from the incomplete closure of the septal wall between the right and left atria (upper chambers) of the heart, a problem that allows blood to partially bypass the natural filtration process provided by the lungs. The new technology was originally developed to target the small group of patients who suffer from strokes related to PFO. However, other exploratory data suggested that PFO closure could potentially help relieve migraine headaches. If confirmed, this would, of course, affect a significantly larger patient population. In the face of this type of new opportunity, the Amplatzer device would not be eligible to address the broader market unless the company obtains approval for a PMA with supportive data for the broader use. **Off-label** promotion by the sponsor of an HDE-approved product for broader use can lead to withdrawal of the HDE.[34]

Costs of FDA submissions

The Medical Device User Fee and Modernization Act (**MDUFMA**), enacted in 2002, established user fees in the medical device industry. Pharmaceuticals and biologic applicants had been paying user fees for some time, but medical devices were historically reviewed "for free." MDUFMA was created to generate resources to help the FDA address increasing review times and to facilitate quicker access to market for medical device applicants. User fees for establishment registration and for covered submissions are published for each fiscal year.

In general, the fees for a 510(k) submission are modest when the time and expertise required by the FDA for review of these applications is taken into account. For fiscal year 2013, the standard application fee was $4,049 and the small business fee (for companies with less than $100 million in sales) was $2,024. PMA applications are considerably more expensive ($220,050 for a standard application; $55,013 for small businesses).[35] There is no PMA fee for the first PMA submitted by companies with gross receipts or sales less than $30 million.

One upside of the fee schedule is that the law now requires increased measurement and accountability of the FDA in terms of its performance and review times. Effectively, this creates a more businesslike model in which there is an agreement between the applicant and FDA to complete a regulatory review. Importantly, however, it does not hold the FDA to specific timelines associated with applications. The FDA is not obligated to clear a 510(k) in 90 days or approve a PMA in 180 days. Yet, Congress measures the FDA's performance against these goals and the agency's continued funding depends on its performance, thereby creating an incentive for the agency to complete timely reviews and approvals. To date, FDA performance seems to have improved as a result of these changes.[36]

A note on FDA regulation of mobile health technologies

With increasing mobile phone usage, a larger number of health-related services are being delivered by mobile devices, a trend known as digital health or mobile health (**mHealth**). In 2013, the FDA issued guidance for the regulation of mobile medical applications (apps) in the US, which developers of these technologies can use to help determine whether or not they will face oversight from the agency.[37] The intended use of the app is central to making this determination, with the FDA focusing its attention on those mobile technologies that are intended to: (1) be used as an accessory to a regulated medical device, or (2) transform a mobile platform into a regulated device. The guidance further clarifies that,

"In general, if a mobile app is intended for use in performing a medical function (i.e., for diagnosis of disease or other conditions, or the cure, mitigation, treatment, or prevention of disease), it is a medical device, regardless of the platform on which it is run."[38]

The guidance outlines three different categories of apps that correspond to FDA's view toward regulating them, as shown in Figure 4.2.3. The category at the base of the pyramid includes technologies that do not meet the FDA definition of a medical device and, therefore will not be regulated (e.g., medical references for physicians or patients, educational apps for medical training, or apps meant to facilitate medical office functions such as scheduling or tracking insurance claims data). In the middle are apps that meet the definition of a medical device, but pose limited risk to patients (e.g., technologies that coach patients or provide them with health-related prompts and reminders). For these apps, the FDA will exercise discretion regarding the need for regulatory oversight. Stated another way, the agency will investigate potential issues on a case-by-case basis, and it reserves the right to enforce the need for regulation after a product is already in the market. At the top of the pyramid are apps that are considered medical devices and have a significant risk of harming patients if they malfunction. Technologies in this category will be classified using the same risk-based scheme that applies to traditional medical devices (Class I, II, or III) and regulated according to the same requirements.[39] Key examples include:[40]

- *Mobile apps that are an extension of one or more medical devices* by connecting to such device(s) for purposes of controlling the device(s) or displaying, storing, analyzing, or transmitting patient-specific medical device data (e.g., remote display of data from bedside monitors, display of previously stored ECG waveforms). These apps are subject to the regulations governing the devices to which the apps serve as extensions.
- *Mobile apps that transform the mobile platform into a regulated medical device* by using attachments, display screens, or sensors or by including functionalities similar to those of currently regulated

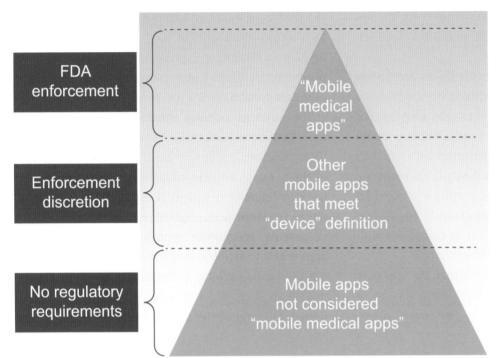

FDA enforcement

Enforcement discretion

No regulatory requirements

"Mobile medical apps"

Other mobile apps that meet "device" definition

Mobile apps not considered "mobile medical apps"

FIGURE 4.2.3

FDA oversight of mobile medical applications (www.fda.gov).

medical devices (e.g., a mobile app that uses a mobile platform for medical device functions, such as attachment of a blood glucose strip reader to a mobile platform to function as a glucose meter; or attachment of electrocardiograph (ECG) electrodes to a mobile platform to measure, store, and display ECG signals). These apps are required to comply with the device classification associated with the transformed platform.

- *Mobile apps that become a regulated medical device (software) by performing patient-specific analysis and providing patient-specific diagnosis, or treatment recommendations* (e.g., apps that use patient-specific parameters and calculate dosage or create a dosage plan for radiation therapy; computer aided detection (CAD) software; image processing software; and radiation therapy treatment planning software). These types of mobile medical apps are similar to or perform the same function as those types of software devices that have been previously cleared or approved. The FDA recommends that the manufacturers of these technologies contact the agency regarding applicable regulatory requirements.

Importantly, the FDA excludes from regulation the hardware on which the mobile medical applications run, as well as the sources (e.g., online app stores) that distribute them.

Summarizing the philosophy underlying the guidance, Geetha Rao, Vice President of Corporate Development at Triple Ring Technologies, Inc., said, "Given the proliferation of mobile medical apps, there's just no way the FDA can exhaustively review everything. So the agency is taking a risk-based approach."[41] The challenge, she continued, is for innovators to interpret the FDA's view of risk. According to Jafar Shenasa, Senior Director of Regulatory Affairs at Proteus Digital Health, one of the most relevant flags is if the app provides patient-specific information. "If the outgoing data from the mobile app becomes patient specific, and that information can potentially lead to clinical decisions, then it's highly likely to be regulated," he stated. "If a technology creates the potential for a patient to bypass the clinician or doctor that is another flag that could lead to an oversight case." The agency will also be on the look-out for companies promoting their apps for use in performing any function of a medical device without having fulfilled appropriate regulatory requirements.

A company called Biosense Technologies was the subject of the FDA's first enforcement action in this space, related to its uChek Urine Analyzer. The company promoted the uChek mobile app to people with diabetes who want to check the amount of glucose in their urine. These individuals download the uChek app to their mobile device. Then, after a "mid-stream collection" of a small amount of urine, they dip an over-the-counter urine test strip into the sample. After a few minutes, the strip changes color to reflect the presence of relevant compounds in the urine, and the patients photograph it with their device's camera. The app assesses the colors and provides the results via email for patients to review and even chart over time.[42]

After uChek had been in the market for some time, the FDA issued the company a letter, with the following message at its core:[43]

Please note that though the types of urinalysis dipsticks you reference for use with your application are cleared, they are only cleared when interpreted by direct visual reading. Since your app allows a mobile phone to analyze the dipsticks, the phone and device as a whole functions as an automated strip reader. When these dipsticks are read by an automated strip reader, the dipsticks require new clearance as part of the test system. Therefore, any company intending to promote their device for use in analyzing, reading, and/or interpreting these dipsticks need to obtain clearance for the entire urinalysis test system (i.e., the strip reader and the test strips, as used together).

After significant back and forth with the agency, Biosense Technologies pulled its original device from the market and made "uChek Lite" available, which did not measure blood or glucose in urine. In parallel, the company began preparing a 510(k) application to pursue regulatory clearance for the next generation of the complete uChek system.[44]

Of course, gaining regulatory clearance from the FDA can be resource intensive. However, it offers certain benefits above and beyond the ability to market the full capabilities of a mobile medical app. The regulatory strategy pursued by AirStrip

Technologies, one of the first app providers to receive FDA clearance, illustrates a few of the advantages. Airstrip's technology allows physicians to monitor mothers and their babies remotely during delivery and make clinical decisions that affect the childbirth experience. According to the company, its efforts to secure FDA clearance gave it the confidence it needed to market the app as a clinical tool and not a simple mobile app. Moreover, executives reported that seeking FDA clearance helped AirStrip develop a better product because of the quality standards required by the agency.[45] As a competitive strategy, FDA clearance also puts a stake in the ground for the rest of the market. In effect, it forces all potentially similar apps to follow suit, giving a company that is first to market a head start. Specifically, if the quality bar is set high and includes a thorough clinical validation, it will be more difficult for competitors to easily market a "**me-too**" solution. Also, if the company wants to test the app in a clinical setting – even if only to establish usability – an approval by an IRB will most likely be required, and having 510(k) clearance will simplify this process. Finally, FDA clearance can open the door to reimbursement since it is often a prerequisite to be considered for **coverage**.

To gain a better understanding of how to develop a mobile medical app that will withstand FDA scrutiny, it is invaluable for innovators to engage a regulatory expert well versed in the mobile medical arena as part of their strategic planning process. Rao also recommended investigating the standards being developed by a joint working group of **ISO** (the International Organization for Standardization) and the International Electrotechnical Commission (IEC).[46] In the US, the American National Standards Institute (ANSI) has initiated a joint working group, AAMI-UL 2800, to address interoperability among medical devices in clinical use situations.[47] The FDA itself is unlikely to develop detailed standards for mobile technologies, but will consult the output of these working groups as a reference for decisions about which new mobile technologies to regulate, as the agency currently does with ISO and IEC standards on medical device quality and **risk management**.

Regulatory approval outside the US

Innovators and companies seeking regulatory approval outside the US should conduct in-depth research to understand the unique regulatory requirements they will face. Working with a seasoned regulatory expert with experience in each target country is also advisable.

Traditionally, the most significant medical device markets outside the US have been in Europe, Canada, Australia, and Japan (based on their size and revenue potential). Regulatory requirements in these countries are generally well understood and have converged through the efforts of a Global Harmonization Task Force, which has now been replaced by the International Medical Device Regulators Forum (IMDRF).[48] However, many developing countries are becoming increasingly important medtech markets, particularly Brazil, Russia, India, and China. The regulatory policies of these nations are also converging towards the regulatory model defined by the IMDRF. The most significant difference between the US and the rest of the world is FDA's requirement for proof of efficacy for medical devices, whereas in most other countries the objective is to demonstrate appropriate performance, with an emphasis on safety.

Europe

Medical devices are regulated in the European Union (EU) by three European Commission (EC) directives.[49] The main directive, which covers the vast majority of medical devices from surgical gloves to life-sustaining implantable devices such as heart valves, is the Medical Devices Directive (**MDD**). Further directives cover active implantable medical devices (the **AIMDD**) and in vitro diagnostic medical devices (the **IVMDD**). These three EC directives have been transposed into the national laws of each EU member state, resulting in a legislative framework comprising literally dozens of medical device laws. The MDD and AIMDD will be replaced, sometime after 2015, by a single European Regulation. Unlike Directives, a Regulation is directly applicable without the need for transposition into national laws.

Regulatory approval in the EU is signified by a "CE" mark of conformity. "CE" is not an official acronym, but may have originated in the terms Communauté Européenne or Conformité Européenne, meaning European community or conformity.

The medical device directives are known as "new approach" directives; that is, their purpose is to ensure the free movement of goods and services. Based on these directives, medical devices bearing the **CE marking** can circulate freely and be sold and marketed according to approved indications throughout 31 European countries. The underlying principle of the new approach is that each of the medical device directives contains a list of "essential requirements" that must be met by any product falling within its scope. In the MDD, the list of essential requirements can be broken down into two groups: (1) a set of general requirements for safety and performance that apply to all devices; and (2) a list of specific technical requirements with regard to design and manufacturing that may or may not apply, depending on the nature of the device. For example, the technical requirement for electrical safety will not apply to a urinary catheter. Compliance with the technical requirements is generally demonstrated by using the relevant harmonized European standard. Most European standards are now identical to ISO standards. A harmonized standard is one that has been verified by the EC to be sufficient to carry the presumption of conformity with essential requirements.

Under the MDD, there are four classes of devices that generally correspond to the US device categories, as shown in Table 4.2.4.[50]

In Europe, a clinical evaluation is needed to demonstrate compliance with the Essential Requirements of the Medical Device Directives (90/385 or 93/42). A clinical evaluation may be a written review and summary of published literature which demonstrate that similar devices meet the Essential Requirements. However, data from clinical trials may be required for Class IIa, IIb, and III devices, depending on clinical claims and risk management outcomes. Data from clinical investigations are generally required for Class III, active implantable and implantable medical devices. The key difference between CE marking and FDA clearance or approval is that the CE standard is based on safety and *performance*, which in practice is generally less rigorous than the safety and *effectiveness* standard applied by the FDA (where the

Table 4.2.4 The four device classes in the EU generally correspond to the three classes defined by the FDA in the US.

EU class	Description	US equivalent
Class I	Devices that present a relatively low risk to the patient and, except for sterile products or measuring devices, can be self-certified by the manufacturer. Typically, they do not enter the human body.	Class I
Class IIa	Devices that present a medium risk to patients and may be subject to quality system assessment. Generally, they are invasive to the human body, but only via natural body orifices. This category may also include therapeutic diagnostics and devices for wound management.	Class II
Class IIb	Devices that present a medium risk to patients and may be subject to quality system assessment, as well as third-party product and system certification. They are usually either partially or totally implantable and may modify the biological or chemical composition of body fluids.	Class II
Class III	Devices that present a high risk to patients and require design/clinical trial reviews, product certification, and quality system assessment conducted by a European Notified Body. In most cases, they affect the functioning of vital organs and/or life-supporting systems.	Class III

data must demonstrate that the device is effective in the indications for use). This difference is a major reason why many companies developing devices in the US pursue CE marking first, before entering an FDA pathway (refer to 5.4 Regulatory Strategy). In order for clinical trials to be carried out in the EU, the studies must comply with EN ISO 14155 "Clinical Investigations of Medical Devices for Human Subjects – Good Clinical Practice" and the national laws in the member states where the trials will be conducted. This involves obtaining ethics committee approval (comparable to an IRB in the US) and making necessary notifications to the relevant Competent Authorities. In some countries, approval must also be obtained from the Competent Authority.

A key aspect of medical device regulation in the EU is that the responsibility for ensuring that devices meet the essential requirements lies with the manufacturer. For low-risk devices (Class I), such as tongue depressors or colostomy bags, the manufacturer is allowed to self-declare conformity with the essential requirements. For medium- to high-risk devices (Class IIa, IIb, III) or for devices supplied sterile or with a measuring function, the manufacturer must call on a third party (a **Notified Body**) to assess conformity. To some degree, the manufacturer may choose among methods for "conformity assessment" of the device and/or manufacturing system.

The end result is a certificate of conformity that enables the manufacturer to apply the CE marking to the product.

Another major aspect of the CE marking process is that, contrary to the US, the "conformity assessment" for medical devices in Europe is not conducted by a central regulatory authority. The CE marking system relies heavily on third parties known as Notified Bodies to implement regulatory control over medical devices. Notified Bodies are independent commercial organizations that are designated, monitored, and audited by the relevant member states via the national "Competent Authorities." The Competent Authority, which reports to the minister of health within the member state, is empowered to act on behalf of the member state government to apply and uphold the requirements of the EC directives. Currently, there are more than 70 active Notified Bodies within Europe, although this number is likely to be significantly reduced in the wake of stricter control by Competent Authorities. A company is free to choose any Notified Body designated to cover the particular class of device under review. After approval, post-market surveillance functions are the responsibility of the member states via their Competent Authority.

The medical devices directives (MDD and AIMDD) were amended in 2007. This amendment significantly increased the requirements for clinical data in the EU in the preapproval and post-market phases, emphasizing the importance of clinical risk/benefit assessment. For more information, see the European Commission's guidance to the Medical Device Directives.[51]

Canada

In Canada, medical devices are regulated by Health Canada's Therapeutic Products Directorate and are subject to the Medical Devices Regulations under the country's Food and Drugs Act. Before they can be sold in Canada, all medical devices must be classified and most must be licensed. Health Canada has enacted four device classes (I, II, III, or IV), which vary based on the level of risk associated with their use. Classification is based on 16 rules or factors, including the degree of invasiveness and whether the device is active or non-active.[52]

As in the US and EU, Class I devices present the lowest potential risk and do not require a medical device license (MDL). Class II devices involve moderate risk and require the manufacturer to submit an MDL application, quality system certification, and a declaration of conformity to Health Canada.[53] Class III and IV devices involve substantial risk and for, this reason, are subject to more in-depth regulatory scrutiny before licensing. Manufacturers of these devices must submit the same information as Class II manufacturers; however, they must also prepare a premarket review document. Among other information, this document must include a summary of safety and effectiveness studies, a risk assessment, and information on labeling and instructions for use. For Class IV devices, a complete package of clinical trial data must also be made available. In some cases, data will be accepted from trials conducted in other device markets.[54] More detailed information about completing an application to secure an MDL are provided in the guidance document on this topic[55] and also on the Health Canada website, the most up-to-date resource on device regulation in Canada.[56] Review times vary from less than a month for Class II devices to up to six months for Class III and IV devices.[57]

Australia

Formal processes for regulating medical devices and medicines in Australia were enacted in 1989 under the Therapeutic Goods Act.[58] The Therapeutic Goods Administration (TGA) is the entity responsible for administering the Act. The Office of Devices Authorization (ODA) oversees premarket regulation of medical devices and the Office of Product Review (OPR) manages post-market regulation. The Australian Register of Therapeutic Goods (ARTG) is the central point of control for the legal supply of devices and other therapeutic goods in the country. Most devices have to be entered in the register before they can be made commercially available. Limited exceptions to this requirement (e.g., devices undergoing experimental use) are outlined in the Australian Regulatory Guidelines for Medical Devices (ARGMD).[59]

Australia maintains five device classes that correspond to their level of risk – I, IIa, IIb, III, and active implantable medical devices (AIMDs). Class I devices (except those that are sterile or have a measuring function) do not require assessment by the TGA before they are included on the ARTG. All other classes of device must undergo premarket regulation procedures. This includes: (1) the manufacturer applying appropriate conformity assessment procedures; (2) the manufacturer making an Australian Declaration of Conformity – a legal declaration that all the required evidence exists and that the device complies with Australian regulatory requirements; (3) the Australian sponsor submitting the manufacturer's conformity assessment evidence to the TGA; (4) the TGA evaluating the available evidence for high-risk devices; and (5) the device being included on the ARTG.[60] If the manufacturer of a medical device resides outside of the country, it must work with a sponsor that is a resident or incorporated body conducting business in Australia to apply to complete step 3 listed above.[61]

Because the Australia regulatory system is modeled after the system used in the EU, devices that have been granted a CE mark from a European Notified Body can more easily substantiate conformity according to TGA requirements.[62] More about the Australian regulatory system can be found on the TGA website[63] or in the ARGMD.[64]

On a related topic, the Australian government has actively sought to develop a strong clinical research industry in the country. Companies from around the world come to Australia to conduct **first-in-human** testing and **pilot studies** to capitalize on the country's well-developed healthcare system. Australia's clinicians have a reputation for being highly skilled and interested in trialing new medical device technologies, and the data generated through these studies is generally well regarded internationally. Additionally, conducting clinical trials in Australia is considered to be relatively cost effective.[65] More information is available on the Australian Clinical Trials website.[66]

Japan

Japan's regulatory authority is the Ministry of Health, Labor, and Welfare (MHLW), which oversees the regulation of medical devices through its technical arm, the Pharmaceutical and Medical Device Agency (**PMDA**). Medical devices are regulated under the Pharmaceutical Affairs Law (PAL), with the regulatory pathway determined by device classification. General medical devices (Class I) require a premarket submission (*todokede*), but do not undergo assessment by PDMA. Some Class II devices – referred to as specified controlled medical devices – can undergo premarket certification (*ninsho*) by working with an independent Registered Certification Body (similar to European CE marking through a Notified Body). Higher risk Class II devices as well as Class III and Class IV technologies (referred to as controlled and highly controlled medical devices, respectively) require premarket approval (*shonin*) through the PMDA. Once a device is certified or approved, the MHLW decides on the reimbursement of the technology and its price based on the documents submitted as part of the regulatory process and advisory panel opinions.

All companies intending to manufacture and/or distribute medical devices in Japan must first seek a business license (*kyoka*) that certifies them as a Marketing Authorization Holder (MAH) before pursuing a regulatory pathway.[67] There are three categories of MAH: Category 1 can distributed all classes of devices; Category 2 can distribute Class I and II devices; and Category 3 can distribute only Class I devices. Overseas manufacturers can appoint an MAH (often a local distributor) or a designated MAH (called a D-MAH, which

is an independent entity) to manage the device registration process and all interactions with the PMDA.[68]

In late 2013, PAL reforms were passed to strengthen safety measures for devices and medical equipment, as well as to more closely link regulatory oversight to device characteristics.[69] One of the primary changes is to allow Japan's third-party Registered Certification Bodies to review a greater number of devices, including some considered controlled and highly controlled, to speed the approval process and make innovative technologies available in a more efficient manner. Additionally, the reforms will require stand-alone software used for diagnostic to undergo premarket submission, certification, or approval depending on its risk profile.[70] (This requirement may be broadened to include software used for treatment purposes, as well.)

The lack of forms available in English and the country's complex registration process can make Japan a more challenging and time-consuming market for device manufacturers to enter than Europe, Canada, and Australia. For this reason, the advice and guidance of a regulatory consultant is particularly important for companies seeking to do business in the country. It is also helpful to seek consultation from PMDA. The MAH and/or manufacture can schedule a meeting with representatives from PMDA during the early stages of device development to discuss specific regulatory requirements and an appropriate approval strategy. PMDA will also provide guidance on the clinical data necessary to support a submission and the preparation of essential documents. Note that through these consultations, companies can also discover if they may be able to take advantage of special regulatory pathways for particularly innovative devices or those that address important diseases. Although regulatory approval for devices in Japan has typically lagged the United States, the international Harmonization by Doing (HBD) task force was launched in late 2003 to move Japan and the US toward greater synchronization in terms of their pre-submission requirements and timing.[71]

The BRIC countries

The large and developing markets of Brazil, Russia, India, and China are becoming increasingly important medical device markets.

Brazil The Brazilian National Health Surveillance Agency (ANVISA) is an independent agency that works in cooperation with the country's Ministry of Health under a management contract. ANVISA has responsibility for the regulation of all medical products and pharmaceuticals. A 2001 resolution pertaining to medical devices outlines the specific documents necessary in order to register devices and equipment with ANVISA before making them commercially available.[72] In contrast to the EU Notified Body system, ANVISA performs its own registration and inspection functions.[73] Only companies based in Brazil can apply for ANVISA registration, so those without a local headquarters or subsidiary must contract with a hosting company or distributor. This representative becomes the only company authorized to sell or distribute the product in Brazil for a five-year period, so choosing the right partner is essential.[74]

In addition to registration, companies bringing medical devices to Brazil must comply with the Code of Consumer Protection and Defense. This code is meant to ensure consumers that equipment is safe and will be used correctly by requiring companies to provide sufficient documentation to demonstrate the safety of their products. As in other countries, medical devices covered by this code are classified into one of four risk-based categories (I–IV),[75] each with different safety requirements. Companies may be required to compile extensive documentation in a technical file, depending on the nature of the device. Defining the specific requirements for each unique device can be complicated,[76] and all documentation must be submitted in Brazilian Portuguese.

One important form of frequently required documentation is INMETRO certification, which applies to many electromedical devices as well as other non-electrical devices as required by ANVISA. Brazil's National Institute of Metrology, Standardization and Industrial Quality (INMETRO) is the body responsible for accrediting organizations to certify medical devices (and other products) for compliance with the Brazilian Conformity Assessment System (SBAC). To qualify for INMETRO certification, medical device manufacturers must have their products tested to SBAC-recognized standards by an INMETRO-accredited testing laboratory.[77]

Another unique requirement is the completion of an Economic Information Report (EIR), which ANVISA requires for some devices. The EIR includes price comparisons for other countries, patient or user information, information on comparable products, and device marketing materials.[78]

Some devices will also require certification by the National Telecommunication Agency (known as ANATEL) if they include a telecommunications component. ANATEL provides oversight of these products to ensure they meet minimum quality and security standards. Products seeking ANATEL certification must be tested by an authorized laboratory in Brazil.[79] Additionally, companies are required to have an import license for each shipment of medical devices into Brazil. These licenses are used to control imports into the country and are separate from the manufacturer's Ministry of Health registration and safety hurdles.[80]

Innovators are encouraged to work with local regulatory experts to help them understand the specific requirements for bringing a product to market in Brazil. Although the process for registration of medical products has been harmonized across the MERCOSUR countries (Argentina, Brazil, Paraguay, and Uruguay) in the past few years, it can still be lengthy, in some cases taking more than two years between filing for registration and final approval by the government.

Russia Two major government entities oversee the regulation of medical devices in Russia. The first is the Ministry of Health, which works to ensure the clinical safety and effectiveness of products through the process of issuing registration certificates. The second is the Federal Committee (government entity) for Healthcare Oversight (Roszdravnadzor) that reports to the Ministry of Health and focuses on registration of medical devices in the Russian Federation. Roszdravnadzor issues certificates of conformity based on established technical and safety standards.[81]

Before pursuing a certificate of conformity, a company must register its product with the Ministry of Health and have it added to the national register. Within three days of submitting an application, Roszdravnadzor appoints one of the two Expert Council Commissions that scrutinizes product documentation, as well as the results of technical, safety, toxicology, hygienic, and clinical tests to make its registration recommendations. If the council determines that foreign products have analogs in Russia, it may deny registration of the foreign product. Once registered, the company can seek a conformity assessment by submitting a declaration-application to the authorized certification organization of its choice. The focus of a conformity assessment is on end-product testing.[82]

Roszdravnadzor maintains a website where it publishes the requirements for the registration of foreign-made medical technologies. However, the contents of the site are only available in Russian. Additionally, Roszdravnadzor officials prefer to conduct any consultations in person rather than by phone or email. For these reasons, it is imperative for companies to have an experienced local representative to assist with registration and other regulatory matters in the country.[83]

India Medical device regulation is relatively new in India, although the Central Drug Standard Control Organization (CDSCO) has regulated pharmaceuticals since 1940. In the absence of a comprehensive regulatory framework for new medical technologies, any form of medical device oversight has consisted of classifying certain devices as "drugs" and then applying drug laws to them under the purview of the country's Drug & Cosmetic Act (DCA).[84]

In 2014, the CDSCO regulated only 14 technologies referred to as notified medical devices. This list includes catheters, cardiac stents, drug-eluting stents, heart valves, intraocular lenses, orthopedic implants, internal prosthetic replacements, bone cements, disposable hypodermic syringes, perfusion sets, scalp vein sets, I.V. cannulae, and some in vitro diagnostics.[85] Medical technologies that do not appear on the notified medical devices list do not require registration or certification prior to their sale in India. Historically, they have been adopted based on the purchaser's evaluation of quality, with FDA- and CE-regulated products receiving preferential treatment. Imported medical devices on the notified devices list that have already obtained clearance or approval in the United States (by the FDA) or the

European Union (by CE marking) are allowed on the Indian market without undergoing separate conformity assessment procedures.

Recently, a bill was introduced that seeks to amend the DCA and modify its name to the Drugs, Cosmetics & Medical Devices Act. The bill proposes a series of changes, including recognizing medical devices as independent from drugs. Under the new Act, a Centralized Drug Authority would replace the CDSCO and have a separate chapter for overseeing the regulation of all medical devices imported, manufactured, and/or sold within the country.[86] The new rules would provide for the more systematic regulation of medical devices in India and bring its requirements in line with those in other established countries.

The newly proposed legislation will create a Central Licensing Approval Authority (CLAA) under the new Centralized Drug Authority. CLAA will manage the licensing and classification of medical devices, based on the classification system shown in Table 4.2.5.[87] Once the medical devices are appropriately classified, assessed, and licensed, they will bear the Indian Conformity Assessment Certificate (ICAC) mark, authorizing their sales in India.[88]

In 2013, India passed a new policy meant to protect patients who participate in clinical research studies performed within the country. While the intent of the new rules has been lauded, companies and researchers have raised concerns about the burden the policy creates in terms of compensating patients for research injuries or death. Critics assert that the definition of what qualifies for compensation is too broad to be reasonable (for example, one condition attributable to research injury is the "failure of investigational product to provide intended therapeutic benefit").[89] The long-term effects of the policy remain to be seen, but some observers of the medtech and pharmaceutical industries anticipate that it will significantly reduce the number of trials conducted in India and, in turn, the availability of new treatments.[90]

China China passed its first set of laws for the regulation of medical devices in 2000.[91] Regulation is managed by the China Food and Drug Administration (**CFDA**) and

Table 4.2.5 The proposed classification systems for notified medical devices in India (www.icac.in).

Class	Risk level	Device examples	Procedures
A	Low risk	Thermometers, tongue depressors	Class A devices, the manufacturer is required to register with the CLAA.
B	Low–moderate risk	Hypodermic needles, suction equipment	For Class B devices, a Notified Body must assess and certify the manufacturing facility quality management system and the manufacturer is required to register with the CLAA.
C	Moderate–high risk	Lung ventilator, bone fixation plate	For Class C devices, certification by a Notified Body is required with regard to the design and manufacture of the device(s); the manufacturer is required to apply for a manufacturing license from CLAA.
D	High risk	Heart valves, implantable defibrillator	For Class D devices, certification by a Notified Body is required with regard to the design and manufacture of the devices; the manufacturer is required to apply for a manufacturing license from CLAA; the manufacturing facility will also be inspected jointly by CLAA and state licensing authority.

its local counterparts, which are the primary regulatory authorities that apply and enforce laws and regulations concerning medical devices. The CFDA oversees the complete lifecycle of medical device regulation, from clinical trials and marketing authorizations to manufacturing, distribution, and post-market surveillance. Other key regulators governing the medical device industry include the State Administration of Quality Supervision, Inspection, and Quarantine (AQSIQ), the Ministry of Human Resources and Social Security (MOHRSS), the National Development and Reform Commission (NDRC), the National Health and Family Planning Commission (NHFPC), and the State Administration of Industry and Commerce (SAIC).[92]

Medical devices are primarily governed by the Regulations on the Supervision and Administration of Medical Devices,[93] which were amended in 2014.[94] To sell a medical device in China, a company must register its product with the CFDA. Devices are organized into three product classes (I–III) based on their risk profile. The CFDA is primarily focused on regulating Class III products, delegating oversight of Class I and II devices to provincial government agencies.[95] A medical device is either registered as a locally manufactured device or an imported device manufactured outside China. The marketing authorization for a medical device is in essence a permit to manufacture or import a device, and is issued to the manufacturer on record. Importantly, devices to be imported must receive marketing approval in their country of origin before the manufacturer pursues their registration in China.[96]

In some cases, devices may also require approval by either the Ministry of Health or AQSIQ. AQSIQ conducts mandatory safety registration, certification, and inspection for certain devices. Once certified, devices are awarded a "China Compulsory Certification" (CCC) mark, which serves as evidence that the product can be imported, marketed, and used in China.[97]

FDA view of clinical trial data from outside the US
Many US companies perform their initial clinical studies and/or seek preliminary regulatory clearance in foreign locations, driven by the less cumbersome regulatory processes found in some other countries (e.g., those in South America). In some cases, regulatory bodies outside the US accept compilations of key literature and a written analysis of those papers to make a case that a device is expected to be safe and efficacious in lieu of expansive clinical data.

As a result, approval can be quicker, easier, and less expensive in other countries than it is in the US. The FDA does not prohibit this practice, but expects to see all patient data obtained from such studies. Many times, companies use these trials to support an IDE application in the US. In some cases, data from another geography may be sufficient to support a 510(k) submission although, depending on the disease state, the agency may have concern about potential differences in the response of American patients to the device. However, almost all PMA applications and the majority of 510(k) submissions that require patient data will require at least some studies to be performed in the US.

The Edwards LifeSciences story provides an example of how one company approached regulatory approval, as well as the collection of clinical data to support its submissions, for a ground-breaking product with global relevance.

FROM THE FIELD ▶ EDWARDS LIFESCIENCES LLC

Navigating global regulatory pathways for a novel device

Beginning in the late 1950s, Edwards Lifesciences established itself as a leader in design and manufacturing of prosthetic heart valves for surgery. Over time, however, the company began to appreciate the **need** for a non-surgical approach to treating aortic stenosis and began working on a solution that became known as Transaortic Valve Replacement (see the CoreValve story in chapter 4.1 for more information on aortic stenosis and TAVR). Progress was accelerated by the acquisition of a start-up company, Percutaneous Valve Technologies (PVT), which had developed and tested a unique valve consisting of tissue leaflets buttressed by a surrounding, expandable stent framework. The valve plus stent could be compressed at the tip of the delivery catheter, so that it could be inserted through a sheath into the femoral vein for delivery to the heart via an antegrade or retrograde approach. Because this technology was deployed without the use of a heart-lung machine and did not require sternotomy (opening of the chest cavity), it promised to be appropriate for patients contraindicated for standard valve replacement surgery and potentially for other patients for whom surgery was challenging.

With its new implant and delivery system, the company anticipated that the new valve (called SAPIEN, see Figure 4.2.4) would be a Class III device and would

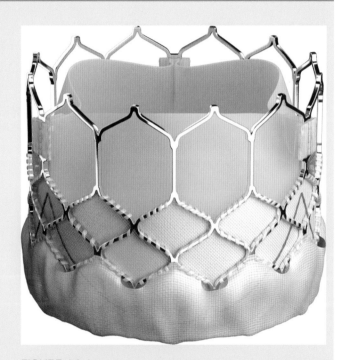

FIGURE 4.2.4
The SAPIEN transcatheter heart valve (courtesy of Edwards Lifesciences LLC).

require clinical data for approval in almost all geographies. The challenge was that Edwards had no specific precedents to follow in terms of what specific studies would be needed to demonstrate the device's safety and efficacy to regulators. Scott Beggins, Edwards' Vice President of Regulatory Affairs, recalled the first steps of the company's approach: "We tried to accumulate knowledge related to the different elements of our device." For instance, he and his team looked at how similar delivery systems had been regulated and studied the guidance documents available for the existing surgical heart valves in the market. "We started mining through all kinds of documents to figure out which ones were potentially relevant to what we were trying to do," he continued. "And we used that information to help us approach regulatory bodies to open a dialogue on how they would expect us to challenge this type of a device from an engineering perspective." Through this approach, Edwards discovered that many well-established tests would be applicable to TAVR devices to validate their safety and durability.

From the outset, the company thought globally about its approach to bringing the device to market. "The US is such a large market opportunity, we always knew we would end up there," Beggins explained. "It was more an issue of timing. Would it be the first market we'd go after, or the second, or the third?" At the time of the TAVR project, he explained, "It was very difficult to get an IDE through the FDA to get a US trial up and running. It was much simpler, faster, and more consistent to initiate clinical trials outside of the US." Once bench testing had confirmed that the SAPIEN valve was safe to move into human trials, Edwards decided to launch parallel clinical trial processes in Europe and the US, with the expectation that the company would make faster progress in Europe and use that accumulated experience to help guide its clinical and regulatory interactions with the FDA.

One early step in pursuing a CE mark in Europe is to choose a Notified Body to work with on making the company's regulatory submission. "At the time, Edwards had a few different Notified Bodies that we were working within our other business units," Beggins said, including the leader in performing conformity assessments for surgical heart valves. "But for this product, we wanted a Notified Body that had more experience with interventional cardiology products. The idea was to proactively identify an organization that would appreciate the characteristics of the device, and would not bring a lot of surgical valve preconceptions into the process," he recalled.

Next, the Edwards team began meeting with the Notified Body to discuss clinical data requirements to support its CE mark submission. One factor central to these interactions was agreeing on the target patient population for the preliminary submission. "We came in with a phased approach," Beggins stated, focused first on inoperable patients with severe symptomatic aortic stenosis (using transfemoral access). This was a population where trials could be relatively aggressive from a risk/benefit perspective, he emphasized, "Because there was no treatment alternative available to these patients and people were dying." Based on this grave situation, Edwards and the Notified Body agreed that the clinical hurdle should be manageable with the hope of providing this population with a treatment opportunity in a timely manner.

Edwards requested ethics committee approval to conduct trials in Europe and made necessary notifications to the relevant Competent Authorities. These trials were run at 2–5 sites, and the results were promising. Edwards initially planned to submit data from these preliminary EU studies, but ultimately supplemented this early experience with data from additional studies prior to making a CE mark submission. Although the early data showed potential for the therapy, it also demonstrated that there was room for improvement. Specifically, the company worked to strengthen its approach to patient selection and design of the delivery systems, as well as improvements to the procedure itself. Edwards then submitted a Class III design dossier to the Notified Body. In turn, the Notified Body took the company through a series of questions and answers. After a total review period of about a year, Edwards received a CE mark for the SAPIEN device.

"Post CE marking, we did a controlled rollout in Europe since this was a brand new therapy," Beggins remembered. "At the same time, we continued to pursue approval for transapical access [through the chest wall into the ventricle wall of the heart], as well as continuing studies for a broader population. The next generation of the device, the SAPIEN XT was also in development." Additionally, Edwards initiated regulatory efforts in parts of Asia, South America, and Eastern Europe.

Meanwhile in the US, Edwards was still in discussions with the FDA regarding requirements for pursuing the PMA pathway with the primary emphasis on clinical trial design. "We were in negotiations for many months with the FDA on trial requirements and getting the IDE protocol finalized," Beggins stated. "We were dealing with something new and novel to the FDA, with people who were used to dealing with surgical valves over the last 5, 10, 15, 20 years of their careers. It became a lengthy process and resulted in all of us becoming educated about the engineering dynamics associated with this technology. There was also some resetting of expectations on both sides about what type of testing is appropriate, what that should look like, what the durations and acceptance criteria should be, and so on. There was a lot of interaction on the test side and a huge amount of interaction on what it took to ultimately design the PARTNER trial," with the FDA taking a relatively conservative stance and requiring a full randomized controlled trial. This decision was likely directed by the current approval process for a new surgical valve, which is fairly prescriptive in terms of the engineering testing needed and requires a single arm clinical trial with defined criteria that is then compared to objective performance criteria (OPC).

In the end, Edwards conducted two randomized controlled trials in the US. The first would target a population of patients for whom surgical valve replacement was not an option, comparing TAVR with medical management, with approximately 350 patients at 22 sites and a one-year endpoint. The second trial would target high-risk surgical patients, comparing TAVR with traditional valve surgery, enrolling over 650 patients at 26 sites with a one-year endpoint.

"Once we started the trial and got the enrollment going, our progress was a little more seamless," Beggins commented. "The hard part was just how long it took to get the actual pivotal trial up and running in the US." Once the trial data had been collected, the PMA review process was also relatively uneventful. "We had some back-and-forth negotiations on what the clinical report would look like," he said. "And we had to go through the panel process, which added time to the overall process." The FDA approved SAPIEN in 2011 – a full four years after the device received its CE mark. By then, the valve was approved in more than 40 other countries.[98] Edwards began a controlled rollout of the product across the United States, limited mostly by the extensive physician training requirements associated with the new technology.

Admittedly, the much longer US regulatory process translated into additional time and expense for Edwards. However, Edwards learned a tremendous amount through these interactions. "FDA was very engaged throughout the process." Beggins stated. "Early EU data was also helpful in discussions with FDA. As the early procedural feedback was obtained in both the EU and the US, our engineering teams were able to iterate the delivery systems as well as incorporate new features into the next generation valve designs."

Looking ahead, Beggins explained that the regulatory landscape is undergoing something of a shift, which could potentially affect the way innovators decide to navigate global regulatory pathways. "From the European perspective, their risk appetite is decreasing. I think they're under some pressures to raise the bar for the safety and efficacy of certain products, and they're moving in that direction," he said. In parallel, the FDA is working to be more responsive, focused and supportive of innovation. "My sense right now is that it's easier to interact with the FDA today than it was when we started with SAPIEN. I think we now get better feedback from FDA and they're a better partner."

In terms of other lessons, Beggins stressed that initiating early discussions with regulatory bodies is essential, regardless of geography and especially if clinical data will be required to support a submission. "The early interactions are critical; you don't want to be guessing what they want," he said. Innovators used to worry, he noted, that these negotiations were relatively informal and non-binding, creating the risk that certain recommendations could change without warning. "But now they tend to be more definitive, the feedback more tangible, and the interactions more valuable to innovators." With novel therapies, Beggins also pointed out that education is essential. "The more you can educate the regulatory bodies about your new therapy, the better the process will work. In some cases, regulatory approval requires a paradigm shift, and they won't support that in the absence of information."

As of early 2014, more than 50,000 patients had been treated with SAPIEN valves around the world.

A final note: using the regulatory pathway to screen and eliminate concepts

To recap, one important reason for gaining a basic understanding of regulatory issues at this stage in the biodesign innovation process is to clarify the risks associated with the **concepts** under consideration. Those with killer risks can be eliminated immediately (e.g., regulatory and related clinical requirements that are way out of alignment with the parameters of a team's strategic focus). For the others, innovators can use the data they gather about the regulatory pathway to help populate a risk scoring matrix (along with IP, reimbursement, and business model factors) as described as part of 4.6 Final Concept Selection.

As should be clear from reading this chapter, regulatory issues can be exceedingly complex: sorting out the final pathway for a new product requires a great deal of time and effort. However, for purposes of screening, some straightforward guidelines can help innovators prioritize which concepts look promising and which are potentially more problematic.

1. **Revisit the team's strategic focus** – Because the path to regulatory approval or clearance is such a major determinant of the time and expense of bringing a technology forward, the level of regulatory risk for different concepts can provide a highly effective screening criterion. The amount of risk a team is willing to take on should correspond directly to its strategic focus. If the team's goal is to create a breakthrough technology platform, it should not be wary of pursuing concepts that require a PMA approval in the US (or a comparable pathway in other geographies).

Alternatively, if the team has a desire to bring a device to market relatively quickly and inexpensively, it should target concepts that can be likely can be cleared via the 510(k) pathway in the US (or an equivalent elsewhere).

2. **Look for a template** – If the team is working in an area where other products have previously achieved regulatory clearance or approval, it can be useful to filter concepts based on a template of how these products achieved regulatory success. In cases where a prior technology has gone through the FDA or another regulatory body reasonably quickly and the trial design is available through a journal publication or other means, the existence of this information can make it much easier for companies to understand what it will take to navigate the regulatory pathway. Of course, there is no guarantee that FDA or another agency will agree to the same testing criteria (regulators sometimes modify their requirements based on new clinical or scientific data, or even political pressure). But, in general, the clarity provided by a preceding product can be helpful for companies taking a fast follower approach. In some technology areas, the regulatory agency may have created guidance documents, which can make the regulatory pathway clearer than for a concept where no guidance exists (with the same caveat that the FDA can always change its approach). Based on this knowledge, if the template suggests a pathway that is attractive to the team, concepts that look like they could use this regulatory approach might be more attractive.

3. **Estimate the cost to clearance/approval** – As emphasized earlier, the main burden that regulatory requirements represent for a team or company is the time and cost required to get a product to market. Time and cost are directly driven by the size of the clinical trial, the type of data to be gathered, and the duration of follow-up required. In most cases, it should be possible to estimate what information the FDA or another regulatory agency is likely to request and, on this basis, perform a "back of the envelope" calculation of the time and expense needed to provide the data. The difference between a 510(k) that requires only bench data and one that will involve a two-year study with hundreds of patients may be tens of millions of dollars or more. These numbers provide a critically important measure for screening different concepts in order to find ones that are consistent with a level of funding that is realistic for the team. The key is to make sure that the benefits (in terms of improvement in clinical outcomes, economic impact, and potential financial return to the innovators, investors, or company) are sufficient to justify whatever level of effort is required to clear the anticipated regulatory hurdles.

↘ Online Resources

Visit www.ebiodesign.org/4.2 for more content, including:

 Activities and links for "Getting Started"
- Confirm the appropriate regulatory branch
- Classify the device
- Determine the regulatory pathway
- Secure a regulatory consultant

 Videos on regulatory basics

 An appendix that shows a sample FDA premarket approval letter

CREDITS

The editors would like to acknowledge Janice Hogan and Nancy Isaac for their extensive editing of this material. Many thanks also go to Scott Beggins of Edwards Life-sciences and Ritu Kamal for their assistance with the case example, and to Geetha Rao of Triple Ring Technologies, Inc. and Jafar Shenasa of Proteus Digital Health for their guidance on the mHealth regulatory section. Further appreciation goes to Milena Adamian of Life Sciences Angels Network; Andrew Cleeland of The Foundry; Julie Delrue and Sarah Sorrel of MedPass International; Ajay Pitre of Pitre Ventures; and Katherine Wang of Ropes & Gray LLP; as well as Robson Capasso, Raj Doshi, Peggy McLaughlin, Jan B. Pietzsch, and Chris Shen for contributing to the information on international regulatory requirements. Darin Buxbaum, Trena Depel, and Asha Nayak deserve recognition for their early assistance with the chapter.

NOTES

1 "The Food and Drug Administration Celebrates 100 Years of Service to the Nation," U.S. Food and Drug Administration, January 4, 2006, http://www.fda.gov/NewsEvents/Newsroom/PressAnnouncements/2006/ucm108572.htm (December 26, 2013).

2 "FDA History," U.S. Food and Drug Administration, http://www.fda.gov/oc/history/default.htm (December 26, 2013).

3 Ibid.

4 Ibid.

5 "What We Do," U.S. Food and Drug Administration, http://www.fda.gov/aboutfda/whatwedo/ (December 26, 2013).

6 Diane Dwyer and Kelly Bauer, "Take the Lead on Safety with Temporary Cardiac Pacing" U.S. Food and Drug Administration, March 2010, http://www.fda.gov/medicaldevices/safety/alertsandnotices/tipsandarticlesondevicesafety/ucm203731.htm (October 17, 2013).

7 "Metal-on-Metal Hip Implants," U.S. Food and Drug Administration, January 17, 2013, http://www.fda.gov/MedicalDevices/ProductsandMedicalProcedures/ImplantsandProsthetics/MetalonMetalHipImplants/default.htm (November 12, 2013).

8 Except for the agency's role in reimbursement categorization for investigational devices. Please see "Implementation of the FDA/HCFA Interagency Agreement Regarding Reimbursement Categorization of Investigational Devices," U.S. Food and Drug Administration, http://www.fda.gov/medicaldevices/deviceregulationandguidance/guidancedocuments/ucm080302.htm (October 17, 2013).

9 If a device is used in a way that deviates from the indication on the FDA-approved label, it is called "off-label" usage. While physicians may elect to engage in off-label use, the manufacturer of the device is not allowed to promote any methods of use other than those cleared by the FDA.

10 "The State of the FDA Workforce," Partnership for Public Services, November 2012, http://www.washingtonpost.com/r/2010–2019/WashingtonPost/ 2012/11/19/National-Politics/Graphics/PEW_FDA_Public_19112012.pdf (November 12, 2013).

11 "FY 2013 Budget Overview," U.S. Food and Drug Administration, 2012, http://www.fda.gov/downloads/AboutFDA/ReportsManualsForms/Reports/BudgetReports/UCM291555.pdf (October 17, 2013).

12 "Is the Product a Medical Device," U.S. Food and Drug Administration, http://www.fda.gov/medicaldevices/deviceregulationandguidance/overview/classifyyourdevice/ucm051512.htm (December 26, 2013).

13 "How to Write a Request for Designation (RFD)," U.S. Food and Drug Administration, April 2011, http://www.fda.gov/regulatoryinformation/guidances/ucm126053.htm (October 17, 2013).

14 "Classify Your Medical Device," U.S. Food and Drug Administration, http://www.fda.gov/MedicalDevices/DeviceRegulationandGuidance/Overview/ClassifyYourDevice/ (December 26, 2013).

15 Ibid.

16 Estimates made by authors based on publicly available information.

17 "CDRH Management Directory by Organization," U.S. Food and Drug Administration, http://www.fda.gov/AboutFDA/CentersOffices/OfficeofMedicalProductsandTobacco/CDRH/CDRHOffices/ucm127854.htm (October 18, 2013).).

18 "Performance Report to Congress for the Medical Device User Fee Amendments of 2007," U.S. Food and Drug Administration, FY 2012, http://www.fda.gov/downloads/AboutFDA/ReportsManualsForms/Reports/UserFeeReports/PerformanceReports/MDUFMA/UCM342644.pdf (October 18, 2013).

19 "Classify Your Medical Device," op. cit.

20 An innovator can check the possibility of achieving exempt status (as well as any limitations that may apply) through parts 862–892 of the code of federal regulations; see http://www.accessdata.fda.gov/scripts/cdrh/ cfdocs/cfcfr/CFRSearch.cfm?CFRPartFrom=862&CFRPartTo=892 (December 26, 2013).

21 See FDA Program Memorandum K86-3.

22 "510(k) Premarket Notification," U.S. Food and Drug Administration, http://www.fda.gov/medicaldevices/deviceregulationandguidance/howtomarketyourdevice/premarketsubmissions/premarketnotification510k/default.htm (October 18, 2013).

23 "Draft Guidance for Industry and Food and Drug Administration Staff – The 510(k) Program: Evaluating Substantial Equivalence in Premarket Notifications [510(k)]," U.S. Food and Drug Administration, December 27, 2011, http://www.fda.gov/medicaldevices/deviceregulationandguidance/ guidancedocuments/ucm282958.htm (October 18, 2013).

24 "Medical Devices and the Public's Health: The FDA 510(k) Clearance Process at 35 Years," Institute of Medicine, July 29, 2011, http://www.iom.edu/Reports/2011/Medical-Devices-and-the-Publics-Health-the-FDA-510k-Clearance-Process-at-35-Years.aspx (October 18, 2013).

25 For more specific guidelines on making a 510(k) submission, see "Screening Checklist for Traditional/Abbreviated Premarket Notification [510(k)] Submissions," U.S. Food and Drug Administration, http://www.fda.gov/downloads/MedicalDevices/DeviceRegulationandGuidance/HowtoMarketYourDevice/PremarketSubmissions/PremarketNotification510k/ucm071369.pdf (October 18, 2013).

26 "How to Prepare a Traditional 510(k)," U.S. Food and Drug Administration, http://www.fda.gov/medicaldevices/deviceregulationandguidance/howtomarketyourdevice/premarketsubmissions/premarketnotification510k/ucm134572.htm (October 18, 2013).

27 See "Guidance for Industry and FDA Staff Format for Traditional and Abbreviated 510(k)s," U.S. Food and Drug Administration, http://www.fda.gov/medicaldevices/deviceregulationandguidance/guidancedocuments/ucm084365.htm (October 18, 2013).

28 See "Screening Checklist for Traditional/Abbreviated Premarket Notification [510(k)] Submissions," op. cit.

29 Julia Post, "FDA Plans to Complete Classifications of Pre-amendment Devices by the End of 2014," Inside Medical Devices, April 12, 2013, http://www.insidemedicaldevices.com/2013/04/12/fda-announces-plans-to-complete-calls-for-pmas-and-reclassifications-for-preamendment-devices-by-the-end-of-2014/ (October 18, 2013).

30 See "PMA Application Contents," U.S. Food and Drug Administration, http://www.fda.gov/MedicalDevices/DeviceRegulationandGuidance/HowtoMarketYourDevice/PremarketSubmissions/PremarketApprovalPMA/ucm050289.htm (October 18, 2013).

31 "PMA Application Methods," U.S. Food and Drug Administration, http://www.fda.gov/MedicalDevices/DeviceRegulationandGuidance/HowtoMarketYourDevice/PremarketSubmissions/PremarketApprovalPMA/ucm048168.htm (October 18, 20013).

32 Ibid.

33 See "IDE Application," U.S. Food and Drug Administration, http://www.fda.gov/MedicalDevices/ DeviceRegulationandGuidance/HowtoMarketYourDevice/ InvestigationalDeviceExemptionIDE/ucm046706.htm (October 18, 2013).

34 See, for example, "NMT Medical Voluntarily Withdraws CardioSEAL(R) PFO HDE and Receives FDA Approval for New STARFlex(R) PFO IDE Study," *Medical News Today*, August 17, 2006, http://www.medicalnewstoday.com/articles/49797.php (October 18, 2013).

35 "Medical Device User Fees Have Been Reauthorized for Fiscal Years 2008–2012," U.S. Food and Drug Administration, September 28, 2007, http://www.fda.gov/medicaldevices/ deviceregulationandguidance/overview/ medicaldeviceuserfeeandmodernizationactmdufma/ ucm109089.htm (October 18, 2013).

36 "MDUFMA Performance Reports," U.S. Food and Drug Administration, http://www.fda.gov/AboutFDA/ ReportsManualsForms/Reports/UserFeeReports/ PerformanceReports/MDUFMA/default.htm (October 18, 2013).

37 "Mobile Medical Applications: Guidance for Industry and Food and Drug Administration Staff," U.S. Food and Drug Administration, September 25, 2013, http://www.fda.gov/ downloads/MedicalDevices/DeviceRegulationandGuidance/ GuidanceDocuments/UCM263366.pdf (December 26, 2013).

38 "FDA Submits Final Guidance on Mobile Medical Apps," Foley and Lardner LLP, October 4, 2013, http://www.foley.com/fda-submits-final-guidance-on-mobile-medical-apps-10-04-2013/ (December 26, 2013).

39 Ibid.

40 "Mobile Medical Applications: Guidance for Industry and Food and Drug Administration Staff," op. cit.

41 All quotations are from interviews conducted by the authors, unless otherwise cited. Reprinted with permission.

42 Ben Richardson, "Why the FDA Took the Piss Out of uChek, a Urine Analysis App," Motherboard.Vice.com, 2013, http:// motherboard.vice.com/blog/why-the-fda-took-the-piss-out-of-uchek-a-urine-analysis-app (December 26, 2013).

43 "Letter to Biosense Technologies Private Limited concerning the uChek Urine Analyzer," U.S. Food and Drug Administration, May 21, 2013, http://www.fda.gov/ MedicalDevices/ResourcesforYou/Industry/ucm353513.htm (December 26, 2013).

44 Shiv Gaglani, "uChek Launches Indiegogo Campaign: Interview with CEO Myshkin Ingawale," MedGadget, August 7, 2013, http://www.medgadget.com/2013/08/smartphone-based-urine-analysis-interview-with-ucheks-myshkin-ingawale.html (December 26, 2013).

45 Pamela Lewis Dolan, "FDA Signals it Will Regulate Medical Apps," amednews.com, May 30, 2011, http://www.amednews. com/article/20110530/business/305309966/5/#cx (December 25, 2013).

46 See "Common Aspects of Electrical Equipment Used in Medical Practice: SC 62A Dashboard," International Electrotechnical Commission, http://www.iec.ch/dyn/www/f?p = 103:29:0:::: FSP_ORG_ID,FSP_LANG_ID:1359,25#1 (January 29, 2014), specifically sections JWG-3, JWG-4, and JWG-6.

47 "AAMI, UL Team Up on Interoperability Standards," AAMI News, October 2012, http://www.aami.org/publications/ AAMINews/ Oct2012/AAMI_UL_Team_Up_Interoperability_ Standards.html (January 29, 2014).

48 See International Medical Device Regulators Forum, http:// www.imdrf.org/ (December 10, 2013).

49 See "Medical Devices," European Commission, http://ec. europa.eu/ enterprise/medical_devices/legislation_en.htm (December 10, 2013).

50 Les Schnoll, "The CE Mark: Understanding the Medical Device Directive," September 1997, http://www.qualitydigest.com/ sept97/html/ce-mdd.html (December 10, 2013).

51 "Guidance MEDDEVs," http://ec.europa.eu/health/medical-devices/documents/guidelines/index_en.htm (December 10, 2013).

52 "Medical Device Registration in Canada (Video)," Emergo Group, http://www.youtube.com/watch?v = i46hyOu5EkQ (January 8, 2014).

53 Ibid.

54 Ibid.

55 "Guidance Document: How to Complete the Application for a New Medical Device License," Health Canada, 2011, http:// www.hc-sc.gc.ca/dhp-mps/alt_formats/pdf/md–im/applic-demande/guide-ld/md_gd_licapp_im_ld_demhom-eng.pdf (January 31, 2014).

56 "Guidance Documents," Health Canada, http://www.hc-sc.gc. ca/dhp-mps/md–im/applic-demande/guide-ld/index-eng.php ((January 8, 2014).

57 "Medical Device Registration in Canada (Video)," op. cit.

58 "Medical Device Regulatory Requirements for Australia," Export.gov, September 25, 2002, http://www.ita.doc.gov/td/ health/Australiaregs.html (January 8, 2014).

59 "Australian Regulatory Guidelines for Medical Devices," Department of Health and Ageing, Therapeutic Goods Administration, 2011, http://www.tga.gov.au/pdf/devices-argmd-01.pdf (January 8, 2014).

60 "The Regulation of Medical Devices: A Life-Cycle Approach to Regulation," Department of Health and Ageing, Therapeutic Goods Administration, November 18, 2011, http://www.tga. gov.au/newsroom/devices-basics-regulation.htm#. Us4AOZ5dV8G (January 8, 2014).

61 "Role of the Sponsor," Department of Health and Ageing, Therapeutic Goods Administration, March 28, 2013,

http://www.tga.gov.au/industry/basics-role-of-sponsor.htm#.
Us3-2Z5dV8E (January 8, 2014).

62 "Medical Device Regulatory Overview for Australia," Emergo
Group, http://www.emergogroup.com/services/australia/
australia-regulatory-strategy (January 8, 2014).

63 "The Regulation of Medical Devices" Department of Health and
Ageing, Therapeutic Goods Administration, November 18,
2011, http://www.tga.gov.au/newsroom/devices-basics-
regulation.htm#.Us4AOZ5dV8G (January 8, 2014).

64 "Australian Regulatory Guidelines for Medical Devices," op. cit.

65 "Australia Clinical Research and CRO Consulting," Emergo
Group, http://www.emergogroup.com/services/australia/
australia-clinical (February 6, 2014).

66 Australian Clinical Trials, http://www.australianclinicaltrials.
gov.au/ (February 6, 2014).

67 "Japan PMDA Medical Device Approval and Certification,"
Emergo Group, http://www.emergogroup.com/services/
japan/medical-device-approval-japan (January 2, 2014).

68 "Japan's Regulatory Process for Medical Devices," Emergo
Group, October 24, 2011, http://www.slideshare.net/
emergogroup/japan-webinar-presentation-oct-2011 (January
2, 2014).

69 Stewart Eisenhart, "Medical Device Regulatory Reform
Advances in Japan," Emergo Group Blog, December 2, 2013,
http://www.emergogroup.com/ blog/2013/12/medical-device-
regulatory-reform-advances-japan (January 2, 2014).

70 Ibid.

71 Ames Gross, "Harmonization by Doing Program Between the
U.S. and Japan," Pacific Bridge Medical, January 1, 2007,
http://www.pacificbridgemedical.com/news/harmonization-
by-doing-program-between-the-us-and-japan/ (January
6, 2014).

72 "Medical Device Regulatory Requirements for Brazil," Export.
gov, September 23, 2008, http://www.ita.doc.gov/td/health/
Brazil%202008%20Medical%20Device%20Profile.pdf
(October 22, 2008).

73 Tara Kambeitz "Medical Device Approvals in Brazil: A Review
and Update," Med-Tech Innovation, November 14, 2011,
http://www.med-techinnovation.com/Articles/articles/article/
30#sthash.Z6XoU4ff.dpuf (February 25, 2014).

74 "Brazilian Medical Device Import Regulations," DRW Research
& Information Services, March 2008, http://www.drw-
research.com/newsletter/Mar%2008.htm (February 25, 2014).

75 Ibid.

76 Kambeitz, op. cit.

77 Ibid.

78 "Brazil's Registration Process: Initial Steps Medical Device
Registration According to ANVISA," Medical Products
Outsourcing, June 11, 2012, http://www.mpo-mag.com/
contents/view_breaking-news/2012-06-11/brazilrsquos-

registration-process-initial-ste/#sthash.8Tp2MENI.dpuf
(February 25, 2014).

79 "Home," ANATEL: National Telecommunication Agency
Telecom Regulatory in Brazil, http://majorevents.anatel.gov.
br/en/ (February 25, 2014).

80 "Latin American Medical Device Regulations," Medical Device
Link, July 2000, http://www.devicelink.com/mddi/archive/
00/07/005.html (February 25, 2014).

81 "Medical Device Regulatory Requirements for Russia," Export.
gov, October 24, 2008, http://www.trade.gov/td/health/
russiaregs.html (October 22, 2008).

82 Ibid.

83 "Overview of the Medical Device Approval Process in Russia,"
Emergo Group, http://www.emergogroup.com/resources/
articles/russia-medical-device-approval-process (February
6, 2014).

84 Ames Gross and Momoko Hirose, "Regulatory Update: Asia's
Largest Medical Device Markets," Pacific Bridge Medical,
January 2008, http://www.pacificbridgemedical.com/
publications/html/AsiaRAFocusDevJan08.htm (January
2, 2014).

85 Frequently Asked Questions on Registration and Import of
Medical Devices in India," Central Drugs Standard Control
Organization, February 21, 2013, http://cdsco.nic.in/FAQ-
IMPORT%20&%20REGISTRATION% 2002022013_DONEE.pdf
(January 2, 2014).

86 "The Drugs and Cosmetics (Amendment) Bill, 2013," PRS
Legislative Research, August 29, 2013, http://www.prsindia.
org/billtrack/the-drugs-and-cosmetics-amendment-bill-2007-
2903/ (January 31, 2014).

87 "India's Central Drugs Standard Control Organization
(CDSCO)," Pacific Bridge Medical, http://www.
pacificbridgemedical.com/indias-central-drugs-standard-
control-organization-cdsco–/ (January 2, 2014).

88 "Indian Conformity Assessment Certificate (ICAC) Mark,"
http://www.icac.in/ (January 2, 2014).

89 Jeremy Sugarman, Harvey M. Meyerhoff, Anant Bhan, Robert
Bollinger, and Amita Gupta, "India's New Policy to Protect
Research Participants," British Medical Journal, July 2013, http://
www.bmj.com/content/347/bmj.f4841 (January 2, 2014).

90 S. Seethalakshmi, "Foreign Companies Stop Clinical Trials in
India After Government Amends Rules on Compensation," The
Times of India, August 1, 2013, http://articles.timesofindia.
indiatimes.com/2013-08-01/bangalore/40960487_1_clinical-
trials-iscr-suneela-thatte (January 2, 2014).

91 "Regulations for the Supervision and Administration of Medical
Devices," Emergo Group, http://www.emergogroup.com/files/
china-medical-device-regulations.pdf (March 22, 2014).

92 Information provided by Katherine Wang, Chief China Life
Sciences Advisor, Ropes & Gray LLC.

93 "Regulations for the Supervision and Administration of Medical Devices,: op. cit.

94 "The State Council Passes the Draft Amendment to the Regulations for the Supervision and Administration of Medical Devices," Healthcare Cooperation Program, February 26, 2014, http://www.uschinahcp.org/state-council-passes-draft-amendment-regulations-supervision-and-administration-medical-devices#sthash.9UlPc6w6.dpuf (March 22, 2014).

95 Ames Gross and Rachel Weintraub, "China's Regulatory Environment for Medical Devices," Pacific Bridge Medical, 2005, http://www.advamed.org/MemberPortal/Issues/International/Asia/china_reg2005.htm (October 22, 2008).

96 "China Medical Device Registration," Export.gov, August 2004, http://www.ita.doc.gov/td/health/regulations.html (October 22, 2008).

97 Ibid.

98 "Less Invasive Heart Valve Replacement Is Approved," *The New York Times*, November 2, 2011, http://www.nytimes.com/2011/11/03/business/fda-approves-less-invasive-heart-valve-replacement.html?_r=2& (December 24, 2013).

4.3 Reimbursement Basics

INTRODUCTION

Patients interested in a new technology or treatment; doctors committed to delivering it. In most industries, the combination of motivated consumers and capable providers would be a formula for success. But in the healthcare field, one critical factor is missing: the role of payers, or those third-party private or public insurance companies that make the decisions whether or not to pay for (or reimburse) a new medical device. With healthcare costs escalating and few patients able to afford their own medical expenses without insurance coverage, payers (and their reimbursement decisions) are exercising unprecedented levels of influence and control over the adoption of new technologies and, in turn, the direction of patient care.

The purpose of understanding the reimbursement landscape in the early phases of developing a new medical technology is to determine whether or not the existing healthcare payment infrastructure will accommodate a new solution to the clinical need it solves. A reimbursement analysis addresses whether there can be adequate payment for the physicians who would deliver the solution and for the facilities where patients would be treated. It also explores whether or not the coverage would be applicable to a large enough segment of the target market to make the development of the solution financially viable.

This is one of two chapters on reimbursement. This first chapter focuses on understanding the basic landscape for reimbursement and the approach a team or company can take when their innovation fits into the existing reimbursement structure for coding, coverage, and payment. The second chapter, 5.6 Reimbursement Strategy, explores how to expand the existing payment infrastructure to accommodate a new technology if the established payment and coverage levels are inadequate. It also addresses how to develop a comprehensive reimbursement strategy that takes into account the evolving healthcare economics environment in the US and abroad.

 See ebiodesign.org for featured videos on reimbursement basics.

OBJECTIVES

- Obtain a high-level understanding of the reimbursement system for medical devices in the US.

- Learn how to identify relevant coverage and billing codes supporting the reimbursement of existing medical devices relevant to a need.

- Understand the status of payment for existing medical device codes, including reimbursement amounts and restrictions on types of patients covered.

- Evaluate differences between US-based private and public payers.

- Survey the reimbursement landscape in select countries outside the US.

- Appreciate how to use reimbursement risks as a screen for prioritizing concepts.

REIMBURSEMENT FUNDAMENTALS

Reflecting on the increasing importance of **reimbursement** in medical device innovation, Thomas Fogarty, the renowned innovator of the embolectomy balloon catheter and dozens of other **medtech** devices, said:[1]

*Regulatory and reimbursement have always been the two big barriers for devices. I used to focus more on **FDA** and the long and unpredictable path to approval. But FDA seems to be getting better – at least they are working to improve. These days, reimbursement is the biggest worry I have for new technologies that are trying to make it through to patient care. The forces of healthcare economics all seem to be working to keep new technologies from coming forward. **CMS** is not the only problem. Private insurers are a bigger threat. They are going to ask for data before reimbursement that will exceed the requirements of CMS. Watch out!*

In the US, reimbursement for medical devices is handled by both public and private insurance programs. The largest public healthcare program is Medicare, which is a health insurance system for the elderly and disabled that is regulated by the federal Centers for Medicare and Medicaid Services (CMS) with headquarters in Baltimore, Maryland and Washington, D.C. The volume of **payment** transactions by CMS is enormous: each year Medicare processes more than one billion claims from over one million providers.[2] Given its large scale, the Medicare **coding**, **coverage**, and payment system exerts a dominant influence on the US healthcare system and is watched closely by the country's private insurers. The US has hundreds of private insurance carriers (depending on how the businesses are defined), with the largest including UnitedHealthcare, Wellpoint, Kaiser Permanente, Humana, and Aetna. US healthcare spending is split roughly equally between Medicare (20 percent) and private insurers (21 percent), with the remaining portions covered by household spending (28 percent), state and local governments, including Medicaid (18 percent), other private sources (7 percent), and other federal spending (5 percent).[3] Total healthcare spending has already reached nearly $3 trillion, on its way to nearly

$5 trillion by 2020 (equivalent to more than 19 percent of the nation's economy).[4]

Against this backdrop, the reimbursement landscape in the US is undergoing fundamental changes that are intended to gradually move away from traditional fee-for-service incentives (that basically reward volume of care) toward new **value**-based programs (that reward quality per unit cost). The passage of the Affordable Care Act (**ACA**) in 2010 was a watershed moment in this transition, mobilizing the healthcare industry to focus on the "triple aim" of improving the experience of care for patients, improving the health of populations, and reducing the per capita cost of healthcare.[5]

Global health insurance systems run the gamut from single-**payer** government systems in many developed countries to predominantly self-pay arrangements that are still common in emerging market countries. While some payment systems are well defined and others are still emerging, each country is characterized by its own unique policies and requirements (see global section later in this chapter).

Whether seeking reimbursement in the US or in any other country in the world, innovators should keep two important points in mind. First, reimbursement policies and procedures are sufficiently complex that innovators almost certainly will depend on experienced consultants to help them develop an understanding of the reimbursement landscape and establish a viable payment approach tailored to specific geographies (see Figure 4.3.1). Second, innovators will almost always achieve reimbursement faster and more easily if they can utilize existing reimbursement pathways for a new technology, rather than having to pursue new coding, coverage, and payment decisions. The determinations of whether or not an existing reimbursement pathway is viable will depend on the nature of the new technology and its match with the existing payment mechanisms, which again may require expert advice and guidance. The main goal of this chapter is to provide innovators with a basic understanding of reimbursement policies and procedures, providing sufficient background that they can collaborate with experts in the field to perform a preliminary reimbursement analysis.

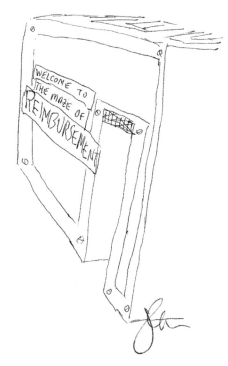

FIGURE 4.3.1
After regulatory clearance, innovators still must navigate a complex maze of challenging reimbursement policies.[6]

Reimbursement in the United States

Both Medicare and private payers in the US follow the same general processes for reimbursing medical services. At a high level, payments are made for medical encounters, which can occur in a wide range of settings: in a hospital (as an inpatient or an outpatient), in an outpatient facility not attached to a hospital, in a doctor's office, or outside of any medical facility. Payments for the physician (or other health provider) and the facility (if the treatment is provided in a hospital or ambulatory health facility) are typically made under separate payment systems. Part of the complexity in the process comes from the fact that there can be multiple payments to the physician and the facility if the encounter involves multiple services.

In its simplest form, the payment process has three main components: Once a service is performed, providers identify one or more appropriate *codes* for the service(s) and provide these codes in a claim that they submit to the payer for evaluation. The payer, in turn, must evaluate *coverage* for the service. If coverage is allowed, the billing codes are translated into appropriate *payment(s)* to the facilities and providers. Each of these

elements – coding, coverage, and payment – warrant further examination.

Coding

In short, coding is the language of reimbursement; it tells a payer exactly what was done, how it was done, and why.

In order for physicians and facilities to be paid, they must submit claims to payers using standardized codes to document the diagnoses and procedures performed. Different codes are chosen depending on the setting in which the care is delivered (see Table 4.3.1).

For the *inpatient* setting, billing for both facilities and physicians is based on an identification of appropriate diagnoses and procedures as specified by **ICD-10** codes (this acronym refers to the International Classification of Diseases, 10th revision). At the time of this writing, the use of ICD-10 codes (rather than the **ICD-9 codes** that preceded them) was expected to become mandatory before the end of 2015.[7] There are two types of ICD-10 codes: diagnoses are specified by ICD-10-CM (Clinical Modification) codes, and procedures by ICD-10-PCS (Procedure Coding System) codes. A typical hospital stay

Table 4.3.1 Different codes are used by different parties for the reimbursement of different procedures, services, and supplies.

Provider	Setting	Procedure code	Diagnosis code
Hospitals	Inpatient	ICD-10-PCS	ICD-10-CM
Hospitals	Outpatient	CPT (HCPCS Level I)	
Ambulatory Surgical Centers	Outpatient		
Physicians	Facility/Office		

will be characterized by a primary diagnosis and any secondary diagnoses (with the appropriate ICD-10-CM codes) and, if a procedure or procedures have been performed, by the ICD-10-PCS code(s).

In the *outpatient* setting, facility claims are submitted using different types of procedures and patient diagnoses. Similar to the inpatient setting, patient diagnoses are described using the ICD-10-CM (diagnosis) codes. For procedures or services, a different set of codes are used, called HCPCS. (This acronym stands for the Healthcare Common Procedure Coding System, and is pronounced "hick-picks.") The HCPCS system consists of two levels of codes. **HCPCS Level I** codes are called Current Procedural Terminology (**CPT**) codes. As the name suggests, CPT codes are used to denote procedures and services provided by medical professionals (e.g., physician claims) – again, these are only used by hospitals in the outpatient setting and for physician services provided in any facility or office setting. CPT codes are established and maintained by the American Medical Association (**AMA**), not CMS. **HCPCS Level II** codes are for products, supplies, and services that are used or provided outside of a physician's office and are not included in the CPT codes. For example, level II codes would be used to submit claims for prosthetics, orthotics, or other supplies used outside the medical office. Level II codes would also be used to bill for ambulance services. The level II codes are established and maintained by CMS.

Coverage

In short, coverage determines *if* a technology or procedure will be reimbursed, and under what conditions.

A properly coded claim, submitted to CMS or an insurance company, does not automatically translate into a payment. To grant reimbursement, the payer must have policies in place which state that the procedures, services, and supplies described in the claim are covered by the patient's health insurance plan. These policies specify conditions under which a procedure is covered and provide details about the codes that should be used to submit and justify the claims. For example, coverage policies can stipulate that specific procedures are only covered for certain patient diagnoses, patient subpopulations, sites of service, or other conditions that the payer may specify. Importantly, coverage policies are not uniform and can vary dramatically across payers. In the US, CMS may set national policies (called national coverage determinations, or **NCDs**) that apply to the whole country, or local policies that cover specific regions. For the local coverage determinations, or **LCDs**, Medicare has multiple jurisdictions across the country, each of which is empowered to make separate decisions administering the Medicare program for its region, as long as those decisions do not contradict an existing NCD. In most situations, a company can decide whether to pursue local or national coverage and there are different strategic considerations for each path (which are discussed further in chapter 5.6). Private payers have their own coverage policies, which may or may not align with the Medicare policies.

Payment

In short, payment describes who is paid, and how much. Payment typically varies depending on the setting where the service is provided (e.g., physician office, hospital inpatient, hospital outpatient).

With appropriate codes and coverage decisions in place, the final factor required for reimbursement is

Table 4.3.2 Medicare payment systems are based on the site of the medical encounter and map to the procedure and diagnosis codes.

Provider	Setting	Payment	Procedure code	Diagnosis code
Hospitals	Inpatient	MS-DRG + MPFS	ICD-10-PCS	ICD-10-CM
Hospitals	Outpatient	APC	CPT (HCPCS Level I)	
Ambulatory Surgical Centers	Outpatient	ASC Fee Schedule		
Physicians	Facility/Office	MPFS		

a payment. With Medicare, payment levels are linked to another set of codes that introduce even more acronyms to the reimbursement landscape (see Table 4.3.2).

Medicare payment to hospitals is made under separate payments systems depending upon whether the patient is admitted as an inpatient or treated as an outpatient. Some services are deemed by CMS to be covered only when performed in the inpatient setting, and this list of "inpatient only" services is published annually by CMS. There are specific considerations that factor into whether patients should be classified by hospitals as inpatients or outpatients. Historically, physicians admitted Medicare patients to the inpatient setting if they were expected to require a hospital stay of more than 24 hours or an overnight stay, whereas hospital outpatients typically were discharged from the hospital on the same day. More recently, Medicare requirements have evolved such that CMS issued a "two midnight rule," stating a presumptive expectation that patients should normally be treated as outpatients unless they require a hospital stay crossing at least two midnights (i.e., three calendar days).

Inpatient hospital payments under Medicare are based on a set of payment category codes called **MS-DRGs** (Medicare Severity Diagnosis-Related Groups). A hospital MS-DRG payment is designed to provide a single, prospectively determined payment amount to reimburse for all hospital services that are provided during the hospital stay (tests, procedures, devices, operating room, recovery rooms, nursing services, etc.) except for physician services, which are paid separately under the Medicare physician fee schedule (CPT coding). All services associated with the hospital stay are assigned to a single MS-DRG payment amount. The appropriate MS-DRG payment category is determined from the ICD-10 codes of the primary diagnosis and any secondary diagnoses. The MS-DRG code also takes into account any complications and/or comorbidities and, in some cases, there are specific MS-DRG payment categories that hinge on complications and/ or comorbidities. Once the appropriate MS-DRG is determined, the actual payment that the hospital receives is adjusted to reflect local labor costs and conditions. In addition, academic teaching hospitals and hospitals treating a large number of indigent patients receive additional reimbursement under Medicare as part of their hospital-specific DRG payments to account for incremental costs associated with providing those services.

For hospital *outpatient* payments under Medicare (but where the care is still delivered within a hospital facility), an Ambulatory Payment Classification (**APC**) code is assigned to the claim based on the procedure(s) or service(s) performed. This APC code is usually translated or "cross-walked" by the payer based on the CPT codes that are submitted by the hospital. Each APC is assigned a specific payment amount, which is updated annually by CMS.

The major distinction between inpatient and outpatient hospital payments is that inpatient payments are primarily based on patient diagnosis (the D in DRG), whereas outpatient payments are based on the procedures performed in the outpatient setting. Whereas hospital inpatient stays, by design, can only result in a single MS-DRG payment, hospital outpatient stays may result in more than one APC payment depending on the specific services performed during the outpatient encounter.

Ambulatory surgical centers (ASCs), are designated facilities separate from hospitals or physician offices that perform only same-day discharge services – examples include colonoscopies or uncomplicated orthopedic procedures. ASCs are paid under Medicare based on still another fee schedule, separate from the hospital payment system.

For *physician payments*, the CPT codes are translated into actual payments using the Medicare Physician Fee Schedule (MPFS). Many private health plans in the US also base their physician fee schedules on the MPFS. For the same CPT code, the payment to the physician may vary depending on whether the service was performed in the physician's office (called a "non-facility setting" by Medicare) or in a facility setting such as a hospital or ASC. Payments to physicians also vary by geographic region, reflecting the different costs of practice (such as staff labor charges).

Information about how to ascertain exact payment rates for specific codes is given in the Getting Started section of this chapter. Online Appendix 4.3.1 provides a summary of all the codes introduced above. The Working Examples provided below are also meant to clarify how, when, and why different codes are used to secure reimbursement payment.

More about physician reimbursement

As described above, physicians use CPT codes (HCPCS Level I) to bill for medical, surgical, and diagnostic procedures, regardless of the setting in which they are performed. In parallel, they use ICD-10-CM codes to describe their patient's medical conditions and diagnoses to support their insurance billings and the chosen CPT code.[8] Medicare and private payers, in turn, use the ICD-10-CM codes to audit physician claims and validate the appropriateness of the billing codes used, based on the diagnosis and treatment performed. The following Working Example based on a medical encounter for atrial fibrillation illustrates how billing will proceed in a moderately complicated case.

> **Working Example**
> Atrial fibrillation: ICD-10-CM and CPT codes
>
> When a patient with atrial fibrillation seeks treatment, the physician may use the default ICD-10-CM diagnosis code I48.91 to indicate that the patient has been diagnosed with this particular heart rhythm disorder. If the patient is more specifically diagnosed with paroxysmal or persistent AF, a more specific diagnosis code should be utilized (I48.0 is paroxysmal AF; I48.1 is persistent AF; I48.2 is chronic/permanent AF; and I48.91 is unspecified AF) with appropriate documentation in the patient's medical record.
>
> The chosen code would serve as a means of justifying the medical necessity of charges made to a payer for any procedures performed, tests ordered, or services provided related to the management of atrial fibrillation. If the patient has multiple other conditions being evaluated or treated, s/he may have many ICD-10-CM codes noted by the healthcare provider. Again, this is because ICD-10-CM codes are descriptors of some or all of a patient's medical conditions relevant for a particular healthcare encounter.
>
> If the physician performs a cardiac ablation procedure for atrial fibrillation (using catheters maneuvered through the blood vessels to the heart), s/he would likely use CPT code 93656 in billing the payer for this procedure. If additional procedures are performed beyond cardiac ablation, the physicians would use multiple CPT codes to receive payment for all of the procedures performed. See Figure 4.3.2 (step 2) for an illustration of the various codes used by the physician.

Physician reimbursement payments under the MPFS are determined using a Congressionally mandated national fee scale developed by a team of researchers at Harvard University under contract with CMS. The fee scale is called the Resource Based Relative Value System (RBRVS) and is maintained through a joint collaboration process between the AMA and CMS. The AMA's Health Care Professionals Advisory Committee Review Board (HCPAC) obtains

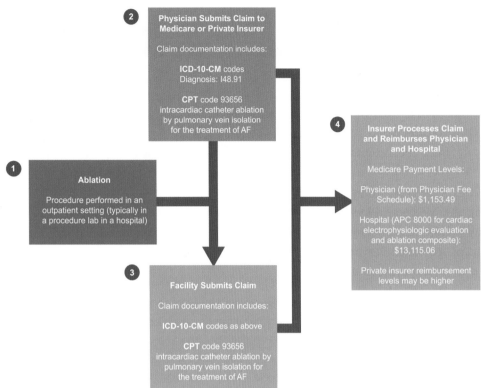

FIGURE 4.3.2

This example for the reimbursement of ablation for atrial fibrillation demonstrates how to trace a procedure through the outpatient reimbursement process. The payment values reflect Medicare national averages across all geographic locations as of 2014.

input from practicing physicians and makes recommendations to CMS on relative value units, or RVUs, for each CPT code. RVUs provide a standardized measure of the resources needed to provide a particular procedure and include three components: (1) a work component, which reflects the time the physician spends on the procedure; (2) a practice expense, which includes nursing time, overhead, and supplies used in the procedure; and (3) a malpractice component that covers liability insurance for the procedure. The number of RVUs is then multiplied by a monetary conversion factor that is determined annually by CMS. The component RVUs may be further multiplied by factors to account for the cost of practicing in different geographic locations (geographic wage index). The result is the final amount Medicare will pay for the procedure or service. The main rationale for using RVUs is that they provide a way of standardizing the measurement of resources and complexity associated with particular physician services across different specialties. Depending on where the procedure or service is

performed, there may be two RVUs: a non-facility one for services performed in the physician's office and a facility one for all others. Changes to CPT codes and to the RBRVS are effective January 1st of each year.

Many critics have pointed to weaknesses of the existing fee-for-service Medicare payment systems that determine payments based on the provider's resource inputs rather than the clinical value or health outcomes of the services being provided to patients or the healthcare system. A variety of new "value-based" payment methodologies are being introduced in the Medicare program and by private payers (sometimes alongside or in addition to the traditional fee-for-service payment systems) that are designed to increase rewards for higher-value services while placing increased financial responsibility on providers and patients for lower-value services. This transition will unfold over a number of years, and it will continue to add extra complexity to deciphering the regulatory pathway for a new technology at any given point in time.

Working Example
Atrial fibrillation: physician payment

As shown in Figure 4.3.2, step 4, the payment for performing a cardiac ablation by pulmonary vein isolation (CPT 93656) in an outpatient setting, such as a hospital catheterization lab, is based on the Medicare physician fee schedule. The 2014 national average payment of $1,153.49 was calculated using total RVUs of 32.20 and a conversion factor of 35.83 (which is the same across the US). For specific locations, geographic modifiers are taken into account such that the reimbursement payment would be, for example, $1,238.68 for a physician working in San Francisco.

Note that for physician services including cardiac ablation performed in an outpatient setting, one or more CPT codes may be utilized, as appropriate, to reflect all of the physician services performed for the patient. Whereas CPT code 93656 happens to be a relatively comprehensive code that reflects the majority of the preparation and procedural work involved with the ablation to treat AF, additional CPT codes may also be billable by the physician to reflect three-dimensional electrophysiological mapping (CPT code 93613) and additional linear or focal ablation of atrial fibrillation remaining after completion of pulmonary vein isolation (CPT code 93657) if these services are performed and medically necessary.

More about hospital inpatient coding and reimbursement

As previously described, hospitals apply ICD-10-codes to cover both diagnoses and procedures for inpatient admissions and then translate these codes into the MS-DRG system to determine the amounts of payment. More than 140,000 ICD-10 diagnosis and procedure codes are grouped into over 900 MS-DRGs. Patients within each MS-DRG category are considered by CMS to be similar in terms of their clinical characteristics and resource use, and the hospital receives a single payment amount that reflects the patient's diagnoses and procedures that were performed. For each admission, only one MS-DRG is assigned, regardless of the number of services provided or the duration of the patient's stay. Each MS-DRG has a unique relative weight, which is then converted into the payment amount. Changes in DRG codes are effective October 1st of each year, along with the updates to the ICD-10 procedure and diagnostic codes.

Working Example
Atrial fibrillation: MS-DRG payment categories

If the patient in the atrial fibrillation example is hospitalized for a complex cardiac surgical procedure to treat his/her disease (e.g., a stand-alone open surgical Maze procedure which requires many days of post-operative recovery in the hospital), the hospital would bill using the Maze ICD-10 code and CMS would assign the hospital stay to MS-DRG 228 if the patient had major complications or comorbidities, or MS-DRG 230 if the patient was without major complications and comorbidities. These codes encompass all of the relevant patient care issues (and, thus, relevant ICD-10 codes) that would come up during the procedure and subsequent inpatient hospitalization.

Alternatively, if the patient has a catheter-based cardiac ablation procedure requiring a brief inpatient stay, the hospital might use the ICD-10-PCS codes 025T3ZZ (destruction of left pulmonary vein, percutaneous approach) and 025S3ZZ (destruction of right pulmonary vein, percutaneous approach) for the description of the inpatient service. If Medicare is the payer, this code can get assigned to either MS-DRG code 250 (percutaneous cardiovascular procedure without coronary stent, with major complication or comorbidity) or 251 (percutaneous cardiovascular procedure without coronary stent, without major complication or comorbidity). (Note that MS-DRG codes can be broad, with individual payment categories encompassing a wide array of procedures.)

More about hospital outpatient coding and reimbursement

Recall that hospitals and free-standing outpatient facilities use CPT codes for claiming outpatient services. Under Medicare, each CPT code is assigned by CMS to an Ambulatory Payment Classification (APC) group with a unique relative weight, which is then converted into a payment amount. The APC payment rates are designed to capture all of the hospital expenses involved in the outpatient service (e.g., medical device costs, costs for nurses and technicians involved in the patient's care, procedure and recovery room costs, bundled services such as fluoroscopy, ancillary supply costs, costs of medications such as regional anesthesia, and associated overhead) except for physician services which are paid under a separate fee schedule. Unlike the MS-DRG system, multiple APCs can be assigned and separately paid for as part of a single outpatient encounter, depending on the procedure(s) performed.

> **Working Example**
> Atrial fibrillation: APC payment categories
>
> As shown in Figure 4.3.2, step 3, both a hospital and physician could bill CPT code 93656 for a cardiac ablation procedure to treat atrial fibrillation if the procedure is performed in the outpatient setting. Whereas the physician would bill this code to reflect the physician's work to perform the procedure, the hospital bill reflects the hospital's own costs associated with the procedure, including medical device and ancillary supply costs, nursing and technician costs, procedure and recovery room costs, etc. The Medicare contractor would assign this outpatient service to APC 8000, as shown in step 4, which includes all cardiac ablation procedures as well as associated electrophysiology studies and mapping services that are "**bundled**" into the APC with a national average payment of $13,115.06 for this service.

Reimbursement by private payers

As noted, private or commercial health insurance companies have historically looked closely at Medicare

reimbursement and have often followed Medicare's lead in the areas of coverage policy and physician reimbursement. In years past, there was an adage in the US healthcare industry that "what Medicare does, the rest will follow." Although this has generally been true for physician and procedure payments, for hospital payments the private health plans have long negotiated specific contracts with individual hospitals or hospital groups using payment methodologies that differ from Medicare. These include a complex assortment of approaches including "per diem" payments, "carve-out" payments for implantable devices, and all-payer DRGs. In addition, private payers are actively deploying a range of new value-based payment methods that place financial risk on providers and tie payments to performance on hospital quality and outcomes-based performance measures. So, while private payers certainly look to Medicare for direction, they are increasingly likely to make independent decisions based on the goals and objectives of their individual plans. This means that reimbursement decisions can differ not only between Medicare and private payers, but among the hundreds of private payers in the US. This has led to a reimbursement environment in which medical device companies must often seek to establish reimbursement coverage (and appropriate reimbursement rates) on a payer-by-payer basis – a daunting undertaking, particularly for start-ups with limited resources.

Unlike the government, most private insurance companies are in business to make a profit. Even non-profit insurers, such as Blue Cross/Blue Shield (**BCBS**), have goals to maintain their financial health through mechanisms such as increasing their cash reserves (the equivalent of profit to a non-profit company). While patients would like to think that health insurance companies have their best interests at heart, the decisions of these companies may actually be driven by a series of interrelated and complex factors. Patient well-being is certainly among the considerations, but other issues such as profitability, efficiency, and **risk management** also come into play. As one professional association stated on its website:[9]

*Insurers are not necessarily in business to assure that everyone receives access to care. Nor are they in business to guarantee that all qualified healthcare providers are fairly and adequately compensated for their services. Healthcare providers often try to assign "moral obligations" to insurance companies, but they are not obligated to accept them. Although the hope is that insurance companies have a basic concern about the health **needs** of the general public and fair payments to practitioners, it should not be expected that this is their primary consideration.*

Generally, as a guiding philosophy, commercial insurance plans state that they cover services that are deemed "medically necessary" by medical doctors. However, the concept of medical necessity is highly subjective and open to interpretation across payers and types of plans.

The two most common types of health insurance plans within the US are health maintenance organizations (HMOs) and preferred provider organizations (PPOs). HMOs bring together healthcare providers (e.g., doctors and hospitals) that have contracted with an insurance company to offer their services as network participants according to a fixed payment schedule and other pre-negotiated terms and conditions. Typically, HMOs are one of the most affordable insurance options available to healthcare consumers. In comparison to PPOs, the premiums for HMOs are relatively low and **copayments** are less expensive (or free). However, in order to offer services at a low, fixed price, many HMOs depend on a high volume of patients and are notorious for being restrictive. HMOs have stringent rules regarding which physician(s) a patient can see and where service can be delivered. From a reimbursement perspective, they tend to place more limitations on what services will be covered, at what rate, and for which patient groups.[10] PPOs, on the other hand, tend not to be as restrictive as HMOs and offer patients a broader range of options.[11] While PPOs contract with medical provider networks, they try to manage medical expenditures through financial incentives (e.g., charging different copayments and deductibles for services performed by network providers versus providers outside the network, charging different

copayments for preventive versus corrective treatments, reimbursing certain medical procedures at rates higher than others). In addition, patients have greater choice in choosing a physician, selecting a facility, and managing their medical care in exchange for higher premiums.

As mentioned, Accountable Care Organizations (**ACOs**) are emerging as another type of healthcare entity, stimulated in part by financial provisions of the Affordable Care Act. There are a variety of different ACO payment arrangements and they are rapidly evolving. These voluntary consortiums of *independent* physician groups, hospitals, and insurers share the responsibility for caring for a defined population of Medicare beneficiaries over a defined period of time, but they can earn financial incentives for saving money through more coordinated care that avoids duplicate or unnecessary procedures and tests if they maintain performance on quality and outcomes measures.[12] In general, ACOs have strong motivation to be careful in evaluating new technologies or procedures that have the potential to escalate costs.

Across HMOs, PPOs, and ACOs, the policies of private insurance providers are not created equal. Kaiser Permanente is one major HMO (emerging now as an ACO) that uses stringent **evidence-based** criteria for adopting new technologies.[13] As a result, Kaiser often approves and covers new technologies only *after* they have been thoroughly studied in the post-approval environment and when there is strong evidence (ideally from **controlled clinical trials**) that the new device improves clinical outcomes. Because the process of conducting studies and publishing the results in reputable medical journals can take years to accomplish, Kaiser physicians may be reluctant to use the latest generation of a technology if the evidence behind it is insufficient.[14] Consider, for example, implantable cardiac defibrillators (ICD) used to treat patients with life-threatening cardiac rhythm disorders. While it is undisputed that ICDs help prevent sudden cardiac deaths, new generations of the device include features for which the clinical benefit is not fully supported by the results from extensive **clinical trials**. An article published in the *Permanente Journal* pointed out that while new features, such as dual-chamber and rate responsive pacing, had driven up the cost of ICDs, most patients do not necessarily benefit from this specialized functionality. For this reason, it advised physicians to consider whether the additional cost associated with the latest technology was justified relative to the potential benefit that each individual patient would receive.[15]

In general, it is necessary for innovators and companies to study their market in determining how much time to invest in understanding the specific reimbursement policies of private payers. If there is a significant Medicare market for the device, understanding Medicare reimbursement practices will be sufficient in most cases. However, if most reimbursement will come from commercial payers, then it will be necessary to invest more time in investigating private payer policies, reimbursement behaviors, and the precedents that have been set by other devices. Some commercial insurers have established groups that specifically evaluate new medical technologies before reimbursement decisions are made. The Blue Cross Blue Shield Technology Evaluation Center (**TEC**) is one well-known example.[16] Many individual plans also post their medical policies regarding coverage and reimbursement on the Internet. Individual payment rates for private insurers, however, can be difficult to obtain from public sources.

The self-pay reimbursement model

Even though government and private payers account for the large majority of healthcare spending in the US, individuals and households finance a full 28 percent of the total.[17] These **out-of-pocket** expenses can include copayments that are a common part of regular insurance coverage, but also self-pay expenditures for elective procedures, such as laser eye surgery or aesthetic and cosmetic interventions, that are typically not covered by insurers. In addition, new types of consumer-directed health savings plans have emerged in which the patient has increased decision-making and financial responsibility for their healthcare decisions.

Innovators should examine carefully whether or not the need on which they are working is appropriate for a self-pay model (see the Miramar Labs story in 5.7 Marketing and Stakeholder Strategy). This approach can be advantageous since it can remove some of the obstacles of reimbursement. However, self-pay also has certain risks and challenges. For instance, the amount of money consumers may be willing to spend out-of-pocket is likely to be much smaller than the amount **third-party insurers** typically pay for procedures. Outside the United States, self-pay is an increasingly important mechanism for reimbursement of new technologies that are not yet covered by government insurance systems, which is a factor that also should be considered (see the global section later in this chapter).

Copayment in reimbursement

As mentioned, a major way that consumers pay directly for healthcare is called "copayment." This is a requirement that patients with a health insurance plan personally pay a certain percentage of the bill, to be collected either at the time of the encounter (e.g., during a clinic visit) or in a subsequent billing. The copay percentage is usually determined by the insurance provider, representing a small portion of the total bill and generally subject to caps. Copayments apply to Medicare recipients, as well as patients covered by private plans. As a rough average, Medicare pays approximately 80 percent of a patient's total bill). Accordingly, many Medicare patients purchase supplemental insurance plans (called "Medigap" plans) that cover most if not all direct copayment responsibilities. With respect to private health plans, copayment levels can differ significantly between insurance plans, with relatively high copayments being one of the major features of less expensive plans. In general, copay levels are becoming an increasingly important factor in reimbursement in the US, and this is contributing to the "consumerization" of healthcare as patients begin paying more attention to the relative costs of procedures and technologies. The implication for innovators is that it is not just the government and private insurance payers who are paying close attention to the cost of new technologies – patients themselves are now comparison shopping with respect to health spending.

Preparing a reimbursement analysis

Before selecting a final **concept**, innovators should have a substantial understanding of the reimbursement environment associated with each of their solution ideas. Such an analysis summarizes the reimbursement landscape for similar or related innovations, including relevant codes, coverage decisions, and payment levels that will potentially serve as precedents for a new technology in the field. The story of Metrika, Inc. exemplifies many of the issues discussed in this chapter, as well as illustrating how basic reimbursement analysis serves as a bridge to the more complex exercise of developing a reimbursement strategy (see 5.6 Reimbursement Strategy).

FROM THE FIELD ▶ **POINT OF CARE DIAGNOSTICS IN DIABETES**

Evaluating the adequacy of reimbursement under established codes

In the care of diabetes, it is recommended that all patients have a hemoglobin A1C (HbA1C) test twice a year. This test shows the average amount of sugar in the patient's blood for the three months preceding the test. The test results are then used by the patient's physician to adjust treatment (i.e., possibly changing medications and/or modifying nutritional guidelines). Traditionally, the HbA1C test involved drawing a patient's blood in the lab. As a result, some patients skipped testing because of the inconvenience. Michael Allen, a California Bay Area innovator and entrepreneur, came up with a vision for a disposable, convenient, hand-held point of care test for HbA1C, which would enable patients to have their HbA1C tested when they visit their physicians rather than having to take the test in a laboratory setting. Being able to perform the test in the doctor's office had several potential advantages: it was more convenient for patients, it would likely increase patient compliance since there was no second step to the testing process (i.e., the visit to the lab), and it would potentially lead to better management of diabetes because the results would be available immediately so that the physician and patient could discuss them face-to-face in the office. Allen founded Metrika, Inc. to commercialize his idea (see Figure 4.3.3), which was later acquired by Bayer.

At the time Allen was developing his technology, Medicare was reimbursing HbA1C lab tests using the CPT code 83036 (Hemoglobin; glycosylated).[18] That is,

labs ("facilities" in reimbursement terms) performing HbA1C tests were using this code to submit their claims to Medicare and were reimbursed at the rate of approximately $13 per test.

The main question in the reimbursement analysis at this point would be whether or not the company could make use of this code to cover its test. The answer (with perfect hindsight) appears to have been relatively straightforward: if the manufacturer could demonstrate that its point-of-care (**POC**) test was equivalent to the existing HbA1C tests performed in the lab, then it would expect to take advantage of the existing code (there is no

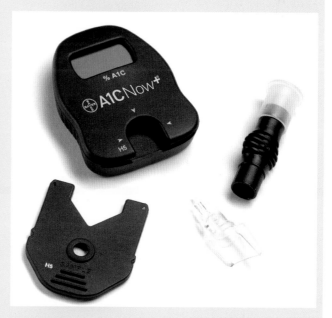

FIGURE 4.3.3

The A1CNow+® point of care HbA1C monitor (courtesy of Bayer HealthCare).

reason why Medicare would not be willing to pay the same amount for an equivalent test performed at a different setting, although careful consideration of any **stakeholder** issues arising from a potential threat to a lab's business would be warranted).

The next question to be addressed would be whether or not the reimbursement level associated with the existing code was adequate. To determine this, the manufacturers would have to calculate a rough estimate for their manufacturing cost per test plus a reasonable markup (i.e., margin on that cost). Without knowing the exact numbers for any manufacturer of an HbA1C POC test (since this information is proprietary), one can perform a quick "back of the envelope" calculation using what is known as the "50 percent rule" of approximation. To be viable, the $13 reimbursement rate needed to be sufficient to cover the end **user's** (physician's) cost of acquiring and performing the test, plus a potential markup for the manufacturer. Applying the 50 percent rule, one can infer that the cost of supplies for the physician should be no more than approximately 50 percent of the total reimbursement rate. This means that the price a physician would pay to purchase the test at wholesale would be approximately $6.50 (including shipping and handling). Applying the 50 percent rule again, one can infer that a wholesaler's markup is typically 50 percent, which means that the price a manufacturer could charge the wholesaler for distributing the device would be about $3.25. Assuming that the manufacturer wants and/or needs to make a 50 percent margin on its costs to justify development of the product, the company's production cost would need to be no

more than $1.62. This means that the manufacturer should be able to produce the test at a cost of no more than $1.62. If such a cost level is not technically feasible, then any manufacturer must consider seriously whether it should seek a new CPT code and a different reimbursement rate. Alternatively, even if this cost is technically feasible, the manufacturer may want to evaluate whether a value-based argument could be created to support a higher reimbursement and a different code to reflect both the innovation inherent to the POC test and the benefit arising from it.

In the case of the HbA1C POC tests, what actually happened seems to suggest that the reimbursement level associated with the existing CPT code was inadequate to cover the manufacturing costs, manufacturer's margin, and distributor's profits (or that the team felt the innovation in the POC test could justify higher reimbursement). In 2005, the retail price for Metrika's test was $24 (including shipping and handling). Assuming a rough 50 percent profit margin at any part of the value chain (as described above), this means the wholesale price was about $12 per test, the price paid by the wholesaler to Metrika was approximately $6, and production costs per test were about $3. In 2006, the AMA issued a new CPT code (83037). As of 2007, Medicare was providing reimbursement for the Metrika device and the products of its main competitors at $21.06 in most states. Importantly, however, this achievement took seven years from the first FDA approval (September 2000) to the time that the new CPT code was approved by the AMA, accepted by CMS, and associated with a standard reimbursement amount.

A comprehensive reimbursement analysis should include information on the topics shown in Table 4.3.3 for each of the concepts under consideration. Innovators can then use this information to compare the concepts in terms of how difficult or simple it may be to obtain reimbursement for them and the financial viability of the business plan in light of economic and reimbursement considerations. They can also use the information gathered through this research to refine and/or validate key market assumptions (see 2.4 Market Analysis), since it results in new data on the number of procedures performed and total reimbursement granted per procedure.

The steps for completing a reimbursement analysis are outlined in the Getting Started section for this chapter.

Table 4.3.3 Reimbursement analysis should provide innovators with an overview of the current reimbursement landscape for technologies related to the need being studied.

Topic	Description
Payer mix	Identify the primary payer(s) and mix of services by payer.
Location of procedure	Describe the setting in which the new device would be utilized and/or procedure would be performed.
Coverage decisions and technology assessment by Medicare, private payers, and/or health technology assessment agencies	Summarize the coverage decisions for comparable devices and/or procedures, including how long it took to achieve reimbursement and any constraints or exceptions that may affect coverage.
CPT code and payment amount	Define the appropriate CPT code and patient copayment under Medicare or determine if a new code is needed.
MS-DRG, APC, and/or other codes and payment amounts	Define the appropriate DRG, APC, or other code (e.g., HCPCS Level II) and patient copayment under Medicare or determine if a new code is needed.
Facility costs	Summarize the anticipated facility-related costs for the device and/or procedure.
Number of procedures	Summarize the number of procedures performed and reimbursed, reimbursement per procedure, and payer mix.

Global reimbursement

Medical reimbursement outside the US has a reputation for being even stricter than in America. Innovators intending to market products in other geographies have an obligation to carefully understand the reimbursement systems and processes in the countries they are targeting. This section provides a general framework that innovators can use to approach global reimbursement, with an outline of key issues and challenges in major developed and emerging markets.

To understand the reimbursement process for a medical device in any country, the innovator first needs to investigate the basis for financing and delivering healthcare. In most developed nations, there is significant public financing of healthcare from taxpayers, so the government is ultimately responsible for deciding how healthcare funds are distributed to providers. Delivery of healthcare could be public (as in the United Kingdom) or private (as in France). Providers may be allocated a fixed budget from which they are to cover the expenses of all medical services, or they may be paid from the government for each service they provide (using payment processes analogous to the ones in the US). Another model is a hybrid system in which capital expenses (e.g., purchases of expensive equipment or capacity expansions) are covered by a budget allocation, with additional payments granted for each service provided (e.g., variable costs such as physician salaries and supplies). Across the globe, the hybrid systems appear to be the model that is most widely utilized.

It is important for innovators to recognize that the method used to allocate funds to providers is critical, since it is directly related to the adoption of a new technology. As one moves from a fixed budget allocation to a per-service system, the incentives of the providers to use a new technology go up, as long as the technology is adequately reimbursed. Another key issue to understand is whether the purchasing decision and the price paid for a device will be determined through direct negotiations with each provider, or through some global purchasing agreement that involves multiple providers and possibly the government.

High-level reimbursement questions that innovators should ask, for each global market they are considering for device sales, include the following:

- Is healthcare financing public or private?
- Is delivery public or private?
- How are providers paid for health services and for capital expansions?
- Is the volume of services delivered by each provider regulated?
- Are the prices for devices regulated?
- Who negotiates the purchase price and reimbursement level for a device and what is the process used (e.g., direct negotiations between providers and manufacturers, government contracting, or contracting with an alliance of providers)?
- Do new devices have to undergo technology assessment before they can be used and reimbursed?
- Is there a list of approved devices that are reimbursed?
- What is the process for obtaining new technology reimbursement?

With a solid understanding of the answers to these general questions, innovators should next consider the unique aspects of the reimbursement systems in the countries being evaluated. A sample of these issues is provided below for three European and two Asian nations. Innovators must carefully understand the local reimbursement policies and practices of the countries in which they seek to do business, particularly if they anticipate the need to negotiate with government agencies. One way to quickly and effectively learn about national practices is to partner with a local distributor. A strong relationship with an established local player can be a valuable asset in securing reimbursement, as well as setting up an effective sales and distribution infrastructure overseas. In addition, local medical specialty societies and physician opinion leaders can provide valuable information needed to assess and plan for reimbursement.

Germany

In Germany, Europe's largest medtech market, the country's Statutory Health Insurance system (SHI) provides health insurance coverage to approximately 85 percent of the population, with the remainder covered through private health insurance.[19]

The process for the reimbursement of medical devices or diagnostics is highly dependent on the setting in which the product should be used. While CE-certified medical devices may be readily used in the inpatient setting (if hospitals are willing to purchase them and cover them partly or fully from their own budgets), reimbursement of devices and care products in the outpatient and home care setting need to be formally approved by SHI based on applications for reimbursement. Companies must provide supporting information with their applications, including detailed product specifications, data demonstrating safety and quality, and, in some cases, medical or nursing care benefits.[20]

Hospital reimbursement[21] Hospital funding in Germany is regulated by the Hospital Financing Act, which has two primary funding mechanisms. First, payment for inpatient hospital care is the responsibility of SHI (or private insurance), with these payments covering facility, labor, and equipment costs. Second, each state is responsible for covering costs related to capital equipment (assets with an economic life longer than three years). These investments are negotiated between the state infrastructure fund and individual hospitals.

The principle mechanism of inpatient reimbursement is the German DRG or G-DRG prospective payment system. The G-DRG system is maintained by the Institute for the Hospital Remuneration System (Institut für das Entgeltsystem im Krankenhaus, Siegburg), or InEK. This entity is responsible for collecting and analyzing hospital cost data, updating the payment rates associated with each code, and maintaining the codes themselves. The diagnostic and procedure codes that support the G-DRG system are controlled by the German Institute of Medical Documentation and Information (Deutsches Institut für Medizinische Dokumentation und Information, DIMDI). The G-DRG employs ICD-10-GM (German Modification) diagnostic codes, which closely resemble the ICD-10 codes maintained by the World Health Organization. Its procedure codes come from the OPS (Operationen- und ProzedurenSchlüssel) system, which is also maintained by DIMDI.

Reimbursement of new technologies in the German healthcare system depends on the availability of appropriate diagnostic and procedure codes, as well as the uptake and correct coding of this new technology by the hospitals that participate in the InEK calculation system. InEK updates the G-DRG annually, both in terms of its overall DRG code structure and the individual reimbursement amounts. However, the updates are determined based on cost data obtained from the previous two years, so innovators can experience delays in receiving new codes and associated payment rates for new technologies. In response to this lag, InEK created an "on-top" funding process for innovative new products that are not appropriately covered under the current DRG structure and payment amounts. Under the NUB (Neue Untersuchungs- und Behandlungsmethoden) process, individual hospitals can file an application for reimbursement of newly introduced devices. If approved, add-on reimbursement for the use of the technology is granted to the hospital(s) that applied for the NUB payment (not to all hospitals using the product in Germany). Importantly, InEK does not decide on the actual amount of the NUB "on-top" payment. Instead, the payment rate must be negotiated between the hospital and the SHIs. Once an NUB has been granted, InEK monitors usage and cost to determine if the underlying DRG payment should be adjusted, thereby increasing the total G-DRG payment on a nationwide basis. While still somewhat complex and time-consuming to navigate, the NUB pathway can help accelerate market access for innovative new technologies.

Ambulatory reimbursement[22]   In Germany, the majority of ambulatory or outpatient procedures are delivered by private practitioners in the community. These physicians are compensated by their regional physician association (Kassenärztliche Vereinigung, KV), which in turn is paid by the SHIs. These payments are generally made on a fee-for-service basis. The physician associations are responsible for distributing payments to their members in accordance with the Uniform Value Scale catalog, or Einheitlicher Bewertungsmaßstab (EBM). The EBM is a fixed budget distribution system, and physicians are only able to invoice services that appear on the EBM. The EBM is maintained by the Kassenärztliche Bundesvereinigung (KBV), the federal association of office-based physicians.

The process to get a new technology for ambulatory use listed on the EBM requires physician support and may involve a health technology assessment by the Institute for Quality and Efficiency in Healthcare (Institut für Qualität und Wirtschaftlichkeit im Gesundheitswesen, IQWiG). The technology must be specifically approved for reimbursement before it can be used at all, which is different than the inpatient setting, where technologies can be used at the discretion of the hospital unless the Federal Joint Committee (Gemeinsamer Bundesausschuss, G-BA) formally disapproves its use. IQWiG will only endorse new technologies and procedures that are "necessary, appropriate, and economically reasonable." IQWiG does not provide standard guidance for the health technology assessments it conducts for medical devices and diagnostics, but has published its methodology for the assessment process. G-BA maintains a separate Valuation Committee that determines the actual payment for newly listed procedures on the EBM.

France

France is the second largest medical device market in Europe. Healthcare is publicly financed and delivered by both public and private providers. All employed individuals in France, as well as their children and spouses, are covered by the national health insurance plan called Securité Sociale. Individuals who are not entitled to participate in this program (e.g., affluent individuals who are not employed) must purchase special coverage, known as Assurance Personelle. Many people covered by the state-run program also choose to purchase additional insurance to supplement their basic coverage.[23]

Outpatient procedures are reimbursed if they are listed on the Liste des Produits et Prestations Remboursables (LPP) found on the Securité Sociale website.[24] Inpatient procedures are reimbursed using a DRG-system referred to as the Groupes Homogenes de Sejours (GHS). Payments may vary according to whether the procedure is performed in a private or public setting, and the list of approved outpatient procedures may vary according to the setting. Expensive devices may receive an add-on payment. Public hospitals have a fixed budget to cover all of their capital expenses, so these do not have to be funded by revenue from fees.

When a new device that is not included in the existing lists for the outpatient or inpatient system enters the market, the manufacturer submits a reimbursement application for the new technology and it undergoes a health technology assessment by a special division of the French National Authority for Health (or Haute Autorité de Santé, HAS).[25] This new entity, known as the Commission Nationale d'Evaluation des Dispositifs Médicaux et des Technologies de Santé (CNEDiMTS), was established in 2010 to perform more rigorous health technology assessment and provide scientific opinion concerning the usefulness, interest, and good use of medical devices and other non-drug healthcare products. Based on the recommendation of CNEDiMTS, the Health Ministry then decides whether a device will be granted reimbursement. The Comité Economique des Produits de Santé (CEPS), also known as the Economic Committee on Health Care Products, then negotiates with the manufacturer to set the price or the tariff. They also may set utilization targets and "clawback" provisions for a new technology that can require manufacturers to provide rebates if usage exceeds pre-negotiated targets. The process may take three to four years (or longer). In some cases, innovative medical devices or procedures can be funded temporarily through a dedicated exceptional pathway.[26] This funding covers partial or total reimbursement related to patient stay, the medical device and/or procedure, and the costs of additional data collection. Admission to this exceptional pathway is decided by the Health Ministry with input from the HAS.

As part of the health technology assessment, CNEDiMTS requires a company to submit a reimbursement dossier that includes a technical description of the technology and its mode of action, specifications for use, for which indication reimbursement is required, the severity of the targeted condition, relevant clinical evidence, comparisons existing treatments, and estimates of the size and characteristics of the target population based on epidemiology and/or market research data. Cost/effectiveness analysis is not required but may be submitted with the dossier.[27]

Similarly, CEPS requires the company to submit an economic dossier that includes the recommended price or tariff for reimbursement, sales forecasts up to the market stabilization, price or tariff justification based on cost minimization versus existing alternatives, anticipated sales, and the status of pricing and reimbursement in other EU countries. In addition, CEPS requests a breakdown of costs for manufacturing and distribution, information about the company (location in France and foreign countries for manufacturing and commercialization, number of employees, turnover, and sales information of any other company products sales covered by reimbursement in France).[28] Again, no formal cost/effectiveness analysis or budget impact data are formally required, but they may be submitted for consideration by CEPS.

A more recent development in French device reimbursement and market access is the "STIC" program (soutien aux techniques innovantes coûteuses).[29] This program provides government funding for innovative medical technologies that have already been initially validated by prior clinical studies, but for which no formal clinical and economic evaluation has yet been completed in France.[30] The program aims to establish evidence necessary to determine eligibility for long-term reimbursement. While the trial funding and reimbursement seems an attractive option for innovative device companies, innovators should appreciate that they cannot take an active role in the design and conduct of the study.

United Kingdom

In the UK, the majority of healthcare is provided by the National Health Service (NHS), a publicly funded healthcare system established in 1948.[31] The NHS is organized in Primary Care Trusts (PCTs) and Hospital Trusts with responsibility in their geographic areas. The Department of Health allocates funds to these trusts, which are used to provide necessary medical services and to invest in infrastructure. Trusts reimburse providers for most services using a payment-by-results system, which establishes fixed payments for hundreds of hospital and outpatient procedures.[32] Medical devices reimbursable in the outpatient setting are listed in the drug tariff list.

Inpatient procedures are reimbursed using a DRG system referred to as HRG (Healthcare Resource Groups).[33]

General practitioners (GPs) play a critical role in the UK health system because they serve as gatekeepers for specialized care. Similarly, the PCTs are central to the system because they are charged with financial responsibility for providing optimal care across primary, secondary, and community healthcare services, staying within a given budget.

As the primary payer across England, Scotland, Wales, and Northern Ireland, the NHS has established a variety of mechanisms for evaluating the cost-effectiveness of new medical technologies. In fact, each region has its own approach to health technology assessment (e.g., in Scotland, this function is performed by the Scottish Medicines Consortium and Wales maintains the All Wales Medicines Strategy Group).[34] However, the National Institute for Health and Care Excellence (**NICE**), which is a government organization chartered to issue national treatment guidance and assess cost-effectiveness of healthcare for England, Wales, and Ireland, has become internationally recognized as a model for health technology assessment. In addition to serving as a gatekeeper to reimbursement in the UK, there is growing evidence that NICE guidance is also referred to by other countries since the organization is perceived as having a robust methodology.[35]

Established in 1999 as a department of the NHS, NICE was tasked with producing national guidance on specific health technologies. The organization issues guidance on the use of medical devices and medicines through its technology appraisal process, and also on clinical practice through its guidelines development process.[36]

In contrast to some other countries, NICE does not evaluate all technologies as they reach the market. Instead it selects technologies for review based on factors such as: (1) how likely the technology is to result in significant health benefits across the NHS population if it is given to all indicated patients; (2) how likely the technology is to result in a significant impact on other health related government policies (e.g., reduction in health inequalities); (3) how likely the technology is to have a significant impact on NHS resources (financial or other) if given to all indicated patients; and (4) how

likely the institute is to be able to add value by issuing national guidance.[37]

NICE performs its health technology assessments via two routes. The first is the Multiple Technology Appraisal (MTA) process, which commonly examines all relevant drugs and devices within a disease area. The MTA process uses evidence provided by any number of sources, including manufacturers, healthcare professionals, and patient/caregiver representatives. The assessment is made a panel of independent, academic experts from one of a number of academic centers that are commissioned by NICE to perform the evaluation and issue an assessment report.[38] The second route is the Single Technology Appraisal (STA) process. Using this approach, an independent Evidence Review Group is commissioned to evaluate a single technology or drug. The STA process tends to be more streamlined because of its limited scope, with the emphasis placed on evaluating the evidence submitted by the manufacturer.[39]

NICE provides clear guidance on its expectations of what should be included in submissions for its health technology assessments.[40] For both the MTA and STA processes, companies must provide background on the technology, available clinical evidence, cost-effectiveness data, and data supporting the impact of adoption on the NHS. Typically, a health technology assessment from NICE can take 12–24 months.[41]

Although it is a mandatory requirement for the NHS to provide funding in England and Wales for medicines and treatments recommended by NICE, sometimes this funding is delayed due to budgetary constraints. In some cases, device manufacturers can negotiate reimbursement on a **pass-through** basis directly with the trusts.[42]

China

China's healthcare sector is growing rapidly and is expected to provide interesting opportunities for biodesign innovators. In recent years, there has been a gradual shift in the healthcare system with more autonomy being granted by the government to local hospitals and healthcare providers. While the vast majority of hospitals are administered by China's Ministry of Health (MOH), they are now expected to generate revenue to cover as much as 90 percent of their operating expenses.[43] In some cases, this can create incentives for hospitals to emphasize services they can charge for, such as dispensing prescription drugs.

China's government provides health insurance coverage to approximately 90 percent of the population.[44] However, the state insurance programs are inadequate to cover basic care and instead focus on protecting patients from catastrophic health events. As a result, the Chinese pay for most basic health services out-of-pocket. This dynamic continues to keep modern healthcare beyond the reach of many of China's citizens.

In general, only medical devices that are approved by China's Food and Drug Administration and put in the government regulated pricing formulary can qualify for reimbursement under medical insurance coverage. However, pricing and reimbursement is complicated since it varies significantly by region and at the provincial/municipal level.[45] The Ministry of Human Resources and Social Security (MOHRSS), the National Development and Reform Commission (NDRC), the National Health and Family Planning Commission (NHFPC) oversee the pricing and reimbursement of medical devices in the country. Medical devices are either directly or indirectly reimbursed by the state's Basic Medical Insurance (BMI) Fund (indirect reimbursement may apply to medical equipment used as part of a medical treatment whereas direct reimbursement covers certain medical consumables and implants). MOHRSS, NDRC, and NHFPC jointly issue the National Scope for Reimbursement of Medical Treatments under the BMI, which includes a Non-Reimbursable Catalogue and a Partially Reimbursable Catalogue. Each province may add to or reduce the items in the Partially Reimbursable Catalogue issued on the national level, to the extent that no more than 15 percent of the total items in the national catalogue are altered. The reimbursement ratio is then decided by each province in accordance with their budgets. Patients must pay for any non-reimbursable items, as well as the uncovered portion of partially reimbursed devices.[46]

Overall, the Chinese system remains in flux. The government is working to reform China's healthcare system, with a goal of making basic care available across the country by 2020.[47] However, given the size of the population, the Chinese government is critically concerned about managing healthcare costs. This results in frequent efforts by the MOH to introduce policies that place downward pressure on medical device prices. For example, since 2005, expensive capital equipment (priced more than the equivalent of $730,000) is purchased through a centralized, government-run bidding process.[48] Other cost containment programs have been the subject of experimentation at the local and regional level.

India

The delivery of healthcare in India is provided by both public and private sector entities, although the system is generally considered to be highly privatized. Private providers are responsible for roughly 90 percent of all hospitals, 85 percent of doctors, 80 percent of outpatient care, and almost 60 percent of inpatient care.[49]

In terms of healthcare expenditures, private financing, mostly in the form of out-of-pocket payments by Indian citizens, accounts for approximately 70 percent of all healthcare spending in the country.[50] Estimates regarding the portion of the country's people covered by some form of health insurance vary from less than 15 percent[51] to over 25 percent,[52] although a much smaller number have full or substantial coverage.

Both government and private insurers are working to increase access to health insurance and the industry is growing quickly.[53] According to some estimates, roughly half of the population should have health insurance coverage by 2020.[54] In India, private health insurance programs, group health (in which employers buy insurance for their employees), and government-sponsored health insurance programs are all available. However, public programs cover the majority of patients with insurance, primarily through schemes available to government employees and individuals living below the poverty line. For example, the Rashtriya Swasthya Bima Yojna (RSBY) is one central government program striving to increase health insurance access for poor families.[55] The Central Government Health Scheme (CGHS) is another example that provides health insurance to central government employees, pensioners and their dependents in CGHS covered cities. Although both of these programs face challenges, they represent a start in addressing the country's need for greater health insurance coverage.

In terms of reimbursement for medical technologies, coverage is inconsistent. As one article described, "a patchwork of different government programs sometimes reimburses for devices, but more to public hospitals than to private ones."[56] This heterogenous approach to reimbursement is viewed as less than favorable by medtech companies. For example, in 2013, CGHS reduced its reimbursement rates for drug-eluting stents by up to 60 percent.[57] It also capped prices for angioplasty procedures and bare metal stents. Although the price reductions only apply to individuals covered by CGHS (approximately three million people), hospitals may follow the government's lead in seeking to negotiate lower prices.[58] More broadly, the Ministry of Health is leading discussions regarding government regulation of prices for patented drugs and medical devices.[59]

A final note: using reimbursement analysis to screen and eliminate concepts

As the material in this chapter demonstrates, the analysis of a reimbursement pathway for a new concept can be an extremely demanding and sophisticated task. However, a few high-level factors can be considered during concept screening to help innovators eliminate solutions with killer risks and better understand the opportunities and challenges associated with the others to help populate the risk scoring matrix described in 4.6 Final Concept Selection.

Because reimbursement has such a profound impact on the success of a new innovation, it is worth the time and effort to make a best-guess estimate of the reimbursement pathway at this point in the process. The key questions to ask are outlined below.

1. **Will an established code work?** Achieving reimbursement for a new technology will almost always be faster, less complicated, and less expensive if a company can utilize existing codes. When evaluating established codes, innovators should determine if the new device and its usage is mismatched in any way to the existing descriptor for the most relevant code(s). If so, the language of the code may need to be expanded or a new code required. Importantly, minor differences in the

indications or procedure associated with the new device relative to the established code descriptor can be grounds for a payer to reject the new device for reimbursement. Even if new codes are required, the presence of applicable existing codes may serve as a "bridge" for the company until new codes, coverage, and higher reimbursement can be secured (this strategy is discussed further in chapter 5.6).

2. **What are the cost implications?** When evaluating the presence of existing codes, another essential factor to consider is if the payment level associated with the code is adequate to cover the new technology and the procedure used to deploy it. As mentioned, codes (and the payments associated with them) are typically issued to describe procedures, not just the technologies. Accordingly, the payment for an existing code has to adequately cover costs to the facility and/or the time of the physician (and other involved care providers) in performing the procedure, as well as the cost of the technology. Innovators sometimes find that an appropriate code exists to describe their new procedure, but the established payment level is insufficient to cover the technology and adequately compensate the facility and/or provider(s) for their time.

3. **When is a new code worth pursuing?** If the concept under consideration could truly represent a breakthrough in clinical care, it may be worthwhile to move forward despite the fact that a new reimbursement pathway is required. In this case, innovators need to be realistic in estimating the time and cost involved in establishing new coding, coverage, and payment, and then be prepared to present this picture to investors. Here it can be useful to look at other technologies with similar regulatory requirements and clinical impact that have made it through the system to achieve reimbursement (and, in so doing, have engaged with relevant clinical experts and medical specialty societies). Even with a precedent of this type, it is important for innovators to take into consideration the fact that the current healthcare environment (with its emphasis on economics) is less friendly to new technology adoption than it has been in past decades.

Online Resources

Visit www.ebiodesign.org/4.3 for more
content, including:

 Activities and links for "Getting Started"
- Identify payer mix
- Confirm location of procedure
- Research coverage decisions and
  ICD-10/CPT codes
- Investigate reimbursement information
  for non-covered devices
- Identify payment categories (MS-DRGs,
  APCs) and reimbursement rates
- Identify number of procedures

Videos on reimbursement basics

 An appendix summarizing
U.S. reimbursement codes

CREDITS

The editors would like to acknowledge John Hernandez of Abbott for his extensive editorial assistance in developing this chapter. Emily Kim of Medtronic also made substantial contributions and guided the organization of the revised chapter. The material in the first edition was written with input from Mitch Sugarman of Medtronic. Further appreciation goes to Jagdish Chaturvedi, Ritu Kamal, Raj Doshi, Darshan Nayak, Jan B. Pietzsch, Chris Shen, and Katherine Wang of Ropes & Gray LLC for assisting with the international sections. Many thanks to Darin Buxbaum, Trena Depel, Asha Nayak, and Eb Bright of ExploraMed for their early assistance.

NOTES

1 All quotations are from interviews conducted by the authors, unless otherwise cited. Reprinted with permission.

2 "Innovator's Guide to Navigating Medicare," Centers for Medicare and Medicaid Services, 2010, http://www.cms.gov/Medicare/Coverage/CouncilonTechInnov/downloads/InnovatorsGuide5_10_10.pdf (March 14, 2014).

3 "National Health Expenditures 2012 Highlights," Centers for Medicare and Medicaid Services, http://www.cms.gov/Research-Statistics-Data-and-Systems/Statistics-Trends-and-Reports/NationalHealthExpendData/downloads/highlights.pdf (February 25, 2014).

4 "CMS Predicts U.S. Health Care Spending Will Increase by 6.1 Percent in 2014," September 19, 2013, California Healthline, http://www.californiahealthline.org/articles/2013/9/19/cms-predicts-us-health-care-spending-will-increase-by-6point1percent-in-2014 (February 25, 2014).

5 "Top Questions About ACOs & Accountable Care," Care Facts, http://www.accountablecarefacts.org/topten/what-are-the-barriers-and-challenges-such-organizations-might-face-1 (March 5, 2014).

6 All cartoons by Josh Makower, unless otherwise noted.

7 ICD-9-CM codes were in effect for more than 30 years, during which time they gradually became outdated and inconsistent with current medical practice. Also, the structure of ICD-9 limited the number of new codes that could be created, and many ICD-9 categories were full. The compliance date for switching to ICD-10-CM/PCS codes was originally October 1, 2013, but this deadline was delayed two times (with the expected implementation date expected to be October 1, 2015 as of the time of this writing). For more information, see "ICD-10," Centers for Medicare and Medicaid Services, http://www.cms.gov/Medicare/Coding/ICD10/index.html?redirect=/icd10 (March 11, 2014).

8 "Classification of Diseases and Functioning and Disability," National Center for Health Statistics, http://www.cdc.gov/nchs/icd.htm (January 22, 2014).

9 American Academy of Physician Assistants, 2007. Reprinted with permission.

10 Eric Wagner, "Types of Managed Care Organizations," in *The Managed Health Care Handbook* (Aspen Publishers, 2001).

11 Ibid.

12 Jenny Gold, "FAQ on ACOs: Accountable Care Organizations, Explained," Kaiser Health News, August 23, 2013, http://www.kaiserhealthnews.org/stories/2011/january/13/aco-accountable-care-organization-faq.aspx (October 2, 2013).

13 Ken Krizner, "Health Plans Apply Scientific Evidence to New Technologies," Managed Healthcare Executive, July 1, 2006, http://www.managedhealthcareexecutive.com/mhe/article/articleDetail.jsp?id=357680&sk=&date=&pageID=2 (January 22, 2014).

14 Mitchell Sugarman, "Permanente Physicians Determine Use of New Technology: Kaiser Permanente's Interregional New Technologies Committee," *Permanente Journal*, Winter 2001, p. 46.

15 Michael R. Lauer, "Clinical Management for Survivors of Sudden Cardiac Death," *Permanente Journal*, Winter 2001, p. 24.

16 See "Technology Evaluation Center," Blue Cross/Blue Shield Association, http://www.bcbs.com/betterknowledge/tec/ (January 22, 2014).

17 "National Health Expenditures: 2012 Highlights," op. cit.

18 This case example was developed by the authors using information from publicly available sources.

19 David Green, Benedict Irvine, Emily Clarke, and Elliot Bidgood, "Healthcare Systems: Germany," Civitas, 2013, http://www.civitas.org.uk/nhs/download/germany.pdf (February 13, 2014).

20 "Germany – Medical Devices," International Society for Pharmacoeconomics, April 2011, http://www.ispor.org/htaroadmaps/germanymd.asp (February 18, 2014).

21 This section is based on "Germany – Medical Devices," International Society for Pharmacoeconomics, April 2011, http://www.ispor.org/htaroadmaps/germanymd.asp (February 18, 2014).

22 Ibid.

23 "Healthcare in France," National Coalition on Healthcare, http://www.nchc.org/facts/France.pdf (August 8, 2007).

24 See Liste des Produits et Prestations Remboursables at http://www.ameli.fr/professionnels-de-sante/gestionnaires-de-centres-de-sante/exercer-au-quotidien/liste-des-produits-et-prestations-lpp/liste-des-produits-et-prestations-lpp.php (February 17, 2014).

25 "France – Medical Devices," International Society for Pharmacoeconomics, October 2011, https://www.ispor.org/HTARoadMaps/FranceMD.asp (February 17, 2014).

26 Ibid.

27 Ibid.

28 Ibid.

29 "Soutien Aux Techniques Innovantes Coûteuses," Wikipedia.org, http://fr.wikipedia.org/wiki/Soutien_aux_techniques_innovantes_co%C3%BBteuses (March 18, 2014).

30 Ibid.

31 "About the NHS: How the NHS Works," NHS Choices, http://www.nhs.uk/aboutnhs/howtheNHSworks/Pages/HowtheNHSworks.aspx (July 2, 2007).

32 "United Kingdom-Diagnostics," International Society for Pharmacoeconomics, May 2010, http://www.ispor.org/htaroadmaps/ukdiagnostics.asp (February 18, 2014).

33 See www.dh.gov.uk and www.ic.nhs.uk for more information on both programs.

34 See "United Kingdom-Diagnostics," International Society for Pharmacoeconomics, May 2010, http://www.ispor.org/htaroadmaps/ukdiagnostics.asp (February 18, 2014) for more information about other programs.

35 "United Kingdom-Diagnostics," op. cit.

36 Ibid.

37 Ibid.

38 "Developing NICE Multiple Technology Appraisals (MTAs)," National Institute for Health and Care Excellence, http://www.nice.org.uk/aboutnice/howwework/devnicetech/developing_nice_multiple_technology_appraisals.jsp (March 18, 2014).

39 "Developing NICE Single Technology Appraisals (STAs)," National Institute for Health and Care Excellence, http://www.nice.org.uk/aboutnice/howwework/devnicetech/developing_nice_single_technology_appraisals.jsp (March 18, 2014).

40 See "Guide to the Single Technology Appraisal Process," National Health Service, National Institute for Health and Clinical Excellence, October 2009, http://www.nice.org.uk/media/42D/B3/STAGuideLrFinal.pdf (February 18, 2014) and "Guide to the Multiple Technology Appraisal Process," National Health Service, National Institute for Health and Clinical Excellence, October 2009, http://www.nice.org.uk/media/42D/8C/MTAGuideLRFINAL.pdf (February 18, 2014).

41 "United Kingdom-Diagnostics," op. cit.

42 Ibid.

43 "Medical Device Reimbursement in China," International Trade Administration, http://www.ita.doc.gov/td/health/medical%20reimbursement%20in%20china%202007.pdf (October 21, 2008).

44 Bradley Blackburn, "'World News' Gets Answers on China: Health Care," ABC News, November 18, 2010, http://abcnews.go.com/International/China/health-care-china-trails-developed-countries-world-news/story?id=12171915&singlePage=true (January 13, 2014).

45 "China Mainland – Medical Devices," International Society for Pharmacoeconomics, January 2011, http://www.ispor.org/htaroadmaps/chinamd.asp (February 18, 2014).

46 Information provided by Katherine Wang, Chief China Life Sciences Advisor, Ropes & Gray LLC.

47 "China's New Health Plan Targets Vulnerable," *Bulletin of the World Health Organization*, January 2010, http://www.who.int/bulletin/volumes/88/1/10-010110/en/ (January 14, 2014).

48 Ibid.

49 "Private Sector in Healthcare Delivery in India," National Commission on Macroeconomics and Health, 2005, http://www.nihfw.org/WBI/docs/PPP_SessionBriefs/PPP%20Course%20sessions/Need%20and%20Scope%20for%20PPP/Private

%20Sector%20in%20Health%20Care%20Delivery%20in%20India.pdf (February 22, 2013).

50 "Health Financing: Private Expenditures on Health as a Percentage of Total Expenditures on Health," World Health Organization, 2011, http://gamapserver.who.int/gho/interactive_charts/health_financing/atlas.html?indicator=i2&date=2011 (October 2, 2013).

51 "India Tries to Break Cycle of Healthcare Debt," *Bulletin of the World Health Organization*, July 2010, http://www.who.int/bulletin/volumes/88/7/10-020710/en/ (March 5, 2014).

52 "Government Sponsored Health Insurance in India: Are You Covered?," The World Bank, October 11, 2012, http://www.worldbank.org/en/news/2012/10/11/government-sponsored-health-insurance-in-india-are-you-covered (February 22, 2013).

53 "2013 Cover Note on Health Insurance," 15th Global Conference of Actuaries, February 17–19, 2013, http://gca.actuariesindia.org/images/docs/15thgca_cover%20note%20for%20gca-health.pdf (March 5, 2014).

54 "Indian Pharma 2020: Propelling Access and Acceptance, Realizing Potential," McKinsey and Company, 2010, http://www.mckinsey.com/~/media/mckinsey/dotcom/client_service/Pharma%20and%20Medical%20Products/PMP%20NEW/PDFs/778886_India_Pharma_2020_Propelling_Access_and_Acceptance_Realising_True_Potential.ashx (February 22, 2013).

55 Nagesh Prabhu, "Rashtriya Swasthya Bima Yojana for BPL Families Too," *The Hindu*, July 8, 2013, http://www.thehindu.com/news/national/karnataka/rashtriya-swasthya-bima-yojana-for-bpl-families-too/article4892098.ece (March 7, 2014).

56 Ames Gross, "Medical Device Market Opportunities in India," Pacific Bridge Medical, March/April 2007, http://www.sona-enterprise.com/pdffiles/medicaldevicemarketopportunitiesinindia.pdf (February 18, 2014).

57 "India Slashes Reimbursement for Drug-Eluting Stents," Pacific Bridge Medical, April 1, 2013, http://www.pacificbridgemedical.com/news/india-slashes-reimbursement-for-drug-eluting-stents/ (February 18, 2014).

58 Ibid.

59 "Patented Drugs Face Price Caps," *The Times of India*, January 27, 2014, http://timesofindia.indiatimes.com/business/india-business/Patented-drugs-face-price-caps/articleshow/29430775.cms (February 18, 2014).

4.4 Business Models

INTRODUCTION

"It seemed like such a good idea, but why did it fail?" asks the frustrated engineer. "Hospitals are capable of purchasing massive million-dollar imaging systems in this specialty, so why were they so resistant to buying disposables?" Even though the total cost per year of this innovator's idea was comparable to (or even less expensive than) the established device solution, there was a fundamental problem that he failed to overcome. The flaw was in designing and executing the business model. Sometimes an idea can be medically compelling, capable of clearing all regulatory hurdles, and manufacturable at a profit, but if the business model does not work, the business will fail. Innovators must consider the business model as one of the key factors that affects an innovation's success, and as an issue that requires as much vetting as the feasibility of the device.

A business model broadly refers to how an offering (e.g., a product or service) is defined and the way it will generate revenue and deliver value to customers. In the medtech field, common business models include disposables, reusables, implantables, and capital equipment, although many others exist. With each of these models, offerings make money in different ways and, in the process, pose different challenges to the innovator in terms of how the company organizes its resources, operations, and business processes. The business model also dictates, to some extent, how interactions with customers and other external stakeholders should be managed to achieve mutually beneficial results. Just as a company needs to assess the intellectual property (IP), regulatory, reimbursement, and technical feasibility of its ideas, it should evaluate the appropriate business model before selecting a final concept.

See ebiodesign.org for featured videos on evaluating and selecting business models.

OBJECTIVES

- Understand the different types of business models that are typically utilized in the medical device field, including their relative advantages and disadvantages.

- Determine how to choose an appropriate business model based on the unique characteristics of the innovation and its customers.

- Appreciate how to use business model risks as a screen for prioritizing concepts.

BUSINESS MODEL FUNDAMENTALS

The business model is often the forgotten axis of innovation, arrived at by default after all other aspects of the product or service are determined. If an innovator recognizes that considering alternative business models is yet another design variable to be utilized in achieving the customer's **needs**, the innovation will have its greatest chance of success.

Regardless of where in the world an innovator is working, the defining characteristics of any business model are:

1. The innovation, which could be a product, service, or blend.
2. The customer.
3. The primary interface between the two, or the way they interact.

The dominant factors that affect interactions between the customer and the innovation are shown in Table 4.4.1.

When innovators choose a business model, they must take these factors into account. The idea is to design an offering that plays to the strengths and capabilities of the company while providing the customer with a desirable interaction.

Choosing a business model

During the **concept** screening stage of the biodesign innovation process, selecting a business model starts with an evaluation of the match between the unique characteristics of the innovation and the defining aspects of the different business models. The revenue stream and manner in which the innovation will get into the customers' hands are two other primary considerations. The information necessary to more fully develop the business model will be extracted from the other steps in the process as the innovation continues to progress.

Table 4.4.1 Each of these factors can change dramatically based on the business model chosen for an innovation. Understanding the impact of the business model choice and its influence on these factors can allow an innovator to determine which business model is the most favorable and compatible with success.

Factor	Explanation
Revenue stream	How revenue is generated and its frequency.
Price	How much the business can charge for its products or services.
Margin structure	The profit to the company from sales (and its adequacy to support the inherent characteristics of the chosen business model).
Sales investment	The required mechanism for getting the innovation into customers' hands.
Customer training requirements	The extent to which specialized training is required to utilize the innovation.
Competitive differentiation	The degree to which the innovation is unique.
IP	The importance of IP protection to the success of the business model.
Other barriers to entry	Factors that could serve as barriers to adoption (e.g., high **switching cost**, brand or customer loyalty, access to distribution channels, etc.).
Clinical/regulatory hurdles	The complexity and duration of clinical requirements (e.g., trials) and the necessary regulatory pathway before commercialization can begin.
Reimbursement	The way physicians, surgical centers, and hospitals are paid.
Financial requirements	The level of investment necessary to develop and commercialize the innovation.
Culture/geography	The extent to which customer needs related to the same clinical area differ across geographic boundaries or different cultural environments.

An appropriate business model must allow the company to extract **value** for its innovation in a way that makes sense to the customer. For example, if a new company tries to market an MRI machine by charging for the equipment and then requiring hospitals to buy a disposable platform for each patient tested, it might have difficulty generating interest.[1] Even if the company offers to service the machine for free (creating an ongoing revenue stream through the sale of the disposables rather than through a service contract, as is typical in this field), it is likely that the buyer would object. Most customers understand the value of paying for a service contract to keep their expensive, high-utilization equipment in good repair. However, it is much harder to convince them to pay for something, such as a disposable in this case, for which they do not understand the need or appreciate the value – especially if the MRI machines from competitors do not have the same associated charge.

Acclarent, a company that manufactures and markets endoscopic, catheter-based tools to perform what is known as *Balloon Sinuplasty,*® a procedure in which a balloon is used to dilate various areas of the sinuses, provides a real-world example. In the ear, nose, and throat (ENT) specialty, companies have traditionally pursued models that are dependent on selling high volumes of inexpensive disposable materials and smaller numbers of moderately priced reusable products. Acclarent sought to change the field's dominant business model by selling higher-value disposables that cost more per patient but lead to improved results. Such a change is difficult, but not impossible if the innovation's performance is substantially better against one or more metrics that are important to the customer. As one board member described (prior to the acquisition of Acclarent by Johnson & Johnson in 2010):[2]

Whenever you try to extract more value from an established market, you have to be sure you can deliver more value through your product or service. The reason Acclarent has been successful in moving the market is that the products provide improved clinical performance and save money in the overall treatment of the disease.

Of course, anytime a company seeks to change a dominant business model, it automatically takes on a market development challenge. Incremental investment (and often a significant commitment of time) is required to adjust customer expectations, modify their perceptions, and/or alter their behavior. However, the possibility of "changing the game" in an industry or field can also be the deciding factor that makes an opportunity interesting. Hypothetically, selling Acclarent's technology as a low-margin disposable might not have made a compelling business proposition. However, through the lens of a different business model (higher margins, lower volume), the opportunity became viable.

Importantly, many diverse business models are relevant within the medical device field. Investors have historically favored models that generate an ongoing revenue stream or an annuity (e.g., disposables, implantables, or capital equipment with an associated service contract), as opposed to pure capital equipment businesses. However, numerous **medtech** innovators have proven that different approaches can be successful. In some cases, making a creative twist to a traditional business model has been the factor that permitted a solution to succeed. For example, innovators have devised approaches where a technology is provided in advance at little or no cost and the **user** is only charged per use (regardless of the consumables required). Or, a device is offered in a "resposable" format that allows a certain number of reuse cycles before it is thrown away. Because some individuals and companies have shied away from novel business models in the past (seeking to attract investor interest more easily), innovative approaches to medical device businesses have inevitably been overlooked. For this reason, as technologies continue to advance, innovators and investors alike should anticipate the emergence of an increasing number of nontraditional medtech business models in the coming years.

Another trend is to play with the factors in Table 4.4.1 that affect interactions between the customer and the innovation in order to make traditional business models work under increasingly difficult circumstances. In fact, this movement is being led by innovators seeking to bring important medical technologies to underserved populations in low-resource settings. Again, the

FIGURE 4.4.1

The first baby treated using D-Rev's Brilliance device in Ogbomoso, Nigeria (courtesy of D-Rev).

fundamental business models are the same in these environments, but the ways in which they are implemented may be different. D-Rev, a non-profit product development company whose mission is to improve the health and incomes of people living on less than $4 per day, is one company experimenting with creative approaches to addressing these factors when defining business models for its products. With the Brilliance device, a reusable treatment for infant jaundice (see Figure 4.4.1), D-Rev used grants and other forms of philanthropic funding to underwrite R&D and operational costs to take pressure off the amount of revenue it would need to generate from product sales. This, in turn, contributed to the organization's ability to price the product affordably for healthcare providers in markets such as India (Brilliance would sell for roughly $400 per unit versus $3,000 for comparable products). To commercialize the device, D-Rev set up a manufacturing and sales partnership with Phoenix Medical Systems, a for-profit medical equipment distributor with well-established channels across the country, so it would have a reliable and sustainable way to get the product to the providers who needed it most. Phoenix licenses the technology from D-Rev in exchange for licensing fees and

royalties. The D-Rev team got especially inventive in thinking about the margin structure that would underpin this agreement. To motivate Phoenix sales representatives to devote time and energy to selling Brilliance, the team knew it needed to offer fair but competitive rates. But to entice Phoenix to target providers in India's public and district hospitals who need the technology the most, rather than primarily higher-end, urban facilities that are easier to reach and better resourced for making equipment acquisitions, D-Rev believed it needed a special incentive. Ultimately, it devised an approach that would make Brilliance available to customers across all locations and market segments at the same price, but D-Rev would take a lower royalty on sales to public and district hospitals. By making interactions with this audience more lucrative, D-Rev hoped that Phoenix would be more focused on reaching these target providers.[3] While the approach appeared promising, only time would tell if it would yield D-Rev's desired results.

Types of business models

As noted, there are multiple types of business models for medical device innovators to evaluate and optimize to their unique innovations. When assessing each one, innovators should be looking for a good fit between the innovation and the business model such that a meaningful, growth-oriented, and profitable business can be envisioned. The 10 business models included in Figure 4.4.2 and described below in more detail are among the most commonly employed medtech models. While certain ones may be more common in different geographies, the fundamentals of the models are the same worldwide. Clearly, a model can only be considered to be functional in any environment once it has proven to sustainably permit the organization to grow profitably.

Disposable products

Disposable products are those goods that are used and then discarded without being reused. Low-cost disposables include items such as paper examination gowns and stopcocks (used with intravenous tubing), both of which might costs pennies per unit. A surgical stapler is an example of a more expensive disposable, which might cost $100 or more for a single per-patient use.

Business model	Margin structure	Sales investment	Importance of IP	Barriers to entry	Customer training	Clinical/regulatory hurdles	Financial requirements
Disposable – High Cost	● High	● High	● High	● High	● High	◐ Neutral	◐ Neutral
Disposable – Low Cost	○ Low	○ Low	○ Low	○ Low	○ Low	○ Low	◐ Neutral
Reusable – Pure	○ Low	○ Low	○ Low	○ Low	○ Low	◐ Neutral	○ Low
Implantable – Mid to High Cost	● High	● High	● High	● High	● High	● High	● High
Capital Equipment – Pure or Combined	● High	◐ Neutral	● High	● High	○ Low	◐ Neutral	● High
Service – Pure or Attached to Product	○ Low	○ Low	○ Low	○ Low	◐ Neutral	○ Low	◐ Neutral
Fee per Use – Pure or Combined	◐ Neutral	● High	● High	◐ Neutral	● High	● High	◐ Neutral
Subscription	○ Low	○ Low	○ Low	○ Low	○ Low	○ Low	○ Low
Over the Counter – Pure or Combined	◐ Neutral	● High	◐ Neutral	◐ Neutral	○ Low	◐ Neutral	◐ Neutral
Prescription – Pure or Combined	● High	● High	● High	● High	● High	● High	● High
Physician-Sell – Pure or Combined	◐ Neutral	● High	◐ Neutral	◐ Neutral	◐ Neutral	◐ Neutral	◐ Neutral

FIGURE 4.4.2

Every medtech business model has an expected set of opportunities and challenges that dramatically impacts the plan for the business.

Disposables can also be attached to major medical equipment, such as ablation catheters used in combination with generators that produce the energy for ablation. Additionally, they can be coupled with reusable devices (used several times before requiring replacement), such as disposable razor blades for reusable surgical shavers.

Whether they are attached to medical equipment or reusables, low-cost disposables:

- Require high sales volumes (to compensate for low margins).
- Must be easy to use.
- Should be marketable through low-cost distribution channels (e.g., medical equipment catalogs).

Higher-cost disposables typically require specialized training to use and, as a result, demand significantly higher margins in order to support a technical sales force and ensure a reasonable level of IP coverage. To justify a higher-cost disposable, the innovator must achieve competitive differentiation (e.g., enabling superior clinical results or establishing key barriers to entry to the competition). Gross margins are usually favorable for most high-cost disposables and can be in the 70–80 percent range (or better). If pricing for a disposable is pegged in such a way that it is unrealistic to achieve margins close to this range, it may be an indicator that this is the wrong business model for a given technology.

In terms of their advantages, disposable products generate a regular revenue stream since customers must acquire them on an ongoing basis. In addition, as the volume of procedures increases (a goal of most healthcare providers), so too does the volume of the disposables used. Often, their value is directly correlated with a specific event, so it is easy for customers to understand – for example, every time a provider draws blood, a disposable syringe is required. Finally, based on their relatively low cost and rapid turnover, there is little risk to the buyer in trying a disposable. This can make it much easier to convince decision makers, such as purchasing managers or physicians, to place an initial order.

On the downside, there are ethical issues to consider that are associated with the environmental consequences of disposable medical devices. Innovators must also be aware that, in some cases, disposable products can easily be displaced by reusable products that meet the same need, if the **value proposition** is compelling.

The Concentric Medical example below highlights some of the issues relevant to a disposables business model.

| FROM THE FIELD | CONCENTRIC MEDICAL |

Pioneering devices in the stroke market using a disposable model

According to Gary Curtis, former CEO of Concentric Medical, stroke affects more than 700,000 people in the US each year. Approximately 85 percent of those patients experience an ischemic stroke (caused by blockage in an artery supplying blood to the brain). Yet, despite the large number of people suffering from ischemic stroke each year, "The **standard of care** for 95 percent of these patients is aspirin and a dark room," said Curtis. Intravenous recombinant tissue plasminogen activator (t-PA), a medication used to dissolve the clots that cause blockages, can be used in some cases of ischemic stroke, but only if it can be administered within three hours from the onset of symptoms. For everyone else, Curtis continued, "they have to wait passively to see how the stroke resolves so that then, a day or two later, a neurologist can consult with the patient's family and tell them their Aunt Martha is lucky enough to go home. For Uncle Harry, on the other hand, his stroke didn't resolve itself and he's going to a nursing home. Dad is kind of in between. He'll need some rehab and then maybe he'll get some of his normal function back."

Concentric Medical is seeking to change that model. "We are trying to change passive to active," explained Curtis when he was at the company's helm. Using the Concentric's Merci Retrieval System™ (see Figure 4.4.3) to restore blood flow in ischemic stroke patients by

removing blood clots in the neurovasculature, "We can now intervene within a reasonable time period. Our trials

FIGURE 4.4.3
The Merci Retrieval System (courtesy of Concentric Medical).

showed that if we intervene and reopen an artery within zero to eight hours, less damage is done. We can change the course of events. It's the first time that's been done," he noted.

When Curtis became Concentric's CEO, the company was working on multiple projects but was having difficulty getting funding. Recognizing the stroke market as the largest unmet need with the most compelling technology in development, he quickly cut all other products in the development pipeline, gained funding for the Merci Retrieval System by promoting the company's improved sense of focus, and led Concentric to the first **FDA** approval of a device to remove thrombus in ischemic stroke patients.

In terms of a business model that would best support Concentric's product, the company's choices were somewhat limited due to the sterilization issues associated with products that come into contact with the bloodstream. "It has to be guaranteed sterile when you use it," said Curtis, which led Concentric to a single-use (disposable) model. "We never even contemplated another model. The device also has to have mechanical performance expectations that are exactly the same every time. Device performance can be compromised as you're cleaning, sterilizing, or repackaging it. If we offered a reusable, all those activities would lessen the ability of the physician to predict how the product is going to perform."

According to Curtis, a business model built around a single-use, disposable device offers many advantages. "Clearly, the recurring revenue is a benefit. Once you get surgeons trained in how to perform the procedure safely, they're going to order the device again and again. We had spent roughly $35 million by the time we completed our first **clinical trial** to get the product approved before we got our first dollar of revenue. We're now trying to recoup that cost. Knowing that there would be this predictable revenue stream is one way we got the venture capital community to invest."

On the downside, he noted that single-use devices are costly to the healthcare system: "A hospital will pay $5,000 to us for every patient treated with our device. If we had a reusable product, they wouldn't have to pay that much." However, he commented, the costs associated with the disposable device are more than offset by the extreme long-term expense of caring for stroke patients "who survive, but survive poorly." That said, competitors are working on reusable technology to address the same need that is targeted by Concentric's device. For example, earlier in Concentric's development, one company was working on a device to break up the clots non-surgically by using focused ultrasound. "People are trying to develop reusable solutions to make treatment less costly," said Curtis, "but no one has proven that you can."

When asked if he worried about the environmental impact of disposable products, Curtis responded quickly: "I don't think two seconds about it. Not when I'm saving a life."

Importantly, Curtis confirmed that there is a difference between the business model used to support high-cost, single-use devices versus low-cost, high-volume disposables. With low-cost disposables, there is a constant, never-ending pressure from investors and customers regarding ways to reduce the cost of the product. "It's all about how you change your manufacturing costs from 5 cents to 4 cents, so you can make that extra margin," he said. "That's not our business model." Instead, Concentric remained focused on what Curtis called a value creation model. "All the pressure I [had was] how to make it more effective. How do I change from 50 percent to 60 percent to 70 percent to 80 percent success rate in restoring flow in these patients?" As a result of these efforts, and the efforts of Curtis' successor, Maria Sainz, Concentric was acquired by Stryker in late 2011.

Reusable products

Reusables are multi-use products with a moderate lifespan, but their cost is orders of magnitude smaller than a capital equipment item. Scalpel holders, laparoscopic graspers, and endoscopes are examples of products that can be reused for a period of time before they eventually wear out and need to be replaced. Reusables can be attached to a disposable (e.g., the surgical shaver mentioned earlier that is used with disposable razor blades) or a service (e.g., servicing of flexible endoscopes).

Business models built solely around a reusable product tend to have no sources of recurring revenue other than replacement. As a result, they generally cannot support a specialized sales force and are commonly sold through medical catalogs and/or through major distribution companies. Although the margins for reusables can actually be quite good – also in the 70–80 percent range – the lack of a sustained flow of cash, as compared to disposables, usually makes the size of the associated business opportunity smaller. Reusable products also carry with them a higher level of risk since customers use them for an extended period. The product is not "factory fresh" each time it is used (as is the case with disposables), so the user faces a greater chance of failure. For example, if surgical scissors become dull or are damaged during a procedure, they may not perform effectively the next time they are used. Many providers of reusables attempt to address this concern by providing service for their products. However, because the products have a finite lifespan (which is much shorter than the lifespan for capital equipment – see below), customers are typically unwilling to pay a great deal for maintenance. This contributes to the factors that make reusables a difficult business model to profitably sustain and grow.

Some entrepreneurs have attempted to force a disposable business model onto a technology that is clearly reusable (and there is no good case for disposability). However, this can be a risky move. If the reusability of an item is discovered, as it usually is, the business model can quickly change from disposable to reusable, outside the innovator's control. This, in turn, can have a negative impact on the business and significantly affect perceptions and expectation of the company. On the other hand, in some cases it may be appropriate for products that were originally designed to be reusable to become disposables. Syringes provide one compelling example. Despite relatively widespread awareness that syringes reuse contributes to the transmission of blood-borne pathogens, including hepatitis B virus, hepatitis C virus, and human immunodeficiency virus (HIV), up to 40 percent of worldwide injections continue to be given with syringes and needles reused without sterilization. This problem is particularly severe in low-resource settings where reuse rates can be as high as 70 percent.[4] The development and dissemination of single-use syringes is one way that manufacturers are helping to address the issue, and it represents an appropriate shift from a reusable to a disposable model.

In terms of advantages, customers intuitively respond to reusable products and often favor them over disposables when the business case makes sense. For example, one would not consider anything other than a reusable weight scale or reusable stethoscope in a doctor's office – the cost/benefit ratio of these devices is essentially unbeatable. Once reference devices such as these are introduced into a marketplace, a new technology which seeks to challenge the business model, even at a modest cost increase, is often highly scrutinized and has a greater risk of failure. This should not deter one from investigating improvements which deliver dramatically better outcomes or a significantly lower cost under these types of circumstances, but the innovator needs to realize that the bar will be set relatively high.

Implantable products

Implantable products are typically mid ($1,000–$5,000) to high (> $5,000) cost items. An example of a mid-cost implantable device is a coronary artery stent. Examples of high-cost implantable devices are pacemakers and artificial joints. Implantable products can be pure, or they can be associated with a service, such as pacemaker follow-up service. Implantable products tend to have high margins – in the 80 percent range or better – and these prices have traditionally been supported due to the high barriers to entry associated with the technology, such as the regulatory pathway, **reimbursement**, and IP requirements. However, a shift is underway in the segment. As these high-margin devices slowly go off

patent, new opportunities emerge for innovators to create generic versions. This causes the implants to become less differentiated and their prices to decline. Companies looking to preserve their margins must, in turn, focus on addressing new needs or creating different offerings that clearly deliver value to their intended users.

Implantable products require the highest level of clinical validation and, as a result, can present significant clinical hurdles to their developers. Because a major investment will be required to support comprehensive, long-term clinical trials, the market for implantable devices must be significant so that investors can recoup relatively sizable returns over a long-term payback cycle. Being able to ensure ongoing IP coverage is another essential aspect to the business model for implantable products.

One benefit of a business model focused on implantable devices is that there is a direct pairing of the value proposition and the procedure – every patient that receives the procedure gets one or more implants. As a result, implantables have an ongoing revenue stream whose growth is linked directly to the increase in the number of procedures performed each year. Additionally, since certain devices eventually wear out due to continuous use (e.g., heart valves) or battery consumption (e.g., pacemakers), patients may need replacement devices and, thus, provide a source of recurring revenue (although this may take many years to capture).

From a risk perspective, however, implantables represent a recurring liability to the company that manufactures and markets them. The challenge to the company is to design an implant that can be replaced or otherwise taken out of service before it malfunctions. Many implants serve a primary function, which is useful for a period of time and then yields to other physiologic processes which persist even after the implant is no longer functional (i.e., drug-eluting stents or resorbable drug delivery systems). However, even in these cases, the residual impact of the presence of the implant may present some risks that need to be managed and accounted for by the company. In addition, implantable products often require a **direct sales** force at the point of care to stay in touch, answer questions, and provide follow-up – a requirement that can be costly for a company to maintain. Although there may be some limited protection offered by legal provisions such as "preemption," under which devices approved in the US via the premarket approval (**PMA**) pathway are exempt from common law claims that impose requirements different from or in addition to FDA's requirements,[5] companies offering implantable products will continue to face liability risks.

FROM THE FIELD ST. FRANCIS MEDICAL TECHNOLOGIES

The challenges of being a pathfinder with an implantable device

St. Francis Medical Technologies focused on the discovery, development, and commercialization of novel treatments for degenerative spinal disorders until the time of its acquisition by Kyphon in January 2007. The company's first product, the X-STOP interspinous process decompression system, was developed to alleviate the symptoms of lumbar spinal stenosis (LSS). LSS is a common spinal problem that primarily affects middle-aged and elderly adults, causing significant pain in the back and legs.

Traditionally, the most common surgical solution to LSS was laminectomy. In this procedure, the surgeon trims and removes part of the bone of the vertebrae to reduce the pressure on the spinal nerve root (which causes the pain and other debilitating effects associated with stenosis). According to Kevin Sidow, former president and CEO of St. Francis, the idea for the solution that eventually became the X-STOP®

FIGURE 4.4.4
The X-STOP implant (courtesy of Medtronic Spinal & Biologics).

emerged when two orthopedic spine surgeons, Jim Zucherman and Ken Hsu, were pursuing the need for *a less invasive way to treat the symptoms of spinal stenosis*. "They had a couple of older patients who had experienced short episodes of dementia as a result of the general anesthesia," explained Sidow. Based on this undesirable side effect of the procedure, "They were hoping to find a way to treat patients under a local anesthetic," he added.

All of the solutions that Zucherman and Hsu conceptualized to address the need involved the use of an implant (see Figure 4.4.4). As a result, when the experienced spinal device executives who co-founded St. Francis began thinking about a business model based on Zucherman and Hsu's leading concepts, they knew that they would be dealing with this type. Highlighting the benefits of the model, Sidow said, "The upside is that implants have a very straightforward revenue recognition process. There are also strong distribution networks that you can easily tap into in order

to market the product, that have great credibility with spine surgeons."

On the other hand, he acknowledged, there are some sizable risks associated with implants that need to be considered early in the biodesign innovation process. Among those he mentioned, "The first is the regulatory hurdle, especially with a brand-new therapeutic option. This is really hard because everyone is incented to say 'no' to new things or extend the pivotal trial timelines," Sidow noted. When he joined St. Francis from Johnson & Johnson, the company expected to receive FDA approval shortly thereafter. In fact, just two months later, St. Francis was notified by the FDA that its PMA application for the X-STOP implant had been turned down. Under Sidow's leadership, the company completely regrouped its US business to address the issues raised by the FDA. Reflecting on the situation, Sidow explained, "It really was a function of a lack of understanding. People at the FDA have an incentive to turn things down if they don't understand them perfectly. Nobody at the FDA gets rewarded for getting products to market quicker to help patients, but people get punished for Vioxx®-like results." When he elevated the matter to a higher level within the agency, "it was a much more objective, straightforward process," and the X-STOP was approved by the FDA.

Sidow emphasized another risk: that new implantable technologies have to anticipate issues associated with getting physicians in the target population to adopt the product. "Laminectomy was a big operation that was well reimbursed," he said. "The surgeons were very skeptical of this little company coming out with this new device. To make matters worse, all the big players in the business – J&J, Medtronic, Stryker, etc. – have very close relationships and tremendous credibility with these doctors. And they each had hundreds of sales reps telling the physicians that the device was a gimmick and that we were a nobody." Extensive efforts were required of St. Francis to overcome this skepticism and resistance in the field. When a new implant hits the market, "the large, incumbent companies have a strong incentive to battle it," Sidow reiterated.

In addition to anticipating regulatory and adoption hurdles, Sidow advised companies with new implants to think carefully about reimbursement before getting too far in the biodesign innovation process. "It is becoming more and more critical to be proactive regarding reimbursement, all the way back to the point that you're designing your clinical studies," he said.

Ultimately, the device that St. Francis brought to market could be surgically implanted via a less invasive procedure performed in under an hour. According to Sidow, in the company's first year following FDA approval, it had $58 million in worldwide sales. At the time of the Kyphon acquisition, the X-STOP device had been implanted in more than 10,000 patients. St. Francis sold the business to Kyphon for $725 million.[6]

Capital equipment products

Capital equipment products are, in essence, another form of reusable products. They require customers to make a capital expenditure in order to obtain a technology that they will use to repeatedly produce/sell a product or provide a service over an extended period of time (i.e., more than one year). Capital expenditures come from funds used by companies or entities, such as hospitals, to acquire or upgrade physical assets to maintain or increase the scope or competitiveness of their operation.[7] In the medical field, capital expenditures are often made to obtain equipment such as MRI or CT scanners,[8] blood analytics equipment, or ultrasound machines. Capital expenditures can also be made for important software programs such as electronic health records systems, as well as facility-related expenses.

A pure capital equipment business model depends on the sale of the equipment, with little or no ongoing interaction between the company and the customer until it is time to purchase new equipment. However, as noted earlier, these products can also be associated with a service (e.g., technical support and maintenance contracts) or disposables. In these scenarios, it is not uncommon for the equipment to be sold at or near cost, with the expectation that greater, recurring revenue will come from the sale of the service and/or disposables.

Capital equipment purchasing decisions usually occur at the administrative level in the healthcare system. For example, while physicians (as the users of the products) may be the targets for the sale of many disposable, reusable, and implantable technologies, hospital purchasing committees are almost always involved in buying decisions for capital equipment. These transactions still require the buy-in and support from doctors in one or more specialties. However, the decision to invest the funds will be made by committee, based on the broader priorities of the facility. As a result, the sales cycle can be long (as much as 18 months) and may require the vendor to support the sale with a careful business case for the purchasing decision (a plan for how the purchaser will recoup the investment in the capital equipment).

An advantage of capital equipment businesses is that sale of the technology usually represents a long-term commitment by the purchaser. The switching costs of moving from one MRI provider to another, for example, are extremely high. As a result, unless there are significant problems with the equipment, customers tend to be loyal to the company once the buying decision has been made.

The example of Gradian Health Systems demonstrates some of the dynamics associated with capital equipment models, as well as creative ways these products can be made more accessible in resource-constrained environments.

FROM THE FIELD ▷ GRADIAN HEALTH SYSTEMS

Blending commercial and philanthropic approaches to make capital equipment accessible in low-resource settings

Globally, 2 to 3 billion people lack access to adequate surgical care.[9] This problem is particularly prevalent in developing countries, where inadequate infrastructure often renders hospitals unable to provide safe surgery. Dr. Paul Fenton, a British anesthesiologist, became frustrated with this situation over his 15 years as Head of the Department of Anesthesia at a busy teaching hospital in Malawi. After observing too many unnecessary deaths caused by surgeries that were interrupted or canceled due to the unreliability of his hospital's anesthesia equipment and infrastructure, Fenton designed a machine that could deliver safe, reliable anesthesia even in the midst of a power outage. The device, which he called the Universal Anaesthesia Machine (UAM), also generated its own oxygen from an integrated oxygen concentrator, which eliminated reliance on expensive cylinder or pipeline gas. If no electricity or other source of oxygen was available, the UAM continued operating, defaulting to room air with integrated oxygen monitoring to ensure the safety of the gas mixture (see Figure 4.4.5).[10]

Fenton began using the **prototype** in his hospital in Malawi. When he saw how well the device performed in this environment, he sought to expand production so that other health facilities could benefit from the device. Unfortunately, Fenton was unable to convince investors to provide funding to further develop the technology. He was also unsuccessful in identifying a buyer or licensee to bring the idea forward. At the time, those he spoke with were either hesitant to get involved in a business based in Africa or they did not perceive capital equipment targeted at low-resource healthcare providers to be commercially viable.

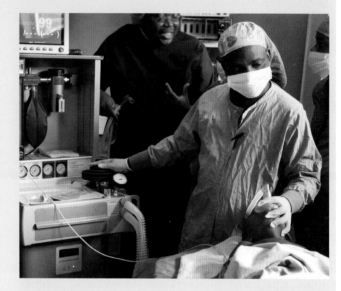

FIGURE 4.4.5

A healthcare provided in Malawi delivers anesthesia using Gradian's UAM (courtesy of Gradian Health Systems).

It was not until years later, after he had retired, that Fenton had the opportunity to bring the idea forward. After meeting with Fenton about a program in Nepal, a private philanthropy called the Nick Simons Foundation (NSF) offered to provide seed funding to develop the UAM into a viable model for use in other hospitals. They identified a manufacturer in the UK who took it through the CE certification process, and then NSF tested it in hospitals the UK and Nepal. Crucial to the UAM's success was its effectiveness in all environments – from resource-constrained health facilities to highly sophisticated medical centers with the highest performance standards. "The UAM was designed very distinctly to not be a 'poor person's machine,'" emphasized Erica Frenkel, who later became involved in the project.

Based on positive results from all testing sites, the Nick Simons Foundation was enthusiastic to bring the device to market. It spun out a wholly owned subsidiary from the

Foundation – called Gradian Health Systems – to dedicate itself to this goal. Foundation leaders established a legal structure for the new organization that would afford them flexibility and protection from the liability risks associated with any medical device company by setting up Gradian as a single member limited liability corporation (LLC).[11] From a tax perspective, the organization was considered a "disregarded entity," which meant the IRS would "disregard" the company as being separate from its owner and roll up its financials into the Foundation's financials so that Gradian would be also tax-exempt.

Gradian's first big challenge was designing a business model to support the UAM. Steve Rudy, a medical device industry veteran, and Erica Frenkel, a global health expert, were hired to run the organization. Together Rudy and Frenkel considered numerous hybrid models for structuring Gradian's approach.

Anesthesia machines are typically priced upwards of $50,000. The target price for the Gradian UAM was significantly less, about $15,000 per unit,[12] but it still qualified as capital equipment and represented a major expenditure for the low-resourced hospitals and surgical centers that comprised the target market. Gradian had to define a blended model for addressing the upfront sale of the UAM, plus the ongoing training, service, and maintenance associated with the device. "Previous thinking suggested that to be a non-profit working in this space you had to just donate machines and all the peripheral services they require – that there was no way to use pricing and sales to build an organization. We felt strongly that a non-profit could incorporate traditional business strategies," Frenkel explained.

Ultimately Rudy and Frenkel determined that Gradian would cover the production and distribution of the UAM by selling it to non-profit organizations and governments at roughly its cost. This would ensure that the company could achieve scale as the team started building demand for the product. "If you're giving away the machines, you can only produce as many as your budget allows. But if you're selling the product and covering your costs,

you can produce and distribute as many as the market demands," Frenkel noted. Funding from the Foundation would then be used to support other aspects of getting the business off the ground during its first few years in operation, as well as to underwrite the expensive but essential work of training anesthesia providers to use the machine and covering the technology's ongoing service and maintenance requirements for customers. Gradian saw this as an imperative differentiator for its offering since donated machines rarely included training or support, and companies selling new machines were notorious for offering only the most basic support, without in-country support representatives, at prices that were largely unaffordable.

With the fundamental business model defined, Gradian still had to address issues related to revenue realization and the sales cycle. As with most capital equipment models, the sales cycle for anesthesia equipment is lengthy. Further, Gradian realized that the timing of anesthesia machine sales is important. "A hospital may buy one anesthesia machine this year and not purchase another for three years or more." Accordingly, the company's revenue for sustaining operations would be uneven and a missed sales cycle could mean a lost opportunity for multiple years.

In addition, Gradian's small team had to market and sell the UAM to four distinct but interconnected **stakeholder** groups: users, hospitals, Ministries of Health, and donor organizations. The direct users of the product are the anesthesia providers who, in the developing world, are often not medical doctors. Instead, they are mid-level healthcare practitioners with specialized (although sometimes minimal) training in basic anesthesia delivery. These providers personally experience the challenges created by the unreliable and unsafe equipment used in low-resource operating rooms. They are perhaps the easiest to convince of the UAM's benefits but, as Frenkel pointed out, "They're not the ones with the resources or decision-making power." Those with the ability to make purchasing decisions resided at the hospital administration or Ministry level. However, a significant percentage of facilities and Ministries in the developing world are severely resource constrained, with operating budgets that cannot accommodate regular capital investments. To acquire new equipment, such as anesthesia machines, many hospitals depend on non-governmental organizations (**NGOs**) or international donor organizations to make these purchases. Decision makers within these organizations are furthest from the problem and sometimes lack adequate information about the field of anesthesia or the needs and constraints of the facilities they are intended to serve. "The users and hospitals that know what they need are often not even involved in the decision-making process for the equipment that they're going to get," Frenkel said.

To operationalize its capital equipment business model, Gradian determined that it needed a coordinated and highly strategic marketing plan that would align the interests of key stakeholders in the purchasing process. Ideally, anesthesia providers would report to hospitals what they required; and hospitals, in turn, would pass this information along to their governments and donors who would use it to make their purchasing decisions. To move the market toward this optimal model, the Gradian team would pursue a multi-part plan that included: (1) publishing meaningful results in **peer-reviewed journals**; (2) building a network of key opinion leaders; (3) connecting with other users through conferences and professional societies; (4) applying for large-scale tenders; and (5) researching donor organizations.

As a top priority, Gradian would continue to conduct field studies of the UAM in collaboration with well-known, respected partners to produce high-quality clinical data demonstrating the efficacy of the device. In parallel, the company would actively seek to expand its network of "champions" – respected users and hospital personnel who had direct experience with the value that the UAM delivered – in Gradian's target geographies. Their advocacy was extremely valuable, particularly for a large capital equipment purchase like the UAM. They could be engaged to answer questions about the UAM and share their experiences with other users and hospitals, either on their own or at Gradian's request. As Frenkel put it, "It's very influential and boosts our credibility in the sales

process when we're not the ones saying that it's a great machine." Gradian would also seek to build a broader base of advocates in the user community promoting the UAM through medical conferences and professional society activities.

Because capital equipment purchases by governments in African countries are often conducted through tender processes, Gradian would also invest significant time and energy into better understanding the needs of the governments and financing institutions issuing the tenders so it could optimally present the UAM in its proposals. "Part of it is understanding how the decision-making processes work for these major organizations," Frenkel said. Another key aspect was raising awareness among these entities that affordable, appropriate technologies like the UAM even existed. To assist, Gradian hired an outside consultant with deep global health experience in Africa "to help us learn how to really speak to these types of organizations," she noted. In terms of reaching high-volume purchasers outside the tender process, Gradian planned to develop a list of the wide variety of organizations that made capital equipment purchasing decisions for individual hospitals. For example, "In Malawi, we targeted a number of organizations that we had worked with, and in Uganda we're starting to get a sense of organizations that train users but also fund equipment," Frenkel said. The Gradian team gathered information about these organizations and also started tracking what it could about their purchasing cycles so that a team member could approach them at an appropriate time. "The idea is to get ahead of the organizations, before they make a decision about anesthesia equipment, so we can make them aware of the UAM, begin a dialog, and answer their questions," she explained.

Reflecting on Gradian's strategy to make capital equipment more accessible in low-resource settings, Frenkel noted that, "It's like a huge knot we're trying to untangle. I wouldn't say by any stretch of the imagination that we've solved it, but we're working on it." The early phases of this plan had enabled Gradian to expand the sites where the UAM was being used from two to over 15 countries over its first two years in operation.

The Gradian story highlights how innovators are devising creative approaches to making new technologies more accessible in low-resource settings – a movement that extends beyond just capital equipment. However, innovators should note that they will face additional risks regarding the long-term sustainability of their organizations if the selling price of their products and/or services does not completely cover the cost of operating the business.

Service

A service is work performed by one person or group for the benefit of another. A long-term care facility is an example of a *pure* service model. Dialysis service centers commonly provide another example of a pure service model. However, in some cases, the companies that provide these services also play a role in providing capital equipment and disposables. As a result, organizations in this market may have a blended product/service business model. A service plan to maintain and support an MRI machine after it has been installed represents one example of a blended product/service model.

A notable characteristic of blended product/service models is the fact that companies may frequently sell a product at or near cost (with very low margins) in order to generate service revenue. It is not uncommon for companies such as General Electric, Philips, and Siemens to maintain relatively slim margins on their capital equipment in order to stimulate the sale of more service contracts, which typically have high margins and provide a recurring revenue stream. Once in place, these contracts tend to be extremely stable given the high switching costs mentioned earlier. In some ways, service contracts on equipment can be thought of as insurance policies. Customers buy them to ensure that their equipment will be properly maintained and quickly repaired in the event of a problem. However, both the customer and the company benefit if little or no service (beyond regular maintenance) is actually required because this means that the equipment experienced no unintended downtime, which can be costly and a source of risk for the customer.

In contrast to their blended counterparts, pure service models can be challenging. They are highly dependent on having the right management capabilities, organization tools, resources, and staff to make the model a success. Furthermore, customers tend to be sensitive to changes in management and company leadership, often valuing their personal relationships with individuals within the company higher than the service the company provides. Additionally, there are few economies of scale that can be realized in a service business. Unlike the production of a physical device (which allows a company to simply "turn up" manufacturing as it adds new customers), service businesses must add staff to keep pace with the acquisition of new customers. As a result, they face increasing risk in managing their costs during periods of volatility. Another substantial source of risk comes from having to recruit, hire, train, manage, and retain human resources.

Although service contracts for capital equipment can be lucrative (with high profit margins), pure healthcare services tend to have lower margins that are often squeezed by **third-party** public and private **payers**. Moreover, services that focus on or utilize technologies or procedures that the company itself does not manufacture or control are further at risk if there is a major change in the external environment. For example, if a company sets up a business servicing some piece of capital equipment (but does not produce the machine itself), its business would evaporate if the technology was suddenly replaced by a new innovation in the market. While this model is relatively uncommon in the device field, it highlights the need for companies using service models to stay attuned to changes in the external business environment and be prepared to adapt their business models.

The Radiology Partners story provides an example of how a service offering can be designed to potentially overcome some of the challenges typically associated with this model.

FROM THE FIELD ▶ RADIOLOGY PARTNERS

Building a national service model in radiology

For nearly a decade, Mohamad Makhzoumi, co-director of the healthcare services and healthcare information technology (IT) investment practice at venture capital firm New Enterprise Associates (NEA) had been watching IT innovations transform radiology. "It started with the advent of teleradiology, followed by increasing subspecialization and shared radiology models," he recalled. More recently, Makhzoumi saw pressures linked to US healthcare reform spawn new uncertainties in the field, especially for the hospitals that relied on radiology groups to diagnose their patients at critical junctures in the **cycle of care**. "Insurance incentives are changing," he elaborated. "Hospitals are starting to be incented for discharging patients sooner, rather than keeping them longer. And radiology is the key to making that happen." Meanwhile, outsourcing businesses had been able to reduce costs, improve quality, and increase service levels in other hospital-based practice areas such as anesthesia, intensive care, and the emergency room. "But we hadn't yet seen anyone achieve all of those goals in radiology," Makhzoumi noted.

Sensing an opportunity, NEA venture partner and veteran healthcare services operator Rich Whitney came up with a strategy for developing a national service business that would leverage scale and technology to deliver more efficient care supplemented by value-added services. His plan was to build the practice, called Radiology Partners, by partnering with leading regional radiology groups that had a culture of physician-led quality and service. Based upon experience in other segments of healthcare services, Whitney believed that scale would offer significant competitive advantages across the dimensions of clinical quality, service to clients and referring physicians, revenue enhancements, cost improvements, and strategic positioning. "By joining a

national group practice, physicians could do better for their patients, have more security, make more money, and avail themselves to leadership and career development opportunities not available in small practices," he described.

Chief among the advantages that Radiology Partners would offer to its partner practices was load balancing. "Radiology groups have to staff their practices at a level that supports the maximum volume of the hospitals that are their clients. However, that maximum volume is only realized periodically," Makhzoumi explained. "As a national group with hundreds or thousands of radiologists, we could load-balance those radiologists across providers using telemedicine technology and sophisticated workflow tools to gain tremendous efficiency and cost-savings."

A second, significant benefit to the radiology groups would be access to the newest and most advanced IT (see Figure 4.4.6). "Building a state-of-the art viewing system, supplemented by clinical decision support tools and intelligent worklists is expensive, and it's not always feasible or practical for a small or even regional

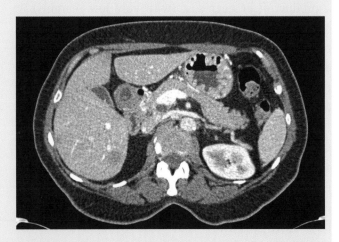

FIGURE 4.4.6

An abdominal CT scan taken with a late-generation scanner (courtesy of Professor Sandy Napel, Department of Radiology, Stanford University).

practice to make the capital investment necessary for such a system," Makhzoumi continued. "But," added Whitney, "these technologies offer the ability to manage the practice and deliver clinical excellence, service, and data in ways that smaller practices aren't capable of achieving."

Another critical leverage point was linked to reimbursement. "Radiology reimbursement rates have traditionally been pretty comfortable," said Christine Guo, an associate on the NEA team. "But with increasing reimbursement pressure in radiology exacerbated by a renewed focus on utilization, many independent radiology groups are hurting for the first time." As a large physician practice with increased negotiating power and resources to focus on revenue cycle management, Radiology Partners could achieve more attractive reimbursement rates from payers and more effectively and efficiently collect the dollars that are due.

To attract hospitals as clients, the team believed it could, first and foremost, improve the quality of care providers were used to receiving. This could be accomplished by implementing state-of-the art technology throughout its network, and giving providers access to a comprehensive range of radiology subspecialties. This benefit would be particularly valuable to regional hospitals and those in remote areas that typically did not have access to specialized expertise. "There's a huge benefit to being able to have the right radiologist read the right scan at the right time," said Guo.

Additionally, since Radiology Partners formed partnerships with local radiology practices rather than competing against them, hospitals and their referring physicians working with the national group would still be able to maintain their existing relationships and clinical interaction with trusted local radiologists.

To help deliver increased efficiency and cost savings to hospital customers, Whitney and team would implement national standards for turnaround time and other important metrics and then carefully monitor the performance of the radiology groups in their network.

Summarizing the model, Makhzoumi said, "We can go to hospitals and say, 'Radiology is a huge pain point for you. It costs you money, it's difficult to manage, and it's the biggest slow-down between getting patients admitted, diagnosed, and eventually discharged. Give us your radiology department – outsource it 100 percent to us – and we'll manage it for you. We'll do it better, more cost-effectively, and we will improve your quality and efficiency.'"

Because the Radiology Partners business model depends on hospitals to outsource their radiology needs to the company, hospital executives are the key decision makers in entering into these integrated service contracts. However, Makhzoumi pointed out that the company actually receives **payment** for its services from the insurance companies and other payers that cover the hospital's patients. Accordingly, the company must ensure that its offering is attractive to both stakeholders for its business model to succeed. For hospitals, this means convincing them that Radiology Partners can more effectively meet their total radiology needs, as described above. For payers, it means persuading them that a national radiology model can provide higher quality care and more timely and accurate diagnoses. Whitney and the NEA team agreed that satisfying this constituent seemed more important than ever as the prospect of new payment models taking hold in healthcare likely will require that radiology practices figure out how to help hospitals and other increasingly capitated payers reduce imaging utilization. According to Whitney, "Many radiology practices in the future are likely to be faced with either delivering a highly valuable clinical, process, and technology-driven approach to reducing imaging volumes while maintaining or improving care, or have it done to them with a resulting decrease in physician incomes, not unlike what happened in the early 90s with radiology benefit managers. With an average US practice size of 16 radiologists, very few groups have the resources to make the massive investments necessary to develop and deliver these kinds of programs, and if they don't offer this value to the system, what payer is going to agree to pay them more to offset the impact of the lower volumes?"

The success of the Radiology Partner business model also hinges on the company's ability to attract the right radiology practices and individual radiologists as partners. As Guo emphasized, one of the key challenges of building and scaling a service business (compared to a product-based business) is the fact that there is typically no proprietary technology, which makes delivering high-quality service essential. This intense focus on patient-focused clinical and service quality necessitates a rigorous screening process to ensure that the practice groups and physicians who join the network are equally committed to these goals. "Before asking a practice to partner with us, Radiology Partners meets with key physicians to make sure that they embrace the values and mission of the Radiology Partners practice. It's not about how many reads you can do in a day," Guo said. The team evaluates potential partners on the caliber and breadth of the physicians they have on staff, as well as the long-terms results they have achieved against critical performance standards and their passion to deliver the best possible service for the benefit of their patients. Those practices that make the cut are invited to become partners. Limited information is publicly available about the details of the partnership agreements, but groups in the network are given the chance to share in the profits of the national Radiology Partners consortium. In fact, a large percentage of the practice's physicians have an ownership interest in Radiology Partners.

In reflecting on other challenges associated with service models, Guo reiterated that, "Operating a service business on a national scale while maintaining consistency and quality is a lot harder than maintaining the quality of a product." Performance has to be carefully monitored and problems must be corrected swiftly before they impact customer satisfaction. And with multiple constituents using the service, keeping everyone happy can be difficult. "You have to pay attention to the underlying patients, but also to the referring physicians, hospital administration, and the health plans." Guo said. Managing these multiple relationships requires "a fully built-out infrastructure that includes layer upon layer of human capital," Makhzoumi said. "Complicating this challenge further is the need to tailor the service offering to each individual stakeholder. "It's not like selling a static device that is always going to be used in the same way," he stated. "You have to iterate, evolve, and amend your service for every customer, and that's a real challenge."

Accordingly, Whitney has recruited a top-notch team with experience in various functional disciplines, including several physicians and radiologists. Many of these individuals have a track record of leading large-scale healthcare organizations distinguished for leadership development. "In the end, developing a culture of service excellence and physician leadership will be our biggest asset," he remarked. "In fact, for Radiology Partners to achieve its full potential we'll have to, of course, be good at many things. But we must be *excellent* at attracting, retaining, and developing extraordinary physician leaders who share our passion for transforming radiology."

For innovators interested in building a healthcare service business, the members of the Radiology Partner team offered the following additional advice. First, be keenly aware of market forces and industry dynamics that yield windows of opportunity. "The timing has to be right," said Guo. "For us, the key factors included the maturation of technology to support this model and market trends converging on lower healthcare costs and increased access and quality." Second, they emphasized the need to be flexible in terms of defining a service business model in order to adapt to market realities. And finally, as with any service business, they reiterated that people are a company's greatest asset. "Seek out, identify, and recruit the best leaders you can for all levels of your new organization," Whitney encouraged.

Fee per use

A fee per use business model can be appropriate for innovations that sit squarely at the intersection of products and services. For example, laser eye surgery requires a capital expenditure to cover the cost of the equipment. However, practitioners are also charged a fee every time they perform a surgery using the machine. The event that triggers a payment to the company is nothing other than use of the machine.

Another form of the fee per use business model in the medical device field is referred to as a capitated model. With this approach, a medical provider is given a set fee per patient, regardless of treatment or equipment required. For example, a company that provides all of the disposables necessary to perform a laparoscopic gallbladder removal may charge a fixed price per patient, independent of which disposables are actually used. Similarly, a company with cardiology products may charge a fixed fee per patient for all of the stents, balloons, and catheters required to perform certain predefined procedures. This approach allows companies with

broad product lines to achieve an advantage over those that offer a smaller number of related products – the bundling of products makes it more difficult for less diverse competitors to penetrate accounts.

This model can be appealing to customers seeking increased certainty in their costs since the per patient payments are fixed regardless of the complexity of the individual cases they perform (e.g., hospitals have traditionally lost money on procedures requiring the use of more than two or three drug-eluting stents in a single patient). On the downside, customers sometimes resist paying the fixed cost for cases that require few devices (e.g., one stent), which can necessitate extra time and effort from the company in defending its business model. Also, unless a business generally has a high degree of IP protection or other barriers to entry, this type of business model is often challenged by other businesses trying to compete in more traditional ways.

The VISX example below illustrates the challenges and benefits associated with a fee per use business model.

FROM THE FIELD ▶ VISX

Developing the fee per use model in the vision correction field

In the late 1980s, Charles Munnerlyn, founder and CEO of VISX, and Allen McMillan, the company's COO, were considering how to commercialize the first system for photorefractive keratectomy (PRK), the original method used for laser vision correction (see Figure 4.4.7). They knew there must be a better approach than the traditional capital equipment model. The cyclical nature of capital equipment sales would leave little money available for research and development since they would sell one model to as many customers as possible and then have to wait for years until the same customers were ready to invest in the next generation. They were also concerned that the high purchase price of capital

equipment could make market penetration quite slow for a new company entering the field.

Seeking a way to realize a more consistent revenue stream and quickly build a commercial presence, Munnerlyn and McMillan conceptualized a business model that would allow the company to sell equipment and service contracts to its customers, but also charge a fee each time the system was used. "The laser vision correction industry was really the first in the medical device field, and certainly in the US, to use a true per procedure model," explained Liz Davila, CEO of VISX in the early 2000s.

In the late 1980s and early 1990s, VISX, and its main competitor, Summit Technologies, were each developing excimer lasers for eye surgery and had patents on different aspects of the technology.

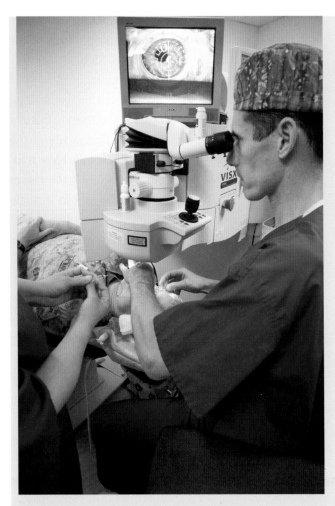

FIGURE 4.4.7

A physician preparing for a laser vision correction procedure (US Navy, photo by Mass Communication Specialist 1st Class Brien Aho via Wikimedia Commons).

In order to avoid patent litigation, they combined their IP in 1992 and created a third entity, Pillar Point Partners, which licensed the combined IP. Pillar Point Partners would then sub-license the patents back to VISX, Summit, and any company that wanted to license the patents. All licensees paid Pillar Point a per procedure royalty. Pillar Point profits were then distributed to VISX and Summit. Since the licensees were paying on a per procedure basis, it was logical that they would charge their customers by procedures. The result was that the US laser vision correction industry adopted the fee per use business model.

VISX's shareholders immediately appreciated the value of the new model, but there was a period of time when the surgeons were resentful and antagonistic toward the company. VISX originally charged upward of $500,000 for the equipment, plus $250 per eye treated. The $250 charge was positioned as a licensing fee. As Mark Logan, the company's CEO (following Munnerlyn but before Davila), explained in an interview for the *Tribune Business News*, "We've spent $52 million bringing this to market, and there's no way we could sell these lasers for $500,000 and ever recoup what we've put into it. If it weren't for the $250 fee, we'd have to sell these systems for $3 million each and have a very small market, which isn't good for anyone."[13] VISX also asserted that these fees enabled the company to sustain its research and development efforts so that it could continue improving its technology to treat new indications.

The surgeons accepted the need to pay for the disposables required by the procedure because they had a utilitarian function, but they did not understand the value of the VISX keycard (which tracked equipment usage and calculated the associated licensing fees). According to Davila, "What eventually got them to quiet down was that they fairly quickly realized how much money they were making." Previously, as corneal specialists, most of the VISX surgeons performed cataract surgeries, which were primarily covered by Medicare and reimbursed at a total of about $1,600 per procedure. In contrast, Davila continued, "By expanding their practices to include vision correction surgery, the surgeons could now charge $1,500–$2,500 *per eye* and were making money as they had never made money before." Eventually, they accepted the $250 per use fee as a necessary expense associated with the more lucrative laser surgery market.

Since VISX and Summit were the first two companies to offer this technology in the US, they exercised a high level of market control (approximately 70 percent and 30 percent market share respectively). Eventually, however, other competitors began to enter the market. Several took licenses to the patents. However, a company called Nidek refused to license the IP from Pillar Point Partners. Despite the fact that the company offered inferior technology, it was able to gain market share because it also did not charge a per procedure licensing fee. To protect itself against substantial market erosion, VISX reduced its per procedure fee from $250 to $110. Surgeons responded positively to the change, and many doctors who had switched to Nidek eventually came back to VISX because of the company's superior technology. Over time, Nidek's business declined as the company lagged technologically further and further behind. As VISX continued to improve its technology, further outdistancing its competitors, the company was able to increase its per procedure fee back to $250.

The Nidek example demonstrates that a fee per use model has the greatest likelihood of success in an industry or market where everyone is playing by the same rules. Alternatively, the company employing the model must sustain such a strong competitive advantage that customers view the license as being worth the additional cost. "When people ask me if the VISX model would work for them," said Davila, "I ask them if their technology is truly revolutionary. Because if

it's not, it will be very difficult to get medical professionals to pay a per procedure fee."

Davila also underscored the importance of having IP that can be protected. Among the reasons that VISX reduced its usage fee in the face of competition from Nidek was the slowness of the patent enforcement process. VISX did sue Nidek for patent infringement, but the court process required several years and VISX decided that it was too costly to wait.

Finally, she pointed out that eye surgery is an elective procedure that is not reimbursed through Medicare and infrequently covered by private insurance providers. The fact that VISX did not have to take third-party payers into account simplified its decision to pursue an alternative business model. The company only had to convince providers and patients that the value provided by its offering would outweigh any non-traditional costs – a somewhat easier hurdle to clear than the cost-effectiveness requirements increasingly imposed by payers.

VISX was acquired by Advanced Medical Optics (AMO) in 2005, and AMO was in turn purchased by Abbott in 2009.

Subscriptions

Subscription models require a customer to pay a fixed price to gain access to a product or service for a defined period of time. The model was pioneered in the publishing field and has now been popularized across industries, including healthcare. The access granted through a subscription can be limited to a certain amount or configuration of a given product or service, or it can be unlimited. Usage can also be given to an individual or group in a transferrable or non-transferrable fashion. These are all variables innovators must consider and define when evaluating a subscription business model. The most important factor is to ensure that there is a clear and accurate way to differentiate subscribers from non-subscribers as they seek to interact with the product or service, especially if some basic portion of the offering will be given away for free.

One example of a subscription-based offering is Welltok, a consumer engagement platform that helps health plans, large self-insured employers, and other population managers connect with their members and reward them for healthy behavior. These population managers subscribe to Welltok's CafeWell on behalf of their individual members. In turn, these consumers gain access to the "social health management platform," which includes health information, competitive wellness and fitness challenges, incentive programs for healthier lifestyle changes, support for chronic conditions from peers and professionals, and anonymous social networking. In exchange for subscription fees, the population managers receive de-identified metrics on member wellness and participation.[14] Rather than targeting consumers to generate income from its offering, Welltok intentionally went after payers. As Chairman and CEO

Jeff Margolis explained, "The insurance companies are at the top of the food chain, in terms of how the dollars flow."[15]

Another example of the subscription model in healthcare can be found in direct primary care offerings. Two of the most common approaches are concierge medicine and direct primary care, through which subscription-based healthcare practices charge patients anywhere from $60 to $30,000 per year to deliver a defined range of health-related services. Patients that sign-up with direct primary care provider like One Medical pay roughly $150–$200 per year for same-day appointments, online prescriptions, and personal email access to doctors. Many patients retain their health insurance to cover acute conditions and emergencies, but pay **out-of-pocket** for the better access and increased convenience that the direct-pay subscription model provides. In parallel, the physicians that belong to these practices maintain a steady income stream without the pressure to see more and more patients each day as they would in traditional medical practices. They also report that they can spend more time coaching patients on wellness and prevention.[16]

On the plus side, subscription businesses have the benefit of being an annuity once a customer-product relationship is established. Growth is driven by maintaining customer satisfaction with the offering and expanding the reach of the business to other customers. On the downside, subscription business models can be put at risk by other models that provide the same or a similar offering for free by leveraging a broad customer base to derive income from advertising or another type of information exchange.

Over-the-counter products

An over-the-counter (**OTC**) business model depends on patients' ability to choose a treatment path and then acquire the product(s) for themselves. OTC products are more typically seen with drugs (e.g., Motrin®, Benadryl®), but can also include devices (e.g., steam machine for treating sinus congestion, home blood-pressure cuff devices, glucose monitors, or the vast array of emerging "quantified self" technologies). OTC products can, at times, be combined with services. For example,

companies selling at-home blood pressure monitoring devices may offer data analysis and feedback services to users who upload their blood pressure readings via a computer.

Because physicians often are not involved in recommending OTC treatments, they must be relatively simple and easy to use. Advertising is usually targeted directly to consumers, and generic retail outlets (e.g., Walgreens) can be used to support sales. On the upside, OTC products rarely require an expensive direct sales force. On the other hand, they require sizable marketing budgets to promote brand recognition, which can be a significant barrier to overcome. In this model, the goal is to make the consumer aware of the product and how it works through conventional advertising channels (e.g., television, magazine, radio). Within this construct, companies at times find it challenging to differentiate their products since consumers spend less time than physicians understanding the clinical benefits of one product over another.

Models that require doctors to specify an OTC product to their patients can be even more challenging. In general, the smaller and more specific a physician population is, the easier it is for a company to reach. However, physician-specified OTC products tend to rely on the recommendations of general practitioners – a vast market that is nearly impossible for any single company to reach effectively with a reasonable investment of time, money, and resources.

Prescription products

Prescription drugs provide a classic example of prescription-based medical products. However, prescriptions can also be used for devices and combined with services (e.g., physical therapy), disposables (e.g., blood glucose monitoring testing strips, drug cartridges to a delivery system), and hardware (e.g., nerve stimulator pain control units, certain inhalers). With this type of business model, the physician selects the treatment and directs the patient toward it, but the patient is still required to act on the physician's instructions.

Prescription products often require more specialized training to understand than OTC alternatives and the process of selling them is more complicated since the physician is directly involved in selecting the treatment. However, a more clinically oriented sales approach can be used to differentiate products, usually through a direct sales force. The size of the sales force is typically proportional to the number of physicians the company is trying to reach. For example, a sales force focused on primary care physicians would be the largest, whereas a smaller, but still significant sales force would be required to support specialty products, such as anti-nausea medications used for patients receiving chemotherapy. To support these requirements, the products must command high enough margins, as is the case with most prescription medications and devices. Direct-to-consumer advertising is sometimes used in parallel with the physician-focused sales effort, because physicians are expected by many patients to respond to their desires and requests.

Given the strength of the advanced marketing expertise amassed by large companies over the last couple of decades, prescription markets can be difficult to penetrate for small start-ups. Companies are often advised to enter the prescription business only if their product is clearly differentiated from the competition and/or if they can enter into a co-development partnership with a large pharmaceutical company that has an established sales force. "Me too" products have little chance of success without the marketing "muscle" and deep pockets of a major partner.

Physician-sell products
Physician-sell products are those treatments that are sold directly through physicians. With this business model, the physician essentially becomes a retailer for the product and usually receives some direct incentive for helping to promote and provide the treatment. Common examples include BOTOX® injections, teeth whitening products, and hearing aids. Physicians can also sell disposables or OTC products, such as contact lenses or solutions.

Typically, physician-sell products are offered on an outpatient basis and are often paid for by the patients (rather than by insurance). Once again, the margins for products sold through this channel must be high enough to cover not only the compensation to the company, but also to the physician for being a distributor. While some physician-sell products can be quite profitable, the primary downside to this model is the potential for ethical conflict. Anytime a physician receives a direct incentive to steer a patient toward a particular treatment, questions may arise about whether the physician is truly keeping the patient's best interests in mind. For this reason, physician-sell products tend to be limited to non-essential, elective treatments that patients may desire but are not necessarily purely medically indicated. Although there are some settings – such as in dermatology or dentistry – where physician-sell models are completely legal and appropriate, recent laws have changed the scope of the products physicians can sell in their offices for a profit. As a result, it is important to consult these regulations prior to utilizing this type of model.

A note on mobile health

Mobile health (**mHealth**) and health information technology (HIT) are rapidly becoming important segments of the medtech industry. New IT-based applications, devices, and services are being launched at a rapid rate, all with the goal of more effectively facilitating the flow of information between patients, healthcare providers, and other stakeholders in the healthcare ecosystem.

mHealth applications and devices, along with their related service offerings, are quickly proliferating in two broad categories: technologies targeted at consumers and those aimed at physicians and institutions (hospitals, employers, insurance companies, etc.). Estimates of the patient/consumer market for remote monitoring devices alone range from $7.7 billion to $43 billion, with 40 percent of consumers surveyed indicating that they would pay for remote monitoring devices and a monthly service fee to send data automatically to their doctors.[17] Within the physician community, doctors surveyed by the Health Research Institute reported that they would like to use mobile health to enable them to remotely access electronic health records, prescribe medications, monitor patients in and outside of the hospital, and better communicate with patients. A full 56 percent believed that mobile health could help them expedite decision making, 39 percent said it would decrease time spent on administrative tasks, 36 percent it had the potential to increase collaboration among physicians, and 26 percent anticipated the ability to spend more time with patients.[18]

Mobile technologies have the potential to create value in multiple ways, including the ability to help deliver less expensive solutions, enable new ways of managing care, and facilitate better health outcomes. The challenge is how to monetize these technologies in the healthcare space. When these technologies first began to emerge, the traditional fee-for-service reimbursement model, which rewards physicians for the volume of patient office visits they conduct, was firmly in place. The Health Research Institute's physician survey revealed that some phone consultations related to chronic disease management were being reimbursed, but insurance payments for remote transactions related to patient wellness and prevention was lagging.[19] As public and private health insurers make the shift toward paying for outcomes as the implementation of US healthcare reform and payment reform efforts accelerate, many more mHealth and HIT solutions will be better positioned to achieve reimbursement. For example, within new **pay-for-performance** healthcare models like Accountable Care Organizations (**ACOs**) and patient-centered medical homes, physicians and the organizations they work for will have direct incentives to improve health outcomes for a defined population. Mobile technologies that enable remote monitoring and provide patients with rapid access to medical staff when questions arise or when changes in the patient's health status are detected are expected to play a key role in enabling these organizations to succeed.[20] In the near term, however, these approaches are being experimented with on a relatively small scale in the US. Other countries are somewhat further along. In the United Kingdom, for example, a telehealth trial conducted by the National Health Service (NHS) yielded a 15 percent reduction in doctor's office visits, a 20 percent reduction in emergency admissions, a 14 percent reduction in the need for planned admissions, and a 45 percent reduction in mortality rates.[21] In February 2012, the NHS, which is run by the government and acts as both payer and provider, began encouraging physicians to prescribe smartphone applications to their patients in an effort to improve monitoring between visits to reduce unnecessary appointments when patients are stable and proactively intervene if they take a turn for the worse.[22]

As they wait for the reimbursement landscape to shift for mobile health applications and online health services, many companies are attempting to have patients pay for their technologies directly through an over-the-counter business model. Numerous health-related consumer products have emerged that depend on an up-front sale of a reusable piece of hardware and/or software, supported by online services made available at no charge. For instance, for approximately $100, the FitBit daily activity tracker allows individuals to automatically record their steps taken, calories burned, and stairs climbed with the purpose of motivating improved fitness and well-being. For roughly $200, AliveCor offers a heart monitoring application and sensors that work with an iPhone to deliver a clinical-quality electrocardiogram. Thousands of other health-related smartphone apps are available online, with worldwide consumers downloading 44 million of them in 2012.[23]

Fee per use and subscription models are also common in mHealth. For example, HealthTap offers a free online knowledgebase of medical information and the ability to get general health questions addressed by practicing physicians in a discussion forum format. However, for about $10 per inquiry, patients can get their more detailed, specific questions answered through a private, one-on-one interaction with a doctor in the appropriate specialty (with follow-up questions charged at $5 each). This approach to acquiring customers by offering basic services at no charge and then trying to monetize their interactions as they seek more premium services is sometime referred to as a "freemium" model.[24] In mid-2013, HealthTap was devising additional ways to charge patients for premium services, like connecting with a specific physician or getting faster responses. HealthTap was reportedly considering additional fee-for-service offerings as well as subscription models.[25]

Some subscription models target physicians, not just consumers. ZocDoc charges doctors a flat monthly fee in exchange for referring patients to them through their online site. This kind of subscription-based model for sourcing leads is also used by companies such as 1-800-DENTIST.

Other mHealth and online health offerings rely on advertising models to support their businesses.

WebMD, a health information knowledgebase, got its start selling ads to pharmaceutical firms and other health-related companies (more recently it has expanded its revenue streams to include selling private health-portal services to employers).[26] Practice Fusion, which offers a free electronic health record system for individual practitioners and small- to mid-sized physician groups, earns the bulk of its revenue from ad sales. However, it too is trying to diversify into other models.[27] One factor driving companies away from advertising is the sensitivity to the economic swings that can dramatically affect advertisers. Another issue is that the traditional display ads that dominate the Internet do not translate well to smartphones and other smaller, more portable devices that are proliferating globally.[28]

On the whole, the mHealth space is changing rapidly. Although many interesting approaches are being developed and tested in the segment, the most effective strategies have yet to be proven. The following example about remote monitoring company Cardiocom and its acquisition by Medtronic illustrates how large and small companies alike are experimenting with business models in the mHealth space.

FROM THE FIELD ▶ CARDIOCOM

Business model evolution in mHealth

Cardiocom, based in Chanhassen, Minnesota, entered the telehealth space in 1999 with the introduction of its Telescale, a remote monitoring device designed to help reduce acute care hospital admissions for heart failure patients. Individuals using the device stepped on to the scale and responded to simple "yes/no" questions to provide information about how they felt, and also about their diet, medications, and exercise. This data was then transmitted to the patient's healthcare provider and Cardiocom via a phone line. The company used data analysis software to proactively detect and act on risk factors that could potentially signal a change in health. For instance, an increase in weight could mean that a heart failure patient was retaining water, a dangerous symptom of the disease. As risk factors were identified, a registered nurse in the Cardiocom call center would telephone the patient to check in and devise an appropriate course of action, from patient education to scheduling a doctor's appointment for prompt and thorough diagnosis.[29]

Over time, Cardiocom expanded from heart failure into other long-term health problems, including chronic obstructive pulmonary disease, asthma, diabetes, hypertension, and obesity. Each of its hardware- and software-based products (along with their related services) was designed to identify symptomatic patients in these disease areas and intervene early to keep them out of the hospital.[30] The Commander Flex was one of the company's core products (see Figure 4.4.8). This device was a table-top display that posed questions to patients about their health. As patients responded, the system used branching clinical logic to provide additional inquiries based on their answers. The device also connected to ancillary vital sign monitoring devices

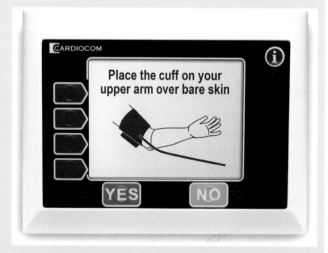

FIGURE 4.4.8

The Commander Flex telehealth system (courtesy of Medtronic).

manufactured by Cardiocom (such as a weight scale, pulse oximeter, etc.) to gather and transmit relevant physiological data to Cardiocom so that nurses in the call center could understand patients' specific symptoms and their severity.

Because reimbursement for remote monitoring services was not available under the fee-for-service paradigm, Cardiocom decided to build its business model around selling to insurance companies interested in reducing the cost of caring for patients with chronic conditions. They also targeted home care agencies that would potentially adopt the technologies as a way to make their services more competitive.[31] The company devised a "per member per month" (PMPM) subscription model for generating revenue. Cardiocom retained ownership of the equipment and charged its customers approximately $100 PMPM for its use. For example, a home health agency with an average of 100 complex, high-risk patients under its care at any given time would purchase 100 subscriptions from Cardiocom (the typical service agreement was 3 years in length). The agency would provide its high-risk patients with the Cardiocom communication hub and appropriate plug-ins for a defined period of time (e.g., someone experiencing heart failure might use the system for 30 days after being discharged from the hospital to help prevent readmission during this critical period). Once the patient stabilized, s/he returned the equipment to the agency, which refurbished it and then redeployed it to the next patient. "This model is attractive," commented Brett Knappe, Executive Director, Strategy & Corporate Accounts, US Region for Medtronic, "because it enables patients to enjoy the benefits of daily monitoring and support, and at the same time helps providers distribute their resources across a population of patients more cost-effectively."

Using this approach, Cardiocom grew its business to $10–15 million in sales by 2010.[32] Then two important events transpired. First, the Affordable Care Act (**ACA**) was signed into law in the United States, which stimulated greater demand for remote monitoring technologies. Specifically, the ACA's Readmission Reduction Program established penalties for hospital

readmission of patients with acute myocardial infarction, congestive heart failure, or pneumonia within 30 days of discharge.[33] Along the same lines, the law also led to the formation of Accountable Care Organizations (ACOs) through which consortiums of hospitals, physician groups, and private payers could share in savings realized from reducing the overall cost of care for a defined population of Medicare recipients.[34] Second, Cardiocom was one of six companies awarded $225 million in health monitoring service contracts by the US Department of Veteran's Affairs (VA) in 2011.[35]

Roughly in parallel with these important changes, another Minnesota-based company, medical device manufacturer Medtronic, appointed a new CEO. Upon assuming his new position, Omar Ishrak, who joined the company from GE Healthcare Systems, outlined a handful of strategic themes for Medtronic, including globalization, an increased emphasis on economic value, and an expansion into healthcare services. "That's where Cardiocom fits in," said Knappe. Medtronic had a strong presence in heart failure, offering a number of implants such as pacemakers, cardiac resynchronization therapy (CRT) devices, and implantable cardioverter defibrillators (ICDs). But it was looking to broaden its footprint in this disease area, and remote monitoring seemed like a good way to do it. "By getting involved longitudinally across the continuum of care and providing services not just devices, Medtronic can be more relevant to a variety of healthcare stakeholders," Knappe noted. "It allows us to have more impact on patient outcomes, not just in the acute setting. And we can also be involved in improving long-term outcomes and reducing costs for healthcare systems." To expand into healthcare services, Medtronic needed to make a strategic acquisition in telehealth; one that would give the company critical mass from which it could continue to build.

Cardiocom surfaced as a strong acquisition candidate, in part, because it was an independent entity. Knappe elaborated: "If you look at the leader board, you'll find remote monitoring offerings tucked under Bosch and Philips and Honeywell – big industrial companies that leverage a position in security, or another connection to the home, to get involved in patient monitoring. As a rare example of a pure play, Cardiocom was an attractive company to pursue." At the time, Cardiocom had been growing at 80+ percent annually for several years and had a run rate of approximately $50 million in annual revenue, making it relatively insignificant compared to Medtronic's $16 billion in net sales in 2012.[36] However, the remote monitoring category was forecast to experience double-digit growth for the next several years.[37] "And Cardiocom was growing even more rapidly," said Knappe. This was quite a contrast to Medtronic's own, single-digit growth rate. "We were also drawn to their commercial success," Knappe said, referencing Cardiocom's VA contract and the fact that the company "was performing very well for this and other blue-chip customers." After a period of due diligence, Medtronic acquired Cardiocom in August 2013 for $200 million.[38]

According to Knappe, who was managing the post-acquisition integration activities, Cardiocom would retain its PMPM subscription business model in the nearterm. But, over time, it would play an important role in Medtronic's strategy to transform its own business model. "It used to be about selling device features to physicians," he explained, "and sometimes these physicians didn't have a vested interest in the financial performance of the institution. That was the old model that the medical device industry grew up with. It depended on relationship-based selling with sales reps who were deeply embedded in the hospital and with the docs." Medtronic's vision for the future was to shift from selling implants through this traditional approach to offering a holistic suite of products and services that help decrease variability of outcomes and reduce risk for its customers. "This new strategy will enable us to engage with payers, integrated payer-providers, ACOs, and even hospitals who are at risk of incurring readmission penalties in new ways that are not based just on the features and functionality of our devices," Knappe said. "We can go to a [healthcare alliance] or a hospital system and say, 'We know you're at risk and we can align ourselves better because we believe that we've got the

solutions to improve patient outcomes. And if we get better than expected results, we're both going to profit from it. If we struggle, we're both going to struggle.'" He continued, "In a sense, the ability to reduce risk becomes part of the product and is a new way that we can help physicians and healthcare providers succeed."

Linking this idea back to Cardiocom, Knappe stated, "For heart failure patients, we have a set of implants that communicate data to our CareLink system. The CareLink system produces reports that are available to electrophysiologists and can be shared with the heart failure docs. The problem is that only 8–10 percent of heart failure patients are indicated for this kind of implant therapy, and only a portion of them use Medtronic devices. So, in the scheme of things, we're saying to the hospital or to the payer, 'We've got really great solutions for some small percentage of your patients with heart failure.' And their response is, 'Thank you, but ho-hum, because it's the rest of the patients that we're having trouble managing.' So with Cardiocom, we now have something that can benefit the vast majority of heart failure patients, who need to have their medications managed but are not indicated for an implantable device." By more closely managing these patients, Medtronic intended to reduce their cost of care. "We can make sure that any complications are treated through scheduled clinic visits, which is a far more cost effective approach than treating them in the emergency room."

While this aligned incentives model remained to be proven, Knappe was optimistic. "To responsibly take risks, you need to have control of the factors that create the risk. We may need to add additional technologies and services to get us to the point where we have our arms around improving outcomes for a broad population of patients. And then we can start reducing risk for payers, integrated delivery networks, and ACOs," he said. Summing up his thoughts, Knappe added, "What we're trying to do is improve outcomes for any patient diagnosed with heart failure, and for the healthcare system in general. And Cardiocom is a great start."

The strategy of bundling multiple offerings to more holistically address a customer's needs in a disease area is becoming increasingly popular as companies seek new ways to increase the value of their products and services. For more information about this approach, see 5.9 Competitive Advantage and Business Strategy.

Validating a business model

From the array of business model choices, one or two will usually seem most immediately viable for a given concept. However, innovators are encouraged to evaluate all available alternatives before making a decision so that opportunities and risks are not overlooked. In the current healthcare environment, where external factors such as reimbursement hurdles are rapidly changing and cost sensitivity is becoming particularly acute, it can be difficult to predict which model will have the greatest likelihood of success. The one that seems to deliver the greatest value to the customer (decision maker) in terms that are meaningful to that audience is a good place to start.

Ultimately, innovators will depend on the strength of their own research and judgment to match a concept to a business model. But they will be well-served to validate the choice by seeking input from other innovators, entrepreneurs, and business advisors. A few important questions to ask include the following:

- Can the technology or therapy be delivered via a different business model?
- If so, and if this was to be done by a competitor, would it be a threat to the business model or is the chosen model still sound?
- If the other business model could be a threat, are there significant barriers to allowing customers or other businesses from executing that business model against the one the innovator has chosen?

If the business model selected by the innovator is the most appropriate and sustainable one available, it should rise to the top regardless of the competitive scenarios outlined above. Because it is so important not to choose a business model in a vacuum, it should be tested with potential purchasers of the solution before a final decision is made. Additionally, the innovators should consider the primary risks associated with the model and develop specific plans for managing them. A framework such as Michael Porter's Five Forces can be helpful when evaluating business models within the medtech industry (see 2.4 Market Analysis).

Operationalizing a business model

Fundamentally, the steps in the biodesign innovation process that come after a business model has been chosen are focused on helping innovators operationalize the business model they selected. As soon as a business model is chosen, it is important to begin thinking about what kind of expertise will be required to implement it. Ideally, a young company will identify a potential hire who has previously built the kind of business it is pursuing. For example, if a disposable model has been chosen, it might be time to augment the engineering team with someone who has successfully pioneered disposable products in the past.

There are many different ways to align a company's expertise with its business model. One approach, of course, is to directly hire individuals with the right experience. Depending on the stage of the company's development, however, consultants can also be leveraged. Another approach is to seek this expertise in the form of advisors who might sit on the company's board of directors. When considering what type of expertise is required, it is important to consider technical and business competencies, both of which will play a critical role in successfully operationalizing a company's business model.

Sometimes, as a company grows, a new product is considered that requires a different model than what is used for the other products in the portfolio. The first and most important step is to recognize this fact before proceeding with development. Typically, for instance, companies that are organized around large capital equipment devices with service contracts have a hard time transforming themselves to act as a disposable company, and vice versa. While there are some notable exceptions, such as Intuitive Surgical which has executed well on both capital equipment and reusable models, the list of companies able to master dual-business models is short. Unless the managers of a business are uniquely prepared for the challenge of executing multiple models, the best choice, at least initially, is to "keep it simple" and focus on one model at a time, based on an evaluation of the company's total overall opportunity.

A final note: using business models to screen and eliminate concepts

Using business models as a concept screen differs substantially from IP, regulatory, and reimbursement. While these other factors require significant, increasingly in-depth research to enable innovators to assess the risks associated with each concept, business model screening is more of a thought experiment. By working through the questions outlined in this chapter, innovators are able to determine the most appropriate business model for a concept. Accordingly, this screen can be applied to every idea under consideration without undue time or effort. If a concept falls squarely into a business model category and there are examples of companies successfully using that model for comparable technologies within or outside the medtech sector, then this idea should move up on the prioritized list of concepts (and will perform well on the risk scoring matrix described in 4.6 Final Concept Selection). On the other hand, if a concept seems out of alignment with typical medtech business models or there are few examples of companies successfully employing the chosen business model for similar offerings, then innovators may be well served to consider how to modify the concept or model. Alternatively, if the business model appears to be particularly difficult or infeasible, they might treat this as a killer risk and set aside that concept, particularly if it is not a top performer relative to the other concept screens.

Online Resources

Visit www.ebiodesign.org/4.3 for more content, including:

 Activities and links for "Getting Started"
- Understand characteristics of different business models
- Choose a business model
- Validate the preferred business model and identify risks
- Determine what new expertise will be required to operationalize the model

 Videos on business models

 An appendix that outlines rules of thumb for choosing a business model

CREDITS

The editors would like to acknowledge the individuals who contributed to the case examples: Gary Curtis, Liz Davila, Krista Donaldson, Erica Frenkel, Christine Guo, Brett Knappe, Mo Makhzoumi, Maria Sainz, Kevin Sidow, and Rich Whitney. Many thanks also go to Darin Buxbaum, Jared Goor, and Joy Goor for early assistance, and Ritu Kamal and Stacey McCutcheon for helping with the updated cases.

NOTES

1 An MRI machine, which stands for magnetic resonance imaging, provides a non-invasive method for visualizing various structures and disease functions within the body.
2 All quotations are from interviews conducted by the authors, unless otherwise cited. Reprinted with permission.
3 See Stefanos Zenios, Lyn Denend, and Julie Manrqiuez, "Brilliance I-From Prototype to Product Company," *Global Health Innovation Insight Series*, 2012, http://csi.gsb.stanford.edu/brilliance-i-prototype-product-company (September 12, 2013) and Stefanos Zenios, Lyn Denend, and Julie Manrqiuez, "Brilliance II-Achieving Impact Through Licensing," *Global Health Innovation Insight Series*, 2012, http://csi.gsb.stanford.edu/brilliance-i-prototype-product-company (September 12, 2013).
4 "Injection Safety," World Health Organization, October 2006, http://www.who.int/mediacentre/factsheets/fs231/en/ (September 12, 2013).
5 Brianne Bharkhda, "Preemption Applies to 510(k)-Cleared Components of a PMA-Approved Device," *Inside Medical Devices*, November 20, 2013, http://www.insidemedical

devices.com/2013/11/20/preemption-applies-to-all-components-of-a-pma-approved-medical-device-even-components-that-were-previously-510k-devices/ (April 22, 2014).

6 "Kyphon Stock up 30 Percent on $725 Million St. Francis Medical Tech Acquisition," *Silicon Valley/San Jose Business Journal*, December 4, 2006, http://sanjose.bizjournals.com/sanjose/stories/2006/12/04/daily2.html (September 12, 2013).

7 "Capital Expenditure," Investopedia.com, http://www.investopedia.com/terms/c/capitalexpenditure.asp (September 12, 2013).

8 CT scanners, which perform computed tomography, provide another type of diagnostic imaging.

9 Jeremy P. Hedges, Charles N. Mock, and N. Cherian, "The Political Economy of Emergency and Essential Surgery in Global Health," *World Journal of Surgery*, 2010, pp. 2003–6, http://www.gradianhealth.org/wp-content/uploads/2012/08/Hedges_Politica-economy-of-EES.pdf (September 13, 2012).

10 "Birth of the UAM," Gradian Health Systems, http://www.gradianhealth.org/case-studies/#case-study-2 (September 17, 2012).

11 An LLC protects the owner's personal assets from the debts and obligations of the company as long as the owner maintains an account of business income and expenses separate from personal [foundation] accounts. However, any personal [foundation] assets used as collateral for business loans are not protected. See Tom Chmielewski, "What Is an LLC Disregarded Entity?," *Houston Chronicle*, http://smallbusiness.chron.com/llc-disregarded-entity-3645.html (December 9, 2013).

12 R.M. Schneiderman, "Reinventing Anesthesia," *Stanford Business Magazine*, March 1, 2013, http://www.gsb.stanford.edu/news/headlines/reinventing-anesthesia (October 30, 2013).

13 "VISX, Incorporated," Answers.com, http://www.answers.com/topic/visx-incorporated?cat=biz-fin (September 13, 2013).

14 Jonah Comstock, "WellTok Raises $18.7 Million for Healthy Social Network CafeWell," *Mobihealthnews*, April 10, 2013, http://mobihealthnews.com/21570/welltok-raises-18-7m-for-healthy-social-network-cafewell/ (September 23, 2013).

15 Michael Davidson, "WellTok Raises $18.7 Million Series B to Gamify Healthcare, Gets New CEO," *Xconomy*, April 10, 2013, http://www.xconomy.com/boulder-denver/2013/04/10/healthcare-it-startup-welltok-closes-18-7-million-series-b-round/ (September 23, 2013).

16 "Clinical Transformation: New Business Models for a New Era in Healthcare," Accenture, 2012, http://www.accenture.com/SiteCollectionDocuments/PDF/Accenture-Clinical-Transformation-New-Business-Models-for-a-New-Era-in-Healthcare.pdf (September 12, 2013).

17 "Healthcare Unwired: New Business Models Delivering Care Anywhere," Health Research Institute and PricewaterhouseCoopers, September 2010, http://www.mobilemarketer.com/cms/lib/9599.pdf (September 16, 2013).

18 Ibid.

19 Ibid.

20 "mHealth: Mobile Health Poised to Enable a New Era in Health Care," Ernst & Young, 2012, http://www.ey.com/Publication/vwLUAssets/mHealth_Report_January_2013/$FILE/mHealth%20Report_Final.pdf (September 16, 2013).

21 Ibid.

22 Brian Dolan, "U.K. to Encourage Doctors to Prescribe Health Apps," *Mobihealthnews*, February 22, 2012, http://mobihealthnews.com/16401/uk-to-encourage-doctors-to-prescribe-health-apps/ (September 16, 2013).

23 "mHealth: Mobile Health Poised to Enable a New Era in Health Care," op. cit.

24 Oliver Roup, "Mobile App Monetization: Think Business Model, Not Ads," *VentureBeat*, June 2, 2013, http://venturebeat.com/2013/06/02/mobile-app-monetization-think-business-model-not-ads/ (September 16, 2013).

25 Parmy Olson, "Medical Device Network HealthTap Surpasses One Million Users," *Forbes*, April 17, 2013, http://www.forbes.com/sites/parmyolson/2013/04/17/medical-advice-network-healthtap-surpasses-one-million-users/ (September 12, 2013).

26 Susan J. Aluise, "WebMD's Business Model May Need Surgery," *InvestorPlace*, January 11, 2012, http://investorplace.com/2012/01/webmds-business-model-may-need-surgery/ (September 16, 2013).

27 Ken Congdon, "The Truth Behind 'Free' EHRs," *Healthcare Technology Online*, January 25, 3013, http://www.healthcaretechnologyonline.com/doc/the-truth-behind-free-ehrs-0001 (September 16, 2013).

28 Roup, op. cit.

29 Chris Newmaker, "Cardiocom Set to Cash in on Health Care Reform," Finance & Commerce, May 2, 2011, http://finance-commerce.com/2011/05/cardiocom-set-to-cash-in-on-health-care-reform/ (October 11, 2013).

30 "About Us," Cardiocom, http://cardiocom.com/about.asp (October 11, 2013).

31 Newmaker, op. cit.

32 Ibid.

33 "Hospital Readmission Reduction Program," Centers for Medicare & Medicaid Services, http://www.cms.gov/Medicare/Medicare-Fee-for-Service-Payment/AcuteInpatientPPS/Readmissions-Reduction-Program.html (October 11, 2013).

34 "Accountable Care Organizations(ACOs): General Information," Centers for Medicare & Medicaid Services, http://innovation.cms.gov/initiatives/aco/ (October 11, 2013).

35 Jonah Comstock, "Medtronic Increases Focus on Home Health with $200M Cardiocom Buy," Mobihealthnews, August 12, 2013, http://mobihealthnews.com/24603/medtronic-increases-focus-on-home-health-with-200m-cardiocom-buy/ (October 11, 2013).

36 "Transforming for Growth: 2012 Annual Report," Medtronic, http://www.medtronic.com/wcm/groups/mdtcom_sg/@mdt/@corp/documents/documents/ar12_annual_report_final.pdf (October 11, 2013).

37 "Concern Over ER Diversions Driving Patient Monitoring System Sales," Kalorama Information press release, July 10, 2012, http://www.kaloramainformation.com/about/release.asp?id = 2877 (October 25, 2013).

38 Comstock, "Medtronic Increases Focus on Home Health with $200M Cardiocom Buy," op. cit.

4.5 Concept Exploration and Testing

INTRODUCTION

If "a picture is worth a thousand words," then in the medtech field, "a prototype is worth a thousand pictures." Fundamentally, there is no substitute for taking all of the conceptual, abstract thinking that has been performed to date and giving it a physical or functional form. Simple concepts can be fraught with problems or result in elegant, effective solutions. Complex ideas can lead to revolutionary results or be impossible to achieve. The only way to find out is to start more deeply exploring concepts through techniques such as prototyping.

The goal of concept exploration is to translate a promising concept from an idea into a rudimentary design, and then into a working form, in order to answer important technical questions. Concept exploration can be thought of as the beginning of the research and development (R&D) process since it is an essential step through which the innovator learns about functionality, gathers preliminary feedback from target users, and tests features that can only be understood and proven through the manifestation of the design.

One of the most important techniques used in concept exploration is prototyping, which is the process of creating early, experimental versions of a product. Prototyping plays a role at multiple stages of the biodesign innovation process. During ideation, the use of crude props to communicate ideas is one form of prototyping. Later, during concept screening, prototyping can be used as a mechanism for quickly and inexpensively evaluating multiple solution ideas against specific design criteria before deciding on a final concept. As innovators move forward, their technical requirements, designs, and models become more advanced. The more robust prototypes that result are used to refine product functionality and features through some combination of user, bench, simulated use, and tissue testing. As these tests are completed, successful prototypes begin to more comprehensively meet important design requirements and innovators begin to transition from prototyping to R&D (or product development).

Importantly, this chapter does not provide innovators with detailed instruction on how to construct different types of prototypes (some preliminary information on this topic is

OBJECTIVES

- Understand how concept exploration is facilitated by techniques such as prototyping and supported by the testing modalities outlined in the biodesign testing continuum.

- Learn guidelines for prototyping, a key technique used in concept exploration, and what questions can be addressed through this approach.

- Become familiar with how to use prototyping in different engineering disciplines, including how to translate concepts into functional blocks.

- Understand how to use prototyping to create design requirements and generate high-level technical specifications related to product feasibility.

- Appreciate how to employ user, bench, simulated use, and tissue testing to transform a concept into increasingly advanced prototypes as the biodesign innovation process progresses.

- Recognize how to use concept exploration as a screen for prioritizing concepts.

provided in online Appendices 4.5.1, 4.5.2, 4.5.3, and 4.5.4). Instead, it outlines an approach innovators can follow to maximize the effectiveness of their concept exploration and testing efforts.

 See ebiodesign.org for featured videos on concept exploration and testing.

CONCEPT EXPLORATION AND TESTING FUNDAMENTALS

Initial concept selection helps innovators narrow their focus from hundreds of **concepts** to a few (see chapter 3.2). In contrast, exploring concepts through prototyping may lead innovators to broaden their focus again as they explore dozens of different ways to technically realize the few concepts under consideration. Subsequently, by employing various testing approaches, innovators once again narrow in on the most promising solutions. However, all of these activities have the same goal: to provide additional data that eventually enables the innovators to choose a final concept and a single way of addressing an important **need**.

In determining which activities to undertake during concept exploration and which questions to answer through prototyping, innovators should strategically prioritize which elements of a concept present the greatest risk to the technical viability of an idea and thus need to be addressed early on. For instance, if a concept revolves around using a certain material in a truly novel way, innovators may seek to prove during concept exploration that the material will perform as hypothesized. If the answer to a limited prototyping experiment (**bench** or tissue test) suggests that the material could be successfully utilized in the intended way, this information helps mitigate significant risk in terms of the concept's feasibility. As risks are mitigated, innovators are able to attract increased interest in their solutions (see 5.2 R&D Strategy and 6.3 Funding Approaches). Accordingly, they should be thoughtful about how they prioritize their work.

At a more tactical level, concept exploration and the testing of the resulting **prototypes** represent the earliest parts of the R&D process. The biodesign testing continuum, shown in Figure 4.5.1, provides a complete overview of the progressively detailed and rigorous tests that are performed as early prototypes are transformed into full-fledged products. However, innovators typically rely on **user**, bench, tissue, and simulated use tests during concept exploration (although this varies based on the type of project and critical risks to be addressed). The later stages of this diagram – from live animal and cadaver testing through **clinical trials** – will be covered in 5.2 R&D Strategy and 5.3 Clinical Strategy.

Guidelines for effective prototyping

Early-stage prototyping is a creative exercise that can be energizing as well as highly informative. The approach described in this section applies to developing all types of prototypes, from mechanical solutions to those based on biomaterials science, electrical engineering, or computer science. As shown in Figure 4.5.2, some prototypes are quick, easy, and inexpensive to develop, while others are more complex, costly, and time-consuming. This is determined primarily by the stage in the biodesign innovation process and the objectives the innovator is seeking to address.

Prototyping is highly iterative, with each successive prototype built to answer questions that arise from the performance successes and deficiencies of previous versions. One way to effectively prototype in an iterative and focused manner is to break a concept down into smaller blocks that correspond to its different functions. Rather than prototyping the whole concept at once, the innovator focuses on proving the feasibility of smaller, essential components before testing them as a system. For example, in developing a new intravascular drug-eluting stent that elutes a commercially available drug over a longer period of time, innovators might explore a block focused on the polymer or mechanism used to allow for longer elution time (a biomaterials science and/or chemical engineering block), one focused on the mechanical properties of the stent itself so that it is

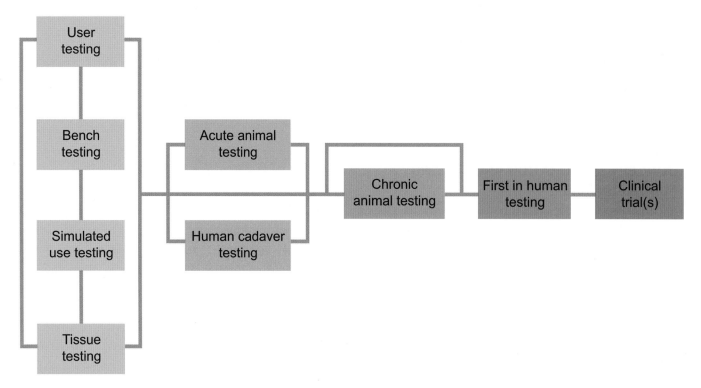

FIGURE 4.5.1

The biodesign testing continuum represents points of evaluation and feedback as innovators move from prototype to product. Note that not all tests are required as part of every project. The specific tests undertaken should be based on the characteristics of the product.

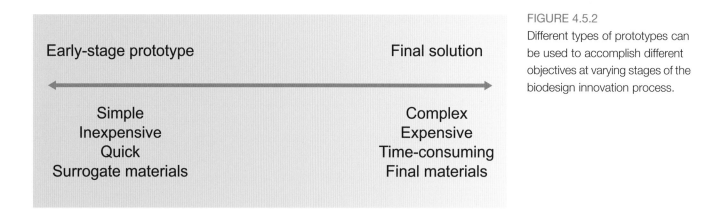

FIGURE 4.5.2
Different types of prototypes can be used to accomplish different objectives at varying stages of the biodesign innovation process.

well-seated within a vessel (a biomaterials science and/ or mechanical engineering block), and another looking at the collapsed shape and pattern of the stent to allow for delivery (a mechanical engineering block). They then might tackle each block separately or in parallel, with several different prototyping exercises early in concept exploration to prove the feasibility of each element and/ or uncover new issues that need to be addressed. This approach makes it easier for the innovators to define

highly specific objectives for each preliminary prototype, and it also helps them determine which aspects of a concept are truly novel and, therefore, may represent the greatest risks that need to be mitigated. Additionally, for large-scale concepts (e.g., a new type of X-ray scanner), smaller blocks may be all that can be prototyped at an early stage of the project.

Through the creation of an iterative series of prototypes, innovators refine their understanding of how the concept will work and gain an appreciation of issues related to the technical feasibility of the concept. As the smaller functional blocks are successfully prototyped, innovators can then start to put them together to understand how the solutions to the various blocks work together. As this process unfolds, they will identify some of the key design requirements that are most likely to satisfy the need in the user's mind. In response, they can continue to iterate on the prototypes until the solution adequately addresses the underlying need.

To take this approach to prototyping, innovators should follow the guidelines outlined below:

- Clearly identify the questions to be addressed through prototyping, focusing on the issues that – if answered with a prototype – will mitigate significant risks.
- Recognize that different types of prototypes may be necessary to address different questions.
- Isolate the most important functional blocks of a solution that need to be explored.
- Understand what is known about each functional block and what must be learned or proven.
- Use what is learned through prototyping to define more detailed design requirements and technical specifications.

To help illustrate these guidelines, an example of innovators working on a solution to the need for *a way to prevent strokes in patients with atrial fibrillation caused by left atrial appendage (LAA) thrombus* is followed throughout the chapter. As described in 2.2 Existing Solutions, the LAA is a small, pouch-like structure attached to the left atrium of the heart in which blood can coagulate to form clots, or thrombi, in patients with an abnormal heart rhythm (such as atrial fibrillation). These clots can dislodge and travel to the brain where they have the potential to create a blockage in an artery, leading to a stroke. In this example, the innovators are focused on the concept of preventing LAA thrombus by filling the LAA with a material that can be delivered in liquid form, after which it will transform into a more solid consistency to eliminate the space where thrombi can form.

Identify the questions to be addressed

When innovators prepare to develop a prototype, their first challenge is to define the specific question or issue to address throughout the prototyping process. Clearly articulating this for each prototype helps to identify what critical elements of the concept should be included in the model. Including "extra" or extraneous features in a model complicates the end results, distracts from testing issues on the critical path, and can unnecessarily increase the time and cost required to build a model. The point is to construct the simplest model possible that will adequately address the key question or issue.

One of the most critical issues that often needs to be addressed during the concept exploration process is whether or not a concept is technically feasible. Most early prototyping work is focused exclusively on proving the technical feasibility of an idea. For instance, it is important to determine whether or not the invention will work with living anatomy. To answer this question, a crude mock-up in conjunction with an animal tissue test could be used to demonstrate that the basic approach is feasible. In the LAA example, the heart could be accessed by means of an existing device, such as a catheter. With only minor modifications, it could be used to inject dye into the LAA of a cow's heart to demonstrate that the general approach of filling the LAA with liquid is feasible. While the functionality of this preliminary model is severely limited, the innovator is able to learn that it is possible to fill the LAA with liquid, and is not distracted by other features that were not necessary to answer this question.

As prototyping progresses, other questions beyond technical feasibility become increasingly important. Prototyping can be equally effective in resolving these issues. A sample of the many different types of questions prototyping can be used to answer is shown in Table 4.5.1.

Continuing with the LAA example, the innovators might want to answer questions about what components are required to make the concept feasible. The crude catheter and cow's heart could be used to try various materials that start as liquids and then solidify, in order to find one that might work. Next, they could seek to demonstrate to key **stakeholders** that this can be done in

Table 4.5.1 There are many different and important questions that can be answered through prototyping.

Issue	Process for resolving issue	Related chapters
Will the concept work? Is it technically feasible and will the product function as designed?	Understand the underlying fundamentals of the clinical problem and then build the prototype and test it.	2.1 Disease State Fundamentals 5.2 R&D Strategy 5.3 Clinical Strategy
Is the innovation novel and unobvious? Can it be patented?	Having sketches and/or prototypes of the idea will help to define the claims of the invention.	4.1 Intellectual Property Basics 5.1 IP Strategy
Will customers adopt and use the product?	Take the prototype to thought leaders and target users in the market. Let them touch and feel it to provide feedback and confirm their interest.	5.3 Clinical Strategy 5.7 Marketing and Stakeholder Strategy
Can it be manufactured?	The development of sketches and/or CAD drawings will help determine manufacturability. It can also be helpful to find precedents and reverse-engineer them. Discuss the prototype with materials and manufacturing vendors to understand their input, ideas, and concerns.	5.2 R&D Strategy
Can it be offered at a price that supports a clear **value proposition** and a sustainable business?	Use vendor estimates/quotes (volume and price) to understand what is the realistic cost to produce the product. That information can then be used to estimate pricing relative to the product's value proposition and the price of competitive offerings. It can also help determine whether the innovators can sustain a reasonable business based on the expected selling price.	5.2 R&D Strategy 5.7 Marketing and Stakeholder Strategy 6.1 Operating Plan and Financial Model
When can it be made available?	Use vendor estimates/quotes (volume and time frame) to develop a project plan that estimates when it is realistic to expect a finished product. Consider hiring a consultant to assist with this exercise.	5.2 R&D Strategy 6.1 Operating Plan and Financial Model

a live animal. They may further develop or modify a catheter, then deliver the previously identified material in an animal study, and later study its pathology post-mortem. Finally, the innovators may want to translate their design and everything that has been learned about the technical specifications to a human scenario, which would include additional design requirements. They also might need to do additional research and make more modifications to the design to ensure that the catheter and material are inert and will not have side effects in the human bloodstream.

As shown by this example, the question or issues that prototyping can address evolve as the design progresses. Each prototype is built on the learnings from previous models. In this fashion, prototyping and its associated questions and issues are highly iterative,

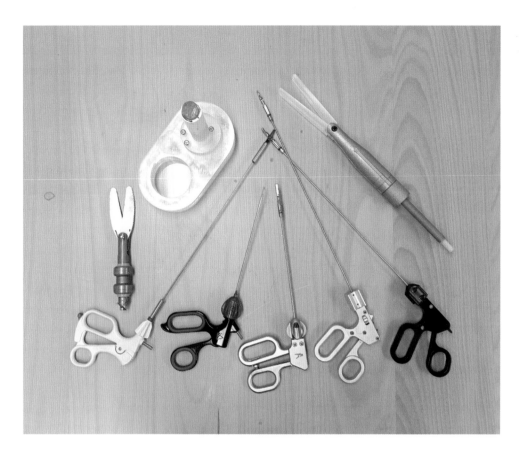

FIGURE 4.5.3
A display of the prototype iterations, from an early works-like prototype (top left) to current day product (bottom right) (courtesy of Miret Surgical Inc.).

similar to much of the biodesign innovation process (see Figure 4.5.3).

Choose the best type of prototype

To answer specific questions or issues that are identified as being important during prototyping, innovators can expect to use a combination of the following types of models:

- **Works-like model** – Demonstrates how the device works. It may not look or feel quite right, but it can be used to demonstrate basic technical feasibility, assess whether the customer would be interested in a device that works in this new way, and/or gather feedback about what they like and/or don't like about the functional aspects of the solution. The bulk of prototyping often is performed using works-like models. Works-like models can apply to mechanical concepts, software programs, applications, or most other types of concepts that have a specific function that needs to be tested.

- **Feels-like model** – Something made of the final material or a surrogate material to demonstrate ergonomics, grip, weight, size, and other tactical factors (e.g., surgical tools). Three-dimensional (3D) printing technologies and other rapid prototyping techniques may be used to create feels-like models, particularly to represent the more physically structured and solid elements of a concept. These models can serve as an important input into human factors considerations. For this reason, they are often used in the early stages of developing wearable technologies as they can be helpful in gathering feedback about **user experience** as it relates to form or design. (See online Appendix 4.5.1 for more information about 3D printing.)

- **Is-like model** – A prototype that performs the desired function and works as intended. An is-like model may not resemble the final form, but could be used clinically. Depending on the nature of the device and the likely regulatory requirements, these models might be used in animal or possibly early human testing. They are often used to transition from design into human testing and manufacturing. For software solutions, an is-like model may represent alpha or beta versions of the software in which the code is functional, but it is still undergoing internal and/or external testing and is not yet ready for general release. These models are also very useful for gathering user experience feedback.

- **Looks-like model** – What the device will look like in terms of its shape, color, size, and/or packaging. Though this is important closer to the end of development when user feedback and marketing decisions are made, a crude looks-like model may be useful earlier to help communicate to others what the innovators are trying to achieve. With 3D printing becoming more accessible and affordable, innovators can more readily produce sophisticated and "finished-looking" looks-like models. These models can be excellent communication tools to a wide variety of stakeholders, including investors and customers. For software applications, innovators can use mocked-up screenshots to gather feedback about the user interface and design for software and applications.

- **Looks-like/is-like model** – A combination of the previous two models. This type of model both functions as and looks like the final device. This step may be undertaken by manufacturing as the is-like model is modified to incorporate the looks-like elements during the technology transfer process from design to manufacturing. This more robust model is a natural candidate to best understand user experience with the product and applies equally well to a range of products from software and applications to mechanical solutions.

Again, it is generally most effective to start with simple works-like prototypes that *only* convey basic information about the concept and are narrowly focused on answering a single, specific question. In the LAA example, the innovators' first step was simply to determine if the LAA could be accessed and filled with liquid. The exact shape, size, weight, look and feel, and complete functionality of the device were irrelevant to answering this basic question. As a result, a simple works-like model was not only appropriate, but relatively easy and inexpensive to construct. It also did not introduce unnecessary complexity into the process of answering the relatively simple question the prototype was designed to address.

The more questions a single prototype seeks to answer, the more risk an innovator faces in understanding the results of the model (i.e., the specific cause of a particular technical problem). This is not to say that advanced prototypes that represent more finalized designs should be avoided. On the contrary, they are essential to proving the total concept. However, they should be built in such a way that they represent the summation of all the work done to date (and all the questions that have been answered by previous prototypes). In this way, advanced prototypes can still be built to answer a single question – it may just be a higher-order inquiry, such as "Will the model work in a living system?"

Before attempting to build complex prototypes, it is essential for innovators to have a deep understanding of how similar solutions currently available in the market function and what their strong and weak points are in terms of design, interface, and user experience. Often referred to as "reverse engineering," the idea of disassembling and analyzing existing products or devices can be helpful in making the prototyping process more efficient and giving innovators a foundational understanding of what does and does not work well within existing solutions. However, it can be dangerous if innovators start to anchor on the "current state" and adopt unnecessary constraints dictated by the existing technology.

Just as developing a prototype that is unnecessarily complex can be distracting to the innovators, it can also be distracting to users when asked to give their feedback. For example, if an early works-like prototype looks too much like a finished product, users may concentrate on how the device looks and feels instead of focusing on critical issues related to its fundamental functionality. The level of the prototype must match the question or issue being considered and only incorporate as much complexity as is needed to find this answer. Crude prototypes, using common materials, can provide a highly time- and cost-effective mechanism for testing the basic functionality of a device. Some examples of rudimentary, but effective prototypes are shown in Figure 4.5.4.

The Oculeve example further demonstrates how different types of prototypes can be used to answer different questions and the value that rough, "low tech" models can provide.

FIGURE 4.5.4

Clockwise from top left to center: an electrical engineering breadboard (iRhythm Technologies); a rough prototype of a device to constrict a body passage and the mold that was used to cast it (Jeremy Koehler); a prototype pump and vacuum system (Calcula Technologies); an early model for a neonatal hearing screening device (Sohum Innovation Lab); and a rough functional prototype for an intravascular tool (InterVene Inc.).

FROM THE FIELD ▶ OCULEVE

Using early models to test technical feasibility and likelihood of adoption

Several years ago, a team in the Stanford Biodesign program began pursuing a way to treat chronic dry eye disease that would be more effective than the prescription eye drops (topical cyclosporine) that are the current **standard of care**. Dry eye disease (DED) is caused by insufficient tear film on the eye, either because the lacrimal gland does not produce enough tears, or because the tears evaporate too quickly from the eye's surface.[1] Without a stable protective tear film, the surface of the eye becomes inflamed, causing patients to experience painful dryness, burning or grittiness, and impaired vision.[2] An estimated 20 million Americans are affected by dry eye disease, with 1.6 million suffering from a form of DED that is severe enough to be debilitating.[3]

As the team vetted the dry eye need, member Michael Ackermann began mulling over ideas for a solution. Ackermann, an electrical engineer who had previously worked for Boston Scientific on neurostimulation devices, believed that this approach might provide an answer to the dry eye problem. "The lacrimal gland, which is innervated by a parasympathetic nerve, produces tears when stimulated by this nerve. So, by delivering stimulation pulses to activate the nerve, we could potentially activate the downstream organ," he explained.[4] Although the idea resonated with the team almost immediately, Ackermann noted that, "True to the process, we didn't chase it down too quickly. We went through the steps in the biodesign innovation process, but as the dry eye need became increasingly compelling, this solution idea kept resurfacing and withstanding the various trials we put it through."

Taking the neurostimulation idea forward into concept exploration and testing, the team began grappling with how to prototype a solution. "It was an interesting phase for us," recalled Ackermann. "The downside of neurostimulation is that it requires a really sophisticated device, so it is hard to build exactly what you want and go and get feedback with it." Faced with this challenge, the team focused on the most pressing questions they had to answer. "We needed to know two things – one, would electrical stimulation produce lacrimation, and two, would the doctors and patients be willing to use it," said Ackermann. Accordingly, they decided to focus on two primary types of prototypes: "looks-like" and "works-like" models.

The looks-like prototypes would be used to gather user feedback and address the adoption question. These "low resolution models," made initially from Play-Doh and cardboard, were used as a visual tool to help explain the concept to doctors and gauge their interest. "The first looks-like prototype was for a device that would be implanted in the chest with a lead up to the ocular orbit," Ackermann described. "The second iteration resembled a cochlear implant with a lead that would go over the ear." Both versions garnered mostly negative feedback from physicians, sending the team back to the design table. For the next version, Ackermann and colleagues evolved their concept significantly, designing a tiny device that could be placed directly into the orbit. "Our first intraorbital prototype, which looks much like the device we have now, was made out of a piece of plastic cut from the packaging of a handtool, and a little piece of modeling clay to simulate the titanium can portion of it," he remembered (see Figure 4.5.5). "This time, the feedback we got was, 'Yes, if you have something that kind of looks like this and it works, then that would be something we'd be excited about.'" Encouraged, the team began testing the looks-like prototype on cadavers in Stanford's anatomy lab to "start getting some sense of how the surgical procedure would work, and how the device would fit in with the orbital anatomy," he said.

confirmatory testing anyway to be extra careful before considering human use," stated Ackermann. For example, in one of its animal tests, the team delivered a 10× dose of stimulation to rabbits and then studied the histology to confirm the safety of the treatment, as well as its efficacy. "We also performed a thorough risk analysis to be sure we had considered and mitigated all of the significant risks we could encounter in the clinic," he added.

Reflecting on the team's early prototyping and testing experience, Ackermann commented that, "Both types of prototypes really pushed us along." He also credited "an extraordinary team of advisors" that included Daniel Palanker, Associate Professor in the Department of Ophthalmology, Stanford School of Medicine, and Mark Blumenkranz, Professor of Ophthalmology at the Stanford University Medical Center. "These guys are savvy, with great reputations and experience. Their feedback and advice as field experts was instrumental in helping us shape the concept," he emphasized. Finally, Ackermann stressed the importance of using rough prototypes to quickly begin figuring out what works. "You can learn a lot by using surrogate devices and early prototypes that execute on the concept, or parts of the concept," he said. The information gathered through concept exploration and testing positioned the team to begin thinking about more robust R&D and clinical strategies to bring the idea forward. (See 5.2 R&D Strategy for more information about this project, which would eventually lead Ackermann and his colleagues to found a company called Oculeve, Inc.)

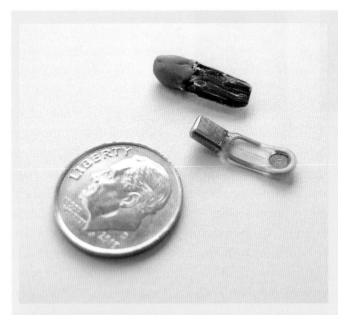

FIGURE 4.5.5
One of Oculeve's look-like prototypes alongside the near-final version of its intraorbital implant (courtesy of Oculeve).

In parallel, Ackermann and his teammates began to develop a works-like prototype to test the concept of electrical stimulation of the lacrimal gland to increase tear production. "We started with a clinical stimulator that was used to identify nerves in anesthesiology for nerve blocks, and we attached some percutaneous fine wires electrodes to it," said Ackermann. Although the prototype "looked nothing like our final product," it replicated the mechanism of action accurately enough for the team to put the prototype through successful bench and animal testing. "The neurostimulator we used was an **FDA**-cleared, off-the-shelf product, but we did

Identifying the functional blocks of a solution

When the initial questions or issues the innovators want to answer are too broad to be easily addressed with a single series of prototypes, they are encouraged to divide the concept into "functional blocks." The original **need criteria**, along with more specific functional design requirements that may have emerged through the analysis performed in chapters 4.1, 4.2, 4.3, and 4.4 can be used to establish the boundaries for the blocks. Each block should represent one aspect of the concept that can be prototyped and is usually tied to a distinct engineering discipline based on its characteristics. Some common examples of the engineering disciplines addressed by different functional blocks include mechanical engineering, biomaterials science, electrical engineering, and computer science/software engineering.

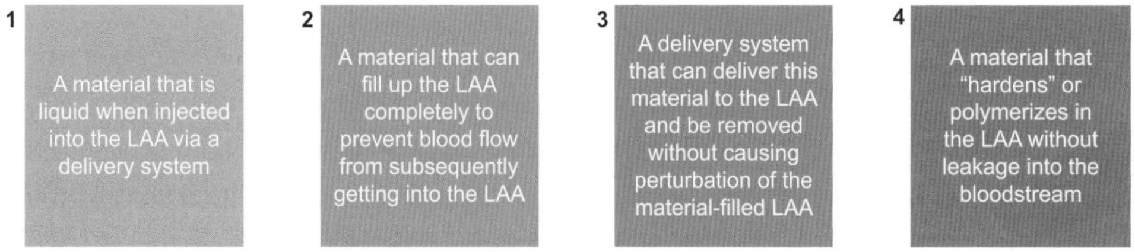

1	2	3	4
A material that is liquid when injected into the LAA via a delivery system	A material that can fill up the LAA completely to prevent blood flow from subsequently getting into the LAA	A delivery system that can deliver this material to the LAA and be removed without causing perturbation of the material-filled LAA	A material that "hardens" or polymerizes in the LAA without leakage into the bloodstream

FIGURE 4.5.6

The primary functional blocks related to the LAA example.

Once the relevant functional blocks for a concept are defined, innovators can prototype each one independently, based on which questions they seek to answer, before bringing the blocks together to prototype the concept as a whole. Importantly, the key elements that could demonstrate the viability of an idea may be represented by only a few blocks (or sometimes a single one).

The primary functional blocks for a prototype for the LAA example might include those shown in Figure 4.5.6. As these blocks demonstrate, the original concept has a mechanical engineering component (#3) and biomaterials science components (#1, 2, and 4). While some concepts reflect a "pure" solution within a single engineering discipline, it is not unusual for innovative concepts to combine different types of engineering science. Instead of breaking down the issues related to biomaterials science into several blocks, another approach might be to classify all of the blocks related to material into a single block with multiple properties. While this may be appealing if the blocks are closely related, by breaking the concept into a more granular level of detail, with multiple blocks for a given engineering discipline, the innovators can more easily determine which of these smaller blocks should be prototyped first, based on what might be novel and what represents the highest degree of risk that needs to be addressed. As the innovator learns about what blocks can be addressed with readily applicable existing solutions, there is no harm in combining them if it makes sense.

The emergence of combination products, such as the drug-eluting stent or biologic coated stent for treatment of coronary artery disease, now demands even more sophisticated prototyping. These new devices often have even more complex functional blocks that can interact with each other in complicated and unpredictable ways. Biologic and pharmacologic blocks add other mechanisms of action to consider that may be mechanical, electrical, or chemical in nature. In this case, expertise in pharmacology and molecular biology would be essential to the prototyping team. In addition, lab resources for both macro and microscopic evaluation of mechanisms would be required. This may not only involve cell and tissue culture facilities, but also new technologies to measure these mechanisms, such as small and large animal imaging.

Understand what is known

Once the functional blocks of a concept have been defined, innovators can evaluate what is already known about each one. This is essential since it helps determine which blocks to focus on first, in order to address the ones that represent the greatest risk. Having a thorough understanding of how similar and competitive solutions function can also help innovators identify which functional blocks are truly novel and which might already have reasonable solutions that can be leveraged to make a concept work. In some cases, many elements of a solution concept may already be well understood, while just one or two aspects still need to be proven through prototype development. For instance, in the LAA example, the innovators can potentially leverage available catheter-based delivery systems, which provide a well-understood mechanism for reaching the heart and are acceptable to certain stakeholders such as cardiologists. Much greater uncertainty exists around the material for filling the LAA and its interaction with the heart. As a result, the innovators might decide to focus their earliest prototyping efforts on the biomaterials science functional blocks #1, #2, and/or #4 before moving on to #3.

Examining the functional blocks further, the innovators may also discover that there are existing materials that can fill an anatomical space (such as the LAA) and be delivered in a liquid form via a catheter. In contrast, they may be uncertain about the existence of materials that can polymerize while in contact with the bloodstream. In this case, they could choose to focus on this latter functional block first to ensure that unknown can be resolved before investing significant time in other areas. Note that under these circumstances, it makes sense for the innovators to divide the biomaterials science aspects of the solution into different functional blocks, rather than considering them all together since many more unknowns exist in this one area versus the others.

Define more detailed design requirements and technical specifications

Just as breaking a solution concept into functional blocks can simplify how to approach the prototyping process, it can also aid with defining more precise design requirements and technical specifications. Once the original need criteria (which are usually fairly high-level) have been considered, additional criteria, relevant to each of the functional blocks, can be defined based on what has been learned to date. For instance, in the LAA example, a more precise design requirement for block #3 is that a catheter-based delivery system should be less than 8 French[5] in diameter and steerable to deliver the chosen material to the LAA. This is based on the determination that cardiologists, not cardiac surgeons, would probably perform the procedure and would prefer a method and tools similar to what they already use. Such specifics can be used at this point to guide prototype development along a pathway that takes into account relevant outside factors. At the same time, detailed design requirements help further define the questions that should be answered through each successive prototyping exercise.

Once a concept or a key element of a concept has been given a working form, it is easier for innovators to solicit specific input to guide improvements affecting the usefulness and marketability of the idea. Real users – members of the target audience for a device – play an important role in directing prototype development as they can identify detailed design requirements that may be unknown to the innovators. Sometime the term "design requirements" is interchanged with "**user requirements**," which emphasizes the importance of the user in generating this information. Users can also raise issues and risks to which the innovators have become "blind" due to their deep involvement with the concept.

Having a crude prototype that meets the need criteria and some basic design requirements of key functional blocks (as understood by the innovators) is the easiest way to start the process of gathering more detailed information. Going back to the LAA example, the innovators can gain important input by showing users a model based on a standard cardiology catheter, a substance such as glue, and a cow's heart. In this scenario, the innovators could demonstrate how the glue can be introduced via the catheter to fill the LAA, after which it polymerizes and hardens within one hour. The cardiologist might then point out, for instance, that the glue would need to polymerize in less than 20 minutes, as one hour could be too long to keep a catheter in the left atrium. Accordingly, the next prototype in the series would have a goal to address this more refined design requirement. Consulting with multiple users can help the innovator avoid the **biases** of any single individual.

Through the process of developing increasingly detailed prototypes and refining design requirements, innovators will naturally develop an appreciation of key technical specifications. Technical specifications represent the important engineering parameters that a solution must satisfy in order to meet the need and important design requirements, provided that these requirements are within the bounds of technical feasibility. Stated another way, technical specifications capture and explain how a solution must function. These specifications typically focus on technical features (e.g., what loads the device must tolerate, what material a device must be made of, how durable it must be, what level of encryption is needed), but may also address "softer" attributes desired by the user which are related to device functionality (e.g., compatibility with other devices, how it is operated, ergonomic features). All such specifications warrant careful consideration during device design as they will likely impact R&D and product development activities later in the biodesign innovation process.

Technical specifications can also emerge from the successes and failures that the innovator experiences in developing different prototypes. The LAA example can once again be used to illustrate this point. Prototyping in a live animal may show the innovator that a catheter has to have a certain rigidity so that it does not collapse as it is inserted into a blood vessel. Through trial and error, the innovators would gradually come to understand the appropriate load factor that it has to withstand. Once this determination is made, this specification would be considered essential for the solution to be technically feasible.

While little emphasis is placed on rationalizing design requirements and technical specifications early in the biodesign innovation process, innovators must begin to acknowledge necessary trade-offs between these parameters as design and development progresses. For example, users might want a device that has many technically complex features but also is miniature in size. When a design is initiated, they may be faced with limits on the degree of miniaturization that is possible while meeting the requirements for technically complex features. If the innovators do not have a clear sense of which requirements are most critical, more user input may be needed to help balance these requirements against relevant technical specifications. Alternatively, multiple designs can be created to emphasize different combinations of key requirements. Target users can then be asked to respond to the specific designs instead of helping to prioritize requirements in the abstract – an exercise that many individuals find more challenging. Regardless of which approach is chosen, innovators must carefully monitor the interplay between design requirements and technical specifications to ensure that a solution can be feasibly engineered.

One approach that can be useful in prioritizing and rationalizing the technical specifications that lead to product features is called Kano model analysis. Innovators can use this method to systematically make feature-level decisions (whether or not to include a single feature in a new product) or to identify an optimal mix of features that generates the most excitement with users and is recognized by key stakeholders

to offer real **value**.[6] More information about Kano analysis is provided in the User Testing section of this chapter.

Timing of prototyping

While there is not necessarily a "right" time to start prototyping, many experts encourage innovators to begin as early as possible so that they learn from the process and apply that learning to iterative design and development. If there is a question that engaging in prototyping can answer, that exploration is usually a worthwhile effort. As Tom Kelley, general manager of design firm IDEO, explained:[7]

Quick prototyping is about acting before you've got the answers, about taking chances, stumbling a little, but then making it right. Living, moving prototypes can help shape your ideas. When you're creating something new to the world, you can't look over your shoulder to see what your competitors are doing; you have to find another source of inspiration. Once you start drawing or making things, you open up new possibilities of discovery [by] doodling, drawing, [and] modeling. Sketch ideas and make things, and you're likely to encourage accidental discoveries. At the most fundamental level, what we're talking about is play, about exploring borders.

The downside of prototyping too early in the biodesign innovation process is that innovators can get swept up in creating a working model before they are certain that the solution meets key need criteria and/or design requirements. Another pitfall is becoming committed to a concept once it takes a working form, despite fundamental flaws that may affect the technically feasible of a final product. Issues such as these may lead innovators to invest money and to spend days, weeks, or months defining and refining a concept that may ultimately be inadequate to satisfy the underlying need (see the OneBreath example in 4.6 Final Concept Selection). Innovators must make a careful judgment about the ideal time for initiating (and sustaining) prototype development, since there are many pros and cons (see Table 4.5.2).

Table 4.5.2 Advantages and disadvantages of working with early models as part of the prototyping.

Advantages of early working models	Disadvantages of early working models
• Flexible; can be relatively easily changed. • Relatively inexpensive (compared to later prototypes). • Can help identify critical design requirements from users early on and incrementally throughout the development cycle. • May provide the proof of concept necessary to attract funding. • Can demonstrate key technical challenges and issues regarding feasibility. • Can help give users an idea of what the final product will look like and how it will operate. • Can encourage active engagement in the development process by users. • Can motivate the innovator(s) and drive an increase in product development speed.	• User needs and design requirements may not yet be adequately understood, rationalized, and reflected in the designs. • Early designs may not adequately address user expectations and may, therefore, disappoint the target audience. • The innovator may become prematurely attached to certain aspects of an early prototype that are counterproductive to upstream (user requirements) or downstream (technical feasibility and manufacturing) considerations. • The innovator might waste time and money prematurely prototyping an incomplete or flawed design.

Concept exploration as part of the biodesign testing continuum

As described earlier, concept exploration is a key early component in the larger R&D effort. In particular, prototypes are used to gather user feedback and prove the technical feasibility of certain concepts through various types of early testing. These testing modalities, as shown in Figure 4.5.7 (excerpted from the biodesign testing continuum), are important to understand as part of concept exploration.

User, bench, simulated use, and tissue tests are usually undertaken multiple times throughout the iterative design process. Live animal testing, cadaver testing, and human testing are frequently performed during the later stages of the biodesign innovation process when more advanced works-like and is-like models are available (see 5.2 R&D Strategy and 5.3 Clinical Strategy). Throughout the testing continuum, the innovator continues to collect information about key design requirements and understand increasingly important technical specifications. Key considerations for user, bench, simulated use, and tissue testing are outlined below.

User testing

User testing involves evaluating a concept by testing it with users to gather their input and observe their interactions with it. At first, this activity may involve showing target users drawings or storyboards of a solution idea and asking for their feedback. Next, it may involve showing users works-like, looks-like, or is-like models to observe their informal reactions and interactions. Over time, it may evolve to be more rigorous, relying on techniques such as usability testing to systematically gather qualitative and quantitative data under controlled circumstances. In a usability test, users are given a relatively advanced prototype and observed interacting with the technology or device to determine how well it meets key requirements and technical specifications. Innovators are unlikely to perform robust usability tests during preliminary concept exploration, but it is a useful method to consider as R&D progresses.

As mentioned, Kano model analysis is another valuable tool for systematically gathering and acting on user feedback. This approach enhances the early work innovators performed in defining need criteria by

Stage 4: Concept Screening

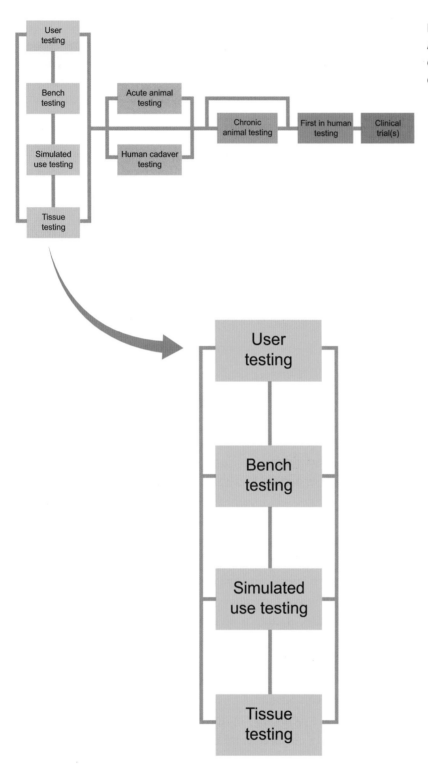

FIGURE 4.5.7
A combination of user, bench, simulated use, and/or tissue tests can enable innovators to effectively evaluate their early prototypes.

providing a framework to put features into four key categories:[8]

1. **Must-haves** – The features that are most critical to customers.
2. **Dissatisfiers or must-not-haves** – The features that would turn people off and drive them away.
3. **Yielders of indifference** – The features that cause a "who cares?" reaction.
4. **Exciters** – The features that delight users, provide unexpected excitement, and contribute to a high level of satisfaction and perceived value.

To classify features according to this construct, innovators use a prototype or model to show users a series of features. For each one, they should ask the users to answer a "functional" and a "dysfunctional" question, such as:[9]

- How would you feel if the new device offered this feature as part of its design?
- How would you feel if the new device did NOT offer this feature as part of its design?

For both questions, users are asked to choose from the following responses:[10]

- It would excite me to have it that way.
- I would require it to be that way.
- It does not matter to me if it were that way or not.
- I wouldn't like it, but I could live with it that way.
- I would not accept it to be that way.

Once the results are gathered, the Kano technique outlines an approach for analyzing the data to understand user priorities and preferences related to specific features, as well for determining an optimal mix of product features. See the Getting Started section for references on Kano Analysis.

Before innovators are ready to engage in formal user testing methods such as usability testing and Kano model analysis, they can still gather useful user input by following some general guidelines:

- **Gather user experience feedback by asking users open-ended questions** – What do they think of this? What could be better? Why don't they like it? Is the reason they do not like it functional, design-related, or practical in nature?
- **Understand user interface and experience issues by observing users as they handle and use the prototype** – What features give them the greatest difficulty as they use it on the bench or in animals (e.g., for a mechanically oriented solution) or interact with it in their typical surroundings (e.g., for an application or software program)? What seems to attract or appeal to them?
- **Gather design input by asking users to imagine using the device in a clinical setting** – If it is not possible to directly observe users trying the device or technology (i.e., due to privacy or regulatory constraints) or it is too early in development to provide them with an advanced model, ask them about the other factors that must be considered in designing the device to integrate it into an operating room, physician's office, catheterization laboratory, healthcare system, etc. Are there other equipment/environment considerations that affect its use? For software and applications, are there compatibility or integration issues that need to be incorporated?

While it is unlikely that all of the feedback gathered from multiple users will be consistent, the opportunity to consider a wide range of opinions can reveal the full spectrum of strengths and deficiencies associated with a prototype. The innovators will then have to determine, based on their own observations, what user feedback to act on, as well as which elements of the prototype to further refine or modify, and which ones to accept.

The Vynca story highlights the importance of gathering user input often and early. Though important across the whole range of biomedical technologies, it can be useful (and relatively easy) to get early input for consumer-oriented products such as software or mobile health applications.

The importance of understanding users and user testing

Rush Bartlett, Ryan Van Wert, and Frank Wang of the Vynca team knew as they launched Vynca that one of the most important elements of developing any **medtech** solution – especially a software-based offering – is making sure it is tailored to the needs of the most essential and influential end users. As they set out to address problems surrounding end-of-life physician orders for resuscitation, they identified numerous users (or stakeholders) to consider. "Stakeholder analysis was the single most important factor that we looked at," said Bartlett. While patients, physicians, nurses, nursing home and hospital administrators, dispatch center personnel, and hospital IT staff were all key audiences that could affect the adoption of a new solution in the space, they prioritized physicians and hospital IT staff as being most central to accepting or resisting whatever they designed.

When conducting clinical observations in pulmonary and critical care medicine as Biodesign fellows, Bartlett and Van Wert were drawn to problems related to acute resuscitation orders and related end-of-life issues. Within the span of a week, they encountered two different cases where elderly patients were put on life support despite having previously filled out a Physician Orders for Life-Sustaining Treatment (POLST) form that clearly specified their wishes to avoid such an intervention. POLST forms are widely accepted in the medical community, but the challenge is ensuring that physicians have reliable and timely access to the information. This is particularly important because the vast majority of patients with terminal illnesses do not want aggressive care at the end of life, but physicians are obligated to provide this care unless instructed otherwise.

In one case, the patient had a POLST form with him (indicating that he did not desire to be put on life support)

when he was transported to the hospital, but it was left behind in the ambulance. Unable to speak for himself, the patient was placed on a ventilator and transferred to the intensive care unit. When the team was validating the need through further observations in the emergency department, the members realized that the POLST form, as with other paper documents, is given to the registration clerk and sent to the hospital's central document management center where it is manually scanned, labeled, and uploaded to the electronic medical record (EMR). This process typically takes one to two days. In effect, the existing workflow for POLST in the emergency department precluded the treating physician from viewing the form to make critical management decisions. Bartlett and Van Wert's physician advisor, Dr. Allen Namath, a critical care specialist, had encountered the problem of recording and acting on patient's wishes on numerous occasions. As a pulmonologist and an intensivist, Namath had the unique perspective of seeing both sides of the problem. He had traditionally filled out POLST forms in his pulmonary clinic, but would unexpectedly see some of the same patients who clearly indicated they never wanted to be placed in the ICU appear there during his ICU shifts. "Very quickly, this need filtered to the top of our list based on patient impact and high cost to the system," explained Van Wert. "One-third of the Medicare budget in the US is spent on patients during their last year of life, much of it for care that patients do not want to receive in the first place." Accordingly, they decided to focus on the need for *a way to rapidly transmit POLST form information to any part of the healthcare setting*.

The team brainstormed a wide variety of concepts for capturing and transmitting end-of-life treatment wishes. However, said Bartlett, "We realized that the POLST form was well accepted and backed by two decades of clinical literature." For this reason, the members decided to first focus on making the form easier to complete in a digital format with added features that would improve security, assist with patient education, and make the captured information more accurate in terms of reflecting patient preferences. The second critical step was then to innovate ways to make the information rapidly accessible to any healthcare providers in any setting.

Physicians were a top-priority user group because they played a central role in completing the forms during an office or hospital visit with the patient. They were also the ones who needed to access the information on a just-in-time basis when administering care. Additionally, as Van Wert noted, "Having a clinical champion is always essential to drive adoption of a new technology in a health system." With this audience in mind, the team defined a series of user requirements that would serve as the basis for its solution, including eliminating paper, integrating the solution with EMR systems, and making the POLST information available via a single click. "I have seen a lot of suboptimal hospital IT software," said Van Wert. "Every extra mouse click and every extra login adds a lot of time to the physician's day, especially when they're doing activities that are repeated over and over again. We were committed to putting ourselves in the healthcare provider's shoes as we built the system because we realized this would be a key driver to adoption."

At this point, the team was ready to begin prototyping. "One attractive thing about software-based solutions is the speed with which they can be prototyped and tested," Bartlett stated. "Relative to traditional medical devices, it's easy to create different types of prototypes of software products to demonstrate them to potential customers and gather more information about what they're looking for in terms of usability."

The team's prototyping approach was based on the development of a series of increasingly robust models. The members began by drawing the workflow of the website with pen and paper under different use case scenarios. Based on preliminary feedback on these mock-ups, they created diagrams and slides with more detail, which they sent out for feedback from a larger group (see Figure 4.5.8). Next, they developed an online

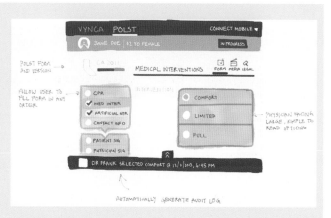

FIGURE 4.5.8

A wireframe sketch of the preliminary website concept that the Vynca team used to gather feedback from users (courtesy of Vynca).

looks-like model that appeared on the screen as it would, but did not have active functionality. Finally they created a "bare bones" but functional website (a looks-like/is-like model) which they used to gather critical input from a wider audience. "Having a functional prototype to show to potential customers was very powerful," Wang said. "We didn't want to build the solid, **HIPAA** compliant, robustly scalable, final site until we were sure. You only want to do that once."

With each model (see Figure 4.5.9), the team gathered increasingly detailed feedback from an expanding group of physician users. They shared the initial drawings within the team and with their close physician advisors. They showed the subsequent storyboard slides and the looks-like model to a slightly larger group of physicians in their network. By the time the looks-like/is-like model was ready, they developed an online user survey to send to an expanded group of doctors that represented potential users of the POLST form. Importantly, when gathering each round of feedback, Bartlett, Van Wert, and Wang made a concerted effort to sit down with the physicians, observe them as they interacted with the prototypes, and ask open-ended questions about how they could imagine the system fitting in with their day-to-day work flow. Although the team members were eager to get input from anyone who would talk with them, Bartlett commented that not all user feedback is equally valuable

(a)

(b)

(c)

and innovators should be somewhat selective in terms of who they approach. "Make sure you're getting the right users because not all feedback you receive is going to be relevant. Sometimes you have to ignore input or take some of it with a grain of salt. Make sure you connect with high-quality users that are directly in the line of fire, so to speak. They'll give you the highest impact feedback. If you send a random email to the followers of a blog, a lot of the feedback is going to be garbage and it's going to dilute your signal." For this reason, Bartlett and his colleagues invested significant time in cultivating relationships with leaders in the palliative care field.

As the team began to deeply understand the requirements of the physicians users of their solution, they also started exploring what it would take to get hospital IT departments to adopt it. "There was a lot of pounding the pavement and getting people to talk to us," said Bartlett. The team used its Stanford affiliation to connect with members of the hospital's IT department. "We asked them, 'What would you be looking for in a solution as an IT director?" noted Van Wert. "IT departments are often negative stakeholders in this process. They sometimes lobby against changes because those changes expose them to risk. Further, they see so many new tools that people want to integrate with the hospital EMR system, but they're not necessarily exposed to the clinical needs behind them. As a result, they're unable to judge the merits of the solutions except by how it increases their risk exposure," explained Bartlett. Accordingly, the team committed itself to differentiating its software by ensuring that its implementation would be as minimally disruptive as possible. The members also took the time to explain the need and garner wide support for the new system across the hospital system.

Elaborating on this approach, Bartlett said, "By working with IT department to ease integration challenges, innovators can mitigate adoption risk." In the process,

FIGURE 4.5.9

These screenshots show how the Vynca website progressively changed with feedback from key stakeholders (courtesy of Vynca).

these conversations helped the team uncovered additional technical and user requirements, especially around HIPAA compliance, privacy, and security issues. To augment this information, said Wang, "We learned a lot from looking at how other third-party tools available in the market integrate with the EMR system."

In 2014, the team, which adopted the name Vynca, was on the cusp of launching an e-POLST project **pilot** with a prestigious academic hospital. This particular institution

was an early pioneer in POLST management in the United States and remains a national leader. Through the pilot, Vynca will be able to test a fully functional, cloud-based, HIPAA compliant version of its software and gather additional user feedback on a large-scale basis. The hospital is also helping the team build connections to other potential hospital customers by leveraging its record of leadership and advocacy in POLST management.

Bench and simulated use testing

Bench testing refers to the testing of materials, methods, or functionality in a small-scale, controlled environment, such as on a laboratory workbench. These early tests typically serve to help identify key elements of a design in terms of materials or functionality. This allows the innovator to make optimal choices about how to build more robust prototypes for later stages of testing. In bench tests, individual components or subassemblies are put into a test loop where all of the associated variables can be independently controlled, measured, and recorded.[11] Bench tests should be designed as simply as possible. For example, for a catheter device with an articulating arm and handle mechanism, the innovator should test the arm mechanism independently of catheter tracking or lubricity to ensure that each of the individual features works independently. This makes it easier to understand the test results when combinations of features are eventually tested together.

In simulated use testing, various features or functions of the prototype are tested as a system on an anatomical model or proxy. Device characteristics and design parameters can be tested in a predictable and reproducible manner to demonstrate actual performance. For instance, a bench top simulated use system that mimics physiological hemodynamics and vascular anatomy can be used to test and refine a vascular implant prototype before going into expensive animal studies. For some devices, simulated use testing can help fulfill validation testing requirements of the FDA or other regulatory agencies.

When developing bench tests or simulated use experiments, innovators should use simple methods and

generic materials or machines (tensile testers, weights, levers, calipers, etc.). They should test in real tissue later in the testing continuum, as many less-costly surrogates are readily available to refine and develop prototypes to a point where the design has a better chance of working in real tissue. During preliminary tests, materials that can provide a substitute for tissue include silicone, chamois, nylon, sponge, foam, or ceramic pieces (for bone).

Niveus Medical, a company started by Stanford Biodesign fellow Brian Fahey, employed simulated use testing to gain valuable feedback on the design and functionality of early prototypes. The company created a novel muscle-stimulation solution to address the problem of weakness, debilitation, and muscle atrophy in patients facing extended periods of immobilization or bed rest. In designing a solution, the team had a series of human factors issues to explore. Some of these dealt with the interaction between the device and patient since the solution included a component affixed directly to the skin. Others dealt with the workflow of the nurses who would equip patients with the device and monitor its usage. Early in the concept generation and screening stages of the innovation process, the Niveus team recognized that the product had to be designed to fit into the complicated and hectic environment found in hospital settings. "Any solution that was viable had to be very easy, quick to deploy, and intuitive to use, with limited training required," Fahey explained.

To address these human factors, the Niveus team created a looks-like prototype, which consisted of a disposable electrode pad, a connection cable, and a mock generator. Then, the engineers set up two different types of simulated use tests. First, they used mannequins to explore different human factors related to fit, sizing, and placement on the skin. Second, they set up focus groups of 8–10 nurses, who were asked to use the looks-like prototypes on mannequins lying on beds with only limited instruction. "So we didn't just prototype the product, we prototyped the workflow," said Fahey. The team observed the nurses through the end-to-end process of finding the product on a shelf, opening the package, interacting with the product components, placing the pad on the mannequin, connecting it to the generator, and then removing the system. "Testing the entire workflow allowed us to validate several assumptions and learn some new things as well," he commented. For instance, the team noticed that different nurses placed the generator component in different parts of the room, which meant that the connecting cable had to be flexible and long enough to accommodate the different use cases. As a result of this insight, the engineers ended up redesigning the connecting cable and changing the position of where it interfaced with the disposable pad. Fahey also stressed the importance observing how users "naturally do things" with the product, such as how they might bend or break it, so that the team can design around possible failure modes of the device. "It is always good to know your limitations early on," he noted.

As a rule, innovators should use tests consistent with the stage of prototyping that is being undertaken. As the prototypes and models grow more sophisticated (through design evolution) and risks are retired, tests can become more complex such that they begin to more closely resemble actual use. Eventually (and often rapidly), the design gets to a point where tissue testing is required to generate the next set of design requirements.

Tissue testing

Tissue testing is an essential component of medical device prototyping for certain types of products that should be undertaken after a concept has been proven in simpler tests. This is not to say that tissue testing cannot be undertaken earlier, but by answering many of the straightforward questions through bench and simulated use testing, results are less likely to be confounded by non-critical data. Innovators can quickly evaluate a device's mechanism of action using animal tissue available from the local grocery store or butcher – for example, substituting the skin on chicken feet to emulate human skin when testing the ability of a new suturing device to penetrate tissue. Electrical mechanisms may be evaluated in much the same way. For instance, a skinless chicken breast can be used to assess the ability of a new ablation catheter energy source to create tissue injury. With chemical mechanisms, a similar approach can be used. As shown in the LAA example, developing a glue that would adhere to the endocardial (inside) surface of the heart is critical. For a preliminary tissue test, an inexpensive cut of beef could be used to evaluate the interaction between the glue and the tissue.

Although mechanisms of action can often be evaluated using simple, widely available animal tissue (like beef tongue, steak, gelatin, or poultry), device feasibility must almost always be confirmed using animal or cadaver tissue that is more representative of the anatomical or physiologic properties that the device intends to address. For example, swine and sheep organs can be obtained from many commercial slaughter houses. These tissues can be shipped fresh or be frozen for later use. When using animal organs or body parts, the innovator's goal is usually to assess the basic functionality of the prototype. While animal parts are good substitutes for many (but not all) prototyping projects, human tissue testing will eventually become essential for products that interact with the human body. Human tissue can be drastically different from animal anatomy for certain organ systems, so more advanced tests are necessary to evaluate a device. In an effort to better understand the size, shape, and contour of the space in which a device needs to operate before a human tissue test is designed, an innovator may find that a trip to an anatomy lab is invaluable. Additionally, with proper licensing and disclosure of intent, human cadaver tissue can be obtained from commercial sources (note that when working with human tissue, proper lab facilities and processes for its safe use and disposal are essential). See 5.3 Clinical Strategy for more information.

Keep in mind that dead tissue has obvious limitations compared to living tissue. However, for purposes of initial anatomical and physiologic properties, it is a good place to start. Once live tissue fed, by oxygenated blood, at body temperature, and with certain intact anatomical relationships is required, then tissue testing may not be adequate. Importantly even with dead animal tissue, there are certain guidelines that need to be followed with respect to proper handling and disposal. For instance, universities and other laboratory facilities typically recommend that innovators wear closed toe shoes, a lab coat or disposable gown, gloves, and potentially eye and/or mouth protection when working with tissue samples.

In-house versus outsourced prototype development

When innovators or companies are ready to begin prototyping, they must decide whether prototypes will be developed in-house or by a third party that specializes in prototype development. This decision depends in large part on the nature of the concept, the cost-efficiency of each approach, and the skills and resources of the team. While third parties can provide access to vast experience and specialized equipment and materials, a team may give up an important learning opportunity if it decides to outsource the development of its prototypes. However, this may be essential for more complicated and sophisticated projects. Innovators, especially relatively inexperienced ones, are encouraged to entrust at least some portion of their prototyping efforts to the company's in-house engineering team to capitalize on the invaluable learning that can take place. The advantages and disadvantages of these two different approaches are summarized in Table 4.5.3. These principles apply across project types, from mechanical engineering solutions to software.

Table 4.5.3 Advantages and disadvantages of in-house versus outsourced prototyping.

	In-house prototyping	Outsourced prototyping
Advantages	• Hands-on prototyping experience contributes to the expertise of in-house engineers. • Company engineers directly learn what works and what does not and can apply those learnings directly to the design to test changes or add/delete features. • Direct prototyping experience allows the iteration process to occur faster. • In-house prototyping can save a great deal of time if the design proves to be not feasible. • Overall, in-house prototyping is usually cheaper than outsourcing (except for complex processes and custom parts). • Intellectual property and proprietary information about what may or may not work can be captured. This information may be used to block competitors.	• Outside specialists possess high levels of expertise on specific processes and equipment. • Outside shops have developed many prototypes for other companies and can offer invaluable advice and design assistance, especially in the prototyping of non-critical, proven components. • In some cases, outsourcing can be less expensive than purchasing the required equipment (and undergoing training) in-house.
Disadvantages	• Some equipment may be too costly for a small start-up to purchase. • Some processes may require expertise that is not available in-house. • The team typically cannot be built up quickly or scaled down easily based on prototyping needs, which may have significant time and money implications.	• Outsourcing can often take more time since the shop has to meet the needs of multiple customers. • An outside shop does not possess unique or vast knowledge about the innovation. • Issues related to intellectual property must be managed carefully in every outsourcing deal. • Design iterations and risk retirement can take longer. • Outsourcing can be costlier than in-house prototyping, especially for simple designs.

Whenever possible, innovators should make an effort to establish a basic prototyping lab of their own to perform early prototyping activities. Depending on the project, this may include tools, materials, computing resources, and software to create mechanical, electrical, material, and software models, or prototypes that combine these different disciplines. Then, as the complexity of the models increases and more specialized equipment and materials are required, specific steps or prototypes can be outsourced on an as-needed basis. This hybrid approach is often the most sensible and affordable way for a lone innovator or young start-up company to proceed. However, there are also times when innovators might choose to establish a more comprehensive prototyping facility. For example, Perclose, a pioneer in the femoral vascular closure space, took this approach and gained vast efficiencies in its design cycle, thereby realizing a significant return on its investment. In this scenario, the cost of setting up the prototyping lab was built into the initial financial model (see 6.1 Operating Plan and Financial Model).

The Evalve story below highlights the advantages that can be realized through in-house prototyping and also demonstrates how one company followed the guidelines outlined in this chapter.

FROM THE FIELD ⟩ EVALVE, INC.

Understanding prototyping as part of the biodesign innovation process

Evalve, Inc. was founded to develop percutaneously delivered devices and tools for repairing the mitral valve, one of the heart's four valves.[12] Mitral valve regurgitation (MR) occurs when the valve's leaflets or flaps do not close completely, resulting in the backflow of blood from the left ventricle into the left atrium and into the lungs, dilation of the left atrium, and the eventual enlargement of the left ventricle. A minor amount of MR occurs in as much as 70 percent of adults.[13] Significant (moderate to severe) MR is much less common, affecting roughly 4 million people in the US.[14] Despite the significant sequelae of untreated mitral valve regurgitation, which can include atrial fibrillation, heart muscle dysfunction, congestive heart failure, and an increased risk of sudden death, just 50,000 individuals undergo surgery in the US each year to correct the problem.[15] Three to four times that number of patients experience symptoms and complications serious enough to warrant the traditional therapy used to treat MR – open heart surgery – but do not receive the procedure, either because they are too sick for heart surgery or are in a position of "watchful waiting." Evalve's founder, Dr. Fred St. Goar, a Stanford-trained interventional cardiologist, was familiar with the **morbidity** associated with open heart surgery and saw a market opportunity for a less invasive solution. After **brainstorming** several different approaches focused on suturing the valve leaflets, St. Goar and other Evalve team members, including vice president of R&D Troy Thornton, hit on the idea of a clip that would stabilize the leaflets and also hold part of the valve closed, allowing the leaflets to function properly.

Evalve's small team of engineers and technicians began building prototypes early in the development process. First, they made rudimentary sketches by hand. Then, technicians and engineers developed crude prototypes in ones and twos in Evalve's lab, using pig hearts from a local butcher to experiment on. According to Thornton, "Eliminating options is an important part of prototyping," and simple experiments with early prototypes were used to rule out certain concepts and ideas that might have consumed significant time and money down the road. Additionally, even though they started working with pig hearts, the team recognized the need to understand the disease in humans as quickly in the process as possible. Many development paths that may seem promising early on can be derailed by the unique intricacies of the human anatomy. "We sent a few of the engineers out to the pathology lab to look at human hearts that have mitral valve disease," said Thornton. "That was enlightening and important. Surgeons can tell you what it looks like and you can see pictures in a book, but seeing the problem in person is quite different." With a strong first-hand understanding of the disease in humans, the team was able to be more efficient in its approach to prototype development, relatively quickly eliminating paths that proved not to be fruitful.

As Evalve pushed ahead, it discovered that there was too much prototyping and testing for a single engineering team to handle. In addition to the clip, Evalve needed to develop a guiding system and a clip delivery device, so it developed three engineering groups to help break down the problem into smaller, more manageable pieces. "From a conceptual standpoint, we realized that the project was too big to tackle all at once," recalled Thornton. "Breaking the product into manageable chunks became critical for a complex project like this." One team, which was focused exclusively on the guiding system, spent many months developing different designs and conducting animal studies just to isolate and understand the functionality required to address this variable. Another team worked on the clip delivery catheter that would be able to actuate the clip once inside the heart. That left one additional engineering group to focus solely on designing the best clip they could to hold the mitral valve leaflets closed and restore normal heart function. While breaking into separate teams was effective in terms of advancing Evalve's understanding of the device's core components, it required a certain amount of additional coordination. "Everything has to come together," said Thornton. "All three engineering teams have to talk to each other constantly. They have to communicate well because if somebody changes something on part A, it could affect system B or C." Due to the interdependencies of the different components, each team also needed to maintain and document specific design requirements for their components to keep the effort synchronized. Adjustments to resources had to be made at various

times during the project to assure that all three teams could move forward at the same fast pace. Despite these challenges, however, the narrower focus of each of the engineering teams allowed Evalve to innovate much faster and more efficiently than when they were a single working group.

Another important input to Evalve's prototyping efforts was the collection of physician feedback as early and often as possible. St. Goar said, "The downfall of more than a few companies has been getting an idea and then having a group of engineers go off to create something very elegant, very sophisticated, and very complex that the physicians just can't imagine using in clinical practice." For Evalve, the downside of physicians potentially being turned off by flaws in early, crude designs was far outweighed by the upside of having their input to take into account in the design process. Having users who were able to be involved all the way through the process, looking at many versions of prototypes and speaking to the positives and negatives of different design features was especially critical. In addition, although the new, less invasive procedure would ultimately be performed by interventional cardiologists out of the catheterization laboratory, St. Goar and Thornton made a point of involving surgeons in their prototyping work. St. Goar explained, "We were trying to match what the surgeons did. If we simply addressed the need from an interventional cardiologist perspective, we might miss the therapeutic mandate. The surgeons knew what needed to be accomplished, and we certainly didn't want to lower their clinical standards. Having surgical input early on was invaluable."

Eventually, after its engineering workstreams started coming together, Evalve had to make the decision about when to begin testing in humans. "There is a little bit of a leap of faith that's required," St. Goar said. "You've got to bite the bullet and say, 'Okay, this is it, we're going to go.' But that's not always easy to do." For Evalve, having positive, longer-term results from animal tests was one factor that helped the team make this decision. It was also important that the design had been shown to meet its critical design requirements. Additionally, executive management played a key role. "Our President and CEO, Ferolyn Powell was very helpful. An engineer by training, Ferolyn had been intimately involved in the development of the technology which allowed her to fully comprehend the risks," said St. Goar. In this case, the board of directors also weighed in. Thornton remembers one director saying, "You can do all the animal studies you want, but you're not going to learn a damn thing until you get into people. You could spend another six months refining this thing, but it's time to move forward."

This decision was closely related to a development challenge that many inventors and companies struggle with: **design creep**, or the tendency for

engineers to spend too much time changing minor aspects of a design or prototype that are unlikely to affect the overall performance of the device. Clearly defining what the "must-haves" versus the "nice-to-haves" are for a device is one strategy that inventors often use to keep the iteration process under control and avoid design creep. Fully satisfying the must-have requirements should be a signal to the team to take the next step forward. According to St. Goar and Thornton, the FDA has particularly strict rules regarding design changes that occur after human trials are initiated, especially for a permanent implant, so design creep can be highly problematic for firms at this stage. "There are always ways to make a device better than it is today, but you really have to be committed, once the decision is made to move into human studies, to stick with your design," said St. Goar.

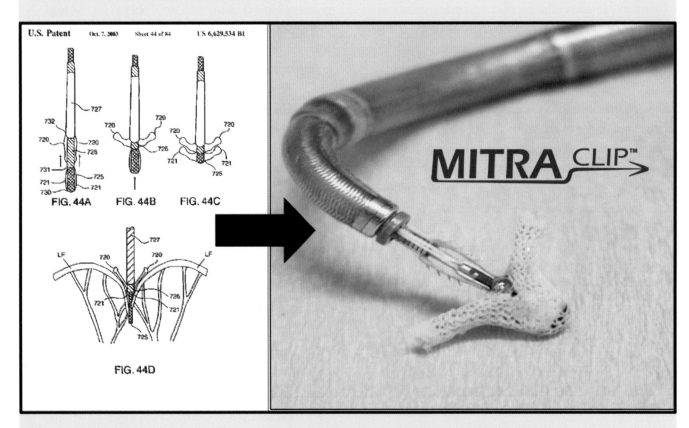

FIGURE 4.5.10
From the early design captured in a patent drawing to the finished product, the Mitra Clip evolved significantly over time (courtesy of Evalve, Inc.).

Although Evalve developed its prototypes in-house, the company also worked with an outside consultant with experience in medical device design, development, and manufacturing. "I think there's sometimes a little hesitancy on the part of the engineers to bring in an outside consultant," said Thornton, "but you've just got to get over it and say, 'Hey, maybe we've got something to learn from somebody else.'" He continued, "We brought in the consultant to work with our top three engineers who were working on the clip design to help them refine it. We used him extensively for brainstorming and developing ideas. He had years and years of experience and knew these specialized vendors who could do micro stamping processes, which we really needed but none of us had known about before. Bringing in outside expertise at the right time in the project can be a pretty critical step."

When offering advice to other inventors and companies, both Thornton and St. Goar stress perseverance as being essential to prototyping, as well as the larger biodesign innovation process. In the case of Evalve, as Thornton described, "There were a couple years of trial and error before it really took off." St. Goar added that patience also comes in handy. "We spent a period of time trying a number of different approaches, none of which looked anything at all like where we eventually ended up," he said (see Figure 4.5.10).

Evalve received a **CE mark** for the Mitra Clip in 2008. A year later, the company was acquired by Abbott Vascular. In 2013, the device received premarket approval by the FDA for use in patients with significant symptomatic degenerative MR who are at prohibitive risk for mitral valve surgery.[16]

Tips for effective concept exploration

As innovators gain experience with prototyping and testing, they build confidence and become increasingly effective in designing and developing models. Those without vast experience, however, will be well served to consider the following advice as they get started:

- **Consider multiple factors** – While exploring concepts, be sure to think about the function, form, material, manufacturing, cost, and feasibility of a prospective device or software solution.
- **Understand how competitive solutions work** – One of the best things to do early in the concept exploration process is to try to obtain examples of similar or competitive devices or solutions to understand how they work and what is good or bad about their design, user interface, or technical underpinnings. Having this knowledge can go a long way in making the concept exploration process more efficient. Just be careful not to anchor on the approaches employed by existing technologies.

- **Play with scale** – When prototyping small devices, consider scaling up $5\times$ to $10\times$ to make the initial exploration process easier during preliminary experiments. Look at things (especially moving parts) under a microscope to be certain to achieve a detailed understanding of how they function. Just remember that it will be necessary to find solutions that can be replicated at the size appropriate for their actual use.
- **Iterate** – Anticipate that concept exploration and specific prototyping exercises and tests will require multiple iterations and plenty of trial and error. Keep ample resources/materials on hand for practicing and correcting mistakes.
- **Consider the effects of reuse** – Construct reusable prototypes and test them for degradation with time and repeated use. Consider computer modeling and/or the use of specialized testing machinery to subject prototype devices to repeated physiologic loads. This process (and the associated test data) is often required for regulatory submissions. For example, devices implanted into the vasculature must endure

tens of thousands of simulated heartbeats without failure or breakdown. Some computer models are commercially available to test this. Others are being developed for specific disease or device applications.

- **Take pictures** – Photograph everything! A great deal can be learned from understanding and keeping a record of how each prototype has progressed based on lessons learned from earlier models. This also serves as compelling history for the company and customers as the product is introduced into the market and the company's story evolves.

- **Save all work** – Keep all prototypes as a physical record of development.

- **Maintain detailed notes** – Keep detailed notes (measurements, techniques, materials, etc.) regarding exactly how the first prototype was constructed so that the process can be replicated. Use a disciplined version control system for all drawings so that the design can be replicated later, as needed.

- **Be cautious regarding confidentiality** – Because there may be a significant amount of invention that occurs during prototyping, be sure to contract with a consultants who agree to relinquish all rights to any intellectual property generated from the design process. Protect all technology that is developed as a result of the prototyping process. Non-disclosure agreements should be signed with any contract suppliers/shops used (see 4.1 Intellectual Property Basics).

For more information and tips for prototyping specifically in the disciplines of mechanical, biomaterials science, electrical engineering, and computer science/software, see online Appendices 4.5.1, 4.5.2, 4.5.3, and 4.5.4.

↘ Online Resources

Visit www.ebiodesign.org/4.5 for more content, including:

 Activities and links for "Getting Started"
  - Identify the questions or issues to be addressed through concept exploration

  - Design the minimal model needed to answer those questions
  - Identify and prioritize functional blocks
  - Build the model
  - Test/refine prototypes to develop design requirements and technical specifications

 Videos on concept exploration and testing

 Appendices that include reading lists and additional guidance on building:
  - Mechanical device prototypes
  - Biomaterials-focused prototypes
  - Electrical prototypes
  - Application and software prototypes

CREDITS

The editors would like to acknowledge Michael Ackermann of Oculeve, Rush Bartlett, Ryan Van Wert, and Frank Wang of Vynca, and Fred St. Goar and Troy Thornton of Evalve for sharing the case examples. Further appreciation goes to Jeremy Koehler, David Myung, David Gal, and Frank Wang for their assistance with the online appendices. Kityee Au-Yeung, Gary Binyamin, Scott Delp, Trena Depel, Brian Fahey, Steve Fair, Mike Helmus, Ritu Kamal, Craig Milroy, Stacey McCutcheon, Asha Nayak, Ben Pless, and Weiming Siow also made valuable contributions to the chapter.

NOTES

1 Johnny L. Gayton, "Etiology, Prevalence, and Treatment of Dry Eye Disease," *Clinical Opthamology*, 2009, pp. 405–12.
2 "Dry Eye," American Optometric Association, http://www.aoa.org/patients-and-public/eye-and-vision-problems/glossary-of-eye-and-vision-conditions/dry-eye (October 31, 2013).
3 According to Oculeve estimates.
4 All quotations are from interviews conducted by the authors, unless otherwise cited. Reprinted with permission.
5 The French catheter scale is used to measure the outside diameter of a cylindrical medical instrument. 1 French (Fr) is equivalent to approximately 0.33 mm.

6 Gerard Loosschilder and Jemma Lampkin, "Using Kano Model Analysis for Medical Device Product Configuration Decisions," MDDI Online, January 7, 2014, http://www.mddionline.com/article/using-kano-model-analysis-medical-device-product-configuration-decisions (February 5, 2014).

7 From Vadim Kotelnikov, "Prototyping," 1000ventures.com, http://www.1000ventures.com/business_guide/new_product_devt_prototyping.html (January 9, 2014). Reprinted with permission from Tom Kelley.

8 Loosschilder and Lampkin, op. cit.

9 Ibid.

10 Ibid.

11 Bill Adams, "Bench vs. System Testing: The Pros and Cons," Overclockers.com, October 16, 2002, http://www.overclockers.com/articles638/ (January 9, 2014).

12 The heart's mitral valve lies between the left atrium and the left ventricle.

13 William H. Gaasch, "Mitral Regurgitation: Beyond the Basics," UptoDate.com, http://www.uptodate.com/contents/mitral-regurgitation-beyond-the-basics (January 9, 2014).

14 "About Mitral Valve Regurgitation," Abbott Vascular, http://www.abbottvascular.com/int/mitraclip.html#about-mitral-regurgitation (January 9, 2014).

15 Ibid.

16 Amanda Pedersen, "FDA Approves MitraClip, But Reimbursement Still an Issue," *Medical Device Daily*, October 28, 2013, https://www.medicaldevicedaily.com/servlet/com.accumedia.web.Dispatcher?next=bioWorldHeadlines_article&forceid=84529 (January 9, 2014).

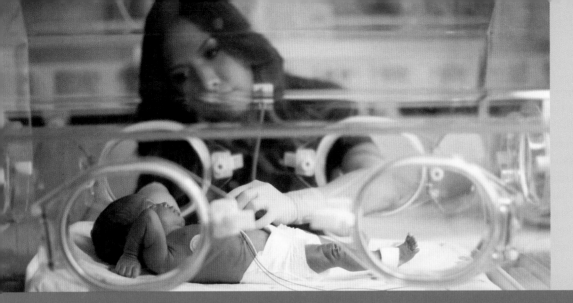

4.6 Final Concept Selection

INTRODUCTION

The team started with duct tape, tubing from the hardware store, and the mesh from a flour sifter. Now it has three working prototypes that prove the solution is feasible. But there is only enough time, money, and manpower to take one forward into detailed development. In a move that can feel a lot like betting everything on a single hand of poker, innovators must decide which solution is most likely to come to fruition, challenge the "gold standard" in the current treatment landscape, align with the needs of key stakeholders in a way that compels them to adopt it, and support the development of a sustainable business.

The purpose of final concept selection is to use everything that has been learned to date about the concepts under consideration to choose the one that will be brought forward into development. This final concept will continue to be refined, tested, iterated, and improved during development and implementation, but by choosing a single concept, innovators are able to focus on optimizing one idea rather than continuing to divide their attention across multiple needs and concepts.

In the chapters leading up to final concept selection – 4.1 Intellectual Property Basics, 4.2 Regulatory Basics, 4.3 Reimbursement Basics, 4.4 Business Models, and 4.5 Concept Exploration and Testing – innovators began screening their solution ideas and eliminating those few with fatal flaws (also called killer risks) that render them truly infeasible. The set of leading concepts that survived these screens can now be researched in more depth, and the data from this additional research combined to select a single, final concept to take forward.

While some experienced innovators are able to make this final decision based on gut instinct, most benefit from a more systematic process, including the incorporation of techniques such as the Pugh method (described within the chapter). While incorporating a structured approach as part of final concept selection may not initially appeal to everyone, it helps ensure that key user and design requirements are kept at the forefront of this important decision. Additionally, it provides a mechanism for reflecting any

OBJECTIVES

- Understand how to combine the data about intellectual property (IP), regulatory, reimbursement, business models, and technical feasibility gathered up to this point through the biodesign innovation process to efficiently identify leading solution concepts.

- Recognize how to apply a structured approach, such as the Pugh method, in developing a concept selection matrix to identify a final concept.

important data gathered through this rigorous research and screening process in the team's increasingly detailed user and design requirements.

 See ebiodesign.org for featured videos on final concept selection

FINAL CONCEPT SELECTION FUNDAMENTALS

Selecting a final **concept** is considered by some innovators to be one of the most crucial steps in the product development process.[1] Teams should take their time and use care in deciding on a final concept to avoid the mistake of anchoring on a single solution too early, without fully exploring a range of potentially viable alternatives to identify the one with the greatest likelihood of addressing the **need**.

Choosing a final concept is essentially an exercise in risk mitigation. Figure 4.6.1 informally depicts how concept selection works. After ideation, innovators identify a group of interesting concepts to evaluate, out of potentially tens to hundreds of solution ideas, by selecting those that best satisfy the need statement and must-have need criteria (see 3.2 Initial Concept Selection). They then begin researching these *multiple* concepts to understand the critical risks associated with each one. In particular, they consider the IP, regulatory, **reimbursement**, and business model implications associated with the potential solutions (as described in chapters 4.1, 4.2, 4.3, and 4.4,), building on what has already been learned about issues such as relevant markets and **stakeholders**. In any one of these areas, the team may find a glaring risk – that is, a problem so profound that it makes clear that the solution under consideration cannot go forward. Those without killer risks, that make it through these screens, are considered *viable* concepts and can then be evaluated for their technical feasibility through concept exploration and testing (as described in chapter 4.5). Again, the team will set aside solutions that prove to be technically impossible, while those that show technical promise advance to become the team's *leading* concepts.

Of course, as is common with so many other activities in the biodesign innovation process, these steps are not nearly as linear as they appear in the diagram – it is typical, for instance, for innovators to initiate their

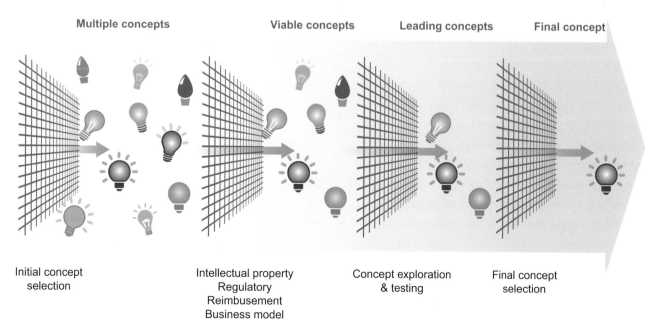

Multiple concepts   **Viable concepts**   **Leading concepts**   **Final concept**

Initial concept selection

Intellectual property
Regulatory
Reimbusement
Business model

Concept exploration & testing

Final concept selection

FIGURE 4.6.1

A visual representation of concept screening and final concept selection (courtesy of John Woock).

assessments of IP, regulatory, reimbursement, and business models in parallel with **prototyping**. The point is that concepts are evaluated – and some are eliminated – on a rolling basis until only a few remain. At that point, innovators have a sufficient understanding of the risks inherent in each solution, and how they can potentially be mitigated, to formally decide which one *final* concept to take forward.

Efficiently getting to a set of leading concepts

During the first part of the concept screening stage, innovators have been on the look-out for killer risks. As counter-intuitive as it may seem, innovators should celebrate the discovery of these fatal issues because it means that they have been successful in avoiding one of the most common traps that teams can fall into – namely, pursuing an expensive and time-consuming development path only to find out much later that the project is doomed. Mark Deem, co-founder of medical device **incubator** The Foundry, acknowledged that he and his team routinely set aside solutions based on little more than the IP, regulatory, reimbursement, or business model risks they entail. As an example, he noted, "There's no sense in going out and spending a lot of time coming up with a great idea if it's something that will require a 1,000 patient study with five years of follow up that's never going to get done with a venture capital financing structure."[2] Similarly, if innovators learn that a great deal of IP has already been filed related to one concept under consideration, whereas the other concepts are relatively clear, they may decide to drop the concept with the IP risk. By necessity, this first round of concept research and analysis, is more broad than deep. As Deem described, "We're not going into great detail when we do this initial assessment, but a top-level analysis gives you a pretty good sense of the risks you're facing," and, in turn, the concepts that should be swiftly eliminated.

While innovators certainly will have identified a number of killer risks through their work to date (and, in turn, eliminated some concepts), they also will have uncovered issues that are troublesome but not necessarily fatal (as initially described in chapter 4.1). As these issues are identified, teams are advised to pay attention to them, but not necessarily to eliminate a concept from

consideration without further research. It takes judgment and a great deal of further work to calibrate how serious these issues actually are. As mentioned, IP issues, for example, seem to wind up in this "intermediate zone" fairly often. In fact, it is almost uniformly the case that once innovators have done a reasonably detailed IP search, they will have serious concerns about freedom to operate and/or patentability.

At this point in the process, innovators can use the information they have gathered about the risks associated with their concepts to create a risk scoring matrix. This can be done by expanding the simple scoring system used initially to assess IP risks: red (troublesome issues), green (looks OK), and yellow (some problems raised). It is important to emphasize that a red rating is different than a killer risk. A killer risk indicates a barrier of sufficient magnitude to grind a project to an immediate halt. In contrast, a red rating flags an issue that is problematic in terms of substantially increasing the uncertainty, complexity, time, and/or expense associated with an effort, but it is not necessarily a "deal breaker." By combining IP, regulatory, reimbursement, and business model risks into a common assessment, the team gains a cohesive (although simplified) picture of the risk profile for each concept.

The team should give itself a realistic yet finite amount of time to complete a risk matrix for all of the concepts across the four primary screening categories (as shown in Figure 4.6.2). If the scoring has been done with reasonable diligence, it is unusual to find a concept that is "green" across all categories. But a combination of mainly greens and yellows means that a concept is viable and probably worth taking further. Based on experience, a team of several fellows in the Stanford Biodesign program is likely to spend a few days to one week generating the information required to complete the matrix for approximately 10 concepts. Since this work can, at times, be frustrating because the team is not "getting to the bottom" of all the key issues, imposing a deadline with the goal of doing the best possible job in the time allotted can help keep the innovators focused and minimize their frustration.

Identifying viable concepts using IP, regulatory, reimbursement, and business model data as described

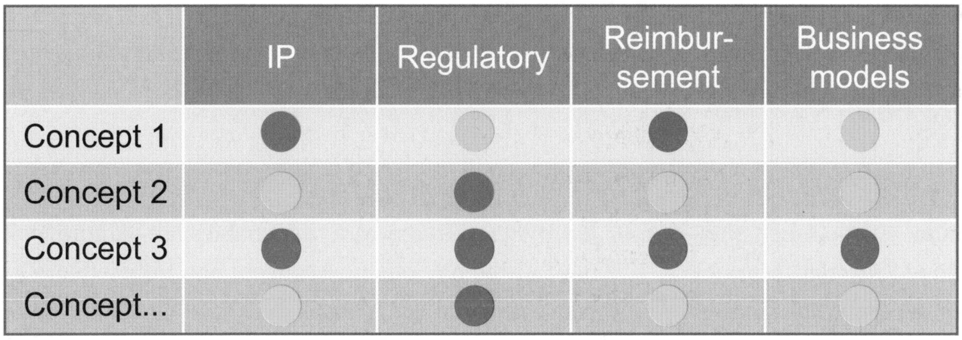

FIGURE 4.6.2
An example of a simple risk scoring matrix for multiple concepts under consideration.

above next blends naturally into the step of concept exploration and testing (as described in chapter 4.5). By **prototyping** different solutions, innovators generate additional data that can help them gauge the technical feasibility of the ideas and the likelihood that they can reasonably be brought to fruition to address the defined need. In this way, technical feasibility serves as an additional screen that enables a team to get to a small set of leading concepts. It is called out separately in Figure 4.6.1 primarily to emphasize the critically important role that prototyping plays in concept screening.

Using a structured approach to select a final concept

For the leading concepts that are left following the screening steps described above, innovators can move toward a final concept by evaluating the extent to which the ideas satisfy important **user** and design requirements. In doing so, it usually makes sense to directly compare the leading solutions (and how they perform against the user and design requirements) because these requirements tend to apply more universally across all (or many) solutions. Innovators typically compare solutions to one another and/or to the current **standard of care** (as described in the section on the Pugh method). Given the comparative nature of this assessment, teams (especially those just starting out) are often well-served to take a structured approach to help ensure the objectivity, accuracy, and thoroughness of the exercise.

Because user and design requirements play a core role in enabling teams to select a final concept, innovators should spend some time assessing these criteria

before comparing solutions against them. The original and most basic user and design requirements come from the **need specification** in the form of must-have and nice-to-have **need criteria**. However, innovators will naturally modify and refine some of these requirements based on the information gathered about IP, regulatory, reimbursement, business models, and technical feasibility. For example, through their research, the innovators may have uncovered new stakeholder requirements – that is, criteria based on the needs of the influential stakeholders who have some important role in the selection, deployment, and use of the technology. Or, as they studied the concept area, they may have identified some features of the technology or design itself that have clearly become important. Often, requirements for new features are related to broader need criteria that were there from the start but, by virtue of the deeper knowledge the team has gathered, it is now possible to put more precise and measurable parameters around them. If time is not taken to incorporate these learnings into user and design requirements, innovators may find themselves with a fundamentally flawed concept under development, a product that does not effectively address critical stakeholder needs, or other problems that can threaten to send them back to the drawing board.

Although this chapter advocates a structured approach to comparing leading solutions and selecting a final concept, innovators should recognize that a wide variety of approaches to final concept selection exist. In simple terms, they can be thought about on a continuum. At one end, innovators depend on extensive research supported by a structured method to help them objectively

FIGURE 4.6.3

General steps of the Pugh concept selection method.

analyze and act on that information. At the other, they rely primarily on intuition, developed through years of experience, to make this important decision. Most innovators usually employ a combination of both types of approaches – structured and intuitive – to get to a final concept.

Seasoned innovators sometimes argue that they do not need to utilize a systematic process in their overall approach to concept selection. However, when probed about their reasons for making certain choices, it often becomes apparent that (consciously or not) they use a semi-structured approach for picking a winning concept. While possibly more informal than some of the well-defined methods used in the medtech field, their favored process has likely been honed through years of experience but is nonetheless rigorous enough for evaluating important factors – such as market size, regulatory pathway, likelihood of reimbursement, match between product and user type, etc. – before deciding on a final concept to pursue.

The Pugh method

The Pugh method is one of the most widely referenced and easily understandable structured approaches for final concept selection, as it focuses on user and design requirements and employs a fairly straightforward ranking system. It was developed in 1981 by Stuart Pugh, a professor at the University of Strathclyde in Glasgow, Scotland. Details of the method can be found in Pugh's book *Total Design*.[3] Since it is relatively simple and generally effective, the Pugh method can be quickly and easily utilized by experienced medtech practitioners and novices. Other benefits of the Pugh method include the following:[4]

- It compels the design team to review the user and design requirements in detail and to understand how the requirements apply to the concept.

- It provides a framework that enables the team to look beyond the obvious first concept and fully explore a wider range of concepts.
- It provides an objective way to evaluate concepts.
- It results in a concise, auditable document for the product's design history file that is easily understood and defensible (see 5.5 Quality Management).

In its most basic form, the Pugh method has four steps, as shown in Figure 4.6.3. Through this process, the innovator creates a two-dimensional decision matrix to facilitate the quantitative evaluation of the leading solution concepts against a baseline concept. The idea is that by comparing each potential solution to a baseline (rather than assessing each one against every other alternative), the analysis is significantly simplified.

Another factor to keep in mind is that, even though this method is useful in helping select a final concept, the process of refining and adjusting a solution is by no means complete once that solution is chosen. In fact, as the concept undergoes development, design modifications will undoubtedly be generated. The Pugh method can repeatedly be used to select from among different variations of the concept, using increasingly specific requirements. Certain early concept-specific technical specifications that have been learned through prototyping can also be incorporated into these later rounds of selection. This point underscores once again how iteration and progressive refinement are hallmarks of the biodesign innovation process.

To illustrate how the Pugh method works, the example introduced in 4.5 Concept Exploration and Testing will continue to be followed. This example involves an innovator working on the need for *a way to prevent strokes in patients with atrial fibrillation caused by left atrial appendage (LAA) thrombus*. As noted, the LAA is a small, pouch-like structure attached to the left atrium of the heart in which blood can coagulate to form clots, or thrombi, in patients with an abnormal heart rhythm (such as atrial fibrillation). These clots can then dislodge and travel to the brain, where they have the potential to create a blockage in an artery that supplies blood to the brain, leading to a stroke.

Step 1 – Identify user and design requirements Innovators should begin final concept selection by referring back to the need criteria outlined in the need specification. These criteria, particularly the must-have criteria, continue to represent the most important, overarching requirements that the final concept must meet in order to satisfy its target audience. Recall that need criteria are defined *before* the innovator begins considering specific solutions to address a need. This is why they play such a central role in **brainstorming** and then screening and refining the list of ideas that is generated (see 3.2 Initial Concept Selection). All of the potential solutions taken forward into concept selection should satisfy the need criteria.

Once innovators have revisited the need criteria to refresh their memories, they should compile the most important user and design requirements. In contrast to need criteria, user and design requirements emerge as individual solutions are investigated through iterative brainstorming, as research is performed related to IP, regulatory, reimbursement, and business models, and through concept exploration. For this reason, they are typically much more detailed than need criteria and may become specific to a particular concept. For instance, in the LAA example, the idea to fill the LAA with glue would eventually have design requirements that are quite distinct from those related to the concept of placing a mesh in the aorta, even though both concepts support an approach to obstructing LAA thrombi. At a fundamental level, user and design requirements begin to resemble **product specifications**.

When choosing which requirements to use for final concept selection, a careful balance is required. Using too many requirements that resemble precise, product-like specifications at this stage can skew the comparison of the concepts toward one particular idea (this is a particular temptation if the team already favors one of the concepts). In the LAA case, for example, specifying that the procedure can be done using a catheter of 7-French caliber or smaller might specifically favor the glue solution. On the other hand, though uncommon, including *only* general requirements that closely resemble the need criteria (e.g., device must be delivered percutaneously) will not allow the innovator to adequately differentiate between concepts. While there is no hard and fast guide to setting the right level of specificity, the goal is to define a well-rounded set of requirements that can be met by more than one of the solutions being considered, yet still illuminate important distinctions between the concepts being evaluated. No more than three to seven user or design requirements should be chosen to keep the assessment manageable and sufficiently encompassing.

Although many user and design requirements are discovered through prototyping, as alluded to earlier, innovators should also take into consideration key facts uncovered through the activities described in chapters 4.1, 4.2, 4.3, and 4.4, both to understand the limitations of certain concepts as well as to help shape how some of the requirements are worded and weighted (see below). For example, if an innovator learns that pursuing a premarket regulatory approval trial is something that certain stakeholders would not support (e.g., investors might not tolerate the typically long premarket approval pathway or patients might not want to participate in a trial if existing treatments are readily available), the concepts that could utilize a **510(k)** pathway might be favored. Thus, intelligence such as this can be used to help shape the way user and design requirements are selected and worded.

Step 2 – Weight user and design requirements Once a meaningful set of requirements has been defined, the team must then assign weightings to reflect their relative importance. At a high level, this process is similar to that performed in 2.5 Needs Selection, although in this case weightings are applied to user and design requirements instead of the important factors used to assess needs. For example, using a scale of 1 to 5, an inventor might assign a "5" to requirements that are "essential" or "must-have" and a "1" to those considered "nice-to-have." Each user or design requirement (along with its weighting) should be captured in the selection matrix. For the LAA example, the requirements might be chosen as shown in Table 4.6.1.

Assigning weightings is a relatively subjective process. However, through developing prototypes and understanding information gathered about IP, reimbursement, regulatory issues, and business models, innovators gain a reasonably good idea of the issues that are important for users that should be translated into design requirements. For instance, in the LAA example, the team may have learned that a comparable procedure can be performed in 30 minutes, which allows physicians to complete a certain number of cases and generate a particular level of reimbursement every week. Accordingly, these issues, related to the physician's business as well as reimbursement, result in a specific user requirement that will play out in the design of the concept ultimately selected. Other issues relating to IP, regulatory issues, reimbursement, or business models may have a different yet distinct impact in helping rule in or rule out certain concepts.

In assigning weightings, innovators should try to ensure that they use a range of different values that actually reflect the true importance or impact of the requirement. If all weightings are high (e.g., all "5s") or low (e.g., all "1s" or "2s"), then it will be difficult to distinguish concepts from one another. In the sample weightings shown in Table 4.6.1, the first requirement that a "device must be delivered percutaneously and not through a surgical incision" may have been weighted so high because it is directly linked to a key need criterion and is also tied to research showing that a percutaneous solution would be more likely to utilize a 510(k) regulatory pathway, which is preferred by key stakeholders. In contrast, the requirement that a "device must be less than 12 French in diameter" may have been weighted on the low side if the innovator's research showed that smaller alternatives are preferred but larger French sizes are routinely used in the need area.

When in doubt, innovators can ask business advisors, physicians, nurses, patients, or other subject matter experts to weigh in on the relative importance of the requirements they have defined. This kind of external input can help keep the process grounded in the real world and safeguard against any **biases** that may have

Table 4.6.1 Sample user and design requirements and assigned weighting for LAA example.

Requirement	Weight
Device must be delivered percutaneously and not through a surgical incision	5
Device must be able to be used in a cardiac catheterization environment	3
Device must not lead to iatrogenic thrombi development	5
Device must be less than 12 French[5] in diameter	1
Device must allow procedure to be performed in less than 30 minutes	3
Device or solution must be reversible at the time of the procedure	4

inadvertently surfaced within the team. Some teams wonder how many experts to approach. Conventional business wisdom is to seek out a broad range of opinions in order to get a diverse and objective set of perspectives. On the other hand, there are advocates for a more focused approach. Todd Alamin, an entrepreneur and practicing spine surgeon, recommends the following: "Find one clinician who you think is smart and creative and knows a lot about his field, and stick with him. If you talk to too many people, you end up creating a cacophony of noise that doesn't make any sense, with some physicians recommending exactly the opposite course from what other physicians recommend. The risk in asking too many people for their advice is that you'll end up getting stuck doing nothing."[6] Ian Bennett, Louie Fielding, and Colin Cahill, who worked with Alamin to found Simpirica Spine, agreed, but acknowledged that there are times when more than one opinion will be helpful. In these cases, Bennett advised learning about the physicians in advance of seeking their input. "You have to know a little bit about their backgrounds, what they like to do, and their natural tendencies so that you can frame their answers with these factors in mind," he said. "This provides a context for their answers and helps you understand them a little bit better," rather than

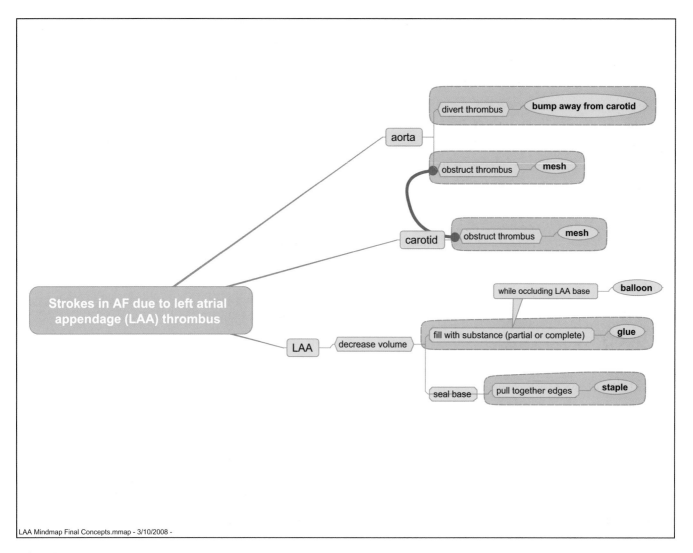

LAA Mindmap Final Concepts.mmap - 3/10/2008 -

FIGURE 4.6.4

Key concepts from a sample mind map for preventing strokes in patients with atrial fibrillation caused by LAA thrombus (provided by Uday N. Kumar).

necessarily taking their input at face value. Once the right experts are identified, "Be prepared to meet them anytime, anywhere. Their schedules are crazy," noted Bennett. Additionally, Cahill emphasized, "Self-educate as much as possible before the first interaction. If you're new to the field, you will not get everything totally right, but at least you'll show that you've done your homework."

Step 3 – Confirm the concepts and choose a baseline
The next step is to confirm which concepts will be evaluated. This should include the most promising solutions that have made it through the concept screening

process and were not eliminated due to undue risks related to IP, regulatory, reimbursement, business model, or technical feasibility. Typically, no more than three to five concepts should be under consideration at this point. For the LAA example, Figure 4.6.4 shows a list of the potential solutions that were chosen via concept screening for further investigation.

It is also essential to choose a baseline concept against which all solutions can be compared. This could be one of the team's solutions that it believes is most likely to achieve success, the "gold standard" among existing treatment options, or a competitor's solution.[7] The important thing is to gain enough information about the

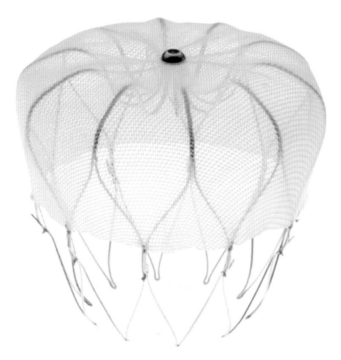

FIGURE 4.6.5

The WATCHMAN LAA device incorporates a permeable fabric over a nitinol frame. It is implanted using a catheter to deliver it into the LAA (courtesy of Atritech).

baseline to be able to effectively evaluate each new solution concept against it. For the LAA example, a baseline could be provided by the WATCHMAN®, depicted in Figure 4.6.5. This device (introduced in chapter 2.2) has been used in US human **clinical** trials, but is not yet **FDA** approved as of the time of this writing.

Step 4 – Assign scores and rank concepts using the selection matrix Each concept should next be rated against the user requirements based on how likely it is to exceed the performance of the baseline concept (+1), lag behind it (–1), or perform at a comparable level (0). Because it serves as the point of comparison, the ratings for the baseline solution should be set at (0) for each requirement. Even if the baseline does not perform particularly well against a certain requirement, it should still be given a (0) rating since this level of performance represents the current standard.

The total score for each concept can be obtained by multiplying the score for each user criterion with its weight, and then totaling the results. The concept with the highest total score is the leader. Graphically, a matrix

can be constructed with each concept listed in its own column, with the baseline solution furthest to the left. For the LAA example, a completed selection matrix could appear as shown in Table 4.6.2. In this example, the LAA glue concept emerges as the clear leader.

In some cases, numerous concepts may score poorly against the baseline. Assuming that the requirements are generally sound, this may suggest that the baseline concept meets both the need and the currently defined user and design requirements relatively well (indicating that the innovator may face a difficult market). Alternatively, it may mean that the new concepts have not been developed to a sufficient degree of detail such that they can possibly satisfy more requirements. Sometimes more than one concept scores well or no clearly superior concept emerges from the exercise. In these cases, the innovators should reexamine the user and design requirements to see if there is adequate specificity and check that the weightings of the requirements are appropriate.

In the case where the requirements and weightings seem adequate, but there is no front-runner emerging from the concepts, it is most likely time to perform additional ideation to be certain that the most promising concepts have, indeed, been identified. In this round of ideation, the team should consider if successful elements of one concept could be applied to other concepts to make them stronger and boost their likelihood of success. Additional prototyping, in an iterative loop, may provide valuable insights about these questions. After taking these steps, and revalidating the requirements and weightings in the selection matrix, the team should assign scores to each concept again, repeating this process until a lead concept becomes clear.

Ultimately, innovators choose their final concept by taking into account all of the information gathered and analyzed, formally and informally, through the entirety of concept screening activities as roughly depicted in Figure 4.6.1. Again, the process in practice is much more iterative than linear, with ideation, research, and prototyping, as well as adjusting requirements and weightings, guiding innovators to a "lead horse." If a team does this cycle only once, there is a chance that the chosen concept is a good candidate, but it may not be as risk-free as it could be with additional research and

Table 4.6.2 Sample selection matrix showing requirements, weightings, ratings, and total rank scores for the LAA example.

Requirement	Weight	WATCHMAN® (baseline)	Aortic Bumper	Aortic Mesh	Carotid Mesh	LAA Glue	LAA Stapler
Device must be delivered percutaneously and not through a surgical incision	5	0	0	0	1	1	−1
Device must be able to be used in cardiac catheterization environment	3	0	0	0	0	0	0
Device must not lead to iatrogenic thrombi development	5	0	0	−1	−1	0	1
Device must be less than 12 French in diameter	1	0	−1	−1	1	1	0
Device must allow procedure to be performed in less than half an hour	3	0	1	0	0	1	0
Device or solution must be reversible at time of procedure	4	0	−1	0	0	−1	−1
Rank Score			−2	−6	1	5	−4

analysis. That is not to say that any final concept is ever risk-free, but an iterative approach to final concept selection can help optimize a solution concept as much as is reasonably possible.

Alternate approaches to concept final selection

As noted, there are many different ways to tackle final concept selection that span a continuum from highly structured and systematic to more informal and intuitive. Some innovators are able to take what looks like (at least from the outside) a more instinctual approach to concept selection. Often these are experts who have had experience in assessing many technologies and have developed "pattern recognition" skills that allow them to short-circuit a more systematic evaluation process. Again referencing the Simpirica Spine team, Fielding recalled that Alamin was one of these intuitive experts who was able to quickly sift through a number of the team's concepts and "use his imagination to see through to something that could be realistic." From the list that the team

presented to him, Alamin focused on one concept for a minimally invasive implant that clinically made sense to him. Even though Alamin's decision-making approach seemed more intuitive on the surface, as with many experienced innovators, there was some structure to his internal process that boiled down to a few key steps. According to Alamin, he routinely assesses if: (1) the device has a clear function; (2) it addresses a clear clinical need; and (3) it offers the simplest possible solution for solving the given problem. "Particularly with the spine," he noted, "anything that's too mechanically complicated isn't going to work because you have to worry about wear debris being created over time and individual pieces malfunctioning. Anything that is designed to repetitively take on axial load over the life of the patient is also a concern because the likelihood of ultimate failure is high." This is pattern recognition in action. Finding an expert like this is, of course, one of the most effective strategies for helping the team decide on the most promising concept to pursue. In this case, Alamin ultimately agreed to join Simpirica as a co-founder, working with the team to bring the concept forward to the market (for more information about Simpirica Spine, see 6.2 Strategy Integration and Communication).

Additional examples, in the form of the short case studies that follow, demonstrate other techniques that innovators use to assist them with the important decision of final concept selection. The first story about medtech legend Advanced Cardiovascular Systems provides a classic illustration of how one set of experienced innovators approached final concept selection in the

founding of ACS. Note that although they did not formally use the Pugh method, there are strong similarities between Pugh's approach and the one they employed – well-defined criteria were used to evaluate the concepts against a baseline solution before making a decision.

A second story about the experience of the team that developed the OneBreath ventilator underscores the benefits of bringing a certain level of structure and rigor to concept selection, especially for innovators who are just starting out in the medtech field.

FROM THE FIELD ▶ **ADVANCED CARDIOVASCULAR SYSTEMS**

Driving concept selection through intuition and an intimate understanding of user requirements

Advanced Cardiovascular Systems (ACS) was the first cardiovascular medical device start-up in California's Silicon Valley. Founded in 1978 by cardiologist John Simpson and entrepreneur Ray Williams, it pioneered the development of percutaneous balloon angioplasty catheters and over-the-wire catheter systems. Carl Simpson, who was working as the lead technician at Stanford's catheterization laboratory when he met John Simpson (no relation), later became the company's first full-time employee. Together, with the help of cardiologist Ned Robert and a team of engineers, they revolutionized the treatment of coronary heart disease and helped create the field of interventional cardiology.

John Simpson, who was training in cardiology at Stanford at the time, became intrigued with angioplasty when he heard a presentation by Andreas Gruentzig, the creator of this innovative new technique. Gruentzig's approach used a fixed-stylet catheter system and balloon to widen a coronary artery that had narrowed or become obstructed through the build-up of atherosclerotic plaque. Although initially skeptical,

Simpson went to Switzerland to learn about the procedure and became convinced of its benefits. Back in the US, he ran into several problems. First, given that the procedure was considered quite radical in its approach, he did not receive much support from colleagues to perform it. Second, as a young cardiologist assumed to be lacking in experience and training, he had difficulty obtaining the parts and tools he needed to perform the angioplasty procedure when he ordered a catheter system from Gruentzig. Nevertheless, with a strong commitment to the therapeutic benefits of coronary angioplasty and faced with no other alternative, John Simpson and his colleague, Ned Robert, began experimenting with materials to build a balloon angioplasty system of their own. In the process, they came to believe that Gruentzig's device was somewhat difficult to use, especially due to the fixed-stylet system. They also recognized that his technique required such a high skill level that it would be challenging for an average cardiologist to replicate.

As John Simpson began developing prototypes and testing concepts, Carl Simpson was managing Stanford's cardiac catheterization laboratory. He helped the young innovator by connecting him with various engineers within the hospital, such as people in the machine shop, before leaving Stanford to join Hewlett-Packard. John Simpson continued to "tinker" and eventually met Ray Williams, a prominent investor, to whom he showed some of his prototypes. Williams saw the market potential of Simpson's ideas and put together a network of angel investors to fund the creation of ACS. At its core, ACS was focused on developing the tools and techniques that would provide a better way to perform coronary angioplasty. After ACS was created, John Simpson contacted Carl Simpson, who then joined the company to accelerate the process of designing, prototyping, and testing device concepts. The team would ultimately have to develop three primary components for its system – a balloon catheter, a guidewire for catheter navigation, and a guiding catheter.

One of the crucial things the ACS team did early in the biodesign innovation process was to define clear user requirements that its balloon catheter would have to meet in order to provide a significant improvement over the available technology. According to Carl Simpson, these requirements included creating a balloon that had a relatively low profile (thin enough to penetrate severe lesions), was relatively non-compliant (did not keep expanding beyond the maximum desired diameter), and could sustain a certain pressure without rupturing. "That was our focus. We knew what we had to do," he said. "You might have 10 ideas for how to make it better, but you can't try to do everything. You have to make a choice on what's most important and stick with it. Because if you don't, that's the kiss of death."

Likewise, in understanding how a guidewire should perform, they realized that it first had to be very fine in order to navigate the small branches of the heart's coronary arteries. Given the complex and tortuous anatomy of the coronary arteries, the guidewire would also have to be flexible, but in a way that could be controlled. Through many unsuccessful attempts to navigate the coronary anatomy with shaped guidewires, they ultimately determined that one of the

guidewire's most important design requirements was to be "torqueable," with the tip responsive to guided control from the other end. This key design requirement (which was integral to the clinical problem) eventually helped make ACS guidewires the standard in the industry.

While ACS initially had an entire "tree" of potential user requirements, Carl Simpson advised innovators to be wary of becoming distracted by "nice-to-haves," especially in the early stages of product development. "You're never big enough to do more than three or four things right at any given time," he noted. ACS identified its top-priority user and design requirements, including the need for a torqueable guidewire, by using its prototypes to facilitate detailed and extensive interactions with cardiologists. ACS routinely sent its engineers to the cath lab to observe catheter-based procedures and study what competitive products, like the Gruentzig system, could and could not do well. "You need to understand better than the physicians what the real needs are," he said. "The company has to have a really high IQ when it comes to understanding what drives user requirements."

Once these requirements were identified, the ACS team was able to weight them to assist the company in making trade-offs during design and prototyping. "We weighted them based on what we learned from being so involved on a daily basis with what the customer needs next," said Carl Simpson. For example, in the development of the balloon catheter, the requirement for a thin balloon profile was given highest priority. "Profile was always number one," he explained, "because if you can't get across the lesion, you can't dilate it." If necessary, he recalled, the team was willing to make sacrifices around other requirements such as inflation pressure, to maximize the desirability of the balloon's profile. Carl Simpson reiterated, "Trade-offs must be driven by the clinical need. That's why the engineers must understand the medical need so well that they can make those decisions intelligently."

Fundamentally, because there was only one competing product in the market, Gruentzig's device was the baseline against which all of ACS's test results were compared. The company worked through multiple concepts before finally deciding on one to take into development. When asked how the team made that decision, Carl Simpson replied, "We just knew it." When the prototype of the balloon catheter concept could perform better than the baseline device on all three of the fundamental user requirements the company had chosen, ACS knew it was ready to move forward (see Figure 4.6.6). At that stage of the process, Carl Simpson said, "If it's not intuitively obvious, you need to get some help. Talk with other successful entrepreneurs that have a track record in this business." He also advised inventors to go back to physicians and get more input until they feel certain that the concept will satisfy their most important requirements.

Carl Simpson acknowledged that developing this kind of intuition was not easy. "You learn by doing it," he said. However, even for new innovators, "After a certain accumulation of knowledge, the right solution should start to present itself." He also issued a reminder that success is often built on a series of failures. "There is no right formula, other than knowing what the patient and the customer really needs." Simpson has personally dedicated himself to helping other inventors develop this sense of intuition. During his tenure at ACS, the company became informally known as the University of Medical Devices for Silicon Valley, training individuals who would go on to found over 100 medical device companies.[8] ACS was sold to Eli Lilly in 1984 and spun off to Guidant in 1994.

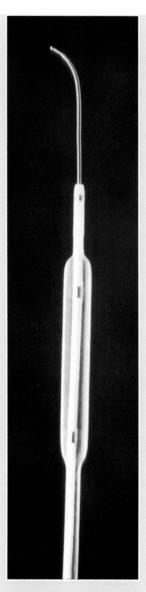

FIGURE 4.6.6
An early Simpson–Robert version of ACS's first catheter (courtesy of Carl Simpson).

FROM THE FIELD ▶ ONEBREATH

Reflections on an alternate approach to concept screening

Matthew Callaghan was a surgery intern at University of California at San Francisco when the avian flu was becoming a global concern. This illness can severely compromise an individual's respiratory system and its treatment often calls for ventilators to help patients breathe. Analyses of hospital preparedness conducted by the US government revealed that, in a pandemic scenario, a shortage of trained healthcare providers would be the greatest challenge in the government's ability to respond. The second biggest constraint would be providing enough mechanical ventilators to treat all those in need. Traditional ventilators are expensive, require specialized training to deploy, and have to be regularly maintained, making them impractical to stockpile for emergencies. As Callaghan described, "Mechanical ventilators were the rate limiting step: the most expensive, the most cumbersome, the most complicated, and the least adaptable equipment."

An engineer by training, Callaghan was intrigued by the idea of designing a better, lower-cost ventilator that could be used for disaster relief in pandemics and mass casualty events. Around this time, he met Bilal Shafi, another engineer-surgeon, who joined Callaghan in seeking to address this challenge. Shafi and Callaghan were both working fulltime in other capacities, but began meeting on the weekends to further investigate the problem. Their initial research pointed to two main target markets for a new ventilator: (1) the disaster relief market, with the US government as the main purchaser of equipment for emergency preparedness; and (2) the general hospital market, particularly resource-constrained facilities that needed access to effective but more affordable ventilator equipment.

Three main types of mechanical ventilation products are currently available in the market. Full-featured ventilators,

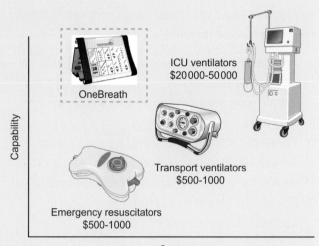

FIGURE 4.6.7
The treatment landscape for mechanical ventilation as depicted by the OneBreath team (courtesy of OneBreath).

used in state-of-the-art intensive care and emergency rooms settings, cost $20,000–$50,000 and require a skilled operator as well as a reliable external power supply. Portable or transport ventilators are less expensive ($8,000–$15,000), but have a short battery life and are not approved for critical care patients. Finally, the most basic devices are low-cost resuscitators that cost approximately $500–$1,000 and are manually operated by a skilled doctor or nurse. Callaghan and his team hoped to position to the new device on the lower end of the cost spectrum, but at the higher end of functionality and effectiveness (see Figure 4.6.7).

Callaghan and Shafi augmented their knowledge of the ventilator space by diving into available clinical literature and guidelines published by the US Health and Human Services Department for required features for pandemic ventilators.[9] Based on their accumulated understanding of the problem, they decided to focus on the need for *a way to affordably ventilate and support patients in acute respiratory distress in low-resource environments*. They further defined low-resource environments as settings with limited infrastructure (power, medical gas, climate control, and repair capabilities), as well as novice users (newly minted RN). The team next created a need specification that included the following need criteria to help guide concept generation:

1. Affordability: Inexpensive enough that hospitals can allocate the device on a one-per-patient basis (the team's preliminary cost target was $3,000 per ventilator).
2. Portability: Ability to operate as a self-contained unit, without electricity or compressed gas, with a long battery life and low maintenance requirements.
3. Operability: Simple enough for a floor nurse to operate without specialized training.
4. Adaptability: Flexible enough to be used on both adult and pediatric patients (unlike most available devices).
5. Compatibility: Universally compatible with currently available breathing tubes and other disposable equipment.

Armed with these need criteria, the team members initially thought they could jump straight into engineering. "Coming from a background in product design, one of the first things we wanted to do was take apart the ventilators that were already out there," Callaghan said. "Our first idea was to take everything that was metal and make it out of plastic to cut the cost in half." But after borrowing existing equipment to dissect, they realized that would not be feasible due to the complex and precise nature of the component parts. "You can't just make them out of cheap plastic," he noted. "We realized that existing devices were built on legacy technology and it would be impossible to make them cheaper in their current form. No one ever went back and started over. It was like layers upon layers of paint, as far as the design went, and no one had peeled off the old paint for decades," Callaghan continued. After weeks of tinkering, they determined that a more creative approach to concept generation would be required. "We probably shouldn't have started with a reverse engineering, tinkering approach," Callaghan admitted. On the other hand, he pointed out, one of the benefits of this exercise

was that it motivated the team to conduct additional research to fully understand the mechanics of respiratory distress – how people breath, how the airways function, and how that potentially can be replicated – which served as critical input to their subsequent ideation efforts.

According to Callaghan, the team conducted multiple rounds of ideation but, because the members were working fulltime, they did not follow a traditional ideation process. "We would meet every night and bat around ideas," he said, recalling that they would sketch ideas on a whiteboard and discuss them as a group. Afterwards, they would divide up the ideas among the team members, based on who had the most experience in the relevant area. Team members would then dig more deeply into promising ideas, independently sketching in their notebooks and working on the ideas that had been discussed so they could bring more information back to the next ideation session.

Through this informal process, Callaghan stated, "We came up with probably 20–30 ideas, including some rather unrealistic ones like modern extra-thoracic ventilation (an updated iron lung). Then we refined these on paper until we couldn't exclude something without seeing if it worked." This approach allowed them to whittle down the list of potential concepts to roughly six promising ideas. At that point, they started prototyping. However, instead of taking their concepts forward into a parallel concept screening process, they decided to pursue one idea at a time and jump right into engineering. "The first prototype made sense on paper in terms of physics, but it didn't work at all when we built it," Callaghan remembered.

The next most promising concepts included a pneumatic electrical solution and an idea for a combined pneumatic and software approach. The team started prototyping the pneumatic electrical idea. "It was closest to our backgrounds [in mechanical engineering], easiest to build, and we didn't have to write any code with this idea," Callaghan said. As they worked on various prototypes, he and Shafi refined their need criteria and added several additional requirements that they derived from analyzing possible use cases of the new ventilator. Unfortunately, after several months, they were forced to admit that the pneumatic electrical approach would never meet the need criteria and user and design requirements that they had defined.

Moving on to the next idea, they started working on models that would use software to control ventilator function via breath delivery algorithms. Roughly six generations of prototypes later, they felt confident that the concept would satisfy the need specifications and had the design that would become the OneBreath v1 ventilator.

Reflecting on this experience, Callaghan acknowledged the limitations of the team's linear approach to concept screening and final concept selection. "We were too quick to jump to fleshing out engineering concepts rather than trying to mate [the concept] to our design target. We would get away from our need specification and hang too tightly to what we were most familiar with. If we'd followed the biodesign innovation process more closely, we would have saved ourselves close to a year in getting to a final concept. This approach cost us time, and we were lucky it didn't kill the project."

According to Callaghan, "We did a good job of understanding the problem, but we didn't give the need spec enough authority." Had the team members more thoroughly and realistically evaluated their solution concepts against the need criteria, they potentially could have anticipated earlier that some of the ideas would not

have been able to meet those requirements. "Concepts that didn't meet the need spec should have been rejected earlier," he acknowledged. As engineers, he added, they had a tendency to believe that with a little more ingenuity or hard work they could overcome fundamental limitations inherent in the idea.

In hindsight, a more objective, comparative process that explicitly enabled the evaluation of potential solutions against key user requirements could have helped OneBreath more efficiently select a final concept. Callaghan likened the optimal process to the rigorous approach used to screen and select needs: "Once you come up with needs, you beat them up to make sure you've got them right. And you validate them across different stakeholders, not just one MD who told you this was a problem for him/her. In other words, you try and 'kill' them. We get attached to things that interest us – which is important; you need passion to keep working on something for long hours with no pay – but this needs to be balanced with whether or not the dog really wants that dog food. The same goes for choosing a design concept."

OneBreath is now focused on developing its low-cost ventilator for the disaster relief market, as well as resource-constrained facilities in developing countries. The company has set up a wholly-owned subsidiary in India, which is its first target market, and closed a Series A financing from international investors in 2013. OneBreath plans to launch its first product in early 2015.

 Online Resources

Visit www.ebiodesign.org/4.6 for more content, including:

 Activities and links for "Getting Started"
- Complete research, prototyping, and preliminary concept screening
- Identify user and design requirements
- Weight user and design requirements
- Confirm the concepts and choose a baseline
- Assign scores and rank concepts using the selection matrix

Videos on final concept selection

CREDITS

The editors would like to acknowledge Carl Simpson of Coronis Medical Ventures for sharing the ACS story and Matt Callaghan for providing the OneBreath case. Many thanks go to Todd Alamin, Ian Bennett, Colin Cahill, and Louie Fielding of Simpirica Spine and Mark Deem of The Foundry for offering examples, as well as Ritu Kamal and John Woock for their assistance updating the chapter. Steve Fair contributed to the original material.

NOTES

1 David Warburton, "Getting Better Results in Design Concept Selection," *Medical Device Link*, http://www.devicelink.com/mddi/archive/04/01/006.html (November 20, 2013).

2 From remarks made by Mark Deem as part of a lecture for the Biodesign Innovation course offered by Stanford's Program in Biodesign, 2010. Reprinted with permission.

3 Stuart Pugh, *Total Design: Integrated Methods for Successful Product Engineering* (Addison-Wesley, 1991).

4 Ibid.

5 The French catheter scale is used to measure the outside diameter of a cylindrical medical instrument. 1 French (Fr) is equivalent to approximately 0.33 mm.

6 All quotations are from interviews conducted by the authors, unless otherwise cited. Reprinted with permission.

7 Pugh, op. cit.

8 Carl Simpson biography, Stanford Biodesign: People, http://biodesign.stanford.edu/bdn/people/csimpson.jsp (November 20, 2013).

9 "Advanced Development of Next Generation Portable Ventilators," Solicitation HHS-BARDA-08-20, April 18, 2008.

Acclarent Case Study

STAGE 4: CONCEPT SCREENING

To determine whether the idea of using balloons to address chronic sinusitis was feasible, Makower and Chang had to take the concept through steps described in the concept screening stage of the biodesign innovation process. Even though they were not directly comparing the balloon solution against another idea, they had to be sure it would clear all appropriate hurdles before making a decision as to whether or not this would be their final concept. "We realized that there were a million questions that needed to be answered once we had cleared the simple hurdle of widening a sinus passageway. But none of these questions were worth asking if we did not know if we could find a technology that worked," Makower said.[1]

4.1 Intellectual Property Basics

To assess the intellectual property (IP) landscape, Makower and Chang spent considerable time researching patents in the ENT field. They were particularly focused on understanding existing patents related to the nose. "We looked at this area in detail," Chang summarized, "and it seemed pretty open." He elaborated:

There was some IP around epistaxis [nosebleeds]. There was a fair amount of IP around turbinate reduction, cutting things, reducing things, and a lot of energy delivery. But the field was generally pretty wide open in terms of interventional devices, as well as methods and different procedures. We found some patents related to drug delivery, but not really specific to ENT. This was important because we recognized from early on that drug delivery could eventually become part of our solution. As we started to hone in on balloons and catheters, we knew that these devices had been around for a long time in cardiology and other spaces, so we focused on aspects of these technologies that were unique

and novel to this application and started to create a massive set of disclosures.

As they performed more research and consulted with an IP attorney, Makower and Chang relatively quickly became convinced that their solution would be patentable. They decided that the first step in this process was to focus on building a picket fence around their unique method of using balloons and other tools designed specifically for the nose. According to Chang, "We recognized from the get-go that the method was where we would really shine." Then, as they learned more about the devices and how they would be used, they began to build increasingly stronger IP protections around what made them uniquely different from existing balloons and other tools for cardiovascular interventions.

"So we filed a series of patent applications," said Chang. "Our first ones were not provisionals. We worked extremely hard and fast with our IP attorney and decided to put in a tremendous effort up-front into our patents." He described using a comprehensive approach to these early applications. "While we clearly described our preferred technology and preferred way of using it, we also put in many other technologies and methods to lay the groundwork for future products. There are multiple approaches to achieve a desired outcome and if you don't cover them, someone else will after they see what you are doing" (see Figure C4.1).

One unexpected outcome from these IP efforts was the decision to operate in what the team called "stealth mode." As they learned more and more about the opportunity, Makower became increasingly concerned about another company learning of the idea and trying to beat them to market with a similar solution. They intentionally decided to disclose their ideas to as few people as possible, and agreed to use carefully constructed non-disclosure agreements (NDAs) when a disclosure was

US007462175B2

(12) **United States Patent**
Chang et al.

(10) Patent No.: **US 7,462,175 B2**
(45) Date of Patent: **Dec. 9, 2008**

(54) **DEVICES, SYSTEMS AND METHODS FOR TREATING DISORDERS OF THE EAR, NOSE AND THROAT**

(75) Inventors: **John Y. Chang**, Mountain View, CA (US); **Joshua Makower**, Los Altos, CA (US); **Julia D. Vrany**, Sunnyvale, CA (US); **Theodore C. Lamson**, Pleasanton, CA (US); **Amrish Jayprakash Walke**, Milpitas, CA (US)

(73) Assignee: **Acclarent, Inc.**, Menlo Park, CA (US)

(*) Notice: Subject to any disclaimer, the term of this patent is extended or adjusted under 35 U.S.C. 154(b) by 524 days.

(21) Appl. No.: **11/037,548**

(22) Filed: **Jan. 18, 2005**

(65) **Prior Publication Data**

US 2006/0095066 A1   May 4, 2006

Related U.S. Application Data

(63) Continuation-in-part of application No. 10/829,917, filed on Apr. 21, 2004, and a continuation-in-part of application No. 10/912,578, filed on Aug. 4, 2004, now Pat. No. 7,361,168, and a continuation-in-part of application No. 10/944,270, filed on Sep. 17, 2004.

(51) **Int. Cl.**
A61M 31/00   (2006.01)
(52) **U.S. Cl.** **604/510**
(58) **Field of Classification Search** 604/509–510, 604/94.01, 103.1, 103.04, 103.05, 96.01, 604/164.01, 164.09, 164.1, 164.11, 164.13, 604/164.08, 171, 173, 101.02
See application file for complete search history.

(56) **References Cited**

U.S. PATENT DOCUMENTS

705,346 A   7/1902 Hamilton

2,525,183 A   10/1950 Robison

(Continued)

OTHER PUBLICATIONS

Strohm et al. Die Behandlung von Stenosen der oberen Luftwege mittels rontgenologisch gesteuerter Balloondilation Sep. 25, 1999.*

(Continued)

Primary Examiner—Kevin C Sirmons
Assistant Examiner—Deanna K Hall
(74) *Attorney, Agent, or Firm*—Robert D. Buyan; Stout, Uxa, Buyan & Mullins, LLP

(57)   **ABSTRACT**

Sinusitis, mucocysts, tumors, infections, hearing disorders, choanal atresia, fractures and other disorders of the paranasal sinuses, Eustachian tubes, Lachrymal ducts and other ear, nose, throat and mouth structures are diagnosed and/or treated using minimally invasive approaches and, in many cases, flexible catheters as opposed to instruments having rigid shafts. Various diagnostic procedures and devices are used to perform imaging studies, mucus flow studies, air/gas flow studies, anatomic dimension studies and endoscopic studies. Access and occluding devices may be used to facilitate insertion of working devices such asendoscopes, wires, probes, needles, catheters, balloon catheters, dilation catheters, dilators, balloons, tissue cutting or remodeling devices, suction or irrigation devices, imaging devices, sizing devices, biopsy devices, image-guided devices containing sensors or transmitters, electrosurgical devices, energy emitting devices, devices for injecting diagnostic or therapeutic agents, devices for implanting devices such as stents, substance eluting or delivering devices and implants, etc.

21 Claims, 44 Drawing Sheets

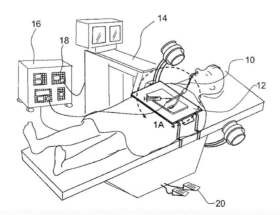

FIGURE C4.1

The first core Balloon Sinuplasty patent was finally issued to Acclarent in December 2008 (provided by Acclarent).

absolutely necessary. As Chang explained, "With the balloon idea, we were basically working with well-understood technology. There are a lot of companies that have experience with balloons, guides, and wires. There are smart people out there who could potentially do this and they might get a hint about what we're doing and invent in front of us. The longer we could keep the potential competition in the dark, the better."

4.2 Regulatory Basics

When thinking about the regulatory pathway associated with the leading solution, Makower referred back to the mantra "no science experiments." This meant that the solution would need to be a candidate for the 510(k) pathway. "We wanted something with a relatively simple regulatory path," said Chang, along with limited requirements for clinical data. "So, we set up a meeting with a regulatory consultant and started talking about how we could get clearance with this kind of a product." They found a regulatory consultant who had some experience in the ENT field. "His impression was that most ENT devices fall within Class I or Class II," Chang recalled. "From there, we pulled out the regulations to make sure we really understood them. After carefully examining this information with the consultant, they felt confident that the decision to pursue a 510(k) pathway was directly in keeping with the guidelines of the Food and Drug Administration (FDA). The key would be to identify a highly effective set of predicate devices upon which to base their submissions.

4.3 Reimbursement Basics

When it came to reimbursement, the team took great care in understanding the codes, coverage, and payment decisions associated with FESS. As described by William Facteau, who joined the team in November 2004, as the CEO of the soon-to-be company, the ENT specialists performing FESS procedures were heavily invested in how their procedure was reimbursed:

Functional endoscopic sinus surgery is the gold standard for the surgical treatment of chronic sinusitis and it is reimbursed by Medicare, as well as all private insurance plans. In its early days, though, there were challenges in getting the procedure reimbursed – this conflict is often referred to as the "FESS mess." The leading physicians in the field used a significant amount of political capital in order to get these codes established. And they are some of the most sacred codes throughout all of otolaryngology – they pay very well, they pay individually by sinus, and physicians also receive additional reimbursement for the follow-up treatments, called debridements [the surgical removal of scar tissue from a wound], which is pretty unique. Due to this favorable reimbursement, sinus surgery is one of the most profitable outpatient procedures for both the physician and the hospital. For this reason, leaders in the field, as well as the [professional] societies, are very protective.

Because the physicians had fought long and hard to establish codes, coverage, and payment levels, the team knew not to proceed if it seemed to be a questionable fit. On the other hand, if this technology were considered just another device to be used in sinus surgery, it could fit comfortably within the existing reimbursement guidelines and the team would be well positioned from a reimbursement perspective. The codes being considered at the time included those shown in Figure C4.2.

Lam Wang performed a detailed analysis of the reimbursement landscape. While the target population of chronic sinusitis patients was largely covered by private insurance, she used Medicare data as a baseline for payment levels and rates since it was readily available, and also because most private payers followed CMS's lead at that time. For physicians, CMS paid 100 percent for the highest weighted code, and then 50 percent for all additional codes. Use of an image-guidance system was addressed separately as an add-on payment at 100 percent. Facilities were also reimbursed for the first sinus at 100 percent, and at 50 percent for all others. However, while image-guidance received an add-on payment at 100 percent in a hospital outpatient setting, no additional payment was granted in an ambulatory surgical center (ASC) setting. CMS data at the time indicated that approximately 70–90 percent of all sinus surgeries were

CPT Code	Description	Hospital Outpatient		Ambulatory Surgical Ctr		Physician Payment	
		APC	APC Paymt	ASC Grp	ASC Paymt	Total RVUs	Ave Paymt
31254	Nasal/sinus endoscopy, surgical, with ethmoidectomy, partial	0075	$1,112	3	$510	7.92	$296
31256	Nasal/sinus endoscopy, surgical, with maxillary antrostomy	0075	$1,112	3	$510	5.71	$213
31276	Nasal/sinus endoscopy, surgical, with frontal sinus exploration	0075	$1,112	3	$510	14.76	$551
31287	Nasal/sinus endoscopy, surgical, with sphenoidotomy	0075	$1,112	3	$510	6.73	$251
61795	Stereotactic computer assisted volumetric (navigational) procedure	0302	$345.20	N/A	$0	7.05	$263.23

FIGURE C4.2

Sinus surgery was reimbursed by Medicare and private payers under a series of existing codes (provided by Acclarent).

performed in a hospital outpatient facility. Importantly, existing FESS codes could not be used in an office setting. If the team wanted to position its solution as an office-based procedure, it would need to apply for new codes (as well as prove that the treatment was safe to administer in an office setting).

To validate what she had learned and secure a formal opinion, Lam Wang contracted with a professional reimbursement consulting firm – Clarity Coding. In its report, Clarity indicated that the team should qualify to use existing FESS codes, since the objective of the procedure when using balloons would be essentially equivalent to that of FESS using other instruments and fell directly within the language of the code. "In other words," said Facteau, "the codes for functional endoscopic sinus surgery are broad enough that they do not specify how you create an opening in the sinus or what tools you use. It leaves it up to the physicians to decide whatever tool, device, or instrument is appropriate to make the opening." The company's technology would be just one more option available to surgeons when performing this procedure.

Encouraged by this outcome, but still concerned about the sacred nature of the FESS codes, Facteau sought the opinion of another reimbursement consultant when he joined the team. He explained:

Reimbursement was a big concern of mine. I worked closely with a legal/reimbursement expert at Reed Smith at a previous job, so one of the first things I did was call her and ask her for her opinion. She did some research and came back with the same opinion as Clarity Coding. In fact, she said, "Bill,

I really enjoyed working with you at Perclose. Unfortunately, we're not going to have to do much work on this one. It's pretty straightforward. The existing codes apply and you guys are all set. So, good luck."

4.4 Business Models

Another issue on Facteau's mind when he joined forces with Makower, Chang, and Lam Wang had to do with the business model. "One of the biggest business risks that I saw was just the fact that it was in ENT. I struggled with whether you could build a sustainable company solely dedicated to ENT. When I researched the space, it was not very impressive. The field just hadn't seen a lot of innovation over the years. It was primarily a reusable and capital equipment environment," he said. Following the path established in interventional cardiology, the devices being designed by Makower and Chang were intended to be disposable. Given the nature of the procedure and the devices themselves, this model made practical sense. More importantly, the team saw this approach as one way to help inject the field with greater innovation. According to Lam Wang, "When you shift from reusable instruments to a disposables platform, you open a field up to much more innovation. That's because the doctors are not stuck with the same tools they used the last year." Facteau elaborated: "Disposables drive innovation. We have seen it in general surgery, with laparoscopy and surgical staplers, orthopedics, spine, and cardiology – they were all driven by implants and innovative disposable technology. With a disposables model, we can innovate very, very quickly. The question was if ENT physicians would be willing to adopt new disposable technology. To some extent, the business model was predicated on our ability to get doctors addicted to innovation and be able to provide them with new, value-added technology every six to nine months."

One obstacle to introducing this kind of a paradigm shift would be in getting the target population of physicians to give up the reusable instrument sets to which they were so attached. The team also anticipated some resistance related to price – many ENT specialists were known for being somewhat cost conscious. "As a general rule of thumb," Lam Wang noted, "disposables command higher margins." However, convinced that the benefits would outweigh the costs, and encouraged by the relatively straightforward pathways identified in the areas of IP, regulatory, and reimbursement, the team felt comfortable taking on the challenge of building a business in ENT based on a disposable model. Basically, said Chang, "This didn't stop us. We just said, 'Okay, remember that. That's going to be a challenge.'"

4.5 Concept Exploration and Testing

In parallel with these other efforts, Makower and Chang had begun developing basic prototypes in earnest (see Figure C4.3).

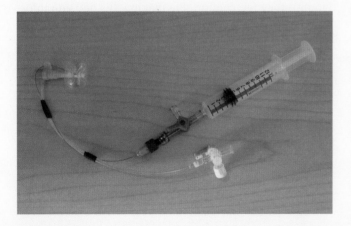

FIGURE C4.3

An early prototype of the stabilized guide access concept (courtesy of Acclarent).

Their goal was to prove the feasibility of the concept. Chang recalled how they got started:

We asked ourselves, "Why can't they treat these sinuses better?" One thing was that they operated via endoscopy – in other words, required direct visualization. So that led us to consider alternative visualization methods since our devices would at some point be placed around and behind structures, out of view of most common endoscopes. We looked at fluoroscopy as the most reasonable way to navigate the anatomy, similar to the way interventional cardiologists guide catheters in the heart. So don't remove the anatomy; look around or through it with fluoroscopy. And then there was the fact that there were no flexible instruments for FESS. When you examined what they were doing, they were just chopping, cutting or ripping the anatomy to gain access and make the opening bigger. We asked ourselves if there were less traumatic ways to accomplish this, ways that wouldn't lead to so much bleeding and the scarring cascade that often brought patients back for a revision surgery. We looked at using wires and flexible instruments. And we also thought, "Well, maybe you could dilate the ostium. Would fracturing the bone via balloon dilation be less traumatic? Was it going to cause the mucosa to necrose and cause more problems?" We didn't know. We wanted answers to these questions. And so that led us to prototyping.

"We'd never read or heard of anyone doing any dilating of bone," Chang continued, "so there were a lot of things that we just didn't know." Using readily available materials, they built a preliminary working model and then went back to the cadaver lab to try it (importantly, they decided that there were no animal models that would serve as an effective proxy for human sinuses). According to Chang, "Not knowing what would work, we were like MacGyver with duct tape and boxes of materials and supplies for cutting, melting, bonding, and shaping existing things. We were ready to make changes right there in the cadaver lab."

Makower described their initial experience using a prototype:

John and I went in with very crude guides, wires, and balloons. We didn't even have a scope for our first study. We went in and poked around. And, to our amazement, we were able to wire all the major sinuses in a cadaver head, and pass balloon catheters up over those wires, and deploy stents, and deliver balloons. It was unbelievable. The first time we tried, we got it to work. So, in terms of the feasibility, we said, "Okay, this is doable!" Now, what we didn't have was any real way of assessing what we had done. And we didn't know what pressures to inflate the balloons to. We saw that at some point the balloons would crack open, and we saw that we were dilating bone. We just didn't know exactly whether we were doing something bad or good.

Encouraged by these early results, Makower and Chang eagerly pressed forward. "We brought Dave Kim in, under a confidentiality agreement, to help with those early exploratory studies. And he brought a scope with him. For the first time, we got to see what it looked like, and it looked pretty good. So, that's when we realized it was time to really talk to some rhinologists and figure out whether this thing had any merit."

They put together a brief presentation on their concept that included the need specification, an overview of the prototyped technology, and the approach, results, and X-rays from the preliminary feasibilities studies. Tapping into a member of their network, they arranged a meeting with Dr. Mike Sillers, who was president of the American Rhinological Society (ARS) at the time. According to a friend who had pitched various ideas to Sillers in the past, he had a reputation for being a tough critic, so they were prepared for the worst. Makower recalled the meeting:

When we sat down with Mike [who agreed to sign an NDA], he was silent for most of the presentation. When we got to the end, he pushed himself away from the table and looked at us and said, "If this works, it's going to change everything we do." So then I asked the question, "What would you need to see before you'd be willing to try it on a patient?"

And he said, "Well, I'd want to do it on a cadaver head." So, we said, "Okay. We'll get you out to the cadaver lab. What next?" He said, "Well, try it on some patients." And then I said, "Well, how many patients would you need to see, and what kind of study would you need to see before you think we could commercialize it?" And his first response was, "You know, this is a just a tool. If I see that with this tool, in a handful of my own patients, I can make a wider opening, I don't need any clinical studies beyond that. I know that in the right patients, if I made a wider opening, it's going to be good for the patient."

4.6 Final Concept Selection

With this feedback from Sillers, Makower and team had reached a decision point. "John and I sat in that lobby after that meeting with Mike," he remembered. "We knew that there were still a lot of other pages to turn, but we realized at that moment that we had hit upon something. If an important potential critic couldn't come up with anything to kill the concept, then we knew we wanted to move forward with balloons. We looked at each other and said, 'I think we found the one. Let's go for it.'"

Because all of the key issues to consider in the concept selection process had panned out relatively favorably (IP, regulatory, reimbursement, business model, and prototyping), they felt comfortable making this decision informally, in lieu of a structured concept selection exercise. However, just to be sure, they took the idea to one more leading specialist in the field – Dr. Bill Bolger, a famous surgeon and anatomical expert in ENT. When Bolger (who also signed an NDA) was equally impressed with their presentation and willing to play an ongoing role in the development of the device, the team committed itself to proceed, initiating the development of the devices that became known as *Balloon Sinuplasty*™ technology.

NOTES

1 All quotations are from interviews conducted by the authors, unless otherwise cited. Reprinted with permission.

PHASES

STAGES

ACTIVITIES

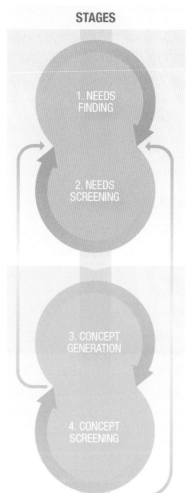

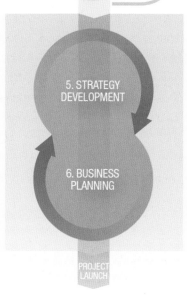

IDENTIFY

1. NEEDS FINDING

2. NEEDS SCREENING

1.1 Strategic Focus
1.2 Needs Exploration
1.3 Need Statement Development

2.1 Disease State Fundamentals
2.2 Existing Solutions
2.3 Stakeholder Analysis
2.4 Market Analysis
2.5 Needs Selection

INVENT

3. CONCEPT GENERATION

4. CONCEPT SCREENING

3.1 Ideation
3.2 Initial Concept Selection

4.1 Intellectual Property Basics
4.2 Regulatory Basics
4.3 Reimbursement Basics
4.4 Business Models
4.5 Concept Exploration and Testing
4.6 Final Concept Selection

IMPLEMENT

5. STRATEGY DEVELOPMENT

6. BUSINESS PLANNING

5.1 IP Strategy
5.2 R&D Strategy
5.3 Clinical Strategy
5.4 Regulatory Strategy
5.5 Quality Management
5.6 Reimbursement Strategy
5.7 Marketing and Stakeholder Strategy
5.8 Sales and Distribution Strategy
5.9 Competitive Advantage and Business Strategy

6.1 Operating Plan and Financial Model
6.2 Strategy Integration and Communication
6.3 Funding Approaches
6.4 Alternate Pathways

PROJECT LAUNCH

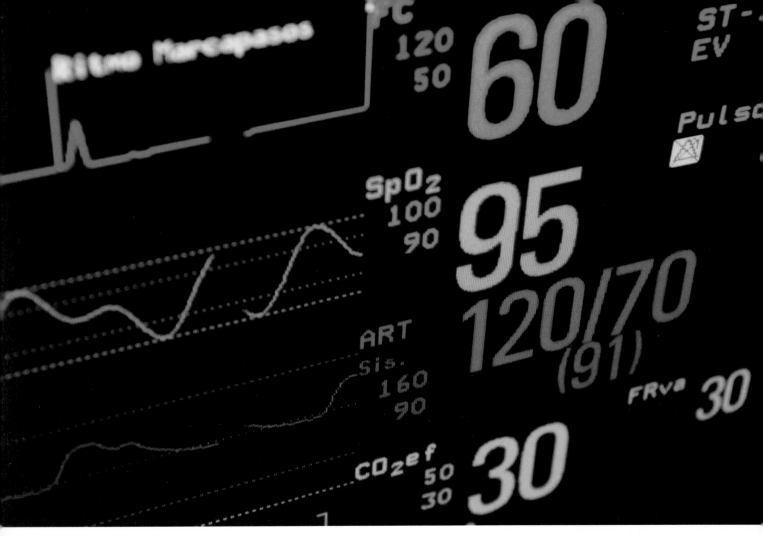

Failing to plan is planning to fail.

Alan Lakein[1]

Always in motion is the future.

Yoda[2]

5. STRATEGY DEVELOPMENT

By this point, you've taken steps to Identify and Invent. Now it's time to consider how to Implement. As Tom Fogarty, surgeon inventor extraordinaire says: "A good idea unimplemented is no more worthwhile than a bad idea. If it doesn't improve the life of a patient, it doesn't count."[3]

This is by far the longest and most complex stage of the biodesign innovation process, with good reason. Regardless of the validity of the need, the ingeniousness of the concept, and the size and scope of the market, at the end of the day, sound business underpinnings are essential if a product is to be delivered to the bedside, in a box, with a toll-free number on the side for sales and service.

Getting to this reality requires a balanced consideration of the rules of the road from Stage 4 (intellectual property, reimbursement, regulatory, and business models) with the addition of a series of overlying and overlapping strategies. These strategies focus more deeply on the following key areas: (1) intellectual property integrated with ongoing research and development and clinical plans; (2) regulatory strategy, including quality management; (3) reimbursement strategy; (4) basic business blocking and tackling – marketing, sales, and distribution; and (5) combining all assets to develop a sustainable competitive advantage.

Heretofore, a small team may have the wherewithal to work independently; now is the time to involve "varsity players." Use them as consultants or mentors, retain them on a part-time basis, or hire them. Without deep technical knowledge, failure is likely.

What to do first? Gordon Bethune's book chronicling the turnaround of Continental Airlines, called *From Worst to First*, espoused the importance of tackling problems and opportunities right away and all at once.[4] The lessons are that few things are simply sequential, that nothing can be considered in a vacuum, and that time waits for no one. Given the fast pace of change in the medtech space, a six-month delay may leave you behind your competitors.

NOTES

1 Unsourced quotation widely attributed to Alan Lakein, author of *How to Get Control of Your Time and Your Life* (New American Library, 1973).
2 From the movie *Star Wars: Episode V – The Empire Strikes Back*.
3 From remarks made by Thomas Fogarty as part of the "From the Innovator's Workbench" speaker series hosted by Stanford's Program in Biodesign, January 27, 2003. Reprinted with permission.
4 Gordon Bethune, *From Worst to First: Behind the Scenes of Continental's Remarkable Comeback* (Wiley, 1999).

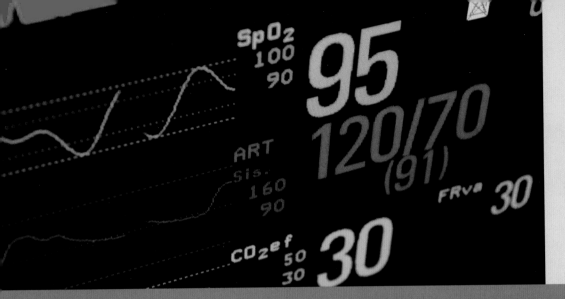

5.1 IP Strategy

INTRODUCTION

The team submitted the original provisional patent six months ago, and a second provisional last month. A venture firm has just completed IP diligence and, although there are no nasty surprises, they have raised concern that a potential competitor is moving in the same direction. The team needs to develop a plan to protect its core ideas and decide where it can afford to pursue patent coverage outside the US. The venture firm is recommending that the team hire the law firm it has used for the diligence, but it could make more sense to recruit an in-house patent council.

Intellectual property (IP) is a particularly important strategic business asset in the medical device field. Not only can a strong IP position help mitigate costly legal settlements, but it can serve as a barrier to entry for competitors, an early source of potential revenue through licensing agreements, and an important overall form of collateral for an entrepreneur or business. Further, an innovator's IP position is a major factor considered by investors in deciding whether or not to fund a technology or its associated company. The best way to develop an effective IP strategy is to establish a solid working knowledge of the patent process, employ experienced IP counsel, work diligently in pursuing patents, and – importantly – understand that patenting is an ongoing activity that requires continual monitoring of the IP landscape and adjustments to one's patent portfolio.

While 4.1 Intellectual Property Basics provides an overview of fundamental IP concepts, the purpose of this chapter is to help innovators understand the value of taking a strategic approach to IP and developing an effective patent portfolio.

 See ebiodesign.org for featured videos on IP strategy.

OBJECTIVES

- Understand strategies involved in filing provisional and utility patent applications.
- Appreciate how to conduct and utilize a freedom to operate (FTO) analysis.
- Understand when and how to hire a patent attorney.
- Review basic strategies of international patent filing.
- Appreciate the importance of an ongoing patent management strategy based on continuous internal and external monitoring.
- Consider different defensive and offensive IP portfolio strategies.

IP STRATEGY FUNDAMENTALS

With more than 2.1 million **patents** in force in the United States alone,[1] effectively navigating the IP landscape is undoubtedly complicated. The number of US-based IP lawsuits filed annually continues to increase, crossing the 5,000 case threshold in 2012.[2] The **medtech** industry is especially litigious, with large teams and significant budgets directed toward pursuing and protecting patents.

Against this backdrop, there are two main categories of IP strategy to consider – first, issues that the innovator should keep in mind while pursing patentability and **freedom to operate** and, second, strategies to establish and maintain an integrated and ongoing IP management process.

Strategies in pursuing patent coverage

There are many strategic issues to consider regarding patentability, but a handful of them are particularly important concerns, especially for first-time innovators. One frequent question is whether to file a **provisional** application or proceed directly to a full **utility patent**. In general, a carefully prepared provisional patent has the advantage of gaining the inventor an effective extra year of protection, comparable to the situation of an inventor filing outside the US. The utility application has a 20-year life (from its filing date) and can be filed up to a year after the provisional. So the provisional can be used to firmly establish the **priority date**; then, once the device is in sales (typically a few years later), the protection offered by the utility patent will last a year longer than if the utility patent was filed immediately. Another important consideration is that, in some cases, it may be faster to file a provisional patent application than a non-provisional application. The US is now a "first to file" patent system, as described in chapter 4.1, so receiving an early priority date with the **USPTO** can be important in order to increase the probability of being the first inventor to file for a patent to a particular invention. Additionally, there can be a cost advantage to a provisional patent compared to a utility application if an attorney is not used to draft the provisional application. This can be a false economy,

however, if the resulting provisional patent application is not thorough and of high quality. As described more fully in 4.1 Intellectual Property Basics, a careless provisional application can create enormous downstream trouble and expense for innovators and their companies.

A second question that comes up frequently is whether it is possible to "improve" a provisional patent application once it is filed. The short answer is that it is not possible to add to or modify a provisional application. If there is a substantial new discovery related to the initial disclosure, the innovators have the option to file a new provisional application or to initiate a utility filing that includes both the original and the new material. However, the priority date of a new provisional patent application will become the date of the second filing. Once an initial provisional patent application is filed, it may be beneficial to file additional provisional patent applications in the year leading up to the first provisional conversion deadline. These additional applications should be filed in a timely fashion to cover new incremental changes in a product or product features as they are developed over the course of the year. By waiting to file a single non-provisional patent application that includes all of the new subject matter invented since the first provisional patent application, innovators may risk losing the right to obtain a patent to the new subject matter if another party files a patent application or otherwise discloses the new subject matter first.

Innovators often wonder whether to bundle all aspects of their invention into a single patent, or try to divide it up into multiple patents. For a provisional application, this issue is usually not critically important. At the time of filing a utility patent application, an IP attorney can generally advise the innovators on whether or not to file more than one application. At this point, there are several considerations that may come into play. On the downside of filing multiple patent applications, there is the issue of the cost (which is magnified greatly if there will be international filings, too). However, multiple filings may be attractive if some piece of the original **concept** is more clearly novel and unobvious than the rest. Creating a separate filing for this piece of the original provisional patent could help the innovators gain approval more quickly for a particular subset of claims, which could be desirable, for instance, in advance of an upcoming product release. Similarly, if the innovators have a possibility of **licensing** some part of their IP to an existing company for **royalties**, it may make sense to isolate that aspect of the technology in its own patent filing to make the licensing agreement more straightforward.

Separating a patent application into parts can be accomplished by filing continuation applications, described in more detail later in the chapter. Another way that carving up a patent application occurs is in response to an **office action** (OA) by the patent office in which there is a restriction requirement. This occurs when the patent examiner deems that the original filing covers more than one invention (patent law allows only one invention per patent) such that it needs to be divided into several applications. The filing mechanism for dividing a patent into parts in this setting is called a **divisional**, which is a new patent application that claims a distinct or independent invention based upon pertinent parts carved out of the specification in the original patent.

A more sophisticated patent filing strategy issue – one that is more often dealt with by companies than individual inventors – is the question of whether to pursue **trade secret** protection rather than a patent. As described in chapter 4.1, a trade secret is information, processes, techniques, or other knowledge that is not made public but provides the innovators with a competitive advantage. In the medtech field, trade secrets often have to do with how a particular device is made (materials, manufacturing process, etc.). There is no application for a trade secret and no official granting of this right by the government. Innovators or companies assert trade secret protection by taking protective measures within their own operations and then suing any entity which wrongfully obtains the protected information. In order to prevail, the company with the trade secret rights must demonstrate that it has taken reasonable precautions to protect the information (for example, having its employees sign an agreement not to disclose, creating a special, high-security room where the trade secret is protected, and so on).

In some circumstances trade secret protection is much more effective than a patent, something which first-time innovators tend not to realize. For a start, if an invention is kept as a trade secret, no competitor should become aware of it. In contrast, a utility patent publishes 18 months after filing. Perhaps more important, a properly kept trade secret can potentially last forever, as long as the company is careful to keep the information secret (the most famous example of this is the formula for Coca-Cola). Although trade secret protection does not involve any filing fees up front, the costs and organizational effort involved in keeping a process secret can be considerable. Commonly, in the medtech field, a company might consider protecting its manufacturing methods as trade secrets, particularly if they are not easily reverse engineered by examining the commercial product.

Relying on trade secrets to protect intellectual property assets can be especially attractive in markets where traditional patent protection is weak and infringement is rife. Yoh-Chie Lu, the chairman of Biosensors International, was faced with such a situation in China when his team began to collaborate with a Chinese medical products manufacturer called Shandong Weigao. The partnership was a 50–50 joint venture called JW Medical Systems (JWMS), which manufactured and sold drug-eluting stents (DES) in the Chinese market.[3] Biosensors, the foreign partner, owned the intellectual property and the technical know-how necessary to manufacture the innovative DES, while Shandong Weigao had a local Chinese presence, deep connections, and a cost-effective manufacturing facility. Although Biosensors had patent protection in China for its novel drug-polymer coating for stents, Lu knew that the expense and time involved in taking an entity to court in China for patent infringement was not worthwhile. Instead, Biosensors chose to implement several strategies to protect its trade secrets, even from its local partner. JWMS set up a separate compound within the Shandong Weigao manufacturing facility and employed its own security team. Engineers from Shandong Weigao were not permitted to enter JWMS facilities and only a few select JWMS employees were authorized to view critical documents. Biosensors' CEO, Jack Wang, was the only JWMS employee who knew the formula to mix the drug and polymer to create the novel stent coating. With such strict trade secret protection strategies in place, Biosensors was able to maintain its position in the Chinese market as the sole manufacturer of the novel drug-polymer coated stent, and JWMS grew to become the third largest DES company in China.[4]

Freedom to operate strategies

It is a common misconception among first-time inventors that gaining a patent on a device means that the device is free and clear to commercialize (see 4.1 Intellectual Property Basics). The device may indeed have a patentable feature, but it may also have other features that infringe on claims of existing patents that are in force. With medical devices, inventors often discover potential freedom to operate (FTO) problems once a careful analysis is performed for a new invention. In certain heavily trafficked areas of medical devices – vascular stents, for example – it is extremely difficult to come up with an idea that has complete FTO.

An FTO analysis begins with a comprehensive search of patents that are currently active (generally in the past 20 years). The inventor is looking for **prior art** that is in the same area as the new invention. Importantly, a thorough search will almost always turn up prior art that is potentially of concern. It is easy to get discouraged by a superficial look at the abstract or drawings of an existing patent. However, inventors should not automatically assume that there is no FTO if a device pictured in a patent or described in the abstract looks or sounds similar to the device they invented. *The key is in the claims*.

The way to understand whether the new invention is distinct from the prior art is to carefully assess whether one or more claims of the potentially infringed patent apply to the new invention. In order for an existing patent to prevent an inventor from going forward with an invention, every part of at least one of the independent (stand-alone) claims has to fit the invention. For example, suppose an inventor has developed an angioplasty catheter (a catheter with a balloon used to dilate a narrowing in a blood vessel) where the key feature is that the balloon is wider in the middle than at the ends. In a

patent search, the inventor might find an existing patent where the claim is: "A tubular member with a balloon at the distal end in which said balloon has a larger dimension in the middle region than at both ends of the balloon and said middle region has a generally convex outward shape." The good news is that if the new invention does not have a convex shape in the middle, it is not covered by this particular claim. In fact, this claim may suggest that these inventors thought that anything besides a convex shape would not work, leaving room for a different shape to be patented. The bad news is that the existing claim covers the general idea and, thus, would potentially make the new invention obvious and not patentable. Furthermore, the inclusion of specific language about the convex shape raises the possibility that somewhere in the patent world is a more general claim covering the broader case (otherwise, it might not have been necessary to make this independent claim so specific). In this situation, it would be helpful to look at the prior art cited in the patent to see where this claim might lie. Another helpful place to look is the prosecution history of the claim. A lot can be learned about a claim and where it fits into the broader patent landscape by reviewing the amendments and arguments that were made to bring the claim in question to allowance. The prosecution history of any published application can be retrieved from the USPTO Public Patent Application Information Retrieval (PAIR) system. In practice, it can be difficult to determine whether or not there is an FTO issue. A patent attorney with medical device domain expertise can be extremely helpful in clarifying the situation. Ultimately, the final decision about FTO may be made in court through the litigation process.

During an FTO analysis, the innovators should keep in mind that under US patent law inventors may "act as their own **lexicographer**" in a patent application. This means they can give common words or phrases meanings that are specific and different from their normal definition. However, all such terms must be described or defined in the description section of the patent specification. If a term is defined in the description, it must be construed in the remainder of the patent and claims in accordance with that definition; otherwise, the plain meaning of the word (as determined by a how a regular person in the field would understand the term) will be used. In medical device patents, it is common to list examples to help clarify the use of a term without limiting it. For example, in the classic Palmaz® stent patent, the key claims referred to an "expandable intra-luminal vascular graft." What this term means for this patent is clarified in the second paragraph by examples: "Structures which have previously been used as intra-luminal vascular grafts have included coiled stainless steel springs; helically wound coil springs ... and expanding stainless steel stents ...."[5] Sometimes patent attorneys will define a key term by saying it "includes *but is not limited* to X, Y and Z" to keep the definition as broad as possible.

There is a range of possible outcomes from an FTO analysis, which vary from low to high risk as shown in Figure 5.1.1. If a thorough analysis suggests that there may be an FTO problem, there are a number of conditions to check that may help the inventor avoid the issue:[6] (1) the patent may not have been applied for, or granted, in the country where the company is seeking to operate; (2) the patent may have lapsed (e.g., if the patent-holder has not kept current with the required maintenance fee payments); (3) the patent may have expired (be sure to check expiration dates); (4) the patent may be invalid (sorting this out will take legal expertise).

If there is a clear, high-risk FTO problem, the innovators still have potentially fruitful options to pursue. First, they can try to "design around" the claims of issue by modifying the invention so that the patentability is maintained but the features that infringe current patents are removed or altered. This is a routine part of the biodesign innovation process, which most inventors go through with their inventions. It is a perfect opportunity to convene **brainstorming** sessions and in particular to bring in fresh advisors to see if there are other approaches that avoid infringement. In many cases, this process of re-inventing leads to important and unanticipated improvement in the fundamental invention.

If the feature revealed in the FTO analysis is absolutely essential to the new invention, the second option is to pursue licensing of the problematic patent from

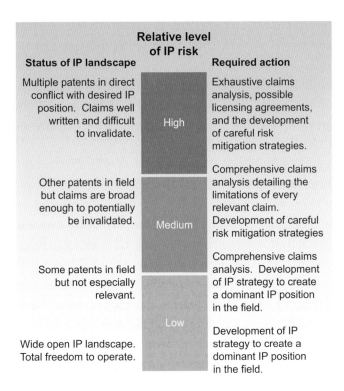

FIGURE 5.1.1

The level of risk associated with protecting a technology has profound strategic implications on the inventors and their company (developed in collaboration with Mika Mayer and Tom Ciotti of Morrison & Foerster LLP).

the entity that owns it. Medtech patents owned by universities may be readily available for an innovator to pick up, since these can be complicated and time-consuming for the university to license to large companies. Licensing of a key patent from a company is generally more challenging. It is sometimes possible to get a license to a limited **field of use** (usually a clinical area) that the company is not interested in pursuing. A company may also be willing to trade IP, such that the innovator gives rights to the new invention (perhaps in a limited field of use) in exchange for rights to the critical patent from the FTO analysis. See 6.4 Alternate Pathways for more information on licensing strategies.

The Spiracur example illustrates how one emerging company managed a complex patent landscape analysis as part of developing its IP strategy.

FROM THE FIELD ▶ SPIRACUR

Building an effective IP strategy

Moshe Pinto, Dean Hu, and Kenton Fong began working together while students in the Stanford Biodesign Program to address the need *to promote the healing of chronic wounds* (see Figure 5.1.2). This team, which eventually became Spiracur, initially developed an IP strategy through extensive prior art searching and the filing of a provisional patent application. However, these steps were just the beginning of Spiracur's activities.

The team also went to work right away on an in-depth patent analysis. Spiracur recognized the importance of working with strong IP counsel to guide its IP activities and the development of a holistic IP strategy. However, the team also felt strongly about performing much of the early work themselves, so they could save money on fees by providing the IP attorneys with a concrete, comprehensive analysis from which to start. This work would also help them become more knowledgeable in the field and increase their credibility with investors and other **stakeholders**. For these reasons, the three entrepreneurs initiated the analysis on their own, performing extensive searches using resources such as the USPTO databases and Free Patents Online.

In terms of performing patent searches, Pinto commented, "You have to be savvy about the search strings you use."[7] While many start-up companies use a funnel approach – starting broad and then getting more specific – Pinto offered another suggestion, which he called "inverting the funnel." The idea is to start with highly specific search strings that yield a relatively small number of relevant patents. After examining these patents, the team can then use the reference charts within the patents to identify other relevant patents. It can also mine this information for clues for developing new, slightly broader search strings that yield a greater number of results. "Through this approach, the inventors quickly gain familiarity with the IP landscape in their space and can more effectively separate the wheat from

FIGURE 5.1.2

Hu, Pinto, and Fong of Spiracur (courtesy of Spiracur).

the chaff as they continue their analysis with the broader search terms," Pinto explained.

To support the patent analysis, the team carefully analyzed each search result. Without any patent law training, the team was at somewhat of a disadvantage in analyzing them. "Dean was an engineering PhD and Kenton was a resident of plastic surgery," explained Pinto. The team, however, worked out a process. Hu and Fong would utilize their scientific training to screen the initial search results for subject matter that might have some threshold interest to them. These interim results were then passed to Pinto, who had more of a business background, and later to their attorneys.

Pinto admitted that the patent analysis could, at times, seem daunting. "If, for example, you run a search on the words 'pressure sores,' you'll get 60,000 patents. There is an initial evaluation of the patents that almost anyone can do just by looking at abstracts. But when it's time to derive conclusions based on what you find, it's important to have someone who is legally savvy to perform this work."

In total, Pinto estimated that the team reviewed the abstracts and drawings of more than 1,000 patents and identified roughly 250 to 300 patents for their IP attorney to evaluate and use to help them develop a patent strategy for their technology. When Spiracur initiated its Series A funding round, the team was prepared for the due diligence performed by the investors' IP attorneys and was able to successfully close the funding round.

In terms of offering advice to others who are building an IP strategy, Pinto emphasized that IP needs to be a primary concern for entrepreneurs. "Inventors need to realize that their exit from a medical device venture is pretty much constrained or determined by the strength of their IP position," he said. "If there is one thing I learned in retrospect, it is that we should have dedicated more time and effort to the entire IP endeavor than we did." When asked to estimate the amount of time a start-up company should invest, he commented: "The investors have it calculated. If we look at it from a budget perspective, they want to see the ongoing time of an IP lawyer at roughly $20,000 dollars a month. That translates to about 25 to 30 hours of an IP expert per month."

Pinto also stressed that one's focus on IP had to be continuous and progressive. "What I suggest is that even after you submit a provisional with claims with the help of a legal expert, revisit it after a month. Don't wait for the end of the 12-month period to say, 'Now, let's see how we can convert this into a utility patent application.' Do the iterations sooner as more information is gathered." Finally, he said, "Remember that quantity does matter." A sound IP strategy must seek to consistently improve the coverage and protection surrounding the company's technology. "When later-stage investors asked what we had done in IP, I told them that we had eight provisionals and three pending utility patent applications," said Pinto. "This kind of ongoing activity assures them that you are paying sufficient attention to IP."

Hiring legal help

While medtech innovators sometimes file provisional patent applications on their own, it can be highly beneficial to have an expert attorney prepare these or at least review them before filing. To file a utility patent application, it is almost always essential to hire a professional. The cost of an IP attorney varies significantly. Usually an initial consultation can be obtained without any charge. Hourly rates thereafter range from $300 to $800. This means that the costs of preparing a utility application on a relatively simple medical device will be in the $8,000 to $14,000 range, including a relatively narrow patentability search. A utility patent application on a complicated technology may run as much as $30,000. A search with a patentability opinion might be $3,000 to $5,000. A complete, formal FTO opinion may cost $25,000 to $100,000 (this expense not only reflects extensive search work, but the fact that, in some cases, the attorneys issue a detailed, written opinion on FTO). Some medtech patent attorneys offer an intermediate approach, consisting of a careful search for prior art and a discussion with the inventors (and potential investors) about the broad scenario for FTO. The cost of this less formal analysis may be in the $5,000 to $10,000 range, depending on the complexity of the patent landscape.[8]

Law firms recognize that inventors are cash-challenged in the early stages of the biodesign innovation process and some will offer alternate payment strategies, including deferring payments or taking equity (stock). Because it is fairly easy to run up a large patent "tab" quickly, many established firms are not willing to defer payments or take equity alone. Another option, which can often be less expensive than working with a patent attorney, is to engage a patent agent. Patent agents are registered to practice *only* before the USPTO and, as such, cannot conduct patent litigation in the courts or perform services, which the local jurisdiction considers as practicing law (for example, providing opinions on patentability, ownership, or validity, or performing FTO analyses).[9] Usually, patent agents are engineers who have successfully passed the USPTO patent office exam, or have served for four years or more as a US patent examiner before entering private patent practice.[10] Patent agents can offer a cost-effective approach when the innovator's IP needs are relatively straightforward, there is broad FTO in the field, and/or the innovator does not anticipate litigation. Note that many law firms employ patent agents and patent attorneys in their IP practices, so that a client company can benefit from a full range of services, for example, by using agents to draft applications and associates and partners to help with IP strategy and litigation.

There are several guidelines to consider in finding a patent attorney or agent.[11] It is important to seek someone with experience filing similar medtech patents (in the same technology and clinical area, if possible). The better medtech patent firms will have previously performed detailed patent landscape analyses in many areas, and these provide attorneys in these firms with substantial background. In some situations, certain attorneys may not be able to work on a filing because their firm has another case that represents a potential conflict of interest. The law firm will make this determination before agreeing to begin the contract. As with engaging any consultant, it is helpful to get recommendations from experienced inventors and entrepreneurs in the field. Note that faculty or student inventors may be steered to a particular attorney by the university's office of technology licensing (**OTL** – see 6.4 Alternate Pathways). However, it is worthwhile doing an independent validation of the expertise of the attorney before proceeding.

Working with an attorney or patent agent who is geographically located near to the innovator is of course convenient. Yet, given the ease of electronic communication, it is generally more important to find someone with expertise in the technical and clinical domain of the invention than someone who is co-located. Make sure there is an explicit understanding of how (and how much) any attorneys will charge for their time to avoid unpleasant surprises. Ask for a face-to-face consultation to gauge the attorney's approach and assess the "chemistry" between the attorneys and the individuals with whom they work most closely (open, clear communication is essential).

In meeting with the attorney, do not ask general questions that can be addressed by reading a book or performing online searches. Invest time in advance to become educated on IP issues so that the attorney's time can be used more effectively to address unique and complex issues (rather than basic questions). Consider creating an initial draft of the application and/or the claims before getting too deeply into the patent drafting process. While the attorney will surely modify the draft, it will likely save time and money by ensuring that everyone has the same understanding of what the patent application will look like.

International filing strategies

An international IP strategy should be developed with the costs and benefits associated with the filings outside the US clearly in mind. Every foreign patent application will require a significant financial commitment (e.g., $250,000–$500,000 over the life of the patent, not including the cost of any potential litigation required to enforce it). This high cost is driven primarily by the need for specialized legal and translation services for each country (the European Patent Organization is an exception, see 4.1 Intellectual Property Basics). To justify such a sizable financial commitment, an innovator or company must be certain that the benefits of the foreign filings will be measurable and central to the overall business strategy. Particularly during the early stages of a company's development, foreign filing should be made when there is a clear strategic rationale (e.g., the product is anticipated to generate a sizable percentage of its revenues internationally). The selection of countries varies from case to case. For many medical devices, it is common to include Europe, Japan, Australia, Canada, and Israel in international filings of US patents. Increasingly the BRIC countries (Brazil, Russia, India and China) are of interest because they can offer attractive emerging markets. Mexico is sometimes considered because it is a cost-effective location for manufacturing.

Other special considerations – and geographies – may come into play depending on the nature of the technology and business goals of the start-up company. Consider Neodyne Biosciences as an example. Neodyne develops and commercializes tissue repair devices that minimize scar formation, which the company has initially targeted at providers in the aesthetic and reconstructive plastic surgery fields. Given the attractiveness of aesthetic surgery markets outside the US, Neodyne began to analyze its international patent protection strategy early. According to Bill Beasley, the president and chief operating officer of the company, "Neodyne looked at two main questions when selecting countries in which to seek patent protection. First, is there a strong market opportunity for our product in that country? And, second, how do those countries value intellectual property?" Another factor that can come into play is the cost and complexity of filing IP in certain countries, which has to be weighed against the risks of the company's technology not being protected in those markets. "Certain investors are more interested than others in comprehensive patent protection, which may guide a start-up's decision to invest in their IP strategy," said Beasley. Neodyne filed patents in China and India due to their enormous market size; in Brazil, Japan, and South Korea because of the large, local aesthetic products markets; and in Israel because "it's a good place to secure a strong patent position," explained Beasley. The company also filed in Europe via the European Patent Office (**EPO**), in Australia for **clinical trial** access, and in Canada to establish access a neighboring geography from its headquarters in the US. Beasley advised innovators to think early about their commercialization strategy, and which markets are most attractive for their technology. "Think of your IP as assets that have value, for instance, to potential acquirers. It is also important to understand your company's overall appetite for investing in IP and whether the technology and market opportunity warrant a large investment."

Innovators should also take into account the fact that they may ultimately want to *manufacture* in a different country. To the extent they can anticipate this, it is critically important to file in the candidate locations (major medical device manufacturing capabilities exist in Belgium, Costa Rica, Ireland, Mexico, Taiwan, and several other countries). If innovators are targeting a particular large company as a potential acquirer of their technology, it is also worth considering where that company has its deepest roots. For example, a German

medical technology company like Maquet might have some hesitation in acquiring a technology for which no patent rights have been pursued in Germany.

Strategies in managing a patent portfolio

Many early-stage innovators put much more emphasis on the filing and obtaining of an initial patent than on developing a comprehensive and ongoing patent strategy. This kind of approach to IP carries with it sizable risks that can negatively affect an innovator's ability to raise funding, not to mention threatening the viability of the company once it gets started. In contrast, innovators and companies that strategically manage their IP portfolios stand to minimize their IP-related risks, differentiate themselves from their peers, and increase the value of their businesses. By taking an active approach to managing IP, the value a company receives from its IP investment can evolve from merely protecting its inventions to controlling (and even preempting) competition, building markets, delivering revenue, and driving business strategy.[12] It is useful to think of these benefits in the form of a hierarchy, as shown in Figure 5.1.3, in which protecting inventions is the base upon which increasingly important benefits are built.

There are two basic features of an effective and valuable patent portfolio. First, the patents must actually cover the products that are developed. This sounds trivially obvious but, in practice, there is a real risk that the development and refinement of a medtech device will outrun the patent coverage. This happens due to a lack of continued attention to IP during the development process (i.e., in response to **prototyping** and testing, the team adds key features to the design that may or may not be patentable). In the hurry and pressure to get a product to market, the team may simply forget that new IP has been created (or that the new features may infringe existing patents).

The second characteristic of an effective patent portfolio is that it creates real barriers to entry. It is not sufficient for the innovators to patent the features of their new device. It is essential to anticipate who might be able to develop a similar device or a next-generation product and then develop patents to block that from happening.

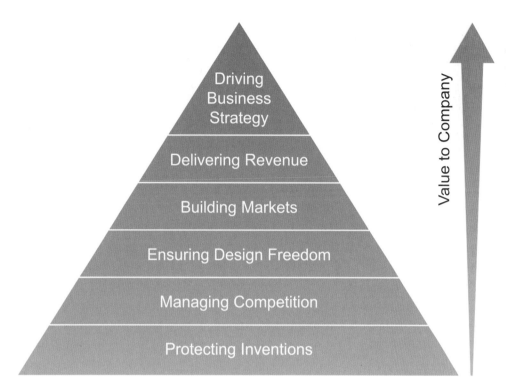

FIGURE 5.1.3
As innovators and companies take a more strategic approach to managing IP, the value they receive from their IP investments increase dramatically (based on Ron Epstein, "Building a Business Relevant IP Strategy," IPotential, LLC; reprinted with permission).

This requires an in-depth understanding of the competitive landscape – not only the patent portfolios and filings of the main competitors, but their business strategies as well.

The mechanism for developing and maintaining a valuable patent portfolio is *continuous internal and external monitoring* of IP. Internally, IP must be integrated into the ongoing research, development, and business planning strategies – that is, it must be an active part of the awareness and work of all team members. Time and attention need to be explicitly allocated to correlate the progress being made in prototyping and testing with the approach to patenting. As new team members are added, it is essential that they be educated in the importance of keeping careful **innovation notebooks** and in the dangers of **public disclosure**. External monitoring means that at least some member(s) of the team must focus on surveying and understanding what competitors and potential competitors in the field are doing on a regular basis. Effective ways to gather this information include talking with physicians and attending symposia, courses, and trade shows. This will also help the team stay up-to-date on the other relevant developments in the field (including the clinical science).

The ongoing monitoring of internal and external IP activity leads to two basic types of proactive patent strategies: *defensive* and *offensive*.

Defensive strategies

Defensive strategies represent the more familiar of the types of approaches innovators can pursue. They refer to developing patents and claims that will exclude others from making, using, or selling the technology that the innovator has invented. The key issue is to understand where the innovation is going as it moves toward clinical practice. This means that in thinking about an invention it is essential to consider not only the device in isolation, but also how it will be used. Is it part of a system (and if so, are there other parts of that system that also need to be protected)? How will it be packaged? Will it come in a kit? Are there certain methods of use that should be patented along with the device claims? Are there other indications (other places in the body or alternate procedures) in which the device could be used? If so, are there other components or aspects that need to be added? Using these considerations to create a series of secondary patents and claims that serve as a barrier to competitors is

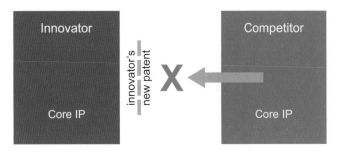

FIGURE 5.1.4

A picket fence strategy gets its name because it uses adjunctive patents to surround the core intellectual property, keeping a competitor outside of the innovator's core patent territory.

sometimes called a picket fence strategy (see Figure 5.1.4).

To illustrate this approach with an example, consider a new vascular stent that has the property of being bioresorbable (it dissolves within the body over time). The inventors are clearly interested in protecting the stent itself – the special, resorbable material and the strut configuration that gives the stent enough radial force to prop open the artery. This is the core IP. But there are other, potentially patentable aspects of using the stent that could also be covered. How is the stent expanded (e.g., with a balloon or some other deployment mechanism)? Does the catheter delivery system have special features, perhaps related to the deployment method chosen? Does the packaging need to be specialized, for instance, to preserve the shelf life of the new stent material? Is there a need for this kind of stent in some other application (e.g., to open the esophagus in children with atresia (narrowing))? Pursuing patents in some (or all) of these different areas would provide the innovators with a defensive shield around the core IP. Just be aware that laws in this area are constantly changing and recent efforts to target "non-practicing entities" may potentially limit a company's ability to protect patents that cover areas that they are not directly pursuing.

One extremely important tool in building a defensive patent strategy is through continuation patent applications, which allow inventors to refine the claims structure during the patent prosecution process. Basically, a continuation is an application that is filed to pursue additional claims to an invention using the identical description or specification from the original patent.

A continuation must list at least one of the same inventors as the original patent. Claims filed as part of a continuation will receive the same filing date priority as the original application.

The power of the continuation process for defensive strategy is significant. As the team gains experience with the original concept, including further prototyping and early-stage testing, important aspects of the invention become clearer. With time, the team is also likely to gain a more informed idea of what potential competitors in the field are doing and what will be necessary to protect the core IP. The continuation process then allows the team to submit new claims that provide a sharper focus on the invention, based on continued experience, thinking, and surveillance of the marketplace. Although obtaining new claims for an invention can seem like "getting something for nothing," keep in mind that claims are meant to clearly define what constitutes the invention. The invention or inventions – as described in the specification – cannot change with continuations, which is why it is important to include multiple well-described embodiments in the original filing. The new claims provide a way to communicate more exactly what the invention is. Keep in mind, also, that the continuation process does not extend the life of the patent – the priority date goes back to the filing date of the original patent specification and the expiration date is based upon the priority date.

In developing a defensive strategy for patent protection, it is essential to promote an active dialog between the patent attorney and the development team. Early teams tend to be worried about the expense of frequent meetings with an attorney. This can be mitigated by having a regular but brief check-in, perhaps by phone, where the team has prepared a succinct update of new test results, development milestones, and the concepts that have come from these. In addition, as the team grows, every member needs to be educated about the basics of intellectual property, either by the attorney or by savvy members of the team. This kind of communication and education is the best way to prevent the loss of rights of a patent either in the US or internationally.

One mechanism to employ the creative talents of the team in building the defensive patent portfolio is to organize invent-around (or design-around) brainstorming sessions. Have the team pretend they are working for a competitor and challenge them to design approaches that will bypass the existing patent claims. These sessions are of course worthwhile for the patent attorney to attend. The results can be used to fill in the gaps in the defensive position, either through new patent filings or new claims through the continuation process.

With respect to claims, an important approach in developing a defensive strategy is to create a range of broad and narrow claims, sometimes called "layered" protection. Broad claims are of course desirable, but keep in mind that they are also at higher risk of being declared invalid in litigation. Narrow claims do not offer as much protection (i.e., they are easier to work around), but they can be granted more quickly. Narrow claims, for example, can include language that is targeted toward specific commercial embodiments. A layered combination of broad and narrow claims provides the best chance to maintain an effective defensive barrier after the patent prosecution process. The timing of product entry into the market can be an important factor in suggesting an appropriate mix of broad and narrow claims. If **FDA** approval is anticipated to be relatively quick (a **510(k)** clearance, for example) and the product can be brought to market easily, it may be strategically important to have some narrow claims already issued on the device to protect its introduction to the marketplace.

To benefit from the defensive protection of a patent, the owners must threaten or take legal action (e.g., filing a lawsuit) against anyone who infringes on their IP. In the realm of medical devices, the company that has been assigned the IP or licensed the patents (from a university, for example) is generally the one that takes on the responsibility of pursuing potential infringers. Occasionally, universities will also take on this role, although in medtech there are a relatively small number of "blockbuster" patents that rise to this level of attention on the part of the university.

There is a cost-effective variation on the picket fence defensive patenting strategy that some innovators and

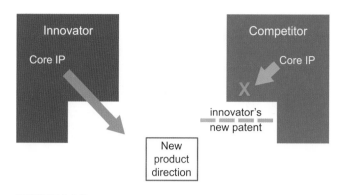

FIGURE 5.1.5

One offensive IP strategy is to preempt a competitor's ability to move toward a new product design by means of a blocking patent.

companies employ – namely, to intentionally publish information on incremental improvements and new uses related to a core IP asset in order to create a buffer of publicly disclosed prior art to keep competitors away. This has the same effect as additional patent coverage in terms of keeping competitors from owning these improvements, but has the advantage of being fast and free.

Offensive strategies

Offensive patent strategies refer to approaches that look outward to the competitive landscape and try to take advantage of the patent position of other companies. These strategies tend to be less intuitive and more difficult than defensive strategies for first-time innovators (who are typically focused on developing and protecting their inventions). However, understanding the direction of competitors in the marketplace and moving in an intelligent and proactive way in reaction to (or anticipation of) their IP portfolios is an absolutely essential means of ensuring that the innovator's product will successfully make it into clinical care.

The first offensive strategy is to anticipate the direction that a competitor may be headed and create IP that blocks that approach (see Figure 5.1.5). Sometimes an innovator can see the competitive pathway clearly enough to create new IP just for the purpose of blocking a particular competitor. In other cases, the block arises from both groups independently seeing the same general technical direction, and one group inventing earlier than

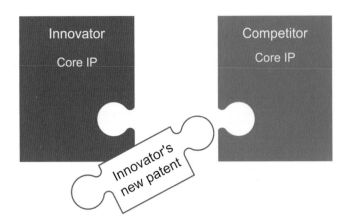

FIGURE 5.1.6

An offensive bridging strategy that complements and/or fills a gap in a competitor's IP portfolio can be effective for an innovator seeking to license or sell a technology to a larger company.

the other. In the case example on Intuitive Surgical, Dr. Fred Moll recounts a classic situation in which Intuitive Surgical, a robotics company, became concerned about patents owned by a competitor, Computer Motion. The patents, which described fundamental software-controlled mechanisms of translating hand movement to instrument tip movement, were directly in the path that Intuitive wanted to pursue in developing its da Vinci® robot. After a great deal of wrangling, this offensive block ultimately led to the **acquisition** of Computer Motion by Intuitive.

A second offensive strategy is also based on an intimate knowledge of a competitor's IP portfolio and business strategy, but in this case involves explicitly linking into the competitor's IP by building a bridge to it with a new patent, as shown in Figure 5.1.6. This strategy should be used specifically when there is a candidate company that the innovators thinks would be a good fit to acquire the new technology. It requires the innovators to have a good understanding of both the basic IP position of the acquiring company and at least some sense of the company's priorities with respect to new technology directions. In practice, this latter information may only become apparent as the inventor has discussions with the company about potential licensing of the core technology.

All of these offensive and defensive strategies require a high level of diligence in pursuing information about competitive companies. With respect to IP, it is important to perform regular and ongoing searches of US and international patent databases using the appropriate keywords, competitor companies, and known inventors in the field. In addition to patent searching, the team must keep up with the clinical literature in the area, stay current on new product releases, and periodically perform Google searches around the technology itself, key companies, inventors, and also clinicians. In an active technology area, it is not uncommon to find 50 or more "hits" a week of potentially interesting new information.

Strategies related to timing of patents

Real patent coverage begins when a utility patent is issued. Between the date of publication and the date of issuance of a utility patent, innovators cannot prevent another company from operating in the space of the invention. Having additional companies operate in the same space, even if just for a short while, can erode prices, thus, impacting long-term revenues even once the other companies are restrained. Ideally, innovators will have patent coverage (under an issued utility patent) at the time their product goes to market.

A relatively new tool from the USPTO is the Prioritized Patent Examination Program, known as "Track One."[13] This program allows applicants, for a fee, to receive a final action from the USPTO on their patent application within twelve months. This expedited process can be helpful in coordinating the timing of patent issuance with a company's entry into the marketplace. It can also be used to ensure the company had at least one issued patent in its portfolio as it approaches other important milestones, such as those related to fundraising. Furthermore, in a space with a competitive IP landscape, Track One can be used beat competitors *through* the patent office and give the company a basis from which to defend its IP position (i.e., patent lawsuits cannot be initiated based on a pending patent application). Track One is best suited for patent applications with narrow claims because they move through the patent office more quickly and stand up better in litigation. The primary reason not to pursue Track One is if the company cannot afford it. Additional fees must be paid to the USPTO, and this expedited approach also condenses the amount of attorney time the company spends on its application into one year, starting very soon after filing.

An alternate approach is for the company to delay its filing to extend the patent term as long as possible. This strategy also can prevent competitors from learning about new devices and methods, since a patent application is not public knowledge until it is published. However, the delay should not be so long that a competitor is able to negatively affect the innovator's desired IP position. There is also a risk that the company may inadvertently disclose the invention prior to filing.

A slightly technical but important timing issue is called "provisional protection" and refers to the fact that if claims are issued in substantially the same form in which they were published, a company can seek back damages starting from the date of publication. That is, if a company B infringes the claims of company A, company A may collect damages dating back to the time those claims were initially published (generally 18 months after filing) rather than the date the claims were granted (which may be several years after filing). This makes it important to write at least some claims in the initial filing that have a strong chance of being granted (as opposed to trying only to get the broadest conceivable claims in the first round).

In some cases, it is possible to extend patent coverage beyond the conventional 20 years when the innovators or company can make the case that the introduction of the product in the marketplace has been delayed for legitimate reasons. This strategy has been relatively widely used in the pharmaceutical industry where development times are so long that the 20-year life of a patent is perceived as a major limitation (by the time a drug or biologic is approved by the FDA, there may be only 10 years or less remaining of patent coverage). Under these circumstances, a company can apply to the patent office for an extension of the term of the patent, which is calculated based on the length of time the development was held up in clinical trials and regulatory approval. However, a company is only allowed to do this for a single patent on the technology and there is a relatively short window once the company receives FDA approval to apply for the extension. In the past, medical devices – which tend to have a much quicker approval process

than drugs or biologics – generally have not had patents extended through this mechanism. Recently, however, with the evolving complexity of medical devices (particularly with drug-device combinations) this approach to extending patent covered is being pursued with increasing frequency in the medtech industry. For instance, Abbott Vascular applied for an extension of the Rapid Exchange™ delivery system for the Xience drug-eluting stent based on the relatively long FDA approval process for this device. Generally, innovators should keep their patent attorneys updated on their FDA timelines so they can help determine if this is a viable strategy (and be prepared to take action once clearance or approval is received).

A related IP timing strategy employed by some innovators and companies is colloquially called "**evergreening**" – a process of introducing modifications to existing inventions and then applying for new patents to protect the invention beyond its original 20-year patent term. Again, this strategy has been much more widely applied in the drug industry compared to medtech. In certain situations, however, devices lend themselves to this approach. For example, during the rapid expansion of the stent market in the 2000–2010 period, companies brought out new stent and stent delivery designs with patentable modifications on a frequent basis, every 12–18 months. This provided ongoing IP protection that in some cases extended beyond the life of the original patents.

Patent litigation

Strategies for patent litigation are beyond the scope of this text, since they generally come into play at a later stage of product development within a company and involve company management and legal experts in the decision making. However, a few basic issues are important for the medtech innovator to understand at the outset of developing an IP strategy.

There are several potential outcomes to be aware of in medical device patent litigation. In rare cases, the courts may perceive a case to be clear enough that they will issue an injunction against the party that is accused of infringement (violating the patent rights of the plaintiff). Consequently, the company accused of infringing the patents must stop the manufacturing and sales of the device. Much more commonly, the case goes to trial or is settled without an injunction. The outcome of a trial may be decided by either a judge or a jury, depending on the nature of the lawsuit. If the patent holder wins (that is, proves that there was infringement), that innovator or company is awarded damages that typically reflect a reasonable royalty based on the sales of the infringing company (e.g., something on the order of 5 percent of net sales). In recent years, there have been a number of multimillion-dollar medical device patent infringement settlements paid by large companies to individual inventors who hold key patents, particularly in the cardiovascular and orthopedic fields. If the courts make a determination of "willful infringement" – that is, the courts believe that the infringing company acted in bad faith, with full knowledge that it was infringing and no mitigating circumstances – then the damages can be tripled (treble damages).

Attorneys who perform patent litigation in court are generally different from the attorneys who write and "prosecute" (pursue the approval of) patents, although both types of attorneys may work for the same firm. It is not unusual for venture capitalists or companies that are doing serious due diligence on an innovator's IP to pay a litigator to evaluate the robustness of the patent protection, particularly in a space that is known to be litigious.

Patent litigation is a time-consuming, expensive, and emotionally draining experience for the participants. The big companies that dominate the medtech industry have large "war chests" for IP litigation. At present, it is part of the culture of the industry sector that important products are, more often than not, embroiled in patent litigation. Practically speaking, if the innovators develop an important medical device they can essentially rely on being involved in significant litigation at some later stage. The burden of litigation, in terms of productivity alone, can be considerable. The inventors and other key contributors to product development are required to participate in depositions in advance of a trial, which are formal opportunities for counsel – from both sides – to discover essential facts about the case. These can be highly adversarial and tricky encounters, which require a great deal of careful preparation. The discovery process may go on for many months before a trial or a settlement occurs, with the process consuming hundreds or even thousands of hours of potentially productive time from inventors and other personnel.

Alternative mechanisms to challenge patents

The America Invents Act (**AIA**) of 2011 introduced a suite of new or improved procedures that companies can potentially utilize to either strengthen their own patents and/or challenge competitive patents. These tools include pre-issuance submissions, supplemental examination, Post Grant Review, and Inter Partes Review.[14]

Preissuance submissions provide a mechanism by which third parties can monitor and potentially affect the course of prosecution of a competitor's patent application by submitting publications, along with an explanation of their relevance to the examination of a pending patent application. The publication and explanation of relevance will be made of record in the prosecution history of the application, which may prove useful in a future infringement action.[15]

Supplemental examination provides a means for a company to strengthen its own patents against future invalidation. Specifically, supplemental examination provides for the post-issuance examination of an issued patent to "consider, reconsider, or correct information believed to be relevant" to an issued patent.[16] For example, supplemental examination may be used to address issues that may relate to an applicant's duty of candor and good faith before the USPTO, including the duty of disclosure.

Post Grant Review and Inter Partes Review proceedings offer third parties an alternative to patent litigation for invalidity issues. The rules for Post Grant Review proceedings permit a party other than the patent owner to file a petition to institute a Post Grant (after issuance of the patent) Review of the patentability of one or more patent claims for any reason (not limited to prior art issues). The petition must be completed and filed within nine months after the date of the issuance of the patent. Therefore, it is important for companies and/or their patent attorneys to closely monitor pending competitive patents of interest to be able to file the petition within this timeline.[17] A petition for Inter Partes Review is similarly used to challenge the validity of patent claims based on patents and printed publications, but it may be filed *after* nine months from the date of issuance.[18] However, there are significant differences between Post Grant Review and Inter Partes Review, including the fact that the grounds upon which invalidity is based and the evidence that may be submitted are limited under the second option.[19]

The Intuitive Surgical story provides an example of the importance of IP strategy in effectively managing

competitive threats. Although, in general, innovators do not face litigation until some years into development of their product (typically after sales begin), the seeds for successful litigation can be planted from the beginning of the biodesign innovation process. Focusing on the basics, as presented in chapter 4.1, provides a good chance of avoiding the biggest problems in litigation. Careful and honest record keeping, engaging a competent attorney early in the process, and diligence in uncovering prior art and publications are the key success factors.

FROM THE FIELD ▶ INTUITIVE SURGICAL

Using IP strategy to manage risk and opportunity

Intuitive Surgical was founded by physicians Fred Moll and John Freund, as well as engineer Robert Younge, to pioneer the field of minimally invasive, robotic-assisted surgery. The company's primary product, the da Vinci® Surgical System, translates the surgeon's natural hand and wrist movements on instrument controls on a console into the corresponding micro-movements of specialized surgical instruments positioned inside the patient through small incisions.[20] By combining the technical skill of the surgeon with computer-enhanced robotic technology, da Vinci enables a minimally invasive

approach to procedures such as prostatectomy, hysterectomy, myomectomy, gastric bypass, and mitral valve repair (see Figure 5.1.7).

Moll, who became the company's first CEO, learned of the technology through a non-profit research organization called the Stanford Research Institute (now SRI International), which had developed an initial prototype under contract to the US Army. Seeing the potential of such a system to accelerate the viability of minimally invasive surgery for a broad range of procedures, Moll and team licensed SRI's core IP surrounding surgical robotics and founded Intuitive Surgical. According to Moll, "We were lucky early on in that we licensed a significant patent portfolio that had

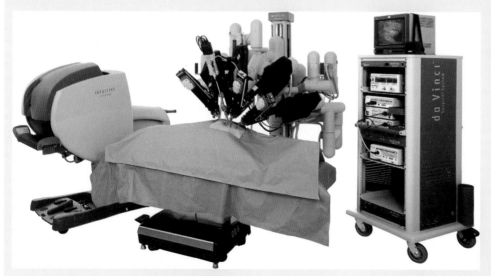

FIGURE 5.1.7
The da Vinci system (courtesy of Intuitive Surgical).

early dates on inventions that were important to the system and the core capabilities that we wanted to develop in surgical robotics." By adding some of its own proprietary technology and collaborating with IBM, MIT, and other institutions and thought leaders in the field, they developed the da Vinci system,[21] which received its preliminary FDA clearance in 2000.

Intuitive Surgical placed great importance on its IP strategy early in the company's existence. Regarding the licensing agreement with SRI, Moll explained the steps the company took to ensure it would maintain adequate control over the IP: "When you license intellectual property from the inventors, you can do it in various forms. But the important points are to get an **exclusive worldwide license**, which we were able to get. We also secured the ability to manage the patents ourselves. By that, I mean that we took over the prosecution of the IP." This allowed Intuitive Surgical to add to the patent portfolio and protect it in the way that best covered the specific product it was developing. "If we didn't have this ability, it could have been a real disadvantage," said Moll.

In terms of expanding the company's IP position, Moll noted, "We were able to do follow-on continuations to the patent portfolio in certain areas that were important to continuing to protect the Intuitive product. By that I mean we used the dates of these early patents to expand the breadth of the filings to cover features and capabilities that were anticipated but not specifically covered in the first filings by SRI." The company also took a systematic and proactive approach to managing its IP position. "You need to have a disciplined way to continuously look for prior art that would somehow modify your view of what patent positions are possible or the strength of your patents, as well as continuing to have a strong system to collect invention disclosures and turn them into a continuous stream of new filings that protect the newer aspects of the technology that you're developing on a weekly or monthly basis," Moll explained. To accomplish this, Intuitive added an in-house IP attorney to its team in the company's early days. "It makes a lot of sense, even for start-up companies, to hire in-house IP counsel sooner rather

than later," Moll said. "In-house counsel is always going to understand better than an outside law firm what's going on within the company and what's most important to the product because they're just a lot closer to it." Relative to the value that medtech companies extract from their patent portfolios five or more years into their existence, "most people tend to under-invest in IP, which can be a big mistake," Moll added.

Although Intuitive Surgical established a leadership position in the surgical robotic market, the field quickly became competitive with the entry of companies such as Computer Motion. While there were important differences between da Vinci and Computer Motion's Zeus® system, they competed head-to-head. Before long, the companies became entangled in a series of patent disputes. Moll described the issue at the core of the conflict: "Computer Motion had some early patent filings with a number of broad claims that had to do with fundamental ways of controlling rigid tools and how you translate hand movement to instrument tip movement if the connection is electromechanical and software related rather than directly mechanical . . . . And, arguably, a couple of these patents had the potential to inhibit our ability to control instruments the way we wanted to control them to make sense to the **user**."

The Intuitive Surgical team believed that it had a strong IP position and a good chance of invalidating the Computer Motion patents. However, "The risk, not only the cost, associated with going through the legal process to find out whether their patents were valid was very significant. If we went through a litigation process and were unsuccessful, meaning that we were judged to be infringing on their very broad claims about how you control an instrument, it would be a difficult situation from a business standpoint and put the company at risk of having to redo a lot of what we had done," Moll said.

Despite this risk, "there was no settlement in sight," recalled Moll. Intuitive Surgical briefly considered cross-licensing, but determined that this was not an attractive solution. According to Moll, "It progressed to the point where it was clear that it would probably end up in a jury trial. As a lot of litigators will tell you, a company's chances of winning a patent dispute that makes it all the way to a jury trial is about 50 percent. Most juries have a difficult time understanding a highly technical dispute. So you don't want to get all the way to a jury because your chance of winning isn't entirely related to the strength of your case."

The team continued to believe it could win the suit, but the risk of losing was so great that it decided to pursue one more option. "It became clear that there was an opportunity to buy the company," Moll recalled. "The upside of an acquisition was not that we would be getting a new product line, because we didn't consider the Computer Motion product line additive to ours. It was competitive and, we believed, not as clinically useful. But what we would be getting was a very strong position with the combined IP portfolios in the area of surgical robotics controlling rigid instruments." Ultimately, he explained, "It came down to a judgment call on how much it was going to cost us to litigate the dispute versus how much it would cost to buy the company, in addition to the certainty of outcomes on each path." Fortunately for Intuitive Surgical, the company's balance sheet was strong enough that acquiring Computer Motion was a viable alternative. The two companies joined forces in March 2003.

The breadth of the combined IP position gave Intuitive strength in the surgical robotics market that Moll believed had deterred some competition. "We always worried that the Japanese would enter the market for surgical robotics. To date, they haven't in a significant way. I think some of that is due to our success in technical development and the know-how associated with building the Intuitive system. And, I think some of it is also due to the perception of a very strong IP barrier, or picket fence, around the system that we developed," he said. Since the sale of its first da Vinci system, Intuitive Surgical

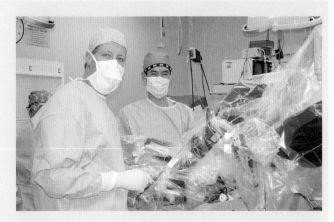

FIGURE 5.1.8
Surgeon Thomas Krummel and surgical resident David Le use the da Vinci system at the Lucille Packard Children's Hospital (courtesy of Thomas Krummel).

has expanded its installed base to more than 2,500 hospitals and has sustained its position as the global leader in the field (see Figure 5.1.8).[22]

Reflecting on his experiences, Moll underscored the importance of a strong IP strategy: "You can't just start a company, file a patent, and hope for the best. You need to file, as early as possible, continuations of the invention that surround the initial technology effort. In other words, you should have a rigorous process of invention disclosure by employees, new filings, and refinements of existing filings that's continuous, throughout the life of the company. The idea of continuously monitoring prior art and other inventions going on within the field is also very important. At almost every company I've been involved with, there are problems with intellectual property that the company doesn't own. If this blindsides you late in the game, it can be devastating."

Finally, he said, "IP gets more and more important every year in the medical device business for principally one reason: the cost associated with clarification of who owns what has gotten so enormous. If you start with and maintain a strong IP strategy, it helps ensure that you don't get into situations where you're spending as much on IP litigation as you are on product development."

Online Resources

Visit www.ebiodesign.org/5.1 for more content, including:

 Activities and links for "Getting Started"
- Understand the IP landscape
- Validate freedom to operate
- Hire IP counsel
- Devise defensive and offensive IP strategies
- Develop a comprehensive IP strategy and implementation plan

Videos on IP strategy

CREDITS

The editors would like to acknowledge Jessica Hudak of the Hudak Consulting Group for her extensive input to the second edition of the chapter. Many thanks also go to Eb Bright of ExploraMed, as well as Ritu Kamal, for their valuable assistance. The original version of the chapter was based in part on lectures by Tom Ciotti and Mika Mayer of Morrison & Foerster LLP and Jeffrey Schox of Schox PLC, who also provided editing and consultation. Further appreciation goes to Gary Guthart and Fred Moll for providing the Intuitive Surgical story and Moshe Pinto for sharing the Spiracur case, as well as Bill Beasley of Neodyne and Yoh-Chi Lu of Biosensors for those examples.

NOTES

1 Dennis Crouch, "How Many U.S. Patents Are In-Force?," Patently-O, May 4, 2012, http://www.patentlyo.com/patent/2012/05/how-many-us-patents-are-in-force.html (January 20, 2014).

2 "2013 Patent Litigation Study: Big Cases Make Headlines, While Patent Cases Proliferate," PricewaterhouseCoopers, 1995–2012, http://www.pwc.com/en_US/us/forensic-services/publications/assets/2013-patent-litigation-study.pdf (January 20, 2014).

3 Pamela Yatsko, "Biosensors International and China (A)," Stanford University Biodesign Case CHINA-01-A, November 18, 2012, http://biodesign.stanford.edu/bdn/resources/casestudies/chin-01-A_biosensors_111812.pdf (January 22, 2014).

4 Ibid.

5 "Expandable Intraluminal Graft, and Method and Apparatus for," Google Patents, http://www.google.com/patents?id = iioBAAAAEBAJ&dq = 4733665 (January 20, 2014).

6 "What Does Freedom to Operate Mean?," *Patent Lens*, http://www.patentlens.net/daisy/patentlens/g4/tutorials/2768.html (January 20, 2014).

7 All quotations are from interviews conducted by the authors, unless otherwise cited.

8 Based on estimates provided by Tom Ciotti and Mika Mayer of Morrison & Foerster LLP, Fall 2008. Updated by Jessica Hudak of the Hudak Consulting Group, Winter 2014.

9 "Attorneys and Agents," USPTO, http://www.uspto.gov/web/offices/pac/doc/general/attorney.htm (January 20, 2014).

10 "Patent Agents and Patent Attorneys," Wilson Enterprises, http://www.wilsonenterprises.org/AgntAtty.htm (January 20, 2014).

11 Sujith Pillai, "Working with a Patent Attorney," Invention Patenting Group, http://www.inventionpatenting.com/patent-article-23.html (January 20, 2014).

12 Ron Epstein, "Building a Business Relevant IP Strategy," IPotential, November 10, 2006.

13 "Track One Prioritized Examination," USPTO, http://www.uspto.gov/patents/init_events/Track_One.jsp (January 20, 2014).

14 Leahy–Smith America Invents Act, Section 6, Post–Grant Review Proceedings," Bitlaw, http://www.bitlaw.com/source/America-Invents-Act/6.html (January 20, 2014).

15 "Preissuance Submissions," USPTO, http://www.uspto.gov/aia_implementation/faqs-preissuance-submissions.jsp (January 20, 2014).

16 "Supplemental Examination," USPTO, http://www.uspto.gov/aia_implementation/faqs-supplemental-exam.jsp (January 20, 2014).

17 "Post Grant Review," USPTO, http://www.uspto.gov/aia_implementation/faqs_post_grant_review.jsp (January 20, 2014).

18 "Inter Partes Review," USPTO, http://www.uspto.gov/aia_implementation/faqs_inter_partes_review.jsp (January 20, 2014).

19 "Inter Partes Review," Fish & Richardson, http://fishpostgrant.com/inter-partes-review/ (January 20, 2014).

20 "Intuitive Surgical's da Vinci Surgical System Receives First FDA Cardiac Clearance for Mitral Valve Repair Surgery," Intuitive Surgical press release, November 13, 2002 http://phx.corporate-ir.net/phoenix.zhtml?c=122359&p=irol-newsArticle&ID=355838&highlight= (January 20, 2014).

21 Gary Singh, "The Robot Will See You Now," *Metroactive*, October 2004, http://www.metroactive.com/papers/metro/10.06.04/scalpelbots-0441.html (January 20, 2014).

22 "Company Profile," Intuitive Surgical, http://www.intuitivesurgical.com/company/profile.html (January 20, 2014).

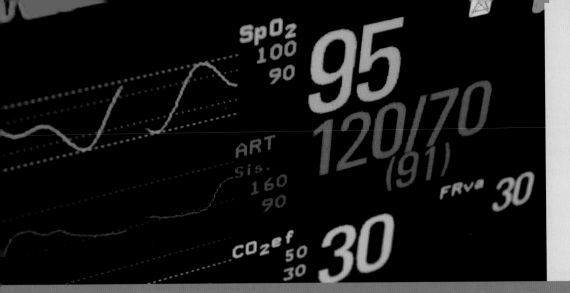

5.2 R&D Strategy

INTRODUCTION

Initial exploration and testing of the chosen concept has shown that it is technically feasible and can address user requirements. But now the solution has to be made real. Successive iterations must lead to a device that is safe, performs effectively in humans, and can be efficiently manufactured. Thousands of hours, scores of raw materials, and varied and expensive equipment will be dedicated to engineering and testing to incrementally reduce project risk and produce a final product. This is R&D.

In the medtech field, research and development (R&D) typically refers to the scientific and engineering work required to take a concept from an early-stage prototype to a user-ready final device. Whereas early prototyping focuses on proving the general feasibility of an idea, the goal of R&D is to develop a series of progressively advanced working iterations until all critical user requirements and core technical specifications have been met. Along the way, bench, tissue, and animal tests are employed to confirm that a product can be safely and effectively used in humans. The entire process can be lengthy and complicated, and it calls upon many different engineering disciplines and skills. But, taken as a whole, the two main outcomes of R&D are to reduce project risk, typically technical, business, or pre-clinical in nature, and in the process, lead to the creation of significant value. As such, it is essential to have a cohesive strategy for approaching R&D.

An R&D strategy defines key milestones that need to be achieved to demonstrate development progress, identifies and prioritizes the technical challenges that must be addressed to achieve each milestone, and calls out the engineering activities, resources, and testing necessary to validate the solutions to these challenges. At its core, an effective R&D strategy, and the tactical R&D plan that supports it, seeks to resolve the greatest risks associated with developing an innovation as early as possible, with the most efficient commitment of capital, time, and effort. Ultimately, the articulation of a sound R&D strategy with clear milestones (and an understanding of how to tactically achieve them) may be one of the most critical factors enabling the successful development of an innovation.

OBJECTIVES

- Appreciate that R&D encompasses a wide range of iterative activities that work together to help innovators retire risk and create value.

- Understand the importance of defining strategic R&D milestones.

- Recognize how to identify and prioritize the key technical challenges associated with each milestone.

- Learn to outline the engineering activities, resources, and tests required to address these challenges as part of a high-level R&D plan.

Importantly, this chapter focuses on providing innovators with an overview of key considerations and a recommended approach for developing an R&D strategy, which includes a high-level R&D plan. Innovators without an engineering background should seek more tactical information about R&D from the additional resources listed in the Getting Started section. Additionally, while the discussion is primarily focused on more traditional medical devices, many of the concepts and approaches integral to R&D strategy apply equally well to a wide range of medical innovations, from software to more complex projects.

 See ebiodesign.org for featured videos on R&D strategy.

R&D STRATEGY FUNDAMENTALS

R&D is sometimes referred to as "engineering" in the **medtech** field, since the design and development of medical technologies is heavily dependent on activities related to one or more engineering disciplines. In the context of the biodesign innovation process, R&D typically includes all engineering and testing activities, beginning as early as **concept** exploration and final concept selection (see chapters 4.5 and 4.6) and then continuing to the point when a product is ready to be released into production. The type of testing used to support R&D varies from project to project, but innovators can draw from all of the types outlined in the biodesign testing continuum (see Figure 5.2.1).

The primary distinction between the development and testing described in 4.5 Concept Exploration and Testing and the activities outlined in this chapter is that now innovators are working on a single concept rather than building and testing models to help them choose a final solution idea to take forward. Typically, after selecting a final concept, innovators may perform additional simple testing (e.g., **user**, **bench**, simulated use, or tissue tests) and then transition to more advanced testing methods (e.g., animal and human studies) as R&D progresses and critical project **risks** are retired. However, R&D is a highly iterative activity that sometimes requires innovators to move backwards in the testing continuum. For example, if innovators have made a long-term implant that fails due to an unforeseen degradation of a material while chronically implanted in an animal, they would need to make important technical modifications. Once these changes had been made, they would then need to revisit bench and tissue tests before resuming animal studies.

Overall, a company's ability to transform an initial concept to a "proof-of-concept" **prototype** into a final product, with many other iterations in between, is central to its viability. However, there are many other reasons why a strategic approach to R&D is important. From a practical, near-term perspective, an effective R&D strategy:

- Plays a role in determining how the original **need** is ultimately addressed.
- Provides an optimized engineering framework for developing a company's technology in order to address the most significant technical risks.
- Establishes an approach for incorporating **user experience** feedback and design input in order to create products aligned for real-world use, thereby reducing **clinical trial** and commercialization risk.
- Helps manage a primary driver of cost early in the company's life in terms of how personnel and other resources are used and managed.
- Lays the foundation (i.e., processes and culture) that helps the company continue innovating to develop new products and product iterations.
- Can lead to important insights related to the firm's intellectual property (IP) position.

In the longer term, a strategic approach to R&D can also help a company:

- Continually increase product differentiation and help mitigate market risk.

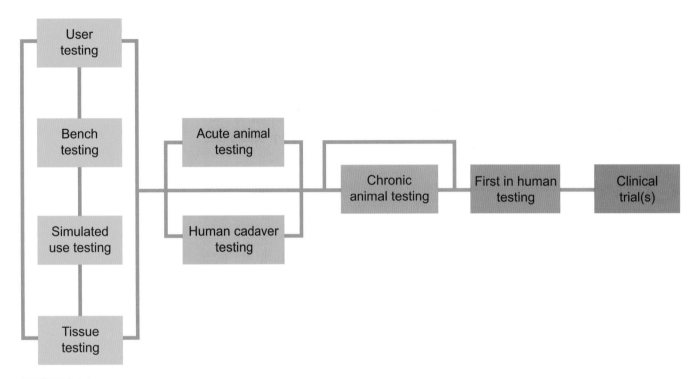

FIGURE 5.2.1

The biodesign testing continuum applies to R&D as well as concept testing and exploration (chapter 4.5), where it was originally introduced. It also extends into human clinical trials (chapter 5.3).

- Drive growth through new product innovation.
- Create a product development pipeline, which can make the company more attractive to investors and/or prospective acquirers.

As these benefits demonstrate, R&D is inextricably linked to other important aspects of the business and must be pursued keeping these other functions in mind. Additionally, the R&D strategy must be built with a clear focus on the clinical need and how it can best be solved from the perspective of the target audience, not just from a technical point of view. As Guy Lebeau, a physician and businessman who became company group chairman of Cordis Corporation, explained:[1]

One big issue is the lack of connection that can arise between the clinical need and the pure engineering desire. And that means you have to tell your engineers, "Be sure you're in contact with people who are using the tool every day."

The linkage between R&D strategy and planning

An R&D strategy can be thought of as a company's overall approach to addressing the key engineering challenges anticipated in the development of the final product. Usually, the innovators already have a general idea of what the final concept should do and how it will solve key **user requirements** and satisfy important technical specifications based on the work performed as part of concept exploration and final concept selection (see chapters 4.5 and 4.6). An R&D strategy picks up from this point by articulating how the team or company will actually develop the product, in particular by considering the obstacles that it will need to overcome and how long that is expected to take. Accordingly, an R&D strategy can be developed by focusing on three key activities:

1. Defining high-level R&D milestones.
2. Using these milestones to identify and prioritize anticipated technical challenges.

3. Creating an early R&D plan that identifies the engineering work, testing methods, resources, and time required to solve these challenges.

The creation of an initial R&D plan can be considered part of R&D strategy since understanding high-level issues related to engineering personnel, resources, and timelines is important to determining when certain milestones can be achieved. More detailed information about R&D planning, building on the information in this chapter, can be found in 6.1 Operating Plan and Financial Model. The specifics of an R&D plan are usually articulated as part of an integrated approach that accounts for all of the personnel and resources required to take a final solution into commercialization. Some of the resources listed in the Getting Started section of chapter 6.1 may also be helpful when developing a more detailed, tactical R&D plan.

Quality (and the implementation of a quality system) is another facet of development that is closely linked to R&D strategy and planning. Quality management systems (**QMS**) play a key role in informing the engineering methods that address the development and manufacturing requirements necessary to receive regulatory approval for a new medtech innovation. Specifically, quality systems lay out processes for properly capturing user requirements and technical specifications. They also detail key testing and validation methods, which are important considerations when thinking about the timeline for engineering R&D work. Finally, quality systems require that detailed risk analyses that articulate potential failures and hazards of a device and how to mitigate them are performed. Risk analysis exercises and their results are a crucial input into R&D in order to ensure that the final product is safe and will pass regulatory scrutiny. However, because R&D strategies vary significantly from innovation to innovation and are usually relatively high level, the requirements governing the implementation of a quality system, which tend to be more consistent across companies and encompass many more aspects than those just related to R&D, are discussed separately in 5.5 Quality Management.

Defining high-level R&D milestones

As outlined above, the first step in creating an R&D strategy is to define essential R&D milestones. Such milestones usually embody significant events in the R&D process and correspond to the retirement of key risks. Many such risks relate to regulatory and clinical requirements that must be addressed on the way to market and, thus, planning to address them is crucial. Importantly, the emphasis innovators place on certain milestones and how they sequence them can be affected by the regulatory pathway for the product. For example, if a device will require a **510(k)** pathway in the US, innovators might choose milestones and anticipate risks based on what is known about the predicate. In contrast, when working on a **PMA** device in the US or a **Class III** device in the EU, for which there might not be precedents to follow, innovators will likely need to achieve additional milestones to retire additional risks.

Accomplishing key R&D milestones is one way that companies can demonstrate value in the eyes of their investors and potential end users. Certain R&D milestones also represent well-accepted **valuation** points to investors. For example, a start-up making an invasive or therapeutic device, is viewed as more credible among clinicians and more attractive to investors after it demonstrates the feasibility of a device on the bench, and then again once it develops a prototype that is effective in animal tests or a pre-production model that is safe and efficacious in early human studies. For other types of projects, particularly related to software or mobile health applications, gathering extensive feedback from many users about the interface and their experience can be valuable. While some of these steps may have been done during concept exploration and testing, additional, iterative work in these areas is often needed to successfully retire risk and achieve a value-creation milestone. By thinking about milestones early, innovators can also use R&D to their strategic advantage by sequencing and timing these types of milestones to correspond with other important events in the company's evolution (e.g., financings, or the formation of clinical advisory boards).

The specific R&D milestones chosen by a company as part of its R&D strategy can vary, but for many medical technologies they may resemble those shown in Table 5.2.1. In the company's overall operating plan, these R&D milestones may be "rolled up" into a smaller number (see chapter 6.1). Regardless, innovators should

Table 5.2.1 Key R&D milestones are typically chosen to demonstrate the retirement of risks.

Important R&D milestones
Proof-of-concept that addresses the scientific and technical feasibility of the concept
First working prototype that performs effectively in a bench model
First prototype that performs effectively in tissue testing
First prototype that performs effectively in live animals
First prototype that results in long-term safety in live animals
First prototype that performs safely and effectively in humans
Pre-production device that demonstrates manufacturing feasibility
Production device that supports scalable manufacturing

recognize their serial, progressive nature – no single milestone can be achieved until those before it have been successfully addressed. That said, sometimes learning related to a later milestone may produce results that require innovators to revisit earlier R&D activities to make changes (recall the example of the long-term implant that fails during chronic animal studies due to an unforeseen degradation of a material and then must be modified and retested on the bench and in tissue tests before animals studies can resume). Furthermore, although milestones such as proof-of-concept and working prototype may already have been achieved as part of concept exploration and testing (before much thought may have been given to an R&D strategy), they are still important R&D milestones to note as having been accomplished, though additional refinement along these paths may still be needed. Given the iterative nature of R&D, it is important to understand that the process of achieving these milestones is likely to take multiple, successive attempts on a development pathway that is not necessarily linear.

To help illustrate how innovators define an R&D strategy, the Working Example introduced in 4.5

Concept Exploration and Testing will be continued in this chapter. This example involved injecting the left atrial appendage (LAA) with a material that can be delivered in liquid form, which then changes to a solid-like consistency to eliminate the space where thrombi can form via a percutaneous approach as a way to solve the need for *a way to prevent strokes in patients with atrial fibrillation caused by LAA thrombus.*

Assuming that the innovators pursuing this need have already used some crude works-like prototypes and bench tests to show that the concept is likely feasible, the next most important strategic milestones to achieve might be to: (1) show acute effectiveness of the material in a live animal; (2) show long-term safety of the material in an animal; (3) show effectiveness using a percutaneous method of delivery in an animal; and then (4) show safety and effectiveness in a human.

Importantly, the milestones chosen and the way they are sequenced reveal information about the project and the innovators leading it. For instance, this particular sequence of milestones would seem to indicate the innovators' belief that demonstrating effectiveness and long-term safety in live tissue is of higher importance from a risk standpoint than developing a percutaneous solution. This is why it is addressed earlier in the development path. As it relates to R&D planning, this sequence further indicates that resources and personnel aligned with developing the percutaneous component of the solution will likely be needed only after or perhaps in parallel with understanding long-term tissue compatibility as part of the R&D effort.

Identifying and prioritizing technical challenges

After selecting relevant R&D milestones, innovators can next consider the key technical challenges that need to be addressed to achieve them. They must think about all of the important engineering issues that they will likely encounter and then prioritize them. Typically, those challenges that involve the greatest uncertainty should be addressed sooner rather than later, so that a minimal amount of time is wasted if a particular challenge cannot be overcome.

Table 5.2.2 Key technical challenges associated with the LAA example.

#	Technical challenge
1	Creating a prototype that can deliver the material to the LAA in less than 1 hour (timeframe based on a user requirement by physicians who would use the device). (Note: The solution to this particular challenge does not necessarily have to be percutaneous, as that requirement is associated with a subsequent milestone.)
2	Ensuring that the substance from which the delivery prototype is made does not react with the material being delivered.
3	Finding a material that can remain in a liquid form for delivery into a living animal.
4	Finding a material that, once released, can reliably and predictably become solid when required within the conditions of a living body (e.g., blood flow, body temperature, clotting factors).
5	Finding a material that will not migrate from the place where it is released or finding a method to prevent migration.

Technical challenges obviously vary greatly by project. Each may have a different focus, depending on the milestone they are associated with in the overall development path. However, identifying engineering challenges is similar to thinking about the functional blocks of a concept during concept exploration and testing (see chapter 4.5). Often, the challenges that need to be solved break down along the lines of different engineering disciplines (e.g., mechanical engineering, materials science, electrical engineering). In this way, technical challenges represent the critical questions that the functional blocks of a prototype seek to address.

Using the LAA example, and focusing on the milestone of showing effectiveness in an animal, the innovators might come up with a short list of key technical challenges, as shown in Table 5.2.2. With these technical challenges identified, the innovator can think about how to prioritize them. From the above list, it is clear that there are three important challenges related to the material intended to fill the LAA, one related to a delivery mechanism, and one related to the interaction of the two. To prioritize them, the innovators can think about which challenges have been solved before and which have not. For example, many innovators working on other projects have accessed the LAA of an animal. In contrast, it is unclear whether anyone has ever found a material that meets the challenges outlined in items 3–5. For this reason, the innovators might sequence the work as follows: 4, 3, 5, 2, 1. This order makes finding a material that reacts in a specific manner with living tissue – which

is likely the greatest challenge with the most inherent uncertainty – their top priority.

As this example demonstrates, innovators often need to speculate which technical challenges will be the most difficult to address. In doing so, they must not forget to draw from foundational knowledge in the various engineering disciplines that exists and, more importantly, leverage the expertise of other engineers. While some of this knowledge would likely have come to bear during early concept exploration related to the feasibility of certain elements of a final concept, it is nonetheless important to tap into this knowledge when considering all of the technical challenges together. Depending on the innovators' backgrounds and level of experience, they may have difficulty making the determination about which technical challenges will actually be the most important (or difficult) to address. For instance, with the LAA concept, an innovator with a strong mechanical engineering background might have a clear understanding of the challenges associated with developing a percutaneous delivery system, but may not be able to predict with great accuracy the challenges involved in finding a material to deliver to the LAA. The best way to address gaps in one's knowledge base and avoid the effect of personal **biases** is to involve individuals from all of the engineering disciplines that play a part in developing the total solution. These individuals can help innovators understand the degree of engineering complexity and uncertainty associated with each challenge. which is critical in the prioritization process. However, even with input from a comprehensive and capable team, unanticipated technical challenges will almost always arise in any R&D effort. By considering known and anticipated challenges at the outset, having plans in place to address them, and staying on the look-out for unexpected problems, the innovators will be in the best possible position to overcome whatever challenges arise.

As innovators and companies start developing an R&D strategy, they must carefully consider how to customize the milestones and testing strategies they choose for their specific project. The importance of thinking critically about these assumptions and testing the most important ones is illustrated by the ArthroCare story.

A second story about Oculeve provides a look at the specific R&D milestones and testing requirements for a product that requires a more rigorous regulatory pathway.

FROM THE FIELD ▸ ARTHROCARE CORPORATION

Tailoring an R&D strategy

Hira Thapliyal and Phil Eggers, the founders of ArthroCare Corporation, originally came up with a novel way to use an electrical current directed to electrodes in contact with coronary artery plaques, often the underlying cause of arterial narrowing and occlusions, to "melt" the plaques away. However, "This was before the field of coronary interventions had exploded with the introduction of stents," Thapliyal recalled.[2] Accordingly, he and Eggers had difficulty raising funding for the idea. To help them understand if there were other needs that would benefit from their solution, they began seeking advice from experienced colleagues in the medtech field. Among others, they talked with Bob Garvey, "a marketing guy who had spent part of his career in arthroscopy." When Garvey heard about their experience with treating occlusions, he suggested that they look at various orthopedic applications since there were still important unmet needs related to tissue removal. At the time, the tools used to perform arthroscopy, a minimally invasive orthopedic procedure involving a small endoscope and tools inserted via small incisions in a joint to examine and treat various joint conditions, were still relatively crude. Taking this advice, the two men spent some time understanding the need for *a better way to remove joint tissue* and tried their device in a chicken meniscus, which he likened to "the gasket between the bones of a joint." "It worked very well," Thapliyal said. "It eliminated the tissues, which just melted away."

Based on the results of these early tests using a prototype of the device, Thapliyal, Eggers, Garvey, and another colleague from Thapliyal's past, Tony Manlove, were able to raise some funding to develop the arthroscopic application. As part of this effort, they realized that they would have to lay out an R&D strategy that would make sense to investors and give the project the best chance to achieve regulatory clearance in the shortest time frame. Thapliyal and his colleagues made a series of new assumptions regarding the sequence of events that needed to occur. "The recipe is the same for all medical devices. Bench-top, some live animals, cadaver types, and then maybe on to humans. But it has to be looked at from the context of the regulatory framework. If you have a PMA device, you structure it differently than a 510(k). If it's a 510(k), then you need to understand more about the 510(k) requirements. Some are straightforward **substantial equivalence**. Many require live animal data. But you don't want to be second-guessed there. By the time you have filed your regulatory submission, you have put in a lot of resources, time, money, and sweat. You don't want to get caught without the right data. Review all your early assumptions."

The team sought advice from various sources and used that information to decide on the best course of action. They first determined that they would likely be able to file a 510(k) based on substantial equivalence to the current electrosurgical cautery tool used in arthroscopy. Their regulatory advisor told them to "keep it a ho-hum 510(k). Don't make a big deal of it. Electrosurgery tools are already in the market. You want to be in the market. So just show, with some live animal data, that your device performs equivalently to what's already available." With this regulatory approach in mind, they next sought to understand which R&D milestones would be most important and how to prioritize them based on the technical challenges they anticipated.

They had already achieved an early technical feasibility milestone through their work in chicken menisci. For their next milestone, they had two options. They could either seek to show efficacy in live tissue through live animal testing or perform cadaver studies to show feasibility with human anatomy. For many devices, especially ones using combinations of electrical and mechanical components, animal studies are done earlier in the R&D path in order to optimize the electrical engineering functionality in live tissue. Once this is achieved, and the risk associated with using the device in live tissue is eliminated, the technology can then be refined for use in human anatomy. However, in Thapliyal's case, he and his team felt confident about the electrical components of their solution and felt strongly that effective arthroscopy was anatomy-dependent. Therefore, they decided to focus on the mechanical components of their device and optimize them for human anatomy using cadavers before performing animal tests.

To accomplish this, they teamed up with a physician they knew from San Diego, Dr. James Tasto, who helped them to work on cadaver samples. "We linked up with him and he said, 'Sure, you can come to where I'm working in the evenings, after hours, and test some menisci with your tool.'" Several months of testing and design iterations on cadavers allowed them to validate the feasibility of using the technology in human anatomy and to optimize the design. "We used the cadaver data to make sure that what we did was correctly sized and designed for human use," Thapliyal said. For example, it was essential to understand the correct angles to build into their probe to make it compatible with human joints.

The next challenge was to find an appropriate animal model. The model needed to provide similarities to

human anatomy and needed to be a widely available resource. Thapliyal recalled, "We settled on a goat model. Goat models were well published by that time and they had important similarities to humans. The size of the joint was a little smaller, but the anatomy and presentation of the meniscus was similar. Plus, goats were available and we could do a lot of them."

Over the course of a few months, the team performed experiments on about a dozen goats, with the objective of generating in vivo data to support regulatory clearance. "We didn't have to do a chronic study; we just had to acutely show that we can remove tissue in a way that was equivalent to what was currently on the market. Our tests were all statistically designed to show non-inferiority. We were not targeting that our tissue damage would be lower, but we were amazed that we saw virtually no tissue damage."

Based on the results of the goat tests, regulatory clearance was granted for the company's arthroscopy device. (See Figure 5.2.2 for a summary of the high-level R&D milestones leading up to regulatory clearance.) Once in the market, the technology caught on relatively quickly because it operated at a lower temperature and was more precise than traditional surgical tools so that damage to healthy tissue surrounding the target area could be minimized.

In terms of its ongoing R&D strategy, the team committed 80 percent of its time, resources, and engineering staff to development opportunities in arthroscopy and 20 percent on exploring new areas. Over time, ArthroCare expanded into the fields of spine and ENT. According to Thapliyal, "What we found is that we have a platform. It's not just arthroscopy. We believed it would be useful in other areas and we wanted to explore the limits of our technology." However,

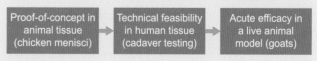

FIGURE 5.2.2
Arthrocare's early R&D milestones, which led to regulatory clearance for its device.

he offered a word of caution to innovators as they manage their R&D strategies: "Guard against projects that can take away from the focus of your company. If they become too strong, do something about it to protect the core business."

Reflecting on his experiences, Thapliyal suggested that innovators should "take a lot of advice." Moreover, he said, "Don't be seduced by the elegance of your assumptions. You have to go back and review them and test them from time to time." This philosophy allowed Thapliyal to effectively recast his focus when he ran into difficulties, and to also keep a complex technical development effort on track through the achievement of important company milestones.

FROM THE FIELD ⟩ OCULEVE

Defining an R&D strategy to retire key risks

Seeking a way to treat dry eye disease (DED) that would be more effective than the prescription eye drops that are the current **standard of care**, Michael Ackermann and his Stanford Biodesign team developed the idea of using neurostimulation to activate the lacrimal gland, which produces natural tears, in order to increase its tear production. Earlier in the biodesign innovation process, they developed a series of "looks-like" prototypes to explore the idea of an intra-orbital implantable device with physicians. They also created a "works-like" model to test the technical viability of the idea. After multiple refinements, the team used the works-like prototype to confirm the basic **mechanism of action** through a progression of bench and animal tests (see chapter 4.5 for more about its experience with concept exploration and testing).

To chart a course forward, Ackermann defined a series of strategic R&D milestones for the team (see Figure 5.2.3). These milestones were intended to sequentially retire the most significant risks in the project and help the team make steady progress toward the market. "Establishing milestones requires innovators to identify what the value creators are," said Ackermann.

"You have to ask yourself, what do you need to prove to yourselves, to your potential investors, and to the public?" He added, "Oftentimes, those things aren't exclusively R&D related, but they require R&D to achieve."

For Oculeve, an early strategic R&D milestone was to confirm that the mechanism of action worked in the team's population of interest. As Ackermann explained, "We needed human clinical data to prove that the concept of stimulating the lacrimal nerve would produce tears in patients with dry eye disease." From a technical perspective, the company already had a works-like prototype (based on a commercially available, 510(k) cleared neurostimulation device) that would allow it to test the concept in the clinic. However, Ackermann and colleagues faced a significant challenge in terms of planning and executing an acute, **first-in-human**, **off-label** study. "We were low on cash and under some financial pressure to meet this milestone quickly," recalled Ackermann. "We had a venture firm interested in providing seed financing, but the partners wanted to see human clinical data before they would commit to coming on board."

Although Oculeve would have liked to conduct its first-in-human study at Stanford, where the team was familiar

R&D Milestones	Critical Technical Challenges	Study Type	Risks to be Retired Based on Results
First works-like prototype that performs safely and effectively in humans	• Confirming that neurostimulation of lacrimal tissue in humans will create tears	First-in-Human Acute Safety & Efficacy Study	Mechanism of action works in humans with DED
First works-like/looks-like model that performs safely and effectively in humans	• Building custom neurostimulation electronics • Embodying all required functionality into a form factor that can be implanted into the ocular orbit • Ensuring hermeticity, biocompatibility, and reliability of the implant • Developing a basic insertion device and technique • Validating the ability to communicate with implant from remote energizer outside the body	Chronic Safety & Efficacy Studies	Mechanism of action can be given a functional form and works in humans with DED over a sustained period of time
Pre-production system that performs safely and effectively in humans	• Optimizing functionality of system for use on a larger scale • Finalizing implant design • Confirming that the implant can be manufactured at a reasonable cost	Pilot Trials	Identify optimal trial design for pivotal trial to collect data for US regulatory submission
Market-ready system that performs safely and effectively in humans	• Optimizing functionality of system for use on a commercial scale • Validating human factors and usability of the insertion device and energizer • Confirming large-scale manufacturability of all system components	Pivotal Trial	Collect data for US regulatory submission

FIGURE 5.2.3

The major milestones that served as the basis for Oculeve's R&D strategy (courtesy of Oculeve).

with the quality of the research and the clinic, "The realities of early stage clinical work made that impossible," Ackermann stated. "It's not realistic for a small company without much money to do a study in the US, especially if it requires an IDE [Investigational Device Exemption] from the **FDA**." He noted that this process is not only arduous and time-consuming, but prohibitively expensive. "Fortunately," he said, "there are plenty of high quality, early-stage clinical study sites outside of the United States."

The team's mentors, Daniel Palanker, Associate Professor in the Department of Ophthalmology, Stanford School of Medicine, and Mark Blumenkranz, worked with one such site in Mexico. "Daniel and Mark had experience with a highly regarded clinical facility in Mexico City, as well as an investigator there who had an excellent reputation and had run sophisticated trials before," Ackermann said. Importantly, the facility had a fairly streamlined process for getting ethics committee approvals and addressing other important requirements, making it possible for the team to launch a ten-person, first-in-human study in just a few months. "Using our works-like prototype, we were able to get up and running pretty quickly to get real clinical data on patients with dry eye disease," he stated.

The study confirmed the general concept that electrical stimulation would produce lacrimation. "By placing needles adjacent to the lacrimal gland, we were able to stimulate tear production acutely in an office setting, proving that the concept did work in human patients," Ackermann reported. Based on these data, he was able to secure several hundred thousand dollars in venture backed-seed funding. With these funds in hand, Ackermann recruited Jim Loudin, a postdoctoral physicist, electrical engineer, and bioengineer, who "had been building electronics for small microstimulators as a hobby and research project," to lead the next phase of electrical engineering work. Together, Loudin, Ackermann, Palanker, and Blumenkranz officially co-founded Oculeve. The seed funding also allowed the two men to hire a deeply experienced mechanical engineer, Janusz Kuzma, to build the housing for Loudin's electronics. "Janusz helped pioneer multi-channel cochlear implants in the 1980s," Ackermann explained. "We actually pulled him out of retirement to spearhead our mechanical efforts." He added, "So we had this great combination of young, smart, hardworking people, and people with lots of experience and gray hair."

With this team in place, Oculeve turned its attention to another major R&D milestone: developing a "works-like/looks-like" model that would perform safely and effectively in humans over time. To achieve this particular milestone, the team had multiple technical challenges to overcome, including "packaging all of these 'smarts' into a tiny device that could fit into the ocular orbit and addressing issues related to developing an implant such as maintaining hermeticity, reliability, and **biocompatibility**," recalled Ackermann. Oculeve also needed to show that it could adequately power and communicate with the device from outside the body. With a long list of related specifications to address, Loudin and Kuzma progressed quickly through 30–50 iterations before deciding on the model they would use in the chronic clinical study. After approximately 15 months of development and testing,

Oculeve was satisfied that its solution was safe for a chronic safety and efficiency study.

The team's next R&D milestone was to develop a pre-production system that performs safely and effectively in humans in a series of **pilot trials**. One important goal of the pilot trials would be to help Oculeve prepare for its subsequent milestone: a **pivotal trial** to support the company's regulatory submission to the FDA. The idea was to conduct two or three different pilot studies to help Oculeve decide on an optimal design for the pivotal trial. According to Ackermann, "Despite its significant regulatory hurdles, the US is still the biggest, most viable market for a device like ours. So we're preparing to go through the premarket approval [PMA] process, and we're doing everything we can to be confident in the outcome of our pivotal trial."

The key technical challenges to address in advance of the pilot studies involved advancing the development of all three components of the Oculeve system: (1) the intraorbital implant, which delivers small electrical pulses that activate the gland and restore tears to the eye; (2) an insertion instrument to place the implant; and (3) an energizer that serves as an external "remote control," for powering and communicating with the implant (see Figure 5.2.4).

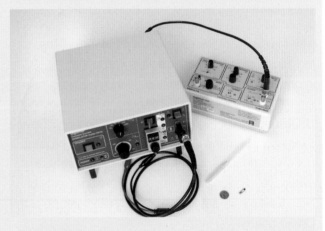

FIGURE 5.2.4
The Oculeve System, including the implant, insertion device, and energizer (courtesy of Oculeve).

In this round of R&D, the team devoted special time and resources to the implant. First, the engineers tried to lock in the design of the implant. As Ackermann described, they could continue to iterate the design of the insertion device and energizer because the testing requirements and cycle time associated with these elements of the systems were much lower. But each time they made significant changes to the implant, they had to complete lengthy and expensive animal tests, biocompatibility tests, and so on before they were able to try the device in humans. "So concerns that might seem down-the-road, like cost of goods and manufacturability, were actually important from the beginning," he said. "The other two components are much less critical and risky. We could afford to put off their productization until we're sure the implant works really well."

However, they did initiate productization of the implantable device, under **design controls**, focusing on long-term manufacturability and ensuring that it could be produced at a price that decision makers and other **stakeholders** would find compelling. Commenting on this transition, Ackermann observed, "I think that bringing in a heavy quality system should be delayed until you're confident that you have good working device. There's a little bit of push-pull in the beginning. You need enough of the quality elements in place so that you are comfortable doing clinical work with the prototype. But not so much that it prohibitively slows down your R&D efforts and your ability to iterate quickly. Once you're really confident that things are working on a clinical level, and on various business and commercial levels as well, then it's time to go in and start developing a formal product."

When asked for advice about building an R&D strategy, Ackermann encouraged other innovators to keep the "big picture" squarely in mind. "Although it is tempting as an engineer to be overly focused on building the actual device, it's more important to pull up and concentrate on proving value," he said. "You have to view your R&D strategy as a means to an end, rather than an end in itself."

Developing an initial R&D plan

With specific R&D milestones and related technical challenges identified and prioritized, innovators can next develop an initial R&D plan. As part of an R&D strategy, this initial plan should be kept at a relatively high level. However, to construct a high-level plan, innovators still must begin with a detailed list of engineering activities required to achieve each milestone, making sure to thoroughly consider the technical challenges and testing that will be required to achieve them. With this understanding, they can then layer in R&D personnel and engineering resources, such as equipment and facilities, which will then help determine overall R&D timelines. Many innovators capture their R&D plans in a Gantt chart, which is a project management tool that displays activities sequenced over time (see Figure 5.2.5). More detailed Gantt charts also can be used to capture interdependencies between activities (see Figure 5.2.8) and the resources/staff they require.[3]

More specific details, such as costs, will be reserved for inclusion in the company's integrated operating and staffing plan (see chapter 6.1). Costs are not directly addressed in this chapter because they differ so significantly from project to project, as well as across geographic areas. For instance, developing a mobile health application will likely be dramatically less expensive than creating a traditional medtech device. And developing a complex system that relies on multiple engineering disciplines will probably be far more costly than engineering a simple mechanical solution. The best way for innovators to gain a benchmark for R&D-related expenses is to network with other innovators working on projects that roughly resemble their own projects.

A71 fx

A	Operating Plan																						Year	
										Quarter														
	2012				2013				2014				2015				2016							
	1	2	3	4	1	2	3	4	1	2	3	4	1	2	3	4	1	2	3	4			6	7
Phase X: Legal																								
Incorporation																								
FTO/Patentability Search																								
IP Filing																								
Phase 1: Proof of concept																								
Initial prototyping																								
Bench/Cadaver Feasibility																								
Phase 2: Final product development																								
Define Design Inputs/Outputs																								
Finalize delivery device design																								
Cadaver Validation																								
Design freeze																								
CEO hire																								
Phase 3: Regulatory approval																								
DV & DVAL Testing																								
CE Mark approval process																								
510k writing & approval process																								
FDA approval																								
Review w/ potential acquirers																								
Phase 4: Clinical Studies																								
Limited pilot study & Follow-up																								
Review w/ potential acquirers																								
Larger randomized study																								
Phase 5: Market Launch																								
Initial sales force hires (EU & US)																								
Larger randomized study																								
EU distributor launch																								
US Sales force growth																								

FIGURE 5.2.5

This relatively high-level Gantt chart, which might be presented to investors or at a board meeting, can be used to communicate important milestones across the company (courtesy of Ciel Medical).

High-level discussions about these issues are necessary as part of the R&D strategy process to allow innovators make more accurate forecasts of when R&D milestones can realistically be met. As noted, once the timing of key milestones is determined, innovators can strategically position R&D as a contributor to other important development milestones, such as funding and regulatory clearance.

R&D activities

The first step in creating an R&D plan is to define the scope of work required to resolve the technical challenges related to each major R&D milestone. This should include listing all of the steps involved in building any required models and then testing them through the most appropriate method(s) in the biodesign testing continuum (revisit Figure 5.2.1). To do so, company or team leaders will need to work closely with the engineers to ensure that the key activities are captured in the plan. Most teams start by making their R&D milestones the major headings in their Gantt chart. Then, under each one, they list the critical technical challenges that have been identified. Next, they outline the work steps required to understand and address each technical challenge. Finally, they add the testing-related activities that will confirm that the technical challenges have been overcome and the milestone has been met.

One important factor to remember is that development and testing is highly iterative. This can sometimes make it difficult for innovators to anticipate all activities that should be included in the R&D plan. While there is no "right" answer, innovators are usually advised to build in multiple rounds of development and testing when addressing a complex technical challenge.

Innovators may also struggle with the right level of detail to include in an initial R&D plan. Again, each team must use its own judgment but it can be helpful to keep in mind that all key activities, as well as the tasks that represent discreet pieces of work that need to be accomplished, should be listed so that specific personnel and resources can be assigned to these activities and tasks. What is typically not included in an R&D plan is how these activities and tasks are performed. This is usually left to individual engineers and/or engineering subteams to determine once the overall plan has been defined.

In addition to being important milestones themselves, completing certain tests is often integral to demonstrating that critical certain technical challenges have been overcome. Accordingly, to create a reasonably accurate R&D plan, the testing activities that will be undertaken need to be thought about carefully, as multiple rounds may be needed. This work can be resource intensive. For instance, when a company has moved into animal studies, each iterative round of testing to achieve a milestone can be costly in terms of time, personnel, and other resource requirements.

Considering the LAA example provides some insight into how to think about technical challenges, how to forecast activities in the initial R&D plan, and how different testing methods may have different impacts on the use of time and resources. It also demonstrates how playing out potential testing outcomes can elucidate important issues relevant to subsequent R&D activities and milestones.

An important early technical challenge in the LAA example is the identification of material that can occlude a simulated LAA using a prototype in a bench-top test. To perform this test, an LAA-like pouch could be created simply by suspending a plastic bag in a chamber filled with moving and heated blood. By attempting to deliver the material into the plastic pouch, the innovator could then determine if it solidified to achieve the occlusion. If it did not solidify, a new plastic bag could be swapped with the old one and testing could be performed again, once a new prototype (e.g., using different material) that addresses a limitation of the failed prototype is created. The failure may have been due to the material, the delivery mechanism, or the artificial set-up. Even though iterative prototyping and bench testing at this stage can be performed relatively quickly, innovators may want to think about allocating a small buffer of time for multiple iterations given the number of failure modes that are possible.

If the material did solidify, in addition to meeting the technical challenge, an innovator might gain an appreciation of the timeline to solidification. While perhaps not initially anticipated, understanding what this finding means could impact other parts of the plan. If it took many hours, for example, this might be problematic if an important user requirement is that the material solidifies in one hour or less. Hence, in the next round, the innovator might need to make repeated modifications until this important design requirement is met. Once a prototype that works adequately is created, additional experiments would be needed to further optimize the solution, especially with respect to integrating it with other components of the solution. Bench and simulated use testing could allow various technical challenges to be tested in parallel or with one variable altered each time. While the innovators would need to scope this activity to take into account how many series of tests might be needed, at this pre-animal stage multiple iterations are unlikely to require significant additional resources or cost.

Taking the example one step further, if achieving effectiveness in a tissue model is an important subsequent milestone, a similar set-up could be created using a cow's heart instead of a plastic bag. While this might be somewhat more complicated, based on the availability and cost of hearts, this type of testing can be performed iteratively and relatively quickly. The basic point is that to achieve milestones that do not involve live animal testing, the testing methods do not have to be time-consuming or expensive, but budgeting some time and resources for iterative rounds of prototyping and testing is important.

In contrast, live animal testing, which is typically another important R&D milestone, takes more time to prepare and plan for, is far more costly, and must be performed in approved facilities. When working with live animals, it is important to keep in mind that the need for multiple rounds of prototype iteration can have a significant impact on the R&D timeline because of ethical and other requirements associated with these tests (see 5.3 Clinical Strategy). If chronic studies are needed that involve many animals, important controls, and specific follow-up protocols to investigate long-term safety and effectiveness, the time and expense to conduct them will be even greater. As such, chronic live animal tests should typically be used only when necessary and be performed as few times as possible to achieve the desired results.

To progress to milestones related to live human testing, a final prototype must demonstrate its safety and effectiveness in animal studies (if animal testing is required) and meet quality requirements (see 5.5 Quality Management). It also will require the transfer of a design to manufacturing. Many innovators make the mistake of thinking about design for manufacturability late in R&D planning. However, to maximize results while controlling time and costs, innovators should consider this factor when developing their initial R&D strategy and plan. Engineers with experience designing products for high-volume production can add tremendous value from the relatively early stages of the R&D life cycle. For one, they can help ensure that the right materials are used so the technology is safe in humans but also compatible with large-scale manufacturing techniques. Additionally, they can help eliminate unnecessary design constraints and process inefficiencies that add time, cost, and complexity when it comes to manufacturing. Further, these engineers can ensure that key components are sourced from vendors and suppliers who are reliable and can scale to handle larger volumes. Finally, they can assist a company in accelerating its time to market by anticipating critical production requirements and initiating long lead-time activities early enough in the R&D plan so they do not become bottlenecks later as development and testing progresses.

The story about 3rd Stone Design and the DoseRight syringe clip illustrates some of the issues related to the transition from prototyping and initial R&D to design for high-volume production. Additional references on manufacturing can be found in the online Getting Started section.

FROM THE FIELD ▶ DOSERIGHT® SYRINGE CLIP

The importance of design for manufacturability

According to Robert Miros, CEO of 3rd Stone Design, a design, strategy, and development consultancy, issues related to design for manufacturing are common among teams of aspiring innovators, especially university-based teams and early stage start-ups. Often, since the focus of these innovators is on early-stage design, concept generation, prototype development, and preliminary bench and field testing, the importance of considering manufacturing and industrial engineering issues early in the process is lost. As a result, while many early engineering teams can successfully develop and produce small numbers of prototypes for testing under controlled circumstances to overcome important technical challenges, they sometimes lack the expertise to design for mass production and create products that perform as intended when implemented outside an R&D environment. "They [university-teams] are coming from a largely theoretical training, so the basics of manufacturing are often not top of mind – that is, to get a factory set up, to have the right sort of drawings and quality documents and inspection criteria," Miros commented.

3rd Stone Design worked with one team of undergraduate students from the Rice 360° Beyond Traditional Borders (BTB) program to help address such challenges. The team had successfully designed and tested what the members dubbed the DoseRight syringe clip to enhance the dosing accuracy of liquid, antiretroviral medications in resource-limited settings. The product was a simple plastic clip that could be inserted into the top portion of a standard oral syringe to control the amount of medication that could be drawn into the syringe. Produced in varying lengths to correspond to different dosing volumes, the clips could be quickly and easily affixed to a standard syringe to ensure dosing accuracy regardless of caregiver literacy or visual acuity (see Figure 5.2.6). Based on field tests that demonstrated that the clips increased dosing accuracy, the team lined up its first customer and a sizable order: the Clinton Health Access Initiative (CHAI) requested 200,000 clips to be distributed in Swaziland,

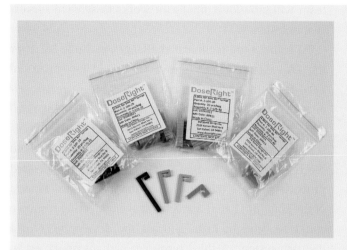

FIGURE 5.2.6
The DoseRight syringe clip (courtesy of 3rd Stone Design).

Africa via the Ministry of Health. However, the students had fabricated their plastic prototypes in-house, using a 3D printer. Although this approach enabled them to complete a proof-of-concept, it was not cost-effective for high-volume production. "Even if they could have made 200,000, which they really couldn't have with that technology, it probably would have cost them a dollar apiece," said Miros. With CHAI eager to procure the clips as quickly as possible at a much lower price, BTB asked Miros for help. 3rd Stone Design **licensed** the technology from Rice and took the project forward.

The primary challenge was figuring out how to transfer the process to manufacturing at scale. Injection molding offers a lower-cost alternative to 3D printing that would produce high-quality results. With this process, heated polymers are forced into a multi-part mold that was essentially a negative form of the desired part. The mold is opened after cooling and the finished part is ejected. Given its simple design, molding the part would be relatively easy. However, the DoseRight clips needed to be produced in 10 different lengths in order to provide dosing volumes from 0.5 mL to 5.0 mL in half mL graduations. Normally, this would require the creation of 10 different molds. The 3rd Stone Design team estimated that 10 hard tooling molds (made of hardened steel with multiple cavities for large-quantity production) would cost $200,000–$300,000 – far too much for a product in start-up mode. Tooling costs had to be held

to a fraction of that in order to keep the capital investment reasonable and to produce the clips at a cost that global healthcare customers could afford to pay while maintaining the quality of the finished clips.

Miros and team solved the problem by designing a limited mold set based on multiple inserts of varying size that enabled them to produce high-quality parts at a reasonable cost. While this solution did not bring the cost down to a few pennies per clip (which was the ultimate goal), it was a good start. As the first customer for the DoseRight clips, CHAI understood that it would have to pay slightly more for the product initially, with subsequent costs likely to decrease as production volume rose. "The customer had been coached, a little bit, on what the initial pricing would be, and we were able to hit that price and still have some profit in it; not a lot."

Using the limited mold set, 3rd Stone cost-effectively manufactured all 10 lengths of the DoseRight clips out of polypropylene in a quality-controlled environment, with all parts bagged and labeled in a manner appropriate for medical disposables. In partnership with CHAI, Swaziland's Ministry of Health began distributing more than 200,000 dosing clips as part of its Prevention of Mother to Child Transmission of HIV/AIDs program. 3rd Stone subsequently went on to pursue additional orders in developing countries, working with large NGOs and local ministries of health that were running, or planning, countrywide vaccination programs.

When asked for advice to help other innovators, Miros recommended that teams develop strong competencies in design for manufacturing, either by adding experienced individuals to the group or by contracting with a seasoned partner. Especially for early-stage start-ups, personnel with a background in manufacturing processes should be brought in well before most technical challenges have been overcome and customers are being lined up. If they are a part of the design process and R&D planning effort, they can help influence important choices made by the R&D team, potentially prevent significant rework, and ensure a more efficient transfer of the product to manufacturing.

R&D personnel

Once the activities in an R&D plan have been outlined, innovators can next consider what R&D personnel are necessary to complete those activities. Thinking about the skills sets and number of engineers needed is critical because it has a direct impact on when R&D milestones can be achieved. Furthermore, R&D staffing costs can be the highest R&D budget expense and, thus, will affect financing events.

To determine what types of engineering and technical resources a company will need, innovators should assess each milestone and associated activities individually, in terms of the relevant engineering skills required to address the specific engineering work. Then, the R&D plan should be evaluated as whole, paying particular attention to the interactions between key aspects of the solution and any unique engineering skills needed to address them.

Assessing the difficulties involved in particular activities can be a challenge in itself, particularly for innovators working outside their area of expertise. However, developing reasonable estimates is important since it directly drives the number (and skill levels) of engineers required to get an activity completed in a reasonable amount of time. The best approach to overcoming this hurdle is to query others with varied backgrounds to get a rough idea of the time and effort needed to address a challenge, assuming that knowledgeable resources are found. As a rule of thumb, the length of time spent on engineering depends heavily on the complexity of the innovation being developed. Even with a clear plan, achieving certain milestones will simply take longer for a complicated solution.

During concept exploration, innovators may have been able to develop working models that demonstrated a concept's basic feasibility with little more than hard work, book research, and some guidance and/or coaching from those with relevant expertise. However, the transition from early-stage prototyping to R&D usually necessitates much deeper knowledge and expertise in each of the relevant functional blocks or engineering disciplines. Unless the company intends to outsource the development of its innovation, it must hire a team of appropriately skilled engineers and map out how that

team will grow over time as it works to achieve successive milestones.

To address the technical challenges in Table 5.2.2 for the LAA example, the innovators would recognize the need for both mechanical engineering and materials science expertise. Based on how these challenges are prioritized, they would also surmise that getting a materials science expert on board first would be advantageous to retire the risk of finding a material with the desired properties before thinking about the mechanical aspects of the solution. Later, a mechanical engineering resource could be brought in to address the development of a delivery system. A third resource, perhaps an expert in how the material and delivery system interact, could be added next, at the appropriate point in time. In a resource-constrained environment, it would not be practical to hire the latter two resources until enough is known about the core material used in the solution, since this will affect the design of the delivery mechanism and the substance from which it is made. Understanding these types of sequential interactions allows the innovators to more effectively manage R&D staffing within prevalent resource and time constraints.

Other projects may lend themselves more readily to parallel development efforts. For example, recall the ACS case example in 4.6 Final Concept Selection, in which parallel development efforts focused on a balloon catheter, a guidewire for catheter navigation, and a guiding catheter are described. In this scenario, the company made a strategic decision to run multiple engineering work streams rather than working in sequence. In some types of project, and where time and money allow, this alternative is available to the team.

The key point is that, in a start-up with scarce resources, it may be possible for a company to stagger its approach to hiring based on the R&D strategy and product development pathway. However, this decision should still be predicated on ensuring that the greatest technical challenges and those with the longest development time are addressed early in the process so that adjustments in the development pathway can be made as soon as possible.

Another reason to think about staffing as part of the R&D plan is that hiring itself poses an element of risk. For example, if a highly skilled electrical engineer with many years of experience in a given field will be needed in three months to help solve a key technical challenge, actually finding and hiring the right resource may be unrealistic in the given time frame, especially if the demand for such engineers is high and availability is low. Understanding this may help an innovator resequence key technical challenges more realistically. This, in turn, may affect the timing of certain strategic milestones, causing further impact on other factors, such as company funding. Furthermore, while having more people on the R&D team does not necessarily lead to faster resolution of critical issues, planning for adequate staff eliminates at least one constraint that can interfere with a company's R&D progress.

Importantly, when hiring R&D personnel, companies face a trade-off. They can spend more money to attract and hire engineers with deep, often specialized experience. Or, they can pay less to add bright but relatively inexperienced staff members to the team. Inexperienced engineers might be easier to find and hire, but they typically introduce more risk from a development standpoint (in terms of needing more time to solve a technical challenge). On the other hand, while experienced engineers can represent less risk from the standpoint of knowledge, they are more expensive and harder to find. That said, it is usually necessary to have at least one experienced manager with product development expertise to oversee the R&D process and ensure that product development is launched in the right direction (e.g., in the role of vice president or director of R&D). Beyond that, a company should consider the type of work it must accomplish in order to achieve its goals, as well as the likelihood of hiring experienced engineers in a reasonable time frame. With complex technologies in treatment areas that are relatively unexplored, it usually makes sense to hire experienced engineers, even if that means the company can bring on fewer people and needs more time to hire. In contrast, when working on incremental technologies in areas where speed to market may be an important driver, a somewhat greater number of less experienced individuals can more effectively swarm the development challenge. As a general rule, however, start-up companies almost always benefit from hiring

the best people they can – particularly in assembling a core team of experienced engineers to ensure that product development is launched in the right direction.

Another approach that can help a company carefully manage its investment in R&D personnel is to use consultants to assist with certain tasks. Hiring engineers on a contract basis can aid the company in several ways. First, it can be an effective mechanism for gaining access to expensive, specialized expertise required to address a specific engineering challenge, but not required on an ongoing basis. For example, in the LAA scenario described above, an expert materials consultant could help guide the materials scientist(s) working at the company to more quickly find an appropriate material, but may not be required once this effort has been completed. In some cases, seeking such expertise on a contract basis is more feasible than trying to hire a full-time employee because many engineers with critical, yet highly specific skills often gravitate to consulting simply due to the nature of their knowledge base (i.e., their value-added skills are not required for long-term projects). Second, hiring consultants can allow the company to respond to temporary or short-term peaks in the engineering workload without having to hire (and then lay off) dedicated resources. For instance, as the LAA delivery system moves toward regulatory submission, there may be multiple, short-term tests that can be accomplished simultaneously. Hiring contract labor may be beneficial to execute the tests without distracting in-house engineers from their other priorities. Finally, consultants can help bring a fresh perspective to the resolution of challenging problems.

According to Hira Thapliyal of ArthroCare, the company used contract manufacturers to develop and produce its prototypes for its animal studies. "Seek out known entities with expertise in a specific area," he said. "In the early stages, you can get a lot of work done very quickly when you outsource. You just need to have a smart group of people inside the organization who can communicate and convey to your vendors what you need." However, as part of its plan, the team knew it would eventually have to bring development in-house. By the time of ArthroCare's FDA filing, Thapliyal recalled, "We had a director of R&D, a couple of engineers, and some quality assurance people. We had started bringing disposable device development in-house because we knew we had to have control of it since this would allow us to iterate more quickly."

When using consultants, just be aware that they may work differently from in-house staff and require different incentives. Occasionally, consultants may not exercise the same sense of urgency as dedicated, full-time employees in performing their tasks. They also may not be knowledgeable about all of the intricacies of the development process that would allow them to move at a fast or consistent pace. Moreover, issues related to confidentiality and IP must be adequately addressed before any contract work is initiated. Finally, consider the fact that when consultants finish their assignments, much of the working knowledge that has been accumulated through their involvement leaves the company with them. At a minimum, the company should orchestrate a proactive knowledge transfer and ensure the completion of relevant documentation prior to their departure.

Engineering resources

In addition to R&D personnel, a company will be required to make an investment in facilities, equipment, and other resources to support R&D (see Figure 5.2.7). While there is less risk associated with the acquisition and utilization of these resources from a timeline perspective (i.e., most of these other resources are readily available), some thought must be given to this from a strategic development standpoint so that the resources obtained match the availability of R&D personnel and both are fully and efficiently utilized. This is particularly important given that decisions related to engineering resources can have significant financial implications.

In the LAA example, if the company decides to develop a delivery system via an in-house engineering effort, it will need to purchase enough equipment to support this activity. However, if finding a mechanical engineer with skills in developing percutaneous systems takes too much time, the development work could instead be contracted to an outside firm. Such a move would obviate the need to purchase the equipment required to develop this portion of its device. In another scenario, the company might be able to hire an engineer, but decide to not

purchase any equipment and have the engineer work with an outside vendor that possesses the machines and equipment to develop the product. All of these choices have an impact on the timeline to develop the delivery system. They also directly affect issues related to resources, which must be taken into account when developing the overall operating plan. Finally, the ultimate decision can affect the level of funding required (as different investments are needed to equip and maintain an engineering shop versus contract with an outside firm).

When the need for specialized equipment, parts, materials, or processes is anticipated in the development of a product, lead-time issues often have to be considered. If a company requires hard-to-get items, this may affect the hiring of personnel, the sequence of addressing challenges, and the time required for milestones to be achieved. For example, creating a certain electronic circuit may require many circuit board revisions to reach a final board design. Outsourcing the fabrication of each of its prototype board designs with every revision may necessitate a substantial lead time of weeks to months for each of the boards to be fabricated. By understanding and anticipating such needs, a company can determine whether a different development pathway is more efficient (e.g., outsourcing design as well as fabrication to the same board house).

Other resource issues to take into account include laboratory space, as well as the specific lab equipment that will be required. These questions are largely driven by the type of engineering work being performed, along with what functions might be outsourced by the company. For example, a company with a mechanical engineering-based device that it plans to develop in-house might need a basic machine shop that includes a mill, lathe, drill, and/or grinder, as well as common hand tools and assorted supplies from a hardware store (nuts, bolts, springs, tubing, etc.). More specialized equipment and test fixtures may also be required based on the specific type of device being developed. For instance, if the engineering team intends to use computer-aided design (CAD) programs (e.g., SolidWorks, SolidEdge), appropriate computer equipment will also be needed. As the cost of 3D printers continues to decrease, this

FIGURE 5.2.7

Clockwise from top left: the team at ExploraMed working on a new concept in the R&D lab (ExploraMed); engineers at Element Science, performing two different types of device tests (Element Science, Inc.); and two views of R&D space and equipment (Singapore-Stanford Biodesign).

equipment is becoming more common in engineering shops as well.

Materials science engineers often require significantly different equipment from that of mechanical engineers.

For example, a wet lab for extensive tissue testing will likely be needed (e.g., sinks, burners, precise scales and measuring equipment, ovens, fume hoods and exhaust, refrigeration), as well as space for analyzing, mixing, and

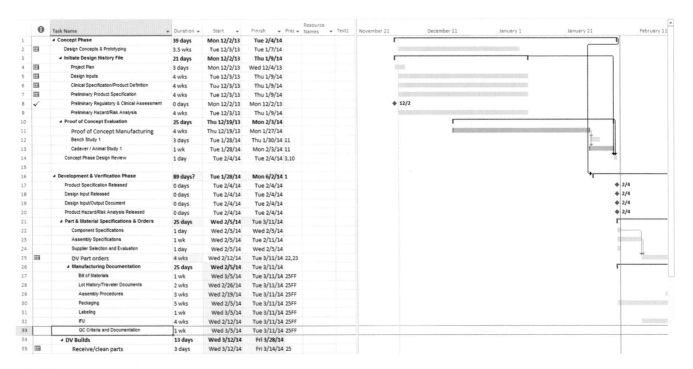

	ⓘ	Task Name	Duration	Start	Finish	Prec	Resource Names	Text1
1		⊿ Concept Phase	39 days	Mon 12/2/13	Tue 2/4/14			
2	▦	Design Concepts & Prototyping	3.5 wks	Tue 12/3/13	Tue 1/7/14			
3		⊿ Initiate Design History File	21 days	Mon 12/2/13	Thu 1/9/14			
4	▦	Project Plan	3 days	Mon 12/2/13	Wed 12/4/13			
5	▦	Design Inputs	4 wks	Tue 12/3/13	Thu 1/9/14			
6	▦	Clinical Specification/Product Definition	4 wks	Tue 12/3/13	Thu 1/9/14			
7	▦	Preliminary Product Specification	4 wks	Tue 12/3/13	Thu 1/9/14			
8	✓	Preliminary Regulatory & Clinical Assessment	0 days	Mon 12/2/13	Mon 12/2/13			
9		Preliminary Hazard/Risk Analysis	4 wks	Tue 12/3/13	Thu 1/9/14			
10		⊿ Proof of Concept Evaluation	25 days	Thu 12/19/13	Mon 2/3/14			
11		Proof of Concept Manufacturing	4 wks	Thu 12/19/13	Mon 1/27/14			
12		Bench Study 1	3 days	Tue 1/28/14	Thu 1/30/14	11		
13		Cadaver / Animal Study 1	1 wk	Tue 1/28/14	Mon 2/3/14	11		
14		Concept Phase Design Review	1 day	Tue 2/4/14	Tue 2/4/14	3,10		
15								
16		⊿ Development & Verification Phase	89 days?	Tue 1/28/14	Mon 6/2/14	1		
17		Product Specification Released	0 days	Tue 2/4/14	Tue 2/4/14			
18		Design Input Released	0 days	Tue 2/4/14	Tue 2/4/14			
19		Design Input/Output Document	0 days	Tue 2/4/14	Tue 2/4/14			
20		Product Hazard/Risk Analysis Released	0 days	Tue 2/4/14	Tue 2/4/14			
21		⊿ Part & Material Specifications & Orders	25 days	Wed 2/5/14	Tue 3/11/14			
22		Component Specifications	1 day	Wed 2/5/14	Wed 2/5/14			
23		Assembly Specifications	1 wk	Wed 2/5/14	Tue 2/11/14			
24		Supplier Selection and Evaluation	1 day	Wed 2/5/14	Wed 2/5/14			
25	▦	DV Part orders	4 wks	Wed 2/12/14	Tue 3/11/14	22,23		
26		⊿ Manufacturing Documentation	25 days	Wed 2/5/14	Tue 3/11/14			
27		Bill of Materials	1 wk	Wed 3/5/14	Tue 3/11/14	25FF		
28		Lot History/Traveler Documents	2 wks	Wed 2/26/14	Tue 3/11/14	25FF		
29		Assembly Procedures	3 wks	Wed 2/19/14	Tue 3/11/14	25FF		
30		Packaging	5 wks	Wed 2/5/14	Tue 3/11/14	25FF		
31		Labeling	1 wk	Wed 3/5/14	Tue 3/11/14	25FF		
32		IFU	4 wks	Wed 2/12/14	Tue 3/11/14	25FF		
33		QC Criteria and Documentation	1 wk	Wed 3/5/14	Tue 3/11/14	25FF		
34		⊿ DV Builds	13 days	Wed 3/12/14	Fri 3/28/14			
35	▦	Receive/clean parts	3 days	Wed 3/12/14	Fri 3/14/14	25		

FIGURE 5.2.8

This Gantt chart shows a more detailed engineering project plan with durations assigned to key activities and critical interdependencies noted between work streams, which can help innovators understand how long the entire R & D process may take (courtesy of Ciel Medical).

testing chemical reagents, polymers, and other compounds. Raw materials and tissue samples can be moderately expensive, especially for certain polymers.

For electrical engineering and computer science-based projects, more advanced computer equipment and different software programs (MatLab, LabView) are needed, but fewer machine tools must be purchased. Additionally, no raw materials are typically needed.

R&D timelines

As described, estimating R&D timelines is an important part of R&D planning because time is a significant driver of both value and risk given its obvious connection to the consumption of capital and resources.

The key inputs to determining a realistic timeline are the necessary R&D activities and tests that must be performed to prove the effectiveness and/or safety of a design, the type of R&D personnel that are needed, and the number and type of engineering resources required. Proper documentation of user requirements, technical

designs, experimental reports, and verification/validation testing (which are all aspects of the quality system – see chapter 5.5), will also affect R&D timelines. Understanding the trade-offs between different scenarios can help innovators articulate a realistic timeline that still takes into account the effort required to address key technical challenges and retire risks. This information also allows the innovator to answer the important strategic question: "How long will R&D take?" (see Figure 5.2.8).

Seasoned innovators and engineers will be able to estimate the time associated with many activities in the R&D plan based primarily on past experience. However, if the R&D plan includes unfamiliar work steps or the innovators are relatively inexperienced, the best way to come up with appropriate timelines is to network with consultants or other innovators working on similar projects.

While R&D timelines vary dramatically across projects, some engineers believe that it generally takes less time to

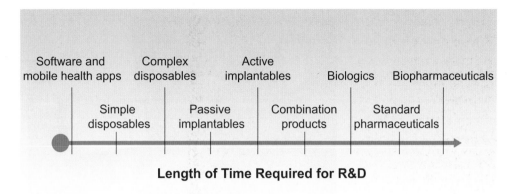

Length of Time Required for R&D

FIGURE 5.2.9

A simplified view of the relative R&D time spent on different types of projects. As combined business models that include products, services, software, and other offerings become more common, innovators can expect to spend increasing time on R&D, as well as testing and integration activities.

achieve key R&D milestones for mechanical devices and solutions that are primarily software based or mobile health applications. In contrast, devices that depend on materials science and/or electrical engineering fundamentals are often more time-consuming (see Figure 5.2.9). For instance, it might take innovators as little as a year to develop a mechanical device that is ready for controlled animal testing. On the other end of the scale, five or more years may be needed to reach that milestone for a high-complexity or combination product, such as a drug-eluting stent. These variations are primarily driven by the greater number of unknowns that are introduced when chemistry, biotechnology, and other scientific disciplines become part of the medical device R&D process.

While creating an R&D plan often may seem to be an isolated exercise done primarily by an engineering team, it must be integrated into the context of the broader organization. R&D plans, as part of R&D strategies, that do not consider the interplay between achieving technical milestones and the larger clinical, regulatory, and marketing needs of the company often lead to misalignment of timelines, create significant inefficiencies and, ultimately, require more time and resources to move ahead and this may affect funding requirements. The case example focused on NeoTract demonstrates how one company, which prioritized a user-focused approach to developing its product, was able to merge this organizational priority with the rigorous R&D milestones they identified as most important.

FROM THE FIELD ▷ NEOTRACT, INC.

Developing and deploying a user-focused approach to R&D

According to Ted Lamson, co-founder and CTO of NeoTract, Inc., a start-up based in Pleasanton, California, there are 4.2 million men in the United States on medication for benign prostate hyperplasia (BPH) and its accompanying urological symptoms. However, each year 1.3 million of these patients stop taking their medication, with only about 250,000 of them seeking surgical intervention to address the disease. The rest, though still

suffering, remain untreated. "We sought to find out through a needs assessment why there is such a huge discrepancy," said Lamson, describing NeoTract's focus when the company was founded in 2005. "We spent our time trying to tease out what it is about the current interventional option that is scary or doesn't fit the risk/benefit profile that these patients are looking for." Over time, the team conceptualized a less invasive, office-based procedure that could be performed relatively quickly and easily, and required only a local anesthetic. To realize this solution would require the development of an implant and the instruments to deliver it.

However, when NeoTract began developing an approach to R&D, it set aside some of its more elaborate designs and defined a first strategic milestone focused solely on proving the fundamental technical concept in the clinic with a strong focus on the end user – the patient. "To develop the complete, more elegant solution would have taken quite a bit of time and money," recalled Lamson. "But we recognized before making this investment that we really hadn't answered the one question that would make it all worthwhile, which is will patients respond well to our approach? So rather than spending time building the perfect device that could potentially do the wrong thing, we asked ourselves what's the quickest way we can answer this question. It turned out that this was to figure out a more invasive, bare bones way to see if the concept would achieve the result we were after and then bring that forward into an operating room setting. We were going for a clinical proof of concept that would deliver results that were better than the surgery that was currently available, but was not necessarily all the way to our end goal."

The NeoTract team developed a somewhat crude set of surgical instruments to deliver the implant and a procedure performed under general anesthesia that more closely resembled the current surgical treatment (see Figure 5.2.10). "Of course," said Lamson, "It was still a lot of work. We had to build everything within the guidelines of a quality system. But the intense design work and engineering to build the more elegant solution could be put on hold." The company also had to act ethically, developing a reversible procedure so that the patients receiving it could still undergo conventional therapy if there were any adverse effects. But in the end, the user feedback was encouraging. "We did ten patients with that device – only then did we confirm that they responded really positively to our idea," Lamson explained. "At that point, we knew we had the right approach, and we were ready to start the program to develop the real device."

"We probably spent about six months on this first milestone, which ultimately delayed the final product," he continued. "But this early clinical experience [and direct

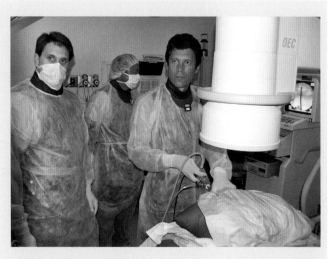

FIGURE 5.2.10
Lamson testing a device in an animal study (courtesy of NeoTract).

user input] gave us the confidence that this was worth doing. More times than not, I've seen projects develop a very elegant and well-engineered device and procedure only to find that there's a critical flaw once they get into the clinic. When that happens late in the engineering process, it's harder to abandon the project. Or, even worse, you've got so much momentum that you just keep trying to tweak the solution when really there's more of a critical flaw in it and you should really just change course. If I had one piece of advice that served us well, it is that a team needs to identify the single biggest risk in the program and emphasize taking that on first."

Another benefit of NeoTract's user-oriented approach, including its rapid R&D work, early clinical experience, and collection of user input, was linked to the fact that the device was focused on improving **quality of life** for sufferers of BPH. "Your ability to eliminate pain and discomfort is not something that you can get from an animal experiment," he said. "These cases allowed us to figure out if what we want to do is feasible and helps patients, but also if it's going to be something patients want or if they're going to say, 'This doesn't feel good. I don't want it,'" Lamson noted.

Although the patient was the primary target of the user-oriented approach, NeoTract also paid close attention to physician feedback. "By getting into the clinic early, we learned a lot about aspects of the therapy that we might have designed incorrectly," Lamson said. "For example, at one point, we had the surgeon performing the procedure holding a sterilized force gauge to measure forces on the instruments. By doing that, we found out what force works and could make that a key requirement." The team also spent considerable time assessing the physician's user experience and then analyzing the data it had collected in the clinic and translating the information into important design requirements and technical specifications to guide the R&D effort.

NeoTract's next strategic R&D milestone was to develop a device and a procedure that more completely addressed all of the team's goals and then get those into the clinic. To accomplish this, Lamson focused on identifying the key technical challenges that needed to be overcome along the way and then set up checkpoints or sub-milestones to make sure the project was tracking to them. "When you have a broad project where people are doing different subassemblies and it's supposed to all come together at the end, you realize that the end is often too far away to make sure everything works. A lot of times we break up the development work into functional elements. But before we get too far down the road on developing a certain configuration, we try to identify the key things we can test in some simple mode to know that we have the right basic design. This approach can keep you from going too far down the wrong path." When an engineer reached a key checkpoint, the entire R&D team would get together to review the work, ask questions, and help identify any fatal flaws. "Often," said Lamson, "the people who aren't working on that element have most valuable insights about what may present a problem."

Another approach that Lamson advocated to help his team members keep the physician squarely in mind during R&D was for them to become "amateur surgeons." "To the extent that you can actually do your procedure and operate your device [in animals or cadavers], it helps you understand what a clinician worries about and doesn't worry about." Lamson and team performed approximately 80–90 percent of their own pre-clinical procedures, enabling them to capture key learnings. That said, "I always bring in at least one clinician to do early experiments, teach us about his or her concerns, and help mark out the procedure," he noted. Then, experts are involved periodically throughout the development process to "make sure we're not being biased in what we're seeing and to help give us a fresh look at our work," he added.

When asked about common pitfalls in the R&D process, Lamson reiterated the importance of "being disciplined in assessing your key unknowns and in being sure you are addressing them." Get into the clinic as soon as possible, he said, and "be clever about answering your key clinical issues." He also underscored the need to "build a bed of testing and quality that applies to everything as it moves forward." He continued, "It's always a delicate balance to assess how well your previous testing applies to the next generation device – what can you take from it, and what you have to re-do. It requires a fairly constant and continuous conversation because enough improvement is going on that we're always weighing whether it creates the need for additional testing, or different testing, or that sort of thing."

Finally, Lamson offered, "I truly believe one of the key strategies to pursue when doing a start-up is to constantly seek expert advice. Ask yourself what you critically need to know and who you can ask. People will generally share information and you usually don't have to disclose much to get your answer. You can say, 'I'm building something that looks like this, this is what I think the team looks like, and here's our proposed development approach.' If you have the rapport with the person, they'll generally challenge you on a few things and give you important information to think about," which can be used to refine and improve the R&D strategy.

In late 2013, NeoTract achieved a major milestone, based on its R&D efforts and thorough clinical testing, when the FDA cleared the UroLift system for the US market through the de novo 510(k) pathway.[4] Shortly thereafter, the company announced that both the US Centers for Medicare and Medicaid Services (**CMS**) and the UK's National Institute for Health and Care Excellence (**NICE**) issued coding and support for the device's routine use by American and British doctors.[5,6] The UroLift system was then awarded a Category 1 **CPT** (Current Procedure Terminology) code by the American Medical Association (**AMA**).[7]

↘ Online Resources

Visit www.ebiodesign.org/5.2 for more content, including:

Activities and links for "Getting Started"
- Determine strategic R&D milestones
- Identify and prioritize key technical challenges
- Develop an initial R&D plan

Videos on R&D strategy

CREDITS

The editors would like to acknowledge Scott Delp, Ron Jabba, and Ted Kucklick for editing the material in the chapter, as well as Michael Ackermann of Oculeve, Ted Lamson of NeoTract, Robert Miros of 3rd Stone Design, and Hira Thapliyal of ArthroCare for sharing the case examples. Many thanks also go to Joy Goor and Stacey McCutcheon for their contributions.

NOTES

1 From remarks made by Guy Lebeau as part of the "From the Innovator's Workbench" speaker series hosted by Stanford's Program in Biodesign, April 5, 2005, http://biodesign.stanford.edu/bdn/networking/pastinnovators.jsp. Reprinted with permission.

2 All quotations are from interviews conducted by the authors, unless otherwise cited. Reprinted with permission.

3 "What Is a Gantt Chart?," Gantt.com, http://www.gantt.com/ (January 30, 2014).

4 "NeoTract, Inc. Receives U.S. FDA De Novo Approval for the UroLift® Prostate Implant," NeoTract press release, September 16, 2013, http://www.prnewswire.com/news-releases/neotract-inc-receives-us-fda-de-novo-approval-for-the-urolift-prostate-implant-223893231.html (March 31, 2014).

5 "NICE Gives Go Ahead to NeoTract's Groundbreaking UroLift Prostate Implant That Can Preserve Sexual Function While Offering Urinary Symptom Relief for Millions of Men Affected by Enlarged Prostates," NeoTract press release, January 29, 2014, http://www.prnewswire.com/news-releases/nice-gives-go-ahead-to-neotracts-groundbreaking-urolift-prostate-implant-that-can-preserve-sexual-function-while-offering-urinary-symptom-relief-for-millions-of-men-affected-by-enlarged-prostates-242576791.html (February 4, 2014).

6 "NeoTract's UroLift® Implant Procedure for Men with Enlarged Prostate Receives Medicare and Medicaid Outpatient Billing Codes for Reimbursement," NeoTract press release, March 13, 2014, http://www.prnewswire.com/news-releases/neotracts-urolift-implant-procedure-for-men-with-enlarged-prostate-receives-medicare-and-medicaid-outpatient-billing-codes-for-reimbursement-250045231.html (March 31, 2014).

7 "CPT® Editorial Summary of Panel Actions," American Medical Association, February 2014, http://www.ama-assn.org/resources/doc/cpt/summary-of-panel-actions-feb2014.pdf (March 31, 2014).

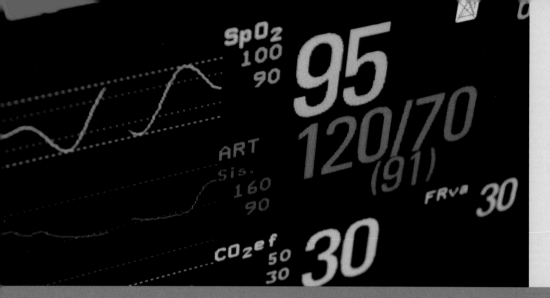

5.3 Clinical Strategy

INTRODUCTION

Human device testing is at the pinnacle of the medtech innovation process. The stakes are high: Two to three years to perform patient recruitment, enrollment, investigation, and follow-up. Costs as high as $100,000 per patient studied. Regulatory approval and reimbursement hanging in the balance. The credibility and commercial viability of the company on the line. And, most importantly, patient lives at risk. These are the types of issues that make clinical device testing a risky and challenging (yet essential) undertaking. An effective clinical strategy not only provides a mechanism for innovators to plan and initiate human testing, but to optimize its overall approach to clinical development.

In the context of the biodesign innovation process, clinical trials are broadly defined as human studies performed to determine specific outcomes based on the use of a new medical technology. Traditionally, the objective of clinical trials has been to demonstrate that a new device offers measurable, clinically important benefits to patients in terms of its effectiveness and safety.[1] This evidence is essential to supporting the regulatory approval of the device. However, the complexity of clinical development has changed dramatically over the last 15–20 years. Trial data are now commonly required to demonstrate clinical value to patients and financial value to payers as part of reimbursement pathways. Similar data are also required to address physician interests and concerns in order to drive market adoption. Because it can be difficult to design a single study that is capable of meeting all necessary regulatory, reimbursement, and marketing endpoints, innovators are often advised to think about study design as part of a larger clinical strategy that includes a progressive series of trials to achieve the company's overarching goals. However, the downside of this approach is that clinical trials are often the most costly and complex activity a start-up will undertake, and conducting multiple studies can strain the resources of even the most promising project. An effective clinical strategy must therefore define a clinical pathway that addresses the company's top priorities in a realistic manner.

OBJECTIVES

- Recognize the importance of establishing an overarching clinical strategy for early preclinical and human clinical studies.

- Appreciate the different types of clinical studies and how their designs relate to the overall goals of clinical development (including regulatory, reimbursement, and market adoption considerations).

- Learn the process of planning a human clinical trial.

- Understand tactical considerations for developing safe, cost-effective, and statistically robust clinical trials.

The creation of a medtech clinical strategy begins in parallel with research and development (R&D) since the clinical development pathway must be tightly integrated with device engineering (see 5.2 R&D Strategy). In addition to maintaining a tight linkage to R&D, an effective clinical strategy must be synchronized with the other strategic work streams in the development stage – see 5.4 Regulatory Strategy, 5.6 Reimbursement Strategy, 5.7 Marketing and Stakeholder Strategy, and 5.8 Sales and Distribution Strategy.

 See ebiodesign.org for featured videos on clinical strategy.

CLINICAL STRATEGY FUNDAMENTALS

Every innovator strives to develop a medical innovation that ultimately addresses the defined clinical **need**. Clinical studies, performed as part of a comprehensive clinical strategy, provide the evidence necessary to ensure that the new innovation is safe and effective.

Clinical strategy defines a prospective approach through which the organization can anticipate and manage the clinical risk associated with the project. A well-developed clinical strategy also helps ensure that clinical activities are tightly coupled with other important efforts underway within the company as part of the biodesign innovation process. No longer can innovators think of products as simply requiring development and testing. Rather, through the clinical strategy, they should seek to integrate testing into the development plan as a series of **value** building steps that ultimately result in a technology that addresses a clear clinical need and satisfies regulatory, **reimbursement**, and market requirements to positively impact patients.

Innovators and companies often feel both excited and apprehensive as they prepare to execute **clinical trials**. In recent years, the requirements for clinical data to address regulatory, reimbursement, and adoption concerns have become increasingly stringent and continue to grow more complex. According to Frank Litvack, former chairman and CEO of Conor Medsystems:[2]

The complexity, size, and expense of clinical trials for important medical devices are going to do nothing but get bigger before they get smaller. And I think that has implications for everybody who's in the start-up business. It means it's going to take longer and you're going to need more money. And there's going to be more risk. That's just the world in which we live.

For innovators developing new medical technologies, clinical trials are often the largest line item (by far) within their preliminary, pre-revenue budgets. Early in the innovation process, generating trial data is absolutely essential to attracting talent and funding to the project, but teams can often struggle to cover study costs, even on a relatively small scale. Consider Respira Design, a team from Stanford University that developed an asthma spacer for use in resource-constrained settings such as Mexico (where the students conducted their needs finding exercise). "We were focused on achieving best possible quality at the lowest possible cost," recalled Santiago Ocejo, one of the company's founders.[3] After benchmarking the make-shift solutions physicians used to help deliver asthma medication to the lungs of their pediatric patients, he explained, "Our cost benchmark was the price of a plastic soda bottle." The Respira team's innovative solution is a spacer produced from a single sheet of paper so it can ship and store flat and then be transformed into a usable device through a series of cuts and folds (see Figure 5.3.1). Despite the simplicity of the design and its low cost to produce and distribute, the team needed to collect clinical data to obtain regulatory clearance for the device. These data would also be used to demonstrate safety and effectiveness to physicians and patients, so the team felt an imperative to understand the extent to which the product impacted the delivery of asthma medication, how many uses each device could sustain, and whether the spacer would function as

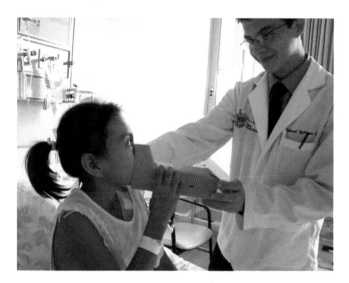

FIGURE 5.3.1

The Respira asthma spacer being used by a young patient in Mexico (courtesy of Respira).

intended in situations of emergency or distress. "This was a medical device that would potentially be used for someone who was having an asthma attack," explained Barry Wohl, another co-founder. "We couldn't put it in the hands of a mother to treat a child without a detailed understanding of how effective the device was in transferring aerosolized particles from the inhaler to the lungs. That was the minimum amount of clinical data we needed to be able to sleep at night." To conduct the necessary tests, the Respira team needed substantial funding. Unfortunately, Ocejo, Wohl, and their third co-founder, Eric Green, quickly discovered that potential donors and investors alike wanted to see clinical data showing that the device worked before making a sizable financial commitment. Ultimately, this quandary was one of the factors that stalled the development of the solution. "Cost was on our radar from the beginning, in terms of materials and distribution," said Green. "But we didn't realize how expensive testing would be." The team cautioned other medical device innovators to plan carefully and proactively for the time and expense associated with gathering user data in a safe and ethical manner, even for relatively simple, straightforward technologies.

For more complex technologies with a higher risk profile, the burden of generating clinical data can be even more daunting. As regulators, **payers**, and practitioners demand data from larger trials conducted over longer periods of time, clinical activities take longer to complete and are more complicated and costly to plan and execute, especially for truly novel technologies. Companies not only must raise the resources to fund the trial themselves, but also must cover their monthly cash expenditures, which can be significant as the technology nears the market. The two-part story on a company called Emphasys Medical illustrates the stress this can place on a start-up organization (see part one at the end of this chapter and part two in 5.4 Regulatory Strategy). A well-constructed clinical strategy helps innovators anticipate and manage the growing challenges associated with clinical testing.

Before diving into the detailed overview of clinical strategy development and tactical trial planning that follows in this chapter, it is important to emphasize the ethical responsibility that innovators assume when initiating clinical evaluation. Individual innovators and companies alike must commit themselves to the ethical treatment of all animal and human **subjects** involved in their studies and use this commitment to guide the development and implementation of a clinical strategy. Care should also be taken to avoid conflicts of interest and other ethical dilemmas that have the potential to negatively affect actual (and perceived) study results.

Clinical study goals

The development of a clinical strategy begins with the definition of the objectives (desired outcomes) of the strategy and of the studies that the company intends to undertake. One of the first questions innovators should ask themselves is, "What is the indication for how the solution will be used in practice?" Far too often, lofty goals result in a strategy that targets an overly ambitious indication before the innovators deeply understand the strengths and limitations of their solution. Although single clinical studies with large market indications are appealing to investors when reviewing a business plan, a clinical trial strategy comprising a series of studies with expanding indications has a far greater likelihood of being successful.

A team's objectives can be achieved progressively – it is not uncommon for a project or company to stage its studies based on its strategy. For example, trials with specific objectives (regulatory, reimbursement, or marketing) are synchronized with the milestones in a company's operating plan such that the data generated by the trials become available as they are most needed (see 6.1 Operating Plan and Financial Model). Implantable cardiac defibrillators (ICDs) provide a good example of a staged approach to clinical testing. Initially, ICDs were studied and approved for use as a form of secondary prevention in patients who had experienced cardiac arrest due to ventricular fibrillation. However, after this approval, the MADIT trial demonstrated that patients with a history of coronary artery disease and heart pump dysfunction could benefit from ICDs as a form of primary prevention (i.e., even if they had not yet experienced a cardiac event caused by ventricular fibrillation).

Within this framework, innovators should next ask themselves questions such as: "What results are needed to support regulatory approval? Are economic outcomes important to support reimbursement decisions? Will data be necessary to help market the device to physicians and/or patients?" Ultimately, the design and execution of each clinical study will be based on the answers to these types of regulatory, reimbursement, and marketing inquiries.

Regulatory considerations

For the large majority of medical device clinical studies, regulatory considerations are the primary drivers of trial design. (Basic requirements for the two primary regulatory pathways in the US – 510(k) or **PMA** – are described in 4.2 Regulatory Basics. More strategic regulatory considerations are addressed in 5.4 Regulatory Strategy.) Usually, a final trial design will result from a negotiation between the company and a given regulatory agency. Whether dealing with a **Notified Body** as part of the European Union **CE marking** process, the US Food and Drug Administration (**FDA**) or another regulator, these regulatory agencies try to base their decisions on prior experience with similar technologies, as well as changes in science and the practice of medicine over time. In certain areas, the interests of the company and the agency are aligned with regard to the details of the trial design; that is, both parties want to achieve valid data that will be publishable in high-quality medical journals, while providing maximal safety to the research subjects who participate in the trial. However, the company has

the additional imperative to achieve an endpoint that is favorable to the product, as quickly as possible, and at the lowest possible cost to deliver value to its investors.

Reimbursement considerations

Increasingly, medical device trials are being designed to provide cost-effectiveness outcomes and economic evidence for attaining reimbursement (particularly from the Centers for Medicare and Medicaid Services (**CMS**) in the US and technology assessment authorities outside the US – see 5.6 Reimbursement Strategy). This may include the incorporation of a formal cost analysis into the trial, along with measures of the financial impact of the outcomes and stricter endpoints than regulatory agencies normally require. On the upside, these requirements can lead to more rigorous trials that result in more robust data and more formidable barriers to entry for competitors. However, they also make trial design more complex since the desired outcome is not only to demonstrate the safety and efficacy of the device (as required by the FDA, for instance), but also to show equivalence or superiority to an existing technology or procedure on the market, as this latter outcome is often necessary to justify reimbursement.

Marketing considerations

A third objective of many clinical trials is to generate data that help give the new technology an optimal launch in the marketplace. Such marketing considerations can have a major impact on trial design. It may be that the physician group targeted by the new technology will be most convinced by a certain type of trial – for example, a **randomized trial** that compares the new technology to a current standard of practice. Or, if there is a choice to test the technology in different patient groups, it may be advantageous from a marketing standpoint to target a certain population (e.g., younger patients who are willing to pay directly for the technology and, thus, provide an early revenue stream for the company). The choice of which investigators to include in the trial can also be important. Companies frequently pursue investigators from among "marquee" physicians and key opinion leaders (**KOLs**) – high-profile practitioners who give talks frequently and are influential among their colleagues (see 5.7 Marketing and Stakeholder Strategy).

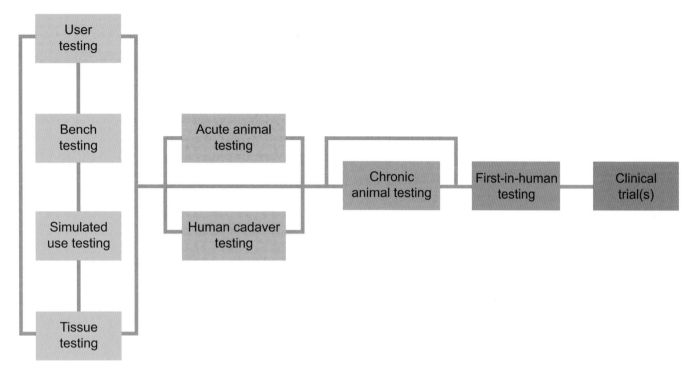

FIGURE 5.3.2

The biodesign testing continuum includes preclinical and clinical studies. Recall that not all tests are required as part of every project and should be determined based on the characteristics of the technology.

Preclinical and clinical evaluation

Prototyping, device R&D, and preclinical animal studies often overlap, with the boundaries for where one begins and another ends varying significantly from project to project. To help make sense of how these efforts interrelate, one can think about the steps in the biodesign testing continuum (see Figure 5.3.2; originally introduced in 4.5 Concept Exploration and Testing) as generally corresponding to two different types of efforts: preclinical and clinical studies. Preclinical studies begin with basic prototyping and continue through chronic animal tests. The focus of early tests is to assess feasibility through user, bench, simulated use, tissue, and cadaver testing or acute animal studies. Later, testing usually involved **hypothesis**-driven studies in animal models that primarily seek to evaluate safety, but sometimes also can provide insights into the potential efficacy of a device. Clinical studies involve human device testing for the evaluation of specific outcomes, which often include both safety and effectiveness.

Preclinical studies

The studies performed during concept exploration and testing (see chapter 4.5) are part of preclinical testing. When a company seeks to answer critical feasibility questions that extend beyond the prototyping of basic device functionality, it typically extends its preclinical efforts into ex vivo human cadaver or live animal testing. Usually, preclinical live animal tests are performed as acute animal studies. In vivo animal tests enable researchers to perform a preliminary evaluation of whether or not the device will function as desired in a living system. For example, an ablation catheter might successfully destroy tissue in simple tissue tests (i.e., using steak as a proxy for heart tissue and/or animal hearts from a butcher). However, until the device is placed in a living system, the inventor cannot be certain regarding fundamental device feasibility (e.g., is the device able to reach the correct anatomic area? Does its interaction with the animal's blood affect the device's ability to effectively ablate tissue?).

Acute animal studies An acute animal study is often an ideal method for evaluating device feasibility. These studies are performed in order to answer a specific, important question and then the animal is euthanized immediately following the test. Acute animal studies allow researchers to assess the device in a small, finite number of living systems prior to accepting the risks and expense of initiating a more extensive, systematic preclinical study. As with any animal study, researchers must be certain to apply for approval in advance of performing the tests, and should adhere to defined protocols for the ethical treatment of animals (see the section "Institutional Animal Care and Use Committees" for more information).

The first step in initiating any in vivo animal testing is to establish if an animal model of the human disease or disorder exists. If multiple animal models have the potential to provide a viable test, then consider secondary factors such as what animal models have been used by similar existing technologies in the past. For instance, there is a long history of using a swine model for evaluating coronary stents. However, not all models are this straightforward. No animal model may be available for an obstetric device due to the unique characteristics of the human pelvis and the high hormone levels experienced during pregnancy.

Chronic animal testing Once researchers have successfully addressed critical issues related to device feasibility through acute animal studies, they can transition into more systematic preclinical studies, such as chronic animal testing. In contrast to the acute animal studies, which are primarily focused on the viability of the technology, chronic animal studies typically involve larger numbers of specimens and seek to prove or disprove a specific predefined hypothesis in small or large animal models (e.g., mice or pigs, respectively). The safety and effectiveness of the device also become important objectives since the overarching purpose of these experiments is to gather evidence to justify research in human subjects.

Given the emphasis on safety and effectiveness, animal survival is almost always included among the important endpoints in a chronic study. Careful attention to analgesia and post-operative comfort and mobility is

> **Working Example**
> An overview of institutional animal care and use committees
>
> According to US federal law, any company or institution that uses laboratory animals for research or instructional purposes must gain approval from an Institutional Animal Care and Use Committee (**IACUC**) prior to beginning research. The purpose of the IACUC is to oversee and evaluate all aspects of the company's animal care and use program. Such a committee is assembled by the institutions performing animal testing and often includes administrators, veterinary experts, and members of the public. All research and teaching activities involving live or dead vertebrate animals must be reviewed and approved by the IACUC before a study is launched. Approval is obtained by submitting a protocol, which establishes the reason for the study, justification for why an animal study is necessary to evaluate the medical problem, and the processes put in place to ensure the ethical treatment of the animals in the study.
>
> Importantly, over the last decade, the intensity and quality of IACUC oversight has evolved substantially. Members of the committee are experts dedicated to both the safety and comfort of the animals, as well as the quality of the science resulting from the trials. The committees themselves are under careful, routine scrutiny by government agencies.
>
> It is worth noting that many members of the public have grave ethical concerns about animal experimentation. Keep in mind that this may include members of the development team. Importantly, all other options for testing a device must be fully explored before any animal is caged, anesthetized, operated on, or potentially sacrificed. Furthermore, it is essential to have a discussion about these issues prior to conducting animal tests to allow any team members with ethical objections to voice their concerns and opt out of the studies.

essential. Methods for monitoring the animals must be established and procedures defined for ethical euthanasia for any animals that are suffering unduly, regardless of how this affects study outcomes.

An important challenge associated with chronic studies of healthy animals is that it can be difficult to get an accurate prediction of device performance compared to the case of a patient with a disease, several comorbidities, and/or advanced age. In some situations there are disease models in animals that have been developed to mimic at least some aspects of the human condition. For example, pig coronary arteries can be pre-treated to create narrowings that are angiographically similar those seen in humans. The tissue composition of these lesions, however, is substantially different than the plaque that builds up in human coronary arteries. As a result, extrapolating device performance based on this animal model has limited utility. In general, there is wide variation of the comparability of disease models in animals to the corresponding human condition – and this can be a major factor in determining the time, effort, and money that a team chooses to invest in this type of testing (refer to the Emphasys case later in this chapter).

Although not common, some preclinical studies look beyond safety by seeking to statistically establish the efficacy of a particular endpoint. **Efficacy endpoints** are selected and then placebo or "sham" procedures (designed to simulate the risk of the device procedure without providing active therapy) are used to analyze the effectiveness of a device treatment. These tests require careful attention to design since the data are only of value if the animal model reflects human disease characteristics and if the endpoint chosen is meaningful for a human study. For example, if the goal of a new device is to improve heart function through the transplantation of stem cells into an injured heart, efficacy could be evaluated by examining the change in pumping function following cell transplant. In this example, an injection of saline could be used to assess the effect of the injection alone, while the difference between the effect of the injection of saline on heart function and the injection of stem cells on heart function could be used to evaluate stem cell efficacy. Having a clearly defined clinical strategy in place prior to embarking on device testing helps ensure that researchers choose the most appropriate endpoint for each preclinical study. This can help them to maximize the value of the data when it is time to transition into human trials.

The demonstration of safety is essential for obtaining approval for **first-in-human** testing. Chronic animal testing, although limited, provides some of the strongest evidence for safety in support of a first-in-human ethics committee submission. These data can also help establish the potential clinical value of the device so that physicians are willing to enroll patients in the study. Additionally, they expand the researchers' base of knowledge so that the probability of causing harm to patients (or of not meeting desired endpoints) is reduced.

Working Example
An overview of good laboratory practices

When conducting preclinical studies, researchers must determine whether or not to follow a research standard called Good Laboratory Practices (**GLP**), as defined by the FDA. In essence, GLP is a set of guidelines that describes in detail how studies should be performed and data collected. The FDA requires that data generated to support a 510(k) or premarket approval (PMA) submission be gathered according to GLP standards. The guidelines specify minimum standards for safety protocols, facilities, personnel, equipment, test and control activities, quality assurance, record keeping, and reports used in conducting the trial. GLP also requires the laboratory to have an extensive written set of operating guidelines (called standard operating procedures or SOPs – see 5.5 Quality Management) for conducting the study. Regulatory agencies outside of the US do not require companies to follow GLP standards; however, they may insist on adherence to **ISO** standards for their study submissions.

Importantly, not every preclinical study needs to be conducted in accordance with GLP guidelines. Although GLP standards ensure robust results, early studies that are performed to assess feasibility or even answer questions of safety do not require such rigorous standards, which can be time-consuming and add increased expense to the project. Synchronization of the clinical strategy with the company's developmental timeline will help establish the need for GLP studies to support key milestones.

Human clinical studies

Progressing from preclinical animal studies to human clinical studies represents a major milestone for most device companies. The opportunity to test and evaluate a new medical device in a human is both an opportunity and risk. Medical devices that fail in a human (or, more importantly, harm a patient) are unlikely ever to be used again. Several different types of studies can be performed for human device testing, depending on the nature of the device, the clinical problem being addressed, and the stage of testing. Each study type may provide an opportunity to further advance the technology (see Figure 5.3.3).

First-in-human studies   Before a company can initiate large-scale clinical studies, it first must complete first-in-human studies. The most important outcome of these small-scale, preliminary human studies is safety, although investigators are also looking to see whether the device performs as intended (even though a specific efficacy endpoint may not be defined). Because first-in-human studies are not designed to establish a clinical benefit, efficacy is anecdotal. Careful thought must be given to the appropriate time and place to perform first-in-human studies, given their significance. Ensuring that the device design has been optimized for human anatomy will improve the researchers' chance of success and minimize the risk of causing harm. There are many reasons that may explain device failure in early human studies, but it is often due to too much reliance on prior animal studies in which the organ or system was overly forgiving and free of disease. This risk can be minimized through the careful evaluation of human anatomy and human cadaver studies as part of the clinical strategy.

Registries and observational studies   A **registry** is a collection of cases performed in a real-world setting (rather than patients treated in a specifically designed comparative study), which may be accumulated either prospectively or retrospectively after a number of cases have been performed. Prospective registries are also called "**observational**" studies and are often used in the pilot phase of testing a new device – typically, the first 10–100 cases following first-in-human tests where the

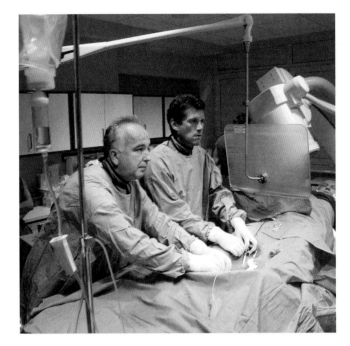

FIGURE 5.3.3

Human clinical trials can take many different forms, but all require careful planning and the highest of ethical and safety standards (courtesy of ExploraMed).

company, investigators, or the regulatory agency may be attempting to learn enough about a device's performance to design a definitive trial. Registries from overseas, for instance, have recently been used in support of 510(k) applications in which the burden of clinical proof is modest (i.e., the device is known to be comparatively safe).

An increasingly important use of registries is to monitor the outcomes of a device after it is approved by a regulatory body and as it is launched into more widespread clinical use. This approach has resulted from recent, high-profile examples of devices that showed unanticipated complications following an approval (e.g., implantable cardioverter defibrillators, drug-eluting stents). Registries are generally much less complex and expensive than randomized trials, but have a lower power to demonstrate important discernable differences.

Case control studies   A case control study statistically compares outcomes of a group of patients treated with a new device or procedure to a matched group receiving no treatment (or a **standard treatment**). Patients in both groups must be well matched on characteristics that are known to influence outcomes associated with treatment for the disease state, such as age or comorbid conditions. Case control studies are often accomplished by searching a large database of patients treated with the standard (or no) device or procedure to find a group of patients that matches the characteristics of those treated with the new device or procedure. A retrospective comparison is then performed to identify which group experienced better outcomes. For example, a new device to treat lumbar disc herniation could be compared to fusion surgery (an accepted technique) by finding patients in a database who have undergone fusion and are matched to patients receiving the new treatment according to age, gender, weight, duration and severity of back discomfort, location of disc herniation, etc. Once a comparable population is identified, researchers could then analyze the medical outcomes of both groups to determine whether the new device or fusion surgery led to more favorable results.

The primary advantages of case control studies are that they are less expensive than randomized studies (see below), can be carried out by smaller teams of researchers, and take less time to complete than prospective studies. The main disadvantage is that the results are not as definitive as prospective, randomized, **controlled** studies and almost always necessitate further research. However, these studies are a valuable step in the development of a technology and can be helpful in revealing both the strengths and limitations of a technology. Case control studies also help innovators understand a potential treatment effect so that large studies can be designed to be successful in meeting statistical and clinical outcomes.

Prospective, randomized, controlled (blinded) trials
Prospective, randomized, controlled trials are considered the "gold standard" for medical device testing. They are increasingly required in the US for both approval by the FDA and for reimbursement by CMS (see 5.6 Reimbursement Strategy), which is one reason why the cost and complexity of clinical testing has skyrocketed over time.[4] The term *prospective* refers to the fact that the trial is

designed before any devices are tested. The patients are divided into treatment and comparison, or control, groups by a statistically *random* assignment. Interpretation of the results is based on the outcome of the group treated with the new device or procedure, relative to a *control* group that may receive no treatment or treatment with a more established approach. The advantage of prospective, randomized, controlled trials is that they have the greatest statistical **power** to discriminate whether or not the outcome and safety profile of the new technology is, indeed, superior to the control group. The main disadvantage of these kinds of trials is the considerable time and expense required to complete the study compared to the simpler trial structures described above.

In some cases, additional rigor can be added to a randomized, controlled trial if it is possible to **blind** the study participants (patients, physicians, nurses, data analysts) to the device being used. The term "double blind" is used when both the patient and the physician are blind to the treatment. For example, a double blind trial can be performed to assess a bare metal versus a drug-eluting stent if it is not possible for the physician or the patient to tell the difference between the two stents based on appearances or deployment. For the trial, both stents would be provided in nondescript packages with code numbers that would eventually be used to determine which type of stent was placed in which patient. Sometimes in device trials, it is not possible to blind the physician to the treatment (e.g., in the comparison of two implants with a different appearance) and only a single blinded study can be performed.

Before deciding on what study(ies) to conduct, researchers are advised to complete a *thorough* review of available literature for the condition being studied and understand *in detail* what has been studied to date. They should also consider specifying outcomes that are similar to those of previously published studies so that the results can be compared. This is particularly important for observational studies, but is also useful in randomized, controlled studies. Being able to demonstrate how an outcome is significantly better in a trial of a new device compared to the same outcome in a previous trial of an older device can have a significant impact.

Consultation with the regulatory agency, users, and/or payers can help generate the best possible study design, with modifications made as input is gathered.

Innovators are also encouraged to investigate emerging trends in clinical trials focused on finding less expensive ways to generate reliable, informative results. Although randomized trials are still among the most powerful tools available to clinical researchers, there is growing awareness in the scientific community about issues related to their excessive complexity and expense, as well as challenges linked to the time required to recruit study participants and the fact that their results cannot always be generalized to a broader population.[5] In response, registries are attracting increased attention as a mechanism for describing practice patterns and trends, identifying outliers, and detecting safety signals. Researchers are also using registries to assess comparative effectiveness, although they acknowledge that observational findings may not be internally valid owing to the absence of **randomization**.[6]

One response to the trade-offs between randomized trials and registries has been the registry-based randomized trial. A recent example is the Thrombus Aspiration in ST-Elevation Myocardial Infarction (TASTE) trial in Scandinavia.[7] As described in an article in the *New England Journal of Medicine*:[8]

The TASTE investigators designed a large-scale trial to answer an important clinical question and carried it out at remarkably low cost by building on the platform of an already-existing high-quality observational registry. With this clever design, which leveraged clinical information that was already being gathered for the registry and for other preexisting databases, the investigators were able to quickly identify potential participants, to enroll thousands of patients in little time to avoid filling out long case-report forms, to obtain accurate follow-up with minimal effort, and to report their findings, all for less than the amount of a typical [individual research grant from the National Institutes of Health]. Their findings may well be broadly generalizable, since they included in the

randomization process the majority of all patients treated for ST-segment–elevation myocardial infarction in the study area.

Although this specific approach will not be applicable to all situations, it is representative of the innovative thinking that can unlock new opportunities in trial design and execution. Another example can be found in the work of the United States' Patient-Centered Outcomes Research Institute (**PCORI**), which is seeking to improve clinical practice through the better implementation of **evidence-based** research. Through its PCORnet program, it is experimenting with models to advance the shift in clinical research from investigator-driven to patient-centered studies.[9]

Other trial nomenclature

As it relates to studies conducted to secure FDA regulatory approval, the term "**pilot**" is used to describe early clinical trials, usually conducted as a registry. The definitive clinical trial conducted for approval (and perhaps reimbursement) is known as a "**pivotal**" trial. "**Post-marketing**" studies refer to trials performed after the commercial approval of the device. These studies are often required as a condition of approval for a PMA device, but also can be used for other purposes as part of a staged clinical strategy. (See online Appendix 5.3.1 for more information about pilot, pivotal, and post-marketing studies and a comparison to trials in the pharmaceutical industry.)

Basic issues in designing a trial

When designing a specific trial, innovators must consider a number of important factors, including the trial's hypothesis, endpoints, and its statistical power and size.

Hypothesis

The design of a clinical trial starts with a hypothesis, which the study will prove or disprove. Generally, this is in the form of a comparison of the outcomes (endpoints) achieved in patients treated by a new device or procedure relative to the control group. For example, a hypothesis for an ophthalmology trial might be that a new intraocular lens will provide superior visual acuity and an equivalent safety profile to an existing lens.

For the purposes of FDA regulation in the US, an innovator or company might define a hypothesis that the device being studied is equivalent or "non-inferior" to an existing standard; that is, the outcomes are statistically indistinguishable. From a statistical standpoint, equivalence can usually be demonstrated in a much smaller trial than would be required to show superiority. A company might undertake an equivalence trial even if it thinks its device is better than the competition's, because it could save a substantial amount of money and still sell effectively against the competition based on the performance of its device.

Endpoints

The efficacy of a new device or procedure is measured in the form of *endpoints*. Endpoints are the prospectively identified and quantifiable parameters that a study is designed to meet. The endpoints relate to the regulatory claim that is being sought for clinical use. Well-designed studies one or more **primary endpoints** which the company, investigators, and regulatory agency agree will be the main criteria on which the device or procedure is evaluated and the one(s) which the study will focus on from a power standpoint (see below). There may be **secondary endpoints**, which are of interest scientifically and clinically, but will generally not lead to a regulatory approval (e.g., endpoints to determine cost-effectiveness for reimbursement). Selection of the proper endpoints for any study is exceedingly important and, is usually a major point of negotiation between the company and the regulatory agency.

To be useful, an endpoint must be measurable and as unambiguous as possible. For instance, even in situations in which the outcome is improved **quality of life** (e.g., a new treatment for tennis elbow), some method of measuring that outcome should be utilized (e.g., a quantitative questionnaire designed to measure the degree of improvement). Another important consideration is that the trial endpoints must be achievable in a reasonable span of time. A new cardiac device may have the potential to prolong life, but the company will not be interested in a trial that takes 40 years to prove this point statistically. In this situation, a **surrogate endpoint** must be selected. For instance, in the cardiac example, the

surrogate endpoint might be a measure of the pumping function of the heart and a nuclear medicine measurement of its blood supply – variables which have been shown in previous studies to correlate with survival.

Importantly, the statistical power of an endpoint is maintained only when it is identified prospectively. Post-trial data analysis ("data dredging") may suggest that some unanticipated subgroup of patients seems to gain special benefit from the device. However, to prove this rigorously, it will be necessary to test this endpoint prospectively in a new trial.

Before deciding on the endpoints for a study, the innovator or company that will sponsor the trial should complete a thorough review of available literature for the condition being studied and understand in detail what has been learned to date. Additionally, it will be helpful to select outcomes that are similar to those of previously published studies so that the results can be compared. This is particularly important for observational studies, but is also useful in randomized, controlled studies. Being able to demonstrate how an outcome was significantly better in a trial testing a new device compared to the same outcome in a previous trial of an older device can have a significant impact.

Statistical power and trial size

One of the most important issues in trial design is the number of patients to include in the study. A properly designed trial uses a "**power calculation**" to determine how many patients are required in the treatment and control groups to adequately test the hypothesis. This calculation is based on a best guess of the impact of the new therapy on the primary endpoint. For example, suppose a new technology for obesity has been tested in a pilot observational study and has been shown to reduce body mass index (BMI) in the study group by 20 percent. The company and investigators might propose a primary endpoint of 10 percent BMI reduction. This means that if the device truly reduces BMI by 10 percent, the results from the trials will prove that. The clinical trial will then collect data to perform a hypothesis test. The null hypothesis is that there is no difference and the alternative is that there is a 10 percent improvement. The null hypothesis will be rejected if the difference between the BMI in the control group and treatment group has a **statistical significance** of 5 percent. Because of the randomness in the data, it is possible, especially if the number of patients enrolled is small, that the null hypothesis will not be rejected even if the device has an effect. The statisticians will help determine the sample size such that the chance of erroneously failing to reject the null hypothesis will be small (typically 10 percent). The endpoints and power calculations will be reviewed by the regulatory agency and approved or modified as its experts see fit.

It may be possible to achieve the goals of the power analysis using other forms of randomization besides a 1:1 model. For instance, when the control group will be treated with a relatively well-understood procedure or technology, it may be possible to perform a 2:1 randomization (that is, the group receiving the new therapy will be twice the size of the control group). This approach reduces the size and expense of the trial without negatively affecting the significance of the results. (See online Appendix 5.3.2 for a basic primer on statistical design.)

Clinical trial planning and operations

Beyond the design of the trial, innovators must address a series of important issues in planning and preparing to launch a trial, as outlined in the sections that follow.

Choosing investigators and centers

An innovator or company, as the sponsor of a clinical trial, will select one or more principal investigator(s) (**PIs**) for a trial based on a number of considerations, including experience in leading trials, track record of publishing trial results, stature in the field, quality of support personnel and – an increasingly important criterion – freedom from financial conflict of interest. The PIs will typically help the company select additional investigators to conduct the trial. These investigators will be chosen based on their technical skill with similar devices or procedures, their ability to enroll patients quickly (generally meaning bigger practices), their prior experience with clinical trials, their reputation among their colleagues, the effectiveness of their research support staff, and their geographic location.

Finding highly productive investigators can be facilitated by searching the literature (looking for contributors who appear on the prior studies that are similar to the one being designed). Polling companies and contract research organizations (**CROs**) that have worked in the specific therapeutic area can also be helpful. However, newcomers (physicians who are new to a particular field) should not be overlooked, as they often can make significant contributions to the trial. Many up-and-coming physicians are hungry to make a mark. Being associated with marquee physicians through a clinical trial can be highly motivating to these new physicians and, as a result, can yield significant enrollment from lesser-known sites.

Typically, all investigators and their institutions involved in a trial will sign a contract with the trial sponsor that specifies their obligations, including accurate recording of data and timely reporting of complications. These agreements also address indemnification and the assignment of ownership rights of new discoveries (IP) made in the course of the study. The investigators and their centers are reimbursed for their costs in conducting the study, which includes their time (and staff time), extra equipment, tests, hospital time, and other expenses. If the investigators are faculty members in a university, the contract is made with the university and an "indirect" charge (typically an additional 25–50 percent) is added to support general university infrastructure. Innovators and companies should expect contracts to take anywhere from 2 to 12 months to finalize, depending on the type of institution with which they are working.

The number of investigators and centers required depends on the size of the trial and the expected rates of enrollment. In practice, enrollment often turns out to be much slower than the investigators believe will be the case. A rough rule of thumb to help manage this risk is for the sponsor to budget for the pace of enrollment to be approximately half as fast as expected and total enrollment from each center to be half as large as projected.

Investigational device exemption

As noted in 4.2 Regulatory Basics, most clinical studies in which there is risk to the patient and in which data will be used to support a regulatory submission in the US, require an investigational device exemption (**IDE**). An IDE allows a premarket device to be tested in humans such that the necessary safety and effectiveness data can be collected to support the FDA submission. For a PMA, clinical studies supported by an IDE are always required. In contrast, an IDE is only needed to support the small percentage of 510(k) submissions that requires clinical data. An IDE can also be used to cover the clinical evaluation of certain modifications and/or new intended uses of legally marketed devices[10] (e.g., off-label usage).

The purpose of the IDE process is to ensure that researchers "demonstrate that there is reason to believe that the risks to human subjects from the proposed investigation are outweighed by the anticipated benefits to subjects and the importance of the knowledge to be gained, that the investigation is scientifically sound, and that there is reason to believe that the device as proposed for use will be effective."[11] Once approved, an IDE clears a device to be lawfully used in conducting clinical trials without the need to comply with other FDA requirements for devices in commercial distribution. All clinical trials that include investigational devices must have an approved IDE *before* the study is initiated, unless the device is determined to be exempt from IDE requirements.[12]

To obtain an IDE, researchers must complete an IDE submission. As discussed in 5.4 Regulatory Strategy, early communication with the FDA and a pre-IDE meeting can help facilitate more efficient approval through the IDE process, as well as assisting with the development of a study design that supports the desired/necessary endpoints for regulatory approval. The FDA is mandated to respond to every IDE application within 30 days. Though an IDE is rarely approved in the first 30-day period, the FDA must provide the applicant with feedback and/or request additional information within this time frame. Often a series of back-and-forth communications with the agency is required before an application is approved. If, after 30 days, the FDA has not responded to an IDE application, a company is authorized by default to proceed with the study.

During the study, the researchers must maintain compliance with specific IDE requirements which include (but are not limited to):

- Obtaining advance approval from the institutional review boards (**IRBs**) where the study will be conducted, and working with the IRB through the execution of the trial.
- Obtaining **informed consent** from all patients involved in the study.
- Labeling the device for investigational use only.
- Carefully monitoring the study.
- Completing all required records and reports.

In rare circumstances, the FDA allows for investigational devices to be used in patients that are not part of an IDE-approved clinical trial. Such usage is allowed under clearly defined conditions of emergency use, compassionate use, treatment use, or continued access[13] (see online Appendix 5.5.3 for more information).

Institutional review board approval

Once an IDE is obtained, researchers must then seek IRB approval before the study is initiated (note that if the study is IDE-exempt, IRB submission is the first step following the design of the trial protocol). It is a federal mandate that each site where a clinical trial will be conducted has an IRB that is responsible for protecting the rights, safety, and welfare of research subjects. The IRB can be developed and managed in-house or contracted from a third-party provider. In either scenario, IRBs are regulated by the FDA and their policies and practices are subject to periodic review and certification. An IRB is generally made up of clinicians, nurses, and one or more hospital administrators. Optimally, the IRB will also include a statistical expert, an expert in medical ethics, and one or more "lay" representatives (such as community advocates, clergy members, or a working professional). The IRB is responsible for monitoring complications of the study and, in many cases, also serves as the screening point for issues of conflict of interest on the part of the investigators or their institutions.

The lead **clinical investigator** is responsible for preparing the application to the IRB at each institution where the study will be conducted. This application describes the device, outlines the proposed clinical study and trial endpoints, and includes a sample patient consent form. The IRB formally reviews this application, often requesting changes as needed, before approving the study. This process usually takes several months to complete. It can require more or less time, depending on how often the members of the IRB meet and how they work together.

It is worth noting that local IRBs often have additional policies and restrictions beyond the general requirements specified by the FDA. IRBs have come under increased scrutiny recently, in part as a result of the 1999 death of a patient enrolled in a gene therapy trial at the University of Pennsylvania. In this case, involving 18-year old Jesse Gelsinger, it was alleged that the IRB did not adequately review the safety of the study and the protections put into place. Furthermore, investigators did not adequately counsel participants on the extent of the risks involved in the study[14] and were not following all of the federal rules requiring them to report unexpected adverse events associated with the gene therapy trials.[15] The Gelsinger case, in conjunction with other events, served as a catalyst for IRB reform and the improved protection of human research subjects. In June 2000, the Office for Protection from Research Risks (OPRR) was officially renamed the Office for Human Research Protections (**OHRP**) and moved from the National Institutes of Health (NIH) to the Department of Health and Human Services or **HHS** (which also oversees the FDA). The move was intended to increase both the visibility and accountability of human research protections, as monitored by the federal system.[16] To carry out this mission, the OHRP has established formal agreements ("assurances") with nearly 10,000 universities, hospitals, and other research institutions in the US and abroad to comply with the regulations pertaining to human subject protections.[17]

Because each IRB maintains its own governance, trial sponsors may have an inconsistent experience from institution to institution. Sponsors and investigators should be prepared to modify their **clinical protocols** (and especially the informed consent document) to meet the requirements of each IRB. It is the sponsor's responsibility to ensure proper document control for each location, a practice that can present complex control/management issues for trials that utilize multiple sites.

It is also important to note that many IRBs charge for their services (i.e., initial review and approval of the study, as well as ongoing reviews). An initial review can cost anywhere from $2,000 to $4,000, while ongoing reviews may range from $1,000 to $2,000 every 6 to 12 months, with these fees paid for by the sponsor.

Patient enrollment

Patients are screened for studies by the investigator and his/her staff. In the modern era of careful trial design, IRB review, and patient advocacy, the ability of researchers to enroll subjects has become the rate-limiting resource in clinical research.[18] Despite this challenge, innovators and companies must carefully adhere to all guidelines and requirements for enrolling their trials (see Figure 5.3.4).

There is no doubt that patients are increasingly cautious about agreeing to participate in trials of new devices and drugs. A recent article summed up the patient's perspective about participating in a cardiac defibrillator trial this way: "Would you sign up for an experimental heart device trial, where there's a

FIGURE 5.3.4
In today's environment, innovators are cautioned not to take chances in performing human research.[19]

50 percent chance your device will be inactive ('control group'), and there's an FDA-approved device already on the market?"[20] For those involved in sponsoring and conducting clinical trials, it is important to keep in mind a key fact: volunteering for a clinical study is an act of generosity and public service on the part of the patients and their families. As a result, study participants should receive all the respect and gratitude they deserve.

The process of enrolling patients into a study must be meticulous and is often highly time-consuming. Subjects must be evaluated carefully and thoroughly to determine if their conditions and health profiles match the targeted audience for the device. This is accomplished through detailed screening using:[21]

- **Inclusion criteria** – Characteristics or indications that subjects must have in order to participate in the clinical trial. For example, if a new spinal disc is being tested, participants might be required to demonstrate a specific site of disc degeneration in the back to be included in the study.
- **Exclusion criteria** – Characteristics or contraindications that eliminate subjects from participating in a clinical study. For example, individuals who have had fusion surgery or advanced spinal arthritis might be excluded because their spines already show significant differences from those that a device is intended to treat.

Meticulous patient screening involves multiple participants in the care continuum. First, patients must be correctly diagnosed. Then, their physicians must be familiar with the clinical studies available for which the patients might qualify. Researchers often impose other requirements and testing procedures on patients to ensure that they fit the trial criteria. At the same time, patients may seek counseling from another healthcare professional, particularly if they are dealing with stressful or life-changing news regarding the illness.

It is worth keeping in mind that patients can benefit from trial participation in several ways. The new technology or procedure may indeed represent a substantial improvement in care. Even if allocated to the control group, the patient may benefit from the increased level of medical attention provided to participants in the trial,

including additional tests and provider visits. Patients also appreciate the chance to learn more about their condition by participating in a study. Some studies will reimburse patients for their participation in the trial, particularly if there is a requirement for the patients to return for further evaluation or testing. Finally, and perhaps most importantly, many patients are genuinely motivated by a desire to serve others. Having suffered themselves, they are inspired to do what they can to reduce or eliminate the medical problem for subsequent patients.

Informed consent and patient protection

Before entering a study, patients must provide written *informed consent*. In practice, this means that someone (generally not the physician performing the procedure) reviews the IRB-approved consent form with the patient, explains the risks and benefits of the study, answers any questions, and obtains the patient or guardian's signature on form. The consent form provides the opportunity for the patient to withdraw at any point in the study and identifies a third party, who can be contacted if there are perceived irregularities or other issues with the trial.

Standards governing the protection of human research subjects, including mandatory informed consent, arose from several shameful experiments conducted in the early part of the twentieth century (most notably in Nazi Germany and in Tuskegee, Alabama). The standard of informed consent emerged in the 1940s as the field of clinical research became codified and well established. Today, requirements for informed consent in research involving human subjects exist in every country, as defined by international standards.

Because new investigational devices have not, by definition, been tested previously in humans, a certain level of risk exists for the subjects involved in any clinical trial. As the level of device invasiveness and procedure complexity increases, so does the level of risk. According to the FDA, "Although efforts are made to control risks to clinical trial participants, some risk may be unavoidable because of the uncertainty inherent in clinical research involving new medical products. It's important, therefore, that people make their decision to participate in a clinical trial only after they have a full understanding of the entire process and the risks that may be involved."[22]

This philosophy is at the heart of the FDA's policy for informed consent, which states:[23]

> No investigator may involve a human being as a subject in research covered by [FDA] regulations unless the investigator has obtained the legally effective informed consent of the subject or the subject's legally authorized representative. An investigator shall seek such consent only under circumstances that provide the prospective subject or the representative sufficient opportunity to consider whether or not to participate and that minimize the possibility of coercion or undue influence. The information that is given to the subject or the representative shall be in language understandable to the subject or the representative. No informed consent, whether oral or written, may include any exculpatory language through which the subject or the representative is made to waive or appear to waive any of the subject's legal rights, or releases or appears to release the investigator, the sponsor, the institution, or its agents from liability for negligence.

Data entry and monitoring

In many medical device trials, the time at which the device is used represents a critical point in the study where intensive data collection is typically performed. Generally, this is accomplished by a study nurse or other trained expert who collects all of the pertinent information about the encounter. The primary data can be kept carefully in files or notebooks (see Figure 5.3.5) or more recently documented by electronic data capture. It is the responsibility of the sponsor to regularly monitor the primary data to ensure quality and to evaluate adherence of the investigators to the protocol. Protocol violations must be carefully documented and tracked. Investigators are often counseled about their continued involvement in the trial if there are problems with the quality of the data or compliance with the protocol. This entire process is also subject to auditing by regulatory agencies.

Core laboratories

Certain endpoints of a clinical study may best be analyzed in a special **core laboratory**, staffed by clinical scientists with specialized equipment and expertise to make the quantitative measurements required. These core labs also add a level of consistency and objectivity to data analysis in that they are independent of the investigators and sponsor. Core laboratories may be in universities or private clinical research organizations. For example, clinical trials of coronary stents routinely use core angiographic laboratories, where

FIGURE 5.3.5
Clinical research nurses and coordinators must keep detailed records on every enrolled patient (courtesy of Todd Brinton).

radiographic images of the arteries (angiograms) before and after stent placement are carefully measured. Core laboratories typically charge on a per-patient basis and are subject to audit by the FDA.

Clinical events committee

Studies with the potential to harm patients must include mechanisms for evaluating clinical events. Clinical events are clinical signs or symptoms that occur during the course of the study. Monitoring of these events is often accomplished through an independent review committee. The clinical event committee (CEC) usually comprises clinicians and statisticians who are charged with independently adjudicating clinical events and reporting the results to both investigators and the study sponsor. Clinical events are typically screened as adverse events (AEs) or serious adverse events (SAEs). The committee must determine if its members believe the event is not related, possibly related, probably related, or related to the procedure or the technology. For example, consider a patient who receives a new cardiac pacemaker but then experiences pain for an extended period following the procedure and ultimately develops an infection of the implant several weeks after the placement. The committee would review the hospital records and reports, including all tests and might deem the pain an AE probably related to the procedure and device. Further it may also determine that the infection is an SAE related to the procedure, but only possibly related to the technology since the infection was most likely a result of contamination of the surgical site and not the device itself. These results are important data points that impact the safety endpoints of a trial. Members of the CEC are compensated by the sponsor but cannot have further financial ties to the company. The reports of the CEC must be available for auditing by the regulatory agencies.

Data safety and monitoring board

Sizable trials will generally have an independent data safety and monitoring board (**DSMB**) to review the preliminary results at prespecified time points during the trial in order to ensure that patients are not being inadvertently harmed by the study. The DSMB may decide that the initial trial results are within the range of expected outcomes and allow the trial to continue; it may terminate the trial based on unexpectedly bad outcomes in the treatment group; or it may terminate early based on unexpectedly good results (when it is no longer ethical to continue the control treatment).

Clinical trial management

Once the trial has been launched, innovators must next address a number of ongoing management issues to ensure the timely and effective completion of the trial.

Research site staff

The resources required at each center to perform the high-quality research necessary for a randomized, controlled trial are formidable. In addition to the resources needed to initiate and manage a clinical trial program at the site (including physicians, facilities, and equipment), the study center must devote research nurses to each study. Research nurses play an instrumental role at all phases of the trial, including general study management, IRB process management, and accurate completion of case report forms. It is the sponsor's responsibility to make sure each clinical site has the necessary resources in place to fulfill the demands set forward in the protocol.

Sponsor personnel

Most sponsors will have at least one or two in-house employees managing a clinical trial, if not an entire team (depending on the level of control and support the company determines to be appropriate). Different models commonly adopted by trial sponsors include the following:

- Companies may engage a CRO to manage the on-site elements of the study, as well as data management. A CRO typically provides the infrastructure required to recruit, qualify, and audit sites. A CRO is particularly beneficial when the scope of the trial is large and the CRO has experience in the type of study being conducted. Also, because of the transient nature of clinical trials, it may not make sense for a company to hire an entire staff of full-time researchers who will not have roles when the trial ends.

- Alternatively, some companies prefer to maintain tight control over a clinical study by managing it in-house. In these cases, certain elements of a trial still may be outsourced [some CROs offer "a la carte" services, such as monitoring only, data management, or regulatory (IRB, IDE, etc.) responsibility], but the overall management of the trial is still performed by the internal team.

The sponsor, or its CRO, has responsibility for monitoring the progress of the trial, compliance with the protocol, and various other requirements. When using a CRO, the external organization should be carefully managed. All contracts also should be explicit with respect to expectations and deliverables. Outsourcing can be risky since the contractor does not have as much at stake as the sponsor. On the other hand, CROs have vast experience and expertise that can be valuable to young, start-up companies. Close, proactive management of any external partners can help minimize this element of risk while capitalizing on its benefits.

Data management

Data management is perhaps one of the most important elements of a clinical trial. However, this factor often is overlooked until the data already have been captured. Traditionally, paper forms are completed for each patient at each visit. These data forms must then be entered into a database to track progress of the endpoints of the study. Increasingly, web-based data capture is being used, which minimizes the back and forth of case report forms through fax or email. Another advantage of these web-based systems is that they can be designed to accept only data that fall within expected limits (i.e., data that make sense), prompting the person entering the data to recheck the values that violate defined expectations.

Clinical trial costs

As noted, the cost of medical device clinical trials can account for a significant portion of a start-up's total budget on the way to market. Based on the purpose and complexity of the study, the costs for medical device

clinical trials can range from hundreds of thousands to tens of millions of dollars. Study costs are driven by:

- The cost of the device(s) being used in the trial.
- The cost of performing the procedure, including physician costs and hospitalizations, if needed.
- The costs of follow-up clinical visits and/or tests to evaluate the safety and efficacy of the medical device.
- The cost of paying investigators and institution study coordinators to perform the clinical studies.
- The cost of conducting the trial, including training, monitoring, and data management.
- Patient recruitment costs, including advertising and potential payment to patients.
- In-house management and personnel costs.
- The cost of trial support and other resources provided by CROs.
- IRB costs.
- Consulting expenses for data safety monitoring boards, physician advisory boards, and core laboratories to independently evaluate trial results.

Clinical trial expenses may range from as little as $2,000 per patient (e.g., for a non-implantable device with a short follow-up period) to as much as $100,000 on a per patient basis (e.g., for an implantable or therapeutic device with a lengthy follow-up period). Investors will show a greater or lesser willingness to fund a device with high anticipated clinical expenses based on factors such as the size of the total market opportunity. As the Respira team (featured earlier) learned with its low-cost asthma spacer targeted at underserved patients in developing countries, raising money to conduct even inexpensive trials can be difficult without a strong commercial opportunity for selling the device. Companies with devices that require trials at the higher end of this range are often expected to have a device with "blockbuster" market potential.

Within start-up medical device companies that expect significant clinical trial program costs driven by a PMA regulatory pathway in the US, financing rounds are often coordinated with clinical trial milestones. It is not unusual for series B or C (early to mid-stage) financing to correspond to milestones for first-in-human or pilot studies. Series C to E (mid to late-stage) financing tends to correspond to milestones related to pivotal trials (see 6.1 Operating Plan and Financial Model and 6.3 Funding Approaches). The close interplay between financing and clinical trial milestones underscores the importance of staging clinical studies and gathering results to build sequential value to a company which can, in turn, justify subsequent rounds of funding. It is also indicative of the significant financial burden that is associated with a clinical program. To efficiently manage this sizable expense, as well as the other important issues related to clinical studies, careful strategic planning is essential.

Location of clinical studies

Researchers also must consider whether or not to conduct clinical trials within the US or outside the US, based on their objectives, particularly with regard to where device approval is ultimately desired. The most common reasons that studies outside the US are favored, particularly for first-in-human testing, are for increased patient access, quicker enrollment, and reduced cost. Locations such as Austalia, Central and Eastern Europe, and Central and Latin America have developed burgeoning eco-systems and a deep bench of qualified investigators to efficiently and cost-effectively support the clinical trials of medtech companies from around the world. Accordingly, a sizable percentage of clinical device testing has shifted overseas. For example, one literature review found that the number of countries serving as trial sites outside the United States more than doubled from 1995 to 2005, while the proportion of trials conducted in the United States and Western Europe decreased.[24]

Another reason that innovators have been encouraged to globalize their clinical strategies is because initiating a US trial has historically been more difficult and time-consuming than launching a study in another country. For instance, a survey of more than 200 medical device companies found that the average time required to obtain an IDE was nearly 14 months. Respondents reported that many delays were linked to disagreements with the FDA regarding the definition of primary efficacy endpoints (27 percent), the definition of primary safety endpoints (15 percent), and other factors such as the use of historical controls (8 percent), size of the trial (12 percent), statistical techniques (6 percent), and/or the need for randomization (5 percent). Once an IDE was obtained, it took these companies an average of 21 months to conduct a pivotal trial that was designed to satisfy FDA's requirements, with no assurance of the adequacy of the FDA-mandated study design.[25] Such delays are one factor that can add significant time and cost to conducting trials in the US. Another is the tendency in the US to demand larger and more extensive clinical trials than regulatory agencies in locations such as the European Union. As the author of one article described, "The European approach places more responsibility on physicians and their clinical judgment rather than on government officials who may have little appreciation of or experience with the exigencies of the clinical circumstance."[26]

That said, regulatory and study monitoring protocols for trials outside the US are gradually becoming more stringent, particularly as developing countries adopt international guidelines on the performance of clinical trials. For example, India recently passed a new policy meant to protect patients who participate in clinical research studies performed within the country. The intent of the new rules has been praised, but some observers have raised concerns about the burden the policy creates for companies and researchers.[27] As described in 4.2 Regulatory Basics, the long-term effects of the policy remain to be seen, although the expectations is that it will significantly reduce the number of trials conducted in India.[28] Nonetheless, depending on the disease state or condition being studied, patients can still be enrolled more quickly in overseas locations where the patient volume is simply higher.

When considering a clinical strategy that involves trials in overseas locations, it is important for companies that eventually intend to target the US market to understand that the FDA may not consider the patient population treated overseas to be equivalent to a population in the US. In order for these clinical data to be relevant for the US regulatory process, the patient population must be comparable (ethnically, as well as in terms of treatment regimens) to that of the US population. As a result, careful guidance should be pursued prior to the initiation of a trial to ensure that data collected outside the US will be accepted by the FDA if that is an assumption within the company's clinical strategy.

Clinical trials performed for FDA PMA regulatory approval are predominantly performed in the US. Companies may include globally dispersed sites in large pivotal studies to increase enrollment and gain footholds into important markets where they intend to sell the device. However, when the US is the ultimate target market, the majority of clinical sites should be based there. The FDA often looks favorably upon data from studies conducted outside the US as a precursor to an IDE to validate initial safety testing of the device. As a result, these studies can be a useful strategy in this regard for companies preparing to conduct pivotal trials in the US.

Working Example
An overview of good clinical practices for conducting clinical trials

When conducting a trial in the US under an IDE, the investigators, sponsors, IRBs, and the devices themselves are all subject to Good Clinical Practices (**GCP**) regulations. These guidelines are analogous to FDA's GLP guidelines for the execution of preclinical studies. They outline specific standards for the design, conduct, performance, monitoring, auditing, recording, analyses, and reporting of clinical trials to provide assurance that the data and reported results are credible and accurate, and that the rights, integrity, and confidentiality of trial subjects are protected. Resources for learning more about GCP regulations can be found in the Getting Started section.

Importantly, IDE-exempt trials are subject to their own regulations (outlined in FDA's exemption regulation), which are not quite as robust as GCP requirements. However, it is worth noting that even exempt trials should comply, at a minimum, with all GCP requirements regarding the protection of human subjects if their credibility is to be maintained.

A note on conflicts of interest in clinical trials

Conflicts of interest in the context of clinical trials refer to a situation where an individual or group has potentially competing interests in the outcome of the trial – for example, wanting results that benefit the patients in the trial but also seeking outcomes that benefit the company that has worked hard to develop the new technology. There are different kinds of competing incentives for different **stakeholders** in the clinical trial process. For instance, the lead physicians may be strongly motivated to have positive study results that can be published in an important medical journal. But, without question, the conflicts that receive the most scrutiny are those related to financial ties to the success of the product.

Financial conflicts of interest are essentially an unavoidable part of the testing of new medical technologies. Clinical trials of new technologies are so expensive that governmental health agencies like the NIH or the FDA can only sponsor a tiny fraction of the trials that need to be conducted. Companies are expected to take on the financial burden of the trials. In doing so, they automatically assume a conflict of interest in the process. As a result, sufficient regulatory checks and balances must be maintained to ensure that these conflicts do not significantly distort trial data.

It is important to understand that conflicts of interest occur at all organizational levels. Any individuals who participate in the trial could have a potential conflict if they stand to benefit financially (or by reputation, publications, etc.) from the outcomes. Their institutions also may have a conflict of interest. For example, a university may receive **royalties** on a new technology that one of its faculty members has invented. In the current climate, many universities would opt not to participate in the clinical testing of the device because of this financial conflict of interest. The national press has recently highlighted a number of examples of physicians and medical centers that have received questionable payments from pharmaceutical and device companies.

A special situation arises when physicians are involved in the invention and early development of a device and, as a result, play an integral role in the early studies. Through this involvement, the clinician/inventor obtains in-depth, first-hand knowledge of the device's performance and its failure modes. Given this experience, it may be most ethical from the standpoint of patient safety for these physicians to perform the first clinical studies. However, because these clinician/inventors often have leadership and/or equity positions in the company developing the device, careful steps must be taken to mitigate and manage conflicts of interest. This situation is recognized in the Association of American Medical Colleges (AAMC) guidelines on conflicts of interest, which allow the conflicted surgeon/clinician with substantial preclinical experience to perform the first-in-human studies, but recommend that these individuals not serve as principal investigators in any definitive, multi-center trials.[29] In addition to processes put in place by the various IRBs at the clinical facilities where the studies are conducted, the FDA's IDE process requires disclosure by any investigators with a significant equity or consulting stake in a company. Investigators are not barred from participating in studies involving their devices, but the nature of the relationship with the company must be disclosed to the FDA, the IRB, and the subjects in the study.

Tips for designing and managing successful clinical trials

A few final words of wisdom can be helpful for innovators and companies as they prepare to engage in clinical trials:[30]

- Collect data early on enrollment patterns from the different centers. This exercise provides a realistic view of the speed of the overall trial and can help to ensure that each clinical research site is screening all patients and is giving the study top priority.
- Statistics only summarize the data. Examine the real data directly to look for early indicators and issues.
- Beware of drawing inferences from small samples over large populations. Design the pivotal trial keeping in mind that results from the pilot might be overly optimistic. Review all data that have been published to date as a "sanity check" of what has been achieved in the field so far.
- Resist the temptation to develop an overly ambitious study design. Review the company strategy and plan a study that seeks to answer important questions reflected in the study endpoints.

Selecting endpoints for trials is a significant challenge, and clinical experts often have different opinions about what endpoints are most appropriate. However, the company and its clinical investigators and advisors must be aligned in their goals for studying a technology. The classic story about the company Devices for Vascular Intervention highlights how interactions between the company, scientific advisory board, and principal investigators can be challenging when designing a clinical trial. It also underscores the importance of thinking about each trial as one part of an overarching clinical strategy.

FROM THE FIELD ▶ DEVICES FOR VASCULAR INTERVENTION

The overall importance of clinical strategy and trial design on business success

In the mid-1980s, John Simpson pioneered the concept of directional coronary atherectomy (DCA). The DCA procedure used a device called an AtheroCath® to cut, capture, and remove plaque from the coronary arteries. The need for the device was created by the ongoing desire in the field for a less invasive, more cost-effective solution to removing obstructive atherosclerotic lesions than cardiac artery bypass grafting (CABG), as well as the perceived failure of percutaneous transluminal coronary angioplasty (PTCA) to yield a durable result for all patients.[31] Simpson believed that the clean removal of plaque through atherectomy would render a larger and less traumatized lumen and, therefore, less restenosis.[32] He founded Devices for Vascular Intervention (DVI) as a mechanism for developing and commercializing this new approach (see Figure 5.3.6).

Allan Will joined DVI in 1986 as vice president of marketing and sales and became the company's CEO in 1987. "At that point in time, most of our work was focused on what we would need in order to translate our success in peripheral vessels to the coronaries and get

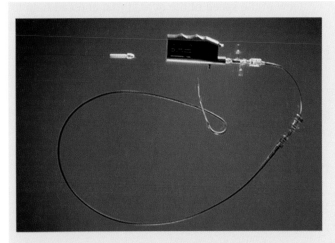

FIGURE 5.3.6
The DVI AtheroCath (courtesy of Allan Will).

the coronary product approved by the FDA," Will recalled. "We knew that eventually we would be involved in a large-scale clinical study." Between 1988 and the time of the company's advisory panel meeting in 1990, clinical testing of the AtheroCath in 873 patients was performed at twelve US hospitals. The value of DCA was assessed by means of a large patient registry that catalogued acute and late clinical outcomes and then compared them to a historical control. "Ultimately, our final PMA registry data included roughly 2,000 patients," Will said. "These data were compared against generally accepted angioplasty results. During this initial clinical study, we used the commonly accepted threshold for a significant stenosis – greater than 50 percent residual stenosis – as our success endpoint." Data from the study indicated that DCA achieved comparable or better results than those achieved by balloon angioplasty and pointed to a lower incidence of serious coronary dissection and abrupt occlusion after DCA.[33] DVI received FDA approval of its device in September 1990 based on the outcome of this study. AtheroCath use grew quickly after its US commercial launch. "Sales grew from approximately $10 million in the year of FDA approval, to $41 million, then $60 million, and finally $84 million in our third full year," Will noted.

In the meantime, however, DVI was acquired by Eli Lilly in 1989. "As part of that **acquisition**," explained Will, "Lilly wanted to include an **earn-out**, so that the value they ultimately paid would be based upon actual versus a speculated value. And they originally proposed that a very significant amount of the earn-out – I believe it was around $25 million – should be based on the results of a prospective, randomized clinical study demonstrating at least a 50 percent reduction in restenosis compared to PTCA. At the time, we didn't even know the full results of our approval study, let alone how the device would perform in a hypothetical, prospective, randomized clinical study." Moreover, these kinds of studies, while common in pharmaceuticals, were relatively new in the medical device industry. According to Will, "I believe this was the first seriously conducted prospective randomized clinical study of a medical device in the field of interventional cardiology."

Will and team "pushed back" on the proposal from Lilly's management. "There was just so much we needed to learn about how to best use this device before we launched a definitive trial," he said. DVI was ultimately able to negotiate the amount of the earn-out that was linked to the clinical study down to $5 million, but Lilly management insisted that the company conduct it.

In an effort to better understand the procedure and thereby produce the most positive results possible, the DVI team intentionally sought to delay the trial until late in the earn-out period. "We hoped to avoid conducting the study until we learned enough to know how best to use the device," said Will. "We needed answers to questions like how much plaque should you remove, do certain lesions respond differently to atherectomy than other lesions, and do certain vessels respond better than other vessels?" For this reason, DVI waited until 1991 to launch what would become known as the CAVEAT trial.

Will and his clinical staff also worked carefully to propose a study design. "We had become convinced that in order to get an optimal atherectomy result, you needed to remove enough plaque to achieve less than 20 percent residual stenosis. However, when we proposed that as a technical endpoint for a successful atherectomy and as a guideline for conducting the atherectomies in the study, a couple of very vocal investigators argued back that requiring clinicians to reach less than 20 percent residual stenosis would **bias** the study toward atherectomy. They told us that if we set that endpoint, it would be the equivalent of saying we're only going to measure the successful atherectomies against all angioplasties. Our argument was that the endpoint for a successful angioplasty was quite well known, because angioplasty procedures had been well proven and accepted for quite some time. All that we were doing was establishing a similar standard for a technique that wasn't yet as widely understood in terms of what the clinical endpoint was."

The need for a clearly defined endpoint was driven, in part, by the fact that DCA was a directional procedure, whereas angioplasty was concentric. When an angioplasty balloon was inflated, "it acted circumferentially on the vessel," explained Will. "With DCA, it was relatively easy to obtain a great angiographic result without performing a complete circumferential atherectomy." To adequately assess a DCA endpoint, the vessels had to be evaluated from multiple angiographic views to ensure completion of the procedure. "Otherwise, the physician might think he or she got a great result even though, when looking axially down the vessel, it had only been atherectomized at 12 o'clock and 6 o'clock, and the result had been assessed by looking at the 'wrong' cross-sectional view."

DVI argued its case aggressively, "But unfortunately, we had selected a principal investigator for the trial who was, if anything, going overboard to ensure that he wasn't favoring atherectomy or the corporation. So any hint in his mind of bias needed to be removed. In the end, that was one of the components to our undoing. We lost that argument, because we had effectively given up control over trial design to the investigator group," Will said. Instead of being able to define an endpoint of less than 20 percent residual stenosis for atherectomy, DVI had to adopt the clinically accepted endpoint at the time for significant stenosis, which was greater than 50 percent. "If you achieved less than 50 percent stenosis, you'd completed a successful procedure as measured by CAVEAT," Will recalled. "We were allowed to 'encourage' clinicians to achieve less than 20 percent, but we were

not allowed to require that as a technical endpoint for success."

DVI's troubles continued when its principal investigator initially refused to share the results of the trial with the company in advance of the 1993 American Heart Association meeting where the data were scheduled to be released. "The chairperson of our clinical trial was determined to maintain control over the clinical study until he released the data. We told him, 'We have an obligation to the patients and the clinicians who use our product to understand what the data is and answer questions. When a clinician comes up to us at our booth, what do we say?'" Ultimately, the DVI team secured the data and was able to quickly prepare for its release. But, said Will, "It wasn't pretty because we had not been able to fully analyze the data prior to its release by our PI."

To make matters worse, the results of the CAVEAT trial failed to show any significant improvement in early or late clinical and angiographic outcomes with DCA.[34] Interestingly, according to an editorial in the *Journal of American College of Cardiology*, "The surprise results of CAVEAT led some investigators to question the conduct of the trial. The CAVEAT was a multi-center study with significant variation in operator experience and skills. The failure of DCA to reduce restenosis could not be reconciled with experience at several of the premier DCA centers. A popular theory arose that DCA fared poorly in CAVEAT because the procedure was not performed optimally in all collaborating centers."[35] Two additional studies (BOAT and OARS) were subsequently completed at hospitals with a strong commitment to DCA as a procedure, as well as substantial experience with the technique. In these trials, aggressive atherectomy, usually combined with adjunct balloon angioplasty, led to significant reductions in restenosis rates when compared to balloon angioplasty alone. However, for many, these results were described as being "too little too late."[36]

Reflecting on the CAVEAT trial results, Will acknowledged that "it threw a damper on the business." However, he continued, "What had a much greater impact on the business was that stents were coming on the scene. The major shortcoming in stenting at that time was recognized to be acute thrombosis. Right around then, Dr. Antonio Colombo had discovered that acute thrombosis post-stenting was caused by an incomplete expansion of the stent. So clinicians began to think, 'Oh, now I can place a stent and not have the acute thrombosis problems that we've been experiencing, and the procedure takes me no longer than an angioplasty would. And maybe I get as good results as an atherectomy, but atherectomy's a lot harder work.' [The average atherectomy took roughly twice as long.] So, it's hard to assess the specific impact that CAVEAT had. Clearly it was negative, and it positioned us to be vulnerable to this finding that came out on stenting. If we had definitively proven in that study that we lowered restenosis . . . perhaps people would have required stenting to have prospective, randomized results before they abandoned atherectomy and ran to stenting. Instead, stenting began to take off even before the release of their prospective randomized clinical studies."

When asked what advice he would offer to entrepreneurs and companies planning clinical trials, Will had this to share: "First, look at trials as incremental building blocks. Don't consider any single trial as the be-all and end-all. Think carefully and strategically about what you need to show in the first study and what you should prove in subsequent studies. Recognize that you're going to learn how the device should be used or optimized and build that into your clinical trial strategy. These days, unfortunately, companies are required by the FDA to prove acceptability in a prospective, randomized study right out of the blocks, with their first large-scale clinical trial. Do what you need to do for the approval study, but don't overreach. Second, be aware that, in order to get paid for these devices, you need to build reimbursement endpoints into your approval study. Third, always opt for likelihood of success and speed of enrollment in your clinical study design. And finally, to as great an extent as possible, maintain control over the design and conduct of your clinical trial while not giving up objectivity. You need to be fair, you need to be honest, and you need to do good science, but it's still valuable to maintain control over the design and conduct of the clinical study."

Will also commented that, "A whole lot of ground has been plowed between when we began the CAVEAT study and today. There's now more than 15 years of experience in the device industry conducting prospective randomized clinical studies. As such, it's a much more accepted practice and there is a raft of principal investigators to choose from – good, bad, and in between. Carefully choosing the right principal investigator is of critical importance . . . one who believes in the potential of the technology, who will conduct a quality, expeditious trial, and who has no agenda other than to fairly and effectively evaluate the technology considering its stage of development." Will also encouraged entrepreneurs to work with experienced consultants in the field and the FDA to design studies that are objective and likely to lead to clearly defined, positive results.

A second story about a company called Emphasys Medical further underscores the challenges of gaining consensus around trial design and execution. In addition, it highlights the fact that the development of a clinical strategy for a new "white space" technology can be a major challenge. Although there are a number of business advantages of being the first to bring a new product forward in a clinical area, one of the disadvantages can be that no roadmap for clinical development exists.

When working in clinical fields where previous technologies have been designed and implemented, innovators have access to guidance concerning knowledge of the disease, the heterogeneity of the patient population, effective animal models, and endpoints for study design. By assessing the clinical strategies of other companies working in the space, teams can save tremendous time, effort, and expense. Where this guidance does not exist, defining a clinical strategy can be characterized by greater than average uncertainty and risk. Seeking input and collaboration from clinical experts, investigators, trialists, statisticians, and regulators can help. But the

demands of defining, executing, and funding a clinical strategy that addresses the needs of these many stakeholders can be difficult.

Navigating complex clinical and regulatory challenges on the path to market – part 1

As described in 2.2 Existing Solutions, Emphasys Medical was launched by medtech **incubator** The Foundry to provide better solutions for patients with advanced emphysema.[37] Emphysema is a form of chronic obstructive pulmonary disease (COPD) in which the alveoli, the tiny spherical air sacs in the lungs, are gradually destroyed. The damaged clusters of alveoli degrade into large, irregular pockets of non-functional lung tissue, reducing the surface area of the lungs available for oxygen/carbon dioxide exchange. In addition, air becomes trapped in these pockets, preventing complete exhalation and crowding the functioning portions of the lung. As a result, patients suffering from the disease must work harder and harder to take in adequate oxygen and feel perpetually short of breath.

Emphysema is irreversible, and there are few treatment options. The **standard of care** is medical management, including the use of oral and inhaled bronchodilators and supplemental oxygen to ease symptoms. For patients with severe emphysema, another option may be lung volume reduction surgery (LVRS), in which the most diseased portions of the lung are surgically removed so the remaining healthy tissue has more room to function. However, because the procedure is highly invasive (requiring the opening of the chest) and the eligible patient population is fragile, LVRS is considered high risk and has limited usage. The only other surgical option is lung transplantation, a last resort that is constrained by a critical shortage of donor organs.

The Emphasys Medical technology was designed to ease breathing in the same way as LVRS but without

surgery. The concept was that tiny one-way valves, placed in the major airways leading to the most diseased portions of the lungs, would prevent air from entering the diseased segments but allow air to escape. The team hypothesized that this would cause the diseased portions of the lung to collapse (called atelectasis), creating more space for the healthier parts of the lung to function and making it easier for the patient to breathe. Emphasys demonstrated the technical feasibility of the approach by placing one-way valves into the bronchial passages of sheep. The valves created immediate and total collapse of the treated lobes.

Encouraged by these results, the company developed a novel one-way valve that could be placed through the airways over a guidewire (the same way a stent is placed within a blood vessel – see Figure 5.3.7). The procedure was non-surgical and completely reversible. Led by CEO John McCutcheon, the team completed early **bench testing** then moved into chronic animal studies, using sheep to study the delivery of the device, deployment of the valve, mechanical success, biocompatibility, and safety.

Based on the positive results of its animal studies, Emphasys filed an application with the FDA for 510(k) clearance of the device. "We used a matrix argument with tracheobronchial stents and some surgical products to establish substantial equivalence," said McCutcheon. "Tracheobronchial stents, one of the few pulmonary devices on the market, were 510(k) cleared, so we thought that this might be a viable approach." The company's submission was reviewed by the Plastics and Reconstructive Surgery branch of the FDA's Center for Devices and Radiological Health (CDRH) because that was where tracheobronchial stents had been assigned. "We expected that the FDA might come back and

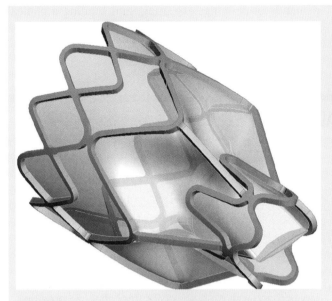

FIGURE 5.3.7
A representation of the Emphasys Endobronchial Valve (courtesy of Emphasys).

require a clinical trial to back up the submission, which we would have done anyway prior to commercialization," said The Foundry partner Hank Plain. "But, they said, 'No, this is definitely a PMA.'"

Knowing that clinical data would be paramount to its success in navigating the PMA pathway, Emphasys considered its next move. Because there was no animal model that replicated human lung anatomy or human emphysema, the team could not measure efficacy through additional preclinical tests. "You can only learn so much on animals," said McCutcheon. "The real learning starts when you get to human clinicals." The team decided to transition into human pilot tests in order to validate the feasibility of the procedure and, more specifically, the safety and effectiveness of the device in humans. Importantly, Emphasys also hoped to learn from the pilot studies which patients were most likely to benefit from the intervention. Emphysema is a complex disease state in that the areas of destroyed tissue can be localized to certain lobes[38] or parts of the lung (heterogeneous), or spread more widely and evenly throughout the lungs (homogeneous). Additionally, the company would study which treatment approach was most effective since the valves could be placed

unilaterally (on only one side of the lungs) or bilaterally (both sides), and could include one or more complete lung lobes and/or partial lung lobes. By testing different patient characteristics and treatment approaches during the pilot studies, Emphasys intended to gather enough information to guide the design of the pivotal trial, which the FDA would review as part of the company's PMA submission.

The first ten patients treated with the Emphasys device were in Melbourne, Australia. All had severe heterogeneous emphysema and received bilateral placement of the valves, which, for the purposes of the study, meant that Emphasys treated the right upper lobe without treating the right middle lobe, and treated part of the left upper lobe without treating the lingula. However, none experienced the desired outcome, atelectasis. "We quickly learned that 70-year-old patients with emphysema are very different than juvenile sheep. Their lungs are a mess, and they're much weaker overall. So we didn't see the same kind of dramatic results," said Hanson Gifford, another partner at The Foundry.

Additional pilot studies proved challenging on a number of other fronts. Enrollment was slower than anticipated, as many pulmonologists used to medically managing their patients were reluctant to try an interventional procedure. This unexpected hurdle led to a broad geographic dispersion of trial sites, as well as a wide range of patients in terms of the characteristics of their disease. As a further complication, the team discovered that the valves were leaking in a way that had not been observed in the sheep. Emphasys engineers modified the design and continued gathering data.

Following the introduction of the modified device, the team began to see some encouraging clinical success. However, as more data from the pilots became available, the company observed significant variability in success rates among the sites without any clear reason for the difference. Reflecting more than the redesign of the valve, the variability clearly indicated either a difference in patient selection or technique between sites. The pilot studies also led to the discovery of a problem that had not been foreseen by Emphasys' clinical experts. In trying to understand why a lobe with complete valve occlusion of all airways sometimes failed to collapse, the Emphasys investigators performed a series of sophisticated tests and determined that the lobes treated with the valves sometimes received sufficient collateral airflow from adjacent lobes to keep that lobe inflated. The team was aware of collateral flow between segments of the lungs (subdivisions within the lobes), but the phenomenon of interlobar airflow had not occurred in sheep, was not anticipated by the pulmonologist advisors, and was not documented in the clinical literature.

In total, investigators treated nearly 100 patients across the pilot sites. In carefully analyzing the resulting data, the statisticians determined that a unilateral procedure worked better than a bilateral one. This finding surprised the group, since a recent national study of LVRS had found that patients did better when the surgery was performed bilaterally. Ultimately, Emphasys determined that a subset of 38 patients who had been treated with unilateral lobe exclusion experienced the greatest volume reduction of the treated lobe. The effect of this could be quantified through two lung function tests, one that measured the amount of air the patient could exhale with force in a single breath in one second (forced air expiratory volume or FEV_1) and a measure of exercise tolerance called the six minute walk test (6MWT).

Despite the fact that the company still lacked clearly defined anatomic/disease markers for successful lobe collapse and a standardized selection algorithm, "Our investigators felt we had enough evidence of success and had shown the therapy was safe and well-tolerated by patients," said McCutcheon. "All we wanted was to get into our pivotal and prove this." Gifford added that, "At this point, we really did feel like we had a device that worked and enough predictability about which patients would benefit that we could design the pivotal trial." Most importantly, as Plain pointed out, "Until you do the randomized trial you don't get definitive data." Finally, external pressures came into play for the young start-up. With a burn rate of close to $1 million per month and a

number of prospective competitors exploring the space, Emphasys felt compelled to move forward.

The first step in initiating a pivotal study was for Emphasys to apply for an IDE with the FDA. According to McCutcheon, "We were the first company to submit an IDE application in this particular space and the Plastics and Reconstructive Surgery branch didn't have any expertise to evaluate our proposed protocol. Consequently, they brought in two medical reviewers from outside their branch. One was a pulmonologist from the drug division and the other was a thoracic surgeon who consulted for the FDA as a clinical reviewer." From the company's perspective, this created some difficulties. The consultant thought Emphasys should randomize its non-invasive treatment versus LVRS. However, recently published clinical data showed that LVRS had an unacceptably high death rate in high-risk patients. As a result of the article, which appeared in the *New England Journal of Medicine*,[39] LVRS fell out of favor "virtually overnight." "So this was a non-starter for us since it would be impossible to enroll anyone in our trial with LVRS as the control," remembered McCutcheon.

While these discussions were taking place, the FDA convened a panel to help establish the appropriate trial design parameters and clinical endpoints for medical device treatments of emphysema. McCutcheon reiterated, "They'd never seen a device like ours before. Plus, they knew that Spiration, Pulmonx, and Broncus would be coming right behind us [with competitive devices]. So they decided to convene the panel of experts to specify how to design trials for devices like ours: the target patient population, endpoints, control group, and length of follow-up."

In the end, the FDA panel made a series of recommendations regarding the trial design for devices intended to treat emphysema (see Figure 5.3.8). "We were thrilled that the panel advocated the use of medical management rather than LVRS as the control group," McCutcheon commented. Emphasys made minor modifications to its trial design to comply with the other

#	Recommendations
1	The trials should include only patients who are candidates for no other procedures or those who have refused other treatments. All patients should have received optimized medical treatment for 3 to 6 months before enrollment. Lung volume reduction surgical patients are not the appropriate control group, and comparisons should be made to patients receiving optimized medical treatment in multi-centered studies.
2	Safety analyses should include an assessment of deaths, bleeding, mechanical ventilation, pneumonia, air leaks hospital days, re-operations, respiratory failure, decreases in FEV_1, 3, 6, 9, and 12 month assessments of device positioning, ease of device removal, COPD exacerbations, intubations, bleeding, and a tabulation of patients who were discontinued due to a lack of benefit.
3	Effectiveness determinations should include exercise capacity, 6 minute walk test, St. George's Quality of Life Assessment, spirometry (FEV_1 increase), decrease in oxygen consumption, and increase in length of life.
4	The duration of follow-up should continue for at least 6 months for effectiveness and at least 1 year for safety.

FIGURE 5.3.8
FDA panel recommendations regarding clinical trial design for devices intended to treat emphysema (US Food and Drug Administration).

panel recommendations, and it received IDE approval to proceed with its pivotal study. "However, the surgeon consultant who had advocated using LVRS as the control group never agreed with the panel's decision. This created problems for us later in the review process," McCutcheon said.

Following the FDA guidelines, there would be two primary efficacy endpoints of the Emphasys study: (1) improvement in lung function measured by FEV_1 (forced expiratory volume: the amount of air that can be exhaled in one second), and (2) exercise tolerance measured by 6MWT (6 minute walk test: how far a patient can walk on flat surface in six minutes). There was also a primary safety endpoint, which consisted of a major complications composite – that is, a combination of unintended consequences, including respiratory failure, pneumonia or severe bleeding associated with the valves. Patients in the treatment group would receive valve treatment of one targeted lobe to achieve complete lobar exclusion followed by optimal medical management. "Importantly, in retrospect, when the right upper lobe was the target for treatment, the right middle lobe was not treated along with it. Conversely, when the left upper lobe was treated, the treatment included the lingula," noted McCutcheon. Patients in the control **arm** would receive optimal medical management. The

individuals in both groups would be given six weeks of pulmonary rehabilitation prior to enrollment to ensure that any post-treatment improvement was not due to the post-procedure rehabilitation regimen. According to the study protocol, the trial was powered to detect a 15 percent improvement in the treatment arm in FEV_1 and a 17 percent improvement in the 6MWT, plus or minus a 33.7 and 41.5 percent standard deviation respectively, based on the pilot results. The Emphasys team recruited centers to participate in both the US and Europe.

Because of site-to-site variability observed in the pilot studies, Emphasys went beyond what was required by the FDA to ensure uniform patient selection and targeting, employing a core imaging lab at UCLA to detect and measure important information such as the completeness of the fissures that divide the lungs into lobes. By analyzing this information, the core lab would look for patterns that would help the team more accurately identify which patients to target. While the use of core labs was established and highly regarded in interventional cardiology studies, Emphasys was the first company to use one for a pulmonary study.

Meanwhile, Emphasys continued to work on product development. The original over-the-wire approach to implanting the valves had been difficult for pulmonologists, who were more accustomed to using a bronchoscope to steer catheters through the lungs. For this reason, the company developed a new variation of its device that could be delivered via the working channel of the bronchoscope. Partway into the pivotal trial, the new system was ready for use. The FDA agreed to the change in the trial protocol but required that Emphasys restart the trial and exclude the 62 patients who had already been treated.

At the end of the pivotal study, 31 trial sites had treated 321 patients with the Emphasys device. However, when the company received the data, some work was required to fully understand the results. "We met the endpoints, but the data were noisy," recalled Greg Bakan, who joined the company as vice president of sales and marketing. While some degree of "noise" was common in all trials, especially one as complex as this, the challenge was in determining its effect on the fundamental outcomes. One of the reasons for this was that emphysema as a disease waxes and wanes, with patients feeling better or worse day-to-day, which could skew their performance on measures such as FEV_1 and 6MWT. As with most studies of this magnitude, there was also missing follow-up information for some patients. The follow-up protocol required the collection of literally thousands of data points for each patient, along with extensive lab work, lung function testing, and X-rays. This proved to be a huge burden for many of the chronically ill patients in the study, as even the lung function test maneuvers could provoke an exacerbation of COPD symptoms. "Ultimately, however, our statisticians determined that we had successfully met all of our primary endpoints, both safety and efficacy," said McCutcheon. "Based on that finding, we filed our PMA and were scheduled for panel review in June of 2010." However, in May of 2010, a reviewer contacted the company to notify it that the surgeon consultant on the panel did not believe that Emphasys had met its endpoints. "The message was that we could go ahead to the panel meeting, but that the FDA would tell the panel

that we had not met our endpoints, and that we would not be allowed to state that we had," said McCutcheon. "We were told that if we voluntarily withdrew from the panel, the FDA would work with us to resolve these issues."

To better understand the data, and these issues, Emphasys subjected the trial results to a rigorous analysis. According to Gifford, "In those patients for whom we didn't have follow-up data, we needed to impute the results. We had all kinds of sophisticated statisticians look at the data and tell us exactly how we should do the imputations and so on. We spent several months doing that, and no matter how we did it, it came out that our data was statistically significant, showing roughly a 6 percent difference between treated and control patients." Finally, the lead FDA statistician agreed that Emphasys had met its endpoints. "We could now resubmit the PMA in time to make the panel meeting in December of 2010. However, the FDA statistician warned us that the surgical consultant on our panel did not believe that the treatment was clinically significant," said McCutcheon. More than four years (and approximately $75 million) after initiating the pivotal trial, the Emphasys team prepared a clinical report and made its PMA submission to the FDA. "We knew the data was not without its flaws, but we firmly believed it was good enough to get approval," said Mike Carusi, an investor with Advanced Technology Ventures (ATV) and a member of the Emphasys board of directors.

In reflecting on the numerous challenges the team faced in its clinical trials, Plain pointed out that many of these issues are common when a company is pioneering a treatment in a relatively untapped disease area. "It's not like working in interventional cardiology where there's the history of angioplasty that informed stenting and then the history of bare metal stents that informed drug-eluting stents. This was new ground, and frankly, we had to overcome a lot of unknowns," he said. Added McCutcheon, "There were so many hidden variables that not even the experts knew what to expect. So we had to constantly challenge ourselves and what we thought we knew. But our team was proud to have developed a technology that provided a clinical benefit to a very desperate patient group. Despite all of the ups and downs and struggles with the FDA, we could see the light at the end of the tunnel with the pending panel meeting." Read more about Emphasys Medical's regulatory experience in 5.4 Regulatory Strategy.

↘ Online Resources

Visit www.ebiodesign.org/5.3 for more content, including:

 Activities and links for "Getting Started"
- Determine the purpose of the clinical strategy
- Determine the overall study strategy
- Identify clinical research specialist(s) to work/consult with the internal team
- Choose a trial design/model
- Determine trial endpoints
- Write research protocol
- Decide where to conduct the trail(s)
- Determine resources required to implement protocol
- Understand and implement GCP

 Videos on clinical strategy

 Appendices that provide:
- A comparison of medtech pilot, pivotal, and post-marketing studies to trials in the pharmaceutical industry
- Further reading on the null hypothesis, type I/type II error, P-values, and sample sizes
- Information about when investigational devices can be used outside an IDE-approved clinical trial

CREDITS

The editors would like to acknowledge Eric Green, John McCutcheon, Santiago Ocojo, and Allan Will for sharing the case examples. Many thanks also go to Trena Depel and Stacey McCutcheon for their assistance with the chapter.

NOTES

1 Carol Rados, "Inside Clinical Trials: Testing Medical Products in People," Vidyaa, http://www.vidyya.com/vol5/v5i262_1.htm (March 31, 2014).

2 From remarks made by Frank Litvack as part of the "From the Innovator's Workbench" speaker series hosted by Stanford's Program in Biodesign, March 5, 2007, http://biodesign.stanford.edu/bdn/networking/pastinnovators.jsp. Reprinted with permission.

3 All quotations are from interviews conducted by the authors, unless otherwise stated. Reprinted with permission.

4 "Creating a Learning Healthcare System: Defining a New Approach to Clinical Research," PCORnet, December 30, 2013, http://www.pcornet.org/2013/12/558/ (March 31, 2014).

5 Michael S. Lauer and Ralph B. D'Agostino, "The Randomized Registry Trial – The Next Disruptive Technology in Clinical Research?," New England Journal of Medicine, October 24, 2013, http://www.nejm.org/doi/full/10.1056/NEJMp1310102 (March 31, 2014).

6 Ibid.

7 Ole Fröbert et al., "Thrombus Aspiration During ST-Segment Elevation Myocardial Infarction," New Englance Journal of Medicine, October 24, 2013, http://www.nejm.org/doi/full/10.1056/NEJMoa1308789 (March 31, 2014).

8 Lauer and D'Agostino, op. cit.

9 "Creating a Learning Healthcare System: Defining a New Approach to Clinical Research," op. cit.

10 "Device Advice: Investigational Device Exemption," U.S. Food and Drug Administration, http://www.fda.gov/medicaldevices/deviceregulationandguidance/howtomarketyourdevice/investigationaldeviceexemptionide/default.htm (March 31, 2014).

11 "IDE Application," U.S. Food and Drug Administration, http://www.fda.gov/medicaldevices/deviceregulationandguidance/howtomarketyourdevice/investigationaldeviceexemptionide/ucm046706.htm (March 31, 2014).

12 One of the most common reasons that a trial is exempt from IDE requirements is if the device being studied has been cleared for commercial use under a 510(k), but the company wishes to

conduct trials to support reimbursement, market adoption, or consumer preference testing.

13 "IDE Expanded/Early Access," U.S. Food and Drug Administration, http://www.fda.gov/MedicalDevices/DeviceRegulationand Guidance/HowtoMarketYourDevice/InvestigationalDevice ExemptionIDE/ucm051345.htm (May 31, 2014).

14 Nicholas Wade, "Patient Dies During a Trial of Therapy Using Genes," *The New York Times*, September 29, 1999, http://query.nytimes.com/gst/fullpage.html?res = 9E06EED8173EF93AA1575AC0A96F958260 (March 31, 2014).

15 Larry Thompson, "Human Gene Therapy: Harsh Lessons, High Hopes," *FDA Consumer Magazine*, September/October 2000, http://archive.is/CeHVZ (March 31, 2014).

16 "A New Broom? From OPRR to OHRP: Transforming Human Research Protections," *Modern Drug Discovery*, November/December 2000, http://pubs.acs.org/subscribe/journals/mdd/v03/i09/html/Clinical1.html (March 31, 2014).

17 "OHRP Fact Sheet," U.S. Department of Health and Human Services, http://www.hhs.gov/ohrp/about/facts/ (March 31, 2014).

18 Nancy Stark, "The Clinical Research Industry: New Options for Medical Device Manufacturers," Medical Device Link, January 1997, http://www.devicelink.com/mddi/archive/97/01/029.html (January 28, 2008).

19 All cartoons drawn by Josh Makower, unless otherwise cited.

20 Stark, op. cit.

21 "Clinical Trials of Medical Products and Medical Devices: What You Need to Know," Spine Health.com, http://www.spine-health.com/research/what/trials01.html (March 31, 2014).

22 Rados, op. cit.

23 "Protection of Human Subjects," U.S. Food and Drug Administration, http://www.accessdata.fda.gov/scripts/cdrh/cfdocs/cfcfr/CFRSearch.cfm?CFRPart = 50&showFR = 1 (March 31, 2014).

24 Glickman et al., "Ethical and Scientific Implications of the Globalization of Clinical Research," *New England Journal of Medicine*, February 19, 2009, http://www.nejm.org/doi/full/10.1056/NEJMsb0803929 (April 1, 2014).

25 Josh Makower, Aabed Meer, and Lyn Denend, "FDA Impact on U.S. Medical Technology Innovation," November 2010, http://www.nvca.org/index.php?option = com_docman&task = doc_download&id = 668&ltemid = 93 (April 1, 2014).

26 Paul Citron, "Medical Devices: Lost in Regulation," *Issues in Science and Technology*, 2011, http://www.issues.org/27.3/p_citron.html (April 1, 2014).

27 Jeremy Sugarman, Harvey M. Meyerhoff, Anant Bhan, Robert Bollinger, and Amita Gupta, "India's New Policy to Protect Research Participants," *British Medical Journal*, July 2013, http://www.bmj.com/content/347/bmj.f4841 (January 2, 2014).

28 S. Seethalakshmi, "Foreign Companies Stop Clinical Trials in India After Government Amends Rules on Compensation," *The Times of India*, August 1, 2013, http://articles.timesofindia.indiatimes.com/2013-08-01/bangalore/40960487_1_clinical-trials-iscr-suneela-thatte (January 2, 2014).

29 "Protecting Subjects, Preserving Trust, Promoting Progress II," Association of American Medical Colleges, Task Force on Financial Conflicts of Interest in Performing Clinical Research, October 2002, http://ccnmtl.columbia.edu/projects/rcr/rcr_conflicts/misc/Ref/AAMC_2002CoIReport.pdf (March 31, 2014).

30 Adapted from Harvey Motulsky, *Intuitive Biostatistics* (Oxford University Press, 1995).

31 Stephen N. Oesterle, "Coronary Interventions at a Crossroads: The Bifurcation Stenosis," *Journal of the American College of Cardiology*, December 1998: 1853.

32 Ibid.

33 David O. Williams and Mary C. Fahrenbach, "Directional Coronary Atherectomy: But Wait, There's More," *Circulation* (1998): 309.

34 Charles A. Simonton et al., "'Optimal' Directional Coronary Atherectomy," *Circulation* (1998): 332.

35 Oesterle, op. cit.

36 Ibid.

37 This story is based on a case study by Stefanos Zenios, Paul Yock, Lyn Denend, Marianna Samson, and Cynthia Yock, "Emphasys Medical: Navigating Complex Clinical and Regulatory Challenges on the Path to Market," Stanford Graduate School of Business, 2010, https://gsbapps.stanford.edu/cases/detail1.asp?Document_ID = 3331 (January 29, 2014).

38 Human lungs are separated into lobes by fissures. The right lung has three main lobes (upper, middle, lower); the left has two lobes (upper and lower), as well as a small appendage to the upper lobe called the lingula.

39 National Emphysema Treatment Trial Research Group, "Patients at a High Risk of Death After Lung-Volume-Reduction Surgery," *New England Journal of Medicine*, October 11, 2001, http://www.nejm.org/doi/full/10.1056/NEJMoa11798 (January 25, 2014).

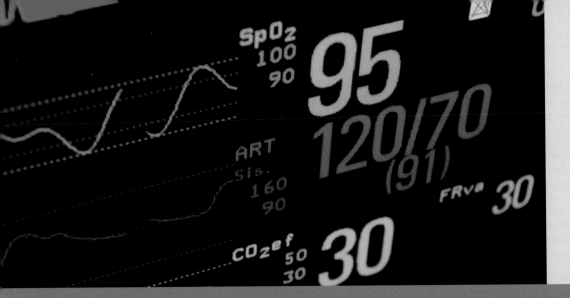

5.4 Regulatory Strategy

INTRODUCTION

The team spent 28 months and more than $45 million on the clinical trial. The PMA submission totaled over 10,000 pages and required three detailed responses to questions from the FDA. The panel meeting was a rollercoaster, with several experts voicing significant initial concern about the safety profile of the device. Now the official letter has finally arrived from CDRH and the team is savoring one short sentence: "You may begin commercial distribution of the device."

Developing an effective, strategic approach to regulation is of critical importance in the biodesign innovation process, not only because this is the gateway to clinical use of the product, but because of the considerable time, cost, and effort associated with this work stream. A sound and thoughtful strategic regulatory plan is tightly coupled with the competitive positioning of a new technology, and it informs the sales and marketing approach, clinical strategy, quality processes, and risk management policies that the innovator puts into place. The regulatory strategy establishes the foundation and sets the constraints within which these interrelated issues must be managed.

While 4.2 Regulatory Basics provides foundational information about the requirements and tactical implications of current regulatory requirements, this chapter addresses the strategic aspects of regulation as part of the broader biodesign innovation process.

 See ebiodesign.org for featured videos on regulatory strategy.

OBJECTIVES

- Understand the strategic risks and opportunities associated with the PMA and 510(k) pathways in the US.

- Explore strategic and tactical issues in dealing with the FDA.

- Consider alternative regulatory approaches outside the US, as well as ways of integrating these strategies with FDA approval.

- Recognize common regulatory mistakes and learn how to avoid them through the creation and implementation of a strong regulatory strategy.

REGULATORY STRATEGY FUNDAMENTALS

When innovators set out to develop a regulatory strategy that includes the US as a target market, there are a few key considerations that should guide this effort. First, although it is possible to help inform the thinking of the Food and Drug Administration (**FDA**) about the regulatory pathway for a **medtech** innovation, ultimately the decisions will be made by the center, division, and branch based on the experience and orientation of the specific group charged with reviewing the submission.

FIGURE 5.4.1

Approaching the FDA can be daunting, especially for first-time innovators. A well thought-out strategy, the right consultant, and effective communication are key elements to success.[1]

As described in 4.2 Regulatory Basics, the overall culture of the FDA is strongly rooted in its primary mission of protecting public health. How this mission is interpreted when making a decision on a particular submission depends on the context of current events (and can be heavily influenced by recent problems with devices or drugs). In general, regulatory processes for devices are becoming more rigorous and the level of evidence required is moving closer to that for drugs (though the size, complexity, duration, and expense of device trials still tends to be lower). There is significant risk to innovators and companies working within this constantly changing context (see Figure 5.4.1) – but this risk can by mitigated, at least in part, by understanding the dynamics in the regulatory environment. At the same time, the FDA is showing an increased willingness to accelerate the regulatory process for truly innovative technologies – and a willingness to work with companies to explore ways of reducing the time and expense in getting new products to patients in general.

A second important "big picture" issue is that FDA clearance or approval, although an important milestone for the company, by itself is essentially worthless if there is not a way for the company to get paid for the new technology. As described more fully in 5.6 Reimbursement Strategy, the Centers for Medicare and Medicaid Services (**CMS**) is responsible for determining whether or not a device will be reimbursed by the US government. This, in turn, can influence the practice of hundreds of private insurers. Therefore, unless the device is intended to be paid for by individual consumers, developing a **reimbursement** strategy must proceed in concert with crafting a regulatory strategy.

Is the device regulated?

One important question to clarify at the start is whether or not the innovation should be regulated as a medical device. In some cases, it is possible to develop claims for the uses of a product that take it outside of an FDA pathway. One example is exercise equipment, where the boundaries regarding whether or not these are medical devices are not completely clear. If a product can be developed and sold as a consumer product and not a medical device it can be advantageous to take a route that bypasses FDA clearance or approval. However, innovators must be careful about how they promote the product so as to not place it under FDA's jurisdiction. For instance, an exercise treadmill could be advertised as a way of increasing the pulse rate, but not as a technology that reduces the incidence of heart disease. In some cases, it may be feasible to market a product directly for consumer use while developing a related product that will be regarded by the FDA as a medical device. An innovator can petition the FDA to obtain a formal determination of whether or not a product is a medical device under the FDA definition (this is known as a 513(g) petition). If, in response to a 513(g) petition, the FDA determines that the product is a device, the agency will generally provide information on the device classification and applicable regulatory requirements.

The tactical question of whether or not to engage the FDA comes up frequently in the case of mobile wellness and other **mHealth** technologies (for example, monitoring devices for ambulation and heart rate). Similar to the treadmill example, there can be a gray zone regarding the degree to which medical claims are being asserted for the technology. The final guidance for mobile medical applications, published in 2013, helps clarify for innovators whether or not their technologies will face oversight from the agency.[2] At its core, the guidance states that if a mobile technology or application is intended for use in performing a medical function (i.e., for diagnosis of disease or other conditions, or the cure, mitigation, treatment, or prevention of disease) using patient-specific information, it is a medical device and will be subject to regulation.[3] Given the rapidly growing number of mobile apps being developed, the FDA will rely on "enforcement discretion" to pursue companies that overextend their marketing claims (see 4.2 Regulatory Basics for more information on the regulation of mobile medical apps).

Innovators and companies are encouraged to think carefully about whether or not to pursue regulation and seek professional advice about an appropriate course of action. Examples like the action against genetic testing start-up 23andMe increasingly underscore the potential risks of deciding to bypass regulation. In late 2013, the FDA issued a warning letter to the company expressing concern that 23andMe did not have adequate scientific evidence demonstrating that its health-related genetic tests provide accurate results. Moreover, they raised the issue that the unsubstantiated genetic information provided by the Personal Geromic Survey (PGS) could cause consumers to make unnecessary or potentially dangerous health-related decisions, consistent with the 23andMe marketing claims that many of its test results offer a "first step in prevention" that enables **users** to "take steps toward mitigating serious diseases."[4] As one article explained, "By definition, these claims classify the PGS as a medical device under the federal Food, Drug & Cosmetic Act, and as such it is subject to FDA regulation including the requirement for FDA approval or clearance prior to marketing. However, 23andMe has never been granted such approval."[5] In response, the company agreed to stop marketing its health-related genetic tests in order to pursue **510(k)** clearance.

Which center and division/branch to target?

If it is determined that a product is a device and will be regulated by the FDA, innovators must make an initial determination of which class of product – device, drug, or biologic – is most appropriate, as well as which center – the Center for Devices & Radiological Health (**CDRH**), Center for Drug Evaluation & Research (**CDER**), or Center for Biologics Evaluation & Research (**CBER**) – should evaluate the technology. Most often for medical devices, this is a straightforward decision (see chapter 4.2 for FDA's definition of devices) and CDRH is the appropriate center to perform the evaluation. One practical point is that CDRH tends to be somewhat easier for innovators to deal with than the other centers and, if clinical data are required to demonstrate safety and effectiveness, the requirements and number of **subjects** can be substantially less. If a technology is intermediate with respect to classification, innovators are often advised to push for regulation by CDRH.

Increasingly, devices are being integrated with drugs or biologics and this has led to the creation of the Office of Combination Products (**OCP**). OCP's charter is to make decisions about whether or not a new technology constitutes a combination product and, if so, which centers regulate it (and which one takes the lead). The most well-known example of a combination product is the drug-eluting stent, in which a basic metal lattice device used to hold open an artery was modified with a coating or surface that releases a drug to inhibit the rest-enosis (tissue re-growth) that occurred with bare metal stents. In this case, the OCP determined that the mechanical effect of opening the artery was the primary mechanism of action and that the drug effect on restenosis was secondary to the mechanical aspect in achieving its intended use. As a result, CDRH took the primary regulatory role, with substantial input from CDER.

All things being equal, the approval pathway for a combination product will be clearer (and the review will be more likely to be based at CDRH) if the drug or biologic that is added to the device is already cleared through its respective center (and the sponsor can argue that it is secondary to the device aspect of the product). However, because combination product submissions are so complex, strategies to address them must be developed on a case-by-case basis. Even if the combined product is evaluated by CDRH, it will likely have a consultation review by CDER or CBER and, in any event, will need to meet requirements for the drug or biologic.

Once a technology is directed to CDRH, there may be latitude in some cases regarding which division or branch reviews the submission. The various divisions and branches tend to have somewhat different standards of evidence (type and length of trial, number of subjects, etc.), based on their traditions and experience. For example, the first company to seek approval for vessel anastomotic devices (devices to attach blood vessels together without requiring a surgeon to suture by hand) was reviewed by the General Surgery Group. Once the devices reached the marketplace, there were some unanticipated cardiovascular complications. As a result, the market leader – along with a number of other companies that had begun the regulatory process in the interim – were shifted to the Cardiovascular Division, with new sets of standards that were much more difficult to meet. In practice, it is unusual for the innovator or company to have much latitude in choosing a division, but it is worth understanding the implications of dealing with one division versus another. Expert regulatory consultants track these dynamics closely and can be extremely valuable as advisors in navigating this process.

Strategies related to IDEs

As described in 4.2 Regulatory Basics, an investigational device exemption (**IDE**) is required before any trials are initiated with US patients. For a device or study with **significant risk**, the innovator must submit an IDE to the FDA prior to initiating the clinical investigation. The FDA will review the protocol and other accompanying information and must grant IDE approval in order for the study to commence. For a device or study that the company considers to pose a non-significant risk to patients, the company is not required to submit an IDE to the FDA as long as the hospital institutional review boards (**IRBs**) approve the study protocol and informed consent documents. In this case, the study can proceed (for example, to generate data for a 510(k) submission). It is important to note, however, that the FDA must be notified if any IRB rejects the study because it poses a significant risk, even if several others have approved it.

It is also worth emphasizing that pursing a non-significant risk study with the expectation of using the data to support a marketing submission without prior discussion with the FDA can be problematic. In this situation, the innovators have no direct perspective on what the FDA will consider to be suitable data to support the ultimate claims, once the study is presented to the agency. For this reason, it is generally advisable to request a pre-submission meeting for an IDE with the reviewing branch for either a non-significant or a significant risk device. The purpose of the meeting is to ensure that the company and FDA are in general agreement prior to executing a clinical study. These meetings are also useful for companies intending to launch **pilot** human studies outside the US. In these meetings, the company, often accompanied by the lead **clinical investigator**(s), presents data to a team of reviewers at CDRH. The company will explain or demonstrate the device and outline the preclinical and clinical development plans. In the pre-submission package, the company also must pose specific questions to the FDA. The agency group members review existing **bench** and animal data, and they make informal, non-binding suggestions regarding the need for additional preclinical data, as well as comment on the suitability of the study design to support the proposed intended use and product claims. The FDA will generally have a statistician in the group, so it is advisable to provide a detailed statistical analysis plan for the study (and bring a statistician along from the company side to answer questions and explain the rationale for the statistical methods chosen). Depending on the complexity of the device and study, it may be useful to request two meetings to occur at different stages in the planning process.

Each meeting will generally last an hour and typically requires scheduling at least four to six weeks in advance. The FDA has specific recommendations for the content of the pre-submission meeting package available on its website.[6] It is critically important that the team making the presentation outlines a study plan with sufficient detail in order to solicit meaningful agency feedback on its proposal and to address the specific questions posed. An open-ended approach to the FDA, asking for guidance in helping to design a study, is not a good idea and will likely result in a more complicated and expensive study than the company wants to undertake (and often can hurt the company's credibility with the agency).

Keep in mind that the FDA is not obliged to "approve" any aspects of the study design at this meeting and has the right to change its advice as the study matures or as new information becomes available. Nonetheless, it is extremely important for the company to formulate questions for the agency that will provide as specific and useful guidance as possible. For example, a company might ask, "Does the FDA agree that 100 patients is a sufficient study cohort for this trial to demonstrate a 20 percent improvement in the clinical endpoint specified?" or "Are nine months of follow-up data sufficient to adequately demonstrate safety and effectiveness or durability of effect?" A designated member of the company team should take detailed notes of the meeting and send a summary to the group leader from the FDA. There is no requirement for the FDA to record notes, although some agency teams will keep minutes. However, the FDA will send responses to the specific questions submitted as part of the pre-submission package.

510(k) versus PMA

The appropriate regulatory pathway within CDRH – 510(k) versus premarket approval (**PMA**) – is generally determined by the risk classification of the device. As emphasized in 4.2 Regulatory Basics, it is unusual that an innovator or company has the opportunity to exercise significant influence over the pathway chosen. However, it is worthwhile to have a clear understanding of the high-level advantages and disadvantages of a 510(k) versus a PMA, as summarized in Table 5.4.1 and in the discussion of the two pathways below.

Strategies for 510(k) clearance

Clearance under a 510(k) relies upon the concept of substantial equivalence to a predicate device, where a predicate device is defined as a device cleared before 1976, a device already cleared by the FDA through the 510(k) process, or a 510(k) exempt device. Substantial equivalence means that the new device is at least as safe and effective as the predicate device or devices. Importantly, according to the FDA, substantial equivalence does not mean the new and predicate devices must be identical. In fact, the predicates used in successful 510(k) applications sometimes appear not to have close similarities to the new device. But, from the standpoint of the agency, they do provide relevant comparisons. FDA's criteria for substantial equivalence are outlined in chapter 4.2.

Strategically, selecting predicate devices and specifying the intended uses for the new device requires a sophisticated understanding of substantial equivalence – and, in practice, requires input from a regulatory specialist. 510(k) clearance limits the use of a device to a finite set of clearly defined indications (which will ultimately be described in the package insert once the product is sold). For expeditious 510(k) clearance, the indications for the new device must be described in language that is the same as that used to obtain clearance for the chosen predicate device(s). There can be no "creative license" in the language about indications. The only way that new language surrounding indications will be acceptable to the FDA is if data are provided to back up any modifications, while still proving substantial equivalence in terms of the technological characteristics and intended use of the device. For these reasons, companies must think carefully about the predicate devices they choose and the clinical uses they

Table 5.4.1 The advantages and disadvantages of a 510(k) versus a PMA pathway have important strategic implications.

Pathway	Pros	Cons
510(k) Substantial equivalence (**SE**) to predicate device(s)	• Quicker route to market • Less expensive submission • Easier to modify the device post-clearance • Clinical data needed only 10–15 percent of the time • Fewer post-surveillance requirements • No facility pre-inspection required • No panel meeting required	• Competitors can more easily follow company to market (claiming substantial equivalence, with the new device as the predicate) • May limit company's ability to market the device as desired (since it must follow the indications of its predicates)
PMA Reasonable assurance of safety and effectiveness established by valid scientific evidence (**clinical trials**)	• Harder and more expensive for competitors to follow (as they are also subject to PMA requirements) • Can be used to allow a company to market a device for a new or different indication than existing devices • Potentially exempt from product liability cases	• Safety and efficacy must be proven • Longer, more complex application/approval process • Expensive submission and approval process • Requires clinical data (often **randomized**, **controlled**) • May require panel review • Requires a pre-approval facility inspection (PAI) • Difficult to make post-approval device modifications • May require post-marketing studies as a condition of approval

intend to promote. Otherwise, they may find themselves in a situation where regulatory clearance of the new device has been achieved, but the new device cannot be marketed to the target clinical population with the desired claims.

It is worth noting that regulatory practice is akin to case law in that the interpretation of substantial equivalence changes according to the accumulated experience of the FDA and the regulatory professionals with whom the agency works. Understandably, the FDA reacts to public and Congressional concerns. For example, the association of silicone breast implants with autoimmune conditions in the 1990s brought into suspicion many devices containing silicone elastomer when, in fact, it was liquid silicone that was thought

to be the risk factor (note that these concerns were never scientifically proven). Clearance of devices containing silicone – of any type – was stalled for a time while the FDA reacted to the events generated by these fears.

A company pursuing a 510(k) strategy must consider its desired speed to market in the context of the indications for which it intends to market its device. If speed to market is particularly important, a company may choose the simplest predicate device(s) and indications for use to clear the device quickly, and then pursue additional indications following initial clearance. This may take less time overall than testing the limits of predicate bundling and undergoing multiple rounds of questioning with an FDA reviewer. One example of this was the 510(k)

Table 5.4.2 A company has different alternatives in the way that it approaches the 510(k) regulatory pathway, depending on its strategic priorities.

510(k) strategy	Pros	Cons
Quickest: Use one established predicate device, if it provides the indications for use that are needed to sell to the initial target market	Facilitates the fastest clearance with fewest question rounds (90 days or less)	May not provide the desired indications for use to sell to the target market
Moderate Risk: Bundle two or more predicate devices to add desired indications	Increases the possible market to which the device can be sold	May be subject to further rounds of questions or additional testing which will increase review time
Riskiest: Push the limits of what is a **Class II** device by using predicate device(s) that have tenuous substantial equivalence arguments	May enable a company to avoid a PMA	Definitely will be subject to increased FDA questioning, additional data, and possible determination that the device is not substantially equivalent to anything currently on the market Caution: Split predicates are not allowed, that is, one predicate for the intended use and another for the technology under a different intended use

strategy pursued by Intuitive Surgical (see case example in 5.1 IP Strategy), which developed a robotic technology to enhance a surgeon's capabilities. As a first step, Intuitive Surgical approached the FDA with the concept of using its device as a surgical assistant to hold tools while the surgeon operated. This required mostly bench data for the 510(k) clearance. In order to obtain clearance for the purpose of actually performing the surgery, the FDA required a randomized, controlled trial. However, the company was able to use the initial 510(k) clearance for the surgical assistant as a predicate for itself in obtaining a second 510(k) clearance for performing the surgery. In the meantime, the initial 510(k) clearance provided the company the opportunity to familiarize surgeons with the technology and gain initial revenues from sales of the robot system and equipment for the more limited use.[7]

The amount of time and risk a company is willing to bear in terms of generating all desired indications in one submission (versus pursuing additional indications following initial market release of the product) is an important issue that requires careful thought early in the regulatory process. Rapid product release enables a company to accumulate early experience with the device and directly understand which alternative clinical applications are worthy of further study. On the other hand, if the device will be a direct competitor to a product already on the market, the company may benefit from taking more time to gain clearance for additional indications or to gather data that substantiate the benefits of its device.

If a company receives regulatory clearance under a straightforward 510(k) based on substantial equivalence to a competing device, it may be difficult for the company to differentiate its new device from the predicate for marketing purposes. When a device is cleared for use under a 510(k), the company is not allowed to make claims above and beyond those of the substantially equivalent device or beyond what was allowed in the 510(k) clearance. Common 510(k) regulatory strategies are summarized in Table 5.4.2.

It is useful to be as broad as possible in describing the features of a device for a 510(k) submission. For instance, if a company makes a type of catheter that comes in a variety of diameters, it should be sure to include a range of sizes in its 510(k) submission that incorporates all diameters of the existing device, as well as those diameters that may be developed in the future. Thus a guide catheter of the same design can be submitted in 5, 6, 8, and 9 French sizes,[8] even if the original released product is only 6 Fr (though the original submission performance data must bracket all of the sizes for which clearance is sought).

Clinical data in 510(k) submissions

Currently, the inclusion of clinical data is only required in approximately 10–15 percent of all 510(k) submissions. In some cases, the FDA issues specific requirements or expectations for clinical data in a guidance document. There are other situations in which submitting clinical data is strategically important. If there are measurable differences between the new device and the predicate device(s), clinical data can be used to demonstrate the safety and effectiveness of the new device. Clinical data can also be submitted to the FDA in order to support the expansion of the device's intended use (over and beyond what is indicated by the predicate device(s)).

There is an ongoing trend for the FDA to require clinical data more often for 510(k) clearance and to require higher levels of clinical proof in the study designs. Clinical data can come in a variety of forms. It is not uncommon for companies to submit small human registry-type trials to satisfy the data requirement. In general, in the 510(k) setting clinical data are used to validate bench data and to establish *safety* rather than effectiveness. Trials should be designed to be as small and as simple as possible. Rarely (if ever) are randomized clinical trials required for the 510(k) pathway, and rarely do studies have to be large (20 to 100 subjects is commonly considered adequate). Often, the necessary clinical data can be collected outside the US and the studies, therefore, are exempt from the formal requirements of an IDE. However, they are still subject to international ethical standards and informed consent, and may require approval by the country's competent authority (the regulatory body charged with overseeing clinical research in that country) (see 4.2 Regulatory Basics and 5.3 Clinical Strategy for more information).

Meeting with the FDA

For a 510(k) submission that does not require clinical data, it is generally not necessary to have any "pre-meeting" with the agency to discuss the application. If the technology and predicate device(s) are reasonably straightforward, a request for such a meeting could demonstrate to the FDA that a company does not possess the expertise necessary to understand their obligations and comply with them (which could be a liability in terms of how the company is perceived by the FDA). However, a 510(k) pre-meeting might be appropriate under circumstances in which: (1) the company knows of problems with the predicate device it plans to use; (2) the company is concerned that the agency might have new questions of safety or effectiveness based on the technology employed; or (3) the company is aware of a competitor using the same predicate device(s), technology, or materials and there has been difficulty with its submission. Similarly, a meeting with the FDA may be appropriate when an innovator or company wants to confirm the appropriate pathway before investing significant time or effort in the development process. As mentioned above, when there is a substantial clinical trial involved, it is wise to meet with the FDA to review the adequacy of the trial design.

Strategies for premarket approval (PMA)

The ratio of PMA to 510(k) submissions ranges roughly from 1:50 to 1:100, reflecting the fact that the time, cost, and risk involved in the PMA pathway is significantly higher than for 510(k) clearance. However, there are some clear advantages to the PMA pathway. An approved PMA is essentially a private license granting the owner permission to market the device.[9] In effect, this is a kind of "regulatory patent." The PMA provides a barrier to entry such that competitors who desire the same type of device and same indications are required to undergo the longer, more costly PMA process as well. In early 2008, the US Supreme Court added further weight to the importance of PMAs in one of its landmark medical device preemption decisions in Riegel versus Medtronic, in which the high court stated that medical device manufacturers could not be sued for complications if the device used was approved by a PMA or PMA supplement.[10] This ruling was generally viewed as a shield for industry from a legal perspective, as it prevents litigants from pursuing tort action on a state-by-state basis in most cases. Companies can still be sued for negligence, however, and the full implications of this decision are subject to further interpretation and clarification.

While it is not common for a company to pursue a PMA if a 510(k) can be justified, a company may decide to pursue a PMA in an attempt to get approval for indications that are beyond the scope of those cleared for predicate 510(k) devices. This would usually only be undertaken by large companies with significant resources and an aggressive desire to erect competitive barriers. If successful, the **first mover** can block followers who simply do not have the resources to conduct the necessary clinical trials and overcome the other hurdles arising from the new bar set by the first PMA. It is also worth mentioning that Medicare and other insurance plans are increasingly requiring evidence from randomized clinical trials that are similar to the scale and complexity of PMA trials to support reimbursement decisions. If the company is obliged to conduct large trials for purposes of reimbursement, the added benefit of PMA approval may come at a relatively small cost. It is also clear that in recent years both the FDA and the Department of Justice are being more aggressive in pursuing medtech companies that promote devices for "**off-label**" uses – that is, uses beyond those explicitly cleared or approved by the FDA (more information about off-label usage is provided below). A PMA approval provides a clear go-ahead for marketing the device for the indications that have been studied.

PMA approval is required for **Class III** (high-risk) devices, even those that are similar to competitors in the field (e.g., stents and pacemakers). Seeking regulatory approval under the PMA pathway as a follow-on device can have both upsides and downsides. For a technology that has a long history of established clinical safety and efficacy (for example, femoral closure devices pioneered by AngioSeal and Perclose), undertaking a

PMA for a similar device leaves little guesswork. The study design and endpoints are well established, and the FDA is familiar with the technology. The PMA process, therefore, is still rigorous, but is significantly de-risked due to the many lessons and shortcuts that can be leveraged as a direct result of the predecessor's experience.

On the other hand, the FDA can impose acquired learning onto the PMA process. This happened with abdominal aortic aneurysm stent grafts (synthetic conduits placed within an aortic bulge to reduce the risk of rupture). The FDA approved two technologies (Medtronic's Aneuryx and Guidant's Ancure) based upon early, promising results. As these two technologies enjoyed wide commercial use and market share, more was learned about performance issues and design problems. Followers in this area, therefore, were hit with significantly increased requirements over and above those faced by the original applicants.

In order to obtain PMA approval, a company is required to validate every indication it seeks with clinical research data. Often, there is temptation to pursue as many indications as possible with the initial submission. However, it is essential to consider how this will complicate clinical trial design. Initial clinical trial design should focus on the simplest, most achievable endpoints possible to maximize the study's chance of success (see 5.3 Clinical Strategy).

A strategy employed by many companies is to target the most important indications for use to capture the initial market, and then add indications following initial approval through a mechanism known as a PMA supplement. A PMA supplement is required by the FDA following PMA approval before a company makes any changes affecting the safety or effectiveness of the device. These changes may include new indications for use of the device, changes in labeling, changes in sterilization procedures or packaging, changes in the performance, technology, or design specifications, changes in operation or layout of the device; and use of a different facility to manufacture or process the device.[11] Approval of a PMA supplement can require anywhere from 30 to 180 days of review time and, in some instances, might require another advisory panel meeting.

The strategy of obtaining PMA approval for core indications first (followed by subsequent PMA supplements) allows the company to earn initial revenue to support the development and testing required for the additional indications. Every year, many more PMA supplements are received and approved by the FDA than PMA applications.

Meeting with the FDA

From a strategic perspective, the PMA process should be highly collaborative, with numerous interactions occurring between the FDA and representatives of the company. One of the most important opportunities for collaboration occurs early in the evaluation and submission process.[12] A meeting with the appropriate FDA branch should occur before significant time and expense is put into device development. While these meetings are optional, they are strongly recommended because they help establish a relationship with the FDA and allow the agency to unofficially buy in to the selection of safety and effectiveness measures in advance of the submission. Companies may request a formal **determination meeting** before the PMA application process is initiated, in which the FDA is obliged to give an official response to the study that is proposed. However, a determination meeting usually does not turn out to be productive for the company (i.e., the agency does not endorse the study). For this reason, it is generally recommended for companies to work within the context of the more informal meetings.

PMA post-approval requirements

Increasingly, the FDA is interested in collecting data on the impact of new technology *after* PMA approval, as the technology is disseminated into more widespread practice. The high-profile issue of stent thrombosis (clotting) following implantation of drug-eluting coronary stents added impetus to this trend. The FDA approved the first two drug-eluting stents based on data that the restenosis rates were lower than with the conventional bare metal stents. Following PMA approval, the stents were deployed in millions of patients, and clinicians began to see a problem with some stents abruptly clotting off completely – a serious event that can occur years after the procedure. This complication was not detected in the PMA trials and led the FDA to require post-market surveillance for the two companies with approved products, as well as mandating larger and longer trials for the approval of new drug-eluting stent products from other companies.

Post-approval requirements, which may include continuing evaluation and periodic reporting on the safety, effectiveness, and reliability of the device for its intended use, are an increasingly common part of the PMA landscape. Evaluation and reporting may be achieved through any of the following measures: a post-market study or registry to track outcomes; reporting on the continuing risks and benefits associated with the use of the device; maintaining records that will enable the company and FDA to trace patients if such information is necessary to protect the public health; inclusion of identification codes on the device or its labeling or, in the case of an implant, on cards given to patients in order to protect the public health; submission of published and unpublished reports of data from any clinical investigations or non-clinical laboratory studies involving the device or related devices; and submission of annual post-approval reports.[13] Another post-approval reporting mechanism under development is the unique device identifier (UDI) system, which will require device manufacturers to assign a distinct number to most medical devices distributed in the U.S so they can be tracked through a publicly available database. The intent of the program is to reduce medical errors, facilitate more accurate reporting of adverse events, and provide an improved understanding of any underlying problems with devices. Class III devices are required to comply with the new system first, with staggered deadlines for Class II and **Class I** devices following soon after.[14]

The Emphasys example, which is a continuation of the case study in 5.3 Clinical Strategy, demonstrates some of the many challenges of devising and executing a regulatory strategy.

FROM THE FIELD ▷ EMPHASYS MEDICAL

Navigating complex clinical and regulatory challenges on the path to market – part 2

When the Emphasys team made its PMA submission to the FDA for its novel endobronchial valve for treating patients with severe emphysema (see 5.3 Clinical Strategy for the first part of the Emphasys story), the company anticipated that it again would be working with the Plastics and Reconstructive Surgery branch of the Center for Devices and Radiologic Health (CDRH). This branch had cleared tracheobronchial stents, one of the few device interventions available in the pulmonary arena. Emphasys had negotiated with the Plastics and Reconstructive Surgery group on its IDE application and the parameters of its **pivotal trial** for the device (as shown in Figure 5.4.2). However, recalled Emphasys CEO John McCutcheon, "A week later we found out our submission had been transferred to the Anesthesiology and Respiratory Therapy Device branch of CDRH."[15] Although this seemed to be a more logical place to evaluate pulmonary devices, the branch had no experience with the Emphasys study or the history leading up to the company's PMA submission. "We had to start all over with developing the trust we had gradually built up," McCutcheon noted. The only reviewer that followed Emphasys to the new branch was the consultant who had disagreed with the team about the control group for the pivotal trial.

Based on the lack of treatment options available to emphysema patients, the FDA granted the Emphasys device expedited review status. Yet, said Hanson Gifford, a partner with The Foundry, which had incubated the device, "Our submission spent a year at the FDA. Their statisticians got into every detail and said, 'No, you're doing the statistics wrong.' We went back and forth and back and forth." The main source of disagreement had to do with how Emphasys imputed outcomes data for roughly 20 percent of patients in the study for which the company did not have complete follow-up data. To resolve its standoff with the FDA,

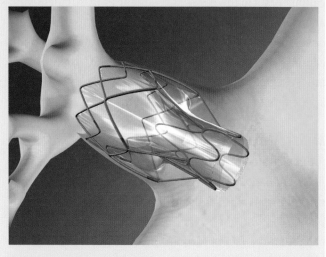

FIGURE 5.4.2

The Emphasys device (courtesy of Emphasys Medical).

the Emphasys team finally hired a leading "imputation guru" in the field to work through the data one more time. "That's when the FDA statistician said, 'Okay, you're right. No matter how you did the imputation, it really is statistically significant,'" Gifford stated. Having reached an agreement on the statistics, the FDA scheduled an advisory panel meeting to review the company's PMA submission.[16]

In parallel, a confluence of events in the external environment created increased public scrutiny of the FDA and device manufacturers alike. First, the withdrawal of the arthritis medication Vioxx from the market[17] and the **recall** of the Guidant defibrillator heightened public concerns about the safety of FDA-approved drugs and devices. Subsequently, in response to criticism that drug and device makers were asserting undue influence over the approval process through panel members, the FDA passed new conflict of interest rules regarding advisory panel participation. Adding more fuel to the fire, a group of CDRH scientists and physicians sent a letter to the then chairman of the House Committee on Energy and Commerce alleging that managers at the device center "knowingly corrupted the scientific review process and approved or cleared medical device applications in gross violation" of the law and agency regulations.[18] These claims led to a House Committee investigation into the matter, which was underway at the time the Emphasys device went to panel. Meanwhile, *The New York Times* launched an article series called "The Evidence Gap," with the purpose of examining how the FDA each year allows "thousands of medical devices onto the market with only cursory review and with no clear evidence of improved clinical outcomes."[19] Unfortunately, as some observers pointed out, the FDA rarely received credit for doing its job well and approving good therapies. Instead, it was only blamed for any problems that occurred with drugs and devices after they were approved. This trend persisted even though it was impossible for the agency (or a company) to foresee all potential problems at the time of approval, no matter how exhaustive the clinical research.

Against this backdrop, the FDA convened the Anesthesiology and Respiratory Therapy Devices Panel to discuss, make recommendations, and vote on a premarket approval application for the Emphasys medical valve system. In addition to the chairman, the panel included 13 voting members, a consumer representative, an industry representative, and an executive secretary. Among the voting members, only three had backgrounds in pulmonology and worked directly with emphysema patients, three were trained in anesthesiology, three in cardiothoracic surgery, two in statistics, one in pediatrics, and one in device research and development.

Prior to the panel meeting, participants were provided with briefing materials prepared independently by Emphasys (referred to as the "sponsor" of the PMA application) and FDA. In the meeting itself, both the sponsor and the FDA were asked to make a presentation to the panel, followed by question and answer sessions. Next, there was a panel discussion, during which the sponsor responded to additional questions from the panelists. Deliberations followed, with the panel considering pre-specified questions about the PMA application put forth by the FDA. Afterwards, the agency and the sponsor were given the opportunity to make summary remarks and the consumer and industry

representatives had the opportunity to make comments. Finally, the panel voted on which of three possible recommendations to put forth to the FDA's Office of Device Evaluation (ODE) and the branch in charge of the PMA regarding the submission: (1) approvable, (2) approvable with conditions, or (3) not approvable. Once a recommendation had been made, each panel member was asked to explain his/her vote. After the meeting, the ODE would consider the recommendation, along with input from the branch, and then render a final decision.

During the meeting, the Emphasys team asserted that the company met its co-**primary endpoints** in a statistically significant way, even when considering only completed cases (without including any imputed data or patients who had inclusion/exclusion protocol violations). Furthermore, by using a **core laboratory** to gain additional insights, the team had identified two clinical subgroups that had experienced even greater improvements in the lung function tests than the general patient population in the trial. These algorithms could be used during future patient screening to facilitate patient selection and treatment targeting. In terms of safety, while there were more complications in the treatment group, the difference was not statistically significant and diminished with time. The team also presented data that the devices could be safely removed when deemed necessary.

In response, the FDA raised four primary issues that it believed affected the estimation of the treatment effect. These included: (1) the lack of blinding in the study, opening the door to the possibility of a placebo effect; (2) the fact that Emphasys had extended the follow-up window without explicit FDA approval; (3) the issue that data were missing for more than 20 percent of patients; and (4) that there were violations in the **inclusion** and **exclusion criteria** for the trial. In addition, the FDA argued that while Emphasys had met the endpoints in a statistically significant way, those outcomes were not necessarily clinically significant. According to the FDA, "The study was designed to show a pre-specified clinical difference between the ... treatment and control arms of 15 percent for co-primary endpoints, FEV_1 and 6MWT as recommended by the General and Plastic Surgery Devices Panel . . . . The 15 percent clinically significant response for an individual was used along with pilot study results to power the trial."[20] So, argued the agency, "An analysis of the co-primary effectiveness endpoints showed statistically significant differences at 6 months. However, at no point in time did either of the endpoints reach clinical significance." Emphasys disputed that the 15 percent threshold had been formally established as a clinical significance requirement for each of the co-primary endpoints in the study. "The FDA went back to our study protocol and pulled our powering number, which had 15 percent in it, and then tried to say, 'We agreed that this is the threshold.' But that just wasn't the case," said McCutcheon. While this difference of opinion was not resolved in the meeting, the panelists spent a significant amount of time discussing the matter. Regarding the subgroup analysis based on the findings from the core lab, the FDA contested the results, stating that the software used by the lab for its analysis had not been validated by the agency.

After a lengthy discussion surrounding these and other issues, the panel arrived at a "not approvable" recommendation by an 11:2 vote. Significantly, the two dissenting members of the panel were those with the most direct experience with emphysema. In statements after the vote, both of these physicians underscored the limited options available to emphysema patients, the fact that the company was well on its way to finding the most appropriate use of the technology, and that the improvement conveyed by the therapy outweighed the limited safety concerns.

The ODE stood by the panel recommendation, requiring Emphasys to do a confirmatory study prior to any possible approval. Faced with this difficult situation, the company evaluated a wide variety of strategic options, including raising additional funds to support another trial. But most of Emphasys' backers had been investors for as much as five to eight years and were not in a position to commit more capital. Speaking as one of these investors, Mike Carusi of Advanced Technology Ventures remembered, "We did the math for putting in

another $50 million and concluded that the return on investment was not going to be there. Plus, the investors were tired. People didn't have budget and the economy had started to get bad. Nobody was going to write those checks – and certainly not for a company with this kind of FDA risk." Another option was to market the device more aggressively in Europe, where it had been approved more than five years earlier based on the results from the Emphasys pilot studies. Ultimately, however, the company was forced to sell its assets at auction to a competitor called Pulmonx.

Reflecting on the Emphasys experience, The Foundry partner Hank Plain said, "In the current environment, fewer and fewer projects like this will be financeable, which I think is really unfortunate for patients and for innovation. The levels of uncertainty, the financing risk, the clinical risk, the FDA risk, all of those hurdles have gotten too high." As Gifford added, "Speaking for The Foundry, as we look at new things, unless they are just clearly going to have a dramatic benefit on huge patient populations, we're unlikely to consider any PMA products."

As for the Emphasys technology, it remained uncertain when it would reach emphysema patients in the US. At the time of the acquisition, the initial strategy of Pulmonx and founder Rodney Perkins was to concentrate on marketing the device in Europe. As Perkins explained, "We didn't want to go back to the FDA too early because if you don't have a period of separation you're just carrying the same baggage." Importantly, the company intended to market the device with its Chartis assessment tools, which could be used to help pulmonologists detect the leaks (collateral ventilation) in the lung airways that could cause the valve to be less effective. Pulmonx would carefully design a European study, selling the approved valves into the trial sites, and then would decide when to go back to the FDA. However, Perkins emphasized, "One of the points to consider is that we didn't have to be in the US to be successful as the market outside the US is larger."

Over the next several years, Pulmonx established a sizable field office in Switzerland and completed close to 8,000 cases with more than 120 hospitals across

Europe. Based on the strength of the results, the Pulmonx team determined that the time was right to once again to pursue FDA approval. "We want to get back to the US because we know we have a good treatment," Perkins said. In 2013, Pulmonx applied for and was granted an IDE by the FDA to initiate a US pivotal trial to support a new premarket submission for the valve.[21]

Strategies regarding off-label device use

One of the more subtle areas of regulatory strategy deals with devices that are used outside of the cleared or approved FDA indications (so-called off-label use). Recall that the FDA does not have jurisdiction to regulate the practice of medicine. Physicians may use a device in any fashion they see fit, provided this use is in the best interests of the patient and is broadly within the **standard of care**. In practice, this means that many devices are used outside of the indications for which they are cleared or approved by the FDA. A classic example is the biliary stent, which was approved to prop open the bile tracts in the intestine. For years, the large majority of biliary stents were, in fact, used by cardiologists to treat blockages in the coronary arteries, despite the fact that these stents were not approved for this indication (coronary stents were approved in Europe several years earlier than in the US, convincing U.S cardiologists that this practice was within a reasonable standard of care).

When companies ultimately sought approval for coronary stents from the FDA, they designed trials strategically with an eye toward proving that the stents were superior to the existing technique of balloon angioplasty. Among other things, this meant targeting vessel sizes that were likely to yield favorable results (smaller coronary arteries, it turns out, have a higher incidence of renarrowing after stenting than vessels greater than 3 mm in diameter). Once the stents were approved for vessels of the optimal size range, however, cardiologists began using the stents widely in the smaller vessels.

Although companies often are tempted by the off-label potential for their devices, a blatant strategy based on this approach is not advisable. The FDA understands this issue well and is on the look-out for companies promoting off-label uses of their products. As a cautionary tale, the 10 or more companies making biliary stents were effectively censured by the FDA for not being more forthcoming about the dominant use of their products.

Integrating US, European, and other regulatory strategies

Innovators can often achieve important advantages by integrating regulatory strategies across geographic areas. In some cases, regulatory processes may be optimized by leveraging clinical data and regulatory approvals obtained in one market to the shorten time to regulatory approval, reimbursement, or market adoption in another. For example, many US medical device start-ups pursue **CE marking** of a device subject to a PMA pathway in advance of US approval (see the Edwards Lifesciences case in 4.2 Regulatory Basics). Conversely, companies may seek the clearance of a 510(k) device in the US before entering the EU market and then use the US clinical data to gain reimbursement abroad. Another common strategy is to use clinical data from a CE marking trial in lieu of a US pilot study to enable the company to start a US pivotal trial earlier. This data can also be used to obtain regulatory approval in other markets, such as Canada. More information about these integrated strategies is provided below.

Early CE marking of PMA devices

For the majority of devices on a PMA pathway in the US, CE marking still can be obtained more quickly than a PMA. This may be helpful to companies for several reasons: (1) it can provide a valuable revenue stream while the product is working its way through FDA approval in the US; (2) it provides the company with early user feedback on device performance and adoption; and (3) it provides early clinical data that may be used in subsequent FDA submissions.

The reason why CE marking of Class III devices is faster than premarket approval is directly related to the differences in the clinical data requirements between the US and EU regulatory pathways. The FDA requires that Class III devices demonstrate reasonable safety and effectiveness, which is typically achieved through prospective randomized controlled trials involving hundreds of patients (see 5.3 Clinical Strategy). In contrast, CE marking traditionally only requires that devices demonstrate safety and performance. Usually, compliance with the EU requirements, even for Class III devices, can be demonstrated with much simpler trials. For instance, the GuardWire® from Percusurge, Inc., which enables debris created during endovascular interventions to be captured to prevent it from embolizing, was awarded CE marking on the basis of a 22 patient single-arm study[22] demonstrating safety and performance (i.e., that debris was aspirated during the interventional procedure). In the US, however, the FDA required an 800 patient multi-center randomized trial[23] for effectiveness (i.e., that compared the device to the standard care to demonstrate a reduction in complications).

Deferring EU market entry of 510(k) devices

Early European approval of a device that is headed for 510(k) clearance in the US rarely provides the same value to a company seeking to penetrate the US market as it does for a PMA device, unless the 510(k) device requires clinical data. 510(k) clearances that do not require clinical trials can usually be obtained much more quickly than premarket approvals, allowing a company to begin providing the device to US physicians and building this market. Furthermore, clinical data generated in the US (before or after 510(k) clearance) may subsequently be used to build markets in other geographies. This strategy was successfully used, for example, by Kyphon® for its interventional device to perform kyphoplasty for the treatment of vertebral body compression fractures.

Using EU clinical data for US and other regulatory approvals

Pre- and post-CE marking trials may be used for US and/or other regulatory approvals, provided that certain aspects of clinical trial design and conduct conform to necessary requirements. These include compliance with all relevant local regulations and any applicable Good Clinical Practices (e.g., set forth by the ICH[24]). All such requirements should be designed into the trial from the outset to ensure that the clinical data will be acceptable by the authorities outside of the EU.

It is important to be aware that the FDA looks critically at data obtained from foreign studies in terms of its applicability to the US patient population. If medical therapy or practice differs significantly from that in the US, or if there are expected or unexplained differences between the patient populations within and outside the US, the data can be deemed not applicable. For example, in coronary interventions, a certain class of blood thinners or anticoagulants (IIb/IIIa inhibitors) is not used as commonly in the EU as in the US. Therefore, trials involving devices for coronary interventions must be carefully designed to ensure that data acquired overseas will be accepted by the FDA if the sponsor is expecting it to be used to support a US approval. Note that the FDA reserves the right to audit foreign clinical sites to confirm whether or not the data are valid for the purposes of obtaining an approval to market a device in the US. The agency has a specialized Bioresearch Monitoring (BIMO) Program for this purpose.

Efforts toward the increasing harmonization of regulatory standards have created a dynamic and complex environment that demands a regulatory consultant with strong international experience. The EU is still the largest market outside the US for most medical devices and is the best understood region in terms of regulatory approvals. Japan is a market that reimburses well but has been notoriously slow to approve new devices, often expecting the company to repeat clinical studies in Japan (the government is working to streamline the approval process). Young companies are becoming increasingly familiar with regulatory agencies in Australia and South America (especially Argentina and Brazil) since "**first-in-human**" studies are often conducted in these countries. As described in chapter 4.2, regulations in other large markets (e.g., China, India) are evolving. When considering overseas regulatory approval as a means of conducting earlier clinical trials, be cautious about the credibility and applicability of the data to ensure a wise investment.

Online Resources

Visit www.ebiodesign.org/5.4 for more content, including:

 Activities and links for "Getting Started"
- Validate device classification and regulatory pathway
- Develop a regulatory strategy
- Modify and monitor regulatory strategy

 Videos on regulatory strategy

An appendix on common regulatory pitfalls

CREDITS

The material in the original chapter was based, in part, on lectures by Howard Holstein and Janice Hogan of Hogan and Hartson. The editors would like to acknowledge Howard Holstein, Jan B. Pietzsch, and Nancy Isaac for their substantial contributions to editing and updating the content, as well as Julie Delrue and Sarah Sorrel of MedPass International for reviewing the information on international regulatory requirements. Further thanks go to John McCutcheon and the Emphasys team for sharing the Emphasys story, and Trena Depel and Stacey McCutcheon for their contributions to the chapter.

NOTES

1 Cartoons created by Josh Makower, unless otherwise cited.

2 "Mobile Medical Applications: Guidance for Industry and Food and Drug Administration Staff," U.S. Food and Drug Administration, September 25, 2013, http://www.fda.gov/downloads/MedicalDevices/DeviceRegulationandGuidance/GuidanceDocuments/UCM263366.pdf (December 26, 2013).

3 "FDA Submits Final Guidance on Mobile Medical Apps," Foley and Lardner LLP, October 4, 2013, http://www.foley.com/fda-submits-final-guidance-on-mobile-medical-apps-10-04-2013/ (December 26, 2013).

4 Joanne Gibbons, "There May Be More to the FDA/23andMe Story Than Meets the Eye," *Xconomy*, December 12, 2013,

http://www.xconomy.com/boston/2013/12/12/may-fda23andme-story-meets-eye/ (December 27, 2013).

5 Ibid.

6 "IDE Approval Process," U.S. Food and Drug Administration, http://www.fda.gov/medicaldevices/deviceregulationandguidance/howtomarketyourdevice/investigationaldeviceexemptionide/ucm046164.htm#pre_ide (December 4, 2013).

7 From in-class remarks made by Howard Holstein as part of Stanford's Program in Biodesign, Winter 2008. Reprinted with permission.

8 The French catheter scale is used to measure the outside diameter of a cylindrical medical instrument. 1 French (Fr) is equivalent to approximately 0.33 mm.

9 "PMA Review Process," U.S. Food and Drug Administration, http://www.fda.gov/MedicalDevices/DeviceRegulationandGuidance/HowtoMarketYourDevice/PremarketSubmissions/PremarketApprovalPMA/ucm047991.htm (December 4, 2013).

10 Janet Moore, "Ruling Shields Medtech Firms," StarTribune.com, February 20, 2008, http://www.startribune.com/business/15801117.html (December 4, 2013).

11 "PMA Supplements," U.S. Food and Drug Administration, http://www.accessdata.fda.gov/scripts/cdrh/cfdocs/cfcfr/CFRSearch.cfm?FR=814.39 (March 3, 2008).

12 "Draft Guidance for Industry and FDA Staff Medical Devices: The Pre-Submission Program and Meetings with FDA Staff," U.S. Food and Drug Administration, July 13, 2012, http://www.fda.gov/MedicalDevices/DeviceRegulationandGuidance/GuidanceDocuments/ucm310375.htm (December 4, 2013).

13 "PMA Postapproval Requirements," U.S. Food and Drug Administration, http://www.fda.gov/MedicalDevices/DeviceRegulationandGuidance/HowtoMarketYourDevice/PremarketSubmissions/PremarketApprovalPMA/ucm050422.htm (December 4, 2013).

14 "Unique Device Identification," U.S. Food and Drug Administration, http://www.fda.gov/medicaldevices/deviceregulationandguidance/uniquedeviceidentification/default.htm (December 4, 2013).

15 All quotations are from interviews conducted by the authors, unless otherwise cited. Reprinted with permission.

16 All PMA applications are reviewed by a special panel, a group of 5 to 15 physicians, statisticians, and other experts (all non-FDA employees) who serve a 3-year term. In addition to the core experts, the panel can add topic experts on a case-by-case basis and also has nonvoting industry and consumer members. After the PMA is submitted, the panel convenes to hear presentations from the company sponsor, its expert consultants, and from the FDA team. The panel votes to recommend whether the technology should be approved, approved with conditions, or

disapproved. The recommendation is non-binding, but generally carries great weight in the approval process. The final decision is based on the analysis of the CDRH branch team, subject to the approval of the director of the Office of Device Evaluation (ODE).

17 In September 2004, drug maker Merck withdrew its arthritis medication Vioxx from the market in the face of new data that showed its sustained use doubled a patient's risk of heart attacks and strokes. In the investigation, which followed, the FDA faced allegations in the press that its relationship with drug companies was "too cozy" and was interfering with public safety.

18 '"Ordered, Intimidated, and Coerced'? CDRH Targeted in Misconduct Probe," *The Gray Sheet*, November 24, 2008.

19 "New York Times Examines FDA Market Approval of Medical Devices," *Medical News Today*, October 28, 2008, http://www.medicalnewstoday.com/articles/127109.php (January 6, 2014).

20 "FDA Executive Summary," Anesthesiology and Respiratory Therapy Devices Panel, December 5, 2008, http://www.fda. gov/ohrms/dockets/ac/08/briefing/2008-4405b1-01-FDA-ExecSummary_Final_Panel%20Pack.pdf (January 6, 2014).

21 Pulmonx Gets FDA Nod for U.S. Trial of Emphysema Therapy," Pulmonx press release, August 27, 2012, http://pulmonx.com/en/news/pulmonx-gets-fda-nod-for-us-trial-of-emphysema-therapy/ (December 30, 2013).

22 J. G. Webb, R. G. Carere, R. Virmani et al., «Retrieval and Analysis of Particulate Debris Following Saphenous Vein Graft Intervention," *Journal of the American College of Cardiology*, 1999, pp. 468–75.

23 D. S. Baim, D. Wahr, B. George et al. "Randomized Trial of a Distal Embolic Protection Device During Percutaneous Intervention of Saphenous Vein Aorto-Coronary Bypass Grafts," *Circulation*, 2002, pp. 1285–90.

24 ICH stands for the International Conference on Harmonization of Technical Requirements for Registration of Pharmaceuticals for Human Use. This project brings together experts and regulatory authorities from Europe, Japan, and the United States to discuss scientific and technical aspects of product registration.

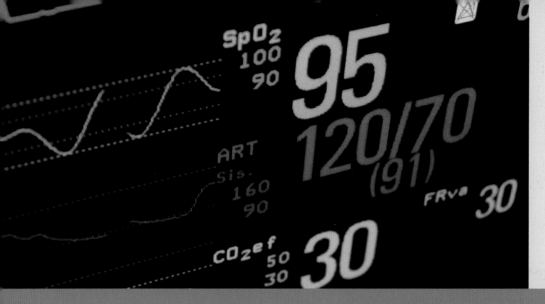

5.5 Quality Management

INTRODUCTION

An engineer looks over the shoulder of a cardiologist and sees something that never should have happened – a broken nitinol stent now embedded in a patient's artery. Shortly thereafter, several other doctors from hospitals across the country start reporting the same finding, although many others indicate that they have experienced no problems at all. After studying all the films, it is confirmed: despite the fact that all cycle testing before commercial launch was successful, some of the stents are now breaking in the field. In an effort to understand why some are breaking and others are not, the engineer thoroughly reviews the causes of failure. By checking the detailed records maintained at the company as part of its quality management system, he determines that only stents from certain lots seem to be involved in the failures, and within these lots, records to support device traceability reveal that the failure is associated only with devices using a particular nitinol. In fact, the failures can be isolated to a specific material source. Stents from this ingot are then recalled and, after communicating the issue and its resolution to the proper authorities, the company is able to overcome this challenge, eliminating the failures and returning to the market to achieve great success.

Scenarios like the one above highlight the fact that a disciplined and rigorous approach to development and production is essential when bringing any innovation to market. If the innovator's end goal is to create a new medical device that can be manufactured according to precise specifications and used safely and reliably in medical care, then rigorous quality processes are central to making this happen. Not only are such processes critical for achieving regulatory clearance for a new technology, but they allow the innovator and team to transition from producing one device at a time to reliably manufacturing batches of products whose performance can be tested and validated. Moreover, patient lives may be put at risk if the quality process guiding development and subsequent production fails to ensure the creation of a product that performs as intended and meets all specified safety requirements.

 See ebiodesign.org for featured videos on quality management.

QUALITY MANAGEMENT FUNDAMENTALS

In the early stages of the biodesign innovation process, there is typically not a "real" medical technology company – that is, an innovator might be working alone or with a small group of engineers to solve problems and produce enough **prototypes** for preliminary testing. But, at some point as testing begins, this informal approach to development must incorporate defined operating procedures to ensure that the product can achieve regulatory clearance. Further, the team must transition into an entity with precise systems capable of producing devices predictably and reliably, delivering them to the end **user** in a timely fashion, and monitoring their performance in the clinical environment.

While essential, this transition is fraught with risks. If the systems put into place are inadequate in any way, they can potentially undermine the success of the venture. Even from an early stage, rigorous processes are needed to govern all of the activities that must be scaled up as a company prepares for commercialization, including those related to initial product development, manufacturing, packaging, labeling, storage, distribution, installation, and service of medical devices. The systems used to manage and monitor these processes are collectively referred to as "quality management systems" (**QMS**) or more simply quality systems.

Quality management in product development and manufacturing

The term "quality" broadly refers to the activities undertaken by the company to ensure that certain regulatory requirements are met and that products delivered to the customer (primarily patients and physicians) are safe and reliable. Relevant regulations require that detailed specifications are developed for the devices, that the devices are tested and manufactured according to these specifications, and that the devices perform according to these specifications once distributed and/ or installed. Their performance must also be monitored so that problems can be reported to the regulatory bodies as they are identified and corrected.

Quality systems can be a source of competitive advantage for a company if they lead to rapid product development cycles, superior quality as measured by extremely low defect rates, and low production costs. Additionally, strong quality systems contribute to the perceived **value** of a product (and, conversely, prevent the value of a product from being undermined by in-market failures). Finally, since quality systems are required by regulatory bodies, such as the US Food and Drug Administration (**FDA**) and other governmental authorities around the world, they are closely considered a part of any regulatory submission. The FDA, for one, demands that "companies establish and follow quality systems to help ensure that their medical devices consistently meet applicable requirements and specifications."[1] The agency has precise regulations, called Quality System Regulations (**QSR**) – called out in the Code of Federal Regulations Title 21 Part 820 (also referred to as simply **21 CFR 820**) – that specify the exact requirements that a company's quality system will need to meet before a product can be cleared or approved for the market. Despite the close ties of quality systems and regulatory processes, it is important to note that quality is much more than just a requirement or checkmark in a regulatory approval or clearance process. When considered more holistically, quality management can be thought of as one specific approach to **risk management** in a medical technology company in that it involves the systematic application of policies, procedures, and practices to the tasks of identifying, analyzing, controlling, and monitoring risk.[2]

Traditionally, quality management has carried the stigma of being a "policing" function within many organizations. For instance, quality measures have been perceived as imposing extra work steps and stringent requirements that necessitate a lot of effort for little measurable return. Busy employees sometimes question why the company needs formal quality processes, especially in young start-ups that may not yet have products in the market. However, recent cases demonstrate that the cost of not having an effective quality system in place can be devastating, no matter what stage of development the company has reached.

For example, consider the case of Boston Scientific. In 2000 and 2004, FDA quality inspectors found hundreds of quality control lapses in six of the company's US-based manufacturing facilities. As a result, the agency issued three warning letters to Boston Scientific. When subsequent inspections in three additional plants revealed quality control and regulatory issues, the FDA issued a broad "corporate warning letter," indicating that the company's corrective actions to address prior violations were inadequate. Such a move by the FDA, which was considered a broad critique of the company's entire quality control systems, is unusual.[3] Disclosure of the letter led to an almost immediate 5 percent drop in Boston Scientific's share price.[4]

Examples of the quality problems uncovered at the company varied from facility to facility, but all related to the procedures, processes, and timeliness of Boston Scientific's corporate quality management system. At one plant, employees were unaware that company headquarters had recalled a needle used to treat tumors in cancer patients. In another location, managers missed deadlines for notifying the FDA of reports linking Boston Scientific devices to serious injuries (federal regulations require notification within 30 days).[5] While the corporate warning was in effect, the FDA informed Boston Scientific that it would not approve any new devices that could be affected by the quality problems. At the time the warning was issued, this had potentially damaging consequences for the company since it was preparing to submit its new drug-eluting stent for FDA approval.

After receiving the FDA's warning, Boston Scientific launched one of the most systematic and extensive quality system enhancement programs in medical device history. The company dedicated two years, millions of dollars, and hundreds of employees to implementing changes in its manufacturing, distribution, and monitoring systems.[6] For example, one action the company took was to consolidate 23 separate processes for tracking complaints into a single system.[7] Based on the company's remediation efforts, the FDA lifted a number of the restrictions it imposed in the warning letter as the company worked with the agency to resolve the remaining issues.[8]

Even in small start-ups, the cost of not having effective quality systems can be high. These costs are known as failure costs, which represent the expenses incurred by a company as a result of having products or services that do not conform to requirements or satisfy customer needs. They are divided into internal and external failure categories by the American Society of Quality (**ASQ**).[9] Internal failure costs occur prior to delivery or shipment of the product (or the furnishing of a service) to the customer. This includes scrap, rework, reinspection, retesting, material review, and downgrading. External failure costs occur after a product or service has been delivered to the customer. These costs include processing customer complaints, customer returns, warranty claims, product **recalls**, and even the risk of being shut down by the FDA or another regulatory body. "Soft" costs must also be taken into account, particularly as a company is seeking to establish itself in the marketplace. These include the negative effect of poor quality on a company's reputation, its ability to attract investors, and its ability to attract and retain valuable employees.

Organizations that proactively address quality early in the design and development of a product have the potential to save significant time, money, and other resources in the long run while reducing the likelihood of devastating product safety issues and/or recalls. Innovators and young start-ups often feel overwhelmed by the amount of work required to implement a quality system, but it is important to note that not all elements of a complete system need to be put into effect right away. Initially, innovators can focus only on those elements that are relevant to their stage of growth. Then, as the start-up expands, so too can the quality system. This staged approach ensures that a solid foundation and good quality practices are established, while not overburdening an early-stage company.

Components of quality management

The concept of quality management is best understood by breaking it down into two components (see Figure 5.5.1). Quality assurance (**QA**) refers to processes that attempt to ensure – in advance – that products will meet desired specifications and perform according to specifications when delivered to the end user. As defined, QA is a broad concept that covers all company-wide activities, including design, development, production, packaging, labeling, documentation, and service, as well as support activities such as employee training and procedures. Quality control (**QC**) refers to activities

Quality Assurance

Improve quality systems

Process oriented

Prospective

Intended to prevent
problems from
occurring

Catch problems

Quality Control

Product oriented

Retrospective

Intended to confirm
that product
meets specification

FIGURE 5.5.1

The concepts of QA and QC are
distinct, but closely interrelated.

performed after these processes have been executed
(e.g., once a product has been produced but before it is
released to the customer) to confirm that the specifications have, in fact, been met. Unit testing, with the intent
of finding defects, is one way that QC is executed.[10] QC
applies to all stages of production from incoming materials, in-process materials/subassembly testing, and finished goods testing.

The primary difference is that QA is *process* oriented
and QC is *product* oriented. QA makes sure a company
does the right things, the right way, whereas QC makes
sure that the results of what the company has done
perform as expected. Stated another way, QA is a
prospective process and QC is a retrospective process
in product development. However, the two are interrelated: QC is a component of a QA system used
to identify problems that can then lead to changes in
QA practices to prevent the same problems from
resurfacing.

Quality management systems

A QMS is the vehicle through which both QA and QC
activities are implemented. A QMS includes policies, processes, and procedures for the planning and execution of
all quality-related activities within an organization. It
also delineates clear responsibilities, starting with the
senior executive level, and helps drive performance
improvement through the measurement and careful
management of core business processes. Importantly,
quality activities were historically the responsibility of
an isolated functional team; however, senior executives
across functions are now considered accountable for
quality activities.

A typical QMS that would satisfy the basic requirements of most regulatory bodies involves as many as
seven components, which are interrelated in specific
ways. The aspects of the QMS shown in Figure 5.5.2
reflect the requirements of FDA's quality system regulation but the underlying principles are also generally
applicable to other countries with well-developed regulatory systems.

At the highest level, **management controls** provide
processes and guidelines for administering the complete
system and are essential from the outset. **Design controls** refer to specific processes used to manage design
specifications and their modifications. Implementing and
spending time on design controls relatively early in the
innovation process is important as these processes are
the cornerstone of any good quality system. Production
and process controls (**P&PC**) ensure that production processes have minimum deviations from their desired performance targets and result in a safe product. Corrective
and preventive actions (**CAPA**) refer to the systems
used to prevent and correct failures. According to FDA's
inspection techniques, these four subsystems make up
the heart of a company's quality system. The three
remaining components – **material controls**; **records,
documents, and change controls**; and **facilities and
equipment controls** – complement the primary subsystems. Of these, records, documents, and change controls
are usually the most important to implement early in the
innovation process as they are central to providing a
solid framework for initial product development.

Importantly, quality practitioners in the field often
point out that there is not a single successful quality
system model. The best way to evaluate whether a

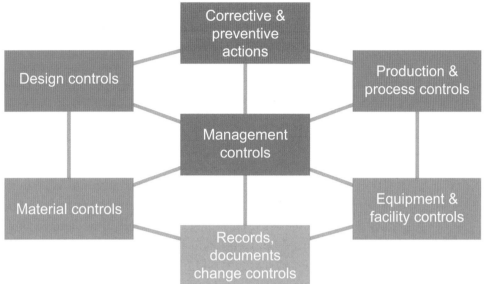

FIGURE 5.5.2

The components of a quality management system (based on FDA's quality system regulation).

company's quality system is adequate is to verify that it achieves certain core objectives:

- It results in the documentation of all product and system requirements.
- Employees are well-trained and follow the documented requirements.
- Records are generated to prove that the requirements are consistently followed.
- It establishes proactive systems to deal with the identification and resolution of problems and improvement opportunities.

Innovators should also understand that quality-related regulations by the FDA and other authorities are written as directional guidelines and must be interpreted appropriately for each company's products and business model. As such, there is no correct sequence in which to implement the various subsystems though, in general, management, design, and records, document, and change controls are often needed as a practical matter before other elements of a QMS. Subsystems such as P&PC and CAPA can usually be more thoroughly implemented a bit later. However, given that these two subsystems are considered core elements of a quality system by the FDA and are assessed as such during audits, putting some basic pieces of these subsystems in place may be useful. How a company interprets and documents its approach then becomes the standard against

which the FDA will audit it for compliance (i.e., the company is audited against its own quality policies and internal documentation).

The seven subsystems of a typical QMS are explained at a high level in the sections that follow. A listing of more detailed references can be found in the Getting Started section.

Management controls

The purpose of the management control subsystem is to ensure that a company's management team provides adequate resources to support effective device design, manufacturing, distribution, installation, and servicing activities. It also establishes mechanisms for ensuring that the quality system is functioning properly, and for allowing management to monitor the quality system and make necessary adjustments on an ongoing basis. According to the FDA, the rationale for this subsystem is that a quality system that has been implemented effectively and is monitored to identify and address problems is more likely to produce devices that function as intended.[11] The key components of the management control subsystem include:[12]

1. Clearly defined, documented, and implemented quality policies and plans.
2. Well-defined quality objectives.
3. An executive in charge of quality management.

4. An organizational structure that includes provisions for resources dedicated to quality management that enable the organization to fulfill stated quality objectives and requirements.

5. Systems for management reviews to monitor the suitability and effectiveness of the quality system and take corrective action where necessary to bring the system into a state of effectiveness.

6. Audit processes to verify that deficient matters are being addressed.

Design controls

From an engineering perspective, design controls are absolutely essential to any quality system and are likely to be the first aspect of a quality system that an early start-up needs to address. The key objectives of design controls are to demonstrate that the design itself is reproducible and traceable and is proven to be both safe and effective. The basis for design controls is initiated in 4.5 Concept Exploration and Testing, as detailed **user requirements** are collected. However, the process for organizing and documenting this information is formalized with the implementation of a design control process.

Design control requirements are implemented to help ensure that every device performs as intended when produced for commercial distribution. To conform with these requirements, device engineers must establish and maintain design plans and procedures that describe design and development activities, define responsibility for implementation, and identify/describe the interfaces with different groups or activities that provide (or result in) input to the design and development process.[13] These plans must be reviewed, updated, and approved as the design and development of a device evolves. These activities are not prescriptive, but provide a method for management to exercise appropriate control over early design work, as well as to assign responsibilities that are consistent with the scope of the design effort.

Risk management is an important part of design controls.[14] Within risk management, hazards refer to potential sources of harm; harm includes physical injury or damage to the health of people, or damage to the property or the environment; and risks represent the probability that harm will occur, as well as the potential

severity of that harm.[15] To assist innovators in formally identifying and addressing risks, QSR advocates for the integration of risk assessment and management activities throughout the design process. With this approach, unacceptable risks can be recognized and mitigated earlier, when changes are less expensive and easier to implement. Again, management has responsibility for applying its experience, insight, and judgment to successfully address this guidance.[16]

Two other important concepts that are central to design controls are **design verification** and **design validation**. Sometimes, these elements are termed "V & V." Design verification confirms that the product meets the specifications laid out by the product development team, while design validation confirms that the design specifications meet customer requirements. Any design specification will need to be validated before it is implemented in a production system and verified before the product is delivered to the customer. A simplistic example that highlights this difference involves a company that has decided to produce a cement lifejacket. Following design input and output, the company can verify that its cement lifejacket meets the **product specification** (i.e., it is made of cement). However, when the company seeks to validate the design against user needs, it encounters problems, since the cement lifejacket clearly will not float (therefore, it will not satisfy user requirements or the lifejacket's intended use).

In the US, premarket approval (**PMA**) and **510(k)** submissions for **Class III** devices typically must include a complete description of the design controls that the company implements. Without this information, the FDA cannot complete its premarket quality inspection (more information about quality inspection is provided later in this chapter). Table 5.5.1 presents guidelines from the FDA that outline the main activities, including those related to risk assessment and management, that must be accomplished to achieve an effective design control system.[17]

Production and process controls

The P&PC subsystem is focused on ascertaining that companies develop processes to ensure that manufactured devices meet their specifications. It also focuses

Table 5.5.1 Understanding and implementing design controls can be valuable even before the innovator has a full-fledged quality system in place.

Activity	Description
Identify design requirements for the device	• Establish and maintain procedures to ensure that the design inputs (or requirements) relating to a device are appropriate and address the intended use of the device, including the needs of the user and/or patient. • The procedures should include a mechanism for addressing incomplete, ambiguous, or conflicting requirements. • Design input is the starting point for product design, providing a basis for performing subsequent design tasks and validating the design. Therefore, development of a solid foundation of requirements is the single most important design control activity. If the majority of design time is spent upfront, doing things correctly, later stages of design can be expedited. • Design input requirements must be comprehensive and include functional (what the device does), performance (speed, strength, response time, accuracy, reliability, etc.), and interface (compatibility user, patient, and other external needs) requirements. Almost every device will have requirements of all three types. • Perform risk analysis to identify all possible sources of failures for different components of the device, acceptable failure rates, consequences of these failures, and corrective actions. The more severe the consequences of a failure, the lower the acceptable failure rates and the more robust the **correction** actions and back-up systems should be.
Develop the design output or specifications for the device	• Establish and maintain procedures for defining and documenting design output (i.e., the physical manifestation of the design planning and input) in terms that allow an adequate evaluation of conformance to design input requirements. Input requirements generally result in what is called a product specification (a document that details the technical and clinical needs the device should meet to satisfy the design intention). • Design output procedures should contain or make reference to acceptance criteria (in the product specification) to ensure that those design outputs that are essential for the proper functioning of the device are identified.
Verify that the design output meets the design input	• Establish and maintain procedures for verifying that the design output meets the design input requirements (i.e., does the device adhere to the design specification?).
Hold design reviews throughout the design process to identify significant problems with the design or the design process	• Establish and maintain procedures to ensure that formal documented reviews of the design results are planned and conducted at appropriate stages of the device's design development. • Multiple reviews of design verifications are not uncommon for complex devices that undergo successive design iterations, and will occur at each

Table 5.5.1 (*cont.*)

Activity	Description
	stage of the design process. Reviews must be conducted by cross-functional teams, thoroughly documented, and approved (signed off) by responsible personnel up through the senior management level.
Validate that the design meets defined user needs and intended uses	• Establish and maintain procedures for validating that devices conform to defined user needs and intended uses. This should include the testing of production units under actual or simulated use conditions.[18] • Design validation should also be performed under defined operating conditions on initial production units, lots, or batches, or their equivalents. • Design validation should include software validation and risk analysis, where appropriate.
Transfer the device design to production specifications	• Establish and maintain procedures to ensure that the device design is correctly translated into production specifications.
Control changes to the design during the design process and changes in the design of products on the market	• Establish and maintain procedures for the identification, documentation, validation or (where appropriate) verification, review, and approval of design changes before their implementation. • This is commonly addressed under a document control or change control system. Especially during the design process, the design teams often manage change and then move the design into the document control systems as part of transferring the design to manufacturing for production.
Document design control activities in the **design history file** (DHF)	• The DHF should be set up to contain or reference the records necessary to demonstrate that the design was developed in accordance with the approved design plan and all design control requirements. • The DHF must be made available to the FDA (or other certified inspectors) for review. The FDA will evaluate the adequacy of manufacturers' compliance with design control requirements in preapproval inspections for class III devices and also during routine quality systems inspections for all classes of devices subject to design control. • A product cannot be legally marketed in the US without a **DHF**.

on validating (or fully verifying the results of) those processes and activities to monitor and control them.[19] It is important to note that validation in this context does not refer to design validation and verification as was discussed under design controls. Rather, in this case, it refers to process validation and verification, in which the focus is on the processes themselves and the operation and use of equipment involved in the enabling these processes. Validation of sterilization processes used to ensure that a product is indeed sterile when delivered to a user provides a good example of process validation. As part of this activity, the sterilization equipment, the sequence of steps followed, and the environment would all be tested in order to establish a threshold for performance and an acceptable operating range in order to meet the specification of consistently delivering sterile products. As a company's R&D efforts move it toward a device that is ready for production, it must consider these types of controls, though implementing some basic elements of this control subsystem may be helpful in preparation for any FDA audits.

In developing P&PC systems it is important to understand when deviations from device specifications could occur as a result of the manufacturing process or environment and to pay special attention to processes with high risk of potential deviations. Management should be on the lookout for potential P&PC problems when a process is new or unfamiliar to the company, is used to produce higher-risk devices, or is used for manufacturing multiple devices. P&PC problems may also be indicated when a problem with a particular process is identified through the CAPA subsystem, when a process has a high risk of causing device failures, employs a variety of different technologies and profile classes, or has never been examined or inspected.[20]

Once processes with a higher than average risk for P&PC problems are flagged based on the presence of one of more of these indicators, methods can be developed and implemented for controlling and monitoring them to minimize production deviations.

Corrective and preventive actions
The purpose of the CAPA subsystem is to establish processes for collecting and analyzing information related to product and quality problems. Referring back to the QA/QC distinction, QC results are a core part of CAPA, but CAPA also incorporates the process used to act upon the QC results. QA is more of a broad, umbrella concept that reflects the combination of all components of the QMS and how they work together to provide the infrastructure for prospectively delivering a high-quality product.

CAPA is made up of processes such as non-conforming raw materials reporting, production process deviations, and customer complaints. Systems are put into place not only to ensure that deviations, complaints, and other

problems are reported and documented, but that appropriate actions are taken to correct and prevent the recurrence of problems and that these actions are verified. Communicating CAPA activities to responsible parties, providing relevant information for management review, and documenting these activities is essential to dealing effectively with product and quality problems, preventing their recurrence, and minimizing device failures.[21] As a result, this subsystem is one of the most critical components of the quality system and is viewed as such by the FDA. However, as product and quality problems often arise a little further downstream in the product development cycle, it is usually sufficient for innovators early in the innovation process to be aware when a CAPA subsystem and its attendant processes need to be fully implemented and focus initially on establishing good management, design, and document controls, after which a general framework for this subsystem can be put into place.

Key activities that should be undertaken in this area include defining, documenting, and implementing robust CAPA processes that ensure the visibility of quality-related problems all the way to the top of the organization. The CAPA subsystem is closely related to medical device reporting (**MDR**). MDR is the mechanism through which the FDA receives information about significant adverse events from manufacturers so they can be corrected quickly. For this reason, CAPA and MDR processes must be tightly integrated (see online Appendix 5.5.1 for more information about MDR).

Equipment and facility controls
Equipment and facility controls are meant to ensure that a company's equipment and facilities are qualified (i.e., they are suitable for their intended purposes), and that standard operating procedures have been designed, implemented, and enforced for all equipment and facilities managed by the company. Qualification of equipment involves its installation, ongoing operations, preventive maintenance, and the overall validation of its outputs and related processes. This subsystem applies to equipment and facilities involved in design, production, and post-production activities. These controls allow the company to determine if any quality events are

related to equipment at a particular facility so that equipment changes can quickly be made (when needed) to address a quality event.[22] On a practical note, early-stage companies often use outside services, vendors, or manufacturers for various aspects of product development. Sometimes choosing contract partners with established equipment and facility controls, and that are familiar with the quality requirements of the medical device space, can help in the long-run with quality assessments related to regulatory and manufacturing processes.

Material controls

Whether they are related to design, production, or post-production activities, all materials used in a medical device must be carefully controlled. Medical device companies must maintain processes to track all materials and their associated suppliers to ensure the quality of those materials and that the final product satisfies the design specifications. All suppliers must be rigorously screened and able to demonstrate that their materials are traceable to qualified and appropriate sources. Detailed records (including traceability of material to suppliers and specific delivery lots) must also be maintained so that any problems that arise can be tracked at the materials level.[23] In addition to thinking about materials from a quality standpoint, innovators should think carefully about the materials they choose for practical reasons, especially if these materials will come in contact with or be implanted or used in the human body. There are key requirements for the **biocompatability** of materials, so choosing materials that are medical grade and may have already been used in other human applications can potentially have a significant impact on the cost and timelines of regulatory processes and product qualification.

Records, documents, and change controls

This subsystem is focused on ensuring that medical device companies maintain a secure, comprehensive, and centralized approach to managing all records and documents related to their quality systems.[24] Along with design controls, this is one of the first, most important subsystems for innovators to implement because it

affects early product development activities, as well as those that come later as product development progresses. For example, maintaining a DHF is essential. Clearly defined protocols should be put into place to track changes to processes, policies, and the products themselves, manage version control, and make necessary documentation accessible to those who need it during design, production, and post-production activities. Another key principle is that all parties involved agree on how work will be done. When changes to these original work agreements are required, they must be formally evaluated and confirmed.

Implementing a quality system

Implementing a quality system can be time-consuming and resource intensive, as detailed processes and procedures must be carefully orchestrated and then documented. While regulatory bodies require that quality systems are in place, these requirements focus on the goals of these systems and not on *how* the procedures must be implemented. Therefore, designing a quality system can be somewhat of a creative exercise (and, as mentioned earlier, the company's interpretation becomes the standard against which it is regulated). On the one hand, this allows a company to iterate and develop a system that works with its particular needs and fits the company culture. On the other hand, the degree of interpretation and analysis required to make "appropriate" decisions that are likely to achieve compliance can be challenging. As quality consultants tend to point out, many "gray areas" exist when it comes to achieving compliance.

Another problem that companies struggle with is getting and keeping management's attention when it comes to the quality system. Too often, over-extended managers have a habit of only devoting their attention to the quality system after a problem has surfaced. This is due, at least in part, to the fact that it is so difficult to quantify the value of a quality system that works, whereas the cost of a quality problem once it emerges is all too obvious. However, given the growing emphasis on designing and implementing quality systems as a part of an overall risk management strategy, executives should recognize that these practices are invaluable to the health of the overall business. Management commitment and attention is essential to making sure a quality system is supported and maintained by employees throughout the organization. Given that executives have to "sign off" on most aspects of the quality system as part of the management controls subsystem and, thus, share responsibility for its outcomes, it is in their best interest to ensure that a sound quality system has been designed and implemented. Careful attention to implementing management controls early in the design of a quality system can help in this regard.

A third issue that many companies face is characterized by a reactive versus a proactive approach to quality management. Especially with start-up ventures, too many medical device companies postpone or overlook the need to develop a quality system and fail to adequately document their work according to the procedures outlined in the quality plan. As a result, they are forced to scramble when problems arise and have to retrospectively "fill in the blanks." The quality system can be viewed as onerous, or as a system that supports the ultimate goal of the company – to design and produce an innovative, safe product. Viewing it as the latter yields the best systems in which compliance is part of everyday work life, rather than an additional burden. Moreover, innovators must remember that it is illegal to bring a medical device to market in the US without having the supporting quality system in place and working properly. Table 5.5.2 summarizes these and other common implementation pitfalls that can jeopardize the effectiveness of a quality system and its return on investment.

Again, senior management sets the tone for quality system design and compliance. If the management team approaches quality as an essential, important, and value-added activity toward the organization's goals, the system is more likely to be valued, followed, maintained, and improved over time to service the company's growth and expansion. By implementing just the most relevant components of a quality system when the company is first starting out (usually management controls, design controls, and records, documents, and change controls), management may be able

Table 5.5.2 Common mistakes to avoid when implementing a quality system.

Common mistakes when implementing a quality system
Viewing the requirements as a burden, rather than a mechanism for increasing firm effectiveness.
Entrusting management of the quality system to employees who do not have thorough training and experience in quality systems.
Making quality an isolated, non-essential function rather than integrating it into the business via cross-functional and senior management involvement.
Being too prescriptive early on and not considering the stage of the product development or the team.
Not including quality professionals in system design AND not including all of the system users in the system design.
Not thinking about how the quality system will need to scale and grow to keep pace with growth in the rest of the business.
Not training all personnel on the overall quality system, as well as their essential role(s) in making the system work.
Designing and implementing a quality system, but not maintaining it (which ultimately renders the system useless).
Ineffective and/or untimely action to deal with problems within the system – poor use of corrective and preventive action systems.

to get greater company-wide buy-in than if it tries to implement everything at once.

When developing and implementing a quality system, companies usually begin by naming an executive to be in charge of quality (often the vice-president of R&D in small companies). This person will sponsor the quality work stream and work with executives and managers in other parts of the business to ensure cross-functional support. A quality engineer, preferably with QSR experience, is also hired to lead the tactical development of the appropriate processes, protocols, and documentation, again with cross-functional involvement. Sessions to educate employees about the quality system, including its importance and specific requirements that affect their work, are another important early step. Then, the company can begin to hold regular review meetings to ensure that it is achieving desired performance levels.

Keep in mind that the US Supreme Court has allowed criminal penalties to be imposed on corporate officers who were in a position to prevent or correct violations, even if they may not have known about or participated in any illegal conduct:[25]

> The [Food, Drug and Cosmetic] Act imposes not only a positive duty to seek out and remedy violations when they occur but also, and primarily, a duty to implement measures that will insure that violations will not occur. The requirement of foresight and vigilance imposed upon responsible corporate agents are beyond question demanding and perhaps onerous, but they are no more stringent than the public has a right to expect of those who voluntarily assume positions of authority in business enterprises whose service and products affect the health and well-being of the public that supports them.

Implications of an increasingly tough enforcement environment

When considering the best approach to implementing a quality system, it is important for innovators to recognize that, in recent years, both the FDA and public have become increasingly conservative and risk-averse when it comes to medical devices. One example of how the FDA has been more vigilant about exercising its authority over companies and individuals that fail to meet QSR requirements can be found in the story of the 2007 FDA raid on Shelhigh Inc., a manufacturer of heart valves and other implantable devices for heart surgery. FDA investigators and US marshals seized all of the company's devices after finding what the FDA deemed to be significant deficiencies in the company's manufacturing processes.[26] According to an FDA statement, the **seizure**

followed an FDA inspection of the Shelhigh manufacturing facility, as well as meetings with the company at which the FDA warned Shelhigh that failure to correct its violations could result in an enforcement action. The FDA also alerted the company to its manufacturing deficiencies and other violations in two warning letters.[27]

Another example that illustrates the potential severity of FDA actions to enforce QSR is the story of C.R. Bard Inc. In 1995, the company was found guilty of unlawfully selling and distributing unapproved heart catheters. The company was forced to pay record criminal and civil fines totaling $61 million. It was also required to implement stringent measures to prevent such illegal activities from occurring again.[28] In addition, three former senior Bard executives were convicted of conspiring to defraud the FDA. The three officials were found to be aware of the serious patient complications that resulted and of the company's efforts to change its products without FDA clearance. These men each received 18 months in prison.[29]

Stories such as these demonstrate why the collective attitude about quality is shifting toward increased prevention. These issues are also driving an increased focus on post-market quality, not just premarket quality requirements.

Quality system regulations in the medical device industry: QSR and ISO

The two most dominant quality systems in the medical devices industry are FDA's QSR and **ISO** 13485. **ISO 13485** was developed by the International Organization for Standardization and is required for devices marketed in the European Union (EU) and other countries recognizing the **CE mark**. Depending on the target market for their products, companies will follow one or both of these standards. The two systems have elements in common, but are fundamentally separate and regulated differently.

When the FDA began to regulate medical devices in 1976, the agency developed what it called good manufacturing practices (**GMP**) to set forth quality requirements for device manufacturers. In 1997, the FDA revised and expanded the device quality regulations under the QSR rubric (although the term "GMP" still lingers in the device field). The FDA moved to QSR in order to enforce regulation that was more focused on prospectively ensuring quality (QA) as opposed to retrospectively catching quality problems (QC). QMS and its subsystems (detailed earlier in this chapter) meet the basic QSR requirements. Of importance to early-stage innovators, one of the notable changes with the new approach was the introduction of design controls into US quality regulations.

The second important change in the shift from GMPs to QSR was driven by the emergence of **ISO 9001** certification in the late 1980s and early 1990s. The ISO 9001 system was considered a best-in-class model for quality management that was applicable to any industry in any country in the world. In 1994, ISO introduced supplementary guidelines (EN46001) to be used in combination with ISO 9001 to address unique quality requirements for medical device manufacturers. As medical device companies began adopting ISO certification on a voluntary basis, along with the required GMP standards, the FDA recognized the merits of taking a systems-based approach to quality. This realization dovetailed with the overhaul of GMP and stimulated the FDA to make its approach more systems-oriented and consistent with the ISO 9001 standard.

In 2001, ISO 13485 superseded the combination of ISO 9001/EN46001 and was launched as a worldwide quality management system designed specifically for medical device manufacturers. ISO 13485 was subsequently updated in 2003 and again in 2012 (though the 2012 update included no changes to the text of the global standard from 2003 and only had revisions to the fore-word and annexes). Based on the same basic principles as ISO 9001, ISO 13485 is often seen as a crucial first step in ensuring design and manufacturing processes consistently produce high-quality products that meet international regulatory requirements.

Importantly, ISO 13485 dictates that risk management must be thoroughly documented and conducted across all stages of a product's lifecycle, but it leaves the specifics to a related standard, ISO 14971: 2001, Application of Risk Management for Medical Devices. ISO 14971 outlines the steps that must be taken by management to fulfill device-related risk requirements. Companies pursuing ISO 13485 certification are not formally required to be 14971 certified. However, compliance with 14971 can aid in the attainment of 13485 certification.[30]

Although not required in the US, ISO 13485 certification is a prerequisite for achieving regulatory approval in the EU, as well as marketing in Canada, Australia, and Japan.[31] In other countries with regulatory systems that are still taking shape or being modernized, such as China or India, specific quality system requirements may be less well defined or in flux. Innovators working in these countries are encouraged to seek out consultants or firms with knowledge of the regulation landscape to understand current requirements. Regardless, by proactively following the principles embodied by ISO and/or QSR standards, innovators can help ensure that they are prepared for whatever requirements they may face.

Primary differences between QSR and ISO 13485

The FDA participated in the harmonization of QSR with the European ISO standards, but stopped short of adopting the ISO standards outright. The biggest difference between the two systems is that ISO is a voluntary standard in the US and QSR is not, which makes the compliance process a very different experience for companies. The conventional wisdom is that ISO standards are more stringent and rigid than QSR, but they do not necessarily ensure QSR compliance. A sample of other high-level differences is provided in Table 5.5.3.

An important requirement for US companies is to determine where they intend to market and manufacture the device – in the short term and the long term. They must consider whether to build a quality system that is both QSR and ISO compliant, or just compliant with one or the other. Any distribution of the device in the US

Table 5.5.3 Important distinctions between the QSR and ISO quality systems.

	QSR	ISO 13485
General Requirements/ Provisions	"Each manufacturer shall establish and maintain a quality system that is appropriate for the specific medical device(s) designed or manufactured, and that meets the requirements of this part."	"The organization shall establish, document, implement, and maintain a quality management system and continually improve its effectiveness in accordance with the requirements of this International Standard."
Link to Regulatory Approval	Required in the US for all **Class II** and **III** medical devices (as well as some **Class I** where general controls are required).[32]	Voluntary in the US but required in the EU and Canada as a prerequisite for regulatory approval and increasingly in other countries combined with other country-specific requirements.
Auditors	FDA inspectors.	Conformity assessment bodies (**CABs**) in EU or other third-party ISO-accredited inspectors elsewhere in the world.
Audit Frequency	Generally, every two years, but this timetable is rarely maintained by the resource-constrained FDA. Safety-related concerns will merit more frequent visits.	Companies generally follow an annual or semi-annual "maintenance" audit procedure.
Audit Scheduling	Audits can be announced or unannounced.	Audits are scheduled in advance.
Cost	There is no cost for an FDA audit beyond the costs the company faces in implementing and maintaining the quality system, dedicating personnel to the audit process, and addressing required corrections.	Costs of initial certification can exceed $30,000 to $40,000, in addition to annual maintenance fees. Cost depends on the size of the organization, as well as the chosen registrar.
Audit outcomes	FDA has enforcement power over the organization and audit findings must be acted upon.	Because ISO is voluntary, an "unsuccessful audit" simply results in postponed certification.

requires QSR compliance and, in general, if a company is to be registered with the FDA it should follow QSR requirements (unless it has an exempt Class I device). However, a company's strategy might involve early clinical work performed outside the US (i.e., in the EU or other foreign countries). If this is the case, ISO compliance may precede QSR compliance. Or, if a company leads with US marketing but intends to later expand to the EU (or beyond), it would be wise to build a system that is both ISO and QSR compliant. Many quality professionals have experience in building such hybrid systems. If an element of one system does not apply at the current time, it is not necessary for all elements to be turned "on" at once. A framework can be established and necessary elements can be brought "online" as needed.

For additional information about ISO 13485 see online Appendix 5.5.2. An interpretive summary of the FDA's QSR is provided in online Appendix 5.5.3.

The story of Sympara Medical illustrates how one **medtech** start-up company approached the process of implementing a quality system.

FROM THE FIELD ▷ SYMPARA MEDICAL

Setting up a quality system

As Sympara Medical, a venture-backed start-up in San Francisco, California, started working on an innovative device to treat hypertension, the founding team realized that it needed to think about quality management early in the development process. Although the company was not sharing the details of its technology at the time of this writing, co-founder Kevin Ehrenreich offered that, "Because we have a much different design compared to most other medical devices, we have the opportunity to move quickly into the clinic, without going through many of the traditional phases of the design process."[33] This accelerated development and testing schedule caused Ehrenreich and team to begin thinking about how to establish a quality system soon after the company's initial founding in order to most efficiently manage risk and prepare the device for human studies.

One of the main goals of Sympara's Series A funding was to launch a First Human Use study in the clinic within one year. To achieve this milestone, the company had to immediately rent manufacturing space and begin preparing to obtain its California Medical Device Manufacturing License, which would allow the team to manufacture the regulated medical products for use in its clinical study. The Cal-State license, as it is commonly known, requires an inspection of the facility and at least one of the following items: an FDA-issued biologics license, an FDA-approved investigational device exemption (**IDE**), a copy of a federal inspection completed in the last two years, or proof of compliance to ISO standards.[34]

As a first step, the leadership team brought in an external quality consultant who spent approximately six months learning about the company and its device so that he could help design and implement a quality system that was tailor-made to Sympara's

requirements in the near-term, but that would also provide a solid foundation from which the company could expand over the next five years. Accordingly, they used a combination of the FDA's quality system regulation and ISO 13485, which have significant areas of overlap. QSR and ISO both provide medical device manufacturers with guidance to help them design and implement a quality system that meets essential regulatory requirements, but the standards also give companies the flexibility to customize their approach based on the unique characteristics of the device and what is most needed to ensure it is safe. "We didn't just want to release a quality system for the Cal-State inspection, we wanted to be actually testing the system to make sure it worked for the type of device we were building," said Ehrenreich.

While the old approach to quality management was grounded in a compliance mentality, Ehrenreich indicated that the new trend is to consider the quality system as a mechanism for risk management. Early on, the Sympara team made a list of all of the **stakeholders** that would interact with its device, including manufacturers, physicians, nurses, pharmacy staff who stock the device, patients, and so on. "We actually put up photos of these stakeholders," he said. Then, with representatives from all functional areas across the organization, the team thought of all the ways its device could potentially create harm for each stakeholder group at every step in the **cycle of care**. Through this exercise, the company identified critical risks and related hazards that it would seek to mitigate through design, development, and quality management. "The earlier a company starts thinking about risk analysis, the earlier it will be able to set the appropriate specifications for the device and focus on achieving them to keep the patient safe," Ehrenreich stated.

In terms of actually designing and implementing the quality system, "The team's key consultant led our initial focus towards developing basic quality procedures and a quality manual," he continued. "The way to get started is to think about what's value added to the company at the given stage of development and what doesn't need to come online until later." Working with the quality consultant, the Sympara team identified 10–12 core aspects of the blended quality system requirements that needed to be in place early in the development process; for instance, management controls, design controls, records, documents, and change controls, validated testing methods, and part identification. Then, as development progressed and the company moved closer to the clinic (and eventually to market launch), they would integrate more advanced features into the quality system. "Having a flexible quality system that can accommodate the development needs of the company is key," Ehrenreich noted. "You don't want to over build the quality system before it's necessary, but many of these requirements add a lot of value during the early stages of development."

With guidance from Sympara's quality consultant, the team was encouraged to write and release quality system components such as standard test methods with the expectation that they would improve and revise them over time. As Ehrenreich explained, "The first time you write it, it's not going to be perfect, it might not even be good. But go ahead and write it and release it, and try to build or test parts using that documentation. Then, don't be shy about editing it. The first time, you may make 50 percent changes to the document or process. But by the third or fourth time, it will be good enough. Remember that the enemy of good is perfect," he added.

FIGURE 5.5.3

"Symparians" at work (courtesy of Sympara Medical).

Of course, this work should be conducted in collaboration with the quality consultant. In fact, Ehrenreich advised innovators to consider working with dual consultants who bring different points of view to the company. "Get two consultants who know the subject matter really well – one who's conservative in their view of quality systems and one that's more liberal – and *work with them* to settle on the best approach for the technology and the company," he elaborated.

Ehrenreich also strongly recommended keeping everyone in the organization involved in building out the quality system by discussing case studies that are grounded in the company's real-world experiences. For example, he said, "Have someone from the clinical team bring a clinical challenge and talk about how the quality system can help address it." Ultimately, the best quality systems are those that are based on the needs and input of the entire cross-functional team (see Figure 5.5.3).

Finally, Ehrenreich commented, "Remember that risk analysis is key. That's what gets you to the clinic and allows you to show that your device is safe, which is the whole reason to have a quality system."

Quality systems audits and what to expect

An important component of the regulation of medical devices is the performance of inspection audits of the quality system. Any company that manufactures or processes an FDA-regulated product may be audited every two years, although the actual interval between inspections is often longer. Preliminary audits are triggered by regulatory submissions and/or applications to use a device in humans. Following regulatory approval or clearance, most inspections are routine, although the FDA also conducts audits for cause as problems come to light.

To economize resources and maximize the value of its audits, the FDA has developed the Quality Systems Inspection Technique (**QSIT**). This process is based on a "top-down" approach to inspections, placing executive management at the core of the quality system. Rather than starting an audit by looking at any specific quality subsystem or potential non-conformance problems ("bottom up"), QSIT begins with an assessment of the company's total system. Then, QSIT sets forth specific guidelines and processes for the evaluation of the four primary subsystems within the company's overall quality system described earlier in this chapter (management controls, design controls, P&PC, and CAPA).[35]

The tone of a quality audit is often set by the company being audited, not the auditor. The FDA inspector has a job to serve the public by ensuring that companies are compliant with QSR. In turn, companies have the obligation to demonstrate to the FDA that they are, in fact, compliant with these regulations. This is largely determined through documented evidence, as well as some facilities inspections. The duration of quality audits depends on the nature and cause of audit, as well as the size of the firm, with brief audits lasting one to two days and lengthy, complex audits requiring one to two months. Auditors typically spend 80 percent of their time during an audit performing records review (looking for proof that the company's quality system requirements are consistently followed). The remaining 20 percent of the time is spent conducting interviews and observations of processes.

When contacting the company for a QSIT inspection, the auditor will generally request a copy of the firm's quality policy, as well as high-level quality system procedures, management review procedures, quality manual, quality plan, or other equivalent documents. (Note that QSIT inspections techniques only apply to preannounced audits, but not all audits are preannounced.) The company is *not* required to supply this documentation. However, the audit will typically progress more quickly if this information is provided in advance and the company assumes a cooperative posture with the FDA. All documentation is returned at the time of the inspection.[36] Within a company that has a well-designed quality system, compiling this documentation should be an efficient, simple, and straightforward process.

Each FDA quality inspection is designed to begin and end with an evaluation of the management control system. Upon the initiation of an audit, the first thing inspectors will likely do is to meet with the executive responsible for quality within the company to get an overview of the quality system and to verify that appropriate management controls are in place. Typically, the inspectors will then select a single design project to evaluate through the end-to-end design control process. At times, the inspection assignment will direct the inspectors to a particular design project (i.e., as part of a "for cause" inspection – an audit triggered by some evidence suggesting non-compliance in a particular area). Otherwise, they will select a project that provides the best challenge to the firm's design controls system. This project will be used to evaluate the process, methods, and procedures that the firm has established to implement the requirements for design controls outlined in Table 5.5.1.[37]

Based on a discussion with management, inspectors will also choose a manufacturing process to evaluate that seems to be a likely candidate for production deviations. They will then review the specific procedure(s) for the chosen manufacturing process, as well as the methods for controlling and monitoring the process. Their objective is to verify that the process is well controlled and actively monitored.

The agency will also seek to verify that CAPA system procedure(s) have been defined and documented, appropriate sources of product and quality problems have been identified, and data from these sources are analyzed to identify existing product and quality problems that may require corrective action. The inspection also seeks to confirm that the defined CAPA processes are followed when problems arise.[38]

During the audit, employees may be coached to answer only the questions asked by the inspector and not to offer up additional information. It is the auditor's responsibility to follow whatever leads it encounters regarding potential quality problems. An audit potentially can be prolonged by an unraveling of issues, and no company wants the FDA on its premises longer than necessary since audits are time-consuming and disruptive to the organization. They can also be intimidating to employees. However, a company that has implemented effective, comprehensive quality systems from its inception should have nothing to hide from the FDA. Similarly, a well-designed quality system should be easily and successfully audited. In no case should the company try to hide anything. It is far worse to be caught trying to conceal information (which is illegal) than it is to receive a finding that will help improve the system.

The second case example illustrates effective practices for managing an FDA quality inspection to achieve a positive end result. For more information about compliance actions and enforcement, including responding to audit results and important information about product recalls, see online Appendix 5.5.4.

FROM THE FIELD ▸ DIASONICS AND OEC MEDICAL SYSTEMS

Strengthening quality systems to improve business performance

Allan May joined Diasonics, as its senior vice president of business development, just as the company was embarking on a major turnaround. According to May, "Diasonics was the company that introduced real-time imaging to the medical device markets. Until then, if you wanted an image, you had to go to the radiography suite or another location where the images could be taken. The idea that you could look at a dynamic image during surgery didn't exist." By introducing innovative new imaging technologies, such as ultrasound and fluoroscopy, the company grew rapidly from its inception in 1978 to its highly successful initial public offering (**IPO**) in 1983. However, shortly thereafter, it faced a series of performance problems that necessitated the formation of a turnaround team.

When May joined the company, Diasonics was a conglomerate of four or five businesses, each with its own technology. One of the issues facing the turnaround team, he said, "was how to harvest more value out of these various pieces to increase returns to shareholders." May continued, "OEC Medical Systems was the crown jewel of this corporate group. The trouble was that Wall Street wasn't giving Diasonics credit for what we felt was the real value of OEC." OEC specialized in making fluoroscopy C-arms – mobile machines that could be moved into the surgical or special procedures suite for real-time imaging during various laparoscopic or endoscopic procedures (see Figure 5.5.4). "That market was very small in the 1970s," he commented, "but it exploded in the 1980s with the proliferation of minimally-invasive procedures."

The OEC product initially had a reputation for being costly and over-engineered with "needless features." "Fluoroscopy was a Volkswagen market, and OEC was building Mercedes," recalled May. "But as these procedures caught on and the market really started to take off, you had longer procedures, like minimally-invasive cardiac or neurology procedures, that placed new demands on the equipment. Those over-engineered features became critical, and OEC's competitors couldn't duplicate them because they didn't have the more advanced architecture. The new requirements couldn't just be added on as a feature set," he explained.

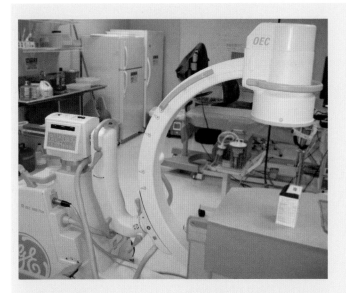

FIGURE 5.5.4

An OEC® C-Arm, similar to the products produced by OEC when it was part of Diasonics. OEC was later acquired by GE (courtesy of Acclarent, Inc.).

This allowed OEC to capture approximately 70 percent market share against competitors such as GE, Siemens, Philips, and Toshiba. "So here was this little dinky company in Salt Lake City, Utah," said May, "competing globally against these major corporations. Every few years, they would appoint vice presidents to study and crush this little start-up, but they never broke it. And we held that market share despite their increasing attempts."

In the midst of OEC's unprecedented growth, the company experienced an unannounced FDA audit. "This was right about that time that a couple of high-profile problems hit the FDA and our sense was that the division offices had been told to step up their inspections," May noted. "The FDA inspector came into the OEC plant in Salt Lake City, inspected it, and said that, in his opinion, it looked like the plant might have to be shut down." During this same period, the US plants of several major competitors had been shut down or their imports restricted.

The FDA's report addressed a series of issues, but primarily focused on complaint handling and the company's corrective actions and procedures. "To be

fair," May said, "the regulations were really just being put into place when this happened.[39] But the concept of an effective complaint handling system didn't exist at OEC. Basically, someone would complain and one of the engineers would look into the problem and try to fix it. There was no senior level involvement or review, few defined processes, and no comprehensive complaint handling system tied to corrective action loops. Most problems were considered 'unable to duplicate or verify' and so little was done to redesign the product or change production processes to prevent the problem from recurring. I also don't think we had manufacturing process instructions on the floor. We were making the best product in the world – number one in its class. Everybody knew how to do their specific job. But it didn't occur to people that, gee, if someone got hit by a bus, no one else would know how to do what they were doing."

News of the inspection results were received by employees in the plant with what May described as "outrage" and the perception of "bureaucratic make-work." "Many employees were in just a complete state of denial, or didn't understand what this was all about. So, our first challenge was to get to the core of the problem. We went to the FDA immediately and got involved with the agency. We wanted to make it very clear to them that we were not going to be confrontational. We intended to be completely cooperative in understanding and fixing our problems and would be inclusive of them in our discussions about what to change. Then, we had to go back to the plant and really patiently explain from the top of the organization down through every layer why we needed to do things differently and what those things were," he said. "The key was getting people's attention and focusing them on understanding the real problem. We couldn't just let them slough it off like, 'Okay, we have to fill out a bunch of paperwork, but we're right and the FDA is wrong.'"

As May and team began to investigate the FDA's concerns within the plant, it became increasingly clear that these quality issues were directly linked to the company's business performance. "Our service calls on the equipment had crept up from one or two a year to 10 to 12 per year," May recalled. "Our customers were seeing our technicians more than our sales people. We recognized that there were some problems with the equipment but, since it had such a tailwind of being the dominant product in this particular sector, we failed to appreciate the extent to which customers were putting up with it. But they were really starting to grumble. On top of that, our margins were decreasing because it was costing us so much to service the equipment."

Over the next 18 months, OEC launched an expansive root-cause analysis to identify all potential sources of product failures. Based on the outcome, the company completely revamped its quality systems, implementing comprehensive new processes and thorough documentation. They reorganized their entire manufacturing facility and changed many of their process and systems, as well. When the FDA reinspected the plant, it was found to be in full compliance. Moreover, said May, "We ended up getting those service calls back down to one or two a year and driving down our costs. We also dramatically improved the quality of the product, our reputation with customers, and the satisfaction of our customers."

When asked for his advice to innovators implementing quality systems, May commented, "You have to completely forget the idea that this is about paperwork. You can't hire a consultant to draft a bunch of documents that you put in a drawer somewhere. Your quality system is totally about the mentality you use to design and manufacture your product. And I believe a hundred percent that if you do what the quality

regulations say, and really understand them, you will design and manufacture a better product with fewer problems that will lead to improved financial performance."

Reflecting further, May added, "Quality should be understood at the top of the organization. When we were going through this, we had a slew of meetings where everyone, right up the board of directors, had to be involved. The higher you get in the organization, the more people may gripe about this sort of thing. But senior executives can't just delegate this and expect it to be done right. Upper management has got to stay personally involved. The phrase we repeated most often was, 'Quality is a journey, not a destination.'"

Overall, May stressed, the strength of a company's quality systems can either help or hinder the attainment of its strategic goals. In this particular case, Diasonics' senior management team had developed a corporate strategy to spin out OEC (along with two other operating companies) to help "unlock" additional value in the market. However, until OEC's quality issues were addressed and its risk of being shut down by the FDA was eliminated, "We had to keep OEC underneath the protection of the corporate shell," May recalled. This delayed Diasonics' ability to implement its significant corporate restructuring plans. However, sometime later, after OEC was found to be in full compliance with the FDA's requirements, Diasonics did, in fact, spin out OEC onto the public markets. Not long after, OEC was able to execute one of the larger corporate exits in the medical device field.

↘ Online Resources

Visit www.ebiodesign.org/5.5 for more content, including:

 Activities and links for "Getting Started"
- Identify quality needs and decide on an approach
- Hire a quality professional
- Engage executive and cross-functional management and define quality policies
- Build a "shell" of a quality system and develop elements as the product progresses
- Assign cross-functional champions to monitor and maintain quality system
- Anticipate and prepare for audits

 Videos on quality management

 Appendices that provide additional information about:
- Medical device reporting
- ISO 13485
- 21 CFR 820
- Compliance actions and enforcement

CREDITS

The editors would like to acknowledge Andrew DiMeo and Ritu Kamal for their help updating this chapter for the second edition of the text. Many thanks also go to Kevin Ehrenreich and Scott Wilson for sharing the Sympara Medical, Inc. story and Allan May for his assistance with the Diasonics and OEC Medical Systems case example. Further appreciation goes to Michelle Paganini of Michelle Paganini Associates for her lectures in the Stanford Biodesign Program and Trena Depel for her contributions to the original material.

NOTES

1 "Medical Devices; Current Good Manufacturing Practice (CGMP) Final Rule; Quality System Regulation," U.S. Food and Drug Administration, http://www.fda.gov/medicaldevices/deviceregulationandguidance/postmarketrequirements/qualitysystemsregulations/ucm230127.htm (February 3, 2014).

2 "Design Control Guidance for Medical Device Manufacturers," U.S. Food and Drug Administration, March 11, 1997, http://www.fda.gov/cdrh/comp/designgd.html (February 3, 2014).

3 Ibid.

4 "U.S. FDA Warns Boston Scientific," CBC News, August 23, 2005.

5 Barnaby J. Feder, "Device Maker Moves to Appease the FDA," The New York Times, http://www.nytimes.com/2006/02/03/business/03device.html?ex = 1296622800&en = 7b669a64d72d2e90&ei = 5088&partner = rssnyt&emc = rss (February 3, 2014).

6 John Chesto, "Boston Scientific Reaches a Crucial Milestone," The MetroWest Daily News, March 3, 2008.

7 Todd Wallack, "A Blockbuster Revisited," The Boston Globe, January 13, 2008, http://www.boston.com/business/globe/articles/2008/01/13/a_blockbuster_revisited/ (February 3, 2014).

8 "FDA Lifts Boston Scientific's Product Approval Ban," USA Today, October 22, 2008, http://usatoday30.usatoday.com/money/economy/2008-10-22-1637289307_x.htm (March 21, 2013).

9 "Cost of Quality," American Society for Quality, http://www.marketwatch.com/story/boston-sci-fda-lifts-some-of-warning-letter-restrictions (February 3, 2014).

10 "Difference Between Quality Assurance, Quality Control, and Testing?," Test Notes, http://geekswithblogs.net/srkprasad/archive/2004/04/29/4489.aspx (February 3, 2014).

11 "Guide to Inspections of Quality Systems," U.S. Food and Drug Administration, August 1999, http://www.fda.gov/downloads/iceci/inspections/ucm142981.pdf (February 3, 2014).

12 Ibid.

13 "Quality System: Design Controls," U.S. Food and Drug Administration, http://www.accessdata.fda.gov/scripts/cdrh/cfdocs/cfcfr/CFRSearch.cfm?CFRPart = 820&showFR = 1&subpartNode = 21:8.0.1.1.12.3 (February 3, 2014).

14 "Design Control Guidance for Medical Device Manufacturers," op. cit.

15 Niamh Nolan, "Quality Risk Management – The Medical Device Experience," Parenteral Drug Association, http://www.pda.org/Chapters/Europe/Ireland/Presentations/Quality-Risk-Management–The-Medical-Device-Experience.aspx (February 26, 2014).

16 "Design Control Guidance for Medical Device Manufacturers," op. cit.

17 Ibid.

18 See Dick Sawyer, "Do It By Design: An Introduction to Human Factors in Medical Devices," U.S. Food and Drug

Administration, http://www.fda.gov/medicaldevices/
deviceregulationandguidance/guidancedocuments/
ucm094957.htm (February 3, 2014).

19 Ibid.

20 Ibid.

21 Ibid.

22 "Medical Device Makers Find Solution to FDA Demands,"
Tooling and Production, December 2007, http://www.
toolingandproduction.com/features/2007_December/
1207_medical_device.aspx (February 3, 2014).

23 Ibid.

24 Ibid.

25 Daniel P. Westman and Nancy M. Modesitt, *Whistleblowing:
The Law of Retaliatory Discharge* (BNA Books, 2004), p. 39. See
United States versus Park, 421 U.S. 658, 672 (1975) for more
information.

26 "FDA Seizes All Medical Products from N.J. Device
Manufacturer for Significant Manufacturing Violations," U.S.
Food and Drug Administration, April 17, 2007, http://www.fda.
gov/NewsEvents/Newsroom/PressAnnouncements/2007/
ucm108893.htm (February 3, 2014).

27 Ibid.

28 Paula Kurtzweil, "Ex-Bard Executives Sentenced to Prison,"
FDA Consumer, December 1996, http://connection.ebscohost.
com/c/articles/9704076198/ex-board-executives-sentenced-
prison (February 3, 2014).

29 Ibid.

30 "ISO 13485," ISOConsultants.com, http://www.iso-
consultants.com/iso_13485.htm (February 26, 2014).

31 "Management System Certification for Medical Device
Manufacturers," www.sriregistrar.com, http://www.
sriregistrar.com/A55AEB/sricorporateweb.nsf/layoutC/
078690568D3D722586257299004F9FC4?
Opendocument&key=Standards (February 3, 2014).

32 For a list of exempt devices, see http://www.accessdata.fda.
gov/scripts/cdrh/cfdocs/cfPCD/315.cfm (February 3, 2014).

33 All quotations are from interviews conducted by the authors,
unless otherwise cited. Reprinted with permission.

34 "Procedure for Obtaining a New Medical Device License,"
California Department of Public Health, http://www.cdph.ca.
gov/programs/Documents/Procedure%20for%20Obtaining%
20MedDevice%20License.pdf (February 25, 2014).

35 "Inspection of Medical Device Manufacturers," U.S. Food and
Drug Administration, June 15, 2006, http://www.fda.gov/
medicaldevices/deviceregulationandguidance/
guidancedocuments/ucm072753.htm (February 3, 2014).

36 "Guide to Inspections of Quality Systems," op. cit.

37 Ibid.

38 Ibid.

39 Recall that the FDA's good manufacturing practices for medical
devices were not issued until two years after the Medical Device
Amendments of 1976 were enacted. It then took quite some
time for these guidelines to have an effect on the processes used
to produce products that were already in the market.

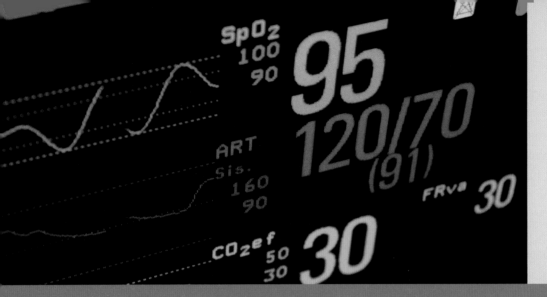

5.6 Reimbursement Strategy

INTRODUCTION

The good news is that the technology the team has developed is a true breakthrough, with the potential to help millions of patients worldwide. And it is likely to save money for the healthcare system in the long run through reduction in hospitalizations and long-term medical management costs. The issue – and it is a big one – is that there is no existing code for reimbursement in the US. The team has a reasonably clear plan for the PMA pivotal trial, but the daunting job ahead is to develop a strategic approach to gaining reimbursement from Medicare and key private payers. The reimbursement consultant is experienced and savvy, but she has been clear in pointing out that the reimbursement environment is evolving in dramatic and unpredictable ways.

The commercial success of any medical technology depends on a company's ability to get adequately paid for providing it to its target customers. Yet, there is no question that the reimbursement landscape is becoming universally more challenging. Across geographies, payers of all types are increasingly asking companies to prove the economic value of their offerings, not just their clinical benefit. In the US, the healthcare system appears to be evolving toward payment systems that explicitly reward value, yet the current environment remains dominated by fee-for-service payments and still features conflicting financial incentives for different stakeholders. The ability to navigate these types of challenges with a successful health economic and reimbursement strategy can be a critical factor determining the success or failure of the technology.

Chapter 4.3 Reimbursement Basics provides foundational information about assessing the reimbursement landscape in the US to determine how a new solution might fit in to existing coding, coverage, and payment paradigms. This chapter focuses on the processes required to expand the payment infrastructure to accommodate a new device if the established reimbursement is inadequate. It also explores the key ingredients for creating a comprehensive reimbursement strategy.

 See ebiodesign.org for featured videos on reimbursement strategy.

OBJECTIVES

- Understand the steps and timing involved in pursuing relevant codes, appropriate coverage determinations, and adequate payment rates for new technologies.

- Learn some of the important strategies for dealing with the Centers for Medicare and Medicaid Services (CMS) and private insurers in the US, and with comparable payment agencies abroad.

- Appreciate the elements of a reimbursement strategy and how to coordinate its development with other functions, including R&D, marketing, and clinical research.

- Recognize the emerging importance of economic value calculations in gaining reimbursement (in the US and abroad).

REIMBURSEMENT STRATEGY FUNDAMENTALS

If a new technology comfortably fits within existing **coding**, **coverage**, and **payment** constructs within its target markets, a company has a distinct advantage. If not, the first step in devising a **reimbursement** strategy is to prepare to systematically pursue appropriate *coding*, a positive *coverage* decision, and a favorable *payment* level for the offering (see chapter 4.3 for an introduction to these three key concepts).

Expanding reimbursement for a new technology

Understanding the steps required to obtain reimbursement approval from Medicare in the US is a useful starting point for determining how to approach this challenge.

Coding

In pursuing coding for a new technology, the company must choose a specific course of action based on the type of code that is needed and its purpose. The two most relevant sets of procedural codes are the **ICD-10 codes** used by hospitals in the inpatient setting and HCPCS codes for services and equipment provided in the hospital outpatient and ambulatory surgery center settings, as well as all physician services. ICD-10 diagnosis codes are used by facilities and physicians to record patient symptoms and diagnoses on health insurance claims in all sites of service.

ICD-10 codes for hospital inpatient claims[1] For *inpatient* procedures, hospital billing is based on the identification of appropriate diagnoses and procedures from the ICD-10 code sets.[2] Responsibility for maintaining the ICD-10 codes is divided between two agencies. The National Center for Health Statistics (NCHS) within the Centers for Disease Control and Prevention (CDC) maintains the classification of diagnoses; **CMS** maintains the classification of procedures. The ICD-10 Coordination and Maintenance (C&M) Committee, co-chaired by representatives from NCHS and CMS, reviews requests to create new ICD-10 codes or to revise existing codes. The

C&M Committee holds public meetings twice a year (usually in March and September), to discuss proposed revisions and solicit public comments. The C&M Committee's role is advisory – no decisions are made at these meetings. The Director of NCHS (for diagnoses) and the Administrator of CMS (for procedures) make all coding decisions.

Requests for coding modifications are accepted from both the public and private sectors.[3] Interested parties are instructed to submit recommendations for ICD-10 modification to the C&M Committee two months prior to a scheduled meeting. Proposals for new codes should include background information on the procedure, patients on whom the procedure is performed, outcomes, and any complications. In addition, the proposals should describe the manner in which the procedure is currently coded, the reasons the existing ICD-10 codes do not adequately capture the procedure, a description of the requested code, and recommended options for a new code title. Supporting references and literature can be included.

The C&M Committee reviews each proposal and decides which ones to include on the C&M meeting agenda. A lead coding analyst is assigned to each selected proposal. This individual contacts the requestor prior to the meeting to discuss the proposal and then prepares and issues a background paper, which includes recommendations on the suggested coding revisions. (Example code papers can be found in the summary reports from previous C&M meetings online.) In the meeting, the requestor is then given the opportunity to make a 20-minute presentation on the clinical nature of the procedure. The lead coding analyst then conducts a discussion of possible code revisions, including alternative suggestions for consideration. CMS makes its final decisions based on this discussion, as well as public comments. Code revisions generally become effective October 1 of the following year.

HCPCS Level I (CPT) codes for hospital outpatient and physician claims As described in chapter 4.3, the **HCPCS** system consists of two levels of codes. The first type, **Level I codes**, are called Current Procedural Terminology (**CPT**) codes, which are used by hospitals in the *outpatient* setting and by medical professionals (e.g., physician claims) in all settings of care. CPT codes are established and maintained by the American Medical Association (**AMA**).

If existing CPT codes are inadequate (or non-existent) for describing a new technology-related procedure for the purpose of physician and physician practice reimbursement, the company can either seek to have a new code created or an existing code modified. In both scenarios, an application must be submitted to the American Medical Association (AMA). However, rather than being submitted to the AMA directly by the company, applications are typically sponsored and managed by an appropriate professional society of physicians willing to advocate for the new code. Fundamentally, the application and supporting data for a new, permanent CPT code (called a category I code) must demonstrate that:[4]

1. All devices and drugs necessary for performance of the procedure or service have received **FDA** clearance or approval when such is required for performance of the procedure or service.
2. The procedure or service is performed by many physicians or other qualified healthcare professionals across the United States.
3. The procedure or service is performed with frequency consistent with the intended clinical use (i.e., a service for a common condition should have high volume, whereas a service commonly performed for a rare condition may have low volume).
4. The procedure or service is consistent with current medical practice.
5. The clinical efficacy of the procedure or service is documented in literature that meets the requirements set forth in the CPT code change application.

Proposals to add, modify, or delete CPT codes are considered by the CPT Editorial Panel, which includes representatives from the AMA, private health insurers, the American Hospital Association, the Health Care Professionals Advisory Committee, and CMS. This group is supported by the CPT Advisory Committee, comprising representatives of more than 90 medical specialty societies and other healthcare professional organizations. The CPT Editorial Panel meets at least three times a year. Applications for new codes are accepted on an ongoing basis but must be received at least four months in advance for consideration at the next meeting. An application form and directions to request CPT changes are available on the AMA website. Category I CPT codes are updated annually to reflect changes in medical technology and practice, with the coding changes taking effective on January 1 of each year.[5]

In the majority of cases, the AMA will not approve a new, permanent CPT code until **pivotal trials** have been performed and the results published, FDA approval/ clearance has been granted, and there is evidence of widespread adoption. For most companies, this means that their code application efforts cannot begin in earnest until the product is launched (although planning of the reimbursement strategy must commence much earlier, as described later in the chapter). Once the product is in the market, it can take considerable time and resources for the company to drive adoption and to perform additional studies that may be required to justify code creation or revision. In total, this process can take from three to six years, depending on the device, study design, follow-up period, and journal publication schedule. After evidence is collected and submitted to the AMA, the company might wait an additional one to two years while the AMA reviews the coding change request and makes a determination. For example, it took Metrika (the company with the point-of-care hemoglobin A1C (HbA1C) test for diabetes patients described in 4.3 Reimbursement Basics) seven years from the time of FDA approval to receive an adequate CPT code and payment level. In contrast, companies producing other technologies, such as transcatheter valves, have successfully worked together with medical societies in parallel with the FDA process to obtain permanent CPT codes less than two years after regulatory approval. Importantly, companies must remember that, despite their decision to invest in this lengthy process, there is still no guarantee that the AMA will approve their coding requests.

The AMA assigns CPT codes in three categories (see Table 5.6.1). Most CPT codes that qualify for reimbursement are permanent category I codes, which fall into one

Table 5.6.1 Different categories of CPT codes are used for different purposes.[6]

CPT code category	Description	Usage
Category I	Codes for procedures and services that are consistent with contemporary medical practice and widely performed.	RVU value and payment is established for procedures and services that have a category I code.
Category II	Supplementary tracking codes that can be used for performance measurement.	No payment attached. Used strictly to facilitate data collection. Not relevant for **medtech** reimbursement.
Category III	Temporary codes for emerging technologies, services, and procedures.	Used to facilitate data collection while an innovation is being evaluated. Associated payments must be negotiated with each individual payer.

of six sections (evaluation and management, anesthesiology, surgery, radiology, pathology and laboratory, and medicine).[7] For category I codes, CMS typically assigns a national Medicare physician fee schedule payment amount for the service and most private **payers** also use this for the purposes of establishing physician payment. Category II codes, applied for and used by health systems to document and monitor quality practices, are not applicable to device reimbursement. Category III codes are for new and emerging procedures. They can be obtained before FDA approval or widespread adoption and are typically used as intermediate codes to track usage and establish the need for a permanent category I CPT code. If a company applies for a category III code, it does so to build a history of widespread usage so that the category III code can eventually graduate to category I.[8] Payers usually do not have fixed fee schedule payment levels for category III codes, but instead determine payments for these codes on a case-by-case basis.

Many companies apply for a category III code as an interim step in obtaining a category I code. While the path to securing a category III code can require less supporting evidence, it is important to reiterate that it does not always have payment linked to it[9,10] and can flag the procedure as experimental, potentially disqualifying it for coverage by some payers. Strategies that include category III CPT codes should be carefully evaluated with the help of reimbursement consultants.[11]

Category III codes usually take up to six years to become category I codes (and thereby trigger payment). Historically, some highly compelling technologies have moved from a category III code to a category I code more quickly. For example, Conceptus Inc. applied for a category III code for its Essure device, which received FDA approval in late 2002 for the minimally invasive treatment of blockages in the fallopian tubes. The company's plan was to use the category III code while it worked to build adoption. However, based on the strong **quality of life** and cost savings data submitted with its application, the AMA instead granted a category I code only two years after FDA approval. More recent examples of technologies that were rapidly converted from category III to category I include transcatheter aortic valve replacement systems, percutaneous mitral valve repair systems, and subcutaneous implantable cardioverter defibrillators.

Another, more common approach to billing for devices and procedures that do not yet have a category I CPT code is to temporarily use a miscellaneous CPT code. Miscellaneous CPT codes, which are grouped by anatomical system and procedure type can be used when "there is no existing national code that adequately describes the item or service being billed."[12] Their advantage is that they can be used as soon as a new procedure is approved by the FDA and while the company is applying for a permanent code. The disadvantage is that claims with miscellaneous codes typically require more detailed

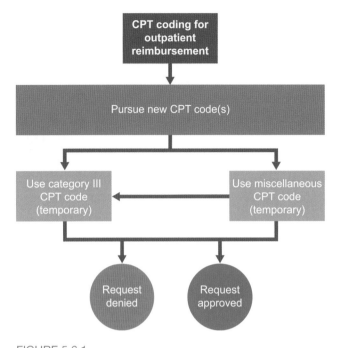

FIGURE 5.6.1

When pursuing a new category I CPT code, companies can seek category III codes and/or use miscellaneous codes to help them build their case.

documentation than usual claims and are often manually reviewed by Medicare and other payers. Therefore, it is to the manufacturer's benefit to closely monitor and support the process of providers submitting these claims so that they are reliably reimbursed at adequate levels. Failure to appropriately support the process may set a bad precedent that affects the reputation of a new device if claims are routinely denied or payments are inadequate.

Often, a company will recommend to hospitals and physicians that they use a miscellaneous code initially, while the company seeks creation of a category III code (with the intent to convert as quickly as possible to a category I code upon FDA approval) or a category I code (after FDA approval). This approach can accelerate reimbursement of the company's device as it seeks to build adoption. However, optimal strategies will depend on the specific technology, and companies are advised to seek input from a qualified reimbursement expert.

Figure 5.6.1 summarizes the CPT coding pathways available to medtech companies.

HCPCS Level II codes for outpatient claims[13] **HCPCS Level II** codes are used primarily to submit claims for products, supplies, and services *not* included in the CPT codes. These codes apply to durable medical equipment, prosthetics, orthotics, and supplies (as well as drugs and biologics). A descriptor is assigned to a code that provides the definition of the items and services that can be billed using that code.

The permanent national HCPCS codes are distributed and maintained by CMS. Permanent codes fall into one of 11 code sets (e.g., E codes are for durable medical equipment; L codes are for orthotics and prosthetics). The CMS HCPCS Workgroup considers each coding request at its regularly scheduled meetings and recommends whether a change to the national permanent codes is warranted based on factors such as the uniqueness of an item's function and operation, its therapeutic distinction relative to existing coded treatments and products, and how it compares to defined volume and marketing criteria.[14] CMS makes the final HCPCS coding decisions. The permanent national codes are updated once a year on January 1.

Temporary national HCPCS codes are also maintained and distributed by CMS. Temporary codes allow CMS the flexibility to establish codes that are needed to meet the national program operating needs of a particular insurer (i.e., Medicare, Medicaid, private insurance sector) before the annual update for permanent national codes or until consensus can be achieved on a permanent national code.

The CMS HCPCS Workgroup has designated certain sections of the HCPCS code set for temporary codes for specific items or types of insurers. For example, G codes are used by Medicare to identify professional healthcare procedures and services that would otherwise be coded in CPT but for which there are no suitable CPT codes. Medicare also uses C codes for certain outpatient items and services, including pass-through devices. (For more information about transitional **pass-through payments**, see the section on hospital outpatient payments later in this chapter.)

Temporary codes do not have fixed expiration dates, and they can be added, changed, or deleted on a quarterly basis. Once established, temporary codes for Medicare are usually implemented within 90 days.

Coding and private payers Codes set up by the AMA (CPT codes) or CMS (ICD-10 and HCPCS Level II) are also used by private payers. If there is a long delay before unique codes are issued, private payers may ask providers to use an existing code to bill services in the interim, agreeing on expanded coverage and/or payment terms under that code. In many cases, private payers would rather have providers use a unique code, even if it is not the permanent code, so that they are able to track and monitor usage and more efficiently process claims. Private insurers use S codes to report drugs, services, and supplies for which there are no permanent national codes, but for which codes are needed to implement policies, programs, or claims processing.

Coverage

In order to obtain reimbursement, a new procedure must gain coverage from payers, either through explicit coverage policies issued by the payer or implicitly based on a payer's determination that it falls within a defined insurance benefit and is "medically necessary" for the insurer's enrollees. Coverage decisions can be favorable, negative, or limited, and may be formalized within a policy or informally addressed on a case-by-case basis.

When making coverage decisions, payers typically seek information about the following key criteria:

- Final approval from the relevant government body (for example, the FDA in the US, **CE mark** in other countries).
- Evidence that proves non-inferiority (and ideally, superiority) on safety and effectiveness compared to currently available treatments that are covered services.
- Data to support favorable health outcomes and clinical improvement in the real world (not just in an investigational setting).
- **Peer-reviewed** evidence published in a (preferably high-impact) journal. For US payers, studies conducted in other countries will be less impactful.
- Published clinical practice guidelines and health technology assessment reviews.
- Specialty society endorsements or positions regarding reimbursement.

When seeking coverage from Medicare, companies can pursue one of two different approaches: (1) a national coverage determination (**NCD**); or (2) a local coverage determination (**LCD**). NCDs are made by CMS and apply to beneficiaries across the country (companies can seek meetings with CMS coverage staff to obtain informal guidance regarding Medicare coverage issues). LCDs are made by Medicare Administrative Contractors (**MACs**) and apply to beneficiaries in their jurisdictions (which typically span one geographic area, although some contractors cover multiple states across the country).[15] Currently, CMS has nine MACs that cover 14 jurisdictions. Because claims adjudication takes place at the local level, providers and companies must negotiate with local contractors, especially in the launch phase of a device (see below) and different LCDs may be in place for the same device under different contractors.

Local coverage determination Requests for an LCD for a medical device are often submitted by local providers, a professional society acting on behalf of the company, and/or reimbursement staff or consultants. Medical professionals among the Medicare contractors tend to have strong personal relationships with the local medical community. These relationships can be leveraged by the company as part of the LCD process. When making an LCD, a local contractor will consider a variety of inputs, including the results of a literature review, medical evidence submitted, physician testimony, and the outcome of existing systematic reviews (such as technology assessments). Once a recommendation has been made, a final review by the contractor-based carrier advisory committees (**CAC**) is mandatory for all proposed local coverage decisions. CACs are made up of practicing physicians representing multiple specialties. Proposed LCDs are posted on the CAC website for 90 days so that public and CAC comments can be collected as input to the final decision. Once a decision is implemented, the coverage policies can be revised or expanded whenever new, sufficiently compelling evidence is presented to justify a change. Furthermore, LCDs are non-binding – so "one-off" rulings are often made at the contractor level. For example, local contractors may allow one-time access and/or deny or pay claims for interventions that preceded local coverage decisions.[16]

Typically, company representatives and/or local physicians pursuing an LCD will visit the appropriate local Medicare contractors at launch in order to secure a coverage commitment for their new technology and associated services using miscellaneous and temporary codes. Until these efforts lead to favorable, published coverage decisions, confusion can exist among providers as claims are potentially denied or underpaid – a common occurrence during the early stages of a local coverage determination. Additional confusion can be caused by the fact that LCDs are not binding, can change at any time, or be highly inconsistent across the country. The rework associated with this uncertainty places a significant administrative and financial burden on the providers that attempt to use the new technology. Reimbursement strategies that include LCDs require the company to build and sustain strong local advocacy and relationships with providers, a relatively large reimbursement team to help manage issues, and persistence in working with local contractors. Success at the local level is most often achieved through a series of small wins, with all favorable policies requiring ongoing maintenance by the company.

National coverage determination CMS issues a limited number of NCDs each year. Most coverage decision making still occurs at the local level by local Medicare contractors. NCDs are primarily used for products and services that may have a significant health and/or budgetary impact on the Medicare program. **Cost-effectiveness** is not explicitly a factor that CMS considers in making NCDs, although CMS informally takes into account the anticipated budgetary impact to the Medicare program in its decision-making process. In order to aid these decisions, CMS sometimes convenes a Medicare Evidence Development & Coverage Advisory Committee (MEDCAC), consisting of outside experts to provide advisory input on specific topics either before or during an NCD coverage review. Companies are strongly advised to seek input from physician key opinion leaders (**KOL**s), professional societies, and/or reimbursement consultants in determining an appropriate coverage strategy.

In some cases, companies may seek an NCD even when the LCD option is possible to avoid the inconsistencies and confusion that can be associated with multiple LCDs. Innovators are strongly encouraged to meet with CMS officials before making a request for an NCD. If a decision is made to pursue national coverage, a written request should be submitted that meets published CMS requirements. This request could be submitted by the manufacturer, a medical society, or initiated internally by CMS.

Although there are many benefits associated with an NCD, this is not a realistic strategy for all companies to pursue. Normally, CMS will only consider NCD requests for a device or intervention that shows breakthrough potential in terms of its clinical benefits and/or cost-effectiveness. CMS may also make an NCD for highly controversial technologies that it deems to be unsuitable for individual rulings by the local contractors (in these cases, CMS can trigger an NCD internally, without a request from the company).[17] Furthermore, CMS typically will only consider an NCD for innovations where there is a large potential budget impact to the Medicare program, significant controversy around the clinical **value** of the treatment, substantial inconsistencies in local coverage by MACs, or some combination of the above. The process of pursuing an NCD can also be extremely time-consuming and resource-intensive, making it infeasible for most small device start-ups. Without the resources of a large company to devote to the pursuit of a NCD, this approach is considered highly risky. A much safer model is to use a series of LCDs to build a basis for national coverage.

NCD versus LCD When considering whether to pursue an LCD or an NCD, there are several other factors that a company should keep in mind. First, even with a published NCD issued by the CMS national coverage group, actual payment rates are established under a separate decision-making process by CMS payment staff according to national payment systems for each setting of service. This means that a sizable effort may still be required, working with CMS payment officials at the national level to secure favorable payment terms. Second, NCDs can be problematic in that they are binding national decisions that can result in unfavorable and often irreversible policies toward the company. For example, in 2005, the FDA approved a vagus nerve stimulation (VNS) device manufactured by Cyberonics for depression in patients who had failed other treatments. Though the device was previously approved and covered for epilepsy, the value of the procedure for depression remained controversial. At a cost of roughly $30,000 for the device and implantation procedure, payers were denying coverage for resistant depression. In 2006, Cyberonics requested that the NCD for VNS be expanded to include resistant depression. In 2007, an unfavorable NCD was issued, citing the lack of compelling evidence to justify such a move. As of 2013, CMS declined a request from Cyberonics to open a reconsideration of this non-coverage decision, so it remains in place.[18] Even though an NCD only formally applies to Medicare, private payers, while not obligated to follow this lead, often can be influenced by decisions made by CMS. Cyberonics had successfully convinced some private payers to reimburse VNS for resistant depression, but still had to instruct providers to seek reimbursement on a case-by-case basis.

A note on parallel review In 2011, the FDA and CMS established a pilot program for the concurrent review of certain medtech premarket regulatory submissions and national coverage reimbursement determinations. The goal of the program is to promote the development of innovative products and shorten the time it takes to bring these products to patients by reducing the time between FDA marketing approval and Medicare coverage. The pilot program, which companies were invited to participate in voluntarily, does not change the distinct review standards for FDA device approval and CMS coverage determination.[19] However, the two agencies are working together more closely throughout the review and decision process.

One of the companies taking advantage of the pilot program is Exact Sciences, which submitted its non-invasive Cologuard™ diagnostic test for detecting colorectal cancer for parallel review in December 2011. During the review period, Exact Science was positive about its experience. Among the main benefits highlighted by CEO Kevin Conroy was the ability to proactively get input on trial design not just from the FDA but from CMS. "One of the most significant inputs [from CMS] was making sure that the patient population for the **clinical trial** was **powered** sufficiently with Medicare patients," he said, noting that the agency also provided other valuable suggestions.[20] For start-up companies without the resources to run multiple trials, being able to conduct a single study for regulatory and reimbursement offers a substantial advantage. As Conroy put it, "I would go so far as to say that I don't know how from an investment standpoint we could have done this without parallel FDA approval and Medicare coverage and payment."[21] Conroy also indicated that the program could save Exact Sciences up to two years in total review time.[22] In October 2014, Cologuard became the first medical device to receive FDA approval and a final NCD as part of the parallel review process.

On the other hand, parallel review is not without risks. Management from other companies have been nervous about being forced into a NCD by the parallel review process. They have also expressed uncertainty about trying to satisfy two agencies with distinct mandates and evidence requirements at the same time.[23]

Building on the parallel review program, FDA's Center for Devices and Radiologic Health (CDRH) announced another pilot to help streamline the link between regulatory and reimbursement approval called the CDRH Reimbursement Program. Under this effort, a Medical Device Reimbursement Task Force will focus on implementing three new processes: (1) to allow a device company to voluntarily request that one or more identified payers participate in a pre-submission meeting to better inform the sponsor of what evidence is necessary to support both FDA approval/clearance and payer coverage; (2) to modify CDRH's current investigational device exemption checklist, as appropriate, and provide it to CMS upon request to support the Medicare agency in decisions of whether to reimburse studies conducted under an approved IDE; and (3) to work with payers and health technology assessment organizations to determine if CDRH can provide them with summary safety or effectiveness information for a device that might reduce the evidence needs to submit to a payer to support coverage.[24] While it remained to be seen how this program would evolve, the effort is indicative of a growing interest in increased coordination between regulatory and reimbursement assessments.

Coverage determinations for private payers There are many similarities between the processes used by Medicare carriers and by private payers to evaluate a new technology and reach a coverage determination, but there are also important differences. Private payers typically have their own internal committees (a medical policy or technology assessment committee), consisting of physicians, plan administrators, health economists, and statisticians. These committees evaluate the evidence supporting the clinical necessity of a new procedure, paying special attention to evidence obtained outside academic and well-controlled settings. Precedence-setting coverage decisions by other major payers are also evaluated. For example, in reaching a coverage position for bariatric surgery, Cigna (a major private payer) based its decision on information provided in a CMS coverage decision, technology assessment reports by the Blue Cross Blue Shield (**BCBS**) Technology Evaluation Center (**TEC**) in the US, the National Institute for Health and Care Excellence (**NICE**) in the UK, and a comprehensive review of the literature.[25]

The final outcome of a private payer coverage determination is either full coverage, no coverage, or coverage with restrictions. A significant difference between private payers and Medicare is that private payers may offer multiple plan designs, with certain plans specifically excluding certain procedures (even if they are medically necessary). Again, in Cigna's case, their coverage determination on bariatric surgery does not apply to health plans where such surgery is explicitly excluded.

Payment

As described in chapter 4.3, payments are structured around different systems for reimbursement to hospitals (inpatient and outpatient), Ambulatory Service Centers (ASCs), and physicians.

Hospital payments[26] For *hospital inpatient payments*, the Medicare-Severity Diagnosis-Related Group (**MS-DRG**) code system assigns a single prospectively determined

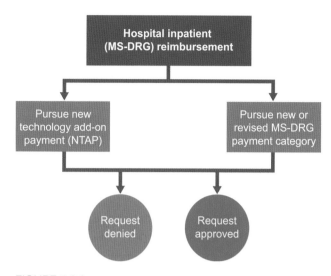

FIGURE 5.6.2

For hospital inpatient payments involving new technologies, companies can apply for new or revised MS-DRG payments or for a new technology add-on payment (NTAP).

payment amount for all hospital services (except physician charges) associated with a given hospital stay. If a new technology is introduced (and the technology adds expense), the hospital could suffer a financial disadvantage even though the care provided is improved. In this situation, a company has essentially two options: (1) to apply for what is called a new technology add-on payment (NTAP); (2) or to seek a new or revised MS-DRG (see Figure 5.6.2).

In order to qualify for an NTAP, the company must submit an application to CMS providing documentation regarding the new technology to demonstrate that it meets three key criteria: (1) it is an innovative new technology that is not substantially similar to existing technologies; (2) it provides a substantial clinical improvement for Medicare beneficiaries; and (3) it involves high costs and is inadequately paid under the MS-DRG system based on thresholds established by CMS. The add-on payment is issued if CMS determines the technology meets these criteria after reviewing public comments and performing its own review. If the data are convincing, the add-on payment is set at up to 50 percent of the average cost of the new device or 50 percent of the overall incremental costs associated with the new technology to hospitals, whichever is less. The add-on payment lasts for two to three years after FDA approval and commercialization. By the end of this period, the company should know whether a new DRG will be issued or an existing DRG will be revised to accommodate the new technology. In some cases, obtaining a new technology add-on payment may comprise just one phase of a larger reimbursement strategy. Companies may apply for an add-on payment, but with the expectation that their device will eventually be reimbursed in the outpatient setting, which would require appropriate APC and CPT codes.

The alternate strategy is to continue to initially seek payment under an existing MS-DRG payment category and apply for a new or revised MS-DRG in the future, after hospitals gain experience using the technology. A company might use this approach if the technology does not meet all criteria for new technology add-on payments upon FDA approval, if there is insufficient clinical evidence to demonstrate differentiated value immediately upon FDA approval, or if it is feasible to commercialize within the existing DRG upon FDA approval during the initial stages while additional experience is collected. An example of this approach involved Boston Scientific's efforts to obtain a new DRG for the Guglielmi Detachable Coil (GDC) for catheter-based treatment of brain aneurysms (see Figure 5.6.3). Basically this procedure involves deploying a tightly nested

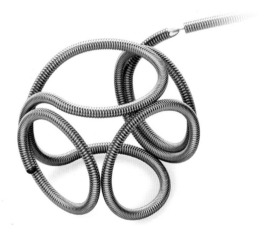

FIGURE 5.6.3

An example of a GDC, the Target® Detachable Coil (courtesy of Stryker Neurovascular).

platinum coil within an aneurysm using a special catheter for access. The coil is left behind and it induces a local blood clot to form which, combined with the structural support of the coil, stabilizes the aneurysm and reduces the chance of rupture. The problem that Boston Scientific encountered was that hospitals were losing large amounts of money using the coils because of inadequate MS-DRG payment levels.

In late 2003, the company presented analysis of Medicare claims data to CMS, showing that hospital costs greatly exceed payments for both coiling and the more invasive surgical procedure of clipping the aneurysm. In its 2004 rulemaking cycle, CMS agreed to the creation of a new DRG for treatment of ruptured aneurysms that doubled average hospital base MS-DRG payments from $17,000 to $34,000 per hospital stay. This successfully expanded access for Medicare beneficiaries to these definitive life-saving treatments.

For hospital *outpatient* payments (but where the care is still delivered within a hospital facility), a company launching a new technology must work within the Ambulatory Payment Classification (**APC**) system. If an existing APC payment category does not provide adequate reimbursement or the descriptors do not match that of the new device, then the company has three choices: (1) seek a pass-through payment; (2) pursue a new technology APC; or (3) seek reclassification of a high-cost new service (typically involving a new CPT code) to a different APC payment category.

The company may apply for a transitional pass-through payment category involving the use of a HCPCS Level II C code to cover the cost of the new device. A pass-through payment is used in the case where there is a new, clinically beneficial, high-cost device being used in an existing procedure (and, therefore, where there are existing APCs that are appropriate). The amount of the pass-through payment is generally calculated as the actual cost for the device, minus the amount already included in the APC payment for the technology that is being replaced. The pass-through provides additional payment for a two- to three-year period while CMS obtains further data on the cost of the technology. For example, in the early 2000s, several companies brought out a new generation of spinal cord stimulation devices that had rechargeable power sources, providing more and longer lasting power. The ability of these systems to last for much longer intervals between generator replacements resulted in substantial potential cost savings to Medicare over the life of the patient. Based on a concerted educational effort from Boston Scientific, Medtronic, and St. Jude, in 2006 CMS approved pass-through payments on the basis that the existing codes did not "adequately describe" the new technology. The rechargeable neurostimulators from all three companies qualified for new technology payments, and the rechargeable systems captured most of the US market. Once CMS retired the new technology payments for 2008, it decided not to create separate APC payment categories for rechargeables and non-rechargeables. However, despite the cost differential in these two technologies, the significant clinical benefit of the rechargeable systems has driven their continued use as the dominant approach in the US.

The second option is to pursue a new technology APC. A new technology APC will only be issued in the case that the new device also warrants an entirely new procedure that cannot easily be described by an existing APC code or combination of codes. Again in this case, the new technology APC provides a period of time for Medicare to gather actual use data before the APC becomes permanent. Applications for a new technology APC can be submitted at any time during the year and are considered for inclusion in the quarterly updates of the codes. With rare exceptions, applications for transitional pass-through payments and new technology APCs cannot be submitted unless the product has received regulatory clearance/approval. A decision from CMS typically takes several months to obtain.

In some circumstances, especially when a distinct CPT code is established for a new technology involving higher costs, then the manufacturer can convince CMS to reclassify the procedure to a different APC payment category if the cost of the new procedure exceeds two-times the calculated cost of the lowest-costing procedure categorized to that same APC (this is referred to by CMS as the "two-times rule").[27] In

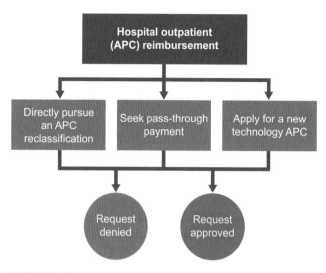

FIGURE 5.6.4

For hospital outpatient payments involving new technologies, companies can seek a pass-through payment or apply for a new technology APC.

such cases, CMS will consider the clinical characteristics and hospital resource costs of the new technology and seek to assign it to the most appropriate APC payment category.

Figure 5.6.4 summarizes the APC coding pathways available to medtech companies.

Payments to physicians As noted in chapter 4.3, payments to physicians for procedures under Medicare are based on a determination of Relative Value Units (**RVUs**) associated with each CPT code for inclusion in the national Medicare physician fee schedule. If a company is seeking new or revised codes associated with a new technology, CMS will look for recommendations from the appropriate AMA committees. The AMA Specialty Society Relative Value Scale Update Committee (**RUC**)[28] recommends RVUs for a new code typically based on an established RUC survey of practicing physicians. In some cases, the company may also provide evidence to CMS that demonstrates the time and effort involved in using the device/performing the procedure. This evidence could include economic studies based on surveys or time and motion studies on physician's time, other practice expenses (such as labor and supplies), and malpractice risk. The RVU recommendations are forwarded by the AMA to CMS, where a final RVU determination is made.

Interestingly, the assignment of payment rates based on RVUs by CMS is budget neutral, meaning that total Medicare funding for all CPT codes remains constant within any given budget year. When a new CPT code is created (or an existing code modified) and a value is assigned to that code, payments made through other CPT codes must be reduced to compensate for the change. Companies must recognize that this creates a disincentive for AMA to approve new codes, especially if they will increase total payment to one specialty at the expense of another. Budget neutrality can raise **stakeholder** issues, especially among specialty societies, as one sees its total payment rates go down while another sees them go up. As a result, these issues need to be identified early and managed proactively.

For CPT codes for diagnostic tests and therapeutics, Medicare establishes a payment level for a new code using one of two approaches. If an existing code is sufficiently similar to the new code in terms of cost, technology, and clinical use, CMS determines the payment amount by "crosswalking" the code – assigning a comparable reimbursement rate to the new code based on its similarities to one or more existing codes. If no comparable code exists, a "gap-filling" method is used to determine a new reimbursement amount. This involves having each local Medicare carrier individually determine an appropriate payment amount for the new code. These recommendations are then shared with CMS, which analyzes the carrier-determined amounts and sets an appropriate payment level for the new code that will be used until the next annual schedule is published. The sponsor will often develop detailed economic studies to support these efforts.

Payment by private payers   As with coverage decisions, the fundamentals for setting payment levels are similar for Medicare and private payers, with a few important differences. All payers place a strong emphasis on **evidence-based** medicine, cost, and the demonstration of value. However, Medicare as the largest purchaser of US healthcare services is responsible for publicly disseminating codes and payment rates, so its decisions can serve

as the pathfinder, influencing the policies of other payers. For the time being, Medicare still cannot explicitly consider cost as a criterion when evaluating coverage for new technologies (though CMS officials have acknowledged that perceptions of value play an informal role in their coverage decision making). Furthermore, officials from CMS, and to some extent local contractor representatives, may be subject to greater political pressures and lobbying. Private payers are less likely to face such pressures since they have no direct links to elected officials and because their policies and payment schedules are not in the public domain. However, private payers are beholden to their providers, and to the employers and individuals that subscribe to their plans. For innovators and companies, there is often a greater opportunity to develop business relationships with private payers, as public payers (and some non-profit plans, such as Kaiser) restrict their employees' participation in industry-sponsored events and activities. Regardless, whatever contacts and relationships a company has with any payers should be leveraged to help secure appropriate payment levels and help facilitate a smooth reimbursement processes when new codes, coverage, and payment levels are being established.

As enrollment in Medicare Advantage programs and health exchanges grows, many private insurers are expected to become more restrictive in their coverage of category III treatments, which tend to be expensive yet still experimental when it comes to delivering proven value. For example, in the absence of strong cost/benefit data, some private payers have retroactively denied coverage of a costly proton beam therapy for prostate cancer that they previously reimbursed under a category III CPT code.[29]

When processing claims from miscellaneous and temporary codes, private payers have different proprietary formulas for calculating payment levels that are based on contractual negotiations with specific providers including physician practice groups and hospital systems. Each private payer generally uses its own formula to reimburse for billed charges on manually submitted claims. With miscellaneous code claims, payment rates usually amount to a percentage of billed charges for services and/or devices or a flat predetermined allowable. Some private payers routinely reject or delay all manually submitted claims until they are elevated for special review within the organization. Companies must understand and anticipate this potential outcome in their reimbursement strategies. For devices and procedures that have been recently approved, private payers will typically set their payment at some percentage (usually greater than 100 percent) of Medicare or at a rate that approximates similar procedures and devices.[30] In the hospital, private payers are more likely to pay for each device separately and not lump them into a per diem payment in order to keep hospitals from losing money. Payment rates vary substantially from one private payer to another and depend on heavy negotiation between providers, payers, and the manufacturer.

Building a reimbursement strategy

Whether a company plans to apply for new coding, coverage, and payment or to fit into the existing reimbursement structure, it must develop a reimbursement strategy. A cohesive, proactive reimbursement strategy allows innovators to anticipate the questions payers and other stakeholders will raise about a new technology, identify reimbursement opportunities and obstacles, and integrate necessary evidence requirements into product development and business planning. Given the complexity of the reimbursement landscape, the initiation of a reimbursement strategy must start early in the biodesign innovation process, well before the offering is nearing the market. Any delays can negatively impact product adoption and can threaten the viability of a start-up company.

When developing a reimbursement strategy, the company should take a multidisciplinary approach. At a minimum, the leaders of the regulatory, clinical, marketing, sales, and reimbursement functions must work together to appropriately integrate and sequence key commercialization activities. As indicated, specialized expertise and accumulated experience may also be needed, which is often best obtained through the involvement of a skilled external consultant.

Developing a reimbursement strategy is particularly challenging given the major changes underway in the health economics environments globally. This is especially clear in the US, where passage of the Patient

Protection and Affordable Care Act (**ACA**) in 2010 signaled a new era of cost containment pressure. A number of features of the ACA are representative of the general policies and strategies around affordability that are surfacing in many other countries. The act mandated a progressive set of cuts in hospital payments, along with the creation of the Independent Payment Advisory Board (IPAB). The IPAB has the authority to make changes in the Medicare payment rates and program rules without prior approval of Congress. The ACA also created the Patient Centered Outcomes Research Institute (PCORI), which is charged, in part, with evaluating the comparative effectiveness of new innovations, such as devices. The ACA authorized CMS to experiment with different mechanisms for containing costs, such as **bundled payments** for certain types of care episodes. A new Shared Savings Program was also created to facilitate coordination among providers, in part through the formation of Accountable Care Organizations (**ACOs**). The Act further expanded the implementation of "**gain-sharing**" mechanisms, through which physicians and hospitals can receive payments linked to cost savings achieved in the delivery of healthcare. (Programs such as gainsharing can influence how innovators choose to price their products as discussed in 5.7 Marketing and Stakeholder Strategy.)

Given the scope and complexity of these changes, innovators and companies have a major challenge in understanding which policy changes have already been implemented, which are still coming, and what the timetable will be for the remaining roll-out. The situation is complicated even further by the intense political nature of healthcare reform, with uncertainty whether major portions of the law will be revoked or changed. Despite this fairly massive uncertainty, however, the general direction of economic reform of healthcare in the US and other global markets is clear: decisions about reimbursement will be driven by assessment of value, with an overall goal of reducing healthcare expenditures. The net result is that, more than ever, reimbursement strategy is becoming a key driver of success for the introduction of a new technology.

A reimbursement strategy certainly cannot account for every possible scenario that might impact the payment landscape. The point is that innovators should take a forward-looking position when developing their strategy and make a best effort to anticipate what changes are most likely to occur. With this framing in mind, an effective reimbursement strategy addresses a combination of decisions and activities, including those related to competitors/proxies, pricing, payer segmentation, clinical trial design, payer **value propositions**, and disseminating critical information to marshal support from physician and patient advocacy groups as well as a potential **payer advisory board**. All of these topics are touched on in the sections that follow.

Competitors and proxies

Identifying competing products and clinical trials underway is an important early step in devising a reimbursement strategy. If two companies are commercializing similar products, understanding the competitor's approach to gaining reimbursement can prevent conflicting or contradictory reimbursement efforts that could jeopardize both parties.

If the technology is completely new, the analysis can be structured around a proxy device – that is, a technology that is already on the market and has relevant similarities to the new product. Proxy devices should have a related function within the same or similar specialty, and should be used by the same type of provider in a comparable site of care. Other comparators to consider include the disease state or mechanism of action.

In most countries, including the US, reimbursement is typically attained for a type of technology, and is not manufacturer-specific. Therefore, a second-to-market company can potentially benefit from understanding the strategy of the first market entrant and save significant resources if the strategy is amenable to its product. Depending on the timing of a competitor's launch, companies may consider this situation similar to one in which they seek coverage under existing codes. However, just because a competitor attains reimbursement first, there is no guarantee that the follow-on technologies will also be reimbursed. Clinicaltrials.gov remains the best public source of clinical trial information, although reimbursement information is typically not provided.

Pricing

Pricing decisions are an essential component of a company's reimbursement strategy. As described in 5.7 Marketing and Stakeholder Strategy, the company will need to set a price that makes the device affordable at the reimbursement levels it expects to secure from payers. To determine the appropriate price, many companies perform what is known as demand curve analysis or value-based pricing. This exercise involves assessing the price sensitivity of providers and key payers to determine the range within which the company can secure adequate reimbursement, which in turn determines how to realistically price its product. The Genomic Health example provided in chapter 5.7 demonstrates how value-based pricing works.

Pricing assumptions directly impact cost models and can exert a substantial influence (positive or negative) on reimbursement. For example, if a new technology substantially improves clinical outcomes but the company sets pricing at a level that does not meet new technology add-on payment cost thresholds, then the technology may not qualify for premium reimbursement. Further, if the company lowers prices before reimbursement rates have been formalized, the final reimbursement may account for the lower price and lead to a reimbursement level that does not cover the cost of the procedure and device, therefore limiting adoption. In contrast, if the price is set at a level that significantly increases the budget impact to the payer without corresponding clinical or economic value, then the payer may respond with negative reimbursement decisions that limit adoption.

Outside the US, pricing is sometimes determined using an approach called **reference pricing** (i.e., setting a price relative to what is charged in another market or geography). When reference pricing is in place, the sequence in which a company releases a product in different markets can be important – it should first be released in countries where a higher price can be commanded.

Payer segmentation

Because early-stage start-ups have limited resources, they typically need to prioritize which payers to approach. At least one year before launch, a company should initiate efforts to segment payers. Some payers will allow immediate use of an approved device (in accordance with product label) at launch, while others will mandate a 6- to 12-month waiting period. Within these more restrictive plans, exceptions may be allowed to varying degrees. Some payers will permit device use after an appeal is made by a physician or physician organization and found to be justified. Often, these waiting periods culminate in an intensive review of the technology (i.e., technology assessment), during which clinical and economic data is scrutinized and the new technology is compared holistically to others in its class or therapeutic area. The outcome of this analysis leads to a final decision about the reimbursement status of the device within that plan. Some payers will wait and follow the lead of assessments performed by TEC or other such committees, while other payers will be more likely to approve more quickly in a manner consistent with the label.

Information about how different payers respond to new technologies (based on primary payer research) can generally be purchased from a reimbursement consulting firm and then used to perform a preliminary payer segmentation. Common criteria used to segment payers include their openness to new technology and the stringency of their review process.

Whenever possible, the number of patients covered by payers in each segments should be estimated to help the company more efficiently allocate the time and focus it will devote to each payer at launch. In general, a small device start-up should follow the 80/20 rule – targeting the top 20 percent of payers that cover 80 percent of patients. At a minimum, companies working in the US should include Medicare and one to three prominent private payers in their initial reimbursement strategy (for the private payers, identify the ones most likely to adopt/cover the new technology).

Design of clinical trials

As mentioned in 5.3 Clinical Strategy, considerations related to reimbursement have an increasingly important influence on the design of clinical trials. Just as company representatives should meet with the FDA or other regulatory agencies early in the biodesign innovation process to understand what evidence will be required to gain regulatory approval, they should do the same with

Table 5.6.2 Evidence to support reimbursement must explicitly be collected as part of a clinical strategy.

Concerns to be addressed by reimbursement-related clinical trial endpoints
What are future significant expenses for patients with this disease or condition?
What expenses could payers avoid as a result of funding the technology?
What inefficiencies exist with the current **standard of care**, and how does the new device address this?
What unacceptable risks are inherent in current treatment standards, and how does the device minimize these?
Do the outcomes that the device will deliver provide patients with a highly compelling reason to seek treatment?
Is the target patient population young, or of working age, and will the device affect their productivity or ability to work?

payers to understand what data will be most likely to convince them of a new technology's value proposition and lead to reimbursement. In particular, payer feedback should be solicited on reimbursement-related endpoints for clinical trials. These endpoints, which are often incremental to endpoints included for regulatory purposes, can be expensive to generate. They can also introduce significant risk, as failure to show a positive effect will work against efforts to support reimbursement.

Gaining feedback from payers will help to determine that the resulting data will have a positive impact. Endpoints should be selected based on the likelihood of addressing important payer concerns (see Table 5.6.2 for a list of concerns to be addressed), for their probability of success, and their ability to support simple, clear value messages that resonate with payers, physicians, patients, and the public.

Outcomes data that address economic value are critical to reimbursement if a company's goal is to market the product outside of the US, particularly in countries with nationalized payer systems such as the United Kingdom, France, or

Australia. It is also becoming increasingly important for establishing private payer reimbursement in the United States. Data from clinical studies serve as direct inputs into the creation of the economic evidence needed to support a strong payer value proposition.

Payer value propositions and building economic evidence
A significant milestone on the path to commercialization is a company's development of one or more value propositions for its offering (as described in more detail in 5.7 Marketing and Stakeholder Strategy). Metrics for demonstrating value can vary widely, but typically seek to express improved outcomes resulting from a product's use relative to the cost of treatment. Such outcomes may be assessed for individuals, populations, institutions, or society as a whole; and may cover clinical, economic, social, or other measurable effects. When it comes to reimbursement, public and private **third-party payers** and technology assessment panels insist that products

are supported by strong value propositions. But, increasingly, such stakeholders want to review data that support those value statements.

Value propositions should be clear and simple, avoiding tenuous projections based on intermediate markers of efficacy. They should also be based on data from **randomized controlled clinical trials** and/or well-defined value models, whenever possible. Using data collected from a patient population and provider type that resembles that of the targeted health plan is also important for maximizing a company's likelihood of success. The Working Example focused on implantable cardioverter defibrillators illustrates a strong, data-driven value proposition.

As mentioned, CMS excludes any explicit cost and cost-effectiveness evidence in its coverage decisions in the US (at least for now). On the other hand, private US payers, such as Blue Cross Blue Shield and Aetna, are increasingly looking at explicit measures of value (health improvement as a function of cost) in assessing the

---

Working Example
Sample value proposition for the cost of treatment with an implantable cardioverter defibrillator[30]

Research shows that implantable cardioverter defibrillators (ICDs) provide an invaluable form of "life insurance" for people most at risk, preventing sudden cardiac arrest (SCA) death 98 percent of the time. Evidence-based medicine has demonstrated that ICDs significantly reduce death among Americans at highest risk:

- 31 percent reduction in death among SCA survivors from a second event.
- 31 percent reduction in death among post-heart attack sufferers.

Despite these statistics, ICDs have been thought to be underutilized, at least from the standpoint of their potential to reduce sudden cardiac arrest and the economic consequences of this deadly condition.

- Fewer than 20 percent of indicated patients receive the benefits of an ICD despite being at high risk for sudden death.
- Although SCA is responsible for more deaths than breast cancer, lung cancer, stroke and HIV/AIDS

combined, spending on SCA prevention is modest when compared to other diseases (AIDS = $19.5 billion; stroke = $6 billion; lung cancer = $1.6 billion; breast cancer = $0.8 billion; SCA = $2.4 billion, including drug and device therapy).

The value of ICDs outweighs their cost to the system:

- An ICD costs approximately $25,000 (including implant costs), which equates to less than $10 per day over the average life of a device (four to five years).
- The cost per day of ICD protection has decreased by nearly 90 percent over the last 10 years from more than $90 in 1990 to less than $10 more than a decade later (equivalent to the cost of optimal medical therapy for these same patients).
- ICD Medicare expenditures are significantly less than for other cardiovascular procedures. In 2002, Medicare reimbursed $1.2 billion for ICD procedures versus $6.4 billion for stent implants and $7.8 billion for bypass surgery.
- The cost of ICD therapy per year is less than 0.2 percent of projected Medicare spending over the next 10 years.

suitability for reimbursement (see online Appendix 5.6.1 for an overview of the Blue Cross Blue Shield Technology Evaluation Center). Outside of the US, health economic data is often required as part of a health technology assessment prior to obtaining reimbursement. For example, in most advanced European nations and many countries in Asia, new devices undergo technology assessment before they are approved for use in nationalized healthcare systems.

Payer value propositions can be significantly enhanced by the inclusion of information from economic models that demonstrate the relation between the health improvement gained and the corresponding costs. Models can be especially helpful for companies seeking to develop economic evidence prior to launch (when limited clinical and cost information is available) in that they allow the company to project outcomes and related costs over time. Which type of model to deploy will depend on the nature of the data available, the purpose of the analysis, and the target country and agency. Companies typically should work with health economists and/or reimbursement experts to help in model development to ensure they meet payer and administrator requirements. Some of the most widely used types of health-economic value models are listed below (see online Appendix 5.6.2 for examples that correspond to each one):

- **Cost analysis** – Cost analysis can be the strongest and most persuasive type of modeling to payers and administrators. It compares the money spent on competing treatments over time from a payer, provider, and/or societal perspective. If a new treatment that is safe and effective (superior or at least non-inferior) to alternative treatments can be shown to produce overall costs savings, then this may be sufficient by itself to warrant positive reimbursement decisions without the need to perform more extensive modeling.

- **Cost-effectiveness model** – Cost-effectiveness analyses evaluate the total incremental cost of the intervention per incremental unit of net health benefit experienced by the patient. In cost-effectiveness analyses, health gains can be measured in a variety of "natural health unit" measures (e.g., treatment successes, cures, lives saved, blood pressure levels, repeat revascularization procedures avoided).

- **Cost-utility analysis** – Cost-utility modeling is a special form of cost-effectiveness analysis where health gains are measured using a generic Quality Adjusted Life Years (**QALY**) construct designed by economists to reflect both quality of life and years lived. Quality of life indexes are multiplied by the number of years gained. Incremental cost-effectiveness is then expressed as a ratio of the incremental costs over the incremental QALYs gained for one treatment when compared to another. This model is often used by national health systems as the standard for evaluating reimbursement outside the US. Commonly used thresholds for determining what constitutes acceptable economic "value for money" in the US are $50,000 per QALY gained or $100,000 per QALY gained when compared to the best available treatment.

- **Budget impact models** – Budget impact models look at the cost and treatable population from the perspective of a particular purchaser of services (e.g., government healthcare program, private health plan, hospital system), as well as the expected annual cost to the plan of covering the device based on the anticipated level of adoption and pricing. The results are normally evaluated in terms of annual cost or per-member, per-month costs. Often, a cost-effectiveness model and a budget impact model will be combined.

When developing a model, companies may produce multiple variations to address the unique interests and requirements of specific stakeholders (e.g., medical specialty societies, government and private third-party payers, hospital systems, ACOs). When data are uncertain or cover a wide range of potential outcomes, they can also experiment with sensitivity analysis to better understand the strength and meaning of varied findings. The approach here is to define a "base case" and then make adjustments to the model's key input assumptions to test the robustness of the base case findings.

Of note, innovators should be sensitive to any price increases or competitor discounts included with a model, as they can "live on forever" once it is distributed, potentially working against the product in the future. The value propositions derived from a model will become important tools to help the company build advocacy and to assist the company's advocates in favorably affecting coverage and reimbursement policies.

The story about a cost-effectiveness study for a new treatment for obstructive sleep apnea illustrates the complexity of creating a value model, but also highlights the benefits that can be realized from the effort.

FROM THE FIELD ▸ WING TECH INC.

Early-stage value modeling to support reimbursement

Early-stage technologies often lack the long-term clinical data or peer-reviewed publications that would ideally form the basis for reimbursement decisions. In many cases, the benefits and cost savings resulting from the use of new diagnostic or therapeutic devices do not become measurable until years after the intervention. In the absence of such data, health-economic modeling can provide quantitative projections of the expected long-term outcomes and costs of new interventions. Accordingly, value modeling has become an important tool for informing the reimbursement-related decisions of both government agencies and private payers. Additionally, the early-stage assessments required to develop such models can also benefit innovators by helping them develop a solid understanding of key value drivers and metrics. In the words of Jan Pietzsch, CEO of Wing Tech Inc., a consultancy specializing in early-stage value modeling, "Health-economic modeling enables innovators to make quantitative, evidence-based statements about a wide range of potential benefits associated with a new technology."[32]

In a recent example of such a study, Wing Tech collaborated with principal investigator John Linehan of Northwestern University to create a health–economic model for evaluating diagnostic and treatment strategies for obstructive sleep apnea (OSA), a chronic condition in which the patient's upper airway collapses repeatedly during sleep. OSA is significantly underdiagnosed and undertreated, and it is associated with numerous adverse consequences, including cardiovascular disease, depression, diabetes, obesity, and stroke, as well as excessive daytime sleepiness that can lead to motor vehicle collisions (MVCs).

The study, "Assessing the Impact of Medical Technology in the Diagnosis and Treatment of Obstructive Sleep Apnea," was supported by a grant from the Institute for Health Technology Studies (InHealth). It was designed to quantify the value contribution of existing medical technologies used in diagnosing and treating OSA and, specifically, to develop an analytical framework for evaluating various diagnostic strategies. An additional objective was the development of an analytical framework for future evaluation of any new types of OSA diagnostic or therapeutic interventions.

The current gold **standard treatment** for OSA is continuous positive airway pressure (CPAP). Physicians use one of three common diagnostic strategies to determine whether a patient has OSA and whether CPAP treatment is indicated:

- Full-night polysomnography (FN-PSG) – The patient spends two full nights in a sleep lab; one for assessment and the other to titrate the CPAP therapy if the diagnosis is positive for OSA.
- Split-night polysomnography (SN-PSG) – The patient spends one night in the sleep lab for both assessment and titration.
- Unattended portable home monitoring (UPHM) – The patient is evaluated with a portable home monitor, followed by unattended CPAP autotitration.

These diagnostics differ in their sensitivity and specificity, as well as their cost to the healthcare system. To determine which method was most cost-effective, the researchers constructed a two-part model for comparing the end-to-end costs and effectiveness associated with each diagnostic strategy when evaluated in conjunction with subsequent CPAP therapy over the course of the patient's lifetime.

They began by identifying published, large-scale cohort studies and other clinical evidence in order to define the diagnostic and therapeutic pathways currently in use and to establish the relative effects of each pathway on health outcomes related to OSA. According to Pietzsch, one of the most significant challenges for model development involves whether published information is available on which to base long-term projections. "When there are established models and long-term epidemiological studies, researchers usually have the evidence they need to define the risk equations and populate the model," he explained. "But where that information is not available, creating a meaningful model is much more difficult."

Existing studies helped the Wing Tech research team understand pertinent elements of the current standard of care, for example, how existing technologies perform, how performance is measured clinically, and the costs of technology-related interventions and clinical events. The team's research also enabled it to quantify and assess the importance of societal outcomes related to OSA, including daytime sleepiness leading to motor vehicle collisions. Measuring non-clinical outcomes such as quality of life, return to work, independence, and productivity can often play an important role in deepening payer understanding about the value of a technology or procedure. When incorporated into a health–economic model, the costs and savings attributable to societal outcomes are readily apparent. Whether or not innovators ultimately go on to develop a health–economic model, Pietzsch explained, this investigational step is essential because "It allows you to identify and pinpoint the characteristics of a technology that really matter – and how you can use this information to make a value statement about that technology."

For the second part of the OSA study, the Wing Tech team built on its analysis of the existing literature to quantify the benefits and cost-effectiveness of CPAP diagnosis and treatment in patients diagnosed with OSA. The team took as its base-case population a hypothetical cohort of 50-year-old male patients with a 50 percent prevalence of moderate-to-severe OSA.[33] The team's analysis focused on adverse cardiovascular events, motor vehicle collisions, and stroke – three areas of high risk for OSA patients – and projected CPAP-related reductions in the average number of such events over a 10-year period, and over a patient's expected lifetime.

In this analysis, the team developed two complementary mathematical models. The first was a multifaceted decision-tree model of current diagnostic pathways (see Figure 5.6.5), which the researchers populated with respective test performance, diagnostic outcomes, and costs as described in the literature. This model enabled the researchers to classify patients according to a starting diagnosis and disease state, and to assess patient outcomes along the pathways during the study timeframes.

The second was a Markov (state transition) model. This type of model is commonly used to illustrate the expected disease progression of a hypothetical cohort through a set of mutually exclusive and collectively exhaustive health states. In the case of the OSA study, health states included the clinically and economically relevant disease states of hypertension; myocardial infarction (MI) and post-MI; and stroke and post-stroke (see Figure 5.6.6). This second model allowed the team to compute the estimated disease progression, costs and outcomes over the patient's lifetime, using the initial results of the diagnostic pathways model as inputs.

The results of the OSA study quantified the health benefits of diagnosis followed by CPAP therapy, demonstrated the cost savings of therapy as a result of reduced illness and death, and showed that the most expensive method of diagnosis under study was actually the most cost-effective of the strategies. Projected over a

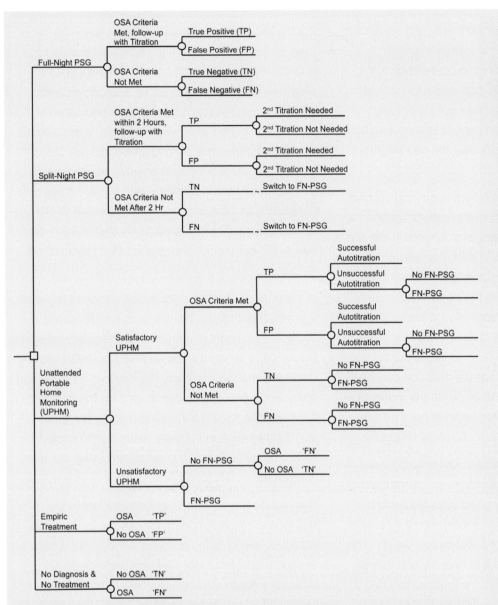

FIGURE 5.6.5

Simplified illustration of the decision-tree structure (diagnosis and titration only). The small square represents the decision to implement a strategy of using a specific diagnostic technology. Circles represent chance events. TN: True Negative; FN: False Negative; TP: True Positive; FP: False Positive (courtesy of Wing Tech Inc.).

10-year period, the team's analysis demonstrated that CPAP therapy can be expected to reduce the risk of motor vehicle collisions by 52 percent, the expected number of heart attacks by 49 percent, and the risk of stroke by 31 percent.[34] When projected over a lifetime, the predicted risk reductions were less pronounced – a result of CPAP-related increases in life expectancy – but they were still significant.[35]

For data projected over a lifetime, the research team quantified the cost savings associated with the reductions in **morbidity** and mortality for OSA patients

who received CPAP therapy. The incremental cost-effectiveness ratio (ICER) of CPAP therapy compared to no treatment was $24,222 per life-year gained, and $15,915 per quality-adjusted life-year (QALY) gained – well below the commonly recognized willingness-to-pay threshold of $50,000–$100,000 per QALY.

Because full-night polysomnography had the greatest diagnostic accuracy, its use significantly reduced the rate of false-positive findings, which meant that, in the long run, it cost less and provided greater health benefits than

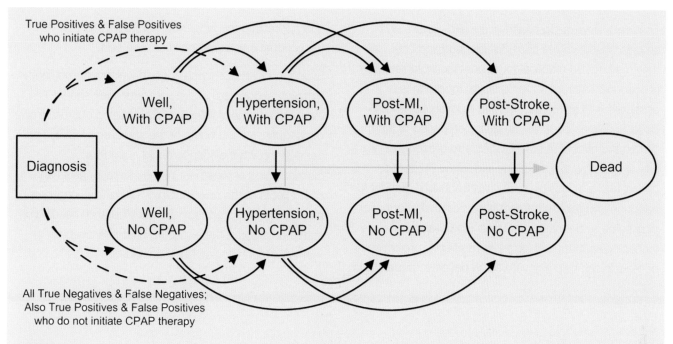

FIGURE 5.6.6

Schematic of the Markov component of the model. Patients are initially distributed based on their diagnosis. Each month, patients can either die, stay in the existing state, or transition into one of the other health states as indicated by the arrows (courtesy of Wing Tech Inc.).

any other approach to diagnosis – even though it was the most expensive approach in terms of upfront costs. At the same time, the study showed that in cases where lab-based diagnosis is unavailable, portable home monitoring can be a viable and cost-effective alternative diagnostic solution. At a time when payers were hesitant to cover or provide reimbursement for portable home monitoring to diagnose OSA, Pietzsch observed that the combination of findings from this study "provided useful insights to guide decision making about coverage and reimbursement of the various diagnostic strategies."

To make the model relevant for evaluating new OSA-related technologies and therapies, the researchers would first incorporate new clinical and economic evidence gathered by innovators during testing or clinical trials of the new technology or procedure, such as therapy compliance or the rates of false-negative and false-positive diagnoses. Using these inputs in the OSA model, the impact of the innovation on relevant health and societal outcomes and costs can be projected.

Next, findings generated by the new model could be used to estimate the relative clinical and economic benefit of the new technology compared to existing technologies.

Reflecting on the value of health-economic modeling, Linehan noted that, "Particularly in complex healthcare scenarios such as those surrounding OSA, where societal implications can be as important as clinical outcomes, it is important to use all available evidence to support decision making. Decision-analytic models like the one completed in this study can help policymakers to understand the clinical and health-economic value of clinical strategies that employ diagnostic and therapeutic technologies." In the absence of data from clinical trials or actual use, healthcare decision makers have come to recognize the value of forecasts developed through health-economic models. "It is increasingly common for innovators to conduct a short-term clinical trial in which some basic evidence is collected and then create a model for projecting the longer-term health-economic profile," Pietzsch added.

That said, developing health-economic models and studies often involves considerable time and cost, so they may not be practical for all new technologies or procedures. According to Pietzsch, modeling is especially well suited for studying innovations whose benefits are realized over an extended period of time. "When the effects of technologies and procedures are fully apparent within the study and follow-up periods of a clinical trial, it's possible to formulate a solid argument for adoption and reimbursement based on that information plus some evidence about costs. However, for technologies that have longer-term effects, projections or models can help innovators and payers to understand and appreciate the longer-term implications that define the product's value proposition," he said.

Even if innovators choose not to develop a formal model, Pietzsch suggested, at the very least they should establish a frame of reference by mapping out the clinical consequences and costs of the standard of care, and considering what evidence could support a distinguishing assessment of the clinical benefits and cost effects of the new therapy. "Ultimately, innovators need to be prepared to make qualified and meaningful statements about the expected cost-effectiveness of their new technologies," he concluded.

As the OSA example highlights, considering whether a value model might be useful in driving reimbursement and adoption can help innovators think critically about the data that is needed to support a compelling payer value proposition. In particular, it helps uncover unknown factors that might make it difficult to formulate a meaningful value statement. As Pietzsch described, "Understanding what information is missing can lead innovators to de-risk their technologies early and in a highly-focused way because they know the key uncertainties that need to be resolved."

Another benefit is that value models can help innovators determine which individuals are most likely to benefit from the treatment. By incorporating data relating to a variety of patient populations with differing characteristics, they can be useful in designing subsequent clinical trials and deciding on inclusion criteria. Additionally, analysis of the dynamics and effects of treatment can help innovators select appropriate endpoints to be studied and ensure that the right clinical evidence is collected for demonstrating the value of the procedure to payers. Study data may also help innovators to devise a rollout strategy for the technology.

Disseminating economic evidence

Providing well-organized economic evidence to important decision makers and other stakeholders is a critically important part of any reimbursement strategy. Documentation is frequently distributed by the company and its sponsor in the form of a **reimbursement dossier** (also sometimes referred to as a "global value dossier") to support the product through this process. A reimbursement dossier can serve as the official source for all key information about the product, including some or all of the topics outlined in Table 5.6.3.

Because the content of a dossier is comprehensive, it is like a roadmap when presenting to payers. Compilation of the data and messages needed to complete the dossier is a lengthy and intensive process that requires the support of specialty medical personnel, as well as experts in cost-modeling, marketing, and managed care. Work should begin up to two years prior to approval to ensure completion before launch. Planning for dossier development should begin even earlier; during **concept** screening the company starts to anticipate the evidence that will be necessary to gain reimbursement. Ideally, a dossier is distributed to payers just after final labeling becomes available from the FDA (or other relevant regulatory body).

For reimbursement decisions outside the US, the dossier (adapted to fit individual country standards) will be essential, as deliberative meetings with industry personnel and advocates are less likely to occur, and a well-documented **need** for the technology will be essential.

Table 5.6.3 A reimbursement dossier can be an important communication vehicle for information related to a company's reimbursement strategy.

Topic	Description
Basic product information	Materials and mechanism of action, indications and product labeling, product cost, associated codes, access, and distribution.
Place of product in therapy	Epidemiology of disease, approaches to treatment, alternative treatment options, product positioning, and expected outcomes of the technology.
Key clinical and economic studies	Treatment populations, number of people studied, study designs, clinical and economic outcomes measured.
Summary of results	All published and unpublished results of clinical studies.
Disease management strategies	Overview of ancillary or disease care management to accompany treatment.
Outcomes studies and economic evaluation supporting data	Results of outcome and economic studies.
Modeling report	Information on cost model(s) and assumptions, results of inputs, parameter estimates, applicable time-horizon and discounting.
Product value and overall cost	Overall justification of cost given data presented.
References	To support all data provided.

Beyond developing and distributing the reimbursement dossier, it is important for the company to publish economic results and key value propositions in peer-reviewed clinical journals and journals targeted to the managed care audience. Often, health economists and/or reimbursement consultants will co-develop and coauthor an article with a physician key opinion leader who was included in the clinical trial. There should also be a publication plan for favorable outcome endpoints from clinical trials as soon as they become available, not just in the clinical journals but in managed care journals as well. For devices that are likely to broaden their indications and usage over time, it is important to plan studies and publications well in advance to support such an expansion.

In order to gain maximum reach with its messages, a company should submit data as abstracts at specialty society and managed care meetings to be published and presented by a KOL. Abstracts and oral presentations are good vehicles for generating positive press, and effective coordination can provide new coverage of the product at the industry, local, or national levels. Scientific symposia at national managed care meetings can be sponsored by the manufacturer to showcase clinical and economic data

to payers. At some meetings, an exhibit booth is also available for sponsorship.

Physician and patient advocacy

Physician advocacy is central to reimbursement success. Developing unique value propositions that specifically focus on physician reimbursement can be useful to get physicians on board with a new technology (see chapter 5.7). Managed care plans are motivated to retain good physicians and, provided the costs are not prohibitive, do not like to deny therapeutic resources that the doctors feel are integral to providing good care. Physicians often see new devices as important innovations that fill an unmet need or have the potential to increase safety and efficiency.

Specialty societies are invested in guiding their profession, helping physicians provide innovative care, gain prominence, and operate financially robust practices. Not only will physician specialty groups be called on to sponsor the creation of new reimbursement codes, but they have the ability to generate positive pressure and influence health plan decision making. Conversely, societies can become powerful enemies in the reimbursement process if the new technology threatens a long-established practice (and/or the revenue stream) for a particular group of physicians (see Stage 6 of the Acclarent case study). In building a case for reimbursement, it can be essential to develop and maintain personal and professional relationships with the physician members and officers of the relevant specialty societies.

Depending on the disease or condition in question, patient advocacy groups can also be mobilized to support the case for reimbursement, particularly if the benefits of the technology are easy to understand and promise significant improvements in quality of life. Patient advocacy groups should be engaged early in the reimbursement process. Strong value propositions are important both to attract patients and to harmonize their advocacy message as they network with others who may have an interest in the new technology. Patient advocacy groups can have an impact on payers through public relations campaigns and direct appeals to health plans. The ability to demonstrate enhanced patient outcomes related to workplace productivity is highly valued by health plans (because they can use this information in their marketing messages to retain and acquire the employers that make up their customer base). Furthermore, private health plans do not want to risk losing members by denying patients access to a technology that is perceived as useful. The threat of negative publicity is a major consideration for payers when deciding whether or not to cover a particular technology.

Payer advisory boards

Establishing a **payer advisory board** (that consists of KOLs and medical directors from select payers in the target payer segment) provides an important way for a company to build relationships with and create awareness and support for the device among payers. Initially, smaller advisory board meetings can be held in order to gather feedback on the product. They can also be used to evaluate the technology's chances for reimbursement success at different prices and/or to validate clinical and economic endpoints. As clinical trials progress and the regulatory submission process begins, the nature of information gathered at an advisory board meeting will shift to the viability of reimbursement given possible

outcomes, coding recommendations, and forward-looking strategies with payers. Just before launch, payers will be able to provide timely information about claims processing, payment amounts, patient prior authorization requirements, provider and facility requirements, and timing of technology assessments. They can also provide the company with feedback on the **financial models**, dossiers, marketing materials, and reimbursement literature that it has developed.

Successful advisory boards can be national, but are more often regional in nature. Their meetings are sometimes planned to coincide with managed care society conventions to make participation easy for payers. They can be held in person or through conference calls and/or webcasts. The typical agenda for a meeting begins with a physician KOL (who is an advocate for the technology and has participated in product testing or a clinical trial) presenting epidemiology data, evidence of an unmet need, clinical data, clinical vignettes, and proposed use to the other members of the payer advisory board. Ideally, this physician will have a strong reputation within the clinical community, good relationships with relevant specialty societies and payers, and will have presented to such an audience successfully in the past. A representative from the company (senior executive, reimbursement staff or clinical-regulatory leader) might next present product positioning, patient population information, progress with FDA, the timeline for launch, proposed code use, value-added services, and the distribution plan. This may also be a good time to vet sections of the product dossier or the validity of economic models. In the event that there is a discrepancy between payer and patient understanding of unmet need, a patient advocate from the clinical trials can be a helpful addition to the meeting. Because an honorarium is typically offered to attendees, not all payer representatives will be allowed to participate in off-site industry-sponsored advisory boards (some health plans limit such employee involvement to avoid conflicts of interest).

The case study on VNUS Medical provides an interesting example of the critical role that reimbursement strategy played for a company that initially thought it could work within established Medicare coding, coverage, and payment policies but ultimately ended up with a different reimbursement pathway.

FROM THE FIELD ▶ VNUS MEDICAL

Navigating the twists and turns of reimbursement strategy

Brian Farley was the first employee of VNUS Medical Technologies, Inc., a company that developed a minimally invasive alternative to vein stripping surgery to treat symptomatic venous reflux disease. This progressive condition occurs when faulty valves in the veins of the legs allow blood to flow backwards and pool, causing leg pain, swelling, skin ulcers, and varicose veins. The VNUS Closure™ technology, also known as radiofrequency thermal ablation, involves threading a catheter into the affected vein and using radiofrequency energy to heat the vein wall. The heat causes collagen in the wall to shrink, collapsing and then sealing the vein. As compared to conventional vein stripping, the standard of care at the time, radiofrequency thermal ablation is equally effective at relieving venous reflux, but causes significantly less post-operative pain and bruising and facilitates a faster return to normal activities. The VNUS Closure technology was cleared by the FDA via the **510(k)** pathway based, in part, on data from a 40-person clinical trial (see Figure 5.6.7).

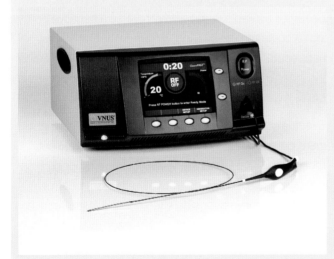

FIGURE 5.6.7

The VNUS radiofrequency ablation catheter and radiofrequency generator (courtesy of Brian Farley).

In preparing its commercialization strategy for the US market, the VNUS team conducted an early investigation of the reimbursement landscape. "Doctors interested in using the product would be able to utilize an existing CPT code for Medicare reimbursement," described Farley, who was President and CEO of the company at the time. CPT code 37204, which covered "trans-catheter occlusion, any method," was not originally designed with radiofrequency thermal ablation in mind, but it provided "a best fit, which according to the CPT coding books, was the guideline for choosing a CPT code," Farley summarized. VNUS initially advised physicians to check with the insurers, many of whom supported the use of this code in conjunction with a corresponding radiologic imaging code, CPT 75894 (a supervision and interpretation or S&I code), since the doctor would be using non-invasive ultrasound imaging to guide the catheter placement in the leg vein. In combination, the payments associated with the two codes adequately covered the costs of the device and the physician's time, which boded well for the early adoption of the technology.

Another positive indication came from the Society of Vascular Surgeons, the primary professional society involved in vein stripping. "The Society passively supported the use of the 37204 code for the new procedure by not suggesting that a new code was needed or intervening for a number of years," Farley recalled.

With regard to establishing coverage from Medicare and private payers, the VNUS team anticipated that this could be a varied and unpredictable undertaking. For example, some private insurers agreed that the treatment was a medical necessity, but did not believe that the description for CPT code 37204 provided a good match. As a result, they asked providers to bill for reimbursement using a miscellaneous CPT code, 37799, designed to cover any vascular procedure without an existing descriptor. In order to realize a reasonable level of reimbursement under a miscellaneous code, doctors had to submit highly detailed operative notes, including a step-by-step account of the activities involved in the procedure and a comprehensive list of all equipment and supplies used in delivering treatment. Although the process was not difficult, physicians disliked the extra time and effort involved in customizing the operative report for each treated patient. Other larger insurers such as Blue Cross Blue Shield asked providers to use a specific S code instead of the miscellaneous code. S codes are temporary codes designated for the private sector that are designed to help an insurer track and monitor usage, often in order to develop a long-term policy for reimbursement.[36] According to Farley, the problem with both of these scenarios was that neither code was associated with an established payment level. "Private payers have their own individual methods for calculating payment levels under miscellaneous codes and temporary codes. And getting a predetermination of a payment level could take an individual provider 6–8 weeks, if it could be obtained at all," he said. Importantly, this uncertainty discouraged some physicians from adopting the new procedure. "Even when the clinical efficacy for your new medical procedure is well established, it's hard to get your business and procedure volume up and running until doctors know precisely how much they're going to get paid," Farley observed.

For roughly three years, providers performing the VNUS procedure billed successfully for reimbursement under the established CPT code, a miscellaneous code, or the S code. During that time, several of the major private payers started to issue negative coverage decisions. "They were not satisfied with the clinical data published to date which was limited to 6 and 12 month follow up of less than 100 patients, and wanted to see stronger evidence that the treatment was safe and efficacious," Farley explained. Fortunately, VNUS had anticipated this issue and voluntarily initiated a large post-market multi-center clinical **registry**. Data from the registry showing elimination of reflux and significant relief of symptoms were published in national and international peer-reviewed journals at a range of follow-up intervals, including six months, one year, two years, and

eventually, five years post-treatment. "In addition to providing additional clinical data," said Farley, "the research helped us develop a stronger understanding of the optimal way to perform the procedure and the likely outcomes it would generate."

Encouraged by the positive research findings and convinced that additional clinical data/publications would continue to strengthen its reimbursement and broader commercialization efforts, VNUS had also launched a randomized controlled trial to directly compare its thermal ablation procedure to vein stripping. The first report from the EVOLVeS randomized trial showed short term advantages of the VNUS Closure procedure compared to vein stripping, along with equivalent efficacy. It was published two years after FDA clearance. "That study was the most important study for obtaining positive coverage policy from large private payers and for obtaining local coverage determinations from Medicare," stated Farley. Once the short-term data from the randomized controlled trial had been accepted for publication, the BCBS Association reviewed the VNUS Closure procedure via its Medical Policy Panel and issued a statement that BCBSA had found the procedure to be medically necessary. Farley explained, "The process is that they review the clinical evidence, speak to experts in the field, and then make a decision. But these decisions aren't always final. If new evidence becomes available, a payer may reverse a negative policy into a coverage policy. And, in our case, that's what happened." After BCBSA issued its findings, nearly all of the independent BCBS plans adopted a formal coverage policy over the next nine months.

Eventually, the two-year results of the EVOLVeS trial were published and conclusions from the 80-patient, multi-center trial found the VNUS Closure procedure as effective as vein stripping at two years post-treatment with significantly fewer side effects and a faster recovery time. The data were published in a peer-reviewed article in the *Journal of Vascular Surgery*[37] five years after FDA clearance and contributed to the company achieving of 100 percent insurance coverage in the US.

Around the same time that VNUS resolved its coverage issues, the company experienced another unexpected twist in its reimbursement strategy. A new vein ablation technology had been developed, called endovenous laser ablation (EVLA). EVLA uses laser energy to thermally damage the vein wall and cause the vein to collapse.[38] Use of this technique became popular among interventional radiologists, and it shifted the balance of which specialist treated saphenous vein reflux. Historically these patients were treated by vascular and general/vascular surgeons who had traditionally performed vein stripping. Accordingly, the Society of Interventional Radiologists took the initiative to apply for new codes to cover energy-based vein ablation procedures. The new codes would separately describe and allow different payment levels for RF vein ablation and laser vein ablation. The Society of Vascular Surgery joined in to co-sponsor the application. Once it became aware that this process was underway, VNUS and the laser companies selling EVLA products worked with the two societies to help ensure that the procedures were properly described and valued, providing supply and equipment cost data, and clinical data that had been published in US peer-reviewed journals.

However, despite these efforts, VNUS remained uncertain whether the establishment of new codes would be a positive development for the company. The primary risk was that the payment level recommended by the RUC and established by CMS for the new code would be less than the payment rates physicians had been receiving to date. "We knew there was some risk because we had a good thing going, with almost 100 percent insurance coverage and current payment levels that were adequate," recalled Farley. At the time, VNUS disposable products commanded roughly a $350 price premium over EVLA in the market, which the company felt was appropriate since it had a more costly catheter that produced better outcomes, as well as less pain and bruising for patient. Plus VNUS had done the "heavy lifting" of pioneering the vein ablation field. "We were first in the market and we had done the randomized trials and published registries. We expected the payment

for the new code for our procedure to reflect a higher cost to the customer compared to EVLA," Farley stated.

Unfortunately for the company, when CMS issued the new codes under which the physician's time and the cost of the device would be covered, it allowed for a premium of only $175 for radiofrequency vein ablation over EVLA. Because this created a situation in which doctors could make a higher profit by choosing the laser procedure, VNUS lost some market share and was forced to reduce the price of its technology. However, in an almost humorous turn of events, several advocates of EVLA systems responded to the $175 payment differential by launching a letter-writing campaign to CMS, detailing the costs of the EVLA supplies and procedure, and requesting that the reimbursement level for EVLA be raised to be on par with the radiofrequency approach. "But what happened," said Farley, "was that CMS took a look at the letters, saw that the actual supply costs cited in the letters were lower than what was in the CMS database of practice expense inputs, and it reduced the payment level for laser ablation. This is a classic example of how reimbursement can be unpredictable." Over the next few years, the payment levels for the two procedures were adjusted until "the payment differential was over $300, close to the way it probably should have been from the beginning," Farley reported. "The doctors could then make a clinically-based decision to choose the technology that was best for the patient and the profit per procedure would be the same."

In parallel, VNUS had to address an issue related to the payment level linked to the initial APC code that was assigned to the procedure when it was performed in a hospital outpatient setting. Most VNUS Closure procedures were conducted in the physician's office, but a reasonable number occurred in the hospital outpatient setting, which necessitated a separate APC code. According to Farley, CMS had two APC codes for vascular surgery procedures and initially placed the VNUS procedure in the lower-paying of the two codes, alongside vein stripping. Aware of the adverse consequences of this lower payment level, executives from VNUS traveled to Washington D.C. with evidence demonstrating the true costs of the procedure. VNUS argued that the CMS method for determining which APC code is appropriate requires the agency to evaluate the cost of the procedure, but that this analysis had not been conducted. While CMS refused to immediately recode the procedure, it did initiate a process for gathering and assessing cost information. "After a year, their representatives agreed with us, and they moved the procedure into the higher-paying APC code for vascular procedures. From start to finish, it took two years for the actual change of assignment," Farley recounted. In contrast to the letter-writing campaign, he added, "It just shows that if there are mistakes made, and payment levels aren't calibrated correctly the first time, it is possible to make a positive change. It's not fast, and it's not a sure bet, but it can happen."

Outside of the US, VNUS was advancing an equally complex portfolio of reimbursement strategies in select geographies. The preliminary focus was on Europe, noted Farley, "where every country is its own challenge." For example, in the UK, the company started off focusing on private insurers, which gave it access to roughly 25 percent of the population. "It was a small percentage of market, but it was a way in," he said. Eventually, VNUS received a positive technology assessment from the National Institute for Health and Care Excellence (NICE). However, the company had to file a business plan with the National Health Service (NHS) Trust to prove that physicians could perform the procedure outside of an operating theater at a significant cost savings compared to vein stripping performed in the operating theatre. Once that business plan was accepted, the company successfully secured national reimbursement coverage in the UK.

At the time VNUS Medical was acquired by Covidien, the company had sales in 38 countries along with full insurance coverage in the US, UK, and the Netherlands and pending reimbursement decisions in Germany, France, and Australia. According to Farley, approximately 1.5 million patients had been treated with the VNUS vein ablation products as of 2014. "This demonstrates the kind of result that can be achieved with strong clinical evidence and an effectively executed reimbursement strategy," he said.

Reflecting on his reimbursement experience with VNUS, Farley summarized, "Reimbursement for a new medical procedure is often a moving target. From coding through payment levels, no decision is final – early success at obtaining coverage and payment may occur because the procedure is very new and payers have yet to review the data or issue negative coverage policies for it. Also, even after coverage is obtained and payment levels are established, those payment levels are adjusted annual by CMS, and reviewed every five years by the RUC."

The key is to remember that regulatory clearance for a new medical product involving a new procedure is "just the tip of the iceberg," he said. "It's a small part of the overall process of being clinically accepted and commercially feasible." In particular, Farley emphasized the need to develop strong clinical data even if it's not required for regulatory approval. "You have to ask yourself from the beginning, 'What clinical studies do we need to prove to the doctors that our technology is better than the standard of care and demonstrate to payers that covering it is in their best interest?'" Additionally, Farley emphasized the importance of having a sophisticated leader who lives and breathes reimbursement every day, responds quickly to changes in the reimbursement landscape, and can engage other senior company executives in the quest. "It's one of the highest priorities in the company," he stated.

Post-launch support for reimbursement strategy

At the time of product launch, reimbursement support begins with educating providers on key reimbursement challenges and giving them access to the support systems that can help them confirm coverage and secure appropriate payment. The company should educate providers about specific payer requirements in their region, as well as identify payers that support reimbursement for the new technology. If health plans have limited coverage of the device to a subset of the indicated population, physicians and their staff should be trained on appropriate patient selection and prior authorization requirements. Billing staff will require training on plan-specific billing, coding, and claim submission procedures and the claim submission package can be very helpful. The product sales force and/or dedicated reimbursement staff will be on the front lines of reimbursement support and must also be trained to field reimbursement-related questions. Monitoring of payer actions should also take place. Furthermore, appropriate reimbursement expectations should be set with providers, internal stakeholders, and the investment community.

Throughout these support activities, oversight by reimbursement experts and skilled legal counsel will be essential. Practice management support is one of the key areas related to reimbursement that can get companies into trouble. Because supportive information helps providers get paid, in some cases from government payers, great care must be taken to avoid illicit activities that involve kickbacks, inducements, or inadvertent counsel to break the law. A misstep in this area can mean huge fines, bad publicity, and added marketing restrictions from which it could be nearly impossible to recover.

In the post-launch phase, a company may have opportunities to develop mutually beneficial partnership arrangements with health plans. For example, it may elect to work with private payers to track usage and performance outcome measures within their plan population. This type of partnership is often an extension of a contract in which the payer receives a discounted price or rebate by reaching predefined share or volume targets agreed upon with the manufacturer. This provides an incentive to both parties to co-promote coverage. For example, Conceptus Inc. issues a press release each time a new payer adds a favorable coverage policy for its minimally invasive procedure to address blockages in the fallopian tubes.[39] Companies can also use in-office pull-through activities, in the form of co-branded sales pieces promoting the device as "preferred" by this payer. Sales representatives can be used as well, to help advertise coverage policies in order to keep providers informed regarding which payers are reimbursing for the device. All of these activities must be carefully integrated and managed through an ongoing reimbursement strategy.

➘ Online Resources

Visit www.ebiodesign.org/5.6 for more content, including:

 Activities and links for "Getting Started"
- Assess the reimbursement landscape
- Perform primary market research with payer decision makers
- Evaluate strategic options
- Develop evidence
- Organize information into a reimbursement strategy

 Videos on reimbursement strategy

 Appendices that provide additional information about:
- BCBS Technology Evaluation Center
- Common cost models

CREDITS

The editors would like to acknowledge John Hernandez of Abbott for his extensive editorial assistance in developing this chapter. Emily Kim of Medtronic also provided substantial review and guided the organization of the revised chapter. The material in the first edition was written with input from Mitch Sugarman of

Medtronic. Further thanks go to Brian Farley of VNUS Medical, Steve Halasey of InHealth, Jan B. Pietzsch of Wing Tech Inc., and Stacey McCutcheon for their assistance with the case examples, as well as Trena Depel, Steve Fair, and Sarah Garner for their early assistance with the chapter.

NOTES

1 Drawn from "Innovators' Guide to Navigating Medicare," Centers for Medicare and Medicaid Services, 2010, http://www.cms.gov/Medicare/Coverage/CouncilonTechInnov/downloads/InnovatorsGuide5_10_10.pdf (March 11, 2014), unless otherwise cited.

2 As noted in 4.3 Reimbursement Basics, the ICD-10 code set was delayed twice but expected to take effect in October 2015 as of the time of this writing.

3 "Process for Requesting New/Revised ICD-10-PCS Procedure Codes," Centers for Medicare and Medicaid Services, http://www.cms.gov/Medicare/Coding/ICD9ProviderDiagnosticCodes/newrevisedcodes.html (March 11, 2014).

4 "Applying for CPT Codes," AMA, http://www.ama-assn.org/ama/pub/physician-resources/solutions-managing-your-practice/coding-billing-insurance/cpt/applying-cpt-codes.page (March 17, 2014).

5 "Innovators' Guide to Navigating Medicare," op. cit.

6 "Understand the Three CPT Code Categories," op. cit.

7 "Understand the Three CPT Code Categories," Advance Health Network, October 12, 2004, http://health-information.advanceweb.com/Article/Understand-the-Three-CPT-Code-Categories-2.aspx (February 24, 2014).

8 "CAD Sciences Announces the Creation of the First Ever Category III CPT Code for Breast MRI CAD," DeviceSpace, January 10, 2006, www.devicespace.com/news_story.aspx?NewsEntityId=6631 (February 24, 2014).

9 "CPT Coding Change Request Form Instructions," AMA, http://www.ama-assn.org//ama/pub/physician-resources/solutions-managing-your-practice/coding-billing-insurance/cpt/applying-cpt-codes/request-form-instructions.page (March 17, 2014).

10 "New Category III CPT Code Created for NeoVista's Novel AMD Treatment," PR Newswire, January 22, 2008, http://www.prnewswire.com/news-releases/new-category-iii-cpt-code-created-for-neovistas-novel-amd-treatment-57115967.html (February 24, 2014).

11 E. R. Scerb and S. S. Kurlander, "Requirements for Medicare Coverage and Reimbursement for Medical Devices," Clinical Evaluation of Medical Devices (Humana Press, 2006), p. 74.

12 Healthcare Common Procedure Coding System (HCPCS) Level II Coding Procedures, https://www.cms.gov/Medicare/Coding/MedHCPCSGenInfo/Downloads/HCPCSLevelIICodingProcedures7-2011.pdf (February 24, 2014).

13 Drawn from "Innovators' Guide to Navigating Medicare," op. cit., unless otherwise cited.

14 "HCPCS Decision Tree," Centers for Medicare and Medicaid Services, https://www.cms.gov/Medicare/Coding/MedHCPCSGenInfo/Downloads/HCPCS_Decision_Tree_and_Definitions.pdf (March 11, 2014).

15 "Intermediary-Carrier Directory," U.S. Department of Health and Human Services, http://www.cms.hhs.gov/ContractingGeneralInformation/Downloads/02_ICdirectory.pdf (February 24, 2014).

16 "Innovators' Guide to Navigating Medicare," op. cit.

17 "Medicare Coverage Determination Process," Centers for Medicare and Medicaid Services, http://www.cms.gov/Medicare/Coverage/DeterminationProcess/ (February 24, 2014).

18 "Vagus Nerve Stimulation (VNS) for Resistant Depression," MLN Matters, Centers for Medicare and Medicaid Services, June 20, 2013, https://www.cms.gov/Outreach-and-Education/Medicare-Learning-Network-MLN/MLNMattersArticles/downloads/MM5612.pdf (February 24, 2014).

19 "FDA-CMS Parallel Review," U.S. Food and Drug Administration, http://www.fda.gov/MedicalDevices/DeviceRegulationandGuidance/HowtoMarketYourDevice/PremarketSubmissions/ucm255678.htm (February 28, 2014).

20 Turna Ray, "Exact Sciences Discusses Benefits of Taking Cologuard through FDA/CMS Parallel Review Pilot," Pharmacogenomics Reporter, April 3, 2013, http://www.genomeweb.com/clinical-genomics/exact-sciences-discusses-benefits-taking-cologuard-through-fdacms-parallel-revie (February 28, 2014).

21 Ibid.

22 Ibid.

23 Ibid.

24 Rebecca Kern, "CDRH Preps New Program To Streamline Approval-To-Reimbursement Path," The Gray Sheet, December 19, 2013.

25 "Cigna HealthCare Coverage Position: Bariatric Surgery," May 15, 2007, http://www.cigna.com/customer_care/healthcare_professional/coverage_positions/medical/mm_0051_coveragepositioncriteria_bariatric_surgery.pdf (February 24, 2014).

26 Drawn from "Innovators' Guide to Navigating Medicare," op. cit., unless otherwise cited.

27 "Two-Times Rule Defined," APCs Insider, HC Pro, May 13, 2004, http://www.hcpro.com/HIM-39495-859/Twotimes-rule-defined.html (March 19, 2014).

28 "The RVS Update Committee (RUC)," AMA, http://www.ama-assn.org/ama/pub/physician-resources/solutions-managing-your-practice/coding-billing-insurance/medicare/the-resource-based-relative-value-scale/the-rvs-update-committee.page (February 24, 2014). The purpose of the RUC process is to provide recommendations to CMS for use in annual updates to the new Medicare RVUs. The RUC is a unique committee that involves the AMA and specialty societies, giving physicians a voice in shaping Medicare relative values. The AMA is responsible for staffing the RUC and providing logistical support for the RUC meetings.

29 Amitabh Chandra, Jonathan Holmes, Jonathan Skinner, "Is This Time Different? The Slowdown in Healthcare Spending," Economic Studies at Brookings, September 2013, http://www.brookings.edu/~/media/Projects/BPEA/Fall%202013/2013b%20chandra%20healthcare%20spending.pdf (February 24, 2014).

30 Barbara Grenell, Debbie Brandel, and John F.X. Lovett, "Obtaining Reimbursement Coverage from Commercial Payers," *Medical Device Link*, http://www.devicelink.com/mddi/archive/06/01/009.html (February 24, 2014).

31 Compiled from available data on the Cost Effectiveness of ICD Therapy in SCD-HeFT, Medtronic's Sudden Cardiac Death in Heart Failure Trial. Reprinted with permission.

32 All quotations are from interviews conducted by the authors, unless otherwise cited. Reprinted with permission.

33 Moderate-to-severe OSA was defined as an apnea-hypopnea index (AHI) of ≥ 15 events per hour.

34 Jan B. Pietzsch, Abigail Garner, Lauren Cipriano, and John Linehan, "An Integrated Health-Economic Analysis of Diagnostic and Therapeutic Strategies in the Treatment of Moderate-to-Severe Obstructive Sleep Apnea," *Sleep*, vol. 34 (2011):695–709, http://www.journalsleep.org/ViewAbstract.aspx?pid = 28141 (March 5, 2014).

35 Ibid.

36 Healthcare Common Procedure Coding Systems (HCPCS) Level II Level II Coding Procedures," Center for Medicare and Medicaid Services http://www.cms.gov/Medicare/Coding/MedHCPCSGenInfo/Downloads/HCPCSLevelII CodingProcedures7-2011.pdf (February 11, 2014).

37 F. Lurie, D. Creton, and B. Eklof, "Prospective Randomized Study of Endovenous Radiofrequency Obliteration (closure procedure) Versus Ligation and Stripping in a Selected Patient Population (EVOLVeS Study)," *Journal of Vascular Surgery*, August 2003, http://www.jvascsurg.org/article/S0741-5214(03)00228-3/fulltext (February 12, 2014).

38 Neil M. Khilnani, "Varicose Vein Treatment With Endovenous Laser Therapy," Medscape.com, http://emedicine.medscape.com/article/1815850-overview (February 12, 2014).

39 "Cigna Now Covering the Essure Procedure," Conceptus, February 15, 2006, http://investor.conceptus.com/common/mobile/iphone/releasedetail.cfm?ReleaseID = 252092&CompanyID = CPTS&mobileid (February 24, 2014).

INTRODUCTION

Two new medical devices with similar clinical data establishing their safety and efficacy. One is hailed as an important advancement, receives broad professional society support, and is quickly adopted upon its commercial launch. Another is dubbed "experimental," blocked by key opinion leaders, and struggles in the market. Although there are many factors that can affect the adoption of a new technology, the difference in its success or failure often comes down to how well the company develops and executes a proactive, forward-looking, multifaceted marketing and stakeholder strategy. In particular, this strategy must effectively communicate the value of the offering in such a way that the improvement–cost equation compels key decision makers to change their behavior and adopt it.

The greatest medtech innovations are disruptive. They deliver breakthrough improvements in patient outcomes and fundamentally change the practice of medicine. Yet, the "do no harm" principle of medical ethics supports a conservative culture that can be at odds with the adoption of new technologies. Similarly, unsustainable growth in healthcare spending in many developed countries and the need to extend health services to millions more citizens in emerging economies both contribute to a cautious attitude toward embracing new medtech products. This tension presents a challenge for innovators and requires them to establish and communicate the value of new device technologies in such a way that the information serves as a catalyst for the desired change. A well-designed marketing and stakeholder strategy seeks to accomplish this by compelling key decision makers and influencers into action. Because there are multiple stakeholders in the medical device field who can sway adoption decisions, an effective strategy must be multidimensional and tailored to the unique perspectives of each primary stakeholder group. The essential components of the strategy are the value proposition, along with the marketing mix, which is used to communicate the value of the innovation to members of the target audience and drive them to adopt.

The decisions made in developing a marketing and stakeholder strategy must be supported by important choices related to 5.3 Clinical Strategy, 5.4 Regulatory Strategy,

OBJECTIVES

- Appreciate the importance of more deeply understanding the perceptions of key stakeholders toward a specific need and/or new solution.

- Learn how to develop evidence-based value propositions that clearly articulate an improvement/cost equation that compels key stakeholders to change their behavior.

- Understand how to develop a marketing mix that enables a company to effectively support and implement its value propositions.

and 5.6 Reimbursement Strategy (these topics are referenced, but not addressed in detail within this chapter). The marketing and stakeholder strategy also informs the activities described in 5.8 Sales and Distribution Strategy.

 See ebiodesign.org for featured videos on marketing and stakeholder strategy.

MARKETING AND STAKEHOLDER STRATEGY FUNDAMENTALS

Early in the biodesign innovation process, through the initial examination of the market, innovators gain a foundational understanding of the **value** associated with a **need** area and the cost/improvement threshold that any new solution would have to meet (or exceed) to cause key **stakeholders** to change their behavior and adopt it. But now that a specific solution has been designed and is under development, the **value proposition** for the new offering can be studied and constructed in much greater detail. Once defined, the value proposition is supported by and implemented via the company's **marketing mix** – the "4 Ps," which traditionally include product/service mix, price, positioning, and place.[1] However, in this text, "place" (or the channels and approach to selling and distributing an offering) is covered in chapter 5.8 Sales and Distribution Strategy. The fourth "P" addressed in this chapter is "promotion" (as shown in Figure 5.7.1).

Revisiting stakeholder analysis

The way in which the value proposition is formulated and then supported by all four elements of the marketing mix directly affects how the new solution is received by the customer. Accordingly, one of the team's first priorities should be to revisit its initial stakeholder analysis. With the new solution squarely in mind (not just the need area that was studied in 2.3 Stakeholder Analysis), the innovators should confirm a solid understanding of the following three factors:

1. Who are the most important stakeholders and who among them will be the critical decision maker(s)?
2. What are their opinions toward the need *and* the new solution; and are these opinions strong enough to cause a change in behavior?

3. Who is most likely to resist the new solution and/or create conflicts between stakeholder groups?

As outlined in chapter 2.3, **cycle of care** and **flow of money** analyses provide a good baseline for identifying relevant stakeholders and beginning to prioritize their importance as decision makers and influencers. Repeating these assessments can be helpful because much more is now understood about the specific solution, how it will be implemented, and by whom. More is also known about regulatory, intellectual property (IP),

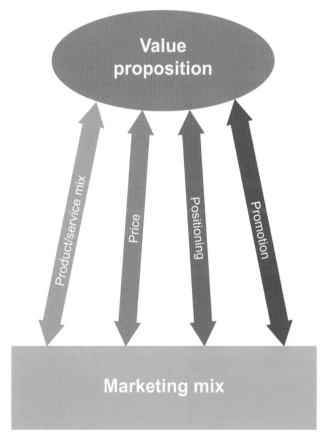

FIGURE 5.7.1

The value proposition is supported by the marketing mix in the creation of an effective marketing and stakeholder strategy.

reimbursement, and clinical issues that will affect adoption decisions. For example, with a proposed solution in hand, the team is now better equipped to assess the impact on workflow with consideration for the detailed practical challenges that may arise. As a result, the cycle of care and flow of money analyses may show different results from the original assessments (which were performed even before the need was clearly specified).

At this point in the biodesign innovation process, innovators are often well served to dive even deeper into stakeholder analysis. Their emphasis should now shift from identifying influencers and decision makers and understanding what drives their behaviors at a high level to gaining an in-depth familiarity with how, when, and why they are most likely to act and in what ways the company can affect their perspectives and decisions. Innovators can choose from any number of different techniques for gathering this more detailed information that range from relatively simple to fairly elaborate. At one end of the spectrum, they can conduct in-person demonstrations of the offering, "walk throughs," and/or interviews with potential **users**. For example, innovators sometimes set a goal to meet with anywhere from 10 to 100 stakeholders before they begin constructing a marketing strategy. They can also conduct surveys of various types. Market research surveys have been effectively used in other industries, such as consumer packaged goods, for decades, but they are largely underutilized in the medical technology field. Carefully designed and executed studies can yield invaluable insights about stakeholder preferences and beliefs, as well as their likelihood of adopting a new technology or taking another desired action in different scenarios. At the other of the spectrum, innovators may choose to conduct pilot studies that involve a limited commercial launch in a specific geography or with a first-generation product (see the Miramar Labs story below).

The key with all of these methods is to ensure that the right questions are being asked, using an approach that minimizes **bias** and other subjective input that can distort the validity of the data in terms of their applicability to a larger population. Closely related is the issue of identifying respondents who will be honest and objective in providing their responses. Working with an expert in marketing research to design a survey instrument or interview guide and define the target audience can add cost to the project, but it is often well worth the investment.

When conducting primary data collection from stakeholders, innovators are looking for indicators that the new solution will matter enough for the target audience to take notice and change its behavior (incremental enhancements and innovations that respondents describe as "interesting" are often not compelling enough to drive a behavior change). They are also looking for direct input to help them craft value propositions and define a marketing mix that optimizes how these stakeholders will respond to the new offering. The story of Miramar Labs illustrates the value of conducting primary stakeholder research. It also previews how this information can be used to shape an effective marketing strategy.

FROM THE FIELD ⟩ **MIRAMAR LABS**

Leveraging market research in developing a marketing strategy

Darrell Zoromski joined Miramar Labs with a background in consumer packaged goods (Proctor & Gamble, General Mills, and S.C. Johnson) and consumer medical devices (Carl Zeiss Vision and Align Technology, maker of Invisalign). At the time, Miramar Labs was in the midst of conducting a **pivotal** clinical study for its novel technology for treating excessive underarm sweat. Axillary hyperhidrosis, a medical condition that causes excessive underarm perspiration, affects 1.4 percent of the US population (or approximately 4 million people). A full 1.3 million individuals report that the condition is barely tolerable and frequently interferes, or is intolerable and always interferes, with their daily activities.[2] Axillary hyperhidrosis is currently treated most frequently with antiperspirants or underarm BOTOX® (botulinum toxin)

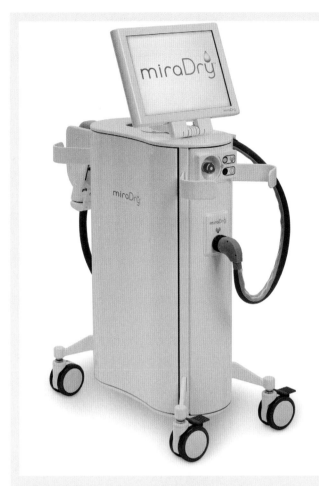

FIGURE 5.7.2

The miraDry system (courtesy of Darrell Zoromski).

injections that temporarily block the chemical signals from the nerves from stimulating the sweat glands. However, both of these interventions have low satisfaction rates among patients. Antiperspirants are not very effective, and BOTOX requires multiple injections and lasts an average of seven months before the procedure needs to be repeated. Miramar Lab's technology, miraDry, uses microwave energy to provide a non-invasive, permanent solution to axillary hyperhidrosis. The miraDry system consists of a capital equipment console and a disposable bioTip used to deliver each procedure (see Figure 5.7.2).

As the company's new CEO, one of Zoromski's first priorities was to focus on a commercialization strategy for the technology, including the creation of a detailed approach for marketing to key stakeholders. At the time, the US was Miramar's primary (although not sole) market

focus due to its large domestic demand for aesthetic procedures. "Our goal was to be sure that we would be successful in the US when we launched because we expected this market to make up a large portion of our sales and be a critical value driver for the company," he explained.[3]

Through a thorough review of the reimbursement landscape, Miramar had determined that the most practical approach in the US would be to make miraDry a patient-pay procedure. Although there was some chance of securing reimbursement based on several favorable characteristics of the treatment relative to available alternatives (e.g., it was permanent while BOTOX injections had to be re-administered every 7 months), Miramar had reasons to believe that the **payment** rate might be set relatively low, making its technology less appealing to physicians.

With a patient-pay model in mind, the company identified the patients who would receive (and fund) the treatment, as well as the physicians who would purchase the technology and deliver the intervention, as its top-priority stakeholders. To jumpstart the creation of a stakeholder and marketing strategy, Zoromski hired a market research firm to survey a representative sample of individuals in both of these stakeholder groups. "These types of studies are common in the marketing toolkit in the consumer packaged goods industry, where consumers are critical stakeholders in the product or service purchase decision," he said. "But medical device companies can benefit from that same approach, particularly given the increasing importance of consumers in making medical care decisions." According to Zoromski, formal market research studies could take roughly three to six months and $75,000–250,000 to complete, depending on the specific scope and approach, but the information could be invaluable to helping shape a company's marketing strategy.

The patient survey sampled 3,000 prospective users of the treatment from the general US population, aged 18–65. Four percent of the respondents indicated that they had previously been diagnosed with excessive underarm sweating, and another 17 percent were severely bothered by the problem. This quantitative research was rounded out with qualitative focus groups and one-on-one interviews. The result was a consumer target profile, indicating that individuals in both groups tended to be younger (18–44 years of age), female, career and relationship oriented, and concerned about how they appeared to other people. Zoromski and his team decided to make the 21 percent of individuals who were diagnosed and/or seriously concerned about underarm sweat the company's first patient targets.

The physician survey similarly helped Miramar Labs identify a target audience within the physician population. This survey went to 300 US-based specialists in dermatology and plastic surgery. The results indicated that physicians in both specialties potentially would have an interest in performing the procedure, but the company should start with dermatologists because they had a higher preliminary purchasing intent. That is, both groups saw the value of the proposed solution, but dermatologists were more inclined to take action on the opportunity in the near term. In particular, the study revealed that the ideal target would be dermatologists who had already been treating axillary hyperhidrosis, had other forms of capital equipment in their offices, and were used to paying/charging for disposable product as part of a procedure.

In addition to helping Miramar Labs determine which stakeholders to focus on, the company used the survey to gather input that would form the basis of its value propositions and marketing mix. One technique used in the survey was to present both audiences with a series of benefit statements and ask respondents to rank the ones that were most meaningful to them. The survey also included questions to help with positioning, understanding pricing, and optimal ways to promote the procedure to the target audiences. Once the company had the results, it was ready to go to work on devising its marketing strategy. As Zoromski explained, "The quantitative and qualitative results gave us a more objective basis for constructing our plans. They provided about 50 percent of the information we needed; the

other 50 percent came from intuition and experience of the team."

To appeal to patients, Miramar defined a value proposition that focused on a series of simple ideas: the procedure was safe, highly effective, and permanent, all of which make it a good value relative to the cost of other solutions. "Even for patients who weren't receiving BOTOX, we tried to communicate that miraDry would provide strong value compared to even antiperspirants, given the high lifetime cost of coping and dealing with excessive sweat. If you look at how much people spend on antiperspirants, keeping one at home, one in the car, one in an office drawer; dry cleaning expenses to try eliminating yellow underarm stains; and the nice clothing that's ruined after four wears," Zoromski said, "it can really add up. And then there's the emotional cost of embarrassing underarm sweat outbreaks, which are just as important.

For physicians, the value proposition centered primarily on the safety and efficacy of the treatment, and secondarily on the fact that it was a leading-edge intervention that would be a lucrative addition to their practice. As Zoromski pointed out, "In the survey, the physicians expressed that profitability wasn't that important to them [normative response], but we knew that it actually would be based on prior experience and candid one-on-one discussions with physicians. So we had to be a little bit thoughtful about how we addressed this key factor."

With both audiences concerned about safety and efficacy, Miramar Labs recognized that its **clinical trial** data would play an important role in supporting its value propositions. When the company completed its pivotal clinical trial to support its **510(k)** application to the **FDA**, the results demonstrated that the procedure had an 89 percent efficacy rate after one month, but then declined and stabilized at 70 percent over time. Together with the R&D team, Zoromski decided that they would work on doing better than 70 percent, even though many dermatologic procedures deliver efficacy in that range. This would ultimately deliver more satisfied patients and physicians, and differentiate Miramar Labs from other "share of physician chair" competition. Diving into the clinical results, they ultimately discovered that physicians in the trial were missing some sweat glands during treatment, which was having a negative impact on the effectiveness of the procedure. By making some adjustments to how the technology worked, Miramar Labs was able to address this issue and provide long-term, stable efficacy of over 90 percent.

In terms of setting a price, Zoromski acknowledged that this decision required some "complex analyses." Specifically, the team had to reconcile (1) patient willingness to pay with (2) physician's willingness to pay and (3) the financial return required for Miramar Labs to become a viable, sustainable business. Using data from the survey as one input into the equation, the company ultimately determined that, for a high level of efficacy, consumers would bear a total price of approximately $3,000 for the two visits necessary to receive the treatment. By comparison, BOTOX injections, which were only covered under certain insurance plans, cost $1,000–$1,500 for up to 30 injections in a single visit to achieve temporary results. For prospective patients not receiving any medical treatment for their condition, Miramar Labs created a calculator that would enable them to estimate the cost of their "coping mechanisms," such as deodorant and dry cleaning bills. "miraDry was a great value if you were using BOTOX and not getting reimbursed, but still a good value for those who were simply coping with the problem," Zoromski said. In the survey, the company had tested different pricing scenarios with physicians that involved either a higher cost for the capital equipment and a lower cost for the disposables or vice versa. In the end, Zoromski and colleagues determined that the "sweet spot" was to charge approximately $50,000 for the capital equipment and $350 for each consumable (for a total of $700 per patient across the two visits). The company also confirmed that this pricing scheme compared favorably to other in-office procedures being performed by specialists in the target audience. With these prices, Miramar Labs would achieve a reasonable gross margin

over time that would allow the company to fund the amount of marketing and sales required to reach its customers while creating a profitable business that is attractive to investors and sustainable in the long run.

Going back to the quantitative and qualitative survey data, the take-aways around how to position miraDry were also helpful to the company. For example, the Miramar Labs team learned that while physicians were excited about the fact that the technology was new and innovative, and that it operated on microwave technology, this same information scared many patients. This caused the company to use the term "microwave energy" only with physicians and refer to "electromagnetic energy" when interacting with patients. Another issue related to the concept that the treatment permanently prevented underarm sweating. To the team's surprise, patients reacted negatively to this terminology, reporting that it made the procedure seem irreversible. However, when the company talked about the treatment being "lasting," this was perceived as a major benefit in patient's eyes, even with patients acknowledging that "lasting" means roughly the same thing as "permanent." "Very subtle differences in wording actually can make a pretty significant difference in how appealing you are to your target group," Zoromski commented.

From a promotion perspective, the survey and qualitative research also revealed important information about the best ways to communicate with patients and physicians in the company's target markets. With patients, establishing a strong online presence would be key. "We knew that the web was going to be our most significant area of focus because it's a cost-efficient way to get a lot of information out. Consumers, especially the younger, female, image-conscious consumers who are our target, start with the web," Zoromski described. The company began by developing its own robust patient-centered website and then established a presence on other sites that its target consumers would perceive as credible, such as WebMD.

To promote miraDry to physicians, Miramar Labs similarly invested in building a strong physician-oriented website. The company also built a strong network of key opinion leaders (**KOL**s) who would act as advocates for the procedure. Zoromski noted that the company's physician advisory board ended up being helpful in this regard, as well. "In addition to advising us on our clinical approach and how physicians will perceive the offering, they also spoke to some of their peers about miraDry," he said. "So they provided another credible way for us to raise awareness." Other aspects of its promotion efforts to physicians included presentations and a strong presence at trade shows and conferences directed at plastic surgeons and dermatologists, publishing clinical data in **peer-reviewed** journals, authoring white papers, and conducting some limited advertising in specialty trade journals. The company also employed a small direct sales force and a team of clinical specialists in the US to help target and train physician customers.

With both consumers and physicians, Miramar Labs realized that early adopters may make a purchase decision quickly, but the majority would require multiple "touches" to move them through the conversion funnel. A web-based lead management program was established for both stakeholder groups to provide topical updates and relevant information to those expressing any interest in the miraDry procedure.

Outside the US, Miramar Labs intended to expand into other countries with large aesthetic products markets, such as Japan, South Korea, and Brazil. Accordingly, it was working on commercialization plans in these other geographies in parallel with its US marketing strategy. However, given the importance of the US market to the company in terms of its overall size and prominence, Zoromski and team decided they would benefit from gaining some fast commercial experience prior to the US launch. They conducted a comprehensive analysis of the other regions of interest and determined that Japan would be the best location to validate their approach to marketing and their commercial assumptions. One of the main reasons they chose this market was that they could move quickly and begin importing the device for direct sale by physicians without first having to seek regulatory approval by Japan's Pharmaceutical and Medical Device

FIGURE 5.7.3
Zoromski (left) with one of the company's early customers in Japan, Dr. Hiroyuki Kanamaru (courtesy of Darrell Zoromski).

Agency (**PMDA**).[4] Over the course of approximately 12 months, Miramar representatives interacted with Japanese physicians and patients and observed how the company's value propositions and marketing mix were received (see Figure 5.7.3). Although the company made a handful of adjustments to address unique factors and preferences in the Japanese market (e.g., placing more emphasis on odor reduction versus excessive sweat; targeting the plastic surgeons who were more likely to treat axillary hyperhidrosis than Japanese dermatologists), the commercial test resulted in rich learning and also helped stimulate early revenues. Most importantly, Miramar Labs confirmed its approach: physicians performed twice as many procedures as the company had originally modeled and the technology was able to command a higher price than initially planned.

Reflecting on the experience in Japan, Zoromski encouraged other innovators to consider running commercial experiments. "Before you move into your primary market, get some fast learning," he advised. "If you're wrong, that's fine. You've only failed on a small scale and can rethink the business model and the marketing approach. And if you're right, that's even better. Just keep accelerating your commercialization."

In terms of creating an overall marketing strategy, he also underscored the value that can come from conducting a well-designed market research study. "Figuring out how to commercialize a B2B2C [business to business to consumer] offering like ours is complex," Zoromski said. "You really need to understand what's in it for each stakeholder, who the influencers are, who the decision makers are, and what are the most important benefits they're seeking. How should you talk about those benefits? And what language should you use to do it?" The advantages of getting this information directly from the source are significant, he added, "Because small differences in your approach can make a big difference in your commercial success. There's no one perfect way to do this, but a combination of quantitative research, qualitative research, and experience-based intuition is a pretty good recipe."

Importantly, not all market research has to be as extensive (or expensive) as the Miramar Labs example, especially during the early stages of a start-up business. Many innovators have successfully used online survey tools (e.g., SurveyMonkey) to collect preliminary user feedback at a fraction of the cost of a professional market study. One Stanford Biodesign team even placed an inexpensive classified ad in a parenting magazine on the topic of children's night terrors and received more than 500 responses. Again, the key is to carefully prepare so that the right questions are asked in a manner that minimizes bias and maximizes the validity of the data in terms of its applicability to a larger population.

Developing value propositions

Value propositions are central to convincing stakeholders to change their behavior and adopt a new technology or offering (recall that value propositions directed at **payers** were introduced in chapter 5.6). In marketing, a value proposition refers to the sum total of the improvement that a company promises to a customer in exchange for payment (or other value transfer).[5] Another way to think about a value proposition is as a marketing statement that summarizes why a customer should adopt a particular product or service. This statement should convince a potential customer that the new product or service will add more value or better solve a problem than other competitive offerings at a reasonable price. Again, in today's value-oriented environment, it is not enough to offer an incremental improvement at a price that it marginally more (or even less) than available options.

Importantly, the improvement offered by a new solution must be perceived as compelling enough to overcome the entire collective cost of making a change. For customers in many situations, this goes well beyond the price of the offering and includes the cost of aligning other stakeholders around the new technology, making a purchase decision, and then implementing the required modifications to standards of care and established workflows to put the new solution into practice.

Companies almost always create value propositions for the key decision makers involved in making a purchasing decision. However, it is important to keep in mind that many health-related solutions involve multiple decision makers and require multiple value propositions as a result. For example, in the Miramar Labs case, the company needed a unique value proposition to influence the physicians who would purchase the technology and perform the procedures, as well as another to target the patients who would elect to receive and pay for treatment. If the company eventually decides to seek reimbursement for the intervention, it will then need to develop value propositions that resonate with the public and private payers.

In addition to decision makers, innovators should pay special attention to the individuals or groups that can help or hinder the adoption of the solution. These secondary value propositions can be devised to help activate other stakeholders as advocates for the offering or to attempt to minimize their resistance to the solution if it may negatively affect them in some way. Again referencing the Miramar Labs example, the company may need to develop value propositions (and a marketing strategy) to anticipate and address potential concerns among the members of physician professional societies in dermatology and plastic and reconstructive surgery.

A value proposition is closely related to the **value estimate** that the team developed as part of 2.4 Market Analysis. However, it is much more specific (addressing the solution, not just the need), comprehensive (taking into account more detailed information about the solution and factors related to its market entry, such as regulatory, IP, reimbursement, and clinical issues), and actionable (based on a more thorough understanding of the target stakeholders and what drives them). As a result, it is a more useful tool for preparing for commercialization and anticipating stakeholder responses.

Recall that the treatment landscape charts developed in chapter 2.4 plotted the available solutions in a need area in terms of their general efficacy and cost (see Figures 2.4.1 and 2.4.2). The development of a value proposition directly builds on this work. First, innovators must choose the existing treatment that their solution is most likely to displace. The cost and efficacy of this treatment becomes the baseline against which they can construct the value proposition for the new solution. Keep in mind that, in this context, the term "efficacy" is used as a proxy for almost any attribute of value not specifically related to cost (e.g., quality, patient satisfaction, waiting time to diagnosis). Similarly, the notion of "cost" can reflect cost to the system, product unit cost, or even time of procedure, which is a proxy for labor cost per patient.

As shown in Figure 5.7.4, there are three viable quadrants in which a value proposition can exist: a solution can offer (1) higher efficacy at a higher cost; (2) lower efficacy at a lower cost; or (3) higher efficacy at a lower

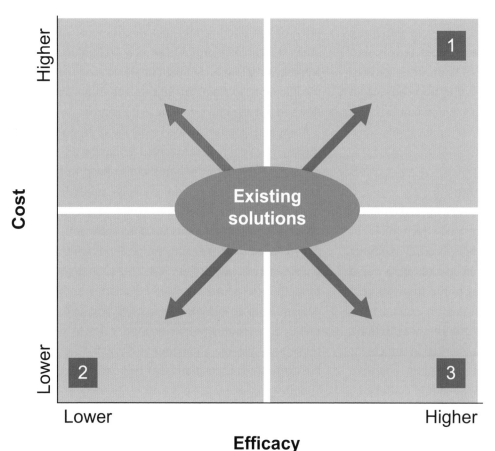

FIGURE 5.7.4

Value propositions help communicate how a new solution is positioned relative to available treatments.

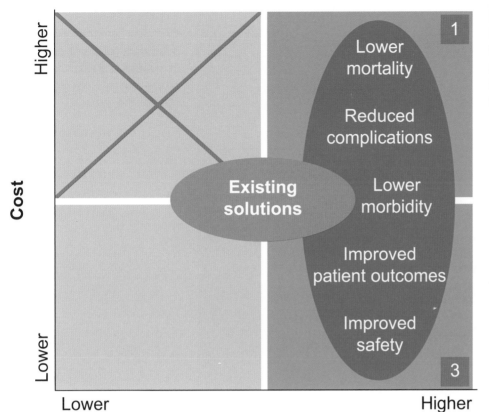

FIGURE 5.7.5

Innovators can differentiate solutions in quadrants 1 and 3 from existing solutions based on the improvements in efficacy that they offer.

cost. Solutions in the upper and lower right quadrants (#1 and 3) compete on the fact they offer better efficacy than existing solutions. Figure 5.7.5 outlines the types of improvements that innovators might highlight in their value propositions for these offerings.

Offerings in the lower quadrants (#2 and 3) compete on their ability to offer a solution at a lower cost than available alternatives. Lower cost offerings tend of reflect one or more of the improvements outlined in Figure 5.7.6.

Importantly, solutions that can be mapped to quadrant #3 offer a particularly powerful dual advantage in that they promise to deliver improved efficacy at a lower cost than the benchmark. Given the current focus of all healthcare stakeholders on optimizing value, solutions in this zone are now often perceived as offering a stronger value proposition (although the extent of the improvement in efficacy must still be sufficient relative to the reduction in cost to warrant a behavior change).

A few examples help demonstrate how this approach translates to actual products (see Figure 5.7.7). Consider the case of stents for treating coronary artery disease, which would appear in upper-right quadrant #1. As described in 2.1 Disease State Fundamentals, angioplasty was introduced as a less invasive, lower-risk alternative to coronary artery bypass (CABG) surgery for treating coronary artery disease. However, restenosis occurred in 30–40 percent of patients within 6 months of the procedure as the body sought to heal the artery. For these reasons, many patients required repeat angioplasty procedures or CABG, which resulted in increased risk for the patient and added cost for the healthcare system. Device manufacturers first began marketing bare metal stents (BMS) in the US in 1992 to hold open the artery, prevent it from recoiling, and reduce the rate of longer-term restenosis. This new technology improved the efficacy of the treatment by reducing complications and the need for repeat procedures, but added approximately $1,600 per stent to the cost of an angioplasty procedure.[6] When drug-eluting stents (DES) were launched in the US 10 years later, they were intended to displace BMS by further improving efficacy rates. For example, head-to-

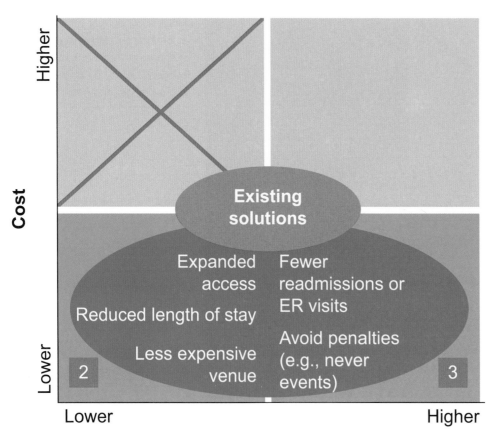

FIGURE 5.7.6
Innovators can differentiate solutions in quadrants 2 and 3 from available alternatives based on the cost reduction (and related benefits) they provide.

head clinical studies demonstrated up to 9 percent in-segment restenosis rates for DES compared to rates up to 33 percent for BMS.[7] By that time, the price of a BMS had dropped to approximately $1,000 per stent, but each DES commanded more than $3,000 per unit.[8] Although payers and facilities were realizing some reduction in cost through the avoidance of repeat procedures linked to restenosis, the additional expense associated with the technology was far greater than the aggregate savings. Value propositions of this type (based on higher efficacy at a higher cost) were common, especially in the US, before unsustainable healthcare spending forced many stakeholders to reconsider the economics related to unrestricted escalations in the **standard of care**.

A second example focused on the HiCARE LIMO team, which developed a new lower-limb splint that is effective, inexpensive, and easy to use, would appear in lower-left quadrant #2. The idea for the new technology stemmed from the observation that many ambulance drivers and other healthcare providers in India had

access to modern splint alternatives, but were hesitant to use them on patients or leave them behind after transport due to their high cost (see chapter 1.2 for more on this story). In this case, the team was willing to trade-off some of the more advanced features of top-quality splints (and the efficacy they delivered) in favor of making a device that could be produced at a fraction of the cost. With a device that was cheap enough to leave behind, healthcare providers would be more willing to deploy it, which would in turn increase patients' access to splints when they needed them. Value propositions of this type (developing a de-featured solution at a significantly lower cost) have become increasingly common among companies working in emerging economies where large portions of the patient population may have limited resources and there is a generally higher portion of private-pay health spending. These value propositions also can be used to target smaller segments of larger traditional markets where a subset of customers is seeking a lower-cost alternative to current solutions. For

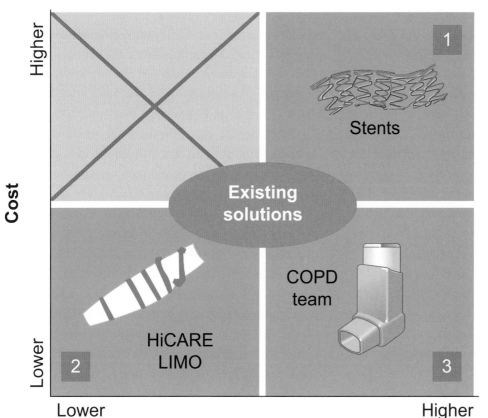

FIGURE 5.7.7
Innovators can devise effective value propositions in quadrants #1, 2, or 3, but those that reduce costs while improving efficacy (as in quadrant 3) can be particularly compelling to stakeholders in today's value-oriented healthcare environment.

example, a company called the Orthopaedic Implant Company (OIC) is seeking to dramatically reduce the cost screws and similar hardware that orthopedic surgeons use in procedures to repair fractures or replace joints. The company's strategy is to identify devices coming off **patent** and then use a lean, outsourced manufacturing model to make comparable products available at 50 to 60 percent less than average prevailing market prices for premium implants.[9,10] Admittedly, these off-patent products do not include all of the latest advancements offered by the current generation of products from major orthopedic implant manufacturers, but certain segments of the total customer base are likely to conclude that these lower-cost alternatives adequately address their basic needs.

A final example focused on the COPD team profiled in chapter 2.4 illustrates a technology that would appear in lower-right quadrant #3. This team went on to design a smart inhaler that could be used to remotely monitor risk factors for exacerbations, allowing the nurses staffed within care management call centers to preemptively intervene when these signals were detected. This device was intended to increase the efficacy (or effectiveness) of the care management call centers by providing them data to help direct their outreach rather than requiring them to perform semi-random outbound calls with the hope of catching patients experiencing exacerbations. In combination, it would reduce the overall cost of care by preventing a greater number of emergency room visits and hospital readmissions, as well as allowing facilities using the system to avoid COPD readmission penalties. This type of value proposition (where efficacy can be still be improved while simultaneously lowering cost to the system) is generally thought to be the most powerful approach to gaining stakeholder support in an era of more limited resources and an increasing focus on the cost of treatment.

Different types of improvements appeal to different stakeholders, which is why value propositions must be highly customized to each stakeholder group. For instance, reimbursement penalties imposed by the

Affordable Care Act (**ACA**) in the US based on quality metrics desired by payers may not capture the attention of individual physicians and patients, but they are likely to be of great importance to healthcare facilities looking to prevent avoidable expenses and negative publicity.

There is no prescribed format that value propositions should take, but they often are constructed as a concise statement or series of statements that articulate the improvement–cost equation of the new offering in terms that the target stakeholder group will find most appealing. At a minimum, this statement should include the audience to which the value proposition is directed, the improvement(s) offered by the solution, and an indication of the cost at which those benefits can be realized. Another approach to creating value propositions suggests that they encapsulate the need, the solution, and the improvement it offers relative to the competition.[11]

Taking the Miramar case as an example, one can speculate how the company might have approached the development of its primary value propositions. Starting with the **need statement**, the company may have framed its focus accordingly: *a way to treat patients with severe hyperhidrosis, which results in a more permanent and effective treatment at or around the cost of current alternatives*. Once the team developed its solution, conducted its early studies, and performed its preliminary market research, its value statements might have read:

- **For patients**: Miramar Labs offers a treatment in the physician's office that is safe, highly effective, and long lasting – with stable efficacy of over 90 percent – that is fast and affordable compared to less effective alternatives such as BOTOX treatments that must be repeated every seven months and the lifetime cost of coping mechanisms that fail to adequately address the problem.
- **For physicians**: Miramar Labs provides a safe and effective private-pay treatment that allows physicians to offer their patients with severe hyperhidrosis a leading-edge solution that is longer lasting than BOTOX – with stable efficacy over 90 percent – while profitably expanding their practice.

As described previously, qualitative data can be used to support a value proposition (e.g., testimonials and anecdotes from key opinion leaders), but they should ideally be augmented with quantitative information that supports those claims to achieve maximum impact. In fact, stakeholders routinely ascribe greater validity to quantified propositions that are supported by high-quality clinical and economic evidence.[12] These data can be collected via surveys, other forms of market research, clinical trials, and/or value modeling (see 5.6 Reimbursement Strategy) and then used as the basis for structuring the value propositions.

Referencing the hypothetical value propositions for Miramar Labs, the company could weave in additional quantitative information by expanding these single value statements into a series of statements for each audience. For example, for patients, the company could highlight the specific cost of recurring BOTOX treatments or cite the average lifetime expense that patients with severe hyperhidrosis devote to dry cleaning and antiperspirants compared to the one-time $3,000 cost of mira-Dry. For physicians, they could demonstrate how with just $700 in disposable equipment to cover for each patient, the $3,000 per-treatment cost would quickly allow them to recoup the $50,000 up-front investment and make the procedure a profitable addition to their practice.

Chris Wasden, the Executive Director of the Sorenson Center for Discovery and Innovation at the University of Utah, recommends five key dimensions across which innovators can create and measure value:[13]

- **Cost** – How much will the innovation decrease cost relative to the customer's available alternatives?
- **Convenience** – How much easier is the solution compared to available alternatives?
- **Confidence** – How much more accurate or better is the solution than available alternatives?
- **Compensation** – How much more money can key stakeholders make over the alternative?
- **Connection** – How much more fashionable, cool, social, emotional, and avant-garde is the solution than available alternatives?

An example involving an automated hearing device called the Otogram demonstrates what a quantified value proposition might look across these dimensions. When the team developing the commercialization plan for this device began experimenting with value propositions, it found little enthusiasm for an approach that stated, "We can decrease your testing costs and improve your testing accuracy." Instead, more quantified and specific statements, such as the following, had more powerful effect on physician stakeholders.[14]

For an average physician practice, the Otogram can:

- Decrease your hearing testing costs by 75 percent (or about $100,000) per year. (Cost)
- Increase the convenience for patients by testing in 11 languages and at any time of the day without an audiologist. (Convenience)
- Deliver the 99 percent accuracy of an expert audiologist, based upon side-by-side clinical trials. (Confidence)
- Increase practice revenue by $30,000 per year in hearing testing and double hearing aid sales to $100,000 per year. (Compensation)
- Demonstrate that the practice is at the cutting edge of the hearing health technology field with an intuitive and elegant design that connects the patient to the clinician. (Connection)

Initially, innovators may not have all of the data required to support quantified value propositions, but this is what they should be working toward as they develop their marketing mix, set pricing, and conduct clinical and economic studies. (For technologies with value propositions focused on delivering improved efficiencies, variations of traditional time-in-motion studies can also be informative.) The important point is that some data must be embedded into the value proposition before the innovators begin promoting their offering. They must work cross-functionally to design studies that will produce the evidence needed to satisfy the data requirements of the many different stakeholder groups they will seek to influence. Of course, studies will need to be prioritized and/or sequenced based on their relative importance and potential impact on the company's ability to achieve its goals, especially in light of early-stage resource constraints. But, over time, as more data are gathered, a company's value propositions will evolve and become increasingly robust. For instance, at market

launch innovators may find themselves working with little more than the clinical data used for regulatory approval and stakeholder testimonials from their market research. Accordingly, their value propositions may leverage this information, but place greater emphasis on appealing to customers that want to be first adopters and pioneers in the field. Later, after additional clinical and/or economic studies have been completed and published, the value propositions can shift to attract customers who want the confidence of knowing the innovation is backed by data from multiple trials. For more information about the types of studies that may be helpful, see 5.3 Clinical Strategy and 5.6 Reimbursement Strategy.

As a team or company determines what studies to conduct, it is helpful to consider who will be involved in those studies from a stakeholder perspective. Study design and management provides an effective way to get key opinion leaders involved with the company and familiar with its offering. Physicians who are considered KOLs have deep experience that can be leveraged to help design and implement studies with a high probability of success. Their expertise and involvement also lends credibility and helps generate interest in the study results among the KOL's peer group. When deciding which KOLs to target, companies must think about factors such as whether they prefer academic or community physicians, and the geographic location where the studies will be performed (see 5.3 Clinical Strategy).

In advance of study design and execution, KOLs can also be engaged to help a company develop and/or refine its critical value propositions. While KOLs are traditionally targeted within the physician population, innovators should remember that influential individuals can be found among facility administrators, biostatisticians, health economists, academics, allied professionals (e.g., nurses and technicians), and patient groups. All such experts can provide invaluable advice regarding which value proposition is most compelling to their associated target audience, and what data are required to effectively support that value proposition.

Sometimes, before a company can convince certain stakeholders of the value of its specific technology, it must persuade them that the need for a solution exists at all. Often referred to as "market development," creating awareness of a need and, in turn, demand for a solution can be a time-consuming, complex, and expensive endeavor – and one that should not be undertaken lightly by a start-up. For example, a company may have to perform or sponsor studies simply to establish the sense of a necessity in the minds of the target audience before it can effectively market its specific technology.

Consider the case of implantable cardioverter defibrillators (ICDs). Before the results of the MADIT-II and SCD-HeFT trial results were released, ICDs were primarily used to treat a relatively small number of secondary prevention patients – those with a prior episode of sudden cardiac arrest. However, MADIT-II and SCD-HeFT demonstrated that ICDs dramatically reduced the mortality rates of primary prevention patients – those at high risk of sudden cardiac arrest who had not yet experienced it. As a result, the FDA expanded the indications for ICD implantation to include primary prevention patients. Similarly, Medicare and private payers modified their policies to provide **coverage** for ICDs used for primary prevention. This increased the market for ICDs by an additional 1 million patients, since this group was much larger than the secondary prevention group. Physicians, payers, and patients just had to be shown in definitive terms that a legitimate need existed within this broader population.

A similar example exists within the treatment of patients with chronic kidney disease (CKD) before their condition progresses to end-stage renal disease (ESRD). While the need to manage heart disease in patients with ESRD was relatively well understood, heart disease was not proactively nor consistently managed in most pre-ESRD patients. However, when a study was published in the *New England Journal of Medicine*[15] that showed pre-ESRD patients have a high risk of heart failure, it was enough to establish the need for a new treatment paradigm for this patient population. It also opened up a new field of treatment to innovators and companies pursuing opportunities related to CKD.

Note that the company's competitive advantage and business strategy can also support the value propositions it has defined. For each value proposition, the team should carefully evaluate what strategies and potential company strengths will enable it to successfully deliver

as promised (see 5.9 Competitive Advantage and Business Strategy for more information). The company's position in the healthcare value chain and its unique capabilities are an important source of leverage in bringing value propositions to fruition.

Marketing mix

With clear value propositions defined for key stakeholder groups, innovators must next make a series of decisions about the marketing mix it will use to support and communicate these value statements. Without an effective marketing mix, value propositions will never reach their intended audience and adoption is likely to falter.

Product/service mix

The first element of the marketing mix presented in this chapter is deciding on the product/service mix. The idea is to define an offering with the greatest likelihood of delivering on the value propositions the company has constructed. In general, the product/service mix is driven by the business model the company has adopted (see 4.4 Business Models). For example, many **medtech** companies with disposable or implantable devices feature a pure product offering, which they may support through ancillary services (e.g., education and training). Capital equipment companies, on the other hand, generally have a more complex product/service mix that involves the bundling of multiple products (e.g., equipment, plus a disposable or reusable component), as well as a service (e.g., maintenance contracts and upgrade agreements). For instance, Accuray (profiled later in this chapter) includes future product upgrades as part of its basic product offering.

The bundling of products and services into cohesive offerings is becoming increasingly common as the providers and payers/purchasers of medical interventions become more value oriented. One emerging business strategy that reflects this shift is called disease management or "**owning the disease**" (see chapter 5.9). With this approach, technology companies are attempting to provide integrated products and services that allow them to solve problems for their customers across the continuum of care. If this trend continues, fewer and fewer medtech companies will remain "pure" product companies and more will bring hybrid product/service offerings to market.

Pricing

Pricing begins with a deceptively simple question: what is the appropriate baseline price for an offering? The price of a product and/or service can play a critical role in encouraging or discouraging adoption, particularly as all stakeholders become more cost conscious. Therefore, pricing not only involves establishing a baseline price for the decision maker who will directly bear the cost of the offering, but recognizing the different ways through which prices (and the related issue of reimbursement) will influence adoption. Reimbursement-related issues are covered at length in 5.6 Reimbursement Strategy. However, they warrant some discussion in this chapter to the extent that they relate to marketing.

Before innovators think about setting a baseline price, they must first understand all of the costs associated with developing, manufacturing, and commercializing the offering. They should also determine what sort of mark-up (profit) the company would ideally earn to support its overhead and ongoing development efforts. Once these factors are clearly understood, the next step is to perform an evaluation of real and perceived value associated with the offering. Survey data, the value propositions defined for the offering, and more advanced analysis such as a cost/benefit model can provide important inputs to this assessment.

Value-based pricing is typically the most effective pricing strategy for a company to support. If the price of a new technology can be directly linked to the value of the improvements it will deliver (with the value exceeding the cost), buyers and payers are far more likely to support the adoption of the offering. The most persuasive value-based pricing argument is related to direct savings in healthcare costs. However, innovators must appreciate that customers will often require a high level of documentation in order to be convinced of a potential savings and justify value-based purchases internally. If a physician or hospital will save money with each device used (relative to the current standard of care), this gives the company a strong argument for justifying its price. Another common pricing argument is related to improved outcomes. If a device leads to improved results such that a payer or provider saves money on follow-up care and/or treatment related to complications, this is frequently a compelling argument. Companies can sometimes encounter resistance, however, when the pricing argument is based on **quality of life**. In some geographies, such as the UK, buyers and payers are generally willing to support the adoption of devices that lead to significant, measurable, **evidence-based** improvements in quality of life. Yet, their standards for demonstrating such a change are growing increasingly stringent, especially for high-end, high-cost devices that represent a sizable potential cost burden to the healthcare system.

Another way that companies can establish a baseline price is to perform a **comparables analysis**. By evaluating the pricing strategies (and associated reimbursement status) of similar offerings in the field, companies can gain valuable information to help them set a price. In general, medical device pricing for established products should give the innovators a strong sense of what the market will bear. Comparables analysis can be accomplished through primary and secondary research. As discussed, performing market research can also be helpful in terms of understanding the price sensitivity of key stakeholders (i.e., exploring different pricing scenarios and identifying the price point at which they will resist the technology).

The story of Genomic Health demonstrates how one company approached the challenge of establishing a price for its product.

FROM THE FIELD ▶ GENOMIC HEALTH

Value-based pricing for a novel diagnostic tool

As described in 2.4 Market Analysis, Genomic Health, a company addressing the need for high-value, information-rich diagnostics based on patient-level genomic testing to predict the recurrence of early stage, N–, ER+ breast cancer and enable personalized treatment decisions, faced an interesting pricing challenge with its product. Because diagnostic companies traditionally charged between $25 and $50 for their tests, commanding margins of just 5–10 percent, the company had to pioneer an entirely different pricing paradigm to support the high cost of R&D and clinical studies necessary to bring its genomic-based test to market.

Three types of analysis suggested that a price in the range of $1,000 to $7,000 per test could be viable. The first was comparables analysis. Kim Popovits, COO of Genomic Health, recalled, "There was another diagnostic in the marketplace at that time, a genetic test that looked at the mutation of the BRAC-1 and -2 genes to assess a woman's hereditary risk of breast cancer." This test was priced around $3,000 and was on its way to being reimbursed on a relatively broad scale.

The second approach was based on the value estimate for the need. Over time, Genomic Health's test had the potential to save money for the overall healthcare system and could, thus, shift the pricing power from therapeutics to diagnostics. Specifically, the total cost of chemotherapy for early-stage breast cancer patients was conservatively $15,000. If the test cut the number of patients undergoing chemotherapy by 50 percent (by predicting low recurrence risk), then the total savings to the healthcare system would be roughly $7,000 per patient. This meant that Genomic Health could potertially command a price of up to $7,000 per test.

Finally, the company performed additional market research, including a market survey with 30 or 40 US medical directors. Through this effort, Genomic Health tested the price sensitivity of payers and discovered that they considered any test over $1,000 to be expensive. However, their reaction was not significantly different between price points of $1,500 and $4,500, assuming the test had clinical value and adequate validation to support high value pricing. Evaluating all of these inputs, the company eventually decided on a price of roughly $3,500 for its product.

Despite the relatively high costs of its test (by traditional diagnostics standards), Genomic Health anticipated that it might take the company approximately 18 to 24 months to gain consistent reimbursement for the product. However, when the company met substantial resistance from payers (driven, in part, by the concern that patients would take the test but then still pursue chemotherapy regardless of the test results), it took an innovative approach to driving adoption.

Genomic Health agreed to enter into a pay-for-performance deal with major payer UnitedHealthcare. The insurer said it would reimburse for the test for 18 months while it monitored the results with Genomic Health. If too many women still elected to receive chemotherapy, even if the test suggested they did not need it, then UnitedHealthcare would seek to negotiate a lower price on the grounds that the test was not having the intended impact on actual medical practice.[16] Genomic Health's management team was confident that women would follow the course of treatment recommended by the test, and believed that the pay-for-performance agreement could be used as a way to advertise the company's confidence in the test. As it turned out, the trial was a success and United Healthcare issued a national payer contract for the technology, which established coverage across all of UnitedHealthcare's plans following the trial period.[17]

As the Genomic Health example suggests, choosing a baseline price has many subtleties. Additionally, the company must decide under what circumstances it might be convinced to deviate from its baseline price. Pricing strategies that require a company to adopt a more complex approach to pricing include **differential** and **bundled pricing**, **gainsharing**, and **pay-for-performance**. Details on each of these alternatives can be found below.

Differential pricing Differential pricing refers to the basic concept of pricing the same product or service differently for different customer segments. For example, in some cases, medical device companies might negotiate discounted pricing with large purchasers (e.g., group purchasing organization, integrated delivery facilities). While this strategy can be effective in driving volume, it has the potential to create conflict in the market among customers, as well as payers. It may also create legal challenges, if it is perceived as creating a financial inducement to physicians.

Differential pricing strategies are becoming increasingly common when companies are working across geographies. In particular, they are growing in popularity with organizations that have a mission to address the needs of underserved populations in resource-constrained settings. For example, Cycle Technologies offers its natural family planning device CycleBeads (developed with researchers at the University of Georgetown's Institute for Reproductive Health) in countries around the world using a differential pricing approach. The company's primary goal is to reach users in developing countries where medical and surgical contraception options are limited or unacceptable; its secondary focus is on meeting the needs of women in developed countries who are seeking effective, non-invasive birth control and proactive family planning tools. To enable the company to achieve both of these objectives on a sustainable basis, Cycle Technologies sells the device at a profit in the US and then uses the proceeds to help it make the product available at close to cost in low-resource settings such as India and parts of Africa.[18]

Bundled pricing Bundled pricing refers to setting a single price for a combination of products and/or services. A medical device manufacturer might bundle service contracts or ancillary products and services with its primary offering to try to drive increased revenue. Bundled pricing can be a way of offering discounts to buyers while incentivizing them to buy a wider range of products and services than they would otherwise. While this works in some cases, it is not successful for all offerings. As classic examples of this is when Guidant offered bundled pricing on its catheter and guidewire products in an effort to drive more widespread adoption. However, because physician preference was so strong in

this particular area, practitioners wanted to choose products à la carte despite the discounts that could be realized by purchasing bundled products.

From a health system perspective, research has shown that bundled payments can align incentives for providers, including hospitals, post-acute care providers, physicians, and other practitioners, and encourage them to work together more closely across an **episode of care**. The Centers for Medicare & Medicaid Services (**CMS**), through the CMS Innovation Center, launched a program known as the Bundled Payments for Care Improvement Initiative to experiment with four new types of payment models. The goal of the program is to determine which models can lead to higher quality and more coordinated care at a lower cost to Medicare.[19]

Gainsharing Gainsharing agreements between a hospital and its physicians represent another subtlety in pricing. Under these agreements, hospitals can negotiate reduced prices with certain manufacturers in exchange for increased volume. Gainsharing differs from differential pricing in that physicians are given direct incentives to adopt certain devices. For instance, these plans often provide physicians with a percentage of the cost savings derived from reduction of waste and use of specific supplies during procedures.[20] While some observers view gainsharing as an effective cost-cutting tool, others perceive it to be laden with inherent conflicts of interest, an obstacle to innovation and proper patient care, and possibly even a violation of anti-kickback statutes.[21] Recently, some hospitals have used gainsharing to help them standardize purchases of high-value orthopedic implants, including hips and knees. Prior to the initiation of gainsharing arrangements, orthopedic surgeons were successful in protecting their individual product preferences, which resulted in a proliferation of products being purchased by each hospital. By selecting a preferred vendor and implementing gainsharing, the hospitals have been able to significantly streamline purchasing and realize cost economies by doing more business with a smaller number of suppliers.

Although gainsharing is still somewhat controversial, the four payment models being tested as part of the CMS Bundled Payments for Care Improvement Initiative allow participating facilities to propose gainsharing arrangements among provider partners.[22]

Pay-for-performance   As the Genomic Health example illustrates, companies may agree to set their prices contingent on the realization of specific results. If the company delivers on its value proposition (as measured by mutually agreed-upon performance metrics), a payer or customer will pay its baseline price. If not, certain discounts will be expected to justify the lower "payback" on the device. As with gainsharing, pay-for-performance arrangements are relatively new and are still somewhat controversial within the industry. However, interest in these types of arrangements seems to be growing as purchasers of all types become more cost and value focused.

Any company should seek legal counsel when creating pricing strategies. The healthcare space is highly regulated and arrangements that may be perceived as creating an inducement for a physician to use a particular device or procedure can run afoul of the Stark Law. This law governs physician self-referrals, or the practice of a physician referring a patient to a medical facility (or form of treatment) in which s/he has a financial interest, be it ownership, investment, or a structured compensation arrangement.[23] While the law remains controversial, innovators should exercise appropriate caution to avoid a potential conflict of interest.

Positioning

Positioning refers to the way a company represents its offering in the market in order to differentiate it from the competition and a make distinct and lasting impression on the customer. Because the medical marketplace is "noisy," with many messages vying for the attention of prospective customers and/or seeking to enlist the support of influential stakeholders, innovators should treat positioning as a selection exercise. They must ask themselves, "What are the most important aspects of our solution that will engage the customer when we begin promoting it?" And, "How can we optimally differentiate the solution from available alternatives (or from the absence of another solution)?" The most compelling answers to these questions then form the basis of the company's positioning messages.

Importantly, not every product feature or aspect of the value proposition rises to the level of a key positioning message. The goal is not to present customers with a "laundry list" of messages – it is to focus on just a few of the most convincing arguments that are most likely to drive a behavior change. These messages can appeal to customers on a rational or emotional level (or ideally some combination of the two).

When defining and customizing their preliminary value propositions, innovators will have considered what is most important to each core stakeholder group. However, now, during positioning, they have more specific information regarding the product/service mix and price for the offering. This enables the team to more clearly and precisely focus attention on what attributes distinguish the solution from the customer's other alternatives and should drive them to try it.

Language is another important factor to consider when crafting positioning messages. As the Miramar Labs story illustrates, small differences in the words and terms chosen to describe a product or service can have a significant impact on whether customers respond favorably (or not) to the messages when they are delivered. Without the benefit of market research with prospective patients, the Miramar team would not have been likely to appreciate the negative connotations associated with the word "permanent" versus "lasting" among target patients.

Keep in mind that positioning is not about telling the story of the innovators and how the innovation came to fruition. It is also not about sharing what the team or company finds most important or exciting about the offering. Positioning should be focused entirely on isolating what is most important to each stakeholder group and devising the optimal way to communicate these factors in a manner that will be meaningful to those audiences.

Intuitive Surgical (introduced in chapter 5.1) provides an example. When this company was pioneering the field of robotic-assisted surgery with its da Vinci® Surgical System, the development team was enthusiastic about the technical capabilities enabled by its device. By translating the surgeon's natural hand and wrist movements on instrument controls on a console into the corresponding micro-movements of specialized surgical instruments positioned inside the patient through small incisions, da Vinci enabled a minimally invasive approach to procedures such as prostatectomy, hysterectomy, myomectomy, gastric bypass, and mitral valve repair. The company might have led with messages about improved precision and reduced complications in positioning the product if it had not discovered that the hospital executives, who would make the purchasing decisions to acquire the system, were more interested in the prestige their facilities could garner from being the first in a geographic area to adopt the technology. Eager to differentiate themselves from competing hospitals, facility executives were actively seeking ways to enhance the reputation of their institutions, and purchasing a da Vinci system was one way to accomplish this. Over time, as more facilities adopted the system and it became less of a competitive differentiator, Intuitive Surgical had the opportunity to highlight the technical benefits offered by the device. However, many of its initial sales were based on the perceived ability of the device to enhance a hospital's reputation. It was their hope that this increase in stature would result in higher patient flow, stronger referrals, and an improved ability to retain key surgeons.

Another interesting example can be found in the TRUE Dilatation™ Balloon Valvuloplasty Catheter. Loma Vista Medical, which was later acquired by C.R. Bard, developed an effective approach to positioning in preparation for marketing its product for use in trans-catheter aortic valve replacement (TAVR) and balloon valvuloplasty (BAV). The product offered multiple benefits over traditional balloons, including the following:[24]

- Highly resistant to ruptures, punctures, and tears.
- Rip-stop fibers to prevent the catastrophic failures seen with other balloons.
- Fiber reinforcement precisely limits the maximum balloon diameter to the labeled size, while allowing conformance to anatomical variation.
- Diameter control within 1.5 percent over its rated pressure range (compared to 15 to 40 percent for competing products).
- Inflates and deflates two to three times faster than competitors' balloons, minimizing pacing time.

- Clean re-wrap and low withdrawal profile after dilatation.
- Able to confirm de-airing before insertion.

The company carefully considered these benefits and how to prioritize them for its primary customer – the physician. Of course, all of this information would be important when answering their questions and helping them understand the total value of the product. But the team had to decide which messages to lead with and how to make an immediate and lasting impression. Ultimately, they narrowed in on four simple, direct ideas, as shown in Figure 5.7.8.

The positioning – "Truly Precise. Truly Tough. Truly Fast. Truly Better" – encapsulated the core ideas from the list of benefits, but articulated them in a way that was compelling to physicians and memorable in its power and simplicity (see 6.3 Funding Approaches for more about Loma Vista Medical).

Promotion and advocacy

The final aspect of the marketing mix covered in this chapter is promotion. Promotion involves how a company communicates with the target audiences. A company generally manages these interactions through a marketing communication and/or public awareness strategy.

Raising awareness through a marketing communication strategy The purpose of a marketing communication strategy is to raise awareness among stakeholders regarding the need and/or solution to prepare them to make a buying decision or change their behavior in some other desired way.

Information that is disseminated by the company is considered a form of *direct awareness*. Information that comes from other sources – such as publications, conferences, KOLs, etc. – is considered *indirect awareness*. An effective marketing communication strategy typically leverages both direct and indirect awareness mechanisms in reaching its target audience.

Usually, a marketing communication strategy outlines which communication vehicles will be used for each key stakeholder group, what the key messages will be for

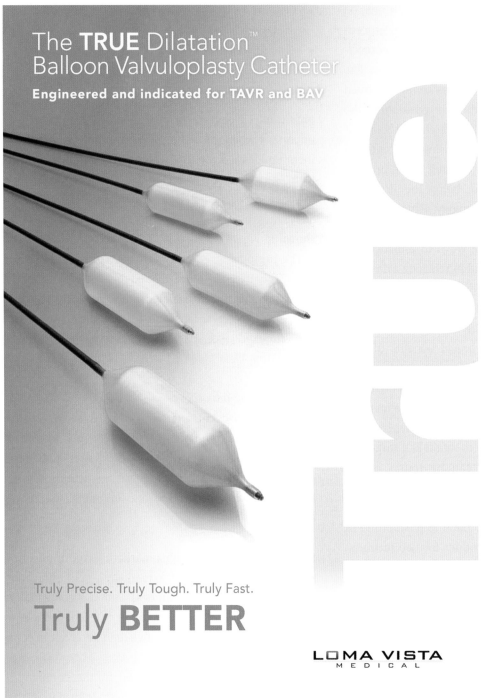

The **TRUE** Dilatation™
Balloon Valvuloplasty Catheter

Engineered and indicated for TAVR and BAV

Truly Precise. Truly Tough. Truly Fast.

Truly **BETTER**

LOMA VISTA
M E D I C A L

FIGURE 5.7.8
Product positioning for the TRUE
Dilatation™ Balloon Valvuloplasty
Catheter (courtesy of C.R. Bard).

each communication, and when each communication will be released. The idea is to create a sequence of communications for each stakeholder group that builds their awareness, introduces and then reinforces key messages (e.g., the value proposition and associated positioning messages), and prepares the target audience to make a decision or change its behavior at the time the device is launched. Table 5.7.1 presents some of the different communication vehicles that can be incorporated into a marketing communication strategy. Note that detailing – when sales representatives and clinical specialists visit physicians or hospitals – is omitted since this

Table 5.7.1 A combination of communication vehicles is the most effective way to reach target stakeholders.

Vehicle	Description	Issues to consider
Peer-reviewed publications	Peer review is the process of subjecting an author's scholarly work or research to the scrutiny of others (peers) who are experts in the same field. Peer-reviewed publications are considered more credible than other sources because they aggregate, filter, and validate author submissions independent of any outside influence or interested third party.	• Submitted data must be based on clear evidence. • Publication can take anywhere from 3 months to 2 years from the time of submission, depending on the publication, topic, and strength of the data; as a result, planning is difficult. • Revisions/rewrites may be required prior to publication. • Publication is not guaranteed. • Any potential conflicts of interest between the company and the authors must be disclosed. • May cost as little as $15,000 but can require hundreds of thousands of dollars to support the research that leads to the publication (conflicts of interest caused by the relationship between the research sponsor and the investigator that may undermine the scientific validity of any studies will need to be managed and disclosed properly).
Peer-reviewed (abstract) conference presentations	An abstract provides a concise statement of the major elements of a research project. Abstracts are reviewed against other submissions, with some subset being chosen for a brief presentation (or a poster) presentation based on their fit with conference criteria.	• Submitted data must be based on clear evidence. • Abstract requirements vary from conference to conference and should be well understood before a submission is made. • The abstract, stating the purpose, methods, and findings of the research project, is submitted to conference organizers to inform them of work-in-progress or completed work that is available to be presented. • Abstracts must be submitted an average of 6 to 8 months prior to the conference.
Technology/clinical talks	Technology talks are given by KOLs, either in the context of a meeting, as an evening session, or pre-conference session.	• Can be used to satisfy CME requirement (see below). • More extensive than the peer reviewed presentation; provides opportunities for more in-depth coverage of a new technology.
Continuing Medical Education (CME)	**CME** is required by physicians in most US states to maintain their licenses. It provides a way for physicians to stay informed and learn about new developments in their field. Content for	• CME programs must be certified by the Accreditation Council for Continuing Medical Education. • Any potential conflicts of interest between the company and the authors must be disclosed. • CME programs can be sponsored directly by companies or through professional societies. • The costs are significant: a professional conference attended by 200 people for 2 days could be $100,000 or more; a 1–2

Table 5.7.1 (*cont.*)

Vehicle	Description	Issues to consider
	CME programs is typically developed, reviewed, and delivered by faculty who are experts in their individual clinical areas.	day academic conference for 50 people could be more than $50,000 (start with speaker honoraria of $5,999 to $15,000 for 10 speakers, then add room, board, travel expenses, etc.).
Reimbursement dossier	**Reimbursement dossiers** typically serve as the official source for all key information about the product, including the place of product in the diagnostic and therapeutic chain, results from key clinical and economic studies, disease management strategies, modeling report, product value and overall cost, and references (see 5.6 Reimbursement Strategy).	• Must be based on well-established and tested facts about the product. • Can be time-consuming and resource-intensive to prepare, requiring specialized expertise and significant lead-time. • Must be customized to meet the needs and interest of each target audience. • Cost can range from $50,000 (for a simple dossier) to as much as $250,000 (for a more complete deliverable).
Direct-to-consumer (DTC) advertising	**DTC** advertising refers to the promotion of medical devices to patients through newspapers, magazines, television, and the Internet. Companies also use brochures, videos, and other materials that are made available to patients in doctors' offices.[33] One recent trend is to focus on direct-to-patient (**DTP**) advertising (i.e., advertising in channels accessed more exclusively by patients, such as the *Diabetes Digest* for diabetic patients).	• DTC advertising is only legal in two developed countries: the US and New Zealand. • Not all medical technologies are well-suited to DTC advertising and this approach traditionally has been used sparingly in the medtech field (exceptions: elective Lasik surgery, BOTOX, drug-eluting stents, Lap Band for bariatric surgery). • However, with the recent proliferation of consumer-oriented medical devices, DTC and DTP advertising is becoming more common. • Online consumer/patient outreach and the use of social media have become the dominant channels for DTC and DTP advertising, especially for consumer-oriented medical devices and health/lifestyle technologies. • DTC advertising is still considered somewhat controversial among regulators, some physicians, and some patient groups, although this varies by product category. • DTC advertising can be complicated and expensive.

is considered part of the sales process, as discussed in 5.8 Sales and Distribution Strategy.

As Table 5.7.1 suggests, the role of marketing in the promotion and adoption of new medical devices is not without controversies. When used properly, relationships between medical device companies and KOLs can be advantageous to a company and can lead to better, more innovative products. However, in some cases, they may be perceived as undermining scientific integrity and unduly influencing medical care. The Advanced Medical Technology Association (**AdvaMed**) has developed a code of ethics on its interactions with healthcare professionals that aims to address concerns about potential conflicts of interest. This code provides guidelines on product training and education, support of education conferences (such as those used for CME), restrictions on gifts and meals provided as part of sales and marketing, and limitations on arrangements with consultants.[25]

More than anything, innovators are advised to protect their clinical activities from the possibility of conflicts and bias by ensuring that those who perform and evaluate clinical studies do not have incentives linked to the success of the company. If research is conducted in this way, the data used to support value propositions and the marketing mix have a much higher likelihood of being perceived as credible and compelling.

Companies should be careful not to create awareness of their technology among patients *before* physicians are fully informed. If a physician is blindsided by patient inquiries regarding a new product, it can cause resentment toward the company and negatively affect adoption or even drive adoption of competing products. This problem, while most common in the pharmaceutical business, is still something to watch out for in the medical device field. And it will only become a larger problem as online and social media-related communication channels continue to proliferate. Traditionally, the best approach was to raise awareness first among KOLs, then expand communications to facilities and payers. Once these stakeholders had a basic understanding of the technology and its value propositions, the company could begin targeting other physicians (who are not necessarily KOLs) and then patients (as appropriate) with its marketing messages. In today's online world, news of

health-related innovations travels rapidly and messaging is more difficult to stage in this manner. Patients are more active than ever before in researching their alternatives and proactively seeking emerging therapies. In some instances, they can be advocates and first-line customers for new medical solutions, such as new methods for parents to monitor chronic conditions in their children.

As noted, professional societies and thought leaders can be especially valuable advocates for a new technology. Societies can create awareness by incorporating new medical technologies in their practice guidelines, sponsoring symposia, and advocating payers for reimbursement (see 5.6 Reimbursement Strategy). KOLs not only influence other physicians, but can guide future generations of product development to create and sustain additional value. Ron Dollens, president and CEO of Guidant from 1994 to late 2005, credits relationships with thought leaders as one of the key factors behind the success of one of medtech's early pioneers, Advanced Cardiovascular Systems (ACS). ACS was acquired by Eli Lilly in 1984 and led by Dollens from 1988 until its spin-off in 1994 to form Guidant. According to Dollens:[26]

ACS had a sales organization that developed great relationships with the thought leaders. So it was associated with the kinds of people that were taking the therapy to another level.

Promoting through a public awareness strategy

A public awareness strategy can be an imperative element of a marketing/stakeholder strategy if the company anticipates strong resistance from one or more stakeholder groups. The concept of public awareness refers to the methods and activities employed by a company to establish and promote a favorable relationship with the public. While public relations (PR) companies are mostly accessible to large companies, focused and shrewd public awareness efforts can also work for smaller companies. For example, athenahealth, a company that initially provided web-based medical billing software and services for physician practices, published a ranking of insurers based on the timeliness of claim processing, which attracted considerable media attention and generated interest among physicians for its products.[27]

Not all device companies require an explicit public awareness strategy. However, if the company is concerned about certain potential barriers to the adoption of its technology, this may be necessary. For instance, payers are more likely to resist a new technology if it requires new reimbursement codes. Yet, if a company launches an effective public awareness campaign that engenders meaningful support from physicians (and their professional societies), payers will have a more difficult time delaying or declining coverage for the device. Similarly, if a physician turf war ensues upon the introduction of a new device, a public awareness campaign targeted at creating patient demand for the new technology can be an effective way to drive physician adoption, despite the conflict surrounding who will administer it.

Guidant employed a public awareness campaign to apply pressure to the Centers for Medicare and Medicaid Services (CMS) while seeking more widespread coverage for its ICDs. Although ICDs had been shown in several studies to prevent sudden death by detecting dangerous abnormal rhythms and shocking a patient back into normal rhythm, CMS was willing to reimburse for the technology in only a fraction of the patients who could benefit from the device. Guidant sought to address this issue by funding a $7 million dollar prospective trial looking at the effect of ICDs on cardiac deaths among those who had suffered at least one heart attack. When the results became available, many private payers recognized the value of the device and agreed to reimburse for all patient types that had shown benefit in the study. However, Medicare took a more conservative approach, limiting its coverage to a high-risk subset of the patients studied in the trial (due largely to the potential billion-dollar price tag associated with expanding coverage to such a large group of patients). To help overcome this resistance, Guidant had advocates for the device place editorials in prominent mainstream newspapers and medical journals and took other steps over a two-year period to rally support among patient groups, specialty societies, and other ICD manufacturers (Medtronic and St. Jude). Eventually, at least partially in response to criticism that it was, to some extent, rationing medical care, CMS finally agreed to expand coverage.[28]

When a company is developing a public awareness strategy, some of the best resources to target include news organizations and journalists, patient advocacy groups, physician professional societies, hospital associations, and even members of Congress (if the value proposition associated with the device is compelling enough to capture the attention of this audience). U-Systems is one organization that has realized positive awareness, as well as legislative benefits through public awareness and patient advocacy. The company developed the first ultrasound imaging device for use in combination with standard mammography in women with dense breast tissue. Nearly 40 percent of women have dense breast tissue, a condition that makes detection of early-stage breast cancer more difficult.[29] In 2003, patient Nancy Capello was diagnosed with stage-three breast cancer after 11 years of "clean" mammograms using standard mammography equipment. Upon learning that a missed diagnosis is not uncommon in women with dense breast tissue, she became an advocate for greater disclosure and better screening practices. Starting in her home state of Connecticut, she championed for a law requiring payers to cover breast ultrasound (not just standard mammography) for women with dense breast tissue, which was passed in 2004. She then fought for legislation mandating that physicians inform patients if they have dense breast tissue, which was passed in 2009. Both were landmark bills that provided a template for other states to follow.[30] In 2011, as the U-Systems technology was nearing the market, the company teamed up with Capello and her organization, Are You Dense, Inc., to continue increasing awareness of dense breast tissue and its significance in the early detection of breast cancer and to push forward similar legislation in other states. These new laws are favorable for U-Systems because they increase the demand for alternatives to mammography for screening patients for breast cancer. As of 2014, 14 states in the US had enacted breast density laws.[31]

The Accuray story below highlights how marketing communication and public awareness techniques and activities have been applied in the past as part of a comprehensive marketing and stakeholder strategy.

FROM THE FIELD ▶ ACCURAY INCORPORATED

Using stakeholder analysis to develop an effective marketing strategy

Accuray Incorporated was founded in 1999 by John Adler, a Stanford neurosurgeon, to develop an image-guided radiosurgical device that could be used to ablate tumors without the need for traditional surgical resection. Adler first conceived of the idea while observing procedures performed with a predecessor device, called the GammaKnife, in Sweden in the 1980s. As Adler described, "Accuray really was developed in response to the shortcoming of the GammaKnife," which was able to treat brain tumors through non-invasive radiosurgery, but had to be anchored to the patient's body via bone screws. Accuray's product, a frameless stereotactic radiosurgical system, simplified radiosurgery by eliminating the need for these invasive, painful anchors while achieving comparable results. The Accuray

system, called the CyberKnife,® also would enable radiosurgery in other less anchorable areas of the body, including the spine, chest, and abdomen. By expanding non-invasive surgical technology, Accuray would give surgeons the option to be more aggressive in treating cancer in multiple anatomic locations or near vital organs, while simultaneously lowering risk and patient recovery times (see Figure 5.7.9).

Starting with his connections at Stanford, Adler put together a team to develop the CyberKnife idea and bring it to the marketplace. Despite difficulties raising enough money to sustain the high, ongoing R&D-related expenses associated with a capital equipment device, Accuray managed to get the CyberKnife ready for human testing under an investigational device exemption by 1994. Although the frameless design of the CyberKnife allowed Accuray to treat tumors in different parts of the body, the company

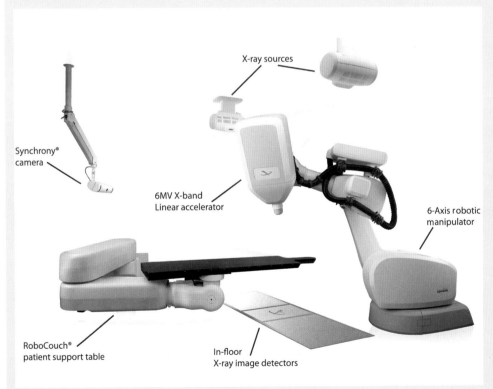

FIGURE 5.7.9
Accuray's CyberKnife system (courtesy of Accuray).

originally focused on brain tumors (much like the GammaKnife), modifying existing procedures but not yet expanding outside the field of neurosurgery. "The first generation was dedicated to replacing many of the GammaKnife's procedures," said Adler, "because we could do it without the frame and modify what had been done to make it biologically a somewhat better procedure."

The decision to begin by targeting neurosurgery was driven, in part, by a series of stakeholder issues. In this market, Accuray believed that it clearly understood the interests of key stakeholders and had the best chance of addressing their concerns. According to the company, the two most prominent stakeholder groups were neurosurgeons and radiation oncologists. Neurosurgeons had the primary relationships with brain tumor patients and had traditionally performed the open operations (to remove solid tumors) that the CyberKnife would replace. This population was receptive to new surgical procedures and innovations. Adler's personal connections with many neurosurgeons also helped generate excitement about CyberKnife within this community, paving the way for the acceptance of the device. With limited resources and funds to attract early adopters to the technology, Accuray became convinced that neurosurgeons would be a receptive audience.

Radiation oncologists, on the other hand, represented a stakeholder group that could be somewhat less receptive to the new device. Yet, Accuray had to earn their support because government regulations required radiation oncologists to participate in any form of radiation delivery, including radiosurgery. Because of various turf wars and financial issues, Accuray anticipated some resistance from radiation oncologists as physicians in different surgical specialties began using the CyberKnife. However, the company believed that it could mitigate these concerns in the neurosurgery field. Adler explained why: "Radiosurgery was developed by neurosurgeons and has been driven by this field for 35 or 40 years now. So, over time, radiation oncologists have come to accept neurosurgeons' role in this type of radiation delivery. Given such a history, the opposition

from radiation oncologists could not be based simply on the fact that they control other forms of radiation." Having arrived at this conclusion, Accuray moved forward with efforts to attract both radiation oncologists and neurosurgeons to the use of CyberKnife in the field of neurosurgery.

When the company launched an improved version of the CyberKnife in 2000, the technology was ready to begin expanding indications for the device beyond neurosurgery. However, with this move, the company recognized that it would face an even greater challenge convincing surgeons outside neurosurgery and radiation oncologists to begin using the CyberKnife to treat patients with tumors in other regions of the body. For Accuray to have success in the long run, the company needed to fully understand the stakeholders involved in the broader radiosurgery arena and how to reach them.

Starting with physician stakeholders, Adler observed that surgeons in other specialties (non-neurosurgeons) tended to be less open to new innovations. Many of them were also threatened in other ways, having already seen their practices reduced by interventional pulmonologists, cardiologists, and gastroenterologists performing less invasive procedures for conditions that had traditionally been addressed through surgery. "What Accuray is doing is pretty heretical," said Adler. "It is anti-surgery in the eyes of true-blue surgeons because we're not cutting people open, there's no general anesthesia, and there's no blood loss – it's not nearly as dramatic as conventional surgery. So it strikes at the heart of what surgeons see themselves as being. It's not a trivial cultural shift to get surgeons to understand that they needn't be defined by blood loss and pain."

To help address concerns within this stakeholder group, Accuray had two primary arguments in its favor. The first, according to Adler, was that 90 percent of patients given the option would choose radiosurgery over traditional surgery. Additionally, surgeons reluctantly recognized that they needed to get involved in such procedures or risk losing this entire portion of their business to the radiation oncologists. Within the current healthcare delivery model, radiosurgery called for a surgeon and a radiation oncologist to be present during the procedure, with both specialists reimbursed separately for their time. Because of the manner in which it is used, CyberKnife procedures could possibly be performed by a single physician, and since most state regulations necessitated the participation of radiation oncologists, surgeons could potentially be seen as expendable participants. Recognizing this fact helped motivate surgeons to involve themselves in the adoption of the CyberKnife and carve out their role in the procedure, beyond just referring patients for treatment. "We achieved a sort of critical mass when surgeons started to realize that this was inevitable," recalled Adler. "They recognized they either had to get in the game, or they were going to be left behind."

Because the use of the CyberKnife in areas beyond neurosurgery would dramatically increase the involvement of radiation oncologists in treating a new class of patient, Accuray hoped that this stakeholder group might be receptive to the company's efforts to expand the use of the device. However, in reality, this turf battle turned out to be what Adler described as one of the company's "biggest problems." Instead of seeing opportunity to expand the use of radiation treatments to more patients "they worried about a new group of physicians playing with what was traditionally their toy. From a regulatory standpoint, it's pretty clear that radiation oncologists have been given complete authority by the state and federal governments to oversee all of this type of work, but they were still threatened." He continued, "They never looked and said, 'This is good because now we're going to be treating 80 or 90 percent more patients.' They don't think that way. Especially the more senior practitioners seemed more committed to creating their little fiefdom and keeping out any interlopers."

With both surgeons and radiation oncologists showing some reluctance to adopt the CyberKnife in areas outside the neurosurgery field, Accuray developed a robust marketing communication strategy to help drive adoption. The activities that the company undertook directly targeted the specific concerns of these surgeons and radiation oncologists, as well as patients, payers, and hospitals where the procedures would be performed.

When asked about the most successful marketing activity that Accuray employed to convince radiation oncologists and surgeons in other specialties of the benefits of the CyberKnife, Adler was adamant that it was the use of publications. "The single most effective strategy is to perform clinical outcomes studies and publish them in peer-reviewed journals," he said. When asked how many studies Accuray had performed, Adler answered, "Maybe a hundred. It never ends." The company's strategy was to perform relevant studies in specific specialty markets and then publish the results in the most prominent peer-reviewed journals in those fields to maximize the credibility of the results. Although it is more expensive to set up multiple studies, showing improved clinical outcome for each condition makes it much easier to market to surgeons, radiation oncologists, and payers. Accuray also presented its data to the targeted specialists at numerous conferences, as well as networking extensively with potential users.

Another interesting move by the company was the development of its own professional society: The CyberKnife Society (now called the Radiosurgery Society). The purpose of this group was "bringing together diverse medical professionals affiliated with radiosurgery worldwide to foster scholarly exchange of clinical information, and to educate the general public with patient information on treating medical conditions, such as cancers, lesions, and tumors anywhere throughout the body using the CyberKnife, most of which are unreachable by other radiotherapy systems." With a strong online presence, Adler said, "The society is focused on providing ongoing professional training and a way to give [surgeons and radiation oncologists] updates and the medical tools they need to use this technology to its latest and greatest capacity." For patients, it is another source of information for learning about their conditions, creating demand for the technology, and networking with other CyberKnife patients.

According to Adler, once patients understand the CyberKnife procedure and how it compares to traditional, open surgery, most of them become intrigued and request more information about the procedure from their surgeons. Accuray seeks to encourage this behavior by sending "evangelists" (patients who have successfully undergone CyberKnife treatment) to patient advocacy and support groups. In their role as evangelists, patients also are vitally important to influencing the next groups of stakeholders: Medicare, private insurance companies, and other payers.

Accuray realized early on that its first adopters would be extremely important in helping the company convince the American Medical Association (**AMA**) and Medicare to create new codes and coverage rates that would allow physicians and facilities to be adequately reimbursed for CyberKnife procedures. The company leveraged its relationship with an early CyberKnife adopter, Georgetown University Hospital in Washington D.C., to help influence the Center for Medicare and Medicaid Services (CMS) regarding the value of its device during the decision-making process. "It's all a matter of persistence," he said, "and of applying enough pressure." To assist in swaying payers, Accuray assisted customers in the formation of the CyberKnife Coalition, a membership of the product's users, to work with lobbyists on issues related to reimbursement.

The last stakeholder group, which is among the most important for Accuray, includes the hospital administrators who make the decision to approve the multi-million dollar purchase of the CyberKnife. To justify such a sizable expenditure, Accuray needed physicians to actively lobby hospitals to buy the equipment. The decision to start with the neurosurgery field proved to be beneficial to the company, as neurosurgeons are often influential within hospitals based on the revenue and profits they generate. However, Accuray also discovered that it needed physicians who would be aggressive in their support for the product and willing to persevere through the lengthy, bureaucratic buying process typical in most hospitals. To help convince hospital administrators to invest in its systems, Accuray developed a customized, detailed business case that outlined how much income each hospital could generate, across specialties, through the CyberKnife product (see 5.8 Sales and Distribution Strategy for more information about Accuray's approach to sales). Although the company had experienced some success in driving adoption among hospitals, this stakeholder group remained difficult to satisfy and could be the primary impediment to more widespread adoption in the future.

Using what Adler referred to as "guerilla marketing techniques," Accuray managed to successfully grow its business. The company went public in 2007.[32]

◥ Online Resources

Visit www.ebiodesign.org/5.7 for more content, including:

 Activities and links for "Getting Started"
- Revisit preliminary stakeholder analysis
- Develop value propositions and collect required evidence

- Define product/service mix
- Develop a pricing strategy
- Decide on positioning
- Develop an approach to promotion

 Videos on marketing and stakeholder strategy

CREDITS

The editors would like to acknowledge John Adler for sharing the Accuray story, Kim Popovits and Randy Scott for the Genomic Health example, and Darrell Zoromski for his assistance with the Miramar Labs case. Many thanks also go to Christopher Wasden and Ron Ho for providing additional examples, as well as Steve Fair and Ritu Kamal for their help with the chapter.

NOTES

1 Stephen N. Silverman, "An Historical Review and Modern Assessment of the Marketing Mix Concept," *7th Marketing History Conference Proceedings*, 1995, http://faculty.quinnipiac.edu/charm/CHARM%20proceedings/CHARM%20article%20archive%20pdf%20format/Volume%207%201995/25%20silverman.pdf (February 24, 2014).

2 D.R. Strutton, J.W. Kowalski, D.A. Glaser, and P.E. Stang, "U.S. Prevalence of Hyperhidrosis and Impact on Individuals with Axillary Hyperhidrosis: Results from a National Survey," *Journal of the American Academy of Dermatology*, August 2004, http://www.ncbi.nlm.nih.gov/pubmed/15280843 (February 11, 2014).

3 All quotations are from interviews conducted by the authors, unless otherwise cited. Reprinted with permission.

4 Japan's regulatory system allows for devices that are approved in other markets or undergoing clinical trials in Japan to be directly imported by a physician into the country without Japanese regulatory approval if the physician receiving the device is: (1) able to name the specific patient that will use the device, and (2) is willing to personally accept the risks associated with importing the product.

5 Kotler and Keller, *Marketing Management*, 14th edn. (Prentice Hall, 2011).

6 Lyn Denend and Stefanos Zenios, "Drug Eluting Stents: A Paradigm Shift in the Medical Device Industry," Stanford Graduate School of Business, February 13, 2006, https://gsbapps.stanford.edu/cases/detail1.asp?Document_ID=2787 (February 26, 2014).

7 Ibid.

8 Ibid.

9 Tiger Buford, "OIC, the Generic Implant Company in Reno, Is Revolutionizing the Price of Implants," May 9, 2012, Orthostreams, http://orthostreams.com/2012/05/oic-the-generic-implant-company-in-reno-is-revolutionizing-the-price-of-implants/ (February 26, 2014).

10 "OIC Releases High Value Proximal Humerus Plate," OIC press release, December 13, 2010, http://orthoimplantcompany.com/category/news-and-events/ (February 26, 2014).

11 "NABC Value Creation," SRI International, September 15, 2010, http://www.slideshare.net/HumInno/nabc-value-creation-5204311 (February 19, 2014).

12 Christopher L. Wasden and Mitchell L. Wasden, *Ride the Innovation Cycle* (expected release date 2015).

13 Ibid. Reprinted with permission.

14 Ibid.

15 A.S. Go, G.M. Chertow, D. Fan, C.E. McCulloch, and C.Y. Hsu, "Chronic Kidney Disease and the Risks of Death, Cardiovascular Events, and Hospitalization," *New England Journal of Medicine*, September 23, 2004, http://www.ncbi.nlm.nih.gov/pubmed/15385656?dopt=Abstract (March 11, 2008).

16 Brandon Keim, "Pay-For-Performance Pharmaceuticals: Satisfaction Guaranteed or Your Money Back," *Wired*, July 16, 2007, http://blog.wired.com/wiredscience/2007/07/pay-for-perform.html (February 5, 2014).

17 "Genomic Health Announces National Payor Agreement With UnitedHealthcare Insurance Company," Genomic Health press release. http://investor.genomichealth.com/releasedetail.cfm?ReleaseID=225085 (February 5, 2014).

18 Julie Manrriquez, Lyn Denend, and Stefanos Zenios, "CycleBeads II: Creating a Dual-Market," Global Health Innovation Insight Series, March 2012, http://csi.gsb.stanford.edu/sites/csi.gsb.stanford.edu/files/CycleBeadsII-CreatingaDualMarket.pdf (February 12, 2014).

19 "Fact Sheet: Bundled Payments for Care Improvement Initiative," Centers for Medicare and Medicaid Services, http://www.cms.gov/Newsroom/MediaReleaseDatabase/Fact-Sheets/2014-Fact-sheets-items/2014-01-30-2.html (March 10, 2014).

20 Lauren Uzdienski, "OIG Approves New Gainsharing Plan," HealthPoint Capital, November 28, 2006, http://www.healthpointcapital.com/research/2006/11/28/oig_approves_new_gainsharing_plan_advamed_critical/ (February 5, 2014).

21 Mark Prodger, "A Primer on Gainsharing," Hospital Buyer, http://www.hospitalbuyer.com/materials-management/cost-savings/a-primer-on-gainsharing-145/ (February 5, 2014).

22 "Fact Sheet: Bundled Payments for Care Improvement Initiative," op. cit.

23 "Stark Law Guidelines," Stark Law, http://starklaw.org/stark_guidelines.htm (February 5, 2014).

24 "TRUE Dilatation™ Balloon Valvuloplasty Catheter," C.R. Bard, http://lomavistamedical.com/wp-content/uploads/2013/05/Datasheet.pdf (March 13, 2014).

25 "Code of Ethics," AdvaMed, http://advamed.org/issues/1/code-of-ethics (February 5, 2014).

26 From remarks made by Ron Dollens as part of the "From the Innovator's Workbench" speaker series hosted by Stanford's Program in Biodesign, March 21, 2005, http://biodesign.stanford.edu/bdn/networking/pastinnovators.jsp. Reprinted with permission.

27 "athenahealth causing Trouble Again," The Health Care Blog, May 7, 2007, http://www.thehealthcareblog.com/the_health_care_blog/2007/05/cigna_ranks_no_.html (February 5, 2014).

28 Thomas M. Burton, "More Are Eligible for Heart Device – Accused of Rationing Care, Medicare Expands Use of Implantable Defibrillators," *The Wall Street Journal*, September 29, 2004, p. D4.

29 "FDA Approves First Breast Ultrasound Imaging System for Dense Breast Tissue," U.S. Food and Drug Administration press release, September 18, 2012, http://www.fda.gov/NewsEvents/Newsroom/PressAnnouncements/ucm319867.htm (February 12, 2014).

30 "Are You Dense, Inc. and U-Systems Announce Partnership to Increase Awareness of Breast Density and Access to Adjunctive Screening Tools," U-Systems press release, April 4, 2011, http://www.businesswire.com/news/home/20110404005186/en#.UvvsJvldV8E (February 12, 2014).

31 "New Jersey Passes Breast Density Law," American Society of Radiologic Technologists, February 5, 2014, http://www.asrt.org/main/standards-regulations/regulatory-legislative-news/2014/02/05/new-jersey-passes-breast-density-law (February 12, 2014).

32 "CyberKnife Frequently Asked Questions," Accuray, Inc., http://www.cyberknife.com/faq/index.aspx (February 5, 2014).

33 "Direct-to-Consumer Advertising," SourceWatch, http://www.sourcewatch.org/index.php?title=Direct-to-consumer_advertising (February 5, 2014).

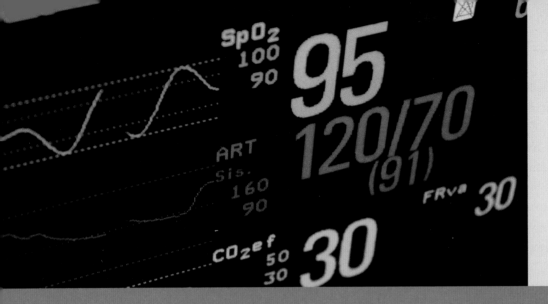

5.8 Sales and Distribution Strategy

INTRODUCTION

When talking with experienced medtech innovators, executives, or investors about commercializing a new medical technology, they all will inevitably ask, "How will you get key decision makers to adopt it?" It is not uncommon for technologies to target specialties with thousands of physician-users or potentially to be focused on reaching millions of patients directly. In either scenario, it can be a formidable challenge to introduce a technology, educate customers about its value, convince them to make a purchasing decision, deliver the offering, and then train them on its use. But all of these steps are necessary – in rapid succession and repeatedly – to create a sustainable business in the medtech field.

At this stage in the biodesign innovation process, a company defines the approach it will use to sell and deliver its offering to customers. This effort focuses on what is known as "the last mile," or the process of educating the key purchasers and users – who may or may not be one and the same – about the technology and its benefits, as well as working with them to make sure that the product or service is appropriately integrated into the care paradigm. For technologies targeted primarily at physicians and facilities, the critical question is whether the offering is best promoted through a dedicated sales force (the direct model), or if it should be offered through a partnership with another entity (the indirect model). Once this is decided, then specific strategies and tactics for executing the chosen approach can be put into place.

 See ebiodesign.org for featured videos on sales and distribution strategy.

OBJECTIVES

- Appreciate the impact that a company's business model has on its options for reaching its customers.

- Understand how traditional direct and indirect sales models work in the medtech field.

- Recognize how shifting forces – including value analysis and purchasing committees, more consumer-oriented products, and global considerations – are affecting medtech sales models.

- Learn how to determine the most appropriate sales and distribution model for a particular offering.

SALES AND DISTRIBUTION FUNDAMENTALS

Great companies are built based on their relationships with customers, and great managers maintain a relentless focus on creating these strong relationships. As John Abele, co-founder of Boston Scientific, put it:[1]

*It always drives me crazy when I walk into a company that posts its daily stock price in the lobby. In my mind, that's the wrong incentive. The incentive is to provide outstanding **value** to your customers. If you do that well, the stock price will eventually follow.*

Abele's statement is a compelling reminder that the success of any **medtech** venture depends on its sales organization and on the choice of the right sales model to deliver a positive customer experience, working within market and financial constraints.

Selecting and implementing an optimal approach to sales and distribution has never been easy. And, unfortunately, this challenge is only becoming more complex. Consider changes underway in the United States as an example. For decades, physicians were at the heart of US medical technology sales. Most of these doctors were self-employed, working alone or in small private medical practices.[2] These independent physicians typically had "privileges" at one or more hospitals that entitled them to provide medical care within the facilities; in exchange, they were expected to provide certain services on behalf of the institutions. Because the hospitals depended on physicians to bring in patients, they traditionally afforded the doctors a relatively high degree of influence over decisions such as which technologies to procure.[3] As one article pointed out, this created a "peculiar economic relationship because physicians benefit financially from the use of hospitals but do not bear direct responsibility for the fiscal health of these institutions."[4] The fact that physicians had little accountability for the consumption of hospital resources is just one example of how incentives for doctors and institutions historically have been misaligned.

But as US healthcare spending has reached unsustainable levels, a shift in these traditional relationships has begun. Hospitals and clinics have been consolidating in an effort to reduce costs and increase efficiencies, and physicians have been joining them as salaried employees. Estimates vary widely, but an American Medical Association survey conservatively estimated that more than 28 percent of physicians were hospital employees in 2012, up from 16.3 percent in 2008.[5] Other sources placed the estimate closer to 50 percent.[6]

Healthcare reform has also contributed to greater consolidation, with providers seeking to take advantage of government incentives by forming accountable care organizations (**ACOs**). The more these ACOs achieve costs saving and quality improvement targets, the greater the financial benefits they stand to gain from Medicare in the form of shared savings.[7] Accordingly, many of these organizations are taking a more active role in standardizing care. This often includes stripping physicians of some of their autonomy in terms of what they do, how they do it, and the devices they use in the process.

This shift is forcing changes in traditional medtech sales models. Rather than being able to work primarily through physicians to sell and distribute new technologies, device companies must deal with fewer, larger, and more powerful hospitals and institutions. Within these organizations, **value analysis committees** or new technology assessment groups are increasingly functioning as gatekeepers to the adoption of new devices. Physicians still play a role, certainly, in deciding which technologies make it onto the shelves of hospitals and clinics; but other **stakeholders**, including representatives from facility administration, purchasing, materials management, finance, and the executive suite are playing a more significant part – and may be the ultimate decision makers in a growing number of cases.[8] Importantly, this new class of decision makers is explicitly motivated to adopt products that help them achieve both clinical and financial goals, which has the potential to change the innovation landscape. As Alex Gorsky, CEO of Johnson & Johnson put it:[9]

> In the United States, buying decisions will shift from surgeons to cost-conscious hospital buyers. And that may create demand for keep-it-simple medical devices – designs that provide 50 percent of the bells-and-whistles of current devices for 15 percent of the cost.

The remainder of this chapter is devoted to explaining basic medtech sales models, the considerations/trade-offs inherent within them, and how they are evolving.

Medtech sales models

When medtech companies prepare to sell their technologies to physicians and/or the provider organizations they work for, they can take one of two fundamental approaches to sales and distribution: the **indirect** and the **direct** models. In an indirect model, one or more sales teams from distributors or third-party manufacturers serve as the primary point of contact with the end

user to manage product sales and delivery. They typically are not dedicated exclusively to any one product from a single company, but instead represent a portfolio of complementary products from multiple companies. That said, while some distributors carry a broad range of products across multiple fields, others may be more specialized, targeting products in a particular field, specific types of physicians, and/or a narrowly defined geographic territory. In a direct model, the company builds its own internal sales force to handle all sales functions, and establishes a separate customer support division to manage distribution of the product to the end user. A third, hybrid model, which is becoming increasingly common in the medtech industry, involves a combination of direct sales force and distributors. For example, a company may decide to use a direct sales force in the US market and local distributors in Asia and Europe; or it may work with a distributor to get its products into hospitals but hire a small team of product specialists to manage training and other important customer interactions.

Certain business models lend themselves better to specific sales and distribution approaches than others. The differences arise primarily from the characteristics of the innovation or product offering, although customer factors also come into play. Chapter 4.4 outlines a total of ten different business models, four of which are most often adopted in the medtech field (disposables, reusables, implantables, and capital equipment). Typical product characteristics that correspond to these four primary business models are summarized below.

Disposables, or single-use products, generally fall into two categories. First, there are low-value, commodity products (as measured by sales price in this context), such as lab supplies, syringes, or gloves. All of these rely on high sales, high volume, and low overhead to be profitable. They also tend to be low-complexity, non-differentiated products (meaning that physicians do not typically express a preference for which brand of the product they use). As a result, low-value disposables often require only limited sales effort beyond the initial product launch, assuming that prices are kept low and quality remains consistent. Marketing and product-related information sharing is generally accomplished via supplier catalogs or online resources, with customer service representatives available by phone.

High-value disposables include products such as ablation catheters used in the treatment of atrial fibrillation or the automated anastomosis systems for cardiac artery bypass graft (CABG) surgery developed by Cardica, Inc. (described in 6.1 Operating Plan and Financial Model). These high-value disposables command relatively high prices, tend to be complex, and may be more appropriately handled through a direct sales model due to the focused and prolonged effort required to sell them.

Reusable products, which are sterilized after each use and have a life expectancy of 10 years or more, tend to be lower-value offerings that provide relatively low margins and depend on moderate to high volumes to be profitable (surgical instruments are a prime example). Like disposables, they can be marketed through supplier catalogs and/or Internet resources. However, if they are sufficiently complex, have a slightly higher value, and/or command somewhat higher prices (such as ambulatory cardiac rhythm monitors), they may be sold and distributed by a third-party distributor that represents complementary products.

When considering the characteristics of *implants*, products that are surgically placed and stay in the body, it is useful to consider them in two categories. High-value, complex implants, such as pacemakers and artificial joints, often require a knowledgeable and specialized sales force in order to complete a sale and ensure proper device usage (through training, etc.). They usually have a somewhat longer sales cycle, as well. Depending on expected sales volume, high-value implants typically lend themselves to either a direct sales force or a specialized third-party distributor. Moderate and lower-value implants, on the other hand, may have characteristics that more closely resemble reusable products, as described above. As a result, it may be more appropriate to distribute them without a direct sales force.

Capital equipment products, or stand-alone machines regarded as fixed assets such as MRI and ultrasound equipment, are often highly complex. They provide high value to the user and, in turn, command high margins. These products tend to have a particularly long sales cycle that requires a prolonged, dedicated effort in order

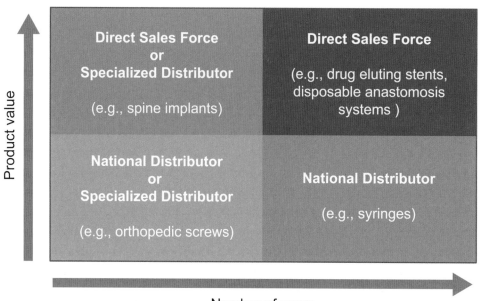

FIGURE 5.8.1

Sales force models are often linked to product attributes.

to make the sale. For this reason, the associated business model favors a direct sales force. Facilities generally act as buyers for capital equipment, with or without broad provider input. Because volume is low, the sales team is usually small, although auxiliary field personnel are required to service the equipment.

The business model, product characteristics, and preferred sales and distribution approach come together as shown in Figure 5.8.1. While this view represents typical scenarios in medtech, the decision regarding an optimal approach is not always clear cut. As mentioned, companies must determine the most effective sales and distribution model based on their chosen business model, their unique product offering, the customer being targeted, and the desired interaction with the customer. Customer accessibility and receptiveness are two important factors that play into the desired interaction that a company seeks to create with its target audience. For example, a direct sales model requires a high-touch relationship with targeted physicians. Such a model is only appropriate if a company can reliably gain access to those physicians, and they are generally willing to engage in the sales process. As noted in 5.7 Marketing and Stakeholder Strategy (see the case example on Accuray), each medical specialty tends to have its own "personality" regarding how receptive physicians are to experimenting with new

innovations, how curious they are to learn about new products, and how loyal they are to established treatments. Understanding the psyche of the doctors in the field where the company is trying to establish a foothold is essential to deciding on an optimal sales and distribution model. Examining the models used by competitors can be a source of invaluable insights.

Another factor to consider is that the skills sets required to sell different types of offerings can vary widely. For example, to effectively sell capital equipment, sales representatives ("reps") must be comfortable with a multiple-step sales process, producing pro forma financial information, amortizing equipment over time, discussing financing options, and negotiating detailed service agreements and installation terms. In contrast, those selling disposables tend to operate on a much more transactional basis, devoting the bulk of their time to conducting product demonstrations and detailing the features and benefits of interest to users in an effort to close sales more quickly. Companies can fail by having a bad match between the types of sales people they hire and the products they ask them to represent. And, if organizations are trying to sell two very different products, they may find that they need two sets of reps with completely different sales competencies, which can quickly become expensive.

The choice of the sales model is also based on a detailed financial analysis. In general, if a company elects to build a direct sales force, it must be prepared to carry all of the overhead and costs associated with hiring and maintaining the sales team. When it chooses an indirect approach, it does not directly incur these costs, but must share a portion of its revenue from product sales with the distributor. One way to evaluate the financial implications of this decision is to determine the largest direct sales force that the potential revenue from the product could sustain, based on the team's market analysis. Then, the innovators can make a judgment as to whether such a sales force could feasibly deliver that amount of revenue. If not, an indirect model may be necessary (or it may even be possible that a stand-alone business may not be viable). The exact financial calculations that can be used to support the choice of the appropriate model are described in 6.1 Operating Plan and Financial Model, while 6.4 Alternate Pathways explains the available options to the inventor when a stand-alone business is not realistic. More information is also provided later in this chapter.

Indirect models

The key advantage of an indirect sales model is that it enables a small or emerging company to enter the market with minimal investment. The primary disadvantage is that the company gives up a portion of its margins, as well as control of the sales process by entrusting it to the representatives of the distributor who, by definition, are paid to promote multiple products. This means that total sales may not be optimized as the product competes for the rep's (and the customer's) time and attention. Additionally, important customer feedback relevant to new product opportunities may be lost since the company is not directly interfacing with its customers.

In the medtech field, common approaches to adopting an indirect model for sales and distribution include entering into a sales and distribution agreement with a national or specialized distributor, or forming a third-party partnership with another manufacturer. The primary differences between these approaches are outlined below (and summarized in online Appendix 5.8.1).

National distributors

National distributors work especially well for disposables, reusables, and simple implants that have a relatively low unit price and low-to-moderate complexity. In these cases, the sales and distribution process is largely managed by a national distributor or wholesaler.[10] The company generally sells to the distributor at a discount. The distributor, in turn, passes along the product at full price, keeping the difference as a form of compensation. The company maintains its own brand, while the distributor maintains the right to represent other products (i.e., the sales and distribution agreement is not exclusive). While this model is almost always adopted for commodity products, such as syringes, surgical gloves, and other such medical equipment, it can periodically be effective for some higher-end disposables and reusables.

In most cases, the customer for the products sold and delivered through a national distributor is not the physician or other end user of the device. Instead, national distributors tend to interact with the purchasing departments of hospitals or clinics, or they interface with intermediaries, such as group purchasing organizations (**GPOs**) or buyers for integrated delivery networks (**IDNs**) and ACOs (more information about these entities is presented later in the chapter).

Specialized distributors

Another indirect model that is widely used in the medtech field involves working with specialized distributors. This approach is best suited to devices that are somewhat differentiated, relatively more complex, and offer at least moderate value to the user (as measured by physician preference for a brand name device, e.g., certain types of implants). Alternatively, specialized distributors work well for low-complexity products that complete a complex product line carried by the specialized distributor.

With a specialized distributor model, a company gains access to sales representatives who are specialists in a particular type of product or therapeutic area. The sales personnel that support this model often work as independent representatives, contracting directly with the company for a portion of the margin made on product sales. These representatives tend to have strong customer relationships within a particular region, and are able to leverage these relationships to sell whatever product they are carrying at the time. Sole source distributors carry one company's products while multiple-source distributors carry a few lines of product, ideally with clear differentiation along product characteristics and price. While it is in the best interest of distributors to carry as many brands as possible, manufacturers prefer exclusivity and can demand it if they are large enough. Invivo Surgical Systems is an example of a multi-source franchise distributor[11] that carries spinal fixation implant product lines and equipment. Invivo Surgical Systems limits its coverage to the Northeast and New England, and recruits its own sales personnel to sell its products.

Specialized distributors can be set up as master resellers or agents. The primary distinction lies in whether the specialized distributor takes ownership of the product. This distinction has important ramifications on economics of the sale. Specialized distributors that are master resellers purchase the product from the company in anticipation of orders coming in from customers, and resell it to the end user. Because master resellers take title of the product, they have more power to negotiate contracts and discounts with the end user. Specialized distributors who act as agents, on the other hand, sell product on consignment and negotiate the sale on behalf of the manufacturer, earning a commission.[12] They have limited pricing flexibility and can only contract and discount at terms that are acceptable to the manufacturer. Although master resellers have more negotiating flexibility, they also take on more financial risk. They purchase goods from the manufacturer at a discount that is roughly equal to their target margin, but have no guarantee that the end user will buy at the retail price, or at all. Agents, on the other hand, receive a 20–30 percent commission[13] on each sale and are not responsible for product that does not sell. However, they do risk losing part of their commission if they accept a lower price from the end user.

Third-party partnerships

A third approach to indirect sales is to manage the process through a partnership with an established medical device manufacturer that has its own direct sales force (this form of partnership is different from partnerships focused on developing the product, which are described in 6.4 Alternate Pathways). Third-party partnerships, while widely used in the pharmaceutical industry, are relatively new to medical devices. Lifescan's agreement with Medtronic to distribute its glucose monitor provides one example of this kind of partnership.[14] A classic example that demonstrates the risks inherent in these arrangements is Israel-based Medinol's decision to sell and distribute its stents through Boston Scientific in the mid-1990s, which ended in a protracted legal battle.[15] The agreement seemed beneficial for both partners: Medinol, a small start-up company with no commercial experience in the US, would benefit from Boston Scientific's size, commercialization, and marketing expertise, while Boston Scientific could enter the stent market more quickly than if it developed its own product.[16] The two companies signed a long-term agreement that included plans to work together on future versions of the devices. Two years after the partnership commenced, in part over concerns about Medinol's manufacturing output, Boston Scientific constructed a secret stent production facility in Ireland. After learning of the facility, Medinol terminated its agreement with Boston Scientific and sued for damages. Ultimately, Boston Scientific agreed to settle the dispute by returning its 22 percent ownership interest in Medinol and paying Medinol $750 million.[17] Sometimes worldwide sales and distribution rights are awarded as part of a third-party deal. In other scenarios, the smaller company maintains local rights (e.g., in the US), while essentially "outsourcing" sales and distribution in other markets (e.g., Europe).

Third-party sales and distribution partnerships are generally only available to companies whose products can be clearly differentiated, have at least moderate perceived value, and/or complete a complex system offered by the larger partner. The most common reason for small companies to enter into these partnerships is to avoid the large financial burden associated with building and maintaining a direct sales force. Companies may also consider such a deal if they lack the breadth in their product lines to sell to entities that wish to purchase a complete line of products. Established manufacturers also have more leverage with GPOs and other important buyers and provide smaller companies with access to these relationships.

There are two primary financial arrangements for forming these third-party partnerships. The distinction is driven by whether or not the distributing company will manufacture the product. If the smaller company will be the manufacturer, a **transfer price** is established to reward the manufacturer each time a unit of product is shipped to the distributing company. If the distributing company will be the manufacturer, then a **royalty** is paid (usually 5–8 percent) to the smaller company based on net revenues. In both scenarios, an upfront payment may be awarded to the smaller company to provide the financial means to bring the product to market.

Although third-party sales and distribution partnerships can provide the resources, infrastructure, relationships, and experience to quickly help get a product into practice, entering into such a relationship can be risky for a small device company, particularly if it is a single-product company. Because the larger company controls all of the customer relationships and manages pricing, training, and usage of the product, the smaller company is essentially putting its fate into the hands of its partner. As the case example on Phoenix Medical Systems demonstrates, this is a lesson that some companies learn the hard way.

FROM THE FIELD ▶ PHOENIX MEDICAL SYSTEMS

When partner sales fall short of expectations[18]

Early in its development, Phoenix Medical Systems, India's leading manufacturer of high-quality, low-cost neonatal care equipment, attracted the attention of a major multinational corporation. The larger organization had a presence in India's medical imaging and patient diagnostics/monitoring systems markets and was actively seeking to broaden its capabilities in the growing maternal/infant care segment. When the multinational approach Phoenix founder V. Sashi Kumar about a partnership deal, he was enthusiastic. "I was very happy to become associated with such a wonderful company," he said.[19] The representatives suggested purchasing Phoenix, but Kumar preferred a **licensing** agreement that would give the company access to Phoenix's technology. The organization was particularly interested in three of Phoenix's products – two infant warmers and a phototherapy unit (see Figure 5.8.2). As Kumar recalled, the multinational team predicted annual sales of $10 million for this equipment within three years.

Ultimately, they entered into a two-year exclusive contract that had two primary parts. First, the multinational would use its established distribution

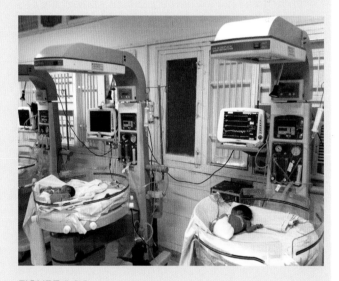

FIGURE 5.8.2
Phoenix equipment in use within a government hospital in Bhiwani (courtesy of Phoenix Medical Systems).

channels to sell all of the products in the Phoenix portfolio, under the Phoenix brand name, exclusively in the Indian market. Phoenix would continue to manufacture the products and the company would purchase the inventory to support Indian sales. Second, the new partner would modify the three products noted above to meet its own international requirements, and then manufacture, sell, and distribute them in markets outside India under the multinational's brand name.

The multinational intended to sell the infant warmers and phototherapy devices into maternity hospitals through the gynecologists who were the customers of its ultrasound machines. "They thought it would be easy to get the products into maternity hospitals because they already had business there," Kumar said. Unfortunately, these sales did not come as effortlessly as expected. "They could sell their ultrasound because their name was well-known in this space. But the company didn't have the same reputation with incubators and phototherapy devices," he explained. In addition, the partner's sales representatives often had to interact with physicians besides the gynecologists. Since the reps were generally unfamiliar with how the new equipment worked, they were not able to effectively position it to these decision makers. "These aren't products that can be sold just like that," Kumar said. Physicians expected the sales reps to deeply understand the benefits of the product relative to other alternatives in the market and answer their technical questions. With no special incentives in place to motivate the sales people to acquire the knowledge they would need to effectively engage in these situations, sales languished.

Meanwhile, Phoenix had made a sizable investment in expanding its manufacturing capacity to meet the multinational's predicted demand. "They expected to move more units in a couple of months than we sold in a year," Kumar recalled. As new production facilities and staff came online, sales remained flat. As a result, inventory started to pile up. "We were left with a large order, which we had to service without any revenue." Phoenix started to feel the financial strain of its new arrangement, but the company's hands were tied. "We had given the total distribution rights to the multinational, so there was no opportunity for us to sell our products into the market directly," Kumar explained. Even if the contract had not prohibited direct sales, Phoenix no longer had its own sales representatives – they had all left the company when the partnership deal was signed. "It was a very difficult time," he added.

Kumar realized that in order for Phoenix to survive, the company would need to play a much more active role in helping stimulate sales – at least in the Indian market. First, he made the multinational aware that Phoenix could no longer afford to sit on the inventory it had amassed. In response, it agreed to purchase some of the equipment in advance. Second, Kumar mobilized his network of service technicians, who remained in the field to maintain the installed base of Phoenix equipment in India. "We got them to begin pulling lots of leads, and we proactively brought them to our partner," he said. These warm leads stimulated enough sales to enable Phoenix to get by. As Kumar put it, "We managed to stay alive."

At the end of the two-year contract, the multinational approached Phoenix about extending the agreement, but Kumar was adamant about regaining control of Phoenix's products. Kumar rehired several of his previous sales representatives and started to rebuild the sales team. In addition, Phoenix again tapped into its network of service technicians to let its customers know that it had reestablished direct control of its product portfolio. "The customers were happy that we were back," Kumar said. Gradually, Phoenix was able to ramp up its sales. The company expected to achieve record turnover of approximately $18 million by late 2013.

Third-party partnerships are usually best employed when some terminal event is spelled out in the partnership agreement toward which the company is working (e.g., an **acquisition**) or the company has another flagship product over which it exercises significantly greater control. Another scenario in which a third-party partnership might be appropriate is when the company is facing financial difficulties that leave it with few other alternatives.

Direct models

For more complex and expensive devices, such as capital equipment and high-value disposables or implants, a

Physicians

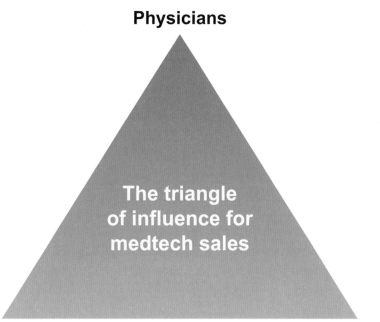

The triangle
of influence for
medtech sales

Purchasing

Administration

FIGURE 5.8.3

The key stakeholders represented in the "triangle of influence" must be aligned on many medtech purchasing decisions in order for a sale to be made.

direct sales model is often required. This high-touch approach may be needed to affect the "triangle of influence" involved in these kinds of sales, which includes physicians, purchasing professionals, and (depending on the dollar amount) "C level" executives representing facility administration (see Figure 5.8.3).

As noted, as the influence of physicians in the purchasing process is waning, the power of purchasing professional and administrators is increasing. All three stakeholders must be considered customers, as they each have a role in the purchase process: physicians evaluate and endorse the performance of a product; purchasers evaluate the economic ramifications of the buying decision; and executives ensure its strategic fit with the direction of the organization (particularly for "big ticket" items). More often than not, convincing these disparate stakeholder groups to buy a product requires information, contacts, and expertise that only a direct sales force can provide. The decision to develop a direct sales force (and how to size it) is complex and must be based on detailed **bottom-up** financial modeling, explained in chapter 6.1.

With a direct sales force, a company hires its own sales people to engage providers (physicians and facility representatives) in a six-step process: (1) account prospecting; (2) relationship building; (3) selling the **value**

proposition to stakeholders; (4) closing; (5) physician user training; and (6) account management, as described below.

Step 1: Account prospecting
Focusing sales efforts on targets that will eventually lead to consummated business is a two-stage process. First, initial sales targets are set by the company based on the amount of calculated potential business a target can generate in the future (6.1 Operating Plan and Financial Model). Next, sales representatives visit or "call on" these targets. Over time, they identify and focus on the subset of providers who are most receptive. Reps may also identify additional providers not included on the original list of targets through networking and other professional activities.

Step 2: Relationship building
Forming partnerships and developing trust with providers and/or facility representatives requires extensive formal and informal relationship-building activity. During this stage of the selling process, sales reps are constantly evaluating providers' **needs** and seeking to convince them of the product's value in addressing those needs (see 5.7 Marketing and Stakeholder Strategy). The off-site social meetings popularized by the drug industry

(such as dinners, sporting events, and company-sponsored weekend advisory board meetings) were initially adopted in the medical device field as one vehicle for gaining access to providers. But since the 2013 enactment of the Physician Payment Sunshine Act, which requires manufacturers of drugs, medical devices, and biologicals that participate in US federal healthcare programs to report certain payments and items of value given to physicians and teaching hospitals,[20] these activities have been significantly curtailed.

In the absence of specific incentives, the focus of this step is primarily on gaining support from physicians so they will advocate for the product through whatever purchasing process is required. Sales reps devote considerable time to giving product demonstrations, sharing **clinical trial** results, and responding to provider questions and concerns. In turn, physicians decide to support a product based on their evaluation of clinical and economic improvements it offers, often based on a product trial. They also may take into account the sales rep's expertise and personal characteristics (such as honesty and integrity), and the commitment of the company to the account (demonstrated through time, attention given to the provider's needs, and the future potential for research partnerships).

Step 3: Selling the value proposition

Ongoing product sales begin once the relationship has become adequately established and the product benefits are clear to the physician. Particularly in a more value-oriented medtech environment, sales representatives must be trained extensively to handle this step in the sales process. Reps use clinical, economic, competitor, and disease state information, as well as an understanding of the decision maker's unmet needs or "pain points," to make a case for their products. Materials are often customized to each stakeholder's circumstances and interests in the triangle of influence. Sales representatives must be prepared to carefully follow the requirements of each institution's purchasing process, which may include presenting to a value analysis or new technology assessment committee. Because these committees typically involve multidisciplinary representatives from all part of the organization, they also must

be equipped to handle objections from stakeholders representing any number of different perspectives. (See the AccessClosure story later in the chapter for a related case example.)

As an integral part of this step in the sales process, especially for higher-cost items sold to large institutions and consolidated provider organizations, companies generally must develop a customized business case model that demonstrates how the product will improve outcomes, generate savings in the buyer's environment, and how the buyer will be reimbursed (or otherwise paid) for use of the technology. This can be a complicated and time-consuming endeavor.

Step 4: Closing

The actual "close" takes place when the sales reps ask for the product to be purchased for a specific patient or within a defined time frame, and the decision maker agrees. Pricing and contract negotiations may occur in parallel with closing the sale; some providers want to address these issues before any final agreement is reached, while others prefer to address them after the product has been tested and found acceptable in a few patients.

While physicians are integral to making the recommendation to procure the product, they participate less frequently in purchasing negotiations, which include agreeing on a price for the product and the delineation of all relevant deal terms (delivery, training, usage, service, etc.). One exception is when technologies are being sold into independent physician practices, where physicians are still likely to serve as the primary point of contact for the end-to-end sales process.

In single hospitals and smaller hospital chains, price and contract negotiations are more frequently conducted between the sales reps and contacts from purchasing and/or facility administration. With strategic customers, such as large IDNs, GPOs, ACOs, and major hospital conglomerates, contract negotiations are usually handled at the corporate level (by the VP of sales and marketing or the chief medical officer). Such negotiations can be complicated and time-consuming. Additionally, they are often strategic in nature and can involve variables that affect large sums of money

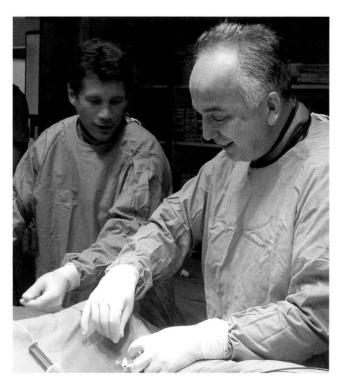

FIGURE 5.8.4
Transvascular, Inc. representative Ted Lamson trains
Dr. Tomasz Siminiak on a new procedures for injecting stem
cells into the heart (courtesy of ExploraMed).

(e.g., if the facility representatives demand pricing concessions in exchange for volume).

Step 5: Physician user training

Once a sale has been made or promised, sales reps help the physicians and facilities use and integrate the new technology into practice. For example, with an implant, they first train the provider on how to select the appropriate model and size of the device and then how to perform the procedure. If appropriate, sales reps are commonly present in the first few procedures to assist physicians in selecting the model or size of device for each individual patient and help troubleshoot problems and issues as they arise (see Figure 5.8.4). For highly complex devices, it is not unusual for companies to restrict the use of a device until physicians have completed appropriate training by a representative of the company (see the Kyphon story in this chapter). The quality of this interaction is critical in determining whether the physicians and facilities will continue using the product. Importantly, the company always bears the cost of delivering this training when it chooses a direct sales model. With an indirect model, the distributor will sometimes, but not always, provide training on behalf of the company. Innovators must carefully consider these kinds of issues when negotiating a deal or they might find themselves covering the cost of training while still sharing their margins with the distributor.

Step 6: Account management

Post-sale account servicing includes everything that happens after the sale to support all servicing included in the contract (e.g., service for a piece of equipment, answering product questions, and assisting with **reimbursement** issues). Additionally, it focuses on customer retention and the development of future sales with the same account. By proactively addressing customer concerns, a sales force can understand what issues might cause a customer to choose another product and, in many cases, make adjustments to the offering to circumvent the problem. Examples of common concerns in the medtech field that must be managed on an ongoing basis include difficulties with the device and/or procedure, patient outcomes that are less favorable than anticipated, and suboptimal customer service.

In addressing these concerns, account management becomes intertwined with the development of new product ideas. As sales representatives assist providers with the use of the device and address product questions, they gather feedback that can be used by engineers and other technical experts within the company to improve and/or expand the company's offerings.

As this brief description of the direct sales process illustrates, there is a sizable investment of time and resources to make each sale. However, once physicians and facilities start reliably engaging the sales representatives and using the company's products, a strong bond is often formed that can be leveraged to make sales easier and less time-consuming in the future and becomes a barrier to entry for the competition.

The AccessClosure story illustrates how the direct sales model, as well as the respective roles of sales representatives, physicians, and facilities are evolving.

FROM THE FIELD ⟩ ACCESSCLOSURE INC.

Direct sales in the changing medtech environment

Serial entrepreneur Fred Khosravi co-founded AccessClosure Inc. to focus on needs in the cardiovascular and peripheral vascular markets. In particular, he became interested in creating a better solution for closing an artery following an endovascular (or minimally invasive) procedure. The current **standard of care** was manual compression, which was used in roughly 55 percent of coronary cases and 80 percent of peripheral cases.[21] This involved a nurse using a hand to put pressure on the wound for 10 to 30 minutes until a patient's bleeding stopped. While effective, compression was time-consuming, resource intensive, and often uncomfortable for the patient. Two primary vascular closure devices were available in the market. They, too, were effective, but Khosravi saw room for improvement. As he described in an article, "The incumbent closure technologies ... are associated with significant complications and, from a **morbidity** standpoint, they are not that patient friendly. They were designed primarily to be physician friendly in terms of being easy to use and simple to operate."[22] Key concerns included bleeding risks and the fact that the solutions remained implanted in the puncture site where they could cause long-term discomfort and create vascular obstructions.[23]

The AccessClosure team received **FDA** approval for its Mynx vascular closure device in May 2007 (see Figure 5.8.5). The new device sealed the femoral artery using a unique hydrogel sealant that dissolved within 30 days, leaving nothing behind but a healed artery.[24] Importantly, the device was still easy for physicians to use, minimized bleeding risks, and was less painful for the patient.[25]

In parallel with pursuing regulatory approval, the team began planning its approach to sales and distribution of the device. According to Gregory Casciaro, who later became the company's President & CEO, "You could go

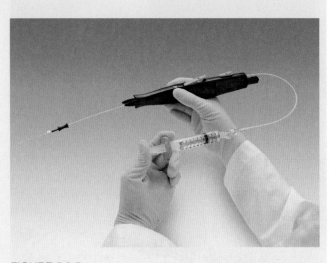

FIGURE 5.8.5

The Mynx Ace vascular closure device is a later evolution of the original Mynx product (courtesy of AccessClosure).

with a network of distributors across the country or you could build a direct sales force. And the company went with the latter for a few reasons." For one thing, he explained, interventional cardiologists, vascular surgeons, and interventional radiologists were used to being called on by direct sales representatives. "You need attention in those areas, meaning that you need to provide the physicians with instruction on how to use the technology." He also viewed the personal relationships that developed between sales reps and physicians as being central to building the trust, reliability, and consistency needed to get the doctors to try the product. Additionally, "Our competitors were using a direct sales force. So, in many ways, there was a standard in place. If we really wanted to compete in this sector, we thought we would be better off using our own sales force rather than relying on a distributor who was going to be distracted with a lot of other things besides our product," Casciaro said. The company also believed that a direct sales force would be better at highlighting the unique characteristics that differentiated the technology from the vascular closure devices that preceded it. In combination, these factors "made our choice relatively easy," he noted. AccessClosure was able to tap into Khosravi's vast network and strong track record to access the funds it needed to hire a sales team, starting off with 15–20 reps to cover key US markets.

The new sales team primarily sold directly to physicians. As Casciaro recalled, "Our reps would go into the hospitals and talk with the doctors. They would get them excited about the technology and they would agree to try the device on their next 10 procedures." The sales people would work with the doctors on those cases, providing them with the necessary training and support they needed to be successful. "After that," he continued, "They'd say, 'I love it. Go tell materials management I want to buy it.'" And this is how AccessClosure would get its device on hospital shelves. Using this approach, the company generated $32 million in sales in its first full year of commercialization.[26] By 2013, annual sales had grown to approximately $80 million.[27]

During this period of time, however, Casciaro acknowledged that a significant shift in the sales process had occurred. "We still do the front end stuff," he said. "But now the physicians often tell us, 'I'm still going to send you to materials management and you can tell them I want the product. But good luck – you better go in with a sharp pencil, and you better have a good economic and clinical argument.'" A direct sales force could still get a new product into a smaller community hospital without necessarily going through a new technology assessment committee. But for larger healthcare conglomerates, it was increasingly common for the sale to be gated by the institution. "The sale still starts and ends with the physician," Casciaro said, "but the institution plays a big role in the middle." On the front end of the process, the technology assessment committees require one or more physicians to advocate for the new product, so the sales reps still have to get them interested and assist them in trialing it. On the back end, the reps work to expand the number of physicians using the technology and make sure they receive the necessary training and support to make their procedures successful. In between these steps, however, representatives from the technology

assessment committee and/or purchasing are the decision makers. "Sometimes the assessment committee meetings happen only once a month or once every two months," Casciaro stated. "Everything takes a little longer in this environment."

Another increasingly common scenario, he explained, is for the company to have to work through a group purchasing organization. "These buyers are seeking economies of scale. They typically ask, 'If we can get our 10 hospitals to use your product, what kind of a deal will you give us?' It sounds great, but getting physicians only to use one product is a big challenge," Casciaro commented. Tracking compliance was another difficult issue, and gaps between the promised and actual usage volume in these accounts were not unusual. "Right now," he continued, "it's more about getting into these institutions and building relationships with them. Then it's up to the sales rep to build volume and growth within the account."

Reflecting on these two emerging sales scenarios, Casciaro emphasized that it has become much more important for the company to make a compelling argument about the cost of the product and the value it delivers. "In our case, that involves demonstrating that we can get the patient in and out of the hospital sooner by using a vascular closure device." He also pointed out that, "It's a more sophisticated process. Sale reps need to expand beyond relationship building at the physician level. They need to be really good at dealing with people at all levels in the organization, from materials management to the executive suite."

In early 2014, AccessClosure was purchased by Ohio-based Cardinal Health, Inc., a major healthcare services company, for $320 million. According to the companies, the acquisition furthered Cardinal Health's strategy of providing products that improve patient care while increasing the cost-effectiveness of hospital services and procedures. Casciaro predicted that Cardinal Health's global reach and significant human resources would help drive rapid business growth while continuing to meet customer needs at all levels.[28]

Direct sales teams are typically organized geographically, with each individual territory manager reporting to a regional manager who, in turn, reports to an area sales director or a VP or sales (depending on number of layers within the organization). Responsibility for the sale is shared somewhat between the regional manager and to the territory manager. They often take a team approach to closing the sale, after which the regional manager handles the contract negotiations and the territory manager handles account maintenance and customer retention in collaboration with the sales representative. Any contracts and discounts are typically managed centrally by the company.[29]

In a direct model, the physical delivery of the device is typically made by the sales representatives or through direct delivery from the company's warehouse (capital equipment requiring a major installation is, of course, one exception to this rule). The provider completes and submits an order form, calls the company's customer service department, or places the order directly through the sales representative. In turn, the product is shipped or hand-delivered. In some cases, multiple sizes or models of the product are left with the provider on consignment so that, during procedures, the device with the correct specifications is always available. Because the physical delivery of a device is not as complex as the underlying sales process, the manufacturer may engage a third-party distributor to take charge of deliveries, even as it maintains a direct sales force.

The role of payers in the direct model

Payers should also be viewed as customers for more complex, high-value medical devices, as they are responsible for the incremental reimbursement of new and expensive products. As addressed in 5.6 Reimbursement Strategy, the reimbursement process is not unlike the sales process described in this chapter; it involves identifying who will be the major payers, developing relationships early to understand what product information and value propositions will be required, lobbying for **coverage** and **payment** policy in anticipation of or in conjunction with launch, contracting and discounting, and payer account maintenance (i.e., fielding product questions or concerns as they arise). However, while "selling" to

payers requires significant effort and focus, it typically must occur only once. After the product is accepted by the payer, less effort is needed to maintain product acceptance, although constant follow-up is necessary to keep the product in a favored position.

Sales force training

Any direct sales force requires extensive training to effectively represent and sell a complex product to the target audience. Such training should be delivered directly by the company that developed the device so that the sales people gain a deep understanding of the product's attributes, the disease state that it treats, how it is used in practice, how it compares to competitive products and/or the standard of care, and its core value proposition to physicians and the facilities where it will be used. In cases where the sales representatives will be asked to educate, train, and/or coach physicians on the use of the device, reps require even more in-depth preparation. Representatives should be prepared to answer detailed and often technical questions related to clinical benefit, economic benefit, and billing and reimbursement, especially if they will face value analysis or new technology assessment committees. At sales meetings, representatives will often be asked to "role play" so they gain experience in addressing the many challenging questions that can arise in the sales process. At times, sales representatives are also given the opportunity to shadow doctors while learning about a new procedure or field of medicine.

It is essential to remember that what a company and its sales representatives can (and cannot) communicate to a buyer about what the product will do or the outcomes it can help the doctor achieve is regulated, to a large extent, by the FDA in the US. For instance, when a device is cleared for use under a **510(k)**, the company is restricted from making claims above and beyond those of the substantially equivalent device. A company can claim new and/or different indications for devices approved under a **PMA**; however, these claims may not extend beyond the intended use for which the device has been approved (see 5.4 Regulatory Strategy).

Furthermore, sales representatives must be trained to adhere to the principles of medical ethics – in particular, the principle to "first, do no harm." As the representatives are closest to the patients on whom a device is used and to the physicians or users of the device, representatives have an obligation to raise any and all concerns to company executives and demand the timely resolution of concerns. Stories surrounding the **recall** of major medical devices, such as implantable cardioverter defibrillators (ICDs), underscore how important it is for a company to mount a timely response to any issues of safety efficacy that arise. If sales representatives have been trained to identify and proactively communicate potential issues, a company may even be able to detect and correct problems early, before they pose a threat to patients, lead to recalls, or affect product and company reputation.

In addition, sales representatives should be aware of potential conflicts of interest that may arise as a result of their interactions with clinicians and proactively manage them. The **AdvaMed** code of ethics, described in 5.7 Marketing and Stakeholder Strategy, provides a minimum set of guidelines. As noted, the Physician Payment Sunshine Act has helped reduce activities that could potentially be construed as creating conflicts of interest. While not unanimously popular across the medtech field, some innovators welcomed the restrictions since they helped to "level the playing field" between smaller technology companies within limited budgets for engaging with key opinion leaders and the larger device manufacturers with more resources at their disposal. As Greg Casciaro, President and CEO of AccessClosure, put it: "Frankly, as a smaller private company that can't match the spending of a larger company, the Sunshine Act is a benefit for us. We've already seen positive changes in physician and company behavior."

Interaction with marketing

Marketing and sales activities must be closely coordinated to maximize the company's efforts to capture and then sustain value. Well-defined marketing activities support the sales effort by providing sales representatives with all of the information necessary to effectively execute the sales process and also to help retain customers once the sale is made. Because of the highly scientific nature of complex medical device sales, promotion/advocacy is one of the most important elements of the marketing mix in terms of supporting sales. Medical education is especially important (based on data that have been published in **peer-reviewed** journals), particularly for devices that seek to change the practice of medicine. The Accuray story demonstrates how marketing and sales activities should work hand-in-hand.

FROM THE FIELD ▸ ACCURAY INCORPORATED

A multifaceted approach to complex capital equipment sales

As described in 5.7 Marketing and Stakeholder Strategy, Accuray was founded by neurosurgeon John Adler to develop a frameless stereotactic radiosurgical system called the CyberKnife® (see Figure 5.8.6). According to Adler, the cost of a CyberKnife system can be upwards of $5 million, including the technology upgrades that were built into the initial deal. Each unit also requires specialized facilities renovations to support the installation, which could add another $1 million to the cost of the system to a hospital. Given the level of required investment, the sales cycle for this product was typically lengthy, challenging, and bureaucratic, taking Accuray anywhere from 18 months to as much as 7 or 8 years to close a deal.

To accomplish its sales goals, Accuray employed approximately 10 full-time direct sales representatives. According to Adler, "We can't use entry-level representatives. These sales go right to the highest levels in a hospital and require signoffs by VPs, purchasing committees, boards of directors, and all kinds of lawyers." In terms of their backgrounds, most of Accuray's reps were seasoned medical device

FIGURE 5.8.6
Adler with the CyberKnife system (courtesy of John Adler).

sales people from heavyweight medtech companies such as Johnson & Johnson or GE.

Accuray's sales process was multifaceted and focused on both initial customer acquisition and customer retention. The primary customer was the hospital, and hospital administrators were the ultimate decision makers in the sales process. However, Accuray learned quickly that physicians, serving as medical champions, were crucial to almost every step in the sales process. "The medical champions are often interested surgeons who are early adopters, who want to have an advantage in the local medical marketplace, or who are just curious and like new toys and gadgets," said Adler. Initially, Adler leveraged his deep professional network to generate sales leads and acquire customers.

As Accuray grew, however, it needed to expand its community of medical champions beyond Adler's contacts. The company sought to persuade additional neurosurgeons of the benefits of the CyberKnife by presenting at neurosurgery conferences and other relevant professional society meetings. Moreover, the company involved key opinion leaders in managing clinical trials that compared the CyberKnife technology to conventional surgery. And, as noted previously, Adler underscored the importance of publishing the results of these studies in peer-reviewed medical journals. Once these physicians became convinced of the benefits of the CyberKnife, they would, in most cases, begin advocating the procedure (and the need for the device) to their hospitals.

After strong demand was fostered within the local physician community, the next challenge for Accuray's sales representatives was to convince multiple layers of hospital administrators to make the required investment. To accomplish this, the company devoted significant time and energy to creating a customized business case and financial models that demonstrated the economics of radiosurgery for each hospital considering the system. Adler further described how the company used revenue-sharing to drive initial sales. "One thing that we did early on to jump-start the business was to implement a shared-revenue model with the hospitals," Adler recalled. "The hospitals were required to build out a room to put the CyberKnife, but we put it in for no charge. And then we treated patients together and split the revenue. We also are linked in with various financing organizations that can provide loans at favorable rates for hospitals." Another factor that enabled Accuray to persuade hospital administrators to purchase its product was the fact that CyberKnife could be used by facilities as a differentiator in the market. Adler explained, "To survive in the medical marketplace, hospitals need to stay ahead of the technology curve. A lot of hospitals [use our system] as a marquee technology to market their institution as a whole."

Once a CyberKnife system was installed at a hospital, it could be used by multiple specialties as long as the practicing physicians had received the appropriate training. Expansion of the device to other specialties not only enabled such hospitals to better exploit the clinical value of Cyberknife, but helped them realize its potential as an investment sooner. In parallel, it could also help the company generate new sales leads as physicians networked in a viral fashion with their colleagues at other facilities (word of mouth among surgical specialists and radiation oncologists, especially within a given institution, remains a powerful marketing tool). As a result, it was in the best interest of both the hospital and the company for Accuray to manage the initial customer acquisition and simultaneously enhance customer value by expanding the clinical indications of the device.

Hybrid models

A hybrid distribution model that combines both a small direct sales force and an indirect approach can be a good choice for some medtech companies, particularly in their early stages. The time, cost, and other resources associated with establishing a full direct sales force can be daunting to a young company. However, if a product is at least moderate in its complexity, value, and anticipated volume, the company could launch a small direct sales team whose efforts are complemented by an independent distributor. The direct sales force can target a narrow segment of the company's highest-value customers, while distributors or third-party partnerships with larger manufacturers are leveraged to give the company breadth in the total available market. The benefits of a hybrid model are greatest when the sales team and the partners(s) work closely together.

Hybrid models are often based on geography, with a company deploying a direct sales force in concentrated markets but relying on regional specialized distributors in diffuse, closed, or global markets. MAKO Surgical is one medtech company that has chosen this approach. MAKO pioneered the use of robotic-assisted surgery in orthopedics with the introduction of its RIO robotic arm. This device enables surgeons to cut through bone with great precision. MAKO employs a direct sales force to sell the RIO systems in the US, but enters into agreements with independent distributors to market, sell, and support its products in international markets. The company has two sales representatives dedicated to defining and executing its global commercialization strategy and maintaining close relationships with its independent distributors.[30]

AccessClosure similarly adopted a hybrid model, using its direct sales force to manage the commercialization of its vascular closure product family in the US and relying on distributors to commercialize the technologies in other geographies. Offering an example, President and CEO Greg Casciaro said, "Each of Europe's countries has its own unique challenges with different buying programs and pricing structures. You adopt a distributor model because you don't have the resources, or because you don't think the selling price warrants putting people on the ground." He continued, "For us, distributors are a more efficient way to get things done. But, you lose the attention and focus you get from a direct sales force. The distributor reps determine what they're going to talk about with a physician. Some might have four products in their bag and some might have 40, and you're competing for their mindshare."

An emerging variation on the hybrid model is for a company to work with a strong distributor in a particular geographic area to help get its product into key hospitals and/or clinics. The distributor reps also manage all logistics related to purchasing and refilling the product. However, to maintain some direct control over important relationships and help expand these accounts, the company will directly employ a small team of product specialists. The members of this team, who are expert on all aspects of the technology, can participate in meetings with significant influencers, deliver all training on behalf of the company, be available to users as questions or issues arise, and offer support to the distributor. In these types of arrangements, the margin given to the distributor is typically lower than if this support from the company was not available. NeoTract provides an example of a company using this approach to commercialize its Uro-Lift system in parts of Europe.

The growing importance of GPOs, IDNs, and ACOs

When it comes to medtech sales, GPOs, IDNs, and ACOs seek to leverage the purchasing power of a group of participants to negotiate for more favorable sales and distribution terms. All three types of organizations have formed in response to rising healthcare costs. GPOs, such as Novation and Premier, organize multiple hospital groups, large clinics, and medical practices into buying cooperatives. Large GPOs may represent as many as 1,000 to 2,000 hospitals.[31] As a result, they have a high degree of negotiating power with companies. Approximately 72–80 percent of all non-labor medical system expenditures go through a GPO. Although estimates vary, providers typically expect to save 10–18 percent on the cost of goods by purchasing through a GPO.[32] In return, GPOs earn a 2–3 percent contracting administration fee (**CAF**)[33] per contract. In the medical device field, this fee is typically paid for by the manufacturer. Many GPOs also provide value-added services to their members, such as product evaluation, training, service, and inventory management.

IDNs take the idea of consolidation one step further by aggregating hospitals, physicians, allied health professionals, clinics, outpatient facilities, home care providers, managed care, and suppliers into a single, closed network. Kaiser Permanente, Tenet, and Intermountain Health are examples of IDNs. IDNs, which include an average of eight hospitals each,[34] are becoming sufficiently large to gain buying clout and act as their own GPO. In addition to size, IDNs also gain negotiating power as they include several healthcare venues (hospital, nursing home, etc.) across the continuum of care, offering an attractive place for companies (or their distributors) to contract a full line of products. The integrated, single-ownership nature of IDNs greatly improves compliance with purchasing contracts, ensuring that the targeted sales volume is reached in exchange for the discount granted.

ACOs, which have emerged in the US in response to the Affordable Care Act (**ACA**), are *voluntary* consortiums of independent physician groups, hospitals, and insurers that act more like an IDN to share the responsibility for caring for a defined population of Medicare beneficiaries over a defined period of time.

Medtech companies may have the opportunity to negotiate exclusive or semi-exclusive contracts with GPOs, IDNs, or ACOs for lower-cost, high-volume products, giving greater pricing and delivery concessions in exchange for this preferred status. This is typically done through a distributor that has an existing relationship with the GPO or IDSN. Manufacturers of high-value devices sold through a direct sales force may or may not enter into contracts with these kinds of organizations.[35] However, healthcare providers sometimes organize themselves into purchasing groups for buying complex, high-value medical devices. Similar to GPOs, these groups take individual physician input into account, but also rely heavily on pooled criteria for making buying decisions that satisfy the interests of all participants, such as the ability of the company to serve as a sole-source supplier of a device, customer service capabilities, pricing, and discounting. Companies faced with the challenge of selling high-value devices to such purchasing groups typically must work hard to sell them on the value of the products. Similarly, they must continue to innovate and improve to retain the business once a buying decision has been made.[36]

The growing importance of the consumer

Another important factor that will affect medtech sales is the growing role of consumers in choosing and using medical technologies. In particular, consumers will be central to the proliferation of preventive care devices and home monitoring systems. In the US, growth in these product categories is being driven by healthcare reform and the move toward rewarding providers for the quality of the care they provide across the lifetime of a patient rather than the volume of service transactions they complete. As people live longer and the prevalence of chronic conditions increases, these factors will further contribute to growth in consumer medical technologies in all geographies. In the US alone, the consumer medical device market is expected to reach more than $10 billion by 2017.[37]

Innovators and companies can use two main approaches for selling consumer-oriented medical technologies. The first is to promote and sell products directly to consumers, with the expectation that they will pay for them out of pocket. The second is to target physicians, provider groups, and/or health plans for the sale, with the products being passed through to consumers. The model that a company chooses is best determined based on which audience is positioned to extract the most value from the offering. In both scenarios, the products must be designed with the consumer squarely in mind to promote an appropriate level of engagement (or compliance) to achieve desired outcomes.

NeilMed® provides an example of a product that is sold and distributed over-the-counter to consumers. This company is the largest manufacturer and supplier of large-volume, low-pressure saline nasal irrigation systems in the world.[38] The NeilMed product family, which includes its Sinus Rinse and a neti pot, helps alleviate common nasal and sinus symptoms. Initially, founders Ketan and Nina Mehta sold the products to regional pharmacists, who would recommend them to customers suffering from sinus-related maladies. They

also promoted them at allergy conferences and sent samples to allergists and ear, nose, and throat (ENT) doctors across the country to try stimulating adoption. But it was not until Dr. Mehmet Oz talked about the benefits of sinus irrigation on The Oprah Winfrey Show that sales took off.[39] Demand soared, with NeilMed selling tens of thousands of its neti pots in a matter of weeks. As Mrs. Mehta explained in an article, "We were well known among doctors and pharmacists before that show, but Oprah Winfrey introduced us to the masses."[40] NeilMed now sells its products throughout the US via supermarkets, drug store chains, and warehouse outlets like Wal-Mart and Costco, working with national distributors including McKesson, Cardinal, and AmerisourceBergen. Its products are also available through regional distributors in Canada, Europe, Australia, New Zealand, Malaysia, Singapore, and India.[41] Additionally, NeilMed conducts direct online sales through its website.

The case example on Propeller Health provides an example of a company that sells its consumer-oriented medical technology through healthcare providers and payers.

FROM THE FIELD ▶ PROPELLER HEALTH

Selling to payers, providers, and patients to improve asthma outcomes

When David Van Sickle was working at the US Centers for Disease Control and Prevention, where he conducted respiratory disease outbreak investigations, he became aware that available information about factors affecting asthma sufferers was not timely enough or geographically specific enough to support targeted interventions. Later, at the University of Wisconsin School of Medicine and Public Health, he explained, "I saw that same information gap plaguing the clinical care and treatment of asthma. Between visits, physicians weren't really aware of how their patients were doing, and we knew from the clinical literature that many weren't doing as well as they could be. The majority of

people with asthma were uncontrolled despite all we know about asthma and how to treat it. So health systems are spending a lot of time and money dealing with preventable exacerbations that result in unnecessary emergency room visits and hospitalizations."

In response to this need, Van Sickle developed a system that includes a sensor and a smartphone application. The sensor, which asthma patients attach to their inhalers, recognizes what type of medication is being used, as well as when and where symptoms are triggered (see Figure 5.8.7). The data is shared with the app, which tracks and triangulates medication use, location, and other information so that patients and their physicians can implement more targeted, personalized care paradigms. The app also prompts users with real

FIGURE 5.8.7

The Propeller Health system.

time tips that can help them more effectively control their disease between physician visits (paper reports are available for users without smartphones). The system received FDA 510(k) clearance as a **Class II** medical device in 2012.

In developing a sales and distribution strategy, the Propeller team initially decided to focus on health plans and healthcare providers as purchasers of the offering. "The primary customer is any organization that has economic risk for the outcomes related to poorly controlled asthma," Van Sickle said. Studies have demonstrated that patients with uncontrolled asthma incur an additional $3,000–$4,000 in healthcare expenses per year. "So there's an immediate opportunity to take those people and help them achieve better control of their disease and, in so doing, significantly reduce the amount of healthcare they consume," he continued. By keeping patients out of the emergency room and hospital, entities such as payers, ACOs, and integrated health systems can directly drive down costs. "Asthma is also widely used in quality metrics to reflect the overall quality of the care being provided by an institution," Van Sickle noted, which gives provider organizations another incentive to adopt the technology.

Expanding on Propeller's approach, Van Sickle was quick to point out, "We're an enterprise technology that

gets picked up by a plan or payer. There's no financial transaction that takes place with individual patients, but there's still a direct to consumer sale that has to happen. The patient has to use the technology and find value in it. If they don't like it or want to use it, we can sell all we want to the enterprise, but it won't make a difference."

To reach providers and health plans in the US (the company's first geographic market), Propeller Health uses a small direct sales force, as well as extensive inbound and outbound marketing. It also maintains an online presence and conducts web-based demos to help convince buyers of the product's benefits, using data from **pilot** studies and clinical trials to help make the case. "It's a consultative sale in many ways," Van Sickle elaborated. "We try to help people understand that the bills they're used to paying for asthma are actually quite susceptible to reduction." Each sale involves multiple stakeholders (that vary based on the type of organization being targeted) and can take anywhere from six to nine months. "In the most common scenario," he said, "a chief medical officer or chief information officer spearheads the investigation of the technology and the vetting of the evidence. But there are always physicians involved. And care management teams have to understand how the product will affect their work flow – they have to find value in it."

When an organization decides to adopt the technology, it subscribes on a per-user, per-month basis. Propeller works with the health plan or provider group to identify which patients will potentially benefit most from the offering. With payers, Propeller usually mails kits with its product directly to the target individuals to help them get enrolled. With providers, the company frequently provides the kits to the physicians and care management teams to distribute during office visits and other regular patient interactions.

At this point, the company's adoption challenge shifts from the institution to the patient. Propeller Health devotes significant time and energy to thinking strategically about how to get patients to use its technology. "In some ways, reaching patients through their insurance company or physicians can be harder than direct-to-consumer sales," Van Sickle commented. "If someone swipes a credit card, you know that they find some threshold of value in your product and have a certain likelihood of being invested in its use. But with our device, it's rolled out across a population, so there's always some uncertainty about its perceived value among patients."

The key, Van Sickle believes, is making it as easy as possible for patients to use the system. "The most important factor that gets patients to adopt the Propeller system is that it only requires passive participation," he said. All patients have to do to begin benefitting from the technology is to attach the sensor to their inhaler, pair it with their phone or base station, and set up a Propeller Health account. "Beyond that, there's really nothing else they have to do to allow us to collect a great amount of valuable information and use it to inform them about how they're doing and provide them with surprising, personalized guidance to help them achieve better control," he said. To make the set-up process simple, the Propeller Health kits are designed to look and feel like something patients would buy at a retail pharmacy or get at any consumer electronics store.

The reason that a passive approach appeals to customers is that they are used to being asked by their doctors to manually log medication use, symptoms, and other factors in daily asthma diaries. Many patients find these diaries burdensome, and they can even be counter-productive if patients forget to keep accurate records or ended up fabricating diary entries. Not only does the Propeller offering automatically fulfill this recordkeeping function, it provides patients with immediate, actionable feedback to improve their health between doctor visits. According to Van Sickle, a depth of knowledge exists for effectively managing asthma,

"But it's locked up in the clinical literature. Our focus is on taking **evidence-based** guidelines, breaking them up, and figuring out what's most relevant for a patient at any given time." This information is provided through the system on a just-in-time basis. "For most people, it's pretty surprising. They learn to accommodate their symptoms and don't realize that they're not doing as well as they should be," Van Sickle stated.

Summing up his approach to consumer adoption, Van Sickle said, "We're trying to help people better manage their asthma with less effort, not more." He continued, "It's really important to think about engagement as a means to an end rather than the objective. To me, it's not about increasing the percentage of time dedicated to the burden of asthma each day. It's about being able to actually delete that time from a patient's calendar."

When asked for advice about selling a consumer-oriented medical device into health plans and providers, Van Sickle cautioned innovators about indiscriminately agreeing to product trials. "Be careful not to become a pilot project where you're considered cute and innovative, but no one is really invested in bringing your technology into the day-to-day work flow. If the program isn't successful, the vendor often gets blamed when, in fact, it was a poorly planned implementation," he said. "You have to build consensus across the organization about how the technology is going to get used and where the budget sits, and you have to make sure that all the right people are invested. Sometimes this means insisting on a longer sales cycle, which feels so wrong when all you want to do is say 'Yes, yes, yes!'"

Propeller Health is growing rapidly, with plans to expand internationally with its asthma offering and launch a product focused on chronic obstructive pulmonary disease.

Determining the most appropriate sales and distribution model

As mentioned, companies must determine the most effective sales and distribution model for sustaining value based on their chosen business model; the complexity, value, and anticipated sales volume of the unique product offering; the customer being targeted; and the desired interaction between the customer, the company, and the device. In practice, however, there are other considerations that must be taken into account, including the size and financial resources of the company, degree of market complexity, and the extent of competition in the therapeutic area (see 6.1 Operating Plan and Financial Model for a more detailed discussion of the economics related to sales and distribution).

If a company lacks the resources and/or access to capital necessary to build a sales force, then it has no choice but to use a distributor or enter into a third-party partnership with another manufacturer, at least in the short term. When an indirect sales and distribution model is determined to be the best option, then the company should assess the characteristics of its product to determine which type of indirect approach makes the most sense. For low-complexity or undifferentiated products, distribution through large national distributors is typically the best strategy (either under a private label or the manufacturer's brand name). For a higher-complexity product with strong competitive attributes, the company can either sell through specialized distributors or via a partnership agreement with a larger, well-established manufacturer.

If the company can access the resources to build a sales force, then a direct sales and distribution model becomes a possibility. A direct sales force is appropriate for complex, high-value products that can command high profit margins, but that will require significant support and training in return. When limited funding for a sales force exists, a hybrid model can be an effective strategy for moving a company from an indirect to direct sales and distribution model as shown in Figure 5.8.8.

Although many medtech companies fit the common approaches described within this chapter, there are notable exceptions. Companies are continuously finding innovative new ways to approach sales and distribution that give their products an edge. They also carefully consider dynamics within their specific medical fields that may cause an alternate approach to be appropriate. For example, St. Francis Medical, a company that developed the X-STOP® technology to alleviate the symptoms of lumbar spinal stenosis, took an approach to distributing its device that might have appeared unusual in any other medical specialty (see 4.4 Business Models for more information about St. Francis Medical). Traditionally, as a

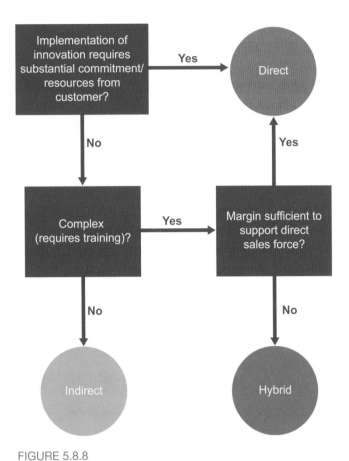

FIGURE 5.8.8
The innovator can follow this simplified decision tree to help determine the appropriate sales and distribution mode.

high-complexity, high-value implant, X-STOP would have been a good candidate for a direct sales and distribution model, especially since it would require a significant change in the established standard of care (shifting treatment of this condition from open surgery to a minimally invasive implant). However, the company instead decided to use an indirect model built around independent, specialty distributors in the spinal field. St. Francis anticipated physician adoption challenges based on the fact that the established surgical procedure was deeply entrenched in medical practice and well reimbursed. Spinal surgeons were also known for having strong relationships with the sales representatives of the major medical device companies. With hundreds of representatives on their sales teams and easy access to the most influential surgeons, companies like Medtronic, J&J, and Stryker had the potential to discredit St. Francis and its device by convincing them that the X-STOP was nothing more than a gimmick (despite its strong clinical data).

To proactively combat this possibility, St. Francis decided to tap into the highly experienced, well-connected network of independent specialty distributors that existed within the orthopedics field. These distributors, who worked as independent agents for both large and small medtech companies, were known for their strong relationships with spinal surgeons (and were potentially perceived to be more objective than the sales representative of the major device manufacturers). Kevin Sidow, former CEO of St. Francis, explained the advantages of this approach: "If you built a direct sales force and had an unknown representative with no relationships come in with some gimmicky product, the surgeon wouldn't see him for six months to a year. On the other hand, if a close friend of his, whose opinion is respected, walks in and says, 'Look, I know this thing looks too simple to be true, but just go to training with me, review the data, and let's take a look,' the physician is far more likely to try the product."[42] Through these independent specialty distributors, St. Francis' product gained relatively rapid adoption, earning close to $58 million in worldwide sales in its first year following FDA approval.

While this approach was successful for St. Francis, another orthopedic start-up, Kyphon flourished in the spinal market by building a direct sales force to market its implants to treat vertebral body compression fractures, as the case example on this company describes. This apparent contradiction underscores the fact that there is no "right" answer when it comes to deciding on a sales and distribution model.

FROM THE FIELD ➤ KYPHON, INC.

Investing in a high-touch direct sales model

When asked at what point in the biodesign innovation process she began to consider the sales and distribution model that would be most appropriate for Kyphon,® Karen Talmadge, company co-founder, director, and executive, said, "I began thinking about sales the second I heard about the product." She continued, "Once I met Mark Reiley, who conceived of the operation that became kyphoplasty, and went into the literature to understand the clinical need, I immediately had to think through every aspect of the business to see if we could build a company. That included the obvious things, like research and regulatory issues, but also how we would get this to the patient."

Talmadge, Reiley, and Arie Scholten, a custom medical device developer, co-founded Kyphon in 1994 to *correct the deformity and stabilize the spine in patients with a vertebral body compression fracture* (VCF). These fractures occur in the large blocks of bone in the front of the spine and most often are found in elderly patients whose bones have been weakened by osteoporosis or cancer. Within the US osteoporosis patient population alone, VCFs affect 750,000 people a year.[43] Treatment for this painful condition traditionally involves bed rest, over-the-counter pain medication, anti-inflammatory drugs, and back bracing.

In the mid-1980s, a surgical procedure was introduced in France, called vertebroplasty, which used X-ray guidance and spinal needles to inject specially formulated acrylic bone cement into the collapsed vertebral body.[44] The cement was intended to strengthen and stabilize the fracture, although the vertebral deformity was not repaired. This procedure, generally performed by radiologists, began to be used widely in the United States in the late 1990s.

The idea of kyphoplasty also began in the mid-1980s, when Mark Reiley, an orthopedic surgeon, thought of performing what became vertebroplasty, but rejected it due to concern about bone cement leaks. He set out to create a cavity in the bone, so the bone cement could be placed inside the bone under low pressure and fine control, with minimal invasiveness. Kyphoplasty was born when Arie Scholten proposed creating the cavity by adapting angioplasty balloon technology to function inside bone, and Reiley instantly recognized that this

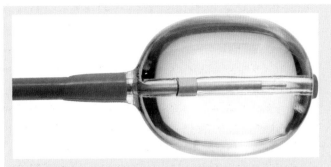

FIGURE 5.8.9
The KyphX Xpander® used to perform balloon kyphoplasty
(courtesy of Medtronic Spinal & Biologics).

would provide the additional benefit of restoring some or all of the lost vertebral anatomy. The surgeon accesses the spine with two small cannulae and delivers orthopedic balloons to the center of the collapsed bone. The balloons are then inflated to compress the inner bone and push the outer bones apart to help restore the patient's spinal anatomy (see Figure 5.8.9). The balloons are removed and the filler (e.g., bone cement) is inserted to stabilize the vertebra and help alleviate the patient's symptoms.[45] Talmadge explained, "Beyond the fracture pain, spinal deformity is a profound problem in these patients. Studies link the vertebral deformity, independent of fracture pain, to loss of lung function, digestive problems, changes in gait that decrease independence and increase the risk of falls, loss of **quality of life**, increased future fracture risk, and increased risk of death. That is why we focus on the deformity along with the pain."[46]

Talmadge, who was trained as a scientist, had more recently spent many years in the biotechnology sector working on business issues before joining the Kyphon team. One such role, in business development for a company called Scios, was instrumental in heightening her appreciation of the importance of sales, as well as aiding her understanding of the orthopedic market. As part of a Scios product development committee, Talmadge attended a meeting of the American Academy of Orthopedic Surgeons (AAOS).

"I went to some of the lectures to understand the clinical landscape, but I also spent a lot of time on the exhibit floor, going from booth to booth and asking people 'How do you sell your products?'" Talmadge recalled. "For the most part, people were unbelievably helpful. They took me under their wing for 10 minutes and said, 'Well, here's how we've been marketing our products and here are some of the issues.' They described the pros and cons of the distributor networks and/or agents used by many orthopedic companies, noting that large companies had at least some direct sales people, requiring more expense and investment upfront, but enabling more attention and focus downstream."

Through her science training and her introduction to the orthopedics field, Talmadge said, "I was fortuitously very prepared to understand the Kyphon concept, as well as the business issues around it." As such, she recognized, "For the product to succeed, we had to change the practice of medicine. Primary care physicians would have to refer patients for an operation they had never had up to that point. In order to do that, we had to do everything right. And getting everything right meant that the basic science, clinical studies, and professional education had to be rigorous." For this, the company worked with a small number of academic spine surgeons and experts in osteoporosis to be sure they were developing "all of the information that the physicians needed, as well as all the information that the future sales people would need," said Talmadge, "including the story of osteoporosis and the fractures, the anatomy, the technical aspects of the procedure, and the appropriate mechanical and clinical studies to document the outcomes. Then, the academic spine surgeons became our expert faculty, along with Mark Reiley, who had conceived of the operation."

But also, it was clear to Talmadge from the very beginning, "Based on everything I had learned about orthopedics, we would need to have a direct sales force." Among other benefits, direct sales representatives would have a vested interest in helping to ensure that the product was used correctly in the field, to help prevent the consequences of product misuse, which could be devastating to spine patients. "Every decision we made as we got the company off the ground was informed by the fact that we were creating a new spine procedure and that we had a responsibility to patients to be careful," Talmadge noted.

To build its direct sales force, the Kyphon team started by hiring a few, select people with a combination of sales and professional education experience. This technical team of professional education managers began supporting the cases that were being performed as part of Kyphon's post-clearance clinical trial in early 1999 (the company's inflatable bone tamp technology was cleared for commercial use by the FDA under a 510(k) in 1998, and it launched a **randomized controlled clinical study** shortly thereafter to collect the clinical data necessary to support product adoption).

In an effort to be certain that only trained surgeons were using the products, and to help prevent the negative outcomes associated with product misuse, Kyphon requested that one of its technical experts be present for every kyphoplasty procedure that was performed, which, according to Talmadge, is "typical for the medical device industry." Other steps were unusual, she noted. The company restricted product use to surgeons participating in **post-marketing trials**, and they did not sell the product to hospitals for inventory. "We did not sell products directly to the hospitals for about two years," Talmadge explained. Instead, the technical expert would bring the product with him/her to the procedure "and a purchase order would be created on the spot in the operating room," she said. As the technical experts observed these procedures, they "provided a resource to the surgeons for information about the product and its use. They also took notes on technical information about the product, such as how the balloons functioned at different pressures and volumes," data that Kyphon then used to refine its instruments and improve its education materials.

"Once we had built a core of knowledge, clinical outcomes data had been collected, and the randomized, controlled clinical trial comparing kyphoplasty to non-surgical management alone was underway, we were ready to develop a direct sales force," Talmadge

recalled. In late 1999, the company hired a sales leader and began recruiting representatives. However, "The role of our sales force was different than a typical sales force because this was still in the pre-inventory days. We needed people who understood that this was market development, in addition to sales. Their role would be to help the physician educate local referring physicians about new treatment options, as well as to support the physician technically."

Talmadge described Kyphon's referral-based approach to market development in a field where most VCFs were diagnosed by primary care physicians: "The sales people would ask the surgeons, 'Who refers to you?' We also got market data on physicians in each area with elderly patients. And then we went and knocked on their doors and asked, 'Would you be interested in hearing about new treatment options for vertebral body compression fractures?' So, the sales people facilitated meetings where the surgeons who were doing the procedure would come and talk to a group of physicians in a lunch or dinner setting. And they would talk to their colleagues about what they were doing, the outcomes they were having, and the clinical outcomes from studies as they became available."

Kyphon expected its sales people to work as a team, which Talmadge explained was "also unusual." This team-based approach was driven, in large part, by the company's ongoing commitment to having a Kyphon representative present at every procedure across its 20 sales territories, if this was requested by the surgeon. "If two cases were going on in the same territory on the same day and one person couldn't cover them both, we flew someone from another territory to bring the product in for the case and observe the procedure," said Talmadge.

In terms of selecting representatives, she recalled, "The sales people that we hired didn't have to come out of spine. In fact, at times we felt that coming from a conventional spine hardware company might be a disadvantage because spine hadn't had many products that really involved a change in clinical practice."

According to Talmadge, many of its first representatives came out of major medical device companies, such as Johnson & Johnson and Ethicon, "where the level of sales training is very, very high, and where they had experience selling devices for new procedures, such as the various laparoscopic techniques of the prior decade." Great emphasis was placed on finding individuals with the right traits. "We went for a series of characteristics that were most important. Intelligence, passion, commitment, and the highest ethical standards," she said. At the same time, representatives had to know how to sell in a hospital setting, including the approval process for buying new devices.

In 2000, the early results of clinical outcomes studies documented the safety and effectiveness of kyphoplasty, while the randomized clinical trial became impossible to enroll. Patients did not want to participate in a study where they might not receive the operation (i.e., in the control group). At the same time, surgeons were frustrated with the policy that product use had to be restricted to clinical study sites. Recognizing this, the company decided to move the randomized clinical trial to Europe and begin a full market launch in the US based on the rest of the clinical data. To support this shift, the company expanded the sales force to 20 representatives. What Kyphon found was that its sales grew proportionately with the number of sales people it

added. As a result, it continued to increase the size of its sales force in advance of sales, along with its continuing investment in clinical studies. As surgeons began to gain more experience with kyphoplasty, the company became more comfortable with procedures being performed without a Kyphon representative present. However, even after this change, physicians continued to demand a high level of professional support from the company. "They love it," said Talmadge. "In fact, for us, it's a little frustrating. You would think after their hundredth kyphoplasty, they wouldn't really want us to be present, but they do. They really want it. Physicians love the technical expertise that the sales person brings. Because the physicians may have done 100 kyphoplasties, but that sales person might have seen 1,000 kyphoplasties. So there's a lot of knowledge in the room."

While Kyphon's high-touch, direct sales model was instrumental to the company's success, Talmadge acknowledged that it was costly and resource intensive. "You have to decide early on that your market is large enough," she said, "because obviously this takes a significant investment."

For Kyphon's shareholders, the investment yielded a tremendous return. In 2007, Medtronic acquired the company in a transaction with a total value of nearly $4 billion.[47]

Online Resources

Visit www.ebiodesign.org/5.8 for more content, including:

 Activities and links for "Getting Started"
- Evaluate the impact of the business model on options for reaching the customer
- Assess the impact of intermediaries on sales and distribution
- Choose a sales and distribution model
- Coordinate marketing, training, and support activities

 Videos on sales and distribution strategy

 An appendix that summarizes indirect sales and distribution models

CREDITS

The editors would like to acknowledge John Adler of Accuray, Greg Casciaro of AccessClosure, V. Sashi Kumar of Phoenix Medical Systems, Karen Talmadge of Kyphon, and David Van Sickle of Propeller Health, as well as Ritu Kamal and Stacey McCutcheon, for their assistance with the case examples. Nick Gaigh provided helpful insights into hospital purchasing decisions and GPOs, and Bill Facteau added great value by reviewing the chapter. Sarah Garner and Jared Goor also made important contributions to the original material.

NOTES

1 From remarks made by John Abele as part of the "From the Innovator's Workbench" speaker series hosted by Stanford's Program in Biodesign, February 2, 2004, http://biodesign.stanford.edu/bdn/networking/pastinnovators.jsp (October 9, 2013). Reprinted with permission.

2 Atul Gawande, "Big Med," The New Yorker, August 13, 2013, http://www.newyorker.com/reporting/2012/08/13/120813fa_fact_gawande (December 3, 2013).

3 "The Relationship of Hospitals and Physicians," New Jersey Commission on Rationalizing Health Care Resources, Final Report, Chapter 8, 2008, http://www.nj.gov/health/rhc/finalreport/documents/chapter_8.pdf (December 3, 2013).

4 Ibid.

5 Carol Kane and David W. Emmons, "New Data On Physician Practice Arrangements: Private Practice Remains Strong Despite Shifts Toward Hospital Employment," American Medical Association, September 4, 2013, http://www.ama-assn.org/resources/doc/health-policy/prp-physician-practice-arrangements.pdf (December 3, 2013).

6 Robert Kocher and Nikhil R. Sahni, "Hospitals' Race to Employ Physicians – The Logic Behind a Money-Losing Proposition," The New England Journal of Medicine, May 12, 2011, http://www.nejm.org/doi/full/10.1056/NEJMp1101959 (December 5, 2013).

7 "Accountable Care Organization 2013 Program Analysis," Quality Measurement & Health Assessment Group, Centers for Medicare & Medicaid Services, December 21, 2012, http://www.cms.gov/Medicare/Medicare-Fee-for-Service-Payment/sharedsavingsprogram/Downloads/ACO-NarrativeMeasures-Specs.pdf (December 5, 2013).

8 Mike Miller, "Evolution of Sales Models in the Medical Device Industry," Alexander Group, 2012, http://www.alexandergroup.com/sites/default/files/images/documents/AGI%20Whitepaper_EvolutionofSalesModelsintheMedicalDeviceIndustry_Final(4).pdf (December 5, 2013).

9 From remarks made by Alex Gorsky as part of the "From the Innovator's Workbench" speaker series hosted by Stanford's Program in Biodesign, March 2012, http://biodesign.stanford.edu/bdn/networking/pastinnovators.jsp (December 5, 2013). Reprinted with permission.

10 The terms wholesaler and distributor are often used interchangeably, although there is a difference between the two: wholesalers sell to another intermediary (such as a pharmacy or a specialty distributor) while distributors sell directly to the end buyer. For simplicity, the focus of this chapter is on distributors since wholesalers are less frequently used in the medtech field.

11 Invivo Surgical System, http://www.invivosurgical.com/product_lines2.htm (April 1, 2008).

12 Healthpoint Capital, "Devices, Distribution and Dollars: Orthopedic Sales Backgrounder and Distribution Model Comparison," 2005, p. 8.

13 Ibid., p. 13.

14 "Medtronic Announces Alliance with LifeScan to Bring Leading Blood Glucose Meter to Its Diabetes Patients in United States," Medtronic press release, August 21, 2007, http://www.businesswire.com/news/home/20070821006155/en/Medtronic-Announces-Alliance-LifeScan-Bring-Leading-Blood (December 13, 2013).

15 Barnaby J. Feder, "Boston Scientific Settles With an Ex-Partner," The New York Times, September 22, 2005, http://www.nytimes.com/2005/09/22/business/22stent.html?_r=0 (December 13, 2013).

16 "The Case of the Medical Stent," Harvard Law School, Problem Solving Workshop, 2009, http://weblaw.usc.edu/assets/docs/contribute/MedicalStent.pdf (April 22, 2014).

17 Barnaby J. Feder, "Boston Scientific Settles with an Ex-Partner," The New York Times, September 22, 2005, http://www.nytimes.com/2005/09/22/business/22stent.html?_r=0 (December 13, 2013).

18 From Lyn Denend and Julie Manriquez with Christine Kurihara, Anurag Mairal, and Stefanos Zenios, "When Partnership Sales Fall Short of Expectations," Global Health Innovation Insight Series, July 2012, http://csi.gsb.stanford.edu/sites/csi.gsb.stanford.edu/files/PhoenixII-PartnershipSales.pdf (December 5, 2013).

19 All quotations are from interviews conducted by the authors, unless otherwise cited. Reprinted with permission.

20 "Toolkit for Physician Financial Transparency Reports (Sunshine Act)," American Medical Association, http://www.ama-assn.org/ama/pub/advocacy/topics/sunshine-act-and-

physician-financial-transparency-reports.page (October 9, 2013).

21 Stephen Levin, "AccessClosure: FDA Clearance – Yes; Product Launch – Not Yet," *IN VIVO*, June 2009, http://inceptllc.com/website/wp-content/uploads/2011/12/4_Article-2009800119.pdf (December 10, 2013).

22 Ibid.

23 Ibid.

24 "Products," AccessClosure, http://www.accessclosure.com/products/ (December 10, 2013).

25 Levin, op. cit.

26 Ibid.

27 Estimate from Greg Casciaro.

28 "Cardinal Health Expands Portfolio of Physician Preference Items," Fierce Medical Devices, April 2, 2014, http://www.fiercemedicaldevices.com/press-releases/cardinal-health-expands-portfolio-physician-preference-items-company-acquir (April 21, 2014).

29 Healthpoint Capital, op. cit., p. 9.

30 "Annual Report," MAKO Surgical, 2012, http://files.shareholder.com/downloads/MAKO/2828377767x0x658217/5F390A66-ED2A-46D3-AA22-3A6AD686F47C/131165_Mako_Annual_Report_Web.pdf (December 6, 2013).

31 L. R. Burns et al., *The Health Care Value Chain: Producers, Purchasers and Providers* (Jossey-Bass, 2002), p. 74.

32 David E. Goldenberg and Roland King, "A 2008 Update of Cost Savings and a Marketplace Analysis of the Health Care Group Purchasing Industry," Locus Systems, Inc., July 2009, http://c.ymcdn.com/sites/www.supplychainassociation.org/resource/resmgr/research/goldenberg_king.pdf (October 9, 2013).

33 Burns, op. cit., p. 48.

34 Ibid., p. 67.

35 Burns, op. cit., p. 43.

36 Richard Cohen, "The Narrowing Distribution Funnel: How to Get Your Medical Device to Market," *Medical Device Link*, February 1999, http://www.devicelink.com/mddi/archive/99/02/003.html (December 13, 2013).

37 Bruce Japsen, "New Study Says Obamacare Will Boost Consumer Medical Device Market to $10 Billion," *Forbes*, September 13, 2013, http://www.forbes.com/sites/brucejapsen/2013/09/13/obamacare-will-boost-consumer-medical-device-market-to-10-billion/ (December 15, 2013).

38 "About Us," NeilMed, http://www.neilmed.com/usa/about.php (December 15, 2013).

39 Matt Villano, "Sinus Sufferer Turns Nasal Spray Project Into Sales Leader," *The New York Times*, November 12, 2008, http://www.nytimes.com/2008/11/13/business/smallbusiness/13wash.html?pagewanted=all (December 15, 2013).

40 Ibid.

41 "About Us," op. cit.

42 From remarks made in an interview with the authors in Fall 2007. Reprinted with permission.

43 "Vertebral Compression Fractures," American Association of Neurological Surgeons, March 2007, http://www.aans.org/Patient%20Information/Conditions%20and%20Treatments/Vertebral%20Compression%20Fractures.aspx (December 13, 2013).

44 Ibid.

45 Steve Halasey, "Growing Up, Globally," November/December 2006, *Medical Device Link*, http://www.devicelink.com/mx/archive/06/11/cover.html (December 13, 2013).

46 Ibid.

47 "Medtronic Buys Kyphon for $3.9 Billion," *Forbes*, July 27, 2013, http://www.forbes.com/2007/07/27/medtronic-kyphon-kyphoplasty-biz-sci-cx_mh_0727kyphon.html (December 13, 2013).

INTRODUCTION

As overwhelming as it may sound, figuring out an approach to intellectual property, R&D, quality, clinical trials, regulatory reimbursement, marketing, and sales is not enough. An innovator must also think about how all of these factors come together to create a compelling competitive advantage and, ultimately, a business strategy. Without an explicit point of view on how a company and its technology will be differentiated from the competition in the complex and dynamic medtech field, a product may never achieve or sustain its market potential.

Fundamentally, a company has a competitive advantage if its competitors cannot replicate its offering in some material way. The value of a competitive advantage is that it can prevent new competitors from entering a market based on the strength of the company's position and the barriers to entry that it has created through the development of key capabilities. It can also dilute efforts by established players to imitate what the company does well. Defining a competitive advantage is highly dependent on external factors within the market (e.g., intellectual property (IP), regulatory, customers, and competitive dynamics), as well as factors internal to the firm (e.g., strengths, weaknesses, and organizational issues). Once defined, the company's business strategy should be built around the optimization of its competitive advantage.

 See ebiodesign.org for featured videos on competitive advantage and business strategy.

OBJECTIVES

- Understand how to apply a fundamental framework for defining a competitive advantage.

- Appreciate how to develop business strategies designed to capitalize on that advantage.

COMPETITIVE ADVANTAGE AND BUSINESS STRATEGY FUNDAMENTALS

A company's competitive advantage is something special about the company that differentiates it from its competitors, creates **value** for its **stakeholders**, and prevents competitors from capturing the value it creates. By definition, a competitive advantage must be unique and/or difficult for others to replicate. No two companies will have exactly the same competitive advantage. In other words, there is no universal source of competitive advantage, even for similar companies in the same industry.[1] A **value proposition** is a description of the improvements a customer will realize (at a defined cost) in terms that are meaningful to that customer (see 5.7 Marketing and Stakeholder Strategy). A competitive advantage, combined with an appropriate business strategy, can help a company to operationalize and deliver on its customer value propositions. There are two primary types of competitive advantages: positional and those based on capabilities.[2]

Positional advantages

Positional advantages come from a company's ability to strategically position itself relative to its competition and stakeholders. Positional advantages are, in general, quantifiable or measurable. A company enjoys a positional competitive advantage if it possesses an asset which is not easily obtainable, such as being first to market, obtaining key **patents**, or having the most widely recognized brand. For instance, companies that have direct access to the customer, through a strong direct sales forces or distributor network, achieve a positional competitive advantage by controlling key customer relationships (e.g., surgeons in the orthopedic market). A company that does not possess those relationships is at a distinct disadvantage when it tries to break into the market. Another way to achieve a positional advantage is to build a strong portfolio of products that offer comprehensive solutions that competitors cannot easily imitate due to technology limitations or other resource requirements. Broad IP portfolios can also be a source of positional competitive advantage. For example, J&J Interventional Systems achieved a strong positional advantage via its powerful Palmaz-Schatz patent portfolio surrounding its first coronary stent. Kyphon® achieved a similar advantage with its intravertebral balloon, as did ArthroCare with its coblation® technology.

Capability-based advantages

Capability-based advantages are built on a company's know-how, or competencies, as well as its ability to leverage its capabilities to become better than its competitors in key areas. Capability-based advantages

sometimes are harder to quantify, and they usually reflect value that is embedded within teams or organizations. For example, a medical device company that develops an exceedingly capable IP team that is able to continually evolve and improve the company's patent portfolio enjoys a competitive advantage based on this capability. Another real-world example is SciMed Lifesystems, which established a capability-based advantage by becoming an expert in rapidly evolving processes to produce catheter and stent products with improved performance characteristics. US Surgical also established itself as a leader in minimally invasive surgical devices by developing a capability in advanced low-cost manufacturing techniques and rapid product iteration.

The interplay between positional and capability-based advantages

In general, positional competitive advantages are driven by a company's relationship to its *external* context. They are based on factors such as timing, size, location, specific assets, and access to important resources. These positional advantages are all relative to the other stakeholders in a field (customers, suppliers, or actual and potential competitors) and are continually changing. In order to understand the external landscape and establish a positional advantage, innovators must revisit the competitive analysis they conducted earlier in the biodesign innovation process (see 2.4 Market Analysis). It is a mistake to assume that the market situation remains static during the many years required to develop and introduce a new solution. Importantly, when performing competitive analysis at this stage of the process, innovators can get much more focused and specific in their assessment since they are now dealing with a **need** and a solution rather than just a need.

Once defined, positional advantages do not necessarily transfer effectively from one field to another. For example, Medtronic is the leader in cardiovascular solutions that utilize active implantable devices (e.g., pacemakers) and enjoys a strong positional advantage in terms of brand name and relationships in this field. However, the company has had a more difficult time creating a durable positional advantage in the field of catheter-based cardiovascular therapies.

In contrast, capability-based advantages are driven by factors *internal* to a company. They are based on a company's ability to do something better or less expensively than its competition and/or customers. These capabilities can be specific to a single process or generalizable across multiple processes. Capability-based advantages also may be transferable from one field to another. For instance, Pfizer's capability of effectively marketing and distributing drugs – frequently enabling the company to achieve the highest sales in a category within weeks or months of obtaining product approvals or acquiring those products – demonstrates how a capability can transfer from one medical field to another.

Sometimes the distinction between a positional advantage and capability-based advantage can be somewhat unclear. For instance, in the case of Pfizer, the company achieves a competitive advantage not only through its marketing capabilities, but also through the position it has achieved in terms of its relationships with physicians. Its broad product portfolio also makes it more appealing for physicians to spend time with Pfizer's sales representatives because they are able to address more products. Similarly, Medtronic has a positional advantage in implantable cardiovascular solutions, but has developed a unique capability for continuously developing complex, closed-loop systems that others cannot easily imitate. Innovators must recognize that multiple strengths often work in combination to create a company's advantage.

The link between competitive advantage and value propositions

Any number of strategies and competitive advantages can be combined to support a company's value proposition(s). For instance, if the value proposition of a company is to continually provide its customer with a constant flow of cutting-edge innovations, it might attempt not only to be first to market with any innovation (a **first mover** strategy), but also support that positional advantage with a strong, constantly evolving IP development capability. Another company seeking to provide its customers with the value proposition of an improved **user experience** might alternatively choose to follow quickly behind the market leader, allowing it to learn from the mistakes and usability issues encountered by the first mover. Alternatively, as a low-cost and low-price leader, a company may choose to offer a "best value" proposition which would de-emphasize market-expanding innovations in favor of a pricing advantage.

Value propositions must always be measured through the perspective of the customer. That is, they represent how a product, service, or company is viewed from the outside. The positional and capability-based advantages that are required to create and sustain a value proposition are what drives the creation of a business strategy for a company. For each value proposition a company defines for its stakeholders, it should carefully consider what strategies and potential advantages will enable it to successfully deliver on that vision and defend itself against imitation from its competitors.

Types of positional advantages in medtech

In the medical technology field, many innovators and young companies initially seek to develop core assets that give them a positional advantage. Another way to think about a positional advantage is as an initial barrier to entry. There are nine main sources of positional advantage.

1. Intellectual property

Establishing an initial patent portfolio (e.g., through a picket-fence strategy – see 5.1 IP Strategy) that is legally defensible can limit other companies' freedom to operate in a field and define a broad positional advantage for the company. J&J and Medtronic are well known for establishing broad, highly effective patent portfolios. Intuitive Surgical, Kyphon, and Acclarent also did this effectively during the early stages of building their companies.

2. Key relationships

The strength of relationships with important stakeholders, particularly physician key opinion leaders (**KOL**s) and powerful facility decision makers, can help a company more easily capture customers. For example, direct sales representatives in orthopedics and pacemakers wield substantial power on behalf of their companies due to the strong relationships they hold with these influential stakeholders. As technologies continue to evolve in the **medtech** space, key relationships with patients (as consumers of some medical products and their related service offerings) can also be a source of positional advantage. Relationships with a wide variety of stakeholders can also be helpful in other aspects of the biodesign innovation process, such as efficiently gaining regulatory approval.

3. Reimbursement coverage

Companies who are first movers into a treatment area can sometimes gain a positional advantage by establishing **reimbursement codes** and **coverage** that favor their unique technologies if they operate in a geography where **payments** from **third-party payers** dominate the healthcare system. Of course, they often must invest significant time and resources into getting reimbursement set up. But once they achieve this goal, it is often difficult for other companies to leverage the same codes and/or justify the creation of additional codes in the same treatment area.

4. Strategic alliances

Partnerships with larger corporations and institutions can allow a new or smaller company to "borrow" reputation and keep other players from accessing the same limited resource (in the case of an exclusive alliance). Angiotech's relationship with Boston Scientific relating to drug-eluting stent technologies provides an excellent example of an effective medtech partnership. Recently, strategic alliances have become increasingly interesting as companies look for innovative ways to not only extend their capabilities but to share risks with other participants in the medtech value chain.

5. Geographic coverage/distribution channels

The ability to gain access to preferred suppliers for distributing products and/or providers for disseminating services to customers in key markets can lead to lower transaction costs and better terms than competitors can achieve. It can also be exceptionally difficult to replicate, as was the case for Stryker when it established strong, early distributor relationships outside the US, most of which continue to hold to this day.

6. Brand

Premium brands often achieve premium pricing. Branding can also increase product awareness, which in turn can lead to higher sales. Companies that establish strong brand awareness within their customer base are at a significant positional advantage with respect to incoming competition. For instance, surgeons are often reluctant to switch brands, especially when the products are life-saving and highly successful, such as the heart valves sold by Edwards Lifsciences or the hip and knee implants sold by Zimmer. As patients continue to amass more power in healthcare, innovators can expect issues of consumer brand awareness to become increasingly important. This is already happening in areas such as diabetes management, where consumers make brand-related choices about which blood glucose testing system to choose.

7. Financial access

Capital is a finite resource. Locking in top-tier venture capitalists, banks, and other funders can keep competitors from accessing the same capital resources, force them to deal with suboptimal terms, and/or leave them with less desirable alternatives. Companies such as Satiety, GI Dynamics, and Endogastric Solutions, which were early pioneers in the area of obesity, captured many of the desired medical device early-stage venture backers, making it much more difficult for later entrants to access capital.

8. Unique, exclusive, or low-cost access to a key resource

Key resources can include hard-to-find materials (such as nitinol, proprietary gene sequencing databases, or limited tissue samples) or exclusive relationships with technology providers in specific fields (such as slippery coatings for catheters, polymer drug-elution platforms, or silicone-polyethylene co-polymer blends). A company that can gain access to key resources exclusively, more easily, or at a lower cost has a positional advantage over its competition.

9. Information access

As healthcare delivery becomes more connected across the **cycle of care** and physicians, facilities, patients,

payers, and other stakeholders continue to expand their efforts to work together more closely and proactively, the company that has the best, most actionable information about key health-related interactions will create a strong positional advantage. Value can be generated by making this data available to health plans and providers to help them improve outcomes and reduce costs. It can also be generated by analyzing the information to understand customer behaviors and identify future customer needs. Medtronic, with its **acquisition** of Cardiocom, provides one example of a company seeking to build a competitive advantage based, in part, on access to information that can be used to optimize patients' experiences across the treatment continuum (see case example in 4.4 Business Models).

Almost every successful medical technology company has, at least at some point in time, developed one or more positional advantages to fend off competition. For example, Fox Hollow, a company focused on the treatment of peripheral arterial disease (which was acquired by ev3), created an almost impenetrable barrier to competition by effectively developing strategic relationships with most, if not all, of the KOLs in the field. The use of key physician relationships has similarly been a fundamental basis for creating an advantage in the orthopedics industry (although this practice has fallen out of favor to some extent following concerns about the potential for these relationships to create conflicts of interest). Strategic corporate relationships offer another common approach towards achieving a positional advantage, such as J&J and Guidant's accord related to balloon catheters and Rapid Exchange (RX)™ technology, which permitted the two companies to cease several lawsuits and refocus efforts into more productive activities.

As noted, positional competitive advantages can be relatively perishable. Just as an army at the top of a hill enjoys a distinct positional advantage in attacking an opposing force in the valley below, that advantage deteriorates as soon as the troops charge down the hill and find themselves on a level playing field with the opposition. In the medical device field, companies that enter the market with an advantageous position must continually seek to improve and/or defend their position to sustain the advantage. For instance, a company that successfully establishes meaningful relationships with key opinion leaders must monitor and invest in those relationships on an ongoing basis or else a competitor may edge in and gradually appropriate them. Similarly, competitors will find ways to design around a company's initial IP position if the market is sufficiently compelling. And, with enough time and money, even new sales and distribution channels can be built to rival an established infrastructure.

Types of capability-based advantages in medtech

Innovators and young companies can also create a competitive advantage by developing a product or delivering a service that is better, cheaper, or more efficient than what is offered by the competition. While doing this creates a competitive advantage in its own right, capability-based advantages are often used to help sustain positional advantages. That is, a positional competitive advantage can be maintained by transforming it into a capability-based advantage and/or can be defended through the development of a strong capability in that area. In most cases, positional advantages can become capabilities if a company can learn to consistently perform better in the given area than the competition. The creation of specific systems and processes focused in the particular area can be the first step in developing a unique capability. Six valuable medtech capabilities are outlined below.

1. Intellectual property management

As noted, if a company seeks to recruit, develop, and retain a world-class IP team, it can transform a strong IP position into a best-in-class IP capability. This is accomplished through systems and processes designed to continually and aggressively monitor the IP landscape, anticipate competitive moves, and act more quickly than the competition to advance, strengthen, and defend the company's IP portfolio.

2. Regulatory, quality, or reimbursement management

Similarly, companies can build a capability-based advantage in regulatory, quality, or reimbursement management by hiring and cultivating talent and building and managing world-class processes in these areas. With capabilities in these areas, companies can more efficiently satisfy essential requirements and more effectively anticipate and mitigate related risks.

3. Relationship management

To create a capability-based advantage focused on key relationships, it is not enough to manage and protect the company's position with established thought leaders. Systems and process must be created for (and resources must be dedicated to) identifying subsequent generations of potential opinion leaders and developing relationships with them before the competition.

4. Alliance management

A company can develop a unique capability when it comes to partnering if it has the right processes in place and resources on board to think about alliances in innovative new ways. Likewise, it can continually monitor the external landscape for mutually beneficial partnerships, and negotiate more exclusive (or otherwise more favorable) terms than the competition.

5. Human resource management

Human resources (HR) management refers to the creation of an optimal team of individuals with the essential competencies necessary to succeed in the market (e.g., technical, clinical, reimbursement). This is a capability-based advantage when it is achieved on an ongoing basis, despite the natural level of turnover inherent in any industry.

6. General management

Simply having good leadership and well-trained managers in key positions within a company could be a strong capability-based advantage. Installing management training and feedback programs can enhance and sustain this advantage.

Capability-based advantages can also be created in other areas where a company has unique process expertise, whether or not it is related to a positional advantage. For example, medtech companies may achieve capability-based advantages related to the following factors:

- **Time to market** – Having the people, processes, and systems in place to consistently get to market sooner than the competition.
- **R&D productivity** – Systematically generating more or better innovations than others in the field.
- **Low-cost manufacturing** – Achieving low-cost manufacturing through the acquisition of low-cost inputs is a positional advantage while achieving it through more efficient processes is a capability-based advantage.
- **Accessing and developing global markets** – Having the talent and experience within an organization to establish and manage global R&D, manufacturing, and/or sales and distribution is becoming an increasingly valuable capability.

Innovators should remember that even the most impressive capabilities are not a source of competitive advantage if one or more competing companies can match them. Capabilities can only become a source of sustainable competitive advantage if they are hard to imitate or the company can continuously improve upon them before others can catch up. When a competitive advantage resists competition, it is said to be sustainable.[3]

Often, capability-based competitive advantages are widely understood in an industry, but still difficult to replicate because no one is certain what causes them. Complex routines, structures, and individual attributes within an organization, combined with a high level of tacit knowledge, can make them difficult for competitors to imitate.[4] For example, Ethicon was widely considered to be the worldwide leader in surgical closure technology and, in particular, suture technology. However, in the 1970s, US Surgical began making inroads into Ethicon's markets with the introduction of its surgical staple technology as a substitute for traditional sutures in minimally invasive laparoscopic procedures. Buoyed by its success, US Surgical decided to enter the suture market and directly compete with Ethicon in its core market by introducing its own line of sutures in 1991. However, US Surgical dramatically underestimated the strength of Ethicon's suture manufacturing capabilities. The company did not realize how challenging it was to produce a high-quality suture that did not break, stayed attached to the suture needle, and performed consistently, package after package. In the end, the product released by the company failed to meet physician quality expectations and, as a result, never claimed the market share that US Surgical expected. Ethicon's proprietary know-how (i.e., trade secrets) in suture manufacturing proved to be a formidable competitive advantage that US Surgical could not overcome, even with its vast resources.

Fundamental business strategies

After a company has identified its potential sources of competitive advantage, it can effectively evaluate the best fundamental business strategies to pursue in order to optimize them. No single strategy can be expected to work for a company indefinitely; instead, innovators should think about choosing the combination of strategies that will enable the company to achieve long-term success.

While some of the strategies described below lend themselves more readily to positional rather than capability-based advantages, they can still be used in combination and/or in sequence to enable both types of competitive differentiation.

First mover

A first mover strategy refers to being the first company to offer a product or service in a market. Being first to market has significant advantages, including the ability to establish key IP that can serve as a barrier to entry, form relationships with important early customers and other resources, and define the most important product attributes and regulatory and reimbursement strategies that future competitors will have to consider. In some cases, a first mover can also create high **switching costs** for customers in order to keep future entrants from selling to the same customer base. This often happens with capital equipment, such as MRI and CT machines, that require a large upfront investment by the hospital that purchases them.

However, first movers incur certain costs for the positional advantages this strategy typically affords them.

These costs include the need to define the regulatory and reimbursement pathways, educate customers to drive adoption of a new product category, and train **users** on new techniques. These pathways can then be used by other companies to more quickly and less expensively create follow-on products. Typically, reimbursement and regulatory pathways paved by first movers are exploited by those that follow. But sometimes the first mover creates barriers by developing those pathways – they set the hurdles higher for those that wish to enter. For example, early developers of certain cardiac implants set a 10-year testing cycle standard that has required all that followed to abide by those criteria.

Importantly, a first mover advantage is only sustainable if the company in the lead continues to find creative ways of exploiting its first mover position. Otherwise, it will fall quickly into the ranks of the competition when other companies arrive in the market, forfeiting any benefits associated with being the first mover.

Another example of a successful first mover is Kyphon. Kyphon focused on treating vertebral compression fractures, a previously overlooked and under-treated market, by creating a new, minimally invasive procedure it called kyphoplasty. By setting up a dominant IP portfolio, the company deterred many competitors from entering the market and remained the leader until it was acquired by the competitor that posed its greatest threat. Guidant, on the other hand, was unable to sustain its first mover advantage in the field of interventional cardiology after several years of being the market leader. Instead, it fell into more of a follower position due to the company's early reluctance to explore new products, such as the area of coronary stents.

Fast follower

Instead of blazing the path to the market, **fast followers** often leverage other advantages to quickly capture market share from the first mover. Large companies in the fast follower position can often leverage their established distribution channels and customer relationships to this end. Both large and small companies may be able to introduce new features that create subtle differences between their technology and the first mover's product.

This can allow the company to overcome barriers to entry created by the first mover's IP position (in that different IP can be pursued). New features can also be used to address known shortcomings of the first mover's technology, further eroding the strength of its position. Fast followers also benefit from the reimbursement, regulatory, and awareness/education campaigns that have already been conducted by the first mover.

Boston Scientific's drug-eluting stents (DES) provides a good example of a successful fast follower strategy. Johnson & Johnson (J&J) created a dominant first mover position in the DES market with its Cypher® stent, holding a monopoly position in the US for nearly a year. During this time, however, J&J ignited tensions with many doctors by pricing its technology at levels that many considered to be unreasonable. The company also experienced significant supply problems, which further angered physicians when they could not obtain an adequate supply of the product to satisfy the patients on their waiting lists. When Boston Scientific launched its Taxus® stent, physicians already understood the benefits of a drug-eluting device and were receptive to the new technology. The company intentionally took advantage of the relationships that had been damaged by J&J's supply and pricing issues by ensuring that it had plenty of supply on hand, and was working with hospitals to develop mutually beneficial pricing strategies. Furthermore, Boston Scientific promoted the product features of its DES that made it more flexible and easier to work with in direct response to criticisms of the Cypher stent, which was perceived as stiffer and more difficult to place. As a result, Boston Scientific captured nearly 70 percent of the DES market from J&J in its first seven weeks on the market.[5]

Me-too

Unlike the products of fast followers, **me-too** products are relatively undifferentiated from the products that are already on the market. Typically, me-too products seek to benefit from the technical and market development of earlier players by creating a product that is cheaper to develop or manufacture and, thus, may be offered at a lower cost or with some very modest improvements. By watching earlier competitors, purveyors of me-too products often manage to avoid costly mistakes and sometimes can go through less arduous regulatory steps. Since me-too companies have lower development costs to recoup than the first movers and fast followers, they often compete based on price.

A prime example of a me-too product is found in SciMed's initial entry into the angioplasty catheter market. The company's first product was not tremendously differentiated from other competitors in the marketplace at the time. Interestingly, as noted previously, SciMed ultimately developed a powerful iteration and rapid development capability that moved it away from this me-too strategy into a first mover strategy.

Another example of a me-too product strategy is seen with generic versions of drugs. Generic drugs can save significant time in getting to market and can incur lower development expenses by using separate regulatory pathways from those used in new drug approvals. Once a generic enters the market, the original product must leverage its brand to retain some advantage. However, a positional brand advantage can be eroded quickly because the less expensive alternatives are favored by price-sensitive payers. This me-too strategy applies not only to generic drugs, but to any follower joining the market on the heels of another company's pioneering product, particularly once that product category has been well established. While fast followers often enter with differentiated products that they believe to be patentably distinct, generic followers enter only once key patents have begun to expire. In both cases, the followers have the distinct advantage of having watched the pioneer in the market make mistakes and adjust its strategies. If they can use these learnings to help them navigate whatever competitive barriers have been erected by the predecessor, they have the opportunity to gain market share much more rapidly.

Niche strategy

Rather than attempting to own a product category, **niche strategies** focus on owning the customer relationships in a specific, focused area of medicine. Companies often pursue a niche strategy when a group of physicians, typically in a smaller subspecialty, remains underserved by the market. To meet the needs of a niche market, a company may tailor a selection of its products to the unique needs of the physician/customer group. The overarching goal is to gain a positional advantage through strong customer relationships as a method for blocking competitors. Another niche strategy is to target a particular geography (or demographic) that may not be compelling to the dominant, established players in the field, and then use this as a springboard into other areas as the company builds momentum. Start-up companies in low-resource environments such as India are using this approach when they believe their products have the potential to be attractive in both developing and more developed settings.

American Medical Systems (AMS) executed a niche play to win the urologist market. While other companies called on urologists with the same sales representative that called on general surgeons, AMS positioned itself narrowly as a urological disorders business with products to treat conditions such as erectile dysfunction, incontinence, and prostate disease. By focusing exclusively on serving urologists and the diseases that urologists treat, the company sought to position itself "not only to benefit from growth in the urology market, but to drive it,"[6] as compared to some of its more broad-based competitors that participated in multiple market segments. This strategy was extremely effective, allowing AMS to build strong relationships with customers that others had trouble accessing.

Another variation on a niche strategy is to pursue "orphaned" drugs and diseases – rare conditions with relatively small populations that previously have been overlooked by medical technology and pharmaceutical companies (e.g., renal cell carcinoma, glioma, and acute myeloid leukemia).[7] One definition of orphaned diseases is that they are diseases where the manufacturer of the vaccine or treatment cannot expect to make a profit (many tropical diseases are also considered orphans because the tens of millions of patients suffering from them are too poor to pay for medical interventions).[8] However, some analysts perceive an opportunity for orphaned diseases to become more interesting to drug and device manufacturers. As medicine becomes more personalized and the discovery of "blockbuster" products gets less and less common, companies in geographies around the world may increasingly take advantage of

the special incentives offered by governments in the regulatory process (e.g., **FDA's Humanitarian Device Exemption** – see 4.2 Regulatory Basics) to try profitably serving these markets[9] and/or using them as a springboard into related treatment areas.

Low-cost provider

An approach that often works well in combination with another strategy, such as pursuing orphaned diseases, is the **low-cost provider** strategy. With this approach, companies must find and sustain creative ways to drive innovation at a significantly lower cost than its competitors so it can pass along lower prices to customers. In mature healthcare markets, innovation has traditionally been stimulated by the desires of advanced health systems to build their infrastructures and expand their capabilities. But some analysts expect this model to become all but outdated as more and more innovations emerge from geographies and other environments faced with severe resource-constraints and infrastructure limitations.[10] Moreover, the incremental approach to making product improvements that traditionally has dominated medtech is becoming obsolete, with the benefits from making incremental improvements to existing devices dwarfed in comparison to the cost of realizing those improvements.[11] Innovations at significantly lower price points are already emerging from companies using a low-cost provider strategy in places like China, India, and Brazil. Companies in established markets such as the US may be at a disadvantage in pursuing a low-cost innovation strategy if they are entrenched in a more traditional mindset. However, over time, government and competitive pressures to lower healthcare costs are likely to make different ways of thinking and acting an imperative for all medtech companies.

Often, companies with a low-cost provider strategy choose to emphasize the "de-featuring" of products over radical innovation. GE's MAC line of electrocardiogram (ECG) machines, which offer basic functionality at a cost of $500–$800 instead of $2,000–$10,000 for conventional ECGs, provides one well-known example.[12] Yet, some companies are able to offer both affordability *and* high-quality innovation, as the Mindray story illustrates.[13]

FROM THE FIELD ▶ MINDRAY MEDICAL

Realizing competitive advantage with a low-cost provider strategy

Mindray Medical International Limited was China's largest medical device company in 2014.[14] Founder Xu Hang originally started the company to serve Chinese hospitals, mostly in rural areas, that could not afford basic medical equipment. The company's initial focus was on providing low-cost in vitro diagnostics and patient monitoring systems of acceptable quality at the lowest possible price. However, over time, Mindray's ambitions expanded. To compete on a larger scale against global medical device manufacturers, the company sought to augment its low-cost provider strategy by making its products more innovative and improving their overall quality.[15] To do so, it would optimize its business around improving what the company calls its "performance-to-price ratio."[16]

In orchestrating this transition, Mindray focused on paying close attention to the needs of mid-tier hospital that traditionally had been ignored by large multinational companies. It then directed its innovation efforts on improving the functionality of its devices based directly on these needs.[17] The company also reported devoting 10 percent of its revenues[18] and 30 percent of its staff[19] to research and development each year. With this investment, Mindray has developed an R&D capability that enables it to launch as many as 13 new products a year.[20]

To hold down costs as it increased innovation and quality, Mindray built a vertically integrated production model that allows it to design, develop, and manufacture its devices in-house. The cost of labor in China is one factor that supports the company's low-cost approach. But other strategies, such as using common resources and modular components within and across product lines,[21] help it maintain a cost advantage. Mindray

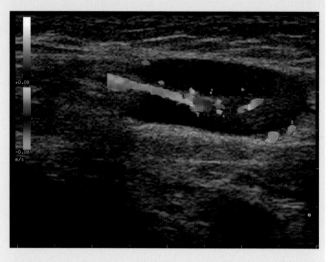

FIGURE 5.9.1

A color ultrasound image of the lymph nodes (by Nevit Dilmen via Wikimedia Commons).

typically offers its equipment at price points as much as 40 percent lower than its multinational competitors.[22]

One of the company's first domestically developed, innovative devices was a digital color ultrasound imaging machine which, when it was launched in China, caught the attention of Mindray's international rivals (see Figure 5.9.1).[23] From there, Mindray has expanded its product portfolio to more than 60 products sold in 140 countries around the world,[24] including Asia, Africa, Europe, and the United States. Roughly half of its revenue is from sales outside of China.[25] In these international markets, Mindray continues to focus on smaller and mid-tier hospitals, as well as private clinics with the intent of expanding from this beachhead.[26]

Relative to its domestic competitors, Mindray garners a 20 percent premium due to its brand name and reputation for quality.[27] However, it is still working to convince some customers in markets like the US that a low-cost provider can deliver high-quality products. Lee Weng, General Manager of Mindray's first overseas R&D center in Seattle, Washington, once referred to the aversion of some international buyers to Chinese medical devices as a "prevailing prejudice."[28] To combat this perception and strengthen its strategy, Mindray remains focused on developing affordable, innovative products that meet the requirements of regulators in the European Union and US.[29]

Distribution strategy

Distribution strategies focus on product breadth and channel relationships rather than on product superiority. By meeting the needs of customers through diverse product offerings, more flexible payment arrangements, or strategies to bundle products and/or services, a business may keep more focused entrants out of the market.

Stryker, a leading manufacturer of orthopedics products, offers a broad product portfolio and thus can be a single solution for all of a hospital's orthopedic technology needs. Theoretically, it would be possible for Stryker to bundle the purchasing of hip and knee implants together and, thus, keep a new knee implant company out of an account. While the bundling of products is regulated by antitrust regulations like the Sherman Act,[30] a company that offers a one-stop solution may win in the market – especially in commodity markets. Baxter and J&J are good examples of companies that use a distribution strategy to maintain an advantage in hospital supplies (e.g., I.V. bags, gloves, bandages) and other lower-cost/commodity markets.

Disease management

A more advanced strategy related to the bundling of products and services that is emerging in the medtech sector is sometimes referred to as a disease management strategy or "**owning the disease**."[31] With this approach, technology companies are:

> *migrating from an episodic or intervention-focused business model to a convergent care model that enables them to provide solutions along the continuum of care. The integrated solutions inherent in this approach ... combine drug, device,*

> *diagnostic, and consumer-centric solutions to establish creative platforms in which a company can dominate the diagnosis and treatment of a disease or condition.[32]*

In simple terms, some companies that once acted as hardware providers, focused on selling medical devices designed for a distinct intervention, are now interested in becoming problem solvers that provide end-to-end solutions for their customers. This strategy is consistent with the anticipated shift from traditional volume-based, fee-for-service health systems to a more cost-conscious, value-driven, and outcomes-focused orientation.

Larger organizations have a certain advantage when it comes to owning a disease due to the scale of their operations, the relationships and resources at their disposal, and the ability to take the time needed to assemble all of the parts of the solution.[33] However, it can be difficult for established organizations to undertake the radical change necessary to dismantle their old strategies and embrace new ones (see the Merck Serono case example). For this reason, start-ups and small companies can sometimes realize greater success in assembling an end-to-end solution. For instance, Well-Doc has created a multi-faceted solution to own type II diabetes that engages prescribing physicians, the company's own trainers and healthcare providers, telecommunications companies, payers, and patients through its BlueStar™ offering.[34] This system, which is FDA-cleared and approved for reimbursement, is sufficiently further along and more disruptive than the offerings of the larger companies with an interest in diabetes management.[35]

FROM THE FIELD > MERCK SERONO

Realizing results through a disease management strategy

To effectively employ a disease management strategy in a large, established organization, Don Cowling, Merck Serono's former Senior Vice President and Managing Director for Western Europe, acknowledged that, "You have to first blow up the current model."[36] His organization did just that in seeking to dominate the market for human growth hormone therapy for endocrine and metabolic disorders. When its drug Saizen® (somatropin) was on the cusp of becoming generic, the company adopted a strategy to transform its UK team from "a power-driven sales organization to a knowledge-driven outcomes organization."[37] The strategy centered on "adding value to the products you already have," said Cowling.[38] As one article described, he and his team determined that growth for drugs (and the devices that support them) is in adherence rather than finding new patients since half of all prescriptions are never filled, and half of those are never taken.[39] Yet, they recognized that stakeholders – including patients, physicians, facilities, and payers – do not want to pay for compliance. "They buy pharmacoeconomic outcomes."[40]

After extensive research and a significant rethinking of its approach, Merck Serono designed the easypod™ injector, a wireless-enabled, electro-mechanical device that resembles an early mobile phone (see Figure 5.9.2). A cartridge of medication is inserted into the side of the device, with a single-use needle attached to the end. The healthcare provider programs easypod with the dose, treatment plan, injection depth, and other specifications. When deployed by the patient, the device records the date and time of the injection, along with other relevant data. A sensor detects the angle of each injection (which is important for preventing unpleasant side effects) and a dose confirmation feature notifies patients when they have properly administered treatment. This information is stored in the device and uploaded to the Serono delivery

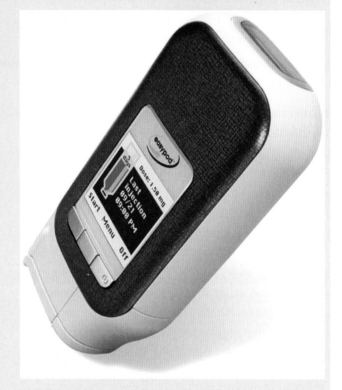

FIGURE 5.9.2
The easypod injector (courtesy of EMD Serono, Inc.).

team, which monitors it for missed doses or other warning signs and then makes outbound contact with patients or their clinics when intervention is needed.

Once the team obtained regulatory clearance for the device, Serono began to gather evidence that demonstrated how the accurate, consistent use of its drugs was linked to desired therapeutic outcomes. It also gleaned insights about patient behavior patterns that allowed it to begin selling its drugs differently to providers and payers. The company was able to document value propositions tailored to customers' particular concerns. If a payer was concerned about the cost of non-compliance and prescriptions that were filled but never used, Serono could tap into patient data, analyze behavior patterns, and then offer per-injection pricing that would enable the insurer to pay only for administered treatment. Alternatively, if a provider was more concerned about how to manage cost overruns, Serono could use its data to devise a capped payment agreement, essentially insuring the hospital or clinic against unanticipated spending.

When Cowling left the company in 2012, after 10 years of building and optimizing the model for Saizen, as well as several other products, he had roughly tripled his division's revenue while dramatically reducing total staff. Few financial results were publicly available, but one report claimed Serono had achieved a 38 percent compound annual growth rate in the mature growth hormone market through this patient-centric, outcomes-based approach to owning a disease state.

Information aggregator

Another emerging strategy in medtech is to become an **information aggregator**. As medical technologies become more and more connected, the value of new devices is no longer solely in the products themselves,[41] but in the information they generate, as well as the insights and solutions that the data enables. However, effectively capitalizing on this information is no easy feat. Issues of connectivity, bandwidth, integration, and security are daunting challenges, not to mention the analytics needed to understand and utilize the data. Companies are needed that can help healthcare stakeholders effectively manage the promise of connected (or mobile) health and mitigate its perils.

In response to this opportunity, companies outside health – mostly technology and telecommunications companies – have been expanding into the sector. For example, Verizon launched its Converged Health Management solution, which is an FDA-cleared remote health monitoring service. This offering provides physicians and patients with real-time access to data from connected biometric devices to enable anywhere, anytime monitoring.[42] Qualcomm is another company that entered the healthcare field through its Qualcomm Life division, as highlighted in the following case example.

FROM THE FIELD ▷ QUALCOMM LIFE

New opportunities in healthcare connectivity and informatics

Mobile health (**mHealth**) and health information technology (HIT) solutions are proliferating at a rapid rate, all with the goal of more effectively facilitating the flow of information among patients, providers, and other healthcare stakeholders to optimize patient wellness, diagnosis, and treatment. Wireless remote monitoring devices for tracking patient blood pressure, blood glucose levels, activity, and other important health indicators provide an example of one growing class of products. The success of this relatively new healthcare sector relies not only on the devices being developed in the space, but on the ability to make the data they gather accessible and actionable.

To help shape the intersection of connectivity, communications, and healthcare data management, telecommunications company Qualcomm launched Qualcomm Life, Inc. This wholly-owned subsidiary has three core initiatives: the Qualcomm Life Fund, a $100 million investment fund managed by Qualcomm Ventures that is focused on start-up wireless healthcare companies; a strategic ecosystem group that collaborates with universities, trade and policy groups, and research institutes to provide resources to developers who want to apply wireless technology in the healthcare field; and a business arm that sells cloud-based communications platforms (specifically, the 2net™ and HealthyCircles™ Platforms) optimized for healthcare customers.

Qualcomm Life's first offering is the 2net Platform, which enables the wireless transfer, storage, and display of medical device data. The platform is designed to be "technology agnostic," meaning that it seamlessly provides wireless connectivity and interoperability between different medical devices and applications from disparate manufacturers that patients and providers use across the continuum of care.[43] Data from these devices and apps reach the 2net Platform through one of four gateways. The first is the 2net Hub, an FDA-listed Class I Medical Device Data Systems (MDDS), stand-alone connectivity device that is roughly the size and shape of a nightlight. As Don Jones, VP of Global Strategy and Market Development for Qualcomm Life, explained, "2net works the way you wish your Wi-Fi actually worked. You open the box, plug it into the wall, and it

recognizes any medical devices in the room that have a cloud-based relationship with our platform." The other gateways include medical data from mobile phones, devices with an embedded cellular component, and server-to-server interfaces between Qualcomm Life and other service platforms.[44] As Jones elaborated, "The 2net Platform gathers the data from these gateways and securely delivers it to healthcare providers [and other stakeholders] in a 'mash-up' format, meaning that it enables the capture of multiple data streams to make one big picture" (see Figure 5.9.3).

Qualcomm Life's second offering, the HealthyCircles Platform, is a hosted enterprise software-as-a-service platform solution that provides a secure communications and record-sharing infrastructure for care team coordination, post-discharge transitional care, and complex condition management.[45] Together, these offerings aggregate device and app data, medication history, labs, care team data entry, and patient

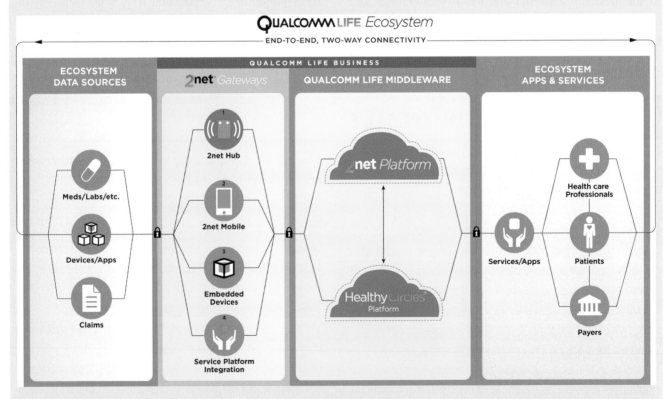

FIGURE 5.9.3
A visual representation of the Qualcomm Life 2net Ecosystem (courtesy of Qualcomm Life).

symptomatic self-assessment to create robust datasets for Qualcomm Life's customers.

According to Jones, Qualcomm's decision to enter the healthcare space was fueled by recognition of a growing unmet need. "Literally hundreds of companies had come to us for advice on how to use wireless, connectivity, and network architecture to create solutions in healthcare," he recalled. "And many of them were trying to reinvent the same wheel." Qualcomm realized that if it took on the challenge of building an enabling platform and connectivity infrastructure, "We could not only do it better and more efficiently, but we could take a lot of cost out of these healthcare solutions and support an ecosystem of companies," Jones stated.

Qualcomm found the healthcare opportunity attractive because it possessed key capabilities that gave it a significant competitive advantage. Specifically, the company's deep expertise in radio and communication protocols, as well as wireless network infrastructure made it possible to cost-effectively develop a robust platform that meets the needs of the multiple user groups involved in mobile health. Additionally, the company had positional assets that enabled it to quickly build a business with a substantial global footprint. "Qualcomm does business in nearly every country in the world and has a broad network of operating centers," Jones said. Further, he noted that Qualcomm is the "hidden name" inside the architecture of almost every 3G and 4G device built worldwide. "We are responsible for either the base system, chips, security, or IP systems that make those devices work." This reach gave Qualcomm Life a head-start on establishing a distribution infrastructure and customer base for the new offering. The company's competitive advantages also enabled it to design a communications platform that met the security, privacy, and regulatory requirements of 43 different countries. "We were in the unique position of understanding how, in an inexpensive way, to design an efficient network architecture that provided the most

robust solution, working within the particular requirements of healthcare," Jones said.

From a strategy perspective, Qualcomm Life was leading innovation in the space as both a first mover and an information aggregator. To maintain its edge, the company's plan is to keep a sharp focus on expanding the business. "The ecosystem we're putting into place increases in value as there are more users and more companies plugged into it," Jones explained. "Patients and providers don't want go to individual portals to get health data from different devices. They want access to it all in one place," he said. "So we're building a scenario in which users don't have to be tied to one manufacturer's bathroom scale or blood pressure cuff or cardiac monitor. As the network grows, we hope they will be able to choose from hundreds of different devices from different manufacturers, because the data these devices generate will be available through a single Qualcomm cloud interface."

Initially, Qualcomm Life will generate revenue primarily by helping stakeholders connect and share data. Using the 2net Platform as an example, Jones said, "2net has as many as six different business models in terms of who pays – from the consumer to a health insurance plan or healthcare provider, to a device manufacturer. However, most customers either pay a one-time fee for access for a defined period of time, or they pay a monthly fee." But, over time, he envisions the company extracting more value from its offering as an information aggregator. "Ultimately, there will be a shift to informatics," he said. "This is the first time in history this kind of health data has been digitized," Jones stated. He perceived that Qualcomm Life would have significant opportunities related to data management and analytics. "Qualcomm is not a healthcare company," he emphasized. "But, most likely, we will build tools that help our customers effectively use this data, anonymized or with appropriate permissions, to develop predictive models that support better clinical decision making."

Original equipment manufacturer/licensing

Using an original equipment manufacturer (**OEM**) strategy, a company provides technology and/or components to another company that then assembles and sells the finished product. This strategy is commonplace in the computer industry where components, such as hard drives, processors, and video cards for a single computer, may all come from different manufacturers. Early-stage medical device start-ups looking to reduce their upfront capital commitments may contract with an OEM (e.g., for the manufacturing of machined or molded components). Or, they may become OEMs themselves by selling their physical products or **licensing** their technologies for use by another company. Some examples of OEM relationships include Gyrus' relationship with Ethicon (involving endosurgery equipment) or Surmodics' relationship with J&J (providing the polymer coating for the Cypher stent).

The contracts involved in becoming an OEM and/or entering into a licensing agreement can range from simple to complex. One of the most strategically important terms is whether the contract is exclusive or non-exclusive. Exclusive agreements can limit the company with the technology from sharing it with any other firms (thereby giving the licensor a strong positional advantage). In turn, the licensor may or may not agree to use only this one technology and not seek alternative suppliers. In non-exclusive arrangements, both parties are free to make deals with other companies.

For example, Wilson Greatbatch is a large company whose primary strategy is to be the OEM of certain battery and electrical component technologies. Specifically, it brands itself as a leader in the development, design, and manufacture of components critical to implantable medical devices.[46] A typical arrangement for Wilson Greatbatch is to supply batteries and electrical components (technologies that are complicated and protected by strong IP positions) for the pacemakers produced by all the large implantable pacemaker manufacturers. As shown by this example, if a company owns a component or service that is considered rare but potentially can be widely used, an OEM strategy is a good way to capitalize on this competitive advantage. Other examples of products/services well suited to OEM strategies include high-volume, five-axis machining used to create orthopedic implants, which is almost completely controlled by the major orthopedic manufacturers, or guidewire technology, which is dominated by Lake Region Corporation.

Partnering

Sometimes two companies become partners so that each one can leverage the competitive advantage of the other (more information is included in 6.4 Alternate Pathways). Typically, two companies form an exclusive, contractual agreement in order to shepherd a product into the marketplace. The most common **partnering strategies** occur in the pharma-biotech field and the pharma-drug delivery space. Usually the smaller entity makes a new technology available and the large pharmaceutical entity brings capital, expertise, and distribution to the partnership. Partnerships are often structured so that the large entity covers portions of the development costs and makes **royalty** payments. In return, the larger entity gains marketing and distribution rights for the product. An example of this arrangement was the partnership between Nektar Therapeutics, a smaller biotech company, and Pfizer, a large pharmaceutical company, to develop and bring inhalable insulin to the market (Exubera).

In the medical device field, partnering between small and large companies is often centered on a distribution, investment, or technology relationship, although more variations are being explored as companies consider disease management strategies. While obtaining a partner may result in fewer strategic options for a company, it also can reduce strategic options for competitors. Consequently, partnering can sometimes be a prudent step, especially when the useful strategic options generated from these relationships are considered relatively scarce. For example, consolidation has resulted in few acquirers of businesses or technologies in some markets. As a result, a key partnership could solidify one of those rare relationships and limit exit options for the competition. Such a scenario can be found in the interventional cardiology field where there are now just four primary players – Boston Scientific, J&J, Abbott Vascular, and Medtronic. If a small start-up is competing against a vast number of companies with similar products or services, a partnering relationship with one of these major players could provide it with a significant advantage against its peers, especially if only three others would potentially be capable of solidifying a similar relationship with the remaining major players. If one of the four major players has a more dominant position than the others, then a relationship with that leader could provide an even more powerful advantage.

Using competitive advantage to define the basis for competition

Beyond choosing the appropriate business strategy(ies) to capitalize on their competitive advantages, innovators and companies should seek to create a basis for competition in its field that plays to their own strengths and attacks the weaknesses of key competitors. The basis of competition refers to the features, benefits, or qualities that become central to the way businesses compete with each other. Price, quality, brand perception, customer support, training, or certain technological features can all serve as bases for competition. If a company is a first mover or early entrant to a relatively new field, it is usually able to exercise more control in this regard and essentially define the "playing field." One example is the way J&J, upon being the first entrant into the stent market, used "stent strut strength" as a basis of competition in the early days of coronary stenting. This forced all subsequent competitors to test their devices to a standard that was difficult to beat while addressing other customer needs. While this basis of competition was later eroded by competitors, it served as a significant barrier in the early days of stent development.

If a company is a later entrant to a well-established field, then it must try to change the basis of competition by developing capabilities and positional advantages in areas where leaders do not have them. For instance, if a small company intends to compete in a medical device market against Medtronic or J&J, it would be impractical to try to challenge these firms on the strength of their

distribution positions and capabilities. However, by interacting with key stakeholders to understand these companies' areas of weakness, the smaller firm may be able to gain a foothold. If physicians are frustrated by the ease of use of an established device, for example, the company might target innovation in this area. Or, if J&J or Medtronic has neglected customer service, a competitor might seek to develop deep capabilities in this area in an effort to win over customers.

In the field of active implantable devices, a completely new basis of competition has evolved to include web-based disease management systems (such as Medtronic's CareLink®) used by doctors and hosted by the companies that provide the implants to ensure improved patient care. The Innerpulse case in 2.3 Stakeholder Analysis provides another example of how a new entrant can seek to differentiate its offerings and redefine the competitive playing field. By choosing to focus on primary prevention of sudden cardiac death, Innerpulse attempted to avoid head-to-head competition with the large, established ICD manufacturers that historically targeted and successfully captured the secondary prevention market.

Importantly, choosing where to invest and what advantages to develop is essential. Otherwise, a company runs the risk of becoming a "jack of all trades, master of none," which can make it difficult to defend any competitive advantage that it may have. It is not realistic to expect that any company can develop all of the sources of competitive advantage discussed in this chapter. Often, as companies are getting started, they first identify gaps in the product offerings of their potential competitors. Then, they seek to determine which advantages to build in the near term and which ones to invest in or develop at subsequent stages of their strategic evolution in order to exploit these gaps. As with any strategy, decisions regarding competitive advantage should be reviewed frequently, based on changes in the external and internal environments, to ensure they remain valid and attainable.

The ev3 story demonstrates how one company explicitly defined and managed its competitive advantage.

FROM THE FIELD ▶ EV3

Building a company on capabilities and business strategy

The formation of ev3, a company focused on delivering products for coronary, peripheral, and neurovascular applications, was somewhat unique in that Warburg Pincus, the private equity investors that backed the venture, funded a capability-based competitive advantage and a supporting business strategy rather than a traditional product. Jim Corbett, ev3's former CEO, described the advantage upon which the company was built. "We had assembled a team of executives experienced in the market who we thought could identify the right technology segments and then acquire or develop products to create a company," he said.

The market ev3 would go after was the endovascular field. "The endovascular market had become very large over the preceding 10 or 15 years, and it had consolidated," Corbett explained. "That consolidation created some rather large global companies – Cordis/ J&J, Boston Scientific, Medtronic, and Abbott Vascular. Those companies, in their success, had developed mega-product categories that they were defending. So, if you're in CRM [cardiac rhythm management] or if you're in drug-eluting coronary stents, you really cannot afford to lose your beachhead because there are billions of dollars of earnings at stake. The consequence of that was that the rate of innovation in the endovascular markets had dramatically decreased, because the big R&D centers were all focused on the defense or preservation of their market position rather than on creating new **concepts**, or new devices, or new market segments." ev3, with its experienced management capability, developed an explicit business strategy to target opportunities that existed within this "innovation gap."

The company's name refers to the three primary types of opportunities it would target in the profitable

endovascular market: coronary, peripheral, and neurovascular devices and interventions. To establish its preliminary product position, ev3 rapidly acquired nine companies and/or technologies that its management team believed could be commercialized in a competitive time frame for a reasonable cost. Because physicians, particularly those in peripheral and neuro, had been relatively underserved by the major medical device players, ev3's strategy was somewhat of a niche play. For example, according to Corbett, the market leader in peripheral stent technology had not introduced an innovation for roughly eight years. In neurovascular, the market leader had been offering the same base product for nearly 10 years. "So there were real opportunities for us to find the chinks in our competitors' armor, so to speak, and create platform technology, and innovate, and bring something new to market," he said.

ev3's idea was to use a niche strategy to establish a foothold and then expand its position. For instance, to become a dominant player in the neurovascular market, which was stronger in Europe than in the US, ev3 added a geographic element to its niche approach, which it reinforced through the development of new positional and capability-based competitive advantages. "Two-thirds of the global market for neurovascular products was outside the United States, which was very atypical for most medical devices," explained Corbett. At that time, "Usually the US market is dominant, or is the largest segment between the two. We made it part of our early strategy to have direct selling operations in Europe. So we put a lot of effort into global distribution which, again, was an uncommon choice for an early-stage medical device company. In fact, a lot of companies these days choose not to do it at all, and rely on improving their product in the United States because they actually plan for consolidation. But that was a very important distribution choice on our part and it paid a lot of dividends." Within a relatively short period of time, a sizable portion of ev3's sales were made outside the

US and it was the fastest-growing segment of the business.

Over time, ev3 recognized the need to further focus its business strategy and its approach to the market. The larger, established companies continued to enjoy a competitive advantage in the coronary field, where Corbett estimated they spent as much as $1.5 billion per year on R&D for DES alone. Staying true to the company's desire to innovate around its major competitors, ev3 decided to drop the coronary market as a focus area. Looking back, Corbett said it was an easy decision to focus on the two fields with the greatest opportunity for innovation. At that time, many peripheral and neurovascular devices were nothing more than repurposed or scaled-down coronary devices, which left ev3 with plenty of room for entirely new devices and intervention innovations. In addition, development of these products was potentially much faster than coronary products and so offered a lower investment risk to ev3.

Another explicit decision made by the company was to invest in developing a strong capability in internal development and innovation. One of the primary reasons for doing so was to help give the company a more sustainable competitive advantage. "Product innovation is the core basis of competition in our industry," said Corbett. By developing a deep capability-based advantage in this area, ev3 would be able to "create a footprint of a company, not just a product line," he added, also noting proudly that, "In 2005 and 2006, we introduced more new products into the market than the four biggest endovascular competitors combined. And that was all internal development." One of the new products the company was most excited about was the EverFlex® peripheral stent (see Figure 5.9.4). Fractures in peripheral stents, which occur in as many as 25 percent of all cases, cause restenosis, surgery, and amputations. ev3's EverFlex stent had been shown to have a fracture rate four to five times lower than the leading product in the category.

FIGURE 5.9.4

The EverFlex peripheral stent (courtesy of ev3).

When thinking about developing a competitive advantage, Corbett had this advice to entrepreneurs: "Resist *not* focusing. Both at the beginning and along the way, there is such an opportunity to take on new tasks. But it will take away from your ability to execute on your core strategy. Focus is the path to success, without question." He also underscored the importance of moving swiftly. "Speed is key. Obviously speed has to be conducted with excellence, and you should never do speed if it compromises quality, or ethics, or any of those types of matters. But the time you spend getting to market burns cash, and you need that cash. There is often a tinkering mentality that emerges in young businesses, especially when they're privately held. But speed is a core capability that you should try to develop from the start."

Ev3 was acquired by Covidien in 2010 for approximately $2.6 billion.[47]

Developing a statement of competitive advantage

When innovators or companies are working on defining their sources of competitive advantage and the business strategies needed to support them, they should consider articulating them in a statement of competitive advantage. In developing such a statement, look at those created by other companies for ideas about how to effectively capture the key ideas (other recommended steps can be found in the Getting Started section).

Unfortunately, detailed information is not always available regarding other companies' statements of competitive advantage because they are often considered proprietary. However, even in the absence of inside insights, innovators can frequently discern the focus of a company's competitive advantage from the information it makes publicly available. For example, when a private company goes public:

- Evaluate its prospectus carefully. Usually there is a section entitled "Our Strategy" or "Business Strategy" that contains relevant information.
- Identify the aspects of this text that are indicative of the company's core assets and chosen fundamental business strategy.
- Use professional judgment and knowledge of the field to translate this information into a statement of competitive advantage.

Information gathered from public sources can be used to reconstruct a statement of competitive advantage, as shown in the Working Example on Kyphon.

Working Example

Kyphon, Inc.: Discerning a competitive advantage from public information

The following information was drawn from a Kyphon prospectus that was released prior to the company's initial public offering.[48] Within this public description of the company's strategy, the elements that correspond to Kyphon's core assets are shown in italics. Kyphon specializes in devices that enable the minimally invasive treatment of spinal fractures caused by osteoporosis or cancer

Business strategy

Our goal is to establish treatments using our proprietary balloon technology as the **standard of care** in orthopedic applications. We are initially focusing our efforts on vertebral body compression fractures. The key elements of our strategy are to:

Penetrate the spinal market using a direct sales force

We believe that a *direct sales force* will allow us to most effectively educate and train physicians in the use of our products. Our products are sold directly to physicians by our experienced sales team, comprising 20 sales representatives, three managers and a vice president of sales. By leveraging their *extensive spinal market experience*, our sales people are able to identify key physicians and provide effective case support to accelerate market adoption of our procedure. Our sales team is supported by two in-house coordinators and four field-based associates.

Educate referring physicians and patients

Patients with vertebral body compression fractures often are not referred to spine surgeons for treatment. Our objective is to *establish referrals from physicians who initially diagnose* vertebral body compression fractures to spine-focused surgeons who perform Kyphoplasty. As a result, we have implemented an *awareness marketing campaign* to educate internists, family physicians, gerontologists, and other primary care physicians about Kyphoplasty and its potential to be an effective therapy. As part of this campaign we provide educational materials to treating physicians, referring physicians and patients, and organize regional market seminars where surgeons trained in performing Kyphoplasty educate referring physicians.

Expand clinical support of the Kyphoplasty procedure

We are conducting *outcome studies* to increase awareness of the procedure within the medical community, to develop additional marketing claims and to support third-party reimbursement. Through our own outcome studies and those of surgeons currently performing the Kyphoplasty procedure, we are gathering data for ***peer-reviewed** journal articles* in support of reimbursement efforts.

Work with opinion leaders

We have obtained the advice and *support of nationally recognized spine surgeons* who are helping us to further develop our products and the procedure, to demonstrate the benefits of Kyphoplasty, and to obtain third-party reimbursement. Because these leading physicians help *set medical policy* in their respective areas of expertise and are experienced in outcome assessments, we believe they will help create patient referrals and advance third-party reimbursement. The reputations of these physicians and their leadership in professional societies help bring recognition and credibility to our products.

Expand surgeon adoption of Kyphoplasty through training

We have implemented specialized training programs and are *rapidly expanding the number of physicians trained in Kyphoplasty*. As of August 31, 2000, we had trained approximately 300 physicians in the United States and Europe and we plan to have trained more than 400 by the end of 2000. We support these physicians through professional development programs, which include funding local seminars, funding travel to national medical conferences and assisting in the preparation of scientific papers for publication.

Expand into Additional Orthopedic Markets

We intend to leverage our *proprietary balloon technology platform* for other applications, including compression fractures of the wrist, knee and hip. These new applications involve refinements of our current products, and we intend to conduct outcome studies in these applications to support market adoption. We believe our *intellectual property position* and our *position within the orthopedic marketplace* will allow us to become the leading provider of minimally invasive medical devices for the treatment of compression fractures.

From this overview, one can infer that Kyphon is developing a competitive advantage that depends on the core assets of:

Intellectual property – To protect its innovative technology in the area of spinal orthopedics, Kyphon defined its initial market narrowly and built a strong and extensive IP barrier to protect its desired position in the market. Through its IP strategy, the company also positioned itself for expansion into other markets.

Owning key relationships – To create additional barriers to entry for competitors, Kyphon has invested heavily in locking up the specialist market (including key opinion leaders in the field) through training and other professional development activities. The company also has gone after generalists (who diagnose the target condition and refer patients for treatment) to increase the strength of its relationships further and make it more difficult for competitors to gain a clinical customer base. To keep the clinical community engaged and convinced of its product's value, Kyphon is also performing ongoing outcome studies to be published in peer-reviewed journals.

Establishing core channels – To support the development and nurturing of its key physician relationships and further strengthen its barriers to entry, Kyphon developed a strong direct sales force. Given the specialized nature of the Kyphon product, this type of channel makes the most sense for sales and distribution. By investing in a team of highly educated sales representatives that can assist in training physicians, the company is positioning itself for easier, more rapid user adoption while also making it more difficult (and costly) for competitors to replicate its sales/distribution capability.

Based on the core assets indicated by the company and a basic knowledge of the industry, one might next infer that Kyphon is pursuing a niche strategy. Rather than initially going after the general orthopedics market, it targeted a small subspecialty and has invested heavily in achieving a leadership position in this area. Kyphon also benefited from a first mover advantage which enabled it to leverage its core assets effectively (i.e., in locking up opinion leaders and creating an impenetrable IP barrier).

Given this positioning, the company's statement of competitive advantage might read something like this:

Kyphon will establish a leadership position in the minimally invasive treatment of spinal fractures caused by osteoporosis or cancer through a strong IP position (from which it can expand beyond this niche), deep relationships with specialists and key opinion leaders in the field developed and maintained by a highly educated, top-tier direct sales force, and widespread general awareness of the efficacy of Kyphoplasty among referring physicians.

Online Resources

Visit www.ebiodesign.org/5.9 for more content, including:

 Activities and links for "Getting Started"
- Understand the competitive advantages of competitors
- Identify the company's competitive advantage(s)
- Create a statement of competitive advantage
- Set a strategy

 Videos on competitive advantage and business strategy

CREDITS

The editors would like to acknowledge Jim Corbett and Julie Tracy for the ev3 story, Don Cowling for the Merck Serono case, and Don Jones for the Qualcomm Life example, as well as Ritu Kamal and Stacey McCutcheon for their help developing these stories. Darin Buxbaum and Jared Goor also made important contributions to the original chapter.

NOTES

1 The following overview is drawn from Garth Saloner, Andrea Shepard, and Joel Podolny, *Strategic Management* (John Wiley & Sons, 2001).

2 Ibid.

3 Ibid.

4 Ibid.

5 Shawn Tully, "Blood Feud," *Fortune*, May 31, 2004, p. 100.

6 American Medical Systems, S-1 Registration Document: Business Description, May 19, 2000.

7 Aarti Sharma, Abraham Jacob, Manas Tandon, and Dushyant Kumar, "Orphan Drug: Development Trends and Strategies," *Journal of Pharmacy and Bioallied Sciences*, October–December 2010, http://www.ncbi.nlm.nih.gov/pmc/articles/PMC2996062/ (December 11, 2013).

8 Ibid.

9 Ibid.

10 "The Race for Global Leadership," PricewaterhouseCoopers, January 2011, http://pwchealth.com/cgi-local/hregister.cgi/reg/innovation-scorecard.pdf (December 11, 2013).

11 "Medtech Companies Prepare for an Innovation Makeover," PricewaterhouseCoopers, October 2013, http://pwchealth.com/cgi-local/hregister.cgi/reg/pwc-medical-technology-innovation-report-2013.pdf (December 11, 2013).

12 "Market-Relevant Design: Making ECG's Available Across India," GE Healthcare, September 30, 2011, http://newsroom.gehealthcare.com/articles/ecgs-india-reverse-innovation/ (December 11, 2013).

13 Developed from public sources.

14 Edward Tse, John Jullens, and Bill Russo, "China's Mid-Market Innovators," Booz & Co., May 29, 2012, http://www.strategy-business.com/article/12204?pg = all (December 12, 2013).

15 Ibid.

16 "Mindray Introduces Enhancements for M Series Ultrasound Platforms at ASA," Mindray press release, October 14, 2011, http://ir.mindray.com/mobile.view?c = 203167&v = 203&d = 1&id = 1617299 (December 12, 2013).

17 Kwaku Atuahene-Gima and Xu Leiping, "Case Study: The Right Focus for Innovations," FinancialTimes.com, June 25, 2012, http://www.ft.com/cms/s/0/104517b4-ba0f-11e1-aa8d-00144feabdc0.html#axzz2nJ2IW9Ij (December 12, 2013).

18 Mena Venu, "A Trend to Low-Cost Imaging?," MorningStar.com, March 12, 2010, http://ibd.morningstar.com/article/article.asp?id = 328275&CN = brf295, http://ibd.morningstar.com/archive/archive.asp?inputs = days = 14;frmtId = 12,%20brf295 (December 12, 2013).

19 Deng Yu, "Mindray Eyes Medical Device Leadership," *China Daily USA*, December 14, 2012, http://usa.chinadaily.com.cn/epaper/2012-12/14/content_16018407.htm (December 12, 2013).

20 Ibid.

21 David Barnes, "Buy Recommendation: Healthcare Sector, Mindray Medical," March 16, 2010, http://www.sba.pdx.edu/faculty/johns/jsaccess/0web/573/Sectors/Healthcare/MR/MR_Mindray%20Medical_Buy_Barnes_Winter10.doc (December 12, 2013).

22 Paul Glader, "Mindray Eyes U.S. as West Looks East," *The Wall Street Journal*, September 14, 2010, http://online.wsj.com/news/articles/SB10001424052748703960004575482031542063628 (December 12, 2013).

23 Atuahene-Gima and Leiping, op. cit.

24 Barnes, op. cit.

25 Yu, op. cit.

26 Atuahene-Gima and Leiping, op. cit.

27 Venu, op. cit.

28 Yu, op. cit.

29 Ibid.

30 "Sherman Anti-Trust Act, (1890)," Ourdocuments.gov, http://www.ourdocuments.gov/doc.php?flash=true&doc=51 (December 11, 2013).

31 This term was dubbed by the Global Healthcare Innovation practice at PricewaterhouseCoopers. See Christopher Wasden and Brian S. Williams, "Owning the Disease: A New Business Model for Medical Technology Companies," *IN VIVO*, December 2011, http://www.elsevierbi.com/publications/in-vivo/29/11/owning-the-disease-a-new-business-model-for-medical-technology-companies (December 11, 2013).

32 Christopher Wasden and Brian Williams, "Owning the Disease II: Adapting Strategy Into Successful Business Tactics," *IN VIVO*, April 9, 2013, http://www.elsevierbi.com/publications/in-vivo/31/4/owning-the-disease-ii-adapting-strategy-into-successful-business-tactics (December 11, 2013).

33 Wasden and Williams, "Owning the Disease: A New Business Model for Medical Technology Companies," op. cit.

34 "WellDoc Launches BlueStar, First FDA-Cleared, Mobile Prescription Therapy for Type 2 Diabetes with Insurance Reimbursement," *BusinessWire*, June 13, 2013, http://www.businesswire.com/news/home/20130613005377/en/WellDoc-Launches-BlueStar-FDA-Cleared-Mobile-Prescription-Therapy (December 11, 2013).

35 Wasden and Williams, "Owning the Disease: A New Business Model for Medical Technology Companies," op. cit.

36 "The Race for Global Leadership," op. cit.

37 All quotations are from interviews conducted by the authors, unless otherwise cited. Reprinted with permission.

38 "Healthcare Unwired: New Business Models Delivering Care Anywhere," Health Research Institute, September 2010, http://www.mobilemarketer.com/cms/lib/9599.pdf (December 11, 2013).

39 Ibid.

40 Ibid.

41 "Medtech Companies Prepare for an Innovation Makeover," op. cit.

42 "Verizon Gains FDA Clearance for Remote Health Monitoring Solution," Verizon press release, August 8, 2013, http://newscenter.verizon.com/corporate/news-articles/2013/08-08-fda-clearance-for-remote-health-monitoring/ (December 12, 2013).

43 "What is 2Net," Qualcomm Life, www.qualcommlife.com/wireless-health (December 12, 2013).

44 Ibid.

45 "HealthyCircle Platform: Care Coordination Engine," Qualcomm Life, http://www.qualcommlife.com/healthycircles (February 20, 2014).

46 Home, Greatbatch, http://www.greatbatch.com/ (December 12, 2013).

47 "Covidien Acquisition of ev3 Inc.," Covidien press release, June 12, 2010, http://investor.covidien.com/phoenix.zhtml?c=207592&p=irol-newsArticle&ID=1446653&highlight= (December 12, 2013).

48 Kyphon Inc. S-1 Form, 2000 http://www.secinfo.com/dut49.54We.htm (December 12, 2013). Reprinted with permission.

Acclarent Case Study

STAGE 5: STRATEGY DEVELOPMENT

As the team began thinking about its next stage of work, it was time to find a name for the venture. After considering a series of different possibilities, they agreed on Acclar*ent*. The decision to embed ENT within the new company name was intentional, as the group began to grow increasingly excited about the many opportunities to address unmet needs in the field.

Their next efforts were focused on transforming the *Balloon Sinuplasty*™ concept into a product and building a business capable of bringing that technology to market.

5.1 IP Strategy

As described earlier, the team recognized intellectual property as an important element of its work early in the life of the company. At this point, it was decided that a full IP search would be conducted to evaluate the complete landscape of patents and study the prior art. While several relevant prior art references were found and added to the information disclosure statement made to the patent office, nothing was discovered that the team believed would stand in the way of proceeding with the chosen technology into commercialization. "At this point we were inventing at a very fast pace and we tried to capture as much as we could in our notebooks," Makower said.[1] "It seemed that wherever we looked we saw more and more opportunities to expand the technology of Balloon Sinuplasty more broadly, along with other product extensions and other ENT needs that still required innovation."

5.2 R&D Strategy

After the meetings with Sillers and Bolger, Chang, who would become the company's vice president of engineering, remembered thinking, "Okay, we need to add to the team. I can't do all of this by myself." His first move was to bring in an experienced engineer with a broad background. "I needed someone who was a jack-of-all trades, executes well, and who I worked well with under pressure and stress." To fill this role, Chang recruited a former colleague, Julia Vrany, with whom he knew he could partner to do everything from "clinical protocol writing, cadaver testing, development, working with vendors, quality, sterilization, validation and verification testing, and packaging – the full gamut of product development to get to first-in-human testing." Chang's next hire was a young engineer, John Morris, who was early in his career. "He was a Stanford grad – smart and hungry for experience," Chang said. As development progressed, Chang recognized the need for more balloon expertise so he identified an OEM vendor that was a specialist in this area. "So we had our super-broad generalist who was great at project management, our worker-bee engineer, and a great OEM partner who really knew balloons," he said. Together with Chang, these three pieces formed the team that would develop the working product used in the company's first-in-human study. "It's a balancing act," he added. "You need enough resources to get you there, but not so many that you sink the ship" (see Figure C5.1).

At this point, they began working on the key technical challenges associated with the "family" of devices necessary to perform a balloon dilation procedure in the paranasal sinuses, which included not just the balloons but the guides, inflation devices, and other key elements of the system. However, not surprisingly, many of the greatest issues to overcome had to do with the balloons. Chang elaborated:

We needed something that was non-compliant, meaning it would keep its shape. We were fracturing bone, so we had to figure out what sort of pressures were required. And then, probably most

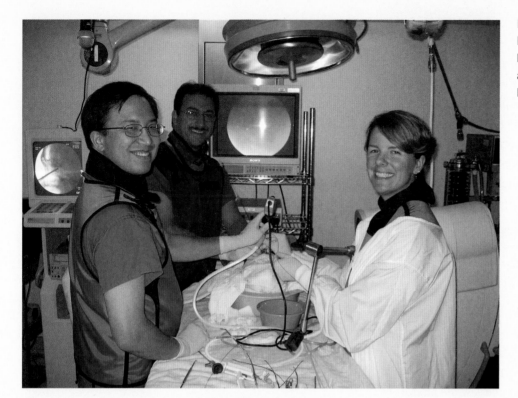

significantly, we saw that we were puncturing balloons. We realized that we were placing more stress on the balloon than typically seen in a coronary or peripheral vascular procedure. The targeted bone was not only strong, but typically sharp. And since we recognized that doctors would likely be inclined to use the same balloon in multiple sinuses, the balloon would need to survive multiple cycles of placement, dilation, and removal.

This durability issue meant that the team could not use available balloon technology – Chang and his engineers agreed they would have to develop something unique. "It turned out that helping with development is exactly what our vendor did very well," he said. The next step was to move into a collaborative development loop, with Acclarent using input from its expert physicians (such as Sillers and Bolger) to refine the requirements it fed to the supplier. According to Chang, "Working with our physicians, we were really able to tune the catheter length, balloon dimensions, pressure requirements, and burst requirements – we quickly iterated a number of times between the doctors and the lab [where the testing occurred] and the balloon manufacturer. It was a great partnership."

Of course, noted Chang, "We were in super-stealth mode at the time. So we didn't really tell the supplier what we were doing, even though we had an NDA. After we had made several iterations and significant progress, they more frequently asked what we were doing. I'd say, 'Look, you've just got to understand that we're not going to tell you what we're doing. But I can tell you what the spec on the balloon needs to be.' It became a running joke. We would have a call about the next iteration and they would close with, 'And you're going to use this on …?' And I would just answer 'Somewhere near, in, or around a body that may or may not be human.'"

Acclarent entered into an exclusive supply agreement with this vendor for balloons in the ENT field. "Our approach as a start-up company was that we didn't need to own every solution or every capability. There are a lot of people who know how to make balloons. Our goal was to find the one that would satisfy our needs. Once we identified the right supplier, we promptly transitioned from a project-based OEM relationship to more of a long-term partnership that we could scale with," Chang explained.

To get a working product, the team used a similar approach for developing the other devices it needed. Basic development, robust testing, and the detailed refinement of all specifications was performed in-house, but then external suppliers were entrusted to produce the actual devices. Under Facteau's leadership, prior to the commercial launch, Acclarent would make the decision to bring all development and manufacturing in-house. But, until then, this model served the company well.

Chang described the engineering environment at the time as requiring "all hands on deck." Everyone on the team was expected to contribute at full capacity. In terms of the philosophy, he said:

Sometimes people fall into the trap of wanting a perfect solution, the perfect prototype. We weren't at all afraid of trying something a little crude and fast – in fact, the faster the better, because you can't replace time. You don't have to be perfect because you're going to learn so much from each prototype. Failure teaches you a ton, right? So just go, go, go, iterate as much as you can. We literally had labs two days apart where we'd go in and test the product to its limits. If something didn't work well, we'd go back, try to fix it, get another cadaver, and try again. It was all about speed. Otherwise, you could spend a month, or two, or three, or more trying to make the perfect prototype. And that just didn't make sense from our perspective.

Before long, the team got to the point where it had a system of devices that allowed physicians to get access to a sinus, demonstrate that they were truly inside it, put a balloon across, dilate the ostium, and show that they had enlarged it via fluoroscopy and endoscopy (see Figure C5.2).

At this point, Chang said, "We felt ready to prove it clinically – to demonstrate proof of concept, safety, and feasibility."

5.3 Clinical Strategy

To prepare for first-in-human studies, Acclarent took a careful approach, working closely with its expert physicians to plan an appropriate course of action. When the team asked the physicians about the risks associated with performing the procedure in humans, they referenced skull-base fractures that could create serious complications. "We needed to determine if a high-pressure balloon

dilation would cause a fracture to propagate in an uncontrolled manner or cause a 'tectonic plate' shift of bone and ultimately lead to a skull base fracture and cerebral spinal fluid leak, or worse," remarked Chang. If the company could demonstrate that this risk was minimal, Bolger, Sillers, and others advised, it would be ready to move into a first-in-human safety and feasibility trial.

To clear this hurdle, they brainstormed with the physicians to come up with a study that would demonstrate that Balloon Sinuplasty devices did not create these adverse events. In the end, they designed a cadaver study that would require a CT scan before and after the ostia were accessed and dilated. In order to test the "worst case scenario," the study would use the largest balloon Acclarent had developed and it would be inflated to its maximum pressure. A comparison of the CT scans would allow the team to assess any adverse events. Concerned about microfractures that might be missed in a CT scan, however, one physician suggested that a complete dissection might then be used to examine the results of the procedure. According to Chang:

So that's what we did. We took six cadaver heads, scanned them, put them in a cooler, jumped in the car to take them to the cadaver lab, dilated them, took them back and scanned them again, and then brought them back again to the lab where we cut them in half. The physician who was there [Bolger] spent three days with us. He flew out and dissected every single cadaver head and said, "Okay, looks good. No fractures." Every single one. We put the results together and the work resulted in a paper and the foundation for our argument to move into live humans.[2] This was the last key piece we needed to get the ethics committee comfortable with the idea that the technology was safe for human use.

Consistent with its stealth strategy, Acclarent decided to perform its first-in-human trial in Australia. This approach would allow the company to keep the study quiet while still generating results that would have a relatively high likelihood of being accepted in the US by the FDA. In addition to having a population that enjoyed reasonably high standards of living and healthcare, Australia had a well-established regulatory agency, the

(a)   (b)   (c)

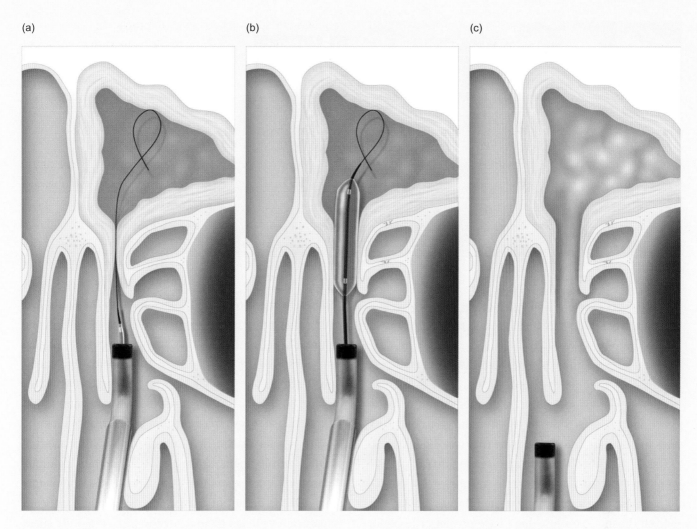

FIGURE C5.2

There are three primary stages of using the Balloon Sinuplasty technology. First, a flexible atraumatic guidewire gains full access to the sinus cavity (a). Second, a balloon catheter smoothly tracks over the wire. Balloon dilation opens the blocked ostia (b). Third, the devices are withdrawn, leaving open ostia with minimal tissue disruption (c) (provided by Acclarent).

Therapeutics Goods Agency (TGA). "You can never be positive how your data will be received, but we were hopeful that it would be received favorably in the US," said Chang. Another advantage of performing the study abroad was to help Acclarent further delay any competitive threat and strengthen its IP position based on the results of the study before word of its approach got out in the domestic market.

In Australia, clinical trials were regulated by the TGA and required ethics committee approval prior to being launched. The process of gaining ethics committee approval was similar to being granted institutional review board (IRB) approval in the US by the hospital or other facility where the procedures would be performed. The first step was to find an investigator willing to work with the company to perform the study. Fortunately, the expert physicians in the team's network were able to help them identify and secure the participation of an appropriate surgeon, Dr. Chris Brown in Melbourne, Australia. Acclarent then put together a package of information that included all the necessary background information, cadaver test data, and other relevant materials for an ethics committee review. "The ethics committee had a couple of questions for us, which we answered to their satisfaction," recalled Chang. "So they gave us a green light for doing the cases in Australia. But they

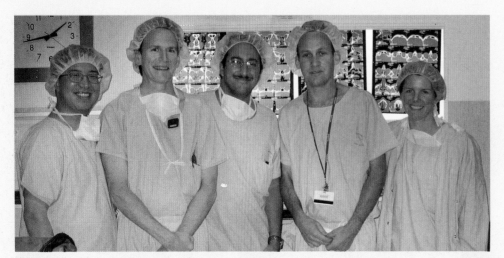

FIGURE C5.3
From left: Chang, Bolger, Makower, Brown, and Vrany after completing the first clinical trial cases using Balloon Sinuplasty technology (courtesy of ExploraMed).

limited us. They said, 'You can only do 10 patients, and you can only treat two sinuses in each patient.' We said, 'Okay, that's good enough.'"

In total, it took the company about two months to get approval to conduct the study. Then, with everything queued up in advance, it took another month or so to enroll the patients, ship all required materials and supplies to Australia, and execute the trial (see Figure C5.3).

Chang described the outcome (see Figure C5.4): "We went in and had unbelievable results – 10 out of 10 patients were successfully treated. It was beyond our wildest dreams. You just don't dare predict or hope for something that good. We were very fortunate."

While an important clinical milestone had been achieved, Acclarent viewed this as just the beginning of its clinical strategy. The company's plan was to begin lining up its next human trials, which it would initiate in the US shortly after receiving FDA clearance for its devices (the next study would become known as the CLEAR study). "We knew we needed more clinical experience and more data that we could share with people," said Chang. "So we started to line up investigators in the US very carefully. The early physicians helped us identify more investigators and called ahead to make introductions. They would tell their colleagues, 'You're going to get a phone call from somebody. Talk to them. They're not going to tell you a whole lot until you sign an NDA.' This process was in anticipation of the need to increase our volume and get data from a mix of academic and private institutions."

There were also important lessons from the Australian study that required the company to change its clinical approach. For example, said Chang, "The outcome measure questionnaire that we used in Australia was much too complicated. People didn't understand it; they couldn't fill it out by themselves. We recognized that and switched to a simpler outcome measure instrument."

The results of the first-in-human trial also affected device development. Chang shared another example: "Based on our experience in Australia, we completely changed the design of our guides. We made significant modifications to get them ready for prime time. We spent a lot of time working on these types of changes before we went commercial."

5.4 Regulatory Strategy

Acclarent needed a regulatory strategy for its entire family of devices. According to Chang, "There were three different balloon sizes, five different guides, an inflation device – just a ton of different things." The team had previously assessed which components of the system would likely fall into each FDA class, as part of its decision to pursue a 510(k) regulatory pathway. The next step was to identify appropriate predicates for each one. For simple components, such as tubing, guidewires, and needles, Acclarent readily found predicates. However, identifying devices upon which it could demonstrate substantial equivalence for the balloons took a little more searching. "From a balloon perspective, we looked hard

Title: Safety and Feasibility of Balloon Catheter Dilation of Paranasal Sinus Ostia: A Preliminary Investigation

Authors: Christopher L. Brown, MD, William E. Bolger, MD

Objectives: Endoscopic sinus surgery (ESS) is an effective option for managing patients in whom medical therapy for rhinosinusitis fails. However, ESS is not always successful, and serious complications can occur. New techniques and instrumentation that improve outcomes and reduce complications would be seriously welcomed. Innovative catheter-based technology has improved treatment of several conditions such as coronary artery disease, peripheral vascular disease, and stroke. Recently, catheter devices have been developed for the paranasal sinuses. Cadaver studies confirm the potential use of these devices in rhinosinusitis. The objective of this investigation was to ascertain the feasibility and safety of these newly developed devices in performing catheter-based dilation of sinus ostia and recesses in patients with rhinosinusitis.

Methods: A nonrandomized prospective cohort of 10 ESS candidates was offered treatment with a new technique of balloon catheter dilation of targeted sinus ostia. The frontal, maxillary, and sphenoid sinuses were considered appropriate for this innovative catheter-based technology. The primary study end points were intraoperative procedural success and absence of adverse events.

Results: A total of 18 sinus ostial regions were successfully catheterized and dilated, including 10 maxillary, 5 sphenoid, and 3 frontal recesses. No adverse events occurred. Mucosal trauma and bleeding appeared to be less with catheter dilation than is typically observed with ESS techniques.

Conclusions: Dilation of sinus ostial regions via balloon catheter-based technology appears to be relatively safe and feasible. Larger multicenter clinical trials are now warranted to further establish safety and to determine the role of this new technique.

FIGURE C5.4

An overview of the results of Acclarent's first-in-human clinical trial, as summarized in the abstract to the clinical journal article in which the results were eventually published (from *The Annals of Otology, Rhinology, and Laryngology*, April 2006).

for predicates in ENT," said Chang. "We actually found something in the eye, a lacrimal duct catheter, that was pretty close." This balloon device was developed for dilating constricted lacrimal ducts which connect the eye to the nasal cavity; thus, the location was similar to Acclarent's intended use. The company would make submissions, with good predicate devices, on all aspects of its system with the intent of having the FDA either exempt or clear them for commercial use.

A strategic decision would need to be made regarding the submission for the balloons. Acclarent had to determine whether or not to submit the data from its Australian study along with its 510(k) application for this device. It was unclear if the data would be required by the FDA as no other device in this category needed clinical data to receive clearance by the agency. For this reason, the decision was made to submit the 510(k) without clinical data, but to be prepared to provide it if the FDA made a direct request.

When Acclarent made its submissions for its class I devices, the FDA responded within 10 days, indicating that all of them were exempt from premarket notification requirements. One week after the first submission, the company filed its 510(k) application for the balloons. Two months later, the FDA responded with a series of questions, including a request for clinical data to demonstrate that there would be no adverse events. "I think they probably expected us to file for an IDE," said Chang, "but we already had the study results in our back pocket." Within a week, the team had prepared a comprehensive response to the FDA's inquiry. "We scrambled that whole week to answer their questions in the most thorough and complete way we could. It was a true team effort. Within a week, we had a thorough and carefully reviewed clinical package with data submitted for the FDA's review. After a few additional phone calls and questions, we were able to provide the appropriate information that eventually led to our 510(k) clearance. The company was now one-step closer to its planned clinical trial in the US, which it intended to launch immediately after clearance.

5.5 Quality Management

In parallel with its clinical and regulatory efforts, the Acclarent team began thinking about quality system and the need to bring some structure to the way the company managed important processes. According to Chang:

When we started getting some traction and realized that we were marching toward a major clinical experience, we recognized that we needed to have a quality system and certain controls in place. Because we were running lean, we had a consultant come in and we worked with him to develop a fledgling, flexible quality system. We knew we needed design controls, processes for supplier audits, and other standard operating procedures so we could prepare for a commercial launch of the products.

The idea was to build a quality system that would initially allow the company to cover the FDA's requirements, but that had the capacity to effectively and efficiently scale with the company as it grew and faced the need for more structure and formality in its approach. "We started with standard operating procedures [SOPs] that were flexible and adaptable," said Chang. "We recognized as we grew, we could then tailor these SOPs and make them more specific. And that's really what has happened over the last four years."

With the plan in place, the team began executing it. Because Acclarent would be using contract manufacturers to produce its first family of products, Chang had particularly vivid memories of performing several supplier audits. "We went out to vendors and audited them with our check list. We examined their quality manual, how they received and inspected materials, as well as their manufacturing processes and methods of keeping lot history records, etc." Based on the outcome of these inspections, the team chose certain vendors to work with, ranging in size from large, established contract manufacturers to smaller "mom and pop" shops with specialized expertise in a certain area. "What we learned over time," said Chang, "was that some vendors were not as good at helping us develop processes, with appropriate quality systems, as others." Gradually, as Acclarent moved closer and closer to launch, it had to replace those suppliers with others that could better support the company's growing needs.

5.6 Reimbursement Strategy

As part of its preliminary reimbursement analysis, Acclarent received two favorable opinions that it would be able to use the existing FESS codes to obtain reimbursement for its Balloon Sinuplasty devices. However, as Lam Wang pointed out, "It's much more political than

that." The company believed that it would be important to get the leading professional society in the field – the American Academy of Otolaryngology (AAO) – to endorse the position that the new balloon technology could be used with these codes. Lam Wang elaborated:

In the end it's the third-party payers who have to pay for the technology. In general, when the volume of claims being submitted by surgeons adopting a new procedure is low, the claims can go under the radar. It's when the volume of claims starts to get kind of high that the insurance companies start looking for reasons to not pay. When you hit this critical mass, they start questioning it. And it's really good if you have a professional society to support you, saying, "We agree." The third-party insurance companies don't pretend to practice medicine – they defer to the physicians. If the physicians say, "We used these new products to perform the FESS procedure because of medical necessity and we agree that they should be covered by these codes," they're more likely to agree.

To solicit AAO support, Acclarent's reimbursement consultant recommended that the company seek a letter from the society indicating that it agreed with the use of FESS codes to cover the Balloon Sinuplasty devices. Acclarent could then take such a letter to the American Medical Association (AMA) and work with that organization to make this recommendation part of the current addendum to its coding policy. Ideally, this letter would be created and provided by physicians familiar and experienced with the technology. Thus, the selection of investigators was an important step in developing this strategy. The good news was that it seemed clear from the feedback that the device's use fell squarely under the existing codes.

5.7 Marketing and Stakeholder Strategy

The company's efforts to secure reimbursement were closely coupled with its marketing and stakeholder strategy. While reimbursement looked at how physicians would be paid for performing the procedure with Acclarent's products, the larger marketing and stakeholder strategy sought to understand the other ways in which Balloon Sinuplasty devices would affect their practices so that Acclarent could promote the benefits and mitigate any downsides. Lam Wang described the in-depth interviews she conducted with a handful of physicians (who were all under NDAs): "My approach was to understand who they are and how their practices work. Fundamentally, I wanted to understand the product positioning of Balloon Sinuplasty within the physician's practice and within the treatment of this patient population."

According to Lam Wang, the responses she received varied significantly. Some physicians viewed the balloon as just another "club in their golf bag" that they would use when it made sense, or in hybrid procedures that used both traditional FESS and Balloon Sinuplasty tools. "One doctor took it on a sinus-by-sinus approach," she recalled. "He said, 'The maxillary sinuses are so fast and easy that I wouldn't bother to use the balloon. But, for the frontal sinuses, this is definitely an awesome solution.'" Other physicians viewed the balloons as a more fundamental leap in technology. "They saw it as a potential paradigm shift in how they would treat patients – no more sharp instruments, no more picking at tissue, but balloons, catheters, and wires." While tools could have utility in most of the sinuses, it was not clear at the time how the technology would apply to the sinus called the ethmoid sinus since it did not have one simple drainage pathway like the others. Some physicians cited this limitation as a reason why the device might have less utility. Comments such as these gave Acclarent a strong understanding of how Balloon Sinuplasty technology would potentially fit into these physicians' practices, how it would be used, and, by extension, how it would affect them financially.

Lam Wang did some additional research into the effect on facilities – primarily hospital outpatient facilities and ambulatory surgical centers. "I also looked at the ramifications of bringing endoscopic sinus interventions back into the office, but it was clear this was not going to be as easy." As the launch approached, Acclarent began revisiting the needs of patients and evaluating the interests of referring physicians (e.g., general practitioners who would refer patients to the sinus specialists).

When it was time to define clear value propositions for each of these stakeholder groups, Lam Wang sought to put these messages in context:

I felt it was important to have a story. I believe that's the backbone of any marketing program, especially for a new company with a new technology. So I spent some time putting together the story for Acclarent. It's not enough to say "new product, unmet need, done deal." You really have to tell a story for why this technology makes sense. The value it provides is the core component of the messages you give out to all of your stakeholders, whether it's patients, surgeons, or facilities.

Once these value propositions were defined and embedded within the context of a story for each stakeholder group (see Figure C5.5), Lam Wang turned her attention to the development of a plan to support promotion and advocacy. "Our next focus was all about getting ready for the commercial launch," she said. The team hired an ad agency to help it establish a brand, as well as a look and feel for the product. This brand was used as the foundation of all marketing communications, including brochures and a website (note that the company's publication strategy was managed separately; see the Sales and Distribution Strategy section for information about professional training and education). Significant effort was also put into planning a major product launch at the American Academy of Otolaryngology – Head and Neck Surgery (AAO-HNS) conference, which would occur in September 2005. This was the meeting at which Acclarent planned to transition out of stealth mode and announce its technology to the world. The launch plan for this event included a booth, where Acclarent staff would show the technology and answer questions throughout the conference, as well as a symposium at the show with presentations by thought leaders in the field (e.g., Bolger and Sillers) and a video demonstrating on a real patient how the procedure worked.

Pricing

Acclarent also needed to decide on a price for its products, an exercise that was closely linked to the company's business model, funding requirements, and its overall viability. "We had to figure out the average price we could sustain per procedure," said Lam Wang. She explained her approach:

Key Messages to ENT Surgeons
- There are over 600,000 chronic sinusitis or recurrent acute patients per year in the U.S. who have unmet clinical need. These patients are underserved by medical therapy; they are surgical candidates, but they are not getting FESS.
- The goals of sinus surgery are to restore ventilation and normal sinus function by opening up blocked ostia while maintaining as much normal anatomy and mucosa as possible.
- However, there are shortcomings associated with today's surgical instruments that make achieving those goals challenging. Straight, rigid tools are used in the tortuous sinus anatomy for a delicate, targeted procedure. As a result:
  - More tissue removal than necessary often required just to gain access (e.g., ethmoidectomy to access sphenoid or frontal areas, uncinate regularly removed during maxillary antrostomy.
  - Procedure hard to tolerate for patients – bloody, packing, general anesthesia, long recovery. Many patients hesitate to undergo surgery.
  - Potential complications can be serious – penetration of eye or brain cavity with these straight, rigid steel instruments. FESS most common reason for litigation against ENT surgeons in the U.S.
  - Frontal procedures are challenging and prone to iatrogenic scarring and stenosis – unrefined instruments make it easy to cut, tear or strip mucosal lining from the recess during ethmoid surgery.
- Improvements to the field have been made – but still same basic problem of using straight, rigid tools.
  - Microdebrider – powered instrument shaves away diseased tissue while sparing normal tissue.
  - Image Guided Surgery – allows surgeon to see instrument location relative to historical CT landmarks.
- There is a need for a new way to accomplish sinus surgery:
  - Devices that conform to the tortuous sinus anatomy.
  - Devices that stretch sinus tissue rather than tear or cut.
- Now, there is a new, less invasive procedure alternative.
- Balloon Sinuplasty uses soft, flexible balloon catheters to gently open up blocked ostia, allowing drainage and return to normal sinus function.
  - Innovative platform change in technology from straight, rigid instruments to soft, flexible catheter-based devices.
  - Devices conform to the tortuous sinus anatomy.
    - Less need to remove sinus tissue in order to gain access.
    - Allow improved access to more remote areas.
    - Less likely to penetrate into the orbit or brain.
  - Balloon catheters stretch sinus tissue rather than tear or cut.
    - Less bleeding during and following surgery.
    - Less likely to stimulate scar tissue.
    - Lower potential of revision surgery.
  - Adheres to physiologic principles and goals of sinus surgery.
  - Safe, low rate of complications.
- FDA cleared.
- Balloon Sinuplasty works.
  - Clinical data to date.
  - Case studies.
- We have thought leader support.

FIGURE C5.5
Acclarent developed a story that included its core value propositions for each key stakeholder group. The example shows some of the messages targeted at ENT surgeons (provided by Acclarent).

For the first pass, I revisited the average payment for FESS per sinus treated, and then by procedure. I understood after this analysis that, in a given FESS procedure, physicians could bill up to nine codes, and that the codes would stack, meaning that they would get paid the full amount for the first procedure done on a certain sinus and then receive 50 percent of each code after that. Next, I did this super in-depth analysis to determine what a typical case looked like at the time. That's when I went back to talk to our physician advisors. And I learned that there was a combination of sinuses that was typical. After that, I looked at an average case from a break-even perspective. If I was going to add cost to the system [by charging for reusable devices], I then had to figure out where I could take cost out of the system.

"I looked at it from a long-term perspective to make an economic justification for this new technology," said Lam Wang. Ultimately, the sum total per case benefit of what a physician or facility would gain from increasing patient flow and decreasing procedure time, taking into account what third-party insurance companies would pay, represented the high end of what Acclarent could charge. She continued, "I completed a 12-page analysis and came up with a range – it was something like $1,200–$2,000 per case. So not per balloon, not per device, not per sinus, but just per patient."

With this range in mind, the Acclarent team again approached its advisors to discuss pricing. For the purposes of conversation, the company used a figure of approximately $1,200 per procedure. Facteau explained what happened next:

When we talked with some of our physician advisors about this, we got some initial push back. At that point, we decided to share some confidential information to help them gain an appreciation for what it really takes to build a medical device company from scratch. We explained to them how and why we came to our decision on price, we shared with them our P&L and cash flow assumptions, and pointed out that in our first year we would lose $16 million. We mentioned that we could do things a lot less expensively, however, we

believed the right thing to do to build a great company is to invest in clinical research, physician training (which included a cadaveric experience), and innovative products. Couple those investments with a direct sales organization and we anticipated the need to raise $75–100 million before we would break even. Therefore, at that point, $1,200 of disposables per procedure really didn't seem to be as big of an issue for them, and they became supportive of the pricing strategy.

5.8 Sales and Distribution Strategy

In terms of sales and distribution, Facteau had a clear vision from the outset regarding the model he wanted to adopt:

I had a strong bias that we should go direct, and that we needed to hire an experienced, clinically savvy sales organization that could execute on our vision. We needed to train physicians really well to preserve the safety of the device and the procedure, and to protect the Balloon Sinuplasty reputation as a safe tool. We thought about how we could leverage distributors, but the direct model made more sense. I was a strong proponent of this. When no one gave me any real resistance, internally or with our investors, we agreed to create a direct sales organization in the US.

Lam Wang underscored the key advantage of this approach. While a direct sales force was more expensive and time-consuming to build, she said, "It allows you to maintain more control."

Deciding on the type of sales representatives that would be best suited to represent Acclarent's technology turned out to be something of a challenge. "We asked ourselves, of all the people in our sales careers that we had interviewed, how many had come from ENT?" recalled Facteau. "And we couldn't point to one." For this reason, the company decided not to target sales people with ENT experience. Instead, they decided to look for people who had worked in interventional cardiology and had experience selling balloons, catheters, guidewires, and other equipment similar to the components of the Balloon Sinuplasty system. "That's where we started," Facteau said, "but that turned out not to be the right model for us."

695

Ultimately, Facteau and team discovered that, while cardiology sales people understood the technology, most had not actually developed the skills to introduce a paradigm shifting technology into the operating room.

Ultimately, Acclarent determined that it needed sales people with deep experience training physicians and who had spent time in the operating room. According to Facteau, representatives who had been in a start-up before also tended to work out well, as did people who had helped develop new markets in the past. "Market development is a lot different than taking share in an established market," he noted. "And I think a lot of people don't really appreciate that. If you have a better mousetrap and there is an established market for it, that's a much easier, predictable sale than if you have to go out and create a whole new market from scratch."

Training

Through its direct sales force, Acclarent intended to invest substantially in the training it delivered to physicians on its technology. "Even though it had never been done before in ENT, we knew that's what we needed to do," said Facteau. The company believed firmly that Balloon Sinuplasty products would lead to improved results if they were used consistently by surgeons – the key was to get each one to use the products and perform the procedure in the same way to achieve comparable results to those achieved in the company's studies. "We had to standardize," Facteau recalled. "What was missing in sinus surgery was standardization. You could go to 10 different hospitals and watch sinus surgery and you'd see 10 different ways to do it." For this reason, the team worked diligently to define best practice processes and protocols that would be taught consistently to physicians and could easily be replicated post-training. Facteau elaborated:

I started my career at US Surgical, and I am of the opinion that they were the first to crack the code on physician training. They spent a significant amount of resources to ensure proper education for new medical procedures. One of the guiding principles to successful patient outcomes and adoption was centered on standardization. No matter what OR

you were in, general surgeons placed trocars in the same location, they held the instruments the same way, and utilized the same retraction techniques to gain better visibility. This ultimately led to better outcomes. We set out to accomplish the same with Balloon Sinuplasty technology. "If you do it this way, we believe you're going to get good results because we know the proper techniques resulted in good outcomes in our clinical studies."

Led by experienced physicians and Acclarent's sales representatives, physicians would be trained on this standardized approach. For those surgeons open to the idea of training, the team was confident that this approach would protect patients and the reputation of the company while enabling desired results. However, as Facteau explained, "We had some concern that this level of standardization – our commitment to standardization – might not necessarily be well received, especially early on because it had never been done before in ENT by a manufacturer. Despite this potential resistance, the company believed strongly that this was an appropriate (and necessary) approach to take when introducing novel technology to a market that had not seen a great deal of innovation in decades (see Figure C5.6).

As far as building the sales organization, Facteau took a somewhat measured approach. For instance, he decided that Acclarent would not begin hiring any sales representatives until the company had received clearance from the FDA on its balloon technology. Other critical milestones were put into place to keep the rate at which the sales organization expanded in alignment with the commercial viability and adoption of the technology. While this made some members of the team a little nervous as the launch date rapidly approached with no sales force in place, "We felt like it was fiscally responsible – and the right thing to do," said Facteau.

5.9 Competitive Advantage and Business Strategy

When the company began thinking about creating a competitive advantage, the Acclarent team relatively quickly

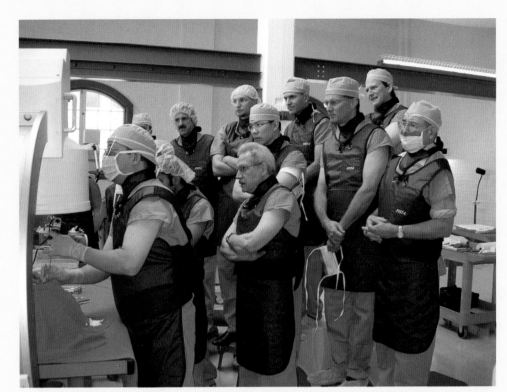

FIGURE C5.6

Surgeons being trained to use Balloon Sinuplasty products in Baltimore, Maryland (courtesy of Acclarent).

FIGURE C5.7

Facteau lays out his vision for Acclarent at a staff meeting (courtesy of Acclarent).

focused on two core capabilities: developing a world-class direct sales force/training organization and establishing a pipeline of innovation. Facteau explained (see Figure C5.7):

The first was around the commercialization. We wanted to build the best ENT commercial organization in the world. There have been some good examples of companies in other specialties that have done that, but no one had tackled it in ENT. We benchmarked companies like Kyphon, Fox Hollow, and Perclose. We performed case studies in an attempt to understand what they did well and to try to learn from their mistakes. In doing this, one of the things we realized was that none of these companies developed a true pipeline of innovation that could drive organic growth for a long time. We decided that we wanted to be great at both, which is no easy feat.

Developing a strong internal R&D capability was also consistent with Acclarent's business model and the company's desire to get ENT physicians "addicted to innovation." Lam Wang added, "Bill's vision was not just to focus on 'N,' but 'E' and 'T,' too. We want to be everything for the ENT surgeon." Sensing that untapped opportunities existed across the ENT field gave Acclarent a source of ideas that it could develop into an ongoing pipeline of innovative products. The decision to focus on developing this capability dovetailed with Acclarent's decision to begin the process of bringing all R&D and manufacturing in-house.

NOTES

1 All quotations are from interviews conducted by the authors, unless otherwise cited. Reprinted with permission.

2 William E. Bolger and Winston C. Vaughan, "Catheter-Based Dilation of the Sinus Ostia: Initial Safety and Feasibility Analysis in a Cadaver Model," *American Journal of Rhinology*, May/June 2006, pp. 290–4.

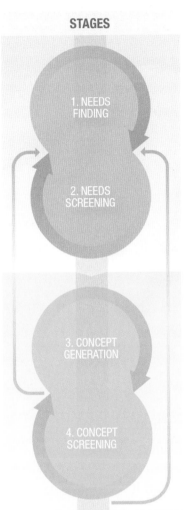

IMPLEMENT ▷ Business Planning

PHASES	STAGES	ACTIVITIES

IDENTIFY

1. NEEDS FINDING

1.1 Strategic Focus
1.2 Needs Exploration
1.3 Need Statement Development

2. NEEDS SCREENING

2.1 Disease State Fundamentals
2.2 Existing Solutions
2.3 Stakeholder Analysis
2.4 Market Analysis
2.5 Needs Selection

INVENT

3. CONCEPT GENERATION

3.1 Ideation
3.2 Initial Concept Selection

4. CONCEPT SCREENING

4.1 Intellectual Property Basics
4.2 Regulatory Basics
4.3 Reimbursement Basics
4.4 Business Models
4.5 Concept Exploration and Testing
4.6 Final Concept Selection

IMPLEMENT

5. STRATEGY DEVELOPMENT

5.1 IP Strategy
5.2 R&D Strategy
5.3 Clinical Strategy
5.4 Regulatory Strategy
5.5 Quality Management
5.6 Reimbursement Strategy
5.7 Marketing and Stakeholder Strategy
5.8 Sales and Distribution Strategy
5.9 Competitive Advantage and Business Strategy

6. BUSINESS PLANNING

6.1 Operating Plan and Financial Model
6.2 Strategy Integration and Communication
6.3 Funding Approaches
6.4 Alternate Pathways

PROJECT LAUNCH

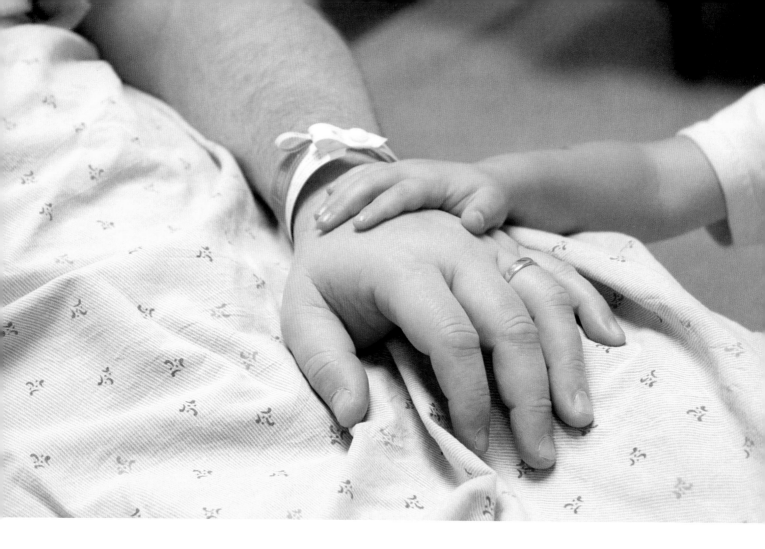

Knowing what to do is not enough. One of our main recommendations is to engage more frequently in thoughtful action.

Jeffrey Pfeffer and R. Sutton[1]

Now this is not the end. It is not even the beginning of the end. But it is, perhaps, the end of the beginning.

Winston Churchill[2]

6. BUSINESS PLANNING

While this is the final stage of the biodesign innovation process, it is in fact the beginning – the beginning of your sustained effort to implement a product or business around the need you've identified and your invention or innovation. Business planning supports the integration and direct execution of all of the strategies plans you've developed in Stage 5.

A very specific focus of this stage, and indeed this book, is around the start-up process. Building and managing a small business, generating the business model, developing a cohesive pitch, and navigating the complicated waters of fundraising are all essential components and are explored in depth. Remember, once outside sources are involved, the enterprise no longer belongs solely to the innovators.

The final chapter looks at alternate approaches to starting a business – that is, the options of partnerships, licenses, or the outright sale of an idea, transferring control to the new owner. If you have a great idea, but there is an existing company with better resources to develop it, one of these pathways can provide a wonderful route to get a solution into practice.

Regardless of how you approach it, the journey should be fun. That's not to say that many lessons won't be learned the hard way, but the optimism and, indeed, idealism of the innovators can profoundly catalyze transformation in healthcare. Good luck.

NOTES

1 Jeffrey Pfeffer and Robert I. Sutton, *The Knowing-Doing Gap: How Smart Companies Turn Knowledge into Action* (Perseus Distribution Services, 1999).
2 According to the Churchill Centre, this statement was made at the Lord Mayor's Luncheon, Mansion House, London, following the victory at El Alamein in North Africa, November 10, 1942.

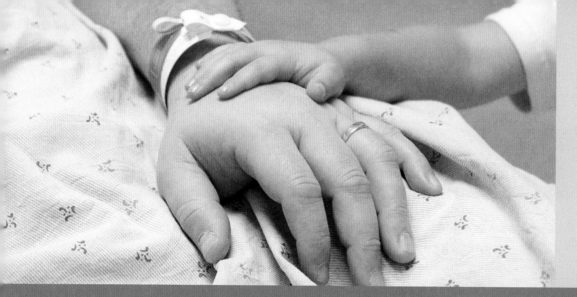

6.1 Operating Plan and Financial Model

INTRODUCTION

Engineers to design, prototype, develop, and manufacture a new product. Clinicians to run clinical trials. Statisticians to analyze the data. Executives to develop marketing and sales strategies. Reimbursement consultants to secure codes, coverage, and payment. Sales people to get the technology into the hands of customers. Key opinion leaders to promote its adoption. All of these individuals are essential to the development and commercialization of a medical device. Yet, without a carefully integrated plan that captures and coordinates these complex, interdependent efforts, innovators cannot fully understand whether the effort required to develop and commercialize the offering will justify the associated investment.

Through the operating plan and financial model, innovators will gain an understanding of the time and money necessary to develop and commercialize their new solution. This information will be used both to assess the viability of pursuing the proposed solution and to provide a blueprint for implementing the company's business strategy, realizing its vision, and monitoring the results it achieves.

To prepare an operating plan, innovators specify precisely who will execute the various strategies defined through the strategy development stage of the biodesign innovation process, when and in what order key activities will be performed, and with what resources. This information is then translated into costs and consolidated into an integrated financial model that also includes a detailed revenue plan for capturing a share of the potential market. By comparing the revenue projections to the detailed cost estimates that stem from the operating plan, the innovators can confirm whether or not the potential market justifies the financial requirements for developing and commercializing the product.

 See ebiodesign.org for featured videos on preparing an operating plan and financial model.

OBJECTIVES

- Understand how to develop an operating plan, cost projections, and revenue model and then integrate them into a unified financial model that can be used to validate an opportunity and support business planning.

- Appreciate how to make and validate medtech-specific assumptions to support the creation of an operating plan and financial model.

- Learn to identify the most important strategic and tactical issues that should be reflected in the operating plan and financial model.

- Understand how to perform a proxy company analysis to validate all components of the operating plan and the financial model against a more established company with attributes similar to the new venture.

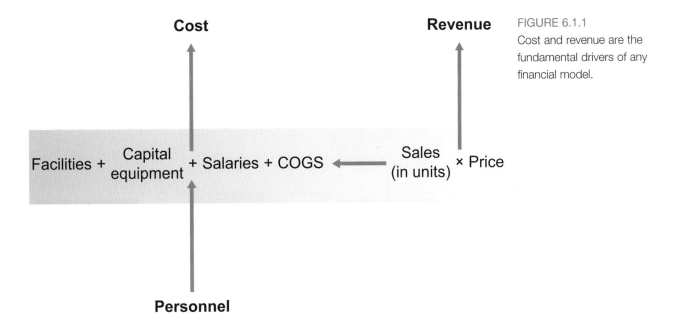

Cost   **Revenue**

FIGURE 6.1.1
Cost and revenue are the
fundamental drivers of any
financial model.

Facilities + Capital equipment + Salaries + COGS ← Sales (in units) × Price

Personnel

OPERATING PLAN AND FINANCIAL MODEL FUNDAMENTALS

Thomas Fogarty, innovator and founder of more than 30 **medtech** companies, highlighted the diversity of the skills required to bring a new device to the market:[1]

> *Particularly in this day and age, you need people from different disciplines – you need intellectual property attorneys, you need corporate attorneys, you need regulatory experts, you need good engineers ... and you need different types of engineers. You need the person who can conceptualize, the one who can **prototype**, the production engineers, and then what I call a "finisher." To bring this whole team together, you have to understand **value** allocation. All of these people create value. They bring something different to the table. If you think, just because you had the idea, you brought all the value, you're not going to be successful. Somebody has to implement your idea, and one individual can't do it.*

A **financial model** for a new venture is a detailed, quantitative articulation of Fogarty's statement. Until this point, innovators largely have been thinking about important activities and costs related to specific steps in the biodesign innovation process, first through landscaping and competitive scanning and then through more focused approaches. Now it is time to bring that information together and integrate it into a cohesive plan so that they can establish a comprehensive understanding of the magnitude of time, effort, and money required to bring the solution to market. In the US, it is not uncommon for a company to require more than three years and $4–$15 million to get a **510(k)** product to market. Technologies on the **PMA** pathway are more likely to take seven to 10 years and anywhere from $50 million to $120 million. Getting a **CE mark** product to market is still significantly more efficient, at roughly two years and $4 million. Of course, these estimates can vary dramatically from project to project, but they underscore the point that medtech development and commercialization is resource-intensive. More than anything, innovators need a realistic, practical understanding of what will be required, and the operating plan and financial model can provide it.

This exercise begins with the company's operating plan, tracking the cost of developing and commercializing the innovation, along with market revenue, over a period of five to seven years. For at least the first three years (and, ideally, the first five years), the model tracks costs and revenue on a quarterly basis and, after that, on an annual basis.

Costs include salaries, capital equipment, supplies, facility expenses, and cost of goods sold (**COGS**), which is the total material cost for manufacturing.[2] Revenue is total units multiplied by sales price. Figure 6.1.1

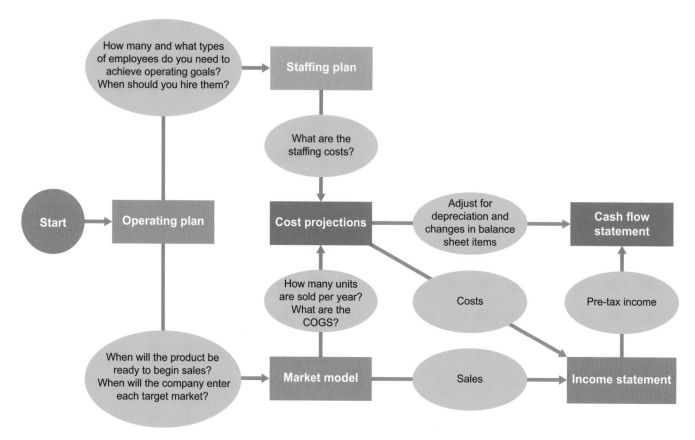

FIGURE 6.1.2

The six primary components of a financial model and how they work together.

provides a simplified view of these basic drivers of the financial model.

The financial model has six components, which correspond to the six steps in its development. The **operating plan** provides an overview of the activities that must be performed to develop the technology, their timing, and the key milestones they support. The **staffing plan** specifies the personnel needed to execute the operating plan. The **market model** includes revenue and market share projections, as well as a preliminary assessment of the sales force required to achieve these forecasts. **Cost projections** then integrate all costs together: salaries, facilities, equipment, supplies, and COGS from product manufacturing. The **income statement** subtracts all costs from revenue to generate an accounting income statement. An important distinction is that, with accounting income, the cost of capital equipment is not subtracted in the year the equipment is purchased. Instead, it is spread across the lifetime of the equipment. However, the capital necessary to acquire the equipment is needed in the

year of the purchase. Cash flows reflect this distinction and are calculated by subtracting the actual cash expenditures from the cash revenue recorded during a given time period. The cash flow analysis then provides the basis for estimating a company's funding needs. Each of these components is explained in more detail in the sections that follow. Figure 6.1.2 provides an overview of how these elements work together.

Throughout this chapter, an example will be referenced to illustrate each of the components of the financial model. The example focuses on a fictitious business developing a relatively traditional medical device. The major components of the operating plan and financial model for other types of offerings should be similar, although details may vary significantly. For instance, for a mobile health (**mHealth**) application, products are typically faster to prototype and can be moved into user testing more quickly. Additionally, while the overall capital requirements for their development is likely to be lower, the corresponding spend on marketing will be higher.

	Quarter																			
	1	2	3	4	5	6	7	8	9	10	11	12	13	14	15	16	17	18	19	20
Phase 1: Proof of concept																				
Hire CEO	▓																			
Initial catheter prototyping	▓	▓	▓	▓																
Generator prototyping			▓	▓	▓															
Clinical proof of concept testing			▓	▓	▓															
Phase 2: Final product development																				
Finalize catheter design						▓	▓	▓												
Finalize generator design						▓	▓	▓												
Complete manufacturing design								▓	▓											
Phase 3: FDA approval																				
Complete animal and bench top testing							▓	▓	▓											
First in human									▓	▓										
510k approval process										▓	▓									
Finalize manufacturing processes											▓	▓								
FDA approval												▓								
Phase 4: Reimbursement																				
Payer engagement						▓	▓	▓												
Preparation of payer materials									▓											
Payer communication										▓	▓	▓	▓	▓	▓	▓	▓	▓	▓	▓
Reimbursement support															▓	▓	▓	▓	▓	▓
Phase 5: Market availability																				
Initial sales force hires											▓	▓								
Design training program												▓	▓							
Limited pilot study and availability													▓	▓	▓					
Product generally available															▓	▓	▓	▓	▓	▓

FIGURE 6.1.3

Sample operating plan for a hypothetical ablation catheter business.

This particular example focuses on a product intended to treat atrial fibrillation (AF), a cardiac rhythm abnormality that causes the atria (the upper chambers of the heart) to contract irregularly. This condition creates a number of undesirable side effects ranging from palpitations and fatigue to the potential for a debilitating stroke. One way to treat AF is to destroy, or ablate, certain areas of the heart responsible for the initiation and perpetuation of electrical signals underlying the disease. Ablation is achieved through the use of an ablation catheter, which can be advanced into the heart via the blood vessels where it can be used to deliver energy to ablate the cardiac tissue (see 2.1 Disease State Fundamentals and 2.2 Existing Solutions for more information).

Operating plan

To prepare an operating plan, innovators should start with the high-level milestones a company chooses to chart its progress. Many of these early milestones are related to research and development (R&D), as explained in 5.2 R&D Strategy. Specific milestones vary from company to company. However, at a high level, they might include:

1. Proof of **concept**
2. Product development progress
3. Manufacturing feasibility
4. **First-in human** studies, **clinical trials**, and **FDA** submission

5. **Reimbursement** progress
6. Development of scalable manufacturing
7. Marketing and sales (US, Europe, Asia)

A detailed operating plan would break down these high-level milestones into additional granularity (e.g., reimbursement initiatives could be divided by **payer** type). Figure 6.1.3 shows a sample operating plan by quarter for the hypothetical ablation catheter business (assuming a 510(k) regulatory pathway and reimbursement under existing codes).

Innovators often would like to think that the operating plan does not change. However, at some point, most ventures face significant technical or clinical challenges that force the team to reconsider its plans and milestones. For example, the company featured in online Appendix 6.1.1 (Cardica, Inc.) faced a formidable set-back when the FDA changed the clinical trial requirements in the middle of its clinical studies as a result of safety problems with a competitor's product. The appendix explains how the company modified its strategy and operating plan in response to the unanticipated change.

Staffing plan

The staffing plan outlines the number of employees that need to be hired over time. The staffing plan is a direct outgrowth of the steps in the operating plan, since staffing requirements should be determined by the company's strategic and operating decisions. Essentially,

	1	2	3	4	5	6	7	8	9	10	11	12	13	14	15	16	17	18	19	20	6	7
Manufacturing team																						
Engineers	0	0	0	0	0	0	0	0	0	1	2	3	4	6	6	6	6	9	9	9	9	12
Manufacturing assemblers	0	0	0	0	0	0	0	0	2	2	2	2	5	5	5	5	10	10	10	10	18	35
R&D																						
Catheter team																						
Engineers	1	1	1	1	1	1	1	1	1	2	2	2	2	2	2	2	2	2	2	2	3	5
Techs	2	2	2	4	4	4	4	4	4	4	4	6	6	6	6	6	6	6	6	6	7	10
Generator team																						
Engineers	1	1	1	1	1	1	1	1	1	1	1	1	1	1	1	1	1	1	1	1	3	5
Techs	2	2	2	2	4	4	4	4	4	4	4	4	4	4	4	4	4	4	4	4	7	10
SG&A																						
Salesforce	0	0	0	0	0	0	0	0	0	0	0	4	4	8	12	12	12	12	12	12	25	50
Marketing, business development, and reimbursement	0	0	0	0	0	1	1	1	1	1	2	2	3	3	3	3	3	4	4	4	5	5
Clinical advisor	1	1	1	1	1	1	1	1	1	1	1	1	1	1	1	1	1	2	2	2	2	2
Regulatory	0	0	0	0	0	0	0	1	1	1	1	1	2	4	4	4	4	5	5	5	5	5
Administrative asst.	0	0	0	0	0	1	1	1	1	1	1	1	1	1	2	2	2	3	3	3	4	5
Management	1	1	1	1	2	2	2	2	2	2	2	4	4	4	4	6	6	6	6	6	8	8

FIGURE 6.1.4

A sample staffing plan.

an innovator needs to figure out the types of employees (in terms of skill sets or function) and overall number of employees required to achieve the milestones set forth in the operating plan. The strategies developed in earlier chapters provide critical input to these estimates.

Figure 6.1.4 shows a sample staffing plan that is driven by the operating plan shown in Figure 6.1.3. The staffing plan is divided into three components: manufacturing, R&D, and **SG&A** (selling and general administrative) expenses. The levels in the staffing plan should directly correspond to the milestones in the operating plan. For instance:

- R&D staffing starts from the first quarter but it increases as product launch approaches in the 12th quarter (Q12). The company hires a second catheter team R&D engineer in Q10 to support R&D activities related to the development of the manufacturing process. This should match the achievement of a specific milestone in the operating plan (e.g., completion of manufacturing design and initiation of the development of manufacturing process).
- The completion of manufacturing design (under Phase 2) is finalized in Q10 and manufacturing processes are finalized in Q11 and Q12. Accordingly, manufacturing employees are hired in Q9, to participate in finalizing the manufacturing process before manufacturing actually begins.
- FDA approval is targeted for Q11, which is why the first members of the sales force are hired in Q12, since it would not make sense to hire them prior to having a

product to sell. In this example, a focused direct sales model (selling directly to the physician or facility customer) is assumed. The peak sales force size assumed (50) is relatively small for a direct sales force and should be rationalized based on a more detailed model (the method used to develop such a model is discussed as part of the **bottom-up market model** later in this chapter). Other sales models, such as using distributors, would require a smaller in-house sales team and these numbers would be reflected in the staffing plan accordingly.

In general, staffing numbers will vary based on factors such as the regulatory pathway, reimbursement environment, and sales model that have been chosen. For example, a direct sales force can require as many as 100 to 200 sales representatives at a peak, while a two- to three-person sales team might be sufficient to support an **indirect sales model**. Additionally, the size of the target physician population (e.g., a smaller specialty as compared to the large numbers of primary care doctors) can significantly affect the required sales force. Reimbursement may require a team of up to ten people if a multinational reimbursement effort is anticipated and a high level of support is desired. On the other hand, one or two consultants may be sufficient if favorable **coding**, **coverage**, and **payment** decisions are already in place.

One way to roughly check if the number of employees reflected in the staffing plan is reasonable is to use commonly accepted staffing ratios and formulas. For example:

- **SG&A to R&D** – Typically, the ratio of SG&A employees to R&D employees is roughly three to one in a mature medical device company. (It is important to note that this ratio may be very different in the early phases of a business as a product is being developed.)
- **Product units per assembler** – If the company intends to assemble or build the product in-house, it must be sure to have enough assemblers to meet demand for the product. Calculate the total number of units produced by each assembler to make sure it is a reasonable match to anticipated demand and a reasonable quantity for an assembler to achieve in a given time period. If either of these are not the case, more or fewer assemblers may be needed. The number of assemblers is linked to the market model, which defines the number of units that need to be produced for a given time period.

Market model

As described in 2.4 Market Analysis, there are two ways to approach market models: top-down or bottom-up. While **top-down models** have been useful to this point in developing "back of the envelope" market estimates, innovators are encouraged to develop a market model from the bottom-up for inclusion in the financial model. This approach helps to ensure that the cascade of assumptions relied upon to develop the numbers are

fundamentally sound and well understood. It lends credibility to the model and provides innovators with confidence when asked to defend it. And, it gives them an opportunity to think through the sales process in depth and develop a better understanding of challenges related to capturing market share. These challenges are not apparent in a rough top-down model. That said, both types of models can be useful in different circumstances and are therefore covered in more detail in the sections that follow.

Top-down model

A top-down market model forecasts projected yearly revenue by outlining the segments of the market that will be addressed and by determining how many customers the company intends to service (the number of patients that will be treated, hospitals/offices that will be sold to, etc.), then multiplying the number of customers by the price of the innovation. This approach is useful for quickly confirming how attractive the market is to pursue. Figure 6.1.5 shows a sample market model; a detailed description of the calculations follow the figure.

As noted, a top-down market model is based entirely on a number of statistical assumptions. Innovators can find many relevant statistics in scientific journals and analyst reports, but they will almost certainly be required to make "educated estimates" when no specific references are available. Developing reasonable assumptions

Key Assumptions		
Assumption	#	Source
# of paroxysmal AF patients	960,000	American Heart Association
Population growth rate	2%	US pop stats - baby boomers
% drug refractory	28%	New England Journal of Medicine
Average selling price (ASP) of ablation catheter ($)	2000	Company research
% annual decrease of ASP	1%	Estimate

US Market Model							
Year	Year 1	Year 2	Year 3	Year 4	Year 5	Year 6	Year 7
Key milestones			FDA Approval				
Paroxysmal AF patients	960,000	979,200	998,784	1,018,760	1,039,135	1,059,918	1,081,116
% drug refractory	28%	28%	28%	28%	28%	28%	28%
Refractory AF patients	268,800	274,176	279,660	285,253	290,958	296,777	302,712
% of refractory patients treated	2.5%	4.5%	7.5%	15.0%	25.0%	30.0%	35.0%
Procedures per patient	1.15	1.12	1.10	1.08	1.07	1.06	1.05
Total ablation procedures performed	7,728	13,818	23,072	46,211	77,831	94,375	111,247
Total market size ($)	15,456,000	27,498,756	45,683,536	91,042,470	152,572,428	184,078,320	215,901,728
% Market share of ablation patients	0.0%	0.0%	0.1%	1.0%	5.0%	12.0%	20.0%
Total units sold	-	-	23	462	3,892	11,325	22,249
Average selling price ($)	2,000	1,990	1,980	1,970	1,960	1,950	1,941
Revenue ($)	-	-	45,684	910,425	7,628,621	22,089,398	43,180,346

FIGURE 6.1.5

A sample top-down market model.

is part of the process of constructing a financial model and something that innovators need to become comfortable doing. Remember that these estimates can always be improved as additional data become available. Where references do exist, cite these sources to enhance the credibility of the projections. Also, draw on the early work completed as part of 2.4 Market Analysis.

To develop this particular top-down market model, the first step was to calculate the number of paroxysmal AF patients per year, starting with 960,000 in year 1 and increasing by 2 percent per year. The number of paroxysmal AF patients per year was then multiplied by the percent of patients who are drug refractory (i.e., not responsive to pharmaceutical therapy) and, therefore, are candidates for ablation (28 percent). This represented the total number of patients who can be treated with ablation. That number was next multiplied by the percentage of drug refractory AF patients who are actually treated to get the number of patients treated per year. These percentages were taken from a Frost & Sullivan research report (it is always a good idea to incorporate published statistics where applicable).

Because patients may need multiple procedures, the number of patients treated was then multiplied by the number of procedures per patient to produce the total number of procedures performed per year. The number of procedures performed per year was multiplied by the average selling price (ASP) per procedure ($2,000) to determine the total market size in dollars. To address the fact that the company will capture a fraction of this market, market share projections were multiplied by the total number of procedures per year to produce the target number of procedures per year, which corresponds to the total number of units sold per year. (Note that if more than one device was used per procedure, this would also need to be taken into account.) Finally, the total number of units sold per year was multiplied by the average selling price to produce the revenue target per year.

When determining market share, bear in mind that market share growth depends on a number of factors, including how innovative or unique the technology is, as well as how likely physicians are to adopt it. Factors such as how aggressive the company's sales plan is and how many sales representatives will be hired should also be

taken into account. Finally, consider the number of competitors already in the market and the extent of their resources.

Experts caution that one should not be overly aggressive when modeling market penetration. In most cases, it is unlikely for a product to gain more than 1 percent market share in the first year unless there is substantial pent-up demand and the sales channel is already well established. Although there are some notable exceptions, innovators should also assume that, in the long-run, competition will join the market (if not already present) and that portions of the market will not necessarily be accessible. As a result, a long-term market share forecast of 15–30 percent is a realistic outcome, even in a successful scenario.

Bottom-up model

The bottom-up model takes a different approach to assessing the market as described in the seven steps listed below:

1. **Determine the fundamental unit of business** – In this first step, the key is to determine what drives business in the market. For a company selling capital equipment, the "unit of business" is the number of facilities that purchase the machine. For other medtech companies, however, physicians may be the key unit of business because they are the ones that directly use a device within a specific procedure. If the company is working on an over-the-counter or fee-per-use business model (see 4.4 Business Models), then the patient may serve as the fundamental business unit.

2. **Consider the sales cycle** – The sales cycle involves all of the time and expense required to sell one unit of business. This includes the time and cost of raising the buyer's awareness of the technology, getting him/her interested, and closing the deal. If the buyer is a facility or hospital, standard purchasing processes and cycles must be taken into account, which can require as much as six months to a year to complete. If the buyer is a physician, the time and cost of training him/her to use the device may also need to be calculated, along with any follow-up

training. For new sales reps, a learning curve must also be factored in to these calculations – it can take three to six months (or more) before a new rep is trained and closes a first sale.

3. **Consider the adoption curve** – The adoption curve refers to the rate at which the buyer will utilize or consume the technology. For devices that present a relatively low risk and/or obvious benefits, the number of devices used over a given unit of time may grow relatively steadily. However, for higher-risk devices or technologies perceived as being more experimental, utilization may grow more gradually. For instance, a physician may perform a first procedure using the technology and then wait for several months to see how that preliminary patient responds before performing another. Certain facilities and medical specialties have a tendency to adopt new technologies faster than others, which should also be taken into account when developing the adoption curve. The key is to derive the anticipated utilization rate for a single unit of business on day one, day two, etc., to get a sense of how quickly the company will be able to build its sales volume. Some typical metrics that can be used to capture utilization include the percentage of the case volume of the business unit that utilizes the venture's products, the number of physicians trained in the use of the product (see the Cardica case example), or the number of individual physicians within the staff who are considered active (repeat) **users** of the product.

4. **Build the commercial effort** – At this point, innovators stop thinking about what is required to make individual sales to a single unit of business and start thinking about how many sales reps should be hired to build a reasonable business. From step 1, the innovators can determine how many units of business exist within the target market. From step 2, they understand how much total effort is required and expected to make each sale. From step 3, they have estimated a realistic adoption rate. Now, the challenge is to pull these factors together to determine how many sales reps are needed to grow the business at an appropriate, realistic pace. The

key is to balance the innovators' desire for "reward" with their tolerance for risk. Hiring too many sales reps at once can be a costly mistake until key assumptions regarding the sales cycle and adoption curve have been tested in the market. On the other hand, the company needs to be able to drive enough sales within a realistic time frame to sustain its operations and keep its commitments to investors. It is common practice to pilot the proof of the sales model – that is, to size the initial sales force to a handful of representatives and postpone expansion (scale up) until the assumptions built into the plan can be validated by actual field experience. Typically, it is advisable to run multiple scenarios (what would it look like to start with 5 sales reps? 10 sales reps? 15 sales reps?), then choose the one that produces the best risk/reward ratio.

5. **Consider market development factors** – The next step is to consider other factors in external environment that have the ability to affect the overall market model. For example, reimbursement coverage can have a major effect on market adoption. If reimbursement has not yet been achieved, this could affect the size of the initial sales force that is appropriate. Similarly, if professional societies have not yet endorsed the technology or key data have not been published, it may be wise to start with a smaller sales force focused on converting early adopters until some of these other factors have been put into place to support more widespread sales. Importantly, the top-down market model does not explicitly take these types of factors into account, which is one of its inherent weaknesses.

6. **Factor in product evolution** – If the innovator anticipates that subsequent versions of the technology will become available within a one-to-three year time frame, this should be reflected in the market model. New versions of a technology have the potential to increase utilization based on the improvements made and/or features added that potentially make the technology relevant to a greater number of procedures, or provide the opportunity to raise the selling price to reflect added value.

7. **Consider other factors** – Finally, an innovator should look closely at the market, the buyer, and the technology to determine if there are any other factors that might affect sales. For example, some medical specialties are more seasonal than others (e.g., orthopedics is busiest in the winter and the summer when people participate in seasonal sporting activities). Buying behavior in certain medical specialties can also be influenced by major medical conferences that occur at a certain time of year (when physicians and hospital administrators convene to check out new technology in the field). These types of considerations should be factored into the overall market model and the timing of key market decisions.

The following example on Cardica, Inc. demonstrates how a bottom-up market model is developed in practice within a medical device start-up. More details about Cardica and a more complete financial model for the company can be found in online Appendix 6.1.1 on **proxy company** analysis.

FROM THE FIELD ▶ CARDICA, INC.

Developing a bottom-up market model for a direct sales force focused on CABG

Cardica, Inc. was co-founded by Bernard Hausen and Steve Yencho, PhD, in 1997 in order to design and manufacture proprietary products that automate the connection, or anastomosis, of blood vessels during coronary artery bypass graft (CABG) surgery. In CABG procedures, veins or arteries are used to construct "bypass" conduits to restore blood flow beyond closed or narrowed portions of coronary arteries. This is typically accomplished by suturing one end of the vein or arterial conduit to the aorta and the other end to the coronary artery at a site beyond the blockage. Cardica's product portfolio included two products by 2004: the C-Port® Distal Anastomosis System (referred to as C-Port; see Figure 6.1.6) and the PAS-Port® Proximal Anastomosis System (referred to as PAS-Port). The C-Port product was used to connect a bypass graft to the coronary arteries while PAS-Port was to automate the connection of the graft to the aorta. In November 2005, Cardica received 510(k) clearance for C-Port, but PAS-Port was deemed to require a PMA and a **randomized clinical trial** (more details are provided in the proxy example analysis in online Appendix 6.1.1). As a result, management expected a potential three-year delay between the US commercial launch of its C-Port and PAS-Port systems.

FIGURE 6.1.6
Cardica's C-Port xA distal anastomosis system (courtesy of Cardica, Inc.).

Early in the biodesign innovation process, Cardica's management team made the decision to develop a direct sales force. As Bob Newell, Cardica's CFO explained, "A key metric we use to measure progress in our business is the number of trained surgeons using the product. In the US there are about 3,000 surgeons performing approximately 250,000 CABG surgeries per year in 1,000 hospitals; but about 225 hospitals perform 50 percent of all procedures. So we can go after these higher volume facilities with a targeted sales force. Our sales force doesn't have to be really huge, like it would be if we were in interventional cardiology. In addition, we do not need to target all surgeons in these hospitals. There are two primary segments: on-pump and off-pump CABG surgery. On-pump is the traditional way of doing bypass surgery where the heart is stopped and the patient is on a bypass machine, which filters their blood and keeps the blood flowing while the heart is stopped. About 75 percent of the market still does bypass surgery that way. A newer method of doing bypass surgery is called off-pump, or beating heart surgery, which means that the surgery is performed on a heart that has not been stopped – it continues to beat and the blood is not bypassed out into a bypass machine. Beating heart surgeons tend to be early adopters, so most of our first customers are likely to be beating heart surgeons."[3] Cardica's management estimated that 225 surgeons performed the overwhelming majority of beating heart surgeries.

Cardica's basic sales strategy was to first train surgeons based on the use of the C-Port device. The company's goal was to have many high-volume beating heart surgeons trained by the time the PAS-Port product was approved and launched. This gave the company three years from the time of C-Port's launch to train its first adopters. Newell estimated that the main expense in educating a surgeon was the time it took the sales person to deliver the training. After a four-to-six hour initial instruction at the company's facilities, which included a brief excursion to the wet lab, a Cardica sales representative would attend the first five to six cases performed by the surgeon. Once a surgeon was trained, it could take anywhere between two to nine months to reach steady-state sales for that customer.

Cardica management estimated that C-Port would have a theoretical total US market of 875,000 units based on an average of 3.5 distal anastomoses per procedure and a total number of 250,000 procedures. Of this amount, the beating heart segment would be about 25 percent or a total available theoretical market of just over 200,000 units. PAS-Port, on the other hand, would potentially have a total theoretical market of 375,000 units based on an average of 1.5 proximal anastomoses per procedure. (The difference between the average number of distal and proximal anastomosis is because some of the arterial bypass conduits, such as the internal mammary artery, are used with their proximal blood flow site intact and thus no proximal anastamosis is required.) The price for C-Port was set at $800 per anastomosis, while PAS-Port was $600 per anastomosis. Using these numbers, management estimated that the total theoretical annual US sales with 100 percent penetration would be $700 million for C-Port (250,000 procedures × 3.5 devices per procedure × $800 per device) and $225 million for PAS-Port (250,000 procedures × 1.5 devices per procedure × $600 per device) The corresponding numbers for the beating heart segment were: $170 million and $56 million (all numbers rounded to the nearest million).

The company then estimated that a sales person would cost between $350,000 and $400,000 per year (including benefits) and each one could generate a maximum of $1.5 to $2.5 million in annual sales. Once a sales person reached that threshold, a new sales person would be needed in the territory to cover additional surgeons. It would take Cardica two to three months to train a sales person, and the company anticipated that each sales person could bring in and train three new surgeons per quarter.

To develop a bottom-up market model based on this information requires two phases of effort. Phase I covers the initial period between C-Port and PAS-Port launch. Phase II covers PAS-Port launch and beyond.

The goal of phase I is to train early adopters at the rate of 25 to 30 per quarter and to have as many as 250 to 300 or more surgeons trained and using C-Port by the time PAS-Port is approved. Assuming a two-thirds retention rate over a three-year period, this translates into roughly

	C-Port® Launch							PAS-PORT® Launch
Quarter	1	2	3	4	9	10	11	12
Direct Sales Force	3	3	3	10	10	10	10	10
Physicians Trained in Current Quarter	10	10	10	10	30	30	30	30
Physicians Trained (and retained) in Prior Q		7	7	7	20	20	20	20
Physicians Trained (and retained) Two Qs Ago			7	7	20	20	20	20
Fully Trained & Retained Physicians				7	80	100	150	140
Total Trained Physicians	10	20	30	40	190	220	250	280
Total C-Port Sales		37,037	111,111	222,222	1,666,667	2,000,000	2,333,333	2,666,667
Total PAS-Port Sales								
Total Sales		37,037	111,111	222,222	1,666,667	2,000,000	2,333,333	2,666,667
Quarterly Sales per Sales Rep		12,346	37,037	22,222	166,667	200,000	233,333	266,667
Total Sales Force Costs	300,000	300,000	300,000	1,000,000	1,000,000	1,000,000	1,000,000	1,000,000

FIGURE 6.1.7

The phase I market model. Notes: (1) The line item "total trained physicians" reflects the cumulative number of physicians trained, ignoring the two-thirds retention rate; (2) the delay in sales following the launch of both products reflects the need for physician training time.

180 to 200 retained surgeons by Q12. This calculation implies that about 10 new surgeons can be trained each quarter in the first year. A reasonable target in Q5 to Q12 is 30 new trained surgeons per quarter (which translates into a direct sales force of 10 reps). The model in Figure 6.1.7 outlines the details (each row in the model represents a quarter and the cost for a fully loaded sales person is $400,000 per year or $100,000 per quarter).

A small sales force of three sales people is in place for the first three quarters, until it increases to ten in the fourth quarter. The model assumes a quarter of training for new sales people; therefore, after training for one quarter, the additional sales people allow Cardica to reach the target surgeon training level of 30 in Q5. The model keeps track of the number of surgeons trained in the current quarter, as well as in the previous two quarters in order to capture the three to nine months it takes for newly trained surgeons to reach steady-state sales. The calculation of total sales assumes surgeons trained in the current quarter do not perform any procedures, while surgeons trained in previous quarters perform increasingly more procedures until reaching 10 percent of their case volume nine months after they complete training. The model indicates that the sales force will generate enough sales to cover its expenses by Q8 (not shown).

The next step is to consider phase II: PAS-Port launch and beyond (see Figure 6.1.8). With an established base of 180 trained and retained off-pump surgeons (140 fully trained with more than two quarters of experience, plus 40 trained and retained for at least one or two quarters) by Q12, one would expect a successful launch and potentially a rapid, "hockey stick" pattern of adoption for PAS-Port. Cardica's management estimated that all retained C-Port surgeons could be trained on the PAS-Port within two quarters after the PAS-Port launch, given their familiarity and use of a Cardiac device and relationship with a Cardiac sales person. New surgeons would continue to be trained at the rate of 30 per quarter on both and C-Port and PAS-Port.

FIGURE 6.1.8

PAS-Port proximal anastomosis system (courtesy of Cardica, Inc.).

Further, management believed the penetration of PAS-Port into the surgeon's case volume would be more rapid than C-Port due to more accepted clinical benefit (one of the benefits of PAS-Port is that it eliminates the need to clamp the aorta, which can lead to a stroke, and is required in manual procedures). They felt that it would be reasonable to expect a 25–50 percent penetration rate within three quarters from launch (the model shown below assumes penetration of 35 percent). Under these assumptions, Cardica's total quarterly US sales would be expected to exceed $8 million by Q17. This sales growth would strain the sales force of 10. Therefore, Cardica would need to build its sales force in anticipation of this growth. The following model assumes that recruitment and training for the sales force happens early in Q11 with a modest increase from 10 to 13 to anticipate the new, additional needs of the existing trained base. The sales force will then increase again to 17 by Q16. The model also assumes that sales territories are split and new sales people are added when the existing sales of a sales person reaches the target $1.5 to $2.5 million per year, and that each new sales person requires one quarter of training. The model in Figure 6.1.9 starts from Q10 and modifies the sales people row to capture the increased needs due to the PAS-Port product launch.

This model demonstrates that by Q18 a sales force of 17 will be able to cover 300 trained and retained surgeons, which represents about 17 surgeons per sales person. Sales activities will include a mixture of supporting existing surgeons and bringing new surgeons on board. Total sales of $8.6 million per quarter will be reached in Q18. The reader will notice that despite the markedly increased sales force, the number of new trained physicians per quarter remains constant. This is because most of the sales people are now spending more of their time maintaining and supporting existing surgeons and a smaller fraction of their time recruiting new surgeons. The assumptions made in the model include the following:

- 3,000 surgeons performing CABG surgeries in the US in 1,000 hospitals.
- 225 hospitals perform roughly 50 percent of all procedures.
- Of the 250,000 CABG surgeries annually in the US, approximately 25 percent are "off pump" or beating heart procedures.
- Surgeons performing off-pump CABG surgeries will be most of the company's first adopters.
- Cardica's goal is to train 300 or more surgeons on the use of C-Port by the time PAS-Port is approved for commercial use.
- The company anticipates that it will take approximately three years (Q1–Q12) from the time of the C-Port commercial launch to the approval of PAS-Port.
- Once a surgeon is trained, it takes two to nine months to achieve a steady-state sales volume.
- Approximately two-thirds of trained surgeons are "retained," or continue to use the product.

		PAS-PORT® Launch						
Quarter	11	12	13	14	15	16	17	18
Direct Sales Force	13	13	13	13	13	17	17	17
Physicians Trained in Current Quarter	30	30	30	30	30	30	30	30
Physicians Trained (and retained) in Prior Q	20	20	20	20	20	20	20	20
Physicians Trained (and retained) Two Qs Ago	20	20	20	20	20	20	20	20
Fully Trained & Retained Physicians	120	140	160	180	180	220	240	260
Total Trained Physicians	250	280	310	340	370	400	430	460
Total C-Port Sales	2,333,333	2,666,667	3,000,000	3,333,333	3,666,667	4,000,000	4,333,333	4,666,667
Total PAS-Port Sales				964,286	2,142,857	3,937,500	3,937,500	3,937,500
Total Sales	2,333,333	2,666,667	3,000,000	4,297,619	5,809,524	7,937,500	8,270,833	8,604,167
Quarterly Sales per Sales Rep	179,487	205,128	230,769	330,586	446,886	466,912	486,520	506,127
Total Sales Force Costs	1,300,000	1,300,000	1,300,000	1,300,000	1,300,000	1,700,000	1,700,000	1,700,000

FIGURE 6.1.9

The phase II market model.

- C-Port would be used in approximately 10 percent of the CABG procedures with an average of 3.5 anastomoses per procedure at $800 each.
- PAS-Port would be used in 25–50 percent of all beating heart procedures and 5–10 percent of on-pump procedures with an average of 1.5 anastomoses per procedure at $600 each.

- The cost of a Cardica direct sales rep would be approximately $350,000 to $400,000 per year, fully burdened.
- Each sales rep has the capacity to generate $1.5–$2.5 million in sales per year.
- Once a sales person reaches that threshold, a new rep would have to be hired and the territory divided.

Discussion of the Cardica bottom-up model

This model is only a starting point. The company can experiment further with its core assumptions, especially with respect to the ramp-up in Q12, training requirements for existing surgeons on PAS-Port, and the training of new surgeons once C-Port and PAS-Port are both in the market. Other considerations that could be included in the model are outlined below.

Reimbursement

A shortfall of this analysis is that it does not factor in the effect of reimbursement, if any, on sales. In reality, product adoption can be significantly influenced by the status of reimbursement coverage and should be incorporated into such models.

Expansion in indications

While the model focuses primarily on adoption of C-Port and PAS-Port among surgeons performing beating heart surgeries, Cardica's management team believes that both devices will eventually be used in non-beating heart CABG surgeries and also believe there may be potential applications in other vascular grafting procedures.

Changes in the business environment

The model can also be used to monitor changes in the external environment and progress within the business to determine if the company's targets remain attainable. Such a model should never be static. Management must actively monitor conditions and then factor them into the model in order to determine the best time to expand the sales force or divide sales territories

The sales organization

The current model does not provide details about how the initial sales territories will be organized. Typically, this kind of analysis should be complemented by a preliminary design of sales territories, since different regions are likely to generate different levels of sales. In the US, it is common for companies to focus their preliminary sales effort in major, high-volume metropolitan areas such as New York, Boston, Chicago, San Francisco, Los Angeles, San Diego, Denver, and Miami. Utilization data from Medicare and the Dartmouth Atlas of Health Care[4] can be used to determine the highest-volume areas for purposes of designing the initial territories.

Reconciliation with the top-down model

As mentioned, a quick back-of-the-envelope calculation shows that the total US markets for beating heart CABG surgeries in which C-Port and PAS-Port could be used were estimated to be $170 million and $56 million, respectively. Assuming a 10 percent penetration for C-Port and a 35 percent penetration for PAS-Port into the total caseload of beating heart CABG surgeries (both are assumptions made in the bottom-up model), this brings the estimated annual US sales to $17 million and $20 million for total annual sales of $37 million, or slightly more than $9 million per quarter. It is reassuring that this quick, back-of-the-envelope, calculation is consistent with estimated quarterly sales of $8.6 million by Q18 in the bottom-up model. It should be noted that both of these sales estimates do not necessarily represent the final steady-state sales for Cardica but rather steady-state sales in the first adopter market segment of beating heart surgeries.

Cost projections

Cost projections (also called the operating statement) calculate the estimated costs of the business, including **manufacturing costs** and operating expenses (**OpEx**). Manufacturing costs include both material costs (COGS), as well as manufacturing labor, facilities, and equipment. They capture how much it costs for the company to make the products it sells. OpEx captures all other costs not included in manufacturing costs, including R&D, sales staff, general and administrative functions, and non-production facilities costs.

A **cost analysis** can be performed at varying levels of detail. It is up to the innovator to decide how much detail is required to satisfy the target audience for the financial model.

Salary analysis

Using the ablation catheter example, the creation of cost projections begins with an employee salary analysis, which includes three main components:

1. Summarize the staffing plan by calculating the average number of hires per year by type.
2. Outline annual salary assumptions, or how much the company will have to pay in salary by employee type, for each position type.
3. Calculate fully **burdened cost** per employee per year by position, taking into account employee healthcare and other benefits, insurance, computers, desk chairs, equipment, etc. for each employee. The fully-burdened cost represents the total annual cost for each employee.

Benchmarks for average annual salaries are widely available. Table 6.1.1 provides some sample data based on a 2012 salary survey.[5]

Rather than figuring out the exact fully-burdened cost of compensation, many experts use a back-of-the-envelope factor of two-times the annual salary for each employee (e.g., the fully-burdened cost of an R&D employee would be $100,500 \times 2 = $201,000). The $2\times$ multiplier, used to account for benefits, insurance, etc., applies to all employees except manufacturing employees, for which a $1.5\times$ multiplier should be used. Additionally, do not forget to take into account annual salary

Table 6.1.1 Medtech salaries can vary by geographic location and stage of the company's development, but generally fall within standard ranges (based on the Medical Device and Diagnostic Industry Salary Survey 2012, unless otherwise noted).

Job title	Median annual salary	Median annual total compensation (including bonuses)
General and corporate management	$145,600	$190,300
Regulatory and legal affairs	$126,600	$163,000
Research and development	$100,500	$120,000
Production and manufacturing	$100,00	$125,000
Product design and engineering	$94,200	$100,000
Quality assurance and quality control	$93,100	$110,000
Medical device sales	$78,489[6]	$150,890[7]
Consultants	$1,000 to $2,500 per day[8]	NA

increases, as well as the need to offer stock to attract and retain high-caliber talent.

Figure 6.1.10 shows a sample salary analysis that builds on the staffing plan in the model. (Note that the salaries reflect prevailing rates at the time the model was constructed for the first edition of this text.)

Manufacturing costs

Manufacturing costs have at least three primary components: manufacturing labor costs, manufacturing facilities costs, and raw materials costs (COGS). Figure 6.1.11 shows a sample model for capturing these data.

In this analysis, to calculate manufacturing labor costs, engineers per year (from the staffing model) were multiplied by their fully-burdened salary cost to produce a

	Year 1	Year 2	Year 3	Year 4	Year 5	Year 6	Year 7
Manufacturing team							
Engineers	0	0	3	6	9	9	12
Manufacturing assemblers	0	0	2	5	10	18	35
R&D							
Catheter team							
Engineers	1	1	2	2	2	3	5
Techs	3	4	5	6	6	7	10
Generator team							
Engineers	1	1	1	1	1	3	5
Techs	2	4	4	4	4	7	10
SG&A							
Salesforce	0	0	1	9	12	25	50
Marketing, business development, and reimbursement	0	1	2	3	4	5	5
Clinical advisor	1	1	1	1	2	2	2
Regulatory	0	0	1	4	5	5	5
Administrative asst.	0	1	1	2	3	4	5
Management	1	2	3	5	6	8	8

Salary assumptions	Factor	Salary ($)
Engineering	2	130,000
Technicians	2	50,000
Clinical, regulatory & QA	2	130,000
Sales, marketing, bus. development	2	100,000
Administration	2	35,000
Manufacturing	1.5	35,000
Management	2	200,000
Yearly salary increase	2.5%	

Fully-burdened salary - per employee ($)	Year 1	Year 2	Year 3	Year 4	Year 5	Year 6	Year 7
Engineering	260,000	266,500	273,163	279,992	286,991	294,166	301,520
Technicians	100,000	102,500	105,063	107,689	110,381	113,141	115,969
Clinical, regulatory & QA	260,000	266,500	273,163	279,992	286,991	294,166	301,520
Sales, marketing, bus. development	200,000	205,000	210,125	215,378	220,763	226,282	231,939
Administration	70,000	71,750	73,544	75,382	77,267	79,199	81,179
Manufacturing	52,500	53,813	55,158	56,537	57,950	59,399	60,884
Management	400,000	410,000	420,250	430,756	441,525	452,563	463,877

FIGURE 6.1.10

A sample salary analysis.

total engineer cost. The total number of labor employees (e.g., assemblers, processors, testers) per year was similarly multiplied by their fully-burdened salary cost to produce total manufacturing labor cost. These two figures were added together, providing the total manufacturing labor cost.

To calculate manufacturing facilities costs, the analysis started with an estimate for the cost of facilities space per square foot. The model assumed $25 per square foot per year, inflated at 2.5 percent per year. However, a more current estimate for Silicon Valley office space is roughly $31 per square foot per year,[9] and R&D space in the area can be leased for approximately $18 per square

foot per year.[10] Many medical device start-ups lease R&D space in order to conserve funds, and because it can accommodate multiple functions under one roof, including business operations, lab space, clean rooms, machine shops, manufacturing, shipping/receiving, and so on.

Next, space per employee was calculated by assuming 250 square feet per employee. This estimate can be reduced over time to 210 square feet per employee as manufacturing processes become more efficient.

Then, the total space required was determined by multiplying the space per employee by the number of manufacturing employees.

Manufacturing labor	Year 1	Year 2	Year 3	Year 4	Year 5	Year 6	Year 7
# of engineers	-	-	2.5	6.0	9.0	9.0	12.0
Fully loaded employee cost ($)	260,000	266,500	273,163	279,992	286,991	294,166	301,520
# of direct labor	-	-	2.0	5.0	10.0	18.0	35.0
Fully loaded employee cost ($)	52,500	53,813	55,158	56,537	57,950	59,399	60,884
Manufacturing labor cost ($)	**-**	**-**	**793,222**	**1,962,633**	**3,162,424**	**3,716,676**	**5,749,180**
Manufacturing facilities	Year 1	Year 2	Year 3	Year 4	Year 5	Year 6	Year 7
Cost per sq. foot ($)	25.00	25.63	26.27	26.92	27.60	28.29	28.99
Inflation	n/a	2.5%	2.5%	2.5%	2.5%	2.5%	2.5%
Sq. footage / employee	250	250	250	240	230	220	210
# of manufacturing employees	-	-	4.5	11.0	19.0	27.0	47.0
Sq. footage required	-	-	1,125	2,640	4,370	5,940	9,870
Projected sq. footage	-	-	3,000	3,000	15,000	15,000	15,000
Manufacturing facilities cost ($)	**-**	**-**	**78,797**	**80,767**	**413,930**	**424,278**	**434,885**
Raw materials	Year 1	Year 2	Year 3	Year 4	Year 5	Year 6	Year 7
Units sold	-	-	23	462	3,892	11,325	22,249
Raw material costs per unit ($)	-	-	600	540	486	437	394
% improvement	n/a	n/a	n/a	10%	10%	10%	10%
Raw material & packaging costs (COGS) ($)	**-**	**-**	**13,843**	**249,539**	**1,891,298**	**4,953,558**	**8,758,685**
Manufacturing costs	Year 1	Year 2	Year 3	Year 4	Year 5	Year 6	Year 7
Manufacturing labor cost ($)	-	-	793,222	1,962,633	3,162,424	3,716,676	5,749,180
Manufacturing facilities cost ($)	-	-	78,797	80,767	413,930	424,278	434,885
Cost of Goods Sold ($)	-	-	13,843	249,539	1,891,298	4,953,558	8,758,685
Total ($)	**-**	**-**	**885,862**	**2,292,939**	**5,467,652**	**9,094,512**	**14,942,750**

FIGURE 6.1.11
A sample model for estimating manufacturing costs.

Additional assumptions were made about projected facilities space. For instance, the model assumes that the company can lease facilities space in blocks and always has more facilities space than needed, but employee moves are limited to a reasonable number. In this example, the first two years of manufacturing (years 3 and 4) are satisfied by 3,000 square feet. A move is made in year 5 to a 15,000 square foot facility. Finally, the projected facilities space was multiplied by the cost of facilities space to produce the manufacturing facilities cost.

To calculate raw material costs, the total units forecast to be sold (taken from the market model) was multiplied by the cost of raw materials per product. In the model, it is assumed that the cost of raw materials drops by 10 percent per year as production volume increases to reflect learning and volume discounts ($600 in year 3 down to $394 in year 7).

Finally, these three cost elements were summed and the result was taken as the total manufacturing cost.

It should be emphasized here that the process of developing a manufacturing cost model is tightly coupled with the company's R&D and manufacturing strategy. Answers to the following questions must be available before an accurate model can be developed: What will the manufacturing process be? Where will manufacturing be done (in-house, outsourced in the US or outsourced outside the US)? What components can be made in-house and what components can be purchased or outsourced? Who will be the suppliers of the outsourced components and who will supply the raw material(s)? What kind(s) of equipment will be needed? What are the cost projections for these components?

OpEx

OpEx refers to the following components in the model: R&D staff costs, clinical trials costs, SG&A staff costs, and non-manufacturing facilities costs. Figure 6.1.12 shows a sample of an OpEx analysis.

R&D	Year 1	Year 2	Year 3	Year 4	Year 5	Year 6	Year 7
Staff costs							
# of engineers	2.0	2.0	2.8	3.0	3.0	14.0	30.0
Fully loaded employee cost ($)	260,000	266,500	273,163	279,992	286,991	294,166	301,520
# of techs	4.5	8.0	8.5	10.0	10.0	28.0	46.0
Fully loaded employee cost ($)	100,000	102,500	105,063	107,689	110,381	113,141	115,969
Staff costs	970,000	1,353,000	1,644,228	1,916,865	1,964,787	7,286,269	14,380,198
Clinical trials costs							
Est. # of patient-years in each year	0	50	100	0	0	0	0
Cost per patient-year	15,000	15,000	15,000	15,000	15,000	15,000	15,000
Total cost of trials	0	750,000	1,500,000	0	0	0	0
R&D costs ($)	970,000	2,103,000	3,144,228	1,916,865	1,964,787	7,286,269	14,380,198
SG&A	Year 1	Year 2	Year 3	Year 4	Year 5	Year 6	Year 7
Staff costs							
# of sales reps	0.0	0.0	1.0	14.0	27.5	70.0	140.0
Fully loaded employee cost ($)	200,000	205,000	210,125	215,378	220,763	226,282	231,939
# of clinical, regulatory & QA	1.0	1.3	2.0	4.5	6.5	9.0	18.0
Fully loaded employee cost ($)	260,000	266,500	273,163	279,992	286,991	294,166	301,520
# of marketing & bus. development	0.0	0.8	1.5	4.0	4.8	12.0	20.0
Fully loaded employee cost ($)	200,000	205,000	210,125	215,378	220,763	226,282	231,939
# of administrative assistants	0.0	0.8	1.0	1.5	2.8	8.0	10.0
Fully loaded employee cost ($)	70,000	71,750	73,544	75,382	77,267	79,199	81,179
# of management	1.0	2.0	2.5	5.0	8.0	12.0	12.0
Fully loaded employee cost ($)	400,000	410,000	420,250	430,756	441,525	452,563	463,877
Staff costs	660,000	1,360,688	2,195,806	7,403,623	12,729,722	27,266,938	48,915,868
Facilities costs							
Cost per sq. foot ($)	25.00	25.63	26.27	26.92	27.60	28.29	28.99
Inflation	n/a	2.5%	2.5%	2.5%	2.5%	2.5%	2.5%
Sq. footage / employee	200	200	200	200	200	200	200
# of employees	8.5	14.0	17.8	38.0	57.8	141.0	256.0
Sq. footage required	1,700	2,800	3,550	7,600	11,550	28,200	51,200
Projected sq. footage	3,000	3,000	15,000	15,000	15,000	60,000	60,000
Facilities cost ($)	75,000	76,875	393,984	403,834	413,930	1,697,112	1,739,540
SG&A costs ($)	735,000	1,437,563	2,589,791	7,807,457	13,143,652	28,964,050	50,655,409
Operating costs	Year 1	Year 2	Year 3	Year 4	Year 5	Year 6	Year 7
R&D costs ($)	970,000	2,103,000	3,144,228	1,916,865	1,964,787	7,286,269	14,380,198
SG&A costs ($)	735,000	1,437,563	2,589,791	7,807,457	13,143,652	28,964,050	50,655,409
Total operating costs ($)	**1,705,000**	**3,540,563**	**5,734,019**	**9,724,322**	**15,108,439**	**36,250,319**	**65,035,607**

FIGURE 6.1.12

A sample OpEx model.

To calculate OpEx, estimate R&D staff spending by multiplying the number of R&D-related employees per year by their fully-burdened salary for each type of employee for a given year. In this example, engineers and technicians are considered R&D employees. All employee data are taken from the staffing plan and salary analysis used in the model.

Next, estimate clinical trial costs as follows. First, review the operating plan and 5.3 Clinical Strategy to determine the length of the clinical trials and the number of patients. The trial length and strategy should be determined early, as part of the overall operating strategy. In this example, the trials will be completed in six quarters: quarters 7 through 12. Therefore, year 2 will include two quarters or 0.5 years of clinical trials (quarters 7 and 8) and year 3 will include all four quarters or a full year of trials. The number of patients participating in the clinical trials is assumed to be 100. Since year 2 includes 0.5 years of clinical trials, it includes 50 patient-years. Similarly, year 3 includes 100 patient-years. Another assumption was made about the cost per patient-year, $15,000, which is a reasonable (but somewhat optimistic) estimate for invasive medical device clinical trials. (Cost per patient-year is the cost of one patient participating in a clinical trial with one-year follow-up.) Multiply the patient-years by the cost per patient-year and sum over the years involved, in this case years 2 and 3, to yield the total clinical trials cost.

Then calculate SG&A staff spending in the same way as R&D staff spending. In this example, SG&A employees include sales reps, marketing and business development, clinical and regulatory employees, administrative assistants, and managers. Determine non-manufacturing or SG&A facilities costs using a method similar to that applied to manufacturing facility costs but using a higher initial cost per square foot. Finally, take the R&D, SG&A, and facilities expenses and sum them to calculate the company's total OpEx. It is worth noting that clinical and regulatory staff is sometimes categorized under R&D staff and sometimes under SG&A. In this example they are allocated under SG&A.

Be aware that clinical trial costs can vary widely, depending on the invasiveness of the device and length of study follow-up. Trials for non-implantable therapeutic devices and diagnostics with a short follow-up period can cost as little as $2,000 per patient-year (or $3,500 per patient on average). However, these costs can climb to as much as $100,000 per patient for an implantable or therapeutic device, which typically requires a lengthy follow-up period. Trial expenses may also depend on whether some of the costs for treating the patient (e.g., physician and facility reimbursement) will be covered by Medicare or private payers.[11] Additionally, the number of patients in the trial should be based on the number of patients statistically needed to establish a particular clinical result. Clinical trial strategy and planning is discussed in more depth in 5.3 Clinical Strategy.

Income statement

The income statement brings together all of the elements of the financial model into a unified view of the company's expected financials. It is also known as an earnings statement, statement of operations, or profit and loss (P&L) statement, and includes the line items shown in Table 6.1.2.

Constructing the income statement is quite simple, since it only requires the innovators to pull together their previous calculations. Figure 6.1.13 shows a sample of an income statement.

When investors examine an income statement, they typically apply a series of guidelines to check and see if

Table 6.1.2 The elements of an income statement.

Item	Description
Revenue (sales)	Total sales for the year.
Manufacturing costs	Total cost of the products *actually* sold by the company. In the case of ablation catheters, this includes cost of raw materials and labor to assemble the device and any other component that went directly into the production of the device. Does *not* include expenses such as marketing, sales costs, management salaries, etc.
Operating (gross) margin	Revenue minus manufacturing costs.
OpEx	All the other expenses associated with running the business that were not incorporated into COGS. Includes items such as R&D, facilities rentals, SG&A, company functions, etc.
Operating income	Operating margin – operating expenses.

	Year 1	Year 2	Year 3	Year 4	Year 5	Year 6	Year 7
Revenue ($)	-	-	**45,684**	**910,425**	**7,628,621**	**22,089,398**	**43,180,346**
Manufacturing costs ($)	-	-	885,862	2,292,939	5,467,652	9,094,512	14,942,750
Gross margin ($)	-	-	(840,178)	(1,382,514)	2,160,969	12,994,886	28,237,595
Gross margin (% of sales)	N/A	N/A	N/A	-152%	28%	59%	65%
Operating expenses							
R&D costs ($)	970,000	2,103,000	3,144,228	1,916,865	1,964,787	3,348,968	5,334,590
% of sales	N/A	N/A	6883%	211%	26%	15%	12%
SG&A costs ($)	735,000	1,437,563	2,458,463	6,165,199	8,408,295	13,492,043	19,708,990
% of sales	N/A	N/A	5382%	677%	110%	61%	46%
SG&A facility costs ($)	75,000	76,875	262,656	269,223	275,953	707,130	724,808
% of sales	N/A	N/A	575%	30%	3.6%	3.2%	1.7%
Total operating expenses	**1,705,000**	**3,540,563**	**5,602,691**	**8,082,064**	**10,373,082**	**16,841,011**	**25,043,579**
% of sales	N/A	N/A	12264%	888%	136%	76%	58%
Pre-tax operating profit ($)	**(1,705,000)**	**(3,540,563)**	**(6,442,869)**	**(9,464,578)**	**(8,212,112)**	**(3,846,125)**	**3,194,016**
Operating margin	N/A	N/A	N/A	N/A	N/A	-17%	7%

FIGURE 6.1.13

A sample income statement for the hypothetical catheter ablation business.

the financial plan is realistic. The following are principles that apply to mature medical device companies:

- A typical gross margin at maturity should be around 70 percent (gross margin equals revenue less manufacturing costs). Many companies target a 60 percent gross margin in their initial financial models.
- R&D is roughly 10–15 percent of sales at maturity.
- SG&A expenses are roughly 15–30 percent of sales (higher at first and then declining over time).
- While not usually highlighted as a separate line on the income statement, SG&A facility expenses are roughly 1 percent of sales.

Double check the income statement against these ratios to help anticipate how it will be received by potential investors, and/or make adjustments. In this example, the only expense category that deviates from medical device norms is SG&A. Therefore, innovators should examine carefully that component of the income statement and either rationalize it or determine ways to reduce it.

In the ablation catheter business example, the business loses money for the first six years due to important early investments in R&D, clinical trials, etc. In general, it takes longer for device companies to become profitable

than it did in the past due to the changing, more competitive environment. The operating margin for the sample company reaches 7 percent in year 7, which is modest for a medical device company.

Cash flow statement

At this point, the innovators are ready to determine the exact cash needs of the business. These are not the same as the net result of the income statement. The discrepancy can be due in part to an accounting concept called depreciation. Sometimes when a business spends cash, it does not record it as a cost on the income statement right away. For example, the company may purchase a computer for $1,000 which has a useful life of three years. On the income statement, the company may record a cost of $333 per year, representing the expended value of the computer each year. However, it still requires $1,000 cash up-front to make the purchase. Similarly, the company will need to spend cash on raw materials to build its product, but it will take time before the products are sold (and therefore recognized as sales on the income statement). In this way, as the company produces more product to support its future sales, it will begin to generate an inventory of finished product and work in process. This, too, consumes cash and needs to be reflected in the calculation as a use of funds. As a result, cash will move out of the hands of the company before the company recognizes the corresponding revenue. A similar adjustment for changes in payables and receivables (or net working capital over time) is not shown in the model but may be needed since it can be another driver of sources and uses of cash in the **cash flow statement**.

When a company's cash is equal to zero, the business is essentially bankrupt. For this reason, having a cash requirements plan is extremely important. Such a plan can also help the company balance the need to be frugal with the need to allow appropriate spending to support the growth of the business – a delicate but essential balance that every company must strike.

Figure 6.1.14 shows a **cash flow statement**, incorporating financing. To complete a cash requirements plan, the innovator needs to begin with an actual pre-tax **operating profit** (or loss) from the income statement. Then, elements representing cash flow "out the door" can be added in, which are not immediately deducted from the income statement. This includes (as shown in the model used here) capital equipment purchased in three categories:

- Cost of capital equipment per employee. This includes the cost of computers, phones, desks, etc. In the model, it is estimated to be $7,500/employee.
- Cost of clean rooms in the years that manufacturing facilities are developed.
- Cost of manufacturing equipment.

Next, consider raw materials costs (taken from manufacturing analysis when developing cost projections) needed to build up inventory. In this example, this is estimated to be 35 percent of the total raw material costs.

Subtract these elements from the pre-tax operating profit (loss) to produce total cash flow per year.

Next, add cumulative cash flow from the prior year (presumably $0 prior to year 1) to total cash flow from the current year to produce cumulative cash flow for the current year. Prior to any financing, the cash balance listed under cash needs will be the same as cumulative cash flow.

Cash balance and financing needs

Once the cash flows are calculated, the innovators can determine the company's financing needs (i.e., funds it needs to raise) and cash balances over time. The cash

	Year 1	Year 2	Year 3	Year 4	Year 5	Year 6	Year 7
Pre-tax operating profit (loss) ($)	(1,705,000)	(3,540,563)	(6,442,869)	(9,464,578)	(8,212,112)	(3,846,125)	3,194,016
Capital equipment purchased							
# of non-mfg employees	8.5	14.8	21.8	41.5	52.8	78.0	117.0
Cost of computers, phones, desks / employee ($)	7,500	7,500	7,500	7,500	7,500	7,500	7,500
Total cost of computers, phones, desks ($)	63,750	46,875	116,250	195,000	200,625	384,375	493,125
Clean room ($)	0	0	300,000	0	0	600,000	0
Manufacturing equipment ($)	0	150,000	150,000	300,000	150,000	300,000	200,000
Total capital equipment purchase ($)	63,750	196,875	566,250	495,000	350,625	1,284,375	693,125
Raw material costs ($)	0	0	13,843	249,539	1,891,298	4,953,558	8,758,685
Factor	0.35	0.35	0.35	0.35	0.35	0.35	0.35
Inventory build ($)	0	0	4,845	87,339	661,954	1,733,745	3,065,540
Total cash flow from the year ($)	(1,768,750)	(3,737,438)	(7,013,964)	(10,046,917)	(9,224,692)	(6,864,245)	(564,649)
Cumulative cash flow ($)	(1,768,750)	(5,506,188)	(12,520,152)	(22,567,069)	(31,791,760)	(38,656,006)	(39,220,655)
Cash needs	**Year 1**	**Year 2**	**Year 3**	**Year 4**	**Year 5**	**Year 6**	**Year 7**
Cash balance ($)	(1,768,750)	(2,506,188)	2,479,848	(7,567,069)	10,208,240	3,343,994	2,779,345
Suggested amount financed ($)	3,000,000	12,000,000		27,000,000			
Post-financing cash in the bank ($)	1,231,250	9,493,813	2,479,848	19,432,931	10,208,240	3,343,994	2,779,345

FIGURE 6.1.14

A sample cash flow statement.

balance will always need to be positive so financing must occur before the cash balance becomes zero. By looking at the cash requirements based on cumulative cash flow (i.e., what is the minimum cash infusion that would make the cumulative cash flow positive), the innovators can input various amounts representing the financing to be raised. The choice of how much money to raise and at what milestones (or points in time) are key strategic decisions (see 6.3 Funding Approaches). Implicit in this decision is also how long each financing round will allow a company to have a positive cash flow.

The spreadsheet in Figure 6.1.14 shows how, starting with the desired financing, the innovator can calculate the post-financing cash in the bank by adding the (pre-financing) cash balance to the current year financing. All subsequent year (pre-financing) cash balances must be adjusted by adding the cash flows to the cash balance. This will need to be repeated for each year until the company has sufficient pre-tax operating profits (based upon revenue) to independently keep cash flow positive.

The financing milestones (i.e., when to raise money from investors) are a critical output from the financial model. In some cases, it becomes evident that not enough important milestones occur within a certain time frame to align investor interest with the funding needs of the business. As a result, it may be necessary to either raise more money upfront to cover the company through these periods, or reconsider the operating plan to allow for a more continuous flow of value-building milestones (milestones that demonstrate the reduction of risk and/ or measurable progress towards the market). See 6.3 Funding Approaches for more information.

When developing the cash flow model, innovators undoubtably realize that the process of building a company involves working on an extremely tight budget. This can be a challenge, but Mir Imran, founder and CEO of InCube Labs, pointed out the silver lining:[12]

> I really believe that the good part of not having sufficient capital is that it really forces you to think through your expenses more clearly and spend the money less frivolously.

A note on profitability

In developing a financial model, it is important to determine a realistic profit goal. This profit goal will directly affect the company **valuation** (i.e., how much the company will be worth) since the valuation is a function of earnings (the higher the earnings, the higher the valuation – see chapter 6.3). Other considerations that impact the business' valuation include:

- **Revenue ramp** – How quickly the company expects to grow its revenue, sometimes referred to as its "growth rate."
- **Time to profitability** – How long it will take the company to break even, and then turn a consistent operating profit.
- **Operating profit percent** – What operating profit the company expects to achieve as a percentage of revenue.
- **Competitive benchmarks** – Benchmarks and comparable data related to the type of business model the company has chosen (e.g., disposable, reusable, capital equipment; see 4.4 Business Models) and the sector/industry within which the company operates (e.g., devices, genomics, pharmaceuticals).

A reasonable operating profit benchmark is approximately 30 percent of pre-tax income in the long term. However, how the company achieves this goal will depend upon its go-to-market strategy. A company might choose a direct sales strategy, in which case it would not have to share revenue with distributors, resulting in high gross margins; but it would have to hire a large sales force, resulting in high sales costs. Alternatively, the company could choose to use distribution partners, resulting in lower gross margins, but lower sales costs. Another option is to blend the two strategies, employing distributors in some markets and direct sales in others. To determine a likely profit model/goal, examine comparable companies to gain insights into reasonable revenue and cost projections. In the example presented here, the operating profit margin in year 7 was relatively low, reflecting the high cost of building a direct sales force. As part of developing the financial model, the innovators should consider the implications of different strategic choices (e.g., direct sales force versus hybrid).

Proxy companies

Proxy companies, also known as comparable companies, are those whose operations resemble what is required to commercialize a product. The concept of **comparables analysis** is that historical precedents can be used as a basis for validating a financial model. In the broadest sense, comparables will include other medical device companies. Further refinement would narrow comparable companies to ones that have, or went through, similar development challenges, clinical and regulatory pathways, are focused on the same disease state, and/or have similar products.

In analyzing a proxy company, there are three primary objectives:

- Determine the milestones that the company selected and/or achieved in developing its business.
- Identify the main risk factors that the company faced in developing its business and devise a plan to mitigate similar risks.
- Reconstruct the company's operating plan and financial model from the details that have been uncovered so that they can be used as a benchmark.

Investors, as well as innovators, use proxy companies to determine appropriate valuations for a new venture using the valuation models outlined in 6.3 Funding Approaches. Therefore, an added benefit of proxy company analysis is that it can help innovators be better prepared in their negotiations with investors since proxy companies play an important role in reaching a fair, market-driven valuation for the company.

This analysis can be performed at a high level or in great detail, depending on the needs of the innovators. A thorough analysis will enable a new enterprise to benchmark and modify its plans against multiple comparable companies, learning from their successes and failures. A high-level or more narrowly focused analysis will provide an innovator with directional information or the answer to a specific question (e.g., how long did comparable companies spend in clinical trials? What are reasonable fund-raising goals and milestones?).

While innovators can learn from the assessment of almost any proxy company, it is most helpful to look for those companies that have been considered a success.

Table 6.1.3 Typical markers for proxy company identification.

Timing	Company focus
Years 1 and 2	- Product development - Intellectual property (IP) filing - Pre-clinical discovery
Years 3 and 4	- Clinical value established - Product defined after multiple iterations - Regulatory approval complete - IP issued
Years 5 and 6	- Revenues ramping up - Franchise established - Pipeline of iterations - Clinical utility established

When the innovators can select from multiple proxy companies they can use the markers in Table 6.1.3 to determine which of these companies will more closely resemble their proposed venture.

See online Appendix 6.1.1 for a sample proxy analysis.

A note on intrapreneurship

Financial models for a new technology that will be developed and commercialized by an established company (this type of development is often referred to as "intrapreneurship") have the same basic components as the model described in this chapter for start-up companies. However, there are at least two key differences to consider.

Leveraging existing resources

Unlike innovators starting a new stand-alone venture, **intrapreneurs** may be able to leverage existing resources within their organization to manage some of the costs. This can involve engaging existing R&D engineers, using existing capital equipment, and/or leveraging the existing sales force. Accordingly, one of the first questions to be resolved is whether any of these existing resources can be accessed for the project. Even if established resources *can* be used, however, they are not cost-free and they should not necessarily always be employed. The finance department of the organization can provide guidance on the cost of existing resources. Another consideration is timing. With multiple projects competing for the attention of limited internal resources, progress may be slower than desired. Therefore, the intrapreneur should develop multiple financial models involving different degrees of reliance on internal resources, and use them to make a recommendation to senior management about the extent to which internal resources should be used.

Complementarities with existing products and company strategy

The financial model should explicitly account for complementarities with existing product portfolios and company strategy and should also provide an assessment of the impact of the project on the company's financial statement. Questions to address include whether the new project will defend existing market share or be used to increase market share. In both cases, the financial benefits to the organization should be articulated. In the defensive case, the market model should explain how market share may decrease without the new innovation. In the offensive case, the market model should highlight how the new innovation will increase revenue without cannibalizing existing sales, and whether the increase in revenue will be significant enough to make a difference.

↘ Online Resources

Visit www.ebiodesign.org/6.1 for more content, including:

 Activities and links for "Getting Started"
- Develop detailed bottom-up financial model
- Identify proxy companies
- Develop proxy company analysis
- Compare and rationalize bottom-up and top-down approach

 Videos on operating plan and financial model

 An appendix that provide an extensive example of proxy company analysis

CREDITS

The editors would like to acknowledge John Cavallaro, Steve Fair, David Lowsky, and Stacey McCutcheon for their help in developing and/or updating this chapter, as well as John White for contributing the core financial models. Many thanks also go to Bernard Hausen, Bob Newell, and William Younger for their assistance with the Cardica story. Special recognition goes to Larry Tannenbaum, who passed away in 2008. Larry was a long-time supporter of Stanford's Biodesign Program and this chapter is based, in part, on his lectures.

NOTES

1 From remarks made by Thomas Fogarty as part of the "From the Innovator's Workbench" speaker series hosted by Stanford's Program in Biodesign, January 27, 2003, http://biodesign. stanford.edu/bdn/networking/pastinnovators.jsp. Reprinted with permission.

2 Walter T. Harrison and Charles T. Horngren, *Financial Accounting* (Prentice Hall, 2007).

3 All quotations are from interviews conducted by the authors, unless otherwise cited. Reprinted with permission.

4 The Dartmouth Atlas of Healthcare, http://www. dartmouthatlas.org/ (February 24, 2014).

5 "Medical Device and Diagnostic Industry Salary Survey 2012: Data Tables," http://www.mddionline.com/article/mddi-salary-survey-2012-data-tables (February 20, 2014).

6 "2013 Medical Device Sales Salary Report," MedReps.com, http://www.medreps.com/medical-sales-careers/2013-medical-device-sales-salary-report/ (February 20, 2014).

7 Ibid.

8 Broad estimate by the authors based on informal input from experienced device executives.

9 "Silicon Valley Office," Kidder Matthews Real Estate Market Review, Fourth Quarter 2013, http://www.kiddermathews.com/downloads/research/office-market-research-silicon-valley-2013-4q.pdf (February 20, 2014).

10 Ibid.

11 "Medicare Coverage: Clinical Trials," Centers for Medicare and Medicaid Services, http://www.cms.hhs.gov/clinicalTrialPolicies/Downloads/finalnationalcoverage.pdf (February 24, 2014).

12 From remarks made by Mir Imran as part of the "From the Innovator's Workbench" speaker series hosted by Stanford's Program in Biodesign, April 28, 2004, http://biodesign. stanford.edu/bdn/networking/pastinnovators.jsp. Reprinted with permission.

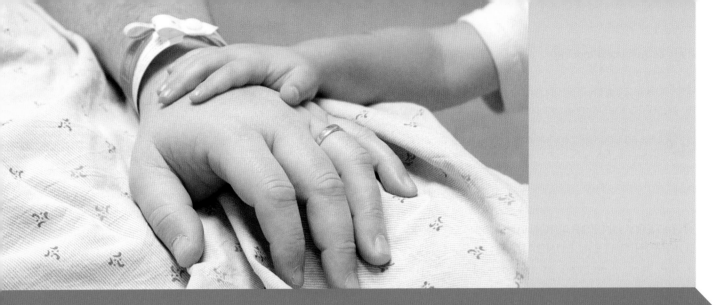

6.2 Strategy Integration and Communication

INTRODUCTION

Having assembled all of the complex components required to build a successful medtech company, at least one essential step remains: being able to explain the vision and plan for the company in a compelling enough way to rally potential investors, partners, advisors, and employees. The process a company goes through to integrate its functional strategies with the operating plan, craft a compelling "story," and prepare to share it with the world was traditionally based on writing a business plan. In today's medtech environment, having a formal business plan is not nearly as important as it used to be. However, the exercise of developing a holistic view of the business, clearly articulating what it is trying to achieve, and outlining how it will execute against its plans is still essential. In most cases, the way a company now communicates this information is through a pitch, which acts as a management tool for the leadership team and provides the rationale for others to commit their time and money to the project.

A company's pitch provides an integrated summary of all of the in-depth strategy and planning work that has been completed to this point in the biodesign innovation process. Among other information, it describes the clinical need, the product's value proposition, the market potential, and the development and commercialization pathway. By putting this material in writing, as part of one cohesive overview, innovators are forced to crystallize their thinking about how they will communicate with internal and external audiences and the story they will tell about the ways in which their business will create and sustain value. They are also forced to identify *key questions* – the few most important unknowns or risk factors that must be addressed in sequence on the way to bringing an innovation to market. Having to articulate a clear, compelling story further helps innovators prepare for important interactions with prospective employees, investors, and customers (whether or not the pitch is directly shared with these audiences), and it can be used as a mechanism for aligning the goals of everyone who is involved in making the business a reality. While the development of the pitch should not take on a life of its own and eclipse the importance of staying focused on the execution of the

OBJECTIVES

- Appreciate how a company's functional strategies come together with the operating plan to form an integrated view of the business.

- Understand the purpose and function of pitch development, both internally in managing the new venture and externally in communicating with potential investors, partners, and employees.

- Learn to identify and develop the main components of a pitch to address the key questions and other information the target audience is most interested in knowing.

- Recognize that a strong pitch does not guarantee funding, but it helps innovators rigorously prepare to present and defend the opportunity to potential investors.

functional strategies and operating plan, the disciplined thinking and integrated view of the business that it imposes on the team can be invaluable.

 See ebiodesign.org for featured videos on strategy integration and communication.

STRATEGY INTEGRATION AND COMMUNICATION FUNDAMENTALS

In the past, innovators were advised to develop a detailed business plan to use for three primary purposes: planning, management, and communication. As a planning tool, the business plan was meant to outline the steps that a company would take in order to bring its product to market and become financially viable. It also integrated the functional strategies behind the venture and the detailed steps that supported the execution of those strategies, along with the required resources. As a management tool, the plan set a short list of important goals to be achieved and the expected timetable, including both near-term and longer-term objectives. It could then be used to monitor progress toward achieving the company's greater goals. Finally, the plan provided a tool for framing and guiding communications, both internally and externally. Internally, it could facilitate cross-functional interactions and help maintain the alignment of objectives and activities across the different departments within the company. Externally, it could be used for recruiting purposes, to convince other businesses to enter into strategic partnership, and to persuade investors to provide capital.

To create the business plan, innovators traditionally started with the operating model and financial plan developed as part of chapter 6.1, and then integrated additional information and materials from the research and planning they completed as part of the strategy development stage of the biodesign innovation process. Typically, all of this information was assembled at a detailed level into a narrative form (e.g., a Word document) that ranged from 40–60 pages in length.

In some cases, this business plan was actively referred to and maintained by the management team and successfully shared with prospective partners and investors to get them interested in the opportunity being pursued by the company. But, with increasing frequency, traditional business plans have been under-utilized (at best) and overlooked (at worst). In parallel, important **stakeholders** in the external environment, such as investors, no longer want to see lengthy, text-heavy descriptions of the business. Rather than full-fledged business plans, they are asking for concise "**pitch** decks" and/or executive summaries that convey much of the same information in a more targeted manner.

Given this important shift, the bulk of this chapter is devoted to what goes into developing an effective pitch. However, there are a few essential points that innovators should understand:

- Companies are no longer required to develop a traditional business plan, but they still must complete the in-depth business planning exercise that integrates the operating plan and the functional strategies.
- Many elements of the traditional business plan are directly relevant to the pitch.
- The pitch is only as strong as the research and analysis that supports it.
- Just as with a business plan, investors, partners, and even prospective employees can all be expected to validate aspects of the pitch through "**due diligence**" before committing resources to the company.

As the author of one article effectively summarized:[1]

I completely agree that old fashioned 30–50 page business plans are relics of the past. But I think that sometimes, when people have this idea that they are only pitching and not planning, they are really just changing the vocabulary. Planning should be nimble. Planning should focus on just the

*information you need to run your business, and
planning should be an ongoing process to help you
understand where you are doing well, and where
you may be headed into trouble – before it happens.
The more information you can gather and
understand about your business and how it relates
to your market, competition, pricing, bottom line,
and cash, the more successful you are likely to be.*

The good news for innovators following the biodesign
innovation process is that most of the required research
and analysis to support the pitch has already been
completed. As noted, the detailed **operating plan** and
financial model serve as the foundation of the business
planning exercise (and will feature prominently in the
pitch). The functional strategies that have been
developed in areas such as R&D, regulatory, **clinical
trials**, **reimbursement**, marketing, and sales, will also
provide important inputs. The primary challenge that
remains is to integrate these strategies and communicate
the company's path forward.

About the pitch

Every invention and every company has a story. The key
is to figure out how to articulate it effectively and learn to
tell it well to attract talent, funds, and other resources to
the project. Generally, innovators must force themselves
to shift their focus from thinking about the solution in
terms of "how it works" to explaining "what it does" to
address the **need** and deliver **value** to important
stakeholders.

At an overarching level, the pitch puts forth a cohesive
argument for why the company is viable. Many innovators mistakenly believe that the primary driver of business success is the underlying technology, especially its
unique aspects. However, there are other considerations
that are equally important to establish in the pitch, and
these should be based on what is most important to the
target audience. For example, consider a pitch developed
to raise funding from investors. The innovators should
recognize that investors are more willing to put their
money into market- or customer-driven companies than
those primarily driven by technology (assuming, of
course, that the underlying technology is viable).

Accordingly, it may be more important to make a compelling case regarding the size and growth of target
markets, potential sales, and expected profitability than
it is to educate the audience on the detailed features of
the innovation. Investors will also want substantial information about the critical financial projections and capital
requirements of the business. On the other hand, if the
pitch is to be used to seek input on the idea from expert
clinicians, then more emphasis might be placed on the
need, how the technology will improve the delivery of
care, and other related subjects. A pitch delivered to
employees may devote more attention to the company's
vision, along with the key milestones outlined in the
operating plan that the team needs to work together to
achieve. In all of these scenarios, the innovators should
explain or avoid terminology or in-depth concepts that
might be unfamiliar or off-putting to the target audience.
The most effective pitch decks demonstrate an understanding of the audience's priorities, issues, and questions, and address them in a straightforward and
comprehensive manner.

It is also important for a pitch to be action oriented. In
addition to communicating the company's vision and
overarching strategy, the pitch should detail the tactical
and operational activities that demonstrate how the
innovation will be developed and commercialized. It also
should instill confidence in the audience that the team
has done its homework, has a clear understanding of
what is required to get to market, and that the members
have the capabilities to make it happen. As Mir Imran,
serial inventor, entrepreneur, and founder of InCube
Labs, put it, "I have learned that good ideas are a dime
a dozen. It's the execution that matters."[2] According to
one guideline, innovators should develop plans that are
10 parts implementation for every one part strategy.[3]

One of the most important ingredients to building
confidence in the target audience relates to the issue of
risk. Most audiences appreciate that any effort to develop
and commercialize a new medical technology is inherently risky. That said, an effective pitch must identify and
address the project's greatest risks head on rather than
trying to side-step them. Innovators should be honest
about the main challenges their company faces, primarily
because these issues almost always become apparent in a

pitch or presentation. Investors will be particularly interested in understanding how the innovators intend to tackle risks in a sequence so that funding can be staged to match the company's requirements as the plan is proven. Again, every project faces risks – the key is to identify them early and outline the decisive steps the team is taking to mitigate them.

Through the development of their functional strategies, innovators identified specific risks in each core area (technical risk, clinical risk, regulatory risk, reimbursement risk, adoption risk, etc.). However, at this point, as they integrate strategies into a single view of the business, they must think more holistically about risk. The idea behind identifying *key questions* is to consider all of the risks facing a company and then organize them in sequence, with an emphasis on determining the one or two most critical issues that must be addressed *first* in order for the project to remain viable. For instance, if the company is still uncertain whether the primary mechanism of action for its technology will function as intended in human anatomy, this risk needs to be retired before substantially more time and effort is invested in the further development of clinical and regulatory strategies and the mitigation of their related risks. Key questions are often, but not always, technical in nature. For example, the team may determine that physician willingness to adopt a new device is on the company's critical path before it invests more time and money into development. The point is that not all risks are created equally, and they do not all require the same level of attention from the team at the same time. Strategic integration is a mechanism through which the innovators explicitly recognize which risks are on their immediate critical path, devise plans for addressing them in the nearterm so that the company's larger goals can be achieved in the longerterm, and think about the optimal way to communicate and manage expectations in these areas.

One more important factor worth considering in pitch development is that the strongest presentation typically makes an emotional connection with the audience and communicates the passion that the team has for addressing its chosen need. Every pitch must display clear logic and a realistic, rational approach, but those that add an emotional element to their vision can often stand out. In many cases, an emotional appeal can be a good way to capture attention and draw in the audience. In healthcare (compared to other industries), demonstrating passion and a desire to help others is usually not difficult for teams, but this aspect of the pitch should not be overlooked. As an example, healthcare pitches will often feature a "day in the life" portrayal of a patient experience as a way to make the problems being solved more tangible and meaningful to the audience.

An effective pitch can range in length from 10 to 50 slides, with 15 to 30 slides often considered optimal. The appropriate number of slides varies based on the target audience and the purpose of the presentation. For example, when using the pitch to recruit a key staff member, 10 slides may be adequate to provide a high-level overview of the business. When sending the pitch to an investor with the hope of securing a first meeting, 10 to 20 slides may be needed to more fully explain the opportunity. In the first meeting with an investor, the innovators may want a 50-slide version of the presentation so they have access to back-up information that can be used to address important questions that arise.

Innovators usually create their pitch decks in Power-Point (or another presentation software program). While the quality of the content is definitely more important than the form the presentation takes, appearances can make a difference. More than anything, innovators should ensure that the pitch appears professional and projects the image of the company they intend to build. The use of features such as interesting graphics, video testimonials, or animations to help communicate complex subjects is certainly not required, but it can enhance understanding and help build excitement about the proposed venture. Just keep in mind that some audiences can respond negatively if a pitch looks too "slick" or as though the team spent an excessive amount of money to prepare it. Striking an appropriate balance is important.

A few additional guidelines for improving the readability and appeal of a pitch are outlined in Table 6.2.1.[4]

Developing the pitch
While each pitch is unique, most cover a similar set of topics. These topics are outlined in the sections that

Table 6.2.1 Paying attention to small details can make a big difference when it comes to the caliber of a pitch (based, in part, on J. Skyler Fernandes, "The 'Best' Start-Up Pitch Decks & How to Present to Angels/VCs," One Match Ventures, July 2013).

Be succinct – Emphasize quality and clarity over length. Don't use too many words on any given slide. Keep each slide uncluttered and easy to read. Make sure text is not too small (font size 20 or larger).

Be specific – Avoid generalizations. Address important issues directly and in sufficient detail to demonstrate preparedness and mitigate major concerns.

Show, don't tell – Emphasize progress and results to date. Include customer input as appropriate. Do not self-aggrandize. Let the facts speak for themselves.

Keep it simple – Make each slide self-explanatory. Avoid content (including graphics) that requires a complex description.

Double check everything – Be sure the pitch is well written (e.g., appropriate in tone and style). Carefully check for typos and the accuracy of all calculations and citations. Be consistent in the use of punctuation and capitalization.

follow.[5,6] To simplify the descriptions, the primary focus is on a pitch targeted at investors for the purposes of raising early-stage funding for a company.

Cover slide

On the cover slide, include the name of the company and/or its logo, the purpose of the presentation (e.g., investor presentation), and the date. Sometimes innovators also include "Copy number ___" in an upper or lower corner of whatever hard copies are produced and shared to help manage and track distribution. This approach can also reassure investors that the plan has not been shared with too wide an audience.[7]

Elevator statement

In an "elevator pitch," innovators try to capture the essence of the business they are developing in just one to two sentences. This concise statement should include the core problem the company is seeking to address, what the solution is that it has developed, and the

benefits it will deliver to its target customers. The time and effort innovators have invested in developing both their **value propositions** and **need statement**s will be useful in crafting the elevator statement for a pitch.

The clinical need

Provide a brief overview of the disease state and the medical need that the innovation is intended to treat. Focus on evidence that confirms the need is real and significant, including opinions of key clinical advisors. Use competitive analysis of existing solutions to make a strong case demonstrating the gap in the existing solution landscape.

Figure 6.2.1 provides a sample from the pitch deck of a start-up called Ciel Medical, founded by Kate Garrett and Dan Azagury. Ciel is developing a series of novel solutions for ventilated patients. Its first product, called the C.L. Bougie, facilitates optimal endotracheal tube placement to prevent complications associated with malpositioning. Its second and third products are targeted at preventing aspiration, which can lead to ventilator-associated pneumonia (VAP). These solutions include the BronchoGuard aspiration barrier and the Sora suction catheter. The Ciel team recognized the importance of clearly and professionally communicating its key messages and worked with consultant Devesh Khanal to develop its pitch.

The solution

Describe the innovation, emphasizing what it does for key stakeholders and the value it delivers to them. Use a short demo to show how the product works, but do not spend too much time here. (Many innovators embed videos in their pitches, but this can create a negative perception if the video fails to work.) Provide evidence that potential **users** find the product compelling by including preliminary results from user studies, quotes from user interviews, or a story from a new or potential user.

The market

Define the initial target market for the innovation. Outline the size of the market, as well as its projected growth and other important trends. Explain key characteristics of

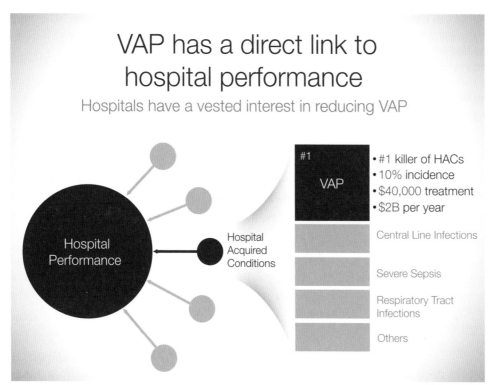

FIGURE 6.2.1
One of the slides used to establish the need for a solution to address VAP (courtesy of Ciel Medical).

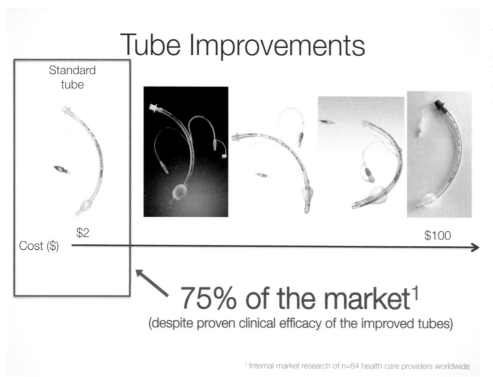

FIGURE 6.2.2
This slide from the Ciel Medical pitch is part of the story describing the competitive landscape for endotracheal tubes (courtesy of Ciel Medical).

the market, including how buying decisions are made, **market segmentation**, important aspects of the competitive landscape (see Figure 6.2.2), and what kind of market position is most appropriate for the innovation.

Describe mitigation strategies for addressing the most critical market risks (e.g., how to differentiate the innovation from existing alternatives or how to defend the innovation against second-generation products if it is the first to market) and the competitive landscape. Touch on potential market expansion strategies.

Intellectual property (IP)

If intellectual property is an important aspect of the proposed business, as it is for most **medtech** start-ups, then it is important to address this issue. However, a sophisticated audience will appreciate that detailed IP discussions should normally not occur without the protection of attorney–client privilege. Despite the fact that most pitches are labeled as "confidential," the innovators should assume that the contents of the pitch could potentially be accessed in the course of IP litigation and keep commentary on claims and prosecution strategy out of the presentation. If the company has issued or licensed IP that is already public it can be listed on the slide together with a summary of relevant facts (for example, the total number of pending applications). A more detailed discussion of IP-related aspects of the plan can be handled during due diligence.

R&D

If the innovation is still under development, outline how much additional research and development is

needed, describe and characterize the technical risks, and cover risk mitigation mechanisms. Timelines are important. Including an R&D Gantt chart can help to rapidly communicate key milestones on the path to commercialization (see Figure 6.2.3). Note, however, that this R&D Gantt is normally a simplified version of what would be used to manage the R&D process itself.

Regulatory strategy

Describe the regulatory pathway for the innovation and other key aspects of the regulatory strategy (see Figure 6.2.4). Provide the current status of all regulatory efforts, outline any regulatory risks, and share planned mitigation strategies.

Clinical studies

Describe what testing has been completed to date, including bench-top validation and animal studies. Detail the clinical studies the company proposes to sponsor. Include information about the endpoints, sample sizes, duration, and location, as well as key investigators. Provide evidence to support integration of studies with regulatory, reimbursement, and marketing efforts, along

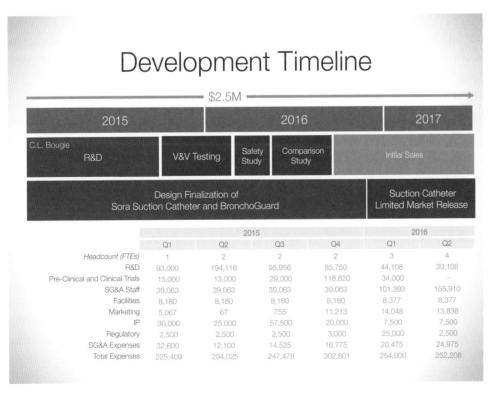

FIGURE 6.2.3

This slide illustrates at a high level the development timelines for the three products in the Ciel Medical portfolio (courtesy of Ciel Medical).

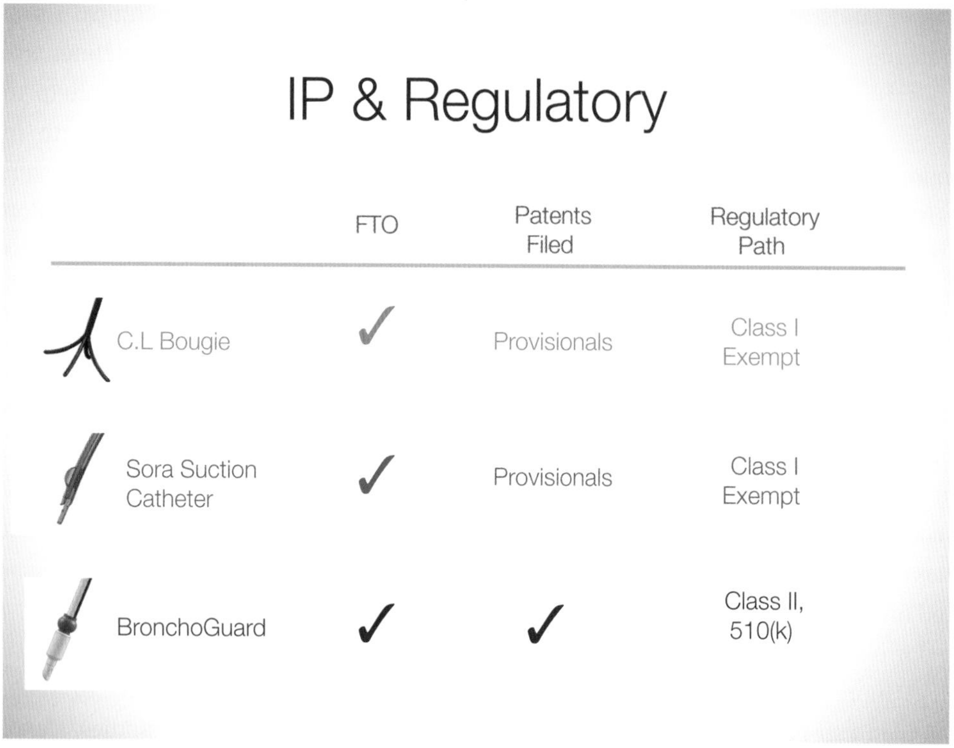

FIGURE 6.2.4
This slide clearly and simply communicates the regulatory pathway for the Ciel products, as well as the status of their IP (courtesy of Ciel Medical).

with information demonstrating that the company has (or can access) the necessary expertise to successfully conduct the trials.

Reimbursement

Describe the reimbursement strategy for the product. Outline the likelihood of (and justification for) receiving reimbursement. Provide examples of the reimbursement codes to be used if they exist, or outline the status of efforts to secure reimbursement to date. Outline reimbursement risks and planned mitigation strategies.

Business model

Highlight how the company will make money, including its anticipated pricing and margins. Justify the type of business model chosen, where the company's key revenue stream(s) will come from, the timing and frequency of revenue realization, and the volume of business needed to make the model viable.

Marketing and sales

Clarify the company's competitive advantage and value propositions for the offering. Outline planned sales and marketing efforts. Be certain to address the costs

associated with sales and promotional activities and provide a convincing explanation of how the chosen methods will result in the greatest return for the most reasonable investment.

Financial information

Include a one-page summary of financial forecasts covering at least the next two to three years of operations, as well as an indication of funds needed and how they will be used. Refer to 6.1 Operating Plan and Financial Model for more detailed information on what specific financial information should be provided.

Exit scenarios

Indicate the pathways through which investors may achieve a **liquidity event** for their investment. Investors are interested in knowing how their financing is most likely to be monetized and in what timeframe that might occur (see Figure 6.2.5). It is highly unusual for a med-tech company to return capital investment to investors through **dividends**. Instead, the two most common paths to liquidity – or an **exit** – are the initial public offering (**IPO**, rare in recent years) and the trade sale. A trade sale, which involves the merger or **acquisition** of the

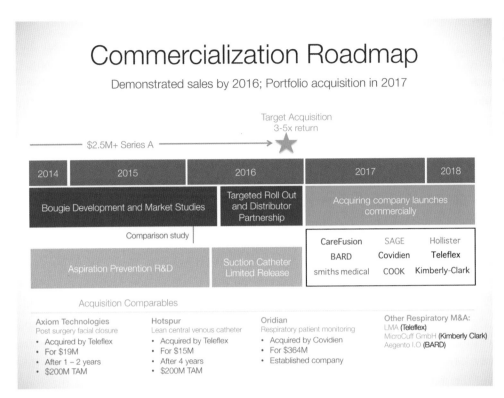

FIGURE 6.2.5

This slide provides an overview of Ciel Medical's commercialization and exit targets, as well as relevant information about comparable exits (courtesy of Ciel Medical).

new venture into a larger company, is by far the most common path. When presenting their opportunity, innovators should be prepared to identify and discuss which of the established medtech companies might be interested in acquiring the business and according to what timeline. If comparable exits have occurred previously, it is helpful to include that data (see 6.3 Funding Approaches for more information about exit scenarios).

The ask

This slide clearly states the objective of the pitch. In many cases, the ask will be for a specific amount of funding. If so, the pitch should outline how much capital is required, specifically how it will be used, and what progress/milestones it will allow the company to achieve.

Management team

Profile key members of the company's management team, highlighting relevant experience, leadership successes, and business accomplishments. It can also be helpful to include advisors or board members if they are well known and respected in the field. Include details about the company's organizational structure and staffing plans, if they are relevant to the team's ability to execute on the strategy outlined in the pitch.

Closing slide

The closing slide should reiterate the company name and provide contact information, even if just the name, phone number, and email address of the CEO. Innovators frequently create pitch decks with too little information about how the company can be contacted for questions and additional information. The closing slide may also propose an actionable and specific next step that drives the process forward (e.g., reconvene within two weeks for a follow-up meeting), although some innovators prefer to leave this topic open for discussion to give them the chance to interact with the investor in deciding what happens next. The process of reaching agreement on next steps can serve as a great preview of future management styles for both parties.

While the sections of the pitch do not need to be presented in this specific order, this organization represents a logical flow that is used by many companies. That said, innovators should feel free to experiment with slide placement, ultimately deciding on an approach that best supports their story.

Importantly, with even one slide on each of these listed topics, a team would already have a 17-page deck. And, it is quite possible that some topics will require more than one slide to clearly capture the team's most important messages and communicate progress in an area to date. One approach is to develop all of the slides needed to clearly tell the company's story and then combine those slides in different ways to address the specific needs and interests of different audiences, using the ones that are most relevant and leaving out those that are not. A list of typical back-up or appendix slides is shown in Table 6.2.2.

To get started in developing the pitch, it can often be helpful for innovators to answer the series of questions captured in Table 6.2.3. These questions can help them think through all of the work they have performed to date and begin identifying some of the most important ideas and information to include within their presentation.

Another approach is to force the "story" onto a single page with no more than four or five main points. This

exercise can help the innovators better appreciate which information is needed to support their argument and which information is interesting, but not as essential. Innovators sometimes make the mistake of assuming that they build credibility based on sharing all of the details of what they have done in each of the functional areas related to building the business. However, in truth, they often receive more "credit" for selectively using the facts they now possess to support a solid argument. If investors are interested in the project, the team will have ample opportunity to impress them with its command of the details during due diligence (see 6.3 Funding Approaches for more information about due diligence).

An alternate approach is to quickly create a crude first draft with no more than 15 slides. With the deck in this "rough draft" form, the team can review the slides and ask, "Why are we sharing these data and what point are we trying to make with them?" Always remember that everything in the pitch deck should support the company's overarching story.

Creating an executive summary

In addition to the pitch deck, many innovators find it useful to create a one- or two-page executive summary that summarizes the most important aspects of the business in a short narrative form. Some investors may request an executive summary as a first step prior to scheduling a meeting, although not all require one. The executive summary can also be helpful in communicating with partners, advisors, and other stakeholders.

Some experts suggest writing a draft of the executive summary to help guide the development of the pitch. Others suggest saving the executive summary for last, after the pitch has been complied. In either scenario, it is essential to make sure the executive summary is concise, well-written, and compelling.

As the innovator begins to prepare an executive summary, the following guidelines may prove useful:

- **Opening** – Start with the elevator statement, which is a compelling statement about the clinical need, proposed solution, and market opportunity.
- **Body** – Provide the most important high-level information about each of the most relevant topics

Table 6.2.2 Innovators can find it helpful to create a series of back-up or appendix slides that are not included in the main pitch but can be referenced when addressing stakeholder questions (J. Skyler Fernandes, "The 'Best' Start-Up Pitch Decks & How to Present to Angels/VCs," One Match Ventures, July 2013).

Timeline: History, milestones, and prior funding
Detailed value propositions for key stakeholders
Detailed financials (revenue and expense breakdown)
Break-even analysis (base-case versus bare bones case)
Pipeline of potential clients (with likelihood of closing each one and potential revenue)
Capital structure (ownership of founders and current investors)
Competitors (capital raises/investors)
Headcount (number of employees, projections, key hires needed)
Proprietary aspects not included in core deck

Table 6.2.3 The major sections of the pitch can be mapped to specific steps in the biodesign innovation processes. Innovators should consider all of the questions listed here; however, not all of the answers will be included in the pitch deck. Focus is key.

Major sections	Questions to address	Relevant activities in the biodesign innovation processes (from which to pull information)
Elevator statement	• What is the core problem the business is trying to solve? • What is the solution it has developed? • What key benefits will the solution deliver to its target customers?	5.7 Marketing and Stakeholder Strategy 5.9 Competitive Advantage and Business Strategy
The clinical need	• What is the need being addressed? • Why is this need important? • How is the need currently being addressed (if at all)? • In what ways are current solutions inadequate?	1.3 Needs Statement Development 2.1 Disease State Fundamentals 2.2 Existing Treatments 2.5 Needs Selection
The solution	• What is the proposed solution to the need? • How will it be used (and by whom)? • How does it better address the need than what is currently available? • What is the value proposition?	5.7 Marketing and Stakeholder Strategy
The market	• Who is the target customer? • What is the market size? • How fast is it growing? • Who are the primary competitors? • How will the company differentiate itself from the competition? • What are the barriers to entry?	2.4 Market Analysis 5.7 Marketing and Stakeholder Strategy 5.9 Competitive Advantage and Business Strategy
Intellectual property	• How will the company protect its IP? • How strong is its IP position?	4.1 Intellectual Property Basics 5.1 IP Strategy
R&D strategy	• How will the company prove (or has it proven) that the solution is technically feasible? • What are the risks that must be addressed and how will they be mitigated? • How will the product be manufactured and where?	4.5 Concept Exploration and Testing 5.2 R&D Strategy 5.5 Quality Management
Regulatory strategy	• How will the company get its product cleared/approved for the market? • What will be required to demonstrate safety and efficacy?	4.2 Regulatory Basics 5.3 Clinical Strategy 5.4 Regulatory Strategy

Table 6.2.3 (*cont.*)

Major sections	Questions to address	Relevant activities in the biodesign innovation processes (from which to pull information)
Clinical studies	• How will the company collect safety and efficacy data? • Where will the studies be performed and who will be the key investigators? • What other endpoints will be studied (and why)?	5.3 Clinical Strategy
Reimbursement	• What is the reimbursement pathway? • What codes will be used? • How much resistance is anticipated?	4.3 Reimbursement Basics 5.6 Reimbursement Strategy
Business model	• How will the company make money, including its anticipated pricing and margins? • What will be the timing and frequency of revenue realization? • What volume of business is needed to make the model viable?	4.4 Business Models 5.7 Marketing and Stakeholder Strategy 6.1 Operating Plan and Financial Model
Sales and marketing	• Why will customers be compelled to use the product? • How will the company close sales? • How will the product reach customers?	5.7 Marketing and Stakeholder Strategy 5.8 Sales and Distribution Strategy
Financial information	• What are the company's financing needs? • What does it intend to do with the money raised (according to what timeline)? • How (and when) will it generate a return for investors?	6.1 Operating Plan and Financial Model 6.3 Funding Approaches
Exit scenarios	• How are investors most likely to achieve liquidity for their investment? • When is an exit most likely to occur?	6.1 Operating Plan and Financial Model 6.3 Funding Approaches
Management team	• Who are the key individuals that make up the company? • What specific qualifications do they bring to bear?	6.1 Operating Plan and Financial Model

outlined in the pitch, focusing on actions that have been taken and results that have been achieved to date. Include evidence to support claims about the extent of the need and the likelihood that the solution will address the need. Describe key risks and propose mitigation strategies. Use headings to make the organization of the summary intuitive and clear. Make sure the text flows smoothly from one section of the overview to the next, using transitions as needed.

• **Conclusion** – Conclude the executive summary with a clear statement of the purpose of pitch (e.g., "The purpose of this presentation is to raise $2 million in funds to establish the technical feasibility of the product and perform animal testing . . ."). If the company clearly articulates its capabilities and needs in the executive summary, it will have a significantly greater chance of engaging the reader.

Just like the pitch, the innovators should always develop the executive summary with the intended reader in mind.

The following case example demonstrates how one team developed a pitch and executive summary and successfully used them with investors to raise preliminary funding.

FROM THE FIELD ⟩ SIMPIRICA SPINE

Putting together a pitch

Colin Cahill, Ian Bennett, and Louie Fielding, three student innovators in Stanford University's Biodesign Program, worked on the need for *a way to stabilize the spine without compromising mobility*. Over time, they invented a minimally invasive implant to help relieve early-stage degenerative disc pain, an area with a vast unmet need (see Figure 6.2.6). The device had a clear mechanism of action, was simple, and would not necessitate large axial loads, reducing the risk of mechanical failure. In combination, these factors caused the team's mentor, spine surgeon Todd Alamin, to recommend that they further develop the idea.

After completing the formal Biodesign Program, the team began an exploratory period to determine how they should proceed. Because Cahill and his colleagues perceived the market to be highly competitive, they decided to work in "stealth mode." This meant they were relatively conservative in terms of sharing information about the device. "At first we were paranoid, because the idea was very raw. We were very careful about which people we talked with about the **concept**," he said.[8]

One top priority during their exploration was proving the technical viability of the idea. "We applied for and were granted some **seed funding**, so we could advance our prototyping and testing," explained Cahill. They outsourced the development of some technical components, but were careful never to reveal how the components would be used or how they fit into the larger design.

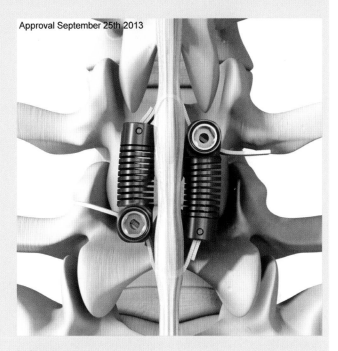

FIGURE 6.2.6

The LimiFlex™ Paraspinous Tension Band (courtesy of Simpirica Spine).

Another priority was ensuring that the solution was truly unique and could be protected with a strong intellectual property (IP) position. Toward this end, the team filed core **patents** early, engaged the services of an outside IP attorney, and used a non-disclosure agreement (**NDA**) to help keep the idea confidential when talking with third parties.

In parallel, they investigated whether the solution was clinically valuable enough to warrant bringing forward. To address this challenge, they leaned heavily on Alamin and also conferred with a small group of trusted spine surgeons to continue to vet the idea. "We didn't want to start a company if it didn't make sense. We had other options – we could have considered it a project and tried to publish a paper, or we could have licensed the idea to a larger company," Cahill stated.

After about a year of hard work, the output of these three work streams started to come together. The data from the team's preliminary tests looked highly promising, the IP position appeared strong, and feedback from advisors was positive. In addition, several events in the external environment related to spinal technologies (such as an IPO filing by St. Francis Medical, a company working on a different type of spinal implant) indicated that the market was ripe for innovative spine-related interventions. "We were also approached by several seasoned entrepreneurs who told us that if we decided not to build a business around the company, they would be interested in taking it on," Cahill remembered. With all of these factors reflecting positively on the opportunity, he said, "We decided to go for it." Cahill, Bennett, Fielding, and Alamin incorporated the company as Simpirica Spine, with Cahill becoming its full-time CEO.

One of their first challenges, once formally committed to pursuing the concept, was to raise additional funding. In preparation, Cahill and team formally began thinking about how to optimally communicate their idea to investors. Instead of developing a traditional business plan, which Cahill suspected would not be fully utilized, they focused on creating a 20-slide overview presentation that covered the company's vision, the need, an overview of the current treatment landscape, a description of the new technology and its value proposition, pre-clinical results to date, and the team, including key project advisors. "We'd been thinking about our 'story' from very early in the project, so what we were really doing was filling in the blanks with more robust data," Cahill described. Much of these data stemmed from the team's operating plan and financial model, the diligence it had completed around IP, the pre-clinical results it had collected, and other related

information. Essentially, the team's exploratory period had equipped Simpirica with a great deal of information that would help it back up its pitch. "If we hadn't felt like it was a plan that could work, or if there had been a massive piece of the story missing, we wouldn't have gone ahead," Cahill noted.

With the slide presentation and a two-page executive summary, Simpirica was nearly ready to begin meeting with investors. However, before taking this step, the team shared its materials with trusted advisors and practiced delivering its pitch. This resulted in valuable feedback that helped the members strengthen their approach. And, although the core messages remained the same, they learned to tailor the presentation to the audience to maximize its effectiveness. "The final pitch was a combination of the dream, the vision for the company, and our ability to execute on that vision," Cahill said.

When asked how they handled concerns about **public disclosure** when talking with prospective investors, Cahill stated that they did not ask these individuals to sign NDAs. "We'd been told that a venture capitalist would never sign an NDA, which isn't necessarily true. But it seemed like a better approach to develop the presentation in such a way that we'd be comfortable if it ended up on the Internet, which it never did," he said.

They did, however, mark CONFIDENTIAL on every slide and, more importantly, made sure all of their IP filings were in place before "going public" with the idea. Reflecting on the disclosure issue, Cahill commented that perhaps he and his teammates were too conservative at first. "As long as you've secured your IP, talking to a lot of people can be a good thing. Use your best judgment about people's character when you meet them, but err on the side of disclosing more rather than being paranoid." Sometimes, he explained, one conversation can lead to another, with new connections and contacts contributing valuable input to the project.

On a related note, Cahill suggested engaging with investors early and potentially talking with them multiple times. "At Simpirica, we didn't want to pitch anyone too early," he said. "But building familiarity with key contacts and showing progress against your milestones before it's time for anyone to make a decision can be valuable, particularly now that there's less capital to be raised in medtech markets. I think the value outweighs the risk."

Finally, he acknowledged that preparing and delivering a pitch and undergoing diligence could be nerve racking. "But if you have all of your ducks in a row and

are well prepared, there shouldn't be too many surprises," Cahill advised.

Using its preliminary pitch, the Simpirica team raised **angel investor** funding, followed rapidly by a venture-backed series A round. In 2014, the Simpirica implant was on the market in Europe where it had been used to treat more than 1,500 patients. In the US, the company was initiating a pivotal trial to support a **PMA** application to the **FDA**.

More about pitching and funding

Some medtech innovators erroneously believe that the most important reason to develop a pitch is to raise funds. Even if this is one of the team's primary objectives, the real value of creating the pitch is that it forces the innovators to carefully think through how they will build a business (in combination with the activities described in 6.1 Operating Plan and Financial Model). That said, the pitch is certainly an important tool that helps facilitates conversations with investors and can lead to funding. Investors expect to see a well-constructed pitch and may rapidly eliminate projects from consideration if the caliber the presentation is lacking.

Of course, funding from external investors will not sustain a new venture indefinitely. A company needs profitable, revenue-generating products to become sustainable. As investors are required in the current investment environment to commit increasingly large sums of money over a longer time horizon to reach an exit, they have higher expectations (and perform more stringent due diligence) related to a company's operational capabilities. A compelling pitch demonstrates how a company will become profitable, as well as the management team's ability to run the business for as long as necessary for investors to generate a return.

One of the most common mistakes innovators make in developing a pitch for investors is to understate important costs (or overstate market estimates). Not only does this cause projections to be blatantly inaccurate, but it can result in the team losing credibility with prospective investors. Studying how other companies of the same

size or in the same field have expanded and how their sales figures have grown over time can provide an effective check on whether the figures outlined in the company's financial projections are truly realistic (refer to online Appendix 6.1.1). Innovators should also recognize that despite their efforts to provide realistic estimates, investors will sometimes discount sales projections by as much as 50 percent and may also inflate costs. By varying sales and/or expenses in the financial model, innovators can study the impact of these assumptions and provide investors comfort that the plan remains viable. Showing investors the results of the analysis over the range of assumptions is called a "sensitivity analysis" and can help build confidence and credibility in the investor audience.

Approaching investors

When seeking funding, it is preferable to use one's connections to gain access to investors. Cold calling, blind mailings, or online submissions of a business plan rarely lead to positive results, and may actually hurt a company's chances of being funded in some cases. Referrals to investors can be gained through any number of different sources. Other innovators, as well as service providers (attorneys, consultants), can be rich sources of leads. If innovators do not have direct contacts in the industry, they should network to find them. For example, if the innovators know a venture capitalist or angel investor who works in the high technology sector, that person might be able to make a referral to another venture capitalist or angel who works in the healthcare or medical device field. Be assertive, but respectful in networking and pursuing referrals. If the innovators have no choice other than to make a cold contact, they should be sure to focus on investors or firms whose criteria seem appropriate to the business. Names and relevant information about potential investors can be found through online directories.

Once potential targets are identified, innovators should next conduct thorough research to understand everything they can about the investors, including their existing portfolios, track records, investment criteria, management styles, and other factors that can help determine if there is a good fit between the company (and its technology) and the interest of the investors. Seek out interviews with founders that the investors have previously supported to gain an understanding of their investment approach and management style. Remember, fit is very important. The questions in Table 6.2.4 can help in this assessment.

Before making contact with prospective investors, the innovators should at least have their elevator statement ready and a sense of the company's funding

Table 6.2.4 Innovators should spend nearly as much time screening investors as investors spend screening investment opportunities.

Questions for screening investors
At what stage does the individual or firm usually invest?
What is the typical investment level over the life of a company?
What is the typical amount invested in each round?
For VCs, when was the current fund started? (If > 5 years ago, then there may not be adequate funds remaining for additional financing rounds.)
For individuals, how did the investor earn his/her money? How many other companies is s/he involved in?
How much time will be spent with the company? What expectations does the investor or firm have regarding involvement?
What prior deals has the individual or firm done in the industry?
Does the company have references (other innovators with whom it has worked) who can be contacted?

requirements. Having the pitch and/or the executive summary available to share is even better.

Preparing to deliver the pitch

In an ideal world, what most investors are seeking is an attractively valued company with a strong, experienced management team that can implement a well thought-out business model to sell a proprietary technology or service in a large, worthwhile market, and grow a substantial, profitable company over a short time frame with a clear exit strategy for the investors. In other words, they are seeking huge rewards with limited risk.[9]

Rodney Perkins, founder of multiple companies including ReSound Corporation, Laserscope, Collagen Corporation, Novacept, and Pulmonx, offered this advice to innovators as they prepare to address investor questions:[10]

> When you're talking with [investors], any complexity makes them nervous. You want to describe a very straightforward, single product; you're going to sell it to a known market, this is the development risk, this is the regulatory risk, and make sure that these are clearly understood.

Many investors make preliminary inquiries in six primary areas: (1) technology or service concept; (2) market size and dynamics; (3) management team; (4) business model and financial requirements; (5) exit scenarios; and (6) **valuation** and deal structure. Table 6.2.5 provides examples of the types of specific questions they may ask in each of these areas when assessing an opportunity. Innovators should develop answers to these questions in advance, in preparation for delivering their pitch. That said, they should also expect that investors will ask at least some questions that they will not be able to answer. During the pitch and diligence process, investors sometimes use their questions as a way to probe whether the team is willing to admit the limits of its own knowledge and how comfortable they are recognizing the never-ending need to seek out ways to get better answers from others. Simply put, every innovation initiative starts off with more assumptions than knowledge and the innovators will be asked to demonstrate their understanding of that reality.

Table 6.2.5 Common questions asked by professional investors (based on Ross Jaffe, "Introduction to Venture Capital," October 6, 2004; reprinted with permission).

Area of inquiry	Technology or service concept
Technology or service concept	• Is the product/service concept clear? Does it make sense? • Is there sufficient proof of principle or evidence of feasibility? • Are there adequate proprietary aspects – patents, trade secrets, or other barriers to entry? • Can it be manufactured at a reasonable expense? • Are there regulatory issues?
Market size and dynamics	• How large is the market, realistically? What is the actual addressable population? • Does the company have realistic potential to obtain substantial revenues in the market? • Is the decision making of purchasers and users well understood? • Are there reasonable marketing and sales costs? Sales cycles? Distribution systems? • Are the business relationships between referral sources, purchasers, providers, and consumers well understood? • Does the company have a strong competitive position? • Is the technology/service consistent with market, regulation, and reimbursement trends?
Management team	• Is the management team smart? Are they knowledgeable about this business? • Does the management team have a proven record, particularly in this business? • Do managers have high levels of honesty and integrity? Can they be trusted? • Do they have reasonable expectations for the business, particularly for the difficulties of product/service development, rate of company growth, capital requirements, ultimate business size and profitability?
Business model and financial requirements	• What are realistic revenue and expense projections for the company? • How much capital will be required to reach positive cash flow? • What are realistic expectations for the timing and sources of this cash? • What are the realistic exit opportunities for investors in this deal?
Exit scenarios	• How are investors most likely to realize an exit? What scenario seems most likely? • For acquisition candidates, what companies are most likely to be interested? What other acquisitions have they made recently? Does the company have an established relationship with one or more of these companies? Why would these companies acquire a technology in the space rather than developing it themselves?[11] • What timeline is most realistic for achieving an exit? • What are the major factors likely to influence whether or not the exit is realized?
Valuation and deal structure	• Will the valuation of the investment in this deal afford a high probability of a substantial return (40 percent or greater internal rate of return)? • Will the deal allow for enough capital to be put to work to make the investment worth the time and effort? • Who will be the co-investors? Are these parties good to work with? • How can this investment be structured to minimize the technological and financial risk?

Even though all investors may be interested in information related to some or all of these questions, different types of investors may have different expectations about what the answers should be to attract their investment. Chapter 6.3 Funding Approaches provides an overview of various funding sources and how their priorities and expectations may vary.

Over and above preparing to address investor questions, innovators are advised to practice their presentations (many times over) to ensure that the flow, timing, transition, and tone are smooth and professional.[12] As suggested in the Simpirica story, it can be helpful to deliver the pitch to advisors and other mentors and actively seek constructive feedback.

The following story highlights other important information related to the investor's perspective and provides an example of how one medtech venture capitalist thinks about the hundreds of business pitches he sees each year.

FROM THE FIELD ▸ VERSANT VENTURES

The role of investors in screening new business opportunities

Versant Ventures is a leading venture capital firm focused on life sciences opportunities, including both diagnostic and therapeutic medical devices. The firm has invested in many of the major medtech success stories since its inception in 1999, including Acclarent (acquired by Johnson & Johnson) and St. Francis Medical (now part of Medtronic). Ross Jaffe, a managing director at Versant, focuses on medical device investments for the firm. He has worked in medtech investing for over two decades, bringing both his clinical training and business experience to bear.

When asked about the medtech funding landscape in 2014, Jaffe said, "We're in one of the most challenging environments for medical device investing I've seen in my career. And it's especially challenging to be an early stage investor in medical technology. We're at the lowest level of medical device **start-up funding** since 1995." Jaffe believes that the current situation is due, in part, to the general contraction in financial markets and longer-term cycles in the venture capital industry, but that is not the whole story. "Compared to the IT industry, healthcare is viewed as more complicated and complex because of the regulatory and reimbursement challenges we face. Fewer people are interested in taking on these additional risks," he explained.

The implications for innovators are stark, but not necessarily dire. Jaffe predicts a "Darwinian environment" where, he said, "Those companies that are truly outstanding and can adapt to the environment will be successful in raising venture capital. This environment will reward the scrappy entrepreneurs who can figure out how to get a lot done on very limited resources." Start-up companies with technologies that are highly attractive to potential acquirers will also do well since they have a clearer path to an exit. **Corporate investments** in early-stage medtech companies are on the rise but, according to Jaffe, this increase in corporate funding will not fully make up for the decline in venture investment in medical devices. "It's also critical to think about how new technologies fit into the evolving healthcare system," he advised. "Technologies that save costs as well as improve outcomes are more likely to succeed in this new paradigm." In contrast, "Incremental improvements that cost more but don't deliver improved outcomes or better economics are no longer compelling," added Jaffe. He also pointed out that the new environment will be challenging for companies developing truly novel technologies with long R&D cycles and no established business parallels.

Jaffe and his partners (see Figure 6.2.7) identify new ideas and technologies from a variety of sources, some higher-yield than others. "There is a hierarchy of where the best opportunities tend to come from," he said. The most promising ideas tend to come from the

FIGURE 6.2.7

Jaffe (center) at a partners meeting. (courtesy of Versant Ventures).

entrepreneurs they have worked with in the past and members of their network (i.e., trusted peers and service providers such as corporate attorneys, IP attorneys, regulatory consultants and the like). "We know their judgment about things, they know how we look at the world, and they often bring to us the highest quality deals. Projects from **incubators** that we generally work with are also in this category," Jaffe stated. Next are the opportunities that deal agents and investment bankers bring to Versant. The largest number of opportunities are the unsolicited proposals that come to Versant because of its reputation as an early-stage investor in the life sciences space, but they are not filtered through other knowledgeable people so the quality of the proposals can be highly variable. He added, "If you don't have direct relationships with venture capitalists yourself, having your proposal vetted by someone knowledgeable in the field who can give you feedback on ways to improve it and then forward the revised plan on to venture capitalists who they know increases the level of review your proposal will receive."

Through these three primary sources, the Versant team reviews 300–400 opportunities annually in the medical device sector. "But keep in mind that we can only invest in two to eight deals each year," Jaffe said. "So we're really trying to find the top one to two percent of

opportunities with the top one or two percent of entrepreneurs that can make these ideas successful."

These days, entrepreneurs rarely send the firm a traditional business plan. Most start-ups submit a PowerPoint-based pitch. According to Jaffe, these presentations should provide a crisp picture of the clinical need, the size and characteristics of the market, the technology and how it solves the need, the key requirements to get to market and drive adoption (clinical, regulatory, reimbursement, and commercialization pathways), the knowledge and experience of the team, the strength of the intellectual property, the operating plan and resulting financial projections, and the financing strategy and exit opportunities for the company.

When he receives a pitch, Jaffe explained, "The first question I ask is: given the market opportunity and the amount of capital required, will the investment generate a return that justifies the risk, the time, and the money?" Importantly, he continued, "I can actually read one of these presentations in about five minutes and tell you whether I'd even consider investing in it or not." As a result, the pitch must be compelling. For Jaffe and many investors at well-known venture capital firms with high-volume deal flows, their time is a more limited resource than their capital. "So this may sounds a little harsh, but my goal is to get to 'no' as quickly as possible so that I can move on to find the most attractive opportunity available," he explained. "And I only need one reason to get to no. The pitch has to communicate why I should be excited about all the elements of the business, otherwise I'll set it aside."

Toward this end, Jaffe shared some tips on how to prepare a compelling pitch. First, he said, "Conciseness and pithiness matter." The pitch should be direct, complete, and clear, without being too wordy or long. The innovator should have thought through the story that needs to be conveyed, and should be able to communicate it in about 20 slides. Second, he continued, "I don't get impressed by presentations that are fuzzy about things." Pitches should be specific and contain enough detail to demonstrate that the innovators have carefully thought through their plan (rather than just preparing a sales document). "I'm always impressed when the presentation anticipates the key questions that I will ask about the specific opportunity," he said.

Innovators should build their business plans as real plans for the business, not just pitches to get funding. The realism and quality of the plan should be of as much concern to the entrepreneur as it is to the potential investors. "As an entrepreneur, you are about to invest your life, your time, your ego, and probably some of your money or your family's money in this opportunity. You should be more hardnosed with yourself than I will be with you about why this opportunity is a valuable thing for you to do and the right way for you to spend the next few years of your life," Jaffe advised.

If the pitch looks intriguing, the Versant team invites the innovators to deliver the presentation in person. This meeting lasts only about an hour, and the determination to take the next step (or not) is made fairly quickly. This decision depends on the strength of the idea, the team, and the overall pitch, as well as the **opportunity cost** of working on this project rather than something else that may be under consideration. "We're in a constant triage mode," Jaffe said. "We're evaluating dozens of opportunities at any point in time and we get together as a group within the firm to prioritize them. A lot of times, proposals come in that are potentially interesting, but they just don't make it high enough on the priority list."

The next step for promising opportunities is diligence. "During the due diligence process, I try to identify the biggest risk areas, the areas where I have the greatest concerns, and dig deep into them. Again, I try to prove to myself why I shouldn't do the deal," Jaffe stated. While addressing the same basic issues, this process is customized to each opportunity and is heavily dependent on the clinical needs being addressed, the market characteristics, type of technology under consideration, and the plans laid out by the team. Importantly, not all risks are created equal. Some can be mitigated or reduced over time, such as technical or clinical risks. While others are more difficult or problematic to potentially affect, such as market size or IP issues.

Once diligence is complete, Versant has the information it needs to make a final decision. Jaffe consults with the rest of his team to make sure that all partners are supportive. The firm then initiates negotiations on a **term sheet** for those opportunities that represent the "highest and best use of our capital at that point in time," said Jaffe.

"Raising funding really depends on how much cash is in the market, what other opportunities are available, and the credibility of the entrepreneurs and how much confidence we have in them," Jaffe summarized. "Great entrepreneurs can still raise money in this environment, but it's not as easy for the average entrepreneur." Accordingly, he advised, "Reach out to people who have built companies and get their feedback early about key challenges. And don't get defensive when they provide negative feedback – they're often shining a light on the areas where you need to look more closely and think more deeply."

Business planning and teams

Most innovators and investors alike acknowledge that it is not new technologies that create new businesses, but rather teams and people. As noted, an effective pitch can help articulate the key team and management structure behind the venture. Even more importantly, it helps the members of the founding and management teams better articulate their goals. Every individual involved in a new venture has his/her own aspirations and objectives. The exercise of building the pitch helps reconcile and potentially find synergies among those expectations. For example, some founders place significant emphasis on ownership and control. If, in the process of developing the pitch, it becomes clear that necessary external funding will dilute management control and ownership, this discovery creates the opportunity for an open discussion about how the issue will be handled.

When it comes to team-related issues, there are several difficult questions that should be addressed as part of the business planning process to ensure that the management structure put in place lays the foundation for a successful venture:

- What will be the equity ownership of each of the founders?
- How will equity and stock options be distributed to key employees?
- What is the process through which these decisions will be made?
- What will be the responsibilities of the founders and early employees?
- What roles will the founder and early employees play as the company evolves and grows?
- Who will be the CEO (and will there be an interim CEO to be replaced by a permanent CEO)?
- What are the desired attributes of early employees?
- How and from where will important hires be recruited?

A thoughtful discussion of these questions should occur early during business planning and pitch development because it can help highlight differences in expectations. Furthermore, the answers may affect the firm's strategic directions and funding options, as well as its overall culture and performance. Do not procrastinate on addressing these issues. Although some of the topics may be difficult for the early team to discuss and resolve, no prospective investor will give a team money before they have been satisfactorily answered.

However, innovators should recognize that while there is some flexibility in how to address these questions, market forces may constrain the range of viable alternatives. Consider, for example, the question of how equity and stock options will be distributed to key employees.[13] The answer depends, to a large extent, on what other companies offer their key employees. If the current trend is to offer two percent of the company to a key senior executive, deciding to offer less may make the

company's offer non-competitive. In contrast, offering more may not be considered acceptable by investors. More details on the question of equity ownership for key employees are presented in 6.3 Funding Approaches.

Business planning and intrapreneurship

Even though most people tend to think of the pitch in the context of entrepreneurship, these documents also play a critical role in intrapreneurship (entrepreneurship within the context of an established company or organization). Importantly, when **intrapreneurs** develop a pitch or executive summary to support an opportunity, they need to consider how to leverage existing resources within the firm and what complementarities exist with existing product portfolios, as described in 6.1 Operating Plan and Financial Model. They should also consider how the opportunity fits within the existing organizational structure and the extent to which it meets defined financial hurdles. Large firms, within which intrapreneurship often occurs, tend to have well-defined organizational structures that may inhibit or support important relationships with internal and external constituencies that are needed to bring an opportunity to fruition. They also have clearly defined capital budgeting processes and requirements for the return on investment (ROI) of any project that is undertaken. So, the financial model should be adapted to reflect internal processes and requirements. Factors such as these can serve as enablers or barriers to intrapreneurs as they develop a pitch.

Innovators operating within larger companies should pay particular attention to three aspects of the pitch, which take on more significance and can influence their ability to gain support. The first is the rationale for what makes the project a strategic fit with the priorities and interests of the organization. Strategic fit is commonly an issue of debate because the intrapreneurs are likely proposing a novel path with inherent risks.

The second issue is the degree of organizational autonomy required for the initiative to succeed. Intrapreneurs often cite their relationship with the core business and the potential for creating conflict as a significant barrier. Corporate innovators should be particularly careful where staffing is concerned. Here, one of the apparent strengths of a larger organization – a large, highly skilled base of human resources – can work against effective implementation. The biodesign innovation process requires a high degree of focus, especially in the early needs finding and needs screening stages. Too frequently, the tendency of the larger company is to believe that the effort can be undertaken by part-time team members "on-loan" from their other core business assignments. Experience has shown that the presence of at least one full-time team member can significantly and positively influence outcomes. That said, large organizations should resist the temptation to make the teams larger than necessary to complete objectives of these early phases. Smaller teams of two to four people are frequently most effective.

Finally, the third factor is to develop an explicit process for characterizing and managing risk. The lower risk tolerance of an established organization can sometimes make undertaking a high-risk initiative difficult. Intrapreneurs can help address this issue by remaining focused on answering their own *key questions* (i.e., sequencing and describing the steps to be taken to reduce risk in as orderly a fashion as possible for the given project). Teams should resist the temptation to use the larger organization's greater financial resources to mitigate risk. It is somewhat counter intuitive but the biodesign innovation process has been proven to run better when resources are constrained, forcing decisions to be made on a more demanding timeline. The corporation must be able to acknowledge up-front that failure rates for intrapreneurship initiatives are generally quite high. Both successes and failures play roles in creating a culture of innovation and should be understood as part of a productive intrapreneurship process.

◥ Online Resources

Visit www.ebiodesign.org/6.2 for more content, including:

 Activities and links for "Getting Started"
- Define the purpose and audience for the pitch
- Identify the key questions
- Develop an outline
- Conduct research and compile supporting documentation
- Write the pitch, seek input, and iterate

 Videos on strategy integration and communication

CREDITS

The editors would like to acknowledge Colin Cahill of Simpirica and Ross Jaffe of Versant Ventures for sharing the cases, as well as Kate Garrett, Dan Azagury, and Devesh Khanal for providing the Ciel Medical examples. Further appreciation goes to Todd Alamin, Tom Goff, John MacMahon, and David Miller for their contributions to the original chapter, as well as Ritu Kamal for her assistance in making updates for the second edition.

NOTES

1 Sabrina Parsons, "Pitching Your Business Versus Planning Your Business," *Forbes*, February 29, 2012, http://www.forbes.com/sites/sabrinaparsons/2012/02/29/pitching-your-business-vs-planning-your-business/ (February 20, 2014).

2 From remarks made by Mir Imran as part of the "From the Innovator's Workbench" speaker series hosted by Stanford's Program in Biodesign, April 28, 2004, http://biodesign.stanford.edu/bdn/networking/pastinnovators.jsp. Reprinted with permission.

3 Tim Berry, "Keys to Better Business Plans," Bplans.com, June 15, 2004, http://www.bplans.com/dp/article.cfm/198 (March 3, 2014).

4 Based, in part, on J. Skyler Fernandes, "The 'Best' Start-Up Pitch Decks & How to Present to Angels/VCs," One Match Ventures, July 6, 2013, http://www.slideshare.net/Sky7777/the-best-startup-pitch-deck-how-to-present-to-angels-v-cs (February 20, 2014).

5 Ibid.

6 "Writing an Effective Business Plan," Deloitte Touche Tohmatsu International, 1993.

7 Stanley E. Rich and David E. Gumpert, "How to Write a Winning Business Plan," *Harvard Business Review*, May 1, 1985, p. 136.

8 All quotations are from interviews conducted by the authors, unless otherwise cited. Reprinted with permission.

9 Ross Jaffe, "Introduction to Venture Capital," October 6, 2004.

10 From remarks made by Rodney Perkins as part of the "From the Innovator's Workbench" speaker series hosted by Stanford's Program in Biodesign, April 14, 2003, http://biodesign.stanford.edu/bdn/networking/pastinnovators.jsp. Reprinted with permission.

11 Fernandes, op. cit.

12 Ibid.

13 "Sharing Equity in a Start-Up or Established Entrepreneurial Venture," The National Center for Employee Ownership, https://www.nceo.org/articles/equity-compensation-startup (March 10, 2014).

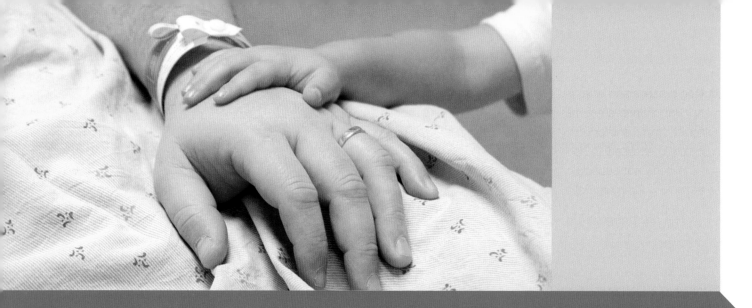

6.3 Funding Approaches

INTRODUCTION

In 2007, US venture capitalists invested $3.9 billion in 365 medical device start-up deals.[1] By 2013, venture activity had declined to $2.1 billion across 308 deals,[2] with only 49 of these investments directed to companies receiving first-time funding.[3] Against this backdrop, getting a new medical device start-up funded is, in many ways, harder than ever before. The US-based venture investors who fueled the industry for decades are increasingly looking to other sectors for less risky investment opportunities, shorter times to exit, and more successful initial public offering (IPO) activity. Meanwhile, capital requirements for medical technology companies continue to grow, driven in part by prolonged regulatory timelines and complex, unpredictable reimbursement processes. In short, the medtech funding landscape is in the midst a historic shift, where "business as usual" has been significantly disrupted and new opportunities are just beginning to take shape. Successful innovators will take advantage of the evolving sources of funding that emerge as the US healthcare industry undergoes restructuring and global markets and businesses continue to expand.

In rare circumstances, medical technology companies can get to market without the involvement of institutional investors. However, in most cases, innovators will require this type of support. A growing number of different funding sources can be leveraged to finance a medtech start-up, and innovators are getting increasingly creative about the sources they tap and the combination of investors they target. When choosing among funding sources, the innovators' focus should be on identifying the best fit between the project's capital needs and the goals of the potential investors. Selecting the right investors is particularly important because the innovators enter into a working relationship with these firms, organizations, and/or individuals that can span multiple years. Accordingly, everyone will be more satisfied if their true financial objectives, timelines, and management styles are aligned.

This chapter explores different types of funding and when they are most relevant to a company, the various sources of funds innovators can consider, and how these sources compare in terms of their advantages and disadvantages. Additionally, it describes the funding process and what innovators might expect when entering into a funding agreement. The chapter's emphasis is on funding sources for companies that can achieve sufficient clinical and/or commercial progress to be acquired, or ultimately generate the cash flows necessary to become self-sustaining entities. If the need addressed by a new solution does not support a stand-alone company, different funding options are presented in 6.4 Alternate Pathways.

 See ebiodesign.org for featured videos on medtech funding.

FUNDING FUNDAMENTALS

Many innovators find the process of raising funds for a new company frustrating and stressful. Similarly, they are often perplexed about why investors fail to grasp the potential of their ideas and decline to fund them. Mir Imran, a seasoned device entrepreneur who has experienced the funding process from the perspective of the innovator and the investor at different times during his career, summarized the situation this way:[4]

> I used to really ponder why these venture capitalists didn't invest in all my companies and give me big checks. I finally got the answer when I started writing checks. When you put on the investor hat, you ask a different set of questions. You're looking at risk – how to measure and gauge risk. So I have sympathy for both sides.

In the **medtech** industry, the costs (and related risks) of forming a new venture can be particularly high due, in large part, to the clinical, regulatory, and **reimbursement** requirements associated with commercializing new products in the field. A survey of more than 200 medical device companies found that the average company expenditure from **concept** just through regulatory approval was $94 million for technologies on **FDA's PMA** pathway and $31 million for those seeking **510(k)** clearance.[5] A sample of assorted medtech companies and their funding requirements to **exit** are shown in Table 6.3.1. Given these substantial capital requirements, funding in the industry was dominated by venture capital (VC) investors who could provide these sizable sums. However, the VC sector as a whole, was hard-hit by the 2008 financial crisis. In the resulting fallout, venture firms began shifting their investments away from start-ups to later-stage, less risky companies.[6]

Medical device companies were impacted by this shift to a greater degree than companies in other industries because it corresponded with other changes that combined to make many medtech investments less attractive. In particular, investors were discouraged by longer regulatory review times, greater clinical requirements, increasing unpredictability surrounding reimbursement, and the implementation of the 2.3 percent medical device excise tax that took effect in the US in 2012.[7] As a result, medical device VC investment declined by more than 40 percent from its peak in 2007 to the end of 2013.[8] Generally, more money flowed into software and, within the life sciences sector, into biotechnology.[9]

In response, the vast majority of medical device start-ups have had a harder time raising money than in years past. On one hand, some observers fear that the current funding environment may be stifling medtech innovation. On the other, aspiring companies have become more rigorous and resourceful in planning to get to market, and a wide range of non-VC funding sources have become more active in the medtech field.[10]

While the downturn in VC funding is almost certainly cyclical, the recent funding trends in medical devices highlight several important points about financing an idea. First, innovators are well served to recognize that the capital requirements of developing and commercializing a medical technology are sufficiently large that they are likely to have to seek institutional funding of some sort – very few device companies are able to get to market by **bootstrapping** alone. Second, deciding on a funding strategy should involve a broad exploration of the financing landscape to determine which type of

Table 6.3.1 Device start-up funding requirements vary widely, but are not insignificant (compiled from Preqin, unless otherwise cited).

Acquisitions				
Company	First funding date	Total known funding (millions)	Exit date	Total known value of exits (millions)
Acclarent	June 2004	$103.5	January 2010	$785
Ardian	February 2005	$64.1[11]	January 2011	$800
BarRx	June 2006	$42.8	February 2012	$325
CoreValve	June 2004	$63	February 2009	$700
Lutonix	July 2007	$35	December 2011	$225
Minnow Medical (now Vessix Vascular)	October 2010	$27.2	November 2012	$425

IPOs				
Company	First funding date	Total known funding (millions)	Exit date	IPO market cap (millions)
Foundation Medicine	April 2010	$89.5	September 2013	$394.2[12]
GI Dynamics	July 2003	$113.6	August 2011	$304[13]
LDR	September 2006	$48.7	October 2013	$339.8[14]
Stentys	January 2007	$22.2	October 2010	$121[15]
Tandem Diabetes Care	June 2008	$158.79	November 2013	$287.7[16]

funding, at what stage of the company's development, and what funding source provide a good match. And, third, in order to raise money from external investors, innovators must be able to tell a compelling story for how those investors will recoup what they put into the business. Importantly, not all sources of funding demand a financial return – but all have specific requirements for what they expect to get out of a deal. By directly taking these requirements into account early in developing an approach to funding, innovators will significantly increase their chances of success.

Topics related to these three important points, along with a series of others, are covered in the sections that follow. The chapter begins with types of funding, stages/uses of funding, and exit scenarios to provide innovators with a context for then evaluating the full range of

funding sources available to them and targeting the one (s) that provide the best strategic fit.

Types of funding

For all practical purposes, equity is the most widely available type of funding for early-stage device start-ups. However, under certain circumstances, innovators can also explore debt and grants.

Equity

Equity refers to a share of ownership in the business received by an investor in exchange for money. This is the most common way that external entities invest in medtech start-ups. The primary advantages of **equity funding** are that equity contributions generally do not have to be paid back (even if the company goes bankrupt), the company's assets do not have to be used as collateral, and no monthly payments are due. On the other hand, equity investments require the innovators to relinquish some ownership of the business and the investors may assert their ownership rights by seeking input into how the business should be run. The company also may be expected to share its profits with its equity investors through the payment of **dividends** (profits can be shared as dividends or reinvested in the business as retained earnings).

Equity investments take two primary forms: **common** and **preferred stock**. Both common and preferred shareholders own a portion of the company, but they are granted different rights in exchange for their investments. Both types of stock give shareholders **voting rights**. However, preferred stock provides shareholders with additional rights, which may include a liquidation preference (meaning that preferred stockholders are paid before common stockholders if a company is sold or its assets are liquidated), or preemptive rights (the option to keep a proportionate ownership of the company by buying additional shares when new shares of stock are issued as they would be in a subsequent financing). Holders of preferred stock are also often given priority over the common shareholders in the payment of dividends when and if declared by the board of directors. Based on the rights associated with preferred stock, it is considered less risky and generally favored by institutional investors, such as VCs and corporations. Individuals as well as founders and early employees, on the other hand, are more often issued common stock. Founders will likely own common stock, while key employees may be allocated stock options for common stock or in rare cases preferred stock as part of an incentive program.

Debt

Debt refers to money that is borrowed by a business. Debt must be paid back by borrowers (usually in monthly payments of principal and interest over a fixed period of time, similar to a mortgage). It is generally obtained from individuals, banks, or other traditional lenders. As noted, debt is primarily available in later stages of the start-up, once the company is generating revenue and has tradable assets that can be used as collateral. Given this later-stage orientation, innovators should keep in mind the effect of debt on other investors in the company. Debt lenders always have the first claim on the company's assets, followed by preferred shareholders and then common shareholders. What this means is that, if the company is sold, the proceeds will first be used to pay any outstanding loans and then distributed to the shareholders according to the rights corresponding to the specific type of shares they hold.

The main advantages of **debt funding** for later-stage companies are that the innovators usually do not have to turn over any ownership in the company or future profits to the lender, the lender does not exercise control over the business, and interest on loans can usually be deducted on the company's taxes. On the other hand, debt financing requires that a company have an adequate cash flow to make loan payments. Loans to start-ups are generally considered risky, carry relatively high interest rates, and may require a co-signer or guarantor. They also may specify that the company's assets be used as collateral (which means they can potentially be seized if the company fails to make its payments). Too much debt can also negatively affect a company's credit rating and impair its ability to raise money in the future. The types of loans most frequently used by start-ups are summarized in Table 6.3.2.

In certain conditions, companies may benefit from using a modest amount of term debt to extend their cash and delay their next round of equity financing.[17] However, from a practical perspective, debt financing only should be considered when the company's prospects appear promising.

Grants

Grants involve funds that do not have to be repaid by the company. They are disbursed to a team or company by a government entity, corporation, foundation, or other grant-making organization to support a particular project. Grants traditionally have been a primary type of funding sought by medtech innovators working in areas without significant commercial potential. Rather than seeking a financial return on the funds that they provide, grant makers instead require recipients to achieve some

Table 6.3.2 Innovators may decide to consider different types of loans, depending on their circumstances.

Loan type	Description	Key considerations
Bridge loan	• A **bridge loan** is the most common form of interim debt financing available to innovators and companies. • Typically used to span a period of time (e.g., before additional financing can be obtained, before the company closes a pending M&A transaction, or until the company achieves positive cash flows). • Amounts can range from $100,000 to several million dollars.	• Commonly extended to companies by existing **angel** or venture capital investors. • Usually can be arranged relatively quickly (with limited documentation), but often commands higher interest rates than conventional debt. • Principal and accrued interest may often be converted into equity at the lender's option, usually at a discount to current price/share. • Points, fees and other costs of obtaining the loan must be amortized over a short period of time. • The term is often short (commonly less than two years) and lenders often require relatively rapid repayment of principal if conversion does not occur.
Venture debt	• **Venture debt** is debt financing available to companies that have at least one professional investor as a significant equity-holder in the company. • Can serve a similar purpose of a bridge loan, but is also used to fund equipment purchases. • Loans typically range from $1 million to $15 million, with the debt maturing in 2 to 4 years. • Sometimes interest-only payments can be made for a pre-defined period.	• Specialized banks and non-bank lenders extend venture debt to companies that do not have positive cash flows or other significant collateral (although they are still likely to place a lien on what assets do exist, including IP). • Lenders can request operating covenants which specify minimum cash balances. • The lenders are compensated for the higher risk of default through a mix of high interest rates and **warrants**, which give them the right to purchase a number of shares of the company's stock at a price per share paid by other investors.
Royalty-backed loans	• With **royalty-backed loans**, the lender extends a loan to the company in exchange for a royalty on the company's product sales. • Lenders will sometimes combine their loan with making an equity investment. • Loans range from $5 million to 10 million.	• Because the lender's source of repayment is based on the success of the product, these loans are usually only available once the company can clearly demonstrate the product's sales potential. • The royalty becomes a cost of selling the product which requires that the product have adequate gross margin to allow payment of the royalty and still remain an economically attractive business opportunity.

other measureable form of "impact" that is aligned with the priorities and interests of the grant-making institution. However, as the number and type of grants and grant makers has proliferated, grants are becoming a viable early-stage funding alternative for more and more medtech companies.

In addition to the fact that grants do not have to be repaid, many innovators like that grant-making organizations often provide other forms of support, such as relevant expertise and resources, connections to other innovators, and increased visibility and credibility for the project among certain audiences. However, grant funding is inflexible from a timing perspective, which means that innovators must work to fixed funding cycles rather than thinking about when funding would be optimal for the project. Innovators also may face stringent restrictions on how grant money can be spent, and this money may come with oversight and reporting requirements that are burdensome to a start-up company.

To receive a grant, teams usually must prepare and submit a grant application, carefully following the guidelines and deadlines set forth by the grant-making organization. Many programs are highly competitive, so there is always uncertainty about whether an award will be made.

Stages/uses of funding

During the earliest stages of its existence, a company's perceived worth is low because there is a great deal of risk and the team has not yet proven itself. This means that the cost of raising money is expensive – a large percentage of ownership in the company will be given up for relatively small amounts of money. As the company accomplishes its major milestones (see 6.1 Operating Plan and Financial Model), it is able to raise

increasing amounts of money to support hiring, manufacturing, marketing, sales and distribution, and other **value**-building activities. As progressively more milestones are met, resulting in lower risk for the investor, funding becomes less expensive to the start-up.[18]

As a company moves through the stages of its development, it will likely seek different forms of investment based on the manner in which the funds will be used (as outlined below and summarized in Figure 6.3.1).

- **Seed funding** – Once a team has defined a compelling need, generated concept(s) to address it, developed a **prototype**, and performed a basic proof of concept, it can consider raising **seed funding**. Seed funding typically comes in increments of $10,000–$100,000. It is used to fuel project progress and help the team reach a stage where more substantial investment can be attracted. Innovators can use equity, grants, or even personal debt as a form of seed funding, but they should seek the type of funding with the most reasonable terms possible since the risk to the innovators and investors at this stage is extremely high.

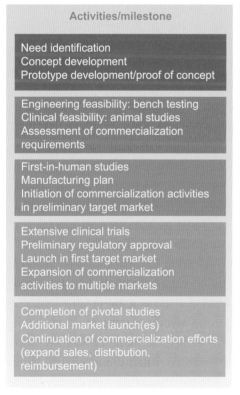

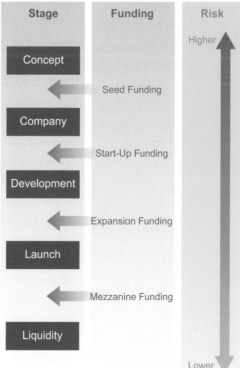

FIGURE 6.3.1

Funding at different stages of the company's evolution and the general activities in the biodesign innovation process to which they correspond (adapted from Leslie Bottorff, "Funding a Medical Device Start-Up," Medical Device & Diagnostic Industry magazine; reprinted with permission).

- **Start-up funding** – Common milestones that lead to **start-up funding** are related to product development and engineering feasibility, animal testing, and early consideration of issues related to commercialization (regulatory, reimbursement, etc.). This type of funding often comes in increments of $1–$10 million, with each round designated with a letter (e.g., Series A, Series B . . .).

- **Expansion funding** – Companies may be able to attract **expansion funding** after they initiate human studies, launch pilot trials, and begin pursuing regulatory/reimbursement and other commercial activities to prepare for an initial launch in a preliminary target market. Expansion funding is then used to accelerate and expand the project's current activities to get the product to market. It is not uncommon for medtech companies to require tens of millions in expansion funding.

- **Mezzanine funding** – At this stage, the company has retired its most significant risks but may not yet be generating sufficient revenue to fuel its growth and be self-sustaining. **Mezzanine funding** is usually available once the product is approved and has been launched. It is used to build distribution channels, fund sales and marketing campaigns, and expand/develop product lines. Commonly, this type of capital is raised in preparation for an **IPO** to ensure the company can demonstrate a strong balance sheet to prospective public investors. Mezzanine financing is sometimes provided by later-stage investors such as VCs, but often comes from private equity firms that specialize in this type of financing (sometimes referred to as "crossover" investors).

Exit scenarios

As explained in 6.2 Strategy Integration and Communication, early-stage investments in any medtech company are **illiquid** – the investors cannot easily sell their ownership stake. But eventually, most investors will want to make their investment liquid to allow them to realize a return. By far, the most common way this occurs in the medtech industry is for the company and/or its technology to be acquired by a corporation. Alternatively, the company can enter the public markets through an IPO.

For perspective, 30 device companies were acquired in 2013 while only four went public.[19] Another option is for a company to enter into a **licensing** agreement as a means of generating revenue and providing a payback to investors, although meaningful licensing deals are becoming increasingly rare. (More information about licensing is provided in 6.4 Alternate Pathways.)

Acquisition

While estimates vary, **acquisitions** in the medical device field account for roughly 80–90 percent of device company exits each year.[20] The sale of a company is usually driven by the strategic fit between the assets or technology of the acquiree and the strategy of the acquirer. Acquisitions can be an attractive exit strategy to investors because they receive either cash and/or a tradable stock that can be acted on immediately, while also avoiding much of the volatility and risk that can be associated with an IPO. They can also provide investors with a reasonable way to exit a troubled company.[21]

An acquisition occurs **outright** when the acquiring company takes full and complete ownership of the business for a single payment at the time of purchase – a point in time also known as the point of "change of control." The primary advantage of this approach is that it can provide a clean break for the founders and current management team, who may be replaced. The major disadvantage is that an outright acquisition typically results in **valuations** at the low end of the typical range because the acquirer assumes the entire risk of future performance. Also, founders and current management team lose the ability to influence future performance and often exit after brief transition periods are completed.

A **structured acquisition** is a variation on an outright purchase, which can allow the selling shareholders to participate in the possible upside based on success realized after the change of control occurs. It normally involves the buyer agreeing to make an initial payment followed by additional **earn-outs** triggered by specified milestones. The founders and employees of the start-up usually continue to be involved under the aegis of the acquiring company to help it meet the defined milestones. They are rewarded when those milestones are achieved and additional payments are distributed to shareholders. One benefit of this approach is that it can be used to resolve differences in perceived value between sellers and buyers by "sharing" future risk and return, often resulting in greater value realization for the founders. Additionally, the founders, as participants, retain some influence on the business through the transition. On the other hand, there may be less certainty regarding the level of value ultimately realized (as it may depend on how much the acquirer invests in developing and promoting the product), the payout milestones may not be under the founder's direct control, and the founders may need to remain with the company, allowing themselves to be directed by the acquiring organization.

Another form that an acquisition can take is an **option to purchase**. This occurs when a buyer receives a future right to purchase a company at a specified price, at a specified time, or following the completion of specified milestones. At the time of exercise, option agreements enable the buyer to use either an outright purchase or a structured purchase using a combination of upfront payments and earn-outs. Although the option holder often participates in governance through representation on the board of directors, there is no actual change of control until the option is exercised. This allows the management team to continue building the business (see the Nanostim case example in chapter 6.4). For the innovator, the biggest downside challenge of entering into an option agreement is making sure that the start-up will be financially strong enough to survive if, for any reason, the acquirer should decide not to exercise the option at the end of the option period.

In general, acquirers target two different types of companies for purchase. First, they will sometimes seek out start-up companies developing new technologies with "blockbuster" potential in large, undeveloped markets. In this scenario, they are seeking revenue potential in excess of $1 billion per year, supported by a novel technology, a strong IP position, and robust **clinical trial** data. Second, it is far more common for corporations to purchase more mature companies that fit well with their current businesses and have the potential to add immediate revenues (ranging from $500 million to $1 billion) and profits to the corporation's product portfolio. With

these deals, the focus is on proof of a sustainable business model, a strong, profitable sales ramp, and a promising product pipeline. More information about acquisitions is provided in 6.4 Alternate Pathways.

Initial public offering

An IPO refers to the first sale of a company's common shares to public (versus private) investors. The main purpose of an IPO is to raise capital for the company. It also provides private investors with a potential exit strategy because they can then trade their shares in the public market. However, the original investors typically face restrictions on when they can sell their shares after the company goes public, with some having to keep their shares for several years before they can start divesting them. IPOs also impose heavy regulatory compliance and reporting requirements on the business (e.g., the Sarbanes–Oxley Act of 2002 in the US), which is one reason the frequency of IPOs by start-ups has decreased. The rash of IPOs in the late 1990s and early 2000s also died down following various economic corrections, with many of the medical device companies that went public during this period failing to deliver on their high valuations. In late 2006 and for some time in 2007, medtech IPOs appeared to be on the upswing, but that trend was relatively short lived. The "window" for IPOs was all but closed for device companies by 2008.[22,23] In contrast, biotechnology companies are experiencing an IPO boom, with 33 public offerings in 2013 alone.[24]

Sources of funding

Company funding can come from a number of different sources. Common funding sources applicable to the medtech industry are described in the sections that follow and then summarized in Figure 6.3.4.

Bootstrapping

Bootstrapping refers to financing a small venture without the use of equity investments, grants, or loans taken by the company. In many cases, innovators use their own savings, personal loans, a second mortgage, or credit card debt to bootstrap a company. For products without clinical or regulatory restrictions, they may also be able to fund preliminary development through small customer advances.[25] Early-stage student teams sometimes enter business plan competitions sponsored by universities, where top prizes can be as much as $100,000. Some universities have also developed internal seed fund programs for faculty and student projects. Bootstrapping can be an effective way to get a project started, but it has obvious limits in terms of how far it can take a team on the path to market.

Friends and family

Friends and family investors refer to members of the innovators' personal networks with adequate means to make an investment in the project. Aside from grant funding, this is one of the least expensive forms of financing since friends and family are usually more flexible than professional investors in terms of their timeline and expected level of return.

Despite their strong personal relationships with friends and family investors, innovators are advised to treat the members of this group as professional investors. Friends and family should be provided with the company's business plan or **pitch** and an investment contract should be put into place.[26] These investors can be given equity in the company at a set price. Or the investment can be made as **convertible debt**. This means the friends and family loan the money, but it can be converted into equity, sometimes at a discount, when the company secures its first round of professional financing.[27]

Getting friends and family on board demonstrates to other potential investors that the innovators believe in the idea and that other people are willing to trust them with their money.[28] The primary disadvantage is that these investors may not understand the level of inherent risk involved in the opportunity. Also, they are not usually able to participate in subsequent rounds of funding, and they often do not have skills, expertise, or connections that can help the company grow.

Incubators

Business **incubators** are for-profit or non-profit entities organized to assist innovators in establishing a new venture and/or accelerating its progress. They accomplish this by providing start-ups with a combination of infrastructure and resources, expertise and connections, and

some amount of initial funding. In some incubator models, funding is provided outright, while in others it is provided indirectly through the provision of services and support.

The amount and type of funding offered by incubators varies widely by organization. At one end of the spectrum, non-profit incubators like StartX provide up to $100,000 in indirect funding from its partners, as well as free office space and legal services. It also offers needs-based stipends for innovator living expenses.[29] Rock Health provides direct capital in the form of an elective $100,000 convertible note or grants ranging from $10,000 to $20,000, in addition to hands-on mentoring and support.[30] Neither organization requires an equity position in its portfolio companies. At the other end of the spectrum, for-profit incubators such as ExploraMed, The Foundry, The Innovation Factory, or Coridea explicitly seek to nurture companies in which they can take a strong equity position and provide substantial support over an extended period of time.

Multinational companies also run incubators with the goal of supporting inventions that may turn into future acquisitions. Some of these, such as such as Johnson & Johnson's Janssen Labs are largely domestically focused, while others explicitly target innovations coming from other geographies. For example, US-based Medtronic has backed an incubator initiative in Europe called MDStart,[31] and Japan-based Sony has launched a medical device incubator called Rainbow Medical in Israel.[32] Many governments have also seized on incubators as a way to encourage entrepreneurship. In Singapore, for instance, the government has put over $30 million towards the Incubator Development Program, which supports local incubators and accelerators in healthcare.[33]

Beyond financial support, the primary advantages of working with an incubator include access to facilities and resources that may be out of reach for a young start-up; the high-quality expertise and guidance that is available; and connections to mentors, other innovators working in the medtech space, and prospective follow-on funders. Many innovators also find it motivating to be part of an incubator community. The main disadvantages vary based on the incubator model, but innovators should consider what "strings" come attached the funding that is provided, possible distractions created by other teams within the incubator, and the potential for micromanagement from incubator leaders.[34]

Governments

One way that governments seek to stimulate economic growth is by supporting research and development that leads to new products and services. With healthcare top-of-mind for governments around the world, innovation in medical technologies has become a high priority area for investment, which many government agencies make in the form of grants.

In the United States, two of the most prominent government grant mechanisms are the Small Business Innovation Research (SBIR) program and the Small Business Technology Transfer (STTR) program. SBIR's mission is to support scientific excellence and technological innovation through the investment of federal research funds in critical American priorities. Through the program, small for-profit or non-profit technology companies (or individual innovators who form a company) can gain access in phases to up to $1,150,000 in early-stage R&D grant funding.[35] STTR has the same mission and funding levels, but it awards its grants to small for-profit or non-profit companies working cooperatively with researchers at universities and other research institutions. Recipients of both types of grants retain the intellectual property rights to the technologies they develop. Funding is awarded competitively, but the process is relatively **user**-friendly. Multiple federal agencies are required to make SBIR and STTR grants on an annual basis. However, each agency designates its own R&D priorities and administers its program separately (according to the same general guidelines), so innovators should identify those agencies with interests similar to their own and approach each application process separately.[36] The National Institutes for Health, for example, makes SBIR and STTR grants for biomedical or behavioral research that supports its mission to improve human health.[37]

Governments outside the US actively fund medical technology innovation, as well. Of course, these programs differ significantly in their focus and details. However, a few examples highlight the range of government grant opportunities available to entrepreneurs and small companies. For instance, the Indian government provides funding for medical device start-ups via several schemes, including the Indo-US Science and Technology Endowment Fund, which awards up to $500,000 in grant financing to US-based companies interested in India as a market for their products.[38] The Irish government is working aggressively to establish the country as a hub for medtech research, development, and manufacturing. Through its Enterprise Ireland agency, the government has established an R&D Fund and a Commercialization Fund to provide early-stage grants to medtech projects based in the country.[39] It also offers a variety of additional incentives to help attract projects, including a sophisticated tax incentive structure for both start-up companies and the investors that back them.[40] Singapore is another country where the government is working to develop a strong medtech ecosystem and is using grants funding and other incentives to draw innovators and small companies to the area.[41]

With government grants, the financial interests of the company's founders are not diluted. Being the recipient of such a grant can also be a source of credibility for a start-up. On the other hand, government grant makers expect innovators to perform diligent research (comparable to what would be required by the most demanding academic institution), competition for funds can be fierce, and the review cycles for awarding funds can be lengthy. Also, with a cap at roughly $1 million per project, the amount of funding awarded through these programs is not likely to be sufficient to meet a medtech company's complete financing needs.

Foundations

Historically, foundation funding was not considered widely applicable to medical technology companies. However, over the last decade, a shift in the focus of some foundations and the nature of the grants they provide has made this a more viable medtech funding source for some device start-ups.

In the broad category of cause-driven foundations, some grant makers are focused on a specific disease state or medical condition. For example, the St. Jude Medical Foundation funds projects linked to device therapies with the potential to change the lives of patients suffering from cardiac conditions or chronic pain.[42] In other cases, a foundation's focus can be defined more generally (i.e., addressing poverty or disparities in global health). For instance, the Bill & Melinda Gates Foundation has an overall mission to help all people lead productive, healthy lives.[43] Within its global health division, it seeks to harness advances in science and technology to save lives in developing countries with an emphasis on vaccines, drugs, and diagnostics to prevent infectious diseases such as HIV, polio, and malaria.[44] Innovators working on projects with the potential to have a positive social impact on a particular population should research cause-driven foundations to identify those that may be in alignment with their goals.

Within the new class of cause-driven foundations that has emerged over the past several years, many will support both non-profit start-ups and those with a for-profit social entrepreneurship orientation (i.e., a commercial approach to driving social change). Additionally, while many foundations were traditionally accustomed to funding programs and services, a growing number are now becoming more open to underwriting product development of technology-based solutions. The Gates Foundation, for example, has awarded close to $1 billion in funding for new technologies through a family of grants known as the Grand Challenge program.[45]

Cause-driven foundations do not expect a financial return on the grants they make, but they do have clear expectations around results. As The Mulago Foundation, which has provided funding to medical technology companies such as Embrace, Mobile Medic, and product development company D-Rev, explained on its website: "We operate like a philanthropic venture fund with proven impact as an analog for profit, and cost-per-impact for return on investment."[46] Innovators working with such organizations must be prepared to think critically about the result their technologies will deliver and build measurement and evaluation processes into their product development and commercialization efforts. Importantly, they should also recognize that the priorities of foundation participants will not necessarily align with those of the other investors that may fund the company at later stages. This disconnect has the potential to create conflicts that may require the ongoing time and attention of the company's leaders.

Another possible funding source for medtech companies is foundations focused more generally on supporting entrepreneurship. Many of these types of programs are targeted at university-based innovators, such as VentureWell (formerly the National Collegiate Inventors and Innovators Alliance – NCIIA). This group provides seed grants to student entrepreneurs developing novel, market-based technologies through its E-Teams program.[47] The Wallace H. Coulter Foundation targets faculty members working on translational research. Given their desire to help spur new products to market, these foundations often have rigorous standards regarding progress against milestones and the realization of other measureable results. The case example on the Coulter Foundation illustrates the expectations one entrepreneurship-oriented foundation places on the university-based projects it funds.

FROM THE FIELD ▶ THE COULTER FOUNDATION

Applying business discipline to de-risk and accelerate university-based translational research projects

Wallace Coulter, a serial inventor and entrepreneur, was passionate about using science to serve humanity.[48] His most renowned invention, the Coulter Principle for counting and sizing particles suspended in a fluid, led to revolutionary developments in industries ranging from medical diagnostics and printing to food and space exploration. The Coulter Corporation was the leader in blood cell analysis equipment and other laboratory diagnostics. During his 40-year tenure as Chairman of this private company, Coulter fostered a culture of entrepreneurship and risk taking among his employees. He routinely invested significant resources in rogue R&D

projects focused on addressing pressing health-related needs. Coulter led the organization until its sale to Beckman Instruments in 1997 (now known as Beckman Coulter, Inc.).

After Coulter's death in 1998, Sue Van started the foundation to continue his life-long pursuits and passions. Van had served as Executive Vice President, CFO, and Treasurer of the Coulter Corporation and was named by Coulter as the trustee of his estate. He had instilled in her a strong commitment to helping patients, and she was determined to sustain his dedication to supporting biomedical innovation. In particular, Van wanted to continue his interest in helping innovators take their inventions from the bench to the market where they could directly benefit patients. Fondly remembering her time working with Coulter, she said, "I wanted to make up for all of the R&D projects I tried to get him to shut down when I was CFO."[49]

Van recognized that, in the changing economic environment, early-stage funding for new technologies was difficult for innovators to obtain. Additionally, she observed that universities, which are the engines for ground-breaking basic research, often struggle to translate interesting discoveries into practical innovations. "Wallace didn't believe in research for research's sake," she said. "He wanted to support research that would save people's lives." With the goal of capitalizing on the best aspects of academia and industry, Van launched a "grand experiment" that eventually led to the formation of the Coulter Translational Research Partnerships in Biomedical Engineering.[50] "But it wasn't easy," she recalled. At the time, many universities were focused solely on basic research and reluctant to broaden their focus to applied research, which was considered the domain of industry. Moreover, biomedical engineering (BME) was the newest of the engineering disciplines. "But I was trained in risk taking, so we pushed ahead," Van said. Ultimately, her effort helped validate translational research in the university setting and enhanced the value of the BME degree. It also positioned the Wallace H. Coulter Foundation as a pioneer in supporting translational research in biomedical engineering with the goal of accelerating the introduction of new technologies to improve patient care.

Elaborating on the Translational Research Partnerships, Elias Caro, the Coulter foundation's Vice President of Technology Development, explained, "The goal is to support collaborative research that addresses unmet clinical needs and leads to commercial products that generate improvements in health care." Specifically, the foundation provides what it calls "bridge funding," coupled with a rigorous project management process to drive projects to the "critical endpoint" of attracting follow-on funding from investors outside the university setting. Over time, the Coulter Foundation established these partnerships with universities across the United States (a complete listing is available online).[51]

Faculty members at participating universities can apply for Coulter funding if the project is translational in nature and multidisciplinary, with at least one principal investigator from the department of bioengineering and another from a clinical department in the school of medicine. The foundation strongly supports partnership between physicians and engineers because the clinicians bring an understanding of the need to a project, along with product ideas and a sense of urgency. The engineers, on the other hand, provide the technical expertise to transform ideas into innovative solutions and then make them a reality.[52]

Each university has a Coulter Program Director and an Oversight Committee, which consists of the bioengineering department chair, other key university representatives, a member of the Office of Technology Transfer, entrepreneurs, local venture capitalists, and industry professionals. Together, these individuals apply a rigorous stage-gate process for screening, selecting, and then managing the projects that seek Coulter Foundation funding (see Figure 6.3.2). The Coulter Process, as the approach is known, provides each academic institution with a disciplined program management approach for de-risking projects,

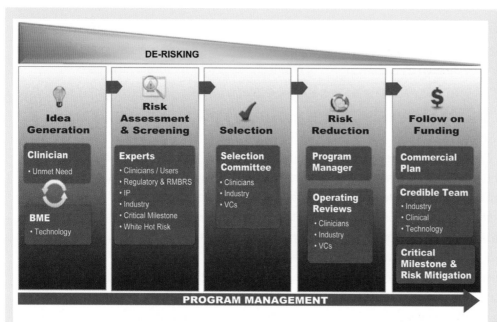

FIGURE 6.3.2
The Coulter Process for de-risking projects and bringing them to the critical endpoint of attracting follow-on funding from investors outside the university.

compressing development timelines, and keeping innovators focused on the most important activities to attract investor interest and increase the likelihood of advancing a product to market.[53]

The first step in the Coulter Process is for clinicians and engineers to work together to come up with a compelling idea. Then, when the project reaches an appropriate stage of development, they can express interest in Coulter funding. A request for proposals is issued each academic year via posters, emails, and presentations by the Coulter Program Director. At most universities, proposals are due in January. "Some universities set up boot camps to coach project teams on how to write the best proposal for Coulter funding," Caro described. Others encourage teams to meet with the Coulter Program Director before submitting a proposal. In these meetings, the Program Director evaluates whether the team is ready for Coulter funding and if the project fits the scope and focus of the award. "A lot of work must be done before the proposal," Caro acknowledged, "including answering questions like, 'What needs to be proved in order to get the next round of funding [after Coulter]?' 'What kind of clinical trials will be needed?' 'What is needed to get the technology to market and eventually to patients?' and so on."

Once submitted, the Coulter Oversight Committee carefully reviews each proposal and assesses the caliber and progress of each project. Criteria for their evaluation include scientific merit, potential healthcare impact, technical feasibility, and the potential for commercialization. Particular emphasis is placed on identifying "white hot risks" that could potentially prevent the project from attracting follow-on funding from angels, venture capitalists, or corporate investors, licensors, and/or acquirers. This involves the rigorous assessment of stakeholder and competitive dynamics, health economics, and other factors that can facilitate or impede adoption. The committee also considers how attractive the opportunity is to those who will fund it beyond the Coulter Process by looking at the most likely exit scenario(s) and potential return-on-investment calculations for the next round of external funders. Finally, the committee evaluates how the team proposes using its Coulter funding and the milestones it will focus on over the next 12 months to ensure that the project has a "laser focus" on retiring critical technical and commercial risks in preparation for graduating the idea beyond the university.

The committee invites about a third to a half of applicants to make an in-person, 20-minute presentation. At some schools, finalist proposals are also reviewed by outside

IP counsel to evaluate any associated risks. The Committee then selects 6–8 proposals to receive an award of approximately $100,000, typically over a one-year period (the money does not have to be repaid and the foundation does not take equity in the projects).

After the selection decisions are made, the focus of the Coulter Program Director and Oversight Committee shifts to providing guidance to teams to help them achieve their defined milestones and build toward the goal of attaining follow-on funding. For example, to address important technical risks, all Coulter-funded teams are advised to conduct "a killer experiment." "Technologies need proof-of-concept data to get seed funding," said Caro. Project teams must talk to domain experts and research clinical precedents to identify and design a killer experiment for their project such that the outcome either increases confidence toward further development of the product or recommends the project be abandoned. To help address commercial risks, each project is assigned an industry mentor with extensive experience in commercializing new healthcare technologies. Teams are encouraged to meet regularly with their mentors.

Importantly, Caro noted, "Projects can be killed and their funding revoked." Most often, this occurs if a project is not progressing to plan and the milestones have not been met for two consecutive quarters, or if the technology simply fails to perform as intended. However, the Oversight Committee occasionally withdraws funding from projects with dysfunctional team dynamics.

Over the life cycle of the funding term, teams prepare brief, written quarterly progress reports and present project updates to the Oversight Committee at specified points in the year. The process culminates at the end of the funding period with projects hopefully securing follow-on funding from angel, venture, or corporate investors outside the university. The Coulter Foundation does not formally engage in a "match making" process, but the Program Director, members of the Oversight Committee, and team mentors informally collaborate to introduce team members to interested parties. By the end of 2012, the Coulter Translational Research Partnerships had funded more than 250 projects. Of those, 45 had led to start-up companies that collectively raised over $840 million; 20 have formed start-ups that are seeking funding; and 29 licensed their technologies to established medtech companies. One start-up sold for $450 million in early 2013, and nine products have already received FDA approval. According to Caro, most university partners achieve a project success rate of more than 27 percent.[54]

Whether or not they pursue Coulter funding, Caro recommended to all innovators that they "begin with the end in mind," emphasizing the importance of focusing on the patient and what it will take to get the technology to market. "Many teams underestimate the importance of commercial risks that can derail a project and often focus solely on the technology. Find business mentors who can help you understand the commercial aspects of the project and help you navigate these challenges," he said. The Coulter Process, with its focus on de-risking projects via a staged approach, "is applicable to all medical technology innovation teams," Caro commented. "Some innovators say that Coulter money is 'expensive' because it comes with so many requirements, but these expectations are often exactly what makes a project successful." And each successful project honors Coulter and his commitment to translational research. As Van concluded, "We want others to see the impact of their research on patients."

Impact investors

Another emerging source of funding that may be applicable to some medtech companies is called impact investing. Impact investors are looking for both social impact and financial return on their investments (sometimes called the "double bottom line"). Specifically, they seek companies with business models that can create a financial return on the investment at (or just below) market rates, as well as deliver measureable social impact. Many impact investors are focused on needs in global markets, such as Asia and Africa.

One pioneer in the space is Acumen Fund. Acumen raises donations from foundations and individual donors that do not expect a financial return on their money but are instead interested in driving social change and impact. Acumen then seeks to identify companies that are finding innovative ways to address core problems affecting the poor in its target geographies. These companies could be non-profit, for-profit, or hybrid organizations. After rigorous **due diligence**, Acumen invests in select projects by providing loans or taking an equity position in exchange for capital. In some circumstances, it also makes guarantees to third-party lenders to facilitate access to local sources of capital, sets up licensing or royalty agreements when assisting with the creation or registration of intellectual property (IP), or provides lab investments to fund innovative but high-risk experiments expected to generate important near-term lessons. For-profit and hybrid companies most often receive equity investments, while non-profits are given loans. Over time, Acumen expects to realize a return on its investments, for example, through the repayment of loans or exit opportunities that allow it to liquidate its equity position. As a non-profit, Acumen invests all proceeds back into the fund so that the original philanthropic dollars can be reinvested many times over.[55]

The amount of capital directed toward impact investing continues to grow, and many of these investors report that their portfolio performance is meeting or exceeding social, environmental, and financial expectations.[56] However, impact investing has been slow to expand into the medical device sector. Acumen Fund invested in a company called Circ Medtech, maker of a non-surgical device for male circumcision,[57] and Khosla Impact, another impact investment firm, backed the Embrace infant warmer.[58] But overall activity in the sector has been light. Mobile health technologies may be one way to draw more impact investor attention to medtech. Both Acumen and Khosla recently funded **mHealth** applications (Sproxil, Inc. and EyeNetra, respectively).

Innovators considering working with impact investors should carefully target appropriate firms and seek to clearly understand their expectations. While delivering results against a double bottom line is appealing to many innovators, it is often not easy to achieve.

Crowdfunding

Crowdfunding, a recent development in fund raising, allows innovators to raise capital from a large number of donors or investors over the Internet. This approach has been successfully used to fund a wide variety of projects in sectors such as technology, consumer products, and entertainment. Crowdfunding across all industries was projected to total $3 billion in 2013,[59] on its way to $93 billion globally by 2025.[60]

Crowdfunding has two main models: (1) donation-based funding, where money is raised without the expectation of financial return, though some perks or rewards may be given to donors; and (2) investment funding, where businesses can raise capital in exchange for equity in the company. Crowdfunding platforms in the medical field, like MedStartr, allow innovators to collect financial contributions from the public in exchange for rewards; for example, pre-orders of devices or services.[61] Others, such as Healthfundr, offer equity investments in health-related companies, but can only allow accredited investors (individuals with a net worth of more than $1 million or income exceeding $200,000 per year) to participate as of the time of this writing.[62]

Raising investment funding (rather than donations) is potentially the more attractive model for medtech innovators, but in the US the Securities and Exchange Commission (SEC) has been hesitant to allow non-accredited investors to get involved. New legislation, called the Jumpstart Our Business Start-Ups (JOBS) Act which was signed into law in 2012, allows start-ups to raise up to $1 million in equity-based crowdfunding from non-accredited investors[63] (although there are limits to how much individual investors can contribute). In order for this provision to take effect, the SEC was required to issue rules governing such investments. The rules were expected in 2013, but were still unavailable as of late 2014 as the agency considered the most appropriate way to loosen long-established investor protections.[64] Until the SEC provides its ruling, non-accredited investors are unable to participate in equity crowdfunding of any type.

The story on VentureHealth.com explores the many issues associated with crowdfunding and outlines some of its benefits and risks.

FROM THE FIELD ▶ **VENTUREHEALTH.COM**

Crowdfunding medtech innovations

Mir Imran, a prolific entrepreneur and seasoned medtech investor, identified a gap in the fundraising landscape. He and his team observed that accredited investors who are looking to deploy capital often lack access to the most compelling biomedical innovations. In turn, promising healthcare opportunities are consistently underfunded, and innovators expend great energy raising capital for the development and commercialization of their technologies. Using his investment firm, InCube Ventures, to run some carefully controlled experiments, Imran and his team allowed a select group of accredited investors to participate in their **syndicated** investments. The response from investors was enthusiastic and the results were promising.

With the passage of the JOBS Act, they decided to initiate a detailed assessment of opportunities related to online financing from individual investors. Imran discovered plenty of online crowdfunding activity, but not many models that appealed to him. One common approach was for crowdfunding portals to post business plans or slide decks from start-up companies and invite investors to contribute funding. But, he discovered, many such websites conducted little or no due diligence on the opportunities. "They make money on the volume of companies on the platform, not the caliber of the opportunities," Imran said.[65] (Typically the portal retains 5–10 percent of the total money raised by the start-up company.) "This is a relatively unsophisticated approach," he continued. "There is no quality control, so investors are exposed to risks that they cannot adequately assess. I would advise innovators and investors to stay away from this approach."

Seeking a better solution, Imran and his team developed a crowdfunding model with much higher standards for the opportunities it presents and the investors it attracts. They founded VentureHealth as an online venture fund platform for accredited investors who want access to breakthrough opportunities in the healthcare sector (see Figure 6.3.3).[66] VentureHealth acts as an aggregator of knowledgeable, accredited investors who are invited to invest in a select group of medtech deals, with full disclosure of benefits and risks involved in the process. "This doesn't mean there is no risk involved, but we have much higher quality control." Investors gain access to disruptive innovations that have been traditionally reserved for venture investors; they invest alongside venture capitalists on similar terms; and the opportunities are rigorously vetted by the VentureHealth team, as with other venture deals.

The VentureHealth team vets investments based on its deep experience in company building. "The start-up companies' business plans go through detailed diligence process, and are selected as a VentureHealth deal offering only if they meet our criteria," Imran emphasized. For example, the team seeks major clinical breakthroughs in a large market, where reimbursement is well-established, and the innovators have devised a thoughtful clinical and regulatory strategy.

In terms of how VentureHealth makes money, "We don't take a fee from the start-up company [for listing them on the site]. Instead, we use a carried-interest model, typical of venture capital firms," Imran stated. (Carried interest refers to the share of profits paid to investors after a company has had a successful exit, which aligns the returns of the investors with the success of company.)

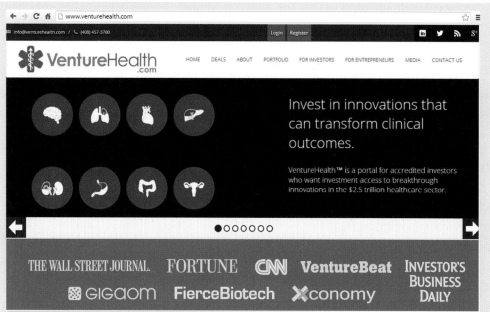

FIGURE 6.3.3
The VentureHealth portal
(courtesy of VentureHealth).

VentureHealth has built a stage-diversified portfolio of companies and deals, which means that it gets involved in early-stage and later-stage investments. "With second and third round deals, investors get to liquidity sooner. This gives them confidence and encourages them to keep investing. Otherwise, long development timelines can cause fatigue in investors not used to deploying capital in healthcare," he said. Recently, VentureHealth closed a $2.6 million Series B round in approximately two weeks for Rani Therapeutics, with InCube Ventures and Google Ventures as co-investors on the deal.[67] Channel Medsystems was also able to raise its $9.7 million Series B round through a syndicate of investors led by Boston Scientific, with VentureHealth participating.[68]

As these examples demonstrate, Imran recommended that medtech innovators undertake crowdfunding as part of a financing round rather than relying on it entirely. "That way, at least the start-up company will undergo some additional diligence," he commented. Bringing in other types of investors also helps validate the valuation of the deal and ensures that innovators have access to some of the benefits offered by other funding sources, such as the connections and expertise that experienced angel and VCs provide.

"We still have more questions than answers when it comes to crowdfunding in medical devices," Imran noted. "For example, in deals with no lead investor, it can be unclear who would set the valuation and draw up **term sheets**." He acknowledged that a certain amount of experimentation and learning was still underway, but at least for now VentureHealth represented the investors on its portal and handled all interactions and negotiations with the company so that the start-up did not have to deal with hundreds of small investors.

When asked for advice for innovators considering crowdfunding as part of an overall funding strategy, Imran said, "Be careful to provide a high-quality opportunity to potential investors. It's the responsibility of the entrepreneur to ensure that less-experienced investors understand the risks and have appropriate expectations about the time and money required to get a medical technology to market." He added, "Crowd-funding is here to stay. I think it will grow into a substantial source of funding for life sciences companies. Both investors and innovators must educate themselves about how best to utilize it."

Angels

Angel investors are experienced, accredited investors who use their personal wealth to invest in start-up companies. With the downturn of venture capital in the medtech sector, angels have become a much more important and prominent source of funding for device innovators. In 2012, angel investors committed more than $3 billion in funding to companies in the broad category of healthcare services/medical devices and equipment (surpassing total VC investment in the space).[69]

On the plus side, angel funding can be less expensive than VC funding. Angels have shown a willingness to take on more risk than some other investors. As a result, they have been a strong source of seed and start-up funding. More recently, these investors have begun to syndicate with VCs to participate in expansion funding. From 2011 to 2012, total angel investment in early-stage funding declined (40 to 33 percent), while their involvement in later-stage deals increased (from 15 to 29 percent).[70] One theory for this shift is the fact that angels may be gaining increased access to later-stage projects through emerging crowdfunding platforms (see the VentureHealth example).

On the downside, funding may need to be raised from several angels to meet the capital needs of a company, and managing the expectations of numerous angels can be daunting. Additionally, angels are traditionally less likely than venture capitalists to invest in multiple rounds of funding for the same company. Typical angel investments range from a few thousand dollars to $1 million. In 2012, the average deal size was just over $340,000.[71]

Corporate investment

Corporate investors have always played an important role in funding new medical technologies and providing companies with exit opportunities. However, as investors, their participation in funding start-ups has been cyclical, rising during times of venture capital shortages and falling during periods of wide capital availability. **Corporate investments** resemble venture investment, but they are usually made from venture funds set up by a sole organization (e.g., a multinational medtech corporation). Unlike private venture firms, large companies make these investments in start-ups primarily for strategic purposes and not strictly for financial returns. Fundamentally, corporate investors are looking for growth opportunities – ways to build their business base and expand their customer impact. Specifically, corporate investors hope to exploit synergies between projects in their internal portfolios and innovation occurring in the external environment.[72] The explicit hope of most corporate venture investors is that the companies they invest in may later become acquisitions capable of helping them to meet corporate growth goals.

Traditionally, corporations got involved in medtech funding during the later stages of a project. But, as Casey McGlynn, head of Wilson Sonsini Goodrich & Rosati's lifesciences practice, explained in an article, "The corporations in general have really stepped up to be major funders of new medtech companies, all the way down to the seed level. The business development people at these large medtech companies are very sophisticated people; they do their homework, they've got huge domain knowledge in their specialist area. They're a bit more targeted than the venture capitalist. I think they're under a tremendous amount of pressure to help find and fund the best new projects."[73]

In terms of their advantages, corporate investments can be less expensive to an innovator than VC funding and can bring with them unique forms of leverage (e.g., access to established sales and distribution networks and complementary technologies). The association with a major corporation can also lend credibility to a young company. Moreover, a strong, mutually beneficial relationship with a corporate investor can provide a smooth and valuable exit strategy.

That said, innovators involved in corporate relationships must recognize the compromises that may accompany this form of funding for their business. Some corporations will only invest in situations where they can attach "strings" to their money (conditions that give them a potential advantage over their competition). These strings can include distribution rights, first rights to negotiate, or even first rights of refusal, which can be triggered if and when the company receives a buyout offer from a competitor. Conflicting agendas may arise as the corporate investor looks out for the corporation's best interests. Issues surrounding the ownership of new intellectual property (IP) that is generated, which may be beneficial to both the start-up and the corporation, can also arise. Additionally, corporate investments may be susceptible to changes in the economy and provide innovators with limited opportunities for follow-on funding. Finally, exiting from a corporate investment can become complicated if there is more than one bidder but the corporate investor has been granted first rights to an acquisition.

A variation on traditional corporate investment occurs when an independent distributor strategically invests in a medical technology that it is potentially interested in adding to the portfolio of products it represents. In addition to creating a reasonably priced funding stream, this type of investment can provide innovators with a rapid path to market once necessary regulatory approval is received. In addition, distributors can add strategic value by connecting the company to physicians and other decision makers. Distributors usually have close relationships with the potential purchasers of the technology and can leverage them to provide insights on the requirements most likely to drive adoption. However, this type of funding can sometimes lead to conflicts if the company chooses to sell the innovation through other channels (e.g., upon acquisition). Innovators considering distributors as investors should think about incorporating a buyout clause into the deal, which allows an acquirer to begin selling the product directly post acquisition. This can help avoid any conflicts and incentivizes the distributor to maximize its selling efforts since it will then participate in the upside of an acquisition through both its equity investment and the buyout calculation. See the Loma Vista Medical story later in this chapter for more information about distributors as investors.

Venture capital

Venture capitalists are professional investment managers who specialize in funding companies with the potential for high returns.[74] Venture capitalists typically raise money from institutional investors (e.g., pension-fund managers, university endowments) or other private and public entities, with those investments put into a fund. The VC firm managing the fund is referred to as the General Partner (GP) and the outside investors are called Limited Partners (LPs). In its role as the GP, the VC firm contributes to the fund (usually 1–2 percent), but the majority of the capital is provided by the LPs. Funds vary in size, but can range from $50 million to $2 billion and beyond.

Each fund has a specific investment focus (medical devices, biotech, information technology) and a time horizon for its investments (usually from three to seven years). The capital in the fund is invested in start-ups that fit the fund's profile with the expectation of reaching a "**liquidity event**" – that is, the company will be acquired or will go public – within the fund's defined investment horizon.[75] Venture capitalists receive a share of any profits realized from investments made out of the fund (approximately 20 percent). In addition, they are compensated through an annual fee, which is a percentage of the total funds raised.

At any given time, VC firms can have several funds under their management, each of which is invested in a different group of companies. Every fund represents a separate pool of capital from which the VC is expected to generate returns over the life of the fund. The normal life of the partnership that supports a given fund is 10 years and the expectation is that most investment returns will be realized during that time interval. For this reason, it is important for innovators to understand the age and investment position of the specific fund from which their investment will be drawn in order to assure themselves that the firm will have the management and financial capacity to support potential future funding requirements.

Venture capitalists are known for having "deep pockets" when it comes to qualified investments, typically investing $3 to $30 million in each of the companies they back over the lifetime of the investment. VCs also have the ability to provide multiple rounds of funding. In addition, they often collaborate with the companies they invest in such that the start-ups benefit from substantial industry experience, as well as extensive networks of contacts (including other potential investors) that can be leveraged to assist the company. Larger financings will often result in VC investors forming syndicates, or groups of venture investors, in order to ensure that the company's future financial needs can be met from internal sources as needed.

On the downside, VC funding is expensive since innovators must be willing to give away significant equity to attract investment, and the due-diligence associated with the funding process can be time and labor intensive. VCs are active investors and will usually insist on having effective control of the company either through percentage ownership or a shareholder's agreement. VCs also have a clear objective for their involvement in a company – a strong financial return on investment – which may be based upon an exit strategy that may or may not be aligned with the company founder's long-term objectives.

Choosing the best investor for the company

When considering funding sources, it is important for innovators to be explicit about what is most important to them. For example, is it essential to find an investor who can make a long-term commitment (and potentially contribute to multiple funding rounds)? Or, is it a higher priority to find an investor that will not restrict the company's strategic options (e.g., by requiring a large ownership stake or by taking majority voting rights that would make the company dependent on the investor's direct support on critical decisions)? The amount of relevant experience and expertise required (or desired) in an investor will vary from innovator to innovator and company to company, depending on the strength of its advisor base and/or board of directors.

Another reason to carefully consider the best investor for a company is because innovators enter into a relationship with their funders. This is generally true, regardless of whether the young company is funded by incubators, impact investors, angels, VCs, corporate money, or distributors, but not necessarily if it is funded by government grants or some types of foundations. Whereas a bank would traditionally give a loan to an established business contingent on the use of its business assets as collateral, investors in a medical device start-up

are handing over millions of dollars in return for a stock certificate and the innovator's promise to build a successful company. For taking on the increased risk, investors in start-ups typically anticipate higher returns, seek to exercise more control, and expect to own more of the company. Not only can the capital carry with it an expected return many times greater than a bank loan, but it may be granted on the condition that the representatives of large investors become board members of the company and have input and voting control over the company's future. For this reason, it is critical to select investors that can add value to the company and ensure that their goals are aligned with those of the company's management team.

Two other important considerations in targeting investors are what type of funding is needed, based on the stage of the company, and how much money is required. Figure 6.3.4 provides a directional representation of when different investors are most likely to get involved. It also provides approximate ranges for the amount of money each funding source typically provides.

At some point, most medtech companies will interact with either angel, VC, and/or corporate investors. Table 6.3.3 provides a general comparison of how their priorities and investment guidelines may differ when evaluating a business opportunity.

Investor due diligence

As referenced in chapter 6.2, the process investors go through to assess how an opportunity measures up to their investment criteria is called due diligence. Due diligence is an iterative exercise that requires ongoing investigation and discovery. Each round of due diligence is different, with the approach and required time varying based on the investor's specific objectives and the characteristics of the company. However, the steps outlined in Figure 6.3.5 are generally followed.[76]

Information collected via this process is continually reviewed. If negative information comes to light at any point, an investor may decide to abandon the due diligence process (and abstain from investing in the company). When dealing with VCs, this process can take from as little as four weeks in rare cases to as long as

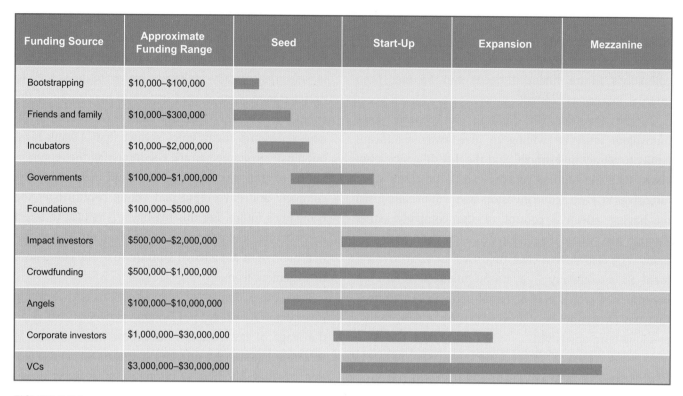

Funding Source	Approximate Funding Range	Seed	Start-Up	Expansion	Mezzanine
Bootstrapping	$10,000–$100,000				
Friends and family	$10,000–$300,000				
Incubators	$10,000–$2,000,000				
Governments	$100,000–$1,000,000				
Foundations	$100,000–$500,000				
Impact investors	$500,000–$2,000,000				
Crowdfunding	$500,000–$1,000,000				
Angels	$100,000–$10,000,000				
Corporate investors	$1,000,000–$30,000,000				
VCs	$3,000,000–$30,000,000				

FIGURE 6.3.4

The timing and amount of money associated with each funding source can vary quite a bit, but this summary gives innovators a directional sense of what to expect.

three years. The typical range is three to nine months. Other types of investors will also conduct due diligence, but they may not devote quite as much time. However, fundraising, regardless of investor type, usually takes much longer and requires more effort than most innovators estimate.[77] Innovators should keep in mind that the typical period between financings can range from 11 months on the low end to 18 months on the high end, with approximately 15 months being relatively common. This is because investors generally like to feel they are financing at least a year or more of "runway," during which time they expect the team to achieve milestones needed to justify raising additional capital under favorable terms. The practical implication of this interval and the time required for diligence is that the innovators are almost always "in the market" actively lining up their next financing.

Milestones

Funding is typically provided to a new venture in a staged manner. As described, rather than committing too much funding up-front, investors provide financing as the business demonstrates its ability to accomplish the milestones laid out in its **operating plan**. This process enables investors to periodically update their information about the firm, monitor its progress, review its prospects, and evaluate whether to provide additional funding or abandon the project. It also enables them to exercise greater control over the direction of the company. It is important for innovators to seek funding in stages, since the attainment of significant milestones in between funding rounds may strengthen their position from a valuation and ownership standpoint during the next round of financing negotiations.

Funding milestones represent significant events in the life of a start-up and should be selected with great care. The operating milestones chosen when creating the operating plan in 6.1 Operating Plan and Financial Model can serve as the starting point for selecting funding milestones. From an investor perspective, funding milestones represent points when a sizable amount of technical, clinical, or market risk has been eliminated from the

Table 6.3.3 Different investors have investment criteria that vary in certain areas.

	Angels	VCs	Corporate investors
Market size	Smaller, emerging markets	Large, established markets with $500 million or more in sales	Same as VCs but with an emphasis on markets in which they already operate; or are of strategic interest for future growth
Investment size	$100,000 to $10,000,000	$3,000,000 to $30,000,000 or more	$1,000,000 to $30,000,000
Expected return	4 to 10 times the first series of capital invested	4 to 10 times initial capital invested	May accept lower returns if investment is aligned with strategy
Capital intensity	Smaller markets with lower requirements; often look for markets untapped by VCs	Willing to enter market with intense competition if potential reward is large enough	Same as VCs
Strategic fit	More likely to be mission-driven	Seeking next blockbuster device, often regardless of specific field	Looking for opportunities that complement their existing portfolio
Investment timeline	Sometimes flexible; could be longer than 8 years	4 to 10 years; tend to prefer devices on 510(k) versus PMA regulatory pathway or early commercialization outside the US	Same as VCs but sometimes may be shorter for corporate investors
Ownership target	Small but will eventually want the start-up to seek funding from VCs or corporate investors	30 to 80 percent in early rounds, rising as high as 95 percent of company by the time of exit	Ultimately may seek to acquire technology
Board representation	Often	Almost always	Sometimes

company. Often these points will coincide with milestones in the operating plan, such as proof of concept, clinical trial initiation, regulatory approval, etc. Ultimately, the final funding milestone for investors is the exit or "liquidity" event, be it through an acquisition or an IPO. Revisit Figure 6.3.1 for sample activities/milestones that investors will expect to be completed between various stages of funding.

As each of these milestones is achieved, the company has the potential to reduce its cost of capital. As described, at the earliest stage of investment, the company's worth is lowest and the cost of raising money is highest because the business has no proven track record

and thus poses significant risk for the investors. Yet, as the company begins to meet its milestones, it is able to raise increasingly large amounts of money at more competitive rates as the risks to investors decline.[78] Three of the most important hurdles for a company to overcome in demonstrating increased value include:

- **Technical feasibility** – Relies heavily on engineering, science, and clinical interactions and is accomplished when the company has proven data regarding in vitro (seed funding), animal use (some seed funding and start-up funding), and human use (start-up funding and expansion funding).

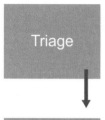

Triage

- Direct or indirect contact between the investor and innovator, often resulting in the business plan or pitch deck being shared.
- Review of the business plan and/or initial meeting between the investor and the innovator or management team.

Initial due diligence

- Initial discussion of concept by investor with some of his/her partners (if appropriate); investor makes initial calls to contacts in the technology/service area to get a general opinion on the concept and identify issues that need to be addressed.
- One or more meetings with innovator/management team to discuss technology/service concept, market, management experience, business/financial model, and valuation expectations.

Heavy due diligence

- Calls or meetings with company references on technology/service concept, market, management experience, etc.
- Calls or meetings with independent experts or knowledgeable individuals among the technology/service area's actual or potential customers, and independent references on management. Investors may hire independent consultants to evaluate aspects of the business, particularly technical, market, regulatory, reimbursement, or IP issues.

Investment decision

- Independent analysis of financial projections and valuation scenarios.
- Further discussion with innovator/management team around specific business issues and company valuation.
- Based on due diligence. If positive, a term sheet is generated (see below).

Investment closing

- If agreement is achieved on the investment terms, legal and patent due diligence is done during the process of closing the investment to ensure that there are no "hidden" issues about which the investors are unaware (e.g., lawsuits, problems in company capitalization, problems in contracts with employees, suppliers, or others that may impede the growth of the company or cause problems at exit).

FIGURE 6.3.5

The due diligence process becomes increasingly rigorous and thorough as the innovators and investors approach an agreement (based on Ross Jaffe, "Introduction to Venture Capital"; reprinted with permission).

- **Product feasibility** – Depends primarily on R&D and clinical and regulatory expertise. Product feasibility demonstrates commercial viability and is proven through the completion of pivotal trials, regulatory approvals, support from key opinion leaders (**KOLs**), and first commercial sales (some start-up funding and expansion funding).

- **Company feasibility** – Relies on continued R&D, sales and marketing, and manufacturing to demonstrate sustainable profitability. Company feasibility can be shown through revenue and profit growth, a full product/technology pipeline with multiple generations of devices being developed, strong brand identification in the market, and defect-free quality (expansion funding, mezzanine funding, and IPO).

When deciding on funding milestones, innovators should select the ones that their target investors are most likely to view as the most significant barriers to the success of a company. Then, as they are achieved, they will yield the greatest increase in company value. That said, it is important to stay somewhat flexible and open to adjusting the company's plans. As Aravind Swaminathan, co-founder of BioTrace Medical, Inc. explained shortly after closing a $3.4 million Series A funding round,[79] "We place a great deal of importance on constructing an ideal operational plan and seeking out an ideal set of investors. I think one lesson we learned is that it's important in this climate to remain flexible in that vision. There are multiple different potentially successful pathways a project can take and any given investor will have its own assessment of how best to handle the milestones and funding. The key is to keep the project moving forward, and flexibility will help achieve that."

Valuation, dilution, and ownership

A company's valuation, or the worth assigned to the business, is directly affected by the following factors:[80]

- The current and expected future valuations of comparable companies in the public and, when available, private marketplace: The higher the valuation of **comparables** and the more optimistic their outlook, the higher the valuation. Innovators should appreciate that active professional investors like venture capitalists develop a strong sense of the market valuation benchmarks due to their participation in multiple financings.
- The supply and demand for capital at the time of financing: Shortage of alternative investment options for investors increases the valuation. Abundance of alternative options decreases the valuation.
- Intangibles unique to a specific company, including the quality of the management team, a company's competitive advantage, and its likely pace of revenue growth and profitability: More experienced teams with proven track records can negotiate higher valuations.

- The nature and timing of an expected exit for the investor: The closer the time to exit and the higher the certainty of the exit, the higher the valuation.
- The implications of future capital raises, as well as needs to expand a company's **option pool** on the company's capital structure going forward: The more future rounds needed, the lower the valuation.

Valuation is important for any company taking on equity investors. Investors often refer to **pre-money** and **post-money valuations**. Pre-money refers to a company's value *before* it receives outside financing (or the latest round of financing), while post-money refers to its value *right after* it gets outside funds.[81] The pre-money valuation reflects the value assigned by investors to the assets the company has developed to date and the promise that the company holds. The post-money valuation is always equal to the pre-money valuation plus the capital raised.

The Working Example focused on Conor Medsystems demonstrates the concepts of pre-money and post-money valuation and introduces the concept of **dilution** to investors, founders, and employees. This particular example was chosen because it provides a rare example, with complete data on changes in valuation as a company went through an entire funding lifecycle that included both private and public investments. It is also especially interesting because Conor Medsystems had two exits; the IPO, which because of the heady medtech market conditions at the time, generated a stunning 28.6x return on the Series A investment, followed by a second liquidity event for founders, employees, and shareholders when the company was acquired by Johnson & Johnson for $1.4 billion.[82] Moreover, the ultimate outcome of the core technology vividly underscores the inherent risk for investors in medtech development. Three months after J&J's acquisition of Conor, the device failed to meet its primary endpoint in late-stage clinical trials. As a result, J&J terminated clinical trials, halted plans to submit an application for premarket approval with the US Food and Drug Administration, and stopped selling the technology in countries in Europe, Asia, and Latin America where it was already approved.[83]

Working Example
Pre- and post-money valuation for Conor Medsystems

Conor Medsystems was founded in 1999 to develop a new generation of drug-eluting stents (DES). Over the next five years it raised more than $78 million before going public in December 2004. In February 2007, it became a wholly owned subsidiary of Johnson & Johnson. Table 6.3.4 presents the history of Conor Medsystems' funding rounds, valuations, and percentage ownership structure until the time of its IPO, as well as a summary of the returns realized by the investors.

The post-money valuation is calculated by adding the amount of funding raised to the pre-money valuation. For example, in the first round of funding on February 1, 2000, the pre-money value assigned by investors to the company based on its progress and plans to date was $2.675 million. The funds raised, $0.325 million, was then added to that amount to give the post-money valuation ($3 million), which represented the value of the company, including its assets, the promise of its plans, and the cash just raised.

Taking the example one step further helps demonstrate the effect of dilution. As more investors provide money in exchange for shares in the company, the increased number of shares outstanding reduces the percentage ownership of existing shareholders.[84] To calculate the percentage of the company owned by investors in the first round of funding, one can divide the amount of funding raised by the post-money valuation:

$$\text{Ownership} = \frac{\text{Investment in current round}}{\text{Post} - \text{money value}}$$

For example, after the completion of the first round of funding, the investors provided $0.325 million to acquire shares in Conor Medsystems. With a post-money valuation of $3 million, this means that the investors acquired approximately 11 percent of the company. The remaining percentage (89 percent) stays with the founders/management and in the employee option pool.

Table 6.3.4 Valuation for Conor Medsystems (compiled from VentureXpert and the Conor Medsystems prospectus).

Date	Funding raised (in 000s)	Pre-money valuation (in 000s)	Post-money valuation (in 000s)	Founder/ mgmt ownership	Investor ownership					
					Series A	Series B	Series C	Series D	Series E	IPO
02/01/00	$325	$2,675	$3,000	89%	11%					
11/01/00	$1,500	$8,500	$10,000	75.65%	9.35%	15%				
06/27/02	$10,200	$17,800	$28,000	48.11%	5.95%	9.54%	36.4%			
10/22/03	$28,000	$29,000	$57,000	24.88%	3.08%	4.93%	18.82%	48.28%		
08/01/04	$38,900	$111,100	$150,000	18.42%	2.28%	3.65%	13.94%	35.76%	25.93%	
12/14/04 (IPO)	$78,000	$409,000	$487,000	15.47%	1.91%	3.06%	11.71%	30.03%	21.78%	16.02%
Summary										
Initial investment					$325 K	$1.5 M	$10.2 M	$28.0 M	$38.9 M	
Terminal value				$75,339	$9,302	$14,902	$57,028	$14,624	$10,606	
ROI					28.6 x	9.93 x	5.59 x	5.22 x	2.72 x	
CAGR					110%	75%	93%	356%	630%	

Numbers may be subject to rounding errors
ROI = return on investment
CAGR = compounded annual growth rate

In each round, the same initial calculation is performed (e.g., $1,500,000/$10,000,000 = 15$ percent in the second round) to determine the percentage of the company sold to the new investors.

However, an additional calculation is needed to compute the effect of each subsequent round of funding on the ownership percentages from previous rounds. For example, in the second round, the ownership of the original investors and founders is reduced by 15 percent (the ownership share of the new investors.) The new investors' share comes from the fraction owned by the original investors and founders. For instance, if the founders originally owned 89 percent of the company, after the new round of funding they will own 89 percent of the fraction of the company retained by the original investors and founder (which is 85 percent). So the founders' diluted share of the company is 89 percent multiplied by 85 percent, which leads to the diluted share of 75.65 percent. Similar calculations apply for investors in series A (their new ownership is 11 percent of 85 percent, which is 9.54 percent).

When Conor Medsystems went public, the terminal value for each investor is simply the value of their shares at the time of the exit (calculated as the number of shares multiplied by the price per share). Then, the **ROI** for each shareholder is given by:

$$\text{ROI} = \frac{\text{Terminal value}}{\text{Initial value}}$$

For example, for A round investors, the ROI is $9,301/$325 = 28.6$. That is, the original investment made by A round investors grew 28.6 times (a return that reflects market conditions at the time, not necessarily the current funding environment).

The compound annual growth rate (**CAGR**) is the annual growth that the initial investment of each investor experienced. This is calculated using the following formula:

$$\text{CAGR} = \frac{1}{\dfrac{\text{Time between original investment and exit}}{\ln\left(\dfrac{\text{Terminal value}}{\text{Initial value}}\right)}}$$

For the funds provided by round A investors to grow 28.6 times from February 1, 2000 to December 14, 2004, this means that the compounded annual growth rate was 110 percent. In other words, the investment grew at an annual rate with compounding of 110 percent per year.[85]

This example provides the funding requirements and valuations for Conor Medsystems, as reported after the fact. However, to derive a company's funding requirements prospectively, innovators should look to the operating plan and **financial model**. Funding requirements should be set such that the company has enough money to reach the next major milestone in the operating plan, enabling it to demonstrate risk reduction and secure the next round of funding.

Although the returns realized by the Conor Medsystems team and its investors do not provide realistic benchmarks for today's medtech environment, the example is instructive in other ways. In particular, it clearly illustrates how the company's valuation determines the ownership percentage of the innovators/founder and employees. As the company progresses and more capital is raised, the individuals should expect their ownership percentage to shrink. However, in parallel, the valuation of the company is expected to increase, which can lead to a higher total value for the owners.

In some cases, it is not uncommon for the valuation of a company to decrease between rounds of funding (a so-called "down round"). While this may be disappointing, it usually reflects a temporary setback that may be reversed in the future. Yet, start-ups are a risky investment, with some of them never generating a return on the capital and time invested. Investors are aware of these risks and seek to mitigate them in two ways: (1) by requiring high ownership stakes in the companies they fund, such that the returns on successful ventures help counterbalance investments in failed start-ups, and (2) by incorporating anti-dilution measures in a deal that prevent their equity investments from losing value (see section on "Term sheets"). Warrants are a vehicle used to protect investors from dilution. These give the investor the option to purchase additional shares of the company's stock at a pre-specified price or else face the dilution of their ownership percentage.[86] Innovators can help retain ownership in the company by carefully and proactively evaluating and managing key risks.

Strategic considerations: how much funding between rounds and at what valuation?

When thinking about the capital and time required to reach each funding milestone, innovators should consider the operating plan (validated relative to a **proxy company** or companies), plus the amount of incremental capital needed to address any deviations from the plan. Each round should include a "cushion" to address these deviations, since running out of cash between valuation points can be incredibly costly and potentially jeopardize the business. However, keep in mind that raising too much capital needlessly dilutes the ownership of the innovator and the previous investors.

Determining the value of the company at each round of funding is both an art and a science. A company can get started by doing simple modeling to develop "back of the envelope" valuations based on expected returns and potential exit valuations. Two common methods of valuing the start-up at each round are: (1) discounting terminal value; and (2) a comparables analysis. The premise behind discounting the terminal value is that investors require a certain return on their invested capital. While not a definitive valuation method, this approach represents a good exercise to understand the drivers of valuation. During earlier rounds of funding when the venture is more risky, investors expect higher returns than in later-stage rounds when risk has decreased. The critical components in determining the company value at each round are listed below:

- **Terminal value** – With the current exit strategy, what amount can investors expect the company to be worth? As an example, a start-up may determine that it could be acquired by a large medical device company for $400 million. The terminal value is often based on what comparable companies received at their exit event but can also be based on the future cash flows that the product may generate after the exit event. For instance, in the Conor example the

terminal value was the IPO pre-money value of $407 million. There could be an alternative terminal value that would represent an acquisition.
- **Duration** – The time frame between the specified round and the exit event.
- **Discount rate** – The discount rate is the return that investors expect to be compensated for putting their capital at risk. Table 6.3.5 illustrates typical discount rates for different types of projects. As a rule of thumb, the discount rates may get smaller with each subsequent round of funding.
- **Calculation** – For each round of funding, discount the terminal value back by the expected duration between that round and the exit using the discount rate. The general form of this equation is:

$$\text{Post money valuation} = \frac{\text{Terminal value}}{(1 + \text{discount rate})^{\text{duration}}}$$

For example, a rough calculation for Conor's valuation on December 1, 2000 could be as follows: assume a terminal valuation of 100 million, a discount rate of 70 percent, and a duration of 5 years (time to IPO); then $100/(1 + 0.70)^5 = \$7$ million. The actual realized valuation at that point was $3 million, which implies that the investors assumed either a longer duration, a lower exit, or required a discount rate in excess of 70 percent.

In addition to using the terminal value, a comparables analysis will help a company target a realistic valuation. This analysis begins with selecting a comparable company based on at least the following criteria: stage of funding, field and application, and founders' experience. The average pre-money and post-money valuation for each round of funding and stage of that company is assessed next. The post-money valuation for the innovator's company should be based on the pre-money valuation of a comparable company, plus the estimated operating expenses to achieve the next major funding milestone from the company's financial model.

The best way to secure a favorable valuation is to have multiple interested investors with multiple deal sheets. While these analyses can help a company target a

Table 6.3.5 Common discount rates for new medtech projects.[88]

Risk level	Example	Expected return
Risk-free project	Build a new plant to make more of an existing product when there is a surge in demand	10–15 percent
Low-risk project	Make incremental improvement in existing products	15–20 percent (above corporation's goals for return to shareholder)
Low to medium-risk project	Develop next generation of existing product	20–30 percent
Medium-risk project	Develop new product using existing technology to address markets served by other products of the corporation	25–35 percent
Medium to high-risk project	Build new product using existing technology to address new markets	30–40 percent
High-risk project	Build new product using new technology to address a new market	35–45 percent
Extremely high-risk project	Build new product using new technology to address a new market when there is an unusually high level of risk associated with one or more of these factors	50–70 percent

reasonable valuation, many other factors can influence the final numbers, including the experience of the team, competitive threats, investor interest in the specific space, and macroeconomic market conditions. Online Appendix 6.3.1 provides another valuation example, using a slightly different approach for the fictitious company analyzed in 6.1 Operating Plan and Financial Model.

Sizing the option pool

Another important decision facing the founders and early investors is how much of the company's stock to reserve for key employees. Stock ownership, typically in the form of stock options, is a crucial incentive that can be used to recruit and retain valuable team members. According to analysis performed by the Silicon Valley-based law firm of Wilson, Sonsini, Goodrich & Rosati, roughly 20 percent of the company's shares should be reserved for the option pool after the first round of funding.[87] Following series A, the amount of stock options reserved for employees (as a percentage of the company's total stock) should follow the rough rules of thumb presented in Table 6.3.6.

Table 6.3.6 Maintaining competitive stock incentive program levels for essential personnel is central to a company's hiring and retention strategy (from Doug Collom, "Starting Up: Sizing the Stock Option Pool," The Entrepreneurs Report, Wilson Sonsini Goodrich & Rosati).

Employees (by position)	Post-series A preferred stock
[Founder] CEO	5–10 percent
Vice presidents	2–3 percent
CFO	1–2 percent
Director level	<0.5 percent

Term sheets

With some types of investors, the funding process culminates in a signed contract between the funder and the company. The important details of the final contract are typically worked out through a term sheet, which outlines the terms for a deal and serves as a letter of intent between the investor providing funds and the company receiving them.[89] The two most important functions of the term sheet are to summarize all of the important financial and legal terms related to a contemplated transaction, and to quantify the value of the transaction.[90] Term sheets are commonly used by VCs, corporate investors, and syndicates of angel investors, among other investor types.

Importantly, term sheets are not legally binding documents because they are put in place before the investors complete legal and **patent** due diligence. However, both parties involved in the deal are expected to interact in good faith, preserving the essence of the agreed-upon terms until the closing of financing. Term sheets are usually prepared by the lead investor and presented to the company's CEO. The term sheet becomes an expression of the investor's interest in a company and outlines the term by which the investor is interested in investing. Following the presentation of a term sheet, a series of discussions between the company and the investor ensue, with the expectation by the investor that the company will accept the proposed terms. If a company is being courted by multiple investors, then the contents of a term sheet are potentially subject to negotiation. Once agreed to, proposing changes to the term sheet can severely undermine the working relationship between the company and the investor. As a result, any potential changes should be considered carefully and initiated only under rare circumstances.[91]

Not surprisingly, term sheets vary significantly from deal to deal – not only in substance, but in style and structure. Because the term sheets set forth all of the details surrounding a funding agreement and may have long-term implications for the business, innovators should confer with top lawyers and trusted advisors when reviewing them. There are typically anywhere from eight to 18 sections within a term sheet, several of the most important of which are outlined in Table 6.3.7. A more comprehensive overview can be found in online Appendix 6.3.2.

Managing the funding process

Whenever possible, companies should proactively manage the funding process and try to create competition among investors. During presentations to investors, anticipate their concerns and be prepared to address them in detail. Also, be prepared to talk "off script" as unanticipated questions arise. Listen carefully before providing answers, and be clear and concise in all

Table 6.3.7 These sections of a term sheets are often among those that innovators find most interesting and important (derived in part from Chapter 3 of *Term Sheets and Valuations*, by Alex Wilmerding, and used with permission. Copyright © 2006 Thomson Reuters/Aspatore).

Section	Contents
Summary of financing	This opening section summarizes the contents of the term sheet and provides an overview of the transaction being proposed. It usually includes: • The name of the investors • The name of the company • The amount of financing being offered • The number of newly issued shares • The purchase price per share • The post-financing capitalization structure (which enables the innovators to calculate the pre-money and post-money valuation). Any milestones that must be met for the release of funds will also be outlined in this section. By linking funding to milestones, investors can more closely monitor company progress and manage their downside risk.
Liquidation preference	The liquidation preference outlines the terms governing the transaction if a company is closed down. While preferred shareholders are given priority over common shareholders, the term sheet often takes this one step further by defining a multiple on the value of their initial investment that preferred and common shareholders will receive. According to these terms, the multiple promised to the preferred shareholders would be paid before any proceeds would be given to other shareholders.

Table 6.3.7 (*cont.*)

Section	Contents
Dilution	One of the single most important issues to investors, as a company grows, is how new rounds of financing will affect the value of their investment on a per share basis. Dilution clauses stipulate how conversion prices will be calculated if future rounds of financing are dilutive to preferred shareholders. If future stock is issued at a price lower than the current round, an anti-dilution clause helps ensure that preferred investors continue to hold an equal (or near equal) percentage of ownership in a company without committing more capital.[92]
Voting rights	Voting rights are included in term sheets to ensure that all shares are treated equally in the event that a shareholder vote is called. Typically, one vote per share is granted for both preferred and common stock.
Board composition	The board composition clause outlines the number of board seats and how they will be filled. Typically, companies seek to build a board with representation by both common and preferred stock holders. Investor-favorable composition would give the preferred shareholders majority control. A more neutral arrangement might give preferred shareholders (investor) and common shareholders (company management) an equal number of seats, with one additional seat granted to a mutually agreed-upon independent participant.

responses. Investors will almost certainly ask some questions for which the innovators do not have the answers. To maintain credibility, innovators should acknowledge where uncertainties and unknowns exist rather than trying to "fake" a response. No investor expects a team to have all of the answers. The importance of paying attention to what investors ask cannot be overemphasized, since it is likely that if one investor asks a question, another may be interested in the same issue. Accordingly, gathering answers to typical questions prior to each meeting is extremely beneficial.

Once innovators receive a term sheet, they may be tempted to seek other investors if the proposed deal does not meet expectations. However, even if there is no exclusivity clause in the term sheet, approaching one investor with the details from the term sheet of another is a strategy that can backfire. Especially in the relatively small and interconnected world of VC and angel financing, "shopping a deal" can potentially undermine an innovator's reputation. It is more professional and effective to do this "shopping" in advance – approaching two or three potential investors simultaneously to try to get multiple term sheets at once, and then deciding who to work with to best meet the company's financing needs and expectations.[93]

The Loma Vista Medical story illustrates how one company defined a diversified funding strategy and managed its fundraising efforts through exit.

FROM THE FIELD ▶ **LOMA VISTA MEDICAL**

Defining and implementing a lean funding strategy

When he became interested in building his own medtech company, Alex Tilson actively set out to find an opportunity that he could pursue using a model that would require fewer resources and less time than traditional device start-ups. His vision was to take a lean, focused approach to developing and commercializing a new technology so that he could get to market faster and would be less dependent on external funding. "The standard model had been to raise $70–$80 million and to take 8 to 10 years to get a product to market," Tilson explained, "But I had seen many problems with that approach. I had also seen fundraising consume one-third of the executive team's time, and I did not want to do that."

Working in his garage, Tilson initially pursued an opportunity in colonoscopy. He was interested in developing a device that would make the procedure less painful and time-intensive for the patient and that also had the potential to dramatically lower facility and personnel costs for providers. Although the initial concept was sound, it was not feasible due in part to the limitations of standard medical materials. Tilson had previous experience with America's Cup sailboats and was familiar with a different class of materials typically not in the lexicon of medical device engineers – thin film flexible composites. Though it was clear that this material had the potential to be superior, it was not optimized to address the unique demands of medical device design. After endeavoring to work with a partner to develop new medical-grade composites, the company (which took the name Loma Vista Medical, LVM) decided to look inward and develop a new material in-house.

As development progressed, the team realized that the new materials opened up a wide variety of opportunities above and beyond colonoscopy. Tilson recognized that the material was not ideal for all procedures, but it was uniquely suited for applications that required exacting dimensional control and toughness. It also enabled balloons with unique shapes, and with localized feature integration. According to Tilson, "We asked ourselves, 'What are we really good at? What would we do if we thought of ourselves as a composite medical balloon company instead of a colonoscopy company?' We looked at all possible applications to find a good fit. It was clear to us that there were multiple unmet clinical needs that we could help solve."

LVM began to actively explore areas where a gap existed between the performance aspirations of an existing technology and the capabilities of available medical balloons. The team uncovered a compelling opportunity in kyphoplasty, which was used to treat painful compression fractures in the spine by inserting a balloon into the patient's vertebrae, inflating it, and filling the resulting cavity with a bone cement to stabilize the bone. "Kyphon had been stunningly successful, despite using what we saw as a balloon technology with distinct limitations," Tilson recalled. LVM produced a prototype of a kyphoplasty composite balloon and initiated discussions with Kyphon's leaders, hoping to accelerate time to market through a partnership.

Another interesting opportunity surfaced in the emerging area of transcatheter aortic valve replacement (TAVR). One of Loma Vista's board members had been a consultant to CoreValve during its early days and suggested that the company explore this area. "They had mastered the valve, but they needed a suite of enabling accessories," said Tilson. "It was clear to us that the procedure needed balloons that had better dimensional control, better puncture resistance, faster inflation deflation, and refolding. We felt that we were out in front of this need, and uniquely had the technology to deliver a solution."

Eventually, LVM held a "bake off" between the kyphoplasty and TAVR opportunities and decided to focus its efforts on TAVR. In this space, the young company found its niche, developing the TRUE Dilatation™ Balloon. The TRUE product received its **CE mark** in December 2011 and began commercial sales in Europe in early 2012. The product was cleared by the FDA via the 510(k) pathway shortly thereafter, in October 2012. In early 2013, LVM introduced the product to the US market through its new direct sales organization, which consisted of two sales reps. Six months later, C.R. Bard acquired the company.[94]

All of these activities were enabled, in large part, by LVM's funding strategy (see Figure 6.3.6). In alignment with his original goal, Tilson took a careful, measured approach to financing the company. "Many entrepreneurs think that VCs fund good ideas. They do, but they typically expect that idea to be backed up by a

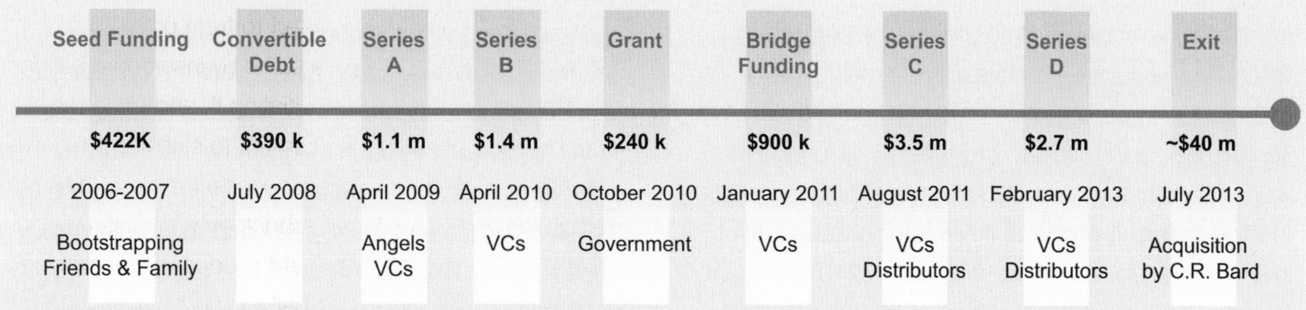

Seed Funding	Convertible Debt	Series A	Series B	Grant	Bridge Funding	Series C	Series D	Exit
$422K	$390 k	$1.1 m	$1.4 m	$240 k	$900 k	$3.5 m	$2.7 m	~$40 m
2006-2007	July 2008	April 2009	April 2010	October 2010	January 2011	August 2011	February 2013	July 2013
Bootstrapping Friends & Family		Angels VCs	VCs	Government	VCs	VCs Distributors	VCs Distributors	Acquisition by C.R. Bard

FIGURE 6.3.6

An overview of the Loma Vista Medical funding strategy, from inception to exit.

long list of milestones. As a first-time entrepreneur, I knew we shouldn't even go knocking on VC doors until we had achieved multiple real milestones," Tilson said. Accordingly, he bootstrapped the company's early development work with his own savings. He also accessed funds through friends and family. "If you're not ready to do that, then you're not ready to be an entrepreneur," he stated.

By mid-2008, Tilson had been working on the company, which was still headquartered in his garage, for almost two years. He had met with investors from De Novo Ventures several times before they finally invited him to discuss a small round of convertible debt. At this point, LVM had achieved several important milestones, including key IP filings, promising bench data, animal data for the colonoscopy system, and prototypes for the kyphoplasty balloon. Moreover, the De Novo team saw that he was persistent and committed to making the company successful. As Tilson put it, "The business plan matters, the size of the opportunity matters, the technology matters, but they are also investing heavily in people. I think that they saw that we were unusually persistent – that we would go to the ends of the earth to make something work." He was able to raise $390,000 in convertible debt in July 2008. Commenting on this type of funding, he said, "Convertible debt is very entrepreneur-friendly. It sets the value in the future and allows you to create value while you're still running. The interest rate is modest and the equity-kicker is reasonable. Importantly, it also shows that real people

have put in real money, which creates a foundation that others often follow."

The next two funding rounds came in relatively rapidly succession. In early 2009, LVM had a working kyphoplasty system and positive cadaver data to support its next fundraising effort. The company raised a $1.1 million Series A in April of that year from an expanded syndicate of small VCs and angels. Over the next 12 months, the company used this money to fuel development of the balloon and also to support its increasing emphasis on TAVR. However, as LVM was getting ready to raise its Series B funding, prospects for using the kyphoplasty balloon to realize an early exit were fading. The team members recognized that they needed to go further with development and potentially prepare to commercialize their products themselves. Tilson began considering a fundraising diversification plan. Through one of his investors, he was connected with Scientific Health Development (SHD), a venture capital firm based in Dallas, Texas. The group liked LVM's approach and was the first to put money down for the B round. "One of the advantages of a smaller firm is that it can make a quick decision and then take action," Tilson stated. With this new investor leading the round, De Novo and several other investors made the decision to commit. LVM closed $1.4 million in April 2010. At that time, the company's burn rate was averaging just over $100,000 per month.

Even though much of LVM's work transferred from kyphoplasty to TAVR, the team needed to complete significant clinical work. Additionally, the composite technology required substantial investment in new manufacturing paradigms – a use of money that sometimes causes VCs to grumble. To help fund this work, the company actively sought alternate forms of funding to help it bridge to its next major round. First, Tilson applied for and was awarded a $240,000 grant from the US government as part of the Economic Recovery Act. LVM also raised almost $1M through one of its investors. "We were lucky," Tilson said. "The investor told us, 'I'll raise almost $1 million dollars for you, commission-free, and I'll be doing it because I'm on your board and I believe in you. And *you* being successful helps *me* be successful.' So he called up other interested investors and helped us raise the bridge funding." LVM used these funds, combined with its Series B money, to advance the TAVR product (see Figure 6.3.7). In particular, the company develop the manufacturing processes and trade secrets that would allow it to scale-up manufacturing, which eventually became key to LVM's acquisition. "By committing to owning the manufacturing, we turned one of our biggest challenges into a valuable strength," Tilson noted.

The company felt ready to raise its Series C after optimizing its manufacturing approach, completing a validation study, moving the product into a design freeze, and making the company's CE mark submission. Always interested in further diversifying his investors, Tilson had

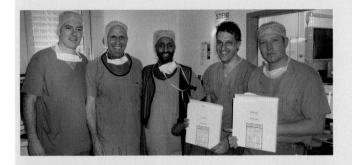

FIGURE 6.3.7

The team after the device's first TAVR case. From left to right: M. Braun, A. Tilson, T. Johnson, Dr. H. Mollmann, Dr. G. Kempfert (courtesy of Alex Tilson).

been speaking with distributors. When Medtronic acquired CoreValve, it also bought out the company's distribution agreements so that it could distribute the TAVR technology through its own channels. This turn of events meant that many distributors with experience in the TAVR space had money on hand to invest in new opportunities. These distributors had developed key relationships with all of the right doctors. Additionally, they had come together to form an investor network, and they expressed interest in investing in LVM. "Suddenly, we had 30 different investors from around the world interested in the Series C round," Tilson remembered. "It was a mind-bender for us, and took some managing, but it worked. We slept well at night because we were very diversified." Loma Vista closed a $3.5 million Series C in August 2011. "Getting the distributors as investors in the company was a big break for us," he said. "It gave us credibility with the VCs because the organizations that knew transcatheter valves as well as anyone on the planet stood up and said they thought we had a good product. Also, when we did get the clearance to launch in Europe, we had an immediate commercialization pathway based on a two-year relationship with people who were ready to go sell the product." Moreover, the distributors provided an invaluable source of feedback, showing a willingness to introduce Loma Vista to key opinion leaders to assist by providing with product and adoption feedback.

The company used the Series C funding to attain its CE mark and launch commercial activities in Europe. The product was well received: Tilson estimated that, within six months of launch, LVM balloons were being used in more than half the CoreValve cases in Germany. LVM had also gained FDA clearance, was manufacturing at scale, and began making progress on a second generation product. Tilson had been actively talking with potential acquirers, but had not yet entered into any definitive discussions. At that point, he faced a decision: "We have half a million in the bank. Do we take on convertible debt? Do we take on venture debt? Do we raise half a million dollars and try to make it to an exit? Or do we raise a big Series D while we can because TAVR is hot and we're firing on all cylinders?" Tilson said. "We were highly cognizant that, as good as things look today, the world can always change quickly." After extended board deliberations, he stated, "We decided to shoot it down the middle." Based on the strength of its commercial milestones achieved to date, Loma Vista raised a Series D in February 2013, capping total investment at $2.7 million. This round brought the company's total investment to $10.6 million raised over approximately six years.

In terms of realizing an exit, Tilson was surprised by the length and difficulty of the process. Over several years, he courted more than a dozen prospective acquirers until, just before the Series D, the company entered into serious talks with C.R. Bard. After eight months of negotiations, Bard purchased Loma Vista Medical in July 2013 for approximately $40 million.[95] Reflecting on the acquisition, Tilson remarked, "As an entrepreneur, I wanted a validated success and the investors were eager for an exit, so we were aligned in this regard. The market at the time was tough for medtech companies. We took the offer." LVM had gotten this far with 19 people, but would have needed to substantially ramp up to continue building the business. "We had a series of next generation products in development, we had new manufacturing paradigms that could take things to the next level, and our sales network could have been substantially expanded. We knew that we could have been worth a lot more, but the activities ahead would be a lot more expensive. There would be a lot of dilution and a lot of risk," he said. Importantly, Tilson added, "Merely existing every day carries risk. Many things are well beyond your control. Opportunities expire. People get exhausted. Sometimes a bird in the hand really is worth two – or three or four – in the bush."

When asked what advice he would offer other innovators as they consider a funding strategy, Tilson recommended his lean approach. "Scarcity breeds efficiency," he commented. "And being lean makes you hard to kill. We were never at risk of getting starved out because we had real technology that we knew that the world needed, and our burn rate was so low."

Tilson also encouraged innovators to think about fundraising as a courtship. "Every interaction matters. Investors are looking for someone they can trust and who they can work with long term." Along these lines, he pointed out that little things can make a big difference. For instance, Loma Vista instituted a policy of sending its investors regular updates about the company's progress. "We were proactive about this even though very few companies do this. Every quarter we would send out a one-page letter about what was happening at the company. And that became part of our brand. People saw us as persistent and inventive and lean, but they also saw us as unusually communicative and honest. Open communication and honesty had a big payback for us, and it made it easier to raise subsequent rounds of funding because our investors were well informed," Tilson said.

Finally, he advised innovators to be realistic about achieving an exit. More often than not, he explained, companies should be prepared to take their products into the market and generate sales before acquirers get seriously interested. This means orchestrating a real and substantive launch that "shows that you're able to ship product, earn real revenue, generate repeat orders, and handle any complaints. It means that your quality system is working, your manufacturing is reliable, and that your product is loved. And it requires creativity to figure out how to do all of this with lean funding," Tilson emphasized.

↘ Online Resources

Visit www.ebiodesign.org/6.3 for more content, including:

Activities and links for "Getting Started"
- Identify comparable companies
- Confirm funding milestones and capital needs
- Determine a company valuation
- Research and select investors
- Approach investors

Videos on funding approaches

CREDITS

The editors would like to acknowledge Elias Caro and Sue Van of the Coulter Foundation, Mir Imran of VentureHealth, and Alex Tilson of Loma Vista Medical for sharing the case examples. Further appreciation goes to Leslie Bottorff, Sami Hamade, Ross Jaffe, Ken Kelley, Richard Lin, and Allan May for presenting in the Stanford Biodesign Program on this topic. Darin Buxbaum, Jared Goor, Ritu Kamal, and Asha Nayak also made important contributions to the chapter.

NOTES

1 "2007 Venture Capital Investing Hits Six Year High At $29.4 Billion," PWCMoneytree.com, January 21, 2008, https://www.pwcmoneytree.com/MTPublic/ns/moneytree/filesource/exhibits/07Q4MT_Rel_FINAL.pdf (March 24, 2014).

2 "Annual Venture Investment Dollars Rise 7 Percent and Exceed 2012 Totals, According to the MoneyTree Report," PricewaterhouseCoopers press release, January 17, 2014, http://www.pwc.com/us/en/press-releases/2014/annual-venture-investment-dollars.jhtml (February 26, 2014).

3 Alex Nixon, "Medical Device Start-Ups Hit by Decline of Venture Capital Investment," TribLive, November 30, 2013, http://triblive.com/business/headlines/5150030-74/companies-medical-capital#ixzz2uTT0xMW9 (February 26, 2014).

4 From remarks made by Mir Imran as part of the "From the Innovator's Workbench" speaker series hosted by Stanford's Program in Biodesign, April 28, 2004, http://biodesign.stanford.edu/bdn/networking/pastinnovators.jsp. Reprinted with permission.

5 Josh Makower, Aabed Meer, and Lyn Denend, "FDA Impact on U.S. Medical Technology Innovation," November 2010, http://nvca.org/index.php?option=com_docman&task=doc_download&gid=668&item=93 (March 1, 2014).

6 Loren Bonner, "Is Venture Capital Funding Shifting Away from Med Tech?," *DOTmed Daily News*, November 20, 2013, http://www.dotmed.com/news/story/22475/ (February 28, 2014).

7 "Medical Device Excise Tax: Frequently Asked Questions," Internal Revenue Service, http://www.irs.gov/uac/Medical-Device-Excise-Tax:-Frequently-Asked-Questions (February 28, 2014).

8 "2007 Venture Capital Investing Hits Six Year High At $29.4 Billion," op. cit.

9 Vlad Lozan, "Venture Capital Fundraising Trends in the Medical Device Industry – The Data," KnobbeMedical.com, October 20, 2013, http://www.knobbemedical.com/medicaldeviceblog/article/venture-capital-fundraising-trends-medical-device-industry-data/9350/ (February 28, 2014).

10 Bonner, op. cit.

11 "Ardian," CrunchBase, http://www.crunchbase.com/organization/ardian-inc (April 25, 2014).

12 "Foundation Medicine," IPOScoop.com, http://www.iposcoop.com/index.php?option=com_content&task=view&id=3277&Itemid=134 (March 16, 2014).

13 "DLA Piper Advises GI Dynamics, Inc. on Largest Float in Australia This Year," DLA Piper press release, September 16, 2011, http://www.dlapiper.com/dla-piper-advises-gi-dynamics-inc-on-largest-float-in-australia-this-year-09-16-2011/ (March 16, 2014).

14 "LDR Holding" IPOScoop.com, http://www.iposcoop.com/index.php?option=com_content&task=view&id=3309&Itemid=148 (March 24, 2014).

15 "Successful IPO of STENTYS on NYSE Euronext Paris," Stentys press release, October 27, 2010, http://www.stentys.com/file_bdd/documents/1287767756_STENTYS_CP_closing_IPO_UK.pdf (March 16, 2014). Note: Market capitalization converted from Euros to U.S. dollars at prevailing conversion rate on access date.

16 "Tandem Diabetes Care," IPOScoop.com, http://www.iposcoop.com/index.php?option=com_content&task=view&id=3383&Itemid=148 (March 16, 2014).

17 Patrick Gordan, "Venture Debt: A Capital Idea for Start-Ups," Kauffman Fellows Press, http://kauffmanfellows.org/journal_posts/venture-debt-a-capital-idea-for-startups/ (March 5, 2014).

18 Leslie Bottorff, "Funding a Medical Device Start-Up," *Medical Device & Diagnostic Industry Magazine*, January 2000, http://www.mddionline.com/article/funding-medical-device-start (March 24, 2014).

19 Timothy Hay, "Medical Device Investing Drops, Though Some VCs Welcome 'Weeding Out' Process," *The Wall Street Journal*, February 7, 2014, http://blogs.wsj.com/venturecapital/2014/02/07/medical-device-investing-drops-though-some-vcs-welcome-weeding-out-process/ (March 4, 2014).

20 Lyn Denend and Stefanos Zenios, "Drug Eluting Stents: A Paradigm Shift in the Medical Device Industry," Stanford University, Graduate School of Business, 2006, https://gsbapps.stanford.edu/cases/detail1.asp?Document_ID=2787 (March 4, 2014).

21 Ross Jaffe, "Introduction to Venture Capital," October 6, 2004.

22 "Medical Device IPOS Are Back ... For Now," MedGadget, November 28, 2006, http://medgadget.com/archives/2006/11/medical_device_2.html (March 24, 2014).

23 "IPO Round Up: Is the Window Slamming for Life Sciences?," Venture Beat, February 2007, http://venturebeat.com/2008/02/07/ipo-roundup-is-the-window-slamming-shut-for-life-sciences/ (March 24, 2014).

24 Leo Sun, "A Year-End Look at 3 High-Flying Biotech IPOs of 2013: PETX, RCPT, STML," The Motley Fool, December 26, 2013, http://www.fool.com/investing/general/2013/12/26/a-year-end-look-at-3-high-flying-biotech-ipos-of-2.aspx (March 4, 2014).

25 Valdim Kotelnikov, "Bootstrapping," Venture Finance Step-by-Step Guide, http://www.1000ventures.com/venture_financing/bootstrapping_methods_fsw.html (March 24, 2014).

26 Michael J. Weickert, "Funding a Medical Device Start-Up in the Current Economy," S.E.A. Medical Systems, June 2009, http://www.slideshare.net/mweickert/funding-a-medical-device-startup-in-the-current-economy (March 5, 2014).

27 Ibid.

28 Ibid.

29 "About Us," StartX, http://startx.stanford.edu/about (March 5, 2014).

30 "FAQs," Rock Health, http://rockhealth.com/about/faq/ (March 5, 2014).

31 "About Us," MdStart, http://www.mdstart.eu/aboutus.php (March 24, 2014).

32 Damian Garde, "Sony Sinks $10M into Israeli Medical Device Incubator," FierceMedicalDevices, May 9, 2013, http://www.fiercemedicaldevices.com/story/sony-sinks-10m-israeli-medical-device-incubator/2013-05-09 (March 5, 2014).

33 Incubator Development Program, Spring Singapore, http://www.spring.gov.sg/Entrepreneurship/FSP/Pages/incubator-development-programme.aspx#.UxUdSfldV8E (March 3, 2014).

34 Bhrigu Pankaj Prashar, "Pros and Cons of Joining an Incubator," *Forbes*, April 12, 2013, http://www.forbes.com/sites/bhrigupankajprashar/2013/04/12/pros-and-cons-of-joining-an-incubator/ (March 6, 2014).

35 "About SBIR," United States Government, http://www.sbir.gov/about/about-sbir (March 6, 2014).

36 Ibid.

37 "Small Business Innovation Research (SBIR) and Small Business Technology Transfer (STTR) Programs," National Institutes of Health, January 3, 2014, http://grants.nih.gov/grants/funding/sbirsttr_programs.htm (March 6, 2014).

38 United States-India Science and Technology Endowment Fund, http://www.indousstf.org/US-India-Endowment-Board1.html (March 3, 2014).

39 Keith O'Neill, "Medical Device Innovation in Ireland," Enterprise Ireland, http://www.bdi.ie/presentations/taiwan_workshop/irish_agencies/Dr_Keith_O_Neill_Enterprise_Ireland_Ireland-_Medical_Device_Innovation_in_Ireland.pdf (March 6, 2014).

40 "Support for Startups," Enterprise Ireland, http://www.enterprise-ireland.com/en/Start-a-Business-in-Ireland/Startups-from-Outside-Ireland/Funding-and-Supports-for-Start-Ups-In-Ireland/ (March 3, 2014).

41 "Pharmaceuticals & Biotechnology," Future Ready Singapore, http://www.edb.gov.sg/content/edb/en/industries/industries/pharma-biotech.html (March 6, 2014).

42 "Foundation Requests," St. Jude Medical Foundation, http://www.sjmfoundation.com/foundation-grants/grant-request-instructions (March 6, 2014).

43 "Who We Are: Foundation Fact Sheet," Bill & Melinda Gates Foundation, http://www.gatesfoundation.org/Who-We-Are/General-Information/Foundation-Factsheet (March 6, 2014).

44 "What We Do," The Bill & Melinda Gates Foundation, http://www.gatesfoundation.org/What-We-Do (March 6, 2014).

45 "Grand Challenges in Global Health," The Bill & Melinda Gates Foundation, http://www.grandchallenges.org/about/Pages/Overview.aspx (March 3, 2014).

46 "How We Fund," The Mulago Foundation, http://www.mulagofoundation.org/how-we-fund (March 6, 2014).

47 "Student Overview," NCIIA, http://nciia.org/students/ (March 3, 2014).

48 "The Grand Experiment: Translating University Innovations to Benefit Patients," Wallace H. Coulter Foundation white paper, September 18, 2013.

49 All quotations from interviews conducted by the authors unless otherwise noted. Reprinted with permission.

50 "The Grand Experiment," op.cit.

51 Coulter Translational Partnership, Wallace H. Coulter Foundation, www.whcf.org/partnershipaward/overview (November 19, 2014).

52 "The Grand Experiment," op.cit.

53 "Translating University Innovation: The Coulter Way," Missouri University, 2011, http://bioengineering.missouri.edu/coulter/Coulter%20Orientation%20for%20MU.pdf (March 9, 2014).

54 "The Grand Experiment: Translating University Innovations to Benefit Patients," op. cit.

55 Lyn Denend and William Meehan, "Acumen Fund and Embrace: From the Leading Edge of Social Venture Investing," Stanford University, Graduate School of Business, 2011, https://gsbapps.stanford.edu/cases/detail1.asp?Document_ID=3457 (March 7, 2014).

56 "Survey Shows Market Growth in Impact Investments and Satisfaction Among Investors," J.P. Morgan press release, January 7, 2013, https://www.jpmorgan.com/cm/cs?pagename=JPM_redesign/JPM_Content_C/Generic_Detail_Page_Template&cid=1320509594546&c=JPM_Content_C (March 7, 2014).

57 "Acumen Fund Invests in Circ MedTech, Developer of Breakthrough Innovation in HIV Prevention," Acumen Fund press release, July 7, 2011, http://www.prweb.com/releases/2011/7/prweb8625404.htm (March 7, 2014).

58 Sainul K. Abudheen, "Infant Warmers Maker Embrace Raises Funding from Khosla Impact, Kiran Mazumdar-Shaw, Others," VC Circle, August 26, 2013, http://www.vccircle.com/news/medical-devices/2013/08/26/infant-warmers-maker-embrace-raises-funding-khosla-impact-kiran (March 7, 2014).

59 "Let's Get Together: Crowdfunding Portals Bring in the Big Bucks," Deloitte, 2013, http://www.deloitte.com/assets/Dcom-Shared%20Assets/Documents/TMT%20Predictions%202013%20PDFs/dttl_TMT_Predictions2013_LetsGetTogeather.pdf (March 5, 2014).

60 "Crowdfunding's Potential for the Developing World," Information for Development Program, The World Bank, 2013, http://www.infodev.org/infodev-files/wb_crowdfundingreport-v12.pdf (March 5, 2014).

61 Mohana Ravindranath, "Crowdfunding the Next Medical Cure," The Washington Post, July 8, 2013, http://www.washingtonpost.com/business/on-small-business/crowdfunding-the-next-medical-cure/2013/07/08/5b3b562c-c871-11e2-9f1a-1a7cdee20287_story.html (March 5, 2014).

62 "Investor FAQ," Healthfundr, https://healthfundr.com/investor_faq (March 5, 2014).

63 Brandon Glenn, "Is Crowdfunding a Viable Option for Medical Technology Start-Ups?," MedCity News, April 11, 2012, http://medcitynews.com/2012/04/is-crowdfunding-a-viable-option-for-medical-technology-startups/ (March 5, 2014).

64 Amy Cortese, "The Crowd-Funding Crowd is Anxious," The New York Times, January 5, 2013, http://www.nytimes.com/2013/01/06/business/crowdfunding-for-small-business-is-still-an-unclear-path.html?pagewanted=all&_r=0 (March 5, 2014).

65 All quotations are from interviews conducted by the authors, unless otherwise cited. Reprinted with permission.

66 "About," VentureHealth.com, http://www.venturehealth.com/about (March 7, 2014).

67 Venture Health Portfolio, http://www.venturehealth.com/portfolio (March 7, 2014).

68 Ibid.

69 "The Angel Investor Market In 2012: A Moderating Recovery Continues," Center for Venture Research, University of New Hampshire, October 16, 2013, http://paulcollege.unh.edu/sites/paulcollege.unh.edu/files/Q1Q2%202013%20Analysis%20Report.pdf (March 7, 2014).

70 Ibid.

71 Ibid.

72 "How Corporate Venture Capital Investing Increases Innovation," Knowledge@Wharton, October 19, 2005, http://knowledge.wharton.upenn.edu/article.cfm?articleid=1299&CFID=7207756&CFTOKEN=25757895 (January 24, 2007).

73 Heather Thompson, "A Near-Term Look at Medtech Investing," Medical Device and Diagnostics Industry, May 16, 2012, http://www.mddionline.com/article/near-term-look-medtech-investing (March 7, 2014).

74 Bottorff, op. cit.

75 Bob Zider, "How Venture Capital Works," Harvard Business Review, November 1, 1998.

76 Ibid.

77 Ibid.

78 Bottorff, op. cit.

79 Chris Walker, "Stealthy BioTrace Medical Reels in $3.4 Million," MassDevice, March 4, 2014, http://www.massdevice.com/news/stealthy-biotrace-medical-reels-34m (March 26, 2014).

80 Alex Wilmerding, Deal Terms (Aspatore Books, 2003), p. 18.

81 "What's the Difference Between Pre-Money and Post-Money?," Investopedia.com, http://www.investopedia.com/ask/answers/114.asp (March 24, 2014).

82 "J&J to acquire Conor Medsystems for US $1.4 billion," Cardiovascular News, February 16, 2007. http://www.cxvascular.com/cn-archives/cardiovascular-news-international-issue-4/jj-to-acquire-conor-medsystems-for-us14-billion (April 22, 2014).

83 "CoStar Stent Fails Head-to-Head Test against Taxus," Medical Device & Diagnostic Industry, May 1, 2007, http://www.mddionline.com/article/costar-stent-fails-head-head-test-against-taxus (April 22, 2014).

84 "Dilution," Investopedia.com, http://www.investopedia.com/terms/d/dilution.asp (March 24, 2014).

85 Ln = the natural logarithm.

86 According to Investopedia, a warrant, like an option, gives the holder the right but not the obligation to buy an underlying security at a certain price, quantity, and future time. However, unlike an option, which is an instrument of the stock exchange, a warrant is issued by a company. Companies will often include warrants as part of a new-issue offering to entice investors into buying the new security. A warrant can also increase a shareholder's confidence in a stock, if the underlying value of the security actually does increase over time. See http://www.investopedia.com/articles/04/021704.asp (March 24, 2014).

87 Doug Collom, "Starting Up: Sizing the Stock Option Pool," The Entrepreneur's Report, Wilson, Sonsini, Goodrich & Rosati, Summer 2008, http://www.wsgr.com/publications/PDFSearch/entreport/Summer2008/private-company-financing-trends.htm#4 (March 24, 2014).

88 Note that discount rates reflect prevailing interest rates. When interest rates are low, it means the cost of capital is lower and, therefore, discount rates (or the expected return) can be commensurately lower.

89 Alex Wilmerding, *Term Sheets & Valuations* (Aspatore Books, 2006), p. 9.

90 Ibid.

91 Ibid.

92 Alex Wilmerding, *Deal Terms* (Aspatore Books, 2003), p. 63.

93 Wilmerding, *Term Sheets & Valuations*, op. cit., pp. 19–20.

94 "Form 10 Q, C.R. Bard," U.S. Securities and Exchange Commission, September 30, 2013, http://phx.corporate-ir.net/External.File?item = UGFyZW50SUQ9NTIyNDQ5fENoaWxk SUQ9MjA3NTc3fFR5cGU9MQ = =&t = 1 (March 19, 2014).

95 Ibid.

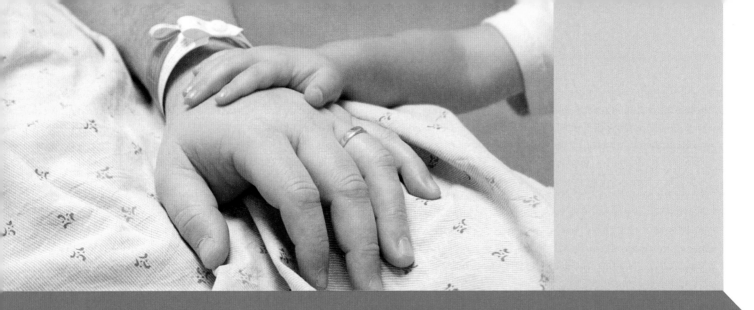

6.4 Alternate Pathways

INTRODUCTION

A doctor who is dedicated to his practice comes up with an innovative concept but has no aspirations to become an entrepreneur. An engineer is advised that her concept potentially represents an interesting product feature but will not support a stand-alone company. In the course of pursuing one device idea, a development team uncovers another compelling opportunity but does not have the bandwidth or resources to pursue both. Under these kinds of circumstances, how can these innovative ideas still be used to change medicine?

While some innovators become entrepreneurs and start their own medical device companies, many have an impact on their fields by licensing or selling their ideas, or partnering with another company to bring a solution to fruition in the market. Every innovator desires to run a company; and not every idea will best be realized through the creation of a stand-alone business. Sometimes the involvement of an outside entity allows an idea or technology to more quickly, successfully, and/or cost-effectively achieve its potential. This chapter describes alternate models for transitioning responsibilities for the development or commercialization of a technology and/or more deliberately moving toward an exit.

 See ebiodesign.org for featured videos on alternate medtech pathways.

ALTERNATE PATHWAYS FUNDAMENTALS

The decision to pursue an alternate pathway is often dictated by the personal situation of the person or team developing the idea. Some innovators may have other priorities and skills that make the entrepreneurial track unappealing or unlikely – such as the desire to avoid being tied up in a single area for too long or other roles that prevent them from diverting their attention to the pursuit of the idea. Others may have the appropriate qualifications and interests, but are not attracted to the risk that is often required in founding a **medtech** business. Still others may be part of a team or company with insufficient resources to pursue a new opportunity or an idea that falls outside of their currently defined focus. In cases such as these, the innovators' circumstances make it undesirable for them to create and sustain a stand-alone business. As described in 1.1 Strategic Focus, the individual or team with the idea should undergo a meaningful self-assessment before deciding whether or not to launch a start-up company.

Factors and constraints related to the idea itself can also affect this decision. To warrant the creation of a business, innovators must ask themselves if a stand-alone company would be able to create and capture adequate **value** from the idea, or if another entity would be in a better position to do so. Preliminary **financial modeling** (as outlined in 6.1 Operating Plan and Financial Model) can expose at least two common obstacles: lack of market access issues or issues related to return, timing, and cost.

Lack of market access

The **bottom-up** market analysis technique described in chapter 6.1 may reveal that the cost of sales for a particular product is greater than the potential margin the product can support. If so, the new product cannot reasonably be brought to the market in a profitable way. In this scenario, there is lack of market access and the idea cannot justify the creation of a dedicated company.

In well-established markets, competitors can "lock up" customers by making significant investments in sales and distribution infrastructures and long-term buyer/supplier relationships that cannot easily be replicated by small, start-up companies. For example, an innovator with a new idea for a syringe product would find it almost impossible to compete against Johnson & Johnson, Baxter, or Becton Dickinson, with their hundreds of sales representatives, deep catalogs of related products, and established connections with the appropriate buyers. While any one of these companies could quickly and inexpensively introduce a new syringe, a start-up company would almost certainly collapse under the strain of the time and effort required to access the buyers, get their attention, and convince them to purchase such a product from a small, unknown entity. In other words, the cost to acquire new customers would be prohibitive to this new market entrant.

Issues of market access can also be problematic when the product in question is generally bundled and sold with other products. For instance, innovators who have developed a new introducer for a guidewire would have difficulty marketing their innovation because introducers and guidewires are usually sold in a set, along with the other elements of a complete catheter system. On its own, buyers see an introducer as a low-value item. As a result, they are unlikely to purchase it separately, even if it performs better than the introducers they have used in the past. The convenience of purchasing the introducer as an integrated part of an overall system usually outweighs any incremental performance benefit, particularly when the bundle is offered at a discount to buying each product individually. These are just a few of the ways market access can support a rationale to license or sell a product **concept** rather than to develop it oneself.

Returns, timing, and cost issues

Product development costs can be another obstacle to the creation of a stand-alone business. The financial burden of developing a new technology *de novo* can be staggering. It is important to complete an assessment of these costs and the timing of expected returns to determine if the potential profitability of the technology will be sufficient to generate a significant payback on the required investment within a reasonable timeframe.

Importantly, not all investors have the same expectations regarding the timing and level of return on their investments. The innovator should consider this before selecting investors to fund a given idea. As described in 6.3 Funding Approaches, venture capitalists may be primarily interested in funding ideas with billion-dollar markets and the potential for a 5–10× return on their invested capital. On the other hand, angels, as well as friends and family, may be willing to explore smaller markets that yield lower returns. Impact investment and grants provide another potential funding source where returns may be measured (in part or whole) in non-financial terms (see chapter 6.3). However, if there is a poor fit between investor expectations and the innovation, this may be another powerful reason to pursue an alternate development pathway.

Types of alternate development pathways

Alternate development pathways leverage other entities to more effectively and efficiently develop or commercialize an invention. In the process, these approaches can allow innovators to transition responsibility for a project or downscale their involvement sooner than if they pursue a more traditional start-up model. There are three primary alternatives available to innovators who decide not to build and sustain a stand-alone business: **licensing**, **partnering**, and sale/**acquisition**. While these options differ in many ways, they are particularly distinct in terms of the degree of innovator involvement and certainty of the payoff upon the completion of a transaction. Regardless of which path is chosen, it is essential to obtain legal advice when pursuing agreements of this type.

Licensing

A license involves the transfer of an idea or invention from the innovator to a licensee in exchange for ongoing royalties or other payments.[1] Licensing deals are sometimes used when the potential payoff from the invention is highly uncertain and/or the innovators do not possess the technical capabilities, experience, time, or money that will be needed to develop the product.

While the prospect of royalties and payments may seem attractive to innovators, they must recognize that it comes with a loss of control. Depending on the other agreements obtained as part of the license, innovators may or may not have a contractual obligation to participate in an ongoing manner. Often, they can add significant value by consulting to the licensee as a technology expert or product advocate within the targeted community. However, even in these scenarios, their ongoing involvement is more limited compared to partnering. Rarely will the licensor be in the position to drive the ongoing management of a technology once it has been licensed. Innovators must be prepared to relinquish control when they enter into a licensing deal. For this reason, the selection of the appropriate licensee should take into account the licensee's ability to deliver on development and commercialization, not just focus on the potential economics of the transaction.

Beyond financial terms, licenses include a description of exactly what rights are being licensed,[2] the field of use and territory for which the rights are being granted, and

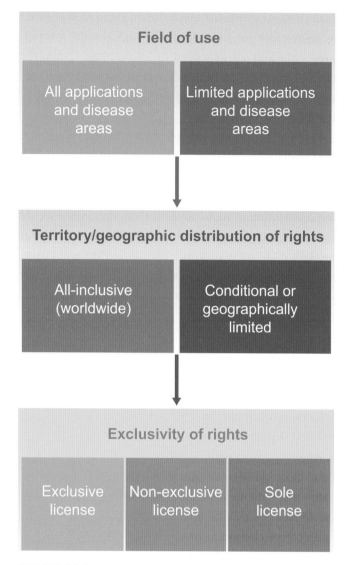

FIGURE 6.4.1

There are different types of licenses that grant different rights to the licensee and licensor.

whether or not the license is to be exclusive, non-exclusive, or sole (see Figure 6.4.1).[3]

An **exclusive license** grants only the licensee (and not even the licensor) the right to use a technology. A **non-exclusive** license allows the licensee to use the rights within the field and within whatever other limitations are provided by the license, but reserves other rights for the licensor or other parties. A sole license grants the same rights to the licensee and licensor, but prevents the licensor from granting rights to anyone else. In general, exclusive licenses tend to engender greater commitment from the licensee and lead to higher payments to the licensor. Exclusive licenses, however, can be higher risk for the licensee because the achievement of milestones (and the flow of royalties) rests in the hands of one party. Non-exclusive and partially exclusive licenses allow for multiple paths and partners to bring a technology to market. On the downside, though, these structures limit a licensee's ability to use the license to differentiate the product. For this reason, exclusive licenses tend to have more value than non-exclusive licenses.

Nothing is more important to the success of a license (or a partnership) than choosing the right entity with which to do business. Although licensing agreements include provisions under which the licensor may terminate the license or license back the technology if the licensee does not commercialize it, there is often a substantial **opportunity cost** associated with the time lost in pursuing those actions.

Partnering

In a partnership, the two entities share responsibility for the development and/or commercialization of an idea or invention. In turn, certain costs and returns are distributed between them. In a partnership, each participant provides something that the other needs. For example, partnerships are a potentially favorable alternative when innovators possess IP and technical capabilities with value that extends beyond a single idea. By partnering, the innovators may be able to explore the development and commercialization of multiple products by parsing each product to a different partner. In cases such as this, innovators are typically a development partner and may choose to retain core IP rights and development responsibilities, while their partners handle sales and marketing.

Partnership agreements can take many forms, and the flow of cash and responsibility between each organization can be tuned to the unique requirements of the two companies. In this way, partnerships provide an effective way to manage risk when necessary technologies, IP, expertise, and resources are spread across more than one organization. Moreover, they can provide a mechanism for large companies to evaluate smaller companies as an intermediate step towards an **outright acquisition**. For instance, when AstraZeneca entered into a five-year research and development alliance with Cambridge Antibody Technology (CAT) to perform monoclonal antibody research,[4] it made a bid to acquire CAT after just two years. This approach allowed AstraZeneca to gain greater insight into the strengths and weaknesses of its target than it would have gleaned during any standard **due diligence** process prior to an acquisition.[5]

In medtech, some skepticism exists regarding the value of partnerships and their ability to be productive over the long term. Most executives recommend them only if the desired endpoints are clearly defined, mutually beneficial to the two organizations, and achievable within a reasonable period of time. If a longer partnership is anticipated, much more flexibility in the agreements are recommended; otherwise, both parties can find it draining as they are continually required to make compromises to satisfy the terms of the original agreement. In general, due to the short product life cycles of most medical devices, partnerships have traditionally played a lesser role than in other medical areas where the timelines for development and product in market are substantially longer.

Because responsibilities are shared in a partnership, innovators and their teams play a critical role in keeping the project moving, working closely with representatives from the other entity. Having a well-defined management structure and open communication channels is important to sustaining a successful partnership. Choosing partners carefully, based on shared cultural values and work styles, also can contribute to desired results. Without alignment in these areas, partnerships can quickly face difficulties.

The challenges of effective partnership management are perhaps one of the reasons why true partnerships (as opposed to more "supplier/service provider" relationships) are relatively rare in the medtech industry. The lack of medtech partnerships provides a stark contrast to the biotech and pharmaceutical industries where they are prevalent and often highly successful. In biotech and pharmaceuticals, partners often come together to share the tremendous cost and risk associated with developing and commercializing their products. Considering that the margins offered by many pharmaceutical products approach 90 percent, there is usually adequate value in most product launches to be shared between one or more partners. In these cases, the decision to partner may be part of a company's fundamental business strategy.

The story of Nanostim and St. Jude Medical provides one example of a successful device-related partnership that led to a reasonably swift exit for the founders and initial investors.

FROM THE FIELD ▶ NANOSTIM

Partnering to advance a ground-breaking product

Medical device engineers Alan Ostroff and Peter Jacobson each had 25 years of experience in cardiac rhythm management, a history of successful collaboration, and an idea for a leadless pacemaker that could potentially revolutionize the industry. For the business expertise they needed to turn their concept into a company, they turned to Allan May, a seasoned medical device executive and investor who had been a founder, board chair, and CEO of numerous medical device and biotech start-up companies. May, Ostroff, and Jacobson founded Nanostim, Inc., with the goal of developing a leadless cardiac pacemaker (LCP) that could be implanted directly into the heart via a catheter (see Figure 6.4.2). The elimination of lead failures, the bane of all existing cardiac pacemakers, would improve device performance, and decrease lead infections and re-operations. At one-tenth the size of a conventional pacemaker, the LCP would also benefit patients by eliminating the constant visual and palpable reminder of the pacemaker's internal, life-saving presence, and facilitate a less restricted lifestyle.

Conventional pacemakers comprise a pulse generator that is implanted into a surgically created "pocket" just below the patient's skin near the collarbone, with leads

FIGURE 6.4.2
The Nanostim™ leadless pacemaker (courtesy of St. Jude Medical).[6]

(insulated wires) that connect the generator to the heart. The pulse generator contains the battery and electronic circuitry (sometimes called the "brain" of the device) that directs the battery to send pulses to the heart when it detects a problem with the heart's rhythm.

One of the main engineering challenges in pacemaker design involves optimizing the trade-off between the lifespan of the implanted battery and the continual quest to reduce size. The greatest drain on the battery comes not from its regular pacing activity, but from the extra power needed when the device is accessed for programming, or interrogated for information. In their early search for a more powerful battery in a smaller package, Ostroff and Jacobson originally reached out to an inventor working on an atomic battery that could deliver tiny amounts of power over very long periods of time. However, this avenue was ultimately abandoned. "It's too risky for investors to stack two revolutionary start-ups together," observed May. "We had to find a way to realize the idea using a conventional battery."[7]

The breakthrough came when the Nanostim team found a way to program and query the device without using anything except pacing pulse power, which allowed them to miniaturize the battery and device while maintaining an average lifespan equal to or better than conventional pacemaker batteries.[8,9] In the design that followed, the pulse generator, electronic circuitry, and battery were all contained in a single, sealed case (smaller than an AAA battery) that would be attached directly to the interior heart wall, eliminating the need for leads. The pacemaker would be implanted through the femoral vein via a steerable catheter and, when necessary, retrieved the same way. The minimally invasive procedure eliminated the surgical incision and the pocket created to house traditional pacemakers. An additional advantage of the technology was that its streamlined design meant that the production cost would be competitive with traditional pacemakers.

To test the market, May and the team arranged focus groups of cardiologists to whom they presented the concept. "The response was overwhelming, with physicians eager to offer their patients a less invasive procedure and a smaller, longer-lasting alternative to conventional pacemakers," he recalled. Encouraged, May began to seek the financing needed to take the concept forward. However, there were two significant obstacles that had to be overcome. First, the development of the pulse generator and circuitry of the device (the ASIC or application-specific integrated circuit)[10] was a complex process that would take an estimated three-and-a-half years to complete. "This was a big hurdle given that, typically, medical device innovators can produce a working **prototype** within six to twelve months," said May. "We were asking investors to commit even though it would be years before we could show them a device that worked." Second, gaining approval from the US Food and Drug Administration (**FDA**) in order to market the pacemaker in the US would require a premarket approval (**PMA**); a long, risky, and expensive process.

Despite these drawbacks, May and his co-founders convinced the members of Life Science Angels to commit $500,000 to the venture. However, the combination of electromechanical complexity and regulatory risk dissuaded other angel firms from following suit. Undaunted, May approached a number of venture capital (VC) firms he knew, where the surprising response to the pitch was that Nanostim "wasn't solving any problem." Recalled May, "Pretty much across the board, the response was, 'Pacemakers are a multi-billion market because we have them, and they work, so you don't have to create a new one.' The lack of vision was stunning." In May's view, pacemaker hardware had not been innovated in decades, leaving a huge existing market with attractive **reimbursement** ripe for disruption via true product differentiation. There were also problems with the conventional technology, including issues around lead failure and replacement, the tendency of patients to play with the bulky implants and dislodge the wires, cosmetic issues around the surgical scar and lump, and occasional

problems with infection. However, "The VCs did not see the disruptive opportunity or agree that these issues were significant enough to justify investing millions of dollars in an idea with a seven year PMA path and no demonstrable prototype," he said. Finally, May turned to Michael Sweeney, a pioneer in the field of cardiac devices who had become a venture capitalist with InterWest Partners. With his electromechanical background and personal experience in cardiac rhythm management, Sweeney "immediately recognized that Nanostim had the potential to reshape the pacemaker industry," May stated. Between InterWest and a **syndicate** of investors who were willing to follow Sweeney's lead, May raised $10 million in Series A financing to fund Nanostim's next phase of development and testing. As important, Sweeney and the venture investors tapped their extensive networks to help attract CEO Drew Hoffman and to assemble an "A team" of cardiac rhythm experts in all relevant functions, optimizing the ability of the team to execute to plan and to budget, despite the usual setbacks and unexpected problems.

Over the next two years, Nanostim worked through the complex process of developing the ASIC. "It just takes that long," reported May. "You have to design it, fabricate it, manufacture it, debug it, and then have all the other attendant circuitry assembled, tested and tweaked." In parallel, because the team knew the precise dimensions (e.g., size, weight, specific gravity, density) of the device the electronics would fit into, they were able to design the housing ("the can") and conduct extensive animal tests to perfect the implantation technique, introducer catheter and retrieval system and explore issues related to free-anchoring a device to the heart wall long before the ASIC was ready. "We did every possible exercise and test to de-risk the actual implant," said May. "We just couldn't yet show that it paced."

Once the ASIC was completed and tested, the next hurdle would be a **pivotal clinical trial** to support a submission for European regulatory approval (**CE mark**). May anticipated that this trial would cost close to $60 million to complete. However, as the company ticked off its R&D milestones, the global economy spiraled into a recession that decimated medtech venture capital funding. Nervous investors were unwilling to commit to projects with substantial clinical and regulatory requirements, and even Nanostim's existing syndicate of VC investors became skittish.

At the time, the pacemaker market was dominated by offerings from several major companies including Medtronic, St. Jude Medical, and Boston Scientific/Guidant. Medtronic had its own leadless pacemaker project and Boston Scientific had just completed the Guidant acquisition. Despite the fact that Nanostim was early stage and still working on the ASIC, St. Jude became interested in the product. In the lengthy partnership negotiations that followed, Nanostim and St. Jude jointly developed a financing arrangement, agreed on an extensive set of key milestones that would de-risk the project going forward, and specified their roles and responsibilities for working together. In terms of highlights from the deal, St. Jude would provide the majority of funding needed to get through the pivotal trial and CE mark submission; Nanostim and its syndicate of investors would provide the remainder. In terms of governance, Nanostim granted St. Jude an observer seat on its board of directors.

"We needed $60 million or more, which would have required two to three sizeable funding raises in a world where everybody was running like hell from PMAs. So our agreement with St. Jude was that we would offer them a better **exit** value on a game-changing technology in return for them committing to put up the bulk of the funds needed to get Nanostim to CE mark. It was a bold step on St. Jude's part to get involved so early in a promising but risky technology. The novel deal was a fair trade-off in which we gave up the theoretical upside of a disruptive technology in exchange for diminishing financing risk in a hostile medical device financing climate," summarized May.

At a more tactical level, Nanostim would continue to lead the development and testing effort over the next two years, with St. Jude providing access to its technical and regulatory expertise.

The ASIC was completed on schedule, making the product real for the first time. As the Nanostim team gained exposure to St. Jude personnel, naturally a trust and bonding occurred which laid a solid foundation for integration after the acquisition decision was made. At the same time, enthusiasm and interest in the device also was growing externally. Whenever the company would share the device (under non-disclosure) at conferences or society meetings, physicians expressed overwhelming interest. "Patients are going to love this" was the oft-repeated response.

The Nanostim device was granted CE mark approval in October 2013. Just days later, St. Jude announced that, per its agreement, it would acquire the company for $123.5 million, with additional payments of up to $65 million based on the achievement and timing of certain revenue-based milestones.[11] Commenting on this outcome, May confirmed that this was a great result for all concerned. "It marks St. Jude as an innovator willing to take calculated risk among large cardiac companies. It represents a solid venture return. It's an excellent angel return. It's quite a good founder return. And it's a great price for the acquirer for a device that is poised to have a major impact on a multi-billion existing market."

Reflecting on the agreement between St. Jude Medical and Nanostim, May said, "St. Jude agreed to fund a risky technology concept at an extremely early stage. So, on balance, escaping the risk of raising $60 million or more and achieving an exit value of almost $200 million on an initial investment of $10 million was an excellent deal for all parties." May also felt that the partnership was invaluable to bringing the technology forward efficiently and effectively. In today's risk-adverse financing climate, this kind of public company-start-up partnership could represent "the future of PMA projects," he predicted. "VCs appear to have shrunk to financing fewer than two dozen PMA projects a year in the United States. If we want more innovation, people are going to have to get creative in how they get their projects funded, and this is a model that could be widely adopted to positively advance innovation for patients and for the US healthcare system."

Importantly, partnerships also can provide innovators with an effective mechanism for bringing a product forward when their primary focus is on achieving a positive social impact rather than a financial gain. The key is to target an entirely different type of partner. In recent years, certain non-governmental organizations (**NGOs**) and foundations have shown an increased willingness to support product development efforts in the medical technology space. PATH, based in Seattle, Washington, is one such entity. As an international, non-profit organization seeking to "transform global health through innovation,"[12] PATH receives a sizable portion of its total funding from the Bill & Melinda Gates Foundation, along with the US government, multilateral organizations, individuals, and a variety of other sources. Through innovative partnerships, the organization believes it can drive sustainable benefits in public health by helping certain companies bring their ideas forward that, in the absence of PATH's involvement, would not be a private-sector priority.[13] One way that it accomplishes this is by providing "significant resources or expertise (such as funding, management, codevelopment, and assistance with clinical studies) to a private-sector collaborator to support the collaborator's development of a product."[14] Take its partnership with PharmaJet as an example. PharmaJet, which makes a needle-free injectors, has a mission to replace needles in the world's poorest countries. The company raised $14 million from investors to develop its product,[15] but faced constant pressure to create a more elaborate device that could be sold at higher prices in developed markets. Through a partnership with PATH, the company is able to continue developing, testing, and evaluating the device for underserved, low-resource environments. As one aspect of the agreement, PATH is funding a study of the device to deliver measles–mumps–rubella vaccinations in Brazil.[16]

Sale/acquisition

In some cases, innovators may choose to sell an idea outright, completely relinquishing control to the acquirer. Acquisitions can happen at any stage in the biodesign innovation process. If the transaction happens early, the acquirer usually obtains only the rights to the technology and IP. If the acquisition occurs later, the acquirer may obtain all of the assets of the nascent company. The reasons why companies acquire IP, technology, and/or companies vary, but they can include the desire to obtain a broad portfolio of **patents** protecting current or anticipated products. Acquirers also may seek to expand into new areas or to block areas from the competition. Or, they may choose to make an acquisition when they do not want to develop the product themselves or to track and pay royalties. Acquiring smaller companies with interesting technologies also can be one way that larger companies extend and enhance their R&D efforts. As Richard Gonzalez, former president and COO of Abbott, described: "We have a strategy that basically says we aggressively invest in internal R&D, but we supplement that investment with opportunities on the outside."[17]

In certain acquisitions, where there is still substantial development or adoption risk, an acquirer may not wish to pay the complete value of the deal until certain milestones have been reached. Such an arrangement is generally referred to as an **earn-out**. An earn-out is a contractual provision stating that the seller is to obtain additional future compensation when certain pre-defined goals are met (otherwise the earn-out will be forfeited). For example, an earn-out might be linked to the achievement of regulatory clearance, positive results from a particular experiment or test, the achievement of predefined gross sales or earnings targets, or the performance of certain individuals for a specified period of time. As described in 5.3 Clinical Strategy, the founding team of DVI faced a sizable earn-out linked to the successful completion of a critical clinical trial when the company was acquired by Eli Lilly. An earn-out can provide the innovators with a higher total value from the acquisition, but the total value is completely dependent on the achievement of milestones, which may or may not be assured or under their control. Overall, although earn-outs carry considerable risk for the seller and are often the source of disputes and litigation, they can provide more value to the innovators if the triggers for the earn-out are achieved.

Equity funding or a loan paired with a put/call option is another structure that is often deployed when substantial risks remain to be resolved in the future, but the buyer believes that the price desired by the seller is not consistent with the current value. In this setting, the parties agree on some level of funding, via a loan or equity investment, and a prearranged acquisition price based on certain milestones that are believed to be achievable within the funding period. This structure allows the seller to potentially eliminate the issues standing in the way of achieving full value, and also allows the buyer to rationalize the price upon completion of the milestones. This structure has its downside, however, as things usually do not go precisely as planned and the two parties can often be left with a tough situation when the money runs out, but the milestones have not been achieved. Further, venture shareholders tend to be negative about such structures because they view them as limiting their potential return and their ability to drive the value to a maximum; thus creating a "capped upside." Regardless, this structure is a viable alternative to consider when there is important strategic value in securing a **corporate investment**, and there is also a need for more predefined outcomes.

In an acquisition, both parties must accept the tenet that the price in such a transaction is rarely a perfect approximation of the value of the asset (i.e., the buyer either overpaid or got a bargain, just as the seller was either paid at a premium or a discount). Value to the innovator from a licensing deal more closely tracks the true value of the innovation since royalties are usually linked directly to revenue. In contrast, with an acquisition, the agreed-upon value of the transaction represents a "best guess" made at a fixed point in time. Once negotiated, the outcome remains the same, regardless of how the asset actually performs in the market (unless there is an earn-out, which makes it possible for the value from the deal to better track the true value of the innovation). Despite this uncertainty, companies often try to acquire assets rather than license them in an effort to gain direct control over them, and also to shift the cost of the asset off the income statement (where it would be if license payments had to be made).

As acquisitions have become the most common exit scenario in the medical device industry, some innovators have begun to think more strategically about designing technologies that will appeal to corporate entities and developing them in a highly capital-efficient manner to maximize their returns at exit. This approach has seen some success in the biotechnology sector, where it is referred to as a "**build-to-buy**" model. In biotech, the approach resembles a partnership agreement with a pre-negotiated **option to purchase** for the corporate entity once the technology is developed. Versant Ventures is one venture capital firm backing such deals. As described in one article, Versant invests in early drug discovery companies and helps orchestrate advanced deals with established pharmaceutical companies for the start-ups to develop specific drugs or devices. The pharmaceutical companies have the right to acquire the companies, but not the talent/teams, which move on to form another new biotech start-up.[18] "Part of what we do is play the middleman," said Jerel Davis, a Versant principal in Vancouver. "We connect great academic science with pharma companies that are looking for that type of program, and we help finance those companies. This is a great way to get pharma's buy-in – there's skin in the game – and also create an exit up front for us."[19]

While in the biotechnology industry, "build-to-buy" models involving structured deals between early-stage companies with pharmaceutical acquirers are gaining traction, this approach is less well developed in the medical device sector. As Ross Jaffe, Managing Director of Versant Ventures in Menlo Park, CA explained, "We and others are exploring these types of relationships for medical devices, but we do not have the models refined yet." The point, he emphasized, is that, "We're in a brave new world right now, where the traditional model of venture capital financing for innovative companies is under pressure, so we've got to continue to evolve to adapt to the new financing environment."

One variation of the pure build-to-buy model is to take a lean approach to designing and developing a technology with the explicit objective of selling it at the first available opportunity. The Vortex Medical example demonstrates how this particular approach to medtech innovation, with the exit scenario clearly in mind, can work.

FROM THE FIELD ▷ VORTEX MEDICAL

A capital-efficient approach for moving from concept to commercialization to exit

Over roughly two decades, Mike Glennon amassed a wealth of medical device industry experience working with corporations such as Guidant, Stryker, and Medtronic. But his career trajectory took a particularly interesting turn when he accepted a leadership role with a company called Accellent, which specializes in providing innovators with comprehensive supply chain solutions on an outsourced basis. "Accellent provides services from developing a comprehensive supply chain solution to warehousing and distribution," Glennon described.

As he worked with this organization, he started thinking about an alternate approach for bringing an innovation to market, which he dubbed a "virtual model." The strategy was to take an invention from concept to commercialization and then to an exit using an incredibly lean, capital-efficient approach. "Most of the great ideas for medical devices come from physicians. The problem is that physicians don't always have the time, skills, and experience to take their ideas and make the products a reality," Glennon said. Rather than putting these individuals into a position where they have to try raising tens of millions of dollars in funding to build a team and a business, he hypothesized that with only one full-time employee it would be possible to usher a solution through development, into commercialization, and to a trade sale using a flexible workforce made up entirely of consultants, contractors, and outsourced service providers.

Glennon's first opportunity to test the virtual model arose when he partnered with Dr. Lishan Aklog, a cardiac surgeon who had devised an approach for the minimally invasive, *en bloc* removal of intravascular material to reduce the risk of complications associated with major open surgery, internal bleeding, and clot fragmentation.[20] "The basic concept was to remove a thrombus using a balloon-actuated, expandable, funnel-tipped catheter and a centrifugal pump. The undesirable

intravascular material would be sucked out in one piece and the shed blood would be re-infused back into the patient in real time so there was zero blood loss," he summarized (see Figure 6.4.3).

Glennon, Aklog, and cardiac surgeon Brian deGuzman co-founded Pavilion Holdings Group and decided to make Aklog's venous drainage cannula its first project. They named the company Vortex Medical, and Glennon became its CEO – and its one and only full-time employee. Dr. Albert Chin, a serial inventor and businessman, served as the sole independent board member.

In this new role, Glennon had several immediate priorities. "First, you must figure out if the idea is going to be novel and nonobvious in nature based on your claim set. Predicting the likelihood of patentability is paramount in the initial phases of funding," he stated. "Next, you have to look at the supply chain solution. How would the device be made? What could you potentially sell it for in the marketplace? It's very important to consider up front whether a device can be made for an overall cost of goods that will support a healthy gross margin that's expected in the medical device industry." Finally, he continued, "You have to focus on a robust development process. Especially with trying to use a lean, virtual model, you have to set a strategy and structure around bringing together all of the different process expertise elements that go into orchestrating an idea from concept to commercialization."

And, of course, the company also had to raise funding. "We raised the money on a PowerPoint presentation," Glennon remembered. "We didn't have a prototype, or a clinically-relevant proof of concept, or even a patent. But we convinced some angels and a venture capital group that we would take this product to market in a capital-efficient manner by outsourcing all of the required functional process expertise. And that was a very different approach than most start-ups use." In total, Vortex raised approximately $350,000 in angel funding and $3.2 million from venture capital firm

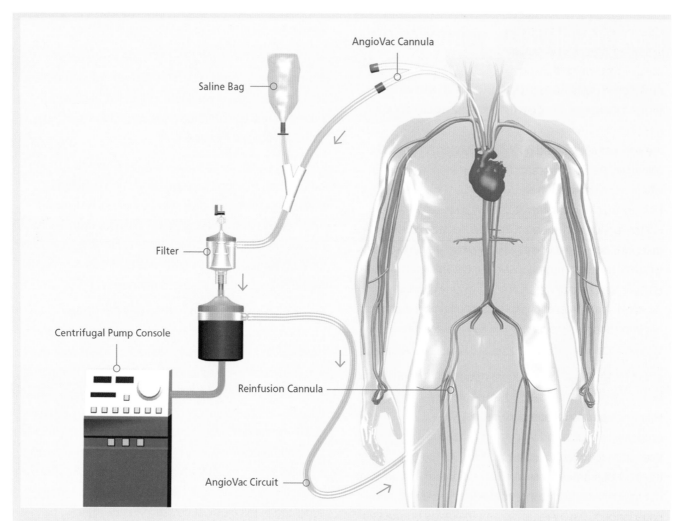

FIGURE 6.4.3
The AngioVac system (courtesy of AngioDynamics).

Catalyst Health Ventures. "We called it a Series O for 'only,'" he laughed, "because we knew we'd have to do everything with that money – developing the technology, all of our regulatory work and clinical testing associated with the **510(k)** filing, manufacturing, preparing for market, marketing, selling, and distributing the device, and ultimately finding an acquirer for the company."

Leveraging his deep relationships with the medical device field, Glennon went to work lining up outsourced service providers and contractors to complete key activities related to the company's major milestones. He acknowledged that managing an entirely virtual workforce required complex coordination and careful oversight. But, he pointed out several important

advantages of working with a network of specialized experts. "With every activity, you know it's going to be done on time and within budget. And you're paying for what you need, when you need it, which is more capital-efficient than maintaining the overhead and the higher burn rate associated with a full-time staff." In addition, he emphasized the availability of top-notch consultants and service providers across a wide range of functional areas: "If you are looking for engineering, operations, regulatory, or quality consultants on an outsourced basis, they are all available. They're all skilled and very experienced. They all want to work with us on a contracted basis, and they understand and embrace my model, so it works well for everybody," Finally, he noted that the company eliminates its investment in property, plant, and equipment when working with a virtual workforce.

Over the course of the next few years, Glennon and his virtual team worked diligently toward Vortex's goals. While the product, which they dubbed the AngioVac, was still undergoing development and testing, Glennon also started thinking about potential acquirers for the technology. "I ran a process to identify companies that might be interested in the product and started talking with the right people. But, ultimately, those conversation didn't heat up until we had a little over a year and a half in the marketplace," he said. Using a virtual distribution network, Vortex established a solid customer base and achieved millions of dollars of revenue. At that point, one of the companies Glennon had been talking with was ready to make an offer. AngioDynamics acquired Vortex Medical for a guaranteed $55 million over the first five years ($15 million up front and at least $8 million per year for five years after that). After the preliminary five-year period, Vortex Medical could receive additional annual earn-out payments, depending on the level of net sales of the AngioVac system.[21]

Glennon admitted that reaching that point on the company's original $3.5 million in funding was no easy feat. "It was a big task, but having deep experience in the business and understanding all of the functions necessary to go from concept to commercialization to exit really helped. We were able to appropriately budget in a very effective and efficient manner and deliver on, not only getting the product to market but ultimately achieving an exit with a very healthy return."

Reflecting on the experience, Glennon stated, "Companies like mine are good at getting products into a box and ready to be sold. And I think we can do that in a more cost-efficient and timely manner than some of the bigger medical device companies. But larger medical device strategics have massive sales networks as well as world-class marketing, professional education, and clinical capabilities. So their teams can introduce and sell products a lot faster and more effectively once they're in the box."

When asked if the virtual model could be replicated by other innovators, Glennon said, "I think it's a risky strategy for young entrepreneurs without a lot of experience. They need to be partnered with folks who have deep industry experience to help them understand the unmet **need**, the market, design, and supply chain solutions as they pertain to the cost of goods and overall gross margin for the product. So there's a lot to do, and I'm not sure if this would be the best strategy for somebody who is just jumping in. But if you built the right team, you could absolutely replicate this model."

He also advised innovators to think carefully about who is most likely to fund this type of approach. "Catalyst Health Ventures is one of a handful of VC firms that invest in early stage start-ups. Our model generated at least an 8x return on the $3.5 million invested. However, that type of return doesn't move the needle for a lot of venture capital firms given the small amount of total capital invested. For many VC firms, this kind of investment may simply not be worth their time." Yet, as the involvement of **angel investors** continues to grow in the medtech sector, Glennon advised that many of these individuals are likely to be interested. Referencing a new company he was working on, he said, "I was able to raise $3 million in two days with all angel investment. That's not a huge amount of money, but using my model, it's more than enough to go from concept to market to exit. For the investors to understand our approach and the potential return, they're excited about investing in our new technology and in our whole team structure."

Building value toward a partnership, license, or sale

Once innovators decide to investigate an alternate development pathway, it may not be clear at the onset which of these paths they will take. In many cases, the actual outcome is determined by the negotiations that lead to the deal. For instance, innovators may approach a company to propose a licensing agreement, but the final deal may take the form of a sale/acquisition. Innovators are advised to stay flexible and consider the advantages and disadvantages of the different financial, ownership, and continued-involvement features of any deal.

Regardless of the development pathway, the innovators must determine the optimal timing for involving another entity. It can be challenging to know how far to develop an idea into a technology or product before licensing it, selling it, or entering into a partnership. Typically, the further one proceeds through the biodesign innovation process, the more risks are eliminated and, therefore, the more valuable the asset should be. This, in turn, should lead to more favorable terms in any transaction. On the other hand, each stage of the biodesign innovation process requires the innovators to invest increasing levels of capital and time. Some innovators may prefer to involve another entity earlier in the process to avoid the need to raise significant funding. Depending on the nature of the invention, licensors, acquirers, and partners may also prefer to get involved sooner rather than later, so that they can help shape the direction of product development. For example, a company seeking to acquire new catheter navigation technology may want to purchase it before the design is finalized so that it can be optimized to work with the rest of its own catheter system.

There is no proven formula for calculating the best time to embark on an alternate development pathway. However, innovators can rely on the evaluation of a number of factors to help them determine the optimal timing for their unique technology. These factors, outlined below, represent value in any licensing, acquisition, or partnership deal. While it may not be desirable (or even necessary) for innovators to address each one, they should understand how these factors contribute to retiring risk and driving value in a transaction.

People

Innovators should start out by assessing the value that they would personally bring to any deal. For example, if a particular innovator with an idea to improve the treatment of kidney disease is a nephrologist with 20 years of experience, that individual will have less difficulty convincing someone about the clinical merits of a solution than someone with a general business or engineering background. If the technology is further along in its development, licensees, acquirers, and partners also may be interested in issues related to

manufacturability or other matters that someone besides a physician might be better suited to address. Rarely can one individual speak to all of these concerns. As a result, innovators should perform an honest self-assessment and then surround themselves with others who can help increase their credibility in other areas of importance.

The quality of one's employees, advisors, consultants, and board members is important because they lend credibility and expertise to the innovator and the project. It can be particularly powerful for physician advisors (and customers) to provide testimonials regarding how they have used a technology in their practice and/or its anticipated impact on treatment in the field. The selection of board members is also important and should receive special attention as this is a common area where innovators make mistakes. Just because someone is a leading physician in the field or a trusted advisor does not mean that they would make a good board member. Board positions should be reserved for only those individuals to whom the innovator would be willing, at times, to turn over control of the company – for example, when a board vote is required for a critical decision. For this reason, it is important for board members to possess a strong general business perspective, as well as specialized expertise in an appropriate field. Ideally, they will also have previous experience serving on a board and will understand the full scope of the associated financial responsibilities.

When the technology being developed is in a highly specialized field that requires hard-to-find skill sets, an outside entity may be enthusiastic about a large, knowledgeable base of employees and an established set of advisors and board members. This is particularly true for partnerships that require the innovator and team to stay engaged in the project's ongoing development. With licenses and acquisitions, however, keep in mind that sometimes outside entities may prefer to pursue technology with limited employee overhead.

Intellectual property (IP)

It is essential for innovators to protect their IP position before initiating discussions regarding a partnership, license, or sale/acquisition. An idea or invention has little value if it has not been protected. Without a patent filing and a confidentiality agreement, the idea can be seen as nothing more than a suggestion, which carries with it no IP rights. In contrast, with an advanced IP position, the innovators can realize significant value from a transaction (see 4.1 Intellectual Property Basics and 5.1 IP Strategy for more information about patent filings, prior art searches, and freedom to operate analysis).

Market assessment

The extent to which a market assessment increases the value of a deal depends, in part, on the target audience that the innovators approach. If they hope to license or sell a technology to the world leader in a given field, that entity surely has more knowledge about that field than the innovators have amassed. In these cases, it may simply make sense to complete a survey of 20 or 30 potential customers to demonstrate that there is demand for the innovation. On the other hand, if the innovators have a technology that could take an outside entity in a new direction or require it to look at a market in a different way, then a more extensive market assessment may be useful. The key is to determine what potential partners, licensees, and acquirers know about a market and then tailor the market assessment to complement and expand that understanding (see 2.4 Market Analysis for specific information about what to include in a market assessment).

Prototype/proof of concept

Prototypes and proof of concept play a central role in eliminating risk, and therefore adding value. In this context, a prototype refers to any non-clinical model that demonstrates the basic feasibility of the technology. Proof of concept generally refers to a more developed version of the device that can be used in a preclinical test to demonstrate its benefits (e.g., in an animal or bench-top study). As noted in 4.5 Concept Exploration and Testing, there is no substitute for being able to *show* someone how an idea will actually work. For this reason, having a working prototype is highly recommended for any innovator seeking to license, sell, or partner. Preclinical and clinical data are certainly beneficial, but is not always needed depending on the type of deal and the interests of the outside entity.

Plan for reimbursement, regulatory, and clinical advancement

Whether or not innovators will have any preclinical or clinical data to use as leverage in a transaction, they should always have a well thought-out clinical plan. Similarly, they should have a strong understanding of the regulatory and reimbursement pathways associated with the device, potentially including an opinion from one or more regulatory and reimbursement consultants. These assessments give the licensee, acquirer, or partner a realistic evaluation of a technology's risks and help them assess its potential worth. More information about the types of information to cover in clinical, regulatory, and reimbursement plans can be found in 5.3 Clinical Strategy, 5.4 Regulatory Strategy, and 5.6 Reimbursement Strategy.

Financial model

Like the market assessment, the value of an extensive financial model varies, depending on the nature of the deal. For a well-developed business with employees, other resources, and infrastructure in place, a financial model can be important (and likely would have been developed before such investments were made). For an early idea that has not progressed as far in the biodesign innovation process (e.g., an innovator and one employee working on early-stage prototypes), it may not make sense to spend significant time creating a detailed financial forecast, especially if the innovator is seeking to license or sell the idea to a company that already has an established infrastructure.

Accurate financial models are determined by the employees, resources, and infrastructure of the buyer (following the transaction) rather than the seller. Accordingly, a model created by the seller may be viewed as relatively useless. For example, an innovator could spend dozens of hours working on a detailed revenue model based on a small team of 20 sales people, but if the acquirer has an established sales force of 150 people, that 20-person model is of little value. When creating a model purely for selling purposes, an innovator should take this issue into account. Ultimately, the financial model has to fit with the business of the acquirer (or licensee), not the innovators' vision of what they would do. Often, the innovators are better served to develop a high-level view of the finances – revenue and cost forecasts – to help justify the opportunity rather than a detailed financial model as described in 6.1 Operating Plan and Financial Model.

Preparation for the process

Before initiating contact with any outside entity, innovators must be properly prepared. The more prepared they are going into negotiations, the stronger their negotiating position will be and the better the result. Innovators who casually enter into negotiations often find themselves feeling "ambushed" by outside entities more experienced at controlling these kinds of interactions.

The first step in preparing is to perform some due diligence on the entities (sometimes called "targets") that are most likely to be interested in entering into a deal. Questions to ask include:

- **Who are they?** – What companies have similar or complementary interests that might be helped by the innovator's technology? What is their current position in the market and how will the technology help them (e.g., increase market share, boost revenue, introduce innovation in a stagnant product area, etc.)?
- **What is their reputation?** – Which ones are most likely to be open to doing a deal? Which would be best to work with?
- **Who will handle negotiations?** – Who within the company has responsibility for licensing, acquisitions, and/or partnerships? What can be learned about this person (or people)?
- **What is their track record?** – What kind of deals have they done in the past? How successful have they been?
- **What is the best way to approach them?** – Do the innovators have people within their networks that could potentially make an introduction? If not, what are the other options for making contact?
- **What opportunities are there to create competition among targets?** – Are some of the targets direct competitors? Who is chasing whom in the market?

Once a handful of appropriate targets are identified, the innovators should next begin thinking about their team – specifically, who should help represent them in the process. It is critical for innovators to have a legal advisor who is closely involved in each step of the process. A business advisor can also be helpful, particularly someone with experience doing licensing, acquisitions, or partnership deals.

Carefully putting together an information packet is also important. Jamey Jacobs, vice president of research and development for Abbott Vascular in Santa Clara, California advises innovators to think about the information packet as a tool that could be used by a potential target to "jump start" an attractive R&D project.[22] The packet should be complete, detailed, and customized to the unique needs and interests of each target. The more the innovators can do to inspire confidence that the technology/idea can be used by the target from the first day to accelerate a project, the higher the odds of a successful transaction.

When deciding what information to include, confidentiality is, of course, a critical issue. One effective strategy is to think about what information should be considered confidential and what can be shared more openly. The information that is shared openly should not reveal anything proprietary about the technology. However, it must be compelling enough to get a target interested in doing a deal. Most companies will not sign a non-disclosure agreement (**NDA**) until they are convinced that an idea is attractive to them, so innovators must strike a careful balance when deciding what to share. For example, the executive summary of one's business plan or pitch might be what is disclosed without an NDA, while the details that sit behind it are shared on a confidential basis. This more detailed information, too, should be organized to support due diligence (actual patents, case-by-case trial results, lists of vendors, copies of contracts signed, etc.), but should not be presented to the target(s) before there has been a formal expression of interest. A letter of intent, which is a written (although non-binding) statement of the partner, licensor, or acquirer's intent to enter into a formal agreement, would be one appropriate way for an outside entity to express interest prior to the initiation of the due diligence process.

Defining a BATNA

With all of this information in hand, innovators can begin thinking about their negotiating position. Specifically, they should invest considerable time determining their best alternative to a negotiated agreement

(**BATNA**).[23] In negotiation theory, one's BATNA is the course of action that will be taken if a negotiation fails to leads to an agreement. Once defined, innovators should never accept a worse outcome than their BATNA.[24] In this way, a BATNA is a safety net and also can serve as a point of leverage in negotiations. This simplified scenario provides an example of a BATNA in action: a team of innovators has an offer to license its technology for royalties that will equal approximately $1,000,000 per year through the remaining 15 years on the patent. With this as a BATNA, the team members would likely not accept a price that is less than the net present value of all those payments to sell the company. If they have no other options available at the time of a negotiation (and would not consider developing the idea themselves), then the BATNA is zero, meaning that almost any prospective deal would be worth considering.

To effectively evaluate alternatives against a BATNA, innovators are cautioned to take into account all considerations, such as relationship value, time value of money, and the likelihood that the other party will live up to its side of the bargain. Considerations such as these are often hard to assess since they are qualitative and uncertain, rather than easily measurable and quantifiable. However, they can make a significant difference when choosing between an offer and a BATNA.

Importantly, in addition to defining their own BATNA, innovators should think carefully about the BATNA of the target. By understanding the outside entity's own best alternative to a deal, they can gain insight into the company's behavior and anticipate how it might react to different demands. Think about what the target's alternatives are, who else could the company work with, and if there is someone else offering a substitute for the technology that could be pursued. Also consider what would happen if the outside entity simply decided to do nothing. Many times, this provides a target with an easier alternative than those available to the innovators.

Managing an effective process

Once the innovators have completed adequate preparation, they can initiate and then proactively manage an effective process. As one innovator put it, "The goal is to drive the process, not to let the process drive you."

The first step is to lay out a timeline for negotiations. This may be driven by external factors (e.g., there is only enough cash to last another six months). But, ideally, negotiations will begin when there is an optimal level of value associated with an idea or invention (e.g., positive results from a small animal study are now available, but it would be good to get a deal done before it is time to launch first-in-human testing). In general, partnership, licensing, or acquisition deals can take anywhere from 6 to 18 months, with smaller companies tending to move faster than larger ones.

Within this broad timeline, it is useful to assign dates to important interim milestones (see Figure 6.4.4). For example, innovators might seek to make contact with all appropriate targets within 30 days, set aside 45 days for presentations, meetings, and arriving at a letter of intent, allocate another 45 days for due diligence, and then allow 60 days for negotiating terms and closing the deal. Establishing this sort of a schedule allows the innovators to engage multiple targets and keep them moving forward at roughly the same pace to maximize their chances of a favorable outcome (assuming the timeline is realistic). Without a timeline, innovators too often find themselves having to decide on one offer before the other targets have had enough time to determine if they are interested. In a best-case scenario, the innovator will receive multiple offers within days of one another, with the targets aware of the competition they are facing in trying to partner, license, or acquire the idea.

With a realistic timeline defined, innovators can begin accessing the targets to determine which ones are interested in engaging. Introductions are usually followed by lengthier, more detailed conversations. Due diligence is initiated once interest has been formally expressed by

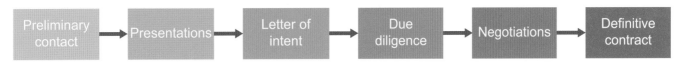

FIGURE 6.4.4
A carefully managed process leads to improved results.

one or more targets. Then, for those that are most serious, formal negotiations ensue, a partner, licensee, or acquirer is eventually selected, and the deal is closed.

Components of a transaction

While acquisitions are treated somewhat differently, there is significant overlap in the common financial deal terms for licenses and partnerships. A typical medtech license or partnership agreement may include some (but not necessarily all) of the terms outlined in Table 6.4.1.

When entering into any agreement that involves equity or **royalty** payments, physician inventors should proceed with caution to avoid conflict of interest concerns. Over the past several years, widespread media coverage has turned a spotlight on the flow of information – and money – between doctors and medtech companies. In response, the US Congress passed the Physician Payments Sunshine Act of 2009. The new legislation requires device companies to track payments or gifts they make to physicians that are worth more than $10 and/or that exceed a total of $100 per year to any given doctor. It also requires companies to annually disclose direct and indirect ownership and investment interests held by physicians and their immediate family members.[25]

Table 6.4.1 The financial terms of licensing and partnership deals can be similar – greater variation between these deals exists in the level of control and ongoing responsibility formally maintained by the innovator (compiled from Shreefal S. Mehta, *Commercializing Successful Biomedical Technologies* (Cambridge University Press, 2008)).

Financial terms	Descriptions
Upfront payment	A lump sum paid by the licensee/partner to the innovator at signing.
Patent prosecution and maintenance fees	The licensee/partner may be asked to pay legal and USPTO fees for maintaining the patent.
Milestone payments	If the technology succeeds at further stages of development, the licensee/partner may be required to make additional payments to the innovator as risk is reduced. Milestone payments may be variable (e.g., if the outcome is good, the innovator receives $100,000; if the outcome is bad, the payment is reduced to $25,000).
Royalties	A percentage of commercial sales paid by the licensee/partner to the innovator. Royalties can scale upwards or downwards based on the volume of sales (e.g., if sales reach 500,000 units per year, the innovator receives a rate of 3 percent; if they reach 1,000,000, s/he received 5 percent). They can also have minimums and caps (e.g., regardless of sales, royalties will not fall below $500,000 per year or exceed $1,000,000 per year).
Equity considerations	Licensees/partners may purchase equity in the innovator's company as part of its payments and/or the innovator may accept equity in a licensee/partner's company in lieu of cash payments.
Sublicensee and sublicense fees	If a patent is sublicensed to another party by the licensee, then a portion of the sublicense payments made to the licensee may be passed through to the innovator as the original licensor. These payments can be structured as a flat fee or a percentage of sales. (Note that this particular term would not be relevant to a partnership agreement.)
Royalty anti-stacking provisions	If a licensee has to license other patents to get a product to market (e.g., a stent that requires the licensing of both a drug and a delivery mechanism), it may be economically infeasible for the licensee to pay full royalties to both the innovator (as the original licensor) and a separate, secondary licensor. **Royalty anti-stacking provisions** provide a mechanism for the licensee to reduce the burden of dual (or stacked) licensing fees. (Note that this particular term would not be relevant to a partnership agreement.)

Under these new rules, physician inventors who want to stay involved in the development, testing, and/or commercialization of their innovations will face increased scrutiny if they accept any sort of licensing, partnership, or acquisition deal that includes equity or royalty payments. The concern is that such payments create the potential for scientific **bias** in the physician's activities. For example, if the physician inventors play the role of **clinical investigators**, some people worry that they may be tempted to interpret or report the data from a trial more favorably with the hope of facilitating regulatory approval and market adoption, thus advancing their personal financial gain.[26] Whether or not this is true, the net result of these requirements is that physician inventors are increasingly forced to choose between playing an active role in bringing their innovations forward and accepting the financial returns that would typically be a part of a licensing, acquisition, or partnership agreement.

The following example on the Rapid Exchange technology highlights the way the concepts presented in this chapter come together to result in a successful deal – in this case, a licensing agreement.

FROM THE FIELD ▸ RAPID EXCHANGE

Licensing a new technology in an established field

Paul Yock was a fellow in angioplasty at Sequoia Hospital in Redwood City, California during the early days of this ground-breaking procedure. In angioplasty, a balloon catheter is inserted through an artery in the patient's groin or arm, carefully moved up the aorta using a long metal guidewire, and then dilated to open a blood vessel that is blocked by plaque. "There was 4½ feet of balloon catheter and, in order to move the balloon in and out of the heart without losing track of its position, the guidewire had to be over two times as long to allow the wire to stay in and the catheter to come out," explained Yock, which made the intervention a two-person procedure. "The more skilled physician was in charge of moving the balloon up and down on the wire, and the less skilled person, positioned down at the patient's feet, had to keep the wire stable. The junior member was supposed to compensate for the movement of the senior person to keep the wire in the same position," he continued. "The trouble was that this was a ten foot long wire and moving the tip even half an inch could be a problem. So it required a dance, a coordination that was very awkward." As a trainee at the time, Yock recalled spending considerable time on what he referred to as "the wrong end of the procedure."

"The senior physicians didn't really perceive it a problem area because they had graduated beyond thinking about it," he said.[27] "But, in retrospect, it was kind of a glaring need." After inevitably losing the guidewire position in a couple of patients during his fellowship, Yock started thinking about better device designs. He made drawings and then developed a prototype rapid exchange catheter. Ultimately, this led him to the development of a technology that became known as the Rapid Exchange (RX)™ balloon angioplasty system (see Figure 6.4.5). RX allowed one physician (using two hands) to perform the entire procedure.[28]

Yock was fortunate to be training under John Simpson, a renowned cardiologist, prolific inventor, and founder of the leading producer of catheter systems at the time, Advanced Cardiovascular Systems (ACS). He shared the RX idea with Simpson who, in turn, referred him to a patent attorney. Thinking back on his days as a young physician, Yock noted that, "Paying for the patent was an issue for me. But I decided to go ahead and do it." He concluded that the best approach for commercialization was to license the technology, and targeted Simpson's company ACS. "I never thought seriously about starting up a new company around RX," Yock remembered. "I was intent on becoming an interventional cardiologist and had no interest in taking a different pathway." Instead, he saw ACS as an exciting company that was

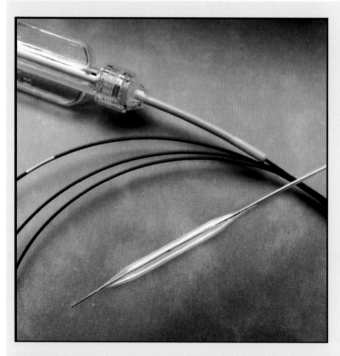

FIGURE 6.4.5

Guidant's ACS RX Comet VP™ Coronary Dilatation Catheter is one example of a balloon angioplasty system that leveraged Rapid Exchange technology through a licensing deal (courtesy of Abbott).

emerging as a powerful player in the angioplasty field. He also viewed it as the ideal candidate to license the RX technology. "With the old style, the over-the-wire systems, you introduced both the balloon catheter and the guidewire together. With my system, the guidewire could go in first, without the catheter. That required really good guidewire performance." Yock had paid careful attention to the other companies in the field and felt that ACS had the best guidewire performance and overall positioning to make to benefit from RX. And, "The fact that they were a local company was a nice coincidence," he added.

Yock had previously met Carl Simpson, the vice president of R&D at ACS (no relation to John Simpson) and arranged a meeting with him to introduce the RX concept. Prior to his initial meeting, Yock made arrangements to put a non-disclosure agreement (NDA) into place. "I didn't have my own NDA," Yock said. "I talked with Carl ahead of time about what the non-disclosure would be and he provided the company's

standard two-way version." Yock also took an extra precaution regarding his disclosure. "I was aware that what they were working on could potentially be influenced based on the discussion. So I came in with a written document that outlined my ideas of examples for how this catheter could be used," he recalled. Carl Simpson signed this document, acknowledging the contents as belonging to Yock in advance of the meeting.

With the disclosures taken care of, Yock demonstrated the RX system using a heart model and his prototype. Simpson was immediately intrigued. According to Yock, "Carl was excited about the system so he called in the VP of sales, who just happened to be down the hall. As soon as the VP of sales saw how it worked he said something like, 'Yep, I can sell that.'"

Just a few weeks later, Yock was invited back to meet with Ron Dollens, ACS's president and CEO at the time. ACS had done some of its own due diligence (including a thorough IP assessment) and was ready to negotiate licensing terms. The company proposed an agreement that would tie Yock's royalty payments to the system's sales volume. As sales increased, the royalty rate would be ratcheted downward, with the increase in volume compensating for the lower royalty percentage at each predefined sales threshold. He asked the company to cover the patent expenses, which it agreed to do. "There was also an indemnification of me, as an individual, for product-related malfunctions. That was extremely reassuring. I didn't think to bring that up, but I was happy to see it in the contract," Yock commented. In exchange, he exclusively licensed his patent to ACS and agreed to cooperate in any RX patent litigation incurred by the company. "There was also language that if they sub-licensed the patent, I would get a royalty stream from it, which turned out to be important," he said, as RX was eventually sub-licensed by multiple companies.

Yock described the fact that the licensing process went so quickly and so smoothly as "remarkable" and admitted that it was "somewhat unrepresentative" of what most innovators face. "These days, companies are really sophisticated in dealing with inventors," he said, which makes it much more difficult for those interested in licensing their technology to enter into mutually favorable agreements. "A lot of power sits with the companies, so the inventor has to create an incentive for them to want to move quickly and aggressively. That will happen spontaneously if the invention is good enough," Yock stated. However, in some cases, it helps if the inventors can orchestrate a competitive situation by approaching more than one company at a time with their technology.

Another difference in the current medtech environment is that companies typically want to have more risks eliminated before licensing an idea. "I was extremely lucky to be able to license RX just off a prototype," Yock said. "Now you have to be prepared to take your invention further." Beyond clinical and device feasibility issues, he underscored the value of addressing other work streams in the biodesign innovation process – such as regulatory, reimbursement, sales – as a mechanism for eliminating risk for the licensee.

Of course, he pointed out, there are also potential risks to the licensor: "The biggest pitfall, I think, is that a company licenses your idea and does nothing about it.

At the time they license it, they may have every good intention of developing the technology. But if it doesn't stay on the priority list, nothing is going to happen. So having some form of milestone – a certain stage reached at a certain time – in the contract is really important, if you can get it." Examples of milestones include FDA approval or first commercial sale. Even in the case of RX, Yock had to stay closely involved through the first year or so of the licensing contract to urge the development process forward. He also acted as an evangelist of sorts to help garner **stakeholder** interest in the technology. According to Yock, "Initially, it just wasn't clear whether people were going to like it or use it. It seems strange in retrospect, but it was a major change of practice. We had push-back from unexpected places. For example, people were worried that if we took the trainees out of the picture, because it was a single operator system, it would be harder to develop the next generation of interventionalists." Yock and ACS were able to overcome these challenges and RX went on to become the dominant balloon angioplasty system used on a worldwide basis.

Specifically for partnership deals, there are other terms that must be considered in the development of the agreement, including those that provide a precise description of the near-term development activities to be undertaken, specific responsibilities and cash flows for each party, a list of milestone events and proposed timing, resources or activities to be provided by both parties, and a management/oversight structure. Usually upon the completion of a phase (e.g., development) or at the termination of the agreement, there may be a decision point for one or both parties regarding whether or not they will continue the partnership and, if so, what the next steps will be. It is possible for partnership activities to continue from one phase to another (e.g., from development into commercialization). However, in most cases, one of the parties assumes responsibility for taking the product into commercialization in predefined markets. Similarly, one of the parties assumes

responsibility for manufacturing. While arrangements such as these are relatively common, each agreement is unique, with the terms defined based on the interests of the partners and the reasons why a partnership makes sense to them.

A note on licensing when development occurs in an academic context

University innovators should keep in mind the implications for licensing when their inventions have been made in the context of their academic work. Typically, the title to any patentable inventions conceived or reduced to practice by faculty, staff, or students of the university is assigned to the university, regardless of the funding source.[29] If the innovator intends to participate in commercialization of the technology (by means of a start-up, for example), it will be necessary to obtain a license from the university.

Every major university has an office of technology licensing (**OTL** – also called the technology transfer office) that is responsible for managing its IP assets. Typically, the goals of an OTL are to promote the transfer of technology from the university into practice, as well as to generate income to support ongoing research and education at the institution. The important issue for innovators in a university setting to know is that they must disclose their inventions to the OTL and work collaboratively with the office to reach a licensing agreement *before* developing a stand-alone business *or* entering into a sublicense, acquisition, or partnership based on the technology in question. The Spiracur example demonstrates how innovators can reach mutually beneficial agreements with an OTL.

FROM THE FIELD ▷ **SPIRACUR**

Negotiating technology licensing with a university

As previously described, Moshe Pinto, Dean Hu, and Kenton Fong joined forces to address the need *to promote the healing of chronic wounds*. Because their invention of the SNaP® Wound Care System (see Figure 6.4.6) was developed while they were still students at Stanford, they needed to work through the university's OTL to submit their preliminary provisional patent application and negotiate terms surrounding the licensing of the invention.

In approaching these negotiations, Pinto emphasized the importance of "preparation, preparation, preparation." The Spiracur team believed that it ultimately achieved a positive outcome by carefully understanding the motivations and needs of the OTL and being prepared to offer mutually beneficial solutions. "To give you an example," Pinto said, "we had to negotiate a royalty rate, which the OTL told us usually ranged from 1–5 percent. Obviously Stanford wanted to start at the high end of this range, but we ended up closer to the 1 percent. What we did to justify that was build a *proforma* model with financial statements that showed the typical margins in this marketplace. We used competitive analysis from established players selling such technologies as comps and made a compelling case that we wouldn't be able to have a viable business if we paid such a high royalty rate. The people at OTL were reasonable. They looked at what we did and felt comfortable based on the data we presented to them that they would be justified in agreeing to a lower royalty rate."

Pinto also underscored the point that a university such as Stanford is not "an economic animal." "It has a broader interest in the community that it factors into its decisions," he said. "One thing that was clear to us was that the folks at the OTL were not interested in maximizing profits and making money for Stanford. They wanted to see the technology disseminated to basically make the world a better place. When we figured this out,

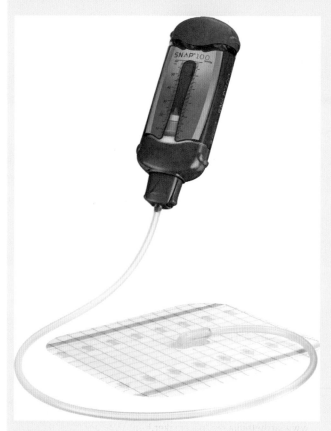

FIGURE 6.4.6
The SNaP® Wound Care System (courtesy of Spiracur).

we started thinking about how we could accommodate this interest while still maintaining the exclusivity of the IP." In the end, Spiracur proposed a creative solution under which it would make the technology available at cost to charitable organizations working in developing countries, such as the Bill & Melinda Gates Foundation, the Ashoka Foundation, and the United Nations Health Program. It would also offer the same arrangement to five African governments. This agreement would take effect in 2012, when the product was fully developed. "We know that every 30 seconds someone has a limb amputated because of a diabetic foot ulcer somewhere in the world, primarily in Africa," Pinto said. "But this is not our natural target market. These folks can't afford to buy the solution. So this agreement would satisfy the altruistic needs of Stanford and feel very good for us, without affecting our anticipated revenue stream for investors."

The key, underscored Pinto, is to recognize that the interests of the university and the interests of the team are not necessarily at odds. If you take a more collaborative approach, he noted, "You can find a middle ground."

⬎ Online Resources

Visit www.ebiodesign.org/6.4 for more content, including:

Activities and links for "Getting Started"
- Determine whether or not to pursue an alternate pathway
- Asses the best timing for a transaction and build value toward that goal
- Identify and evaluate targets
- Prepare to initiate contact
- Manage the process

Videos on alternate pathways

CREDITS

The editors would like to acknowledge Allan May for sharing the Nanostim story, Mike Glennon for providing the Vortex Medical example, and Moshe Pinto and Ken Wu for supporting the Spiracur case. Many thanks also go to Fred Khosravi of Incept LLC for his contributions in reviewing and editing the original chapter, and also to Ritu Kamal and Stacey McCutcheon for their assistance with the updates.

NOTES

1 Shreefal S. Mehta, *Commercializing Successful Biomedical Technologies* (Cambridge University Press, 2008).

2 While many licensing deals focus on patents and patent rights, it is important to recognize that technology development rights, product distribution rights, and brand names can also be licensed.

3 Mehta, op. cit.

4 "Cambridge Antibody Technology and Astrazeneca Announce Major Strategic Alliance to Discover and Develop Human Antibody Therapeutics in Inflammatory Disorders," Cambridge Antibody Technology press release, November 22, 2004, http://www.prnewswire.com/news-releases/cambridge-antibody-technology-and-astrazeneca-announce-major-strategic-alliance-to-discover-and-develop-human-antibody-therapeutics-in-inflammatory-disorders-75510212.html (March 22, 2014).

5 "AstraZeneca: CAT's in the Cradle," DataMonitor Research Store, May 16, 2006, http://www.datamonitor.com/store/News/astrazeneca_cats_in_the_cradle?productid=B4A56667-EBAA-45BD-AE84-12A7CABB362E (January 16, 2014).

6 Nanostim and St. Jude Medical are trademarks of St. Jude Medical, Inc. or its related companies. Reprinted with permission of St. Jude Medical, ©2014. All rights reserved.

7 All quotations are from interviews conducted by the authors, unless otherwise cited. Reprinted with permission.

8 "Treatments and Devices: Pacemakers," Arrythmia.Org, http://www.arrhythmia.org/pacemaker.html (January 20, 2014).

9 "St. Jude Announces Implant of World's First Leaderless Pacemaker in Germany," St. Jude Medical press release, http://www.sjm.com/~/media/SJM/corporate/Media%20Kits/

nanostim/Nanostim%20First%20Implant_Germany_SJM_
FINAL.ashx (January 9, 2014).

10 An application-specific integrated circuit (ASIC) is an integrated
circuit (IC) customized for a particular use, rather than intended
for general-purpose use; such as a chip designed solely to run a
cell phone. The one-time cost to engineer an ASIC can be in the
millions of dollars. For more information see: http://www.
princeton.edu/~achaney/tmve/wiki100k/docs/Application-
specific_integrated_circuit.html (January 20, 2014).

11 "St. Jude Medical Announces Acquisition and CE Mark
Approval of World's First Leadless Pacemaker," St. Jude
Medical press release, http://www.sjm.com/~/media/SJM/
corporate/Media%20Kits/nanostim/NanostimAquisition_
CEMarkApproval_FINAL_WEB.ashx (January 9, 2014).

12 "About PATH," PATH.org, http://www.path.org/about/
(January 16, 2014).

13 PATH"s Guiding Principles for Private-Sector Collaboration,"
PATH.org, http://www.path.org/publications/files/
ER_gp_collab.pdf (January 16, 2014).

14 Ibid.

15 Jeff Bailey, "PharmaJet Finally Gets Unstuck," Inc.com, October
1, 2010, http://www.inc.com/magazine/20101001/pharmajet-
finally-gets-unstuck.html (January 16, 2014).

16 Ibid.

17 From remarks made by Richard Gonzalez as part of the "From
the Innovator's Workbench" speaker series hosted by
Stanford's Program in Biodesign, June 9, 2006, http://
biodesign.stanford.edu/bdn/networking/pastinnovators.jsp.
Reprinted with permission.

18 Nelson Bennett, "American Venture Capital Firm Building New
Biotech in Vancouver," *Business Vancouver*, December 9, 2013,
http://www.biv.com/article/20131209/BIV0112/131209937/-
1/BIV/american-venture-capital-firm-building-new-biotech-in-
vancouver (March 3, 2014).

19 Ibid.

20 "AngioDynamics to Acquire Vortex Medical," AngioDynamics
press release, October 8, 2012, http://investors.angiodynamics.
com/releasedetail.cfm?releaseid=711973 (February 27, 2014).

21 Ibid.

22 From in-class remarks made by Jamey Jacobs as part of
Stanford's Program in Biodesign, Winter 2008. Reprinted with
permission.

23 Roger Fisher, William Ury, and Bruce Patton, *Getting to Yes:
Negotiating Agreements Without Giving In* (Houghton Mifflin
Books, 1991).

24 Ibid.

25 "Physician Payment Sunshine Act: Reporting Physician
Ownership," Policy and Medicine, July 25, 2013, http://www.
policymed.com/2013/07/physician-payment-sunshine-act-
reporting-physician-ownership.html (January 16, 2014).

26 "Guidelines for Avoiding Conflicts of Interest in Multi-center
Clinical Trials," National Heart, Lung, and Blood Institute,
National Institutes of Health, 2000, http://www.nhlbi.nih.gov/
funding/policies/coi-res.htm (December 8, 2011).

27 Jeffrey S. Grossman, *Innovative Doctoring* (Innovative
Doctoring, 2006).

28 Burt Cohen, "Four Minus Two Equals Two – Hands, That Is,"
The Voice in the Ear, October 5, 2005, http://www.ptca.org/
voice/archives/2005_10.html (January 16, 2014).

29 "Our Policies," Stanford University Office of Technology
Licensing, http://otl.stanford.edu/inventors/
inventors_policies.html (January 16, 2014).

Acclarent Case Study

STAGE 6: BUSINESS PLANNING

With strategies defined to address the many different aspects of Balloon Sinuplasty's development and readiness for commercial launch, Acclarent faced the challenge of integrating all of these plans into a unified vision of the business.

6.1 Operating Plan and Financial Model

At the core of Acclarent's planning efforts was the creation of an operating plan and financial model. According to Lam Wang, a company's operating plan and financial model "has to do with the culmination of all of your assumptions, and then how you intend to go about executing your plans."[1] In developing Acclarent's model, "we based a lot of it on prior experience," she said. However, the team also tried to benchmark proxy companies to validate its approach. "When you're trying to do something that's never been done before, like what we have set out to do in ENT, there really aren't a lot of models to look at," explained Facteau. "We had no idea what the adoption of the technology would be. In the early days, we struggled with how fast we should be training doctors, how fast we should expect them to turn into customers, and at what rate they should be reordering. We really had no peer data to look at."

Investigating companies outside ENT turned out to be more fruitful. "We probably looked at 10 companies: Fox Hollow, VNUS, Kyphon, Perclose, and others. Our CFO did a phenomenal job of laying out all their operating expenses and revenues," Facteau remembered. The team assessed these benchmarks from a time perspective (e.g., 0 to 3 years), but then also in terms of revenue thresholds ($20 million and under, $21–$50 million, etc.). These revenue bands allowed Acclarent to determine where (and when) the companies made major infrastructure investments and incurred certain costs. "Then we drilled down one more layer and looked at their business model. Is it similar? Is it disposables? Is it a market development type situation or is it a market share stealer?" Facteau added. "It wasn't perfect, but it gave us a way to plan and also to keep a report card of how we were doing as we moved forward. We also used this information to justify our operating plan and financial requirements to the board. We said, 'If this is the type of growth you're expecting from our company, these are the types of companies that have done it, and this is the investment that it took in order for them to get there. So if you believed those companies were a success, then you should be prepared to fund us at this level and we'll execute.'"

In terms of developing the model, Lam Wang described key aspects of the approach:

> We assumed it would be a surgeon training model, meaning that the way the product would be introduced to the field and generate revenue was through surgeon training. So I modeled that there would be national, regional, and local courses, meaning large, medium, small, and at each of these courses there would be so many doctors per course, per month. Once they were trained, we assumed their business would increase at a certain case ramp rate. The first month they were trained, we assumed they would do no cases, but then it would start to build until they reached some sort of steady state. To calculate a ramp rate, I looked at what were they doing at the time with FESS and then made some more assumptions, like they would be partly converting cases from conventional FESS and partly drawing from the pool of nearly 600,000 new patients to get to some steady state. After 12 months, they would reach a steady state. So that's how the model would grow. But then I added some reality factors, like surgeon productivity. Even in the best

case scenario, maybe only half of the surgeons would be actually performing the way they should. There was also some seasonality in chronic sinusitis that had to be taken into account. Then there was the competition factor. So for all the cases we thought we were going to get, we had to consider that we might lose some portion to new technology in the future. Taking all of these factors into account got us to a monthly number of cases, then we multiplied that by the $1,200 price to get to revenue per month. From there, we could estimate the expenses associated with growing that top line and the staffing that would be required.

Among other things, the financial model was used to determine how much money Acclarent needed to raise to get through the product's preliminary commercialization (approximately $14.5 million). Key near-term milestones in the company's operating plan included receiving FDA clearance for the balloon technology, launching a post-marketing clinical study, and building up its commercial sales force once it verified safe and efficacious results from this study.

Reflecting on the experience, Facteau commented that, "The model is only as good as the information that goes into it. Therefore, you really have to hone in and focus on the assumptions you're making. Challenge yourself to benchmark other companies. And be honest with yourself and your team's ability to execute."

6.2 Strategy Integration and Communication

To guide all important communications with investors, stakeholders, and other contacts, the Acclarent team members set about assembling and integrating information that would help them cohesively articulate the company's clinical need, the market opportunity, technology, and chosen IP, regulatory, reimbursement, and sales/training pathways, as well as how all of these pieces fit together. "These presentations included the main guts of a business plan," Facteau said. "But we never formally developed a plan or sent it out. Remember that we were in stealth mode this entire time, so all of our external communications were really targeted."

6.3 Funding Approaches

When it came to funding, Acclarent was in an enviable, although rare position. As part of ExploraMed II, the entity had received all of its seed funding from Explora-Med to get its efforts off the ground. When it was time to raise additional capital, Makower was able to tap into his established relationship with venture capital firm NEA relatively quickly. He and Facteau easily closed a Series A round of funding ($14.5 million) in January 2005. This was the only fundraising the company completed prior to the commercial launch of its technology. Investors Versant Ventures and NEA shared the round.

Importantly, the company's fundraising efforts corresponded to the completion of its first-in-human clinical study. The positive outcome of that trial was used to help pique investor interest and build confidence in the viability of the technology. Despite this careful timing, Facteau emphasized how fortunate the company was in so readily gaining access to the funds it required: "The credit has to go to Josh on this one." He continued, "At the end of the day, you need to find people who know what they're doing in the fundraising world. Whether you put them on your board, bring them into your company, or have some other type of mentor relationship with them, it's invaluable. You need someone to help you understand what it's going to take to raise money and what investors are going to look for. Josh played this role for us and it was really beneficial."

Into Commercialization: What Happened?

Approximately one month after speaking with representatives from the FDA by phone, Acclarent received notification that its balloon technology had received 510(k) clearance (see Figure C6.1).

Achievement of this important regulatory milestone set into motion a whirlwind of activities as the team began aggressively pressing forward toward a September 2005 commercial launch at the upcoming AAO-HNS conference.

One of the first things Acclarent did was formally launch its next clinical trial, the CLEAR study. Based on

DEPARTMENT OF HEALTH & HUMAN SERVICES

Public Health Service

APR 5 2005

Food and Drug Administration
9200 Corporate Boulevard
Rockville MD 20850

FIGURE C6.1

In April 2005, Acclarent received FDA clearance on the core components of its Balloon Sinuplasty system (provided by Acclarent).

ExploraMed II, Inc.
c/o William M. Facteau
President & CEO
2570 W. El Camino Real
Suite 310
Mountain View, CA 94040

Re: K043527
    Trade/Device Name: Relieva Sinus Balloon Catheter
    Regulation Number: 21 CFR 874.4420
    Regulation Name: ENT manual surgical instrument
    Regulatory Class: Class I
    Product Code: LRC
    Dated: March 24, 2005
    Received: March 25, 2005

Dear Mr. Facteau:

We have reviewed your Section 510(k) premarket notification of intent to market the device referenced above and have determined the device is substantially equivalent (for the indications for use stated in the enclosure) to legally marketed predicate devices marketed in interstate commerce prior to May 28, 1976, the enactment date of the Medical Device Amendments, or to devices that have been reclassified in accordance with the provisions of the Federal Food, Drug, and Cosmetic Act (Act) that do not require approval of a premarket approval application (PMA). You may, therefore, market the device, subject to the general controls provisions of the Act. The general controls provisions of the Act include requirements for annual registration, listing of devices, good manufacturing practice, labeling, and prohibitions against misbranding and adulteration.

the legwork it initiated immediately following its first-in-human results, the company already had several IRB approvals lined up and could immediately begin enrolling patients. According to Facteau, by the time of the commercial launch, "We already had phenomenal results, having treated 75 patients at nine sites with zero complications and a significant improvement on quality of life."

Facteau also gave the green light for hiring a sales team. His first move was to bring on a VP of sales and marketing, who joined in July 2005. "I really put him in a very difficult situation," Facteau recalled. "I said, 'Alright, go hire a training organization, a marketing organization, a sales management team, and a sales organization by September.'" By the launch, the company had added nine sales representatives, but was still in the process of assembling a complete team. To help these new members rapidly come up to speed, said Facteau, "We challenged everyone in the organization to observe cases [being performed as part of the CLEAR trial], to learn from each other, and to commit to the standardized approach so we could train the same way, every single time."

As the AAO-HNS conference approached and the team's launch plans began to come together, Facteau, Makower, and the team had to evaluate their stealth strategy and decide when (or if) the time was right to break their silence. Ultimately, they decided to contact a targeted list of key opinion leaders in the field prior to the launch. Makower explained:

We identified an important group of physicians in the specialty that we didn't want to surprise by our launch. We wanted them to be able to say to their peers that they knew about what we were doing if they were asked and ensure they wouldn't be caught off guard on the floor at the conference. We approached them a month in advance to explain the technology and share our clinical experience. Unfortunately, many of them were insulted by the fact that they hadn't been involved sooner and somewhat in disbelief that we had moved so quickly to be in a position to show up with clinical and cadaver data.

By the time the conference arrived, Acclarent was hit with "a tidal wave of interest." Chang described the first morning of the show:

The exhibit floor opened at 9:30 so we met around 8:30 at the booth to give a pep talk and to make sure we were all prepared. Our booth was in the very back of the hall, literally by the bathrooms and the concessions. We were placed as far away from the entrance as possible. At 9:30, we finished our huddle and I picked up my briefcase to move it out of the way. By the time I put it down and looked up there were at least 40 people at the booth already. It just hit me. Clearly these people had sought us out. They had not meandered through the exhibit hall. They wanted to come and find out exactly what was going on and made a bee-line for our booth.

The team faced crowds of interested physicians throughout the entire conference. It was not until the company was formally approached by representatives of the AAO-HNS that the team's excitement turned to other emotions. "There were people from the Academy and certain committees that came over to our booth and basically said 'Your products are not covered by existing codes. This is different, and we will publish a position statement recommending the use of miscellaneous codes,'" recounted Facteau. Under the use of miscellaneous codes, physicians and facilities would not receive any reimbursement for performing sinus surgery that involved the use of the Balloon Sinuplasty products.

Facteau described what happened next:

After that, the process just took on a life of its own. Within 60 days there was a dedicated task force that was formed by the president of the AAO-HNS at the time to evaluate our technology. Dr. David Kennedy – who was very influential and well connected in the field – headed up this task force, which included four or five other people. Dr. Kennedy had been one of the individuals responsible

for establishing the FESS codes and their values. In parallel, members of the Academy took a very aggressive approach and began calling all of the trained doctors on our website and telling them that they did not recommend the use of existing codes, which was unprecedented in ENT at the time. In response, we did everything that we thought we could do. We really tried to communicate with some of the leaders in the Academy to better understand their concerns and explain our position. We immediately set up a meeting with CMS to get its opinion on the matter. Fortunately, CMS said "We agree with you. This is a device. It's a tool that's used in sinus surgeries. This technology has been around in other specialties. We don't create new codes for devices and tools, per se, we do for new procedures. Whether you create an opening using a shaver, or back biter, or your balloon catheter, it doesn't matter. The codes are broad enough to interpret the use of any of those tools."

Acclarent also invested significant time in formulating responses to refute the concerns and educate its strategic advisors and trained physicians that their position on coding was sound (see Figure C6.2). "We spent a lot of money and resources on experts reviewing the language of the codes and the vignettes to make sure we weren't overlooking anything. After multiple confirmations that the existing codes apply, we were able to respond with conviction and not take 'no' for an answer," said Facteau. We could not allow a few politically motivated individuals to dictate our future, because it could have fundamentally killed the company. So we persevered and did what we believed was the right thing to do for our physicians, patients and the company.

At the crux of the Academy's argument was the concern that using FESS codes to reimburse sinus surgery performed using Balloon Sinuplasty products would potentially lead to the devaluation of FESS reimbursement. Some individuals perceived that FESS would be significantly easier to perform with the new tools, which would affect reimbursement work values. Opponents also speculated that the new technology would cause

FIGURE C6.2
Facteau (left), Bolger (center), and Dr. Fred Kuhn (right) review documents and discuss ideas on how to respond to the coding debate that ensued at the AAO-HNS in Los Angeles, 2005 (courtesy of Acclarent).

surgeons to operate on patients sooner than they really should. This could, in turn, affect utilization and cause the codes to be revalued.

Another common concern was that radiologists and cardiologists would begin performing sinus surgery. "Could that happen?" asked Facteau. "Maybe. But the likelihood is very, very small." He elaborated:

Our counter position was, "Listen, guys, this is great technology that's going to benefit a lot of patients. If you don't embrace it, others will." There are many examples where that has happened in medicine before. Cardiology is one. Cardiovascular surgeons could have been the ones performing angioplasty in the early days, but they made a conscious decision that they were surgeons, they cut tissue, they do surgery, and they opted not to embrace interventional techniques. As a result, there was an opportunistic group that developed – called interventional cardiologists – who were ready to take advantage of their decision. So we thought it was a valid argument, and so did our advisors. They said, "Sinus surgeons deserve to use this technology. We're the ones that need to embrace it. We need to research it. We need to study it. And we need to determine where it fits, not someone else." So, this

coupled with the fact that we were trying to build a dedicated ENT company eventually got us through that concern.

"We had to respond to every one of these arguments," Facteau continued. "We had to organize our thoughts so that our answers were consistent, well thought out, and well articulated as to why we didn't think these things weren't going to happen. We also spent a lot of time saying, 'This is technology that is going to make a positive difference in patient lives. We have to figure out a way to make it stay here.'"

Through this process, Acclarent's advisors and trained physicians became important evangelists – a role that Acclarent appreciated but did not initially expect them to play. "Our advisors probably found themselves defending Balloon Sinuplasty and Acclarent more than they ever wanted to. But they were just trying to do the right thing for the patient and the specialty as a whole."

According to Facteau, the task force to investigate Balloon Sinuplasty was formed in November 2005. Its position statement, dubbing the procedure as experimental and recommending the use of miscellaneous codes, was released in the Academy's January 2006 bulletin. "It had some pretty disparaging comments in there,

Sinus balloon catheterization is a surgical technique for the treatment of sinusitis, during which a wire-guided balloon catheter is inserted into the paranasal sinus and inflated in order to dilate the targeted ostium.

The board-approved policy declares that the evidence regarding the safety of sinus balloon catheterization has been supportive, and that balloon catheterization is a promising technique for the treatment of selected cases of rhinosinusitis. These include those without polyposis involving the frontal, sphenoid or maxillary sinuses either in conjunction with or in place of conventional instrumentation.

FIGURE C6.3
From the revised AAO-HNS position statement on Balloon Sinuplasty, issued March 19, 2007 (from the American Academy of Otolaryngology – Head and Neck Surgery).

as well, about the company from a marketing perspective," Facteau said. "It was pretty alarming and obviously very, very difficult to deal with at the time." Circumstances worsened for the company when *The New York Times* published a critical article in May 2006. In October 2006, the American Rhinologic Society (ARS – a subdivision of AAO) published its own unfavorable position statement on the technology.

The struggle and debate raged on, in both open and closed forums, at every sinus surgery meeting and committee meeting from that point onward. Despite the controversy, Facteau, Acclarent, and the doctors who believed in the technology and its place in sinus surgery did not give up. It was not until March 2007 that the physician users of Balloon Sinuplasty (numbering in the hundreds at that point) were finally successful in getting the AAO-HNS to overturn its position (see Figure C6.3).

The ARS followed in May 2007 with a positive statement of its own. These reversals in opinion were primarily driven by several thought-leading physicians who took it upon themselves to fight for the technology, as well the persistence of Acclarent and the strength of the clinical data that it continued to amass. In September 2006, the company presented the results of its CLEAR study at the AAO-HNS conference. It also hosted a symposium at that meeting, which included a live case demonstration (more than 200 doctors attended). Through ongoing clinical work, a focused publication strategy, and the sharing of information between physicians, Acclarent slowly began changing surgeons' minds.

However, the increasing acceptance of Balloon Sinuplasty among sinus surgeons had little impact on the attitude of insurance companies. While the professional societies initially had failed to support the balloon as medically necessary, most private payers flagged the technology as investigational, which essentially meant it would not be covered on grounds that there was insufficient clinical data to support it. Eventually, as the clinical evidence mounted, many began reimbursing sinus surgery performed with Acclarent's tools, but this was not without tremendous hours of effort, physician lobbying, and more painful persistence. But, said Facteau in 2008, "To this day we are still fighting to get certain private insurance companies, like the Blue Cross Blue Shield National Association and Wellpoint, to eliminate their investigational status. We've done over 50,000 procedures. 150,000 sinuses have been treated with our technology with virtually no complications associated with the technology, yet they still say it's investigational."

Reflections on Acclarent's past and future

Acclarent progressed through the biodesign innovation process and into the market at breakneck speed. According to Facteau, "We moved from concept to commercialization in 18 months, which is pretty unusual in the device industry." Reflecting on the company's experience, he said:

Being in the medical device business, I had anticipated that we would face challenges albeit I thought they would come from clinical, regulatory, or other technical workstreams – would the devices work, could we make them, and other issues that typically come with getting a device to market. But these things we sailed through surprisingly easily. The challenges came from other places – politics, adoption, reimbursement, and the societies. I just didn't anticipate those issues. The key was to respond to them with facts and data and stay focused on overcoming them. These issues definitely slowed us down. As a result, we had to raise more money than we initially anticipated, and make investments in areas that we hadn't contemplated.

When asked how the company's decision to adopt stealth mode may have affected the company's launch, he replied:

None of us had been in ENT, we had no relationships, we had really no idea what process we should follow to work with the American Academy of Otolaryngology and who the key players were. So, I think a lesson learned here is to understand the politics, understand the Academy and society's positions and process, and communicate. Get them involved, even if you're in stealth mode. Put some NDAs in place and be able to engage the key players. It could make your life a lot easier.

At the time the first edition of this text was published, approximately three years after Acclarent's commercial launch, the company had trained more than 4,000 physicians to use its technology and was rapidly building sales momentum. Fifteen peer-reviewed clinical publications had validated that Acclarent's technology was safe, effective, and offered significant quality of life benefits over common alternatives. Both AAO-HNS and ARS had endorsed reimbursement of the technology under existing FESS codes, and CMS as well as many private insurance companies were consistently reimbursing the tools when used in sinus surgery. Beyond the US, the company was selling into 45 other countries around the world. With a team of nearly 300 employees, Acclarent also had developed a significant pipeline of products to support its organic growth for years to come.

Despite these promising initial results, "We still have a long way to go to get this to be the standard of care," said Facteau. As of 2008, Balloon Sinuplasty had achieved less than 10 percent penetration into the existing surgical market. "There are still a lot of physicians that we need to train and that we need to move along the adoption continuum to get them to use this more frequently and think about this as the standard of care for treating chronic sinusitis." The company was also still working to become profitable.

However, Facteau, Makower, and Chang were confident about Acclarent's ability to overcome these challenges and become a leader in the ENT field. Looking back across the arduous journey of the biodesign innovation process, Facteau commented:

Do not underestimate the importance of people. At the end of the day, it is probably the most important thing that will make or break a company. My father used to tell me, "You can't coach speed. You can either run fast or you can't." I think you can apply a similar principle with people and business. You can't teach people to get out of bed in the morning, they either have the drive or they don't. You have to find the ones with the right work ethic and the right mind set. If you get people on your team that have conviction, believe in the product, and have the desire, the work ethic, and the heart, to do extraordinary things – that's the piece that makes great companies.

With a strong team intact, Acclarent was moving boldly toward its goals.

In fact, just two years later, in December 2010, Johnson & Johnson's surgical care company, Ethicon, announced that it would acquire Acclarent in a transaction valued at $785 million. The deal closed in January 2011, after receiving approval from the US Federal Trade Commission. At the time the second edition of this text was published, more than 500,000 patients around the world had been treated with Balloon Sinuplasty and more than 1,000,000 sinuses had been dilated using the system. In addition, Balloon Sinuplasty was documented in 77 peer-reviewed publications, including 32 clinical studies (eight Acclarent supported), 19 technique/cadaver case studies (three Acclarent supported), and 29 reviews/commentaries (two Acclarent supported).[2]

CREDITS

Many thanks go to William Facteau, John Chang, Sharon Lam Wang, and Earl "Eb" Bright for their substantial contributions to the development of this case.

NOTES

1 All quotations are from interviews conducted by the authors, unless otherwise cited. Reprinted with permission.

2 Figures exclude publications with off-label use (ET dilation, choana, fracture reduction, etc.) and include foreign-language papers only when an English translation is available.

About the Author Team

This book carries the fingerprints of literally hundreds of contributors. Although the list of individuals who have helped shape the Biodesign Program, our process, and this book has become far too long to include within the text, we hope you all know how much we appreciate your expert contributions and unwavering support. Our heart-felt appreciation goes out to the entire community of students, fellows, staff, faculty, industry professionals, and sponsors – many of whom are listed online – that have so generously shared with us their time, resources, and wisdom.

Writing the second edition of the book turned out to be a much bigger project than we originally anticipated. But so much had changed within the medtech industry since the first edition was published that we ultimately found the rewrite to be both essential and rewarding. The experts who had large roles in the revision of specific chapters are included in the credits at the end of each chapter. In addition to these folks, we wanted to mention some of the people who assisted with content across multiple chapters. First, we'd like to thank the members of the informal working group that helped us shape our position on healthcare value: Laurence Baker, Ronnie Chatterji, Vic Fuchs, John Hernandez, Doug Owens, Jan Pietzsch, Bob Rebitzer, Gordon Saul, and Chris Wasden. We'd also like to acknowledge the core team that helped with the greater globalization of the text: Robson Capasso, Raj Doshi, Fumi Ikeno, Ritu Kamal, Anurag Mairal, Jan Pietzsch, and Chris Shen. And, we offer our great appreciation to those who helped extend our capacity and/or capabilities: Ritu Kamal and Stacey McCutcheon (case writing), Erika Palmer (chapter review), and Jean Zambelli (graphics).

Finally, the core team behind the second edition (see Figure A1) would like to thank the Biodesign staff for keeping things afloat while we devoted so much of our time and energy to the book: Mary Gorman, Justina Kayastha, Maria Kelly, Linda Lucian, Andrea Mattison,

and Chris Queen. The 2013–2014 Biodesign Fellows and visiting faculty also deserve recognition for their suggestions, inspiration, and willingness to play a "starring role" in the creation of the new Biodesign video library: Varun Boriah, Tiffany Chao, Pranav Chopra, Nicolas Damiano, Rena Dharmawan, Prusothman Sina Raja, Deevish Dinakar, Ryan Crone, Sohail Gupta, Neeraj Kumar, Yujiro Maeda, Shreya Mehta, Kathryn Olson, Andy Rink, Holly Rockweiler, Jonathan Steinberger, Benjamin Chee Keong Tee, Celia Yao Wang, John Woock, and Masakazu Yagi.

Thank you again to the hundreds of people who truly comprise the *Biodesign* "author team." We couldn't do it without you!

Author biographies

Paul Yock is the Weiland Professor and Founding Co-Chair of the Stanford Department of Bioengineering, with a joint appointment in Cardiovascular Medicine and courtesy appointments in Mechanical Engineering and Operations, Information, and Technology in the Graduate School of Business.

Dr. Yock is internationally known for his work in inventing, developing, and testing new devices, including the Rapid Exchange™ angioplasty/stent system, which is now the primary system in use worldwide, and the Doppler-guided access system known as the Smart Needle™ and PD-Access.™ He authored the fundamental patents for intravascular ultrasound imaging and founded Cardiovascular Imaging Systems, acquired by Boston Scientific. Dr. Yock has co-founded several other medical technology companies.

In his academic career, Dr. Yock has authored over 300 peer-reviewed publications, chapters and editorials, two textbooks, and over 45 US patents. Recent awards include the Transcatheter Therapeutics (TCT) Career

The core team behind the second edition textbook. From left to right: Paul Yock, Tom Krummel, Uday Kumar, Josh Makower, Christine Kurihara, Todd Brinton, Jay Watkins, and Lyn Denend (photo by Justina Kayastha).

Achievement Award and the American College of Cardiology Distinguished Scientist Award. Dr. Yock is a member of the National Academy of Engineering. He also founded and directs the Program in Biodesign, a unit of Stanford's Bio-X initiative that focuses on invention and technology transfer related to biomedical engineering.

Stefanos Zenios is the Charles A. Holloway Professor at the Graduate School of Business, Stanford University and the director of its Center for Entrepreneurial Studies. An innovative educator, he was the first to introduce courses on the interface between medicine, engineering, and management in the MBA curriculum, and he is the lead architect of Startup Garage, a popular experiential elective on forming new start-ups. His pioneering research on maximizing the benefits of medical technology to patients when resources are limited has influenced policies in the US and Europe. Dr. Zenios is the co-founder of Konnectology.com, a website funded by the National Institutes of Health to help kidney patients find transplant centers. He received a BA /MA in Mathematics from Cambridge University and a PhD in Operations Research from MIT.

Josh Makower Josh Makower has dedicated his life to the creation of medical technologies that improve the quality of life for patients. He is the CEO and Founder of ExploraMed Development, LLC, a medical technology incubator based on the West Coast. He is also a Venture Partner with New Enterprise Associates, where he supports investing activity in the medical device arena. He serves as a Consulting Professor of Medicine at Stanford University Medical School and co-founded Stanford's Biodesign Innovation Program.

Dr. Makower has founded several healthcare companies through the ExploraMed incubator that have achieved successful M&A transactions, including Acclarent, Inc., a company focused on developing novel therapies in ENT, which was acquired by J&J in 2010, TransVascular, Inc., a company focused on the development of a completely catheter-based coronary bypass technology, which was acquired by Medtronic, Inc. in 2003, and EndoMatrix, Inc., a company focused on the development of a novel therapy for urinary incontinence and GI Reflux, which was acquired by C.R. Bard in 1997. Up until 1995, Dr. Makower was Founder and Manager of Pfizer's Strategic Innovation Group, a group chartered to create new medical device technologies and businesses for Pfizer's medical device businesses.

Dr. Makower serves on the Board of Directors for NeoTract, Inc. (urology), Moximed, Inc. (orthopedics), Intrinsic Therapeutics, Inc. (spine), ExploraMed Development, LLC, Ceterix, Inc. (sports medicine), and Coravin, LLC (wine technology). He holds over 200 patents for various medical devices in the fields of orthopedics, ENT, cardiology, general surgery, drug delivery, and urology. He holds an MBA from Columbia University, an MD from the New York University School of Medicine, and an SB in Mechanical Engineering from the Massachusetts Institute of Technology.

Todd J. Brinton is a Clinical Associate Professor of Medicine (Cardiovascular) and Consulting Associate Professor of Bioengineering at Stanford University. He is also an interventional cardiologist at the Stanford University Medical Center and the Palo Alto VA Hospital. His clinical practice focuses on general cardiovascular disease and complex coronary interventions. He is also a clinical investigator for new interventional-based therapies for coronary disease and heart failure. He has served as the Fellowship Director for the Biodesign Program since 2006, through which he has mentored numerous innovators in the biodesign innovation process. He also serves as the co-director of the graduate courses in Biodesign Innovation and the Biodesign Executive Education Program at Stanford. Dr. Brinton completed his medicine, cardiology, and interventional training at the Stanford Medical Center. He holds an MD from the Chicago Medical School and a BS in bioengineering from the University of California, San Diego. He is co-founder of BioParadox and Shockwave Medical, both venture-backed medical device companies, in which he serves on the board of directors and directs clinical development & strategy. He is a board director for both Infogard Laboratories and Qool Therapeutics, and on the advisory board for a number of other early-stage medical device companies. Prior to medical school, he was the Clinical Research Director for Pulse Metric, Inc., a medical device start-up company.

Uday N. Kumar is the Founder, President, and CEO of Element Science, Inc., a company focused on the treatment of sudden cardiac death, that he started during his time as an Entrepreneur-in-Residence at Third Rock Ventures, LLC. He also is a co-founder and board member of Qurious.io, Inc., a healthcare information technology company, a co-founder and board member of Sympara Medical, Inc., a company developing a novel therapy for hypertension, and the founder of iRhythm Technologies, Inc., a company focused on developing cost-effective new devices and systems for cardiac rhythm monitoring. He served as a board member and Chief Medical Officer of iRhythm from founding through broad commercialization of its Zio® Patch cardiac monitoring device.

Dr. Kumar currently serves as a Consulting Associate Professor of Bioengineering at Stanford University and is the Fellowship Director of Stanford's Global Biodesign Programs. He completed a Biodesign Innovation fellowship at Stanford, a cardiology and cardiac electrophysiology fellowships at the University of California, San Francisco (UCSF), internal medicine residency at Columbia University, and his medical and undergraduate education at Harvard University. Prior to his medical training, he was also Chief Medical Officer and Vice-President of Biomedical Modeling Inc., a company applying 3D printing to healthcare.

Jay Watkins has extensive experience founding and funding healthcare companies. He is a Managing Director with De Novo Ventures, and an active individual investor and advisor to emerging medtech companies. He serves as a Lecturer in Management at the Graduate School of Business, Stanford University.

Mr. Watkins was co-founder and founding CEO of Origin Medsystems, a venture funded medical technology start-up acquired by Eli Lilly & Company. After Lilly formed Guidant, he joined the Management Committee and served as president of several divisions, including the Minimally Invasive Surgery Group, and Heart Rhythm Technologies. While President of the Cardiac and Vascular Surgery Group, he initiated the development of a minimally invasive vein harvesting technology, which has been used to treat almost two million patients worldwide. He also co-founded Gynecare, which was acquired by Johnson & Johnson. At Guidant, he formed and led Compass, Guidant's corporate business development and new ventures group where he initiated venture investments in fourteen companies, including Impella (acquired by Abiomed) and Intuitive Surgical (NASDAQ: ISRG).

Prior to founding Origin, Mr. Watkins held management positions in several start-ups, including Microgenics

Corporation (acquired by Boehringer Mannheim), and was a consultant with McKinsey & Company. He has served as a board member for both private and public medtech companies and has been an advisor and faculty member for the Kauffman Labs program. Mr. Watkins received his MBA from Harvard Business School and his undergraduate degree from Stanford University.

Lyn Denend is the Associate Director for Curriculum of the Stanford Biodesign Program. In this position she is responsible for the development of written and multimedia curricular materials to support Biodesign courses and initiatives.

Previously, Ms. Denend worked at the Stanford Graduate School of Business (GSB) as the staff director for the school's Program in Healthcare Innovation, where she specialized in research related to the challenges of global health innovation. She also authored a variety of papers and teaching materials as part of the GSB case writing office. Prior to joining the Stanford community, Ms. Denend was a management consultant with Cap Gemini Ernst & Young. She has an MBA from Duke's Fuqua School of Business and a BA in Communications from UC Santa Barbara.

Thomas M. Krummel is Professor/Chair, Department of Surgery at Stanford University. He has served in leadership positions in all of the important surgical societies, has mentored over 200 students, residents, and post docs, and is the recipient of more than $3 million in research grants. He is Co-Director of the Stanford Biodesign Program.

Dr. Krummel has been a pioneer and innovator throughout his career. He has received two Smithsonian Information Technology Innovation Awards for work in the application of information technology to simulation-based surgical training and robotics. He remains an active start-up consultant with three successful exits and nine in the pipeline.

Christine Q. Kurihara, Senior Associate Director of Global and Communication, oversees the Biodesign Global Fellowship Programs at Stanford University and is the primary point of contact for all global relationships. She is also responsible for efforts that support Stanford students, fellows, and faculty working on device projects based on global needs.

Ms. Kurihara oversees IT and web projects and is responsible for communication and marketing for the program. She manages several websites for Biodesign including biodesign.stanford.edu; ebiodesign.org, the companion website for this textbook; bme-idea.org, for the BME Academic community and indiabiodesign.org, a social networking site for persons interested in the development of the medical device industry in India.

Ms. Kurihara joined Biodesign after an 11-year career with Stanford in the area of media services. In her previous role, she spearheaded media development efforts for an on-campus service unit, where her team produced websites, online courseware, video and broadcast products. Prior to Media Solutions, she was the first coordinator of the Stanford University website. In 1997, Ms. Kurihara was co-chair of the Sixth International World Wide Web Conference. Prior to Stanford, she worked for The Aerospace Corporation for twelve years managing Computer-Aided Engineering development. She has a Bachelor's in Mathematics from the University of California, Los Angeles.

Image Credits

Glossary

21 CFR 820   The US Code of Federal Regulations, Title 21, Part 820 that sets forth FDA's Quality System Regulation.

510(k)   One of several pathways for medical devices through the regulatory process at the FDA. This pathway is used when similar devices are already in use.

ACA   The Patient Protection and Affordable Care Act, which was signed into law in 2010 in the United States. Among other reforms, the law expands Medicaid eligibility, establishes health insurance exchanges, and prohibits health insurers from denying coverage due to preexisting conditions.

Acceptance criteria   Conditions used to determine whether a project offers a good fit for the innovator. Used as part of choosing a strategic focus.

ACO   Accountable Care Organization. ACOs are groups of US doctors, hospitals, and other healthcare providers who voluntarily coordinate care for a defined population of Medicare recipients. ACO payment models seek to link provider reimbursement to quality metrics and reductions in the total cost of care.

Acquisition   A transaction in which the seller of the property (technology, IP, company) relinquishes complete control of the property to the acquirer.

AdvaMed   Advanced Medical Technology Association. The advocacy group for medical device companies in the US.

AIA   The Leahy-Smith America Invents Act. AIA is a federal statute signed into law in 2011. Among other changes, it moved the US from a first-to-invent to a first-to-file system for patent applications.

AIMDD   Active Implantable Medical Device Directive 90/385/EEC. One of the key regulatory approval directives used in the European Union.

AMA   American Medical Association. The primary association of physicians in the United States. The AMA controls the issuance of new CPT codes.

Angel investor   Experienced individual investors who use their own wealth to fund start-up companies. Angel investors may be organized in groups.

APC   Ambulatory Payment Classification. A method of paying for facility outpatient services for Medicare recipients in the US. APCs are part of the Outpatient Prospective Payment System and applicable only to hospitals.

Arm   Any of the treatment groups in a randomized trial. Most randomized trials have two arms, but some have three or even more (see Randomized trial).

ASQ   American Society of Quality.

BATNA   Best alternative to a negotiated agreement. The course of action that will be taken if a negotiation fails to lead to an agreement.

BCBS   Blue Cross Blue Shield. Health plans that operate in various regions in the US. The BCBS Association is a trade group that, among other activities, helps establish guidelines for reimbursement.

Bench testing   Testing prototypes (materials, methods, functionality) in a controlled laboratory environment (not in animals or humans).

Bias   When a point of view prevents impartial judgment on issues relating to the object of that point of view. In clinical studies, blinding and randomization control bias.

Biocompatibility   The property of a material that indicates that it is suitable to be placed in humans.

Blind trial   A trial in which neither the members of the patient group nor any participating doctors, nurses, or data analysts, are aware of which treatment or control group the patients are in.

Blue-sky need   A large-scale need that would require major new medical or scientific breakthroughs and/or significant changes in practice.

Bootstrapping   Financing a venture without the use of equity investments, grants, or loans taken by the company. In many cases, innovators use their own savings, personal loans, a second mortgage, or credit card debt to bootstrap a company.

Bottom-up model   A sales forecast model that uses a series of detailed sales factors, including sales cycle, adoption curve, hiring effort, commercial effort, etc. to predict future sales.

Brainstorming   A cooperative problem-solving technique that involves the spontaneous contribution of ideas from all participants in the session.

Bridge loan   An interim debt financing option available to individuals and companies which can be arranged relatively quickly and spans the period of time before additional financing can be obtained.

Budget impact model   A model for determining product value that examines the cost and treatable population within a health plan, as well as the expected annual cost to the plan for covering a device.

Build-to-buy When a company intentionally designs a technology that will appeal to one or more corporate entities and develops it in a highly capital-efficient manner to maximize the opportunity for achieving an exit (as well as maximizing investor returns).

Bundled pricing Setting a single price for a combination of products and/or services.

Burdened cost The total cost of an employee, including salary, benefits, associated overhead, and fees.

CAB Conformity Assessment Body. An entity that determines compliance to ISO 13485.

CAC Carrier Advisory Committee. A committee that performs a review of all US local coverage decisions through Medicare.

CAF Contracting Administration Fee. The fee that a Global Purchasing Organization will charge for managing the purchasing contracts for many end users, paid by the manufacturer.

CAGR Compound Annual Growth Rate. The annual growth rate for an investment.

CAPA Corrective and Preventive Actions. One subsystem of a quality management system. The system to implement corrections upon and to avoid future problems in quality control.

Capability-based advantages An advantage over competitors that is driven by a company's capabilities. This type of advantage is based on the ability to do something better or less expensively than the competition and/or customers.

Cash flow statement An accounting statement that shows the cash that flows to the company in each period (typically quarterly) minus the cash that flows out in the same period.

CBER Center for Biologics Evaluation & Research. The part of the FDA that approves biologics.

CDER Center for Drug Evaluation & Research. The part of the FDA that approves drugs.

CDRH The center within the FDA responsible for medical device regulation.

CE Mark Resulting "mark" that is given to a device in the EU to indicate regulatory approval.

CFDA China Food and Drug Administration. The CFDA and its local counterparts are the primary regulatory authorities in China that apply and enforce laws and regulations concerning medical devices. Sometimes also referred to as the State Food and Drug Administration (SFDA).

Class I Classification of a medical device by the FDA that indicates low risk to a person.

Class II Classification of a medical device by the FDA that indicates intermediate risk to a person. Class II devices are typically more complex than Class I devices but are usually non-invasive.

Class III Classification of a medical device by the FDA that indicates highest risk to a person. Class III devices are typically invasive or life sustaining.

Clinical investigator A medical researcher in charge of carrying out a clinical trial protocol.

Clinical protocol A study plan on which all clinical trials are based. The plan is carefully designed to safeguard the health of the participants, as well as answer specific research questions. A protocol describes what types of people may participate in the trial; the schedule of tests, procedures, medications, and dosages; and the length of the study.

Clinical trial A research study performed to answer specific questions about diagnoses or therapies, including devices, or new ways of using known treatments. Clinical trials are used to determine whether new treatments are both safe and effective.

CME Continuing Medical Education. Additional training required to maintain a license for physicians and others in healthcare-related fields.

CMS Centers for Medicare and Medicaid Services. The primary government payer of healthcare charges for the elderly and disabled.

Coding The process of assigning a specific, identifiable code to a medical procedure or process.

COGS Cost of goods sold. Raw materials costs for a product.

Common stock Equity in a company that confers on shareholders voting and preemptive rights (the right to keep a proportionate ownership of the company by buying additional shares when new stock is issued).

Comparables analysis Evaluating the pricing strategies (and associated reimbursement status) of similar offerings in the field.

Concept An idea or solution for how an unmet need can be addressed.

Concept map A diagram that depicts relationships between ideas. Concept maps are often used to organize and structure information generated through ideation. Also called a mind map.

Controlled trial A trial that uses two groups: one that receives treatment, and a second, control group, that does not, in order to compare outcomes.

Convertible debt A hybrid debt-equity alternative for companies seeking financing and their investors. A type of debt instrument that can be converted into shares of stock of the issuing company, usually at some pre-determined rate.

Copayment Out-of-pocket payments made by patients every time they access medical services, as stipulated in their health insurance policy. Copayments usually represent a small portion of the actual cost of the medical service and are intended to prevent people from seeking unnecessary medical attention.

Core laboratory   Laboratories that analyze data from a clinical trial; these laboratories often have specialized equipment and expertise.

Corporate investment   When corporations invest in new companies by: (1) the purchase of equity in support of a research and development or a licensing agreement, or (2) traditional venture investments.

Correction   Repair or modification of a distributed product while it is still under the control of the manufacturer.

Cost-effectiveness model   A model for determining product value where cost is expressed per unit of meaningful efficacy, usually used comparatively across interventions.

Cost-utility analysis   A model for determining product value where cost is assigned for quality of life and years lived. It is based on clinical outcomes measures related to quality of life and/or disability and mortality.

Cost analysis   A model for comparing the cost of a new treatment to the money spent on competing treatments over time.

Cost projections   Provides an integrate view of all of a company's projected costs, including salaries, facilities, equipment, supplies, and COGS from product manufacturing.

Coverage   Establishes whether or not a technology or procedure will be reimbursed, and under what conditions.

CPT codes   Current Procedural Terminology codes. Codes used to classify medical procedures in a standard way so that the same procedure is reimbursed in the same way across all facilities. Also known as HCPCS Level 1 codes.

CRO   Contract (or clinical) research organization. An independent organization that provides management services for clinical trials.

Cycle of care   A description of how a patient interacts with the medical system.

Debt funding   Funding that is repaid with interest. A loan.

Design controls   One subsystem of a quality management system. Controls that ensure the device being designed will perform as intended when produced for commercial distribution.

Design creep   Ongoing, minor changes in developing a device that can lead to significant delays and issues with intellectual property and regulatory clearance.

Design history file   Also known as a DHF, the design history file is the complete collection of documentation that describes the design history of a finished medical device.

Design validation   Ensuring that a design does what it has intended to do.

Design verification   Ensuring that a design meets product specifications.

Determination meeting   A formal meeting with the FDA to request approval of the design for a clinical study.

Differential pricing   Pricing the same product or service differently for different customer segments, e.g., discounts for large buyers.

Dilution   Section of a term sheet that stipulates how conversion prices will be calculated if future rounds of financing are dilutive to preferred shareholders' holdings (i.e., they reduce the total value of the shareholder's ownership stake in a company).

Direct sales model   Hiring a sales force within a company to sell to customers directly.

Distribution strategy   A strategy that focuses on product breadth and channel relationships rather than on product superiority.

Dividend   A payment to the shareholder that is proportional to the shareholder's ownership of a company.

Divisional   A type of patent application that claims a distinct or independent invention based upon pertinent parts carved out of the specification in the original patent.

DRG   Diagnosis Related Group. A set of codes that are grouped together by diagnosis; used specifically for coding hospital-related billing for patient encounters. Also referred to as MS-DRG.

DSMB   Data Safety and Monitoring Board. An independent body that reviews results of clinical trials in the US.

DTC   Direct-to-consumer. A type of marketing that targets the end user of a product, as opposed to the physician or other medical professional.

DTP   Direct-to-Patient. A type of marketing that targets patients directly, as opposed to physicians or other healthcare professionals.

Due diligence   An iterative process of discovery, digging into detail about the various elements of a start-up company's business plan or licensing opportunities.

Earn-out   An acquisition in which additional payments are made to the seller after the sale day if the acquired company reaches pre-specified milestones. See Structured acquisition.

Efficacy endpoints   A result during an animal or clinical study that demonstrates efficacy (i.e., a therapeutic effect). Endpoints are what a study is designed to prove.

Empathy   The ability to share someone else's feelings or to vicariously understand their thoughts, feelings, attitudes, and beliefs.

Epidemiology   Study of factors affecting the health and illness of a population that are used as the basis of making interventions in the interest of public health.

Episode of care   All clinically related services provided to treat a selected condition over a defined period of time.

EPO   European Patent Office. The office that provides unified patent filing for 38 European countries.

Equity funding   Funding in which the investor provides a cash infusion to a company and in exchange obtains equity in the company.

Ethnographic research   Understanding a particular culture or way of life by studying the members of that culture or group.

Evergreening   The process of introducing modifications to existing inventions and then applying for new patents to protect the original device beyond its original 20-year term.

Evidence-based   Treatments, guidelines, and processes based on the results and outcomes generated from experiments and observation, which use specific evidence of outcomes and suggest treatment or processes based on such evidence.

Exclusion criteria   Characteristics or contraindications that eliminate subjects from participating in a clinical study.

Exclusive rights   The rights of the inventor or group of inventors, who have been issued a patent on an invention, to be the sole person(s) creating and marketing that invention.

Exclusive license   A license that grants only the licensee (and not even the licensor) the right to use a technology.

Exit   When a company is either acquired or has an IPO.

Expansion funding   Funding required to ensure completion of clinical trials, initiation of additional trials, or initial product launch. Such funding is often acquired through VCs or corporate investment.

Facility and equipment controls   One subsystem of a quality management system. It ensures, in part, that standard operating procedures have been designed and implemented for all equipment and facilities.

Fast follower strategy   A company that leverages its own corporate advantages to quickly capture market share from the first mover.

FDA   Food and Drug Administration. The primary US regulatory agency, whose oversight includes drugs, diagnostics, and medical devices.

Field of use   A licensing option that allows an existing patented device to be used within a restricted domain, such as one clinical area.

First-in-human   FIH. The first time a device or technology is used in a human subject. Also sometimes called first-in-man (FIM).

Financial model   A detailed numerical articulation of a company's costs and revenue over time. It tracks both the cost of developing the innovation and bringing it to the market as well as market revenue, and it follows these costs and revenue over a period of five to seven years.

First mover strategy   An attempt by a company to be first to market with any innovation.

Flow of money analysis   Analysis aimed at identifying key stakeholders that is focused on payments to providers of healthcare services.

Freedom to operate   The ability to commercialize a product, without infringing on the intellectual property rights of others.

Gainsharing   When hospitals negotiate reduced prices with certain manufacturers in exchange for increased volume.

GCP   Good Clinical Practices. Guidelines from the FDA that outline specific standards for holding clinical trials.

GLP   Good Laboratory Practices. A system of management controls for laboratories that assures consistent and reliable results.

GMP   Good Manufacturing Practices. Formerly used by the FDA to promote quality; replaced by Quality Systems Regulation (QSR).

GPO   Global purchasing organization. An organization that brings together multiple hospital groups, large clinics, and medical practices into buying cooperatives.

HCPCS Level I   Common Procedure Coding System Level I. Codes used to classify medical procedures in a standard way so that the same procedure is reimbursed in the same way across all facilities in the US. Also known as Current Procedural Terminology (CPT) codes.

HCPCS Level II   Common Procedure Coding System Level II. Coding for supplies and services obtained outside the physician's office that are not covered by a CPT code in the US.

HDE   Humanitarian Device Exemption. An exemption to the normal regulatory pathways for a medical device that is intended to benefit patients in the treatment or diagnosis of a disease or condition that affects or is manifested in fewer than 4,000 individuals in the US per year.

HIPAA   Health Information Portability and Accountability Act. Ensures comprehensive protection of patient health information (PHI) in the US.

HCUP net   Healthcare Cost and Utilization Project. A website with data about healthcare cost and utilization statistics (e.g., hospital stays at the national, regional, and state levels).

Hypothesis   A supposition or assumption advanced as a basis for reasoning or argument, or as a guide to experimental investigation.

IACUC   Institutional Animal Care and Use Committee. A committee that institutions must establish in order to oversee and evaluate animals used for trials.

ICD-9 codes   International Classification of Diseases, 9th edition codes. Codes for classifying patient diagnoses, developed by the World Health Organization and then adapted for use in the US in the form of ICD-9-CM.

ICD-10 codes   International Classification of Diseases, 10th edition codes. The US adaptation of this code set has two parts-ICD-10-CM for diagnoses and ICD-10-PCS for procedures – effective as of October 2014.

IDE   Investigational Device Exemption. An exemption granted to a hospital or doctor by the FDA that allows the hospital or doctor to use a device prior to its regulatory approval, usually as part of a trial.

IDN   Integrated Delivery Network. An organization that aggregates hospitals, physicians, allied health professionals, clinics, outpatient facilities, home care providers, managed care, and suppliers into a single, closed network.

IFU   Indications for Use. Instructions on how to use a device. Mandated by the FDA. Typically a package insert.

Illiquid   Not liquid (e.g., stock or other property that is not easily sold or converted to cash).

Incremental need   A need is focused on addressing issues with, or making modifications to, an existing solution, such as the function of a device or other technology.

Inclusion criteria   Characteristics or indications that subjects must have in order to participate in a clinical study.

Income statement   Subtracts all costs from revenue to generate an accounting income statement.

Incubator   Small companies that specifically serve to develop a need or concept at the early stages. An incubator may incubate multiple device concepts for a significant period of time. Successful products may result in the spin out of a company from the incubator into a stand-alone entity.

Indirect sales model   A sales and distribution agreement with an existing distributor, or forming a third-party partnership with another manufacturer.

Information aggregator   A company that seeks to connect solutions across the cycle of care, amass relevant information, and make the analysis of that data possible for healthcare stakeholders.

Informed consent   Consent by a research subject that indicates they are fully aware of all aspects of the trial prior to participating, including both the risks and potential benefits.

Innovation notebook   A notebook in which an innovator documents each aspect of the invention. This notebook may be used in infringement trials to prove inventorship.

Intrapreneur   A person within a company who is tasked to develop new products or business models – an internal entrepreneur.

IPAB   The Independent Payment Advisory Board (IPAB) has the authority to make changes in the Medicare payment rates and program rules without prior approval of Congress.

IPO   Initial Public Offering. The first offering of a company's stock for public sale.

IRB   Institutional Review Board. An internal committee that monitors clinical trials to ensure the safety of human subjects.

ISA   International Searching Authority. The organization that performs patent searches as part of an international patent filing.

ISO   International Organization for Standardization. A non-governmental network of national standards institutes that establishes standards of quality. The name ISO is not an acronym but rather based on the Greek word *isos* meaning equal.

ISO 13485   International Standards Organization Certification 13485. The European Union's Quality System (compare to QSR).

ISO 9001   International Standards Organization Certification 9001. A quality certification in use around the world between the 1980s and 1990s.

IVMDD   In Vitro Diagnostic Medical Device Directive 98/79/EEC. One of three regulatory approval directives used in the EU.

KOL   Key Opinion Leader. Physicians and others in the Medical Device arena who are often consulted when new devices are readying for the market.

LCD   Local coverage determination. One of two types of reimbursement determinations made by Medicare that provides guidance on national reimbursement coverage. Typically applies to payments for inpatient services. LCDs are decisions made by one of 28 Medicare contractors and apply only to the contractors' area of coverage.

Lexicographer   An inventor may use his/her own language and definitions in a patent application, thus becoming a lexicographer.

Licensing   One option in getting a technology to market by transferring the rights to the technology from the innovator to a licensee in exchange for ongoing royalties and/or other payments.

Liquidity event   The transaction that enables an investor to receive cash in exchange for its equity stake in a company. Also referred to as exit events.

Longitudinal data   Data collected in studies that take place over several years, often decades or more.

Low-cost provider   A company with a strategy to find and sustain creative ways to drive innovation at a significantly lower cost than its competitors so it can pass along lower prices to customers.

MAC   Medicare Administrative Contractors. These entities are multi-state, regional contractors responsible for administering both Medicare Part A and Medicare Part B claims in the US.

Management controls   One subsystem of a quality management system. Controls that ensure adequate management support and participation in quality systems.

Manufacturing costs   Costs for material (COGS), manufacturing labor, facilities, and equipment.

Market model   Outlines revenue and market share projections, as well as a preliminary assessment of the sales force required to achieve these forecasts.

Market segmentation   Using specific parameters to partition the market into identifiable, homogeneous segments in order to understand sales and marketing needs.

Marketing mix   A combination of factors, including product/ service mix, price, positioning, and promotion, that a company uses to support and implement its value proposition and help drive adoption of a new technology.

Material controls   One subsystem of a Quality Management system; controls that ensure material quality and consistency.

MDD   Medical Device Directives 93/42/EEC. One of three regulatory approval directives used in the European Union.

MDR   Medical Device Reporting. The reporting vehicle through which the FDA receives information about significant medical device adverse events that was established by the Safe Medical Devices Act.

MDUFMA   Medical Device User Fee and Modernization Act. The federal act that established user fees in the medtech industry.

Me-too products   Products that are relatively undifferentiated from products already on the market.

Mechanism of action   The specific biochemical or biomechanical interaction through which a drug or device produces its effect.

Medtech   Medical Device Technology. A short form to allow comparisons to Biotech, for instance.

Mezzanine funding   Funding that is required when some of the most significant risks have been resolved but the company has yet to generate sufficient revenue to be self-sustaining.

mHealth   An abbreviation for mobile health, which is the delivery of health-related services supported by mobile devices. Also referred to as digital health.

MHRA   Medicines and Healthcare Products Regulatory Agency. The organization that approves devices and drugs for Europe (including the UK).

Mixed need   A need with features that are easily achievable (more incremental to existing approaches) and other elements that introduce significant technical or clinical risk.

Morbidity   When a human is harmed in some way (short of death) by infection, decreased quality of life, extended hospital stay, physical impairment, etc.

NCD   National coverage determination. One of two types of reimbursement determinations made by Medicare that provides guidance on national reimbursement coverage. Typically applies to payments for inpatient services.

NDA   Non-Disclosure Agreement. An agreement between two parties such that the party receiving confidential information from another party will not disclose the information to anyone for a fixed period of time.

Need   Opportunities for innovation.

Need criteria   Requirements that any solution must meet to address the need as defined in the need statement. Usually need criteria are divided into two groups: must-have and nice-to-have requirements.

Need statement   A characterization of a need that communicates the problem or the health-related dilemma that requires attention, the affected population, and the targeted change in outcome, against which solutions to the problem will be evaluated.

Need specification   A document that synthesizes all of the important data gathered about a need through observations, research, and needs evaluation, including the need criteria.

NGO   Non-governmental organization. Non-profit organizations working for a cause. These organizations provide resources and assistance to parties when the governments will not or cannot provide them.

NICE   The National Institute for Care Excellence. A government organization chartered to issue national treatment guidance and assess cost-effectiveness of healthcare for England, Wales, and Ireland, which has become internationally recognized as a model for health technology assessment.

Niche strategy   A strategy whereby a company seeks to own the customer relationships in a specific, focused area of medicine.

Non-exclusive license   A license that allows the licensee rights of use within a given field and within whatever other limitations are specified by the license, but allows the licensor to grant similar rights to other parties.

Notice of allowance   The notice from the USPTO to indicate the patent has been accepted.

Notified bodies   Third-party organizations recognized by the European Commission to carry out conformity assessments as part of the medical device regulatory and CE marking process in the European Union.

NSE   Not Substantially Equivalent. A determination by the FDA that a new device is not equivalent enough to a predicate device and therefore cannot use the 510k pathway.

Observational studies Studies that make conclusions about the efficacy of a treatment or device on a group of subjects where the assignment of subjects into the treated versus control groups is outside the control of the investigator.

OCP Office of Combination Products. The section of the FDA that reviews medical technology comprising a combination of drugs/device or drugs/biologics to determine which center of the FDA will regulate it.

OEM strategy Original Equipment Manufacturer strategy. When a company provides technology and/or components to another company that then assembles and sells the finished product.

Off-label use The use of a treatment for conditions other than those approved by the FDA.

Office action A document issued by the USPTO that outlines objections or necessary changes to an application or claim due to finding prior art.

OHRP Office for Human Research Protections. A federal agency that helps assure the protection of humans participating in clinical research.

OIPE Office of Initial Patent Examination. The first agency that examines patent applications for completeness.

Operating plan Provides an overview of the activities that must be performed to develop the technology, their timing, and the key milestones they support.

Operating profit The difference between income and the expense incurred during operations.

OpEx Operating Expenses. Costs considered not to be manufacturing costs, including R&D, sales staff, general and administrative functions, and non-production facilities costs.

Opportunity cost The opportunity forgone by choosing a different opportunity.

Option pool The total number of stock options available for a company to grant, typically to employees.

Option to purchase A future right given to a potential acquirer to purchase a company at a specified price, at a specified time, or following the completion of specified milestones.

OTC Over the Counter. Drugs or devices that are sold directly to the end consumer.

OTL Office of Technology Licensing. The office within a university that manages its IP assets. Also called the Office of Technology Transfer (OTT).

Outright acquisition When a company acquires another by taking full and complete ownership of the business for a single payment at the time of purchase.

Out-of-pocket A healthcare payment model that requires patients to personally finance their medical expenses. Also referred to as an OOP model.

Owning the disease A company with a strategy to dominate the diagnosis and treatment of a disease or condition across the entire cycle of care through the integrated delivery of relevant products and services.

P&PC Production & Process Controls. One subsystem of a quality management system. Requires that production processes be controlled and monitored to ensure product conforms to specifications.

Partnering strategy One option in getting an idea to market – joining with another company to help develop a device.

Pass-through code Also called a c-code. A code that is issued to cover the cost of a device that is incremental to the services provided under an existing APC code, or set of codes in the US. The cost of the device may be bundled into this transitional APC code, or may still be billed separately under a temporary pass-through code.

Pathophysiology Study of the change of the normal mechanical, physical, and biochemical functions of a human due to disease or other interruption to normal function.

Patent A grant from the government to the inventor for exclusive rights to make, use, sell, or import an invention.

Patient towers Segments of patient populations based on epidemiological factors, market size and other important factors to determine the most favorable target for a device.

Payer Insurance companies (both public and private) who pay on behalf of the patient. Also called a third-party payer.

Pay-for-performance Setting prices contingent on the realization of specific results from a product or device. A new, if somewhat controversial, approach to healthcare fee structuring.

Payer advisory board A board consisting of key opinion leaders and medical directors from select payers in a target payer segment. They advise start-up companies in issues of reimbursement.

Payment Describes who is paid, and how much. In a reimbursement context, payment typically varies depending on the setting where the service is provided (e.g., physician office, hospital inpatient, hospital outpatient).

PCORI The Patient Centered Outcomes Research Institute (PCORI), which is charged, in part, with evaluating the comparative effectiveness of new device innovations.

PCT Patent Cooperation Treaty. A treaty that establishes unified patent filing for foreign countries. Issued by the World Intellectual Property Organization (WIPO).

PDP Product Development Protocol. A contract that describes agreed-upon details of design and development activities, the outputs of these activities, and the acceptance criteria for these outputs.

Peer review Review of a clinical trial or study by experts. These experts review the trials for scientific merit, participant safety, and ethical considerations. Peer-reviewed literature refers to scientific articles published in credible academic journals where a panel of physicians has reviewed the trial report.

PHI Protected Health Information. A patient's personal health information, which is protected by HIPAA.

PI Principal Investigator. The person responsible for conducting a clinical trial. May also be someone who manages a grant or contract for a particular research project.

Pilot trial Early clinical trial, usually conducted as a registry.

Pitch A tool used to communicate a company's integrated strategy and plans to an audience to engender support.

Pivotal trial Typically larger, controlled studies designed to test specific hypotheses when significant clinical data are necessary. Often used to support the submission of a PMA application for a new device.

PMA Premarket Approval. The most stringent pathway through the regulatory process for a device. A PMA is necessary when a new device is not substantially equivalent to any existing devices that were approved before 1976.

PMDA Pharmaceutical and Medical Device Agency. The PMDA is the technical arm of Japan's Ministry of Health, Labor, and Welfare, which is authorized to oversee the regulation of medical devices in the country.

POC Point of care is the place in which the actual surgery, medical intervention, or other medical procedure is done.

Porter's five forces Five forces, identified by Michael Porter (Harvard), that influence competition within an industry: rivalry, new competition, substitute products, suppliers' bargaining power, and buyers' bargaining power.

Positional advantage An advantage that comes from a company's ability to strategically position itself relative to its competition such that the position is not easy for others to replicate.

Post-marketing trial Trials performed following the commercial approval of a device in order to gain acceptance of the device in the field.

Post-money valuation A company's value after it obtains outside funds.

Power Statistical power refers to the probability of detecting a meaningful difference, or effect, if one were to occur. Ideally, studies should have power levels of 0.80 or higher – an 80 percent or greater chance of finding an effect, if one exists.

Power calculation A calculation used in clinical trials to determine the optimum number of patients to include in order to demonstrate efficacy.

PPI Physician preference items. Medical technologies that tend to be complex and differentiated, which physicians request by brand.

Pre-money valuation A company's value before it receives a subsequent outside round of financing.

Preferred stock Equity in a company that gives shareholders a liquidation preference, meaning that preferred stock holders are paid before common stock holders if a company is sold or its assets are liquidated. Holders of preferred stock are also given priority regarding the payment of dividends. Investors, such as venture capitalists, often are granted preferred stock in return for funding.

Primary endpoints Criteria that the company, investigators and FDA agree are required to be met to prove device efficacy during a clinical trial. For the device to receive approval, these endpoints must be met. The number of patients needed in a clinical trial is based on finding a statistically meaningful difference in these endpoints.

Primary prevention patients Patients who are risk of experiencing a particular episode of a disease state.

Prior art Any subject matter that is previously published or known generally in the field prior to an invention.

Priority date The first filing date of a patent application. This date is used to help determine the novelty of an invention compared to prior art.

Product specification A statement of what a solution needs to do and how it is intended to perform in order to address the needs of its intended users.

Prototype An early, often crude or rough sample or model of an idea to test some aspect of the concept. Prototypes are often used to give form to a solution idea and to help determine if it should be taken forward into more formal product development.

Provisional patent A patent filing that is less costly and rigorous to file but that only ensures a filing date prior to filing for a regular patent. Provisionals expire in 12 months if the provisional patent is not converted to a regular patent during that time.

Proxy companies Companies whose operations resemble what is required to commercialize an innovation. Also known as comparable companies. Used to validate financial models.

Public disclosure The moment an inventor describes an invention in a public setting; discussions with a single individual may be considered public settings.

QA Quality Assurance. The process that ensures that the product will operate according to its specifications.

QALY Quality-adjusted life year. A measure to estimate how much additional time (in months or years) of a reasonable quality a patient might gain as a result of treatment.

QC Quality Control. The activities designed to catch defective products in the manufacturing before they are released to the customer.

QMS Quality Management System. A system through which quality assurance and quality control are implemented.

QSIT Quality Systems Inspection Technique. A guide describing the way in which FDA investigators conduct an inspection of quality systems.

QSR Quality System Regulation. Regulations regarding the manufacturing, design, material handling and product testing of devices.

Quality of life The degree of well-being felt by an individual.

Randomization A method based on chance by which study participants are assigned to a treatment group. Randomization minimizes the differences among groups by equally distributing people with particular characteristics among all the trial arms.

Randomized trial A trial that uses random assignment of patients to the treatment and control groups.

Recall A method of removing all of a particular product from the market in an effort to minimize the risk to patients.

Records, documents, & change controls One subsystem of a Quality Management system. Controls for managing documentation and records for all aspects of the quality systems in place.

Reduction to practice Taking an invention beyond the concept stage. Showing that the invention actually works, as opposed to merely outlining its theoretical workings.

Reference pricing Using the prices of comparable products used in healthcare systems outside the US to make decisions on payments within the US, without necessarily considering any compensation for economic and regulatory differences.

Registry A collection of cases that have been performed in real-world settings outside the scope of a formal comparative protocol. Also called observational studies.

Reimbursement The act of paying for a medical device, procedure, visit or other element of patient care by a third-party payer.

Reimbursement dossier A set of documents that describes all aspects of a device (including basic information, studies' results, modeling reports, etc.). Used by companies introducing devices to payers.

Revenue ramp The degree to which a company expects to grow its revenue over time.

Risk management The systematic application of policies, procedures, and practices to the tasks of identifying, analyzing, controlling, and monitoring risk.

ROI Return on Investment. The amount of profit made on a given investment.

Royalty A percentage of commercial sales paid by the licensee/partner to the patent or rights holder.

Royalty anti-stacking provisions Provisions in a licensing deal that provide a mechanism for the licensee to prevent the licensor from reducing the royalty when other royalties are required for the same product.

Royalty-backed loans Loans for which the lender extends funds to the company in exchange for a royalty on the company's product sales. Lenders will sometimes combine a loan with an equity investment.

RUC AMA/Specialty Society Relative Value Scale Update Committee. The committee within the AMA that assigns relative value units to a new code.

RVU Relative Value Unit. Units that measure the resources required for a new procedure; used to determine the payment for a new CPT code.

SE Substantial equivalence. Demonstration that a new device is nearly the same as a predicate device (in terms of safety and effectiveness) in order to use the 510k pathway.

Secondary endpoint Additional criteria that may be met during a clinical trial, but that are not required to obtain a successful positive clinical trial result. The FDA rarely uses secondary endpoints to gain approval or clearance.

Secondary prevention patients Patients who have experienced a particular episode of a disease state and are candidates for preventing a recurrence of that episode.

Seed funding Early-stage funding that supports company creation through prototypes and proof of concept. Typically, this round is often funded by friends, families, and/or angels due to low capital requirements.

Seizure When the FDA takes action against a specific device, taking control over inventory or materials.

SG&A Selling and General Administrative Expenses. Expenses that include salaries, commissions, and travel expenses for executives and sales people, advertising costs, and payroll expenses.

Significant risk A term used by the FDA to indicate that a device is: (1) an implant and presents potential for serious risk to the patient; (2) is used to support or sustain life; (3) has a significant use for diagnosing or treating a disease; or (4) otherwise presents serious risk.

Staffing plan Specifies the personnel needed to execute the operating plan.

Stakeholders All parties with some interest in the delivery and financing of medical care for patients with a specific medical need.

Standard of care   A treatment process that is well-supported by evidence and that a doctor or medical facility should follow for a particular type of patient, disease or procedure.

Standard treatment   A treatment that is currently in wide use and approved by the FDA and considered to be effective in the treatment of a specific disease or condition.

Start-up funding   Funding that is required to make substantial investments in a company. This type of funding often requires millions of dollars and comes from Angels or VCs.

Statistical significance   The probability that an event or difference occurred (or did not occur) by chance alone. In clinical trials, the level of statistical significance depends on the number of participants studied and the observations made, as well as the magnitude of differences observed.

Structured acquisition   An acquisition in which the buyer makes an initial payment followed by additional earn-out payments, triggered by specified milestones. See Earn-out.

Subject   A human who participates in an investigation as an individual on whom an investigational device is used or who participates as a control. A subject may be in normal health or may have a medical condition or disease.

Superseding need   A need that is proximal or upstream of the need under consideration and, if solved, would make it superfluous.

Surrogate endpoint   Substitute criteria that are used to prove efficacy in lieu of actual endpoints that may take too long to prove.

Switching costs   The costs of a hospital or medical provider to change from one piece of capital equipment to another or from one treatment process to another; these costs are often a deterrent to trying a new vendor's equipment or new procedures.

SWOT analysis   An analysis, identified by Albert Humphrey (Stanford), of factors affecting a company: strengths, weaknesses, opportunities and threats.

Syndicate   When more than one investor is involved in funding a company.

TEC   Technology Evaluation Committee (or Center). A committee, established by a payer (such as Blue Cross Blue Shield) that evaluates a medical device or procedure to assess whether to pay for the device or procedure and how much.

Term sheet   A document that outlines the terms for a deal and serves as a letter of intent between the investor providing funds and the company receiving them. Terms may include percentage ownership in the company, intellectual property division and other financial and legal concerns.

Third-party payer   Insurance companies (both public and private) who pay on behalf of the patient. Also called a payer.

Time to profitability   The amount of time it takes a company to break even, and then turn a consistent operating profit.

Top-down model   A sales forecast model that forecasts projected yearly revenue by multiplying the number of customers by the price of the innovation.

Trade secret   Information, processes, techniques, or other knowledge that is not made public but provides the innovator with a competitive advantage.

User   The stakeholder who will ultimately use/apply a new technology.

User experience   The way in which the application of a technology affects the user's behaviors, attitudes, and emotions, as well as his/her perception of the product.

User requirements   Criteria that influence the design of a technology that reflect the needs and preferences of the user.

USPTO   United States Patent and Trademark Office. The government organization that issues patents and trademarks.

Utility patent   The most common type of patent for medical devices, which describes an invention in sufficient detail to determine that it is novel, useful, and unobvious.

Valuation   The worth assigned to the business.

Value   An expression of the improvement(s) a new technology offers relative to its incremental cost.

Value analysis committee   A multidisciplinary group from within a healthcare organization that meets to review and make purchasing decisions for some categories of medical technologies.

Value estimate   A directional representation of the value associated with a need based broadly on understanding who the real decision makers are with respect to adoption/ purchasing decisions, how significant they perceive the need to be, to what degree available solutions are effectively addressing the need, and therefore how much "room" there is to get them to change their current behavior by offering a new technology with a different improvement/cost equation.

Value exploration   Seeking need areas where improved economic outcomes can potentially be realized. Typically performed as part of needs exploration.

Value proposition   The sum total of benefits that one party promises to a second party in exchange for payment (or other value-transfer).

Venture debt   Debt financing available to companies that have at least one professional investor as a significant equity-holder in the company. Can serve a similar purpose of a bridge loan, but is also used to fund equipment purchases.

Voting rights   Section of a term sheet that spells out how voting will be orchestrated when shareholder approvals are required.

Warrants   A form of equity that gives the holder the right to purchase a number of shares of a company's stock at a predetermined price per share, sometimes subject to other conditions as well.

WIPO   World Intellectual Property Organization. The organization responsible for the Patent Cooperation Treaty.

WHO   World Health Organization. The direct coordinating authority for health within the United Nations system.

Index